ION IMPLANTATION TECHNOLOGY

To learn more about the AIP Conference Proceedings, including the
Conference Proceedings Series, please visit the webpage
http://proceedings.aip.org/proceedings

ION IMPLANTATION TECHNOLOGY

16th International Conference on Ion Implantation Technology

IIT 2006

Marseille, France   11 – 16 June 2006

EDITORS

Karen J. Kirkby
Russell Gwilliam
Andy Smith
*Surrey Ion Beam Centre, ATI,
University of Surrey, England*

David Chivers
*Ion Links International, Ltd.,
West Lothian, Scotland*

SPONSORING ORGANIZATIONS

*Advanced Beam Technology - AIBT
Air Products & Chemicals, Inc.
Applied Materials - AMAT
Atmel Rousset S.A.S.
ATMI Inc.
Axcelis Technologies, Inc.
Cascade Scientific
Matheson-Trigas/Linde-Nippon Sanso

Nissin Ion Equipment
Plansee Metal GmbH
Praxair Electronics
Semequip Inc.
SEN Corporation
SOITEC
ST Microelectronics
Varian Semiconductor Equipment

Conseil General des Bouches-du-Rhone
Marseille Provence Metropole
Pays D'aix Development
Provence Promotion
University de Provence
University Paul Cezanne/TECSEN
Ville de Marseille

* Sponsors of the Conference Proceedings CD-ROM

Melville, New York, 2006
AIP CONFERENCE PROCEEDINGS ■ VOLUME 866

Editors:

Karen Kirkby
Russell Gwilliam
Andy Smith

Surrey Ion Beam Centre
Advanced Technology Institute
School of Electronics and Physical Sciences
Guildford, Surrey GU2 7XH
England

E-mail: k.kirkby@surrey.ac.uk
r.gwilliam@surrey.ac.uk
a.j.smith@surrey.ac.uk

David Chivers
Ion Links International, Ltd.
St. Mary's Place
Bathgate
West Lothian EH48 1DS
Scotland

E-mail: ionlinks@btinternet.com

Authorization to photocopy items for internal or personal use, beyond the free copying permitted under the 1978 U.S. Copyright Law (see statement below), is granted by the American Institute of Physics for users registered with the Copyright Clearance Center (CCC) Transactional Reporting Service, provided that the base fee of $23.00 per copy is paid directly to CCC, 222 Rosewood Drive, Danvers, MA 01923. For those organizations that have been granted a photocopy license by CCC, a separate system of payment has been arranged. The fee code for users of the Transactional Reporting Services is: ISBN/978-0-7354-0365-9/06/$23.00.

© 2006 American Institute of Physics

Permission is granted to quote from the AIP Conference Proceedings with the customary acknowledgment of the source. Republication of an article or portions thereof (e.g., extensive excerpts, figures, tables, etc.) in original form or in translation, as well as other types of reuse (e.g., in course packs) require formal permission from AIP and may be subject to fees. As a courtesy, the author of the original proceedings article should be informed of any request for republication/reuse. Permission may be obtained online using Rightslink. Locate the article online at http://proceedings.aip.org, then simply click on the Rightslink icon/"Permission for Reuse" link found in the article abstract. You may also address requests to: AIP Office of Rights and Permissions, Suite 1NO1, 2 Huntington Quadrangle, Melville, NY 11747-4502; Fax: 516-576-2450; Tel.: 516-576-2268; E-mail: rights@aip.org.

L.C. Catalog Card No. 2006935319
ISBN 978-0-7354-0365-9
ISSN 0094-243X

CD-ROM available: ISBN 978-0-7354-0366-6

Printed in the United States of America

CONTENTS

Preface xiii
Committees xv
Sponsors and Session Chairs xvii

DOPING PROCESSES IN SEMICONDUCTORS

Germanium ... The Semiconductor of Tomorrow! 3
    A. R. Peaker, A. Satta, V. P. Markevich, E. Simoen, and B. Hamilton

The Role of Ion Beam Technology in the Development of Integrated Optical Monitors Suitable for Applications in Silicon Photonics 9
    A. P. Knights, J. D. B. Bradley, P. J. Foster, S. H. Gou, and P. E. Jessop

Influence of Energy Contamination at S/D-Extension Dopant Implantation Using Ultra Fast Annealing 13
    M. Herden, D. Gehre, T. Feudel, and L. Herrmann

Formation of Ultrashallow Junctions Less than 10nm with the Combination of Low Energy B Ion Implantation and Laser Annealing 17
    R. Yamada, S. Seto, S. Sato, Y. Tanaka, S. Matumoto, T. Suzuki, G. Fuse, T. Kudo, and S. Sakuragi

Impact of Fluorine Co-Implant on Boron Diffusion during Non-Melt Laser Annealing 21
    T. Noda, S. Felch, V. Parihar, C. Vrancken, T. Janssens, H. Bender, B. Van Daele, and W. Vandervorst

Characterization of B_2H_6 Plasma Doping for Converted p^+ Poly-Si Gate 25
    J.-G. Oh, J. K. Lee, S. H. Hwang, H. J. Cho, Y. S. Sohn, D. S. Sheen, S. H. Pyi, S. W. Lee, S. H. Hahn, Y. B. Jeon, Z. Fang, and V. Singh

Fluorine Profile Distortion upon Annealing by the Presence of a CVD Grown Boron Box 29
    P. López, L. Pelaz, R. Duffy, P. Meunier-Beillard, K. van der Tak, F. Roozeboom, and G. Maas

Deactivation of Low Energy Boron Implants into Preamorphised Si after Non-Melt Laser Annealing with Multiple Scans 33
    J. A. Sharp, N. E. B. Cowern, R. P. Webb, D. Giubertoni, S. Gennaro, M. Bersani, M. A. Foad, and K. J. Kirkby

Co-Implantation for 45 nm PMOS and NMOS Source-Drain Extension Formation: Device Characterisation Down to 30 nm Physical Gate Length 37
    E. J. H. Collart, B. J. Pawlak, R. Duffy, E. Augendre, S. Severi, T. Janssens, P. Absil, W. Vandervorst, S. Felch, R. Scheutelkamp, and N. E. B. Cowern

Control of Phosphorus Transient Enhanced Diffusion Using Co-Implantation 41
    A. Vanderpool, A. Budrevich, and M. Taylor

Integration of Advanced Source and Drain Extension Process Using Carbon/Fluorine Co-Implants and Spike Anneal in 65nm PMOS Devices 46
    C. I. Li, H. Y. Wang, C. C. Chien, M. Chain, C. L. Yang, S. F. Tzou, H. Graoui, M. A. Foad, and R. Ting

Controlling Dopant Diffusion and Activation through Surface Chemistry 50
    K. Dev, C. T. M. Kwok, R. Vaidyanathan, R. D. Braatz, and E. Seebauer

Strain-Enhanced Activation of Sb Ultrashallow Junctions 54
    N. S. Bennett, L. O'Reilly, A. J. Smith, R. M. Gwilliam, P. J. McNally, N. E. B. Cowern, and B. J. Sealy

Phosphorus Implant for S/D Extension Formation: Diffusion and Activation Study after Spacer and Spike Anneal 58
    S. H. Yeong, B. Colombeau, F. Benistant, M. P. Srinivasan, C. P. A. Mulcahy, P. S. Lee, and L. Chan

The Concept of LDSI (Locally-Differentiated-Scanning Ion Implantation) for the Fine Threshold Voltage Control in Nano-Scale FETs 62
    M.-Y. Lee, S. W. Jin, Y. S. Sohn, S. K. Na, K. B. Rouh, Y. S. Joung, Y. J. Ki, I. K. Han, Y. W. Song, and S. W. Park

Time Dependence Study of Hydrogen-Induced Defects in Silicon during Thermal Anneals 65
    S. Personnic, A. Tauzin, K. K. Bourdelle, F. Leterte, N. Kernevez, F. Laugier, N. Cherkashin, A. Claverie, and R. Fortunier

Layer Transfer of SOI Structures Using a Pre-Stressed Bonding Layer 69
    L. Vorrada and N. Cheung

Optimal Preamorphization Conditions for the Formation of Highly Activated Ultra Shallow Junctions in Silicon-On-Insulator 73
    J. J. Hamilton, E. J. H. Collart, M. Bersani, D. Giubertoni, S. Gennaro, N. S. Bennett, N. E. B. Cowern, and K. J. Kirkby

Ion Implantation: A World of Innovations 76
    M. Bruel

Junction Stability of B Doped Layers in SOI Formed with Optimized Vacancy Engineering Implants 84
    A. J. Smith, N. E. B. Cowern, B. Colombeau, R. Gwilliam, B. J. Sealy, E. J. H. Collart, S. Gennaro, D. Giubertoni, M. Bersani, and M. Barozzi

Effects of Hydrogen Atoms on Redistribution of Implanted Boron Atoms in Silicon during Annealing 88
    K. Yokota, S. Nakase, and F. Miyashita

Investigation of Amorphous Layer Formation Using Applied QUANTUM X Single Wafer and QUANTUM Batch Implanter 92
    R. Doherty, B. McComb, R. Ting, and Y. C. Cheng

High Dopant Activation and Low Damage P+ USJ Formation 96
    J. Borland, S. Shishiguchi, A. Mineji, W. Krull, D. Jacobson, M. Tanjyo, W. Lerch, S. Paul, J. Gelpey, S. McCoy, J. Venturini, M. Current, V. Faifer, R. Hillard, M. Benjamin, T. Walker, A. Buczkowski, Z. Li, and J. Chen

A Study of Carbon Effects in Implantation Process for Non-Silicide Contact Formation 101
    T. H. Huh, S. Kim, G. J. Ra, R. N. Reece, S. I. Kondratenko, Y. S. Kim, K. I. Shin, and W. H. Jeon

Pre-Annealing Effects of n^+/p and p^+/n Junction Formed by Plasma Doping (PLAD) and Laser Annealing 105
    S. Heo, S. Baek, D. Lee, M. Hasan, and H. Hwang

Spike Annealing of Shallow Arsenic and Phosphorous Implants in Different Gaseous Ambient 109
    S. Paul, W. Lerch, and D. Bolze

The Effect of Flash Annealing on the Electrical Properties of Indium/Carbon Co-Implants in Silicon 113
    S. Gennaro, D. Giubertoni, M. Bersani, J. Foggiato, W. S. Yoo, R. Gwilliam, and M. Anderle

Local Arsenic Structure in Shallow Implants in Si Following SPER: An EXAFS and MEIS Study 117
    G. Pepponi, D. Giubertoni, S. Gennaro, M. Bersani, M. Anderle, R. Grisenti, M. Werner, and J. A. Van Den Berg

Well Design in a Bulk CMOS Technology with Low Mask Count 121
    M. P. M. Jank, C. Kandziora, L. Frey, and H. Ryssel

Boron Redistribution during Crystallization of Phosphorous-Doped Amorphous Silicon 125
    R. Simola, D. Mangelinck, A. Portavoce, J. Bernardini, and P. Fornara

Ultra-Shallow Junctions Formed by Sub-Melt Laser Annealing 129
    S. B. Felch, A. Falepin, S. Severi, E. Augendre, T. Hoffman, T. Noda, V. Parihar, F. Nouri, and R. Schreutelkamp

Defect Behavior in Bf_2 Implants for S/D Applications as a Function of Ion Beam Characteristics 133
    N. Cagnat, C. Laviron, N. Auriac, J. Liu, S. Mehta, L. Frioulaud, and D. Mathiot

Impact of Dose Rate Effects and Damage Engineering On Device Performance 137
    K. Shim, Y. Hwang, Y. Lee, J. An, S. Ryu, S. Hahn, C. Cho, N. Hur, B. Guo, J. Liu, and Y. Erokhin

Germanium Ion Implantation to Improve Crystallinity during Solid Phase Epitaxy and the Effect of AMU Contamination 140
    K. S. Lee, D. H. Yoo, G. H. Son, C. H. Lee, J. H. Noh, J. J. Han, Y. S. Yu, Y. W. Hyung, J. K. Yang, D. G. Song, T. J. Lim, Y. K. Kim, S. C. Lee, H. D. Lee, and J. T. Moon

CLUSTER IMPLANTATION AND DOPING

20 Years History of Fundamental Research on Gas Cluster Ion Beams, and Current Status of the Application to Industry 147
    I. Yamada

Dose Retention Effects in Atomic Boron and ClusterBoron™ ($B_{18}H_{22}$) Implant Processes 155
M. A. Harris, L. Rubin, D. Tieger, V. Venezia, T. J. Hsieh, J. Miranda, and D. Jacobson

Universal Ion Source™ for Cluster and Monomer Implantation 159
T. N. Horsky

Investigation of Converted p^+ Poly-Si Gate Formed by $B_{18}H_X^+$ Cluster Ion Implantation 163
S.-H. Hwang, D. S. Kim, Y. H. Joo, J. G. Oh, J. K. Lee, T. W. Jung, H. J. Cho, Y. S. Sohn, D. S. Sheen, S. H. Pyi, S. Kim, T. H. Huh, W. A. Krull, and H. T. Cho

A Beam Line System for a Commercial Borohydride Ion Implanter 167
H. F. Glavish, T. N. Horsky, D. C. Jacobson, F. Sinclair, N. Hamamoto, N. Nagai, and M. Naito

Ultrashallow P^+/n Junction Formed by $B_{18}H_{22}^+$ Ion Implantation and Excimer Laser Annealing 171
S. Heo, D. Lee, H. T. Cho, W. A. Krull, and H. Hwang

Productivity Enhancements for Shallow Junctions and DRAM Applications Using Infusion Doping 174
J. Hautala, M. Gwinn, W. Skinner, and Y. Shao

A Vaporizer for Decaborane and Octadecaborane 178
D. Adams, T. N. Horsky, G. Gilchrist, R. Milgate, J. Sweeney, and P. Marganski

Simplifying the 45nm SDE Process with ClusterBoron® and ClusterCarbon™ Implantation 182
W. Krull, B. Haslam, T. N. Horsky, K. Verheyden, and K. Funk

Implantation Characteristics by Boron Cluster Ion Implantation 186
T. Nagayama, N. Hamamoto, S. Umisedo, M. Tanjyo, and T. Aoyama

Sputtering and Chemical Modification of Solid Surfaces by Water Cluster Ion Beams 190
G. H. Takaoka, K. Nakayama, D. Takeda, and M. Kawashita

Cross-Sectional TEM Observations of Si Wafers Irradiated with Gas Cluster Ion Beams 194
H. Isogai, E. Toyoda, T. Senda, K. Izunome, K. Kashima, N. Toyoda, and I. Yamada

Boron Beam Performance and in-situ Cleaning of the ClusterIon™ Source 198
T. N. Horsky, G. F. R. Gilchrist, and R. W. Milgate III

P-Type Gate Electrode Formation Using $B_{18}H_{22}$ Ion Implantation 202
D. Henke, F. Jakubowski, J. Deichler, V. C. Venezia, M. S. Ameen, and M. A. Harris

ClusterBoron™ Implants on a High Current Implanter 206
D. R. Tieger, W. DiVergilio, E. C. Eisner, M. Harris, T. J. Hsieh, J. Miranda, W. P. Reynolds, and T. N. Horsky

Cluster Size Effects of Gas Cluster Ion Beams on Surface Modification 210
N. Toyoda and I. Yamada

High-Speed Nano-Processing with Cluster Ion Beams 214
T. Seki and J. Matsuo

Characterization of Molecular Clusters in the Supersonic Gas Jet 218
T. Kagawa, F. Sato, Y. Kato, Y. Ito, and T. Iida

PIII AND PLASMA DOPING

PLAsma Doping for P+ Junction Formation in 90 nm NOR Flash Memory Technology 225
D. Bigarella, V. Soncini, D. Raj, V. Singh, and S. Walther

Deep Trench Doping by Plasma Immersion Ion Implantation in Silicon 229
S. Nizou, V. Vervisch, H. Etienne, M. Ziti, F. Torregrosa, L. Roux, M. Roy, and D. Alquier

Nitrogen Plasma Ion Implantation of Al and Ti Alloys in the High Voltage Glow Discharge Mode 233
R. M. Oliveira, M. Ueda, J. O. Rossi, H. Reuther, C. M. Lepienski, and A. F. Beloto

Plasma Immersion Ion Implantation with a 4kV/10kHz Compact High Voltage Pulser 237
M. Ueda, R. M. Oliveira, J. O. Rossi, H. Reuther, and G. Silva

Effects of Ion Energy on Nitrogen Plasma Immersion Ion Implantation in UHMWPE Polymer through a Metal Grid 241
M. Ueda, R. M. Oliveira, J. O. Rossi, C. M. Lepienski, and W. A. Vilela

Modified Phasor-Particle Model of Treating a Blocking Capacitor as a Phasor Element in Simulation of Plasma Coupling with an External Auto-Matching Network 245
D. T. K. Kwok and P. K. Chu

B_2H_6 PLAD Doped PMOS Device Performance 249
    Z. Fang, T. Miller, E. Winder, H. Persing, E. Arevalo, A. Gupta, T. Parrill, V. Singh, S. Qin, and A. McTeer

Plasma Immersion Ion Implantation Applied to P+N Junction Solar Cells 253
    V. Vervisch, D. Barakel, F. Torregrosa, L. Ottaviani, and M. Pasquinelli

Ion Behaviour in Pulsed Plasma Regime by Means of Time-Resolved Energy Mass Spectroscopy (TREMS) Applied to an Industrial Radiofrequency Plasma Immersion Ion Implanter: PULSION® 257
    M. Carrere, F. Torregrosa, and V. Kaeppelin

Boron Ion Implantation into Silicon by Use of the Boron Vacuum-Arc Plasma Generator 261
    J. M. Williams, C. C. Klepper, D. J. Chivers, R. C. Hazelton, J. J. Moschella, and M. D. Keitz

MATERIALS—NOVEL TECHNIQUES AND APPLICATIONS

Micro-Patterned Porous Silicon Using Proton Beam Writing 269
    M. B. H. Breese, D. Mangaiyarkarasi, E. J. Teo, A. A. Bettiol, and D. Blackwood

Grazing Incidence Angle X-Ray Diffraction of Implanted Stainless Steel: Comparison between Simulated Data and Experimental Data 275
    J. Dudognon, M. Vayer, A. Pineau, and R. Erre

Investigation of the Impact on Device Parameters of Fluorine Enhanced Oxide in a Power Trench MOSFET 279
    J. H. Rice and C.-T. Wu

Rare Gas Ion Implanted-Silicon Template for the Growth of Relaxed $Si_{1-x}Ge_x$/Si (100) 283
    G. Regula, M. Raissi, J.-L. Lazzari, F. Chevrier, N. Burle, and E. Ntsoenzok

High Temperature Implantation of Aluminum in 4H Silicon Carbide 287
    M. Rambach, A. J. Bauer, and H. Ryssel

Annealing of TiO_2 Films Deposited on Si by Irradiating Nitrogen Ion Beams 291
    K. Yokota, Y. Yano, and F. Miyashita

Thermal Diffusion Barrier for Ag Atoms Implanted in Silicon Dioxide Layer on Silicon Substrate and Monolayer Formation of Nanoparticles 295
    H. Tsuji, N. Arai, N. Gotoh, T. Minotani, T. Ishibashi, T. Okumine, K. Adachi, H. Kotaki, Y. Gotoh, and J. Ishikawa

Water Splitting and Hydrogen Emitting Catalytic Function of Hydrogen-Implanted Oxide Ceramics Studied Using Ion Beam Technology 300
    K. Morita, B. Tsuchiya, S. Nagata, K. Katahira, M. Yoshino, J. Yuhara, Y. Arita, T. Ishijima, and H. Sugai

Structural Changes in Polymer Films by Fast Ion Implantation 304
    M. A. Parada, R. A. Minamisawa, C. Muntele, I. Muntele, A. De Almeida, and D. Ila

Hydrogen Ion Implantation Mechanism in GaAs-on-Insulator Wafer Formation by Ion-Cut Process 308
    H. J. Woo, H. W. Choi, G. D. Kim, J. K. Kim, W. Hong, and H. R. Lee

High Dose Hydrogen Implant Blistering Effects as a Function of Selected Implanter and Substrate Conditions 313
    R. Eddy, C. Hudak, P. Bettincurt, and S. Delgado

Germanium Nanoparticle Formation into Thin SiO_2 Films by Negative Ion Implantation and Their Electric Characteristics 317
    N. Arai, H. Tsuji, N. Gotoh, T. Okumine, T. Yanagitani, M. Harada, T. Satoh, H. Ohnishi, T. Minotani, K. Adachi, H. Kotaki, T. Ishibashi, Y. Gotoh, and J. Ishikawa

Size Analysis of Ethanol Cluster Ions and Their Sputtering Effects on Solid Surfaces 321
    G. H. Takaoka, K. Nakayama, T. Okada, and M. Kawashita

Tuning of Etching Rate by Implantation: Silicon, Polysilicon and Oxide 325
    R. Charavel and J.-P. Raskin

The Influence of Ion Implantation on Cell Attachment to Glass Polymeric Carbon 329
    R. Zimmerman, I. Gurhan, F. Ozdal-Kurt, B. H. Sen, M. Rodrigues, and D. Ila

IMPLANT TECHNOLOGY

Chicane Deceleration—An Innovative Energy Contamination Control Technique in Low Energy Ion Implantation 335
    N. White, J. Chen, C. Mulcahy, S. Biswas, and R. Gwilliam

Characterisation of the Beam Plasma in High Current, Low Energy Ion Beams for Implanters 340
    J. Fiala, D. G. Armour, J. A. van den Berg, A. J. T. Holmes, R. D. Goldberg, and E. H. J. Collart

Next Generation Medium Current Product: VIISta 900XP 345
    A. Renau

Implant Angle Control on Optima MD 349
    R. D. Rathmell, B. Vanderberg, A. M. Ray, D. E. Kamenitsa, M. Harris, and K. Wu

Process Transferability from a Spot Beam to a Ribbon Beam Implanter: CMOS Device Matching 353
    V. Kaeppelin, Z. Chalupa, L. Frioulaud, S. Mehta, B. Guo, K.-H. Shim, H. Lendzian, and Y. Erokhin

Application of Stencil Mask Ion Implantation Technology to Power Semiconductors 357
    T. Nishiwaki, H. Saito, K. Hamada, K. Tonari, and T. Nishihashi

Ion Implanter Cross Contamination and Maintenance Safety Considerations with High Dose Phosphorus 361
    R. Eddy, B. Ostrowski, M. H. Yang, and D. Huntington

Source Life Improvement for Germanium Implant 365
    A. Allen, P. Banks, S. Biswas, and C. Mulcahy

Ion Sources for High and Low Energy Extremes of Ion Implantation 369
    A. Hershcovitch, V. A. Batalin, A. S. Bugaev, V. I. Gushenets, B. M. Johnson, A. A. Kolomiets, G. N. Kropachev, R. P. Kuibeda, T. V. Kulevoy, I. V. Litovko, E. S. Masunov, E. M. Oks, V. I. Pershin, S. V. Petrenko, S. M. Polozov, H. J. Poole, I. Rudskoy, D. N. Seleznev, P. A. Storozhenko, A. Ya. Svarovski, and G. Yu. Yushkov

An Electron Cyclotron Resonance Ion Source with Cylindrically Comb-Shaped Magnetic Field Configuration 373
    Y. Kato, H. Sasaki, T. Asaji, T. Kubo, F. Sato, and T. Iida

Advantages of Dual Magnet Ribbon Beam Architecture for Particle Control in Single Wafer High Current Implant 377
    C. Campbell, G. Redinbo, J. Blake, P. Kellerman, E. Moore, and N. Variam

Optimized Autotuning for Single Wafer High-Current and Medium-Current Implanters 381
    J. T. Scheuer, A. Cucchetti, M. Welsch, W. Callahan, K. Luey, and J. C. Olson

Backing Up Medium Current Implanters Using Single Wafer High Energy Implanter for Manufacturing Efficiency 385
    H. L. Sun, W. Lee, K. Xu, H. Y. Tsun, K. T. Peng, L. S. Juang, and H. P. Tseng

Rising Microwave Frequency of a Broad-Ion-Beam ECR Source with Cylindrically Comb-Shaped Magnetic Field Configuration 389
    T. Asaji, Y. Kato, H. Sasaki, T. Kubo, F. Sato, T. Iida, and J. Saito

Profile and Angle Measurement System of SHX 393
    Y. Kikuchi, M. Kabasawa, M. Tsukihara, and M. Sugitani

Implant Angle Deviation Reduction in Batch-Type High Energy Implanter 397
    N. Suetsugu, T. Yamada, M. Tsukihara, and M. Sugitani

Stencil Mask Ion Implantation Technology for Realistic Approach to Wafer Process 401
    K. Tonari, T. Nishihashi, M. Ishikawa, and J. Fujiyama

Enhanced Dosimetry for Single Wafer High-Current Implanters 405
    J. T. Scheuer, J. Dzengelaski, D. Distaso, D. Timberlake, J. Cummings, and J. C. Olson

Using Multiple Implant Regions to Reduce Development Wafer Usage 409
    S. R. Walther, S. Falk, S. Mehta, Y. Erokhin, and P. Nunan

Advanced Modeling Techniques for Analysis of High Current Ribbon Beam Transport and Control 413
    P. Kellerman and F. Sinclair

Increase of Beam Current Mass-Separated by Long Gap Dipole Sector Magnet for S/D Process in FPD Manufacturing 417
    S. Dohi, Y. Ando, Y. Inouchi, Y. Matsuda, M. Konishi, J. Tatemachi, M. Nukayama, K. Nakao, K. Orihira, and M. Naito

Nissin Ion Equipment Indirectly Heated Cathode Ion Source 421
K. Tanaka, S. Umisedo, K. Miyabayashi, H. Fujita, T. Kinoyama, N. Hamamoto, T. Yamashita, and M. Tanjyo

Improved Ion Beam Incident Angle Control for Varian E220 and E500 Implanters 425
D. Hendrix, Z. Zhao, R. Liebert, K. Gifford, and P. Mitchell

Development of an Ion Beam Aligner for Liquid Crystal Displays 429
T. Matsumoto, Y. Matsuda, M. Tanii, M. Konishi, and Y. Andoh

Real-Time Optimization Method for Optical Parameters of Ion Implanters 433
S. Ogata, T. Nishihashi, K. Tonari, H. Yokoo, H. Suzuki, T. Hisamune, and M. Araki

Implant Angle Monitoring—A Comparison of Channeling Features 437
M. A. Rathmell

Defectivity Reduction and Control in Ion Implant Systems through Hardware and Process Optimization 441
D.-W. Franke, F. Hundt, T. Guenther, M. Schmeide, R. N. Reece, C. E. Ferrell, B. Krimbacher, F. Haerting, and J. Grant

Advanced Charge Control for Single Wafer Implanters 445
P. F. Kurunczi, A. S. Perel, E. Wright, S. Kikuchi, and J. T. Scheuer

APC Implementation on VIISta Ion Implanters 449
Y. K. Kim, B. Adams, N. Parisi, S. Mehta, and J. Hamilton

Optimization of High-Energy Implanter Beamline Pumping 453
M. LaFontaine, M. Pharand, Y. Huang, I. Pokidov, and J. Ferrara

Reduction of the Wafer Pattern Damage on the Batch-Type High Current Ion Implanters 457
E. Oga, H. Izutani, G. Fuse, and M. Sugitani

Charging Mechanism during Ion Implantation without Charge Compensation 460
S. Sakai, H. Fan, E. Chen, and M. Tanjyo

The Impact of Mass Resolution on Molybdenum Contamination for B, P, BF_2, and As Implantations 464
V. Häublein, L. Frey, and H. Ryssel

Maximizing Productivity for Well Implantation 468
Q. Zhai, Y. K. Kim, D. Rodier, and N. Variam

Efficient High Current Process Transfer and Device Matching Strategies for Sub-90nm Manufacturing 472
D. Lee, N. Loh, M. C. Tan, K. Shim, S. Falk, B. Guo, S. Jillson, B. Wong, K. Loh, S. Mehta, and Y. Erokhin

IMPLANT TECHNOLOGY (MATERIALS)

The Development of In-Situ Ion Implant Cleaning Processes 477
S. Bishop, R. Kaim, S. Yedave, J. Arnó, F. DiMeo Jr., and M. Wodjenski

Qualification of the GASGUARD® SAS GGT Arsine Sub-Atmospheric Gas Delivery System for Ion Implantation 481
J. P. Dunn, J. L. Rolland, J. S. Grim, R. M. Machado, and C. L. Hartz

Manufacturing Assessment of SDS3® Gas Upgrade 485
J. P. Dunn, J. Rolland, J. Grim, and B. Brown

ATMI's Ion Implant Process Efficiency Research Laboratory (IIPERL) 489
S. Yedave, J. Arnó, S. Bishop, F. DiMeo Jr., R. Kaim, and L. Wang

A Safe Solution to Dopant Gas Desorption from Metal Surfaces 493
T. Nakanoya and M. Egami

Dopant Cylinder Lifetime Monitor 497
S. Bishop, M. Wodjenski, R. Kaim, S. Lurcott, J. McManus and G. Smith

PROCESS CONTROL & YIELD

Real World Experience with Ion Implant Fault Detection at Freescale Semiconductor 503
D. C. Sing, T. Breeden, H. Fakhreddine, S. Gladwin, J. Locke, J. McHugh, and M. Rendon

Angle Measurement and Control in High Current Ion Implantation 507
B. S. Freer, D. E. Hoglund, and M. A. Graf

Process Control in Production-Worthy Plasma Doping Technology 511
    E. J. Winder, Z. Fang, E. Arevalo, T. Miller, H. Persing, V. Singh, and T. M. Parrill

Gate Dielectric Damage Due to High-Tilt Implant 516
    S. B. Felch, R. Hung, B. Ninan, M. Smayling, N. Toshiyuki, H. Chen, and C.-P. Chang

High Current Implant Precision Requirements for SUB-65 nm Logic Devices 520
    Y. Erokhin, T. Romig, E. Kim, J. Xu, B. Guo, J. Liu, K. Shim, and P. Nunan

Production-Worthy USJ Formation by Self-Regulatory Plasma Doping Method 524
    Y. Sasaki, H. Ito, K. Okashita, H. Tamura, C. G. Jin, B. Mizuno, T. Okumura, I. Aiba, Y. Fukagawa, H. Sauddin, K. Tsutsui, and H. Iwai

PROCESS CONTROL & YIELD (METROLOGY)

Surface Charge Profiling—An Advancement in Ion Implant Monitoring 531
    C. Krueger, C.-H. Ng, Z. Zhao, and G. Krytsch

Metrology and High Resolution Mapping of Shallow Junctions Formed by Low Energy Implant Processes 534
    E. Don, A. Pap, P. Tutto, T. Pavelka, C. Wyon, C. Laviron, D. Sotta, R. Oechsner, and M. Pfeffer

Metrology Requirements for Single Wafer Ion Implanters 538
    J. C. Olson, G. Angel, A. Gupta, R. Mollica, D. Distaso, and J. Liu

Local Resistance Measurement on Polycrystalline Silicon Layer in Low-Temperature Poly-Si Thin Film Transistor Using Scanning Spreading Resistance Microscopy 542
    S. Abo, H. Yamagiwa, K. Tanaka, F. Wakaya, T. Sakamoto, H. Tokioka, N. Nakagawa, and M. Takai

Study on Chemical Binding States of Silicon in Conjunction with Ultra-Shallow Plasma Doping by Using Hard X-Ray Photoelectron Spectroscopy (HX-PES) 546
    C. G. Jin, Y. Sasaki, K. Okashita, H. Tamura, H. Ito, B. Mizuno, T. Okumura, M. Kobata, J. J. Kim, E. Ikenaga, and K. Kobayashi

Characterization of Parasitic Transistor Phenomenon in Nano-Scale NAND Flash Device by Blanket Tilt Implantation and Scanning Capacitance Microscopy 550
    D.-H. Lee, S.-W. Shin, C.-K. Ryu, M.-K. Lee, N.-Y. Kwak, H.-S. Shon, B.-S. Lee, S.-K. Park, and K.-D. Kwack

Direct Measurement of Beam Angle in a High Current Ion Implanter 554
    B. S. Freer, L. M. Rubin, M. A. Graf, D. E. Hoglund, D. Newman, K. Ditzler, K. Elshot, and T. Romig

Photoelectric Measurement Method for Implanted Silicon: A Phenomenological Approach ... 558
    K. Steeples and E. Tsidilkovski

Detection and Reduction of the Yield Impact of Particle Induced Structure Defects at Batch Ion Implanters 562
    M. Schmeide, M. Kokrot, D.-W. Franke, and B. Sauter

Non-Contact, Image-Based Photoluminescence Metrology for Ion Implantation and Annealing Process Inspection 566
    A. Buczkowski, Z. Li, T. Walker, S. G. Hummel, and J. O. Borland

Superior Dose and Energy Monitoring Capability of the Therma-Probe System 570
    M. Bakshi, D. Shaugnessy, L. Nicolaides, and P. Mitchell

Mapping Leakages of USJ Test Wafers 574
    J. T. C. Chen, T. Dimitrova, and D. Dimitrov

Improved Techniques for Characterization and Optimization of SIMOX Implantation 578
    R. Dolan, C. McKenna, S. Richards, Y. Aoki, T. Nakai, S. Nakamura, and M. Walden

Non-Contact Sheet Resistance and Leakage Current Monitoring of Multi-Implant, Ultra-Shallow Junctions: Doping and Damage Effects for MS-Anneals 582
    M. I. Current, V. N. Faifer, T. M. H. Wong, T. Nguyen, and A. Koo

Characterizing Dopant Contamination Using Ion Implantation 586
    J. J. Naughton and J. M. Towner

Manufacturing Precision Polysilicon Resistors Using Ion Implantation 590
    J. M. Towner

Automated Dose and Dopants Level Monitoring by SIMS 594
    H. Maul, N. Loibl, U. Ehrke, A. Merkulov, P. Peres, and M. Schuhmacher

MATERIALS

Optima HD: Single Wafer Mechanical Scan Ion Implanter 601
    P. Splinter, M. Graf, C. Godfrey, Y. Huang, D. Polner, J. Danis, and K. Ota

High Performance Medium Current Ion Implanter System EXCEED3000AH-G3 605
    S. Sakai, M. Tanjyo, N. Hamamoto, S. Umisedo, T. Kobayashi, T. Yamashita, T. Matsumoto, T. Ikejiri,
    K. Tanaka, Y. Koga, S. Yuasa, M. Naito, and N. Nagai

Down to 2nm Ultra Shallow Junctions: Fabrication by IBS Plasma Immersion Ion Implantation Prototype PULSION® 609
    F. Torregrosa, H. Etienne, G. Mathieu, and L. Roux

Advanced Single Wafer High Current Beamline Architecture for SUB-65nm 614
    G. Redinbo, C. Campbell, and J. Mullin

Applied Quantum X Implant System: Technology Enhancements to Enable Production-Worthy Performance at the 45 nm Node 618
    A. Murrell, P. Edwards, R. Goldberg, P. Banks, B. Mitchell, E. Collart, S. Morley, G. Ryding, T. Smick,
    M. Farley, T. Sakase, D. Hacker, and P. Kindersley

Enhancing the Ibis i2000™ SIMOX Oxygen Ion Implanter 622
    C. McKenna, R. Dolan, J. Blake, and S. Richards

Indium Performance on the V810 626
    R. J. Low and Q. Zhai

Understanding the Calibration Methodology for the Axcelis GSD/HE Final Energy Magnet and a Means for Manipulating the Calibration Curve 630
    R. Johnson and J. Schuur

New Medium Current Ion Implanter SOPHI-200 633
    H. Yokoo, H. Suzuki, R. Fukui, T. Hisamune, M. Tomita, T. Nishihashi, K. Tonari, and S. Ogata

Optima MD: Mid-Dose, Hybrid-Scan Ion Implanter 637
    K. W. Wenzel, A. M. Ray, B. H. Vanderberg, and R. D. Rathmell

Optimised Charging Performance on Quantum X Ion Implanters 641
    D. A. Kirkwood, T. Sakase, R. Miura, R. D. Goldberg, and A. J. Murrell

65nm Device Characteristics Matching on Single and Batch System Ion Implanter 645
    K. Okabe, R. Miura, and M. Kase

An Adaptive Knowledge-Based Ion Source Automation Methodology to Improve Beam to Beam Switch Performance on Applied Materials Quantum® X Single Wafer Ion Implanter 649
    C. Burgess, M. Keane, and R. Oliver

Conference Photographs 653
Author Index 657

Preface

IIT 2006 was held at the Palais du Pharo, in the Vieux Port of Marseille, France, from 11 to 16 June 2006.

More than 240 abstracts were submitted with over 400 attendees from 26 countries. Traditionally IIT is held in 3 major centres, Europe, USA and Asia. Attendees figures from this year's conference consisted of Europe (55%), USA (30%) and Asia (15%). IIT is still the premier world meeting for the presentation of the latest advances on all aspects of ion implantation technology.

The conference covered a wide range of topics including Doping Processes, Implant Technology, Materials Science, Process Control and Yield and Novel Applications. Ultra Shallow Junctions were a major feature of the 2006 conference, with major advances in covering recent advances in implantation, Cluster Beams, Plasma Immersion and thermal processing being reported.

This Conference, as is the custom of IIT, was preceded by a School featuring Ion Implantation Science and Technology. Once again, the School was an outstanding success due largely to the efforts of Jim Ziegler.

There was a strong Social Program enabling participants to sample the history and culture of Provence. The conference banquet was held amidst the outstanding natural beauty of the Camargue. Partners of conference attendees enjoyed a memorable program, organised by Pam Chivers, including an opportunity to explore Marseille and the surrounding areas.

There was also a strong Trade Exhibition, consisting of 25 booths, featuring Ion Implantation equipment and services.

We are extremely grateful for the very high level of sponsorship that we have received, both from the world-wide semiconductor industry and from local government bodies and also the strong participation of the Exhibition. Without this support it would not be possible to host IIT2006.

In such a rapidly advancing field, speedy publication of the conference proceedings is essential. We are therefore indebted to Karen Kirkby and her team and would like to thank her for taking editorial responsibility for these proceedings. We also wish to acknowledge our gratitude for the help and support provided by Sabine Kessler of AIP and would like to thank all authors and referees for their fast processing of papers.

Finally we would like to thank most of all the conference participants, organisers, invited speakers and authors, for making IIT2006 such a rewarding and memorable occasion.

We now look forward to IIT2008, the 17th conference in this series, to be held in Monterey, USA in the summer of 2008. This conference will be co-chaired by Sue Felch and Ed Seebauer.

Russell Gwilliam
David Chivers

International Committee:

Robert Brown (USA)
Lih-Juanni Chen (Taiwan)
John Chen (Taiwan)
Dave Chivers (UK)
Paul Chu (Hong Kong)
Michael I. Current (USA)
Russell Gwilliam (UK)
Jozsef Gyulai (Hungary)
Yoshitomo Hidaka (Japan)
Kevin S. Jones (USA) (Chairman)
Masataka Kase (Japan)
Renée Koudijs (USA)
Lawrence A. Larson (USA)
Mike Mack (USA)
Koji Matsuda (Japan)
Charles M. McKenna (USA)
Adrian Murrell (UK)

John Poate (USA)
Tony Renau (USA)
Peter Rose (USA)

Geoffrey Ryding (USA)
Heiner Ryssel (Germany)
Robert B. Simonton (USA)
Thomas C. Smith (USA)
Mikio Takai (Japan)
Mitch Taylor (USA)
Anatoli Vyatkin (Russia)
Wesley Weisenberger (USA)
James Stanislaus Williams (Australia)
Andrew Wittkower (USA)
Isao Yamada (Japan)
James Ziegler (USA)

Local Organizing Committee

Isabelle Armand (Atout Organisation Science)
Marcel Carrere (University of Provence)
Pam Chivers (Ion Links Int. Ltd.)
Bernard Pichaud (Paul Cezanne University)
Luc Jeannerot (Atmel)
Corinne Joachim (ARCSIS)

Jean Claude Nataf (SCS Cluster)
Christelle Neveu (Atout Organisation Science)
Robert Ronchi (ST Microelectronics)
Georges Terzian (University of Provence)
Russell Gwilliam (University of Surrey)
Jean-Louis Lazzari (University of Mediterranee)

Scientific Program Committee

Russell Gwilliam (UK)
Dale Jacobson (USA)
Leonard Rubin (USA)
Mitch Taylor (USA)
Tony Renau (USA)
Michael Current (USA)

Karen Kirkby (UK)
Roger Webb (UK)
Paul Chu (Hong Kong)
Fuccio Christiann (France)
Marcel Carrere (France)
Gilles Mathieu (France)

Editorial Committee

Karen Kirkby (UK)
Russell Gwilliam (UK)

Andy Smith (UK)
Justin Hamilton (UK)

Finance Committee

Celine Auger (ARCSIS)
Philippe Mollard (ARCSIS/Atmel)
Laurent Roux (ARCSIS/IBS)
Russell Gwilliam (University of Surrey)

Exhibition And Sponsorship Committee

Dave Chivers (Ion Links Int. Ltd.)
Alain Gargani (Atout Organisation Science)
Corinne Joachim (ARCSIS)

Corporate Sponsors

AIBT/Advanced Ion Beam Technology, Inc. (sponsoring the CD of the conference proceedings
AIR PRODUCTS & CHEMICALS, Inc.
AMAT/Applied Materials
ATMEL ROUSSET S.A.S.
ATMI Inc.
AXCELIS TECHNOLOGIES, Inc.
MATHESON-TRIGAS/LINDE-NIPPON SANSO
NISSIN ION EQUIPMENT
PLANSEE Metal GmbH
PRAXAIR ELECTRONICS
SEMEQUIP Inc.
SEN Corporation, an SHI and Axcelis Company
SOITEC
ST Microelectronics
VARIAN SEMICONDUCTOR EQUIPMENT

Local Sponsors

ARCSIS
Conseil General des Bouches du Rhone
Marseille Provence Metropole
PAYS D'AIX DEVELOPEMENT
PROVENCE PROMOTION
University de Provence
University Paul Cezanne/TECSEN
Ville de Marseille

Session Chairs

Session 1 Doping Processes in Semiconductors I –	Laurent Roux
Session 2 Implant Technology I –	Paul Chu
Session 3 Implant Technology (Clusters)–	Bob Brown
Session 4 Process Control and Yield I (Metrology) –	TC Smith
Session 5 Doping Processes in Semiconductors II –	Adrian Murrell
Session 6 Materials 1 SOI –	Karen Kirkby
Session 7 Implant Technology II –	Sandeep Mehta
Session 8 Process Control & Yield II –	Michael Current
Session 10 Doping Processes in Semiconductors III –	Lenny Rubin
Session 11 Novel Techniques and Applications–	Heiner Ryssel
Session 12 Implant Technology III –	Russell Gwilliam

DOPING PROCESSES
IN SEMICONDUCTORS

Germanium ... The Semiconductor of Tomorrow?

A.R. Peaker[1,a], A. Satta[2], V.P. Markevich[1], E. Simoen[2] and B. Hamilton[1]

[1]*Center for Electronic Materials Devices and Nanostructures, University of Manchester, M60 1QD, UK*
[2]*IMEC, Kapeldreef 75, B-3001 Leuven, Belgium*

Abstract. Germanium is being investigated as an alternative channel material for mainstream CMOS. Although aspects of processing technology have similarities to silicon, ion implantation, and particularly the activation of dopants and removal of implant damage appear to be fundamentally different. In this paper we review some of the issues and present results that are encouraging in the context of PMOS devices.

Keywords: germanium, CMOS, ion implantation, semiconductor defects, dopant activation
PACS: 71.20.Mq, 72.80.Cw, 73.61.Cw, 79.60.Ht
[a])Electronic mail: peaker@manchester.ac.uk

INTRODUCTION

Extreme scaling of MOS devices has stimulated a search for channel materials exhibiting higher mobilities. The reasons for this are manifold but a key issue is that gains from conventional scaling are yielding less significant returns and many issues contribute to reduce the effective mobility in nano channels eg surface scattering, high halo doping, coulombic scattering from dielectric charge and remote phonon scattering. Strained silicon using either global or local strain is already used to provide some device advantage [1]. However in the longer term more radical solutions appear to be necessary. The introduction of high-k dielectrics and metal gates have led people to question whether semiconductors other than silicon might be considered as channel materials. The issue of hole mobility has always been a problem in silicon CMOS, the disparity between the hole and electron mobilities has necessitated a larger PMOS transistor than NMOS so as to balance drive currents.

This issue has been accentuated with the advent of extreme scaling and it is now evident that the concept of channel mobility is only a partial descriptor of carrier motion in strained nano-materials [2].

However it can be seen from Table 1 that germanium offers particular advantages in relation to hole mobility while also benefiting electron mobility and so quite obviously is a serious candidate as an alternative channel material. Although III-V materials offer much larger benefits in terms of electron mobility they offer no advantage in terms of hole mobility and their more complex processing and volatility of the group V component is a strong discouragement in relation to CMOS. Historically a strong argument in favor of the III-V materials has been the high electron saturation velocity. In early generations of devices this is a key parameter but in extreme scaling (below 100nm channel length) it is now evident that it is the low-field effective inversion-layer mobility that is most significant in device performance [3].

TABLE 1. Semiconductor Properties

	Si	Ge	GaAs
Electron mobility (cm^2/Vs)	1350	3900	8500
Hole mobility (cm^2/Vs)	450	1900	400
Band gap (eV)	1.12	0.67	1.42
Electron saturation velocity (x10^7 cm^2/s)	1	0.6	2
Thermal conductivity at 300K (W/cm/°C)	1.3	0.58	0.55

FUNDAMENTALS

Although germanium was the dominant material in semiconductors in the 1960s this was a time that ion implantation was in its technological infancy. Because the motivation for device related work disappeared with the advent of silicon not much work on the implantation and activation of dopants in germanium existed until quite recently. In this paper we review some of the fundamental issues that distinguish implantation in germanium from silicon.

There are of course major problems other than ion implantation in utilizing germanium for CMOS the principal one being the prime reason for the original move to silicon, namely the difficulty in forming a germanium dielectric interface suitable for MOS devices. The native oxide is soluble in water and the interface state densities are extremely high when silicon dioxide or silicon nitride is deposited on the semiconductor. A considerable volume of literature is now accumulating on the use of high-k dielectrics on germanium and it is evident that although there are many problems to be solved there are some promising results

From the point of view of ion implantation perhaps the most important fundamental difference between silicon and germanium is that the energy needed to create a vacancy in germanium is rather low. Reported values are as low as 1.9eV for the creation of the neutral vacancy compared to 3.3eV in silicon. This results in self-diffusion and the diffusion of most impurities being controlled by the vacancy population rather than interstitials as in the silicon case.

Although there is increasing evidence that boron is an exception to this [4] there is strong evidence that the diffusion of other technologically important dopants is vacancy mediated. This is rather important in relation to transient-enhanced diffusion which would then be expected to result from metastable vacancy clusters unlike the case of silicon where interstitials dominate.

In addition it would be expected that because of the ease of formation of vacancies that the doping limitation mechanisms in n-type germanium may be more dramatic than in silicon as these effects tend to be vacancy driven.

The available data on diffusion in germanium has been reviewed recently by Bracht [5], if the normal procedure of normalizing the temperature to the melting point is adopted in the plot of diffusivity as a function of temperature, then self diffusion has comparable values in germanium and silicon. The diffusion of the Group V impurities is also similar but there is an enormous difference in the behavior of boron in germanium compared to silicon. Using the normalized temperature it is evident that in germanium boron has a diffusivity that is five orders of magnitude smaller than for the silicon case. Boron also seems anomalous in terms of its solid solubility, the accepted data which were measured by Trumbore in 1960 [6] show that the solubility of boron is anomalously low at 600°C at about $5 \times 10^{18} cm^{-3}$, this is two orders lower than gallium at the same temperature.

Taking bulk values of mobility we would expect for equilibrium boron in a 10nm implant the resistance would be 30kΩ-square, this compares very unfavorably with the ITRS requirement for silicon at the 30nm node which is 300Ω-square for a 12nm implant. Essentially this means that boron would have to activate at a concentration two orders more than the 600°C solid solubility to be usable as a dopant in PMOS at the 30nm node.

ION IMPLANTATION

We have investigated diffusion and activation of boron implanted into germanium at a maximum concentration well in excess of the solid solubility and then studied the results of thermal annealing up to 600°C. The germanium was supplied by Umicore and was Czochralski grown 100mm diameter wafers, {100} orientated and antimony doped with a resistivity of ~20Ω-cm. Prior to implantation 10 nm of SiO_2 was deposited by plasma enhanced chemical vapor deposition. Implantation into crystalline germanium was studied but also implantation after pre-amorphization using germanium ions at 100keV at a

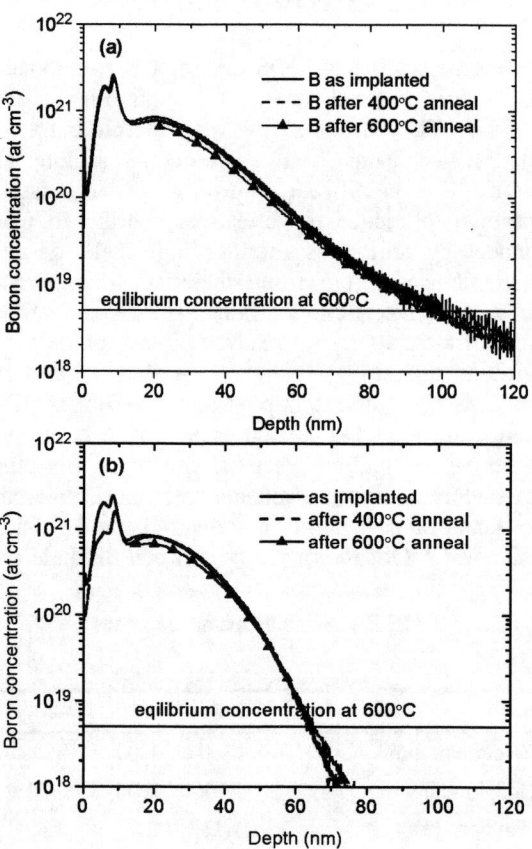

FIGURE 1. SIMS profiles of 6keV 3×10^{15} atom/cm^{-2} boron implantation into a) crystalline germanium and b) pre-amorphised germanium before and after annealing at 400°C and 600°C The pre-amorphisation was effected by a 120 keV 10^{15} cm^{-2} germanium implant

dose of 10^{15} atoms cm^2 with a 7° tilt. The wafers were annealed in a rapid thermal processing system (Heat Pulse 610) in dry nitrogen with a ramp up-rate of 24°C s^{-1} with holding times between 1 and 60 seconds. The chemical profiles were studied by Secondary Ion Mass Spectrometry (SIMS) and the electrical activity with a variable probing spacing technique (VPS). Full details of the experimental procedures are reported elsewhere [7-10].

Figure 1 shows the SIMS profile for an implant into crystalline germanium for the as implanted case and after 60 second anneals at 400°C and 600°C.

Although a channeling tail is clearly visible in the case of implantation into crystalline germanium it is clearly evident there is no significant diffusion on this scale. In the case of implantation into pre-amorphised germanium the channeling is totally suppressed and again no diffusion is discernable up to anneal temperatures of 600°C. This is important ... there is no transient-enhanced diffusion at this temperature and there is no concentration enhanced diffusion making boron look like an ideal dopant for extremely scaled germanium devices in this respect.

When the activation of the boron is examined by measuring the resistance and calculating the resistance per square of implant layer it is apparent that rather high levels of activation can be achieved. This is in contrast to the predictions from the accepted solid solubility data. Fig 2 reports results for 6keV implants into crystalline and into amorphous germanium. When boron is implanted into a crystalline matrix a 3% activation is present immediately after implantation and does not change significantly with anneals up to 400°C, at higher temperatures in some cases the activation decreases. However in the case of the implantation of boron into amorphous germanium, although there is very little electrical activity immediately after implant 30% activation is achieved after a 500°C anneal for one second and the ITRS figure for the 30 nm node of the silicon roadmap is achieved, i.e. the electrical activation is almost two orders above Trumbore's figure for the equilibrium solid solubility at 600°C.

Similar experiments have been undertaken using n-type dopants. In this case the results are not as promising. Work reported elsewhere [8-10] indicates that for high dose implants the active phosphorous concentration is limited to 5×10^{19} atoms cm^{-3} and phosphorous forms large inactive precipitates in the germanium. Out diffusion also occurs and there is evidence of concentration enhanced diffusion, a much more complicated scenario than for the case of boron. Undoubtedly this is the reason for early device successes with Ge PMOS in contrast to NMOS [11]

RECOMBINATION-GENERATION CENTERS

Quite extensive studies have been carried out in relation to radiation damage in germanium. Historically this focused on high resistivity material for radiation detectors but more recently studies of doped material appropriate for IC use [12-14] have been undertaken. Studies of radiation damage are quite important in relation to ion implantation. In the case of silicon our knowledge of implantation defects and their removal stems from early radiation studies of silicon, although of course there are important differences. The objective is to remove all midgap states which will contribute to recombination generation currents, during the dopant activation anneal.

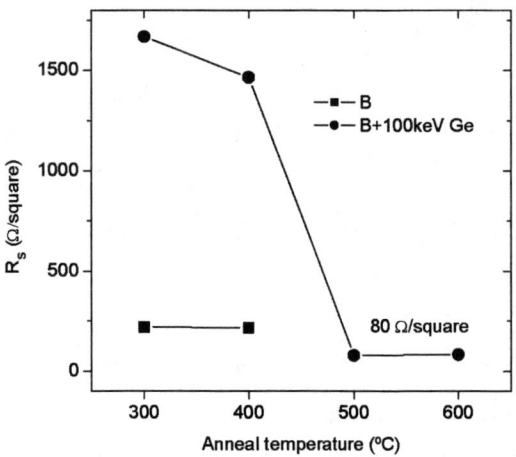

FIGURE 2. Activation of a 6keV 3×10^{15}cm^{-2} boron implant into crystalline and amorphous germanium as a function of 1s anneal temperature in the range 300 to 600°C

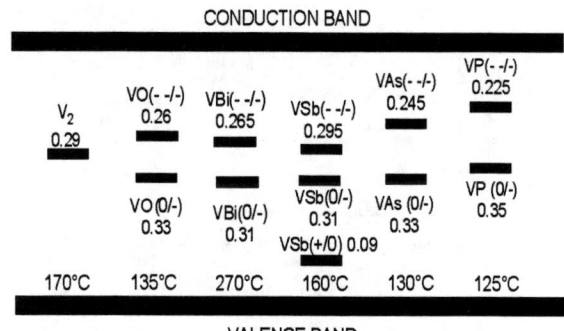

FIGURE 3. Commonly observed radiation defects in gamma irradiated germanium. The figures adjacent to the identity of the defect are the activation energy for hole emission or the enthalpy for electron emission from the state. The temperatures are those at which the defect anneals to half its original concentration after 30 minutes

Fig 3 summarizes the dominant radiation induced defects in germanium. The identity of the defects is given, its energy and the temperature at which they anneal out to half their original value using 30 minute isochronal anneals.

Unlike silicon, Czochralski grown germanium contains very low concentrations of oxygen and so the vacancy oxygen complex which dominates the irradiation defect population in Czochralski silicon is only present in germanium in special oxygen enhanced material. The dominant centres are vacancies combined with the donor (referred to as E centers). In the case of p-type material the divacancy is more significant.

Fig 4 shows DLTS spectra of irradiated n-type (antimony doped) germanium. The samples were irradiated with gamma rays from ^{60}Co source. The samples had an antimony concentration of 2×10^{14} cm^{-3}. In order to produce the test samples Schottky diodes were fabricated by thermal evaporation of gold onto surfaces etched in a 1HF + 10HNO$_3$ acid mixture, characterization was effective with conventional DLTS and Laplace DLTS techniques.

The Laplace method [15] gives much higher resolution and enables us to state categorically that all the defects observed were simple point defects i.e. no large complexes were observed in the gamma irradiated material.

The key issue here is that all the defects were removed after a 30 minute anneal at 400°C and there is no indication of any large complexes forming.

IMPLANTATION DEFECTS

In the case of implantation damage there are important differences when compared to the gamma irradiated case. In silicon the differences between gamma or electron irradiated and implanted material have been extensively studied and documented [eg 16] but despite this the defect population and its evolution are not understood in detail. The principal reason for the difference compared to gamma irradiation is the high rate of damage generation near the end of range of the implanted particle which is likely to result in clustering of self-interstitials and nearer the surface in clusters of vacancies.

Reverse modeling of annealing [17] in silicon and ab-initio calculations [18, 19] predict stable clusters of small numbers of vacancies and/or interstitials. These anneal out at temperatures around those at which dopant diffusion is important and hence play a key role in mediating the diffusion via a transient enhancement. In both the silicon and the germanium cases the difference between gamma and implant damage arises primarily from the generation *rate* of interstitial vacancy pairs.

During implantation the creation of these primary defects is extremely high in the region close to the peak of the implanted profile. This means that the primary defects have a much higher probability of reacting with themselves, which in the case of an interstitial vacancy reaction simply results in annihilation but in the case of vacancies can result in

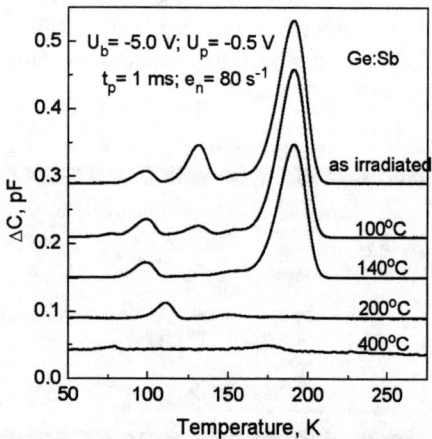

FIGURE 4. DLTS spectra of electron emission from Sb doped n-type germanium irradiated with gamma rays and subsequently 30 minute anneals at the temperatures shown The total radiation dose was 1.1×10^{17} cm^{-2}. The spectra have been vertically displaced for clarity.

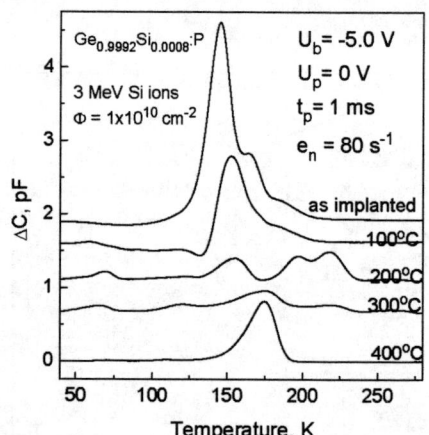

FIGURE 5. DLTS spectra of electron emission for P-doped Ge$_{0.9992}$Si$_{0.0008}$ crystals after implantation with 3 MeV Si ions and a subsequent 30-min anneal at temperatures as shown. The total implantation dose was 10^{10} Si ions cm^{-2}.

the production of the divacancy or larger vacancy agglomerates. In the case of interstitials there is, in the silicon case, some evidence for the formation of clusters. A number of configurations of vacancy and of interstitial clusters are believed to be quite stable thermally and only dissociate on high temperature annealing.

Previous work using very high energy heavy ions (up to ~5000MeV) provides strong evidence for multi-vacancy complexes in germanium after such implants [20]. There has long been evidence of surface erosion and buried void production after large doses of heavy ions implanted into Ge [21, 22]. Recently it has become evident that even after low energy implants the driving force for vacancies to accumulate is so strong that buried voids can form under conditions that might otherwise be considered as technologically important [23].

We have investigated electrically active ion implantation defects in Ge and SiGe alloys after very low ion doses. Fig. 5 shows a germanium rich SiGe sample (0.08% Si), which has been implanted with 10^{10}cm^{-2} of 3 MeV silicon ions. Considering first the as implanted DLTS spectrum used to quantify the electrically active defects, it is apparent that this is significantly different to the γ irradiated. The main peak is similar in emission behaviour and annealing to the P related E centre in γ irradiated germanium. However the minor peaks present in the γ irradiated samples and tentatively associated with interstitials are absent.

Additional features are evident on the high temperature shoulder of the main peak. From association with previous results reviewed elsewhere [13] of electron emission characteristics and annealing behaviour, the shoulder at ~160K is thought to be due to the divacancy. The generation of this defect is much higher during implantation in contrast to gamma irradiation because of the large generation rate of vacancies during implantation. This gives them the opportunity to react together and form a relatively stable entity.

During annealing it can be seen that the DLTS spectrum undergoes a series of complex transformations. These damage related defects are much more difficult to remove than the E centre or divacancy with the result that a substantial concentration of electrically active defects remain after the 400°C 30 minute anneal and although in work not not shown here we observe some defects which remain after 500°C anneals. There is as yet no clear evidence as to the identity of these defects but it is likely that both multi-vacancy and multi-interstitial complexes are involved which as for the silicon case of interstitials seem to be stable up to relatively high temperatures.

DIODE LEAKAGE

As discussed in the last section the presence of these generation recombination centres is potentially an important source of diode leakage current. Fig 6 illustrates the effect of defect removal for a p$^+$n diode made by a boron implantation at 13 keV with a dose of 4×10^{15} ions cm^{-2}, the substrate doping was 10^{16} cm^{-3}. The surface was passivated with GeON/HfO$_2$. It can be seen that there is a significant reduction in leakage current with increasing anneal temperature as would be expected from the DLTS results. In a p$^+$n junction it is the defects in the n type region that will give rise to the defect generation currents. The intrinsic leakage current density for a Ge p$^+$n junction at room temperature with n = 10^{16} is ~10^{-6}A cm^{-2}. However even with a 500°C anneal the leakage current from the device in Fig 6 is three orders greater than this intrinsic leakage. From the deep state concentration determined by DLTS after a 500°C anneal we calculate that defect related generation current will be ~8×10^{-5}A cm^{-2}. It is evident that there is another very significant current component.

Satta et al [24] have recently separated out the surface leakage component in ungarded diodes from the bulk leakage by studying diodes of different area to periphery ratios and using different dielectrics. It is evident that surface leakage is very significant and could account for the excess current reported in Fig 6.

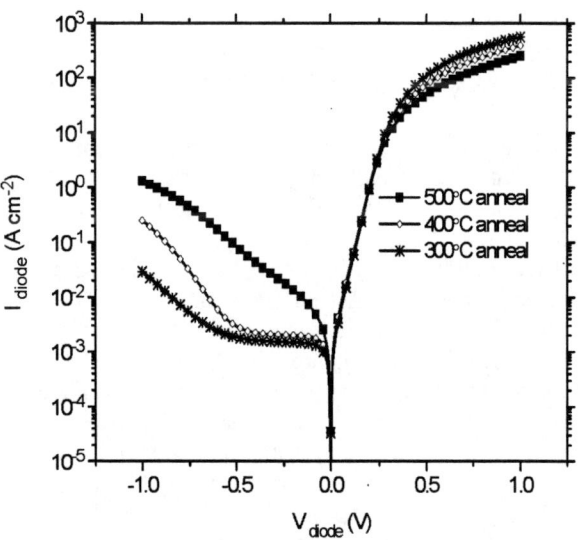

FIGURE 6. Current voltage characteristics measured at room temperature of a p$^+$n diode after annealing for 5 minutes at the temperatures indicated The diode was made by implanting 13keV B$^+$ ions into n-type germanium with a carrier concentration of 10^{16} cm^{-3}

CONCLUSION

Ion implantation of boron into germanium seems to be a well behaved process presenting no major difficulties when using conventional ion implantation and annealing techniques. Using pre-amorphisation techniques shallow implantations with low resistance per square can be obtained. It appears that this aspect of germanium technology is well suited to PMOS well beyond the 45 nm node. Although there are still some questions about the removal of cluster like implant damage and its role as generation centres, early indicators are that the leakage due to bulk generation centres can be reduced to acceptable levels with relatively low thermal budgets.

The situation with regard to NMOS and particularly in relation to the implantation of phosphorous is less hopeful. The preliminary experiments conducted by us indicate that a great deal of work and possibly new techniques need to be used to achieve adequate activation of phosphorous in germanium. However the issue of generation centres and removal of damage appears to be fundamentally the same as in the PMOS case. Surface leakage is a problem, orders of magnitude more severe in germanium than in silicon, and much research on the true nature of germanium interface states is undoubtedly necessary not only in relation to producing a suitable gate dielectric stack but also in the more mundane issue of leakage from the periphery of devices.

ACKNOWLEDGMENTS

We would like to thank Umicore Belgium for germanium materials, Nikoli Abrosimov of the institute of Crystal Growth Berlin for germanium rich SiGe, Gareth Nicholas, Michel Houssa, Marc Meuris Wilfried Vandervorst and Cor Claeys of IMEC for valuable discussions, Arne Larsen and Knud Bonde Nielsen (Århus University Denmark) for collaboration on the electron and gamma irradiation work, Ian Hawkins, Mal McGowan, Ivana Capan and M.Kalid Khan at The University of Manchester for help with defect studies and Russell Gwilliam (UK National Ion Beam Centre at Surrey University). We thank the UK Engineering and Science Research Council and European Commission (INTAS) for funding.

REFERENCES

1. S.E. Thompson et al *IEEE Transactions Electron Devices* **51**, 1790-1797 (2004)
2. M.L. Lee, E.A. Fitzgerald, M.T. Bulsara, M.T. Currie, A. Lochtefeld, *J. Appl. Phys.* **97**, 11101-1-27 (2005)
3. A. Lochtefeld I.J. Djomehri, G. Samudra, D.A. Antoniadis, *IBM Journal of Research and Development* **46**, 347-357 (2002)
4. S. Uppal, A.F.W. Willoughby, J.M. Bonar, N.E.B. Cowern, T. Grasby, R.J.H. Morris, and M.G. Dowsett, *J. Appl. Phys.* **96** 1376 (2004)
5. H. Bracht, Proceedings of EMRS 2006 Symposium T Germanium based semiconductors ... from materials to devices in press *Materials Science in Semiconductor Processing*
6. F.A. Trumbore, *Bell System Technical Journal* **39**, 205-233 (1960)
7. A. Satta, E. Simoen, T. Clarysse, T. Janssens, A. Benedetti, B. De Jaeger, M. Meuris, and W. Vandervorst, *Appl Phys Lett* **87**, 172109 (2005)
8. A. Satta, E. Simoen, T. Janssens, T. Clarysse, B. De Jaeger, A. Benedetti, *Journal of The Electrochemical Society*, **153** G229-G233 (2006)
9. A. Satta E. Simoen, R. Duffy, T. Janssens, T. Clarysse, A. Benedetti, M. Meuris, and W. Vandervorst, *Applied Physics Letters* **88**, 162118 (2006)
10. A. Satta, T. Janssens, T. Clarysse, E. Simoen, M. Meuris, A. Benedetti, I. Hoflijk, *J. Vac. Sci. Technol.* **B 24** 494 (2006)
11. Chi On Chui, K. Gopalakrishnan, P.B. Griffin, J.D. Plummer, K.C. Saraswat, *Appl. Phys Lett.* **83** (2003) 3275-3277.
12. V.P. Markevich, A.R. Peaker, V.V. Litvinov, V.V. Emtsev, L.I. Murin, *J. Appl. Phys.* **95** (2004) 4078-4083.
13. J.Fage-Pedersen, A. Nylandsted Larsen, A. Mesli, *Phys. Rev. B* **62** (2000) 10116-10125
14. V.P. Markevich, I.D Hawkins, A.R. Peaker, K.V. Emtsev, V.V. Emtsev, V.V. Litvinov, L.I. Murin, L. Dobaczewski, *Phys Rev B* **70** (2004) 235213 – 1-7.
15. L. Dobaczewski, A.R. Peaker, K. Bonde Nielsen, *J. Appl. Phys.* **96** (2004) 4689-4728.
16. B.G. Svensson, B. Mohadjeri, A. Hallen, J.H. Svensson, J. W.Corbett, *Phys. Rev. B* **43** 2292-2298 (1991)
17. N.E.B. Cowern, G. Mannino, P.A. Stolk, F. Roozeboom, H.G.A. Huizing, J.G.M. van Berkum, F. Cristiano, A. Claverie, M. Jaraiz, *Phys. Rev. Lett.* **82** 4460-4463 (1999).
18. T.E.M. Staab, A. Sieck, M. Haugk, M.J. Puska, Th. Fraunheim, H.S. Leipner, *Phys. Rev. B* **65** 115210 – 1-11. (2002)
19. J.L. Hastings, S.K. Estreicher, P.A. Fedders, *Phys. Rev. B* **56** 10215-10220 (1997).
20. A. Colder, M. Levalois, P. Marie, *European Physical Journal, Applied Physics* **13** 89 (2001)
21. O. W. Holland, B. R. Appleton, and J. Narayan, *J. Appl. Phys.* **54**, 2295 (1983)
22. I.H. Wilson, *J. Appl Phys* **53**, 1698-705 (1982)
23. T. Janssens, C. Huyghebaert, D. Vanhaeren, G. Winderickx, A. Satta, M. Meuris, and W. Vandervorst *J. Vac Sci & Tech B* **24** 510-14 (2006)
24. A. Satta et al Proceedings of EMRS 2006 Symposium T Germanium based semiconductors ... from materials to devices in press *Materials Science in Semiconductor Processing*

The Role of Ion Beam Technology in the Development of Integrated Optical Monitors Suitable for Applications in Silicon Photonics

A P Knights, J D B Bradley, P J Foster, S H Gou and P E Jessop

Department of Engineering Physics, McMaster University, 1280 Main Street West, Hamilton, Ontario., L8S 4L7, Canada.

Abstract. This paper outlines one aspect of the enabling role that ion implantation has played in the continuing development of silicon photonic devices. It describes work on defect mediated, silicon-based integrated optical monitors, suitable for detection of light at wavelengths around 1570nm. The detection mechanism in such devices relies on the introduction of mid-gap energy levels via inert ion implantation into the volume of the device coincident with the optical mode. The defect-related absorption of a controllable fraction of the optical signal leads to the generation of carriers and a photocurrent is extracted via an integrated *p-i-n* diode. The sensitivity of the device to the defect concentration and the need for shallow junctions for the *p* and *n*-doped regions which form the detector contacts ensures the requirement for well-controlled and characterized ion implantation processes. Further, in an analogous process to that used in the electrical isolation of III-V devices, a combination of defect introduction and evolution via low temperature annealing is found to remove shallow defects which contribute to the dark current. To date, the best devices produce around 80μA of photocurrent for an on-chip optical power of 2.8mW from a broadband source with a central wavelength at 1570nm.

Keywords: Silicon, Defect Engineering, Detector, Photonics, Waveguide
PACS: 07.57.Kp, 61.72.Tt, 42.79.G

INTRODUCTION

Silicon photonics is rapidly emerging as an important technology for future use in microelectronics, telecommunications and even the chemical and biological sensor industries [1]. A great deal of research is currently in progress with the aim to develop a range of photonic functionality using standard CMOS processing. To date, a large range of silicon photonic devices has been demonstrated including low-loss waveguides, optical attenuators and (de)multiplexers, with a large on-going research effort aimed towards the development of efficient silicon-based optical sources [2].

The motivation for the use of silicon for the fabrication of planar optical devices (an indirect bandgap semiconductor and hence seemingly unsuitable as a substrate for photonics) is the potential for seamless integration with standard micro-electronic devices and processes. The ability to call upon the many decades of research and manufacturing experience invested in the silicon-based micro-fabrication industry should then yield significant cost-savings in a high-volume manufacturing environment. There also exists the tantalizing prospect for the introduction of photonics as a solution to ITRS roadblocks, the most discussed such application being optical interconnects.

As a ubiquitous process technology, ion implantation has a significant role to play in the field of silicon photonics. It has already been shown as an enabling technology in the development of the first optically-pumped, silicon waveguide, continuous Raman laser [3], and is key to a number of attempts to coax light from silicon-based light emitting diodes [4].

In this paper we present work on the development of silicon photo-detectors, monolithically integrated with optical waveguides. Sensitivity to sub-bandgap light at wavelengths around the important optical communication window at 1570nm is induced through the use of mid-gap defect levels, introduced through self-ion irradiation. The devices are fabricated using current standard processing technology and are fully compatible with silicon waveguide technology and integrated operational amplifier circuits.

FABRICATION OF OPTICAL WAVEGUIDE DETECTORS

Optical Waveguides

To produce an optical waveguide one requires a variation in refractive index (R.I.) to provide confinement based on total internal reflection. Silicon-on-insulator (SOI) is a natural candidate for the fabrication of waveguides. Strong optical confinement is observed in the vertical dimension due to the large difference in R.I. of silicon, SiO_2 and air for input wavelengths around 1570nm (~3.5, 1.45 and 1 respectively). A symmetric waveguide may be produced through straightforward oxidation of an SOI substrate surface. For practical application, optical confinement in the lateral direction may be provided with a rib structure (fabricated using a dry or wet etch process), producing a small difference in the effective index between the rib and the adjacent slab regions of the SOI overlayer.

In this work, low-loss optical waveguides in the rib geometry were fabricated using a SOI wafer consisting of a low-doped (<$10^{15}cm^{-3}$), 5μm thick overlayer on a 1μm thick BOX. The rib widths, nominally 3μm at the base, and the rib heights, measured using a surface profilometer to be close to 1μm, ensured that all the waveguides would be single mode.

Inducing Optical Absorption in Ion Implanted Waveguides

The bandgap of silicon effectively renders it transparent for wavelengths >1.1μm. This is ideal then for guiding IR light at wavelengths around 1570nm, however, it provides significant challenges for the monolithic integration of detectors at these sub-bandgap wavelengths.

It is well-known that when crystalline silicon is irradiated with energetic ions mobile defects such as interstitials and vacancies are produced. At room temperature, most of these defects rapidly reincorporate into the crystal structure. However, some combine to form complex, stable defects such as the silicon divacancy. A stable room temperature defect disrupts the normal energy band structure of the material imposed by the periodic crystal and introduces mid-bandgap levels. The photon energy required to excite an electron from such levels is lower than the energy required for valence to conduction band excitation, allowing the absorption of lower energy photons. For example, a prominent absorption band positioned at 1.8μm, has been conclusively associated with excitation of the silicon divacancy [5].

Importantly, attached to this absorption peak is a tail which extends to the silicon band absorption edge passing through the communication wavelengths around 1570nm.

For some defects (other than the divacancy) the excess absorption close to the band-edge is observed regardless of the presence of a specific absorption band; for example in the case of silicon which has been fully amorphized by high dose ion irradiation.

In order to utilize defect induced absorption for sub-bandgap detection, we have designed and fabricated a *p-i-n* diode for integration with the waveguide structure outlined above.

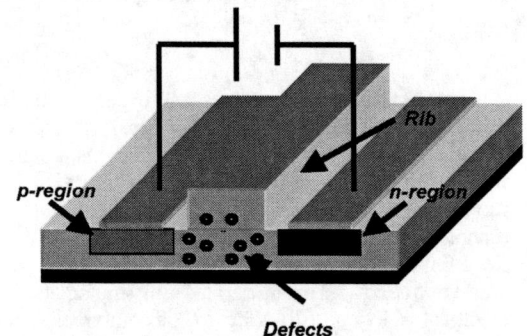

FIGURE 1. Schematic representation of the waveguide detector.

Using masked ion implantation, phosphorus (*n*-type) and boron (*p*-type) doped regions were symmetrically defined on either side of fifteen waveguide structures formed on a 1inch^2 sample. The separation between the doped regions (*x*) and the length of the doped regions (*l*) were varied for each waveguide. Following dopant activation, the samples were capped using a PECVD deposited SiO_2 layer of 200nm thickness, through which contact vias were subsequently defined before the deposition and selective lift-off of an aluminum metallization layer. The final photolithography step defined areas for inert silicon implantation that were centered on the rib, 50μm in width and of length coincident with the doped regions on either side of the waveguide. The silicon ion energy and dose were 1.5MeV (equivalent to a range of approximately 1.5μm in silicon) and $1 \times 10^{12} cm^{-2}$, respectively. One waveguide, completely masked from the silicon implantation, was subsequently used as a control device. Optical quality waveguide end facets were prepared using a Loadpoint Microace dicing saw. This procedure does not require subsequent polishing or post-dicing facet preparation of any kind and has been found to result in facets of quality comparable with those produced via cleaving or dicing and polishing. A schematic representation of the waveguide detector is shown in figure 1.

Following fabrication, the photo-detectors were characterized for optical loss and photoresponse to a broadband source (wavelength range 1530-1610nm). Measurements were repeated following isochronal annealing for 10 minutes up to 350°C in intervals of 50°C.

RESULTS

Electrical Characterization

For each waveguide, current-voltage (IV) characteristics were measured before and after silicon irradiation to ensure diode operation. Figure 2 shows a plot of IV for the device with x=16μm and l=6mm after silicon ion implantation and following a subsequent anneal at 300C for 10 minutes. Also shown is an IV curve for the control device. The control diode exhibits a well-defined turn-on at 0.6V and a forward resistance of approximately 30Ω. A significantly increased forward resistance of around 190Ω, with a greatly softened turn-on, is observed in the as-implanted sample. Following annealing, this resistance is reduced to approximately 90Ω. The comparative change in the IV curve following implantation may be attributed to the mechanism of Shockley-Read-Hall (SRH) recombination at the introduced defect centers, which we assume here to be dominated by point defects such as the silicon divacancy [6]. The decrease in resistance indicates the partial removal or evolution of these defects.

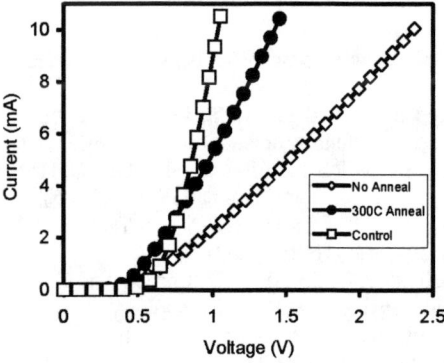

FIGURE 2. IV plot for a device with separation of doping regions x = 16μm and device length l = 6mm following 1.5MeV self-ion implantation to a dose of $1 \times 10^{12} cm^{-2}$ (open diamonds); implantation plus annealing for 10 minutes at 300°C (closed circles); and for an unimplanted control device (open squares).

Optical Characterization

Figure 3 shows a plot of unbiased photocurrent in response to the waveguide coupled, broadband optical source, observed for an as-implanted device (l=6mm and x=16μm); the same device following annealing at 300C; and the control diode. The broadband source output was fixed at 11.5dBm (equivalent to an on-chip power of approximately 1.5dBm), with the coupled optical power being varied by an external variable attenuator. The photocurrent is enhanced by more than an order of magnitude, up to a maximum of 5μA, in

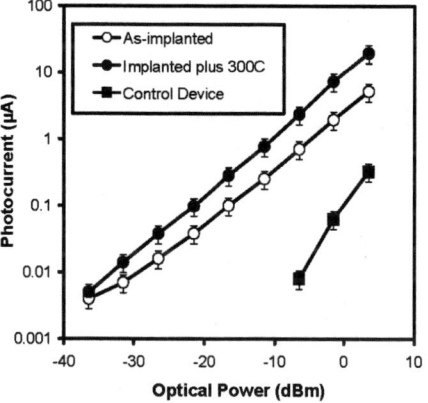

FIGURE 3. Unbiased photo-current versus on-chip optical power for device with separation of doping regions x = 16μm and device length l = 6mm following 1.5MeV self-ion implantation to a dose of $1 \times 10^{12} cm^{-2}$ (open circles); implantation plus annealing for 10 minutes at 300C (closed circles); and for an unimplanted control device (closed squares).

the silicon implanted sample compared to the small intrinsic photocurrent measured in the unimplanted sample. This enhancement is a direct result of carrier excitation by SRH generation via the defects introduced by the silicon implantation. The dynamic range of the detector covers four orders of magnitude, with a sensitivity limit less than −30dBm. The dark current of the device was determined to be <10nA.

A further increase in responsivity is seen for the sample implanted and then annealed at 300C for 10 minutes where the photocurrent increases to 19μA for the unattenuated input. This data set demonstrates the complex nature of defect engineering in silicon device fabrication. The annealing induced increase in photoresponse may be indicative of the competing mechanisms of carrier generation and deterioration of the electrical integrity of the photo-detector. Alternatively, the post-implantation annealing may benefit the device performance by removing carrier recombination centers which do not contribute to the carrier generation process.

The beneficial effects of post self-ion implantation annealing on the forward bias electrical characteristics of the integrated *p-i-n* diode are also observed when the device is reversed biased. For the device with $x=5\mu m$ and $l=6mm$, the dark current at a reverse bias of 5V is almost 4µA compared to a signal of 28µA for input light power of 1.5dBm, immediately following

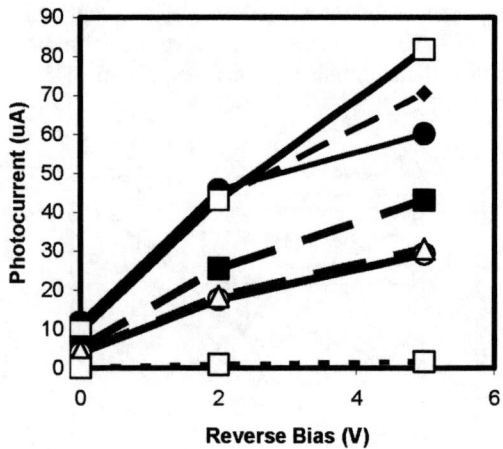

FIGURE 4 – Photocurrent versus reverse bias for on-chip signal of 1.5dBm, broadband source centered at 1570nm. Post-implantation annealing temperature 25C (open circles); 150C (open triangles); 200C (closed squares); 250C (open squares, solid line); 300C (closed diamonds); 350C (closed circles); 250C, dark current (open squares, dashed line).

ion implantation. After subsequent annealing at 250C, the equivalently measured photocurrent is increased to 80µA, while the dark current is reduced to <1.5µA. The complete evolution of photocurrent for a reversed bias of zero, 2V and 5V is shown in figure 4 for an input optical power of 1.5dBm.

SUMMARY

We have demonstrated operation of a SOI rib waveguide, *p-i-n* photo-detector which has response to sub-bandgap wavelengths around 1570nm. The photosensitivity of the device results from the introduction of defects via Si^+ ion implantation beneath the rib in a volume partially coincident with the guided optical mode. Following implantation of 1.5MeV ions to a dose of $1 \times 10^{12} cm^{-2}$, for a device of length 6mm an on-chip signal of 1.5dBm induces a photocurrent of 5µA. Post-implantation annealing at 300°C results in an increase of unbiased photocurrent to 19µA. The post-implantation annealing also reduces significantly the dark current observed under reverse bias. For instance, following annealing at 250C the photocurrent collected under a reverse bias of 5V approaches 80µA for on-chip power of 1.5dBm, while the dark current is only 1.5µA.

The fabrication of these detectors is completely compatible with current CMOS processing strategies. These results then indicate the potential of these devices for use as waveguide integrated, optical power monitors.

ACKNOWLEDGMENTS

The authors wish to thank Ian Mitchell and Jack Hendriks at Interface Science Western, University of Western Ontario, for assistance with the development of the implantation process. We also thank the staff members of the Centre for Emerging Device Technolgies at McMaster University for help with device fabrication.

This work is funded by a combination of sources – Ontario Photonics Consortium, Canadian Institute for Photonic Innovation, Ontario Centres of Excellence, and the National Science and Engineering Research Council of Canada.

REFERENCES

1. S Koehl and M Paniccia, *Photonics Spectra*, **39**, 53 (2005).
2. G. Reed and A. P. Knights, *Silicon Photonics-An Introduction,* (John Wiley & Sons, Chichester, 2004).
3. H. S. Rong, R. Jones, A. S. Liu, O. Cohen, D. Hak, A. Fang, and M. Paniccia, Nature, 433, 725 (2005).
4. L. Pavesi and D. J. Lockwood, *Silicon Photonics* (Springer-Verlag, Berlin, 2004).
5. L. J. Cheng, J. C. Corelli, J. W. Corbett, and G. D. Watkins, Phys. Rev. **152**, 761 (1966).
6. P. G. Coleman, C. P. Burrows, and A. P. Knights, Appl. Phys. Lett. **80**, 947 (2002).

Influence Of Energy Contamination At S/D-Extension Dopant Implantation Using Ultra Fast Annealing

Marc Herden, Daniel Gehre, Thomas Feudel, Lutz Herrmann

AMD Saxony LLC & Co. KG, Wilschdorfer Landstr. 101, 01109 Dresden, Germany

Abstract. Ion implantation at ultra low energies (<1keV) using deceleration of ions adjacent to the wafer surface appear to show energy contamination based on process parameters or design of the beamline of the implanter. Previously energy contamination was overwhelmed by diffusion of dopants when using soak or spike RTA processes. Use of ultra fast anneal techniques (e.g. flash/laser) for ultra shallow junction formation provides diffusion less activation of implanted dopants. Thus energy contamination is becomming a severe issue for precisely positioning pn-junctions. By use of deceleration modes, front of the wafer, fractions of the dose penetrate the wafer at non-decelerated energy and introduce undesired move of the junction depth. The fraction of decelerated to non-decelerated ions typically needs to be less than ~0.1% so that those amounts will not influence device parameters.

Keywords: Ion Implantation, ultra-fast annealing, energy contamination, USJ.
PACS: 61.72.Tt, 85.40.Ry

INTRODUCTION

Scaling of CMOS devices towards smaller geometrical dimensions to allow higher packaging densities also requires scaling of dopant distributions within the device. For upcomming technology generations of 45nm and smaller, the diffusion of dopants must be minimized during the annealing process. Where actual CMOS technologies use conventional spike RTA processing causing dopant diffusion, this amount of diffusion needs to be minimized for smaller device geometries. Scaling of dopant diffusion and gradients is linked to scaling of physical gate length. The projections for parameter scaling are collected in the ITRS roadmap requirements for extension junctions [1]. To minimize diffusion at conventional RTA processing anneal times and temperatures needs to be reduced. Since this is causing increased parasitic resistivities their applicability is no longer sufficient for high performance devices.

Advanced ultra fast annealing (UFA) techniques like laser or flash annealing provide formation of shallower junctions with higher activation and steeper junction abruptness due to reduced process time in millisecond range and increased actual anneal temperatures up to ~1400°C. By this, dopant diffusion is minimized or not apparent during annealing.

When using UFA processing placing of the dopants in the device is given by the actual ion implantation process where the location of the dopants is preserved after the diffusion-less UFA process. For S/D-extension dopant implantation typically energies of 5keV or less for As or 1keV or less for B are used for advanced CMOS technologies. To allow high productivity of the implantation process ion implanters may provide deceleration modes for ions. If the deceleration occurs in front of the wafer energy contamination shows up due to neutralization of ions in the deceleration region if energy filtering is not implemented. The amount of neutralized and non-decelerated ions depends on the design of the deceleration region, of apertures in the beam path, or of the vacuum performance. The neutralized ions cause dopants that are implanted deeper than the desired ones. By this the pn-junctions are moved into the depth of the substrate if the amount of energy contamination is higher than the substrate concentration. Laterally uncontrolled and non-repeatable shift of overlap occurs at gate edges.

Substrate dopant concentration control by optimized well and halo implants is mainly done for Vt adjustment and for overlap control. So the amount of energy contamination from S/D-extension implantation in decel modes needs to be lower than the well doping level. For spike RTA processing the amount of energy contamination was surpassed by dopant diffusion and pn-junction movement was not apparent. For ultra fast annealing where the pn-junctions are given by the implanted dopant profile and no diffusion occurs energy contamination must be less than the well doping. This must be true for all implantation parameters used in conjunction with UFA.

EXPERIMENTAL

Ion implantation of S/D-extension dopants was done in drift mode or in decel mode at various decel ratios. The implants were done on commercially available ion implanters. Ge or Xe pre-amorphized silicon substrates were used in order to avoid channeling effects. After ion implantation annealing was performed by ultra fast anneal processing, i.e. laser annealing or flash annealing. Those UFA tools provide anneal times at typically about 1 msec, and at anneal temperatures between 1100°C to 1350°C.

RESULTS AND DISCUSSION

In figure 1 the simulated dopant distribution of a CMOS device using spike RTA processing is shown. In this process the dopant distribution was given by the RTA process showing graded junction profiles. In the insert of figure 1 a similar device structure is shown using ultra fast annealing which shows diffusion free dopant distributions [2]. This process shows much steeper junction dopant profiles and allows further gate length scaling to improve CMOS transistor performance.

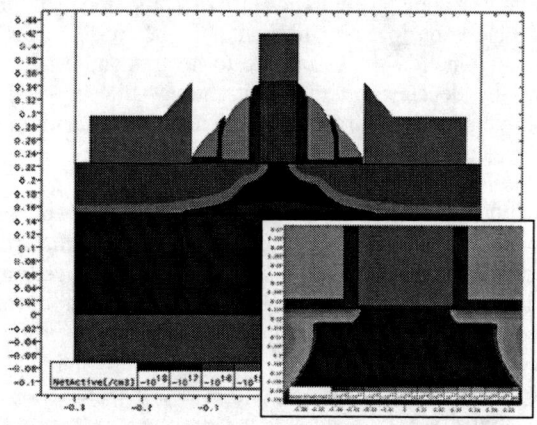

FIGURE 1. Simulated device structure of an advanced CMOS transistor using RTA processing or using ultra fast annealing (insert).

In figure 2 dopant profiles of a B implantation in decel-mode are shown after implantation and after a spike RTA process. The amount of energy contamination in the tail of the dopant profile was surpassed by dopant diffusion during RTA processing. In figure 3 As dopant profiles after implantation in 3:1 decel-mode are shown having non-detectable energy contamination. After UFA processing the as-implanted dopant profile is preserved, showing no additional diffusion during the UFA process.

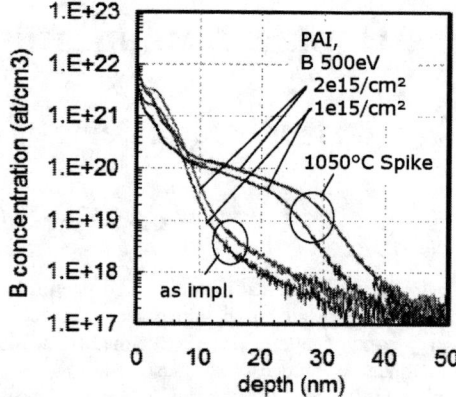

FIGURE 2. SIMS depth profiles of B extension implantation in decel-mode with apparent energy contamination and after RTA processing.

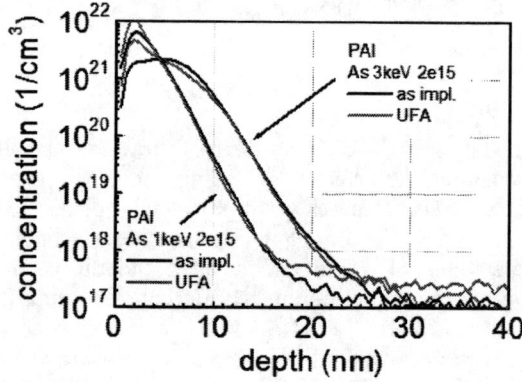

FIGURE 3. SIMS depth profiles of As extension implantation in 3:1 decel-mode showing non apparent energy contamination and after ultra fast annealing.

Because UFA processing showing no diffusion, this process can be applied to allow reducing physical gate length and allow smaller devices for higher packaging densities in future technology generations [2]. Roll-off characteristics from simulations, shown in figure 4, suggest applicability of the UFA process down to 25nm gate length or less.

Application of UFA techniques also offers higher thermal activation of dopants. Thus higher implant doses can be used to reduce parasitic resistances in CMOS devices. In figure 5 sheet resistivities for various B implant conditions are shown using UFA processing. In figure 6 sheet resistivities for As implantations are shown. Activation of extension implant doses of up to $2*10^{15}/cm^2$ for B, or $4*10^{15}/cm^2$

for As is possible without reaching a temperature limited activation level.

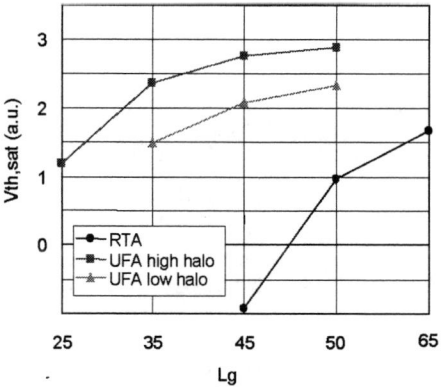

FIGURE 4. Simulated Vt roll-off characteristics of a CMOS transistor for 45nm technology nodes using diffusion less ultra fast annealing.

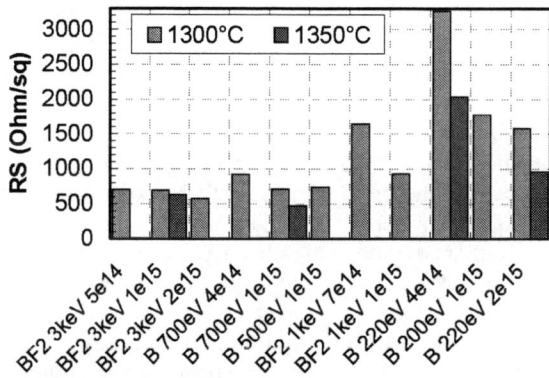

FIGURE 5. Sheet resistance of B or BF2 implanted S/D-extension dopant profiles into pre-amorphized substrates after ultra fast annealing.

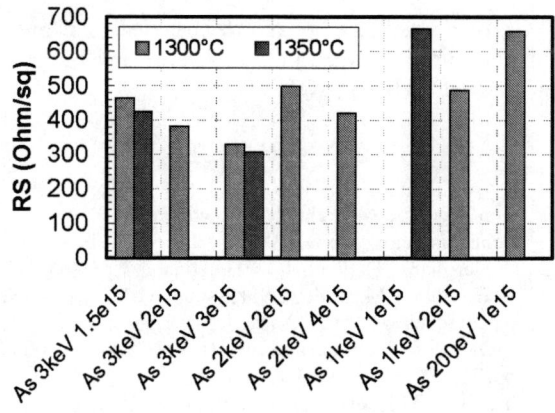

FIGURE 6. Sheet resistance of As implanted S/D-extension dopant profiles into pre-amorphized substrates after ultra fast annealing.

To allow introduction of that amount of implant doses into an advanced CMOS technology the amount of energy contamination has to be lower than the well dopant concentration. In case of the simulated device structure the integrated amount of dose as a result of energy contamination must be below $1*10^{13}/cm^2$ for As, or $5*10^{12}/cm^2$ for B when 3:1 decel-modes are used and which represent about 0.25% energy contamination.

In figure 7, B dopant profiles from SIMS measurements after implantation in decel-mode are shown having noticable energy contamination together with profiles after ultra-fast annealing. For well doping levels of about $10^{18}/cm^3$ a movement of the pn-junction of about 5nm into the depth can be expected. Figure 8 shows As dopant profiles with energy contamination above a $10^{18}/cm^3$ level. There, also the amount of energy contamination leads to a move of the pn-junction of about 5nm into the depth.

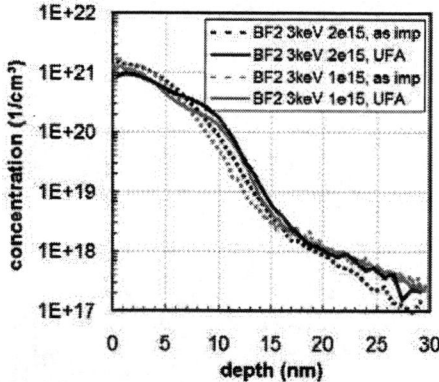

FIGURE 7. SIMS depth profiles of BF2 implanted extension doping in 3:1 decel-mode with 1% energy contamination and after ultra fast annealing.

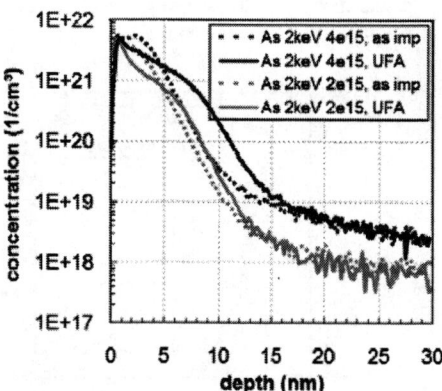

FIGURE 8. SIMS depth profiles of As implanted extension doping in 4:1 decel-mode with 1% energy contamination and after ultra fast annealing.

Simulations of various amounts of energy contamination at extension dopant implantation are shown in figure 9 and 10 relative to various well dopant levels [3]. For low well dopant levels and about 1% energy contamination a move of the pn-junction of about 8nm occurs. With respect to a 25nm physical gate length process this causes a 60% reduction in electrical gate length and causes severe short channel degradation. Additionally parasitic junction capacitances increase. To avoid short channel effects and controlled Vt roll-off characteristics variations of the electrical gate length of about 5% must be ensured. Simulated Vt degradation for various amounts of energy contamination are shown in figure 11 of the device in the insert of figure 1. For 0.1% energy contamination about 10-20 mV Vt shift show up. Where the effect is more pronounced when using shallower extension profiles or higher extension dopant doses. For this an ion implantation process in decel-mode must provide a level of energy contamination that is less than 0.1% of the nominal implant dose, and that is always lower than the well dopant concentration at the pn-junction.

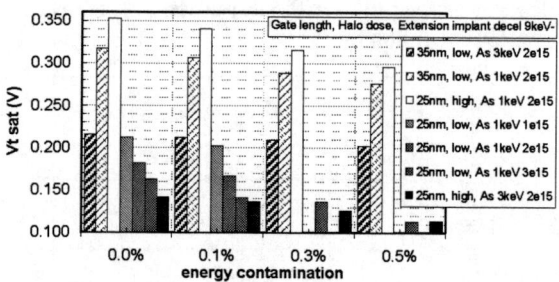

FIGURE 11. Simulated n-type extension dopant implantion in 3:1 decel-modes with varying amount of energy contamination after ultra fast annealing.

CONCLUSION

By using ultra fast anneal techniques for S/D-extension dopant annealing the as-implanted dopant profile is preserved during annealing due to anneal times in millisecond range. To avoid undesired movement of pn-junction depth occuring from uncontrolled energy contamination during implantation the amount of energy contamination must not be higher than 0.1% of the nominal dose of the actual extension implant. The design of an implanter using deceleration in front of the wafer must be adequate to apply this mode into repeateable high performance device manufacturing at 45nm technology nodes or smaller. Implanters for this application have to guarantee less than 0.1% energy contamination for all conditions, i.e. varying vacuum performance due to actual pump conditions, varying beam setups at all times in volume production modes.

ACKNOWLEDGMENTS

This work was funded by the German Federal Ministry of Education and Research, registered under funding number 01M3167B (SWITCH).

REFERENCES

1. The International Technology Roadmap for Semiconductors, www.itrs.org.
2. T. Herrmann, Th. Feudel, M. Horstmann, J. Hoentschel, L. Herrmann, M.Herden, W. Klix and R. Stenzel, „Novel Approaches to Improve Laser Annealed SOI-MOSFETs", Mat. Sci. and Eng. B, Vol. 124-125, 2005, pp. 223-227
3. T.Feudel, M.Horstmann, M.Gerhardt, M.Herden, L.Herrmann, D.Gehre, Ch.Krueger, D.Greenlaw and M.Raab, Temperature Scaling for 35 nm Gate Length High-Performance CMOS, Mat. Sci. in Semicond. Processing, Vol. 7, Issues 4-6, 2004, pp. 369-374

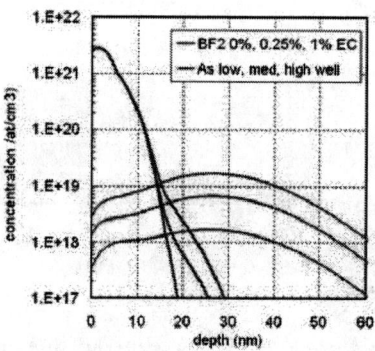

FIGURE 9. Simulated p-type extension dopant implantion in 3:1 decel-modes with varying amount of energy contamination after ultra fast annealing.

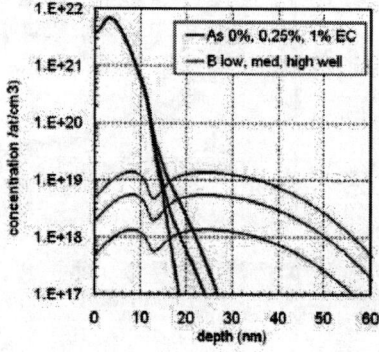

FIGURE 10. Simulated n-type extension dopant implantion in 3:1 decel-modes with varying amount of energy contamination after ultra fast annealing.

Formation of ultrashallow junctions less than 10nm with the combination of low energy B ion implantation and laser annealing

Ryuta Yamada [1], Singo Seto [1], Sosi Sato [1], Yuki Tanaka [1], Satoru Matumoto [1], Toshiharu Suzuki [2], Gensyu Fuse [2], Tosio Kudo [3], Susumu Sakuragi [3]

1 Keio University, 3-14-1 Hiyosi, Kouhoku,Yokohama, 223-8522, Japan
2 SEN Corporation an SHI and Axcelis Company SBS Tower9F, 10-1 Yoga 4Chome, Setagaya-ku, Tokyo 158-0097 Japan
3 Sumitomo Heavy Industries Ltd., 19 Natsushima-cho, Yokosuka-shi, Kanagawa 237-8555, Japan

Abstract. Formation of ultrashallow active junctions will be required in keeping the ongoing miniaturization of ULSIs. We reported the formation of p^+/n ultrashallow junctions less than 10nm with the combination of low energy B ion implantation and non-melt laser annealing. First a Ge pre-amorphization implant was performed at energies of 3 keV, 6 keV with a dose of $3E14$ /cm^2. After the pre-amorphization implant, a B implant was performed at 0.2 keV and 0.3keV with doses of $8E14$ /cm^2 and $1.2E15$ /cm^2. Double-pulsed laser annealing was adopted at annealing process. B depth profiles are measured with SIMS analysis. In case of 0.3 keV B I/I with 3 keV Ge pre-amorphization, ultrashallow junctions less than 10nm was successfully formed with the double-pulse laser irradiation of 760 mJ/cm^2. A reasonable sheet resistance of ~550 Ω/□ was obtained in such a ultrashallow junction.

Keywords: ultrashallow junction
PACS: 71.55.Cn, 61.72.Tt

INTRODUCTION

While LSIs have been getting smaller [1], short channel effect (SCE), i.e., increase of off current, lowering of threshold voltage and so on, has become more serious, especially in p-channel MOS transistor. Ultrashallow junction (USJ) is considered to be one of the effective technologies to suppress the SCE [2-4]. In this article, we report the formation of USJ with the combination of Ge pre-amorphization, low energy B ion implantation and non-melt laser annealing.

EXPERIMENTAL DETAILS

First a Ge pre-amorphization implant was performed at energies of 3 keV and 6 keV with a dose of $3E14$ /cm^2, respectively. After the pre-amorphization implant, a B implant was performed at 0.2 keV and 0.3 keV with doses of $8E14$ /cm^2 and $1.2E15$ /cm^2. All implants were into n-type Si (100) wafers with more than 10 Ωcm. The implantation conditions and sample numbers are summarized in Table 1 for the laser irradiation of 300 ns interval time of double pulse.

For the activation of implanted boron, double pulsed green laser with the same energy density was used in order to slow the cooling rate and to enhance the re-crystallization of amorphized layer. Line shaped laser beam with the size of 0.1 mm × 17 mm was used. Pulse width of the laser was ~10 nsec and irradiation frequency was 1 kHz. The interval of the 1st pulse and the 2nd pulse was varied at 0 nsec, 300 nsec and 600 nsec. For the 0 nsec case two laser beams were irradiated simultaneously. Energy density of both pulses was varied from 560 mJ/cm^2 to 940 mJ/cm^2.

Sheet resistance was measured by a four points probe method and B depth profiles were measured with SIMS analysis. In this SIMS analysis, primary ions were O_2^+, and its energy was 350 eV. The surface morphology of annealed wafers was evaluated by AFM to confirm whether the surface melts or not. Additionally, for measuring the carrier density profile by differencial Hall measurement, we attempted the removal of thin Si layer by the use of O_3 water oxidation, as it will be explained in the latter section of Results and Discussion.

			Ge	
			3e14	
			3keV	6keV
B	0.3keV	8e14	1	5
		1.2e15	2	6
	0.2keV	8e14	3	7
		1.2e15	4	8

Table 1. Implantation conditions and sample number

RESULTS AND DISCUSSION

-1 SIMS profiles

Figures 1 and 2 show SIMS profiles of sample 2 (B:0.3 keV, 1.2E15 /cm^2, Ge:3 keV), and 4 (B:0.2 keV, 1.2E15 /cm^2, Ge:3 keV), respectively. These samples were annealed at an energy density of 760 mJ/cm^2 with 300ns of delay time. It is clearly shown that ultrashallow junction less than 10 nm is formed for sample 4. B SIMS profile of sample 2 does not change from that of the as-implanted, though slight movement of B profile is observed in sample 4. Both samples were annealed with the same laser energy density. There is a possibility that boron atoms diffused in amorphous Si before recrystallization in sample 4 because the as-implanted profile is shallower than amorphized thickness. This must be clarified more in detail.

As far as the junction depth is concerned, ultrashallow junction less than 10nm can be realized as shown in these figures. However, it is important to discuss whether the doped layer is active or not. In the latter section, we tried to perform the electrical estimation in the very shallow doped layer.

-2 Sheet resistance

Figure 3, 4 show sheet resistance as a function of laser energy density after the laser annealing for samples of 1 to 8 in Table 1 (interval, 300 ns). In the case of Ge implantation energy of 6 keV, sheet resistance was very low (100 Ω/□) at the energies lower than 740 mJ/cm^2 and it increased rapidly at 760 mJ/cm^2 and then decreased as the laser energy increased. When the laser energy was low, crystalline regrowth was not enough and the measured sheet resistance seems to be that of the bulk Si. In fact, sheet resistance measured on the back side of the wafer, which was considered to be same as that of the bulk Si, 105 Ω/□.

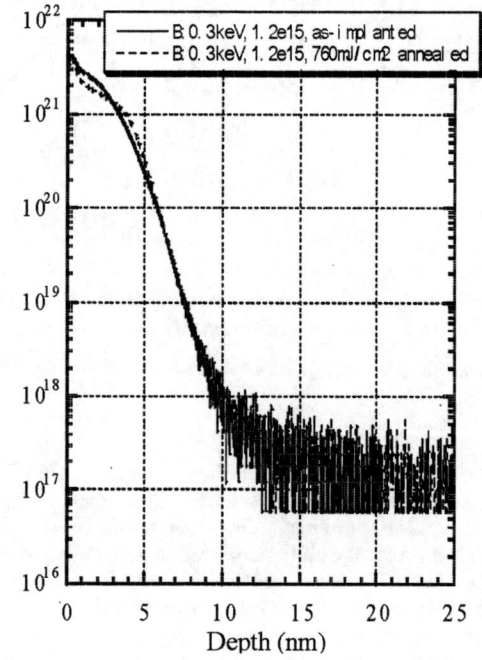

Fig.1 SIMS profile of sample 2 annealed with the energy density of 760mJ/cm^2

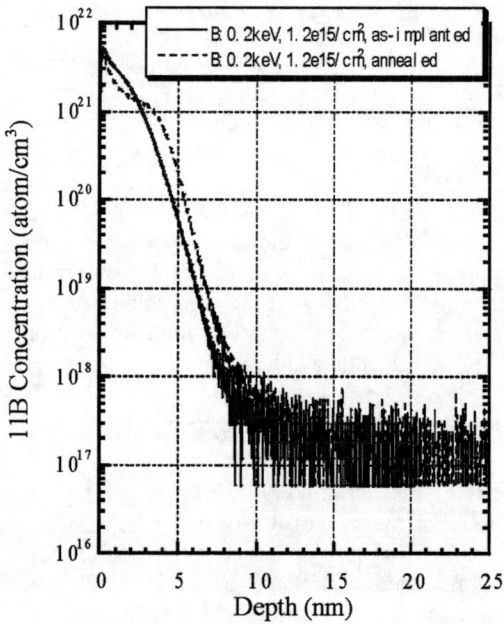

Fig.2 SIMS profile of sample 4 annealed with the energy density of 760mJ/cm^2

On the other hand, in the case of Ge implantation energy of 3 keV, sheet resistance decreases gradually with the increase of laser energy density. If the laser energy density is much lower than 720 mJ/cm^2, similar situation of incomplete regrowth will be observed. Difference in sheet resistance behavior may be related with the difference in pre-amorphous depth. It is considered that much higher energy density is required to regrow the deeper amorphous regions. However, it is necessary to analyze its complex behavior in Fig. 3 with a model based on non-equilibrium thermal conduction.

When the interval was 0 ns, that is, the 1st pulse laser and the 2nd pulse laser are irradiated at the same time, surface temperature becomes higher as compared with that of interval of 300 ns and 600 ns. Therefore lower energy density is enough to regrow in 0 nsec.

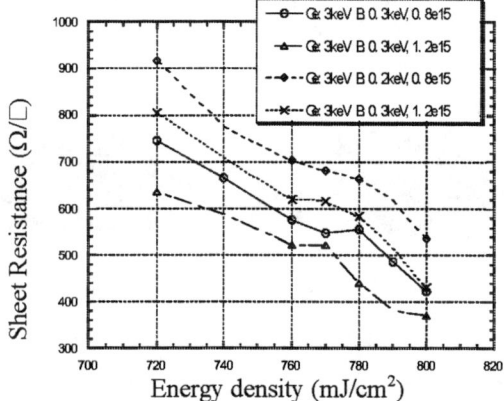

Fig.3 Sheet resistance vs. energy densities with the delay time of 300 ns. (Ge: 3 keV)

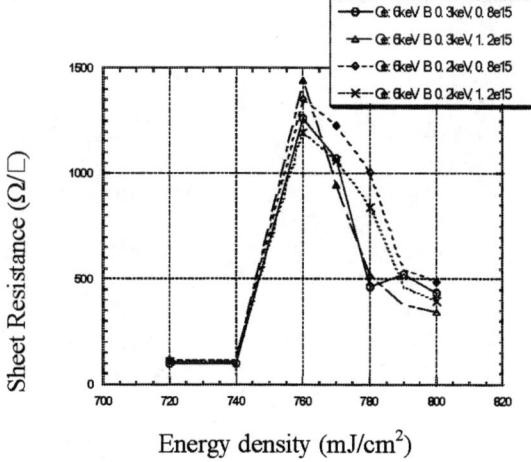

Fig.4 Sheet resistance vs. energy densities with the delay time of 300 ns. (Ge: 6 keV)

-3 AMF analysis

Figures 5 (a), (b) and 6 (a), (b) show the result of AFM analysis of sample 1 annealed with the energy density of 720, 790 mJ/cm^2, respectively. Figure 5 (a) and 6(a) are two-dimensional images and Figure 5 (b) and 6(b) are three-dimensional images. While surface morphology is smooth in case of lower energy (720 mJ/cm^2), the surface of the wafer annealed with high energy density was very rough as shown in Fig.5. Such a rough surface is considered to be due to the melt of the surface. Surface roughness becomes more severe in this wafer with the energy density more than 790 mJ/cm^2. Thus we judge the criterion of non-melt or melt from the surface roughness obtained from AFM images. Therefore it is noted that the values shown in Table 2 are not rigorous ones.

Table 2 shows the laser energy densities of non-melting states for various wafers. As shown in Table 2, in the case of sample 6 (B:0.3 keV, 1.2E15 /cm^2, Ge:6 keV), the melt energy was low in comparison with sample 2 (B:0.3 keV, 1.2E15 /cm^2, Ge:3 keV). This is because amorphous region of sample 6 was deeper than sample 2 and laser absorption is higher in amorphous region than in crystal region.

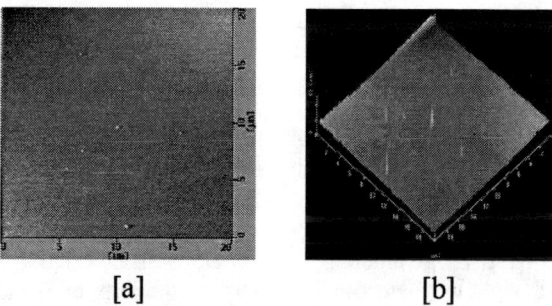

[a]   [b]

Fig.5 AFM analysis for the sample annealed with the energy density of 720mJ/cm^2 ([a]:two-dimensional image [b]:three-dimensional image)

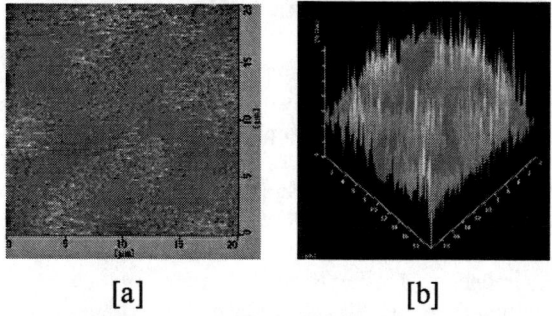

[a]   [b]

Fig.6 AFM analysis for the sample annealed with the energy density of 790mJ/cm^2 ([a]:two-dimensional image [b]:three-dimensional image)

When wafer surface melts, enhanced diffusion of B occurs and junction depth becomes very deep. To form ultrashallow junction, it is necessary to choise non-melt energy density.

Sample number / Delay time	0ns	300ns	600ns
2	640mJ/cm^2	760mJ/cm^2	
6	620mJ/cm^2	740mJ/cm^2	800mJ/cm^2

Table.2 Comparison of melting energy density.

-4 Differential Hall measurement

Figure 7 shows oxide film thickness gained in the O_3 water as a function of dip times in it. It is found that the growth of the oxide film was saturated after about 60sec and the thickness was about 14-15 Å. We also studied thickness of removed Si with step measurement after HF solution etching. Figure 8 shows the image of step measurement, schematically. In this experiment, we found that the thickness of removed Si at one O_3 water oxidation process was about 4.8 Å. Therefore it was possible to measure the electrical characteristic of the USJ of 10nm with the repetitive use of O_3 water oxidation as differential Hall measurement.

Carrier density of the top surface of sample 2 obtained by the differential Hall measurement is about 5×10^{20} /cm^3. This value is different largely from the atomic boron density appeared in the SIMS profile in Fig. 1. Large amount of boron atoms in this region is thought to be electrically inactive in the form of cluster or others. The state of the boron atoms in this region must be clarified precisely in the next step.

CONCLUSION

We succeeded in forming ultrashallow junction less than 10nm and of very low resistivity with the combination of Ge pre-amorphization implantation (3 keV, dose of 3E14 /cm^2), low energy B implantation (0.3 keV, dose of 1.2E15/cm^2) and double pulsed non-melt laser annealing. In this case, low sheet resistance of ~550 Ω/□ was obtained. We also determined the threshold laser energy density between a non- melting and a melting state. Further we performed Si oxidation in O_3 water and its thickness of Si removal at one process was about 4.8Å.

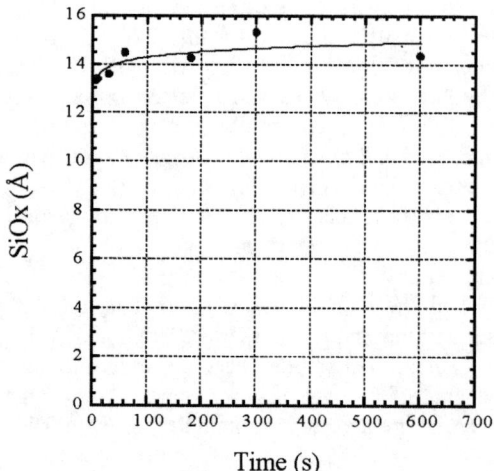

Fig.7 Thickness of oxide formed by O_3 water oxidation

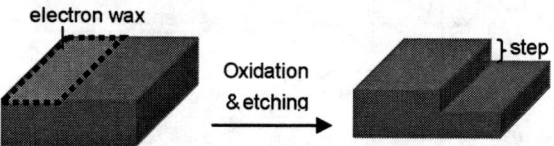

Fig.8 Image of step measurement

REFERENCES

[1] H.Dennard, F.H.Gaensslen, H-N Yu, V.L.Rideout, E.Bassous, A.R.Leblanc, *IEEEJ. Solid-State Circuits*, SC-9, 256 (1974)

[2] T. Kuroi, Proc. of Workshop on USJT, p.4 (2005)

[3] H. W. Kennel, P. H. Keys, M. Armstrong, A. Budrevich, M. D. Giles, M. Liu, Proc. of Workshop on USJT, p.211 (2005)

[4] T. Yamamoto, T. Kubo, T. Sukegawa, Y. Wang, L. Feng, S. Talwar, M. kase, Ext. Abs. The 5th Workshop on Junction Technology, S1-4 (2005)

[5] K. Yamazaki et al., AM-LCD2002 Digest, p.175 (2000)

[6] T. Kudo, S. Sakuragi, and K. Yamazaki, Mat. Res. Soc. Symp. Proc. Vol. 862 (2005) p.269-274

Impact of Fluorine co-implant on Boron Diffusion during Non-melt Laser Annealing

T. Noda[1,2], S. Felch[3], V. Parihar[3], C. Vrancken[4], T. Janssens[4], H. Bender[4], B. Van Daele[4], and W. Vandervorst[4]

[1]*Matsushita Electric Industrial Co., Ltd.,* [2]*Matsushita assignee at IMEC, Kapeldreef 75, B-3001, Leuven, Belgium*
[3]*Applied Materials, 974 E. Arques Ave., M/S 81280, Sunnyvale, CA 94085, USA*
[4]*IMEC, Kapeldreef 75, B-3001, Leuven, Belgium*

Abstract. The impact of F co-implant on B diffusion during Non-melt Laser Annealing (NLA) is investigated. The wafers were implanted with Ge PAI at 30 keV, $1\times10^{15}/cm^2$ and F was implanted at energies ranging from 0.89 keV to 20 keV, and then B was implanted at 0.5 keV, $1\times10^{15}/cm^2$. After the implants, the wafers were annealed by sub-ms NLA. In the presence of F, the B diffusion length during NLA is dramatically increased.

Keywords: Transient enhanced diffusion, Non-melt Laser annealing, Millisecond annealing, Fluorine, co-implant
PACS: 61.72.Tt, 66.30.Jt, 62.72.Cc

INTRODUCTION

For the formation of ultra shallow junctions for the 45 nm CMOS technology node and beyond, sub-millisecond (ms) non-melt laser annealing (NLA) technology is a promising diffusion-less annealing candidate.[1-6] Boron diffusion behavior during the laser annealing process is a 2-step diffusion (SPER + Laser). In the viewpoint of X_j/R_s trade off, ms annealing shows a large advance in comparison to spike-RTA. But it has been shown that the thermal budget is too small for full defect evolution.[6]

F co-implant is often used for the control of boron transient enhanced diffusion (TED).[7-10] It is also known that F can reduce the solid phase epitaxial regrowth (SPER) velocity.[11] Recent study of F co-implant with spike-RTA shows the reduction of B TED and B deactivation.[12,13] For spike-RTA, deep F co-implant is effective because it can stop the back flow of interstitials from end-of-range (EOR) defects. It is also noted that F can affect the EOR defect formation.[13,14]

In this work, the F co-implant impact on boron diffusion during NLA is shown.

EXPERIMENTS

The n-type, 200 mm, (100) wafers were implanted with Ge at 30 keV, $1\times10^{15}/cm^2$ for pre-amorphization implant (PAI) and F co-implant at F implant energies ranging between 0.89 keV and 20 keV, at a dose of $2\times10^{15}/cm^2$. The Fluorine co-implant energy was varied in order to obtain different fluorine distributions around the original amorphous/crystalline (a/c) interface. Subsequently boron was implanted at 0.5 keV, $1\times10^{15}/cm^2$. After the implants, the wafers were annealed by non-melt laser annealing at a peak temperature of 1100 °C or 1300 °C.

Secondary ion mass spectrometry (SIMS) measurements with a quadruple mass spectrometer and 4 point probe sheet resistance measurements were used for the analysis of boron diffusion and activation. Transmission microscopy microscopy (TEM) measurement was also used for the observation of the amorphous layer depth and defects.

RESULTS AND DISCUSSIONS

Fig. 1 shows XTEM images for two as-implanted samples. F co-implant at 0.89 keV shows an amorphous layer depth of 46 nm. This is the same depth shown by the No F co-implant sample. So, F co-implant with lower implant energy does not move the original a/c interface to a deeper position. In the case of low F co-implant energy, the depth of the original a/c interface is determined by the Ge PAI. In the case of deep F co-implant energy, F can move the original a/c interface deeper than 46 nm.

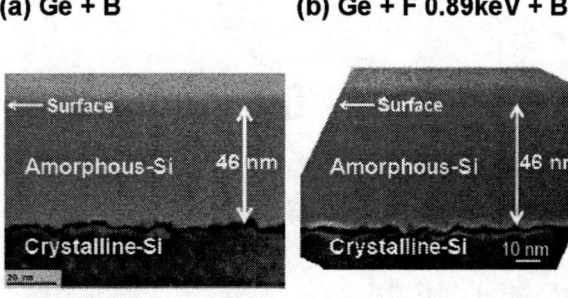

FIGURE 1. The XTEM images of as-implanted samples, (a) Ge 30 keV PAI + B 0.5 keV, (b) Ge 30 keV PAI + F 0.89 keV + B 0.5 keV, respectively.

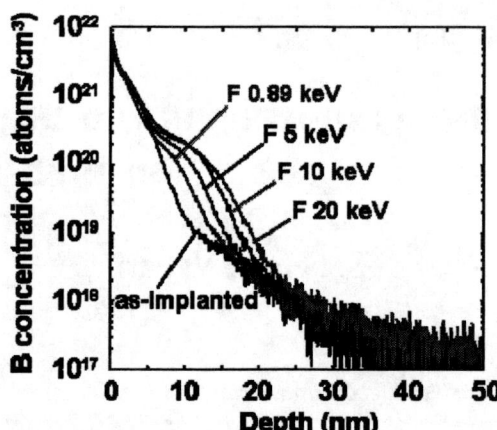

FIGURE 3. Boron SIMS profiles after NLA at 1100 °C as a function of F co-implant energy.

Fig. 2 shows the simulation of F as-implanted profiles with 30 keV Ge PAI. The Monte Carlo (MC) model can take into account the damage accumulation. Therefore MC model can give good predictions of the F implanted profiles with Ge PAI. The F co-implant energy was chosen so as to obtain different fluorine distributions around the original a/c interface. The lowest F co-implant energy (0.89 keV) has R_p close to that of the 0.5 keV B implant. So, this condition creates a similar situation as BF_2 implantation. In the case of 5 keV F, almost all of the implanted fluorine is located inside amorphous-Si (aSi). 10 keV F implant creates a F as-implanted profile beyond the original a/c interface. In the case of 20 keV F, the R_p of fluorine is close to 46 nm, and most of the fluorine is located inside crystalline-Si (cSi). The original a/c interface is also moved deeper than that of 30 keV Ge PAI. Therefore it is considered that the situation is more complicated for the 20 keV F co-implant case.

It is known that dopant in amorphous-Si changes the Solid Phase Epitaxial Regrowth (SPER) velocity.

For instance, boron increases the SPER velocity, while fluorine slows it down. In this work, both boron and fluorine are implanted into an aSi layer. Therefore the consideration of SPER velocity change due to the dopants is important for the understanding of boron diffusion during sub-ms annealing with a pre-amorphizing layer.

It is known that boron diffusion during NLA is very limited, and increasing the laser peak temperature does not increase the junction depth very much.[6] We determined the B diffusivity in aSi in this work. The plot in fig. 5 shows that this B diffusivity is smaller than that in cSi.[15]

Fig. 3 shows that boron SIMS profiles after NLA with F co-implant. As the F co-implant energy increases, the boron diffuses deeper. In the presence of fluorine, the B diffusion length during sub-ms NLA is dramatically increased.

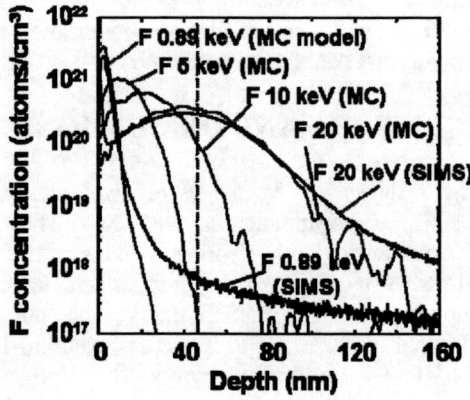

FIGURE 2. Monte Carlo implant simulation results of fluorine as-implanted profiles with 30 keV Ge PAI. F SIMS profiles are also shown in this figure.

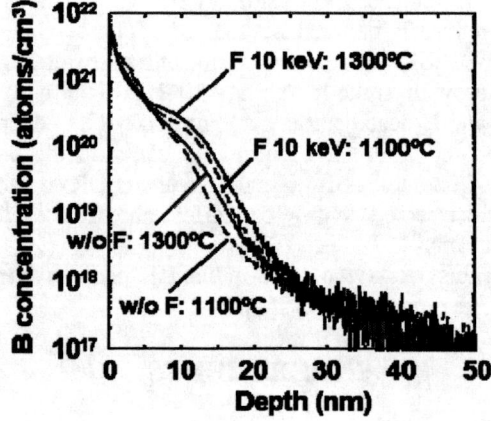

FIGURE 4. Boron SIMS profile comparison between implants with F co-implant and without F co-implant.

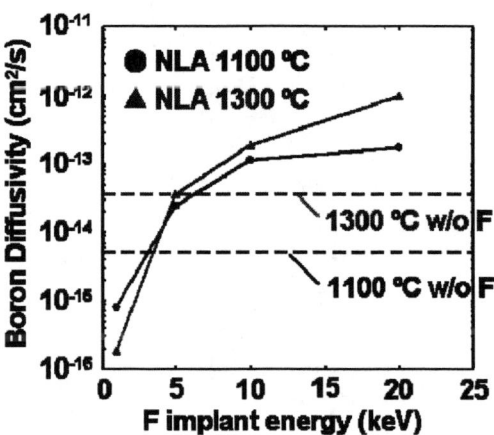

FIGURE 5. Boron diffusivity during sub-ms NLA as a function of F co-implant. The Broken line shows the B diffusivity during NLA without F co-implant.

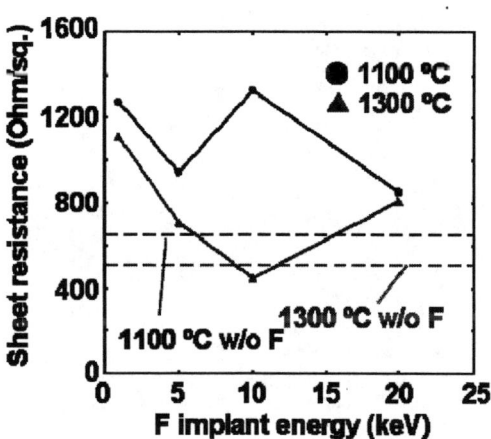

FIGURE 6. Sheet resistance after sub-ms NLA as a function of F co-implant. Broken lines show the sheet resistance after NLA without F co-implant.

Fig. 4 shows a SIMS profile comparison between implants with F co-implant and without F co-implant. F co-implant raises the B shoulder position up to 4×10^{20} atoms/cm^3 ~ 5×10^{20} atoms/cm^3 and creates a more box-like shaped B profile. The junction depth at a B concentration of 1×10^{18} atoms/cm^3 does not change much as a function of F energy. Fig. 5 shows the B diffusivity enhancement as a function of F co-implant energy. In this figure, the broken line shows the B diffusivity during NLA without F co-implant. B diffusivity enhancement is observed for F co-implant energies higher than 5 keV. Here, the time averaged diffusivity is the combination of the diffusivity in aSi and that in cSi. It is also known that fluorine can slow down the SPER velocity. Nevertheless the portion of the sub-ms anneal, when SPER is completed and boron diffuses in cSi is also important.

Fig. 6 shows the sheet resistance as a function of F co-implant energy. F co-implant with NLA increases the sheet resistance. The lowest F implant energy (0.89 keV) shows a significant increase of sheet resistance. The chemical interaction between fluorine and boron is one possible explanation for this increase of sheet resistance. The data for NLA at 1300 °C show that sheet resistance decreases as a function of F co-implant energy up to 10 keV and then increases again at a F energy of 20 keV. Regarding the 20 keV F implant, it must be noted that the depth of the aSi layer is changed, and a more complicated situation occurs for the boron diffusion/activation process. From this plot, the optimum F implant energy seems to be around 10 keV. The fluorine SIMS profile indicates that after sub-ms annealing, many F atoms still remain in the Si-substrate with no significant dose loss of fluorine.

The sample with the lowest F co-implant energy (0.89 keV) shows that the fluorine profile almost completely overlaps boron profile.

Fig. 7 shows XTEM images for EOR defects with F co-implant at energies of 0.89 keV and 5 keV. XTEM shows that EOR defects are formed clearly below the original a/c interface. The depth of the EOR defects is almost identical for the two samples. This indicates that 5 keV F cannot push the original a/c interface formed by the 30 keV Ge PAI. Regarding the defect evolution, previous work with spike-RTA showed that the additional F co-implant greatly increases the defect density at the EOR region.[13] In the case of our sub-ms NLA, for 0.89 keV and 5 keV F implant conditions, the defect size is not very different from that without F co-implant.

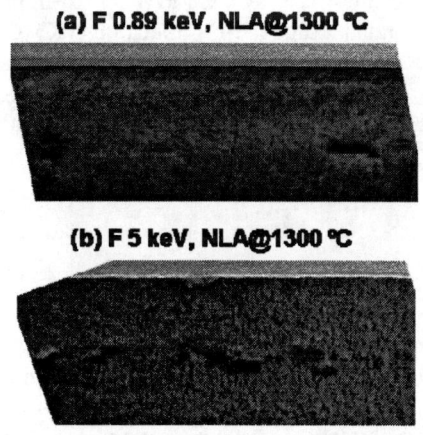

FIGURE 7. XTEM images of samples annealed by NLA at 1300 °C with F co-implant energies at (a) 0.89 keV, and (b) 5 keV, respectively.

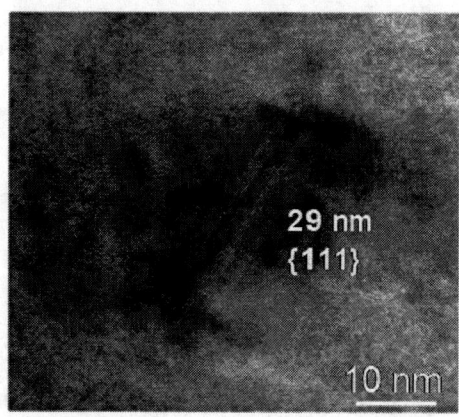

FIGURE 8. XTEM images of EOR defects on the {111} plane.

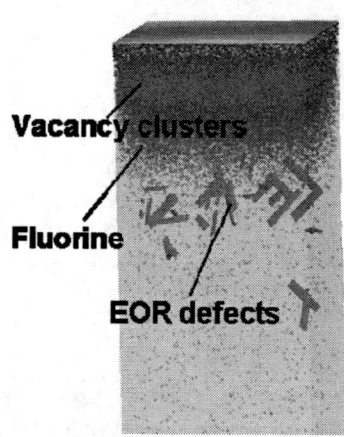

FIGURE 9. An atomistic KMC simulation result for Ge PAI + F co-implant + 0.5 keV B implant and annealing with sub-ms NLA.

However, in the case of F co-implant, dislocation loops on the {111} plane are more clearly seen than those of without F co-implant. Fig. 8 shows a XTEM image for the EOR defects on the {111} plane, which correspond to the dislocation loops. It is known that fluorine can form fluorine-vacancy complexes. The Fluorine-Vacancy interaction is more stable than fluorine-boron or fluorine-interstitial interactions.[16] During SPER, vacancy small clusters (V_m) are formed, and then fluorine-vacancy (F_nV_m) complexes are formed. F_nV_m complexes can capture and annihilate the free interstitials. It is considered that the existence of F_nV_m complexes has some influence on the EOR defect evolution behavior.

In order to clarify those behaviors, an atomistic kinetic Monte Carlo (KMC) approach[17] model with the F_nV_m effect is used. Fig. 9 shows an atomistic KMC simulation result for 30 keV Ge PAI + F co-implant + 0.5 keV B implant and sub-ms annealing. KMC with the F model predicts that F_nV_m complexes are possibly remaining after sub-ms annealing. It is considered that those F_nV_m complexes can have an influence on the EOR defect evolution.

CONCLUSIONS

In this work, the impact of fluorine co-implant on boron diffusion during NLA is shown. In the presence of fluorine, the B diffusion length during sub-ms NLA is increased dramatically as a function of F co-implant energy. B diffusivity enhancement due to F co-implant is observed at F implant energies higher than 5 keV. This diffusivity enhancement is possibly a B diffusivity enhancement in cSi in the presence of high concentrations of fluorine. F_nV_m complex formation is a realistic scenario for the F impact on EOR defect evolution. An atomistic KMC model predicts that F_nV_m complexes are still remaining after sub-ms annealing.

REFERENCES

1. A. Shima, *et al.*, Symp. VLSI Tech. Dig., p. 174 (2004).
2. S. K. H. Fung, *et al.*, Symp. VLSI Tech. Dig., p. 92 (2004).
3. K. Adachi, *et al.*, Symp. VLSI Tech. Dig., p. 142 (2005).
4. A. Poudebasque, *et al.*, in Tech. Dig. IEDM, p. 663 (2005).
5. S. Severi, *et al.*, Mater. Res. Soc. Symp. Proc. Vol.912, 0912-C01-07 (2006).
6. T. Noda, *et al.*, Mater. Res. Soc. Symp. Proc. Vol.912, 0912-C05-06 (2006).
7. D. Downey *et al.*, Appl. Phys. Lett. **73**, p. 1263 (1998).
8. A. Mokhberi *et al.*, Appl. Phys. Lett. **80**, p. 3530 (2002).
9. J. Jacques *et al.*, J. Appl. Phys. **98**, 073521 (2005).
10. T. Noda, J. Appl. Phys. **96**, p. 3721 (2004).
11. S. Mirabella *et al.*, Appl. Phys. Lett. **86**, 121905 (2005).
12. N. Cowern, *et al.*, Appl. Phys. Lett. **86**, 101905 (2005).
13. S. Paul, *et al.*, J. Vac. Sci. Technol. B **24** (1), p. 437 (2006).
14. F. Cristiano, *et al.*, 2006 MRS spring meeting (2006).
15. Y. Haddara, *et al.*, Appl. Phys. Lett. **77**, p. 1976 (2000).
16. G. Lopez, *et al.*, Phys. Rev. B **72**, 045219 (2005).
17. M. Jaraiz, *et al.*, Mater. Res. Soc. Symp. Proc. Vol.532, p. 43 (1998).
18. T. Noda, J. Appl. Phys. **94**, p. 6396 (2003).

Characterization of B_2H_6 Plasma Doping for Converted p^+ Poly-Si Gate

Jae-Geun Oh, J.K. Lee, S.H. Hwang, H.J. Cho, YS Sohn, D.S. Sheen, SH Pyi
S.W. Lee*, S.H. Hahn*, Y.B. Jeon**, Z. Fang** and V. Singh**

Hynix Semiconductor Inc., San 136-1 Ami-ri, Bubal-eub, Ichon-si, Kyoungki-do, 467-701, Korea
Varian Korea Ltd., 433-1 Mogok-dong Pyeongtaek-si Kyoungki-do 459-040 Korea
Varian Semiconductor Equipment Associates Incorporate, 35 Dory Road, Gloucester, MA 01930 USA
E-mail: jaegeun.oh@hynix.com

Abstract. We have investigated the characteristics of B_2H_6 plasma doping (PLAD) process used to convert the n^+ doped poly-Si gate to the p^+ poly-Si gate for pMOS. The throughput of the PLAD process is much higher than a conventional beam line implantation process at low energy and high dose ranges. The B_2H_6 plasma counter-doping on the n^+ poly-Si were performed in the energy range of 5kV ~ 9kV and dose of ~E16#/cm^2. The B_2H_6 Plasma doped poly-Si layers were characterized by TDS, SIMS, AFM, and TEM.
The TDS analysis showed hydrogen desorption from the B_2H_6 plasma doped p^+ poly-Si layer at a low temperature. The surface concentration of PLAD doped boron was much higher compared to the conventional beam line implantation. However, a serious loss of surface dopant was also observed during photoresist strip and post cleaning. The surface dopant loss could be suppressed by 10% with optimization of the cleaning condition, leading to improve characteristics of PLAD doped p^+ poly-Si pMOS, compared to the beam line implantation. Moreover, flat band voltage (V_{FB}) shift was not observed in the C-V curves and there was no significant difference in I-V characteristics between PLAD and the conventional ion implantation. Deeper and higher dopant profile will be helpful to decrease required dose gap between PLAD and beam line implantation.

Keywords : B_2H_6 Plasma Doping, P^+ poly Si
PACS : 79.20.R 85.30.Tv 85.40.Ry

Introduction

Boron ion implantations at low energies have been traditionally used for the formation of dual poly gate in pMOSFETs. However, the increase of implantation dose brings about a low throughput issue. Although the deceleration mode implantation is more effective to increase the beam current, it can cause the energy contamination problem with high energy ions. Plasma doping is one of the very attractive technique to overcome above issues [1, 4~7]. The PLAD technique has been developed to achieve ultra-shallow junctions and high dose doping processes. Another approach is the boron cluster implantation technique [2-3].

In this work we used B_2H_6 plasma doping in order to fabricate the p^+ poly-Si gated MOS capacitors with Varian VIISta PLAD system. We compared results of B_2H_6 plasma doped and B^+ implanted in the viewpoint of capacitor performance. The hydrogen ion issue, one of the concerns for B_2H_6 plasma source, was also investigated. In fact, a large amount of hydrogen ions can degrade device performances, which is simultaneously incorporated into the p^+ poly-Si gate with boron ions. In addition, the sheet resistances and boron profiles in poly-Si / oxide / Si wafers were also analyzed.

Experimentals

In this experiment, we pay attention to the principal doping step for the formation of p^+ poly-Si gated MOS capacitors. Figure 1 shows a process sequence for the p^+ poly-Si gate fabrication. After the plasma nitridation onto the gate oxide, either B^+ ion, BF_2^+ ion implanted or B_2H_6 plasma was doped into the p^+ poly-Si layer. For the control, the samples implanted with conventional B^+ ion, BF_2^+ ions were also prepared. Conventional poly implant annealing (PIA) was performed at 950°C, and spike poly implant annealing at 1075°C with lamp-based equipment. We analyzed doping characteristics and surface morphologies using secondary ion mass spectrometry (SIMS), transmission electron microscopy (TEM), atomic force microscopy (AFM), and a conventional four-point probe. The hydrogen contents in the poly-Si films were examined by TDS (Thermal Desorption Spectroscopy) and SIMS. The electrical properties, such as the leakage current(not shown) and capacitance-voltage (C-V) characteristics of p+ poly-Si gate, were measured by use of 4155B and HP4284A.

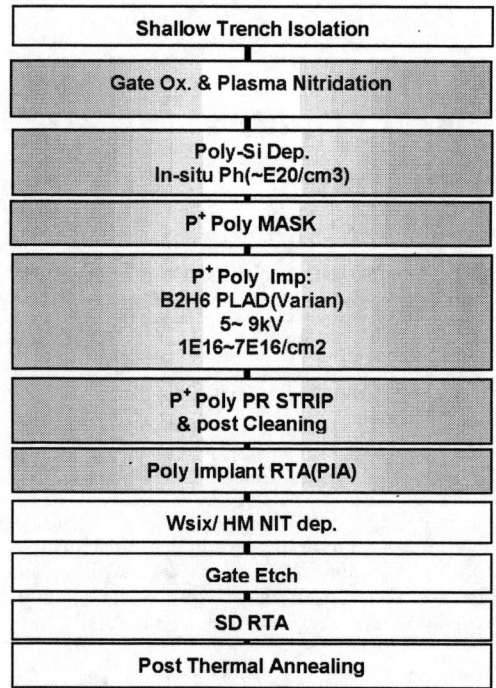

FIGURE 1. Process sequence for p⁺ poly-si gate fabrication

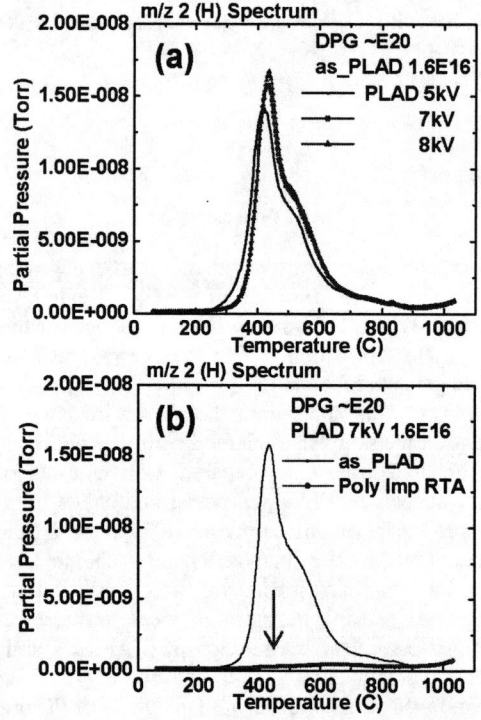

FIGURE 2. M/Z2 spectra of hydrogen ; (a) B_2H_6 as-PLADed and (b) after poly implant annealing.

Results and Discussion

Figure 2 shows M/Z 2 spectra of hydrogen for B_2H_6 PLAD before and after poly implant annealing. In B_2H_6 PLAD case (a), occurrence of H_2 desorption in 200~800°C range must be derived from B_2H_6 plasma source. As the PLAD energy increases, H_2 desorption from poly-Si increases. H_2 desorption was suppressed after poly implant RTA(b). Based on TDS result, desorption of hydrogen can be thought to be also occurred by RTA following PLAD.

Figure 3 shows surface roughness of the samples doped by B_2H_6 PLAD as a function of dose(a), energy(b), and after poly implant annealing(c). As the PLAD dose increases, roughness of PLADed poly-Si film increases. As the PLAD energy increases and after poly implant RTA, roughness of PLADed poly-Si film decreases.

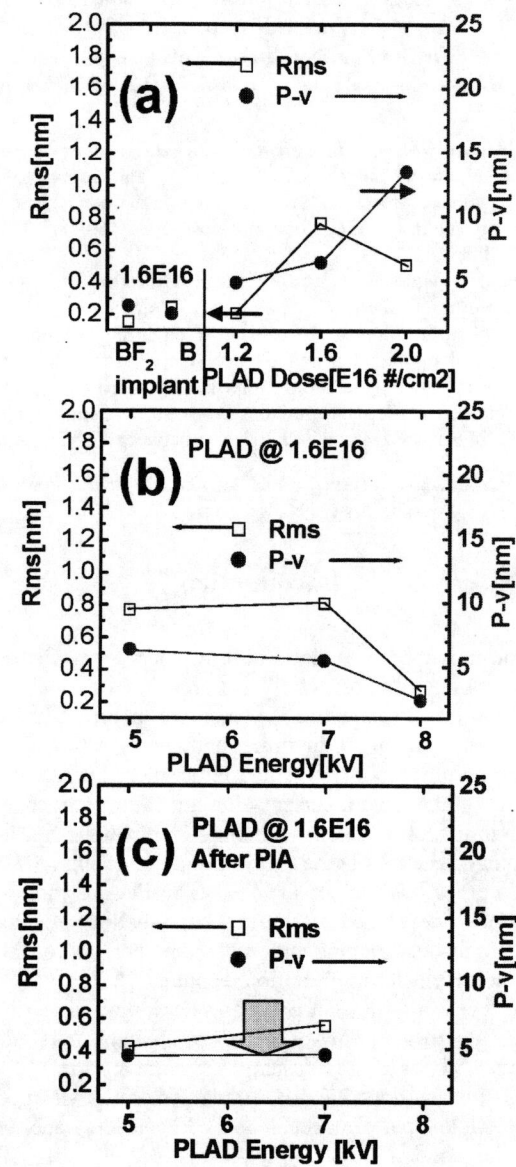

FIGURE 3. Surface roughness of the samples doped by B_2H_6 plasma; (a) dose effect, (b) energy effect, and (c) after poly implant annealing

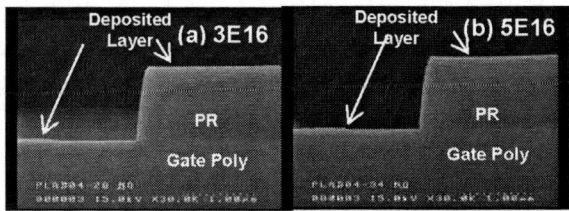

FIGURE 4. Cross-sectional SEM photographs of the Deposited Layer on B_2H_6 PLADed Poly Si and Photo Resist (a) 3E16 and (b) 5E16 dose.

As seen in figure 6, boron concentration for PLAD was still high at surface. after poly implant RTA. Boron concentration of PLAD was lower than that of B implanted Poly Si

As shown in figure 7, sheet resistance of PLADed poly-Si was higher than that of B implanted poly Si. As the RTA temperature increases, sheet resistance of PLAD decreases

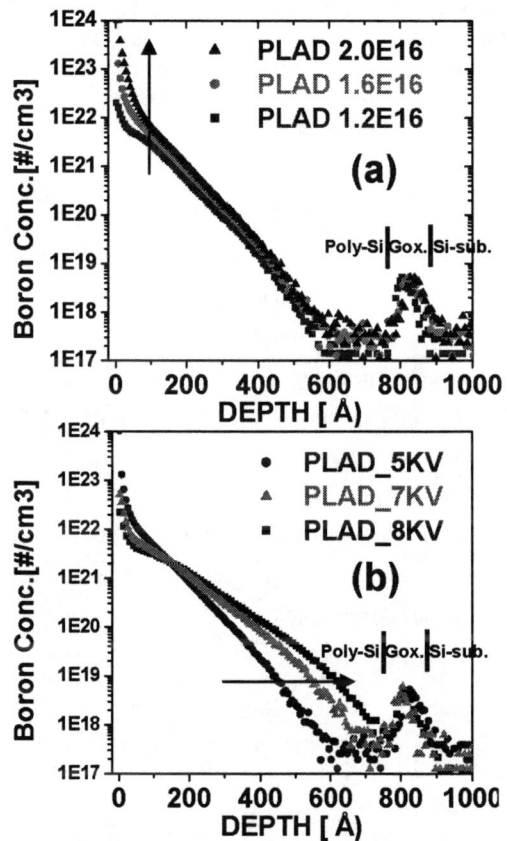

FIGURE 5. SIMS profile of boron in B_2H_6 PLADed poly Si ; (a) dose effect at 5kV and (b) energy effect at 1.6E16

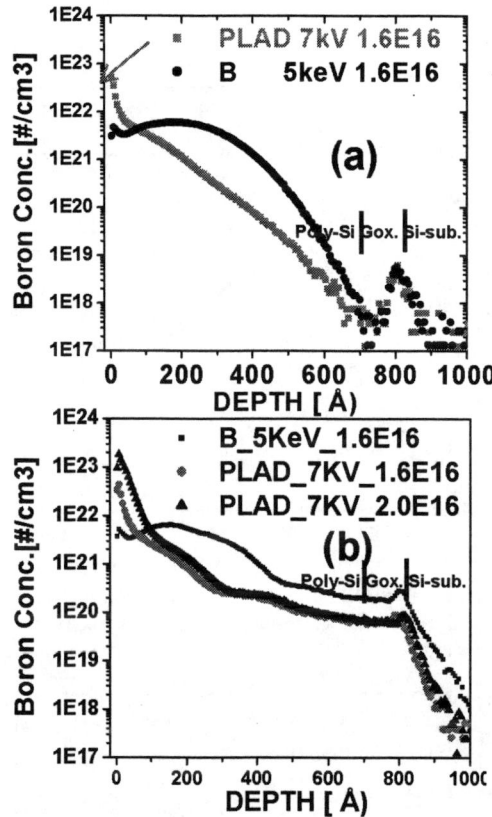

FIGURE 6. SIMS profile of boron in B_2H_6 PLADed poly Si (a)before and (b)after poly implant annealing

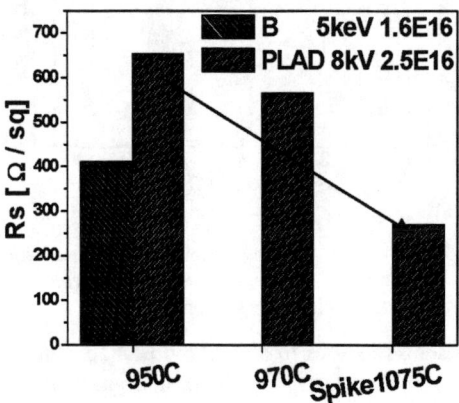

FIGURE 7. Sheet resistance characteristics of p+ poly-Si gate converted by $B_{18}H_x^+$ and B+ counter doping from n^+ doped poly-Si gate.

Figure 4. shows Cross-sectional SEM photographs of the deposited Layer on B_2H_6 PLADed Poly Si and Photo Resist. As the PLAD dose increases, the deposited layer thickness also increases. Even after PIA, the doposited layer was not disappeared(not shown).

Figure 5 shows SIMS profile of boron, of B_2H_6 PLADed poly Si (a) dose effect at 5kV and (b) energy effect at 1.6E16). As PLAD dose increases, boron concentration increases at surface. As PLAD energy increases, boron profile becomes deeper. The PLAD showed an exponential-function-like profile due to both multi-species and multi-energy components.

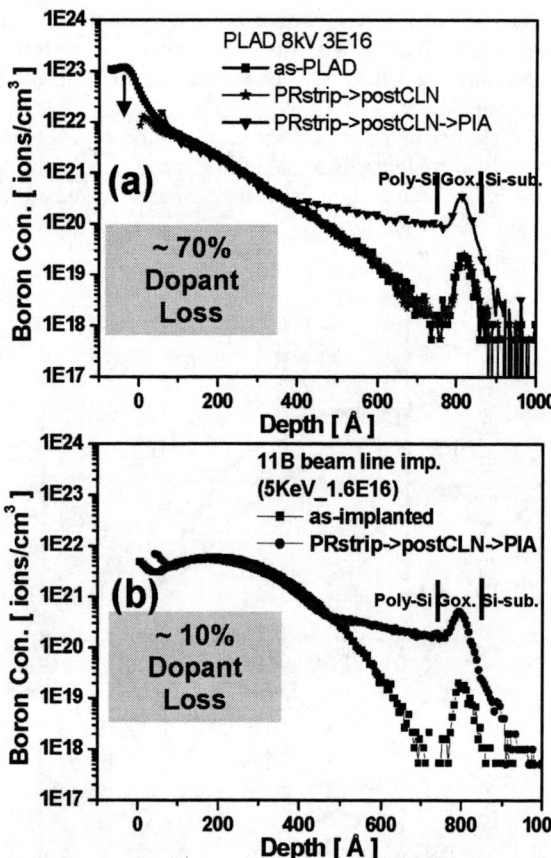

FIGURE 8. SIMS shows dopant loss during the photoresist strip and the post cleaning (a) PLAD (b) conventional B implantation.

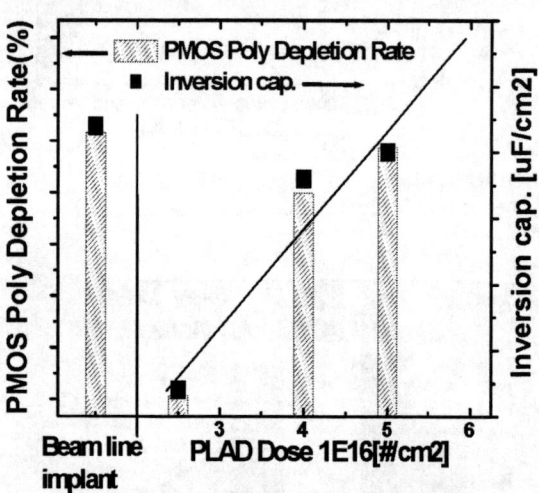

FIGURE 9. Poly depletion rate and inversion capacitance from C-V characteristics of p+ poly-Si gate converted by B_2H_6 and B+ counter doping from n^+ doped poly-Si.

Figure 8 shows serious loss of boron near the surface during the photoresist strip and the post cleaning. About 70% boron loss on the surface and about 40 Å surface loss was observed after cleaning. Deeper boron profile is also required although cleaning method should be optimized.

Figure 9 shows poly depletion rate and inversion capacitance for PMOS. As the PLAD dose increases, both pMOS poly depletion rate increase

Conclusions

We demonstrated B_2H_6 Plasma Doping Process for converted p+ poly-Si Gated MOS capacitors.

Based on TDS result, desorption of hydrogen can be thought to be also occurred by RTA following PLADed poly Si. About 3 times higher dose is required for PLAD compared with beam line implantation in order to obtain the equivalent poly depletion rate by compensating surface boron loss during PR Strip and Post Cleaning. Deeper and higher dopant profile will be helpful to decrease required dose gap between PLAD and beam line implantation

References

1. Y. Sasaki et al, Symp. On VLSI Technology (2004) p. 180
2. K. Goto et al, Tech. Dig. Of IEDM (1996) p. 435
3. D.C. Jacobson et al, International Conference on Ion Implantation Technology (2000) p. 300
4. S. Qin et al, J. Vac.Sci. Technol. B 23(6), 2005, p.2272
5. S. Walther et al, Nucl. Instr. and Meth. In Phys. B237(2005), p.126
6. J Ha et al, Surface and Coatings Tech. 136(2001) p.157
7. S Felch et al, International Conference on Ion Implantation Technology (2000) p. 488

Fluorine Profile Distortion upon Annealing by the Presence of a CVD Grown Boron Box

P. López[1], L. Pelaz[1], R. Duffy[2], P. Meunier-Beillard[2], K. van der Tak[3], F. Roozeboom[3] and G. Maas[3]

[1] *Dpto. de Electricidad y Electrónica, Universidad de Valladolid. E.T.S.I. Telecomunicación. Campus Miguel Delibes s/n. 47.011 Valladolid, Spain*
[2] *Philips Research Leuven, Kapeldreef 75, 3001 Leuven, Belgium*
[3] *Philips Research Laboratories Eindhoven, Prof. Holstlaan 4, 5656 AA Eindhoven, The Netherlands*

Abstract. We provide experimental evidence of the distortion of a F profile by the presence of a CVD grown B box. After annealing, a depletion in the F profile is observed fitting the position of the immobile part of B profile. To study this phenomenon further experiments were designed and atomistic simulations were done using a recently developed F model. In this model F complexes with both Si interstitials (F-I) and vacancies (F-V) are included. The formation of Boron-Interstitial Clusters is found to reduce the local Si interstitials defect concentration. This feature may be responsible for the reported F distortion by the presence of a B box.

Keywords: Fluorine implantation; Boron clustering; Extended defects; Fluorine complexes; Modeling.
PACS: 66.30.Jt, 61.72.Cc, 61.72.Tt.

INTRODUCTION

The continuous scaling down of new generation devices has been for many years the driving force of microelectronics evolution. The reduction of dimensions may give rise to the appearance of short channel effects, which highly degrade device performance. The fabrication of ultra shallow junctions (USJ), in which a high concentration of electrically active dopants is confined in a very shallow area, prevents short channel effects. Co-implant of fluorine with boron, mainly in the form of BF_2, has been widely used for USJ formation. The implantation of BF_2 molecular ions has two main advantages: equivalently low-energy implantation can be achieved due to the large mass number of BF_2, and the BF_2-implanted silicon surface may become amorphous reducing channeling and favoring B activation during solid phase epitaxial regrowth (SPER). The introduction of F has been proved to drastically reduce B diffusion [1, 2, 3] and to increase B activation [2]. The mechanisms underlying these effects have not been fully elucidated yet. Several theories have been proposed: a chemical interaction between B and F [4], the formation of F-Vacancy (F-V) complexes acting as annihilation centers for Si interstitials and hence reducing B diffusion [5, 6], or the trapping of Si interstitials by F atoms [2, 7]. In an amorphous layer the presence of F-V complexes after SPER seems to explain B reduced diffusion [8] but in crystalline Si the scenario is more complex due to the large variety of defects formed.

In this work we report some intriguing experimental results regarding F behavior in the presence of B: the distortion of a F profile by a B box grown by CVD. We combine new experiments and atomistic simulations to gain insight of this particular phenomenon. A good understanding of the interactions of F with B and defects in Si is a key factor to explain F beneficial effects on B.

EXPERIMENTAL

F, P and Sb were implanted in p-type <100> oriented Czochralski wafers. In some of the samples a B box approximately 60 nm wide and a peak concentration of 2×10^{19} cm^{-3} had been grown by CVD. The B box is placed always beyond the mean projected ranged (R_p) of the implant. All samples were then annealed in a N_2 inert ambient using rapid thermal annealing. Secondary ion mass spectrometry (SIMS) was used to determine the concentration profiles of the implanted species. In the first set of experiments, 200 keV 2×10^{15} cm^{-2} F was implanted at Room Temperature (RT) in wafers with and without a B box centered around 650 nm deep. The samples were annealed at 900 °C for 6, 60 and 600 s. In the second set of experiments 75 keV 1.8×10^{14} cm^{-2} P and 215 keV 5.2×10^{13} cm^{-2} Sb were implanted at RT over a 2×10^{19} cm^{-3} CVD grown B box centered approximately at 180 nm. Implant dose and energy were chosen to avoid amorphization and get similar damage distribution for both implants, according to TRIM simulations. All wafers were annealed for 60 s at 850 °C.

SIMULATION MODEL

In our simulation scheme implantation cascades are simulated with the binary collision approximation computer code MARLOWE [9], and the defect evolution with the kinetic nonlattice Monte Carlo diffusion code DADOS [10]. After each implant cascade the coordinates of the Si interstitials (I's), vacancies (V's) and implanted ions are transferred to DADOS. Point defects may interact and form several kinds of defects ranging from small clusters to extended defects. Amorphization is modeled by using the defect known as *Bond defect* or *IV pair* [11, 12]. Dopant diffusion, precipitation and activation are also considered in DADOS. For boron, a detailed boron-interstitial clustering model is included [13].

We have implemented a F model which includes the main features of F regarding diffusion and clustering. The experimentally reported fast diffusion of F is modeled by considering the interstitial F (F_i) diffusion. According to ab-initio calculations the activation energy for F_i diffusion is 0.4 eV [14]. F_i is annihilated at the surface, what generates a gradient from the bulk to the surface. This fact leads to an effective surface-oriented F diffusion and produces F out-gassing.

Our model also includes the formation of immobile F-V and F-I complexes, whose energies are based on those of V's and I's clusters and in the case of F-V complexes also based on ab-initio calculations [15]. The observed F accumulation at V's rich regions can be modeled by F-V complexes [16, 17], whereas F-I complexes may account for the experimental reports of F trapping at end of range defects and around Rp [18, 19]. In short, a complete F model has been developed, including F_i as the diffusing species and the formation of F-I and F-V complexes, reproducing the main features of F behavior.

RESULTS AND DISCUSSION

200 keV 2×10^{15} cm^{-2} F was implanted over a 2×10^{19} cm^{-3} CVD grown B box. F and B profiles after annealing at 900 °C for 60 s are shown in figure 1. The as-implanted and the as-grown profiles for F and B respectively are also included for reference. Several things are remarkable in this figure. There is a F accumulation at the shallow part of the profile and also around R_p. The formation of F complexes with V's or I's is considered to be responsible for F retention. The shallow peak of F in the plot may be related to F-V complexes formation, since this region is known to be rich in V's [16, 17]. Si I's are predominant around Rp and beyond. The high concentration of F atoms and Si I's at this zone may favor F-I complexes formation and the existence of the F peak at R_p [19].

But the most intriguing feature is the distortion of the F profile by the presence of B: F atoms seem to have been ejected from the B box. A closer look at the figure reveals that this phenomenon seems to be restricted to the immobile region of B profile. This region is assumed to be related to the existence of boron-interstitial clusters (BIC's). Thus, there may be a connection between the depletion in the F profile and the existence of BIC's.

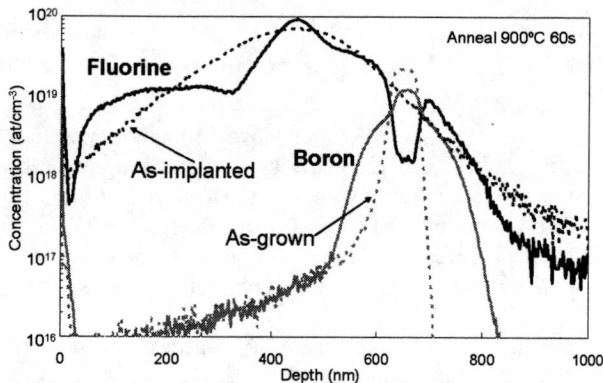

FIGURE 1. Experimental B and F profiles after 900 °C, 60 s annealing. 2×10^{15} cm^{-2} 200 keV F was implanted over a 2×10^{19} cm^{-3} CVD grown B box. As-implanted and as-grown profiles also shown for reference. A depletion in the F profile at the position of the immobile B region is clearly observed.

To check whether BIC's formation distorts the profiles of other implanted species we designed a new set of experiments. The layout is similar to the previous one: an ion implantation is done over a CVD grown B box and then the sample is annealed. Two different species were implanted: P, which diffuses via I's, and Sb, which diffuses via V's.

SIMS data of the 75 keV 1.8×10^{14} cm^{-2} P implant after annealing at 850 °C 60 s are plotted in figure 2. There is a slight P accumulation at the immobile part of the B profile but it is very small and it is difficult to say at this stage whether this is significant.

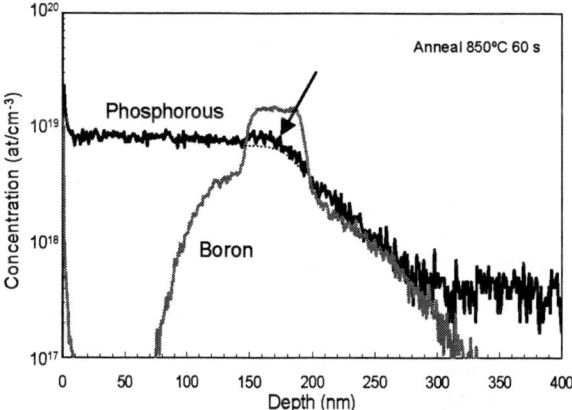

FIGURE 2. Experimental P and B profiles after annealing at 850 °C for 60s. 1.8×10^{14} cm^{-2} 75 keV P was implanted over a 2×10^{19} cm^{-3} CVD grown B box.

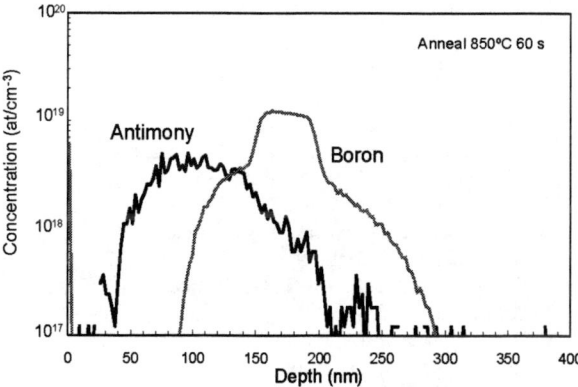

FIGURE 3. Experimental Sb and B profiles after annealing at 850°C for 60s. 5.2×10^{13} cm^{-2} 215 keV Sb was implanted over a 2×10^{19} cm^{-3} CVD grown B box. Sb profile is not altered by the presence of B in these experimental conditions.

5.2×10^{13} cm^{-2} Sb was implanted at 215 keV and annealed at 850°C for 60 s. Concentration profiles of Sb and B are plotted in figure 3. As we can see Sb is not influenced by the presence of B. Sb is a heavy ion and diffuses slowly. The annealing conditions may not be the optimal ones to observe some effect on Sb distribution.

Since experiments are not conclusive we use simulations to try to gain insight of this phenomenon. For this purpose we use a kinetic Monte Carlo code, called DADOS, in which we have implemented a F model. As kinetic Monte Carlo simulators do not solve differential equations we can consider many different interactions for F-I and F-V complexes. Other models are also included in the simulations, such as damage generation and defect evolution and a detailed B diffusion and clustering model.

In our simulations we have reproduced the same conditions as in the experiment of Figure 1. Impurities and defects profiles after temperature ramp-up at 900 °C are plotted in figure 4. It can be seen that voids (V_m) and F-V complexes are present close to the surface whereas 311 defects (I_m) and F-I complexes are predominant around Rp and deeper. However in BIC's region there are no I's defects.

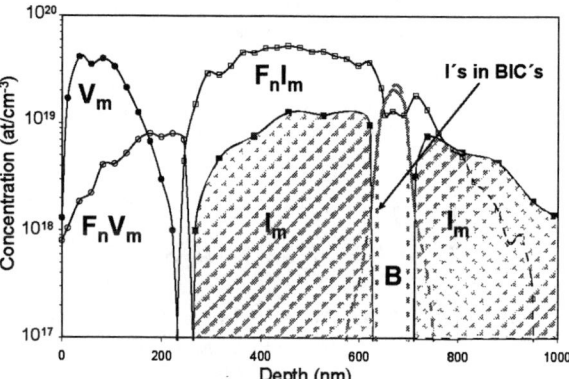

FIGURE 4. Simulated B, F and defects profiles after ramp-up at 900 °C, in the same conditions as Figure 1. Solid symbols represent extended defects and open symbols F complexes. B profile and the I's concentration inside BIC's are represented by a solid and a dashed grey line, respectively.

Upon annealing most of V's and I's generated during the implant recombine, leaving an excess of V's at shallow regions, whereas I's are predominant over the rest of the profile. F complexes need point defects to grow and stabilize. Thus, F-V and F-I complexes are present where V's and I's are predominant, respectively. In the B region, B cluster formation traps Si I's. In fact, BIC's and I's extended defects (such as 311's) compete for I's. BIC's formation has been found to suppress the growth of 311's defects [20]. This may explain the lack of extended defects at the immobile part of the B box in our simulations.

Figure 5 reports simulated F and B distributions after annealing at 900°C for 60s and the as-implanted and as-grown profiles for comparison, respectively. We can see that simulations reproduce the depletion in the F profile at BIC's position. During annealing F atoms are released from F complexes in the B box region. As there are no defects inside BIC's region some F atoms diffuse out and are trapped at extended defect outside BIC's region. This may cause the distortion in the F profile.

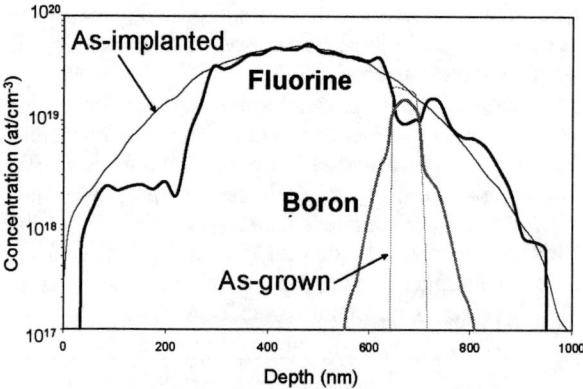

FIGURE 5. Simulated F and B distribution after annealing at 900°C for 60 s. As-implanted and as-grown profiles also shown for reference, respectively. A depletion in the F profile at BIC's position is clearly seen.

CONCLUSIONS

We report experimental data showing a distortion in the F profile by the presence of B, when F is implanted over a CVD grown B box. A depletion in the F profile is formed fitting the position of the immobile part of B distribution. P and Sb profiles after ion implantation in similar conditions were not noticeable altered by B. We used atomistic simulations to analyze the influence of the B box on the F profile. A F model was implemented that includes F_i diffusion and the formation of F-I and F-V complexes. Simulations revealed that BIC's region has a low I's defect density as a result of B clustering process. The trapping of F atoms released from F complexes at I's defects outside this zone may explain the F profile distortion by the presence of the B box.

ACKNOWLEDGMENTS

This work has been funded by the Spanish DGI under project TEC2005-05101 and the JCyL Consejería de Educación y Cultura under project VA070A05.

REFERENCES

1. R. G. Wilson, J. Appl. Phys. 54, 6879-6889 (1983).
2. K. Ohyu, T. Itoga, N. Natsuaki, Jpn. J. Appl. Phys. 29, 457-462 (1990).
3. H. A. W. El Mubarek, P. Ashburn, Appl. Phys. Lett. 83, 4134-4136 (2003).
4. A. Mokhberi, R. Kasnavi, P. B. Griffin, J. D. Plummer, Appl. Phys. Lett. 80, 3530-3532 (2002).
5. H. A. W. El Mubarek, J. M. Bonar, G. D. Dilliway, P. Ashburn, M. Karunaratne, A. F. Willoughby, Y. Wang, P. L. F. Hemment, R. Price, J. Zhang, P. Ward, J. Appl. Phys. 96, 4114-4121 (2004).
6. M. Diebel, S. Chakravarthi, S. T. Dunham, C. F. Machala, S. Ekbote, A. Jain, Mater. Res. Soc. Symp. Proc. 765, D6.15.1-D6.15.6 (2003).
7. G. Impellizzeri, J. H. R. dos Santos, S. Mirabella, F. Priolo, E. Napolitani, A. Carnera, Mater. Res. Soc. Symp. Proc. 810, C5.9.1-C5.9.6 (2004).
8. G. Impellizzeri, J. H. R. dos Santos, S. Mirabella, E. Napolitani, A. Carnera, F. Priolo, Nucl. Instrum. Methods Phys. Res. B 230, 220-224 (2005).
9. M. T. Robinson and I. M. Torrens, Phys. Rev. B 9, 5008 (1974).
10. M. Jaraiz, L. Pelaz, J. E. Rubio, J. Barbolla, G. H. Gilmer, D. J. Eaglesham, H. –J. Gossman, and J. M. Poate, Mater. Res. Soc. Symp. Proc. 532, 43 (1998).
11. L.A. Marqués, L. Pelaz, J. Hernandez, J. Barbolla, and G.H. Gilmer, Phys. Rev. B 64, 045214 (2001).
12. M. Tang, L. Colombo, J. Zhu, and T. Diaz de la Rubia, Phys. Rev. B 55, 4279 (1997).
13. M. Aboy, L. Pelaz, L. A. Marqués, P. Lopez, J. Barbolla, R. Duffy, J. Appl. Phys. 97, 1-7 (2005).
14. G. M. Lopez, V. Fiorentini, G. Impellizzeri, S. Mirabella, E. Napolitani, Phys. Rev. B 72, 045219 1-7 (2005).
16. H. A. W. El Mubarek, J. M. Bonar, G. D. Dilliway, P. Ashburn, M. Karunaratne, A. F. Willoughby, Y. Wang, P. L. F. Hemment, R. Price, J. Zhang, P. Ward, J. Appl. Phys. 96, 4114-4121 (2004).
17. M. N. Kham, H. A. W. El Mubarek, J. M. Bonar, P. Ashburn, Appl. Phys. Lett. 87, 011902 (2005).
15. M. Diebel, S. T. Dunham, Mater. Res. Soc. Symp. Proc. 717, C4.5.1-C4.5.6 (2002).
18. M. Y. Tsai, D. S. Day, B. G. Streetman, P. Williams, C. A. Evans, Jr., J. Appl. Phys. 50, 188-192 (1979).
19. X. D. Pi, C. P. Burrows, P. G. Coleman, Phys. Rev. Lett. 90, 155901 (2003).
20. T. E. Haynes, D. J. Eaglesham, P. A. Stolk, H. -J. Gossmann, D. C. Jacobson, J. M. Poate, Appl. Phys. Lett. 69, 1376-1378 (1996).

Deactivation of low energy Boron Implants into Pre-amorphised Si after Non-Melt Laser Annealing with Multiple Scans

J.A Sharp[1*], N.E.B Cowern[1], R.P Webb[1*], D. Giubertoni[2], S. Gennaro[2], M. Bersani[2], M.A Foad[3] and K.J. Kirkby[1*]

[1]Advanced Technology Institute, *Surrey Ion Beam Centre, School of Electronics and Physical Sciences, University of Surrey, Guildford GU2 7XH, UK.
[2]ITC-irst Centro per la Ricerca Scientifica e Tecnologica, Povo (Trento) 38050, Italy.
[3]Applied Implant Technologies, Applied Materials Inc., 974 E. Arques Avenue, Sunnyvale, CA. 94086, USA.

Abstract: Activation/deactivation of 500eV B implants in pre-amorphised Si after non-melt laser annealing with multiple scans at 1150°C and isochronal rapid thermal post-annealing has been investigated. Under the thermal conditions used for non-melt laser at 1150°C, a substantial residue of end-of-range defects remained after 1 laser scan, evidenced by end-of-range defect decoration by B atoms after 700°C post-annealing and by transient enhanced diffusion after 800°C post-annealing. Dramatic boron deactivation is also observed after post-annealing the 1-scan samples. Most of these features were not present in samples receiving 5 or 10 laser scans, indicating that the end-of-range defects had been stabilised or dissolved within 5 and 10 scans. The results show that the detrimental effects of end-of-range defects can be removed during non-melt laser annealing and is therefore an achievable method for stabilisation of highly activated B profiles in pre-amorphised Si.

Keywords: Boron, Laser Annealing, Deactivation, Silicon.
PACS: 61.72.Tt; 61.72.Hh; 61.72.Ji; 81.15.Np

INTRODUCTION

The continued down scaling of CMOS architecture requires ultra-shallow source/drain extension regions with a low sheet resistance for future devices [1]. Among other thermal processes, non-melt laser annealing of ion implanted silicon has gained attention as a means of achieving these requirements by its short process time and high annealing temperature resulting in high dopant solubility [2-5]. Low thermal budget annealing of pre-amorphised silicon (such as non-melt laser annealing) has a problem of subsequent deactivation of the boron during further annealing [6-12]. The deactivation in pre-amorphised silicon is driven by the release of silicon interstitials from end of range defects (EOR). These evolve through non-conservative Ostwald ripening during the annealing [13]. The interstitials flow towards the surface and deactivate the boron atoms in the profile, by producing boron interstitial clusters [14-17]. In this paper, single and multiple laser scan annealing at a temperature of 1150°C, followed by isochronal rapid thermal post-annealing at lower temperatures is used to investigate the role of end-of-range defects in the redistribution and deactivation of 500eV B implants in pre-amorphised silicon.

EXPERIMENTAL

An n-type (100) Cz-silicon wafer was pre-amorphised with 5keV Ge$^+$ to a dose of 1×10^{15} Ge cm^{-2} producing a surface amorphous layer to a depth of ~15nm. 500eV B$^+$ was then implanted into the amorphous layer to a dose of 1×10^{15} B cm^{-2}. Both implants were made using an Applied Materials Quantum X implanter. The wafer was held at room temperature and exposed to a scanning diode laser source operated under non-melting conditions. Three strips were annealed across the wafer, corresponding to 1, 5 or 10 scans. The surface of the wafer was monitored and a maximum temperature of 1150°C was observed. By using multiple laser scans to anneal the implant, it was possible to study the defect evolution as a function of increasing the thermal budget. The amorphous layer re-grew by solid phase epitaxial regrowth (SPER) during the anneal. Samples were taken from these strips and annealed in dry N$_2$ for 60s at temperatures ranging from 700°C to 1000°C, using a Process Products Corporation rapid thermal annealing system operating with a 50°C/s heating ramp rate. The van der Pauw technique was used to measure the sheet resistance of the samples after laser and deactivation annealing. Secondary ion mass spectrometry (SIMS) was also carried out on selected samples using a Cameca Wf SC-ULTRA tool running an 500 eV O_2^+ primary beam (68° angle of incidence with respect to the surface normal) with oxygen leak. The sample was loaded on a rotating stage in order to prevent the formation of ripples on the crater bottom [18].

RESULTS AND DISCUSSION

The sheet resistance results for the 500eV B^+ implant after laser and then post annealing can be seen in figure 1. The results show an initial low sheet resistance value, which increases as the post anneal temperature is increased (deactivation). Above 900°C the sheet resistance drops considerably to values lower than that observed for laser anneal only (reactivation). The samples that received only 1 scan show a much greater deactivation than the samples that received multiple laser scans. This higher sheet resistance is consistent with deactivation resulting from the

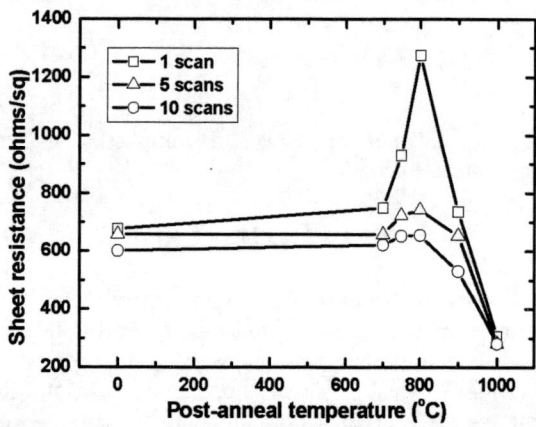

FIGURE 1. Sheet resistance against post-annealing temperature showing deactivation /reactivation of 500eV B in pre-amorphised silicon after laser anneal at 1150°C and isochronal post-annealing for 60s. The 0°C points correspond to activation by the laser only.

dissolution of end of range (EOR) defects [12, 15, 17]. The lower amount of deactivation during the post-anneals in the 5 and 10 scan cases is similar to that observed by Lerch et al [19] who found that flash-lamp annealing (which is on a similar time scale to the scanning laser annealing used in this study) was sufficient to evolve the EOR defect band into perfect and faulted dislocation loops. The supersaturation of interstitials at this level of evolution is much lower than when {311} defects are present and hence the amount of deactivation during post anneals is reduced.

Figure 2a shows SIMS boron profiles of three samples that have been laser annealed with 1, 5 or 10 scans. There is some degree of diffusion from the as-implanted profile during the laser annealing, which increases with the number of laser scans. A kink is observable in the 1 scan sample at a concentration of 10^{19} cm^{-3} which does not show up in the 5 or 10 scan profiles. This kink develops into a boron peak at 16 nm that can be clearly seen after the sample has undergone a post-anneal at 700°C for 60s (figure 2b). Since the amorphous layer is predicted to be about 15 nm thick, this peak is a clear indication that B is being trapped in EOR defects that are still present after 1 laser scan. Moreover, the redistribution to form this peak implies a significant amount of diffusion at concentrations below that of the original kink - which would only occur in the presence of a high interstitial supersaturation. In contrast, the 5 and 10 laser scanned samples do not show any substantial diffusion after the 700°C post-anneal, nor do they exhibit any other anomalous feature.

Further evidence for the presence of EOR defects after 1 laser scan at 1150°C can be seen in figure 2c, where the samples have been post-annealed at 800°C for 60 seconds. Here, anomalous transient enhanced diffusion (TED) is observed in the low-concentration tail region of the profile. This effect is probably caused by a very high supersaturation of self interstitials, which could only be driven by the dissolution of EOR defects. An approximate estimate of the supersaturation can be obtained from the diffusion length in the tail region, which is of the order of 15 nm. Based on standard B diffusivity values under equilibrium point-defect conditions, this implies a supersaturation of ~1000, consistent with the presence of {113} defects, which are known to form in the EOR band. No such enhancement is seen in the samples that received 5 or 10 laser scans. In these samples, post-annealing at 800°C gives rise to diffusion with a much higher kink level of about 1.8×10^{20}/cm^3, with a diffusion profile consistent with near-normal steady-state diffusion.

The presented results consistently point to the conclusion that for the single laser scan, there was not enough thermal budget to break up the EOR defects, leading to strong deactivation as a result of the release of interstitials during post-annealing. By using 5 or 10 laser scans at 1150°C, the EOR defects were largely or completely removed or stabilized, resulting in a greatly decreased deactivation and relaxation towards equilibrium diffusion behavior during post-annealing.

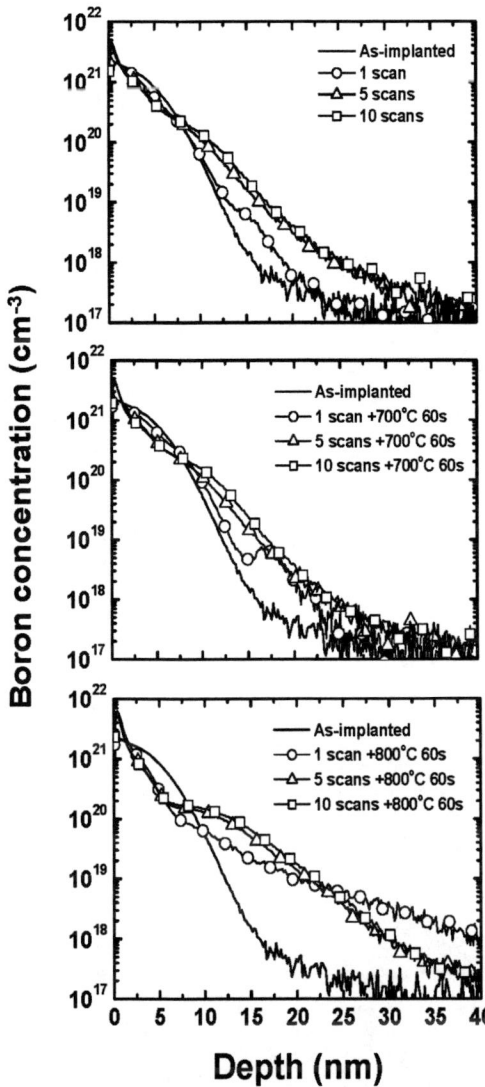

FIGURE 2. SIMS profiles of 500eV B in Ge pre-amorphised silicon showing as-implanted and (a) activated with 1, 5 or 10 laser scans at 1150°C, (b) laser annealed at 1150°C then post-annealed at 700°C for 60 seconds and (c) laser annealed at 1150°C then post-annealed at 800°C for 60 seconds.

CONCLUSIONS

In summary, we have studied non-melt laser annealing as a tool for the formation of highly active ultra-shallow boron implants in pre-amorphised silicon. The deactivation of the boron after annealing using a non-melt scanning laser with multiple scans was examined, by studying the boron profiles after rapid thermal post-anneals over a range of lower temperatures. Under conditions where the laser anneal does not remove (or stabilize) the end-of-range defects, subsequent low-temperature processing leads to strong deactivation and transient enhanced diffusion. In addition, when the post-anneal thermal budget is too low to remove the end-of-range defects, B migrates and decorates these defects. In contrast, when using multiple laser scans to anneal the implant, the EOR defects are evolved into stable loops such that both deactivation and diffusion are strongly reduced.

ACKNOWLEDGMENTS

The authors would to acknowledge the UK Engineering and Physical Sciences Research Council (EPSRC) and the Surrey Ion Beam Centre for their support. J.S. is supported by an EPSRC Doctoral Training Award. N.E.B.C. is supported by an Applied Materials/Philips/Royal Academy of Engineering Research Chair.

REFERENCES

1. International Technology Roadmap for Semiconductors, http://public.itrs.net/. 2005.
2. S. Earles, M. Law, K. Jones, S. Talwar and S. Corcoran, Mater. Res. Soc. Symp. Proc. **669**, J4.1.1 (2001).
3. S. Earles, M. Law, K. Jones, R. Brindos and S. Talwar, Mater. Res. Soc. Symp. Proc. **610**, B10.5.1 (2000).
4. S. Earles, M. Law, R. Brindos, K. Jones, S. Talwar and S. Corcoran, IEEE Trans. Electron Devices, **49**, 1118 (2002).
5. S. Earles, M. Law, K. Jones, J. Frazer, S. Talwar, D. Downery and E. Arevalo, *Proceedings of the 12th IEEE International Conference on Advanced Thermal Processing of Semiconductors, 28-30 Sept. 2004.* (IEEE, Piscataway, NJ, USA 2004) p143.
6. B.J. Pawlak, W. Vandervorst, A.J. Smith, N.E.B. Cowern, B. Colombeau and X. Pages, Appl. Phys. Lett. **86**, 101913 (2005).
7. Y. Takamura, S. Jain, P.B. Griffin and J.D. Plummer, Mater. Res. Soc. Symp. Proc. **669**, J7.3.1 (2001).
8. Y. Takamura, S.H. Jain, P.B. Griffin and J.D. Plummer, J. Appl. Phys. **92**, 230 (2002).
9. Y. Takamura, P.B. Griffin and J.D. Plummer, J. Appl. Phys. **92**, 235 (2002).
10. W-E. Hong and J.-S. Ro, J. Appl. Phys. **97**, 13530 (2005).
11. R. Murto, K. Jones, M. Rendon and S. Talwar, *Proceedings of the 14th Ion Implantation Technology Conference* (IEEE, Piscataway, NJ, 2000) page 155.
12. B.J. Pawlak, R. Surdeanu, B. Colombeau, A.J. Smith, N.E.B. Cowern, R. Lindsay, W. Vandervorst, B. Brijs, O. Richard and F. Crisitano. Appl. Phys. Lett. **84**, 2055 (2004).
13. C. Bonafos, D. Mathiot and A. Claverie, J. Appl. Phys. **83**, 3008 (1998).

14. B.Colombeau, A.J.Smith, N.E.B.Cowern, W.Lerch, S. Paul, B J.Pawlak, F.Cristiano, X.Hebras, C.Ortiz, and P.Pichler, International Electron Device Meeting 2004, Technical Digest IEEE, (2004), p971.
15. M. Aboy, L. Pelaz, L.A. Marqués, P. Lopez, J. Barbolla, V.C. Venezia, R. Duffy and P.B. Griffin, Mat. Sci. Eng., B, **114-115**, 193 (2004).
16. M. Aboy, L. Pelaz, L.A. Marqués, J. Barbolla, A. Mokhberi, Y. Takamura, P.B. Griffin and J.D. Plummer, Appl. Phys. Lett. **83**, 4166 (2003).
17. F. Cristiano, N. Cherkashin, P. Calvo, Y. Lamrani, X. Hebras, A. Claverie, W. Lerch and S.Paul, Mat. Sci. Eng., B. **114-115,** 174 (2004).
18. M. Bersani, D. Giubertoni, E. Iacob, M. Barozzi, S. Pederzoli, L. Vanzetti and M. Anderle, Applied Surface Science in press.
19. W. Lerch, S. Paul, J. Niess, S. McCoy, T. Selinger, J. Gelpey, F. Cristiano, F. Severac, M. Gavelle, S. Boninelli, P. Pichler and D. Bolze, Mat. Sci. Eng., B. **124-125**, 24 (2005).

Co-Implantation for 45 nm PMOS and NMOS Source-Drain Extension Formation: Device Characterisation Down to 30 nm Physical Gate Length.

E.J.H. Collart[1], B.J. Pawlak[2], R. Duffy[2], E. Augendre[3], S. Severi[3], T. Janssens[3], P. Absil[3], W. Vandervorst[3], S. Felch[4], R. Scheutelkamp[5], and N.E.B. Cowern[6]

[1] Applied Materials UK Ltd, FoundryLane, Horsham, W-Sussex RH13 5PX, Uk
[2] Philips Research Europe, Leuven, B-3001, Belgium
[3] IMEC vzw, Kapeldreef 75, Leuven, B-3001, Belgium
[4] Applied Material Inc., 974 E. Arques Avenue, Sunnyvale, CA 94086, USA
[5] Applied Materials, Leuven, B-3001, Belgium
[6] Advanced technology Institute, School of Electronics and Physical Sciences, University of Surrey, Guildford, Surrey, GU2 7XH, UK

Abstract. We have optimised co-implantation schemes for NMOS and PMOS USJ formation down to 30 nm physical gate length. These schemes included Ge or Si pre-amorphisation steps, followed by C and/or F and dopant implants of P and B for NMOS and PMOS, respectively. Junction depth and sheet resistance optimisation on blanket wafers was complemented with electrical device data. Blanket wafer results show junction depths as low as 15 nm at 5×10^{18} cm^{-3}, abruptness around 2.5 nm/decade and Rs in the 400-600 $\Omega/\square$ range. Device data show very good Vt roll-off behaviour down to 30 nm physical gate length, and good Ion/Ioff curves. Leakage currents are higher than in reference devices, but within acceptable limits for general purpose applications. The leakage has been found to be a very sensitive function of dopant and non-dopant species placement. The main cause for the $I_{off,leak}$ is trap-assisted tunnelling through C clustering with residual damage in the depletion layer, rather than band-to-band tunnelling. Optimisation of extension implant conditions as well as halo and spacer may improve leakage characteristics and device performance further.

Keywords: ultra-shallow junction formation, co-implantation, carbon, NMOS, PMOS
PACS: 61.72.-y, 61.72.Cc, 61.72.Ji, 61.72.Ss, 66.30.Jt, 66.30.Lw

INTRODUCTION

Until recently the accepted method to meet the ITRS 65 and 45 nm ultra-shallow junction requirements included advanced annealing strategies such as full-melt or non-melt laser annealing or solid phase epitaxial regrowth (SPER). The option of decreasing ion implantation energies beyond those used for 90 nm technology node and to below a few hundred eV was not considered production viable. This perceived imminent requirement of advanced anneal strategies has recently changed with the realisation that an optimised implant 'cocktail' of different species can drastically improve activation and at the same time reduce dopant diffusion [1, 2]. The 'cocktail' is also attractive from an integration point of view as conventional rapid thermal processing (RTP) spike annealing can be extended for one or more technology nodes.

EXPERIMENTAL

Blanket wafer and device experiments were carried out on 200 mm wafers. In order to form the ultra-shallow extension implants, a first step of Si or Ge pre-amorphisation (PAI) was used. This was followed by a C or F implant and finally a dopant implant. These were B for PMOS extensions and P for NMOS extensions. On the device wafers these extensions have been implemented into a conventional transistor flow with poly gates on SiON gate dielectric. The only significant modifications were the PAI step and C and F co-implantation. Channel and poly doping remain almost unchanged. Damage annihilation and activation was is done by a 1050°C spike anneal. The implants were carried out on an Applied Materials Quantum X ion implanter and the activation anneal on the Applied Materials Centura® Radiance*Plus*® Radiance. Chemical profiles were measured by secondary ion mass spectrometry (SIMS) using an Atomica 4500 instrument with a 500 eV O2 analyzing beam.

RESULTS AND DISCUSSION

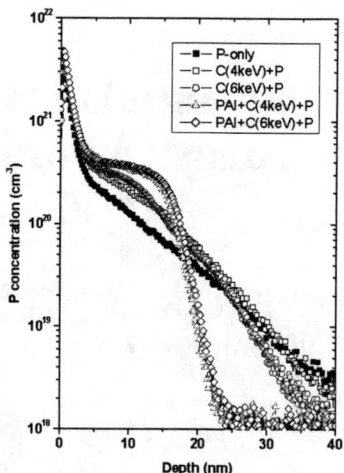

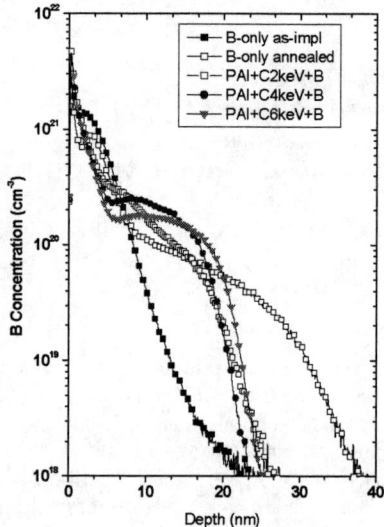

FIGURE 1. Diffusion behaviour of a 500 eV B implant with different co-implantation implant conditions and a 1050 C spike RTP.

Figure 1 shows a 500 eV B implant where a combination of Ge pre-amorphisation and C implantation is used prior to the B implant. The Ge implant conditions were 20 keV at a dose of 5×10^{14} cm^{-2} resulting in an amorphous-layer thickness of around 35 nm. The C implants were all 1×10^{15} cm^{-2} at 2, 4 and 6 keV. The difference between the Ge+C+B profiles and the C-only or with Ge-only is striking. For all three C implant conditions the B profile is now much steeper and considerably less diffused. The best result in terms of steepness is obtained with 6 keV C, where the bulk of the C profile is deeper than the B profile. This condition results in an Rs=468 Ω/□, an Xj=25 nm and abruptness=2.5nm/decade.

Figure 2 shows P SIMS profiles for a 1 keV, 7×10^{14} cm^{-2} dose implant and different combinations of co-implanted species. The PAI is a 25 keV, 1×10^{15} cm^{-2} Si implant, resulting in an initial amorphous layer of around 60 nm.

FIGURE 2. Diffusion behaviour of a 1 keV P implant with different co-implantation implant conditions and a 1050 C spike RTP.

The sheet resistance R_s of the P-only was 411 Ω/□. For the C+P the R_s decreased to 392 Ω/□ and 349 Ω/□ for 4 and 6 keV C respectively, showing that the extra diffusion, compared to P-only, is at least partially activated. The PAI+C+P profile displays a box-like shape up to about 15 nm followed by a very steep gradient between 15-20 nm. The steepest slope was 3 nm/dec for the 6 keV C case. The sheet resistances were 357 Ω/□ and 318 Ω/□ for the 4 and 6 keV C cases, indicating further activation of the diffused profiles.

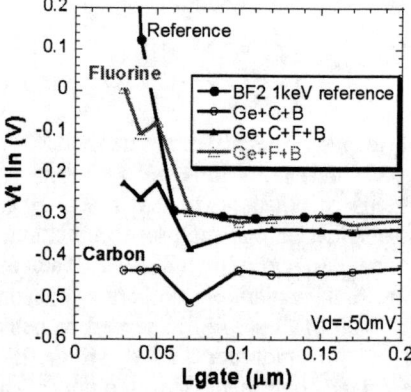

FIGURE 3. V_T roll-off behaviour for a 500 eV B implant with different co-implantation conditions and a 1050 C spike RTP. A reference 1×10^{15} cm^{-2} 1 keV BF$_2$ implant is plotted for comparison.

Figure 3 shows the threshold voltage (Vt) roll-off curve versus gate length for a number of co-implantation conditions. Combinations of PAI, C and

F together with B are compared to a 1×10^{15} cm^{-2} BF$_2$ reference implant at 1 keV. The B conditions are 7×10^{14} cm^{-2} at 500 eV. The Vt roll-off is a well-known short channel effect. With smaller device dimensions the lateral and vertical dimensions of the junctions will encroach more onto the channel. The shorter the channel, the smaller (in magnitude?) Vt becomes. As can be seen, the Vt roll-off behaviour of the Ge+C+B case is superior to all other conditions investigated, remaining essentially flat down to 30 nm gate length. This is a remarkable result, considering that the reference BF$_2$ implant has an equivalent B energy of only 225 eV. It illustrates the effect of the very abrupt, vertical and lateral profiles on device performance. The Ge+C+B is also superior to the Ge+F+B and Ge+C+F+B cases. Although the addition of F is also known to improve junction depth and abruptness[3], it is not as effective as C.

compared to 200 eV. This correlates well with a deeper junction and worse VT roll-off for 500 eV.

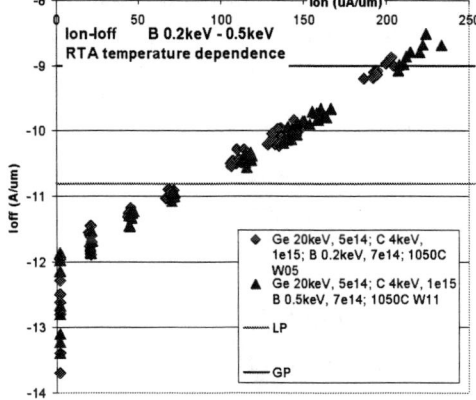

FIGURE 5. I_{on}-I_{off} comparison for 200 eV and 500 eV B implant

Leakage current behaviour is a major concern for junctions formed with C co-implantation. It has been known for a long time that C or C-clusters can cause leakage when located in the depletion region of a p-n junction. Figure 6 shows the cumulative junction leakage curve obtained from diode measurements.

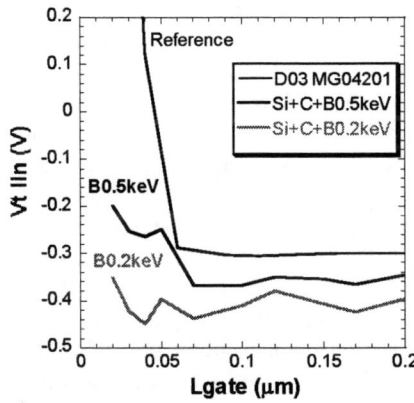

FIGURE 4. V_T roll-off behaviour as a function of boron energy. The reference is a 1×10^{15} cm^{-2} 1 keV BF$_2$ implant.

Figure 4 shows the effect of reducing the B implant energy from 500 eV to 200 eV on V_t roll-off. Results on blanket wafers have shown that a 200 eV implant results in an even shallower profile (~15 nm@5×10^{18} cm^{-3}, not shown), less gate overlap, and compared to the 500 eV implant energy a slightly more negative V_t. Note also that here a 1×10^{15} cm^{-2}, 25 keV Si PAI was used rather than a Ge PAI step. As far as the V_t roll-off characteristics are concerned, Si and Ge PAI give comparable results. From a process integration point this is beneficial, as now a single flavour PAI can be used for both PMOS and NMOS.

Figure 5 shows the Ion/Ioff curves for a Ge+B+C scheme with 500 eV and 200 eV B implants at 7×10^{14} cm^{-2} dose and after a 1050 °C spike anneal. The 500 eV curve is shifted to higher drive currents, i.e. for a given Ioff a 500 eV implant results in a higher Ion

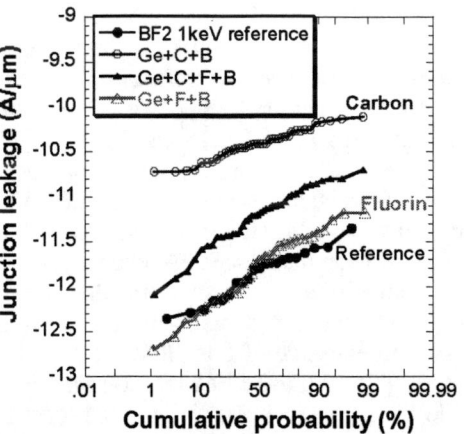

FIGURE 6. Leakage current for different co-implantation conditions.

As expected, the C-containing scheme shows the highest leakage currents. A distinction can be made between these by measuring the thermal behaviour of the diode leakage current. At the higher temperature range (100-200 °C) an activation energy E_a of 1.2 eV was found, indicating thermal excitation across the Si bandgap as main contributor. At lower T (20°C-100°C) an E_a of 0.6 eV is evidence of a midgap defect and trap-assisted tunnelling. There is no evidence of band-to-band tunnelling. It has been shown that the C atoms cluster on the residual end of range defects from the PAI [1,4,5]. If located in the junction depletion region, these will contribute to the leakage. It also suggests a way of minimizing the leakage by careful optimisation of the placement of PAI and C implants. Although these leakage currents are too high for low-power devices, they are compatible with other applications.

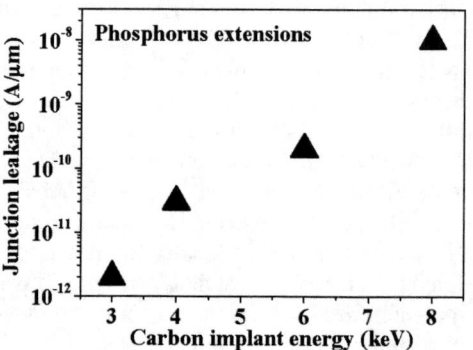

FIGURE 8. Diode leakage current for a 7×10^{14} cm^{-2} 1 keV P implant and different C implant energies.

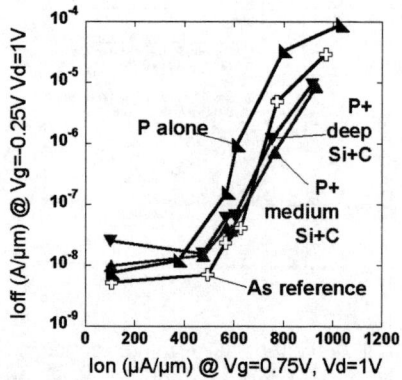

FIGURE 7. I_{on}-I_{off} behaviour for a 1 keV P implant with different co-implantation conditions (see text). The reference is a 1×10^{15} cm^{-2} 1 keV As implant.

Turning to NMOS devices, Figure 7 shows Ion/Ioff curves for different co-implantation schemes and a comparison with an As reference. The co-implantation scheme consisted of a 1 keV P implant at 7×10^{14} cm^{-2} and Si PAI at 15, 25 and 35 keV, all at 1×10^{15} cm^{-2} dose. The P implant on it own is the worst. The co-implantation with C and different Si PAI conditions have a significant impact on the performance, with the medium Si PAI energy (25 keV) showing a better Ion/Ioff characteristic than the As reference. Diode measurements showed, as in the PMOS case, no evidence of band-to-band tunnelling and a Si midgap level (E_a=0.53 eV) responsible for the observed leakage. Figure 8 shows the effect of the C implant energy on the leakage current for NMOS extensions formed with Si+C+P. The leakage current increases by four orders of magnitude, from 10^{-12} A.µm^{-1} to 10^{-8} A.µm^{-1} when varying the C implant energy from 3 keV to 8 keV. This illustrates again the sensitivity and the ability to fine-tune the junction characteristics by careful optimisation of implant conditions.

CONCLUSIONS

We have shown that a combination of pre-amorphisation, C or F and dopant implants can lead to very abrupt, shallow and well-activated extension formation. The improvements seen on blanket wafers also translate to improved device behaviour down to 30 nm physical gate length. Although junction leakage current is a concern, careful optimisation of implant conditions may minimise these issues without much loss of the benefits offered. Tuning of related parameters such as halo implant and spacer thickness may further improve electrical device behaviour. Finally and most importantly, this route of extension formation allows for further scaling of traditional ion implantation and spike anneal to at least the 45 nm CMOS technology node.

REFERENCES

1. E.J.H. Collart, D. Kirkwood, R. Lindsay, W.Vandervorst, T. Janssens, B.J. Pawlak, Proceedings of the 15th International Conference on Ion Implantation Technology Part II, Taipei, Taiwan 2004, edited by L.J Chen, J Poate, T.-F. Lei, pp.11-14.
2. E.J.H Collart, S.B Felch, B.J. Pawlak, P.P. Absil, S. Severi, T. Janssens, and W.Vandervorst, *J. Vac. Sci. Technol. B* **24(1)**, 507-509 (2006).
3. E.J.H. Collart, S.B. Felch, H. Graoui, D. Kirkwood, S. Tallavarjula, J.A. Van den Berg, J. Hamilton, N.E.B. Cowern, K.J. Kirkby, Material Science and Engineering **B 114-115**, 118-129 (2004).
4. B.J. Pawlak, T.Janssens, T.Brijs, W.Vandervorst, E.J.H. Collart, S.B. Felch, N.E.B. Cowern, accepted for publication in *Appl. Phys. Lett.*, June 2006
5. B.J.Pawlak, R.Duffy, T.Janssens, W. Vandervorst, S.Felch, E.J.H. Collart, N.E.B. Cowern, submitted for publication to *Appl. Phys. Lett.*

Control of Phosphorus Transient Enhanced Diffusion using Co-implantation

Aaron Vanderpool[1], Andre Budrevich[1], Mitch Taylor[2]

[1]Intel Corporation, [2]Applied Materials, Inc.

Abstract. The production of Ultra Shallow Junctions (USJ) in silicon devices requires controlling the Transient Enhanced Diffusion (TED) of electrical dopants. USJ development has focused on boron because hole mobility is lower than electron mobility in silicon and because arsenic has such excellent diffusion properties. However, the advent of strain enhanced mobility in P-type silicon has created the need to study higher solubility N-type dopants like phosphorus and find methods to control their diffusion. Co-implants have proven effective in controlling the interstitial diffusion mechanisms of boron TED. In this work the effectiveness of some co-implants on phosphorus to form high performance USJ is reported. It has been found that carbon and fluorine co-implants reduce phosphorus diffusion. As work with boron has shown, this is due to the carbon Kick-out mechanism and Fluorine-Vacancy clusters, both of which consume the interstitials driving TED. It has also been found that record levels of phosphorus diffusion control can be obtained if boron and carbon are co-implanted. In this junction diffusion control increases as the boron implant energy decreases; even as low as 0.5 KeV. However, this may be activating Uphill diffusion. The data also shows that the carbon implant energy has very little effect on phosphorus diffusion. The boron and carbon co-implants also produce the steepest phosphorus USJ yet reported at 2.5nm/decade with a solubility $>1.0E21$ atoms/cm^3. Counter intuitively it has been found that the boron and carbon USJ is shallower with a higher solubility if the phosphorus implant energy is increased from 2 to 3 KeV. These boron and carbon co-implant findings are quite novel even if they are not technologically useful. They strongly support the widely held model that phosphorus TED occurs via an interstitial diffusion mechanism and that techniques to block this mechanism can control it. The boron implanted below the phosphorus is probably consuming interstitials very efficiently in Boron Interstitialcy clusters (BI$^+$) causing boron TED rather than phosphorus TED. The electrical characteristics of a phosphorus USJ with high doses of boron below it may be undesirable; but, it does demonstrate that phosphorus diffusion can be well controlled if the right co-implants are found.

Keywords: Ultra Shallow Junction, Transient Enhanced Diffusion, co-implantation, Carbon, Fluorine, Boron.
PACS: 85.40.Ry

INTRODUCTION

The development of Ultra Shallow Junctions (USJ) in Complementary Metal-Oxide-Semiconductor (CMOS) devices has always focused on controlling the Transient Enhanced Diffusion (TED) of boron in p-type Silicon. While boron is significantly affected by TED, the n-type USJ dopant arsenic is not. This makes the formation of an USJ with arsenic straightforward but very difficult with boron. Phosphorus is an alternative n-type dopant and could be a better choice than arsenic because of its high solubility in silicon [1]. However phosphorus is affected by TED to an even larger extent than boron [2]. For this reason arsenic is used as the n-type USJ dopant and has exceeded the electrical performance of boron USJ's in both sheet resistance (Rs) and junction depth (X_j).

In addition the focus upon boron USJ is important because the hole mobility from p-type dopants is much lower than electron mobility from n-type dopants. However, the development of Strained Silicon [3] in P-type Metal-Oxide-Semiconductors (PMOS) has significantly enhanced the mobility of holes and created the need for higher performance n-type USJ. One way to improve the electrical performance of n-type USJ is to increase the solubility of the dopant and lower the Rs. By using phosphorus instead of arsenic this can be accomplished if the TED of phosphorus can be controlled.

Since the TED mechanisms between phosphorus and boron are very similar many of the methods developed to control Boron TED can be employed; however, additional control techniques are needed because of the stronger bonding of phosphorus to silicon interstitials (I) than boron [2, 4].

THEORY

Transient Enhanced Diffusion (TED) of boron and phosphorus in silicon occurs via the interstitialcy mechanism and is written as:

$$I + A \Leftrightarrow AI \qquad (1).$$

Cowern [5] calculated that TED could only be explained through an interstitial mechanism. He observed the diffusion tails of boron δ doped layers followed an exponential distribution upon annealing rather than a Gaussian distribution. This distribution could only be modeled by an interstitial reaction like either the Interstitialcy reaction in Equation (1) or the Kick-out reaction in Equation (2);

$$A + I \Leftrightarrow A_i \qquad (2).$$

Later Sadigh [4] was able to use total energy and molecular dynamic calculations to show that of all the interstitial paths possible the AI^+ interstitialcy reaction is the lowest energy configuration and that this pair is the fast diffuser responsible for TED. Christensen [2] went on to carefully measure the activation energy of Phosphorus Interstitialcy (PI) and Boron Interstitialcy (BI) clusters in silicon and found that the PI activation energy was lower. He concluded that the PI binding was stronger which resulted in longer diffusion lengths before the pair broke up and the phosphorus became substitutional again.

This characterization of TED suggests that a method to interfere with the creation or motion of the Interstitialcy pair will reduce diffusion lengths. To this end carbon and fluorine co-implants have been utilized to reduce TED.

When carbon is substitutional in silicon it undergoes the Kick-out reaction in Equation (2), and will consume the Interstitials created by implantation damage. These interstitials drive TED, and their reduction significantly limits TED. Carbon was observed to have this effect and then characterized during the growth of Epitaxial SiGeC films. Rucker [6] observed that boron TED from implantation damage in δ doped layers was eliminated when carbon was Epitaxialy grown into the layers. Further, Stolk [7] found that amorphized silicon with implanted carbon would significantly reduce TED upon Solid Phase Epitaxial Re-growth (SPER).

Fluorine has long been observed to reduce TED and in recent years there has been a great deal of progress made in understanding the mechanism by which this occurs. Using calculation methods similar to Sadigh [4], Diebel [8] was able to determine that of all the point defects in implanted silicon (interstitials, vacancies, or dopant impurities) fluorine forms the strongest bonds with vacancies, creating F_nV_m clusters (F-V). Calculation suggests that when these clusters interact with interstitials the fluorine is kicked-out, the vacancy is annihilated, and the interstitial is consumed. This reduction of the interstitial population acts to reduce TED. Pi [9] presents some interesting experimental observations of these F-V clusters using Positron Annihilation Spectroscopy (PAS) in high energy, high dose fluorine implanted silicon. Also, the presence of an implant generated vacancy distribution before fluorine co-implantation in boron USJ has been observed to retard TED [10]. This suggests that the creation of F-V clusters to reduce TED is enhanced when the vacancy distributions are higher. When a vacancy distribution is created after fluorine co-implantation, it is thought that the formation of F-V clusters is reduced because there is not as much energy available to the point defects for diffusion and formation as there is during fluorine implantation. In addition the retardation of SPER and the differences in TED between amorphous and crystalline silicon that Jacques [11] observed fits well with the F-V cluster model now that it is better understood.

RESULTS

In Figure 1 a $2.0E15$ cm^{-2} dose of phosphorus is implanted at 3 KeV and spike annealed above 1050C to form an USJ with Fluorine and or Carbon of co-implants.

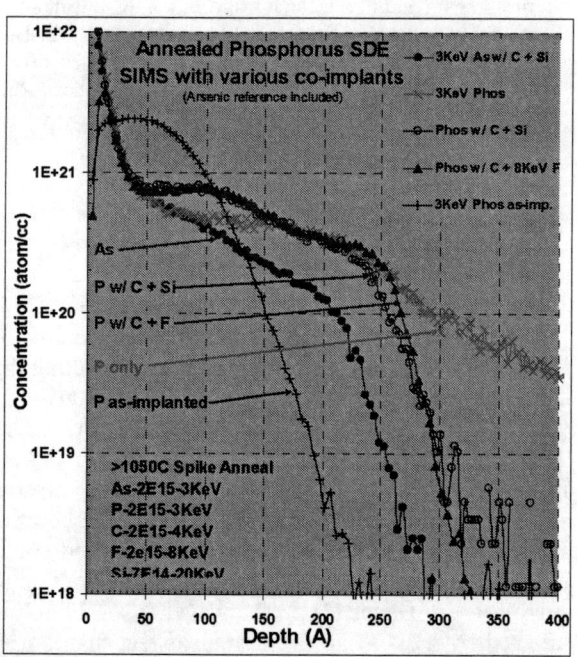

FIGURE 1. 3 KeV Phosphorus USJ SIMS comparing the diffusion effects of Fluorine or Carbon co-implants.

A 3 KeV arsenic implant is displayed for comparison as well at the as-implanted phosphorus profile. A phosphorus implant without any co-implants is also displayed to define the magnitude of TED the co-implants need to control and to demonstrate the increased solubility of phosphorus over arsenic. Carbon is co-implanted into silicon amorphized with a 20 KeV 7.0E14 cm^{-2} Si implant is able to reduce phosphorus diffusion to about the same extent that the fluorine and carbon co-implants can together although the Fluorine and Carbon implants appear to produce a slightly steeper junction. This may indicate that the fluorine TED reduction mechanisms are not as important for phosphorus USJ as they are for boron USJ. It is possible that the primary role of fluorine here is to amorphize the lattice for the carbon mechanism.

forming very efficiently below the phosphorus dramatically reducing the interstitials in the near surface region.

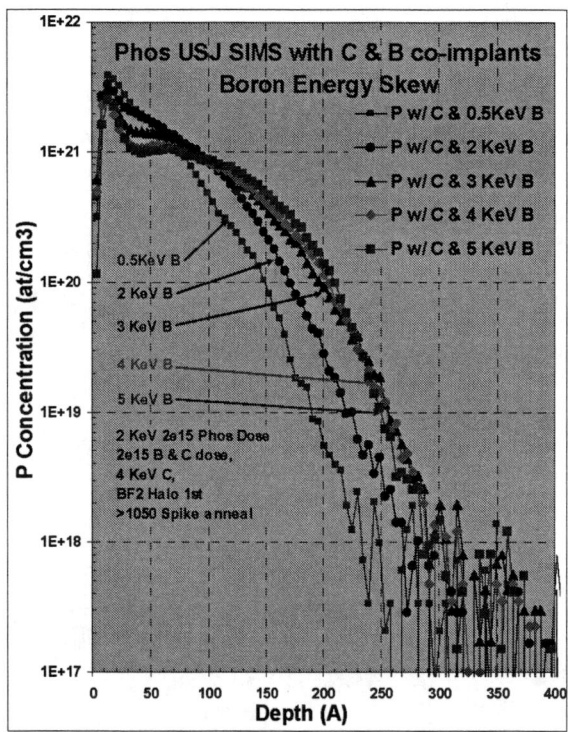

FIGURE 3. 2 KeV Phosphorus USJ's with C & B co-implants. Reducing the B energy reduces B TED.

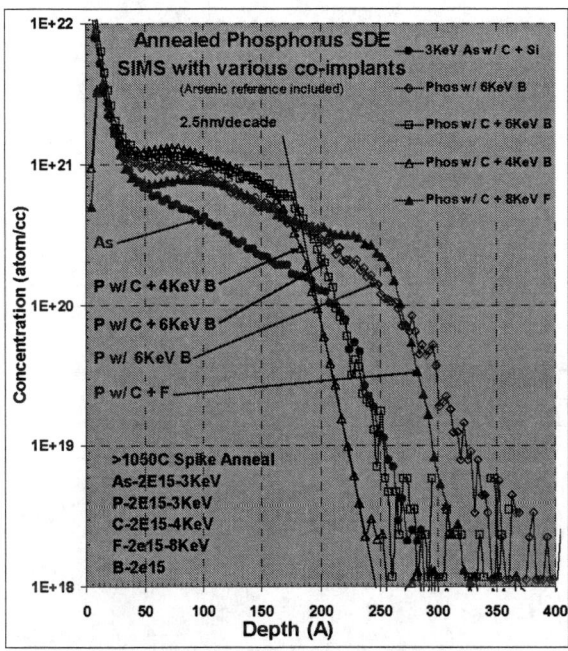

Figure 2. 3 KeV Phosphorus USJ SIMS comparing the diffusion effects of B, C and F co-implants

In Figure 2 Boron co-implantation alone is even less effective at reducing TED. However, when boron and carbon are co-implanted together phosphorus TED is significantly reduced. Junction depths shallower than arsenic can be obtained by reducing the boron implant energy below 6 KeV. In addition the solubility of this junction increases to more than 1.0E21 cm-3. The carbon and 4 KeV boron co-implants also demonstrate a very steep spike annealed junction at 2.5 nm/decade. This is very close to the lateral abruptness resolution of the SC Ultra SIMS tool used to measure it. In addition to the Kick-out mechanism of the substitutional carbon, it is possible that the TED reduction mechanism in this junction is the consumption of interstitials by the co-implanted boron which is deeper than the phosphorus. It is probable that Boron Interstitial Clusters (BIC) are

In Figure 3, it is observed that the reduction in TED continues to improve as the boron co-implant energy decreases; however, the junction begins to loose its steepness. The changes in the profile shape are likely a manifestation of the transient Uphill diffusion that Duffy [12] observed with (1.5 KeV) phosphorus. It is interesting that this effect becomes lager as the boron co-implant energy is decreased. One possible reason for this may be that at lower implant energies the high boron dose is probably starting to cluster, forming the defect structures similar to those observed by Duffy in the early stages of annealing. Perhaps the concentration of interstitials in these clusters increases with decreasing boron energy which enhances the Uphill diffusion instead of reducing the interstitials below the phosphorus.

Part of this change in these profiles is due to the reduction in the phosphorus implant energy from 3 to 2 KeV. This change is better illustrated in Figure 4 where the higher energy implant has a much better profile that would normally be associated with a lower energy implant. It is very counter-intuitive that this would occur. This may also be another aspect of phosphorus Uphill diffusion which suggests that there is a minimum implant energy for high solubility junctions some where around 3 KeV. The higher Rs of the 2 KeV profile suggests that some of phosphorus is inactive and may be trapped in clusters.

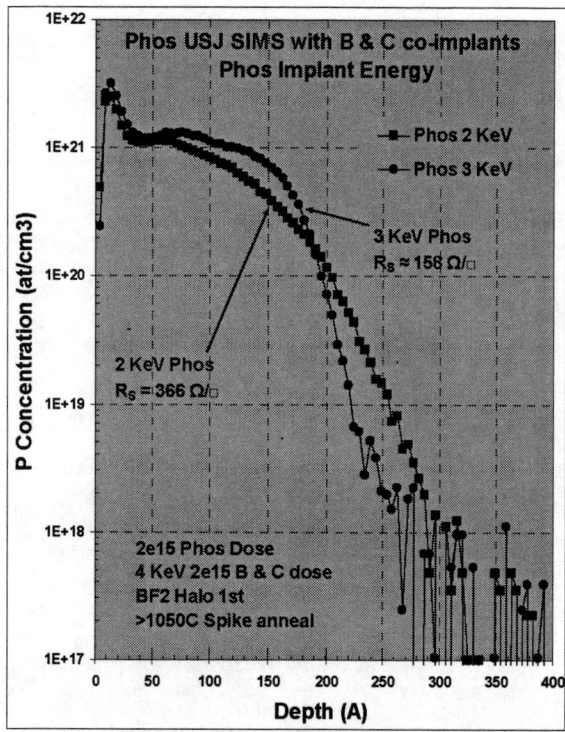

FIGURE 4. Phosphorus USJ SIMS plots with C & B co-implants. Counter-intuitively a higher Phosphorus implant energy reduces TED and increases solubility.

The Rs results shown in this work are speculative because of the high dose boron implants in these junctions but can be used as a relative comparison to one another rather than as an absolute value.

In Figure 5, amorphizing the silicon and increasing the carbon implant helps to slightly improve the solubility of the junction without changing the TED. This indicates that the SPER of the junction only plays a small role in TED control and solubility enhancement of the co-implants. It suggests that phosphorus, boron and carbon implants are completely amorphizing the silicon or that the carbon may be playing another role than consuming interstitials via the Kick-out mechanism. It also explains that the profile differences in Figure 4 are not due to better or deeper amorphizaiton with a 3 Kev phosphorus implant.

In Figure 6, skewing the carbon implant energy has no real effect on the TED or the solubility of the junction; however, reducing it significantly increased the fraction of activated dopants. Conventional wisdom says that high doses of shallow carbon degrade electrical performance. This suggests there is another role for carbon in this junction as well.

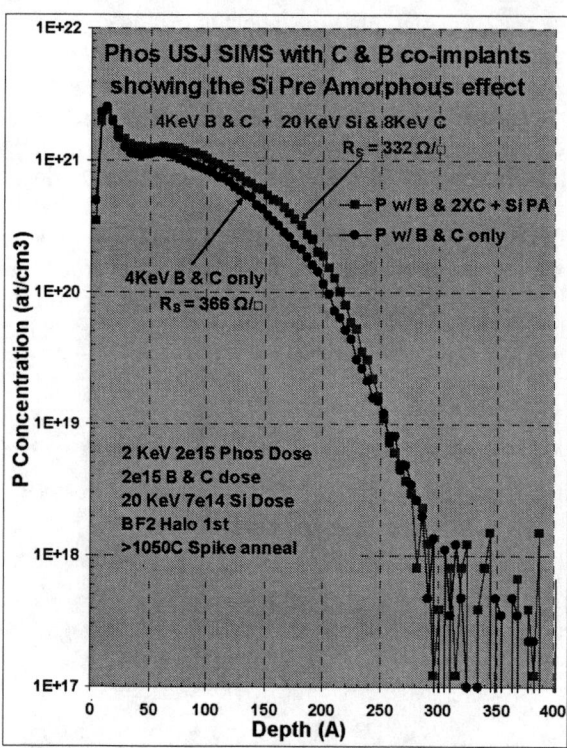

FIGURE 5. 2 KeV Phosphorus USJ with C & B co-implants. A Si pre amorphous implant and an extra C implant do not reduce TED rather they increase solubility.

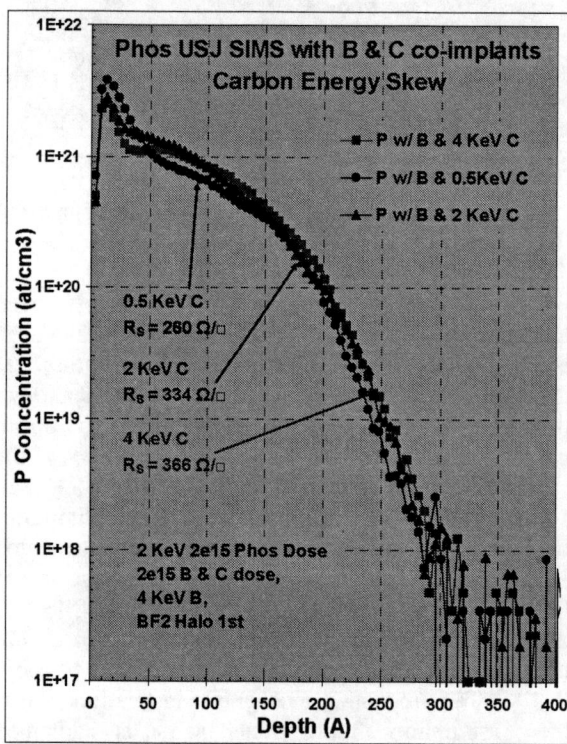

FIGURE 6. 2 KeV Phosphorus USJ's with B & C co-implants. Carbon implant energy does not affect TED but decreasing it dose help Rs.

CONCLUSIONS

Carbon and fluorine have demonstrated the ability to reduce phosphorus TED but not to the degree possible with carbon and boron co-implants. This technique is not applicable to real devices because there probably is a parasitic capacitance that the boron layer forms with the phosphorus layer. This junction revealed more aspects of phosphorus Uphill diffusion and suggests that the high solubility levels may not be scalable down in energy without further modifications to prevent Uphill diffusion.. The boron in this junction is probably forming BI clusters below the phosphorus significantly reducing the interstitials driving TED. It is observed in this work that carbon assists boron with this mechanism. If another non-electrical co-implant or method can be found to reduce the interstitial population in a similar manner, phosphorus can replace the arsenic in n-type USJ.

REFERENCES

1. T. Fiory, et al., presented at the *9th International Conference on Advanced Thermal Processing of Semiconductors-RTP 2001*, p. 227. See also Figure 1.
2. J.S. Christensen, et al., *Appl. Phys. Lett.* **82**, p.2254 (2003).
3. T. Ghani, et al., IEDM Tech. Dig., p.197-200, (2003).
4. B. Sadigh, et al., *Phys. Rev. Lett.* **83**, p.4341 (1999).
5. N.E.B. Cowern, et al., *Phys. Rev. Lett.* **67** p.212 (1991).
6. H. Rucker, et al., *Appl. Phys. Lett.* **74** p.3377 (1999).
7. P.A. Stolk, et al., *Appl. Phys. Lett.* **66** p.1370 (1995).
8. M. Diebel, et al., *MRS Symp.* **765** p.D6.15 (2003).
9. X.D. Pi, et al., *Phys. Rev. Lett.* **90**, p.155901 (2003).
10. A. Vanderpool, M. Taylor, *Nucl. Instr. & Meth. in Phys. Res. B* **237** p.142 (2005).
11. J.M. Jacques, et al., *MRS Symp.* **717** p.C4.6 (2002).
12. R. Duffy, et al., *Appl. Phys. Lett.* **86** p.81917 (2005).

Integration of Advanced Source and Drain Extension Process Using Carbon/Fluorine Co-Implants and Spike Anneal in 65nm PMOS Devices

C. I. Li[*], H. Y. Wang[*], C. C. Chien[*], M. Chain[*], C. L. Yang[*], S. F. Tzou[*], H. Graoui[+], M.A. Foad[+], Richard Ting[+]

[*] *CRD Advanced Modules, UMC, Tainan, Taiwan, ROC, 300*
[+] *Applied Materials, 974 E Arques Ave, Sunnyvale, CA 94086*

Abstract. Carbon and fluorine co-implantation have shown encouraging junction formation improvement, especially for P-type junctions. In this paper, Xj of 20 nm, Rs of 730 ohms/sq and abruptness of 3.5 nm/decade were obtained using carbon co-implantation at 6 keV, 2×10^{15} ions/cm^2, BF$_2$ implant and spike annealing. With LSA, the sheet resistance decreases to 640 ohm/sq. Rs decreased 7% at 1050°C RTP and decreased 13% at 1000°C RTP combined with laser spike annealing. We implemented germanium, carbon, and fluorine co-implanted junctions for SDE fabrication for 65 nm node devices. Results indicated that both decrease in overlap capacitance and junction leakage have proportional correlation with C co-implant dosage. However, Vt needs to be optimized. Device optimization by combining C co-implantation with LSA can yield better control of short channel effects due to the co-implantation and better activation due to the LSA.

Keywords: Carbon; co-implantation; ultra shallow junction; LSA
PACS: 68.55

INTRODUCTION

To meet the 45nm technology node requirements, source and drain extension engineering is faced with a challenging goal. Recently, carbon or fluorine co-implantation, has shown encouraging junction formation improvement to reduce boron TED, especially for P-type junctions [1,2]. S/D extension junctions as shallow as 20 nm at 5×10^{18} atoms/cm^3 with sheet resistance and abruptness of 573 ohms/sq and 2.5 nm/decade respectively have been developed [3,4]. This process was achieved by using optimized carbon implantation and standard spike annealing. In this paper we implement germanium and carbon co-implanted junctions for source and drain extension fabrication in a 65 nm node process flow and compare the co-implant performance of carbon/fluorine co-implants. We also implement a laser spike anneal (LSA) step after spike RTP in device fabrication to study the device performance variation when combining co-implants and LSA.

EXPERIMENT

300 mm wafer size has been used both for device and blanket test wafers. The wafers are N-type doped with a resistivity of 0.02 ohms-cm. The wafers were first pre-amorphized with a 12 keV germanium implant. Carbon was implanted at a range from 3keV to 12keV and dosage from 5×10^{14} ions/cm^2 to 2×10^{15} ions/cm^2. BF$_2$ was implanted at 5×10^{14} ions/cm^2 to 2×10^{15} ions/cm^2. Electrical characterization of overlap capacitance, Vt roll-off, and leakage current was carried out. Secondary Ion Mass Spectrometry (SIMS) at 5×10^{18} atoms/cm^3 level defined Xj, and Rs measurements were done by four-point probe sheet resistance measurements (Rs) on the corresponding blanket wafers.

RESULTS

Carbon co-implant energy/dosage effect

Wafers were pre-amorphized with germanium at 12 keV, 3×10^{14} ions/cm^2 then implanted with carbon at energies of 3 keV, 6 keV and 12 keV, to a dose of 1×10^{15} ions/cm^2 and with BF$_2$ implant. Wafers were annealed at high temperature spike RTP. Boron SIMS analysis on blanket wafers is presented in Figure 1. For carbon 3 keV, the SIMS profile shows the shallowest junction depth. The abruptness for carbon was approximately 4 nm/decade. The junction depth was decreased when adding the lower energy carbon co-implant. Besides, the abruptness is not sensitive to different carbon energies.

Figure 2 illustrates the boron SIMS profile for different carbon energy. The carbon dosage split from 5×10^{14} ions/cm^2, 1×10^{15} ions/cm^2 and 2×10^{15} ions/cm^2. The results show higher carbon dosage can efficiently inhibit boron diffusion and get shallower junction depth. Besides, there is only slight difference for junction depth when carbon dosage is larger than 1×10^{15} ions/cm^2.

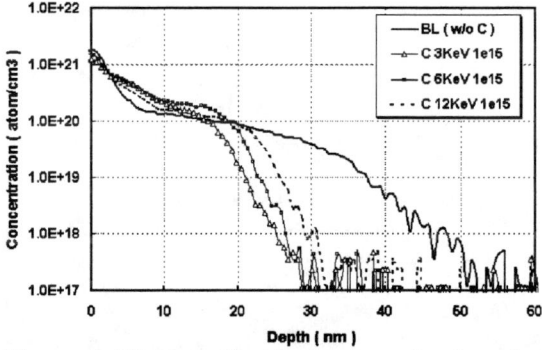

Figure 1. SIMS profiles for boron co-implanted with different carbon energy show junction depth has strong correlation with carbon energy.

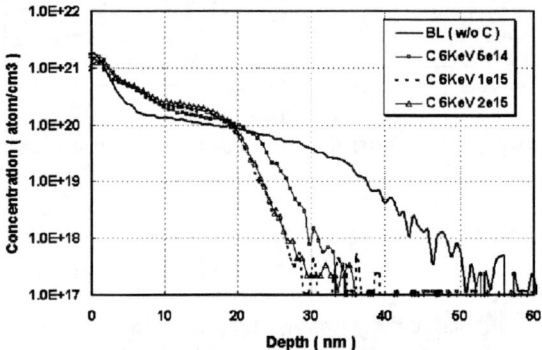

Figure 2. SIMS profiles for boron co-implanted with different carbon dosage show higher carbon dosage gets shallower junction depth.

The sheet resistance (Rs) versus junction depth (Xj) performance is presented in Figure 3. The RTP condition was split from 1080°C to 1000°C. With C co-implants, the diffusion coefficient is not sensitive to RTP temperature. The junction depth variation is only 4.4 nm from 1000 °C to 1080 °C RTP while without C co-implant, the Xj variation is 13 nm. It is worth noting the junction improvement as the energy of carbon decreases from 6 keV to 3 keV, which translates to 7% reduction of Xj but 60% increase of sheet resistance for 4keV BF$_2$ with 2×10^{15} ions/cm^2 dose C co-implant at 1050°C RTP. Lower C co-implant energy will decrease Xj but increase Rs. How to optimize the condition depends on the profile of B/C/Ge. For low-energy BF$_2$ implantation, combining 6 keV, 2×10^{15} ions/cm^2 carbon can yield the shallowest junction depth and lowest sheet resistance.

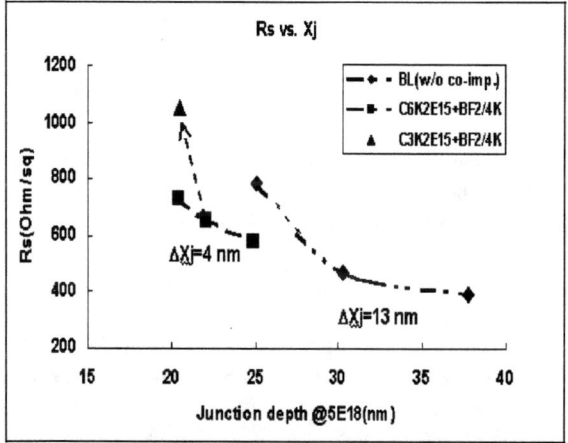

Figure 3. Carbon co-implant sheet resistance versus junction depth

Carbon and fluorine co-implantation comparison

Two co-implant species, fluorine and carbon, were studied to understand the effect of inhibiting B diffusion. Wafers were pre-amorphized with germanium implanted at 12 keV, 3×10^{14} ions/cm^2, creating an amorphous layer. Fig. 4 is a plot of the junction depth (X$_j$) versus the sheet resistance (R$_s$) for annealed wafers. At the same Rs, fluorine reduces boron diffusion 12% over no co-implants and carbon reduces boron diffusion 29%. For lower temperature spike RTP, the abruptness is not good because of serious tailing. Adding fluorine co-implantation can improve vertical abruptness 23% and C co-implantation can improve abruptness 70% to 3.5 nm/decade. Carbon co-implants show significantly sharper junctions and better junction depth (X$_j$) versus the sheet resistance (R$_s$) than fluorine co-implants.

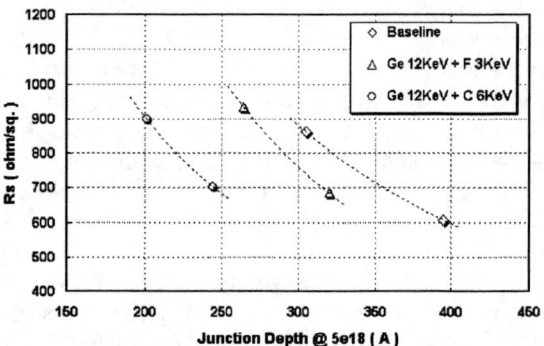

Figure 4. Carbon and fluorine sheet resistance versus junction depth performance comparison

Carbon co-implantation with laser spike anneal

To meet the 45nm Xj requirement, RTP temperature might be decreased to suppress diffusion, and co-implants could even be combined. To have better activation, combining an *m*-sec anneal is an interesting topic. PAI with carbon co-implants (dose = 2×10^{15} ions/cm^2) and BF$_2$ implants were activated by different temperature RTP and a 1300°C laser spike anneal (LSA). With LSA, the BF$_2$ implant has better activation and gets lower sheet resistance without additional diffusion. The SIMS results are shown in Fig. 5. Besides, for carbon co-implants, Rs decreased 7% at 1050°C RTP and decreased 13% at 1000°C RTP combined with LSA. Adding carbon co-implants make boron profile have higher concentration near the surface (> 1×10^{20} atoms/cm^3), and *m*-sec anneal has higher solid solubility (~ 4×10^{20} atoms/cm^3) to activate boron implants and get lower sheet resistance.

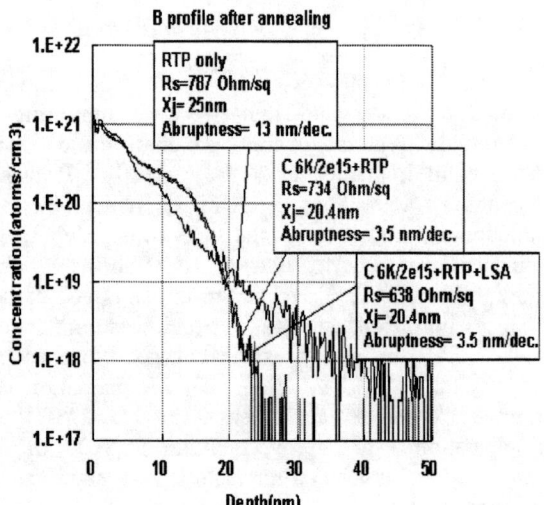

Figure 5. SIMS profile of C co-implant+ RTP+1300°C LSA.

Device performance

Device performance of the studied junctions is presented in Fig. 6 through Fig. 7. About 12%~24% decrease in overlap capacitance (C$_{ov}$) was obtained when 6 keV carbon was co-implanted, which represents an electrical signature of carbon effectively reducing lateral boron TED. Carbon dosage is proportional to the decrease of overlap capacitance. Carbon implanted at 1×10^{15} dose decreases overlap capacitance 12% and 2×10^{15} dose decreases overlap capacitance 24% (Fig. 6). Vt roll-off behavior of implanted carbon also shows better Vt roll-off than the baseline because of better short channel controlling (Fig. 7). However, Vt shift has strong correlation with carbon co-implant dosage (50 mV~120 mV shift). Modifying the channel doping could probably optimize this.

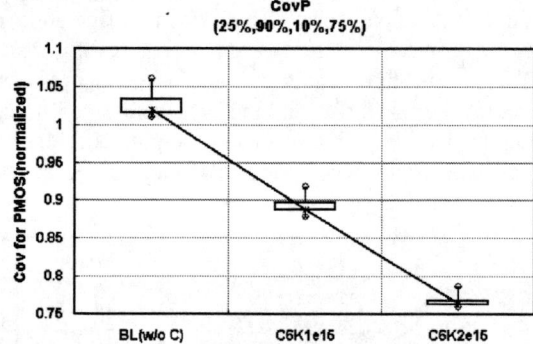

Figure 6. Overlap capacitance change with different dosage

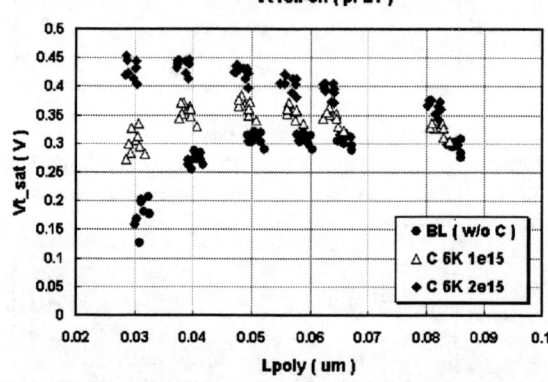

Figure 7. Vt roll-off behavior for PFET with different carbon dosage

The 6 keV carbon at 2×10^{15} ions/cm^2, however, exhibits two orders of magnitude higher junction leakage compared to the baseline. When decreasing carbon dosage, the junction leakage performance stays the same as the baseline (Fig. 8). The leakage current has strong correlation with carbon dosage, which might be due to abrupt and shallow boron profile inducing band-to-band tunneling. Some intrinsic defects also need further investigation. Vanderpool et al. [5] suggested that a carbon co-implant is a likely defect center.

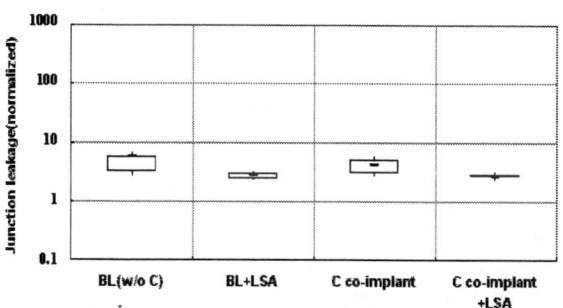

Figure 8. Leakage current comparison with and without co-implant + LSA

Device performance (with laser spike anneal)

Laser spike annealing (LSA) was implemented after S/D RTP for device fabrication. The overlap capacitance decreased 10% with 5×10^{14} dose carbon co-implantation and LSA. Adding LSA only increased overlap capacitance ~1%. The junction leakage behavior is also not changed regardless of using LSA (Fig. 8). Adding LSA also delivers the same Vt roll-off behavior compared with only co-implantation and shows better Vt roll-off than the baseline (without co-implant) (Fig. 9). Device optimization by combining carbon co-implants with LSA can yield better control of short channel effects due to co-implants and better activation due to LSA.

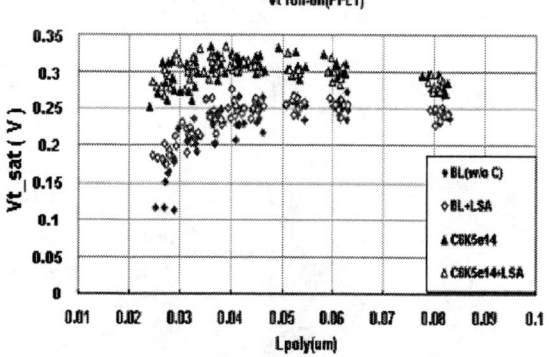

Figure 9. Vt roll-off behavior comparison with and without co-implant + LSA

CONCLUSION

We report successful junction formation on blanket wafers with Xj of 20 nm, Rs=730 ohms/sq and abruptness of 3.5 nm/decade using carbon co-implantation at 6 keV, 2×10^{15} ions/cm^2, BF_2 implant and spike annealing. With LSA, the sheet resistance decreases to 640 ohm/sq. At the same Rs, there was 29% junction depth reduction by the carbon co-implant and 12% junction depth reduction by the fluorine co-implant. For the carbon co-implant, there was 13% Rs improvement with LSA at 1000°C RTP. For device performance, low overlap capacitance is achieved by carbon co-implant and better Vt roll-off than the baseline. Adding LSA only increases overlap capacitance ~1% and Vt roll-off behavior stays the same. After device optimization, junction leakage is not changed with or without carbon co-implant and LSA.

REFERENCES

[1] J. M. Jacques, L. S. Robertson, K. S. Jones, Mat. Res. Symp. Proc. **717**, 175 (2002).
[2] R. Lindsay, B. J. Pawlak, P. Stolk, K. Maex, Mat. Res. Symp. Proc. **717**, 65 (2002).
[3] H. Graoui and M. Foad, Presented in USJ workshop 2005
[4] H. Graoui, M. Hilkene, B. McComb, M. Castle, S. Felch, N.E.B. Cowern, A. Al-Bayati, A. Tjandra and M.A. Foad, Mat. Res. Soc. Symp., Vol. **810** (2004).
[5] A. Vanderpool, M. Taylor, Nuclear Instruments and Methods in Physics Research B **237**, 142-1 (2005)

Controlling Dopant Diffusion and Activation through Surface Chemistry

K. Dev, C. T. M. Kwok, R. Vaidyanathan, R. D. Braatz and E. G. Seebauer

Department of Chemical & Biomolecular Engineering
University of Illinois
Urbana, IL 61801, USA
Email: eseebaue@uiuc.edu

Abstract. The degree of chemical bond saturation at surfaces can affect dopant activation and transient enhanced diffusion (TED) in silicon during annealing for ultrashallow junction formation. Point defects such as interstitial atoms can more easily combine with unsaturated dangling bonds than with saturated ones. Thus, maintaining an atomically clean surface during annealing greatly increases the annihilation probability. Statistical arguments show that such a surface removes Si interstitials from the underlying solid much faster than dopant interstitials. Simulations for boron and experiments for arsenic show that this effect leads to large and simultaneous improvements in dopant activation and TED. This surface-based method of defect engineering is relatively easy to integrate into a process line, and offers benefits over and above what can be obtained with methods used in parallel such as carbon co-implantation and laser annealing.

Keywords: Ion implantation; Ultrashallow junctions; Transient enhanced diffusion; Defect engineering
PACS: 61.72.Tt; 61.72.Cc; 61.72.Ji ; 66.30.-h

INTRODUCTION

Forming increasingly shallow *pn* junctions in silicon-based microelectronic logic devices is critical to accomplish as device dimensions continue to diminish. The amount of dopant in these regions that is electrically active simultaneously needs to increase. Various approaches to defect engineering intended to accomplish these purposes are now under active consideration. These approaches include millisecond-scale annealing methods such as laser and flash annealing [1], and co-implantation of elements such as C and F [2]. However, these methods suffer from limitations such as process integration, removal of implantation damage and foreign atom incorporation.

As junctions move closer to the surface, the possibility arises for using the surface itself for defect engineering. Surfaces should differ markedly in their ability to annihilate defects. For example, adsorption should exert significant effects, as suggested by recent quantum calculations [3,4] by Hwang *et al*. An atomically clean surface can annihilate interstitial atoms by simple addition of the interstitials to dangling bonds. Such a process resembles adding a hydrogen atom to a free radical in the gas phase; there is essentially no activation energy, and the addition is very facile. However, if the same surface is saturated with a strongly bonded adsorbate, annihilation requires the insertion of interstitials into existing bonds. Such insertion should have a higher activation barrier and a much lower probability of occurrence. A schematic diagram of this idea appears in Fig. 1.

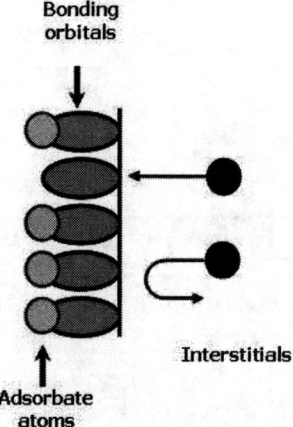

FIGURE 1. Diagram showing how bulk interstitials can react easily with surface dangling bond sites, but less easily with sites saturated by a strongly bonded adsorbate.

Figure 1 suggests that the annihilation probability can be controlled and varied simply by adjusting the amount of adsorbate bonded to the surface in submonolayer quantities. But can such adjustment lead to improved dopant behavior and ultimately device performance? As this paper shows below, modeling suggests the answer is an emphatic "yes," and early experiments bear out that prediction.

BENEFICIAL EFFECTS ON BORON DIFFUSION AND ACTIVATION

Based on the premise that the activity of the surface toward diffusing point defects might be controllable, we performed simulations of boron diffusion and activation in response to implantation and conventional spike annealing. We employed a model for boron diffusion developed previously [5]. The model utilizes continuum equations to describe the reaction and diffusion of boron, self-interstitial atoms, and related defects in silicon. These equations have the general form for species i:

$$\frac{\partial C_i}{\partial t} = -\frac{\partial J_i}{\partial x} + G_i \quad (1)$$

where C_i, J_i, and G_i denote the concentration, flux, and net generation rate of species i, respectively. The model was implemented using the process simulator FLOOPS (by Mark E. Law of the Univ. of Florida and Al Tasch of the Univ. of Texas at Austin) [6]. The values of activation energies were determined by methods drawn from systems engineering, including Maximum Likelihood estimation and Maximum *a Posteriori* estimation based on comptational and experimental results from the literature [7, 8, 9]. The model has no adjustable parameters, and has demonstrated predictive, rather than merely correlative, capabilities.

The ability of surface to annihilate interstitials can be quantified in terms of a "surface annihilation probability", S ($1 \geq S \geq 0$). S represents the likelihood that an interstitial approaching the surface is permanently removed from the underlying solid. For example, with S=0 all interstitials simply "reflect" back into the solid. The annihilation probability is incorporated in the boundary condition as

$$-D\frac{\partial C}{\partial x}\bigg|_{x=0} = J_{total} S \quad (2)$$

where J_{total} denotes the total impinging flux of interstitials. The actual flux at the surface is the product of the total impinging flux and the surface annihilation probability. In the present simulations, the annihilation probabilities for silicon and boron interstitials were simply set equal to each other (since the value for boron is not known, though boron and silicon share many features of their chemical reactivity). Various values of S were put into the model to determine the amount of profile spreading and electrical activation yielded by a conventional spike anneal. Experimental as-implanted boron profiles (0.6 keV, 2×10^{15} ions/cm^2) were used as an initial condition. The degree of activation was calculated by integrating the active boron concentration throughout the entire profile. Figure 2 shows the effect of S on profile spreading and dopant activation. The annealing-induced change in junction depth decreases sharply for S above 10^{-6} and eventually becomes very small. Activation begins to rise substantially for S above 10^{-3}, and increases by about 50% by the time S reaches unity. Separate simulations and comparisons to experimental data (not shown here) indicated that S normally lies around 10^{-5} when a screen oxide is in place. Separate measurements showed that an atomically clean surface gives $S \approx 0.05$ for Si [10]. Thus, using a clean surface has the potential to substantially improve dopant activation and diffusional spreading simultaneously.

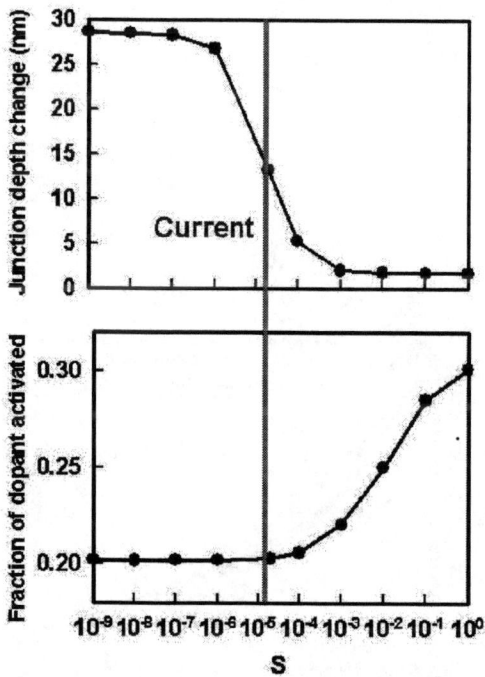

FIGURE 2. Simulated change in junction depth for boron resulting from a typical spike annealing procedure (1050°C, 150°C/s) as a function of surface annihilation probability S (set equal for B and Si). The lower panel shows the corresponding degree of dopant activation integrated throughout the entire B profile. Increasing S from its typical value with screen oxide near 10^{-5} to near unity improves both junction depth and dopant activation. Use of an atomically clean surface would accomplish this purpose.

Figure 3 shows the beneficial effects of a clean surface from a different perspective. In this case, a conventional spike anneal temperature program was used with a ramp rate of 400°C/s and a variety of peak temperatures. Fig. 3a shows that for $S = 0.05$ (the experimental value for Si on a clean Si(100) surface), the fraction of boron that is activated increases with peak temperature. Spike anneal to 1200°C gives 100% activation, since essentially all interstitial clusters dissociate by that temperature. Such large degrees of activation are no surprise at high peak temperatures, but one would normally expect huge amounts of profile spreading from TED. Indeed, Fig. 3b shows that for 1050°C and $S = 0$, the junction deepens considerably. However, for $S = 0.05$ at this peak temperature, the profile freezes with no junction deepening. Even more surprisingly, Figure 3b shows that with $S = 0.05$, even a peak temperature of 1200°C induces minimal spreading. Thus, the simulations predict 100% activation with essentially no TED.

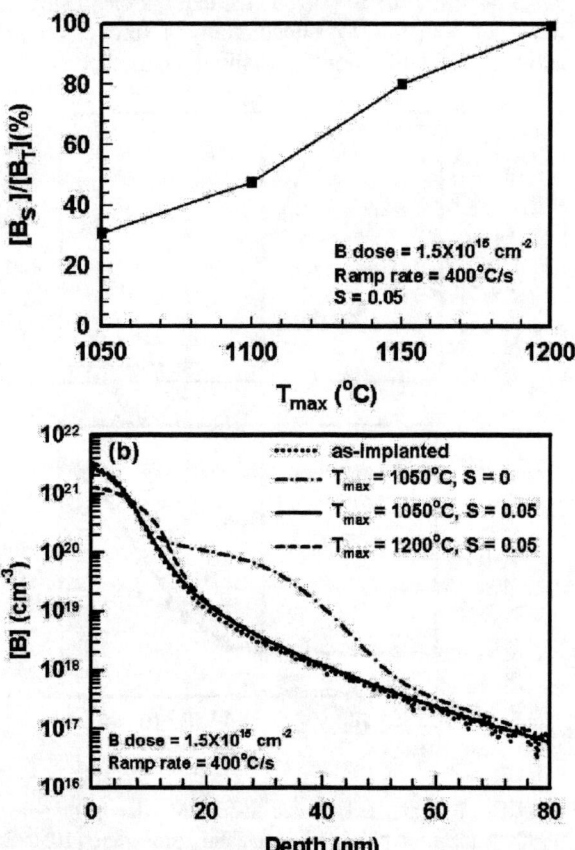

FIGURE 3. (a) Estimated degree of dopant activation integrated throughout the entire B profile as a function of peak temperature in conventional spike anneal (ramp rate = 400°C/s). (b) Simulated B annealed profiles with different peak temperatures and surface annihilation probabilities, S. Annealing with T_{max} = 1200°C and S = 0.05 gives 100% B activation and little profile spreading.

MECHANISM FOR SELECTIVE REMOVAL OF SILICON INTERSTITIALS

How does the presence of a surface that is very active for interstitial removal yield the dual benefits of increased electrical activation and reduced TED? The answer is that the surface selectively removes of silicon interstitial atoms from the solid, even when the values of S for Si and the dopant are identical.

Behavior of dopants such as boron during annealing is determined primarily by the interplay among lone interstitials, interstitial clusters, and the lattice. During implantation, numerous lone interstitials are created that diffuse quickly and accrete into clusters. Subsequent annealing dissociates these clusters. Low annealing temperatures leave most of the dopant locked within clusters, rendering it useless. Higher temperatures dissociate more clusters and release more dopant that can become active by kicking into lattice sites. However, cluster dissociation also releases interstitials of Si that kick the dopant back out of the lattice, rendering it mobile. Thus, standard annealing protocols that yield good dopant activation also lead to substantial junction deepening.

An additional "sink" that removes Si interstitials selectively over dopant interstitials would solve this problem. An atomically clean surface serves this purpose. The reason is statistical. A dopant interstitial diffusing toward the surface periodically kicks into the lattice (and becomes immobile and electrically active). The kick-in process almost always releases an interstitial of Si, and the immobilized dopant atom must wait for another Si interstitial to come along in order to become mobile again. Thus, the lattice serves to impede the motion of dopant interstitials toward the surface. Silicon interstitials also exchange with the lattice. However, at typical doping levels near 1%, a lattice exchange event simply yields another Si interstitial atom. (The remaining 1% of the events results in kickout of dopant.) Thus, the lattice does not impede the net motion of Si interstitials nearly as much as for dopant. In fact, based on crude statistics assuming equal Si-Si and Si-dopant exchange rates, the surface will extract Si interstitials about 100 times faster than dopant interstitials, even if S is the same for both species.

Note that this mechanism does not depend upon the particular type of dopant used, and operates in parallel with other defect engineering methods that are commonly employed. The method is compatible with the use of strained Si, Si-Ge, any kind of annealing protocol (spike, flash, laser), and co-implantation of species such as C and F.

SURFACE CONTROL OF ARSENIC DIFFUSION AND ACTIVATION

To verify the basic outlines of this mechanism, we studied the effect of the surface on arsenic diffusion in silicon. P-type Si (100) wafers (5 Ohm.cm) implanted with arsenic at 2 keV and dose of 2×10^{15} ions/cm^{-2} using Applied Materials Quantum X high current ion implanter were used. Specimens were annealed through resistive heating in an ultrahigh vacuum chamber (to monitor surface cleanliness). Specimens were soak annealed at 750°C for 1 hr with different surface conditions – one with an atomically clean surface and the other with one monolayer (ML) of adsorbed nitrogen created by exposure to NH$_3$ (99.99%) at 700°C before annealing. Control experiments showed minimal spreading of the profile due to this exposure procedure. At this temperature for the low exposures employed, it is well known that only atomic nitrogen remains on the surface; the hydrogen desorbs. Auger electron spectroscopy was employed to determine atomic nitrogen coverage. Diffusion profiles were measured *ex situ* with secondary ion mass spectroscopy (SIMS) using Cameca 4f, with a 700eV Cs beam incident at 50°. Sheet resistance (R_s) was measured with a standard four-point probe.

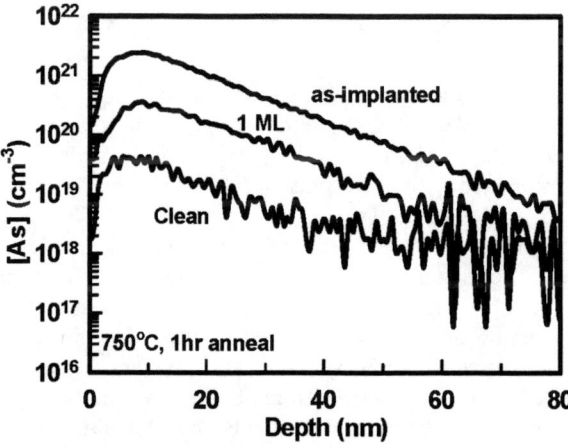

FIGURE 4. SIMS profiles of annealed arsenic for atomically clean and 1ML nitrogen adsorbed surface.

Figure 4 shows the annealed arsenic profiles. There is significant arsenic dose loss for the atomically clean surface. However, for the 1 ML surface, much more arsenic is retained, but the profile spreads considerably more as well. Also, for the clean surface, the sheet resistance was nearly 50 times lower than for the 1 ML case, even though much less total dopant remained in the sample for the clean surface anneal. Adsorption clearly exerts a large effect on R_s, and hence on the degree of dopant activation.

CONCLUSIONS

Simulations and experiments suggest that an atomically clean surface with many dangling bonds preferentially removes silicon interstitials from the bulk. This phenomenon keeps the dopant atoms in an immobilized and electrically active form by inhibiting the "kick-out" reaction.

The mechanism offers promise for simultaneous reduction in TED and increase in dopant activation. Obtaining a relatively clean surface in the processing environment does not need ultrahigh vacuum. It can be simply achieved by cleaning the surface by a procedure that leaves the surface H-terminated (such as an HF-last etch). Most adsorbates such as hydrogen then desorb early during annealing, resulting in an atomically clean surface at the peak temperature where most diffusion and activation take place. Strong surface annihilation may help reduce the dopant deactivation in subsequent processing steps by dissociating most clusters.

The beneficial effects of this method should be observed for any dopant atom whose diffusion is mediated by interstitials and annealing scheme (e.g. spike, flash or submelt laser anneal). It should also work in the presence of strain (for SiGe) as well as co-implanted species such as C and F. Moreover, the effect of surface sink should be felt strongly in 3-dimensions (i.e., under the gate), though only 1-dimension profiles have been studied in this work.

REFERENCES

1. R. Lindsay, B. Pawlak, J. Kitl, K. Henson, C. Torregiani, S. Giangrandi, R. Surdeanu, W. Vandervorst, A. Mayur, J. Ross, S. McCoy, J. Gelpey, K. Elliott, X. Pages, A. Satta, A. Lauwers, P. Stolk, and K. Maex, *Materials Research Society Symposium* **765**, 261 (2003).
2. H. Graoui, M. Hilkene, B. McComb, M. Castle, S. Felch, A. Al-Bayati, A. Tjandra, and M. A. Foad, *Nucl. Instrum. Meth. Phys. Res. B* **237**, 46 (2005)
3. T.A. Kirichenko, S. Banerjee and G.S. Hwang, *Phys. Rev. B* **70**, 045321 (2002).
4. T.A. Kirichenko, S. Banerjee and G.S. Hwang, *Phys. Status Solidi B* **241**, 2303 (2004).
5. R. Gunawan, M. Y. L. Jung, E. G. Seebauer and R. D. Braatz, *AIChE J.* **49**, 2114 (2003).
6. See Mark Law, http://www.swamp.tec.ufl.edu/
7. M. Y. L. Jung, R. Gunawan, R. D. Braatz and E. G. Seebauer, *AIChE J.* **50**, 3248 (2004).
8. R. Gunawan, M. Y. L. Jung, R. D. Braatz, and E. G. Seebauer, *J. Electrochem. Soc.* **150**, G758 (2003).
9. C. T. Kwok, K. Dev, E. G. Seebauer and R. D. Braatz, Proc. of the Joint IEEE Conf. on Decision and Control and European Control Conference 2058 (2005).

Strain-Enhanced Activation of Sb Ultrashallow Junctions

N. S. Bennett, L. O'Reilly[1], A. J. Smith, R. M. Gwilliam, P. J. McNally[1], N. E. B. Cowern, B. J. Sealy

Ion Beam Centre, Advanced Technology Institute, University of Surrey, Guildford GU2 7XH, UK
[1]*Nanomaterials Processing Laboratory, Research Institute for Networks and Communications Engineering (RINCE), School of Electronic Engineering, Dublin City University, Dublin 9, Ireland*

Abstract. Sheet resistance (R_s) reductions are presented for antimony and arsenic doped layers produced in strained-Si. Results show a modest R_s reduction for arsenic layers, a result of a strain-induced mobility enhancement, whereas for Sb, a superior lowering is observed from improvements in both mobility *and* activation. Tensile strain is shown to enhance the activation of dopant Sb whilst creating stable ultrashallow junctions when low-temperature processing is employed. Our results propose Sb as a viable alternative to As for the creation of highly activated, low resistance ultrashallow junctions for use with strain-engineered CMOS devices.

Keywords: Dopant activation, Strained-silicon, Rapid thermal annealing, Junction stability.
PACS: 61.72.Tt, 61.73.Ji, 81.15.Np.

INTRODUCTION

The ability to create highly activated, low resistance ultrashallow junctions is a limiting factor for future silicon-based devices. In parallel, devices seem set to rely increasingly on strain-engineering to enhance device performance and relieve the reliance on aggressive dimension downscaling that has been the industry norm for over forty years.[1]

Although strain-engineering has been employed in devices since 2003 as a mobility enhancer in the channel, little is known about the impact of strain on neighbouring areas of the device. The source/drain extension regions are one such area requiring the creation of increasingly ultrashallow junctions in the presence of strain. The ability to further optimize the use of strain in these regions is still to be realized and limited scientific literature exists: Arsenic is the conventional n-type dopant species in Si and literature discussing As implants in bulk- and strained-Si layers have shown a reduction in sheet resistance in the presence of tensile strain. This is attributed directly to a strain-induced mobility enhancement, with *no* evidence for any improvement in activation.[2-4]

Due to its greater mass Sb is considered a feasible alternative to As as inherently it forms shallower, more abrupt junctions for same implant conditions. Similarly, we have previously shown Sb to be highly activated and free from out-diffusion in Si following low-temperature rapid-thermal processing without the need for more sophisticated anneals.[5,6]

EXPERIMENTAL DETAILS

Experiments were performed on IQE tensile strained Si wafers grown on constant-composition $Si_{0.83}Ge_{0.17}$ relaxed buffer layers. Each wafer was implanted with a low energy (2keV) Sb or As implant to a dose of $2\times10^{14}cm^{-2}$ to create junctions to 13 or 15nm respectively, taken at the $3\times10^{18}cm^{-3}$ level. Control samples were prepared using conventional p-type Si wafers for comparison. The as-implanted junction depth was confirmed by secondary-ion mass spectrometry (SIMS), as was implant dose by Rutherford back-scattering. Following implantation dopant activation was achieved by annealing wafer pieces for 10s in an N_2 ambient. A traditional rapid-thermal process was employed for temperatures in the range 600-800°C.

Van der Pauw sheet resistance measurements combined with sheet Hall-effect measurements were used to measure active carrier density and Hall mobility throughout the implanted region.[7] R_s measurements were confirmed by four-point probe. UV micro-Raman spectroscopy was employed to measure strain in the top 9nm of the wafer and ensure that strain remains present after processing.

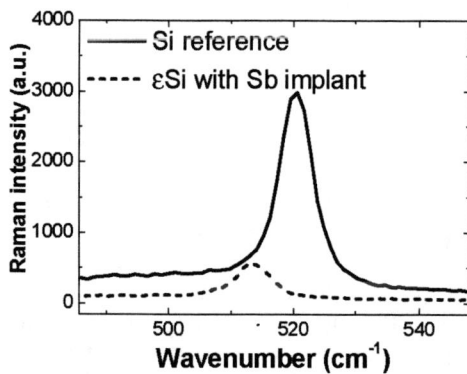

FIGURE 1. UV micro-Raman spectra comparing the signal of a Si reference with that of a strained-Si sample implanted with 2keV Sb at a dose of 2×10^{14}cm^{-2}.

RESULTS AND DISCUSSION

Micro-Raman Spectroscopy

By comparing the UV micro-Raman spectra of samples at various stages of sample preparation the strain content throughout processing can be acquired. The Raman spectra of the strained-Si and reference Si samples are dominated by the signal of the degenerate Si-Si mode at a Raman shift of ~520cm^{-1}. Figure 1 shows the red-shift in the Si-Si mode Raman peak position between a reference Si sample and that of a strained-Si sample implanted with Sb. This shift of 6.5±0.15cm^{-1} implies a strain of 1.62±0.05GPa is present in the strained sample after implantation.[8] Although the peak intensity is reduced indicating some implant damage, it is clear that ion implantation has not relaxed the layer. Additional Raman measurements were made to confirm that strain was present prior to implantation and also to ensure that the Si-Si mode peak height is restored following thermal anneal, showing that the layer is repaired.

Electrical Characterization

Van der Pauw sheet resistance measurements are shown in figure 2 for ultrashallow junctions created in strained and unstrained material. Sb implants into conventional silicon show similar resistance characteristics to those found by Alzanki et al.[5] where the lowest Rs is seen for processing at 600 and 700^{0}C and increased Rs is apparent as the anneal temperature is raised to 800^{0}C and above.

Additionally, figure 2 reveals the fraction of the implanted dose that is measured as electrically active by the Hall-effect technique. For Sb implants in bulk Si, the best levels of activation are seen after processing at 600 and 700^{0}C, in good agreement with Alzanki et al.[5]

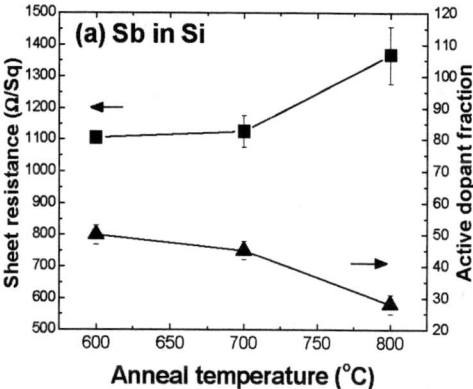

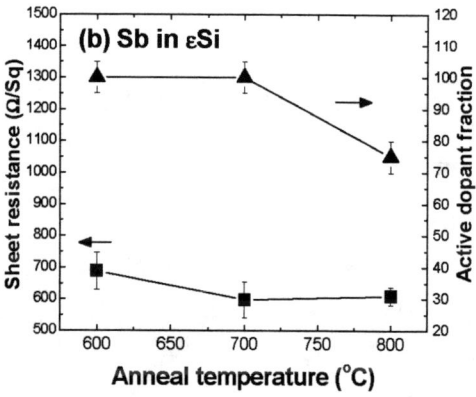

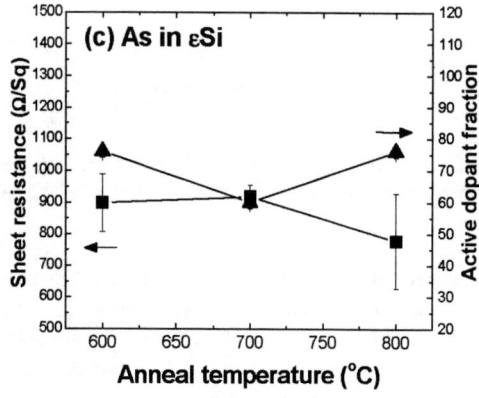

FIGURE 2. Sheet resistance (squares) and active dopant fraction (triangles) as a function of anneal temperature for (a) Sb in Si, (b) Sb in strained-Si and (c) As in strained-Si. Implants were carried out at an energy of 2keV with a dose of 2×10^{14}cm^{-2}.

By comparing figures 2a and 2b the considerable reduction in Rs for Sb implants made in strained material compared to unstrained is clear to see. An improvement factor in Rs of up to x2.5 is evident. Comparing the measured active fraction, the figure also demonstrates that strain has produced a doubling in the best levels of activation i.e. at 600 and 700^{0}C, activation is improved from 50 to 100%. Although deactivation is evident after processing at 800^{0}C the improvement in active fraction is even greater rising from 28% to 75% activation.

Evaluating figures 2b and 2c allows comparison between same-condition Sb and As implants. The reader can see that for lower anneal temperatures Sb produces lower resistance junctions. A shallower junction depth will accompany this, both because the Sb implants are shallower, and because Sb implants do not undergo the transient enhanced diffusion (TED) that occurs with As. For processing at 800^{0}C the Rs for As implants tends to improve towards the Sb case. Here, however, increased junction depth will arise from strain-enhanced TED that becomes significant for As, as demonstrated by Dilliway et al.[3] Additionally, although similar proportions of As and Sb dopants are activated after annealing at 800^{0}C, following 600 or 700^{0}C anneal significantly more (up to 40%) implanted Sb is activated compared to As.

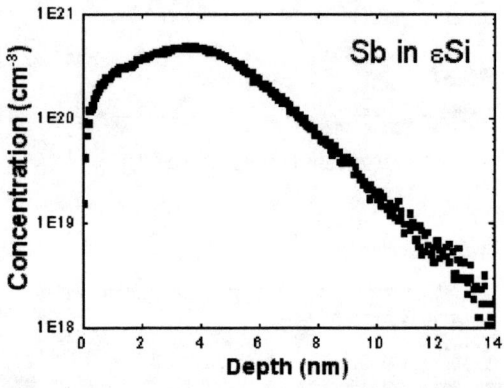

FIGURE 3. SIMS profile of 2keV, 2×10^{14}cm^{-2} Sb implant into strained-Si, followed by 10s rapid-thermal anneal at 700^{0}C.

Previous work on As has shown strain to have no effect on dopant activation. Although a modest amount of Rs lowering is reported, this is clearly attributed to mobility improvements alone.[2-4] We propose that for implanted-Sb the Rs lowering seen in strained-Si is a result of an enhancement in electrical activation in addition to a mobility improvement. Although the carrier concentration is subject to a currently unknown correction factor due to scattering mechanisms inherent in the Hall-effect measurement,[9] it seems unlikely that this will discount such a large activation effect. The reduction in Rs demonstrated earlier appears too great to be simply a result of mobility enhancement.[10]

Secondary-Ion Mass Spectrometry

Figure 3 demonstrates an atomic profile of implanted-Sb into strained-Si followed by 700^{0}C anneal. Experiments investigating the diffusive nature of Sb in conventional silicon have shown it to be free from dopant out-diffusion for processing at temperatures <800^{0}C.[5,6] Likewise, literature predicts that the diffusion of vacancy-mediated diffusers such as antimony, should actually be reduced in the presence of tensile strain.[11] Figure 3 shows the as-implanted junction depth to be conserved at ~13nm following anneal. It can therefore be seen that following implantation and anneal at 700^{0}C (the condition that produces the lowest resistance junction), dopant Sb remains stable against out-diffusion.

CONCLUSION

Sheet resistance improvements for antimony and arsenic doped layers produced in strained Si have been presented. The mobility-induced Rs reduction for As has been compared to that for Sb, where we observe a much stronger reduction in Rs caused by an additional improvement in dopant activation. Tensile strain has been shown to enhance the activation of dopant Sb atoms. Additionally, Sb junctions created with the best resistance characteristics are also seen to be free from dopant out-diffusion effects.

Results propose Sb as a viable alternative to As for the creation of ultra-shallow junctions for source/drain extension regions in strain-engineered CMOS devices.

ACKNOWLEDGMENTS

The authors wish to thank IQE Silicon Compounds for providing the strained material used in these experiments. N.S.B. is supported by an EPSRC Doctoral Training Award and N.E.B.C. is supported by an Applied Materials/Philips/Royal Academy of Engineering Research Chair of Nanoscale Materials Processing.

REFERENCES

1. The International Technology Roadmap for Semiconductors, 2005.

2. N. Sugii, S. Irieda, J. Morioka, T. Inada, J. Appl. Phys., **96** (1), 261-268 (2004).
3. G. D. M. Dilliway, A. J. Smith, J. J. Hamilton, J. Benson, L. Xu, P. J. McNally, G. Cooke, H. Kheyrandish, N. E. B. Cowern, Proc. IIT NIM-B, **237**, 131-135 (2005).
4. N. S. Bennett, A. J. Smith, C. S. Beer, L. O'Reilly, B. Colombeau, G. D. Dilliway, R. Harper, P. J. McNally, R. Gwilliam, N. E. B. Cowern and B. J. Sealy, Mater. Res. Soc. Symp. Proc. **912,** C2.3 (2006)
5. T. Alzanki, R. Gwilliam, N. Emerson, B. J. Sealy, Appl. Phys. Lett., **85,** 11, 1979-1980 (2004).
6. T. Alzanki, PhD Thesis, University of Surrey UK, 2003.
7. N. S. Bennett, A. J. Smith, B. Colombeau, R. Gwilliam, N. E. B. Cowern, B. J. Sealy, Mat. Sci. Eng. B, **124-125**, 305-309 (2005).
8. B. Pichaud, M. Putero and N. Burle, Phys. Stat. Sol. (a) **171**, 251 (1999).
9. Y. Sasaki, K. Itoh, E. Inoue, S. Kishi, T. Mitsuishi, Solid-St. Electron., **31**(1), 5-12 (1988).
10. M. L. Lee, E. A. Fitzgerald, M. T. Bulsara, M. T. Currie, A. Lochtefeld, J. Appl. Phys., **97**, 1-27 (2005).
11. N. E. B. Cowern, P. C. Zalm, P. van der Sluis, D. J. Gravenstijn, W. B. de Boer, Phys. Rev. Lett. **72**, 2585 (1994).

Phosphorus Implant For S/D Extension Formation: Diffusion And Activation Study After Spacer And Spike Anneal

S.H. Yeong*[¶], B. Colombeau[¶], F. Benistant[¶], M.P. Srinivasan*, C.P.A. Mulcahy[¥], P.S. Lee[ϵ], L. Chan[¶]

*Department of Chemical and Biomolecular Engineering, National University of Singapore, 9 Engineering Drive 1, Singapore 1175776
[¶]Chartered Semiconductor Manufacturing Ltd., 60 Woodlands industrial Park D, Street 2, Singapore 738406
[¥]Cascade Scientific Ltd, ETC Building, Brunel Science Park, Uxbridge UB8 3PH, U.K.
[ϵ]School of Material Science and Engineering, Nanyang Technological University, 50 Nanyang Avenue, Singapore 639798

Abstract. Formation of highly activated S/D extension is one of the key issues to meet the requirements for further downscaling of CMOS devices. Germanium-preamorphization implant (Ge-PAI) followed by solid phase epitaxial regrowth (SPER) is capable of forming abrupt and shallow junctions with activation levels well above solid solubility. In this paper, we demonstrate a possible alternative by using phosphorus (P) with the Ge-PAI and boron (B) Halo implant to form the S/D extension of NMOS. The anomalous diffusion and activation of P after the conventional spacer and spike anneals were studied. We observed that a highly activated and shallow junction can be formed via Ge-PAI after spacer anneal which is equivalent to a low temperature SPER. The level of P activation is enhanced even more when B Halo is considered. It is postulated that this is due to the competing interactions of B and P with the emitted interstitials of the end-of-range (EOR) defects. However, improvement in junction depth and electrical properties due to B halo implant decreases when annealing temperature increases.

Keywords: Ge-PAI, SPER, phosphorus, Halo implant, spacer anneal, spike anneal
PACS: 61.72.-y, 61.72.Cc, 61.72.Ji, 61.72.Ss, 66.30.Jt, 66.30.Lw

INTRODUCTION

The main issues for the fabrication of ultra shallow junctions (USJ) in advanced nano-CMOS technology are how to achieve high dopant activation, shallow penetration dopant profile and good junction steepness by ion implantation. Among the various proposed techniques, Ge-preamorphizing implant (Ge-PAI) is one of the most promising methods to fulfill the stringent requirements of future USJ [1]. This technique is performed by amorphizing the Si substrate with Ge implant prior to low energy dopant implant, followed by a low temperature solid expitaxial regrowth (SPER).

A general understanding of Ge-PAI with B is well established in recent years [2-3]. However, there are not many studies applying such a technique for the n-type dopants. Arsenic (As), the favored dopant for NMOS, has heavy atomic mass, low diffusivity and high solubility appears to be suitable for USJ formation. Meanwhile, high junction capacitance and polysilicon depletion are some of the issues associated with As source/drain (S/D). Such problems can be solved by the formation of hybrid (arsenic + phosphorus) S/D [4]. Therefore, the potential of P is remarkable. The most severe problem of P dopant is the transient enhanced diffusion (TED). Its diffusion mechanism is not as straightforward and fully interpreted as other dopants (such as B and As). From the early model of vacancy-mediated diffusion [5] to the more recently consensus of interstitial-mediated diffusion [6,7], P is always one of the controversial dopants for diffusion mechanism. More efforts are required to expand its application in NMOS devices.

In this work, we investigate the anomalous diffusion and activation of P with Ge-PAI after spacer and spike anneals. In addition, we also show the influence of B-Halo implant on the junction characteristics.

EXPERIMENTAL DETAILS

Czochralski-grown, p-type, <100> oriented silicon wafers with a resistivity of 10~25 ohm.cm, were used in this experiment. Ge implant conditions varied between 15 keV and 30 keV with doses ranging from 3×10^{14} atoms/cm^2 to 5×10^{14} atoms/cm^2 were performed to create amorphous layer with different thickness. The standard wafer (non-preamorphized) was implanted with P at energy 2 keV to a dose of 1×10^{15} atoms/cm^2 together with those preamophized wafers. One of wafers was also pre-implanted with B at 9keV to a dose of 7×10^{13} atoms/cm^2 for Halo formation prior to P implant. All implantations were performed at 7° tilt and 27° twist in single quad-mode. The wafers were then subjected to a series of anneal conditions in an inert N2 ambient. Particularly, low thermal budget scheme from 700°C to 750°C, typical of a spacer anneal, and high temperature conventional spike were studied.

The dopant profiles were analyzed by secondary ion mass spectrometry (SIMS). The primary beam source and condition were chosen accordingly to the implant dopant type and energy, so optimum depth profiles would be obtained. In order to reduce the 30SiH interference for P in the mass spectrum, high mass resolution setting was applied. Four point probe and Hall effect measurements were performed to obtain sheet resistance (Rs) and active carrier concentration (Ns) of the samples. A unity Hall scattering factor is assumed. Rutherford backscattering (RBS) was used to measure the amorphization depth and it was verified by cross-sectional Transmission Electron Microscpy (XTEM).

RESULTS AND DISCUSSION

The SIMS P profiles for the as-implanted and annealed samples with different amorphous thickness are presented in Figure 1. Note that there is no significant variation for all the as-implanted profiles. Typical two step spacer anneal scheme, 700°C and 750°, were performed on the samples. Table 1 shows the depth of amorphization from RBS channeling measurements. It was observed that continuous amorphous layers were obtained not only on the Ge-PAI samples but also on P implanted standard sample. This is because the implant dose (1×10^{15} atoms/cm^2) of P is high enough to cause self-amorphization with a thickness approximately 8nm.

TABLE 1. Thickness of amorphous layer at various implant conditions.

Implant condition	Thickness
P 2 keV, 1×10^{15} atoms/cm^2	8 nm
Ge 15 keV, 3×10^{14} atoms/cm^2	21nm
Ge 30 keV, 5×10^{14} atoms/cm^2	44nm

As shown in figure 1, the P dopants moved preferentially toward the surface. This applies to both non-Ge-PAI (P standard) and Ge-PAI cases. Such observation was reported recently [7] and it is closely similar to the phenomenon observed for boron uphill diffusion with the PAI [8]. The phenomenon is explained due to the supersaturation of interstitial and defect evolution which leads to the interstitial-mediated transient enhanced diffusion (TED) mechanism. As the depth of amorphous layer becomes thicker, the impact of surface-driven phosphorus movement is weaker. In term of sheet resistance (Rs), a similar trend was seen, at which the Rs value can be inversely correlated to the amorphization thickness. The P movement length (at $C_p = 1\times10^{20}$ cm^{-3}) and Rs value correspond to the PAI depth was summarized in figure 2. Based on the data, we can conclude that the deeper the end of range (EOR) defect band, the lesser the effect on the surface-preferential diffusion of P and higher level of activation can be achieved under the same anneal condition at low temperature range.

It was also worth to mention here that a higher dopant concentration was noticeable near the surface (~3nm) for P standard profile (non-Ge-PAI) as compared to Ge-PAI profiles. It is speculated that this might due to the difference in surface proximity effect between P self-amorphization and Ge-preamorphization. The further investigation is beyond the scope of this work, but it should not rule out the consideration of SIMS resolution limit near to the surface.

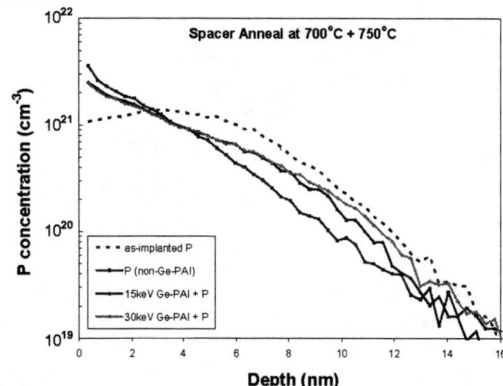

FIGURE 1. SIMS profiles of phosphorus, as-implanted P 2keV (1×10^{15} atoms/cm^2) and after spacer anneal at 700°C +750°C for non-Ge-PAI, 15 keV (3×10^{14} atoms/cm^2) Ge-PAI and 30keV (5×10^{14} atoms/cm^2) Ge-PAI

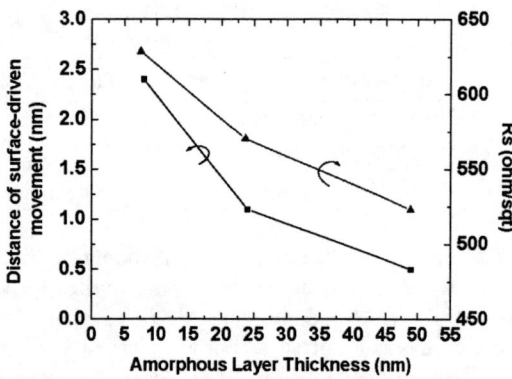

FIGURE 2. The amorphous layer thickness dependence of surface driven distance and sheet resistance (Rs)

Same spacer anneal condition was applied to the 15 keV, 3×10^{14} atoms/cm^2 Ge-PAI sample with B Halo implant prior to the doping of P. Figure 3 shows the comparison of SIMS profiles for P with and without the Halo implant as well as the B Halo profile. P trapping around the EOR defect band can be clearly identified for both cases. However, the concentration of P right before the EOR band is higher for the case which B Halo implant is performed. One possible explanation is that B atoms interact preferentially with the emitted interstitial atoms released by the EOR defects and thus decreasing the P interaction with the interstitial atoms. For the B Halo profile, as expected, we can see the trapping of B atoms around the EOR band. The observed "kink" of B within the n-type P profile is due to the ion pairing between P$^+$ and B$^-$. This ion pairing decreases the B diffusivity in the n-layer.

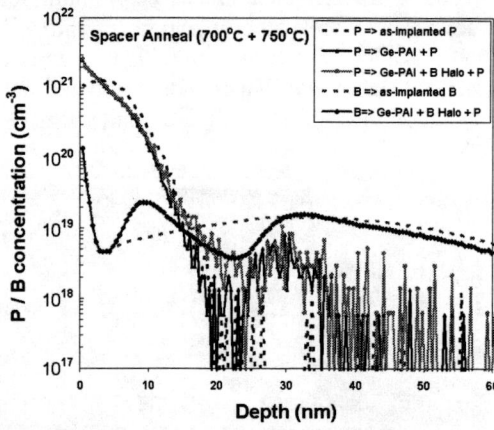

FIGURE 3. SIMS profiles of P (2keV, 1×10^{15} atoms/cm^2) and B (9 keV, 7×10^{13} atoms/cm^2), as-implanted and after spacer anneal for Ge-PAI (15 keV, 3×10^{14} atoms/cm^2) with and without B Halo implant

Interestingly, the Rs was found to decrease from 628.6 ohm/sq to 264.6 ohm/sq when the B Halo implant was added on the Ge-PAI sample. The reduction is more than 50%. Meanwhile, Hall effect measurement verified that active dopant is of n-type and its concentration level is greatly enhanced. Combining the results of Rs and SIMS profiles, it suggests that B atoms from the Halo implant is competing with P or preferentially interacting with self-interstitials released from EOR and hence increases the phosphorus activation under the low temperature anneal process.

So far low temperature scenarios have been discussed; the impact of B Halo implant on phosphorus diffusion and activation after high temperature spike anneal was studied as well. As clearly dictated in figure 4, there is no great difference in term of the metallurgical junction depth (~55nm) for non-Ge-PAI, Ge-PAI and B-Halo coupled Ge-PAI samples. However, it can be clearly seen that the shape of P profiles is different from each other after spike anneal. There is noticeable variation in between the non-Ge-PAI sample and the Ge-PAI samples in term of "kink" and abruptness of the concentration profiles; while the impact of B Halo implant on the Ge-PAI case is greatly reduced. Unlike the spacer anneal, no "kink" can be observed at the tail of the profiles. As one would expect, this is because both the trapped B and P atoms which located around the EOR dissolve fully after high temperature spike treatment. However, the tail of the profiles shows significant P transient enhanced diffusion (TED).

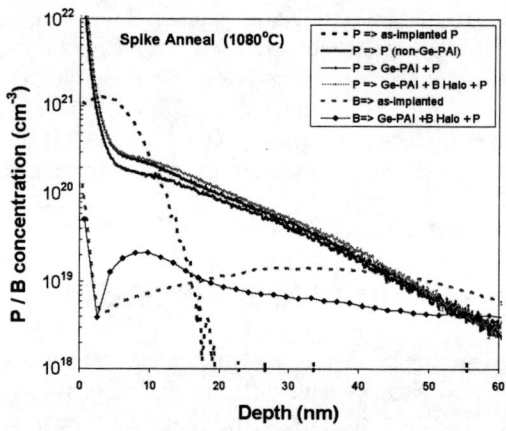

FIGURE 4. SIMS profiles of P (2keV, 1×10^{15} atoms/cm^2) and B (9 keV, 7×10^{13} atoms/cm^2), as-implanted and after spike anneal for non-Ge-PAI, Ge-PAI (15 keV, 3×10^{14} atoms/cm^2) with and without B Halo implant

Figure 5 presents the comparison of Rs values under the effect of Ge-PAI and B Halo implant after the low temperature spacer and high temperature spike anneals. Generally, the spacer annealed samples offer larger Rs value than the spike treatment. The results also reveal the gradual reduction of sheet resistance with respect to amorphization depth for both low and high temperature schemes.

Significant reduction of Rs caused by the B-Halo implant on Ge-PAI samples was observed after the spacer anneal. This is speculated due to the interactions of B atoms with the interstitials which emitted from EOR band. However, the extent of Rs improvement for the spike anneal is decreased to approximately 20% from 50% of the low temperature treatment. The improvement of Rs under spike anneal condition is also attributed to the same mechanism previously mentioned. With higher thermal budget, it is believed that the phosphorus-vacancy clusters start to dissolve and thus decreasing the B halo impact on the final level of P activation. This is in agreement with the slightly higher abruptness and "kink" level as shown in the SIMS profiles of figure 4.

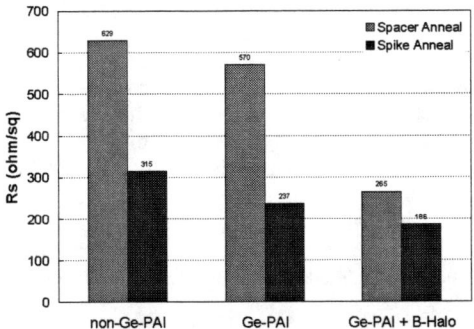

FIGURE 5. Comparison of sheet resistance (Rs) for non-Ge-PAI, Ge-PAI and B-Halo coupled Ge-PAI samples after the spacer and spike anneals

Although the Rs indirectly indicates the level of dopant activation, Hall effect measurement provides a more precise activation information by offering the total active carrier concentration (Ns) and mobility. The same set of samples which used in Rs measurement was re-analyzed by Hall and plotted in Figure 6. When we correlate Ns of this figure with Rs of figure 5, we can suggest that reduction of Rs is due to the higher level of active dopant concentration (Ns). When B-Halo implant is carried out on the Ge-PAI sample, the increment of Ns for the spacer anneal is more drastic than high temperature spike anneal. This in fact is in agreement with the Rs trend. Similarly, the Hall mobility is showing a decreasing trend as the Ns increases for the two anneal conditions. This further verified the electrical measurement by showing theoretically Rs is equal to the inverse product of Ns and mobility.

In summary, we have investigated phosphorus anomalous diffusion and activation after low temperature spacer and high temperature spike anneals. The surface-driven P movement was correlated to the location of EOR defect band as well as the level of activation. It has also been shown that activation level of P can be increased significantly when the B-Halo implant is performed on the Ge-PAI P implanted samples after low temperature treatment (SPER). An ultra-shallow and steep profile can be achieved under such condition. Although the impact of Ge-PAI and B-Halo implant on Rs reduction is noticeable after spike anneal, the enhancement of P activation is decreased to certain extent. Additionally, anomalous diffusion of P is dominant over the surface-driven movement under high temperature range. This study provides an important insight and physical basis for modeling and optimization of future NMOS technology.

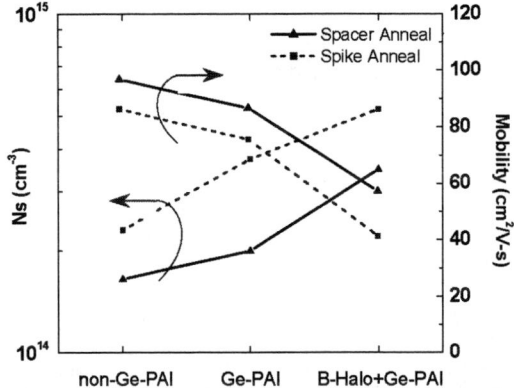

FIGURE 6. Comparison of active carrier concentration (Ns) and mobility for non-Ge-PAI, Ge-PAI and B-Halo coupled Ge-PAI samples after spacer and spike anneals.

ACKNOWLEDGMENTS

This work was carried out as part of National University of Singapore (NUS) – Chartered Semiconductor Manufacturing (CSM) research collaboration program with input from TCAD group (CSM). The supports from IMRE and MSE, NTU are also gratefully acknowledged.

REFERENCES

1. R. Lindsay, R. Severi, B.J. Pawlak, K. Henson, A. Lauwers, X. Pages, A. Satta, R. Surdeaunu, H. Lendzian and K.Maex, *in the Fourth Intternational Workshop on Junction Technology (IEEE, 2004)*, pp.70-75
2. B. Colombeau et al., *Mat. Res. Soc. Proc.*, 810, C3.6 (2004)
3. B.J. Pawlak, R. Surdeaunu, B.Colombeau, A.J. Smith, N.E.B. Cowern, R. Lindsay, W.Vandervorst, B. Brijis, O. Richard, and F. Cristiano, *Appl. Phys. Lett.* 84, 2055 (2004)
4. K.K. Young, et al., *IEDM Tech. Dig.*, 2000, pp.563-566
5. R.B. Fair and J.C.C. Tsai, *J. Electrochem, Soc.* 124, 1107 (1977)
6. P.M. Fahey, P.B. Griffin, and J.D.Plummer, *Rev. Mod. Phys.* 61, 289 (1989)
7. R. Duffy et al., *Appl. Phys. Lett.* 84, 081917 (2005)
8. Ray Duffy, V.C. Venezia, A. Heringa, T.W.T. Husken, M.J.P. Hopstaken, N.E.B. Cowern, P.B. Griffin, C.C. Wang, *Appl. Phys. Lett.* 82, 21 (2003)

The Concept of LDSI (Locally-Differentiated-Scanning Ion Implantation) for the Fine Threshold Voltage Control in Nano-Scale FETs

Min-Yong Lee[*], S.W. Jin, Y.S. Sohn, S.K. Na, K. B. Rouh, Y.S. Joung, Y.J. Ki, I.K. Han, Y.W. Song, S.W. Park

R&D Division Hynix Semiconductor Inc., San 136-1, Ami-ri, Bubal-eub, Ichon-si, kyoungki-do 467-701, Republic of Korea.

Abstract.
New Concept of ion implantation technology, namely Locally-Differentiated-Scanning Ion Implantation (LDSI), is suggested for the first time in this report. The Vth variation caused by process non-uniformity in each process variables such as gate Critical Dimension (CD), spacer deposition Uniformity and spacer etch CD, etc, becomes big huddle in scale down and larger diameter wafer. This LDSI technology has been developed for reducing the Threshold Voltage (Vth) variation within a wafer in nano-scale FETs. The LDSI technology has the capability of implanting of the locally different dose at the desired region in wafer by controlling scan speed of X- or Y-direction or by combining them with step-wise rotation. By applied LDSI on S/D and halo implantation on 90nm era DRAM, the remarkable improvement in Vth variation has been achieved as much as 47%. Also, the Improvements in device yield and Idd fail rate by LDSI on 512M DDR3 product.

By the evidences of this study, new implantation technology concept, LDSI, is believed to be a key doping technology for future nano-scale FETs device fabrication.

Keywords: Locally-Differentiated-Scanning Ion Implantation, Process variables, Threshold Voltage.
PACS: 79.20.R, 85.30. Tv, 85.40.Ry, 85. 40-e

1. Introduction

Even though the current ion implanter can make sure the very good dose controllability and uniformity [1,2], the most of device engineers in semiconductor industry has been suffered from the difficulties on Vth control. Both by the device shrinkage down to sub 70nm era and by the change of market needs to low voltage/high speed devices, the problem of insufficient Vth margin will be more serious in larger Diameter [3].

For a long time, the many reasons of Vth variation have been studied. Recently, the Vth variation caused by process non-uniformity in each process step becomes very critical because of too narrow gate CD. As shown in Fig. 1, the map of Vt variation, typically, has some types of specific shape caused by the map trend of process variables such as gate CD, spacer deposition uniformity and spacer etch CD, etc.

If we implant the differentiated dose on the local region within wafer and they can be controlled, we can compensate the Vt variation caused by process variables.

2. Experimental

The various examples of locally different dose map obtained by LDSI are shown in Fig.2. The concept of LDSI is a controlled non-uniformity ion beam process with a wafer.

Dose depends on staying time of ion beam per the unit area. When the ion beam staying time of the desired region of wafer comes to be shorter than normal beam condition time relatively, it becomes a lower dose region. In the other

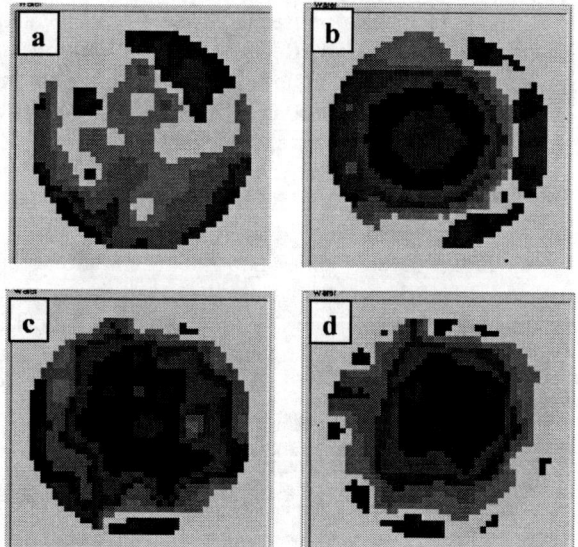

Fig. 1. Wafer map trend of process variables: (a) Gate FICD, (b) Gate Spacer Deposition thickness, (C) Gate Spacer Etch FICD and Device Characteristics: (d) Vth within wafer Trend. (All same Wafer)

hand, when the ion beam time comes to be longer, it becomes a higher dose. It can be realized by altering scan speed of ion beam sweep direction (X direction) and mechanical scan direction (Y direction) at the desired position or by combining them with step-wise rotation of wafer as shown in Fig.2.

The optimum dose map of LDSI can be decided by referencing the measured Vt and/or inline CD & spacer thickness within map trend (Fig. 1.)

TW values were measured using THERMA-WAVE TP630XP. CD and thickness were measured by KLA-Tencor.

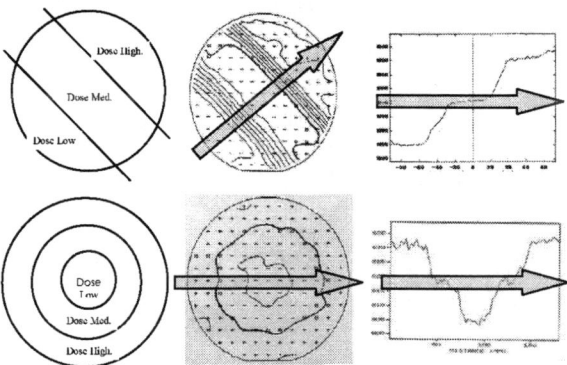

Fig. 2. The various examples of locally differentiated dose map obtained by LDSI.

3. Results and Discussion

In this work, we have first demonstrated that the LDSI can compensate the Vt variation of FETs and resulting the uniform device characteristics including Vth and Idsat in FETs.

For the improvement of 90nm DRAM tech. PMOS Vth variation, we have challenged LDSI on P+S/D implantation. Before trying it, the typical PMOS Vth distribution within wafer should be predicted to choose an appropriate P+SD LDSI implantation condition. Because we misunderstand exact vth variation, level and implanting region, it is hard to improve vth variation, even though it was used by an appropriate LDSI.

More or less dose rate at each desired region in wafer was calculated by In-line Data Feed-back to Electrical Characteristics Prediction System (IFEPS). IFEPS was to find out the way to expect the required post-gate LDSI condition by measuring in-line value of process variables to have effects on Vth variation. (Fig. 1.) So, Vth map trend and LDSI condition predicted a circular shape and Vth value at center region was lower than we expected. Also, Vth value at edge region was higher than we expected in a circular shape.

Finally, an appropriately calculated P+SD LDSI dose condition was 15% less at center and 15% more at edge than 100% of middle within wafer.(Fig. 3(b))

It is clear that the lower Vt region is reduced in center region. Also the higher Vth region is reduced in edge region. The cumulative plot of PMOS Vth exhibits the resulting uniformity resides within the range of only 88mV. This means that 47% (167 to 88mV) improvement has been achieved as indicated with solid circle in Fig. 4. Fig. 3(a) and (b) show the different TW map of conventional Implant vs LDS implant, we confirmed improvement of Vth variation (Fig. 4.) as well as the change of Vth map trend in Fig. 3(a-1) and (b-1). We recognized to increase in number of middle

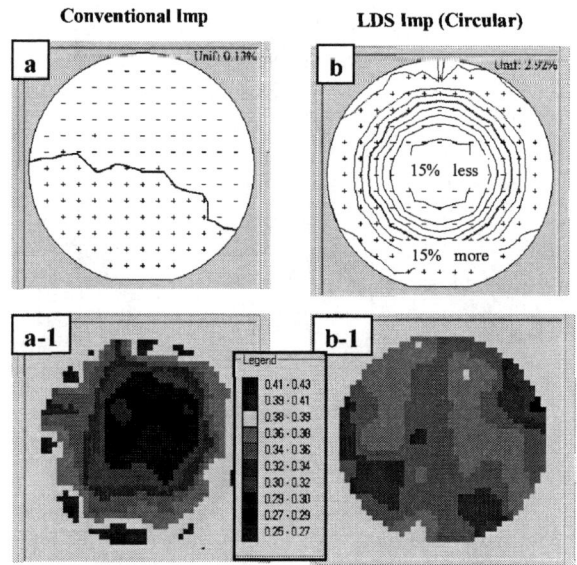

Fig. 3. As TW map obtained by conventional implant (a) and LDS implant (b), Vth map change from circular shape and non-uniform map(a-1) to uniform map (b-1).

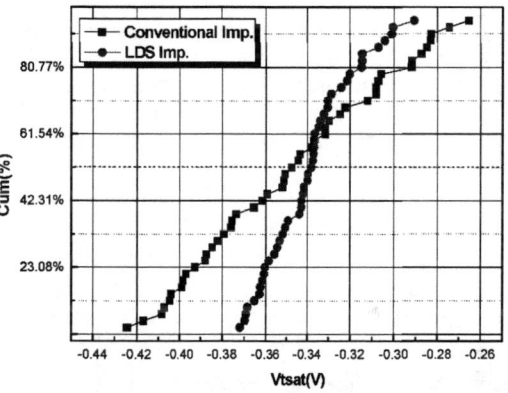

Fig. 4. Comparison with Vth Uniformity of Conventional Implant indicates that LDSI application is improvement of Vth uniformity on P+S/D implantation (15KeV, 49BF_2^+, 2.1E15)

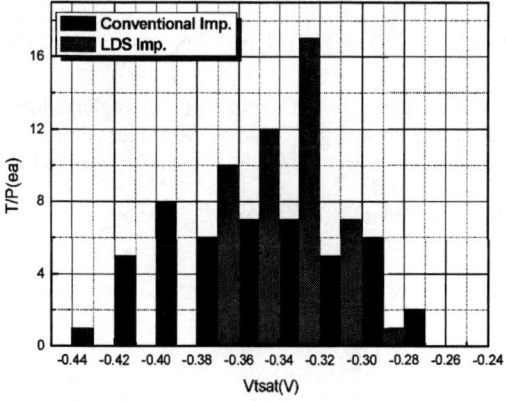

Fig. 5. Comparison with middle Vth value between conventional implant and LDS implant.

Vth value as shown in Fig. 5. It means that Middle Vth value portion increased by 95% (39 % to 76%) within wafer.

By applying the P+ S/D LDSI and P-Halo LDSI on 512M DDR3 SDRAM production wafers, there's a remarkable improvement in Idd fail rate which is the results of uniform Vth and Idsat of pMOSFETs.[Fig. 6.]

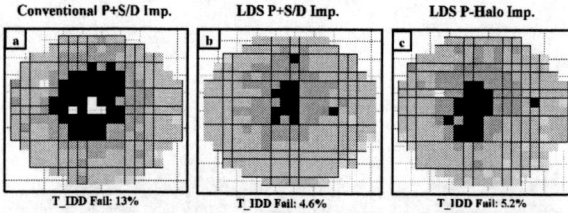

Fig. 6. Idd fail map and fail rate for yield in case of (a) conventional I/I, (b) LDS P+S/D I/I, (c) LDS P-Halo I/I on 512M DDR3 product wafer.

Through the LDSI, especially center region of low Vth was compensated and then Idd Fail dies of center region were reduced as shown in Fig. 6.

4. Conclusion

Locally-Differentiated-Scanning-Ion-Implantation (LDSI) is the new ion implantation technology that is a controllable non-uniform ion beam process within a wafer. We have developed, for the first time, LDSI and demonstrated its remarkable improvement on FETs and device fabrication.

From the results of this study, LDSI is believed to be a key doping technology for fine control of threshold voltage in future FETs

Acknowledgments

Authors would like to appreciate both Varian Semiconductor Equipment Association (VSEA) and Nissin Ion Equipment Company for their cooperation on developing LDSI function in ion implanter.

REFERENCES

1. Hans Glawisching et.al, Chapter 8 of Ion Implantation Science and Technology edited by J.F.Ziegler, 2004
2. M. D. GILES Chapter 8 of USLI Technology edited by S. M. SZE, 2002
3. 2005 International Technology Roadmap for Semiconductors (ITRS) 2005

Time Dependence Study Of Hydrogen-Induced Defects In Silicon During Thermal Anneals

S. Personnic[1], A. Tauzin[2], K. K. Bourdelle[1], F. Letertre[1], N. Kernevez[2], F. Laugier[2], N. Cherkashin[3], A. Claverie[3] and R. Fortunier[4]

[1] *SOITEC, Parc technologique des Fontaines, Bernin 38926, Crolles Cedex, France.*
[2] *CEA-DRT-LETI, CEA-GRE, 17 rue des Martyrs, 38054 Grenoble cedex 9, France.*
[3] *nMat Group, CEMES/CNRS, BP 4347, F-31055 Toulouse, France.*
[4] *CMP "Georges Charpak", Avenue des Anémones, 13541 Gardanne, France.*

Abstract. Hydrogen implantation in silicon and subsequent thermal anneal result in the formation of a wide range of point and extended defects. In particular, characteristic two-dimensional extended defects, i.e. platelets, are formed. The growth of these defects during thermal anneal, related to H migration, induces the development of micro-cracks in Si. In this paper, a time dependence study of H defects during isothermal anneals is performed using SIMS, FTIR and TEM techniques. We calculate the kinetics of H_2 formation based on SIMS depth profiling and FTIR measurements. We show that the splitting is determined by H migration and rearrangement of hydrogenated defects.

Keywords: hydrogen, implantation, platelets, migration, splitting kinetics.
PACS: 6172Cc, 6172Qq, 6180-x

INTRODUCTION

The discovery of the way to fabricate silicon-on-insulator (SOI) structures with the Smart Cut™ technology [1] has significantly enhanced the scope of applications of industrial ion implantation. The Smart Cut™ is based on ion implantation and wafer bonding. The SOI substrates are in widespread use in the microelectronic industry [2], and the technology was shown to be applicable to many other materials.

It is well known that H is mobile and has a strong chemical activity in crystalline Si [3,4]. In the absence of defects, H moves rapidly through silicon. During ion implantation a large variety of H-related defects is created. The reported general form can be written as Si-H_n, and more specifically I_mH_n and V_mH_n where V and I are, correspondingly, a vacancy and an interstitial. In addition, two-dimensional extended defects, i.e. platelets, are formed in a region near R_p, the ion range. The growth of these defects by means of the H migration and defect rearrangement during subsequent thermal anneal, induces the formation of large gas-filled micro-cracks in the implanted Si. It has been shown that these defects undergo a ripening process in which they exchange H atoms [5]. The anneal is one of the possible technological option to obtain the layer transfer of thin films in Smart Cut™ technology in the specific case of thermal transfers. Despite significant efforts, many physical aspects of implanted H evolution in Si are still unclear.

In this paper, we present a time dependence study of the evolution of hydrogen induced defects in Si after isothermal treatment using secondary ion mass spectrometry (SIMS), Fourier transform infrared spectroscopy (FTIR) and cross section transmission electron microscopy (XTEM) experiments. In addition, splitting kinetics measurements are performed.

EXPERIMENT

The (100) p-type Czochralski (Cz) silicon substrates were H implanted at an energy of about 40 keV and a dose of a few 10^{16} H^+/cm^2, through a 140 nm thick thermally grown oxide layer. After the implantation and the hydrophilic bonding to a base substrate, the structures were cut into pieces. The samples underwent anneals at temperatures ranging from 350 to 500°C for different times ranging from 30 seconds to 50 hrs. SIMS technique was used to obtain the depth distribution of implanted H in the as-implanted and annealed samples. FTIR measurements were performed in the MIR (Multiple Internal Reflection) mode to determine the as-implanted state of Si and the evolution of H-complexes during anneals. Finally, the XTEM study provided additional information about the platelet defects distribution, evolution and morphology.

For each annealing temperature we measured the critical splitting time, denoted $t_s(T)$, needed to obtain the layer transfer. The corresponding data depicted in Fig. 1 show an Arrhenius-type dependence with an activation energy (E_a) of (2.3 ± 0.1) eV. In this study, the splitting kinetics results are used to establish the "effective" thermal budget for different annealing temperatures.

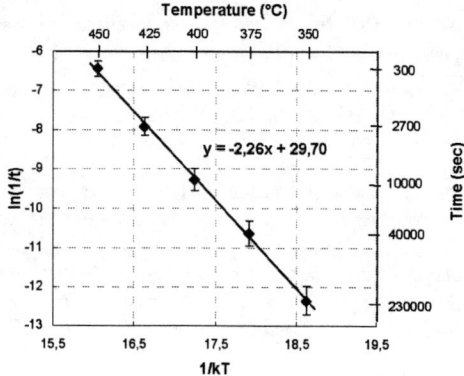

FIGURE 1. Splitting kinetics of Si implanted with an energy in the range of 40 keV and a few 10^{16} H$^+$/cm².

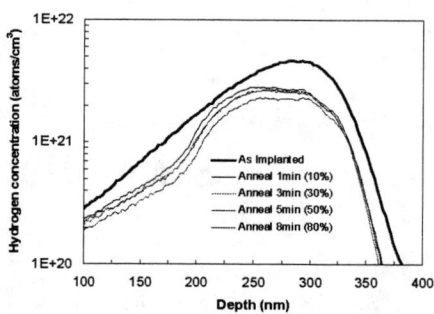

FIGURE 2. Comparison of SIMS profiles from Si implanted with H-ions, before and after isothermal anneals at 450°C.

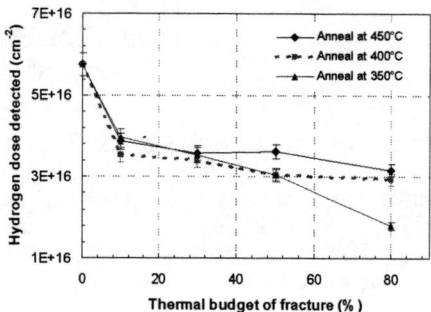

FIGURE 3. Time-evolution of hydrogen dose measured by SIMS after isothermal anneals at 350, 400 and 450°C as a function of the thermal budget of fracture.

RESULTS AND DISCUSSION

Time-Dependence Evolution Of H Implanted Si

We determined the kinetics of evolution of hydrogen profiles from SIMS measurements. For this purpose the samples underwent isothermal anneals at three different temperatures: 350, 400 and 450°C. As an example, figure 2 shows the H-profiles obtained after annealing at 450°C for different times. It is seen that the measured H-concentration decreases with increasing annealing time. This implies that under thermal treatment, the implanted H migrates and evolves in such a manner that it becomes no longer detectable by SIMS, e.g. by forming H_2.

Figure 3 shows the time-evolution of the H-dose measured by SIMS during isothermal anneals at the three investigated temperatures. A strong time dependence during annealing, with two specific domains, is observed. Most of hydrogen SIMS dose loss occurs during the first 10% of the $t_s(T)$ at each temperature. For the time intervals corresponding from 10-30% to 100% of the $t_s(T)$, SIMS measurements show a weak time dependence.

A similar time dependence is observed using FTIR-MIR technique where a fast rearrangement of H-induced defects occurs during initial stages of anneal, the first 10% of the $t_s(450°C)$, with slow evolution from 30% of the $t_s(450°C)$ (Figure 4). The total amount of hydrogen observed by FTIR decreases with the annealing time. Indeed, the attenuation of the spectrum integrated intensity reaches ~50% after a long annealing at 450°C. We believe this is due to conversion of trapped hydrogen to an unbound form, i.e. either atomic H or H_2. The net conversion of Si-H_n into H_2 is more likely since the existence of isolated H atoms is highly improbable (unstable state) [3]. For the three investigated temperatures, we observe the same behavior with a "fast-rate" decrease domain for annealing < 10% of the splitting time at the related temperature, and a "slow-rate" decrease domain in the interval 10-30 to 80%.

The "fast-rate" decrease domain is associated with the fast dissociation of some specific hydrogenated defects. The attenuation of the pair of modes observed at 1833 and 2049 cm^{-1} in the as-implanted sample, assigned to the H_2^* defect, and the weakening multivacancy signature (i.e. the major part of the Si-H_n related features in the sub-2050 cm^{-1}, Figure 4) confirm this assumption. We believe that the dissociation of these defects results in the formation of

H_2 that is not directly detectable by SIMS or FTIR, which will fill the microcracks located close to the R_p region. The "slow-rate" decrease domain seems to correspond to another phenomenon with a slower rate of H_2 formation.

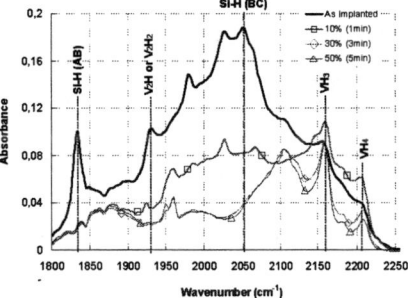

FIGURE 4. FTIR spectra of the Si-H stretching modes of H implanted Si, before and after isothermal anneals at 450°C.

Platelets Nucleation And Growth

A XTEM study was performed for the same samples and provided additional information about the platelets distribution, evolution and morphology. The damage zone is located in the region 170-370 nm. Two types of platelet habit planes are detected: a (001) plane parallel to the wafer surface and a (111) plane. (111) platelets are mostly concentrated in the deeper region of the damage zone in negligible density as compared to (001) platelets.

The platelet nucleation process is finished at the initial stage of annealing, at 5-10% of the $t_s(T)$. To illustrate this phenomenon, figure 5(b) shows the evolution of the platelet volume fraction per unit area as a function of the annealing time, which slightly increases during the first 5-10% of the $t_s(T)$. A signature of this nucleation can be seen with the FTIR experiments. Indeed, during the first minute of annealing at 450°C, a slight increase of the modes assigned to hydrogenated monovacancy defects (VH_n) appears between 2120 cm^{-1} and 2220 cm^{-1} (Figure 4). According to the authors [3, 4, 6], the VH_n defects are supposed to be the precursors of the H-extended defects. The increasing contribution of VH_n in the FTIR spectra and the platelet volume fraction (Figure 5b) measured by TEM are additional evidence that tends to confirm the VH_n precursors theory. Thus, during the first stages of annealing, there is a strong formation of H_2, associated with the end of platelet nucleation (end of VH_n creation).

The following decrease of the VH_n contribution in the FTIR spectra (from 10-30% of the thermal budget of fracture, Figure 4) indicates that the nucleation is finished. This is confirmed by the TEM study showing a stagnation of the platelet volume fraction. This behavior is also seen in FTIR spectra where the intensity of the stretching modes assigned to hydrogen passivating the internal surfaces around 2100 cm^{-1} becomes constant. So, while the duration of annealing increases (from 10% of the $t_s(T)$) the mean diameter of the platelets increases, their density decreases and the volume fraction per unit area occupied by platelets stays constant. These are the typical characteristics of a conservative Ostwald ripening mechanism [5]. This ripening mechanism seems to begin after fast H_2 formation and the slight VH_n creation show up before which takes place during the initial stages of annealing.

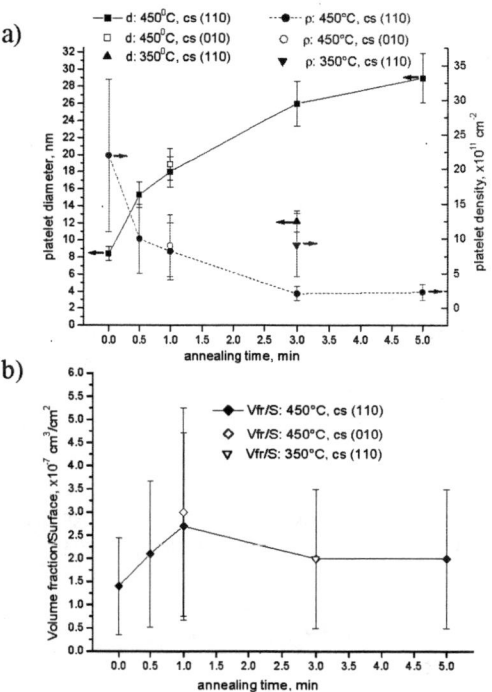

FIGURE 5. Evolution of a) mean platelet diameter and surface density and b) platelet volume fraction per square unit as a function of annealing condition.

Finally, based on the platelets volume fraction per unit area of annealed samples, the dose of H within platelets was found to be in a close agreement with the SIMS dose. That confirms that there is little out-diffusion of H from the damage zone during thermal anneals used in this study.

Kinetics Of H_2 Formation

In order to investigate the rate-limiting mechanism related to the splitting phenomenon occurring during the thermal treatment, we calculated for each temperature the rate of bonded-H decrease, during the entire anneal from as-implanted state until layer transfer. This rate was calculated based on the SIMS profiles (figure 2) and is maximum around 300nm i.e.

in the area where the fracture occurs (presumably due to the formation of H_2 within the microcracks). Hydrogen after implantation exists in the form of Si-H_n defects. The formation of H_2 in a microcrack can be divided in three mechanisms : the dissociation of the Si-H_n complexes, the migration of H towards microcracks and the spontaneous formation of H_2 by association of two H atoms. Thus, we can consider the following general chemical reaction [7]:

$$2\ Si\text{-}H_n \longleftrightarrow 2\ Si + nH_2$$

The kinetics of reaction can be written after development,

$$v = -\frac{1}{2}\frac{d[N_H]}{dt} = \frac{d[N_{H_2}]}{dt} = K_0 \exp\left(-\frac{E_a}{kT}\right) N_H^\alpha \quad (1)$$

Where N_H is the number of H atoms in the Si-H_n form, N_{H2} is the number of molecules formed, α is the order of reaction and K_0 the rate constant (sec^{-1}) mainly defined by the implantation conditions, the experimental procedures and the substrate nature. The order of reaction represents the empirical dependence of the kinetics of reaction in comparison with the reagents concentrations. For our particular case, it is not equal to zero because the rate of H-measured decrease is not constant during the entire anneal. Figure 6 shows the rate-temperature data obtained, plotted as ln(v) as a function of 1/kT, with v the rate of bonded-H decrease and T the annealing temperature. The function is written below as equation (2).

$$\ln\left(\frac{1}{N_H^\alpha}\frac{d[N_H]}{dt}\right) = \ln(K'_0) - \frac{E_a}{kT} \quad (2)$$

To compare efficiently this function corresponding to the formation of H_2 with the splitting kinetics reference (figure 1), the calculation was made on the entire thermal treatment, from 0 to 100% of the $t_s(T)$. The comparison is shown on figure 6. We fixed the order of reaction equal at 1. Indeed, the kinetics of reaction is dependent to the concentration of hydrogen in the Si-H_n form (two rate decrease domains).

The two types of kinetics coincide very well. The critical fracture time given by the splitting kinetics matches the H_2 formation rate found for the investigated depths near the R_p of implantation. We obtained the same activation energy. The E_a of (2.3 ± 0.1) eV found with the classical splitting kinetics is generally related to H-migration in the presence of gettering centers [8], or, according to some results [9], the rupture of the remaining Si-Si bonds in the implanted region. Our results tend to prove that H migration towards microcraks is the rate limiting step in the splitting kinetics.

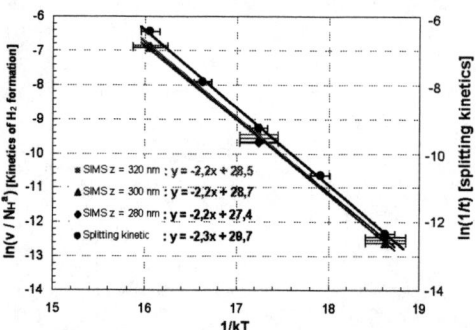

FIGURE 6. Comparison between the splitting kinetics reference and the calculated kinetics of H_2 formation (i.e. decrease of H-measured by SIMS) during the entire thermal treatments.

CONCLUSION

The migration behavior of H obtained from SIMS measurements shows strong time dependence during anneals. Most of H dose loss measured by SIMS occurs during the first 10% of the critical splitting time at a given temperature. SIMS results are in a good correlation with FTIR data where fast rearrangement of H-induced defects has been also observed in the same time. This first stage corresponds also with the end of the platelet nucleation, FTIR results indicating the major role of monovacancy defects in the platelet nucleation and growth. The typical characteristics of Ostwald ripening mechanism were found by TEM after a rapid initial formation of H_2 and the end of the platelet nucleation. For the time intervals corresponding from 10-30% to 100% of the $t_s(T)$, SIMS and FTIR measurements show only slight evolution of the hydrogen-induced defects. Kinetics of H_2 formation indicate that the fracture phenomenon is limited by the H-migration in the presence of gettering centers. Further studies are in progress to identify the thermo-mechanical phenomena occurring at the later stages of microcrack formation and interactions.

REFERENCES

1. M. Bruel, *Electron. Lett.* **31**, 1201 (1995).
2. A. J. Auberton-Hervé, M. Bruel *et al.*, *IEICE Trans. Electron.* **E80-C**, 358 (1999).
3. M. K. Weldon *et al.*, *J. Vac. Sci. Technol.* **B 15(4)**, 1065 (1997).
4. M. K. Weldon *et al.*, *Appl. Phys. Lett.* **73**, 3721 (1998).
5. J. Grisolia *et al.*, *Appl. Phys. Lett.* **76**, 852-854 (2000).
6. F. A. Reboredo and S. T. Pantelides, *Sol. State Phen.* **69-70**, 83-92 (1999).
7. A. Tauzin, *Internal CEA Report*.
8. B. Aspar *et al.*, *Microelectronic Engineering* **36**, 233 (1997).
9. Q.-Y. Tong *et al.*, *Appl. Phys. Lett.* **70**, 1390 (1997).

Layer Transfer of SOI Structures Using a Pre-Stressed Bonding Layer

Vorrada Loryuenyong[1] and Nathan Cheung[2]

[1]Department of Materials Science & Engineering, University of California-Berkeley, Berkeley, CA 94720
[2]Department of Electrical Engineering and Computer Sciences, University of California-Berkeley, Berkeley, CA 94720

Abstract. Using SU-8 polymer as the bonding and pre-stress layer, we have performed an edge-initiated mechanical transfer of Si using bare Si and hydrogen implanted Si as the donor wafers. 40 keV H+ were implanted to donor wafers to three doses; 2, 4, 6 × 10^{16} cm^{-2}. Non-implanted and implanted thermally-oxide grown on Si donor wafers were also investigated. Experiments show that the present layer transfer mechanism is correlated to the interfacial decohesion of residually-stressed SU-8 bonding layer and the mixed-mode loadings. It has been observed that the cut location shifts toward a deeper region with higher hydrogen implantation dose and with thermal annealing. This cutting mechanism involves the effect of shear constraints on the opening mode crack propagation along the weakest interfacial plane. The crack trajectory depends on the proportion of shear mode to opening mode. The objective of this study is to obtain a more refined understanding of layer transfer based on crack initiation and layer cleaving methods such as ion-cut process.

Keywords: SOI, layer transfer, heterogeneous integration, ion cut, shear-constrained crack propagation
PACS: 68.60.-p, 61.72.Tt, 85.40.Ry, 68.55.Jk

INTRODUCTION

Recent advances in Si-based CMOS, optoelectronics, micro-electrical-mechanical systems (MEMS), and nanotechnology necessitate the incorporation of dissimilar material components, which is currently be accomplished with the heterogeneous integration approach. One key component of the heterogeneous approach is the layer transfer technique, which is generically based on a "paste-and-cut" concept. An example is the implanted ions-induced surface layer cleavage technique by either thermally-induced layer separation (thermal ion-cut[1]) or mechanical cleavage (mechanical ion-cut[2]) processes. Although ion-cut processes are currently used in manufacturing, the understanding of fundamental mechanisms that drive the delamination process is still phenomenological.

A General Model for Crack Trajectory

In general, crack will follow the path that is most energetically favorable. For static crack growth, crack propagates in a direction following

$$\underset{\Gamma}{Max}\,[G(i) - G_C(i)] \geq 0 \qquad (1)$$

where G is the energy release rate at the crack front along path Γ and G_C is the resistance to crack propagation along path Γ. The subscript "i" denotes the loading mode. Based on Equation (1), the crack will follow the path that maximizes the energy dissipation, which can be controlled either by reducing the G_C (plane weakening) or by increasing the driving force G. The driving force or energy release rate also depends on the loading mode. When two or more modes of loading whose magnitudes are compatible to each other are present, the energy release rate is contributed from each mode. The opening crack, as a consequence, has a tendency to propagate by kinking and extending along a new trajectory such that Eq. (1) condition is maintained at the crack tip. It is found that the pure mode I crack growth, $K_{II} = 0$, is usually preferred under mixed mode conditions.[3-5]

In layer transfer technology, the controlled crack propagation direction is normally achieved by the introduction of localized damages or defects to be served as a Mode I cleavage plane to the materials system. For ion-cut process, the in-depth weakened layer is defined via ion implantation-induced damage layer. Thermal or mechanical energy is then applied internally or externally to provide the splitting force for the fracture surface.

In this work, we demonstrate a special case of mechanical cut based on the mixed mode crack propagation, which does not even need ion implantation but instead a pre-stressed substrate (e.g. SU-8 on Si).

EXPERIMENTAL PROCEDURES

The donor wafers used for this study were: (A) (100) oriented p-type Si wafers with resistivity 1-20 Ω–cm and (B) 100nm thermally oxide grown on (100) oriented p-type Si wafers. Some wafers were implanted with 40 keV H^+ ions to various doses. The polymer SU-8 was used as the bonding and stress-inducing material between the donor wafer and a glass substrate (i.e., the handle wafer).

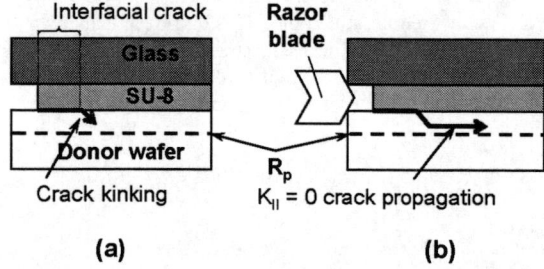

FIGURE 1. Scheme of mechanical cleavage transfer of Si onto glass with SU-8 as a bonding and stress-induced layer: (a) crack kinking, (b) cleavage propagation.

The surfaces of all donor wafers were first cleaned, following a standard cleaning, and dehydrated on a hotplate. Some wafers were pre-annealed before bonding. All donor wafers were taped around one side of their edges to prevent the coating of SU-8 during spinning. This gap will be used later for razor blade insertion to induce mechanical layer transfer. The wafers were spin-coated with SU-8 at spin speed of 1500 rpm for 1 min. The bonding processing steps started with pre-baking to remove all the solvents, followed by UV exposure through the backside of glass and thermal curing. The donor and handle substrates were then bonded together at 120°C. The exposure was done using a UV lamp with an intensity of ~10 mW/cm^2, integrated over a broadband 320-400 nm range of wavelength. The bonded wafers were then annealed at 170°C for 1 hr as a final curing treatment to initiate the crack kinking into Si substrates. The layer transfer step is induced by insertion of a 100 μm-thick razor blade between the bonded pairs from the gap at edge regions, causing the crack to propagate across the whole area. The cleavage experiments are schematically illustrated in Figure 1.

The thickness was measured by profilometry for Si transferred layers and by NanoSpec for SiO$_2$ transferred layers. AFM technique was then used to obtain the root mean square (RMS) values of the 5μm × 5μm transferred surfaces.

RESULTS AND DISCUSSION

A. Fracture Mode

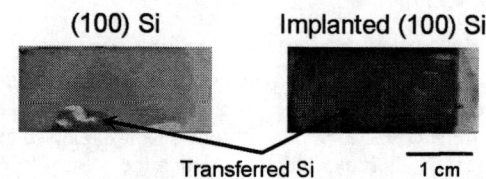

FIGURE 2. Optical micrographs of non-implanted and implanted transferred Si layers.

Figure 2 show optical micrographs of the transfer layers onto the glass substrate. For the glass/SU-8/(100) Si system, cracks propagated mostly or entirely through the bonded interface with partial or none advancing into the substrate. For the Glass/SU-8/implanted (100) Si system, on the other hand, cracks initiated at the sample edge first traverse along the donor wafer/SU-8 interface and then gradually deviated away into the Si substrate, rendering crack propagation across almost the entire sample. These experimental results suggest that the cleavage can be enhanced by ion implantation-induced weakening layer. The lower fracture strength of implanted Si subsurface allows further stress relaxation during crack growth and hence results in large-area transfer.

B. Dose Dependence on Cutting Depth

Layer transfer was carried out for three different implantation doses: 2, 4, and 6 × 10^{16} cm^{-2} into Si (100) substrates. Figure 3 shows the experimental results based on residual-stress induced mechanical cut process. It has been observed that the cutting depth is significantly shallower than the depth of the hydrogen profile peak, and the cut location shifts toward a deeper region with higher hydrogen implantation dose. According to Cho et al.[2], these results specified that the greatest energy release rate or stress relaxation occurred at the area other than the implanted layer and the bonding interface.

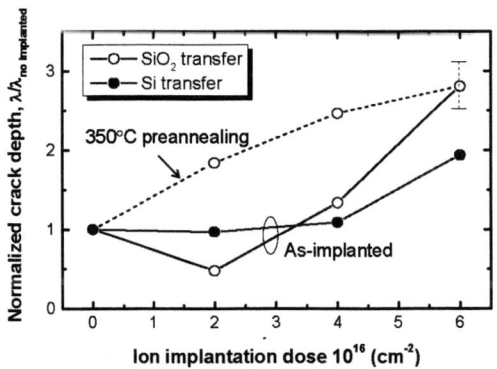

FIGURE 3. Transferred thickness of the Si donor wafers as a function of implantation dose based on residual-stress induced mechanical cut [●] and thermal ion-cut [○].

In general, when the implantation dose is below a threshold value, the residual compressive stress increases linearly with the doses.[6] It is speculated that when a crack propagates under the compression in implanted substrate, the opening crack is suppressed by the shear stresses, and Mode-II crack growth will take over at a certain depth.[7] As a result, the cracks are then assumed to extend further along the plane of maximum shear stress. The analysis agrees very well with our experimental results, except for the data point with a low implantation dose of 2×10^{16} cm^{-2} where the transferred depth is shallower than non-implanted sample. When the compressive stress is small, the shear-mode crack propagation cannot take place but instead prohibits crack opening of the growing crack. As a consequence, shallower crack propagation is observed for the samples implanted at the low dose, e.g. 2×10^{16} cm^{-2}.

The study was also extended further for the 100nm SiO$_2$/Si(100) samples with or without pre-annealing at 350°C before bonding. The entire cleavage occurred within the oxide layer, suggesting that the fracture mechanism involves the shear constraints. A strong dependence of the transferred thickness on the implantation dose as well as the pre-annealing temperature was also observed.

C. Surface Roughness Characterization

The implantation dose-dependent surface roughness is summarized in Figure 4. The roughness of implanted Si samples was larger than non-implanted sample, which is speculated to be a result of shear constraint effects. There might be several reasons to induce these effects: (a) buckle-driven crack propagation (e.g. due to any possible unstiffened blisters) and (b) crack propagation in compressive materials. The shear effects, however, are not significantly large between implanted samples. The AFM micrograph shows a very rough surface with RMS roughness of about 3.5-4 nm for implanted Si samples.

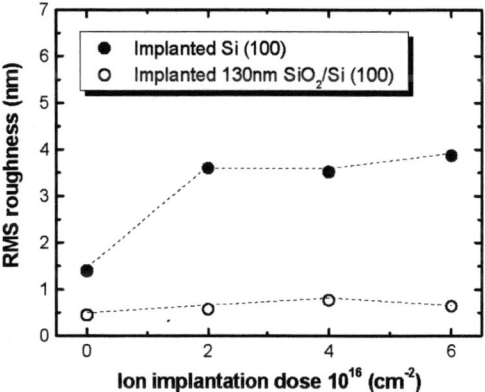

FIGURE 4. RMS roughness of the transferred Si and SiO$_2$ surfaces by stress-cut technique. Dotted lines are intended to guide the eye along the data points.

In summary, when assisted with residual mechanical stress to define a cleavage plane, we have demonstrated that a thin layer of Si and SiO$_2$ can be transferred by mechanical cleavage to a handle wafer without the use of ion implantation. The results indicate that crack propagation is not necessarily confined within the crack plane, and the growth direction can be diverted by the presence of mechanical stresses.

ACKNOWLEDGMENTS

This work is supported in part by the University of California Discovery Grant through the Small Feature Reproducibility and the Feature Level Compensation and Control Projects (Grant no. ELE04-10189) and the SRC Marco Program (Grant no. B-12-M06-S8).

REFERENCES

1. M. Bruel, *Electron. Lett.*, **31**, pp. 1201 (1995).
2. Y. Cho and N. W. Cheung, *Appl. Phys. Lett.*, **83**, 3827-3829 (2003).
3. M. D. Thouless, *IBM J. Res. Dev.*, **38**, pp. 367-377 (1994).
4. M. D. Drory, M. D. Thouless, and A. G. Evans, *Acta Metall.*, **36**, pp. 2019-2028 (1988).
5. J. W. Hutchinson and Z. Suo, "Mixed mode cracking in layered materials," in *Advances in*

Applied Mechanics edited by J. W. Hutchinson and T. Y. Wu, Academic Press, Inc., New York, **29**, pp. 63-191 (1992).
6. B. M. Paine, N. N. Hurvitz, and V. S. Sperious, *J. Appl. Phys.*, **61**, pp. 1335 (1986).
7. P. Isaksson and P. Ståhle, *Int. J. Solids. Struct.*, **39**, pp. 2281-2297 (2002).

Optimal preamorphization conditions for the formation of highly activated ultra shallow junctions in Silicon-On-Insulator

J.J. Hamilton, [1]E.J.H. Collart, [2]M. Bersani, [2]D. Giubertoni, [2]S. Gennaro, N.S. Bennett, N.E.B. Cowern, and K.J. Kirkby

Advanced Technology Institute, University of Surrey, Guildford, Surrey, GU2 7XH, UK
[1]*Parametric and Conductive Implant Division, Applied Materials UK Ltd, Foundry Lane, Horsham, West Sussex, RH13 5PX, UK*
[2]*ITC-irst Centro per la Ricerca Scientifica e Tecnologia, ITC-irst, 38050-Povo, Trento, Italy*

Abstract. Preamorphising implants (PAI) in Silicon-on-insulator (SOI) compared with bulk silicon substrates have been shown to improve junction properties. This paper studies the optimization of electrical behavior of this process in SOI. We will show that the deactivation caused by end-of-range (EOR) defects is vastly reduced in SOI by positioning the EOR band as close as possible to the buried oxide (BOX) interface while still allowing crystal regrowth to occur. Results show a 3% deactivation in SOI compared to 10% in bulk Si.

Keywords: Pre-amorphisation, Silicon-on-Insulator, Rapid thermal annealing, Dopant activation.
PACS: 61.72.Tt, 81.15.Np.

INTRODUCTION

Formation of highly activated, ultra-shallow and abrupt profiles is a key requirement for the next generation of CMOS devices, particularly for source-drain extensions.[1] For p-type dopant implants (boron), a promising method of increasing junction abruptness is to use Ge preamorphizing implants (PAI) prior to ultra-low energy B implantation. Advantages of this method are a reduction in the B channeling and an increase in B electrical activation due to the solid-phase-epitaxial (SPE) regrowth process.[2] In future technology nodes, bulk silicon wafers may be supplanted by Silicon-on-Insulator (SOI) since it can offer many advantages over bulk silicon.[3]

Unfortunately after SPER interstitial defects remain located at depths greater than the initial amorphous/crystalline (a/c) interface. These excess interstitial atoms agglomerate into 'end-of-range' (EOR) defects situated just below the former a/c interface.[4] During higher temperature annealing these defects undergo a series of transitions from self-interstitial clusters, to {113} defects to dislocation loops depending on the initial PAI and subsequent processing conditions.[4] These defects dissolve during annealing, the self interstitials so released migrate to nearby sinks such as the silicon surface.[4] Transient Enhanced Diffusion (TED) and B electrical de-activation are caused by self interstitials released from the EOR band, which diffuse towards the silicon surface during annealing. These self interstitials are also implicated in the formation of boron interstitial clusters (BICs).[4]

Previous results have shown that electrical deactivation and diffusion are reduced in SOI compared to bulk Si.[5] For SOI it is also important to minimize lateral dopant diffusion, this should also be somewhat reduced compared with Bulk Si. This is due to the upper buried oxide (BOX) interface effectively "soaking up" the interstitials from the EOR defect band that would ordinarily move toward the surface. This is explained by the fact that the oxide interface acts as a sink for point-defects coupled with the knowledge that B deactivation and diffusion are driven by self interstitials released from EOR defects.[6]

This work investigates the optimal conditions for B activation using Ge PAI in SOI material and shows the limitations of this process for this material type in terms of SPE process.

EXPERIMENTAL

Experiments were performed on n-type <100> Cz silicon wafers with a resistivity of ~10-25 Ω.cm, and on SOI wafers with a 145 nm-thick buried oxide and a 55 nm-thick p-type Si over layer. Silicon and SOITEC© SOI wafers were implanted with Ge^+ at energies of 8, 20, 32 and 36keV to a dose of 1×10^{15} Ge cm^{-2}, amorphising the silicon to depths of ~20, ~40, ~55, and ~60nm respectively (determined using RBS). The implants were performed using an Applied Materials Quantum X high current Implanter, with an implant orientation of 0° tilt and 0° twist. Boron was subsequently implanted at an energy of 500eV to a dose of 2×10^{15} B cm^{-2}.

The SPE re-growth process was performed at a temperature of 570°C for consistency with previous studies[7] using annealing times from 30s to 180s in N_2 ambient. Normal and glancing-exit Rutherford Backscattering Spectrometry (RBS) measurements were used to monitor the re-growth of the amorphous silicon. The amorphous depth was measured using a 1.5MeV He beam at a glancing angle of 45° to the sample, and the resulting spectra were analysed using the IBA Data Furnace software developed at the University of Surrey.[8]

To study the electrical (de)activation of the boron, isochronal annealing (60s) was carried out using a Process Products Corporation 18 Lamp rapid thermal processing (RTP) annealer, over a temperature range of 700-1000°C, pre-regrowth. The samples were analyzed by Hall Effect measurements using an Accent HL5500 Machine.

RESULTS AND DISCUSSION

Rutherford Backscattering was performed in order to check that the samples had recrystalized during annealing, results for the 32keV and 36keV Ge PAI in SOI are shown in figure 1. A selected portion of the RBS spectra (channel 235 to 265) was chosen to show and emphasize the Si amorphous peaks seen for both PAI conditions. The top figure shows spectra for the 32keV Ge PAI normal random and channeled spectra and also channeled spectra for samples annealed at 570°C for 76 and 180s, the lower figure is the same but for 36keV Ge PAI condition.

In order to determine these regrowth anneal times at 570°C, simulations were made using activation energies from G.E. Olson.[9] Annealing for 180s would be enough time for the 36keV Ge PAI samples to fully recrystalize, according to the simulations. The regrowth velocity would be further increased with the presence of boron in the samples.[5,9] For the 32keV Ge PAI case (top of figure 1) there is some re-growth after 76s anneal time and full re-crystallisation is after 180s, this is shown by a decrease in the amorphous peak size. However the RBS shows that 36keV Ge PAI in SOI does not recrystallise as a single crystal, this is shown by the fact that the channelled and amorphous spectra are coincident for all annealing times.

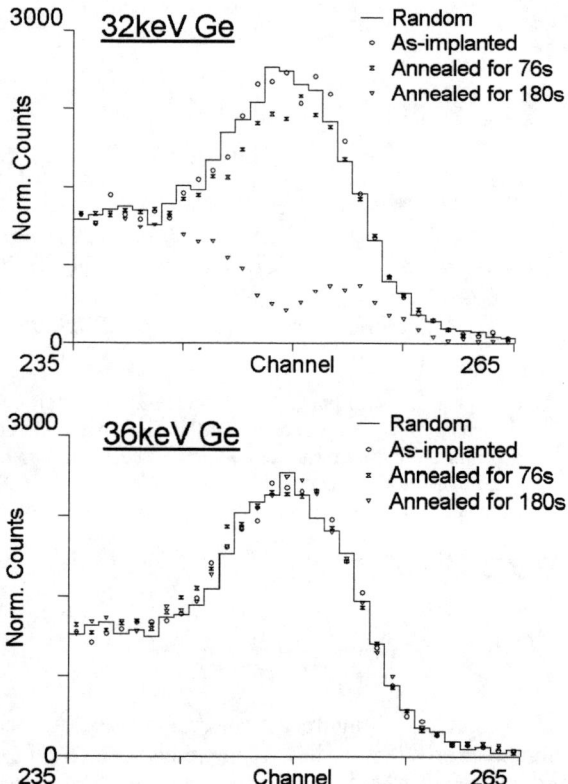

FIGURE 1. RBS spectra for 32keV (top) and 36keV (bottom) Ge PAI condition in SOI material. Both figures show the as-implanted random (solid line) and channeled spectra (circles), channeled spectra for annealed samples at 570°C for 76s (hourglass) and 180s (inverted triangle).

Since there is not enough of a Si crystal seed (the amorphous layer interface overlaps the BOX interface leaving no single crystal between the two) in the Si overlayer of the 36keV Ge PAI SOI, SPER can not occur. This presents a limit for the PAI technique in SOI.

Since the 36keV SOI sample did not regrow as a single crystal the boron was found to be electrically inactive, therefore electrical measurements were only made on the SOI samples where the PAI had been carried out at 8keV, 20keV and 32keV. Figure 2 shows the percentage electrical activation for isochronal anneals using 60s rapid thermal annealing in the temperature range 700° to 900°C. These temperatures were chosen in order to reveal the deactivation and reactivation of B for bulk silicon and SOI, only the

32keV Ge condition for bulk Si is shown (considered as best case scenario in bulk Si). The graphs show electrical deactivation occurring between ~750°C to 850°C and then reactivation after 850°C – the well known 'reverse annealing' effect.[4]

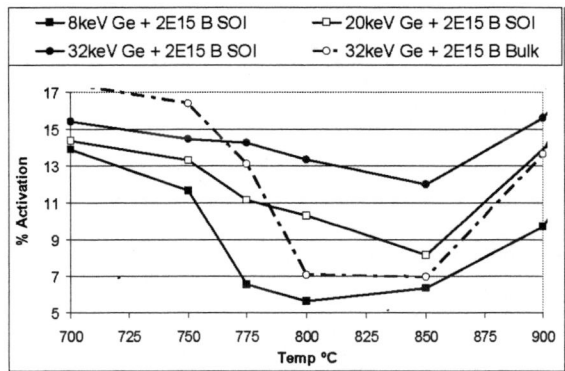

FIGURE 2. Electrical activation after 60s isochronal anneals. Solid lines are SOI, dotted line is Bulk Si, black squares, white squares and circles represent 8, 20 and 32keV Ge PAI respectively.

The results show as in previous work[6] that deactivation is dramatically reduced in SOI compared to bulk Si where the EOR defect band is close to the upper BOX interface. This can be seen by a higher activation level in SOI at temperature ranges from ~775°C to 850°C. The decrease in activation level corresponds to the ripening of EOR defects and the release from the EOR defect band of self interstitials which diffuse towards the surface, driving clustering and surface segregation of the B.[4] The increase in activation after the minima arises from the subsequent dissolution of the boron interstitial clusters (BICs) and the diffusion of B.

As the Ge energy in SOI is increased, i.e. decreasing the distance of the EOR defect band to the upper BOX interface, further improvement in electrical activation is seen. Results show very little deactivation for the 32keV Ge PAI in SOI only a 3% drop in activation between 700°C to 850°C (this represents a maximum increase in R_S of ~80 Ω/□) as opposed to a 10% activation drop in bulk Si. This shows that the junction is much more stable in terms of its electrical properties for the SOI and means that many more interstitials released from the EOR band are absorbed by the upper BOX interface than those by the Si surface. Therefore the dissolution of the EOR band is faster in SOI compared to bulk and is fastest as the EOR approaches the BOX interface.[10]

CONCLUSION

In this paper we have optimized the PAI conditions in 55 nm-thick Si over layer in SOI. For the highest Ge PAI condition 36keV, there is no SPE re-growth since there no longer remains any single crystal silicon after the PAI. The results show a vast reduction in electrical deactivation can be achieved in SOI with a 32keV Ge PAI condition, compared to bulk Si for the same conditions and also compared to pre-amorphising implants at lower energies.

ACKNOWLEDGMENTS

The authors wish to thank the UK Engineering and Physical Sciences Research Council (EPSRC) and the Ion Beam Centre University of Surrey for their support. J.J.H. is supported by EPSRC Doctoral Training Awards. N.E.B.C. is supported by an Applied Materials/Philips/Royal Academy of Engineering Research Chair of Nanoscale Materials Processing.

REFERENCES

1. The International Technology Roadmap for Semiconductors, 2005.
2. R. Lindsay, et al., Mat. Res. Soc. Symp. Proc. **765** D7.4.1 (2003).
3. C.K. Celler, Sorin Cristoloveanu, J. Appl. Phys. **93**, 9 4955 (2003).
4. B. Colombeau, et al., Mat. Res. Symp. Proc. **810**, C3.6 (2004).
5. J.J. Hamilton, E.J.H. Collart, B. Colombeau, C. Jeynes, M. Bersani, D. Giubertoni, J.A. Sharp, N.E.B. Cowern, and K.J. Kirkby, Nucl. Instr. and Meth. in Res. B **237**, 107 (2005).
6. B.J. Pawlak, R. Surdeanu, B. Colombeau, A.J. Smith, N.E.B. Cowern, R. Lindsay, W. Vandervorst, B. Brijs, O. Richard, F. Cristiano, Appl. Phys. Lett. **84**, 12 (2004).
7. G.L. Olson, Mat. Res. Soc. Symp. Proc. **35** (1985) 25-38.
8. C. Jeynes, N.P. Barradas, P.K. Marriot, G. Boudreault, M. Jenkin, E. Wendler, R.P. Webb, J. Phys. D: Appl. Phys. **36**, R97-R126 (2003).
9. GL Olson, JA Roth, Mat. Sci. Rep. **3**, 1 (1988).
10. J.J. Hamilton, E.J.H. Collart, B. Colombeau, M. Bersani, D. Giubertoni, A. Parisini, J.A. Sharp, N.E.B. Cowern, and K.J. Kirkby, Appl. Phys. Lett. **89**, 5 (2006).

Ion Implantation : A World Of Innovations

Michel Bruel

CEA/LETI 17 rue des Martyrs 38054 Grenoble France

michel.bruel@cea.fr

Abstract. Ion implantation is shown to have been at the basis of major innovations, especially in the field of microelectronics. The most significant of which are reviewed and analysed. It is also emphasized that ion implantation could be in the future the source of new and significant innovations.

Keywords: Ion Implantation, doping, SOI, channeling, innovation.

PACS: 34

INTRODUCTION

Ion implantation and Ion implanters were brought to the world of microelectronics by people involved in nuclear physics with accelerators and also by people of research teams using or developing isotopic separation. Ion implantation techniques and the scientific associated domain were in the infancy in the years 60' .i.e. forty years ago.

This domain has grown up since and presents itself to-day as very mature one, and it has been long since it has cut the relationship with its parent scientific area (nuclear physics and isotopic separation).

Ion implantation has become one of the most reliable technological step in factories to-day. Ion implantation is routinely used in advanced technological process on a huge scale.

This scientific and technical domain has been, up to now, at the basis of major innovations. Most significant of them will be described.

Does it mean that everything possible has already been attained, that it is the end of the road for Ion implantation ?. Most probably not. New applications may arise in the future based upon Ion implantation ; some discussion about possible domains will be presented ...

INNOVATION -PAST AND PRESENT

Microelectronics is, par excellence, the domain of Ion Implantation ; other applications like metallurgical ones representing only a niche market. Two domains of microelectronics are addressed by Ion Implantation : doping and SOI

Ion Implantation for doping

The MOSFET circuit technology has dramatically changed over the last three decades. Starting with a ten-micron pMOS process with an aluminum gate and a single metallization layer around 1970, the technology has evolved into a nanometric gate-length, self-aligned-gate CMOS process with up to seven metallization levels.

With the evolutions represented by switching from a metal gate to a poly-silicon gate, from wet chemical etching to dry etching, from aluminium alloys to copper wiring and damascene technology, from oxide reflow to full planarization, the transition from dopant diffusion to ion implantation represents one of the major improvements which have made possible the development of microelectronics as we know it to day.

The primary problem at the time was threshold voltage control. Positively charged ions in the oxide decreased the threshold voltage of the devices. p-type MOSFETs were therefore the device of choice despite the lower hole mobility, since they would still be enhancement-type devices even when charge was present.

The self-aligned poly-silicon gate process was introduced before CMOS and marked the beginning of modern day MOSFETs. The self-aligned structure is obtained by using the gate as the mask for the source-drain implant. Since the crystal damage caused by the energetic ions must be annealed at high temperature (600-1000°C), an aluminum gate could no longer be used. Doped poly-silicon was found to be a very convenient gate material since it withstands the high anneal temperature and can be oxidized just like silicon. The self-aligned process lowers the parasitic capacitance between gate and drain and therefore improves the high-frequency performance and switching time. The addition of a silicide layer on top of the gate reduces the gate resistance while still providing a quality implant mask. The self-aligned process also reduced the transistor size and hence increased the density. Using local oxidation isolation structure (LOCOS), where a Si3N4 layer is used to prevent the oxidation in the MOSFET region. provides an implant mask and contact hole mask yielding an even more compact device.

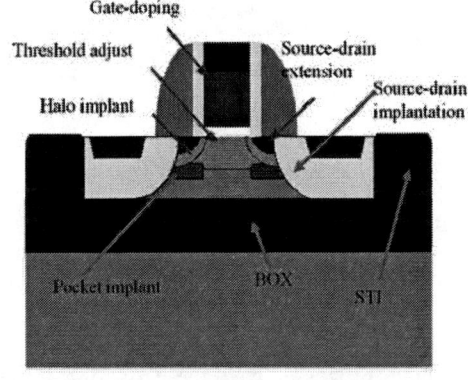

FIGURE 1. Ion Implantation for a CMOS/SOI technology

To-day for a basic advanced CMOS /SOI technology (see fig 1) one needs the following steps

– Drain source doping   2 steps
– Drain source extensions   2 steps
– Dual gate doping   2 steps
– Threshold adjust   2 steps
– Pockets   2 steps

For bulk devices one must add other implantations steps
- well doping   2 steps
- anti-punchthrough   2 steps

Even these figures may vary a little bit depending on the chosen technology, a CMOS device requires around between ten and fourteen different implantation steps (not taking into account multiple energy-dose for one operation).

Knowing that within a given product different kinds of devices are implemented, for example devices with standard oxide thickness and devices with thick gate oxide , I/O devices which require special optimization .

Generally this differentiation has no consequence for drain source engineering, nor for dual gate doping, but affects strongly channel engineering .

So each kind of new device in the technology may bring 4 implantation steps in addition .

Globally a CMOS technology requires at least 10 implantation steps , possibly up to 20 or more .

We are far from the beginning when Ion implantation major but sole role was to adjust the threshold potential of MOS devices !

Looking at the ITRS 2005 road map one can see that Ion implantation and associated equipments are not on the critical path towards new generations of technologies . In fact, even if the doping issue is

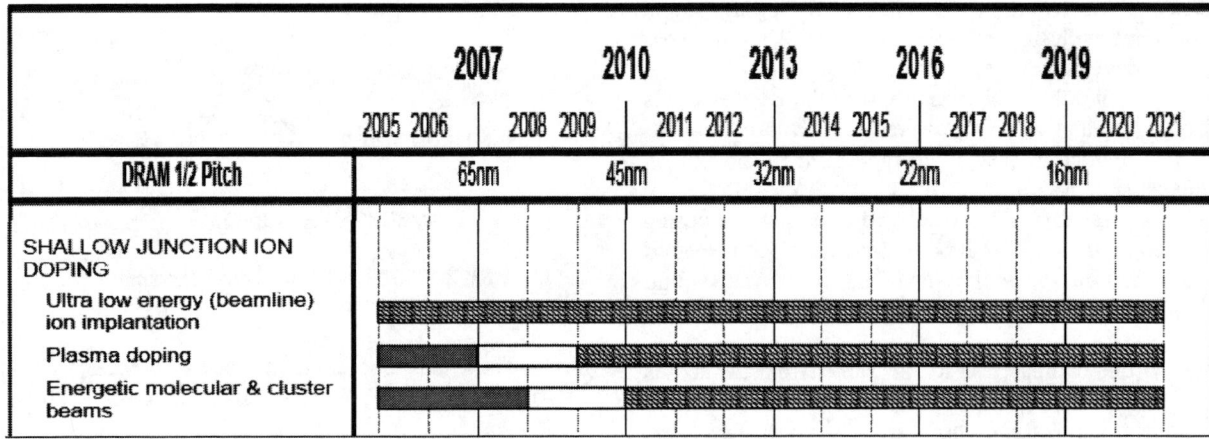

TABLE 1. Extract from ITRS 2005

largely treated, only very few sentences address directly the issue of ion implantation in itself. That confirms that ion implantation equipments are considered as mature and considered as capable of adapting their characteristics and performances in pace with the evolution of the required implantation criteria.

There are two domains where a special attention must be brought :
- implantation at very low energy in relation with ultra shallow junctions
- doping fluctuations in relation with the parametric dispersion of the device enhanced by the quantized number of introduced atoms

The issue of low energy implantation is addressed either by:
-extending mainstream ion implantation technology down to the 500eV range or below.
This makes implanter design more and more difficult, but the manufacturers claim they will cope with these stringent requirements
-developing plasma ion immersion. PII seems promising especially when thinking of productivity and operation cost, however some points have still to be worked out. If in classical implantation Ion implant's precision is taken for granted (the dose is precisely defined), this is not yet the case in PII. The impossibility of tilted implantations strongly limits the usability PII for CMOS technology and the issue of beam purity may represent another kind of limitation compared to classical II where the beam is isotopically selected. However it seems that Pulsed Plasma implantation could possibly offer a solution to the dosimetry issue and lessen the contamination beam purity issue..
-Developping cluster ion sources [1]
Using cluster ions makes possible the implanter to be operated at roughly n times the process energy, where n is the number of doping atoms within the cluster. For example, a 500eV B+ process is achieved by operation at 10keV using B18Hx.
Noteworthy the beam space charge density is only 1/n that of an equivalent atomic ion implant, thus reducing space charge issues and beam expansion issues.

A session is entirely devoted to cluster ion beams during this IIT 2006 conference. Detailed information is to be found in the corresponding proccedings.

If one compares an ion implanter from the sixties to an ion implanter of nowadays, one will find the same functions and the same sub-assemblies (ion source, magnet, acceleration column, end-station etc..). However these sub-assemblies are for sure quite different and much more sophisticated (it is not necessary to comment about that for people who practiced ion implantation in the seventies). To-day implantation machines result in fact from continuous improvements on all aspects of this type of equipment based upon introduction of advanced technologies (vacuum, computers, cybernetics) as well as optimization through modelling and simulation

Is there still room for innovation in the field of implantation machines ? Certainly yes ; and even there is place and need for breakthrough innovation when thinking, for example, of machines dedicated to very low energy implantation. Why could'nt we dream of a machine meeting all the advantages of the classical and of the PII implantation : that is to say
-ion purity and dose accuracy of classical machines
-machine footprint, throughput, Cost of Ownership of PII.

SOI technology

It is noteworthy that ion implantation was the key process which made possible the existence to-day of a SOI industry suited for mass market applications.

SIMOX:
This process was the first based upon ion implantation invented, developed and brought to industrial scale process. It uses ion implantation as a way of introducing foreign atoms into bulk silicon in order to obtain a buried oxide layer beneath a silicon single crystalline layer, in combination with very high temperature annealing.

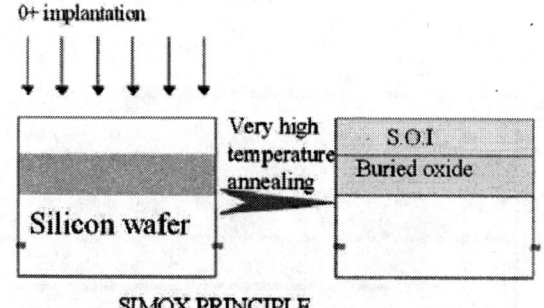

FIGURE 2. SOI/SIMOX material technology.

This ion implantation based process (fig.2) triggered a lot of physical studies as well as innovation in equipments.

First a very high temperature annealing (around 1300°C) was shown to be very effective in the buildup of a good quality buried SiO2 layer and a high quality top silicon layer ; the driving force being a mechanism leading to dissolution of small radius SiO2 precipitates for the benefit of the buried oxide layer which is a precipitate of infinite radius .

In the same time pieces of equipments had to be developped : dedicated Oxygen Ion implanters for doses in the range of some 1E17 to 1E18 Cm-2, with currents in the range of some tens mA,capable of making the implantation at high and controlled temperature (around 600°C), avoiding metallic contamination and capable of handling to-day 300 mm wafers .What a huge evolution, if you think of the first implanters which where devoted to MOS threshold adjust and where mainly suited for implantations in the 1E11- 1E12 Cm-2 dose range .

It was also necessary to develop annealing systems for treating wafers at around 1250°C - 1300°C during many hours . A particular key issue being to avoid contamination, at this high working temperature .

These two pieces of equipments have been brought to the industrial stage . However, due to the small size of the corresponding market, this has given rise to limited industrial commitment .

SMART-CUT

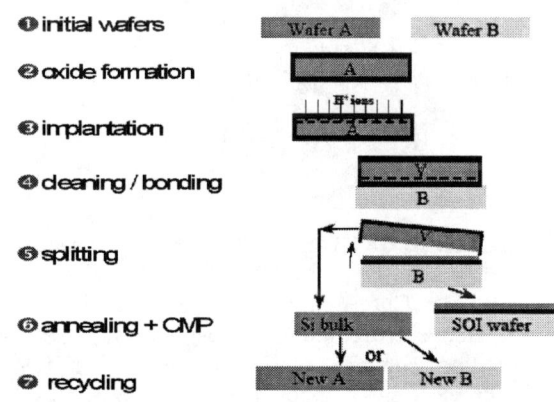

FIGURE 3. The basic steps of the Smart-Cut process.

The Smart-cut technology (fig. 3) was invented and developed during the 90'

In this process ion implantation is used to create a defective layer at the end of range of the implanted ions which will become a splitting layer .

The process slices the wafer lifting off a thin layer from the donor substrate and transferring it onto a new substrate.

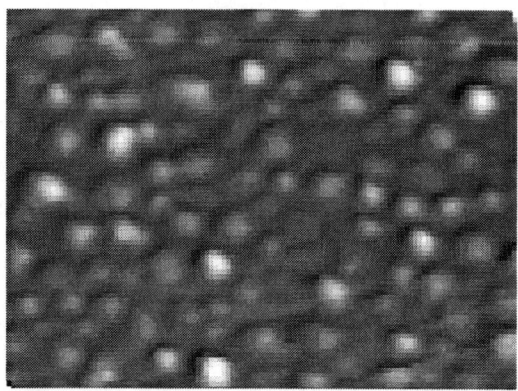

FIGURE 4 Blisters at the silicon surface induced by ion implantation.

The blistering phenomenon (see figure 4) induced by ion implantation, so to say a parasitic effect, has been turned into a powerful generic technology whose range of applications extends beyond the microelectronics world to applications in photonics, opto-electronics, high frequency and high power devices, thanks to the development of composite substrates.

The basic application is the UNIBOND SOI Wafers . Due to the intrinsic versatility of the process, mainly thanks to the intrinsic properties of ion implantation, the thickness of the top Si layer and as the thickness of the buried insulating layer can vary to cover the full range of applications., particularly to cope with the requirements of partially as well as of fully depleted CMOS applications.

This versatility is used for applications, such as displays, micro-wave devices where a thin silicon film onto an insulating substrate is required . Fused silica, sapphire, high resistivity silicon wafers are among the required insulating substrates .

Silicon-on-Quartz (SOQ), a thin singe-crystal silicon film onto a quartz (fused silica)wafer provides in the same time a substrate with excellent dielectric and optical properties.

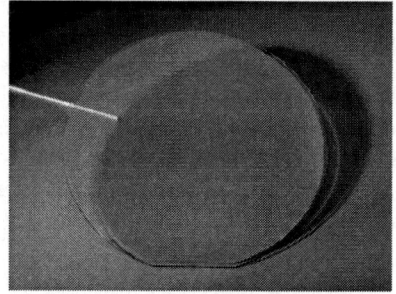

FIGURE 5. Silicon on quartz wafer

SOI layers on top of high resistivity silicon wafers is an attractive choice . Simulations and experimental results have shown that resistivities beyond 1 kohm-cm offer significant gains for substrate loss in the 1-30 GHz range .

It is possible to –day to make single crystalline silicon over-layers on top of a large variety of insulating layers (e.g. SiO2 and Si3N4 multi-layers) and large size substrates. For instance, 200nm thick Si films have been successfully transferred onto 400nm thick insulating SiO2 and Si3N4 layers by the Smart Cut technique referenced as Silicon On Insulating Multi-layers (SOIM) structures .

SOIM structures are attractive because of the high thermal power dissipation of Si3N4 films, as confirmed by its high thermal conductivities (30 W m-1K-1 instead of ~1 W m-1K-1 for SiO2).

The flexibility of the Smart Cut techniques can be used to stack more complex structures. For instance several (Si-SiO2) bi-layers can be subsequently transferred onto the same structure as illustrated in figure 6., where [Si/SiO2] bilayers were stacked up, four times, onto 150mm silicon substrates . The SEM view of the final structure reveals sharp interfaces and perfect intra-layer insulation..

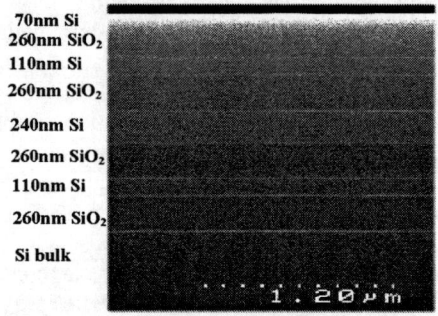

FIGURE 6. Four (Si-SiO2) bilayers subsequently transferred onto the same structure.

Wafer level strained silicon is a very attractive option to further boost the performance of CMOS ICs.

The smart-cut technology makes possible such engineered substrates (Fig.7)

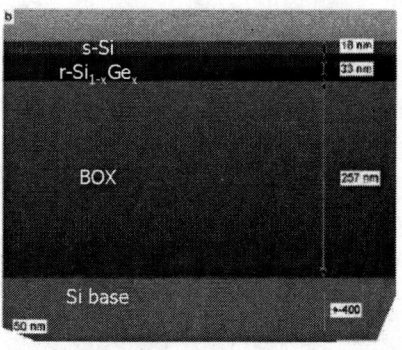

FIGURE 7. .. Strained Si films as observed by TEM cross sections when epi-grown onto transferred relaxed-Si1-xGex films.

Germanium on a silicon substrate is a very promising development as Ge offers a higher mobility than Si and should be better suited for the use of high-k gate oxides. Smart-cut offers also the possibility of such substrates (fig.8)

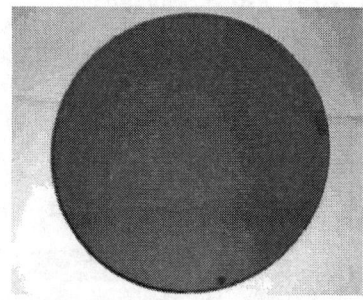

FIGURE 8. Ge 200mm / Si 200 mm hetero-structures

Smart Cut has significantly enlarged the field of engineered composite substrates and opens new exciting application perspectives in the field of microelectronics, photonics, opto-electronics, high frequency and high power devices.

INNOVATION – THE FUTURE

Looking at the existing applications of Ion implantation, it is obvious that they do take advantage of the intrinsic properties of this technique, i.e. accurate dose control, homogeneity, ion purity and possibility of implanting quite any kind of isotopic species, ease of choosing energy and thus depth of the applied treatment .

They use the physical effect of introducing foreign atoms in a matrix (allowing if needed to obtain concentrations well above the

thermodynamical limit).Introduction of foreign atoms is the heart in :
- doping
- Simox technology

They use the defects generation through nuclear stopping in a positive way :
- to get amorphization making possible dopant atoms incorporation during solid phase epitaxial regrowth induced by a subsequent annealing treatment at around 600°C .
- to get a brittle in-depth layer resulting from interaction between hydrogen and defects in the smart-cut process.

However some physical phenomena related to ion implantation,emphasized here below, have not yet been really used for actual applications .

Channeling

Channeling is generally considered as a drawback which must be avoided . for it is rather difficult to control . That's the reason why it is generally considered as absolutely necessary to choose parameters out of channeling conditions (i.e. the quasi canonical 7° tilt angle). Otherwise the junction depth could be strongly affected due to an extended tail in the implantation profile .

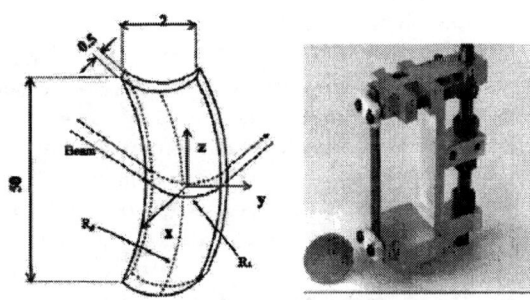

Principe and experimental set-up for deflecting GeV proton beams . *Extracted from : Review of Scientific Instruments 73 , 9 , 3170-3173 , V.M. Biryukov et al.*

FIGURE 9. From ref [2]   principle and apparatus for deflecting energetic beams

Surprisingly, there is a common use of ion channeling . Research in the field of sub-nuclear physics with very high energy particles (The energy unit being the GeV) use channeling in a slightly bent crystal as a way of deflecting the trajectories of the particles . See for example ref [2]

One of the interesting new features which could lead to major innovation is described in the paper of Demkov and Meyer [3].

They simulated le transport of particles within a channel (planar or axial) .In brief they showed that, provided the impinging beam divergence would be smaller than 1E-4, the particles are focused in the center of the channel axis into a spot of roughly a radius below   0.0078 nm which is the thermal vibration amplitude of the atoms .Which is even lower than the amplitude of the thermal vibration of the atoms of the lattice .The phenomenon is quite periodical , that is to say there are convergence spots distributed along the channel at a quasi periodic distance .

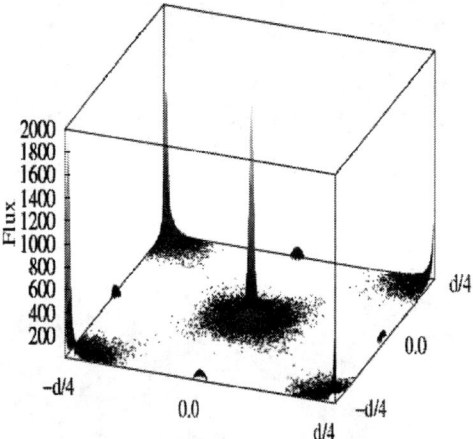

FIGURE 10. From ref [3]   : Focused flux within the channel

This can be related to the phenomenon of flux picking which has been known for long . However what is new here is to show what is possible when paraxial conditions are choosen .

In the paper the authors open some perspectives about the possible use of this discovery One of them being to enhance nuclear reaction probability of a incident ion and the nucleus of an atom situated within the channel at one of the focusing spots . It is thus possible to think of a possible use for thermonuclear reaction ; however, even with this probability enhancement the   target seems not reachable as it stands, because the reaction cross-section lie at best in the barn range (1E-24Cm2) which means a surface of roughly 1picometre diameter.

Anyway it is possible to use this phenomenon to create some kinds of ion probes featuring some tens picometres size, regularly distributed in a planar lattice with atomic spacing . Most probably many applications may arise from these remarkable properties in the field of nanotechnologies or picotechnologies .

Channneling in nanotubes and most probably in the near future in crystals made from close packed nanotubes opens very attractive perspectives which should be looked out carefully.

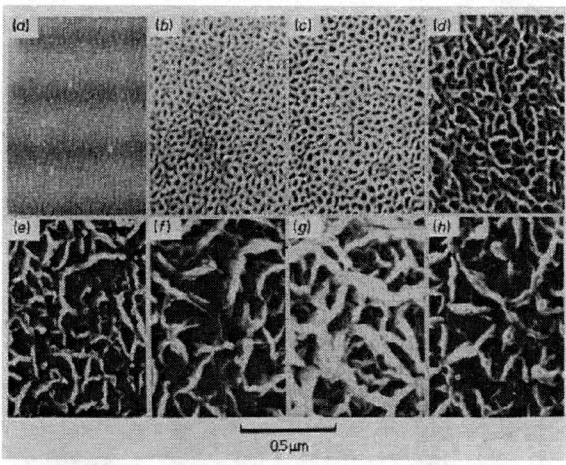

FIGURE 11. from ref [4] : nanostructuration induced in germanium by germanium ions implantation for different conditions

FIGURE 12. Vaporous structure obtained by implantation of heavy ions in a Sb/In bi-layer

Nano-structuring

The effect of low energy ion bombardment on the material surface has been known for long, including the possibility of creating or enhancing surface reliefs.

Ion implantation is also known as a means for creating in -depth structures. Fig. 11 and fig. 12 show two examples of what can be achieved through simple bombardment of material, provided that a good combination between ion implantation conditions and treated material are choosen.

It is noteworthy that the effect of "nano-porosification" of bulk crystalline germanium by heavy ion bombardment has been known for long [4]. It is something which should be carefully looked at a time when doping of germanium by ion implantation is coming to the stage . That could explain at least partly why the behaviour is so different from the one of silicon : quite no boron diffusion and large diffusion of phosphorous and arsenic (the heavy ones which are able to induce the formation of this nano-porous structure).

It is possible to think of different applications of this structured material , for example electrode material in Li-Ion batteries (to allow to accommodate the volume increase).

The fig. 12 displays SEM observations which where obtained at LETI more than twenty years ago . On the right of the picture one sees the layer which has been protected from implantation and on the left the result of implanting heavy ions like heavy inert gases , into the layer which was obtained by evaporation onto silicon of indium and then antimony . The obtained vaporous and filamentary structure , seemingly composed of indium antimonide material , is quite impressive . This is to relate to other experiments which have shown a porosification effect in bulk InSb . However in our case the effect seems to be much stronger.

The starting layer is composer of sub-micrometric islands of indium within a thin matrix of antimony . What has be seen also is that the effect (in terms of step height for example) is related to the average size of this nano-islands . Remembering that InSb has a extremely hign electron mobility, it could be interesting to realize experiments with isolated Indium island and look at the obtained structure (wires ?, two D membranes ?,).

Implanting ion by ion

The title of this sub-paragraph may seem provocative , and yes indeed it is a little . Yet this is a topic which must be looked out carefully because due to the extremely fast progress of electronics, micro and nano-fabrication what seems to-day a simple curiosity may one day come to a standard technology .
Shinada, Okamoto, Kobayashi and Odhomari [5] present a very interesting work in this field . They have succeeded in implanting ions regularly spaced on a square lattice of 100nm pitch . In each spot they implanted the same precise number of ions . Doing that they have been able to show that the resulting threshold voltage is much better centered with a smaller dispersion compared to the random case .

The conclusion which arises from this work is that, not only the extremely precise control on ions number, but also controlling their locations is of prime importance.

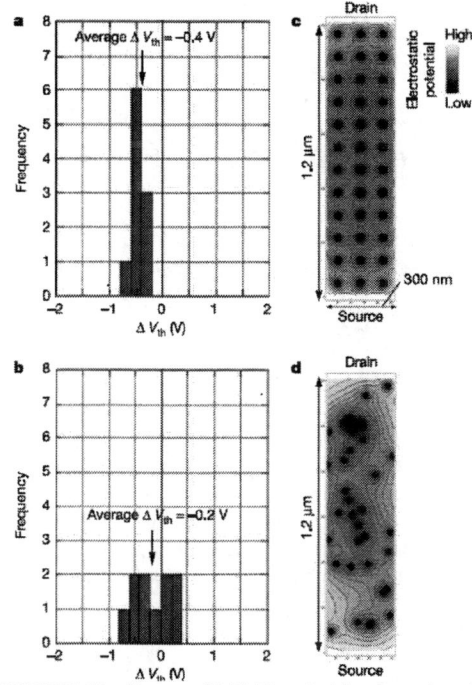

FIGURE 13. From ref [5] Threshold voltage in relation with the distribution of implanted dopant ions

Yet further studies should be conducted to evaluate the limits of the method, especially when the lateral straggling becomes of the same order as the pitch between implanted ions

CONCLUSION

Obviously Ion implantation and innovation are closely associated. We have shown through examples that is has been the case up to know. We have also brought to mind some new trends which could give rise in the future to new applicative developments. These are only examples and there are also many other ideas which could participate in the future of innovative ion implantation.

ACKNOWLEDGMENTS

I had the opportunity to spend a part of my career in the field of ion implantation and it is really a great satisfaction for me ; so I would like to thanks all the persons who at any level have contributed and still continue to the development of this wonderful domain .

I would also thank H. Moriceau from whom I borrowed some figures and Michel Brillouet for interesting discussion and advice .

REFERENCES

1. D.Jacobson, "Using Boron Cluster Ion implantation to fabricate Ultra-shallow junctions," in Extended Abstracts of the 5 th International Workshop on junction technology 2005

2. V.M. Biruykov, Yu.A. Chesnokov, V.Guidi, V.I. Kotov, C.Malagu, G.Martinelli, M.Stefancich, D.Vincenzi, *Review of scientific instruments*, **73**, 3170 (2002).

3. Yu.N.Demkov, J.D. Meyer, *Eur.phys.J.B* **42**, 361-365 (2004).

4. I.H. Wilson " The surface topography of germanium after bombardment with germanium ions " in " *Low energy ion beams 1980*" Conference series Number 54 The Institute of physics Bristol and London, 262-270

5. T.Shinada, S.Okamoto, T.Kobayashi, I.Ohdomari, *Nature* **437**, 1128-1131 (2005).

Junction Stability of B Doped Layers in SOI Formed with Optimized Vacancy Engineering Implants

A.J. Smith[1], N.E.B. Cowern[1], B. Colombeau[2], R. Gwilliam[1], B.J. Sealy[1], E.J.H. Collart[3], S. Gennaro[4], D. Giubertoni[4], M. Bersani[4] and M. Barozzi[4]

[1] *Ion Beam Centre, Advanced Technology Institute, University of Surrey, Guildford, GU2 7XH, UK*
[2] *Chartered Semiconductor Manufacturing Ltd, 60 Woodlands, Industrial Park D, Street 2, Singapore 738406*
[3] *Applied Implant Technologies, Foundry Lane, Horsham, W-Sussex RH13 5PX UK*
[4] *ITC-irst, via Sommarive 18, 38050 Povo (Trento), Italy*

Abstract. Forming highly stable, low resistive, ultra shallow p-type junctions is well known to be a challenge for future transistor devices. This paper investigates the junction stability of boron layers formed with an optimized 160keV silicon vacancy engineering implant in SOI. It is demonstrated that when the electrical activation is well above the solid solubility a combination of diffusion and possible boron precipitation, during prolonged annealing at 850°C, drives the boron to return to an equilibrium level of electrical activation, which is compensated by the carrier mobility to maintain a constant Rs. Reducing the anneal temperature to 700°C shows it is possible to create highly stable p-type junctions in terms of diffusion and sheet resistance.

Keywords: Vacancy Engineering, Junction Stability, Boron, SOI
PACS: 61.72.Ji, 61.72.Tt, 66.30.-h.

INTRODUCTION

The formation of highly conducting ultra-shallow p-type junctions is a key component of the p-type Metal Oxide Semiconductor Field Effect Transistor (MOSFET), especially for the source/drain extension regions which help suppress the detrimental short channel effects. However, as technology evolves the junction requirements become increasingly more difficult to achieve due to the fundamental limitations of the silicon substrates and/or process induced phenomena, such as, Transient Enhanced Diffusion (TED) and dopant-defect clustering mechanisms which increases the junction depth and sheet resistance, respectively. For Boron, it is the remaining silicon interstitials from the ion implantation process (+1 model [1]) which hinders its activation and causes enhanced diffusion, therefore, clever techniques have to be employed to decouple the interstitial-boron interactions.

The most commonly researched technique in an attempt to achieve the junction requirements is a form of pre-amorphisation and solid phase epitaxial regrowth which is proven to result in the incorporation of boron within the silicon lattice well above the solubility limit [2]. However, the inherent formation of an End Of Range (EOR) defect band limits the junction stability whilst increasing device leakage [3].

It has also been suggested that the traditional down scaling of transistor devices, which has been improving the device performance for over 40 years, has come to an end and the transition to new starting materials or device geometries is inevitable [4].

The main replacement for bulk silicon is thought to be high quality ultra-thin Silicon On Insulator (SOI), and such a transition has divided the semiconductor industry. However, many companies are already using SOI as their starting substrate for a range of device architectures. Therefore, any new doping, or co-implant process must be fully compatible with not only standard CMOS processing but also SOI material.

An alternative technique already shown to be compatible with standard CMOS techniques, and uses one of the many advantages of SOI, is known as 'vacancy engineering' [5,6]. In this process a silicon co-implant is used to engineer a surface rich vacancy region due to a number of processes intrinsic to ion implantation. Firstly; the momentum transferred to the host atoms from the impinging ions cause a spatial separation of the generated Frenkel pair population [7], secondly; sputtered silicon atoms out of the surface region leave behind vacant lattice sites and thirdly; missing primary recoils from the surface region leaving behind vacancy defects as they are not replaced as they would be if the silicon substrate was infinite [6]. These generated vacancies help improve the attributes of a subsequent boron implant through a vacancy-interstitial annihilation mechanism.

This paper investigates the effect of an optimized vacancy engineering implant [8] on the thermal stability of boron doped ultra shallow junctions within a thin SOI substrate in terms of diffusion and electrical activation.

EXPERIMENTAL DETAILS

A 55/145nm (silicon overlayer/BOX thickness, respectively) SOI structured material was implanted with a vacancy engineering silicon implant to a dose of either $8 \times 10^{14} cm^{-2}$ or $1.1 \times 10^{15} cm^{-2}$ at an ion energy of 160keV, which results in the peak of the interstitial damage being beneath the buried oxide, and therefore, physically isolate the excess silicon interstitials from the top silicon layer. After the vacancy engineering implant an ultra low energy boron implant followed at 500eV to a constant dose of $10^{15} cm^{-2}$. A schematic of the implantation layout is illustrated in figure 1, showing the defect distribution (Monte Carlo simulation) created by the highest dose silicon co-implant with respect to the ultra shallow boron implant (SIMS analysis) and the SOI structure.

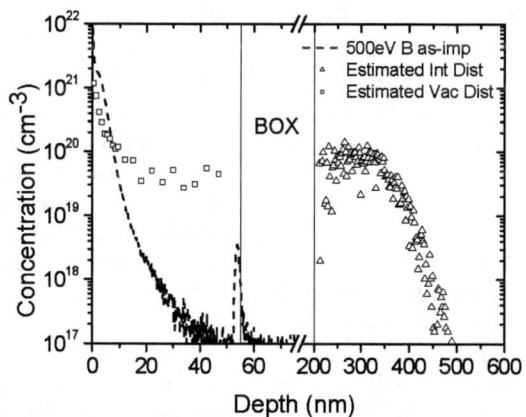

FIGURE 1. Schematic representation of the defect distribution generated by the silicon co-implant, with respect to the low energy boron implant and SOI structure.

It must be noted that the highest dose silicon co-implant forms a buried amorphous layer which on annealing above 700°C completely regrows via a reverse solid phase epitaxial regowth leaving only an interstitial rich EOR defect band beneath the BOX [9].

An 850°C and 700°C isothermal annealing scheme (10s to 2430s) in an N_2 ambient was employed to investigate the effect of the diffusion and electrical properties of the boron doped layer with prolonged annealing via SIMS analysis and sheet resistance/Hall measurements, respectively. For this comparative study the Hall scattering factor has been left at unity.

RESULTS AND DISSCUSIONS

The Rs measurements obtained using the Van der Pauw technique are presented in figure 2, illustrating the Rs of the boron layers, with and without the silicon co-implants annealed from 10s to 2430s duration at a temperature of 850°C.

The boron-only curve as a function of time (□), starts off high at ~3200 Ohms/sq (10s) and reduces exponentially to ~1100 Ohms/sq (2430s). In comparison the effect of the vacancy engineering implant is clearly seen via the $8 \times 10^{14} cm^{-2}$ dose silicon co-implant (○) which shows at an anneal time of 10s the Rs is reduced by a factor of 2. As the anneal time is increased the Rs still reduces, but at a much slower rate, so that eventually at a time of 2430s the Rs is near identical to the boron-only curve. By increasing the silicon dose to $1.1 \times 10^{15} cm^{-2}$ (△) an improved effect on the reduction of the boron Rs is observed at 10s, in the order of ~x4, compared to the boron-only curve. As the anneal time is increased the Rs stays roughly constant until 2430s. Therefore, in terms of Rs it can be seen that boron doped layers formed with this optimized vacancy engineered implant are highly stable compared to traditional preamorphisation and SPER techniques which show a dramatically increasing/decreasing Rs curve as a function of time [10]. As there is no EOR defect band within the top silicon layer of the SOI structure there is no reservoir of silicon interstitials to cause such a deactivation..

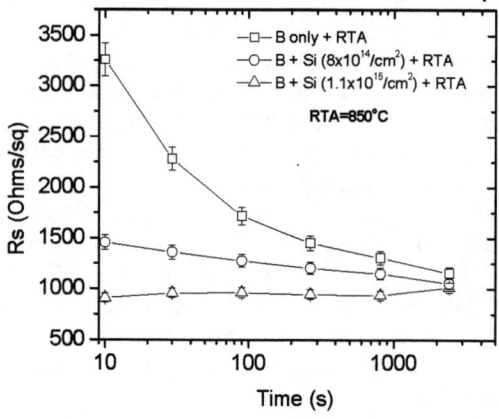

FIGURE 2. Rs as a function of anneal time of the boron layers with and without (□) a vacancy engineering implant performed to a dose of $8 \times 10^{14} cm^{-2}$ (○) and $1.1 \times 10^{15} cm^{-2}$ (△).

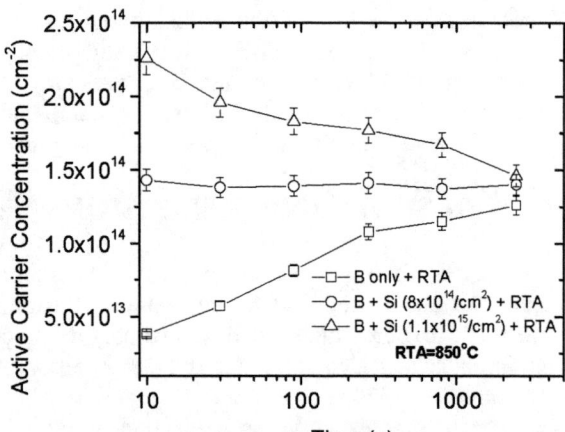
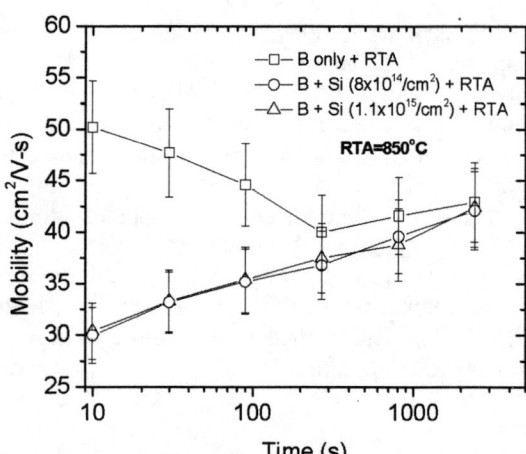

FIGURE 3. Ns measurements of the boron layers with and without a vacancy generating implant, using the same notation as figure 2.

FIGURE 4. Mobility results for the boron layers with and without the vacancy engineering implant, using the same notation as figure 2.

To complement the Rs measurements the corresponding Hall analyses are presented within figures 3 and 4, illustrating the active carrier concentration (Ns) and mobility, respectively, using the same notation as figure 2.

In terms of Ns the boron-only curve shows a rapid rise of ~x3 from 10s to 2430s which is reflected in the corresponding mobility curve which monotonically decreases up to 270s due to the increase in the ionized impurity scattering. However, the rise in mobility from this point onwards can be attributed to the boron layer diffusing which results in a decrease in the local ionized impurity scattering.

With the lowest dose vacancy engineering implant the Ns stays roughly constant, which means that the slight reduction in Rs is actually as a direct result of an increase in mobility which can also be attributed to the movement of the boron profile.

In contrast the highest dose co-implant demonstrates a high Ns at short anneal times which reduces, and tends towards the lower dose co-implant and reference values as time advances. This could be due to two reasons, firstly; the boron could be out diffusing. Secondly; the boron layer is starting to precipitate and the silicon lattice is forcing the level of substitutional boron atoms to return to equilibrium, namely the solid solubility limit.

SIMS analyses were performed on boron samples with and without the highest dose co-implant from 10s up to 810s and are presented in figure 5. By integrating the profiles of the boron layers with a co-implant (b) it is possible to rule out the first possibility as the retained dose is roughly constant as a function of anneal time.

Comparing the effect of the co-implant on the diffusion of the boron profiles, it is possible to observe two different processes. The boron profiles without a co-implant (a) show a 'kink' point which occurs at ~$1 \times 10^{19} cm^{-3}$ after 10s, rising to ~$3 \times 10^{19} cm^{-3}$ at longer times.

This transient behavior is consistent with previously reported results at higher boron implant energies [11]. In contrast, with a co-implant, the 'kink' point starts much higher at ~$2 \times 10^{20} cm^{-3}$. Assuming this point corresponds to the boron fraction which is electrically active, at 850°C this is ~x4 the solid solubility limit ($4.9 \times 10^{19} cm^{-3}$) of boron in silicon according to Armigliato et al [12]. As the anneal time is increased this kink point lowers whilst the boron profile diffuses. As a result, after a 810s anneal the boron profiles with and without the co-implant show very similar distributions. In fact it is interesting to note, that by taking a reference point at $3 \times 10^{18} cm^{-3}$ the tail of the boron distributions occur at near identical depths for all 850°C anneals. Therefore, it can be seen that for a boron-only implant which is contained within an interstitial rich region due to the very nature of the implantation process (+1 model), the electrical activation starts low and comes up to the solid solubility limit. However, the boron contained within a vacancy rich region has an electrical activation which starts off well above solubility, returns towards equilibrium as the anneal time is increased. Here we report the equilibrium value to be around $3 \times 10^{19} cm^{-3}$, just slightly lower than Fair et al. [13] and Kim et al. [14]. However, this should possibly not be treated as a true figure for solubility, as both the boron second phase and the surface could be playing a role in the uncertainties each at long anneal times.

It is speculated that the low dose vacancy engineering implant results in the boron being incorporated within the silicon lattice at around the equilibrium value, and therefore, the Ns does not have to increase or decrease to reach this value.

At 850°C a significant amount of diffusion is seen, However, as boron layers formed with vacancy engineering co-implants do not require a high temperature anneal to achieve a high degree of electrically activation [8,15], the process can be conducted at lower temperatures.

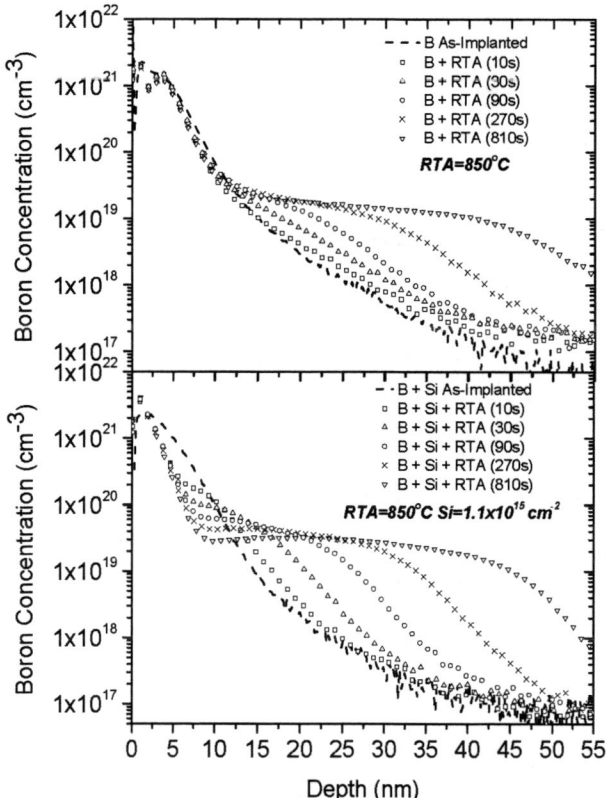

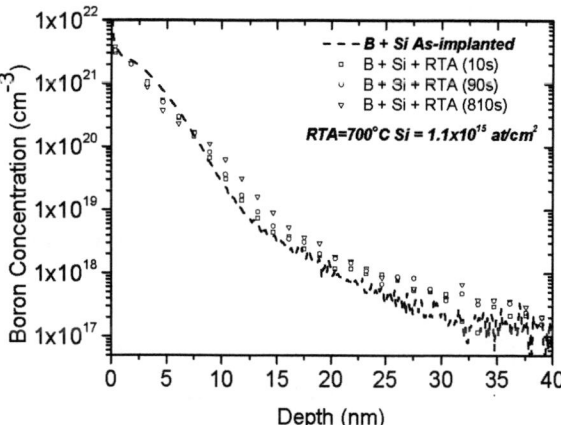

FIGURE 5. SIMS analyses for the boron samples with and without the high dose silicon co-implant as a function of anneal time from 10s to 810s, annealed at 850°C.

The effect of the boron junction depth was studied at 700°C for 10s, 90s and 810s anneal times, and illustrated in figure 6 via SIMS. Taking a reference at 3×10^{18} cm^{-3}, after 10s the junction depth is nearly 'diffusionless' at ~16nm and even after 810s the junction has only moved a further 2nm. In terms of Rs the corresponding values for 10s, 90 and 810s are 890, 876 and 972 Ohms/sq, respectively. The Hall measurements (not shown) indicate the exact same processes already discussed in the 850°C study, although they are less pronounced due to the reduced thermal budget.

SUMMARY

The electrical activation and diffusion of boron layers with and without an optimized vacancy engineering implant have been investigated with prolonged annealing. It has been shown that in terms of Rs the boron layer with a co-implant shows a high thermal stability and only produces a comparable value to a boron-only sample after an 850°C anneal for 2430s. However, Hall measurements clearly show that with a high dose co-implant (Ns well above solubility) a reduction in Ns is compensated by an increase in mobility to keep the Rs constant. SIMS analyses show that after a long anneal the reduction in the kink point combined with dopant diffusion into the substrate

FIGURE 6. SIMS analyses for the boron samples with and without the high dose silicon co-implant as a function of anneal time from 10s to 810s, annealed at 700°C.

results in the boron profile returning to the solubility limit, and therefore, very similar to the reference sample. However, reducing the thermal budget still results in low Rs but also increases the junction depth stability.

REFERENCES

1. M.Giles, J.Electrochem.Soc. 138, 1160 (1991).
2. R.Lindsey et al. Mat.Res.Soc.Symp.Proc. 717, C2.1 (2002).
3. B.Colombeau et al. Mat.Res.Soc.Symp.Proc. 810, C3.6 (2004).
4. B.Meyerson, Soli.Stat.Tech.Online. May 22 (2006).
5. A.J.Smith et al. Appl.Phys.Lett. 88, 082112 (2006).
6. N.E.B. Cowern, A.J.Smith, B.Colombeau, R.Gwilliam, B.J. Sealy, and E.Collart, IEDM Tech.Digest 2005 (IEE, Piscataway, NJ), 39.1.1 (2005).
7. K.Winterbon, Rad.Eff. 46, 181 (1980).
8. A.J.Smith, B.Colombeau, N.Bennett, R.Gwilliam, N.E.B. Cowern, and B.J. Sealy, Mat.Res.Soc.Symp.Proc. 864, E7.1 (2005).
9. A.J. Smith, B.Colombeau, R.Gwilliam, N.E.B. Cowern, B.J. Sealy, M.Milosavljevic, E.Collart, S.Gennaro, M.Bersani, and M.Barozzi, Mat.Sci.Eng.B. 124-125, 210 (2005).
10. B.J. Pawlak, R.Surdeanu, B.Colombeau, A.J. Smith, N.E.B. Cowern, R.Lindsey, W.Vandervorst, B.Brijs, O.Richard, and F.Cristiano, Appl.Phys.Lett. 84, 2055 (2004).
11. N.E.B. Cowern, K.Janssen, and H.Jos, J.Appl.Phys. 68, 6191 (1990).
12. A.Armigliato, D.Nobili, P.Ostoja, M.Servidori, and S.Solmi, J.Appl.Phys. (2004).
13. R.Fair, J.Electrochem.Soc. 137, 667 (1990).
14. Y.Kim, H.Massond, and R.Fair, J.Elec.Mat. 18, 143 (1989).
15. R.Kalyanaraman, V.Venezia, L.Pelaz, T.Haynes, H.-J Gossmann, and C.Rafferty, Appl.Phys.Lett. 82, 2, 215 (2003).

Effects of Hydrogen Atoms on Redistribution of Implanted Boron Atoms in Silicon during Annealing

Katsuhiro Yokota*, Shuusaku Nakase*, and Fumiyoshi Miyashita**

*Faculty of Engineering and HRC, Kansai University, Suita, Osaka, 564-8680, Japan
**Faculty of Informatics, Kansai University, Takatsuki, Osaka, 569-1095, Japan

Abstract. Silicon dual-implanted with B and H ions were annealed at temperatures of 700 - 900 °C for 30 min. On Si annealed at temperatures below 800 °C, the redistribution profile of B atoms was the same as that on the as-implanted Si because transient-enhanced diffusion of implanted B atoms was restricted by the presence of H atoms. On highly doped Si annealed at 900 °C, the diffusion of B atoms was independent of B atom concentration although it was significantly retarded in comparison with that on singly B-implanted Si.

Key words: silicon, boron, hydrogen, diffusion, activation, ion-implantation
PACS: 85.40.Ry, 73.20.Hb, 81.65.Rv, 81.05.Cy

INTRODUCTION

Low-energy implantation [1], molecular ion-implantation [2], and rapid thermal annealing [3] are effective techniques for fabricating shallow junctions in silicon. However, the ion-implantation of impurity atoms generates point defects such as Si self-interstitials and Si vacancies in the silicon, which may make undesirable transient-enhanced diffusion of implanted impurity atoms into deeper regions in the silicon during annealing [4]. The diffusion coefficient of B atoms in transient-enhanced diffusion is about three orders of magnitude larger than that for thermal-equilibrium diffusion [5], yet the transient-enhanced diffusion becomes negligible at longer annealing times [6].

In this study, we examine use of atomic H to restrict the transient-enhanced diffusion of B atoms during annealing at low-temperature, assisting by point defects generated by the ion-implantation of B atoms.

EXPERIMENTAL PROCEDURE

B^+ ions with energy of 10 keV were implanted into P-doped 10-20 Ω cm (100) CZ-Si wafers with a dose of 5×10^{15} cm^{-2} (the sample was named as Si(B)), followed immediately by implantation of H^+ ions with energies of 2 ~ 25 keV. (this sample was named as 2 keV-12-Si(B), e.g. 2 keV H^+ ions were implanted into Si(B) at a dose of 12×10^{15} cm^{-2}). The B and H dual-implanted Si was annealed at 700-900 °C for 30 min in flowing Ar gas. Carrier concentration profiles were obtained through differential Hall and resistance measurements after the removal of subsequent silicon layers by anodic oxidation and oxide stripping. The distribution profiles of B and H atoms were analyzed in a 60 μm-diameter area using a secondary ion mass analyzer (SIMS) with magnetic prism.

RESULTS and DISCUSSION

Figures 1(a) and (b) shows B atom and carrier concentration profiles on Si(B) and 1 keV-12-Si(B) annealed at temperatures of 700 - 900 °C. On Si(B) annealed at 700 °C for 30 min, the B concentration was 1×10^{18} cm^{-3} at a depth of 0.19 μm, slightly greater than 0.16 μm in the as-implanted Si. The thermal-equilibrium diffusion length ($2(Dt)^{1/2}$, where D is diffusion coefficient and t is time) of B atoms on Si annealed at 700 °C for 30 min is only 2.5 nm [7]. The transient-enhanced diffusion of B atoms occurred significantly when annealed at 800 °C for 30 min: the B concentration of 1×10^{18} cm^{-3} occurred at a depth of 0.23 μm. The concentrations of B atoms on the annealed Si(B) samples varied exponentially with depth. Ion-implantation can generate the concentration of Si self-interstitials greater that of the thermal-equilibrium state in Si [4]. The diffusion of B atoms in Si is enhanced by interaction with the Si self-interstitials: transient-enhanced diffusion [6].

The distribution profile of B atoms in Si(B) annealed at 900 °C for 30 min can be approximated by a Gaussian distribution function in regions except the near surface. The diffusion coefficient of B atoms in the 900 °C-annealed Si(B) is obtained as 1.7×10^{-14} cm^2 s^{-1} by fitting a Gaussian distribution function for ions implanted into Si, substituting a straggle of ΔR_p^2 to ($\Delta R_p^2 + 2Dt$) [10], to the measured distribution profile.

The measured diffusion coefficient is about one order of magnitude greater than that for thermal-equilibrium diffusion (2×10^{-15} cm^2 s^{-1}). The diffusion coefficients of B atoms also were calculated using the Matano's method [8], and were the same at all depths on the 900 °C-annealed Si(B). Thus, the fast diffusion of B atoms did not occur by vacancies varying concentration with the Fermi level corresponding to carrier concentrations [9]. Implantation-induced defects such as vacancies and silicon interstitials in high-dose implanted Si are much more abundant than thermally produced vacancies. The defects serve to assist the diffusion of B atoms in the sample before they were annealed out.

The profiles of carrier and B atom concentrations on all samples are very similar to each other in regions greater than the near-surface region. In the near-surface region, the carrier concentrations were limited to about 0.6×10^{20} cm^{-3} at 700 °C and 1.3×10^{20} cm^{-3} at 800 and 900 °C, lower than the equilibrium solubility [10].

Figure 1(b) shows B atom and carrier concentration profiles for 1 keV-12-Si(B) annealed at temperatures of 700 - 900 °C. On the samples annealed at temperatures up to 800 °C, B atom concentration profiles were similar to that for the as-implanted silicon. The diffusion of B atoms was not transiently enhanced in a part of region, but it was slightly enhanced in regions with B concentrations lower than 10^{18} cm^{-3}. The plots for carriers were well corresponding to the B atom concentration profiles: many B acceptors were not neutralized by atomic H, but the atomic H acts to retard the transient-enhanced diffusion of B atoms in the 1 keV-12-Si(B).

The concentration of ionized H is given by $n^+/n_o = K(T)\exp((\varepsilon_D - \varepsilon_f)/kT)$ and $n^-/n_o = K(T) \exp((\varepsilon_f - \varepsilon_A)/kT)$,

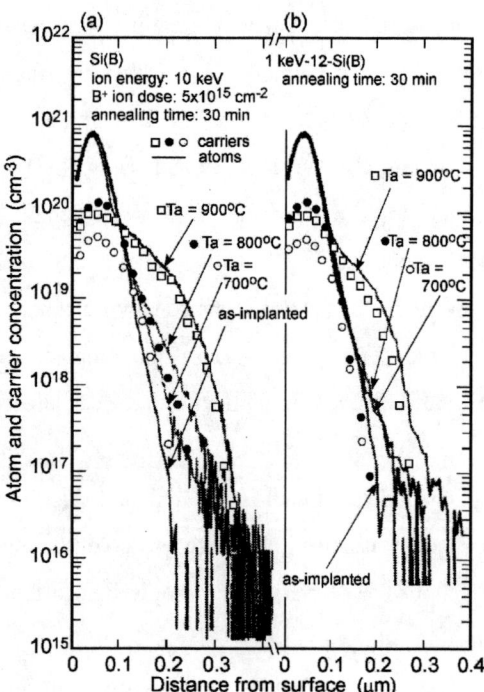

FIGURE 1. (a). B atom and carrier concentration profiles on annealed Si(B) and (b). B atom and carrier concentration profiles on annealed 1 keV-12-Si(B).

where n_o is the concentration of neutral hydrogen, K(T) is a constant at temperature T [°C], ε_D is the hydrogen donor level, ε_A is the hydrogen acceptor level, ε_f is the Fermi level, and k is the Boltzmann constant [11]. On heavily doped p-type Si, the concentration of positively-ionized H is greater than that of negatively-ionized H since $\varepsilon_A - \varepsilon_D > 0$ for normal order. The positively ionized H reacts preferentially with Si self-interstitials via the reaction such $H^+ + I_{Si} \rightarrow SiH + h^+$ rather than the negatively charged B acceptors. This seems to be supported by our experimental result that the plots for carriers fell almost on the B atom concentration profiles below the near-surface region. In near intrinsic regions, the concentration of positively-ionized H is close to that of negatively-ionized H, they can compensate each other,

and neutralized silicon self-interstitials are less than those in heavily-doped p-type regions. Thus, the diffusion of B atoms in the intrinsic region is significantly enhanced by interaction with Si self-interstitials. On this experiment, the transient-enhanced diffusion of B atoms was observed in regions having B concentrations below the intrinsic carrier concentration of 1×10^{18} cm^{-3} at 800 °C.

The boron concentration profile on the 900 °C-annealed 1 keV-12-Si(B) can be approximated by a Gaussian distribution function with a diffusion coefficient of 1.2×10^{-14} cm^2 s^{-1}, which is smaller than 1.7×10^{-14} cm^2 s^{-1} on 900 °C-annealed Si(B). The implanted H atoms serve to reduce the fast diffusion of B atoms by passivating defects generated by B ion-implantation. Complexes of interstitials or vacancies and H atoms have been found on 80 keV proton-implanted Si annealed at room temperature [12] and on ion-implanted silicon annealed at low temperatures in H atmosphere [13].

Figure 2 shows B atom concentration profiles as a function of H ion energy on dual-implanted silicon annealed at 800 °C for 30 min. Annealed 12.5 keV-12-Si(B) and 25 keV-6-Si(B) samples had the same B concentration profiles as annealed Si(B), while the B atom concentration profile on annealed 1 keV-12-Si(B) was the same as the as-implanted 1 keV-12-Si(B). The projected ranges of 193 nm for 12.5 keV H$^+$ ions, and 436 nm for 25 keV H$^+$ ions were greater than the projected range of 29 nm for 10 keV B$^+$ ions [17]. On the annealed 12.5 keV-12-Si(B) and 25 keV-6-Si(B), a large fraction of the implanted H atoms distributes in regions deeper than the region containing implanted B atoms. On the other hand, the projected range of 1 keV hydrogen ions is only 14 nm, being about half the

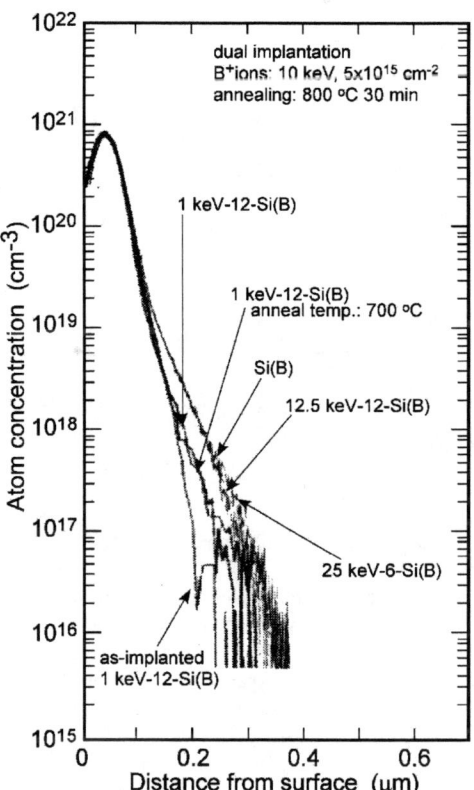

FIGURE 2. B atom concentration profiles on 800 °C-annealed dual-implanted Si as a function of H ion energy.

projected range of 10 keV boron ions [14]. The large fraction of H atoms in the as-implanted 1 keV-12-Si(B) is implanted into the same region as the large faction of boron atoms. The implanted H atoms can interact with B atoms in the 1 keV-12-Si(B) elsewhere. That is, the transient-enhanced diffusion of B atoms is reduced in regions from the surface to about 0.2 μm in where the implanted B and H distribution profiles on the as-implanted sample are overlapped each other.

CONCLUSION

The implanted H atoms react with implantation-induced defects such as Si self-interstitials arising undesirable transient-enhanced diffusion of implanted B atoms. The transient-enhanced diffusions of B atoms in Si(B) annealed at temperatures of 700-800 °C can be restricted by implanting a large dose of H ions with energies such that the populations of B and H ions overlap each other in the silicon.

REFERENCES

1. A. Bousetta, J. A. van den Berg, and D. G. Armour, Appl. Phys. Lett. **58**, 1626(1991).
2. M. Y. Tsai and B. G. Streetman, J. Appl. Phys. **50**, 183(1979).
3. N. C. Tung, J. Electrochem. Soc. **132**, 914(1985).
4. W. K. Hofker, H. W. Werner, D. P. Oosthoek, and N. J. Koeman, Appl. Phys. **4**, 125(1974).
5. E. Vandenbossche and B. Baccus, J. Appl. Phys. **73**, 7322(1993).
6. S. Solmi, F. Baruffaldi, and R. Canteri, J. Appl. Phys. **69**, 2135(1991).
7. R. B. Fair, J. Electrochem. Soc. **122**, 80(1975).
8. D. Shaw, Atomic Diffusion in Semiconductors, Plenum, London, 1973, p.45.
9. T. Y. Tan and U. Gosele, Point Defects, Diffusion, and Precipitation, in K. A. Jackson and W. Schroter, ed., Handbook of Semiconductor Technology, Wiley-VCH, Weinheim, 2000.
10. T. H. Ning, Properties of Silicon, INSPEC, London, 1988, p.384.
11. JC. Herring, N. M. Johnson, Hydrogen Migration in silicon, Hydrogen in Semiconductors, ed. J. I. Pankov and N. M. Johnson, Academic. Boston, 1991, p. 235.
12. K. H. Irmscher, H. Klose, and K. Maas, J. Phys. C **17**, 6317(1984).
13. J.I.Pnkove and C.P.Wu, Appl.Phys.Lett. **35**, 937(1979).
14. J. F. Ziegler, TRIM ver.92, IBM-Research, Yorktown, 1994.

Investigation of Amorphous Layer Formation Using Applied QUANTUM X Single Wafer And QUANTUM Batch Implanter

Roisin Doherty[1], Brendan Mc Comb[1], Richard Ting[2], Yu Chin Cheng[3]

[1] *Applied Materials, 974 E. Arques Ave, M/S 81280, Sunnyvale, CA 94085 USA*
[2] *Applied Materials Taiwan, Hsin-Chu, Taiwan ROC*
[3] *United Microelectronics Corporation, Tainan, Taiwan ROC*

Abstract. With the rapid acceptance of single wafer high current implanters for 90nm technology node and beyond, the challenge that customers face is to device match the single wafer system to their existing batch production line. Due to the differences in scanning architecture, the damage rates will differ significantly between tools and thus can cause variations in amorphous layer thickness and residual defects which impact the post anneal junction depth and activation level. In this study we investigate the damage differences between single wafer and batch implanters for Germanium pre-amorphising implants over a range of energies. Implants were performed on an Applied Quantum® X single wafer implanter and a Quantum® III batch implanter. Amorphous layer thickness was measured using TEM and activation was monitored using sheet resistance.

Keywords: Germanium, pre-amorphisation, dose rate, damage.

I. INTRODUCTION

Germanium pre-amorphisation implants (Ge PAI) have become an industry standard in the formation of highly doped boron (B) ultra shallow junctions at 90nm and below. It is well established that enhanced activation and shallower junction formation are observed in B implanted amorphous silicon over crystalline silicon [1]. The amorphous layer (a-layer) produced by PAI reduces the random channeling of B and therefore the penetration depth of the profile. In addition it overcomes the limits of solid solubility as it places the B atoms directly onto the substitutional sites during solid phase epitaxial regrowth [2,3,4]. The disadvantage of the PAI step however is that it leaves a damage band rich in interstitials just beyond the amorphous - crystalline (a/c) interface. Upon annealing these interstitials will form dislocation loops and {311} defects known as end of range defects which drive transient enhanced diffusion (TED) and B deactivation [5]. The separation of the EOR and the implanted B profile is crucial for minimizing TED. By formation of a deeper a-layer the post implantation damage can be separated from the shallow B implant [1].

The a-layer thickness and residual defects created by the PAI step will differ depending on the implanted dose, energy and dose rate.

The dose rate is dictated largely by beam current but also by the exposure rate of the wafer through the beam (implant scanning architecture). Dose rate is described as the rate of arrival of ions per unit area. From a damage accumulation perspective, this is the time between collisions cascades into the same region. Quantum X (QX) utilizes a two dimensional XY scanning system to scan the wafer in front of a spot beam. The Quantum batch implanter has the same beamline however doping of the wafers is achieved by motion of a spinning wheel containing 13 x 300mm wafers, back and forth in front of the spot beam. The dose rate and damage accumulation differences between these two types of scanning architectures for the same beam current conditions can be quite significant (>>10X) depending on selected scanning parameters.

II. EXPERIMENTAL

Ge implants were performed into 300mm N-type, 10-30ohm-cm resistivity wafers, on both QX and Quantum batch implanters at energies of 2, 10 and 20keV at a dose of $1E15 at/cm^2$. The same beam current settings were applied between tools for same energy condition. Since the beamline optics are the identical between Quantum and QX, the same beam spot size and densities were used.

Thermawave (TW) measurements were taken on the Ge PAI as-implanted wafers using a TW TP630 model. Samples were subsequently analyzed with TEM to compare the amorphous layer thickness and residual defects. The TEM images were taken on a JEOL 2010 FEG TEM using two different magnifications (150KX and 400KX) and on-axis multi-beam high resolution imaging conditions.

A selection of Ge PAI wafers from each system was further implanted with B 1keV 1E15at/cm^2 on a common tool and annealed using a 1050°C 1sec spike anneal in 100% N2. These wafers were measured for sheet resistance (Rs) using a KLA four point probe, Rs-100 Omnimap. Secondary ion mass spectrometry (SIMS) measurements were performed on the annealed B doped wafers to compare the profile matching, and junction depth. The SIMS was performed on a Cameca 4FE7 tool using O2+ beam of 700eV profiling energy. The beam angle was ~ 40 degrees.

III. RESULTS AND DISCUSSION

THERMAWAVE RESULTS

A review of the TW results in Fig. 1 from the as-implanted wafers shows that the QX has higher TW readings associated with a higher level of damage. The difference in TW values is less at the higher Ge energy (Δ6% at 2keV reduces to Δ0.34% at 20keV) as damage saturation has been achieved at the higher energy. A reference at 5E14 dose level is also included from QX for calibration purposes. The lower dose on QX shows comparable level of damage to the Quantum system.

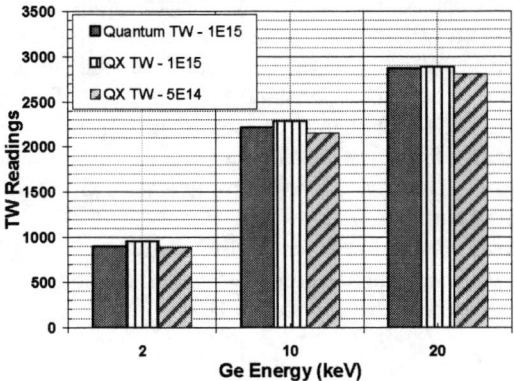

FIGURE 1. TW results for as-implanted Ge PAI wafers

TEM RESULTS

TEM micrograph images of the 2keV and 20keV Ge as-implanted wafers are displayed in Fig 2. The TEM a-layer thickness and a/c interface roughness is reported in Table 1. It is observed that the QX system produces a deeper a-layer for the same implant conditions and a smoother a/c interface roughness.

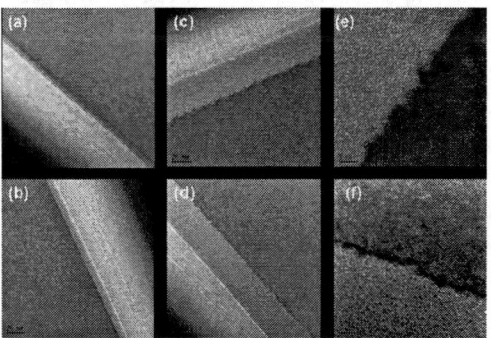

FIGURE 2. Ge PAI as-implanted TEM images from Quantum (a,c,e) QX, (b,d,f) Ge 2keV(a,b) @ 150kX mag. Ge 20keV (c,d) @ 150kX mag. Ge 20keV (e,f) at 400kX mag.

TABLE 1. TEM Measurements			
Implant Conditions	Tool	Amorphous layer thickness	A/C Interface roughness
Ge 2keV 1e15	Quantum	62Å ± 2Å	4Å ± 2Å
Ge 2keV 1e15	QX	72Å ± 2Å	4Å ± 2Å
Ge 20keV 1e15	Quantum	338Å ± 5Å	36Å ± 3Å
Ge 20keV 1e15	QX	370Å ± 5Å	24Å ± 3Å

The linear dependence of the Ge PAI energy on a-layer is as expected due to deeper penetration depth and is consistent with data previously reported by Pawlak et al [1].

The dose rate affects the a-layer thickness by modifying the time of the dynamic anneal. This parameter specifies the temporal separation of the implant cascades. If the dose rate is low, generated damage may anneal out (interstitial-vacancy recombination) during the long time before the next cascade arrives into the same region. When the dose rate is high, damage generated by a cascade may overlap with the damage from previous cascades favoring the formation of more stable structures and the accumulation of the damage [6,7].

A reduction in EOR defect density has also been reported with increase in dose rate [6,7]. During annealing the interstitials and vacancies combine and only the excess interstitials generated from the implant remain. During the regrowth the excess interstitials are swept to the surface and only those beyond the a/c interface remain. A deeper a-layer implies the removal of a larger amount of interstitials, and hence a reduced residual damage [7].

Post anneal TEM measurements have not yet been completed for this study.

SHEET RESISTANCE RESULTS

The Rs data is presented in Fig. 3. It is observed that the Rs decreases with increasing PAI energy for each tool. Since a deeper a-layer is formed with higher PAI energy the lower Rs may be attributed to better activation, deeper B junction formation or less B deactivation.

Lower Rs performance is achieved for the same PAI implants performed on QX system compared to Quantum. The percentage Rs offset between the systems is of the order of 6-9%. All B 1keV implants were performed on the same tool to eliminate any dose matching or dose rate variability for this step. Hence the lower Rs achieved on QX is driven by the PAI step partitioning.

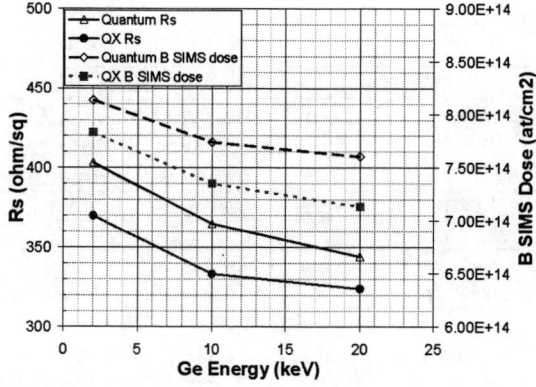

FIGURE 3. Rs results for Ge + B implanted wafers on QX and Quantum. Rs wafers were annealed using 1050°C 1sec spike anneal in 100% N2.

An explanation for the lower Rs with deeper a-layers is that more of the B atoms are contained within the a-layer region. There may be more B activated as a result during the SPER or more damage is eliminated.

SIMS RESULTS

The annealed B SIMS profiles for Ge 2keV and 20keV are plotted in Fig 4. QX produces a 10-20A shallower junction depth as measured at the 1E18 concentration level, for the same PAI condition. The junction depth values are tabulated in Table 2.

The Ge 2keV PAI condition shows a pile up of B dopant at ~ 60A which is the location of the a/c interface confirmed by the TEM measurements. The B profiles for the Ge 20keV PAI condition on both tools have a deeper junction depth compared to 2keV PAI condition which may be attributed to TED. Liu et al. [8] reported the effect of implant energy on diffusion enhancement. It was found that the diffusion enhancement decreased with energy corresponding to the reduction in supersaturation of interstitials.

The integrated B dose as measured by SIMS is also plotted in Fig. 3. A 3-5% lower B dose was measured on QX wafers post anneal which may also be attributed to damage rate differences and migration of the B towards the surface during anneal.

TABLE 2. Differences in junction depth @ 1E18 conc.

Ge PAI Energy (keV)	Quantum junction depth (Å)	QX junction depth (Å)	Delta (Å)
2	476	465	11
10	490	472	18
20	520	500	20

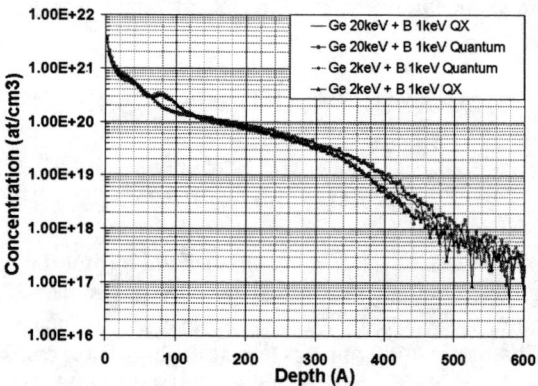

FIGURE 4. B 1keV SIMS profile post anneal on Ge 2keV, 20keV PAI conditions.

IV. CONCLUSIONS

This study confirms that deeper a-layers are formed on QX for the same Ge PAI condition. Since the implant and beam conditions (beam current, density) were kept constant between tools for the experiment the higher dose rates results from slower exposure of the wafer through the beam. A 3.2nm deeper a-layer was achieved formed with Ge 20keV which is quite significant.

Lower Rs performance is also achieved on the QX PAI wafers with shallower B profiles indicating enhanced activation and less TED.

IV. ACKNOWLEDGEMENTS

The authors wish to thank Kevin Jones of University of Florida for TEM measurements and Dimitry Kouzminov and Buzz Morgan of Materials Analysis Services for the SIMS measurements.

V. REFERENCES

1. B.J. Pawlak, R. Lindsay, R. Surdeanu, P. Stolk, K. Maex, and X. Pages, *Proc. 14th Int.Conf. on Ion Implantation Technology*, New Mexico (2002).
2. R. Lindsay, B.J. Pawlak, P. Stolk, and K. Maex, Mat. Res. Soc. Symp. Proc. Vol. 717, C2.1 (2002)
3. J-Y. Jin, J. Liu, U. Jeong, S. Metha, K. Jones, J. Vac. Sci. Tech. B 20(1) 2002, p422
4. K. Tsuji, K. Takeuchi, T. Mogami, IEDM Tech Dig 1999, p9
5. A. Claverie, B. Colombeau, G. BenAssayag, C. Bonafos, F. Cristiano, M. Omri, B. de Mauduit, *Mat. Sci. Semicond. Process.* **3**, 269 (2000)
6. L. Robertson, A.Lilak, M.Law, K.Jones, Appl. Phys. Lett. Vol 71. No. 21 (1997)
7. P. Lopez, L.Pelaz, L.Marques, J. Barbolla ,H.-J.L. Gossmann, A.Agarwal K.Kimura, T.Matsushita , Mat. Sci. and Eng. B 124–125 (2005) 379–382.
8. J. Liu, V. Krishnamoorthy, H.-J. Gossman, L. Rubin, M, Law, K. Jones. J. Appl. Phys, 81, 1657, (1997).

High Dopant Activation And Low Damage P+ USJ Formation

John Borland[1], Seiichi Shishiguchi[2], Akira Mineji[2], Wade Krull[3], Dale Jacobson[3], Masayasu Tanjyo[4], Wilfried Lerch[5], Silke Paul[5], Jeff Gelpey[5], Steve McCoy[5], Julien Venturini[6], Michael Current[7], Vladimir Faifer[7], Robert Hillard[8], Mark Benjamin[8], Tom Walker[9], Andrzej Buczkowski[9], Zhiqiang Li[9] and James Chen[10]

[1]J.O.B Technologies, Aiea, HI; [2]NEC Electronics Corp., Sagamihara, Japan; [3]SemEquip, North Billerica, MA; [4]Nissin Ion Equipment, Kyoto, Japan; [5]Mattson Technology, Fremont, CA; [6]Sopra Optical Solutions, Bois-Colombes, France; [7]Frontier Semiconductor, San Jose, CA; [8]Solid State Measurements, Pittsburgh, PA; [9]Accent Optical Technologies, Bend, OR; and [10]Four Dimensions, Hayward, CA

Abstract. High dopant activation and low damage p+ ultra-shallow junctions (USJ) 15-20nm deep for 45nm node applications have been realized using $B_{10}H_{14}$ & $B_{18}H_{22}$ implant species along with flash, laser or SPE diffusion-less activation annealing techniques. New USJ metrology techniques were employed to determine: 1) dopant activation level and 2) junction quality (residual implant damage) using both contact and non-contact methods.

Keywords: ultra-shallow junction, boron, decaborane, octa-decaborane, flash annealing, laser annealing, SPE annealing, USJ metrology
PACS: 85..40.Ry

INTRODUCTION

For the 45nm node, the p+ USJ for extension varies between 15nm to 20nm deep depending on the device application and trade-offs between dopant activation, junction depth (Xj) and junction quality for high performance (HP), low operating power (LOP) and low-standby power (LSTP) logic devices. To minimize boron dopant diffusion, high temperature >1300°C (flash or laser) or low temperature 650°C SPE annealing are available resulting in <5nm of dopant movement. To eliminate dopant channeling pre-amorphization implantation (PAI) is usually used. PAI and/or co-implantation can lead to higher dopant activation, however, the residual end-of-range (EOR) damage can also result in high damage junctions when using these advanced dopant activation techniques with minimal dopant diffusion [1,2]. For this reason we investigated alternative p+ dopant species to B and BF_2 such as $B_{10}H_{14}$ and $B_{18}H_{22}$ because of their self-amorphization effects avoiding PAI and EOR damage to achieve low damage high quality junctions [3].

EXPERIMENTATION

Boron 500eV/1E15/cm² dose equivalent implants were performed on one hundred, 200mm n-type wafers using B, BF_2, $B_{10}H_{14}$ and $B_{18}H_{22}$ implant species. The implants were performed into crystalline silicon or 11.5nm deep amorphous silicon layer using Ge 5keV/5E14/cm² for PAI. Both batch and serial implanters were used for implant signature comparison.

Dopant activation was achieved using: 1) spike annealing at 1080°C or 1000°C at Mattson/Germany, 2) msec flash annealing at 1300°C at Mattson/Canada, 3) 200nsec sub-melt laser annealing at Sopra/France or 4) 5sec 650°C SPE annealing at Mattson/Germany. Sheet resistance (Rs) was measured with non-contact junction photo-voltage (JPV) at Frontier and contact non-penetrating 4 point probe (4PP) using elastic material (EM) probes at Solid State Measurements (SSM) and mercury (Hg) probes at Four Dimensions (4D). The electrically active surface dopant level/density was measured by a C-V technique (Nsurf) at SSM. SIMS analysis at NEC was used to determine the boron chemical density depth profile and X-TEM to evaluate the amorphous layer depth and after anneal residual implant EOR damage. After anneal junction quality was determined by junction leakage measurement using JPV at Frontier. Silicon crystal lattice damage levels were measured by photo-luminescence imaging (PLi) at Accent on as-implanted and after annealed wafers.

RESULTS & DISCUSSION

Figures 1 and 2 shows SIMS boron depth profiles for B and $B_{18}H_{22}$ after each anneal. Without Ge-PAI, evidence of channeling for B starts at 8E18/cm³ as shown in Fig. 1a and for $B_{18}H_{22}$ it starts at 4E18/cm³ as shown in Fig. 2a. Table 1 shows the SIMS determined chemical junction depth (Xj) defined at 1E18/cm³ for all the conditions studied. Channeling for B and BF_2 is about 9nm, for $B_{10}H_{14}$ is about 4nm and for $B_{18}H_{22}$ is about 3nm. PAI results in greater boron dopant diffusion with lamp based annealing.

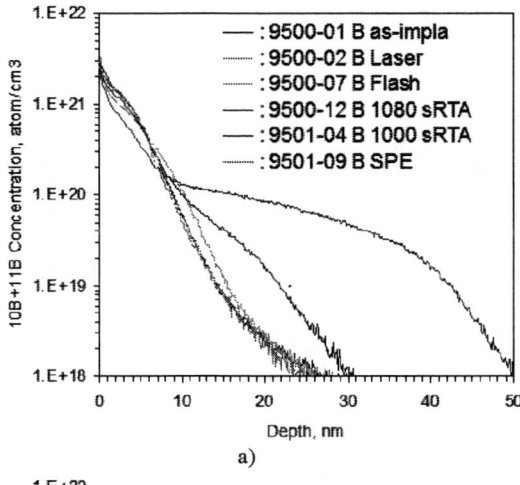

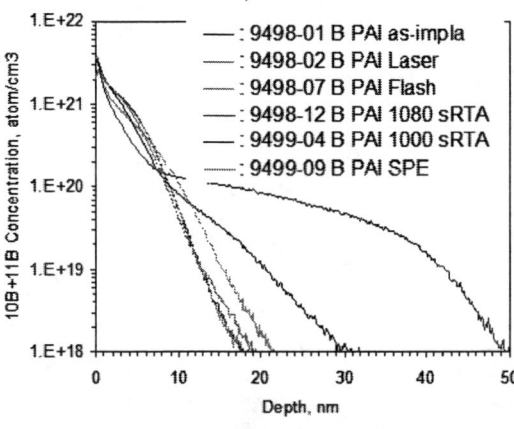

Figure 1. a) B and b) PAI+B SIMS results.

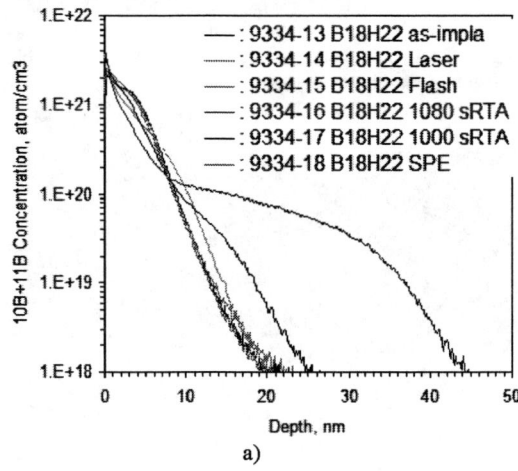

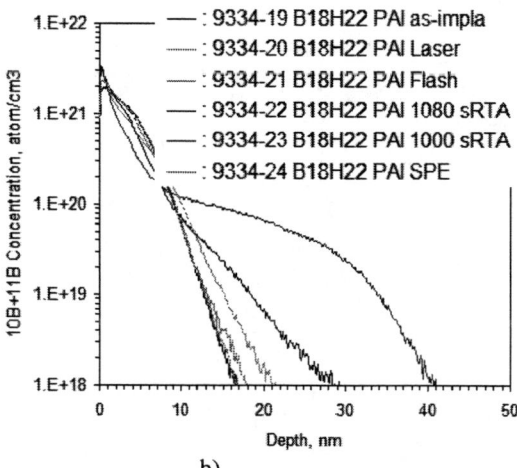

Figure 2. $B_{18}H_{22}$ & PAI+$B_{18}H_{22}$ SIMS results.

TABLE 1. SIMS determined Xj depth in nm at $1E18/cm^3$ and dopant movement.

Dopant	Control	Spike/1080	Spike/1000	Flash	Laser	SPE/650
B11	26.9	50.1+23.2	30.4+3.5	25.9-1.0	26.8-.1	24.1-2.8
+PAI	17.9	49.3+31.4	30.4+12.5	21.5+3.6	18.0+.1	19.4+1.5
BF2	25.1	40.6+15.5	27.6+2.5	25.2+.1	24.0-1.1	22.0-3.1
+PAI	16.6	39.6+23.0	26.9+10.3	21.0+4.4	17.0+.4	17.7+1.1
B10-serial	18.7	36.5+17.8	23.2+4.5	17.5-1.2	19.8+1.1	17.4-1.3
+PAI	14.8	36.5+21.7	29.3+14.5	19.8+5.0	15.2+.4	16.8+2.0
B10	20.0	36.6+16.6	22.5+2.5	18.6-1.4	19.2-.8	18.0-2.0
+PAI	15.5	36.6+21.1	28.5+13.0	20.9+5.4	15.9+.4	16.9+1.4
B18	19.7	44.7+25.0	25.4+5.7	20.0+.3	21.0+1.3	19.4-.3
+PAI	16.6	41.1+24.5	28.8+12.2	21.6+5.0	16.8+.2	18.2+1.6

The 1080°C spike anneal resulted in 15.5-31.4nm of dopant diffusion. When the temperature was reduced to 1000°C only 2.5-14.5nm of diffusion occurred. With flash annealing -1.4 to +5.4nm of diffusion was observed. With laser annealing -1.1 to +1.3nm of dopant movement and with SPE -3.1 to +2nm of dopant movement.

PLi analysis was used to get a full wafer image mapping of the as-implanted damage and after annealing damage recovery (residual implant damage). The as-implanted PAI wafers all had high PLi values between 72-75 due to the 11.5nm deep Ge amorphous layer while the non-PAI wafers had PLi values between 41-46. For the wafers receiving the 1080°C spike anneals and flash anneals complete damage recovery was detected as shown in Fig. 3a&b with PLi values below 14 but the BF_2 samples always had the highest PLi values suggesting a F effect that dominates even over Ge-PAI. The PLi results for 650°C SPE annealing is shown in Fig. 4. The EOR damage from PAI and BF_2 implants resulted in PLi values between 25-29 while B residual implant damage values were 31. The molecular dopant species without PAI had the lowest PLi values between 14-18 suggesting complete damage recovery of the self-amorphizing layer.

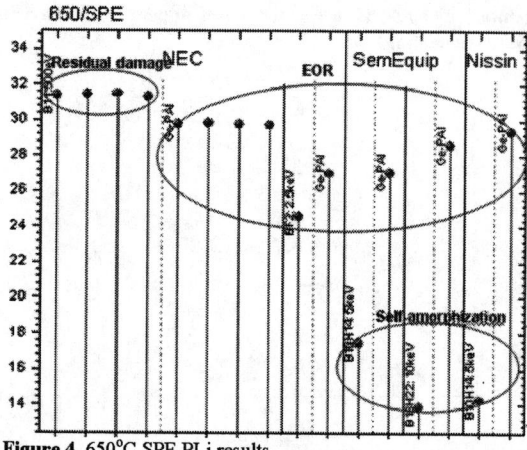

Figure 4. 650°C SPE PLi results.

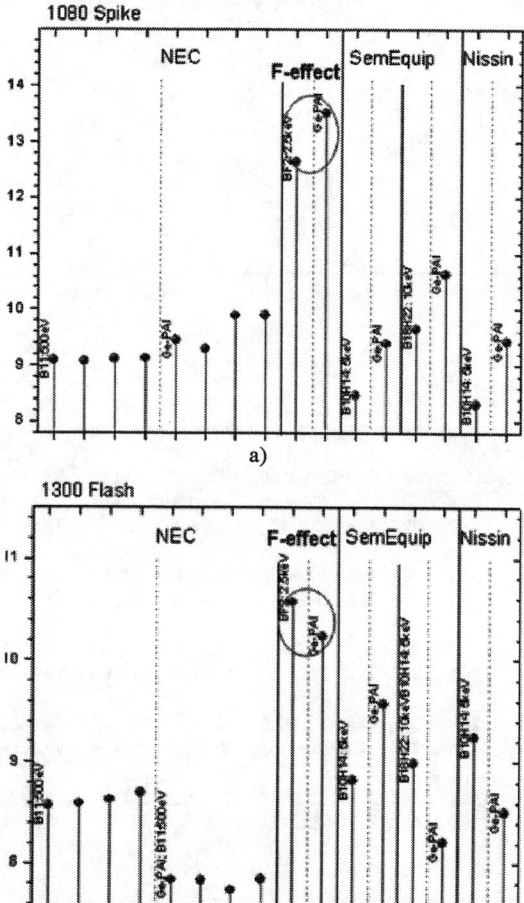

Figure 3. a) 1080°C spike PLi results and b) Flash PLi results.

Figures 5-7 shows X-TEM results for the Flash, SPE and laser annealing respectively. With the high temperature flash anneal all the samples were clean with no evidence of residual implant damage nor EOR damage from the amorphous layer. The PLi values were 8 to 9 in Fig. 5 with JPV leakage all <1E-7A/cm2. With the SPE anneal as shown in Fig. 6 residual implant damage within the top 6nm of the surface could be observed for the B implant as reflected by the PLi value of 30 and leakage of 1E-5A/cm2. The BF_2 sample showed well defined EOR damage resulting in a PLi value of 25 and leakage of 1E-6A/cm2. $B_{18}H_{22}$ was clean with a PLI value of 14 and leakage of 2E-7A/cm2 while with Ge-PAI well defined EOR damage could be seen with a PLi value of 29 and leakage of 2E-5A/cm2. After laser annealing random residual damage could also be seen for the B sample with a PLi of 19 and leakage of 3E-7A/cm2 while the BF_2 sample shows EOR damage as well as a 3nm surface amorphous layer with PLi of 27 and leakage of 3E-6A/cm2. The B18H22 sample was again clean with a PLi of 13 and leakage of 1E-7A/cm2 while with Ge-PAI an 11.5nm of amorphous layer remained suggesting no recrystalization and the PLi value was high around 55 and leakage of 2E-2A/cm2. In fact all the Ge-PAI samples remained amorphous after laser annealing with PLi values between 55-65 and remained n-type, no electrical activation/conversion to p-type.

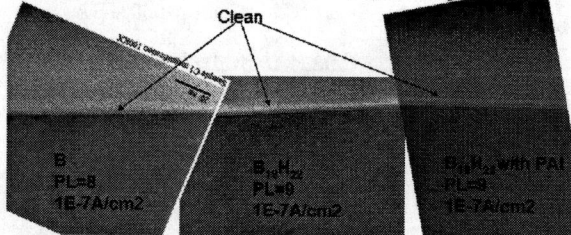

Figure 5. X-TEM of flash annealed samples.

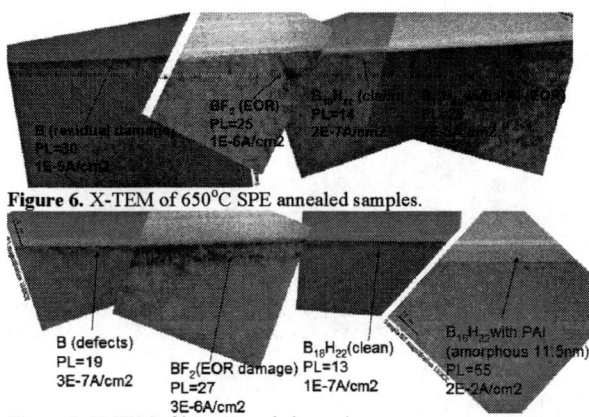

Figure 6. X-TEM of 650°C SPE annealed samples.

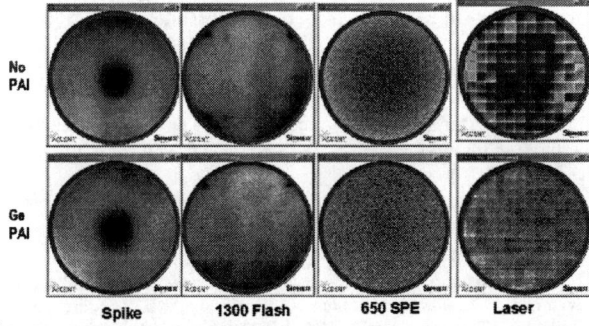

Figure 7. X-TEM of laser annealed samples.

The unique signature of each annealing technique can be seen by full wafer PL imaging as shown in Fig. 8. The gradient are magnified in these figures; the actual variations are quite small. Spike annealing shows a center to edge gradient with the highest PLi of 10.3, in the center and 9.2 at the edge; flash annealing shows dark spots (8.1PLi) where the wafer lifters are located (an artifact of the earlier version of the tool used to process the wafers) compared to the other bright areas (7.7PLi); the SPE signature is slightly darker towards the center (35.9PLi) compared to the edge (32.4PLi); and laser annealing shows a step and repeat checkerboard pattern.

Figure 8. PL full wafer map imaging of annealing signatures.

Junction quality/damage recovery was characterized by JPV R_sL leakage measurement. The results are shown in Table 2 (A/cm^2) and all the spike and Flash annealed samples with or without Ge-PAI had junction leakage <1E-7A/cm^2 (measurement sensitivity limit). The Ge-PAI wafers with laser annealing remained amorphous and the R_sL measured leakage was in the E-2 to E-3A/cm^2 range. While without PAI $B_{18}H_{22}$, $B_{10}H_{14}$ and B were in the 1–3E-7A/cm^2 range and BF_2 was 3E-6A/cm^2 due to EOR damage. Results for the SPE anneal showed that an excellent junction leakage current of 2E-7A/cm^2 measured for the $B_{10}H_{14}$ and $B_{18}H_{22}$ wafers suggests high quality junctions. The Ge-PAI wafers were in the E-5A/cm^2 level. The B and BF_2 wafers were in the E-5 and E-6A/cm^2 range due to residual damage and EOR damage after SPE annealing.

Table 2. JPV junction leakage measurements (A/cm^2)

Dopant	Spike/1080	Spike/1000	Flash	Laser	SPE
B	1E-7	1E-7	1E-7	3E-7	1E-5
+PAI	1E-7	1E-7	1E-7	2E-2	6E-6
BF2	1E-7	1E-7	1E-7	3E-6	1E-6
+PAI	1E-7	1E-7	1E-7	1.5E-3	1E-5
B10-serial	1E-7	1E-7	1E-7	2E-7	2E-7
+PAI	1E-7	1E-7	1E-7	1.8E-3	3E-5
B10	1E-7	1E-7	1E-7	1E-7	2E-7
+PAI	1E-7	1E-7	1E-7	2.3E-2	3E-5
B18	1E-7	1E-7	1E-7	1E-7	2E-7
+PAI	1E-7	1E-7	1E-7	2.3E-2	2E-5

For shallow junctions <25nm deep, an accurate 4PP sheet resistance (R_s) measurement for dopant activation is very difficult to obtain due to probe penetration of any or all of the probes. Therefore, we compared several new alternative methods to measure R_s (ohms/square). Non-penetrating contact EM-4PP and Hg-4PP R_s results, as well as JPV R_s results are compared to standard 4PP with blunted probe tips and are shown in Table 3. For most of the conditions, good agreement between all the various R_s metrology techniques were verified, however, for some of the conditions, a wide range of R_s values were observed, especially for the B laser and SPE diffusion-less activation anneals even though SIMS analysis detected deep junctions of >24nm. The true electrical junction depth could be much shallower than the SIMS determined junction depth which is based on the B chemical (elemental) depth profile.

Table 3. Comparison of various Rs (ohms/sq.) metrology results (Hg/std/EM/JPV).

Dopant	Spike/1080	Flash	Laser	SPE
B	342/340/313/315	527/525/466/475	2200/2224/1209/1100	20403/19410/62K/7500
+PAI	349/340/342/315	538/512/473/470	no p/n	1227/1135/1235/1080
BF2	465/453/466/430	661/641/682/610	998/806/1378/996	4818/4856/4335/3600
+PAI	634/526/719/488	674/675/685/637	no p/n	2183/1746/2954/1872
B10-serial	590/579/535/536	905/895/831/791	844/988/857/706	2680/2676/2456/2175
+PAI	656/650/614/600	860/842/842/759	no p/n	1678/1025/1799/1488
B10	566/553/538/516	809/781/765/732	877/832/1494/511	2940/2961/3521/2377
+PAI	617/601/589/560	780/791/955/703	no p/n	3368/--/1531/1590
B18	405/393/378/368	539/526/503/484	583/580/561/728	1781/1785/1682/1493
+PAI	484/452/640/423	613/595/725/555	no p/n	1384/1099/1400/1239

From the R_s vs. X_j plot, the dopant activation level was determined [5]; however, since there is always uncertainty in the true electrical junction depth, as well as the measured R_s value by each of these techniques, there is uncertainty in the true activated level. The determined dopant activation level also known as the boron solid solubility limit (B_{ss}) is listed in Table 4 [a) Nsurf and b) Bss derived from Rs-vs.-Xj]

Table 4. Carrier density /cm^3 determined by Nsurf or Bss (Rs/Xj)

Dopant		Spike/1080	Spike/1000	Flash	Laser	SPE
B	Nsurf	1.9E19	1.8E19	4.4E19	1.2E20	9E18
	Bss	7-8E19	5E19	1E20	2-4E19	<8E18
+PAI	Nsurf	5.5E19	7.2E19	9.5E19	no p/n	4.6E19
	Bss	7-8E19	5E19	9E19	no p/n	6E19
BF2	Nsurf	3.9E19	3.7E19	6.3E19	1.3E20	1.8E19
	Bss	8E19	3-4E19	6E19	4-7E19	1.5E19
+PAI	Nsurf	4.4E19	4.6E19	8.6E19	no p/n	1.4E19
	Bss	8E19	3-4E19	1E20	no p/n	3-4E19
B10-serial	Nsurf	4.8E19	6.3E19	9.2E19	1.6E20	7.4E19
	Bss	7E19	4E19	0.9-1E20	7-9E19	3-4E19
+PAI	Nsurf	4.2E19	5.2E19	8.9E19	no p/n	4.8E19
	Bss	7E19	3-4E19	8E19	no p/n	6E19
B18	Nsurf	4.2E19	4.7E19	9.7E19	1.4E20	6.3E19
	Bss	8E19	4-5E19	1.3E20	0.9-1.3E20	4E19
+PAI	Nsurf	5.2E19	5.3E19	1.2E20	no p/n	3.9E19
	Bss	7-8E19	3-4E19	0.9-1.2E20	no p/n	6E19

The wide spread in B_{ss} values for a specific annealing technique is due to the wide range in R_s values determined by the various metrology techniques listed in Table 3. PAI+B with a 1000°C spike anneal R_s determined B_{ss} varied from 0.5-1E20/cm^3 and PAI+BF$_2$ with SPE anneal B_{ss} varied from 2.5-5E19/cm^3. For this reason, a new technique to directly measure the near surface electrically active dopant density (Nsurf) within the top 3nm of the surface was developed using an EM-probe CV based technique. Using this technique, we could directly measure the surface activated dopant density, and therefore compare each implant species and annealing conditions without having to know the electrical junction depth.

Table 4 also shows Nsurf results. The highest Nsurf dopant activation levels were seen with laser annealing (1.6E20/cm^3) followed by flash (1.2E20/cm^3), and then spike annealing (7.5E19/cm^3), and SPE (7.4E19/cm^3) as shown in Fig. 9. For each annealing technique, the highest dopant activation was always detected for the molecular dopant species without Ge-PAI, while the opposite conclusion would be made using B_{ss} determined from the R_s vs. X_j data in Table 4. Except for the SPE annealing case, the B_{ss} values were similar, with or without Ge-PAI for the spike and Flash anneals. For SPE anneals, the Ge-PAI wafers always had higher B_{ss} activated levels. For most of the cases good agreement was observed between Nsurf and Rs/Xj determined Bss as shown in Table 4. However, for some cases like the B spike annealed at both 1080°C and 1000°C the difference between Nsurf and Bss was as much as 4x (1.9E19/cm^3 versus 8E19/cm^3). The SIMS profile in Fig. 1a showed B diffusion at approximately 1.5E20/cm^3 with a surface pile-up of 2E21/cm^3 of electrically inactive B. Spreading resistance depth profile (SRP) was conducted on this sample by beveling. A drop in the electrical active dopant level towards the surface is clearly observed by SRP in agreement with the lower Nsurf measurement shown in Table 4.

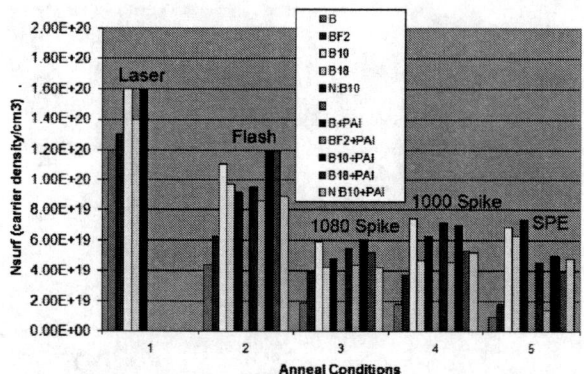

Figure 9. Nsurf for various anneals.

CONCLUSION

High quality and high dopant activation p+ junctions ~15-20nm deep can be achieved using $B_{10}H_{14}$ or $B_{18}H_{22}$ with high temperature (>1300°C) flash or laser annealing, as well as low temperature SPE annealing at 650°C. These anneals enable the extension of beam-line implantation to beyond 32nm node with energies at 5-10keV. Residual implant damage when using B, BF$_2$ or Ge-PAI implants make them undesirable with laser and SPE activation due to degradation in junction leakage current. Therefore, molecular dopant species are very attractive for SiON gates using fast (msec) or ultra-fast (200nsec) annealing, or high-*k* Hf-oxide gates requiring low thermal budget processing (low temperature spike or SPE annealing). New USJ metrology techniques were evaluated and found to be critical in determining the junction quality and dopant electrical activation level.

REFERENCES

[1]. F. Ootsuka, H. Ozaki, T. Sasaki, K. Yamashita, H. Takada, N. Izumi, Y. Nakagawa, M. Hayashi, K. Kiyono, M. Yasuhira and T. Arikado, *IEDM 2003*, section 27.7, p. 647.
[2]. R. Surdeanu, R. Lindsay, S. Severi, A. Satta, B. Pawlak, A. Lauwers, C. Dachs, K. Henson, S. McCoy, J. Gelpey and X. Pages, *SSDM 2004*, section B-3-1, p. 180.
[3]. J. Borland, M. Tanjo, N. Nagai, T. Aoyama and D. Jacobson, *Semiconductor International*, Jan. 2005, p. 52.
[4]. T. Aoyama, M. Fukuda, Y. Nara, S. Umisedo, N. Hamamoto, M. Tanjo and T. Nagayama, *IWJT 2005*, June 2005, section S2-2, p. 27.
[5]. J. Borland, T. Matsuda and K. Sakamoto, *Solid State Technology*, June 2002, p. 83.

A Study of Carbon Effects in Implantation Process for Non-Silicide Contact Formation

T.H. Huh *, S. Kim *, G.J. Ra *, R.N. Reece *, S.I. Kondratenko *,
Y.S. Kim °, K.I. Shin °, and W.H. Jeon °

Axcelis Technologies Inc., 108 Cherry Hill Drive, Beverly, MA 01915, USA
°*Hynix Semiconductor Inc., Memory Manufacturing Center, Ichon-Si, Kyungki-Do, Korea*

Abstract. With a continuous semiconductor devices shrinking approaching to a nanoscale area, the ion implantation processes used for formation of ultra shallow electrically active areas become more sensitive to the material properties at the surface and near surface region. Active control of surface condition and contamination, which can occur during the process of ion implantation, is very important and helps to improve implanted layer parameters and device characteristics. Surface properties modification and mitigation of contaminants by introducing a small amount of gases directed precisely near the target surface during ion implantation is investigated. It is shown that introducing of reactive gases containing oxygen and water vapor into a process chamber results to a significant reduction of surface carbon contamination. Typically, the carbon particles are generated by ion beam striking carbon-based surfaces such as graphite, which is a commonly used material within ion implantation systems. Additionally, photoresist material, which is usually used as a mask for ion implantation, contains carbon, which can then be released as gaseous by-product during ion implantation. Surface adsorption is considered as a potential mechanism of carbon contamination mitigation. It was demonstrated that a suggested method can be applied to control contact resistance of p-MOS transistors and improves DRAM characteristics.

Keywords: Ion Implantation; Carbon; Adsorption, BF_2; Contact Resistance.
PACS: 52.77.Dq, 81.05.Bx

INTRODUCTION

Contemporary technology nodes for semiconductor devices manufacturing include extensive applications of low energy ion implantation. As the ion energies become lower and the dopant profiles shallower, parameters of implanted active areas are getting more sensitive to the condition and chemical composition of the silicon surface undergoing ion bombardment. One common contaminant element that can be introduced into the surface area during the implantation process is carbon. There are two major sources of carbon contamination. First, ion implanter itself can be a source of carbon. Carbon particles are generated by ion beam striking and sputtering carbon-based surfaces such as graphite, which is a commonly used material within ion implantation systems. The second source of carbon is photoresist material used as a mask on silicon wafer surface. During ion implantation photoresist surface layer undergoes chemical bond breaking and release gaseous byproducts such as H_2, CO_2, CO, CH_2, CH_4 and C_2H_2 [1]. Carbon or carbon contained particles either sputtered from graphite parts of the implanter or released as byproducts from the photoresist layer can become adsorbed on the silicon surface and driven into the silicon lattice by the energetic ion beam, thereby altering the chemical composition at the top surface layer of silicon.

It is known that bulk or implanted carbon affects on diffusion and activation of boron in Si-wafer during annealing [2, 3]. The increased carbon concentration can result in suppression of the transient enhanced diffusion (TED) for boron by trapping and reducing the number of available silicon interstitials. In [4, 5] it is shown that surface carbon contamination from photoresist byproducts during implantation of low energy BF_2^+ and As^+ ions can modify boron and arsenic concentration profiles after annealing. The amount of surface carbon added during implantation is difficult to control as it depends on many different factors including implanter beam line design and integrity, ion beam density, photoresist material and wafer surface coverage. It results in higher variability and degradation of device parameters. Traditional methods of surface contamination mitigation using sacrificial oxide layer or chemical surface treatment after implantation before anneal cannot be applied for ultra shallow layers without a significant dopant loss.

In this paper, we discuss the method of surface carbon contamination mitigation and control by introducing of reactive gases containing oxygen and water vapor into the process chamber during ion implantation [6].

EXPERIMENTAL

200 mm, n-type, <100> cz-silicon test grade bare wafers were implanted with BF_2^+ ions with a dose of 3×10^{15} ions/cm^2 and energy 15 keV in a batch, high current ion implanter with a beam current 8-10 mA. Before the implantations the system was checked for meeting base pressure requirements: $P \leq 5.0 \times 10^{-7}$ Torr. The reference test wafers were implanted with standard beam tuning conditions. Two ways were used for introduction of gases containing oxygen and/or water vapor into the process chamber. For the first set of experiments the process chamber was vented, the process disk was exposed to atmosphere, and the chamber was then pumped down to 1×10^{-5} Torr immediately followed by the wafer implantation. The process disk temperature during implantations was 10°C and 40°C. For the second set of experiments the reactive gases were introduced into the process chamber through a gas bleed control system using a mass flow controller (MFC). P-type test wafers with patterned device structures and photoresist on the surface were implanted together with bare wafers. For the device wafers the above BF_2^+ implantation was used for source-drain and poly-Si gate contact areas formation of PMOS transistors (Fig.1). Contact resistance, R_C, quality evaluation were then performed on the finished device structures.

SIMS concentration profiles for boron and surface carbon were measured on the bare test wafers after implantation and annealing at 810°C, 20 seconds in nitrogen. As-implanted boron profiles were measured on selected wafers for implanted dose accuracy check.

RESULTS AND DISCUSSION

A. Effects of Process Disk Exposure to Atmosphere

The SIMS profiles for boron and carbon after test wafers anneal at 810°C are shown in Fig.2 and Fig.3, accordingly. The as-implanted boron profiles (not shown) were practically identical and within the repeatability of the SIMS measurement. This confirms the accurate dose control of the ion implanter, independent on process disk temperature and outgassing effects after the atmospheric exposure. Two disk temperatures, 10°C and 40°C, were compared. As it is seen in Fig.2, boron profiles show a difference in concentration within the surface layer of ~10 nm. For the standard implantation processes the highest boron

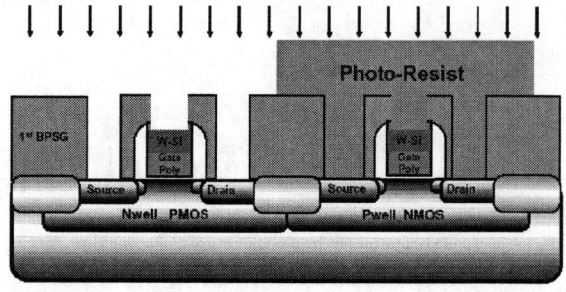

FIGURE 1. Test device structure cross-section during BF_2^+ implantation.

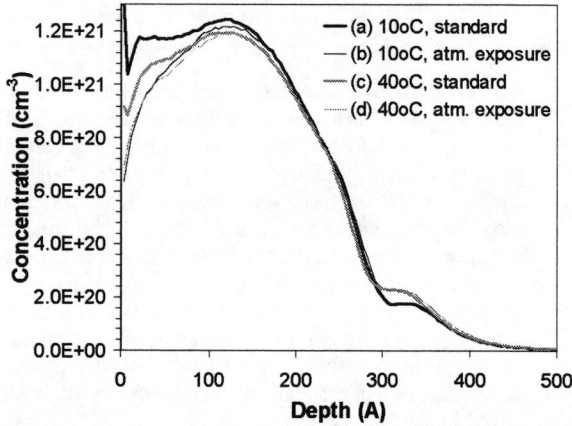

FIGURE 2. Boron SIMS profiles after BF_2^+ implantation and anneal at 810°C.

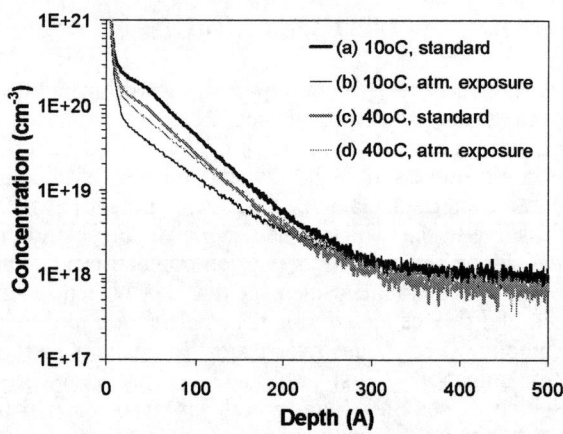

FIGURE 3. Surface carbon SIMS profiles after BF_2^+ implantation and anneal at 810°C.

concentration at the surface area is observed for the lower disk temperature of 10°C. Surface concentration decreases with the disk temperature increased up to 40°C. For the implantations performed right after process disk atmospheric exposure boron depletion at the surface is higher compared to the standard processes. The profiles in this case are very close to each other at the surface area for both implantation temperatures (Fig.2b and d). The second peak observed on boron profiles is related to the amorphous layer end-of-range defects [5], which thickness depends on wafer temperature during implantation.

Carbon SIMS profiles presented in Fig.3 show that after the standard BF_2^+ implantations surface carbon concentration is higher for the wafer processed at lower temperature. During implantation energetic ion beam continuously produces silicon dangling bonds at the surface. Carbon atoms presented near the substrate can become adsorbed on the silicon surface and driven into the silicon lattice by the energetic ion beam. The higher carbon concentration observed at lower substrate temperatures is likely to be related to the higher adsorption rate of carbon. For the implantations performed after process disk exposure to atmosphere surface carbon contamination reduced. The highest reduction of carbon is noticed for the lower temperature of 10°C. We suggest that atmospheric gases containing oxygen, nitrogen, and water vapor were adsorbed on the surface of the process disk and then released during ion implantation altering the carbon partial pressure in residual gas content. Si-O, Si-OH, and Si-N bonds are more ionic compared to Si-C covalent bonds, which electronegativity difference is considerably lower. We believe the dangling bonds formed at the silicon surface during implantation, will be preferably passivated by adsorbed O, OH, and N rather than C. That explains the observed carbon contamination reduction. Table 1 shows the carbon dose integrated from the SIMS profiles (Fig.3) including absolute value and normalized dose relatively to the un-implanted reference wafer. The observed results indicate an important role of surface adsorption in contamination effects during ion implantation. The adsorption mechanisms are under further investigation.

TABLE 1. Carbon Surface Concentration.

Implantation Conditions	Carbon Dose (atoms/cm^2)	Normalized Carbon Dose
Un-implanted wafer	5.8x10^{13}	1.00
10°C, Standard	1.5x10^{14}	2.52
10°C, Atm. Exposed	4.7x10^{13}	0.81
40°C, Standard	9.2x10^{13}	1.59
40°C, Atm. Exposed	7.7x10^{13}	1.33

B. Effects of Reactive Gas Introduction during Ion Implantation

A reactive gas containing oxygen and water vapor was introduced into the process chamber during ion implantation close to the location of the processing wafers. Bare wafers and wafers with device structures were implanted together at different gas flow rates. The partial pressure of the introduced bleeding gas was varied from 0 to 3x10^{-6} Torr, which corresponded gas flow rate changing from 0 to 2 sccm. Carbon SIMS profiles measured on bare wafers after the implantation of BF_2^+ ions are shown in Fig.4. It is seen from Fig.4 that introduction of reactive gas during ion implantation near the wafer surface results in a significant reduction of carbon contamination. Surface carbon dose integrated from SIMS profiles was 1.7x10^{14}, 1.1x10^{14}, and 9.9x10^{13} atoms/cm^2 consequently for the gas flow rate 0, 0.2, and 0.6 sccm. The higher gas flow rate and, consequently, the higher partial gas pressure is not considered to be a practical application due to the higher beam neutralization effects and associated potential implant dose error.

The results of the source/drain contact resistance measurements for PMOS transistor as a function of reactive gas flow rate are shown in Fig.5. As it is seen in Fig.5, the contact resistance value is steadily decreasing with increasing of the gas flow rate. Even a small addition of the reactive gas with a partial pressure of ~3x10^{-7} Torr (corresponds to the flow rate of 0.2 sccm) results to a significant reduction of Rc value about 15%. The relative surface carbon concentration derived from as-implanted carbon SIMS profiles (Fig.4) is ~1.6 times lower in this case compared to the standard implantation without reactive gas. Comparing Rc data with boron and carbon SIMS profiles after annealing it was noticed that the higher contact resistance value corresponds to the higher carbon contamination and higher boron concentration within the surface layer of 10 nm. We believe that the carbon atoms accumulated at the silicon surface during implantation are forming clusters with silicon and boron atoms during a consequent annealing step. As a result, the resistance of ohmic contact formed on the modified silicon surface increases. It is not clear if the implanted fluorine plays any role in the cluster formation and surface modification mechanism during annealing.

In Fig.6 it is shown a comparison of PMOS transistor source/drain contact resistance distribution before and after a reactive gas bleed method was implemented into DRAM manufacturing process. The higher average Rc value with a large distribution range was superseded by the tighter distribution with ~20% lower average. These results indicate that the reactive gas introduction during ion implantation helps to mitigate surface carbon contamination and provides a better control of silicon surface composition before the non-silicide ohmic contact formation.

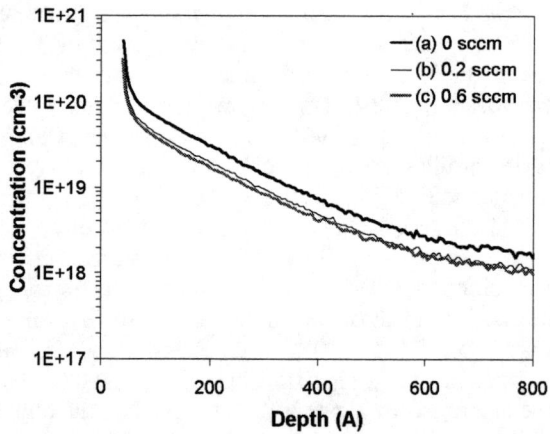

FIGURE 4. Surface carbon SIMS profiles after BF_2^+ implantation with bleeding gas flow rate 0.2 sccm (b), 0.6 sccm (c), and no gas (a).

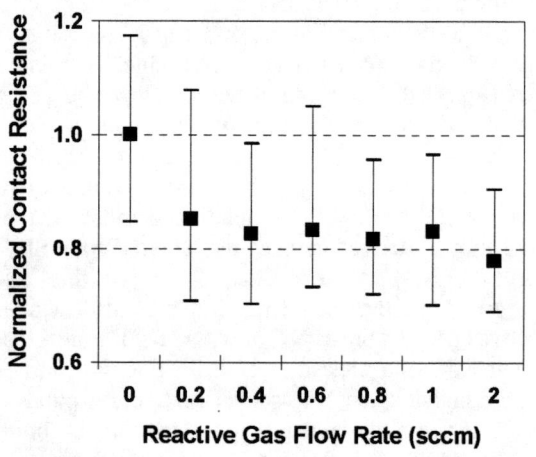

FIGURE 5. PMOS transistor source/drain contact resistance, Rc as a function of reactive gas flow rate.

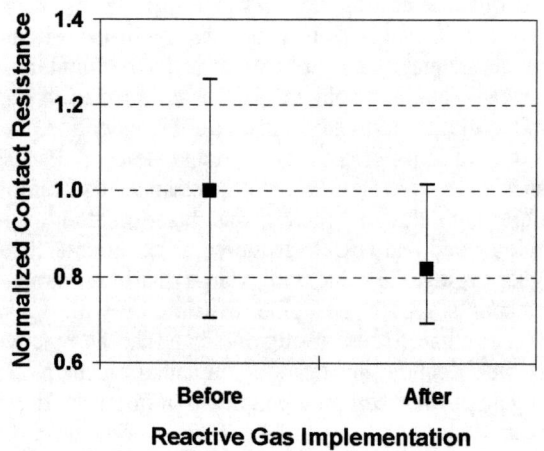

FIGURE 6. PMOS transistor source/drain contact resistance distribution range before and after reactive gas bleed implementation for DRAM process.

CONCLUSIONS

The effects of silicon surface carbon contamination on non-silicide contact formation have been studied. It is shown that a higher surface carbon concentration results in a higher source/drain contact resistance value of PMOS transistor formed by the implantation of BF_2^+ ions. The source of carbon contamination during ion implantation is sputtering of implanter graphite parts and/or release of byproducts from the photoresist layer presented on the wafer surface. Carbon becomes adsorbed on the silicon surface and driven into the silicon lattice by the energetic ion beam, thereby altering the chemical composition at the top surface layer of silicon. It was observed that the higher carbon concentration is accompanied with the higher accumulation of boron at the surface after annealing. As the amount of carbon contamination during ion implantation is difficult to control it results to deterioration of contact resistance repeatability. The suggested method of introduction of a small amount of reactive gases containing oxygen and water vapor into the process chamber during ion implantation helps to mitigate and get better control of surface carbon contamination. It is shown that non-silicide contacts formed on the silicon surface have better characteristics if the reactive gas bleed was used during the implantation. Lower contact resistance value was achieved, which is more stable and less depends on the process parameter variations.

ACKNOWLEDGMENTS

The authors would like to thank Axcelis Korea Field Service Team, especially EG Kim and YC Seo, for reconfiguring the implant system for this study.

REFERENCES

1. T. N. Horsky, *IEEE Proc. of the XII Intl. Conf. on Ion Implantation Technology*, Kyoto, Japan, June 1998, p. 654.
2. P. A. Stolk, H.-J. Gossmann, D. J. Eaglesham and J. M. Poate, *Mat. Sci. Eng.* **B36**, 275 (1996).
3. H. Rucker, B. Heinemann, D. Bolze, D. Knoll, D. Kruger, R. Kurps, H.J. Osten, P. Schley, B. Tillack, and P. Zaumseil, *IEDM*, 345-348 (1999).
4. P. Kopalidis, S. Kondratenko, G. Kay, R.N. Reece, and J. Xu, *Proc. of the XV Intl. Conf. on Ion Implantation Technology, Part II*, Taipei, Taiwan, October 2004, pp. 1-5.
5. P. Kopalidis and S. Kondratenko, *J. Electrochem. Soc.* **152**, (5), G375 (2005).
6. R.N. Reece, S.I. Kondratenko, G.J. Ra, L. Wainwright, and G. Cai, U.S. Patent pending (21 September 2005).

Pre-annealing effects of n$^+$/p and p$^+$/n junction formed by plasma doping (PLAD) and laser annealing

Sungho Heo, Sungkweon Baek, Dongkyu Lee, Musarrat Hasan, and Hyunsang Hwang

Department of Materials Science and Engineering, Gwangju Institute of Science and Technology, #1, Oryong-dong, Buk-gu, Gwangju, 500-712, Korea

Abstract. In this paper, we demonstrated ultra-shallow junction formed by plasma doping (PLAD) and laser annealing. PLAD may be considered as an alternative doping method for the sub 45 technology node due to the possibility of low energy doping and high throughput. However, PLAD has various problems due to the incorporated hydrogen or fluorine. Incorporated hydrogen generally increases damage in the Si substrate and junction depth. Incorporated fluorine also retards dopant activation and increases deactivation behavior after post-annealing. In order to improve the effect of incorporated ions, we applied pre-annealing prior to laser annealing in PLAD samples. By employing low temperature pre-annealing, we can improve electrical characteristics such as low sheet resistance and high activation rates, and also reduce the junction depth after laser annealing.

Keywords : B_2H_6, BF_3 and PH_3 Plasma doping (PLAD), Laser annealing, Low temperature pre-annealing.
PACS : 85.30-Kk

INTRODUCTION

Below the sub-45 nm complementary metal-oxide-semiconductor (CMOS) technology node, conventional low energy (< 0.5 keV) ion implantation faces a process limit. It has a low throughput due to low beam current[1]. As an alternative method, plasma doping (PLAD), which has the advantages of high throughput, and a large implanted area, has been studied in the last 25 years[2]. However, PLAD has problems of incorporated hydrogen (H) or fluorine (F) and other ion contaminations. These have affected the degradation of dopant activation and the increase of dopant diffusion during activation annealing[3-5]. In the case of conventional high temperature spike rapid thermal annealing (RTA), incorporated ions of H and F easily diffuse out to surface regions[6]. However, the spike RTA could not prevent the enhanced diffusion during activation annealing. In order to reduce diffusion, various annealing methods such as solid phase epitaxy (SPE), flash lamp annealing (FLA), and laser annealing (LA) have been developed. LA is the most attractive annealing method because it has a shorter annealing time (~ nanoseconds) and higher activation rate[7]. Despite these advantages, laser annealing could not remove incorporated H or F after laser irradiation. So we need additional annealing prior to laser annealing.

In this study, we demonstrated the pre-annealing effects prior to laser annealing in p$^+$/n and n$^+$/p junction formed by B_2H_6, BF_3 and PH_3 PLAD. In order to evaluate electrical characteristics, we also studied deactivation behavior in laser annealed samples after post-annealing.

EXPERIMETAL DETAILS

The p$^+$/n and n$^+$/p junction were fabricated by B_2H_6 (3%, H_2 dilution) or BF_3 (100%, No dilution) and PH_3 PLAD (3%, H_2 dilution). The bias voltages were between 400 V and 1 kV and the dosages were between 5×10^{15}/cm^2 and 1×10^{16}/cm^2. For dopant activation, RTA or KrF excimer laser annealing was performed from 500 ~ 650 mJ/cm^2 with 1 or 10 pulses. Laser annealing pulse duration was 25 nanoseconds at repetition of 1 Hz. Compared with one-step laser annealing, pre-annealing was conducted at temperatures of 300 ~ 600 °C for 5~10 min prior to laser annealing. Post-annealing was conducted at 500 ~ 800 °C for 5 min. The sheet resistance (R_s) and active carrier concentration (N_s) were measured by van der Pauw pattern and Hall measurement. A cross-sectional transmission electron microscopy (XTEM) analysis was conducted in order to confirm the crystallization rate. The depth profiles of various ions (B, H, and F) were obtained by secondary ion mass spectrometry (SIMS). In order to evaluate junction leakage current, diodes were fabricated.

RESULTS & DISCUSSION

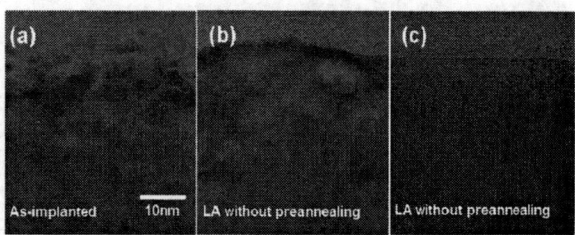

FIGURE 1. XTEM images of (a) as-implanted, (b) LA without pre-annealed and (c) LA with pre-annealed (at 500 °C) B_2H_6 PLAD samples.

Figure 1 shows XTEM images of B_2H_6 PLAD samples followed by LA with or without pre-annealing. The as-implanted sample of B_2H_6 PLAD shows significant damage because of a large amount of co-implanted hydrogen ions ($> 10^{16}/cm^2$). This co-implanted H makes a damaged or defect-rich layer in the Si substrate[4][8]. When pre-annealing is used prior to LA, the damage was effectively removed. Thus, LA with pre-annealed samples shows the better crystallization.

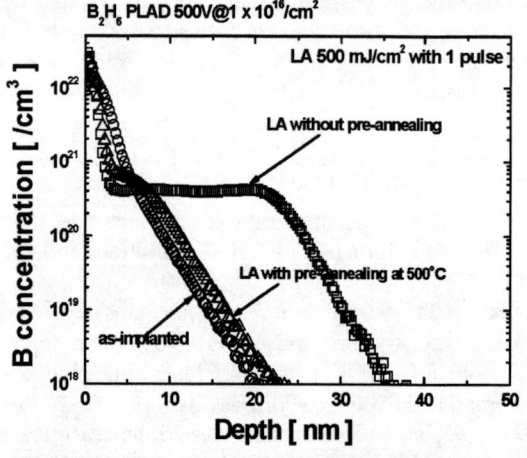

FIGURE 2. Boron SIMS depth profiles of the as-implanted sample, LA with pre-annealed, and LA without pre-annealed samples.

Figure 2 shows B SIMS depth profiles followed by LA with and without pre-annealing. The sample of LA with pre-anealing shows the shallower junction depth because residual defects were reduced after pre-annealing.(Fig. 1.(c)) One-step laser annealing was not enough to reduce the implanted damage due to its short annealing time.(Fig. 1.(b)) Additional pre-annealing offers time to recover incorporated H damage and also incorporated H is easily diffused out to the surface at temperatures of above 500 °C. So, pre-annealing is needed to reduce B diffusion.

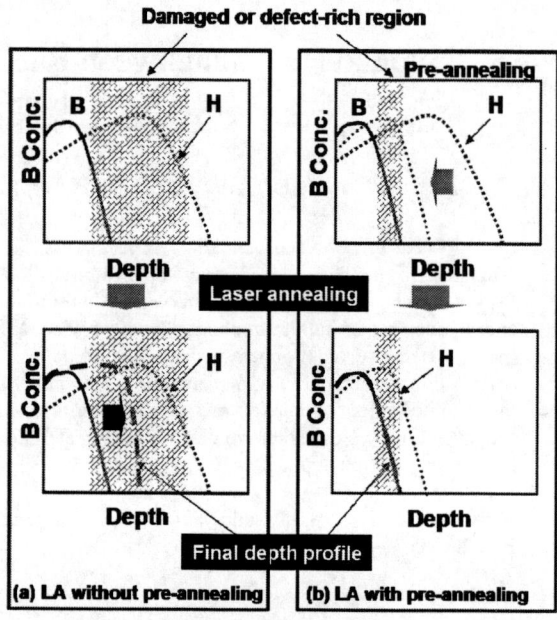

FIGURE 3. Schematic diagram of B diffusion model of (a) LA without pre-annealing and (b) LA with pre-annealing.

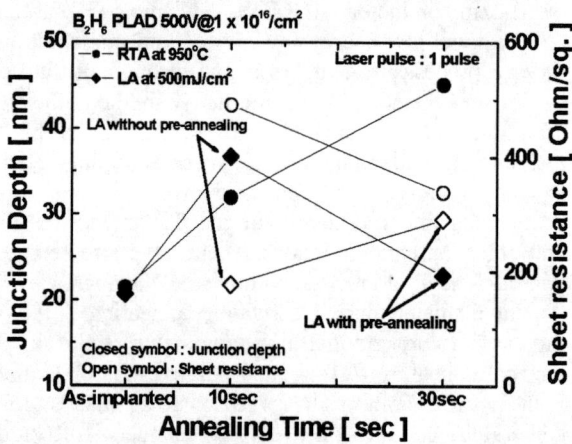

FIGURE 4. Junction depth and sheet resistance followed by various annealing conditions. (RTA and LA with or without pre-annealing)

Figure 3 shows the B diffusion model followed by LA with or without pre-annealing. The damaged or defect-rich layer could be produced by B_2H_6 PLAD. If LA was applied to the sample without pre-annealing, the LA would affect the defect-rich region, resulting in the increase of junction depth.(Fig. 3.(a)) If pre-annealing was performed, the defect-rich region would be reduced near the surface region due to the annihilation and diffusion. It was reported that the implanted hydrogen began to diffuse deeper into the undamaged bulk silicon above 200 °C[6]. Thus, LA with pre-annealing effectively reduces B diffusion.(Fig. 3. (b))

Figure 4 shows the results of junction depth and sheet resistance followed by various annealing conditions. Compared with RTA, LA shows a shallower junction depth and lower sheet resistance. By employing pre-annealing, the junction depth was dramatically reduced but sheet resistance was slightly increased due to the shallow junction depth. Therefore, to remove the incorporated hydrogen effect, a B_2H_6 PLAD process requires pre-annealing prior LA.

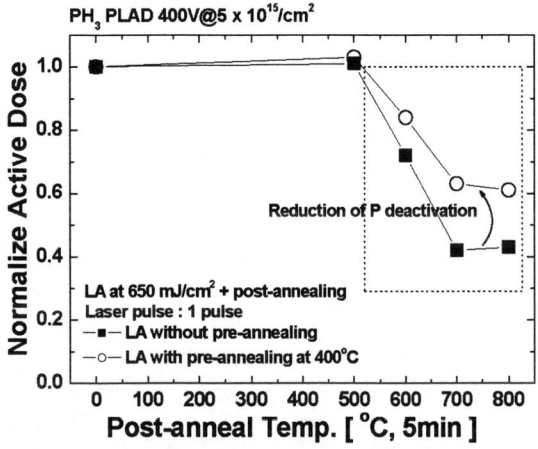

FIGURE 5. Variations of phosphorus deactivation as a function of post-annealing conditions.

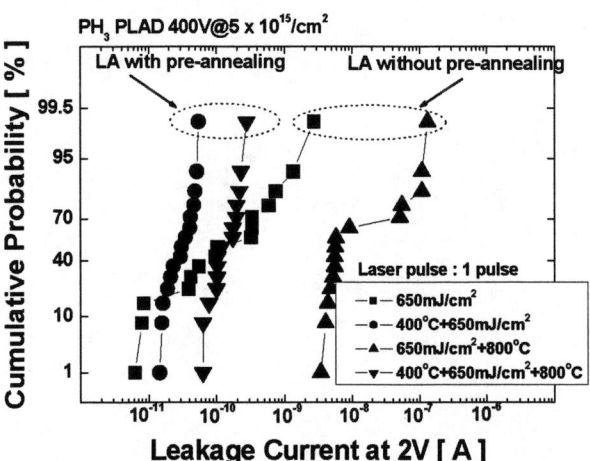

FIGURE 6. Cumulative probability of reverse leakage current at 2V in n^+/p junction followed by various annealing conditions.

Figure 5 shows phosphorus deactivation rates of LA with and without pre-annealing followed by post-annealing conditions. In spite of the various advantages of LA, such as a high activation and low diffusion rates, it has the disadvantage of dopant deactivation after post-annealing due to the meta-stable process. The P dopant shows greater deactivation behavior than stable B dopant because of its unstable nature[9]. In order to improve the deactivation behavior, we applied pre-annealing prior to LA. The samples of LA without pre-annealing shows the P deactivation behavior after post-annealing (> 500 °C). However, by employing pre-annealing, P deactivation behavior was improved. It might be that pre-annealing helps to reduce residual damage remaining after LA.

Figure 6 shows the cumulative probability characteristics of reverse leakage current. Diode reverse leakage current is related to residual defects in depleted regions. The sample of LA with pre-annealing shows lower leakage current. This means that LA with pre-annealing acquires less residual defects and P deactivation behavior. It also shows uniformity of leakage current distribution. Therefore, in order to improve electrical characteristics, PH_3 PLAD needs pre-annealing prior to LA.

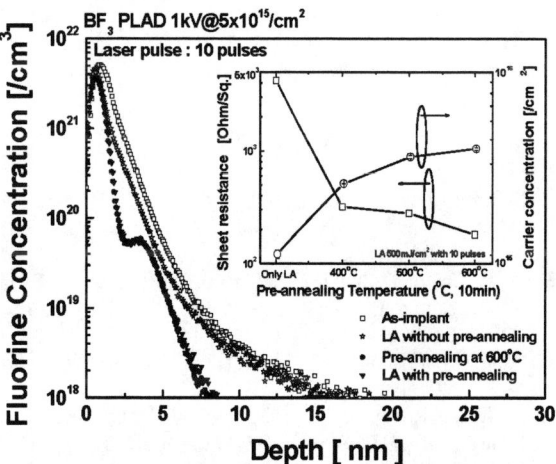

FIGURE 7. Fluorine SIMS depth profile followed by samples of LA with or without pre-annealing. The inset shows the results of sheet resistance and carrier concentration followed by samples of LA with or without pre-annealing.

Figure 7 shows fluorine SIMS depth profile of samples of LA with or without pre-annealing. In BF_3 PLAD, the incorporated F retarded B diffusion but degraded the B activation rate due to the F effect. So, incorporated F should be removed in p^+/n junction to improve electrical characteristics. Incorporated F is easily diffused out by high temperature annealing (> 1000 °C) but LA could not be diffused out of F to the surface region and most F remained. However, the residual F is greatly removed but not completely removed by pre-annealing. The inset figure shows that LA with pre-annealing can achieve low sheet resistance and high active carrier concentration because of the reduction of the F effect. As the pre-annealing temperature increases, electrical characteristics improved due to the reduction of residual F concentration.

CONCLUSIONS

In summary, we demonstrated the pre-annealing effect prior to LA in p^+/n and n^+/p junction formed by B_2H_6, BF_3, and PH_3 PLAD. PLAD degraded the electrical characteristics due to the incorporated H and F. To remove the incorporated ions effect, additional pre-annealing prior to LA was applied in PLAD samples. Pre-annealing acquires higher dopant activation and more stable electrical characteristics. Therefore, pre-annealing prior to LA can be considered as a promising process for the sub-45nm CMOS technology node.

ACKNOWLEDGMENTS

This work was supported by the Center for Distributed Sensor Network at GIST.

REFERENCES

1. X. Lu, L. Shao, X. Wang, J. Liu, W. Chu, J. Bennet, and L. Larson, and P. Ling, *J. Vac. Sci. Technol. B* **20**, 992-994 (2002).
2. S. B. Felch, Z. Fang, B. W. Koo, R. B. Liebert, S. R. Walther, and D. Hacker, *Surface Coating Tech.* **156**, 229-236 (2002).
3. H. Kakinuma, and M. Mohri, *Jpn. J. Appl. Phys.* **34**, 1325-1328 (1995).
4. L. Wang, R. Fu, X. Zeng. P. Chu, W. Y. Cheung, S. P. Wong, *J. Appl. Phys.* **90**, 1735-1739 (2001).
5. S. Paul, W. Lerch, B. Colombeau, N. E. B Cowern, F. Cristiano, S. Boninelli, and D. Bolze, *J. Vac. Sci. Technol.* B **24**, 437-441 (1999).
6. R. G. Wilson, *Appl. Phys. Lett.* **49**, 1375-1377 (1986).
7. B. Yu, Y. Wang, H. Wang, Q. Xiang, C. Riccobene, S. Talwar, and M. Lin, *Tech. Dig. Int. Electron Devices Meet.*, 1999, pp509-512.
8. P. Chen, S. S. Lau, P. Chu, K. Henttinen, T. Suni, I. Suni, N. D. Theodore, T. L. Alford, J. W. Mayer, L. Shao, and M. Nastasi, *Appl. Phys. Lett.* **87**, 111910 (2001).
9. Y. Takamura, S. H. Jain, P. B. Griffin, and J. D. Plummer, *J. Appl. Phys.* **92**, 235-244 (2002).

Spike Annealing of Shallow Arsenic and Phosphorus Implants in Different Gaseous Ambient

S. Paul[1], W. Lerch[1], D. Bolze[2]

[1] Mattson Thermal Products GmbH, Daimlerstr. 10, 89160 Dornstadt, Germany
[2] IHP, Im Technologiepark 25, 15236 Frankfurt (Oder), Germany

Abstract: Spike annealing of shallow arsenic (500 eV and 1 keV, $1 \cdot 10^{15}$ cm^{-2}) and phosphorus (500 eV and 1 keV, $1 \cdot 10^{15}$ cm^{-2}) implants was done in different gaseous ambient. The wafers were spike annealed at a peak temperature of 1050 °C. The gaseous ambient used was pure nitrogen, different oxygen concentrations as well as pure oxygen. It is well known that annealing in oxygen ambient results in injection of silicon self-interstitials. In another set of experiments the wafers were annealed with the same thermal budget in NH$_3$ ambient which is known to inject vacancies. The wafers were spike annealed in different NH$_3$ concentrations and in pure NH$_3$. Resulting from these experiments the minimum sheet resistance of shallow arsenic and phosphorus implants is achieved either with an ambient of 1-10 % oxygen in nitrogen or with 100 % NH$_3$. In terms of limited diffusion, annealing in inert ambient (100 % N$_2$ or Ar) is beneficial. A spike anneal with 1050 °C peak temperature of As 500 eV in inert ambient gives the shallowest junction (16.3 nm) in combination with acceptable sheet resistance (636 Ω/sq.). Furthermore it can be concluded that with the right choice of the gaseous ambient the dose loss due to outdiffusion of dopants can be effectively suppressed.

Keywords: arsenic, phosphorus, RTP, spike anneal, ms-anneal, dopant loss, diffusion, vacancies, silicon self-interstitials
PACS: 66.30.-h, 61.72.Ji

INTRODUCTION

In recent years a lot of work has been done on boron implanted ultra shallow junctions due to the stringent requirements of the ITRS [1]. Nevertheless the requirements for NMOS are also relatively tight and equally hard to fulfill. In this work we investigated the activation and diffusion behavior of shallow arsenic (500 eV and 1 keV) and phosphorus (500 eV and 1 keV) implants with a dose of $1 \cdot 10^{15}$ cm^{-2} that were spike annealed in ambients containing oxygen or ammonia that either introduce silicon self-interstitials or vacancies besides the capping of the surface with an oxide or thermally grown silicon oxynitride (SiO$_x$N$_y$) layer. Arsenic, especially, is known to outdiffuse during annealing at elevated temperatures. This can be effectively suppressed through the addition of oxygen during annealing.

EXPERIMENTAL DETAILS

For the experiments, p-type, 200 mm prime Si wafers of (100) orientation and 1-20 Ωcm were used. All the implants were performed on an Applied Materials Quantum batch implanter at a tilt angle of 0 ° and twist angle of 0 °. To avoid any influence of the native oxide, a wet chemical etch was performed in a diluted HF solution prior to the low energy implantation process. The implants were As$^+$, 500 eV and 1 keV, $1 \cdot 10^{15}$ cm^{-2} and P$^+$, 500 eV and 1 keV, $1 \cdot 10^{15}$ cm^{-2}. After ion implantation the wafers were processed in a Mattson 3000 Plus RTP system equipped with Mattson's Ripple-based pyrometer, a temperature controller for spike anneals (FAC) and wafer rotation. The ramp-up rate to the pre-stabilization step was 50 K/s. The pre-stabilization was at 650 °C for 10 s followed by a spike with a ramp-up rate set to 250 K/s. The peak temperature was 1050 °C, with a peak width of 1.65 s at T$_{(peak-50 K)}$ with a maximum cooling rate of about 80 K/s. The wafers were processed in an ambient of pure nitrogen and in 100, 500, 1000 ppm and 1 and 10 % oxygen in nitrogen as well as in pure oxygen. Furthermore wafers were processed in 5 and 20 % NH$_3$ in argon and in pure NH$_3$. The sheet resistance was measured by a KLA-Tencor RS-100 four point probe using probe type D with a circular 121 site pattern and a 3 mm edge exclusion. Selected samples were analyzed regarding their junction depth and dose using a quadrupole SIMS4550 at CAMECA's laboratory in Munich. Prior to SIMS measurement a 10 nm layer of amorphous silicon was sputtered on top of the samples to reach equilibrium sputtering conditions before approaching the SiO$_x$N$_y$/silicon interface. For phosphorus the dip in the O$^+$ signal was used to align the samples to the interface, for arsenic the dip in the Si$_2^-$ signal was used. Depth profiling was performed with 500 eV primary oxygen beam with normal incidence (0°) for phosphorus and with 500 eV cesium ion beam with 60° for arsenic. Implant standards were used for concentration calibration, depth scale was determined by a CVD-grown delta sample.

RESULTS AND DISCUSSION

Sheet Resistance

The sheet resistance results of arsenic and phosphorus are shown for various oxygen and ammonia concentrations in Figure 1 and 2. The data points for an ambient with 100 % nitrogen are shown at 0.001 % on the x-axis. As can be seen in Figure 1, for arsenic there is a minimum sheet resistance for an oxygen concentration of about 1 %. With higher oxygen concentrations the sheet resistance is increasing again for both implant energies. As expected for the 500 eV implant the sheet resistance is higher than the 1 keV implant for an ambient of 100 % nitrogen and 100 ppm oxygen in nitrogen. But with a further increase in oxygen concentration the sheet resistance values of both energies are similar and the lower energy case even shows a slightly lower sheet resistance. The sheet resistance of the wafers annealed in an ambient containing ammonia decreases continuously with increasing ammonia concentration. Again the sheet resistance of the 500 eV implant is very similar to the 1 keV implant with a tendency to be slightly smaller. For the pure nitrogen and ammonia ambient in all cases surface etching (dose loss) due to residual oxygen concentrations can be neglected.

In Figure 2 the sheet resistance graph of the phosphorus implants is given. Similar to the arsenic implant a minimum sheet resistance is seen for oxygen concentrations between 1 and 10 % oxygen in nitrogen. In contrast to the arsenic implants the sheet resistance of the 500 eV phosphorus implant is always higher than the sheet resistance of the 1 keV implant. For the ammonia ambient the sheet resistance is decreasing with a minimum sheet resistance at 100 % NH$_3$ concentration independent of dopant and implant energy.

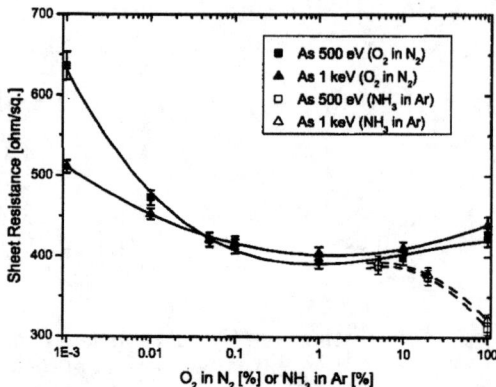

FIGURE 1. Sheet resistance results of arsenic implants in dependence of oxygen and ammonia concentration (lines are to guide the eye)

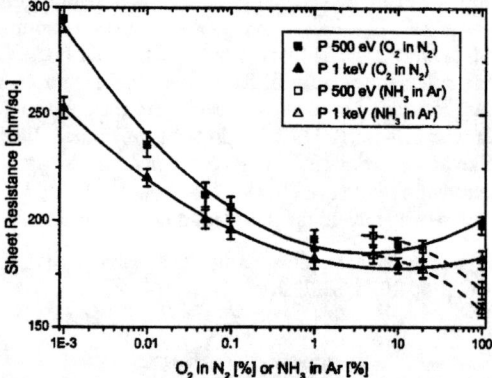

FIGURE 2. Sheet resistance results of phosphorus implants in dependence of oxygen and ammonia concentration (lines are to guide the eye)

SIMS Profiles

Figure 3 shows SIMS profiles of the 500 eV arsenic implants. Up to an oxygen concentration of 10 % O$_2$ in N$_2$ the arsenic profiles are deeper with increasing oxygen concentration but the profile with 100 % O$_2$ is shallower and to a certain extent steeper in the tail region. It is also worthwhile noting that the shoulder as well as the peak concentration increase continuously with increasing oxygen concentration indicating a reduced dose loss with higher oxygen concentrations and therefore higher oxide thickness (not shown here). In general the arsenic profiles are much shallower after annealing than the phosphorus profiles.

In Figure 4 SIMS profiles of 500 eV phosphorus as-implanted and spike annealed wafers are shown. The profiles exhibit the typical characteristics of phosphorus with a kink and fast diffusing tail. With increasing oxygen concentration the profile tail gets deeper. The lowest kink concentration is seen after the spike anneal in 100 % N$_2$. The profile in Figure 4 reveals that part of the phosphorus is accumulated in the oxide layer that is formed during annealing.

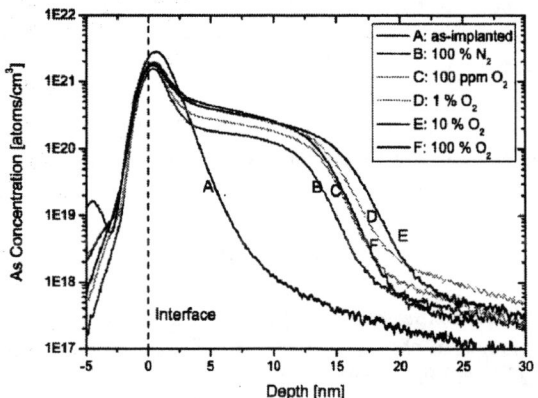

FIGURE 3. SIMS profile of arsenic 500 eV before and after spike annealing with various oxygen concentrations

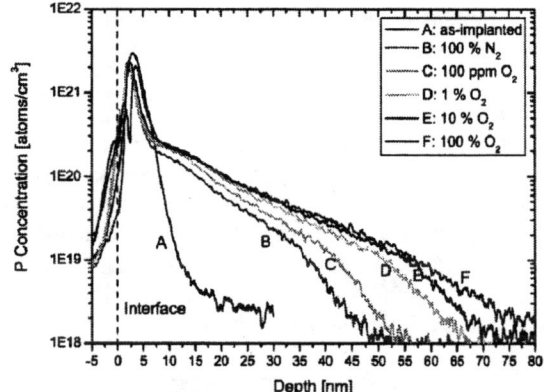

FIGURE 4. SIMS profile of phosphorus 500 eV before and after spike annealing with various oxygen concentrations

Diffusion Length

The diffusion length (Ld), i.e. the difference in junction depth (determined at a dopant concentration of $5 \cdot 10^{18}$ cm^{-3}) between the annealed and the as-implanted dopant profile, is shown in Figure 5 for arsenic and in Figure 6 for phosphorus implants in dependence of oxygen and ammonia concentration. With the arsenic implants and oxygen containing ambient a maximum in diffusion length is seen for an oxygen concentration of approximately 1 %. In case of the 500 eV implant the diffusion length is slightly smaller below 100 ppm oxygen and then slightly higher above this concentration (similar to the sheet resistance behavior). The diffusion length of arsenic in the ammonia ambient nearly doubles from 8.4 nm (500 eV) and 9.4 nm (1 keV) in inert conditions to 19.2 nm (500 eV) and 17.3 nm (1 keV) in 100 % ammonia. Also with the ambient containing ammonia we see the effect that the 1 keV As diffuses less than the 500 eV As. In contrast to arsenic the diffusion length of phosphorus in oxygen ambient nearly doubles from 26 nm to 50 nm for 500 eV and from 28.5 nm to 48 nm for 1 keV implant energy. The diffusion length of phosphorus in the ammonia ambient is slightly smaller than with the oxygen ambient but it also nearly doubles its diffusion length when the ambient changes from inert Ar to pure NH$_3$ ambient. In general, arsenic shows a significantly smaller diffusion length than P in agreement with [2]. In an inert ambient the diffusion length of phosphorus is about 3 times higher than that of arsenic and in 100 % oxygen the diffusion of phosphorus is nearly 5 times higher than that of arsenic. As in the case of the oxidizing ambient, arsenic shows a significantly smaller diffusion length than phosphorus in an ambient containing ammonia. In inert and 100 % NH$_3$ ambient the diffusion length of arsenic is about 3 times lower than that of phosphorus. The diffusion length results from the spike anneals in different gaseous ambients clearly show that arsenic at high concentrations mainly diffuses via vacancies whereas the degree of phosphorus diffusion at high concentrations is nearly equally dependent on interstitials and on vacancies. For a reduced thermal budget, as in spike annealing, it is obvious that phosphorus shows about three times higher diffusivity than arsenic in inert ambient (see [2] ~5 times higher diffusivity for P than for As). A similar effect of the reactive gaseous ambient independent of the thermal budget can be expected for ms-anneals like flash annealing, but on a much shorter length scale. This data in combination with data from earlier soak anneal experiments [3, 4, 5] also suggest this behavior.

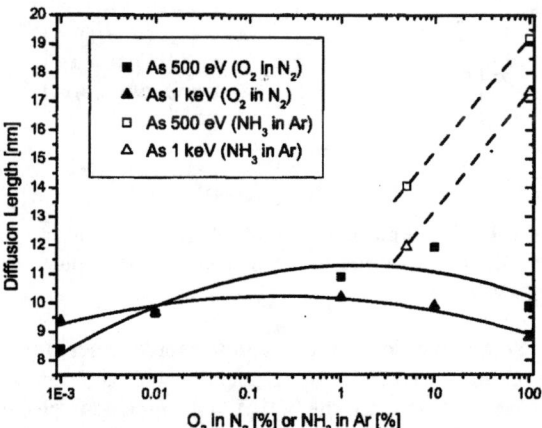

FIGURE 5. Diffusion length of arsenic implants at $5 \cdot 10^{18}$ cm^{-3} in dependence of oxygen and ammonia concentration (lines are to guide the eye)

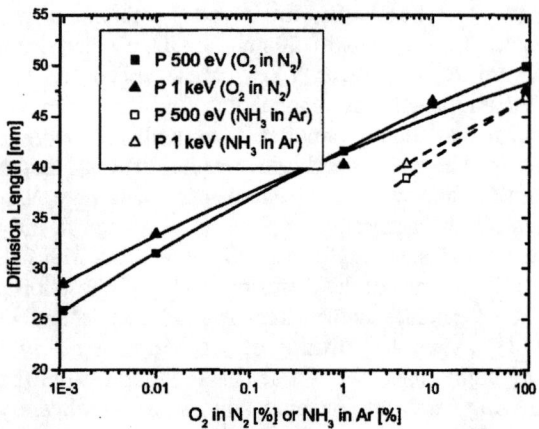

FIGURE 6. Diffusion length of phosphorus implants at $5 \cdot 10^{18}$ cm^{-3} in dependence of oxygen and ammonia concentration (lines are to guide the eye)

Retained Dose

All the wafers were implanted with a nominal dose of $1 \cdot 10^{15}$ cm^{-2}. Figure 7 shows the retained dose, i.e. the dose after annealing in relation to the as-implanted dose, for the phosphorus implants with 500 eV and 1 keV.

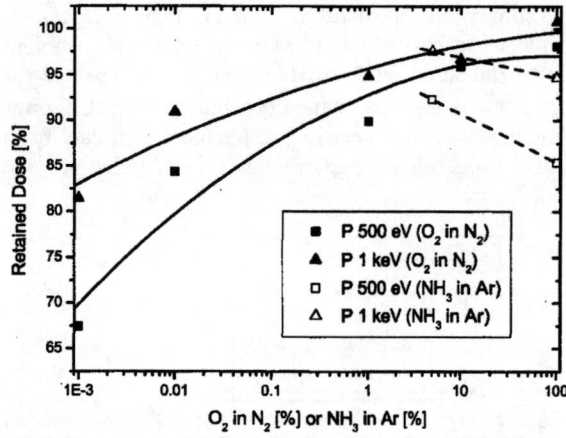

FIGURE 7. Retained dose of phosphorus implants in dependence of oxygen and ammonia concentration (lines are to guide the eye)

Significant dose loss is seen for spike annealing in an ambient of 100 % N$_2$, especially with the 500 eV implants. In general, the 500 eV implant that is closer to the surface suffers from increased outdiffusion due to the proximity of the wafer surface relative to the region of high dopant concentration. With increasing oxygen concentration the retained dose becomes higher and finally reaches nearly the same dose level as for the as-implanted wafers. Thermal oxidation of the silicon surface during the annealing therefore effectively prevents the dopants from outdiffusion. For ambient containing ammonia, the retained dose decreases again for a concentration of 100 % NH$_3$. The retained dose is smaller for 100 % NH$_3$ than for 5 % NH$_3$. Again this effect is more pronounced for the shallower 500 eV implant than for the P 1 keV implant. Due to issues in quantification of the absolute concentration of the as-implanted 500 eV arsenic implant the equivalent data for arsenic are not shown in this paper.

CONCLUSION

These experiments show that the minimum sheet resistance of shallow arsenic and phosphorus implants is achieved using either 100 % NH$_3$ or 1-10 % oxygen in nitrogen. In terms of limited diffusion annealing in inert ambient (100 % N$_2$ or Ar) is beneficial. A spike anneal with 1050 °C peak temperature of As 500 eV in inert ambient gives the shallowest junction (16.3 nm) in combination with still acceptable sheet resistance (636 Ω/sq.). The thermal budget for the phosphorus implants can be further reduced to minimize diffusion with acceptable sheet resistance. Furthermore it can be concluded that with the right choice of the gaseous ambient the dose loss due to outdiffusion of dopants can be effectively suppressed. In case of an undefined annealing ambient a barrier layer needs to be deposited or grown which insures that dose loss by surface etching or outdiffusion is minimized.

ACKNOWLEDGMENTS

The authors would like to thank Zsolt Nényei, Paul J. Timans, Peter Pichler and Jeff Gelpey for inspiring, helpful discussions and critical reading of the manuscript.

REFERENCES

1. International Technology Roadmap for Semiconductors 2005 Edition, http://www.itrs.net/Common/2005ITRS/Home2005.htm
2. P. Pichler, "Intrinsic Point Defects, Impurities, and Their Diffusion in Silicon", edited by S. Selberherr, Wien, Springer, 2004
3. W. Lerch, N. A. Stolwijk, S. D. Marcus, D. F. Downey, M. Schäfer, *Electrochem. Soc. Proc.* **10**, 141-150 (1999)
4. W. Lerch, M. Glück, N. A. Stolwijk, H. Walk, M. Schäfer, S. D. Marcus, D. F. Downey, J. W. Chow, *Journal of The Electrochemical Society,* **146** (7) 2670-2678 (1999)
5. D. F. Downey, J. W. Chow, W. Lerch, J. Niess, S. D. Marcus, *Mat. Res. Soc. Proc.* **525**, 263-271 (1998)

The Effect Of Flash Annealing On The Electrical Properties Of Indium/Carbon Co-Implants in Silicon

S. Gennaro[1], D. Giubertoni[1], M. Bersani[1], J. Foggiato[2], W.S. Yoo[2], R. Gwilliam[3] and M. Anderle[1]

[1]*ITC-irst, Centro per la Ricerca Scientifica e Tecnologica, Via Sommarive 18, 3805 Povo (TRENTO) Italy*
[2]*Wafermasters, Inc., 246 East Gish Road, San Jose, California 95112, USA*
[3] *Surrey Ion Beam Centre, Advanced Technology Institute, School of Electronics and Physical Sciences, University of Surrey, Guildford GU2 7XH, Surrey, UK*

Abstract. Shallow Indium implants and Indium-Carbon co-implants have been subjected to flash anneals and a combination of furnace treatments in order to evaluate the electrical properties of the implant and differentiate the behavior between low temperature and high temperature ultra fast thermal treatments. It is found that by using "flash" anneals, higher levels of electrical activation are achievable for the given experimental conditions. This behavior is related to the indium dose and to the dopant diffusion within the layer and its interaction with the carbon.

Keywords: Ion implantation, Indium, shallow junctions, annealing, electrical activation.
PACS: 81.05.Cy, 85.40.Ry, 61.72.Tt

INTRODUCTION

The ITRS roadmap of semiconductor technology describes the path to follow to produce devices with increased integration and performance [1]. One of the main challenges is the requirement to obtain dopant distributions in the source and drain (S/D) extensions that have to be shallow, abrupt and highly activated to obtain the sheet resistance values lower than 500 Ohm/sq in order to overcome Short Channel effects (SCE) problems. With each new device generation, with shallower S/D extensions, the Source/Drain junctions are expected to be as shallow as 13 nm for the 65 nm technology node. The present technology uses ion implantation as the principal tool for accurately positioning the dopant species within the active layer of the devices and providing good control of dopant concentration profiles. The following step for junction fabrication is a thermal treatment to recover repair the damage induced by the implantation process and to activate the dopant. During this process dopant redistribution and in- and out- diffusion occur. These phenomena lead to undesired changes in the dopant distribution and affect the electrical properties of the produced layers. Therefore, the study of processes which minimize the diffusion have become very important. Among them numerous methods have been tried with the use of new dopant species, such as indium and gallium as alternatives to boron for the production of *pn* junctions [2,3] plus the use of suitable annealing techniques, such as laser and flash anneal [4-8]. Recently, by investigating the use of flash anneal on indium and indium/carbon co-implants in silicon, we reported values of retained dose levels higher than previously reported for low temperature-long time or high temperature-RTA treatments [9]. In this paper we further describe the effects of the flash anneals on the electrical properties of indium and indium/carbon co-implants in silicon.

EXPERIMENTAL

Indium and carbon were implanted in 4-inch CZ silicon (100) wafers, with n-type background doping of P, with resistance of 5-10 Ω-cm. The energies and doses used are given in Table 1. The implantation conditions were selected in order to obtain indium distributions with $R_p \approx 20$ nm and peak concentration of about 2×10^{20} at cm^{-3}. In the co-implant case, the indium distribution was chosen to be completely confined within the carbon profile.

The pre-amorphisation dosage levels (Si$^+$ 70 keV 2e15 cm^{-2}) were chosen to produce an amorphous layer about 100 nm thick.

TABLE 1. Implantation conditions.

#	In	C	c-Si	a-Si
#1	25 keV 3e14 cm^{-2}		v	
#2	25 keV 3e14 cm^{-2}	5 keV 5.5e14 cm^{-2}	v	
#3	25 keV 3e14 cm^{-2}			v
#4	25 keV 3e14 cm^{-2}	5 keV 5.5e14 cm^{-2}		v

The implants were performed using a 200 kV Danfysik 1090 accelerator, at room temperature, using 7° tilt and 22° twist angles. Following implantation, 1cm square specimens were cut from the wafers and were submitted to either flash or furnace anneal. The conditions for annealing are summarized in Table 2.

TABLE 2. Variations of performed anneals

	Flash Anneal				Furnace anneal
	A	B	C	D	
Anneal Preheat Temperature (°C)	350	500	350	500	
Flash Power Level (%)	75	75	100	100	
Annealing Temperature (°C)					650, 750
Annealing time (s)					30, 300, 900

The furnace treatments were performed on a Process Products Corporation RTP system in flowing nitrogen. The flash anneals were performed using a WaferMasters Flash Anneal system. Under these conditions, the temperatures experienced from the samples during the flash are expected to range between 1150 and 1200 °C.

The electrical properties were evaluated by sheet resistance and Hall Effect measurements on a clover leaf pattern defined by photolithography and wet etch. The measurements were performed using an Accent HL5500 system.

The carbon atomic profile were measured using secondary ion mass spectrometry (SIMS) with a Cameca Wf mass spectrometer. A 0.5 keV Cs$^+$ primary ion beam was used and negative secondary ions ^{12}C$^-$ and ^{28}Si$^-$ were collected.

RESULTS AND DISCUSSION

Figures 1 and 2 depict the electrical characteristics, in terms of sheet resistance and carrier concentration, of the processed samples for implants performed in c-Si. As reported in the literature, the indium electrical activation is enhanced, especially when annealed at low temperature, with the presence of carbon [2,11]. Our results replicate those reported in the literature.

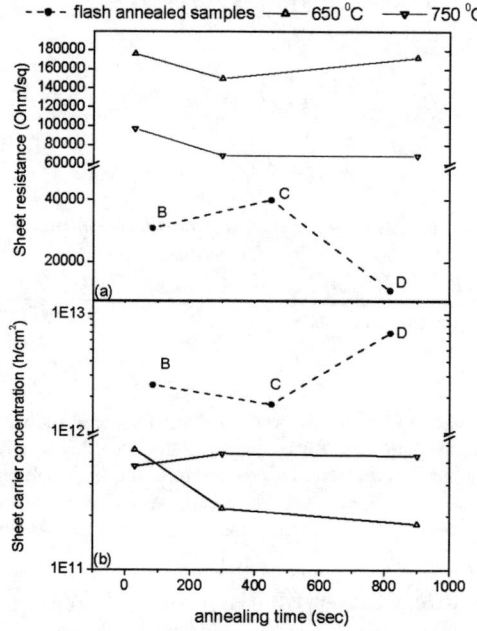

FIGURE 1: Electrical characteristics of the implant #1 in terms of a) sheet resistance b) carrier concentration. The dashed line connecting the data related to flash anneal is used for comparison only.

By comparing the different annealing techniques, it appears that the flash anneals result in lower values of the sheet resistance indicating a higher degree of activation of the doping species in terms of the carrier concentration. This behavior is observed for both the indium single implant (Fig. 1) and the indium carbon co-implant (Fig. 2). Nevertheless the observed sheet resistance is still quite high limiting the electrical analysis of the implants performed in a-Si due to leakage problems

The obtained results, however, present some difference from what was previously reported. In a previous work, in fact, we reported that the carbon presence negatively affect the electrical activation of indium when using high temperature spike RTA [12]. In the present work, where the annealing temperature is expected to hit 1150-1200 °C, instead, this effect does not occur. We assume that this happens because of the extreme rapidity of the treatment which would prevent the formation of carbon based precipitates within the layer.

When relating these data with the values of the indium retained dose previously reported [9], it is possible to obtain a value of activation of the retained dose, which approaches 36% for In/C co-implants.

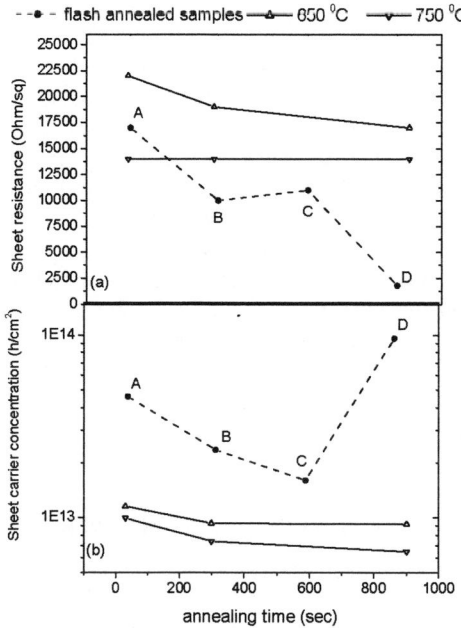

FIGURE 2: Electrical characteristics of the implant #2 in terms of a) sheet resistance b) carrier concentration. The dashed line connecting the data related to flash anneal is to facilitate comparison only.

Previously, for slightly deeper implants annealed in the furnace we reported a value of retained dose activation close to 50% [9]. The difference between these levels can be explained by noting that both the implantation and the flash anneals conditions resulted in creating shallower distributions with higher peak concentrations of the doping species, hence making its activation more challenging.

The effect of the carbon distributions on the electrical properties of the indium implants has been thoroughly explained in several publications [2,10,11,13]. It is mainly due to the formation of In_s-C_s pairs which become the actual dopant species within the active layer. Another effect is to take into account for the implants are performed in c-Si. In fact, we reported that for analogous samples, the indium implantation on c-Si resulted in the formation of a buried amorphous layer which, upon flash annealing, re-grew producing two damaged areas in proximity of the previous amorphous/crystalline (a/c) interfaces. In those areas we observed peaks of indium segregation. This feature was far less evident with the presence of carbon and was not evident for implants performed on a-Si [9]. In what follows we aim to present the behavior of carbon when submitted to flash annealing and its effect on the indium distributions and electrical activation levels. Figs. 3-4 depict the carbon distributions achieved for the given implantation and flash annealing conditions used.

In figure 3, the profiles obtained for wafer #2 show that, after flash annealing, carbon tends to segregate in the damaged area at about 40 nm, competing, hence, with the indium atoms for the segregation points. This can explain the lower degree of indium segregation previously observed. This effect, in turn, may help the dopant species be available for electrical conduction.

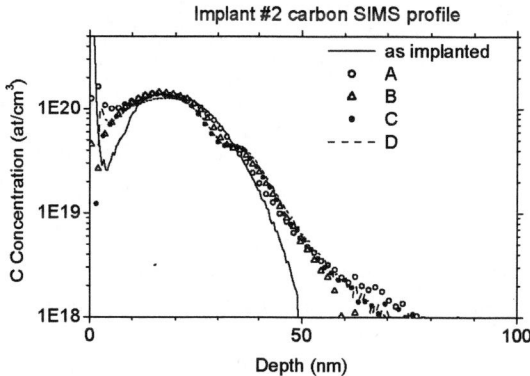

Figure 3: SIMS atomic profiles for the implant #2 (in c-Si) of the carbon distributions as implanted and flash annealed according to Table 2 conditions.

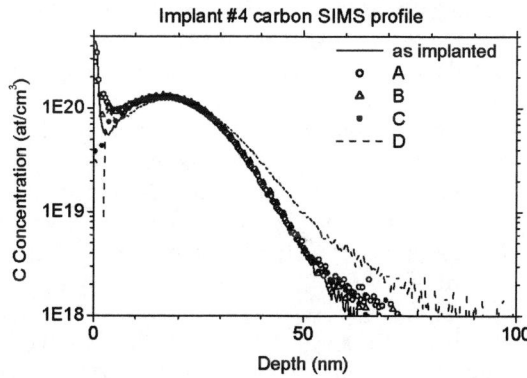

Figure 4: SIMS atomic profiles for the implant #4 (in a-Si) of the carbon distributions as implanted and flash annealed according to Table 2 conditions.

This effect obviously is not present when considering implants performed in a-Si (fig 4) where the as-implanted a/c interface lies far beyond the carbon and indium distributions. In such case any observable effect on the electrical activation of the implant would solely be due to the interaction between carbon and indium atoms.

The experimental set up limited the full electrical characterization of the implanted samples in a-Si. Preliminary results, on only furnace annealed (750 °C) samples, are reported in Table 3.

TABLE 3. Preliminary electrical results for implanted samples in a-Si (#3-4) reported in comparison with data for implanted samples in c-Si (#1-2). Annealing temperature is 750 °C.

#	Annealing time (s)	ρ_s (Ω/sq)	N_s (h/cm^2)	μ_H (cm^2/V-s)
#1	300	69000	4.4×10^{11}	202
	900	69000	4.3×10^{11}	209
#2	300	14000	7.4×10^{12}	60
	900	14000	6.5×10^{12}	67
#3	300	31000	3.5×10^{12}	57
	900	69000	1.6×10^{12}	54
#4	300	11000	1.9×10^{13}	30
	900	6000	2.2×10^{13}	43

It is shown that indeed the pre-amorphisation treatment increases the indium activation when single implants are considered (#1 and #3). It is particularly evident that in this case the decrease of the resistance is driven mainly by the enhancement of the number of carriers available for conduction, which goes up about one order of magnitude.

However, when considering co-implants, the same degree of enhancement is observed when comparing the c-Si and a-Si. When a comparison of the data obtained for indium single implant in a-Si (#3, where the enhancement of activation can only be due to crystal-related effects) and In/C co-implant in c-Si (#2, where any enhancement of activation can only be due to the presence of carbon) the number of carriers available for conduction in the case of the co-implant is higher. This indicates that the carbon couples with indium in substitutional pair defects tending to overcome its effect in inhibiting indium segregation, thus the produced carrier levels are even higher despite the presence of EOR defects which are not present in the a-Si implant.

Hence, we can conclude that any electrical effect of the diminished indium segregation in the EOR damage due to the carbon presence may be not readily evident and actually may be overshadowed by the other influences of carbon on the indium distribution and activation. Although this effect does not clearly show any influence on the electrical activation of the indium implant, it cannot be definitely ruled out.

The mobility values are clearly affected by carrier scattering. By comparing the case on Indium single implant in a-Si and In/C co-implant in c-Si, despite the above mentioned higher carrier density in the co-implant, mobility values are nearly equal. This indicates that, despite the presence of EOR damage in the active part of the profile, the crystalline quality of the environment surrounding the indium atoms is improved, at the specified annealing conditions, by the presence of carbon. This leads towards much lower values of resistivity.

CONCLUSIONS

The effect of a combination of annealing techniques and conditions and of the carbon content on the electrical properties of indium implants on c-Si and a-Si has been presented. It is reported that using flash anneals results in achieving a higher degree of electrical activation within the experimental conditions investigated. The carbon atoms compete with the indium for the segregation points in the areas with defects produced during annealing. The effect of these phenomena on the electrical characteristic of the indium implants is still to be resolved. Although it is presumably covered by the other well known carbon-related mechanisms leading towards enhancement of the indium electrical activation, it may be disregarded, but not totally ruled out.

ACKNOWLEDGMENTS

The PAT (Provincia Autonoma di Trento) is acknowledged for its financial support to the project "65-pMOS".

REFERENCES

1. ITRS, The International Technology Roadmap for Semiconductors, FET, edition 2005. *http://public.itrs.net/*
2. H.Boudinov, J.P. de Souza and C.K. Saul, *J. Appl. Phys.* 86, 5909 (1999).
3. S. Biesemans, S. Kubicek and K. De Meyer, *Jpn.J.Appl.Phys* 35, 1037 (1996)
4. R. Murto, K. Jones, M.Rendon, S. Talwar, *Proc. Ion Implantation Technology-2000*, (2001) 155.
5. W. S. Yoo and K. Kang, Proc of *Ion Implantation Technology 2004, Nucl. Instr. And Meth. In Phys. Res. B* 237 (2005) 18
6. R. A. Camillo-Castillo, M. E. Law, K. S. Jones, L. Radic, R. Lindsay, and S. McCoy *Appl. Phys. Lett.* 88, 232104 (2006)
7. S. H. Jain, P. B. Griffin, J. D. Plummer, S. Mccoy, J. Gelpey, T. Selinger, and D. F. Downey *J. Appl. Phys.* 96, 7357 (2004).
8. Yi-Chao Wang, Ci-Ling Pan, Jia-Min Shieh, and Bau-Tong Dai *Appl. Phys. Lett.* 88, 131104 (2006)
9. S. Gennaro et al in *J.Vac. Sci. Technol. B* 24(1) 473-477 (2006).
10. R. Baron, J.P. Baukus, S.D. Allen, T.C. McGill, H. Kimura, H.V. Winston and O.J. Marsh, *Appl. Phys. Lett.*, 34 (1979) 257
11. S. Gennaro, C. Jeynes, B. Sealy, E. Collart, A. Licciardello, R. Gwilliam, *Proc. Ion Implantation Technology 2002*, 552.
12. S. Gennaro, B.J. Sealy and R.M. Gwilliam *IEE Electronics Letters*, 41, (2005) 1302
13. S. Gennaro, E. Collart, Y. Wang, B.J. Sealy, R.M. Gwilliam *Nucl. Instr. And Meth. In Phys. Res. B* 209 (2003) 136

Local Arsenic Structure in Shallow Implants in Si following SPER: an EXAFS and MEIS study

G. Pepponi, D. Giubertoni, S. Gennaro, M. Bersani, M. Anderle, R. [1]Grisenti, [2]M. Werner and [2]J. A. Van Den Berg

ITC-irst, via Sommarive 18, 38050 Povo (Trento), Italy
[1]*Università degli studi di Trento, Dipartimento di Fisica, via Sommarive 14, 38050 Povo (Trento), Italy*
[2]*Institute of Materials Research, University of Salford, Salford M5 4WT, UK*

Abstract. Solid phase epitaxial regrowth (SPER) has been investigated in the last few years as a possible method to form ultra shallow dopant distributions in silicon with a high level of electrical. Despite the interest for this process, few investigations were related to arsenic. Apart from the fact that it is easier to form shallow distribution with arsenic than with boron, it is also well known that at the moderate temperatures implied by SPER (500-700°C) arsenic easily deactivates, probably by forming inactive clusters around point defects in silicon. In order to have a better understanding of the SPER process for arsenic implanted silicon in shallow regime, an EXAFS (extended x-ray absorption fine structure) and MEIS (medium energy ion scattering) study is reported in this paper. Silicon samples were implanted at 3 keV with arsenic ions (dose was 2E15 at/cm^2 producing a 11 nm amorphous layer) and then annealed in nitrogen at temperatures ranging from 500 to 700°C to have different levels of recrystallisation. From the comparison of the recrystallised fraction as measured by MEIS with the electrical activation measured by Hall effect it results evident that a full regrowth of the lattice is not reflected by a high electrical activation. The activated arsenic corresponds to less than one third of the apparently substitutional dopant for all the samples analyzed. This lack of activation was further investigated by EXAFS: the samples that according to MEIS are fully recrystallised do not reveal a clear local order around As atoms suggesting that either the As atoms are not yet completely relocated within the lattice sites or a deactivation occurred resulting in a more disordered local structure.

Keywords: arsenic, activation, SPER, MEIS, SIMS, EXAFS.
PACS: 81.15.Np, 81.20.-n, 87.64.Fb, 85.40.Ry, 68.49.Sf, 82.80.Yc

INTRODUCTION

Solid phase epitaxial regrowth (SPER) is a process to form ultra shallow distribution of active dopant in solids by exploiting the crystalline regrowth of an amorphous layer from a crystal seed [1]. In fact the dopant implant itself can amorphise the surface (e.g. As at high fluence [2, 3]) or a pre-amorphisation implant is performed to avoid high diffusion by channelling phenomena (e.g. a Ge high fluence implant followed by a B ultra low energy implant [4, 5]). The subsequent medium temperature annealing (600 ÷ 800 °C) recrystallise the layer and relocate the dopant in lattice sites allowing a high level of electrical activation together with very sharp distributions. The main drawbacks are represented by the high level of defect left at the end of range region and by the usually metastable nature of the reached activation [1].

SPER formed As USJ in Si have been less investigated than B in the recent past. One difficulty of applying SPER to As USJ is given by the fact that arsenic is known to easily deactivate at the typical temperatures implied by this process forming presumably clusters around a vacancy (As_nV with n ≤ 4) [6-8]. Whether this clustering starts after the complete re-crystallization of the amorphous layer [9, 10] or is simultaneous to it [2, 11], has been a point of controversy. In this work we combined secondary ion mass spectrometry (SIMS), medium energy ion scattering (MEIS), Hall effect measurements and extended x-ray absorption fine structure (EXAFS) information in order to gain new insight about the SPER of As ultra shallow implants in silicon. In particular, EXAFS was already used in the past for investigating the local order around the As atomic site [12-14] and allowed the first hypothesis of As_nV clusters showing a decrease of the dopant coordination when it is deactivated [13]. In this preliminary work the same technique was applied to the SPER ultra shallow distributions in order to reveal the possible reasons of a poor electrical activation.

EXPERIMENTAL

A Si (100) wafer was implanted normally by As$^+$ ions at 3 keV energy at a fluence of 2×10^{15} cm^{-2}. Some samples were cut from this wafer and were thermal annealed at temperatures ranging from 600 to 700 °C for 10 up to 1200 seconds in a N_2 / 5% O_2 atmosphere. The detailed description of the samples is reported in Table 1.

Sheet resistance (Rs) values, electron mobility and active dose were determined by Hall effect measurements. The measurements were performed on clover leaf structures defined by photolithography and etch using an Accent HL5500 system. SIMS analyses were performed with a Cameca Wf/Sc-ultra instrument in order to obtain the dopant depth distribution using a 0.5 keV Cs$^+$ primary ion beam (45° angle of incidence with respect to the surface normal) and collecting negative secondary ions ^{28}Si$_2^-$ and ^{28}Si^{75}As$^-$ in high mass resolution [15].

MEIS spectra were obtained using a 100 keV He$^+$ ion beam and the double alignment configuration, in which the [-1,-1,1] channelling direction was combined with the [1,1,1] blocking direction: these conditions make possible the separation of the masses of As and Si and provide an excellent depth resolution (better than a nanometre [16]). The quantification procedure has been described in detail elsewhere [17].

EXAFS analyses have been carried out at the GILDA (BM08) beamline of the European Synchrotron Radiation Facility. The bending magnet beamline is equipped with a double crystal sagittaly focussing Si(111) monochromator which was used to select the energy in the range between 11600 and 13000 eV. High harmonic rejection was obtained by total reflection on a Pd-coated mirror with cut-off energy set at 21 keV. The wafer pieces were positioned horizontally with the implanted surface upwards and were illuminated in grazing incidence geometry. Sample alignment as well as choice and setting of angle was carried out by evaluation of the reflected beam and fluorescence intensity curves with respect to the incidence angle. The energy scan was performed with variable steps, the smaller being 1 eV close to edge, the larger being 10 eV in the pre-edge region and 5 eV in the far region. The fluorescence radiation was acquired by means of a side-looking 13-element Ge detector. The acquired data was elaborated using the software ATHENA in the package IFEFFIT. The Fourier Transform of the EXAFS function reported in the Figures was carried out on the 3.5-9 Å^{-1} range, using a Kaiser-Bessel window, k-weight 2, and was not phase corrected.

RESULTS

The electrical data reported in Table I show that increasing the time of the 600 °C annealing Rs value decreases towards a saturation level of around 750 Ohm/square whereas the activated fraction reaches the 2.1×10^{14} cm^{-2} value, well below the 2×10^{15} cm^{-2} implanted dose. The active dose value reaches its maximum for sample F annealed at 650 °C for 10 seconds (2.4×10^{14} cm^{-2}) whereas the sample annealed for the same time interval at 700 °C (G) shows a lower value (2.2×10^{14} cm^{-2}) resulting in a higher Rs.

The SIMS results for all the samples are reported in Figure 1. The curves of the 600 °C isothermal series reveal that the increase of the annealing time affects neither the total retained dose nor the tail of the As distribution but only the first ~12 nm of the profile, where an As peak is progressively "snow-ploughed" to the SiO_2/Si interface.

For the 10 s annealed samples SIMS revealed As distributions quite similar between samples F and G annealed at 650 and 700 °C, respectively, showing an As peak below the surface oxide as already observed on sample E (600 °C 120 s). Instead the sample annealed at 600 °C has a peak at ~5 nm for SIMS, a little deeper for MEIS [15], and the crystalline re-growth is not yet complete but a 7 nm amorphous layer is still present. Also in this case the results can be interpreted arguing that the SPER relocate the dopant in Si lattice sites and at the same time a fraction of it is snow-ploughed by the amorphous/crystalline interface

TABLE 1. Sample description, sheet resistance, electron mobility and active dose as determined from Hall Effect Measurements. All the samples were implanted with As at 3 keV and 2x1015 cm-2 dose

Sample	Annealing T (°C)	Time (sec)	R$_s$ (Ω/sq.)	μ (cm^2/Vs)	Ns (e/cm^2)
A	As implanted		-	-	-
B	600	10	948	41.8	1.58E14
C	600	20	786	47.2	1.68E14
D	600	30	746	45.8	1.83E14
E	600	120	754	39.8	2.08E14
F	650	10	610	42.3	2.42E14
G	700	10	684	41.9	2.18E14

towards the surface ending in a segregated peak below the oxide.

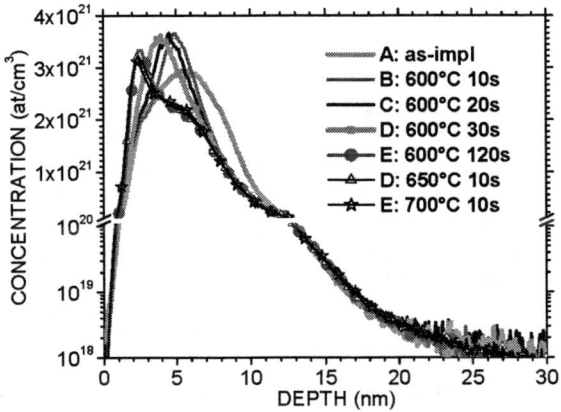

FIGURE 1. SIMS profiles of the samples indicated in Table I. The scale of the axis of ordinates is logarithmic for values up to 10^{20} and linear for higher values.

The MEIS results reported in Figure 2 show that when annealing at 600 °C, increasing the duration of the treatment the amorphous layer thickness (energy 75 ÷ 84 keV) is reduced. Starting from 11 nm on the as implanted sample the final thickness becomes 4 nm including the 2.7 nm of surface SiO_2. The latter is nearly constant with respect to all the annealing processes according to MEIS, indicating that the considered temperatures do not result in the oxide growth. Arsenic is progressively located within the lattice sites and thus it becomes invisible to the ion beam, as shown in the part of the spectra comprised between channel 85 and 95 keV.

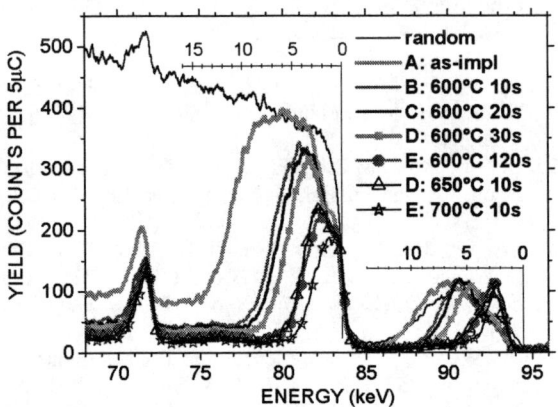

FIGURE 2. MEIS measurements. The axis running from right to left, placed on the graph, indicate the depth calibration for As(85-95keV) and Si(75-84keV).

MEIS observed the visible As to decreases with the annealing time and an As peak is progressively snowploughed up to the interface with the surface SiO_2 as already revealed by SIMS. The difference between SIMS and MEIS can give an indication of the active dose on each sample. In Figure 3 the Hall effect measured active dose is plotted against the amount of substitutional arsenic as measured by MEIS. The comparison with the theoretical active dose value (e.g. the whole substitutional fraction) reveals that a substantial fraction of the dopant is not activated even if it lies on lattice sites and therefore not detectable by MEIS [18].

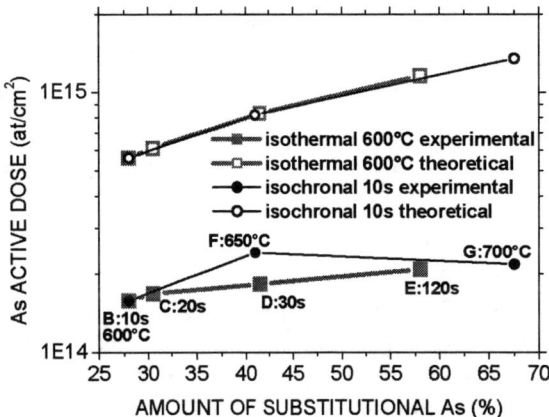

FIGURE 3. Electrical activation: experimentally determined and extrapolated form the MEIS data.

The plot of the active dose against the substitutional one reveals again a relevant electrically inactive fraction of the relocated dopant. Moreover the active dose of sample F is notably higher than the G one even if the latter was re-crystallised at a higher temperature, i.e. 700 as against 650 °C.

In order to understand the reason of the electrical activation behaviour as a function of the substitutional dopant fraction the samples were analyzed by EXAFS to study the development of the local order around the As atomic site. In Figure 4 the Fourier Transform of the EXAFS function is shown. The measurements were performed at an angle of about 0.7 mrad above the critical angle for total reflection, gathering information from the whole implant.

The peaks at $R \cong 1.9$Å, 3.4Å and 4.1Å (the values do not correspond to actual bond lengths since no phase correction was applied) are formed respectively by the contributions of the first second and third shells of neighbouring atoms around arsenic. If we assume a 4-coordinated substitutional arsenic for the activated reference sample and arsenic in an amorphous Si matrix for the as implanted sample, we can qualitatively judge the local order around the arsenic for the other samples.

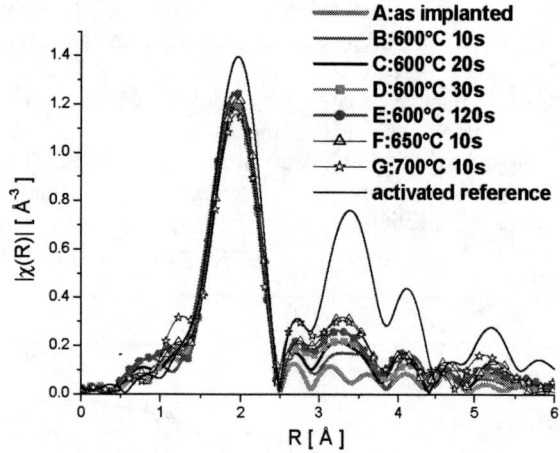

FIGURE 4. Fourier Transform of the experimental EXAFS function.

The 600 °C isothermally annealed samples show higher peaks for increasing annealing time, a trend well explained by the MEIS measurements shown in Figure 2. Still the curves are far from the reference one indicating a poor order as confirmed by the low electrical activation measured. The 650 °C seems to have slightly higher first and second shell peaks than the 700 °C curve. This is in accordance with the electrical data, but not in agreement with the MEIS data which give for the 700 °C sample a better reconstructed crystal lattice. This result could be explained by As beginning to deactivate after the partial readjustment of the crystal.

CONCLUSIONS

MEIS and EXAFS are valuable tools for the understanding of the behavior of arsenic implanted in Si and subject to SPER annealing to achieve maximum electrical activation. The apparent disagreement of the two techniques suggests that the regrowth of the crystal is not a sufficient condition to achieve a satisfying electrical activation and that deactivation of arsenic might play an important role for temperatures above 650°C competing with its relocation in substitutional position.

ACKNOWLEDGMENTS

The authors thank S. Milita for the reference sample, L. Capello for the fruitful discussion, F. d'Acapito and the BM08-GILDA team of the ESRF for the assistance in the EXAFS measurements and R. Gwilliam and the University of Surrey for the opportunity to run the electrical measurements. The authors acknowledge the CLRC Daresbury Laboratory for provision of the MEIS facility.

REFERENCES

1. R.Lindsay, B. Pawlak, J. Kittl, K. Henson, C. Torregiani, S. Giangrandi, R. Surdeanu, W. Vandervorst, A. Mayur, J. Ross, S. McCoy, J. Gelpey, K. Elliott, X. Pages, A. Satta, A. Lauwers, P. Stolk, and K. Maex, *Mat. Res. Soc. Symp. Proc.* **765**, 2003, D7.4.1.
2. B.J. Pawlak, R. Duffy, T. Janssens, W. Vandervorst, K. Maex, A.J. Smith, N.E.B. Cowern, T. Dao, and Y. Tamminga, *Appl. Phys. Lett.* **87**, 2005, 31915
3. C. Tsamis, D. Skarlatos, C. Ben Assayag, A. Claverie, W. Lerch, and V. Valamontes, *Appl. Phys. Lett.* **87**, 2005, 201903.
4. B.J. Pawlak, R. Surdeanu, B. Colombeau, A.J. Smith, N.E.B. Cowern, R. Lindsay, W. Vandervorst, B. Brijs, O. Richard, and F. Cristiano, *Appl. Phys. Lett.* **84**, 2004, pp. 2055.
5. J.J. Hamilton, E.J.H. Collart, B. Colombeau, C. Jeynes, M. Bersani, D. Giubertoni, J.A. Sharp, N.E.B. Cowern, and K.J. Kirkby, *Nucl. Instr. And Meth. In Res. B* **237**, 2005, 107.
6. A. Lietoila, R.B. Gold, J.F. Gibbons, T.W. Sigmon, P.D. Scovell, and J.M. Young, *J. Appl. Phys.* **52**, 1981, 230.
7. P.M. Rousseau, P.B. Griffin, W.T. Fang, and J.D. Plummer, *J. Appl. Phys.* **84**, 1998, 3593.
8. M.A. Berding and A. Sher, *Phys. Rev. B* **58**, 1998, 3853.
9. A. Kamgar, F.A. Baiocchi, and T.T. Sheng, *Appl. Phys. Lett.* **48**, 1986, 1090.
10. M. Orlowski, R. Subrahmanyan, and G. Huffman, *J. Appl. Phys.* **71**, 1992, 164.
11. N.D. Young, J.B. Clegg, and E.A. Maydell-Ondrusz, *J. Appl. Phys.* **61**, 1987, 2189.
12. A. Erbil, W. Weber, G.S. Cargill III, and R.F. Boheme, *Phys. Rev. B* **34**, 1986, 1392.
13. K.C. Pandey, A. Erbil, G.S. Cargill III, R.F. Boheme, and D. Vanderbilt, *Phys. Rev. Lett.* **61**, 1988, 1282.
14. J.L. Allain, J.R. Regnard, A. Bourret, A. Parisini, A. Armigliato, G. Tourbillon, S. Pizzini, *Phys. Rev. B* **46**, 1992, 9434.
15. D. Giubertoni, M. Bersani, M. Barozzi, S. Pederzoli, E. Iacob, J.A. van den Berg and M. Werner, Appl. Surf. Sci. in press.
16. J.A. van den Berg, D.G. Armour, S. Zhang, S. Whelan, L. Wang, A.G. Cullis, E.H.J. Collart, R.D. Goldberg, P. Bailey, T.C.Q. Noakes,, *J. Vac. Sci. Technol. B* **20**, 2002, 974.
17. M. Werner, J.A. van den Berg, D.G. Armour, W. Vandervorst, E.H.J. Collart, R.D. Goldberg, P. Bailey, T.C.Q. Noakes , *Nucl. Instrum. and Meth. In Phys. Res. B* **216**, 2004, pp.67-74
18. L. Capello, T. M. Metzger, M. Werner, J. A. van den Berg, M. Servidori, M. Herden, T. Feudel, *Mat. Sci. Eng. B* **124-125**, 2005, 200

Well Design In A Bulk CMOS Technology With Low Mask Count

M. P. M. Jank[1], C. Kandziora[1], L. Frey[2], and H. Ryssel[1]

[1] Universitaet Erlangen-Nuernberg, Lehrstuhl fuer Elektronische Bauelemente, Cauerstr. 6, 91058 Erlangen, Germany
2) Qimonda Dresden GmbH & Co. OHG, Koenigsbruecker Str. 180, 01099 Dresden, Germany

Abstract. Modifications in the implantation sequence offer wide potential for simplification in integrated circuit manufacturing. We present a novel concept for a CMOS process with only three front-end mask layers for fabrication of bulk CMOS devices. Process and device simulations demonstrate the manufacturability and scalability of the presented approach.

Keywords: CMOS, Process Simplification, High-Energy Ion Implantation.
PACS: 61.72.Tt, 81.05.Cy

INTRODUCTION

The past decades have seen a steady increase in the effort necessary for integrated circuit manufacturing. Ongoing scaling leading to elaborate transistor architectures, an increasing number of metallization levels, advanced power management solutions, and integration of different technology options into Systems on Chip will result in a further increase of complexity [1], expressed in the number of mask levels or process steps. To counteract this trend at least partially, significant interest is spent on options for process simplification [2, 3, 4].

In this work, we investigate the combination of two concepts for process simplification published earlier by Yabu [3] and Kerber [4] and their co-workers. Both approaches are based on a modification of the doping sequence.

Yabu [3] describes an option for distinction between high energy ion implanted n- and p-well using only one mask level. The n-well doping is implanted into the opened mask areas while the p-well doping is implanted through the resist mask. This implantation scheme simultaneously leads to the formation of a highly doped buried p-layer below the n-well (fig. 1).

Kerber [4] demonstrated the combination of high energy well implantation with the corresponding S/D-implant using only one masking level for each transistor conduction type. This approach is obvious as the n$^+$ S/D regions are directly related to the p-well and the p$^+$ S/D regions are found in the n-well only. However, the well contacts should be defined simultaneously to the S/D doping performed in the complementary well region. This problem can be circumvented considering the strong differences in the lateral straggle between high energy well implants and low energy S/D or contact implants. Narrow resist pillars or bars in the opened, i.e. implanted, areas and narrow holes or trenches in the resist above covered, i.e. non-implanted, areas are nearly invisible for high energy implants while being an effective mask or open area regarding low energy high dose contact implants. For example, the resist mask layout and implantation steps used for simultaneous n-well, p$^+$ S/D, and p-well contact doping is shown(fig. 2). The phosphorus or arsenic n-well implant is carried out through the resist openings and is completely masked in the resist covered areas. Due to the strong lateral straggle of the deep implantation, the 2-dimensional ion distribution extends laterally below the narrow resist feature at the n-well contact. Therefore, the n-well is not discontinued in the n-well contact region. Regarding the p-well contact, the total number of ions implanted through the narrow open resist area during n-well doping is very low. In conjunction with the high lateral straggle, the concentration of implanted dopants stays below the substrate doping or the p-well doping to be implanted at a later stage of the process. The high dose boron-S/D implants are carried out in the openend mask regions. Due to the low lateral straggle of this low energy implantation step, a highly p-doped region forms in the PMOS S/D and p-well contact areas while the n-well contact area is not influenced. The doping scheme is finalized by employing the complementary mask and implantations for formation of the p-well and the n$^+$ regions.

Employing the approach of Kerber, a complete CMOS plocess flow can be conducted using only 4 front end mask levels (isolation, poly patterning, NMOS, and PMOS). Solutions for the integration of S/D extensions have been discussed elsewhere [2, 4] and will not be addressed here.

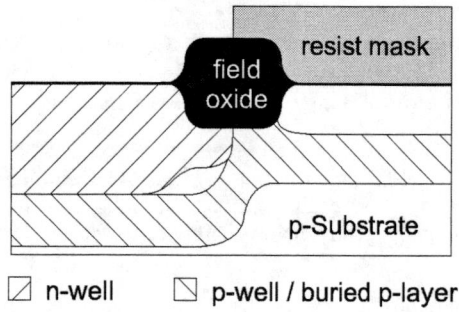

FIGURE 1. Self-aligned process used for high energy implantation of n-well and p-well using a single resist mask. p-well dopants are implanted through the resist. After Yabu [3].

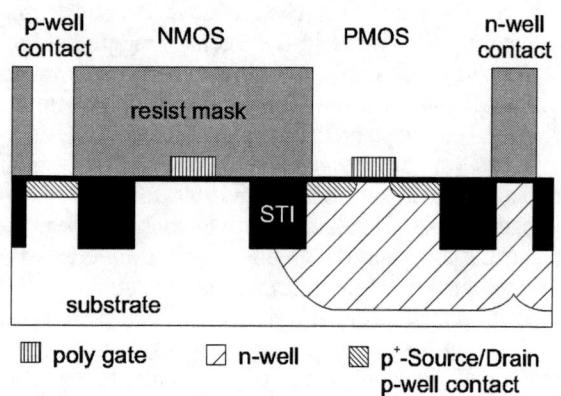

FIGURE 2. Combination of S/D and well implants using only one mask level per transistor conduction type. The implantation scheme utilizes the different lateral straggle between high energy and low energy implantation for additional formation of well contacts.

INTEGRATION CONCEPT

We propose a process flow that combines the two concepts described in the last section (fig. 3). In this process flow, one mask is used for n-well and p-well definition. The n-well is implanted into the mask openings whereas the p-well dopants are implanted through the resist mask. Comparable to the concept shown in fig. 2, the p$^+$ dopants related to the openend n-well S/D and p-well contacts can be introduced with the same mask.

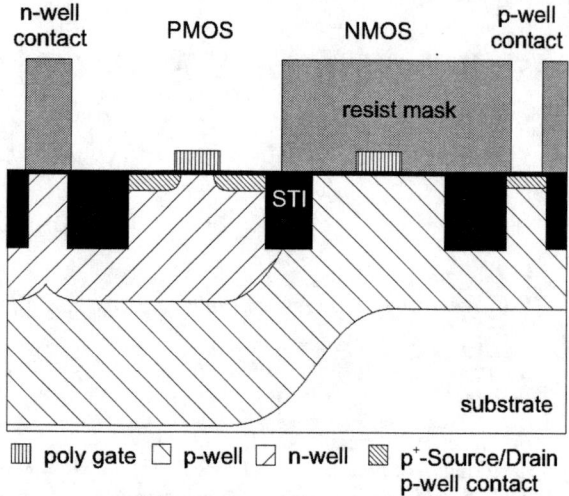

FIGURE 3. Combination of sequence for doping of complementary wells by high energy ion implantation using a single mask and simultaneous well and S/D doping.

Compared to the approach published by Kerber, an additional feature (p-well) can thus be defined without using additional mask levels. Assuming that n$^+$ doping can be carried out without the expense of an additional mask, this doping scheme promises an additional reduction of one mask layer leading to a total of 3 front end mask layers for a complete bulk CMOS technology. The S/D processing will not be addressed here further. In this work we will discuss the well formation with special focus on the well contact regions.

Regarding the n well contact, the influence of the additional p-well implantation has to be taken into account in comparison to previous approaches. The deep implant shows an even higher lateral straggle than the n-well implant does. Accordingly, the deep buried p-layer will also not be discontinued below the resist mask above the n-well contact. A minor portion of the implanted ions is decelerated in the narrow resist pillar or bar above the n-well contact and is implanted in the contact region. If the resist feature is narrow enough, the number of penetrating ions is very low and in conjunction with the strong lateral straggle, the effect of these ions on the contact region can be neglected.

The doping of the p-well contact area is performed by the high energy p-well implant and the subsequent low energy p$^+$ implant. The p-well in the contact region will not be disrupted if the resist opening, i.e. hole or trench, is small enough. In this case, the straggled dopants from the resist covered areas next to the opening form the well doping in the contact area. Ions reaching through the narrow opening are implanted deeply into the substrate. Again, if the opening is small enough the total number of ions is very low and in conjunction with the strong lateral straggle, the ions will be widespread in the substrate and do not effect the electrical behavior of the well and the p-well contact.

In the previous two paragraphs, we noted that the dimensions of the resist features are a critical issue for the success of the presented concept. In the following we present a detailed investigation of the contact design rules for a reference technology. Furthermore, we discuss the application to other technology generations.

PROCESS AND DEVICE SIMULATION

For reference, we employed a triple well CMOS process deduced from a 0.7µm diffused well technology. The optimum resist thickness for the triple well process was 2.5µm. For n-well formation, $4 \cdot 10^{13} cm^{-2}$ phosphorus ions were implanted at an energy of 1.2MeV and for p-well implantation, we used $4 \cdot 10^{13} cm^{-2}$ boron ions with an energy of 2.0MeV. Both implants were performed with a 7° tilt.

Simulation parameters were adjusted to make simulation match to SIMS data and 2-dimensional profile cross-sections prepared with an etching technique based on concentration dependent etch rates [5]. Fig. 4 shows good agreement between the iso-concentration lines detected by SEM in the p-well region of the prepared sample and the process simulation performed with DIOS-ISE [6]. The n-well region can not be analyzed by this technique simultaneously.

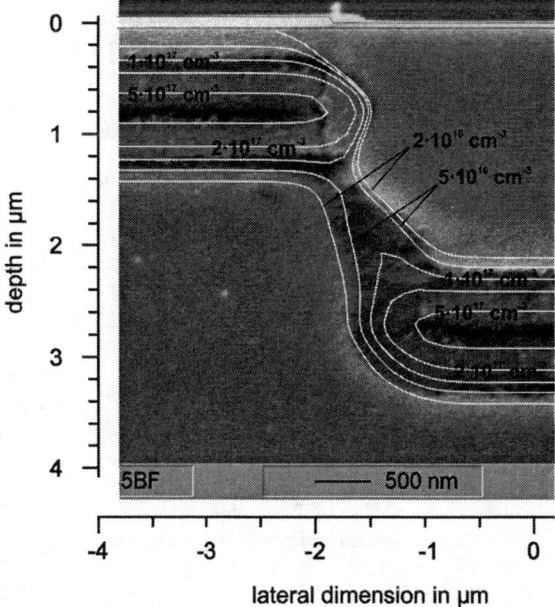

FIGURE 4. Overlay of cross-cut sample prepared by etching with concentration dependent etch rate and simulation

Based on the calibrated simulation environment we conducted a detailed 2-dimensional simulation study on the influence of resist feature sizes and implantation rotation angle with respect to simulated trenches and bars.

Regarding the p-well contact, a sufficient connection between the p^+ contact and the p-well can only be established when the projection of the incident beam onto the wafer surface is in parallel with the trench direction, referred to as a rotation of 0°. At feature sizes below 0.7µm to 0.8µm the p-well extensions formed by scattered atoms from outside the resist opening start to overlap (fig. 5) and give rise to an increase in p-well to substrate breakdown voltage as can be seen in fig. 7a).

Complementary, the n-well contact below a narrow resist bar can only be formed when the projection of the ion beam onto the wafer surface yields a line perpendicular to the resist bar. In this case, the n-well extensions due to scattered atoms start to overlap for resist widths below 0.8µm. A further relaxation of the requirements can be achieved by implanting half of the dose at -90° rotation and the remaining dose at +90° rotation. The n-well extensions will overlap below resist widths of about 1.0µm (fig. 6). A device simulation of the latter configuration shows a strong increase of the n-well to p-well breakdown voltage for bar widths below roughly 1.0µm (fig. 7b).

DISCUSSION

The simulations indicate that the proposed process flow can be realized in the respective technology generation as the maximum allowable feature size is met by the critical dimension.

Regarding the p-well contact, the breakdown voltage of about 60V is sufficient for the projected 5V technology and below, although it is significantly less than the value of about 100V reached by a planar p-well-to-n-substrate junction. Additionally, every p-well corner in the design shows a reduced breakdown voltage, so the p-well contact area is not the critical feature for the determination of breakdown voltage. In the case of the n-well contacts, a sufficient breakdown voltage of about 25V is reached for bar widths below 1.0µm. As this value is defined by the planar n-well-to-p-well-junction the proposed implantation scheme does not add critical features to the existing technology.

Scaling to lower critical dimensions leads to a reduction of implantation energies. With reduction of the implant energy, the lateral straggle is less reduced than the projected range of the ions. Thus scaling generally leads to relaxed requirements for the maximum allowable feature size of resist trenches and bars. This will also counterbalance partially the transition to 0° tilt angle used for high energy implants in technology generations below 100nm. Furthermore, S/D extensions can be integrated into the presented concept without or with only little additional mask count [2, 4].

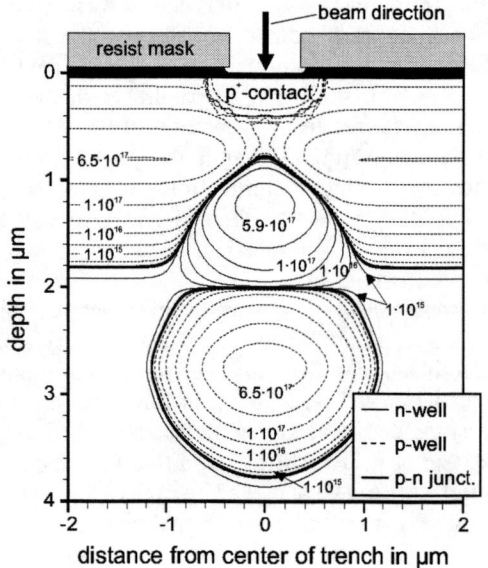

FIGURE 5. Simulation of the p-well contact area shows overlapping p-well extensions for resist openings smaller than 0.7 to 0.8µm. Beam projection onto the surface yields a line in parallel with the trench direction.

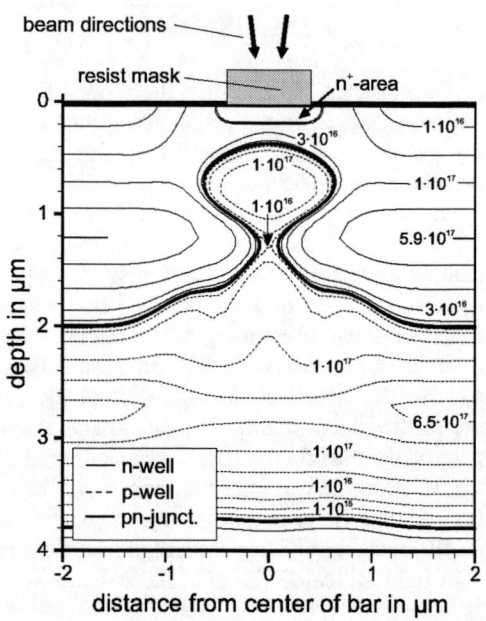

FIGURE 6. Simulation of the n-well contact area below a resist bar 0.8µm in width. Beam projection onto the surface yields a line perpendicular to the bar direction. Overlap of n-well extensions can be optimized performing two implants with half the dose and changing the rotation angle by 180°.

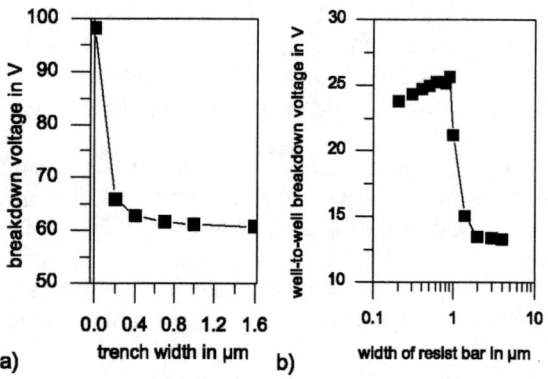

FIGURE 7. Breakdown voltage between p-well and n-substrate depending on the width of the resist trench used for formation of p-well contact (a). Breakdown voltage between and p- and n-well depending on the width of the resist bar used for formation of p-well contact (b). The breakdown voltage of the overall technology is not influenced by the contact regions when resist feature sizes are below the maximum allowable widths.

CONCLUSION

Combining an approach for triple-well formation via a single resist mask with a concept for integration of well and S/D doping using one mask per transistor type, we demonstrate a manufacturable and scalable implantation scheme for a highly simplified CMOS process. With an appropriate formation of n^+ contacts the front end mask count can be reduced to three.

REFERENCES

1. International Technology Roadmap for Semiconductors. http://public.itrs.net
2. J. O. Borland, "Improved Device Scaling & Process Simplification Through Advanced Ion Implantation Techniques", in *Second international symposium on ULSI Process Integration*, Electrochemical Society Proceedings Volume 2001-2, 2001
3. T. Yabu, S. Odanaka, H. Umimoto, N. Shimizu, T. Ohzone, "An Advanced Half-Micrometer CMOS Device With Self-Aligned Retrograde Twin-Wells and Buried p+ Layer", in *1989 Symposium on VLSI Technology*, Digest of Technical Papers, pp. 35-36.
4. M. Kerber, U. Schwalke, R. Heinrich, " Low-Cost CMOS Process with Complete Post-Gate Implantation Scheme", in *ESSDERC '97*, Proceedings, Editions Frontières, 1997, pp. 400-403.
5. L. Gong, L. Frey. H. Ryssel, S. Petersen, "Improved delineation technique for two dimensional dopant profiling", in *Nuclear Instruments and Methods in Physics Research B*, 96 (1) 1995, pp. 133-138
6. *ISE TCAD Release 10.0 – Manual (DIOS, MDRAW, DESSIS)*, Integrated Systems Engineering AG, Zurich, Switzerland, 2004

Boron Redistribution During Crystallization of Phosphorus-Doped Amorphous Silicon

R. Simola[a], D. Mangelinck[a], A. Portavoce[a], J. Bernardini[a], P. Fornara[b]

a) L2MP CNRS UMR 6137, case 142, Faculté des Sciences de Saint Jérôme,, 13397 Marseille cedex 13, France
b) ST Microelectronics, 77 avenue Olivier Perroy, 13790 Rousset, France

Abstract. The redistribution of boron has been studied during solid phase crystallization (SPC) of a homogeneous phosphorus-doped amorphous silicon layer deposited by low pressure chemical vapor deposition, for different thermal annealing. We show that for the lower temperature annealing (T = 586 °C, 1h) boron diffuses without changing the P profile, while for the higher temperature annealing (T = 800 °C, 3h), the initially homogeneous P profile is modified, showing two concentration peaks.

Keywords: Boron, Phosphorus, amorphous Silicon, Crystallization, Diffusion
PACS: 78.40.Fy

INTRODUCTION

Dopant redistribution in silicon is of great importance in microelectronics. While boron diffusion has been extensively studied in crystalline silicon (c-Si), fewer studies exist concerning boron redistribution in poly-crystalline silicon [1-4] as well as in amorphous silicon [5-8]. In this work, we investigate the boron redistribution in a thin silicon film during a typical non volatile memory (NVM) process flow. In this process, the Si floating gate is first deposited on the Si oxide by low pressure chemical vapor deposition (LPCVD) as an amorphous homogeneously phosphorous-doped layer, and then implanted with boron. This layer crystallizes later in the process during the formation of the oxyde-nitride-oxide (ONO) layer. The crystallization phenomenon is expected to greatly influence the redistribution of the dopants, and may lead to unusual variation of their concentration profiles versus the annealing time. For instance, it has been shown that both the nature of the dopant and its concentration can influence the silicon crystallization [9-10]. Furthermore, dopant diffusion depends on the crystalline state of the Si matrix, being faster in polycrystalline and amorphous Si than in monocrystalline Si.

EXPERIMENTAL

The samples were made of a 100 nm thick amorphous Si (a-Si) layer deposited by low pressure chemical vapor deposition (LPVCD) on a 12 nm thick SiO_2 layer, which has been thermally grown on Si(001) substrates. The a-Si film was in-situ co-deposited with phosphorous at a temperature (T) of 530 °C, in order to obtain a homogeneous P concentration of 7×10^{19} atoms / cm^3 in the whole layer. The a-Si film was then implanted with $^{11}B^+$ ions, with a beam energy of 7 keV and an ion fluence of 3.5×10^{15} atoms / cm^2.

The samples were annealed under secondary vacuum (pressure $< 10^{-6}$ mbar) at temperatures ranging from 585 to 800 °C for time between 1 and 50 hours. The temperature was measured using an in-situ thermocouple located at a few millimeters from the samples. Transmission electron microscopy (TEM) measurements and X-ray Diffraction (XRD) measurements have shown that all the samples were polycrystalline after these different thermal treatments.

The dopant profiles were analyzed by secondary ion mass spectroscopy (SIMS). The B profiles were measured using a 3 keV O_2^+ primary ion beam with an incidence angle of 45° while the P profiles were acquired under oxygen atmosphere (oxygen leak) using a 3 keV O_2^+ primary ion beam with an incident angle of 25.2°.

Complementary experiments were performed in order to estimate the crystallization time (time which is needed to completely crystallize the a-Si layer at a given temperature). In-situ XRD measurements were performed on a-Si undoped, phosphorus-doped and boron-doped layers made following the same process as previously described. During isothermal annealing at T = 565 °C under vacuum (p < 10^{-4} mbar), the variations of the (111)Si X-ray peak intensity was recorded versus time, in the classical Bragg-Brentano geometry. The temperature was monitored using a thermocouple in contact with the surface of the samples.

RESULTS AND DISCUSSION

The boron concentration profiles measured before and after a low temperature thermal annealing (1h at 585 °C) are shown in Fig. 1. Despite an annealing at low temperature, a non negligible part the B atoms did diffuse and an additional shoulder appear on the profile.

It is important to notice that for these experiments the concentration of boron in the as-implanted sample is notably higher than the boron solubility limit in monocrystalline Si (~ 4 × 10^{17} cm^{-3} at 600 °C [12]). Thus the fact that a part of the B atoms did not diffuse can be explained considering that during the annealing the immobile atoms formed boron-interstitial-clusters (BICs), whose properties have been the object of theoretical [13-15] and experimental [16-20] studies.

Fig. 1 shows that boron atoms which have diffused during the thermal treatment correspond to a concentration lower than 2 × 10^{20} cm^{-3}, which is higher than the common solubility limit of B in c-Si at 600 °C. This higher value can be understood considering that atoms could have diffused while the layer was still amorphous. In this case the solubility limit of B is expected being higher than in c-Si (a-Si has a lower density than c-Si [21]) and is not precisely known as it may depend on the nature of the amorphous layer. Also, it has been shown that in the case of implanted pre-amorphized Si layer the solubility of B can be greatly increased after solid phase epitaxial crystallization (3.5 × 10^{20} cm^{-3} in the temperature range T = 800-1000 °C [22] and up to 2 × 10^{20} cm^{-3} in the low-temperatures range T=500-600°C [7]). The same mechanism can take place in our experiments.

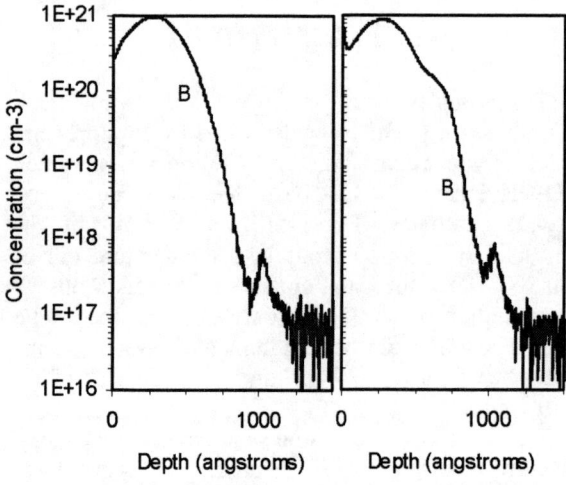

FIGURE 1: Boron profile measured by SIMS, before thermal annealing (left) and after T = 585 °C 1h annealing (right).

One can clearly observe on Fig. 1 that this shoulder located in the region between 50 nm (C_B = 2 × 10^{20} cm^{-3}) and 70 nm (C_B = 8.5 × 10^{19} cm^{-3}) is followed by an abrupt slope which ends at a depth of ~ 94 nm (C_B = 4 × 10^{17} cm^{-3}). The peak at a depth of ~ 100 nm, located in the SiO$_2$ layer zone, is certainly a SIMS artifact due the so-called matrix effect [11], which originates here from a higher ionization rate of B atoms in the SiO$_2$ matrix compared to the Si matrix.

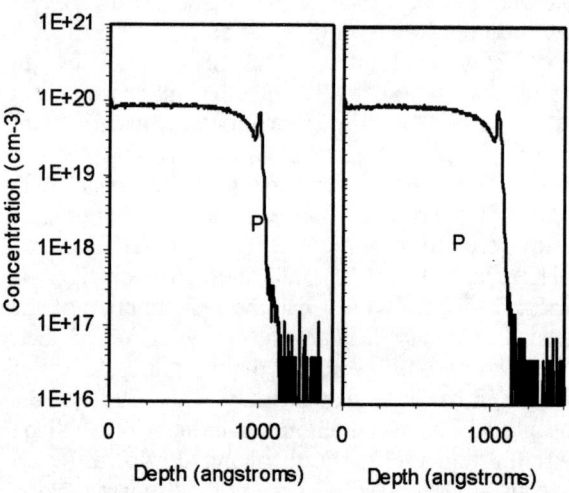

FIGURE 2: Phosphorus profile measured by SIMS, before thermal annealing (left) and after T = 585 °C 1h annealing (right).

The width of the extra-shoulder observed on the SIMS profile measured after annealing can be considered as the diffusion length of the B atoms during the treatment. It is about 20 nm. This distance is somewhat greater than the one expected for boron diffusion in polycrystalline Si (poly-Si), as according to [2] this distance should be only of ~ 7 nm for our annealing conditions. This may be explained by a faster diffusion of boron in a-Si.

Furthermore, the abrupt slope observed after the shoulder (depth greater than 70 nm) cannot be simulated with a diffusion model using a constant diffusion coefficient (even if the presence of boron clusters is considered). Different phenomena that could explain this abrupt slope include a concentration dependence of the boron diffusion coefficient and/or the influence of the mobile interface between the amorphous and the crystalline parts of the Si layer during the crystallization.

Figure 2 shows that the phosphorus profile is found unchanged after the lower temperature annealing (T = 585 °C, 1h). This is what is reasonably expected in the case of a homogeneous material, as the SIMS profile of phosphorus does not present any concentration gradient. Neither the presence of boron, nor the crystallization is thus found having a significant effect on the phosphorus distribution.

As already cited, at sufficient high temperatures crystallization of doped a-Si takes place relatively quickly. Normalized (111)Si X-ray diffraction peaks (2θ=28.4°) for undoped a-Si (250 nm-thick), phosphorus-doped a-Si (150 nm-thick) and boron implanted a-Si (250 nm-thick) as a function of annealing time are shown in Fig. 3. All the samples where annealed at T=565 °C.

In all three cases, the intensity peak does not start to increase immediately; it takes some time, a latent time, for crystallization to start. This latent time is maximum for undoped a-Si and minimum for the phosphorus-doped a-Si film. The phosphorus doped a-Si sample is the fastest to crystallize; total time for crystallization is about 360 minutes at T=565 °C. At higher temperatures, crystallization time should be much shorter because of the exponential dependence of crystallization rate on temperature. Through these measurements we were able to: i) confirm that boron and phosphorus enhance crystallization; ii) set a lower bound for crystallization time for all the experiments that we have performed.

Figure 4 shows the boron (left) and the phosphorus (right) profiles measured after annealing at 650 °C for 50 hours. In contrast with the previous annealing (T = 585 °C, 1h) diffusion in poly-Si cannot be neglected anymore [2]. As the Si film is found being polycrystalline after a thermal treatment at 585°C for 1 hour, it is reasonable to consider that at 650 °C for 50 hours the crystallization is faster (diffusion in the a-Si is shorter), and thus boron atoms diffuse in a poly-Si matrix during at least 49 hours. The boron profile, at a depth below 70 nm is found to be similar to the one obtained during the lower temperature annealing.

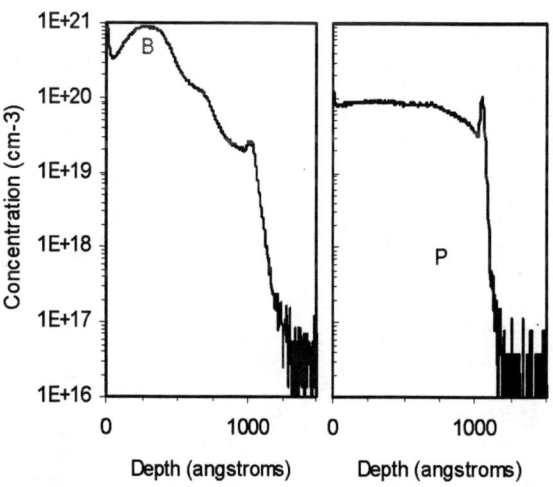

FIGURE 4: Boron (left) and phosphorus (right) profiles after T = 650 °C 50h thermal annealing.

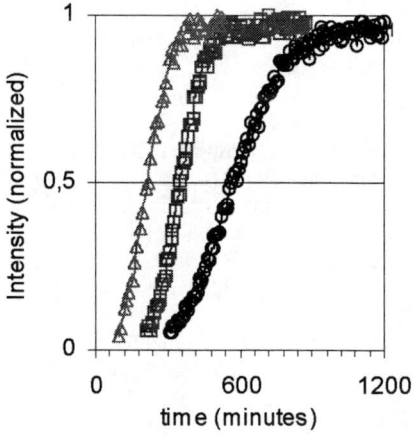

FIGURE 3: XRD normalized peaks for non doped (circle), phosphorus doped (triangle) and boron doped (square) a-Si films, as a function of annealing time at T=565 °C.

The significant changes in the boron profile, compared to the boron profile obtained after annealing at 585 °C for 1 hour, start from a depth of 70 nm with a concentration about $C_B = 1 \times 10^{20}$ cm^{-3}. We conclude that the left part of the profile (depth below 70 nm) is mainly due to the boron redistribution before and during crystallization, while for greater depth the concentration profile of boron results from the diffusion of atoms in poly-Si. Boron solubility limit in phosphorus-doped poly-Si seems thus to be 1×10^{20} cm^{-3} consistent with the previous study [13]. The right-hand phosphorus profile in Fig. 4, shows an almost constant concentration after the annealing, with small variations at depth around 33 nm and 70 nm where little local maxima appeared.

High-temperature crystallization annealing (T=800°C, 5h) results are shown in Fig. 5 for boron (left) and phosphorus (right). Boron profile is quite flat between 75 nm and 94 nm ($C_B=1.2\times10^{20}$ cm^{-3} and 9.3×10^{19} cm^{-3} respectively) due to boron diffusion in polycrystalline silicon. The rest of the profile, namely for depth below 70 nm, is similar to the one obtained after low-temperature annealing (T=585°C, 1h) despite a significant change in the 30 nm peak which concentration diminished and reached the value of 6×10^{20} cm^{-3}. We conclude that at T=800°C 5h, boron diffusion length in polycrystalline is about 20 nm. Atomic Force Microscopy measurements (not shown in this paper) yield an average diameter of 40 nm for the grains. Phosphorus profile displays two concentration peaks at 30 nm and 70 nm; since concentration profile is constant before annealing, these peaks are certainly due to boron-phosphorus interactions; such interaction may include electrical and/or chemical ones.

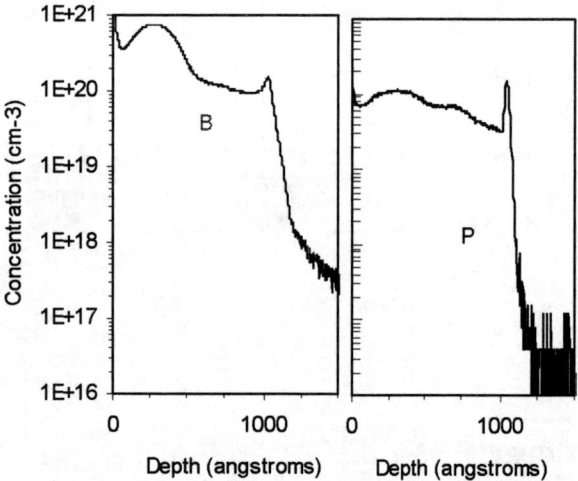

FIGURE 5: Boron (left) and phosphorus (right) profiles after T=800°C 5h thermal annealing

ACKNOWLEDGMENTS

The authors would like to thank K. Hoummada for the help concerning the XRD measurements.

REFERENCES

1. T.I. Kamins, J. Manoliu, R.N. Tucker, Tucker, *J. Appl. Phys.* **43**, 83 (1972)
2. K.Sakamoto, K. Nishi, T. Yamaji, T. Miyoshi and S. Ushio, *J. Electrochem. Soc.* **132**, 2457-2462 (1985)
3. A. Merabet, C. Gontrand, *Mat. Science and Eng.* **B102**, 257-261 (2003)
4. B. Semmache, A. Merabet, C. Gontrand, A. Laugier, *Mat. Science and Engin.* **B38**, 41 (1996)
5. J. M. Jacques, L.S. Robertson, K.S. Jones, *Appl. Phys. Lett.* **82**, 3469-3471 (2003)
6. R. Duffy, V.C. Venezia, A. Heringa, T.W.T. Husken, M.J.P. Hopstaken, N.E.B Cowern, P.B. Griffin, C.C. Wang, *Appl. Phys. Lett.* **82**, 3647-3649 (2003)
7. R. Duffy, V.C. Venezia, A. Heringa, B.J. Pawlak, M.J.P. Hopstaken, G.C.J Maas, Y. Tamminga, T. Dao, F. Roozeboom, L. Pelaz, *Appl. Phys. Lett.* **84**, 4283-4285 (2004)
8. V.C. Venezia, R. Duffy, L. Pelaz, M.J.P Hopstaken, G.C.J. Maas, T. Dao, Y. Tamminga, P. Graat, *Mat. Science and Engin.* **B124-125**, 245-248 (2005)
9. R. Bisaro, J. Magarino, K. Zellama, S. Squelard, P. Germain, J.F. Morhange, *Phys. Rev. B* **31**, 3568-3575 (1985)
10. H-C. Cheng, F-S. Wang, Y-F. Huang, C-Y. Huang, *J. Electrochem. Soc.* **142**, 3574-3578 (1995)
11. V. R. Deline, William Katz, C. A. Evans, Jr., and Peter Williams, *J. Appl. Phys.* **33**, 832 (1978)
12. P. Pichler, *Mat. Res. Soc. Symp. Proc.* **717**, 103 (2002)
13. L. Pelaz, M. Jaraiz, G. H. Gilmer, H.-J. Gossmann, C.S. Rafferty, D. J. Eaglesham, and J. M. Poate, *Appl. Phys. Lett.* **70**, 2285 (1997)
14. X.-Y. Liu, W. Windl, and M. P. Masquelier, *Appl. Phys. Lett.* **77**, 2018 (2000)
15. P. Alippi, P. Ruggerone, L. Colombo, *Phys. Rev. B* **69**, 125205 (2004)
16. G. Mannino, N. E. B. Cowern, F. Roozeboom, J. G. M. van Berkum, *Appl. Phys. Lett.* **76**, 855 (2000)
17. A. D. Lilak, M. E. Law, L. Radic, K. S. Jones, M. Clark, *Appl. Phys. Lett.* **81**, 2244 (2002)
18. S. Mirabella, E. Bruno, F. Priolo, D. De Salvador, E. Napolitani, A. V. Drigi, A. Carnera, *Appl. Phys. Lett.* **83**, 680 (2003)
19. A. M. Piro, L. Romano, S. Mirabella, M. G. Grimaldi, *Appl. Phys. Lett.* **86**, 81906 (2005)
20. L. Romano, A. M. Piro, S. Mirabella, M. G. Grimaldi, E. Rimini, *Appl. Phys. Lett.* **87**, 201905 (2005)
21. J.S. Custer, M.O. Thompson, D.C. Jacobson, J.M. Poate, S. Roorda, W.C. Sinke, F. Spaepen, *Appl. Phys. Lett.* **64**, 437-439 (1994)
22. S. Solmi, E. Landi, F.Baruffaldi, *J. Appl. Phys.* **68**, 3250 (1990)

Ultra-Shallow Junctions Formed By Sub-Melt Laser Annealing

S.B. Felch[1], A. Falepin[2], S. Severi[2], E. Augendre[2], T. Hoffman[2], T. Noda[2,3], V. Parihar[1], F. Nouri[1], and R. Schreutelkamp[1]

[1]*Applied Materials, 974 E. Arques Ave., M/S 81280, Sunnyvale, CA 94085 USA*
[2]*IMEC, Kapeldreef 75, B 3001, Leuven, Belgium*
[3]*Matsushita Electric Industrial Co., Ltd.*

Abstract. This paper presents an overview of blanket and device wafer studies with sub-melt laser annealing as the sole dopant activation technique. Blanket wafer studies of the activation and diffusion of n- and p-type junction implants have focused on the influence of the pre-amorphization depth and laser annealing temperature. In addition, deep sub-micron NMOS and PMOS transistors with effective gate lengths down to 40 nm have been fabricated. With optimized implant conditions several critical transistor parameters, such as gate/extension overlap and polysilicon dopant depletion, can be improved when sub-melt laser annealing is used instead of conventional spike anneal.

Keywords: Laser anneal, dopant activation, diffusion.
PACS: 61.72.Tt, 61.80.Ba, 61.72.Cc

INTRODUCTION

Since the requirements for the source/drain extensions of future devices become more and more severe with respect to dopant activation, junction depth, and abruptness, ultra-fast annealing techniques, such as flash and sub-melt laser annealing, are being explored [1]. Blanket wafer studies of the activation and diffusion of n- and p-type junction implants are helpful to understand the influence of the pre-amorphization depth and laser annealing temperature. However, the integration of diffusion-less junctions creates challenges related to contact and overlap resistance [2,3], so that the junction implant parameters optimized for spike anneal need to be redesigned to address these issues. Nevertheless, with optimized implant conditions several critical transistor parameters, such as gate/extension overlap and polysilicon dopant depletion, can be improved when sub-melt laser annealing is used instead of conventional spike anneal.

EXPERIMENT

Blanket n- and p-doped Si (100) wafers were first implanted with 20 keV B^+ to a dose of 5×10^{12} at/cm^2 or 70 keV As^+ to a dose of 3×10^{12} at/cm^2, respectively, and annealed at 1100°C for 1s. This forms a deep, highly resistive p- or n-well so that during sheet resistance measurements, the probe is not in contact with the counter-doped Si substrate, allowing reliable sheet resistance measurements.

All ultra low energy implants have been performed on an Applied Materials XR80. Prior to USJ implantation, some of the Si wafers have been pre-amorphized by a Ge^+ implantation (PAI) at 8 or 30 keV and a dose of 5×10^{14} at/cm^2. This results in amorphous layers with a thickness of approximately 15 nm and 40 nm, respectively. For the p-type junctions, B^+ ions have been implanted at an energy of 1 keV and a dose of 1×10^{15} at/cm^2, whereas for the n-type junctions, As^+ was implanted with an energy of 2 keV and a dose of 1×10^{15} at/cm^2.

Activation of the implanted ions was achieved using sub-melt laser annealing from Applied Materials. During this process, the wafer is loaded onto a heated chuck at a temperature of < 500°C, and different peak temperatures at the Si surface can be obtained by varying the laser power. For all conditions, prior to laser annealing, an absorber layer (AL), typically grown at 550°C, has been deposited onto the implanted Si wafers. Several wafers received a solid phase epitaxial regrowth (SPER) anneal at 650°C for 1 min. before the laser anneal.

The junction depth (X_j) and sheet resistance (R_s) were measured with secondary ion mass spectrometry (SIMS) and four-point probe, respectively.

The process flow description for most of the NMOS and PMOS transistors studied is shown in Fig. 1. After conventional V_t adjust implants, 1.4 nm SiON gate dielectric was formed. The NMOS and PMOS transistors were processed on separate wafers with PVD TaN gate electrode [4] and Ni Fully Silicided (FUSI) gate [5], respectively, or with polysilicon gates. After halo implantation, several source/drain extension (SDE) splits on Ge and B implants for PMOS and As implants for NMOS have been performed. A low temperature 60nm nitride spacer was formed prior to the deep junction implants (HDD). All implantation energies were chosen to obtain similar junction depths after spike or laser annealing. After formation of NiSi, back end of line (BEOL) processing completed the flow.

	NFET		PFET	
	Reference	Laser	Reference	Laser
	Gate stack: TaN / 1.4nm SiON		Gate stack: NiSi FUSI / 1.4nm SiON	
	Pockets and Extensions:		Pockets and Extensions:	
	B 10keV	B 10keV	As 40keV	As 55keV
	—	Ge PAI	—	Ge PAI
	As 1keV	As 5keV	BF$_2$ 1keV	B 0.5/1keV
	HDD:		HDD:	
	—	—	—	Ge PAI
	As 25keV	As 30keV	B 3keV	B 5keV
	1050°C spike	1300°C Laser	1050°C spike	1300°C Laser

FIGURE 1. Process flow conditions for spike RTA and laser annealed transistors.

RESULTS

Blanket Wafer Studies

Figure 2 shows the SIMS profiles for 1 keV B implants into crystalline and amorphous Si, followed by SPER and laser annealing at 1300°C for the deep Ge PAI and followed by 1300°C laser annealing only for the shallow Ge PAI [6]. To obtain ultra-shallow B-doped junctions, the implementation of a Ge PAI step is essential to eliminate channeling of the B ions in the crystalline silicon, which leads to a deeper junction. Using deep (30 keV) or shallow (8 keV) Ge PAI result in the same ultra-shallow X_j, although higher activation and a more abrupt profile are achieved for a PAI implant of 30 keV. Since the thicker amorphous layer requires a longer time for complete regrowth, more boron can diffuse in the amorphous phase, leading to these observations.

The sheet resistance values for the 1 keV B implants as a function of laser annealing temperature are plotted in Fig. 3 for both PAI energies. R_s decreases as the laser annealing temperature increases for both PAI cases, highlighting one of the key benefits of laser anneal. In addition, higher PAI energy leads to lower sheet resistance. This could be due to an interplay between several phenomena. First, the thinner amorphous layer produced with the low Ge PAI energy leads to a higher flux of Si interstitials towards the surface originating from the EOR region, which cause more B deactivation. Second, a thicker amorphous layer results in more B diffusion and hence more activated boron.

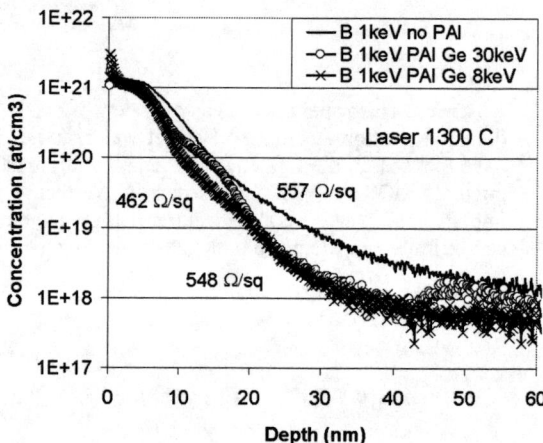

FIGURE 2. SIMS profiles for 1 keV B implants into crystalline and amorphous Si, followed by laser annealing at 1300°C.

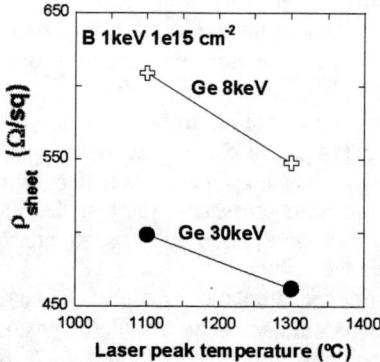

FIGURE 3. Influence of the pre-amorphization depth on the sheet resistance after laser annealing at different temperatures.

The diffusion behavior of As, the typical n-type SDE dopant, is illustrated in Fig. 4. The annealed profiles are identical, regardless of laser anneal temperature or whether an SPER anneal was only used, and are the same as the as-implanted profile (not shown here). The sheet resistance values for these junctions decrease from 790 ohms/sq. with a 1000°C anneal to 670 ohms/sq. with a 1300°C anneal. Since the As junction depth remains constant during the anneal, decreasing R_s corresponds to an increasing electrically active concentration. So, sub-melt laser anneal is truly diffusionless for As and can produce junctions with high dopant activation.

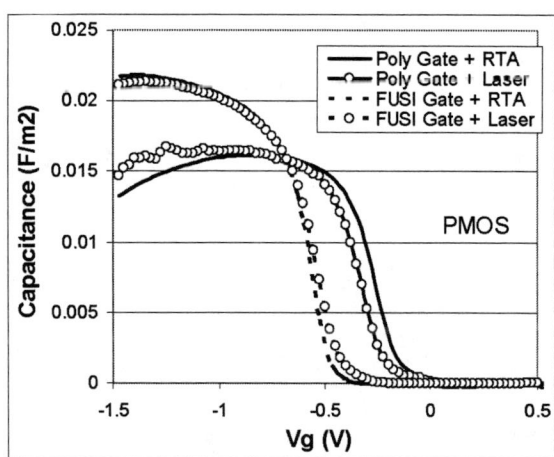

FIGURE 5. Inversion RF C-V curves of PMOS devices with polysilicon or FUSI gates and spike and/or laser anneal.

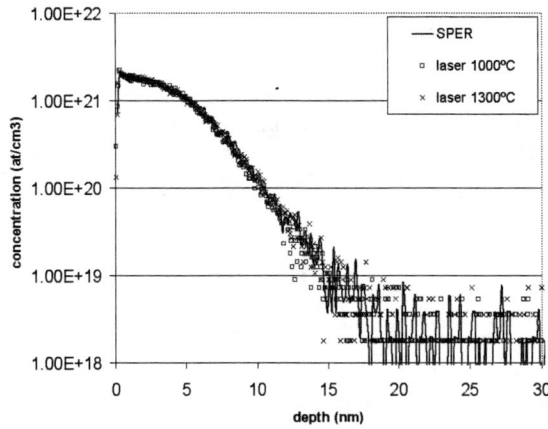

FIGURE 4. SIMS profiles for 2 keV As implants into crystalline Si followed by laser annealing at 1000 or 1300°C or by an SPER anneal at 650°C for 1 min.

Device Studies

One of the primary benefits of using laser anneal with polysilicon gates is the higher dopant activation in the polysilicon and the resultant reduced depletion region near the gate dielectric interface. Figure 5 displays inversion capacitance vs. voltage curves of PMOS devices with polysilicon or FUSI gates and spike and/or laser anneal. For gate voltages below -0.5 V, the capacitance with spike plus laser anneal is significantly higher than that with spike anneal only for polysilicon gates, confirming that the improved dopant activation has produced a smaller depletion region. On the other hand, FUSI gates do not rely on dopant activation for their conductivity, so laser anneal has no impact on the gate capacitance but the FUSI gate capacitance is much larger than that with doped polysilicon.

The extracted gate/SDE overlap length [7] is an electrical measure of the SDE lateral diffusion under the gate, which has a major influence on the electrical channel length and the device performance. Figure 6 shows that the overlap length shrinks from 7 nm to 2 nm moving from spike RTA to laser annealed transistors implanted with 0.5 keV B. A SDE implant of 1 keV B produces an electrical overlap of only 3 nm, which is in agreement with that measured by a scanning spreading resistance microscopy (SSRM) profile at an active doping concentration of 1×10^{19} cm^{-3} [7]. So, laser anneal alone results in less lateral diffusion, even though the 1 keV B implant conditions were chosen to match the vertical SIMS junction depth of the BF_2 reference. In addition, SSRM profiles have confirmed that the HDD diffusion is reduced with laser anneal alone [8].

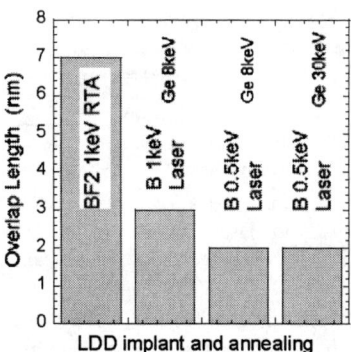

FIGURE 6. Overlap length for the laser annealing implant splits compared to the overlap obtained with spike anneal.

The reduced lateral diffusion of both the SDE and HDD lead to reduced short channel effects (SCE), as exemplified by a flatter change in threshold voltage as a function of gate length with laser anneal. Figure 7 shows that the V_t roll-off continually improves as the SDE implant energy (and gate/SDE overlap) decreases. However, the lowest energy of 0.5 keV also results in a very shallow junction with higher sheet resistance and an I_{on}/I_{off} performance that is worse than that obtained with 1 keV B (see Fig. 8). Thus, the best performance is achieved with a compromise between low sheet resistance (deeper junction) and better short channel effects (shallower junction), as the 1 keV B SDE produces a 15% improvement over the spike-annealed baseline.

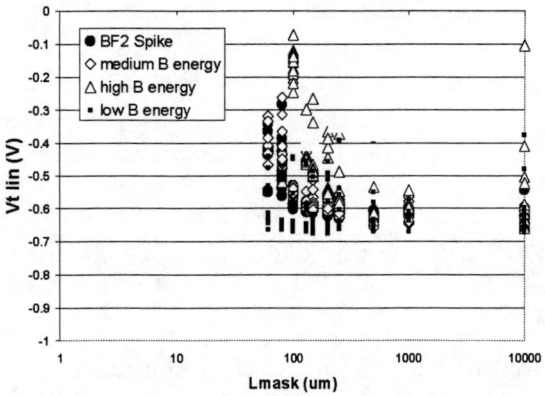

FIGURE 7. Linear threshold voltage as a function of gate mask length for various PMOS SDE implant conditions with laser anneal. The reference devices had BF_2^+ implants activated by spike anneal.

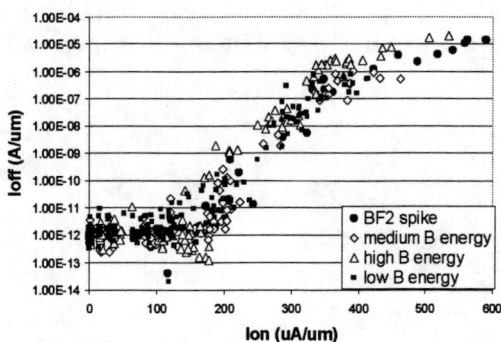

FIGURE 8. I_{on}-I_{off} characteristics for various PMOS SDE implant conditions with laser anneal. The reference devices had BF_2^+ implants activated by spike anneal.

I_{on}-I_{off} characteristics for NMOS transistors with spike or laser anneal are shown in Fig. 9. The good electrical characteristics of the NMOS devices with TaN gates demonstrate that sub-melt laser anneal is compatible with metal gates. In addition, smaller transistors can be fabricated with laser anneal and the same B halo implant. However, during laser annealing the B halo does not diffuse, and the dopant activation is much higher than in the case of spike RTA. So, laser anneal produces a device with a relatively high V_t, and yet good transistor performance is still achieved. Lowering the halo dose should allow further improvement in performance without significant SCE degradation.

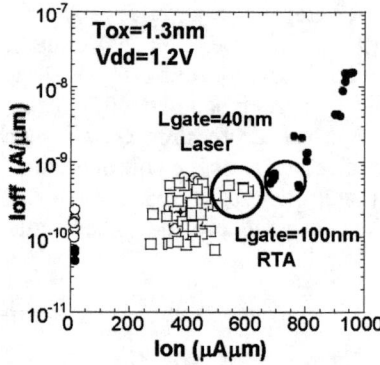

FIGURE 9. I_{on}-I_{off} characteristics comparing spike RTA (dark circles) with laser annealed (open squares) NMOS transistors with the same implanted B halo dose.

CONCLUSIONS

The results of blanket and device wafer studies with sub-melt laser annealing as the sole dopant activation technique have been presented. Laser annealing produces improved dopant activation in crystalline Si and polysilicon, where reduced polysilicon depletion has been observed. Minimal dopant diffusion leads to reduced SDE lateral diffusion under the gate and improved short channel effects. Enhanced PMOS and NMOS device performance can also be obtained with proper optimization of SDE and halo implants. Finally, laser anneal has been shown to be compatible with metal gates and does not appear to limit junction depth scaling.

REFERENCES

1. The International Technology Roadmap for Semiconductors, 2005.
2. S. Severi et al., *IEDM 2004 Tech. Dig.*, IEEE, 2004, pp. 99-102.
3. A. Shima, Y. Wang, D. Upadhyaya, L. Feng, S. Talwar, and A. Hiraiwa, *2005 Symp. on VLSI Technology Digest of Technical Papers*, IEEE, 2005, pp. 144-145.
4. K. Henson et al., *IEDM 2004 Tech. Dig.*, IEEE, 2004, pp. 851-854.
5. K. G. Anil et al., *2004 Symp. on VLSI Technology Digest of Technical Papers*, IEEE, 2004, pp. 190-191.
6. A. Falepin, T. Janssens, S. Severi, W. Vandervorst, S.B. Felch, V. Parihar, and A. Mayur, *Proc. of 2005 RTP Conference*, IEEE, 2005.
7. S. Severi et al., *Proc. of 2006 VLSI Taiwan Symposium*, IEEE, 2006.
8. S. Severi, E. Augendre, and K. De Meyer, to be published in *Proc. of MRS Symposium C*, Spring 2006.

Defect Behavior in BF$_2$ Implants For S/D Applications as a Function of Ion Beam Characteristics

Nathalie Cagnat[1], Cyrille Laviron[2], Nicolas Auriac[3], Jinning Liu[4], Sandeep Mehta[4], Laurent Frioulaud[1] & Daniel Mathiot[5]

[1]*STMicroelectronics, 850 rue Jean Monnet, 38926 Crolles Cedex, France*
[2]*CEA-LETI, 17 rue des Martyrs, 38054 Grenoble Cedex 9*
[3]*Freescale Semiconductor, 850 rue Jean Monnet, 38926 Crolles Cedex, France*
[4]*Varian Semiconductor Equipement Associates, 35 Dory Road, Gloucester, MA 01930-2297*
[5]*InESS, 23 rue du Loess, BP 20 CR, F-67037 Strasbourg Cedex 2*

Abstract. This study investigates characteristics of defects induced by BF$_2$ implantation with 2 different types of ion beams - a spot beam and the other a parallel ribbon beam – on blanket wafers. The effect of dose rate on defect behaviour was also investigated. Transmission Electron Microscopy (TEM) and Ellipsometry measurements were made to understand the damage characteristics. Secondary Ion Mass Spectroscopy (SIMS) analysis was conducted to understand dopant distribution and segregation. In addition, activation of Boron was characterized by sheet resistance measurements. We conclude that it is primarily the inherent difference in dose rate of batch versus single wafer implanters that affects induced defect behaviour in Si.

Keywords: Ion implantation, ion beam, defects.
PACS: 81.05.Dz

INTRODUCTION

Spinning disc batch implanters have been the traditional workhorse for many years to meet the low energy and high dose implant requirements. However, at 90nm node and beyond, the associated gate lengths shrink significantly, making the devices extremely vulnerable to even small variations in process steps. To address the precision needs of advanced scaled devices, the industry is fast transitioning to single wafer high current implanters. At the gate edge where several implants such as Source-Drain Extension (SDE), Source-Drain (SD), Pre-amorphization Implantation (PAI), co-implanted species, halos and pockets all come together, small variations in any of these can lead to unacceptable fluctuations in device performance. The precision required to perform these implants therefore becomes critical. As an example, diffusion in SDE and halo is strongly influenced by damage generated during SD implantation. So the understanding of defect generation as a function of beam characteristics of the ion implanter is critical in the process integration of advanced devices such as 65nm and beyond.

This study investigates characteristics of defects induced by BF$_2$ implantation from two different ion implanters, one featuring a spot beam and the other a parallel ribbon beam (single wafer tool). Effect of dose rate was also investigated. Transmission Electron Microscopy (TEM) and Ellipsometry measurements were made to understand the damage characteristics, sheet resistance measurements were used to characterize activation of Boron, and Secondary Ion Mass Spectroscopy (SIMS) analysis was conducted to study dopant distribution and segregation.

EXPERIMENTAL

Samples Preparation

All samples employed in this study were oxidized to grow a 31Å oxide layer to prevent dopant out-diffusion during annealing.

Two successive room temperature implants were done to mimic SD implantation in a typical CMOS device. The first implant featuring an ion beam of BF$_2$, 15keV, 3.6E15cm² was done either on a spot beam, batch implanter or on a single wafer tool featuring a ribbon beam. For the latter, three different dose rates were investigated by means of adjusting the beam current. The highest one is the one typically used the two others are reduced to reach values ranging between the batch and the single wafer tools values. The second implant featuring an ion beam of B^+, 13keV, 6E13 at/cm² was done on a single wafer tool for all samples.

Finally, two different anneal conditions were used - a furnace anneal at 800°C for 10min and a spike anneal at 1113°C with RTA. While the first anneal was chosen to agglomerate implant induced defects to simplify their characterization, spike anneal was used to simulate the S/D activation anneal step.

The amorphous layer was characterized by Ellipsometric measurements. Following the furnace anneal step, defect analysis was done by TEM. Activation of Boron after spike anneal was analyzed by 4-point probe measurements.

Sample preparation steps are described in Table 1.

TABLE 1. Sample preparation steps

Step	1	2	3	4	5	6	7	8
Oxidation (31Å)	x	x	x	x	x	x	x	x
Implant 1 : BF₂ 15keV 3.6E15at/cm²	Batch tool D_R = 0.02E13 at/cm²/s		Single wafer tool D_R = 0.67E13 at/cm²/s		Single wafer tool D_R = 1.64E13 at/cm²/s		Single wafer tool D_R = 6E13 at/cm²/s	
Implant 2 : B 13keV 6E13at/cm²	x	x	x	x	x	x	x	x
Annealing	10min, 800°C	spike 1113°C	10min, 800°C	spike 1113°C	10min, 800°C	spike 1113°C	10min, 800°C	spike 1113°C

Dose Rate Calculation

A key parameter of this study is the dose rate (D_R) calculation expressed by the equation (1).

$$D_R = \frac{Q}{t \times n}. \qquad (1)$$

Q : dose in at/cm²,
t : implant duration,
n : number of wafers implanted (batch tool : n = 13, single wafer tool : n = 1)

Table 2 summarizes the associated values for each sample.

RESULTS

Ellipsometry

Ellipsometry was used to estimate the amorphous layer thickness by fitting experimental results to theoretical spectra. The model uses a stack of oxide/α-Si/crystalline Si. Thickness of the oxide and α-Si are optimized to reach the maximum goodness of fit between experimental and theoretical spectra without modifying the materials optical properties. Thickness of the α-Si layer after the second implant are given in Table 2.

Results show that amorphous layer thickness increases with increasing dose rate.

TEM

After annealing at 800°C for 10min, point defects agglomerate into dislocation loops. Cross sectional images by TEM are depicted in Figure 1 and Figure 2. The dislocation counts (summarized in Table 2) are expressed in "number of dislocation/μm length". Even if this number depends on the sample thickness, they were all prepared the same manner, so the values can be compared.

It is observed that the number of dislocation/μm in length decreases with increasing dose rate (see Fig 4).

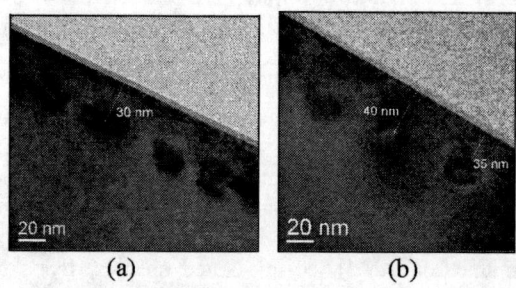

FIGURE 1. TEM cross sectional view after annealing for 10min at 800°C of (a) sample 1 and (b) sample 3

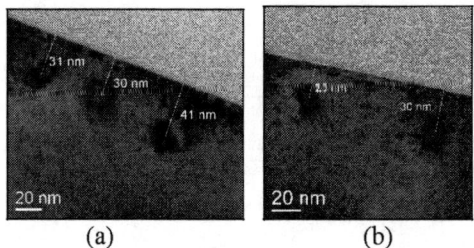

FIGURE 2. TEM cross sectional view after annealing for 10min at 800°C of (a) sample 5 and (b) sample 7

SIMS

Depth profiles of Boron and Fluorine after spike anneal are depicted in Figure 3. Although similar diffusion is observed for all samples, we observe two peaks of Boron and Fluorine segregation - one located at approximately 20nm corresponding to BF_2 implant R_p and another at approximately 30nm corresponding to the End-of-Range (EOR) of the implant in question. This is consistent with works of Ohno and al [1].

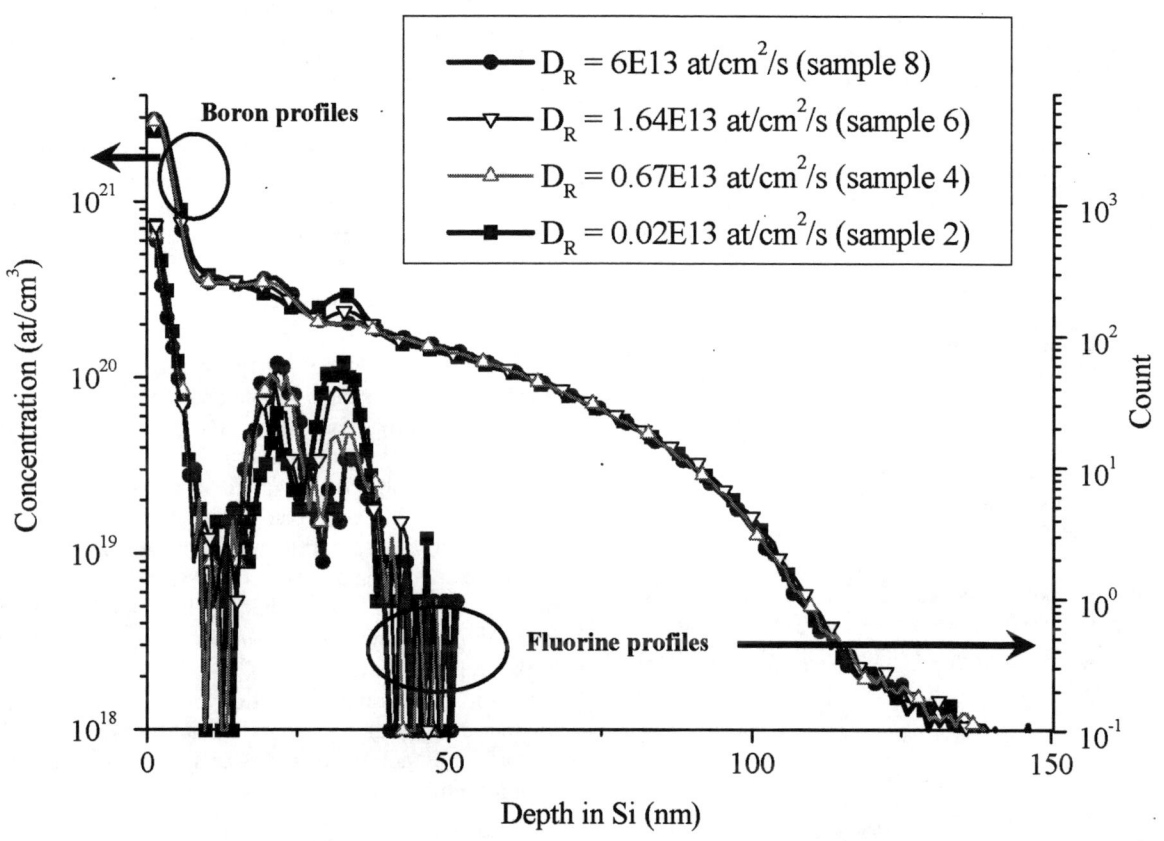

FIGURE 3. Boron and Fluorine depth profiles after spike anneal

TABLE 2. Analysis results summary

Samples	Dose rate (xE13 at/cm²/s)	Ellipsometry α-Si thickness (nm)	TEM estimated number of dislocation/μm length (see comments in the text) after furnace anneal	R_s after spike anneal (Ohm/sq)
1 & 2	0.02	30.3	30	106.9
3 & 4	0.67	31.1	20	103.2
5 & 6	1.64	31.9	10	100.7
7 & 8	6	32.5	5	99.4

DISCUSSION

It is observed that the thickness of amorphous layer is influenced by dose rate. Indeed, a low dose rate allows more time for Frenkel pair recombination before the next collision, leading to a thinner amorphous layer [2]. On the other hand, the implantation depth does not depend on the dose rate so the excess interstitials depicted by the "+1 model" [3] remains at the same depth whatever the dose rate. That means that the defects below the α-Si/crystalline interface are more numerous with a low dose rate. This is clearly revealed by the furnace annealed sample - the number of dislocation loops decreases with increasing dose rate (see Table 2 and Figure 4).

In addition, Moroz & Martin-Bragado [4] have demonstrated that high beam density results in a higher peak temperature for the spot beam, so more damage is annealed during the implant. This leads to a lower amorphization depth and more defects below the α-Si/crystalline interface.

As a result, after spike anneal the concentration of Boron and Fluorine segregated at the EOR location grows with the amount of defects and thus the dose rate as depicted in Figure 5. Moreover, the amount of Boron and Fluorine trapped by the EOR defects stem from the excess concentration located at R_p of the BF_2 implant. Indeed, the height of this hump in the profiles increases as the other one decreases (see Figures 3 and 5).

We can note, from the SIMS profile, that the bulk diffusion is similar for all samples. This suggests that interstitial Si segregates into EOR defects before Boron diffusion.

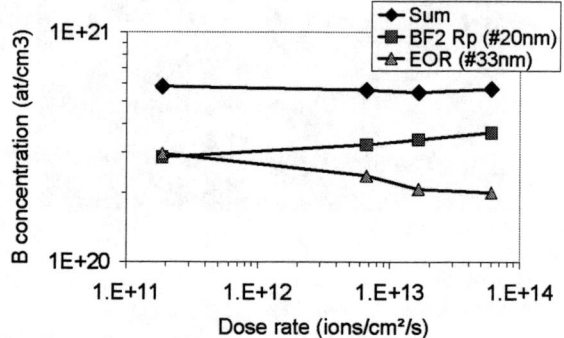

FIGURE 5. Estimated correlation between dose rate and concentration of segregated Boron at BF_2 R_p and at the End-of-Range. The sum of both concentration for each sample is also represented.

Implant activation is lower for lower dose rate (see Rs values in Table 2). This may be due to the fact that the amount of Boron trapped by the EOR defects is inactive unlike when it remains at the BF_2 R_p.

CONCLUSION

Defects and dopant diffusion for two different types of ion beams have been investigated.

Results show that dose rate is an important parameter to fully understand the physics of defect and dopant behaviour. At lower dose rates, the induced Frenkel pairs can annihilate before the next collision, thus reducing the amorphous silicon depth. However, the implant projected range is the same for every dose rate investigated. As a result, the number of Si interstitials below the α-Si/crystalline interface is higher for a lower dose rate.

During subsequent activation anneal, Si interstitials agglomerate leading to trapping and deactivation of Boron. Thus activation is lower for a low dose rate than for a high dose rate.

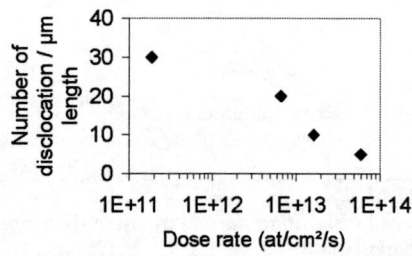

FIGURE 4. Estimated correlation between dose rate and number of dislocations/μm length after furnace anneal at 800°C by TEM cross sectional analysis.

REFERENCES

1. Ohno, N., Hara, T., and Current, M.I., *Journal of the electrochemical society* **152** (11), G835-G840 (2005).
2. Claverie A., Laânab L., Bonafos C., Bergaud C., Martinez A. and Mathiot D., *Nucl. Instr. And Meth. In Phys. Res B* **96**, 202-209 (1995).
3. Mathiot D., Claverie A. and Martinez A., *Defect and diffusion forum* **153-155**, 11-24 (1998).
4. Moroz, V., and Martin-Bragado, I., *MRS Spring Proceeding*, 2006.

Impact of Dose Rate Effects and Damage Engineering on Device Performance

Kyuha Shim[1], Yeonsang Hwang[2], Yongseung Lee[2], Jungsoo An[2], Seonho Ryu[2], Seungho Hahn[3], Changjune Cho[3], Namhae Hur[3], Baonian Guo[1], Jinning Liu[1], Yuri Erokhin[1]

[1]*Varian Semiconductor Equipment Associates, Inc., 35 Dory Road, Gloucester, MA, United States*
[2]*Samsung Electronics Co., Ltd., San#24 Nongseo-dong, Kiheung-ku, Yongin-city, Kyungki-do, Korea*
[3]*Varian Korea, Ltd., 433-1, Mogok-dong, Pyeongtaek-city, Kyungki-do, Korea*

Abstract. Traditional implant conditions during source/drain formation process, such as dopant, dose, energy and incident angle have been known as key parameters determining device electrical characteristics. As devices scale down, instant dose rate of BF_2 ion implantation, however, should be considered as an important factor to control buried channel PMOS characteristics since fluorine and boron diffusion behavior can be varied depending on implant damage and results in change of effective channel length. Ribbon beam single wafer high current implanters enable ion beam density to be modulated. By changing beam size with same beam current during source/drain implantation, PMOS electrical characteristics of 70nm Flash memory have been investigated. With achieved results, device matching to spot beam batch ion implanter has been demonstrated.

Keywords: Buried channel PMOS, BF2, damage, dose rate, 70nm Flash
PACS: 61.72.Tt, 85.40.Ry

INTRODUCTION

Since the first single wafer high current tool was introduced at production fabs, lots of device matching to existing spot beam batch high current have been executed and investigated to find out key variables for successful process transfer. Due to fundamental architecture design difference and beam optics between spot beam batch high current tool and parallel ribbon beam single wafer high current tool, traditional dose matching on bare wafer is not directly applicable to device level any more. A major difference is that effective dose difference on desired areas of the device due to angular properties of ion beam which cannot be detected by sheet resistance(Rs) or 1-D dopant depth distributions(SIMS). Especially, post-gate processes, such as Halo implant, source/drain(S/D), and source/drain extension(SDE) etc, are sensitive to ion incident angle and can cause device electrical parameters shift by shadowing effect or encroachment under the gate [1].

In buried channel PMOS transistors of peripheral circuits, however, BF_2 S/D process by single high current implanter shows an anomalous behavior, lower contact Rs but deeper(more negative) Vt as compared with spot beam batch tools. In order to get both target device parameters matching, implant dose of S/D process is adjusted for contact Rs and also threshold voltage control recipe dose should be changed. [2], [3], [4].

Without modification of other process conditions, this paper discusses beam density modulation of BF_2 ion on single high current implanter to achieve device matching simply.

EXPERIMENTAL

Bare wafers were implanted on VIISion 80 spot beam batch tool and VIISta HC parallel ribbon beam single high current implanter. In order to modulate dose rate on VHC, Drift Expand(DEX) mode is used by applying D1 suppression voltage and D2 suppression voltage in drift region.

Figure 1 shows illustrations of the dose rates from the spot beam batch tool and single high current implanter. Since batch tool loads 13 wafers on a disc, implant time might be 13 times longer than single implanter at given beam current. There are two approaches for single high current implant to mimic dose rate of spot beam batch tool, just to decrease beam current or increase beam size for reducing instant dose rate.

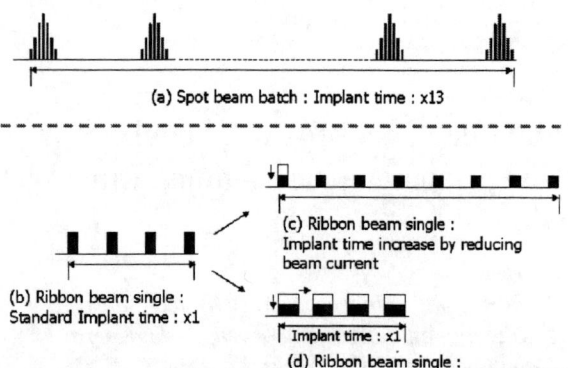

FIGURE 1. Illustration of dose rate (a) spot beam batch tool, (b) ribbon beam single tool, (c) ribbon beam single tool with reduced beam current, (d) ribbon bean single tool with increased beam size

Implant conditions are BF_2 40keV 3E15at/cm^2 with 0° tilt and 45° twist common for both implanters. TW signal was used for estimating implant damage as well as residual damage after anneal at 1000°C for 10 seconds in N_2 ambient. Boron and Fluorine profiles are detected by secondary ion mass spectroscopy(SIMS) and spreading resistance profiling(SRP) was performed to obtain electrically active dopant profile which could not be determined by four point probe or SIMS. 70nm Flash memory technology device was run at selected conditions. 2-D doping profile was investigated by scanning capacitance microscopy.

RESULTS

Figure 2 shows beam size measurement result. Since VIISta HC has parallel ribbon beam, only vertical direction beam size governs overall beam

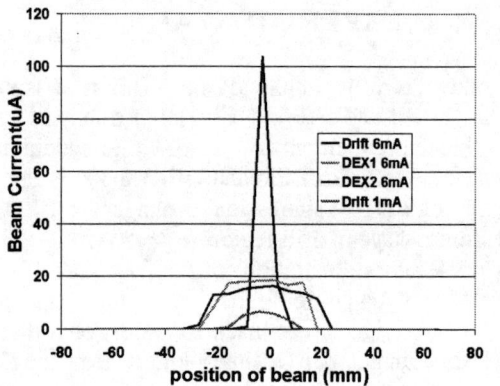

FIGURE 2. Beam size measurement on VIISta HC

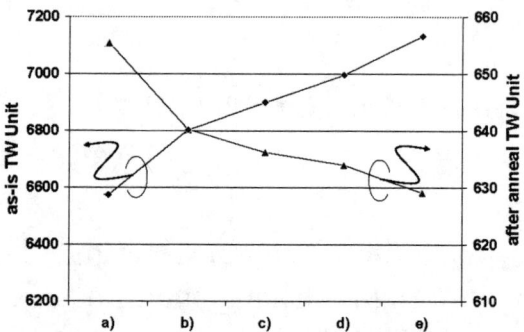

FIGURE 3. TW value from various implant condition on VIISta HC and VIISion80 a) VIISion80 Drfit 6mA, b) VIISta HC Drift 1mA, c) VIISta HC DEX2 6mA, d) VIISta HC DEX1 6mA, e) VIISta HC Drift 6mA

density. Drift Expand(DEX) mode by applying electric suppression field in beamline region creates relatively bigger beam size than standard drift mode.

Figure 3 shows TW measurement result. VIISion80 had lowest as-is TW value compared to any implant condition of VIISta HC. Probably longer implant time leads to self anneal effect and consequently implant damage relaxed. DEX2 mode on VHC looks effective to decrease TW unit closing VIISion80. In case of reduced beam current at 1/6 on VIISta HC, as-implanted TW is much closer than DEX mode. These wafers were annealed all together and checked with TW again to compare residual damage. As shown in Figure 3, totally opposite trend from high to low of as-implanted TW. The reason is fast dose rate implantation creates thick amorphous layer(higher TW value) and less residual dislocations(lower TW) at the end of range after anneal.

Figure 4 shows Boron and Fluorine diffusion behavior. The higher TW value is corresponding to higher Boron and Fluorine segregation at the end of range.

To verify Boron activation in segregation layer, SRP is shown in Figure 5. It is clear that Boron atoms

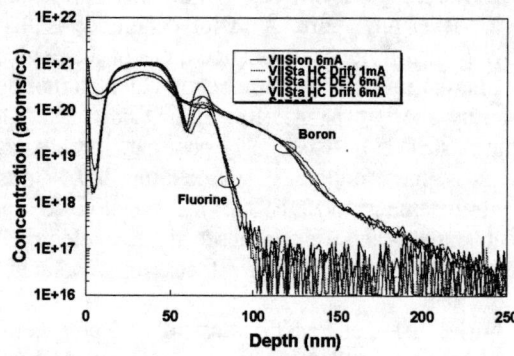

FIGURE 4. SIMS profile of BF_2 40keV 3E15at/cm^{-2} tilt 0° twist 45° after anneal

in higher than 2.0E20 atoms/cc concentration were not activated. However, junction depth of low dose rate was slightly deeper than high dose rate case.

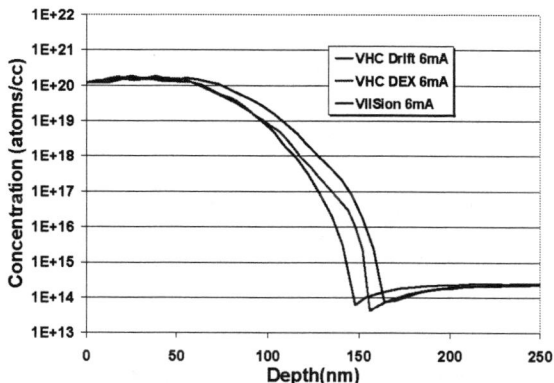

FIGURE 5. SRP of BF$_2$ 40keV 3E15at/cm^{-2} tilt 0° twist 45°

Based on fundamental experiment results, 70nm technology Flash memory device was implanted on VHC with DEX2 mode for P+S/D. Table 1. is summary of key device electrical parameters. Originally, ribbon beam single tool showed 7% shift of drive current at same contact Rs. By implementing DEX mode, PMOS Idsat was recovered to target level without compromising contact Rs.

TABLE 1. Buried Channel PMOS Electrical Data

Implant Conditions	Contact Rs (a.u.)	Idsat (a.u.)
VIISion80 Drift 6mA	100	100
VIISta HC Drift 6mA	100	93
VIISta HC DEX2 6mA	101	101

In order to verify drive current gain by DEX2 mode, lateral diffusion into channel area was examined by 2-D profiling on the device. Figure 6 shows SCM pictures for device split test. It proved that effective channel length by DEX2 was well matched to VIISion80.

CONCLUSIONS

Buried PMOS device matching by BF2 P+S/D implantation was investigated. DEX mode on VIISta HC ribbon beam single high current implanter was tried to mimic low dose rate of spot beam batch tool. TW value and dopant profile from DEX2 is approaching to spot beam batch tool tendency. After all, key device parameters were well-matched to target of spot beam batch tool without other process modification such as Vt control or Halo processes etc.

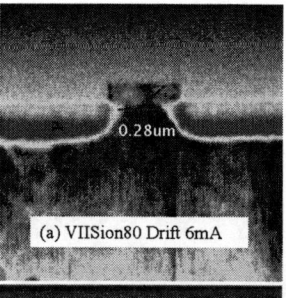
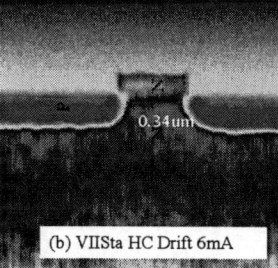

FIGURE 6. Scanning Capacitance Microscope(SCM) on Buried channel PMOS implanted by (a) VIISion80 Drift mode (b) VIISta HC Drift mode (c) VIIsta HC DEX2 at same beam current.

ACKNOWLEDGMENTS

The authors thank VSEA demonstration team for running implants and Genise Bornacorsi for arranging specific analyses.

REFERENCES

1. U. Jeong et al. "Devices dictate control of implant-beam incident angle", Solid State Technology, Oct. 2001
2. Dirk-Wito Franke, "Tool matching by using single wafer implanter Varian VIISta80 and batch implanter Axcelis HC3 for high current applications, IIT 2004
3. Jinning Liu, "Fluorine Effect on Boron Diffusion: Chemical or Damage ?", IIT 1998
4. Jinning Liu, " A damage and diffusion study on ribbon beam versus spot beam system", VSEA internal technical report.

Germanium ion implantation to Improve Crystallinity during Solid Phase Epitaxy and the effect of AMU Contamination

K. S. Lee[*], D. H. Yoo, G. H. Son, C. H. Lee, J. H. Noh, J. J. Han, Y. S.Yu,
Y. W. Hyung, J. K. Yang[a], D. G. Song[b], T. J. Lim[c], Y. K. Kim, S. C. Lee,
H. D. Lee, J. T. Moon

Process Development Team, [a]Process Analysis & Control Group, [b]Samsung Austin Semiconductor, [c]Manufacturing Technology Team 2, Samsung Electronics Co.,LTD., San #24 Nongseo-Dong, Giheung-Gu, Yongin-City, Gyeonggi-Do,446-711, Korea. +82-31-209-3037;fax:+82-31-209-3319;e-mail:daehan.yoo@samsung.com

Abstract. Germanium ion implantation was investigated for crystallinity enhancement during solid phase epitaxial regrowth (SPE) using high current implantation equipment. Electron back-scatter diffraction(EBSD) measurement showed numerical increase of 19 percent of <100> signal, which might be due to pre-amorphization effect on silicon layer deposited by LPCVD process with germanium ion implantation. On the other hand, electrical property such as off-leakage current of NMOS transistor degraded in specific regions of wafers, which implied non-uniform distribution of donor-type impurities into channel area. It was confirmed that arsenic atoms were incorporated into silicon layer during germanium ion implantation. Since the equipment for germanium pre-amorphization implantation(PAI) was using several source gases such as BF_3 and AsH_3, atomic mass unit(AMU) contamination during PAI of germanium with AMU 74 caused the incorporation of arsenic with AMU 75 which resided in arc-chamber and other parts of the equipment. It was effective to use germanium isotope of AMU 72 to suppress AMU contamination, however it led serious reduction of productivity because of decrease in beam current by 30 percent as known to be difference in isotope abundance. It was effective to use enriched germanium source gas with AMU 72 in order to improve productivity. Spatial distribution of arsenic impurities in wafers was closely related to hardware configuration of ion implantation equipment.

Keywords: Pre-Amorphization Implantation(PAI), Solid Phase Epitaxial, AMU contamination, Enriched ^{72}Ge
PACS: 81.15.Np, 61.72.Tt

INTRODUCTION

As scaling of CMOS faces physical limit, three-dimensional (3D) integration of silicon devices has attracted intensive attention world-wide. Silicon-on-insulator (SOI) with lateral solid phase epitaxy (L-SPE) is one of the candidate structures in achieving future 3D-device. Single crystalline seed window is defined in the edge or middle of patterns, and amorphous silicon (a-Si) layer is deposited and finally appropriate thermal annealing is followed in order to achieve epitaxial re-alignment of silicon atoms coherent to seed region. Wafer-bonding technology is also under investigation. Practical convenience for realization of process, researchers have reported on important L-SPE subjects ; the role of polycrystalline in initial amorphous silicon layer, the influence of SiO_2, orientation dependence of the L-SPE on seed layer, and impurity effect[1-4].

Ion implantation has been usefully introduced in order to enhance crystallinity since it can break the native oxide layer at the original a/c interface and remove micro-crystalline seeds in a-Si which induce random crystallization[5-6]. In this paper, germanium, isoelectric element of silicon, ion implantation was investigated for crystallinity enhancement confirmed with EBSD measurement, during solid phase epitaxial regrowth. Its relatively high mass is useful for creating non-buried amorphous layers as thin as a few nm through as thick as several hundred nm with less end of range damage than many other implant species[7]. In this paper, the dependence of the crystallinity of SPE layer on germanium ion implantation condition is studied with EBSD measurement. Electrical characteristics are investigated as well.

EXPERIMENTAL

P-type Si (100) wafers were used and 250nm of high density plasma CVD oxide was deposited. Windows were patterned as a hole type with diameter of 100nm. Selective epitaxial growth through windows was conducted and a-Si layer was deposited by LPCVD to be 50nm thick. The ^{74}Ge ions were implanted with various doses and energies in order to amorphize a-Si, followed by thermal annealing for solid phase epitaxial regrowth in <100> direction. The crystallinities of SPE layer with various Ge PAI splits were quantified by EBSD, which was able to represent the fraction of <100> in channel silicon layer with SPE. The module for EBSD measurement was equipped with conventional SEM, and EBSD value was extracted at confidence index of 0.2. Measured area was 2 x 5 um^2 and 5 samples were provided in a wafer. NMOS transistors were realized on abovementioned 3D-structure using conventional process in Figure 1. Boron was implanted for channel dopants. SIMS measurement was fulfilled to analyze the distribution of implanted atoms in SPE-treated silicon layer.

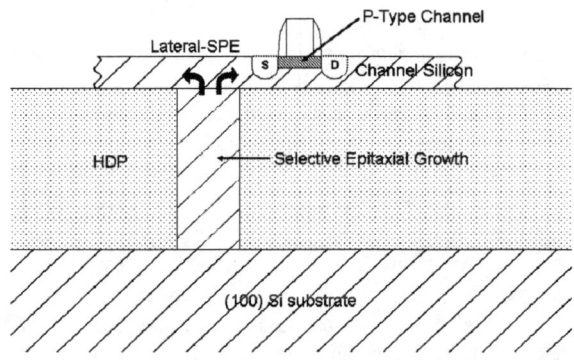

FIGURE 1. Schematic diagram of a NMOS transistor on SOI structure by lateral solid phase epitaxy.

RESULTS AND DISCUSSION

Ge PAI effect for SPE

Table 1 shows ^{74}Ge PAI split conditions used in this paper. With the same dose of 2.0E14/cm2, it is shown that EBSD value to quantify crystallinity of SPE was improved until 40keV, but it began to decrease with further increase of energy. It is also obvious that a certain amount of germanium was required for optimum condition. For the energy of 40keV, EBSD value increased as dose increased to be 2.0~5.0E14/cm2, above the dose, however it began to decrease. For 60keV, similar tendency was observed, optimum dose was to be around 2.0E14/cm2. Dose dependency of crystal quality is thought to be due to amorphization efficiency. However, excess amount of germanium in a-Si layer can cause lattice distortion of SPE layer or prevent silicon atoms in a-Si layer from realigning to epitaxial position toward crystalline seed, which is reflected to the reduction of EBSD value. Energy dependency of crytallinity can be interpreted with dopant stopping phenomena. For high energy, physical damage is dominated by electron stopping, which causes a little damage in silicon lattices. On the other hand, physical damage for low energy is influenced by nuclear stopping, which gives much damage to silicon lattices[8]. It implies that the effect of Ge implantation to knock out micro-crystalline in a-Si layer is more effective at lower energy range. However, reduced Rp with lower energy is insufficient to sweep away micro-crystalline seeds in whole a-Si layer, as a result, crystallinity of SPE layer was not improved for 20keV. Optimal condition was obtained at 40keV, 2.0~5.0E14/cm2 for the crystallinity enhancement by 19% for 50nm a-Si layer with SPE process.

Electrical characteristics

While SPE-treated silicon layer using ^{74}Ge ion implantation showed better crystallinity, electrical property such as off-leakage current was observed to be abnormally increased after fabricating typical NMOS transistor with conventional process. It is noticeable that off-leakage current of NMOS transistor was degraded in specific region, that is, left side of wafers, as shown in Figure 2(a). SIMS analysis revealed that ^{75}As atoms were incorporated into silicon layer during the implantation of ^{74}Ge ions.

TABLE 1. Ge PAI split conditions and EBSD results : the dark region indicates the crystalline in <100> direction.

ENERGY (keV)	20	40				60			SKIP
DOSE (Ions/cm2)	2.0E14	5.0E13	2.0E14	5.0E14	1.0E15	5.0E13	2.0E14	5.0E14	
Crystallinity (%)	41.1	38.1	54.1	57	48.1	37.5	44.7	39.4	39.2
EBSD IMAGE									

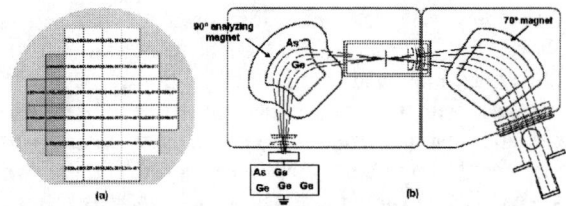

FIGURE 2. Off-leakage current wafer map ; wafer was loaded at 180° rotation. (a), Hardware configuration for implanting Ge and As. (b).

Figure 3 shows that ^{75}As doping level in left region of wafer was higher than that of center region which was supposed to be noise level. On the other hand, ^{74}Ge concentrations for two regions were similar level. It is reasonable that AMU contamination between ^{75}As and ^{74}Ge occurred during ^{74}Ge ion implantation in p-type channel silicon. AMU contamination and spatial distribution of ^{75}As in wafers turned out to be closely related to the hardware configuration of the implantation equipment and mass resolution. The equipment used ribbon-like wide beam and arc-chamber was shared with other gas sources such as BF3 and AsH3. As shown in Figure 2(b), poor mass resolution during ^{74}Ge PAI is able to cause the incorporation of ^{75}As which resided in arc-chamber.

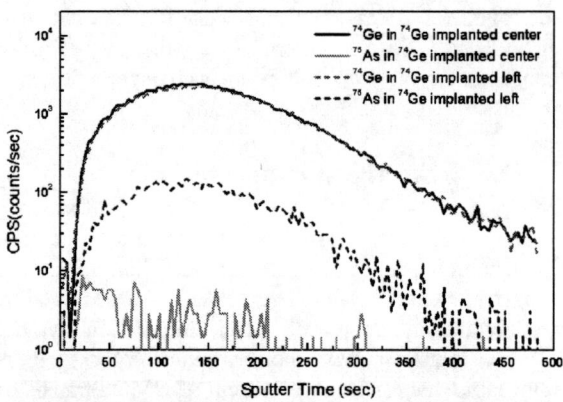

FIGURE 3. SIMS profiles at wafer of figure 2. ; Ge, 60keV, 1.0E15/cm^2.

After extracted to ribbon-like beam from arc-chamber, incorporated ^{75}As ions in ^{74}Ge beam tends to bend less than ^{74}Ge in 90 degree analyzing magnet because of heavier mass, so that ^{75}As ions are more abundant in outer region. Then, the beam is focused at the resolving aperture where the position of ^{75}As-rich region within the beam is converted to opposite direction. Finally ^{75}As ions are implanted in left region of a wafer, while ^{74}Ge ions are relatively uniform within a wafer.

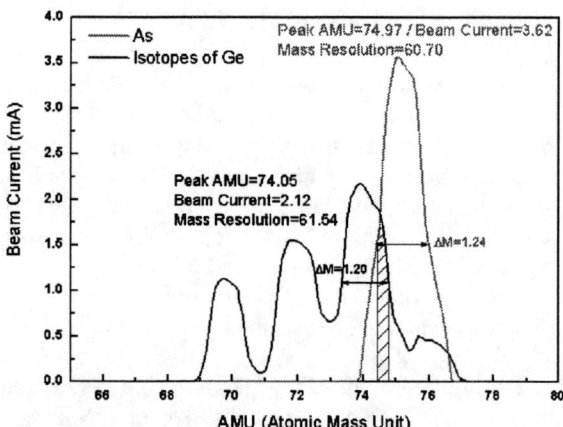

FIGURE 4. The As and Ge mass spectrums used to calculate mass resolution as the ratio of M/ΔM, where M is the AMU at the peak beam current and ΔM is the spread of the distribution at 1/2 of the peak beam current.

Mass spectrum for Ge with As was compared in Figure 4. When Δ M for both ^{74}Ge and ^{75}As was calculated by mass resolution of about 60 degree which the equipment guaranteed at minimum level, they overlapped from 74.28 AMU to 74.61 AMU in dash region of Figure 4. It was impossible to separate ^{75}As from ^{74}Ge and much higher mass resolution was demanded to prevent such AMU contamination.

After running ^{75}As implantation for 30minutes to coat the ion source with ^{75}As deposits, BF$_2$ flushing was used for 1 hr to eliminate ^{75}As residue in arc-chamber. A fluorinated gas such BF$_3$ is known to have such etching effect as to eliminate ^{75}As residue deposited in the chamber wall. However, Figure 5 shows that BF$_3$ gas was insufficient to clean up ^{75}As residue. It is likely that several hours of BF3 flushing is required to remove ^{75}As contamination, which leads serious reduction of productivity.

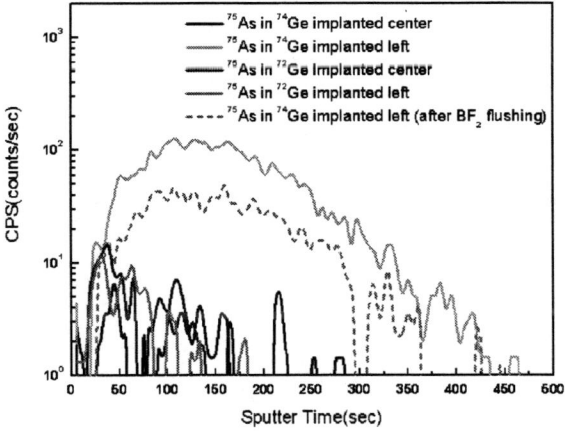

FIGURE 5. SIMS profiles of various AMU cross contamination suppression methods ; implanting ^{74}Ge, ^{72}Ge, ^{74}Ge after BF$_2$ flushing ;Ge, 50keV, 1.0E15/cm^2

^{72}Ge was chosen to improve the mass resolution for ^{72}Ge and ^{75}As. Because of AMU difference between ^{72}Ge and ^{75}As, ^{75}As could be separated from ^{72}Ge by the mass resolution of 60 as Figure 5 indicated. Unfortunately, the beam current for ^{72}Ge decreased from 1.89 to 1.33mA when compared with ^{74}Ge, as known to be difference in isotope abundance. It also led serious reduction of productivity by 30 percent, consequently. Isotope abundance of of Ge in natural GeF$_4$ and enriched ^{72}GeF$_4$ source gas is depicted in the Table 2.

TABLE 2. Isotopic abundance for natural GeF$_4$ and enriched ^{72}GeF$_4$ gas.

AMU	Natural gas	Enriched gas
70	21.1	16
72	27.3	50
73	7.9	10
74	37.1	20
76	6.5	4

Because of ^{72}Ge abundance, it was applied to use enriched ^{72}GeF$_4$ source gas in order to improve productivity, which could increase by 80 percent due to beam current increasing from 1.33 to 2.39mA. SIMS profile was analyzed to identify chemical distribution for natural GeF$_4$ and enriched ^{72}GeF$_4$. Figure 6 shows that enriched ^{72}GeF$_4$ was in a good agreement with natural GeF$_4$. Off-current of NMOS transistors with ^{72}GeF$_4$ is compared in Figure 7, which shows that no spatial degration was observed. Other electrical properties of NMOS transistor such as on-current, threshold voltage and swing were confirmed to be identical, which are not shown in this paper. The on-current for enriched ^{72}GeF$_4$ gas was identical or slightly superior to that for natural GeF$_4$ gas. EBSD value of SPE-treated silicon layer with 40keV, 5.0E14/cm2, enriched ^{72}Ge was 57% which is in the range of measurement tolerance.

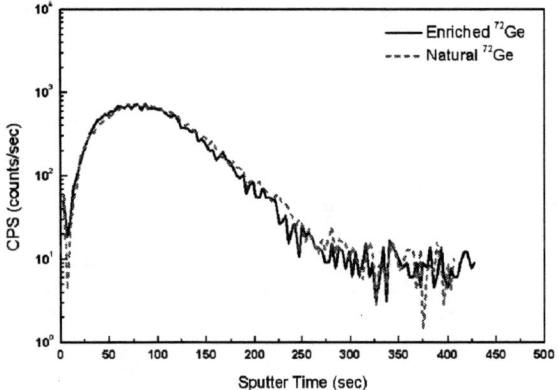

FIGURE 6. SIMS profiles for natural and enriched ^{72}Ge. ; Ge, 40keV, 5.0E14/cm2

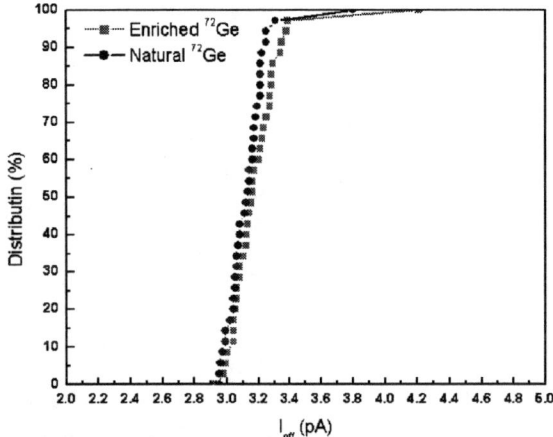

FIGURE 7. Distribution of off-current for natural and enriched ^{72}GeF$_4$ implanted NMOS transistor.

CONCLUSION

The role of Ge PAI during SPE was studied in 3D structure in order to improve crystallinity of channel silicon layer. For a-Si layer of 50nm, the maximum EBSD value 57% was obtained with ^{74}Ge, 40keV, 5.0E14/cm2.

It was confirmed that the incorporation of ^{75}As in P-type channel Si layer during ^{74}Ge PAI caused the off-leakage current increase of NMOS transistor in left regions of wafers. Such distribution of ^{75}As was closely related to hardware configuration. To prevent cross contamination between ^{74}Ge and ^{75}As, it was chosen to apply enriched ^{72}GeF$_4$, which demonstrated the superior improvement of electrical properties and productivity as well.

REFERENCES

1. E. Murakami *et al.*, *J. Appl. Phys.* **63**, 4975-4978 (1988)
2. H. Yamamoto *et al.*, *Jpn. J. Appl. Phys.* **25**, 667-672 (1986)
3. M. Sasaki *et al.*, *Appl. Phys. Lett.* **49**, 397-399 (1986)
4. H.Yamamoto *et al.*, *Appl. Phys. Lett.* **46**, 268-270 (1985)
5. M. von Allmen *et al.*, *Appl. Phys. Lett.* **35** 280-282 (1979).
6. L. Csepregi *et al.*, *Appl. Phys. Lett.* **29**, 92-93 (1976).
7. M. Minondo, J. Boussey, G. Kamarinos, H.C. de Graaff. H. Van Kranenburg, *"Preamorphization induced defects and their effects on the electrical properties of shallow p+/n junction,"* Proceedings of the 25th European Solid State Device Research Conference. 1995, pp. 370-382.
8. J. F. Ziegler, "The Stopping and Range of Ions in Solid." in *Ion Implantation Science and Technology*, edited by J. F. Ziegler, Academic Press, 1984.

CLUSTER IMPLANTATION
AND DOPING

20 Years History of Fundamental Research on Gas Cluster Ion Beams, and Current Status of the Applications to Industry

Isao Yamada*

Laboratory of Advanced Science and Technology, University of Hyogo
3-1-2 Kouto, Kamigori, Hyogo, 678-1205, Japan

Abstract. This paper reviews the development of gas cluster ion beam (GCIB) technology, including the generation of cluster beams, fundamental characteristics of cluster ion to solid surface interactions, emerging industrial applications, and identification of some of the significant events which occurred as the technology has evolved into what it is today. More than 20 years have passed since the author first began to explore feasibility of processing by gas cluster ion beams at the Ion Beam Engineering Experimental Laboratory of Kyoto University. Processes employing ions of gaseous material clusters comprised of a few hundred to many thousand atoms are now being developed into a new field of ion beam technology. Cluster-surface collisions produce important non-linear effects which are being applied to shallow junction formation, to etching and smoothing of semiconductors, metals, and dielectrics, to assisted formation of thin films with nano-scale accuracy, and to other surface modification applications.

Keywords: Cluster ion beams, Shallow implantation, Lateral sputtering, Surface smoothing,
PACS: 61.72T;79.20R;81.15J;3640

HISTORICAL MILESTONES IN GCIB TECHNOLOGY

In 1950, Becker et al first studied cluster formation for thermonuclear fuel applications using gaseous materials passed through supersonic nozzles [1]. The supersonic expansion approach was successful in producing cryogenic beams containing large numbers of clusters. This original work opened the way to employ gas clusters for materials processing.

During the late 1970's and 1980's, an ionized cluster beam (ICB) technique which employed metal vapor clusters from heated Knudsen cells for thin film formation was studied at Kyoto University and elsewhere. Kyoto University investigations of metal vapor clusters ended when collaborative work with W.L.Brown at Bell Laboratories showed the cluster ion intensities within the metal vapor streams to be too low for most practical purposes [2,3]. Subsequent work at the Kyoto University Ion Beam Engineering Experimental Laboratory then focussed upon cluster beam formation employing gas expansion through simple supersonic nozzles.

Initial research on gas cluster beam formation showed that supersonic nozzles having converging-diverging shapes operating at room temperature could produce intense beams of gas clusters. This then led to research and development of gas cluster ion beam (GCIB) techniques [4] and to investigations of new ion-solid interactions produced by gas cluster ion impacts. These studies demonstrated that GCIB produces unique ion/solid interactions and offers new atomic and molecular ion beam process opportunities in areas of implantation, sputtering, and ion beam assisted deposition. Most of the original technical results through to the year 2000 have been summarized in a monograph. [5].

Over the first 10 years of GCIB studies, low energy surface interaction effects, lateral sputtering phenomena and high chemical reaction effects were observed experimentally and were explained by means of molecular dynamics (MD) modeling. Japanese government funding through JST (Japan Science and Technology), MEXT (Agency, Ministry of Education), METI (Ministry of Economy, Trade and Industry) and others provided long term support for the research at Kyoto University. Difficulty in developing GCIB equipment within Japan resulted in development of commercial GCIB equipment by Epion Corporation in the U.S. beginning in 1995 [6].

In 2000, a four year R&D project for development of GCIB industrial technology began in Japan under funding from the New Energy and Industrial Technology Development Organization (NEDO). This project involved subjects in areas of semiconductor surface processing, high accuracy surface processing and high-quality film formation. The project was supported by the formation of a new Collaborative Research Center of Cluster Ion Beam Technology at Kyoto University and University of Hyogo.

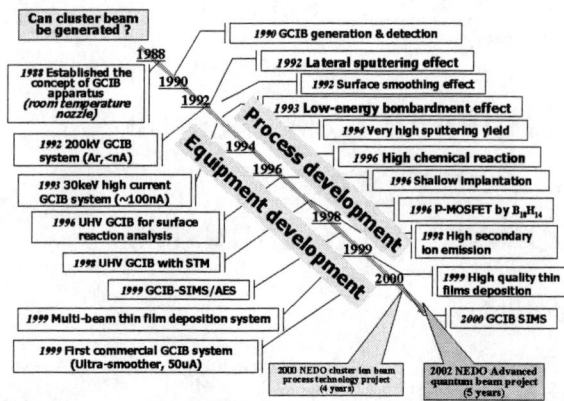

FIGURE 1. Historical milestones of GCIB equipment and process development.

In 2002, another major GCIB project which emphasized nano-technology applications was started under a contract from the Ministry of Economy and Technology for Industry (METI). This METI project currently involves development related to size-selected cluster ion beam equipment and processes, and development of GCIB processes for very high rate etching and for zero damage etching of magnetic materials and compound semiconductor materials.

Figure 1 shows historical milestones of GCIB equipment and process development.

GAS CLUSTER FORMATION AND GCIB EQUIPMENT

GCIB processing of materials is based on the use of electrically charged cluster ions consisting of a few hundred to a few thousand atoms or molecules of gaseous materials. A beam of neutral clusters is first formed from individual gas atoms by expansion of the gas through a nozzle at room temperature into vacuum. The clusters are subsequently ionized and accelerated.

The typical configuration of GCIB equipment is as shown in Figure 2. A small aperture, or skimmer, transmits the primary jet core of gas clusters emerging from the expansion nozzle. Forward-directed neutral clusters are then ionized by impact of electrons

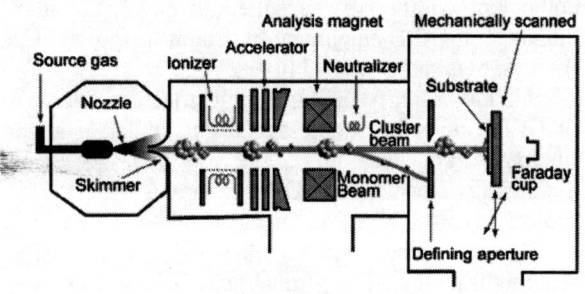

FIGURE 2. Typical configuration of GCIB equipment

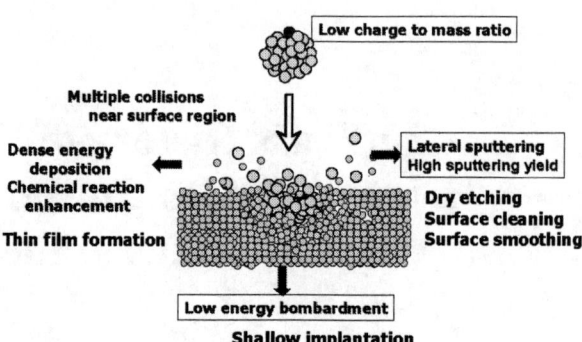

FIGURE 3. GCIB-solid surface interactions and their applications

accelerated from a filament so as to form positive ion gas clusters with nominally one charge per cluster. The ionized clusters are extracted and accelerated through typical potentials of between and 2 and 30 kV using a series of electrodes. Electrostatic lenses are utilized to focus the cluster ions, and monomers are filtered out by means of a strong transverse magnetic field. Usually the cluster ion beam is kept stationary and material to be processed is scanned mechanically through the beam so as to obtain uniform and complete coverage. The cluster ion fluence is measured by means of a Faraday cup.

When an energetic cluster ion impacts upon a surface, it interacts nearly simultaneously with many target atoms and deposits high energy density into a very small volume of the target material. The concurrent energetic interactions between many atoms comprising the cluster and many atoms of the target result in highly non-linear implantation and sputtering effects [7]. These effects, which are fundamentally different from those associated with the more simple binary collisions occurring during monomer ion impacts, include low energy bombardment phenomena, lateral sputtering effects and high chemical reaction effects. Figure 3 summarizes GCIB-solid surface interactions and their applications.

In early 1988, cluster formation was confirmed by electron diffraction. A strong Debye-Scherrer ring, as shown in Figure 4(a), was observed when electrons were diffracted through the beam emerging from a gas nozzle. Time of flight (TOF) measurements were subsequently used to show that the clusters contained several hundred to many thousand atoms [8]. The clusters were believed to be held together by van der Waals forces.

Figure 4(b) shows typical size distributions found for Ar gas clusters. Clusters of the sizes which are normally produced by room temperature nozzles were found to be particularly useful for materials processing. More recent experiments conducted with size-selected cluster beams have confirmed the fortuitous nature of the cluster size distributions which are typically

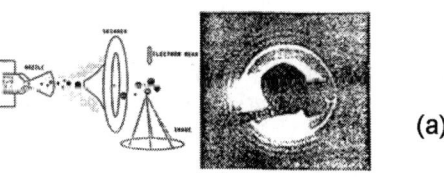

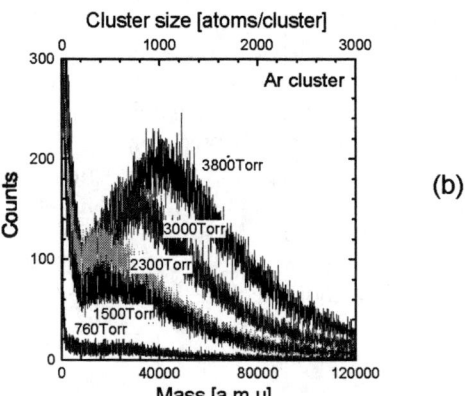

FIGURE 4. Cluster detection by e-diffraction (a) and TOF (b).

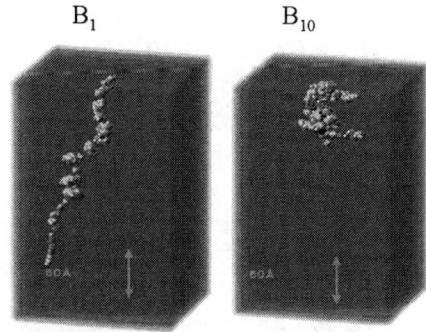

FIGURE 5. MD simulations of B monomer and B_{10} cluster ion implantations into Si at 5 keV

produced [9]. It has been found that clusters can be formed from nearly all gases and gas mixtures, including rare gases such as Ar and Xe, most diatomic gases such as O_2 and N_2, and molecular compound gases such as B_2H_6, BF_3, CH_4, NF_3, SF_6, etc.

One of the advantages associated with the cluster ion is a very low charge to mass ratio. Cluster ions containing up to several thousands of atoms typically become only singly or doubly ionized. Consequently, a cluster ion beam at any given current density can transport up to thousands of times more atoms than a monomer ion beam at the same current density. For example, a 1µA beam of cluster ions with average size of 1000 atoms per cluster can transport the same number of atoms as a 1mA monomer ion beam. These characteristics make it possible to use GCIB for very low energy ion beam processes which are normally difficult by traditional ion beam technology. Available GCIB equipment is now able to produce cluster ion beam currents of hundreds of microamperes or more from gases such as Ar [6].

LOW ENERGY EFFECTS AND SHALLOW JUNCTION FORMATION

From the beginning of cluster ion beam investigations, it was expected that clusters should produce low energy bombardment effects since the kinetic energy of each atom in a cluster ion is roughly equal to the total energy of the cluster divided by the number of atoms contained in the cluster. As an example, within a 20 keV cluster ion consisting of 2000 atoms, each of the individual atoms has energy of only 10 eV. Low energy effects were predicted by MD simulations and were confirmed by experiments, for example by comparing B monomer ion and B cluster ion implantation [10]. The results showed low-energy individual atomic interactions even when the total energy of the clusters was high. Figure 5 shows MD simulations of B monomer and cluster ions into Si illustrating the low energy effect of cluster ion bombardment relative to monomer ion bombardment. Important differences in range and density of the displacements produced are apparent. In the cluster ion case, the penetration range is extremely shallow and the displacements that are produced remain tightly concentrated within the impact region at the target surface.

While, due to space charge effects, it is exceptionally difficult to transport monomer ion beams at energies as low as 10 eV, equivalently low energy ion beams can be realized by using cluster ion beams at high acceleration voltages. The standard configuration of GCIB equipment inherently results in highly directional parallel cluster ion beams which are extremely well suited for ultra shallow doping applications and are known to produce thin film transistors which exhibit operational characteristics superior to those which result when conventional ion implantation is used for the shallow doping [11].

POLYATOMIC CLUSTER IMPLANTATION

The concept of performing polyatomic cluster implantation into Si by using decaborane ($B_{10}H_{14}$) molecules was first investigated in 1996 by Kyoto University researchers working together with researchers at Fujitsu [12]. Effects upon range and damage distributions resulting from implantation of different sizes of B polyatomic clusters have been studied by MD simulations employing B_1, B_2, B_5, B_{10}, B_{18} molecular models shown in Figure 6(a) [13]. The results of these simulations indicate that cluster-like

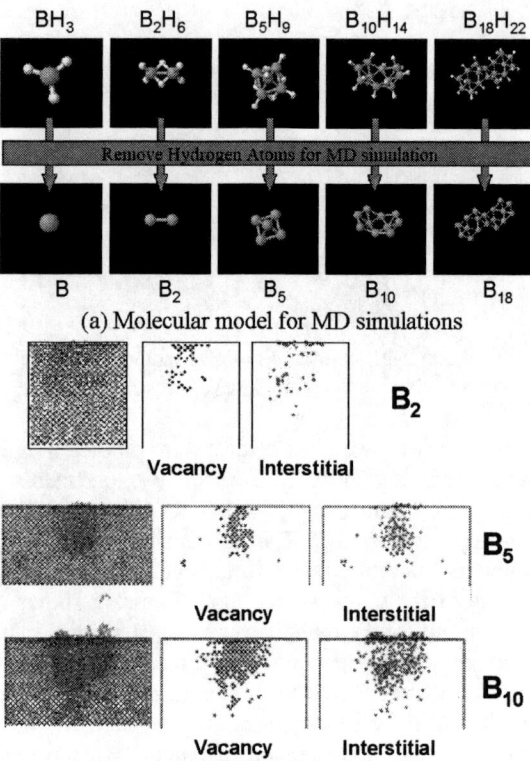

FIGURE 6. MD simulation models of B clusters (a) and typical implant damage distributions for B_2, B_5 and B_{10} ions (acceleration energy: 500eV/atom)

bombardment phenomena, the nonlinear effects which are typical of cluster impact, begin to be observed with clusters containing at least five or more atoms. Figure 6(b) shows the implant damage distributions resulting from the model clusters of 2 atoms, 5 atoms and 10 atoms. From MD simulations of bombardment by even larger clusters, it has been shown that the displacement damage increases with increasing cluster size and very clusters, as in the case of GCIB, cause complete self amorphization [5].

P-MOSFETs with 40nm gates, as in the device shown in Figure 7, were successfully fabricated by Fujitsu in 1996 using $B_{10}H_{14}$ implantation for ultra shallow junction and source/drain formation [12]. Decaborane implantation at 30 keV to a dose of 1×10^{13} ions/cm^2 followed by annealing at 1000°C for 10 seconds was reported to have resulted in a junction depth of 20 nm. For source/drain extensions, $B_{10}H_{14}$ ion implantation at 2 keV was carried out to a dose of 1×10^{12} ions/cm^2 followed by annealing at 900°C for 10 seconds.

Under a 2002 contract from JST (Japan Science and Technology Agency) Nissin Ion Corporation has successfully developed equipment for $B_{10}H_{14}$ ion implantation. A beam current of 3 mA at an acceleration energy of 3 keV has been achieved [14]. Using this equipment, Nissin and Fujitsu have worked to advance the decaborane process technology. Recently they have reported devices with significantly reduced threshold voltage deviation and higher forward currents than similar devices made by conventional B implantation [11]. Other polyatomic cluster implantation, such as that using $B_{18}H_{22}$, has also been developed because of the decaborane success [15].

CLUSTER SIZE EFFECTS AND LOW DAMAGE PROCESSES

The influence of cluster size upon surface damage production has been studied experimentally using GCIB apparatus which incorporates a strong permanent sector magnet for cluster size selection [9]. Displacement damage within the Si surfaces has been evaluated by ellipsometry using a two-layer analysis model which assumes that oxide and amorphous layers are formed by the GCIB bombardment. From ellipsometry measurements, the intensity ratio Ψ and phase difference Δ of p and s waves can be determined. An increase in amorphous layer thickness is indicated by an increase of Ψ and an increase in the oxide layer thickness is indicated by a decrease of Δ. From the Ψ and Δ behaviors, estimates can be made of the damage formation due to Ar-GCIB bombardment.

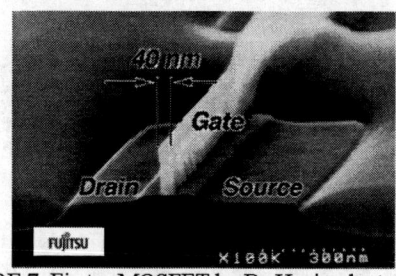

FIGURE 7. First p-MOSFET by $B_{10}H_{14}$ implantation

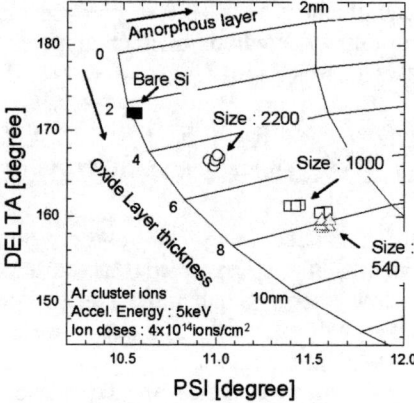

FIGURE 8. Cluster size dependence of Ψ and Δ after 5keV Ar-GCIB irradiation with ion dose of 4×10^{14} ions/cm^2.

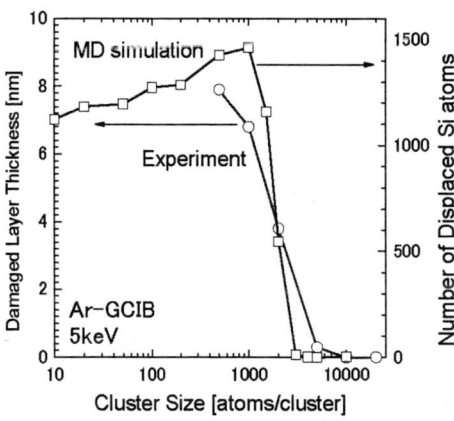

FIGURE 9. Cluster size dependence of number of displaced Si atoms by MD simulations and damage layer thickness

Studies have been made of Si surfaces bombarded by size-selected 5 keV Ar cluster ion beams to dose levels of 4×10^{14} ions/cm^2. Mean cluster sizes utilized were 540, 1000 and 2200 atoms per cluster, resulting in average energies of 9.2, 5.0 and 2.3 eV per atom respectively. At the fixed 5 keV beam energy, both the oxide thickness and the amorphous layer thickness were found to decrease monotonically with increasing cluster size, ie., with decreasing energy of the cluster atoms. Figure 8 shows Ψ and Δ plots after Ar-GCIB irradiation of the Si.

Figure 9 shows plots of estimated damaged layer thickness and total displacements versus cluster size as determined experimentally and also by MD simulations [9]. From the MD simulations, the number of Si atoms displaced from their lattice cites increased with Ar cluster size from 10 to 1000, showed a peak at around 1000 atoms/cluster and then rapidly decreased with further increase of cluster size. At cluster sizes above 4000 atoms per cluster, MD simulations showed almost no displaced atoms after Ar-GCIB irradiation. The experimental results have shown almost the same trend as the MD simulations. These results suggest that shallow implantation and doping can be possible by using large cluster ions. From MD simulations, it is predicted that there is no damage formation in Si when the energy per atom in Ar cluster ion is below 1eV even though large cluster ions having such conditions can still deposit approximately 30% of their acceleration energy into a target Si substrate.

LATERAL SPUTTERING AND ATOMIC LEVEL SMOOTHING

An important characteristic of large gas-cluster ion bombardment is an effect known as lateral sputtering. Angular distributions of surface atoms ejected by cluster ions are considerably different from the distributions produced by monomer ions. Figure 10 shows experimentally measured angular distributions of sputtered atoms by Ar_{2000} cluster ions at (a) normal incidence and (b) oblique incidence. The angular distribution produced by monomer ions, which indicates the usual cosine distribution, is also shown. The angular distribution of sputtered atoms produced by the Ar cluster ions illustrates the lateral ejection. [8].

Sputtering yields due to cluster ions are very high relative to those associated with monomer ions at similar energy. Very high sputtering yields on metal, semiconductor and insulator surfaces due to bombardment with cluster ions have been observed experimentally. Experimentally measured sputtering yields of various materials due to 20 keV Ar (physical) and SF_6 (chemically reactive) cluster ions and monomer ions are shown in Figure 11.

Lateral sputtering produces surface smoothing behavior which does not occur with monomer ions.

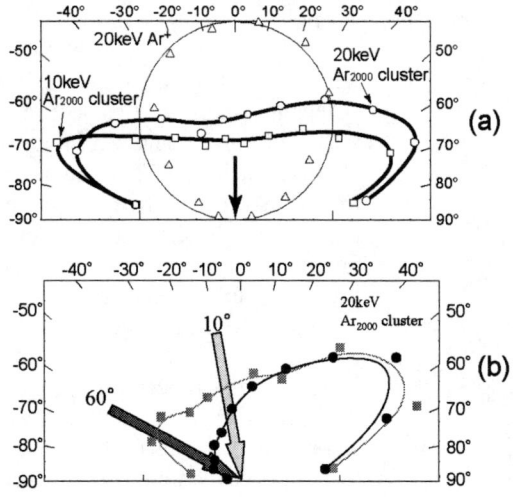

FIGURE 10. Angular distribution of sputtered atoms by Ar monomer and Ar cluster ions. (a) normal incidence (b) oblique incidence.

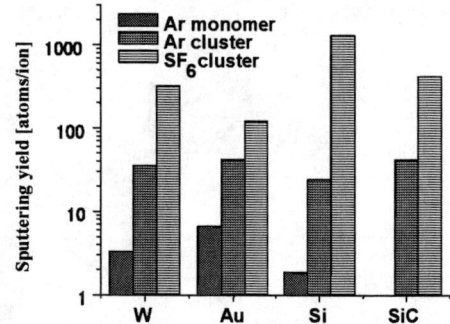

FIGURE 11. Reactive and physical sputtering for various materials at 20keV acceleration energy.

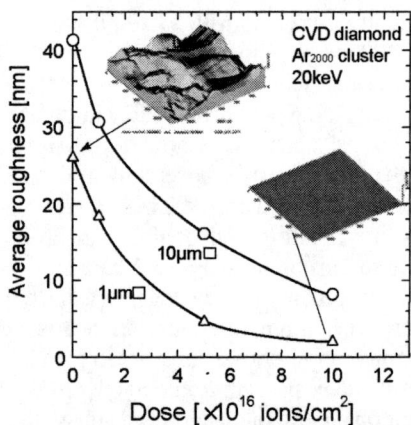

FIGURE 12. Ion dose dependence of the average roughness of a CVD diamond surface bombarded with a 20keV Ar cluster ion beam at normal incidence

Smoothing of surfaces to atomic levels has been the first production use for cluster ion beam processing. Figure 12 shows typical ion dose dependence of the average roughness of a CVD diamond surface bombarded with a 20 keV Ar cluster ion beam at the normal incidence. The average roughness decreased monotonically with increasing ion dose from the initial value of 26 nm to a value of 1.3 nm after an ion dose of 8×10^{16} ions/cm^2. In the case of monomer ion irradiation at normal incidence, surface roughness typically increases with ion dose due to erosion or bubble formation inside the target.

GCIB APPLICATIONS IN INDUSTRY

Epion Corporation in the U.S. has developed GCIB equipment for industrial applications under a license from Japan Science and Technology Agency (JST). As is shown in Figure 13, concurrent with the development of increasingly capable commercial GCIB equipment has been the development of applications for GCIB process technology in the manufacturing of semiconductor devices and other advanced technical devices [16].

A. GCIB applications in semiconductors

As is suggested in Figure 14, a number of candidate applications are being developed for GCIB in the manufacturing of coming generations of semiconductor devices. These applications include ultra shallow junction doping, SiGe alloy formation, film deposition, silicon-on-insulator thinning and uniformity correction, etching of dielectrics, ashing of photoresist, and surface modification of metals and dielectrics for integration of improved Cu interconnects.

GCIB offers excellent characteristics for producing ultra shallow junctions with pre-activation depths ranging from less than 10 nm to a maximum of approximately 30 nm. The mechanism of doping by GCIB, which is referred to as "infusion", depends upon the intense localized temperature/pressure transient which is created at the point of cluster impact. During the moment of impact, molecular gases such as B_2H_6 contained within the cluster undergo dissociation and solid species such as B which they contain then undergo mixing into the target within the region of the thermal transient while gaseous atoms escape from the surface. Unlike ion implantation, the junction depths resulting by GCIB infusion depend upon the 1/3rd power of the cluster ion energy. Although the cluster beams used for ultra shallow doping are formed using dilute concentrations of dopant gas within an inert carrier gas host, for example 1% B_2H_6 in 99% Ar, the GCIB infusion processes are very efficient and throughputs are very high.

GCIB processes employing reactive gases such as halogen compounds can be used to produce very uniform and reproducible chemical etching. An optional mode of operation of GCIB equipment allows very precise but intentionally nonuniform "corrective etching" processes to be performed. One example application which has been demonstrated for this controlled nonuniniform processing capability of GCIB is in corrective etching and smoothing to reduce the thickness and improve the uniformity of active silicon layers on silicon-on-insulator (SOI) materials. As an example, a 20 keV 5 % CF_4 with 95 % O_2

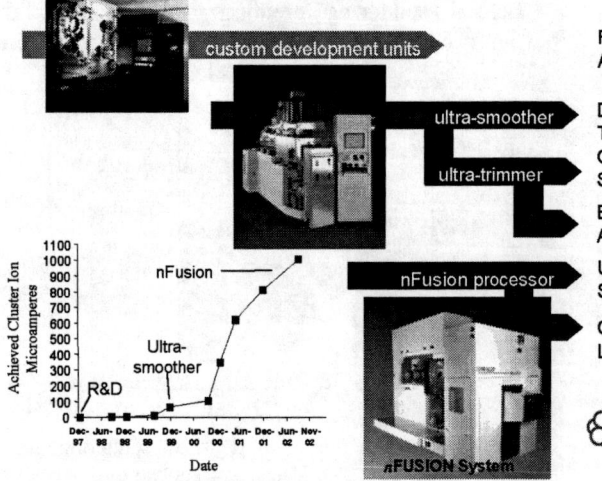

FIGURE 13. Industrial GCIB application areas

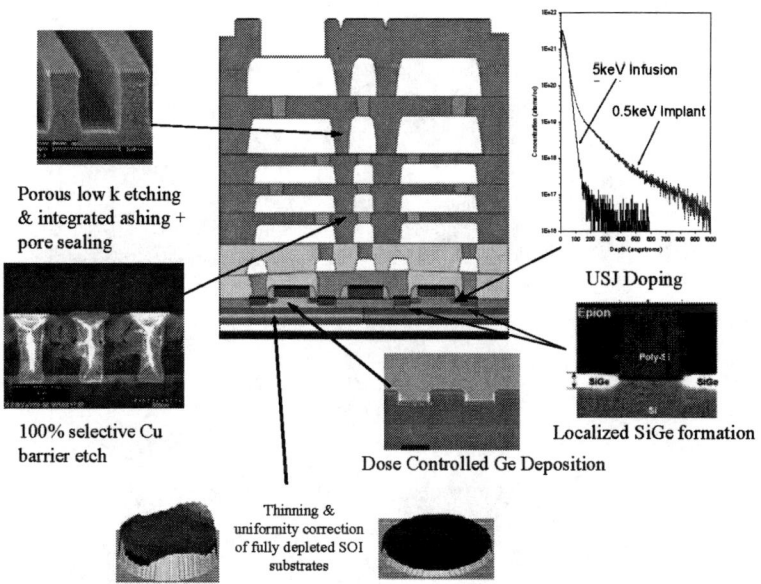

FIGURE 14. GCIB applications in semiconductors

halogen-GCIB process and a 3 keV O_2-GCIB process were used to provide rapid corrective etching and smoothing of a 150 mm SOI wafer. The 145 nm thick Si surface layer with standard deviation of 0.85 nm was reduced to 50 nm with standard deviation of 0.4 nm. In order to fully realize the benefits of SOI substrates for fully depleted MOS devices, Si layers only 20 nm or less thick with thickness nonuniformity of <5 % are desired. A 50 nm thick Si layer having greater than 3 nm nonuniformity over a 300mm SOI wafer can typically be reduced to 20 nm thickness with less than 0.4 nm of nonuniformity.

GCIB densification of porous materials is a new ion beam process which would not be expected to be possible by monomer or molecular ion beams. Argon GCIB processing of porous low-k dielectric materials was investigated on blanket spin-on methyl-silsesquioxane (p-MSQ) films of k~2.2 on 200 mm Si wafers by Epion Corporation and International Sematech. The results showed that a GCIB-densified surface could be formed without alteration of the dielectric material composition. The densified surface produced on the low-k dielectric was able to prevent penetration into the dielectric by Ti from a subsequently deposited PECVD TiSiN film, whereas low-k materials without the surface densification showed Ti penetration under the same conditions.

The highly reactive yet substrate sensitive properties of GCIB have allowed it to be very effective for ashing with high rates of polymer removal without any etching or degradation of other exposed materials such as porous low-k dielectrics. When reactive gases such as O_2, N_2, or C-containing molecules are included in the beams, surface reactions can take place to form thin films, and these films can be used in devices. Chemical GCIB can also be used for cleaning surfaces such as Cu or for etching of thin films in normally hard to reach surfaces within a semiconductor device, such as at the bottom of a trench or via. A GCIB high energy directed chemical beam can perform tasks that are difficult or impossible by other technologies such as plasma processing.

B. GCIB applications to non-semiconductor fields

Currently, industrial applications in several non-semiconductor fields are being developed by a number of Japanese companies under the nanotechnology program called "Advanced Nano-Fabrication Process Technology Using Quantum Beams" of NEDO /METI (New Energy and Industrial Technology Development Organization /the Ministry of economy and Technology Industry). Figure 15 summarizes the project.

GCIB surface smoothing processes are being applied for surfaces of magnetic materials used for HDD sensor heads (Hitachi, Ltd.), for surfaces of polycrystalline SiC wafers which are used as monitor wafers for CVD processes (Mitsui Engineering & Shipbuilding Co., Ltd), for laser annealed poly-Si surfaces which are used for flat panel displays (Mitsubishi Electric Corp.) and for Si nano-structure surfaces for photonics (Japan Aviation Electronics Industry, Ltd.)

One of the more remarkable applications under development involves surface smoothing of the side walls of high aspect ratio Si pillar structures for photonic devices. The successive etching and deposition by inductively coupled plasma – reactive ion etching (ICP-RIE) used to fabricate the tall pillar structures results in side walls which are excessively rough. A GCIB smoothing process has been used to smooth the side wall surfaces to an Ra value of 0.1 nm.

SUMMARY

The history and present status of research and development in the field of gas cluster ion beam processing has been reviewed. Non-linear and non-equilibrium effects due to bombardment of large cluster ions are now attracting attention as new

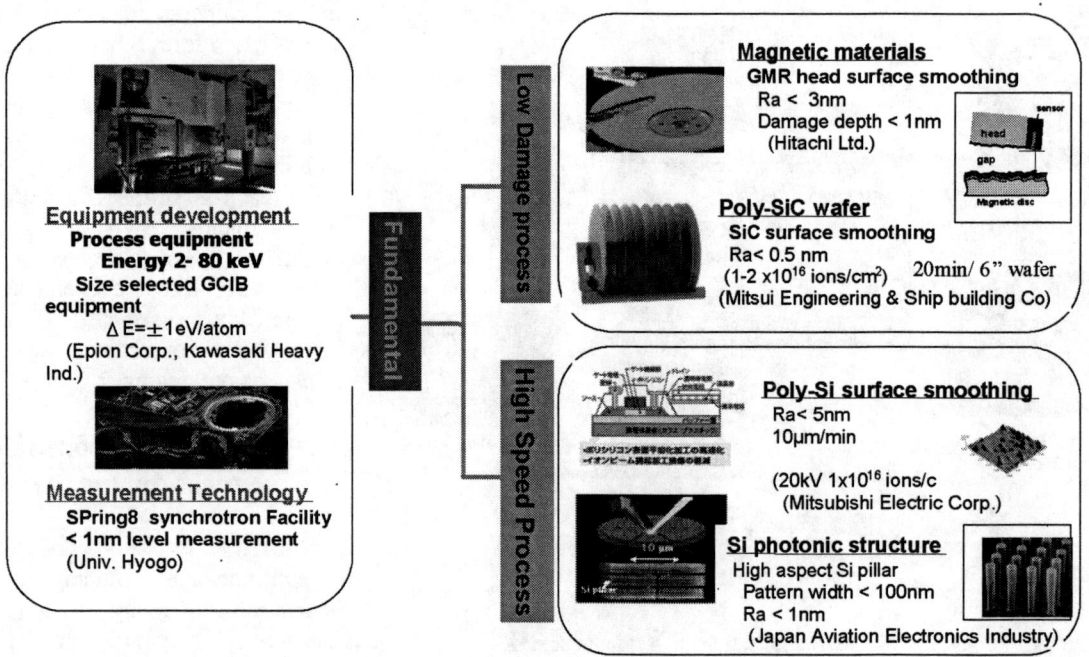

FIGURE 15. Summary of NEDO/MITI project.

technology in the area of ion beam processing. GCIB processing is an advanced approach which will contribute to further progress in the ion beam technology field. Some industrial applications of GCIB which are being developed have been discussed.

ACKNOWLEDGEMENTS

The author wishes to thank to the members of the Collaborative Research Center of Cluster Ion Beam Technology in Kyoto Univ. and Univ. of Hyogo, especially J.Matuso, N.Toyoda, T.Aoki, and T.Seki. He also thanks A.Kirkpatrick of Epion Corporation for long collaboration in the development of commercial GCIB equipment and applications. This work is partially supported by New Energy and Industrial Technology Development Organization (NEDO).

REFERENCES

* Emeritus Professor, Kyoto University

1. E.W.Becker 'On the history of cluster beams', in F.Trager and G. zu Putlitz, eds, Proc. Internat. Symp. On Metal Clusters- 1986, Springer Verlag, Berlin (1986) p. 1
2. R.L.McEachern, W.L.Brown, M.F.Jarrold, M.Sosnowski, G.H.Takaoka, H.Usui and I.Yamada, J. Vac. Sci. Technol. A, 9 (1991) 3105.
3. W.L.Brown, M.F.Jarrold, R.L.McEachern, M.Sosnowski, G.H.Takaoka, H.Usui and I.Yamada, Nucl. Instr. and Meth. B, 59/60 (1991) 182.
4. I.Yamada, Radiation Effects and Defects in Solids, 124 (1992) 69.
5. I.Yamada, J.Matsuo, N.Toyoda, and A.Kirkpatrick, Materials Science and Engineering R34, (2001) 231.
6. A.Kirkpatrick Extended abstracts, Workshop on cluster ion beam process technology, Kyoto International Community House, Kyoto October 12-13, 2000, Osaka Science & Technology Center, (2000) p.17.
7. J.Matsuo, Kyoto University PhD thesis (2000) and T.Seki Kyoto University PhD thesis (2000)
8. N.Toyoda, Kyoto University PhD thesis (1999).
9. N.Toyoda, S.Houzumi, T.Aoki and I.Yamada, Mat. Res. Soc. Symp. Proc. 792, (2004) p.623.
10. T.Aoki, Kyoto University PhD thesis (2000)
11. T.Aoyama, S.Umisedo, N.Hamamoto, T.Nagayama, M.Tanjyo, Extended Abstracts of the Sixth International Workshop on Junction Technology, Shanghai, May 15-16, 2006 , IEEE Press, (2006)p.88.
12. K.Goto, J.Matsuo, T.Sugii, H.Minakata, I.Yamada, IEDM Tech Dig. 1996, IEEE (1996) p. 435. and K.Goto, J.Matsuo, Y.Tada, T,Tanaka, Y.Momiyama, T.Sugii, and I.Yamada, IEDM Tech. Dig., IEEE (1997), p. 471.
13. T.Aoki Private communication
14. Japan Science and Technology Agency (JST) Project completed on 2005-06-18.
15. D.Jacobson, Extended abstracts of the fifth international workshop on junction technology, June 7-8 2005, Osaka Japan, Japan Society of Applied Physics/Silicon Technology division. IEEE, (2005) p23
16. A.Kirkpatrick Extended abstracts, 6th workshop on cluster ion beam and advanced quantum beam process technology, KKR Hotel Tokyo, September 26-27, 2005, Osaka Science & Technology Center, (2005) p.26.

Dose Retention Effects in Atomic Boron and ClusterBoron™ ($B_{18}H_{22}$) Implant Processes

Mark A. Harris[1], L. Rubin[1], D. Tieger[1], V. Venezia[1], T.J. Hsieh[1], J. Miranda[1], D. Jacobson[2]

1 - Axcelis Technologies Inc., 108 Cherry Hill Drive, Beverly, MA 01915 USA
2 - SemEquip Inc., 34 Sullivan Rd. North Billerica, MA 01862 USA

Abstract. Dose control is often assumed to be a function of tool design and calibration. Fundamental interactions, including sputtering and backscattering of ions from the surface of the wafer, modulate the equipment effects. For high dose and low energy processes, such as those necessary for poly-gate doping, these effects can be significant. The use of octadecaborane ($B_{18}H_{22}$) implantation enables production-worthy throughput but can impact these surface interactions. In this work, we have investigated the dose retention in silicon after high dose, low energy $B_{18}H_{22}$ and B implants. A wide range of effective energies and doses was studied. This analysis provides insight into the physical mechanisms and can be used to guide process control development.

Keywords: Molecular Implantation, Octadecaborane, $B_{18}H_{22}$, sputtering, poly-gate doping
PACS: 73.30.+y, 61.72.Tt, 36.40.–c

INTRODUCTION

Dual poly-gate doping for p-poly gates and source drain extension (SDE) implants are driving the need for low energy, high dose boron implants as technology nodes continue to shrink[1]. The high dose boron implants cannot be run productively at the energies required on current beam line implanters using monatomic boron or BF_2 in drift mode. In post-decel operation, very low levels (ppm) of energy contamination can impact device performance[2]. One method of overcoming this challenge is the use of molecular implant species such as decaborane ($B_{10}H_{14}$), or octadecaborane ($B_{18}H_{22}$).

Retained dose for low energy implantation is the ratio of the dopant in the silicon after implant compared to the actual implanted dose. Sputtering and backscattering from the surface cause dopant loss during implantation. This effect has been reported on in the past for low energy boron implants[3]. This work evaluates some of the surface effects and compares retained dose for B_{18} implants as a function of energy and dose as compared with monatomic boron implants. The goal is to better understand the physical interactions for these implants to assist in process development.

Other topics covered include quantification of the surface sputtering rate by high resolution SEM (HR SEM), and discussion of maximum retained dose using extrapolation from existing data. Also, measurement of surface roughness for boron and B_{18} will be compared with that reported in previous work[4,5].

Advantages of Molecular Implantation

Use of molecular implantation, such as octadecaborane, allows higher effective beam currents at lower energies as each molecule carries 18 boron atoms at ~$1/20^{th}$ of the implant energy for the cluster. This allows for low energy, high dose implants in drift mode operation with reduction in space charge by extracting and transporting the beam at higher energies. The use of molecular implantation can allow for productive implantation of the poly-gate and SDE applications using traditional implant architectures without using decel mode operation which can cause energy contamination[1].

Low Energy High Dose Implantation

Figure 1 provides an illustration of the primary mechanisms for dopant loss, backscattering and sputtering. For ultra shallow implants, the probability of a backscattered atom leaving the silicon surface increases when compared to higher energy implants. The differential scattering cross section is given by

$$\frac{d\sigma}{d\Omega} = k\frac{Z_1^2 Z_2^2}{e^4 E^2} \quad (1)$$

therefore, as the energy of the implant decreases, the probability of a backscattering event increases quadraticly. However, as the B concentration increases, the probability of a backscattering event decreases because of the Z_2^2 dependence in the cross section. The value of Z for B is less than that for Si.

Sputtering is the mechanism where near surface atoms leave the target during implantation due to recoil collisions. For a sputtered boron or silicon atom to leave the target surface, the vertical component of the recoil energy must be greater than the surface binding force. As boron concentration and sputtering at the surface increases, a steady state will be reached leading to a maximum retained boron dose.

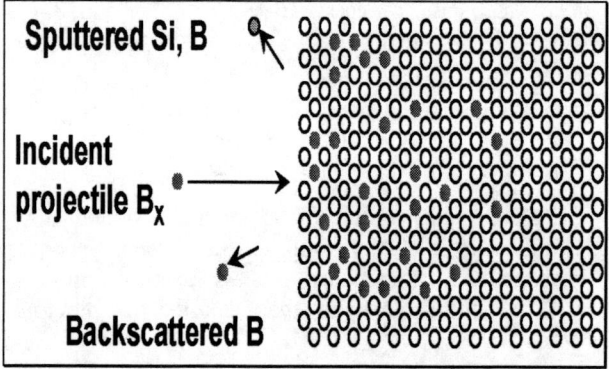

FIGURE 1. Illustration of Sputtering and Backscattering of Si and B from Silicon Surface during Ion Implant.

This effect was previously reported for monatomic boron implants, BF_2 implants, and $B_{10}H_{14}$ implants at effective boron energies <1keV[3]. A measured loss of 10% by nuclear reaction analysis, B(p,α)Be (p-alpha) at 0.5keV and 1×10^{15} ions/cm² boron dose was reported (20% at 200eV). As dose increases and/or energy decreases, the fraction of dopant loss increases due to the above mentioned mechanisms.

Physical Description: B_{18}

Borohydride molecules have been used for decades as source feed material. B_2H_6 has been commonly used. In the last 10 years $B_{10}H_{14}$ has been investigated[6,7,8] for use as a path to low energy high dose p-type implants. In the last couple of years $B_{18}H_{22}$ has become a prime candidate for low energy high dose implants. This large molecule has a molecular weight of 216 AMU. This is of course an average weight because there are two isotopes of boron, mass 10, 20% and mass 11, 80%. There is a binomial distribution of the 18 boron atoms in the molecule. The lightest possible weight would be 202 AMU, which would occur if all 18 B atoms were mass 10. The heaviest possible weight would be 220 AMU; this would occur if all 18 B atoms were mass 11.

Figure 2 is a ball and stick model of $B_{18}H_{22}$. It is of interest that there are six divalent hydrogen atoms in this molecule. These are known to boron chemists as *Bridging Hydrogen Bonds*.

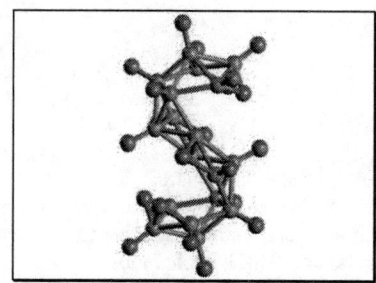

FIGURE 2. Octadecaborane ($B_{18}H_{22}$) model. Note the six bridging hydrogen bonds. They are 3-centre, 2-electron B-H-B bonds containing rare divalent bridging hydrogen atoms.

DOSE RETENTION AND SURFACE EFFECTS FOR B_{18} AND ^{11}B IMPLANT

Surface Sputtering

Figure 3 illustrates the sputtering on Si surface for a high dose B_{18} implant. Sputter yield was measured in silicon for 1.25keV, 9×10^{16} B/cm² B_{18} implants, and the sputter depth was measured by high resolution SEM (HRSEM). The sample was patterned resist covered with a vertical stripe removed. After implant, all the photoresist was stripped off the wafer, and a sample was sent for SEM to measure the thickness of the sputtered layer in Si and SiO_2.

Based on the SEM data and implanted boron dose, a calculated sputtering yield (Y_B) per boron atom of 1 was calculated (Y_{B18}~18). This value is slightly higher than the estimated value[9] for equivalent boron of ~0.3 to 0.4. This could be due to sputtering from an amorphous surface as dose increases or molecular effects from B_{18}. For typical SDE implant doses, high e14 and low e15 range, this would equate to a couple of angstroms surface loss, which is still in the native oxide. For p-poly gate implants, there is potential to sputter on order of 20-30 Å for low e16 doses. This effect leads to lower retained dose as discussed below and must be considered when designing a process, regardless of implementation (monatomic, or molecular doping).

FIGURE 3. HRSEM picture - Si surface of partial PR coated wafer after high dose B_{18} implant to illustrate sputtering by implant at PR interface.

Surface Roughness

Li et al reported at IIT2002 that high dose (1×10^{17} at 5keV and 12keV) implants using decaborane molecules[4] had an effect of smoothing the silicon surface compared to an argon implanted sample. Figure 4 compares doses typical for SDE and P-poly applications. $B_{18}H_{22}$ at 5×10^{15} and 1×10^{16} ions/cm^2 and ^{11}B at 1×10^{15} and 5×10^{15} ions/cm^2 were implanted into polished silicon wafers. The surface roughness was determined by atomic force microscopy (AFM) measurement on an un-implanted sample, and on samples implanted at the above doses.

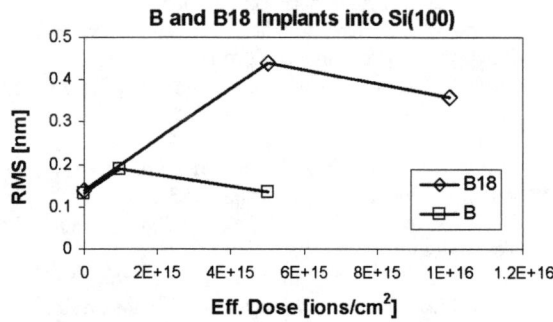

FIGURE 4. Surface Roughness Measurement (RMS) by AFM for ^{11}B and B_{18} implants at various doses

The data indicate a 3x higher RMS roughness for the B_{18} implants at 5×10^{15} ions/cm^2 dose. Both boron and B_{18} implants show greater roughness at the lower dose (1×10^{15} and 5×10^{15} respectively) followed by turnover and then smoothing as dose increases. This supports the findings of smoothing with molecular implant at 1×10^{17} ions/cm^2 dose reported in earlier work[4].

Dose and Energy Dependence for Retained Dose

Dose dependence is illustrated at 1keV effective boron energy in Figure 5. At 1keV and above, monatomic boron shows a 1 to 1 ratio at doses studied.

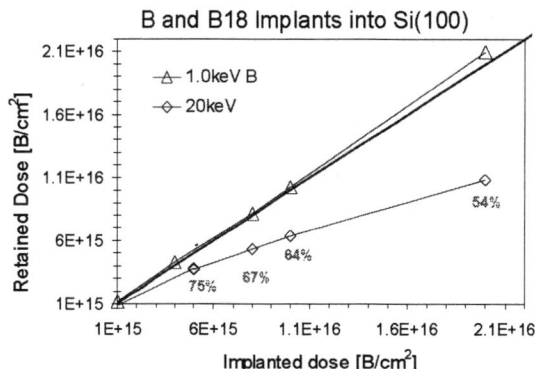

FIGURE 5. Dose Retention of 1 keV B by B_{18} implant in silicon as measured by SIMS and verified by p-alpha

A decrease in retention was quantified with SIMS and p-alpha as implant dose increases at a given energy. As shown the effect can exceed 50% as the maximum retained dose is approached. The energy dependence is illustrated in Figure 6.

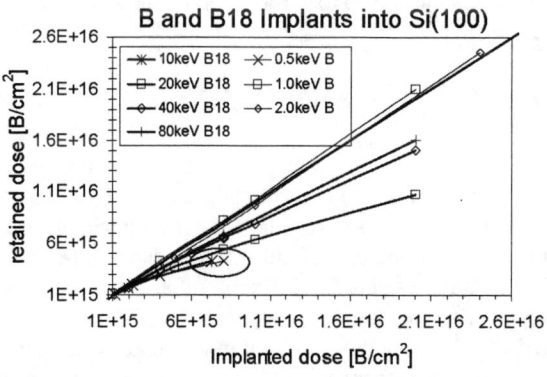

FIGURE 6. Dose Retention by B_{18} energy of B implant in silicon as measured by SIMS and verified by p-alpha

Higher energy implants result in better retention as dose increases, and a higher maximum retained dose. Again, the effect approaches 50% as the maximum retained dose is approached.

Maximum Retained Dose

Based on the above data, there will be a maximum dose of boron in a sample for a given energy. Figure 8 extends the dose to 1×10^{17} for 30keV and 60keV B_{18} implants. The plot shows the retained boron dose at 30keV as leveled off somewhere before 7×10^{16} B/cm^2. This is the point where the dopant loss from surface sputtering is equivalent to the dose entering the surface. The 60keV B_{18} implant doses are still increasing at 1.2×10^{17} B/cm^2, although retained dose is decreasing as implant dose rises.

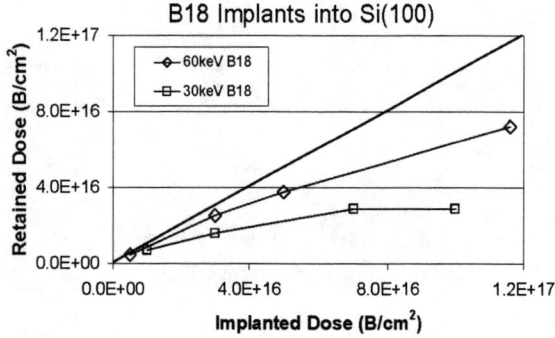

FIGURE 7. Maximum Retained Dose illustrated for 30 keV B_{18}, and Compared with 60keV B_{18}

The above data were collected by SIMS with validation by p-alpha measurement at select points. Both techniques measure the same effect for maximum retained dose in the samples. In addition, the effect was the same at 30keV for implant into 1000Å poly silicon.

SUMMARY

Dose loss for low energy high dose implants results from sputtering and backscattering. The various doping methods (monatomic implant, cluster implant, and plasma doping) will all see this depending on the implantation conditions. The amount of retained dose depends on the doping technique, the effective energy, and the effective dose. This work discusses the effects for octadecaborane implants compared with monatomic boron implants for doses and energies typical for dual poly-gate implants. Topics discussed were surface sputtering loss, surface roughness in relation to dose, retained dose comparisons for B_{18} and boron, and extrapolation to maximum retained dose using existing data for B_{18}.

ClusterBoron™ implants have been shown to match current processes for poly-gate doping[10] and source drain extension[11] applications, and are a viable alternative for significantly improving productivity for energy-pure boron implants using well-known, existing implant architectures. The goal of this work was to document the surface effects and describe some of the physical interactions that will allow a more thorough understanding for guiding process development. It is also pointed out that there is some variability based on how the process is being integrated and some experimentation is required for optimizing the process to fit each case.

ACKNOWLEDGMENTS

The authors wish to thank Dr. Vincent Venezia and Dr. Frank Sinclair for valuable discussions and early contributions to this effort. The authors would also like to thank Dr. Mike Ameen for discussions related to sputter yield measurements and general process topics.

REFERENCES

1. M. Taylor, *et al*, "Material Challenges for CMOS Junctions," in MRS Symposium. Proc. Vol C.1.1. pp. 810
2. H.-J.L. Gossman, "Ion Implantation in Advanced Planar and Vertical Devices," in Nuclear Instruments and Methods in Physics Research B237 (2005). pp 1-5.
3. A. Agarwal, "Ultra-Shallow Junction Formation Using Conventional Ion Implantation and Rapid Thermal Annealing," in proceedings 2000 International Conference on Ion Implantation Technology. pp. 293-299.
4. C. Li *et al*, "Sputtering of Si with Decaborane Clusters," in proceedings 14th International Conference on Ion Implantation Technology. pp 583-586
5. S. Heo, *et al*, "Characteristics of Ultra-Shallow P$^+$/n Junction Prepared Cluster Boron (B18H22) Ion Implantation and Excimer Laser Annealing," in Proceedings 6th International Workshop on Junction Technology. pp. 48-53.
6. D. Jacobson *et al*, "Decaborane, an Alternative Approach to Ultra Low Enrgy Ion Implantation," in proceedings 2000 International Conference on Ion Implantation Technology. pp 300-303
7. A. Perel *et al*, "Decaborane Ion Implantation," in proceedings 2000 International Conference on Ion Implantation Technology. pp 304-307
8. A. Agarwal *et al*, "Transient Enhanced Diffusion from Implantation of Molecular Decaborane Ions," in proceedings 1998 International Conference on Ion Implantation Technology. pp 857-860
9. R. Behrisch, Sputtering by Particle Bombardment III. Topics in Applied Physics Vol. 64. 1991.
10. D. Henke *et al*, "P-Type Gate Electrode Formation Using $B_{18}H_{22}$ Ion Implantation," in these conference proceedings.
11. Y. Kawasaki *et al*, "Ultra-Shallow Junction Formation by $B_{18}H_{22}$ Ion Implantation'" in Nuclear Instruments and Methods in Physics Research B237 (2005). pp. 25-29.

Universal Ion Source™ for Cluster and Monomer Implantation

Thomas N. Horsky

SemEquip, Inc., 34 Sullivan Road, North Billerica, MA 01862 USA

Abstract. Energetic cluster and molecular ion beams have become useful for performing high dose implants at very low implantation energies. The ClusterIon® source, when installed on a conventional beam line implanter, can routinely produce >1 mA of electrical $B_{18}H_x^+$ ion current on the wafer, enabling boron dose rates in excess of 18 mA at an effective implant energy of 1 keV per boron atom implanted. However, this very important application represents only a fraction of the implant processes required for leading edge device fabrication. For example, both p-type and n-type source/drain structures, as well as halo, punch through, poly gate, and threshold adjust implants are required. To address this need for a true multi-species capability, a new ion source has been developed. The source is a novel dual-mode design which operates in a cluster formation mode and a monomer formation mode, such that the cluster formation mode is mediated by electron impact ionization and the monomer formation mode by an arc discharge.

Keywords: High current implantation; molecular implantation; ion sources
PACS: 61.72Tt

INTRODUCTION

Ion sources, for example sources of the Bernas type [1], use an arc discharge to efficiently dissociate molecular feed gases into their constituent elements, enabling dopant monomer ions to be generated, *e.g.*,

$$e^- + BF_3 \rightarrow B^+ + F^+ + BF_2^+ + 4e^-. \quad (1)$$

Here, B^+ is the dopant ion of interest. There are, however, also important applications for the implantation of molecular ions which require the preservation of the parent molecule during ionization. To efficiently produce molecular ions without dissociation, a "soft" ionization method is preferred over the arc discharge method. An electron impact source has been developed [2, 3] for this purpose, the ClusterIon® source, which can produce a beam of several milliamperes of molecular ions. As is typical with electron impact ionization sources, the ion temperature and ion density are low, producing a quiet, low-emittance ion beam which lends itself to ready inclusion in conventional medium-current (MC) and high-current (HC) implanter platforms.

A boron-containing material of interest is the large borohydride molecule *octadecaborane* ($B_{18}H_{22}$), which can be ionized according to the reaction:

$$e^- + B_{18}H_{22} \rightarrow B_{18}H_x^+ + (22-x)H^+ + (24+x)e^-. \quad (2)$$

In eq. (2), x is an even integer [4], and $x \leq 22$. The utility of implanting beams of $B_{18}H_x^+$ ions can be appreciated through the observation that at a given accelerating voltage, the increased mass of the ion reduces the implant energy per boron atom by about twenty-fold, while at the same time producing an eighteen-fold increase in dose rate [5,6,7]. Thus, large conglomerates or "clusters" of like dopant atoms are useful for both ultra-low energy and ultra-high dose applications. For example, the p+ drain extension implant for 45nm CMOS devices requires a 300eV boron implant at a dose of about 1E15 cm^{-2}, which can be achieved at high throughput in drift mode with a conventional beam line implanter by substituting a 6 keV $B_{18}H_x^+$ implant at a dose of 5.6E13 cm^{-2} [8,9]. In addition to increased throughput, it has been reported [8,9] that $B_{18}H_x^+$ implantation provides other important advantages, such as: *i)* obviating the need for a pre-amorphization implant to reduce channeling, *ii)* reducing or eliminating end-of-range defects and reducing the device leakage currents they cause, *iii)* enabling improved dopant activation, *iv)* eliminating energy contamination, and the high-energy tail which is a characteristic of deceleration-mode implantation, *v)* dramatically improving low-energy beam quality (reduced angular divergence, and increased angular and spatial uniformity) on wafer, and *vi)* achieving ultra-shallow junctions with conventional spike anneal technologies. These results have been described in detail elsewhere [5—10].

In addition to borohydride vapors, the ClusterIon® source runs conventional feed gases such as AsH_3 and PH_3 to produce several milliamps of AsH_x^+ or PH_x^+ ions, and indeed can produce molecular beams of any gaseous or vapor feed materials which can be ionized by electron impact. Multiply-charged currents are small, however, as this is a "soft" ionization source.

There has been considerable interest in incorporating borohydride and other molecular implantation capabilities in conventional beam line implantation platforms to enable a wider range of both N- and P-type implants, *i.e.*, a "hybrid" of HC and MC applications [7]. Given the industry-wide move to serial wafer processing for HC as well as the traditionally serial MC systems, it is clearly advantageous to cover the applications spaces of both generic platforms within one serial hybrid implanter. Such hybridization is greatly simplified by the incorporation of an ion source which delivers molecular ions and ionized clusters, in addition to conventional ion species.

ION SOURCE DESCRIPTION

The Universal Source™ is based on the ClusterIon® source [3], but is designed to support a broader range of applications. While the electron impact source has the great utility of providing high currents of molecular ions and ionized clusters, its monomer currents are suitable for medium-current implants only. High currents of N-type dopants, increased currents of As and P clusters, and the formation of multiply-charged ions are features of the Universal source. Its guiding operating principle is that it is "dual mode", that is, it can operate in either an electron impact mode (to produce clusters and molecular ions), or in an arc discharge mode (to produce high currents of monomers and multiply-charged ions). This is accomplished by incorporating a separate electron gun and an arc emitter in the source; for cluster operation, only the electron gun is used; to produce monomers and multiply-charged ions during arc operation, the arc emitter is energized, striking a plasma discharge, as in a Bernas-type source. As shown in Fig. 1, the source incorporates an externally-mounted electron gun and an indirectly-heated cathode (IHC). Although not shown in Fig. 1, the source incorporates both vaporizer and process gas inlets.

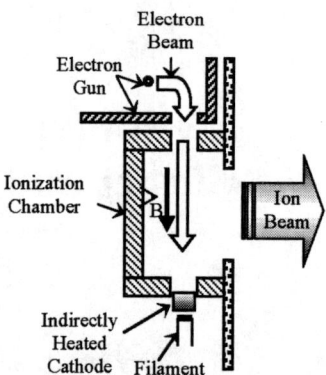

FIGURE 1. Ion Source Schematic.

Electron Impact Operation

Referring again to Fig. 1, we describe the electron impact mode of operation. An externally generated electron beam creates a stream of ions just behind the long rectangular slot from which ions are extracted by the implanter optics.

The electron gun creates an energetic electron beam of between 1 mA and 100 mA, which is then deflected through 90 degrees by a magnetic dipole field. Since the electron gun is remote from the ionization chamber and has no line-of sight to the process gas, it resides in the high vacuum environment of the implanter's source housing, resulting in a long emitter lifetime. The deflected electron beam enters the source ionization chamber though a small entrance aperture. Once within the ionization chamber, the electron beam is guided along a path parallel to and directly behind the ion extraction slot by a uniform axial magnetic field of about 100 Gauss produced by a permanent magnetic yoke surrounding the ionization chamber. Ions are thus created along the electron beam path and adjacent to the extraction slot. This serves to provide good extraction efficiency of the ions, such that an ion current density of up to 1 mA/cm^2 can be extracted from the source. The beam current dynamic range thus achieved is comparable to other sources; by varying emission current and also the flow of feed material into the source, a stable on-wafer electrical beam current of between 5 μA and 2 mA is achieved. Fig. 2 below shows the dependence of extraction and wafer current on the rate of vapor flow into the ionization chamber, as measured by the inlet pressure to the ion source. $B_{18}H_{22}$ or $B_{10}H_{14}$ vapor is typically introduced into the source from an externally mounted vaporizer, through a pressure control device which regulates the flow of vapor into the source ionization chamber [11,12].

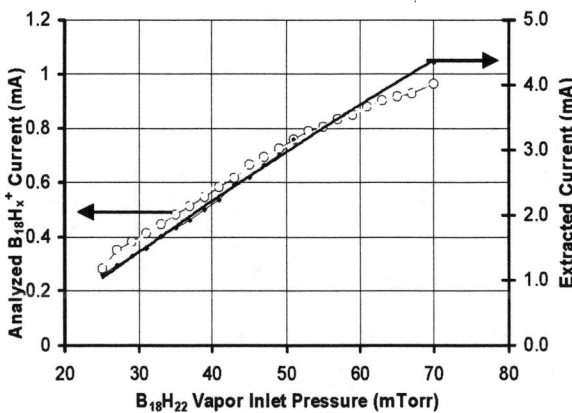

FIGURE 2. Current versus vapor inlet pressure. Solid Line: extracted current. Open circles: analyzed current.

Arc Discharge Operation

In arc discharge mode, the electron gun is not used. Instead, the IHC is energized, creating a plasma column along the magnetic field direction, which provides plasma confinement. Fig. 3 and Fig. 4 show arsine and phosphine spectra, respectively, from the Universal source operating in arc discharge mode. As further indicated in Fig. 5, up to 12 mA of P^+ and As^+ current were delivered to the wafer position at an arc current of less than 0.5A. These values were achieved at an arc voltage of 70V and 80V, respectively. SDS gas flows of 4.3sccm of PH_3 and 5.8sccm of AsH_3 were used to generate the currents shown in Figs. 3—5. The relatively high gas flows are due to the large area of the ion extraction aperture (arc slit), which was 48mm tall by 8mm wide, *i.e.*, an extraction area of about 4 cm^2.

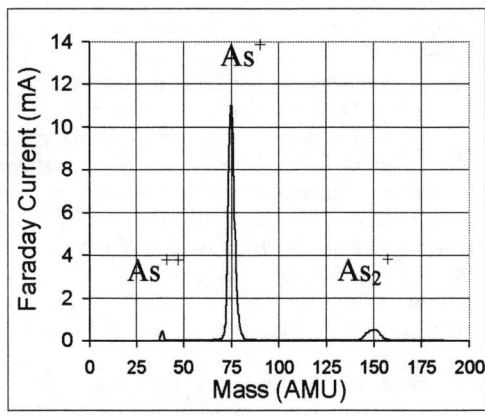

FIGURE 3. Arsine spectrum in arc discharge mode.

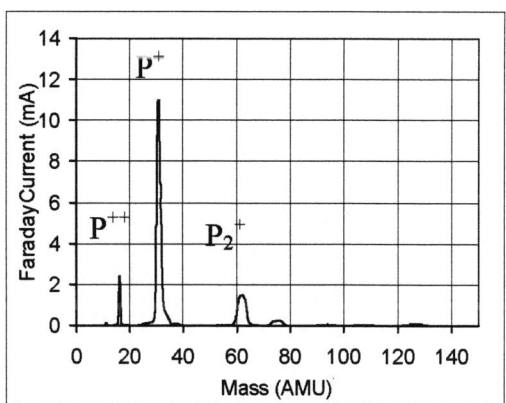

FIGURE 4. Phosphine spectrum in arc discharge mode.

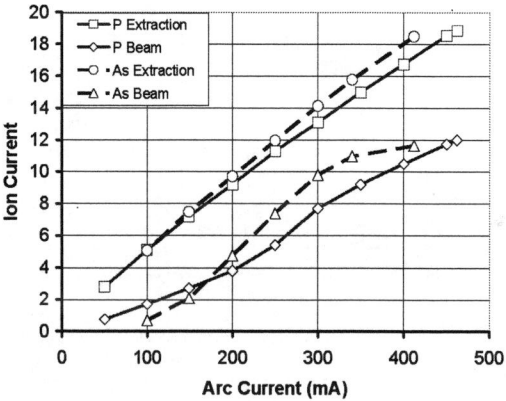

FIGURE 5. Extraction and post-analysis As (dotted line) and P (solid line) currents versus IHC arc current.

N-Type Dopant Clusters

The N-type dopants As and P are important for doping silicon to form NMOS transistors. Substantial increases in effective ultra-low energy beam current for these species can be achieved by implanting the dopant clusters consisting of the dimer (As_2^+, P_2^+) [13] and tetramer (As_4^+, P_4^+). In general, the implantation of a cluster of n dopant atoms can provide dose rates (in atoms/s) a factor of n^2 higher than the implantation of single atomic ions. This follows from the Child-Langmuir law, and so assumes a condition in which the transport of the monomer ion beam is fully space charge-limited, a condition often satisfied when implanting ultra-low energy ions such as 1or 2 keV As, for example, for creating source/drain extensions. The tetramers As_4^+ and P_4^+ do not typically survive as ions in plasma discharges since they readily dissociate via the reactions:

$$e^- + As_4^+ \rightarrow 2e^- + 2As_2^+, \qquad (3)$$

and

$$\gamma^0 + As_4^+ \rightarrow e^- + 2As_2^+, \quad (4)$$

Where γ^0 is an ultraviolet (uv) photon, for example. Since the ions produced within the ionization chamber are also shielded from the uv radiation produced by the electron emitter filament, the tetramer ion is largely preserved, as can be seen in the As and P spectra of Fig. 6 and Fig. 7. Note how large the tetramers peaks are, containing fully half of the As and P atoms of the overall spectra. These spectra were produced in electron impact mode.

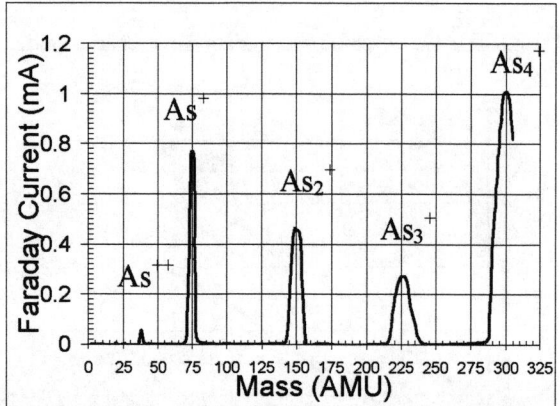

FIGURE 6. As spectrum generated in electron impact mode. Note that half of the As atoms are in the As_4^+ peak.

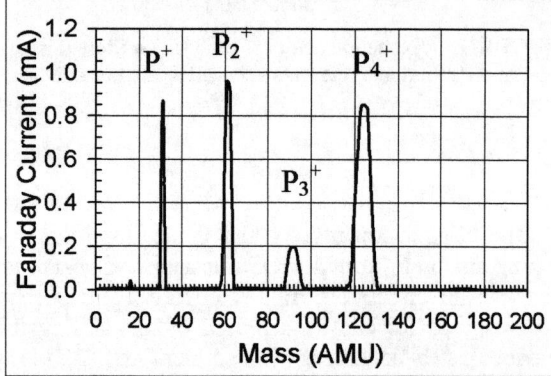

FIGURE 7. P Spectrum generated by electron impact.

SUMMARY & DISCUSSION

A new multi-species ion source has been developed which has two discrete modes of operation: an arc discharge mode (mediated by an indirectly-heated cathode) for the production of monomer and multiply-charged species, and an electron impact mode (mediated by an electron beam) for producing molecular ions. In addition to $B_{18}H_x^+$ ions, the electron impact formation of significant currents of the N-type tetramers As_4^+ and P_4^+ have not, to my knowledge, been previously observed. These species are immediately useful for creating ultra-shallow source/drain extensions, and also for PMOS halo implants. It is expected that, just as has been demonstrated for $B_{18}H_x^+$ and $B_{10}H_x^+$ implantation, the large mass of these tetramers provide self-amorphization, resulting in reduced channeling [13].

ACKNOWLEDGMENTS

The author wishes to thank Dr. Frank Sinclair for his critically valuable analytical work, and the excellent design engineering work of Frank Trueira. Special thanks go to Bob Milgate and Dr. Glen Gilchrist for their wonderful laboratory work and technical leadership. The expertise of George Sacco and Bill Reynolds were also invaluable in generating the results reported in this paper.

REFERENCES

1. N. White, *Nucl. Instr. Meth. Phys. Res.* **B37/38**, p. 78 (1989).
2. T. Horsky, D. Jacobson, W. Krull, and H. Glavish, "Performance of the SemEquip Ion Source", 14th International Conference on Ion Implantation Technology, 22—27 September 2002, Taos, NM (unpublished).
3. T. N. Horsky, "ClusterIon™ Source for Cluster Implantation", XVth International Conference on Ion Implantation Technology, October 25—29, 2004, Taipei, Taiwan, ROC (unpublished).
4. D. Jacobson, T. Horsky, W. Krull, and R. Milgate, *Nucl. Inst. Meth. Phys. Res.* **B 237** (2005) pp. 406—410.
5. Y. Kawasaki, T. Kuroi, K Horita, Y. Ohno, M. Yoneda, T. Horsky, D. Jacobson, and W. Krull, *Ibid.*, pp. 25—29.
6. J. O. Borland, *Ibid*, pp. [ ].
7. J. O. Borland, M. Tanjyo, N. Nagai, T. Aoyama, and D. Jacobson, "Applying Equivalent Scaling to USJ Formation", *Semiconductor International*, January 2005, pp. 52—55.
8. S. Heo, H. Cho, and W. Krull, these proceedings.
9. J. Borland, S. Shishiguchi, A. Mineji, W. Krull, D. Jacobson, M. Tanjo, W. Lerch, S. Paul, J. Gelpey, S. McCoy, J. Venturini, M. Current, V. Faifer, R. Hillard, T. Walker and S. Hummel, IWJT2006, Japan.
10. S. Heo, H. Cho, and W. Krull, these proceedings.
11. D. Adams, G. F. R. Gilchrist, R. W. Milgate III, T. N. Horsky, J. Sweeney, and P. Marganski, these proceedings.
12. T. N. Horsky, R.W. Milgate III, and G. F. R. Gilchrist, these proceedings.
13. A. Agarwal, A. Stevenson, M.S. Ameen, B.S. Freer, and J.M. Poate, *Nucl. Inst. Meth. Phys. Res.* **B 237** (2005), pp. [ ].

Investigation of Converted p$^+$ poly-Si Gate Formed by B$_{18}$H$_X^+$ Cluster Ion Implantation

Sun-Hwan Hwang [a], D.S. Kim [a], Y.H. Joo [a], J.G. Oh [a], J.K. Lee [a], T.W. Jung [a],
H.J. Cho [a], Y.S. Sohn [a], D.S. Sheen [a], S.H. Pyi [a],
Steve Kim [*], T.H. Huh [*], W.A. Krull [**], H.T. Cho [**]

[a] Hynix Semiconductor Inc., San 136-1 Ami-ri, Bubal-eub, Ichon-si, Kyoungki-do, 467-701, Korea
[*] Axcelis Technology, Ltd., 1024-16 Youngtong-dong, Youngtong-gu, Suwon-si, Kyunggi-do, 442-443, Korea
[**] SemEquip Inc. 34 Sullivan Road, N. Billerica, MA 01862, USA
E-mail: sunhwan.hwang@hynix.com

Abstract. Conventional B$^+$ or BF$_2^+$ implantation has a limitation in terms of throughput and energy contamination. Boron cluster implantation is one alternative to solve this problem. We have investigated the characteristics of B$_{18}$H$_X^+$ cluster ion implantation using p$^+$ poly-Si gated MOS capacitors. The B$_{18}$H$_X^+$ cluster was implanted to n$^+$ poly-Si in order to convert to p$^+$ poly-Si with energy of 2.5keV and dose of 1.6E16 ~ 2.0E16 ions/cm^2 The improvement in the inversion capacitance (more then 3%) was observed in the case of cluster ion implantation. This result indicates less poly-silicon depletion effect. It was verified by TDS and SIMS that the hydrogen level of the by p$^+$ poly-Si gate implanted by the cluster ions, and then annealed, was similar to that of a sample implanted by conventional B$^+$ ions. Thus, boron cluster implantation was proven to be more beneficial than the conventional B$^+$ implantation for the capacitor performance.

Keywords : B$_{18}$H$_X^+$, Cluster Boron, P$^+$ poly
PACS : 79.20.R 85.30.Tv 85.40.Ry

Introduction

Boron ion implantation at low energies has been traditionally used for the formation of dual poly gate in pMOSFETs. However, the increase of implantation dose brings about a low throughput issue. Although deceleration mode implantation is more effective to increase beam current, it can cause energy contamination problems with high energy ions.

Plasma doping is one of the very attractive technique to overcome above issues [1]. Another approach is boron cluster implantation [2-3]. Boron cluster implantation, regaining extraction and transportation of ion beams at much higher energies than the desired implant energy, has been proposed as a solution for the both throughput and energy contamination issues. Boron cluster implantation has been investigated using source materials such as B$_{10}$H$_{14}$ (decaborane) for many years. Recently, the beam current of novel B$_{18}$H$_x^+$ cluster ions from B$_{18}$H$_{22}$ has been substantially improved using the SemEquip ClusterIon® Source and becomes more practical for use for the formation of p$^+$ poly-Si gates.

In this work we used B$_{18}$H$_X^+$ implantation to fabricate the p$^+$ poly-Si gated MOS capacitors. The beams were generated in an Axcelis GSD200 ion implanter modified with a SemEquip ClusterIon® Source and vaporizer. We compared results of B$_{18}$H$_X^+$ and B$^+$ from the viewpoint of capacitor performance. The hydrogen ion issue, one of the concerns for cluster implantation, was also investigated. In fact, a large amount of hydrogen ions can degrade device performances, which is simultaneously incorporated into the p$^+$ poly-Si gate with boron ions. In addition, the sheet resistances and boron profiles in poly-Si/oxide/si wafers were also analyzed.

Experimental

In this experiment, we pay attention to principal doping step for the formation of p$^+$ poly-Si gated MOS capacitors. Figure 1 shows a process sequence for the p$^+$ poly-Si gate fabrication. After the plasma nitridation onto the gate oxide, either B$^+$ or B$_{18}$H$_X^+$ was implanted into the p$^+$ poly-Si layer. Conventional poly implant annealing (PIA) was performed at 950°C, and spike poly implant annealing at 1075°C with lamp-based equipment. We analyzed doping characteristics and surface morphologies using secondary ion mass spectrometry (SIMS), transmission electron microscopy (TEM), atomic force microscopy (AFM), and a conventional four-point probe. The hydrogen contents in the poly-Si films were examined by TDS (Thermal Desorption Spectroscopy) and SIMS. The electrical properties, such as the leakage current and capacitance-voltage (C-V) characteristics of p+ poly-Si gate, were measured by use of 4155B and HP4284A.

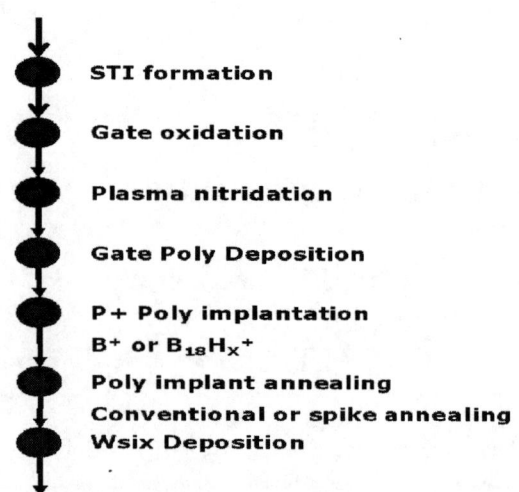

FIGURE 1. Process sequence for p+ poly-si gate fabrication

Results and Discussion

The ion beams for boron cluster implantation were generated from solid $B_{18}H_{22}$. Figure 2 shows the SIMS profiles of boron in an as-implanted poly-Si gate/plasma nitridation/gate oxide/Si substrates. The implantation was performed at 2.5keV-equivalent energy and 2.00E16-ions/cm^2-equivalent dose, corresponding to the extraction energy of 50keV and a dose of 1.11E15 ions/cm^2 of cluster ions. During $B_{18}H_x^+$ cluster implantation, 11B and 10B was implanted simultaneously. The composition ratio of 10B and 11B was 27% and 73%, respectively.

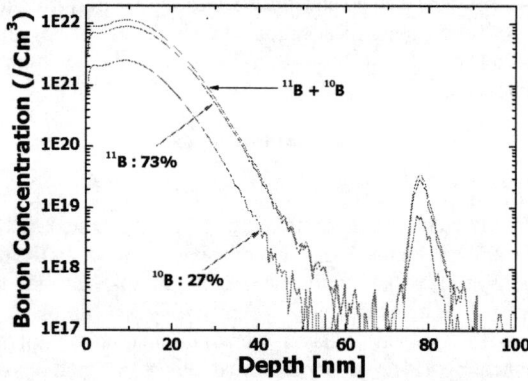

FIGURE 2. Typical SIMS ^{10}B and ^{11}B depth profiles of as-implanted poly/oxide/Si substrates. The implants were done with a $B_{18}H_x^+$ beam extrated at 50keV (2.5keV equivalent implant energy). The composite profile was obtained by a liner sum of the ^{10}B and ^{11}B profiles.

Figure 3 shows boron profiles in 50keV $B_{18}H_x^+$ implanted or 2.5keV B^+ implanted poly-Si films both before and after poly implant annealing. The implantation was performed at 50keV-equivalent energy and 1.60E16-ions/cm^2-equivalent dose corresponding to extraction energy of 50keV with an electrical dose of 8.88E14 ions/cm^2. The implanted poly-Si/oxide/si substrates undergo an identical conventional anneal at 950°C. The SIMS profiles of $B_{18}H_x^+$ and B^+ implants represent only 11B dopant concentrations.

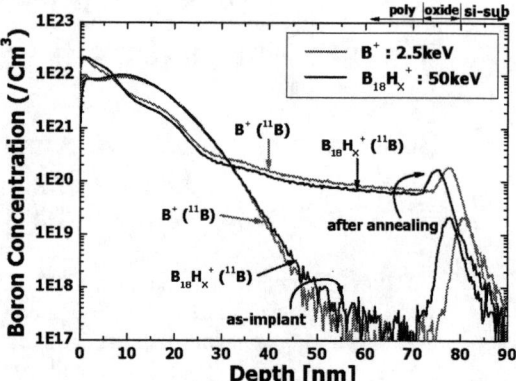

FIGURE 3. SIMS profile of boron, of $B_{18}H_x^+$ at 50keV and B^+ at 2.5keV before and after poly implant annealing.

The SIMS profile of as-implanted $B_{18}H_x^+$ is broader than that of B^+. Due to the fact that poly-Si is an amorphous layer, a channeling phenomenon of boron dopants from clusters or conventionals is hard to consider. Therefore, the broader as-implanted profile can be thought as an effect of atomic mass mixing. After PIA, $B_{18}H_x^+$'s boron concentration at the poly-Si/oxide interface decreased more than B^+'s one. Due to this reason, Rs increased more than 6%.

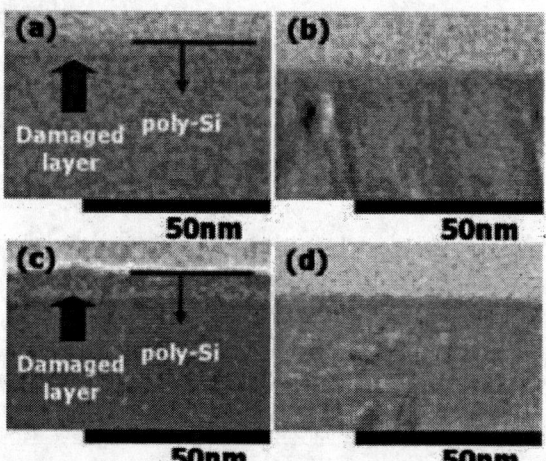

FIGURE 4. shows cross-sectional TEM photographs of 50KeV $B_{18}H_x^+$ implant (a) before and (b) after PIA and 2.5keV B+ implant, (c) before and (d) after PIA. The PIA was performed by conventional-RTA at 950°C.

Figure 4 shows cross-sectional TEM photographs of a 50KeV $B_{18}H_x^+$ implant and a 2.5keV B^+ implant both before and after poly implant annealing. The PIA was carried out by was conventional-RTA at 950°C. In the as-implanted case, the ratio of $B_{18}H_x^+$'s damaged layer is about 3 times of the ratio of B^+'s one.

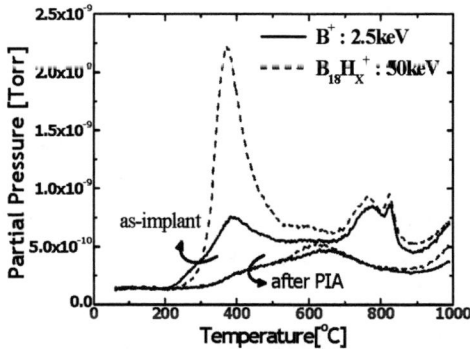

FIGURE 5. M/Z2 spectra of hydrogen for $B_{18}H_x^+$ and B^+ implants both before and after poly implant annealing.

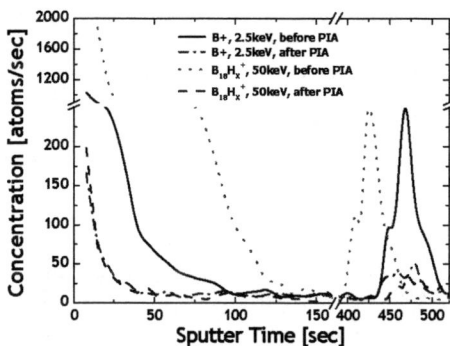

FIGURE 6. SIMS profiles of hydrogen for $B_{18}H_x^+$ and B^+ implant both before and after poly implant annealing.

Figure 5 shows M/Z 2 spectra of hydrogen of $B_{18}H_x^+$ and B^+ implant both before and after poly implant annealing. In the $B_{18}H_x^+$ case, occurrence of H_2 desorption in 200~800°C range must be due to the solid $B_{18}H_{22}$ source. After PIA, H_2 desorption characteristics of $B_{18}H_x^+$ become similar to those of B^+. This result can be verified by SIMS profiles of hydrogen shown in Fig. 6.

Figure 7 shows C-V characteristics of p^+ poly-Si gate converted by $B_{18}H_x^+$ and B^+ counter doping from n^+ doped poly-Si. $B_{18}H_x^+$ and B^+ implant were progressed by a equivalent energy (2.5keV) and dose (2.0E16/Cm^2). Inversion capacitance for $B_{18}H_x^+$ implant increased about 3% and 5% at gate voltage of -1V for thick oxide (55Å) and thin oxide (25Å) capacitors.

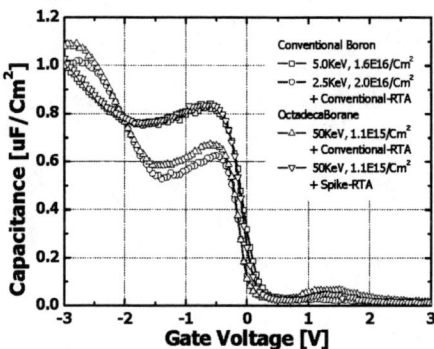

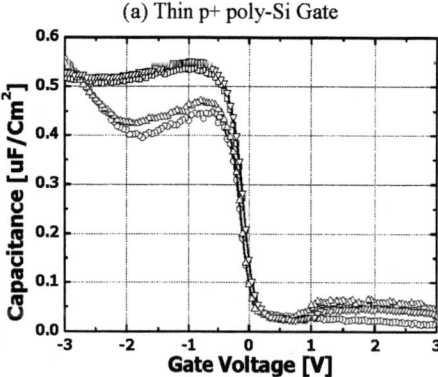

(a) Thin p+ poly-Si Gate

(b) Thick p+ poly-Si Gate

FIGURE 7. C-V characteristics of p+ poly-Si gate converted by $B_{18}H_x^+$ and B+ counter doping from n^+ doped poly-Si gate.

After $B_{18}H_x^+$ implantation, spike annealing was performed at 1,075°C in ambient O_2. The inversion capacitances obtained by present experiments are improved more then 25% and 40% compared with the conventional PIA $B_{18}H_x^+$ and the conventional PIA B^+ respectively. In the case of B^+ implant at 5keV and dose of 1.6E16/Cm^2, inversion capacitance increases only about 3%. These tendencies can be explained by increasing of boron's activation ratio in poly-Si considering spike-RTA application and suppression of boron out-diffusion in ambient O_2.

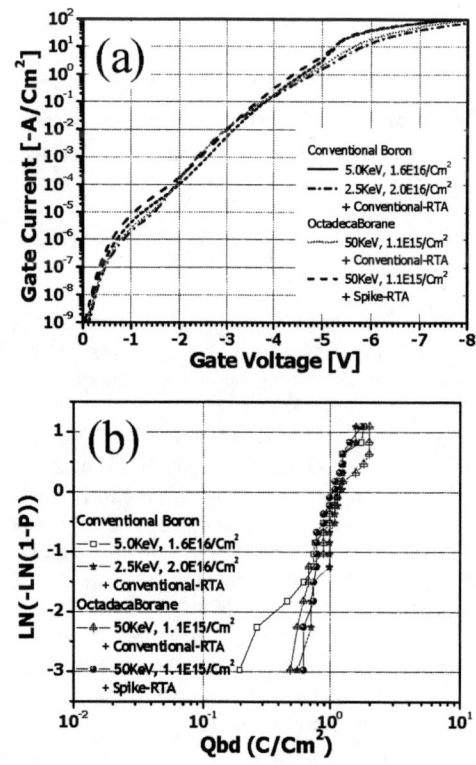

FIGURE 8. (a) leakage current-voltage and (b) cumulative distribution of charge-to-breakdown with p+ poly-Si gate

converted by $B_{18}H_x^+$ and B^+ counter doping from n^+ doped poly-Si gate.

Figure 8 shows leakage current-voltage and cumulative distribution of charge-to-breakdown (Q_{bd}) of gate oxide with p^+ poly-Si gate converted by $B_{18}H_x^+$ and B^+ counter doping from n^+ doped poly-Si. Since conventional poly implant annealing was performed, any difference between $B_{18}H_x^+$ and B^+ can not be observed. In the case of Spike PIA progression, increase of leakage current can be thought as the increase of electrons trapped by boron penetrated-oxide. In the case of Q_{bd} characteristics, there is no significant difference in the annealing methods and types of boron.

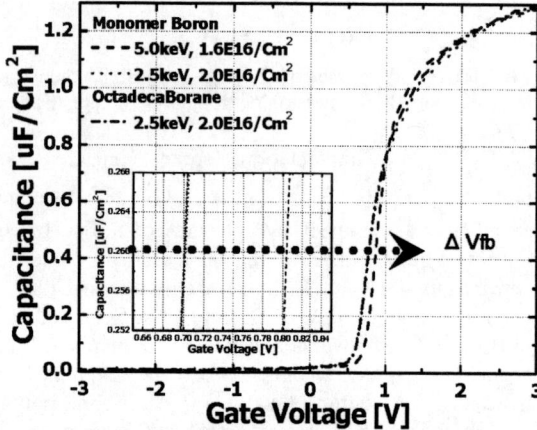

FIGURE 9. Boron penetration characteristics of $B_{18}H_x^+$ and B^+ after conventional PIA at 950°C.

Figure 9 shows Boron penetration characteristics of $B_{18}H_x^+$ and B^+. Conventional PIA was performed at 950 °C. When $B_{18}H_x^+$ and B^+ counter doping were implanted into n^+ doped poly-si, all flat band voltage shifts(ΔV_{fb}) were observed to be almost, same.

Conclusions

We have demonstrated octadecaborane molecular ion implantation for the formation of p^+ poly-Si gated MOS capacitors and examined the characteristics of a high-performance p^+ poly-Si gate.

$B_{18}H_x^+$ cluster implantation consisted of 27% 10B, and 73% 11B in the dose ratio. $B_{18}H_x^+$ shows broader profiles in as-implanted SIMS compare to conventional boron. In the as-implanted case, the $B_{18}H_x^+$'s damaged layer observed in poly-si gate is about 3 times of B^+'s. After poly implant annealing, $B_{18}H_x^+$'s boron concentration at poly-Si/oxide interface decreased. Any damaged layer was not observed in both $B_{18}H_x^+$ and B^+ implanted poly-Si. The hydrogen level of the annealed p^+ poly-Si gate implanted with $B_{18}H_x^+$ was similar to that of sample implanted with B^+.

Improvement in the inversion capacitance (more then 3%) was observed in the case of the cluster ion implantation. It was shown that the flat band voltage shift and leakage current for the cluster ion implantation was at a same level as those for the conventional sample.

References

1. Y. Sasaki et al, Symp. On VLSI Technology (2004) p. 180
2. K. Goto et al, Tech. Dig. Of IEDM (1996) p. 435
3. D.C. Jacobson et al, International Conference on Ion Implantation Technology (2000) p. 300

A Beam Line System for a Commercial Borohydride Ion Implanter

H. F. Glavish*, T. N. Horsky*, D. C. Jacobson*, F. Sinclair*, N. Hamamoto[†], N. Nagai,[†] and M. Naito[†]

*SemEquip Inc., 34 Sullivan Road, Suite 17, Billerica, MA 01862, USA
[†]Nissin Ion Equipment Co., Ltd, 575 Kuze Tonoshiro-cho, Minami-ku, Kyoto, Japan 601-8205

Abstract.
We describe the features of a beam line and ion source system which is being developed for a commercial ion implanter capable of meeting the challenges of high dose, low energy implants needed to fabricate integrated circuits with critical dimensions of 60 nm and less. Intense borohydride ion beams of $B_{10}H_x^+$ or $B_{18}H_x^+$ generated from the source are used to achieve commercially acceptable wafer throughputs for the low energy, high dose applications such as poly-gate and source drain extension implants. The beam transport elements, from ion source to wafer, are designed to achieve wafer boron currents of greater than 30 pmA at an implant energy of 2-4 keV, and greater than 3 pmA at an energy as low as 200 eV. These high currents are obtained at low energy without the need for deceleration just prior to the wafer. Consequently, the beam impinging on the wafer is very pure with respect to energy, and is free of high energy components that can generally degrade shallow junction implants.

After magnetic analysis the beam is parallel magnetically scanned across the wafer at a frequency in the range of 100-200 Hz. In conjunction with a serial end-station, implants with high quality dose and angle uniformity can be achieved using a wafer mechanical scan rate as low as 0.5 Hz in a direction orthogonal to the beam scan direction.

As well as the high wafer throughput performance obtained by using borohydride ions, the beam-line is also capable of transporting conventional monatomic ions up to a maximum energy of 80 keV with a mass-energy capability of 12.6 amu.MeV.

Keywords: Ion Implanter, Analyzer Magnet, Cluster Ion, Decaborane, Octadecaborane, Shallow Junction>
PACS: 41.75.Ak, 41.85.-p, 41.85.Ar, 41.85.Ja, 412.85.Lc

SUMMARY

The potential advantages of using molecular ions containing multiple dopant atoms have been well recognized for several years and can be summarized as follows:

1. The dose received by a wafer for a given ion beam current is increased in proportion to the atomic multiplicity of the dopant contained in the molecular ion.
2. For a given dose, the wafer receives a lower electrical charge, inversely proportional to the atomic multiplicity of the dopant ion.
3. The ions can be extracted from the ion source and transported to the wafer at a much higher energy in proportion to the ratio of molecular ion mass to dopant atomic weight. Consequently, the limitations associated with space charge forces and the intrinsic thermal ion temperature within the beam are much less.

However, there are a number of drawbacks when either decaborane $B_{10}H_{14}$ or octadecaborane $B_{18}H_{22}$ is used as a source vapor in a conventional ion implanter. Firstly, the ion source of a conventional ion implanter has a relatively high density plasma which breaks up most of the borohydride molecules before singly charged molecular ions are produced. Secondly, referring to Figure 1, a range of different ion masses exist in the extracted beam arising from the different number x of hydrogen atoms (and also the different mixtures of the boron isotopes ^{10}B and ^{11}B) in the generated ions $B_nH_x^+$. Different mass ions describe different paths on passing through the analyzer magnet and other mass dispersive elements that might be present in the beam line. In turn, this can introduce undesirable angular and/or dose variations across the wafer surface. Finally, the relatively high mass of the molecular ions limits the single atom implant energy to just a few keV because of the limited field strength and size of the analyzer magnet (and other magnetic elements if used).

To work around these difficulties SemEquip [2] has now developed a commercial very high current borohydride ion source that uses a formed electron beam rather than a high density plasma to ionize the borohydride vapors. This is used on the beam line shown in Figure 3 which has been jointly developed by SemEquip Inc., and Nissin Ion Equipment Co., Ltd. It includes an analyzer magnet, a magnetic triplet quadrupole beam focusing element, and a magnetic scanner and collimator combi-

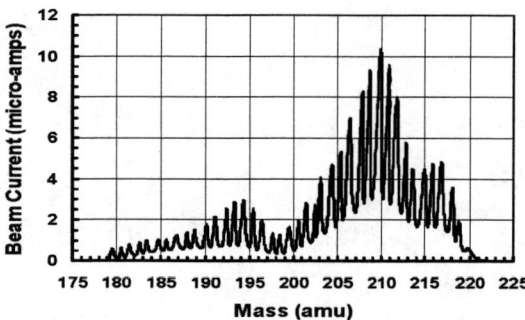

FIGURE 1. High resolution mass spectrum of ionized octadecaborane $B_{18}H_{22}$.

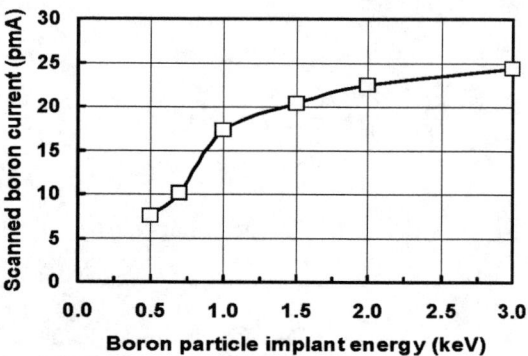

FIGURE 2. Magnetically scanned $B_{18}H_x^+$ beam currents.

nation to parallel scan [1],[5] the ion beam across the wafer. The analyzer magnet has a large working aperture to accept the beam from the large aperture (12.5 mm wide × 100 mm high) of the SemEquip ion source. It is able to analyze 80 keV octadecaborane ions, corresponding to a 4 keV particle boron implant energy (or 7 keV in the case of decaborane). Even for a source slit width as large as that used in the SemEquip ion source, a mass resolution of $m/\Delta m \geq 60$ can be realized by the analyzer magnet system, which is sufficient to properly filter conventional dopant ions. If fitted with a universal ion source, capable of providing both conventional and borohydride ions and expected [3] to be commercially available by mid 2007, the ion delivery system will meet the fab requirements of a fully utilizable tool that, in effect, minimizes device manufacturing costs for 60 nm and below.

Figure 2 shows the magnetically scanned boron particle currents, derived from octadecaborane, measured at the exit port of the collimator vacuum housing labelled in Figure 3. The beam current was essentially unchanged over the entire scan sweep frequency range from dc to 170 Hz. The measured particle beam currents are very much higher than hitherto reported from conventional fixed beam, high current ion implanters. Furthermore, these beam currents have been achieved in drift mode - i.e. without the need to use deceleration just prior to the wafer - therefore avoiding undesirable implantation of higher energy particles neutralized prior to or during deceleration.

ANALYZER MAGNET

The analyzer magnet has a bending angle of $\phi = 120$ degrees and a center bending radius of $R = 500$ mm. The nominally uniform gap of 118 mm and pole width of 166 mm is sufficient to accept the beam emerging from the SemEquip source [2] and allow graphite liners to cover the pole faces, generally needed to avoid beam sputtering of heavy metallic ions from the iron pole surfaces. The entrance and exit pole edges are normal to the beam axes and there are no significant first order field gradients in the working gap of the magnet. Consequently, in the dispersive plane (i.e. the horizontal plane for the beam line arrangement shown in Figure 3), the conjugate image points for the source object and mass resolving aperture are simply determined by Barber's rule [4]. Specifically, the object source point is set at 400 mm prior to the effective entrance field boundary and the mass resolving aperture is at $b = 195$ mm from the effective exit field boundary of the magnet. The object distance of 400 mm is the minimum space needed to accommodate high speed vacuum pumping, an in-line vacuum isolation valve, and a wide energy range extraction optics system.

On passing through the magnet the paths of ions having the same energy but differing mass become spatially separated. For a mass deviation $\Delta m/m$ the spatial separation Δx at the mass resolving aperture is given by [4]

$$\Delta x = D \frac{\Delta m}{2m} \quad (1)$$

where the quantity D, called the dispersion, is given by

$$D = R(1 - \cos\phi) + b\sin\phi. \quad (2)$$

The dispersion D and mass resolving power $m/\Delta m$ increase with bend radius R and bend angle ϕ, in the regime $\phi \leq 180°$. Referring to Figure 4, which applies to the present case of $b = 195$ mm, it may be seen for example that increasing the bend angle from just 70° to 120° increases D and hence $m/\Delta m$ by a factor of 1.7. For the 120° magnet here, Equation (2) yields a dispersion value of $D = 919$ mm.

As for the mass resolving power, we observe that the ion beam is typically formed with a waist near the source aperture region and with a horizontal width X_o equal to

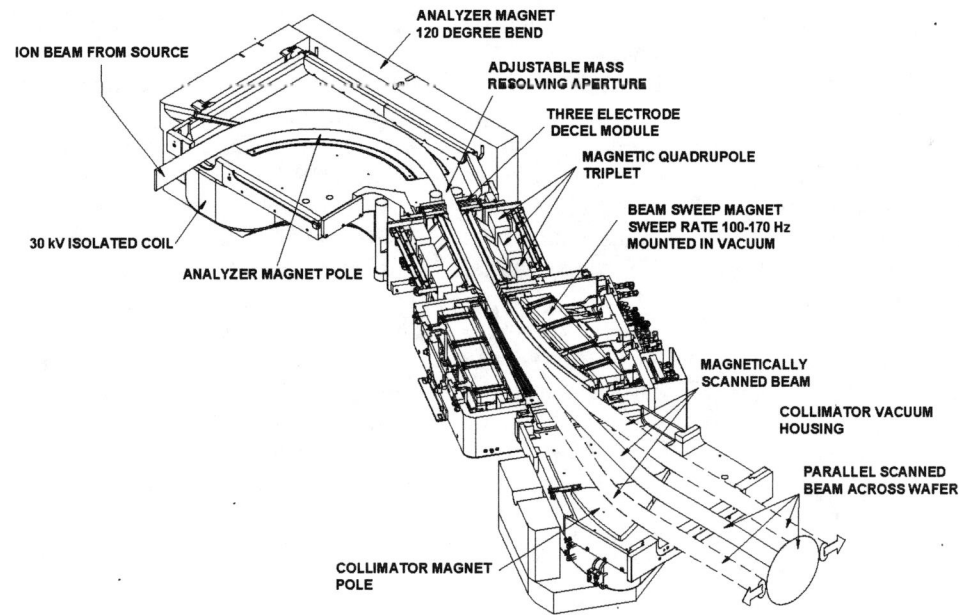

FIGURE 3. Beam line components.

about 60% of the source aperture width - i.e. $X_o \simeq 8$ mm. Therefore, in order to transmit all the ions in the beam of a given mass m through the horizontal mass resolving aperture, the width X_i of the latter is set at a minimum value of $X_i = |M| X_o$ where M is the conjugate image magnification of the beam waist X_o. The realizable mass resolution is therefore

$$\frac{m}{\Delta m} = \frac{D}{2X_i} = \frac{R(1-\cos\phi) + b\sin\phi}{2|M|X_o}. \quad (3)$$

For the present magnet $M = -0.838$ and therefore $m/\Delta m \simeq 68$ which is entirely adequate for the case of conventional ions and corresponds to a mass resolving aperture width of $X_i \simeq 6.7$ mm.

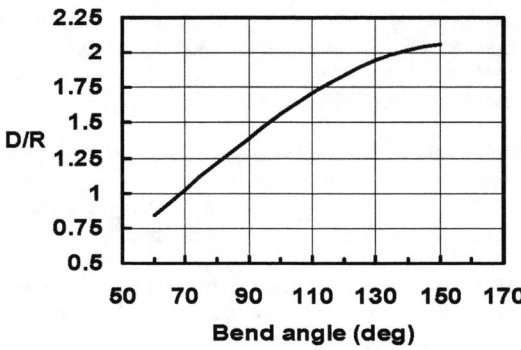

FIGURE 4. Variation of dispersion with bend angle.

In principle it is possible to increase the dispersion and mass resolving power by increasing R. Also, for a given

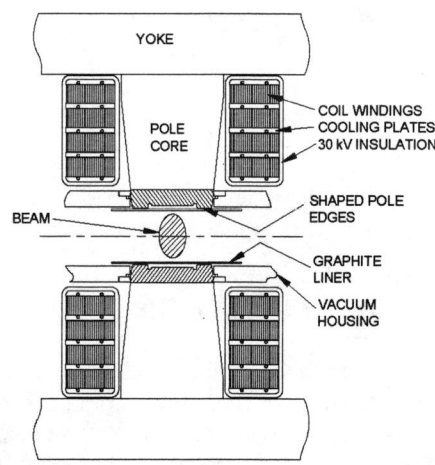

FIGURE 5. Analyzer magnet pole and coil detail.

bending power, increasing R reduces the magnetic field strength needed in the gap, the coil power, and finally the overall weight of the magnet. However, from another standpoint, it is generally preferred to keep $R\phi$ at a minimum because this reduces the path length through the magnet. The value $R = 500$ mm selected in the present magnet requires a gap field of 1.2 tesla in order to bend 80 keV octadecaborane ions. Up to this value of field, the fringing field shapes do not vary significantly with field strength. Consequently the poles can be kept reasonably narrow and shaped as shown in Figure 5 to provide appropriate second, third, and fourth order field components that minimize image aberrations at the mass resolving aperture over the entire magnetic field range.

In the case of borohydride ions, the mass resolving aperture width X_i needs to be set at a wide enough value to accept several of the borohydride peaks in order to achieve a high beam current. For example, in the case of octadecaborane, mass peaks from about 205 amu to 218 amu, i.e. $\Delta m/m = 0.061$, account for most of the useful beam and are all selected by setting the resolving aperture width at about $X_i = (D/2)(\Delta m/m) = (919/2)(0.061) = 28$ mm. In the case of decaborane the usable mass peaks lie between 108 amu and 115 amu which requires a similar resolving aperture width of 29 mm.

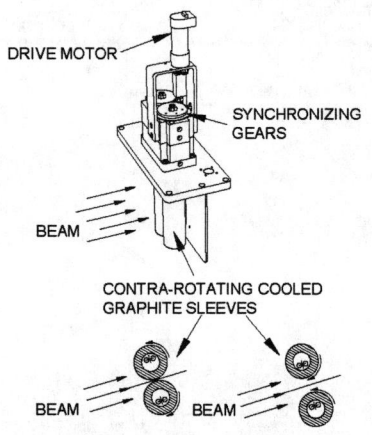

FIGURE 6. Adjustable (6-30 mm) mass resolving aperture.

DECELERATION MODULE

A three electrode deceleration module is located immediately after the mass resolving aperture as shown in Figure 3. Its purpose is to boost low energy beam currents with up to 30 kV deceleration. An adjustable focus voltage applied to the center electrode compensates for the strong divergence generated in the horizontal plane by the internal space charge forces of the beam. Although the applied deceleration voltage biases the entire magnet body from ground potential, a 30 kV insulating cocoon built around the coil windings and cooling plates, as shown in Figure 5, enables the coil power supplies and cooling water to remain at nominal ground potential, avoiding the need for costly high power transformer isolation and high volume deionized water cooling.

It is important to emphasize that high energy particles remaining in the beam after deceleration do not reach the wafer because they are filtered out of the beam by the 30 degree deflection produced by the scanner and collimator combination.

QUADRUPOLE TRIPLET

The magnetic quadrupole triplet shown in Figure 3 is configured as a DFD (i.e. defocus-focus-defocus) array for ion motion in a horizontal plane and, correspondingly, FDF for motion in a vertical plane. The beam size and angular divergence at the wafer are controlled by differentially adjusting the strength of the individual quadrupole elements in the triplet. In the case of borohydride ions the triplet also compensates for the collective mass dispersion introduced by the analyzer magnet, beam scanner, and collimator. In fact, by appropriately setting the quadrupole element strengths, the angular deviation arising from the multiple mass components can be reduced to less than 0.15 deg over the entire scan range.

CONCLUSIONS

This joint development by SemEquip Inc., and Nissin Ion Equipment Co., Ltd. clearly shows the practicality of and the tremendous improvement in drift-mode beam current that can be realized by using borohydride molecules. The results pave the way for a new generation of ion implanter tools and have put to rest previous and somewhat widely held concerns that such beams could turn out to be difficult to transport, and even more difficult to scan, in the vacuum system and general beam line architecture commonly used in ion implanters. Even with the long beam path through the scanning and collimator magnets, gas attenuation measurements show that the beam loss from gas scattering, neutralization, and ion break-up, is only a few percent.

REFERENCES

1. H. F. Glavish, *System and method for producing oscillating magnetic fields in working gaps useful for irradiating a surface with atomic and molecular ions*, US Patent 5,311,028, May 1994
2. SemEquip Inc., Billerica, MA 01862, USA, *Model 350 Ion Source*, 2005.
3. SemEquip Inc., Billerica, MA 01862, USA, *Private communication*, 2006.
4. H. A. Enge, *Deflecting Magnets*, Focusing of Charged Particles, Vol II, Ed A. Septier, Academic Press, New York, p203-264, 1967,
5. H. F. Glavish et al, *System and method for unipolar magnetic scanning of heavy ion beams*, US Patent 5,393,203, August 1995.

Ultrashallow P$^+$/n junction formed by B$_{18}$H$_{22}^+$ Ion Implantation and Excimer Laser Annealing

Sungho Heo[1], Dongkyu Lee[1], H.T. Cho[2], W.A. Krull[2], and Hyunsang Hwang[1]

[1]*Department of Materials Science and Engineering, Gwangju Institute of Science and Technology, #1, Oryong-dong, Buk-gu, Gwangju, 500-712, Korea*
[2]*SemEquip, Inc, 34 Sullivan Road, Billerica, Massachusetts, USA, 01862*

Abstract. Conventional ultrashallow junction processes require two-step implantation such as pre-amorphization by Si$^+$ or Ge$^+$ implantation and ultra-low (<0.5 keV) energy B$^+$ implantation. In this report, we investigate B$_{18}$H$_{22}$ molecular ion implantation. Due to the heavy mass of cluster ions, one step ion implantation at equivalent implant energy of 0.25 keV readily forms a 5 nm-thick a-Si layer and an ultrashallow junction without boron channeling. By employing excimer laser annealing (ELA), we have obtained a shallow junction depth (<10 nm) and low sheet resistance (~830 ohm/sq.). When applied pre-annealing step prior to ELA, junction depth can be further scaled down with reduced residual damage.

Keywords: B$_{18}$H$_{22}^+$ cluster ion implantation, Excimer laser annealing, Pre-annealing effect.

INTRODUCTION

In order to satisfy the formation of ultrashallow p$^+$/n junction in sub-45nm complementary metal-oxide-semiconductor (CMOS) technology node, conventional B$^+$ and BF$_2^+$ implantation is facing its scaling limit. Because low energy B$^+$ implantation has a low throughput and large boron channeling due to the low beam current and small atomic mass of boron.[1] In order to reduce the boron channeling, conventional ultrashallow junction processes require two-step implantation such as pre-amorphization by Si$^+$ or Ge$^+$ implantation and ultra-low (<0.5 keV) energy B$^+$ implantation. However, this implantation makes a larger end-of-range (EOR) damage in amorphous/crystal (a/c) interface than one-step implantation.[2] For BF$_2^+$ implantation, it can reduce the boron channeling but the dopant activation rate normally degrades due to the presence of fluorine.[3]

As an alternative method, molecular boron ion implantation has been studied in the last 20 years, but B$_{10}$H$_{14}^+$ (decarborane) usually showed limited improvement.[4] So, we have studied the advantages of B$_{18}$H$_{22}^+$ (octa-decaborane). B$_{18}$H$_{22}^+$ can effectively make an amorphous-silicon (a-Si) layer without additional ion implantation and reduce boron channeling due to its 20 times higher mass than monomer boron. It has various advantages such as high throughput, low beam divergence, and high dose implantation. However, like B$^+$ implantation, B$_{18}$H$_{22}$ molecular ion implantation also shows transient enhance diffusion (TED) during conventional rapid thermal annealing (RTA), which increases the junction depth due to the long annealing time (~seconds).[5]

To solve this problem, we apply excimer laser annealing (ELA) in B$_{18}$H$_{22}^+$ implanted samples. ELA can achieve a high activation rate above the solid solubility limit and reduce TED due to the meta-stable process and short annealing time (~nanoseconds).[6] In addition, we evaluate the pre-annealing effect prior to ELA.

EXPERIMETAL DETAILS

B$_{18}$H$_{22}$ molecular ions at an energy of 5 ~ 20 keV were implanted to an actual boron dose of 3×10^{14} ~ 1×10^{15} /cm^2 into n-type Si substrates. The equivalent implant energy is 0.25 ~ 1 keV according to the ratio of boron per cluster boron mass (1 / 20 = B$_{mass}$ / B$_{18}$H$_{22\,mass}$).^{8} ELA was conducted with an energy of 300 ~ 600 mJ/cm^2 at one pulse. For comparison with ELA, RTA was performed at 950 °C for 10 sec. Pre-annealing was conducted at 400 ~ 500 °C for 5 min in N$_2$ ambient. The sheet resistance (R$_s$) and active carrier concentration (N$_s$) were measured by van der Pauw pattern and Hall measurement. A high-resolution cross-sectional transmission electron microscopy (HR-XTEM) analysis was conducted in order to confirm the crystallization of the sample and to assess defect evolution. The boron dopant depth profiles were obtained by secondary ion mass spectrometry (SIMS) at Charles Evans and Associates. In order to obtain an accurate dopant profile near the surface region, a SIMS analysis was carried out without O_2 leak.[7]

RESULTS & DISCUSSION

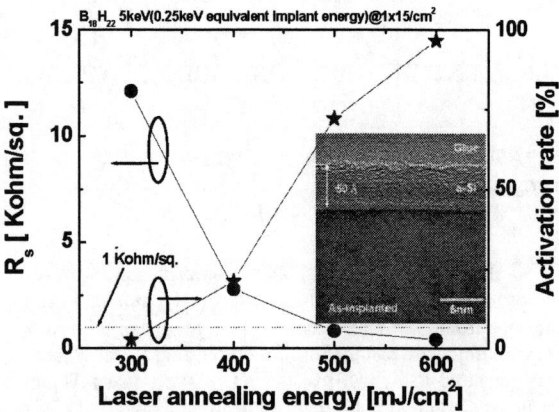

FIGURE 1. Variation of sheet resistance (R_s) and activation rate (%) as a function of ELA energy. The inset of the HR-XTEM image shows the $B_{18}H_{22}^+$ as-implanted samples

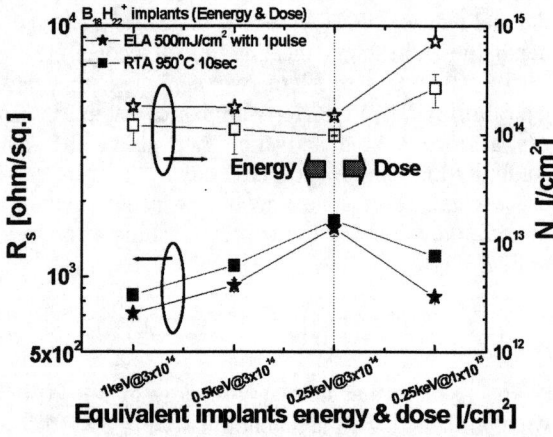

FIGURE 2. Variation of R_s and N_s as a function of implanted conditions (energy & dose) with annealing methods (RTA & ELA). Boron equivalent implantation energy was calculated by cluster ion extraction energy.

Figure 1 shows R_s and dopant activation rate (%) followed by ELA. The inset shows an HR-XTEM image of an as-implanted sample. Up to 500 mJ/cm² of laser energy, the R_s decreased below 830 ohm/sq, and the dopant activation rate was increased above 72 %. This indicates that laser energy above 500 mJ/cm² is sufficient for dopant activation, perhaps owing to re-crystallization of an a-Si layer (inset of HR-XTEM image). Given that high energy ELA above 600 mJ/cm² is not applicable for the conventional CMOS process due to the gate deformation and deeper junction down (X_j),[8] the proposed approach of $B_{18}H_{22}^+$ implantation with ELA at 500 mJ/cm² is expected to mitigate the problem of process integration.

The critical ELA energy is directly related to the a-Si thickness and implanted dose for dopant activation. Hence, we prepared various implant samples with ELA at 500 mJ/cm² and RTA, as shown in Fig. 2. Compared with RTA, ELA shows improved R_s and N_s results, which are attributed to dopant activation above the solid solubility limit. The sample implanted at energy of 1 keV shows lower R_s and higher N_s values after ELA than that of the control sample implanted at 0.25 keV, which can be explained by the deeper junction depth. The high dose implanted sample shows a lower R_s value than that of the control sample without any increase in the implantation energy due to the increase of dopant activation.

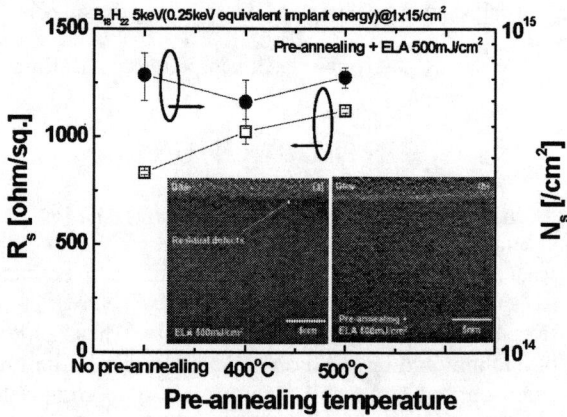

FIGURE 3. Variation of R_s and N_s as a function of ELA with and without pre-annealing. The insets of the HR-XTEM images show the sample without pre-annealing (a) and that with pre-annealing (b).

In order to control a-Si thickness and residual defects, we applied a pre-annealing step for solid phase epitaxial regrowth (SPER) of a-Si prior to ELA. Figure 3 shows the effect of low-temperature annealing prior to ELA at 500 mJ/cm². With increasing pre-annealing temperature, R_s also increased. This increase is attributed to the reduction of a-Si thickness, as shown in Fig. 4. In spite of lower X_j, N_s is similar to that of the ELA sample due to a reduction of residual defects, as shown in the inset of the HR-XTEM image. The pre-annealing step prior to ELA serves to control the junction depth and reduce residual defects.

Figure 4 shows the boron SIMS depth profile after various process conditions. For comparison, the as-implanted boron profile was calculated by Monte Carlo simulations. Compared with the simulated boron depth profile, the $B_{18}H_{22}^+$ implanted sample shows a dramatic reduction of the boron channeling tail and improved vertical abruptness. After ELA at 500 mJ/cm^2, diffusion of the boron profile was almost negligible, which can be explained by non-melting laser annealing. For a channel doping concentration of 5×10^{18}/cm^2, X_j was 9 nm and the vertical abruptness was ~2.6 nm/dec. In the case of a low-temperature pre-annealed sample, the boron depth profile was slightly shifted to the surface region (~0.6 nm). This is attributed to the SPER of the a-Si layer and out-diffusion of incorporated hydrogen.

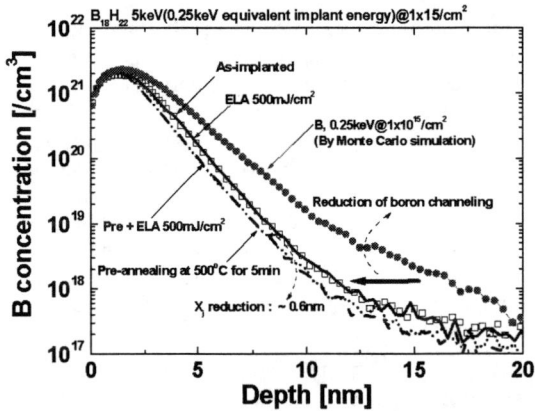

FIGURE 4. SIMS depth profile of boron before and after ELA with or without pre-annealing. The Monte Carlo simulation results show the as-implanted SIMS depth profile of B^+ implantation.

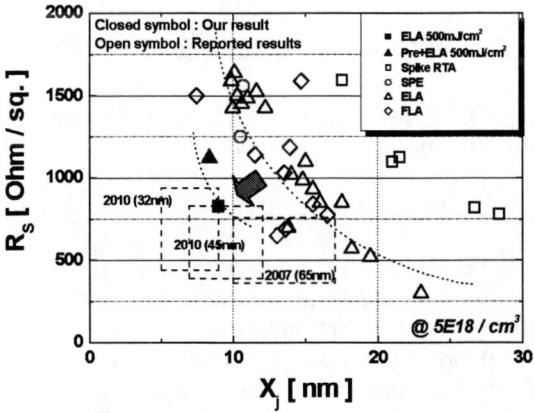

FIGURE 5. Relationship of R_s vs. X_j. The closed and open symbols denote this experiment and reported results, respectively. The reported results were obtained by various annealing methods such as RTA, Spike RTA, SPE (Solid phase epitaxy), ELA, and FLA (flash lamp annealing).

Figure 5 shows the relationship of R_s and X_j obtained in this experiment as well as reported results.[9] Implantation of various dopants such as B^+, BF_2^+ and various annealing methods were evaluated. Compared with the reported results, our samples show shallower X_j and lower R_s. In addition, we can further scale down X_j by applying low temperature annealing prior to ELA.

CONCLUSION

In summary, a sub-10nm ultrashallow p^+/n junction formed by $B_{18}H_{22}$ molecular ion implantation and ELA is demonstrated. The ELA yielded shallow X_j (~9 nm) and low R_s (~830 ohm/sq) due to the none-melting laser annealing process. By employing low-temperature pre-annealing prior to ELA, we can further scale down X_j through a reduction of the a-Si thickness. Considering the various advantages noted here, the proposed approach of $B_{18}H_{22}$ molecular ion implantation with ELA appears to be a promising process for the sub-45nm CMOS technology node.

ACKNOWLEDGMENTS

This work was supported by the Center for Distributed Sensor Network at GIST and Brain Korea 21 project.

REFERENCES

1. X. Lu, L. Shao, X. Wang, J. Liu, W. Chu, J. Bennet, and L. Larson, and P. Ling, *J. Vac. Sci. Technol. B* **20**, 992-994 (2002).
2. J. Borland, S. Shishiguchi, A. Mineji, W. Krull, D. Jacobson, M. Tanjyo, W. Lerch, S. Paul, J. Gelpey, S. McCoy, J. Venturini, M. Current, V. Faifer, R. Hillard, M. Benjamin, T. Walker, A. Buczkowski, Z. Li, J. Chen, *6th international Workshop on Junction Technology, Shanghai*, 2006, pp. 4-9.
3. K. Goto, J. Matsuo, Y. Tada, T. Sugii, and I. Yamada, *IEEE Trans. Electron Devices* **46**, 683-689 (1999).
4. D. Takeuchi, N.Shimada, J. Matsuo, and I. Yamada, *Nucl.Instrum. Methods Phy. Res. B* **121**, 345-348 (1997).
5. D. Jacobson, *5th international Workshop on Junction Technology, Osaka*, 2005, pp. 23-26.
6. G. Fortunato, L. Mariucci, A. L. Magna, P. Alippi, M. Italia, V. Privitera, B. Svensson, and E. Monakhov, *Appl. Phys. Letters* **85**, 2268-2270 (2004).
7. T. H. Buyuklimanli, C. W. Magee, J. W. Marino, and S. R. Walter, *J. Vac. Sci. Technol. B* **24**, 408-413 (2006).
8. M. Hernandez, J. Venturini, D. Zahorski, J. Boulmer, D. Débarre, G. Kerrien, T. Sarnet, C. Laviron, M. N. Semerica, D. Camel, and J. L. Santailler, *Appl. Surf. Science* **208**, 345-351 (2003).
9. K. Suguro, T. Ito, K. Matsuo, T. Iinuma, and K. T. Nishinohara, *4th international Workshop on Junction Technology, Shanghai*, 2004, pp. 18-21.

Productivity Enhancements for Shallow Junctions and DRAM Applications using Infusion Doping

John Hautala, Matt Gwinn, Wes Skinner and Yan Shao

Epion Corporation, 37 Manning Road, Billerica MA 01821

Abstract. The inherent challenges for traditional ion implanters to deliver sufficiently high beam currents at sufficiently low energies required for shallow doping applications such as USJ and polysilicon doping for DRAM and has motivated the development of alternative methods for shallow doping. These include large molecule implantation, plasma doping and infusion doping with ionized gas clusters. Recent advancements in infusion doping have led to a very high flux (>100mA) of boron at low equivalent monomer implant energies (<1keV). In addition to producing uniquely high equivalent beam currents, the infusion process is physically distinct from implantation (beamline, molecular, or plasma) in the following important ways : (1) the doping depth is independent of species mass; (2) the doping depth is proportional to the one third power of the acceleration energy; (3) infusion doping is fully self-amorphizing with no evidence of channeling or end of range (EOR) damage; (4) there is no evidence of any self-sputtering behavior at any energy and no B build up on the surface even for very high doses (>5E16/cm2), and (5) the gas cluster beam operates at very low power requiring no wafer cooling and showing good compatibility with photo resist. In addition to discussing these differentiations, the manufacturability of the process will be discussed.

Keywords: infusion doping, ultra shallow junctions, USJ, poly doping, gas cluster, GCIB
PACS: 61.72-Tt; 41.75-I; 39.10+j; 36.40-c; 36.40-Cg; 36.40-Jn; 36-40-Wa

INTRODUCTION

As device dimensions decrease, new methods of efficiently doping to shallow depths are required. Similar manufacturing issues exist for both the precision doping of ultra shallow junctions (USJ) in the source drain extensions (SDE) and the p+ compensation doping of dual line DRAM devices. In both applications conventional beamline implantation may be reaching fundamental limitations, and many companies are exploring alternatives such as plasma doping, large molecular doping, and gas cluster infusion doping. Each of these relatively new technologies have advantages and disadvantages; here the differentiation of the infusion process from the other techniques will be outlined.

The physics of ion implantation has been studied in detail and is relatively well understood. Although they have substantially different methods of delivering the doping species to the substrate, traditional monomer and BF_2 beamline, plasma, and molecular doping all appear to follow traditional monomer ballistic style implantation. The behavior and properties of infusion doping with gas cluster ion beams (GCIB); however, is an entirely distinct physical process with significantly different behavior from implantation. There are several physical property differences which are relevant to shallow doping applications and will be discussed below.

INFUSION DOPING PROPERTIES

Infusion processing utilizes (GCIB) technology, which has been described in detail elsewhere [1,2,3], The GCIB shallow doping applications use gas mixes with a small percentage (1-5%) of a doping gas, (i.e. B_2H_6) or a mixture of gases (i.e. $GeH_4+B_2H_6$) diluted in an inert gas such as Ar or He. The gas is delivered to the machine at ~10atm and through adiabatic expansion clusters of typically >5000 atoms are formed, ionized, accelerated up to 60 keV and transported to the substrate. The transfer of the cluster energy to the surface results in a rapid heating of a semispherical volume and the molecules in the cluster are disassociated into its constituent atoms and intermixed or infused with the substrate. The gas atoms (H, Ar, He, etc.) will instantaneously leave the surface and all soluble species will infuse to the exact same depth in the substrate independent of atomic mass. A typical 5 keV infusion doping profile is shown in Fig. 1. The extreme abruptness and lack of channeling or energy contamination compared to a standard 500eV beamline implant is a consequence of the different mechanism of the infusion doping process. The penetration depth of the doping species into the substrate is not determined by the energy of individual atoms or molecules (<5eV/atom for 5keV GCIB), but rather exclusively by the collective energy of the cluster.

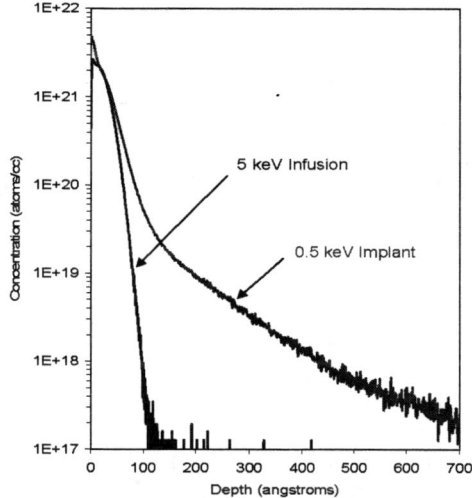

FIGURE 1. Comparison of 1E15/cm2 B 5keV infusion and 0.5 keV traditional beamline doping.

The abruptness of the doping profile results in improved leakage and short channel effects (SCE) for USJ, and the lack of energy contamination and very high equivalent beam currents also make infusion doping useful for p+ poly doping.

Mass Independent Doping

Perhaps the most straightforward and intuitive example of the useful distinction between infusion and all the other doping techniques is the complete independence on mass for doping depths. The profiles and depths are exclusively determined by the collective energy of the cluster which directly determines the super-heating and intermixing depth into the substrate. Any species contained within the same cluster will infuse the same depth as shown in Fig 2. This unique capability of doping all species to the exact same depth allows for a 'cocktail' of multiple gas species such as C, F, As, P, etc. to be used in a single step (no 'chained implants' required) with the assurance of all species going to the exact same depth. The infused Ar dilution gas it typically below the SIMS sensitivity; however, when it has been observed it follows the same profile as the dopant species and is not measurable post annealing. The ratio of the retained doses is determined by the

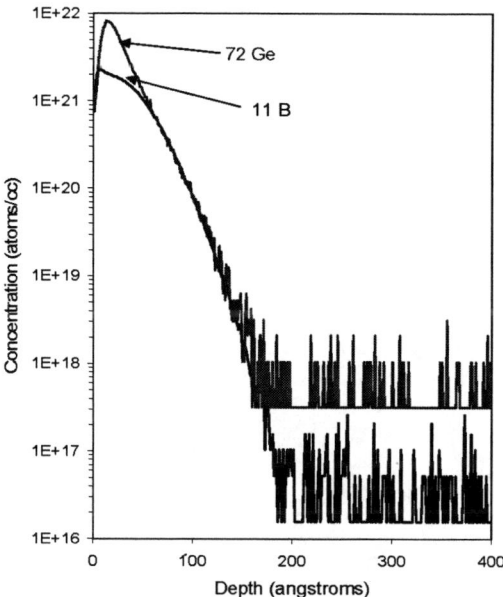

FIGURE 2. SIMS profiles for the simultaneous infusion of B_2H_6 and GeH_4.

gas mix into the nozzle of the machine, and thus the doping ratios or stoichiometry of the resulting compound materials such as SiGeB can be easily controlled by mass flow controllers and/or the ratio of gas species pre-mixed in the bottle [4,5].

$E^{1/3}$ Depth Dependence

At the lower implant energies required for USJ and DRAM poly doping, all implant techniques are a ballistic process and thus have essentially a linear relationship between depth of the implant (determined at $1E18/cm^3$ concentrations) and the acceleration of the ion. The slope of this relationship is mass dependent; however, this strong relationship results in a problematic susceptibility to energy contamination and thus a problem with tight control of a junction depth. This strong dependence on energy also means all shallow doping processes are necessarily restricted to very low beam energies where productivity becomes problematic. This is especially true for beamline tools operating without the benefit of deceleration. On the other hand, for infusion the hemispherical volume of the inter-mixing of the doping species and the substrate is exclusively and directly determined by the collective energy of the cluster, and thus the depth follows a very distinct energy to one third power dependences as shown in Fig. 3. This energy dependence makes GCIB highly efficient for shallow doping, but also restricts the infusion process to doping applications of <40nm regardless of doping species.

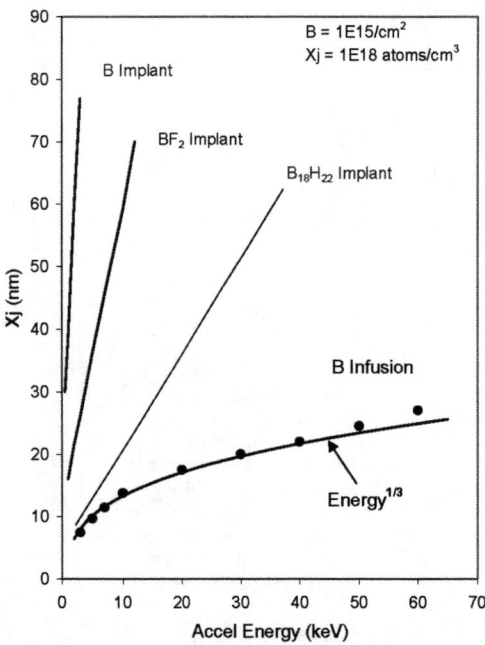

FIGURE 3. Doping depth as a function of beam acceleration energy for increasing molecular mass beamline implantation and GCIB infusion doping. The B, BF_2 curves were taken after [6,7] and $B_{18}H_{22}$ curve was generated from .[7,8].

Self-Amorphizing with no EOR Damage

When each cluster arrives at the substrate surface, the combination of extremely high pressure (>Mbar) intermixing of gas species and the Si substrate and a high temperature (>5000K) thermal spike effectively produces a fully amorphous layer of the approximate depth of the B infusion as shown in Fig. 4. The three dimensional cooling of each of these hemispheres is too quick (<10ps) for the Si surface to re-crystallize, and thus all infused surfaces remain fully amorphous independent of process conditions such as dose, energy, gas species, substrate material, etc. The fact that there are no high energy ballistic ions means there are no knock-on effects and thus no pile-up of interstitials at the end of the doping range. The images in Fig. 4 show an example of a 60 keV as-infused amorphous layer, here ~5nm thick, and the restoration of the crystal after the anneal where no evidence of the EOR damage typically observed in pre-amorphizing implants (PAI). This lack of the EOR is likely responsible for the superior leakage performance of infusion doped USJ devices.[9]

No Self-Sputtering Limits

It is a well known phenomenon for monomer and molecular implantation that the competing effects between implantation/deposition and etching become more problematic at lower implant energies and higher doses. This becomes important for producing both junctions shallower than 15nm for USJ and for >E16/cm2 compensation doping for poly thicknesses <80nm where low energy implants are required. As shown in Fig. 5, the infusion process has a linear relationship between the GCIB dose and the retained B dose in Si. This linear relationship extends from low doses (<1E15/cm2) up through very high doses (>1E17/cm2). This plot is for 60 keV infusion (Xj ~ 30nm), but the linear relationship holds for all cluster beam energies. For the 30nm deep infusion process, B begins to saturate the Si at a dose between 0.5 and $1E17/cm^2$. At doses higher than ~ $1E17/cm^2$ a pure fully dense B film will begin to be deposited on the surface whose thickness is also linearly related to GCIB dose. Analysis of as-infused samples with $5E16/cm^2$ doping levels were analyzed by high resolution XPS and SIMS, and clearly showed no B pile-up on the surface where the concentration of Si on the surface is >30% and all B bonded to Si or Si-O and no B-B bonds were detected.

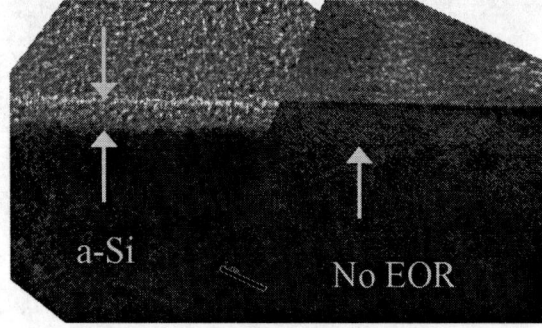

FIGURE 4. Two TEM photos showing the uniform amorphous layer of an as-infused surface (left), and the same sample after a 900C anneal showing no EOR.

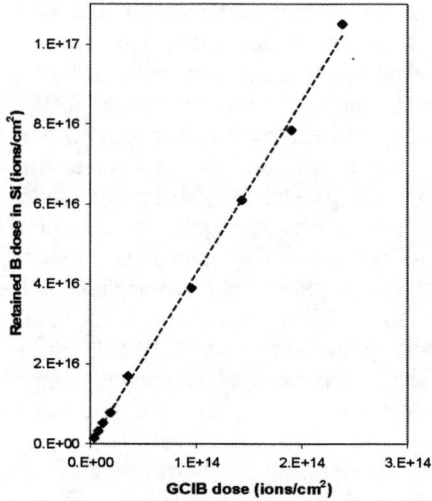

FIGURE 5. Retained B dose vs GCIB dose. Plot demonstrates no evidence of a self-sputtering limit for doping levels up to >1E17/cm2.

Low Beam Power

Since there is less than one ion per 1000 atoms or molecules, the typical maximum beam power delivered to the substrate is <6W for the >100mA boron flux at 60keV used for the DRAM poly doping, and < 1W for the shallower, lower energy USJ doping applications. For both cases there is no appreciable heating of the wafer, and thus no wafer cooling is required or deleterious effects to photo resist. This in combination with the lack of B pile-up on the surface described earlier, allows for a photo resist friendly process requiring no special or aggressive stripping steps, even for the very high doses (>1E16/cm^2).

MANUFACTURABILITY

Significant advances have been made at Epion to demonstrate the readiness of infusion processing for high volume manufacturing. A comparison of equivalent beam currents as a function of equivalent monomer beam energies is shown in Fig. 6. Here assumptions were made that equivalent monomer beam energies of 0.1, 0.2, 0.4, and 0.6 keV result in Xj values of approximately 7.5, 12, 20 and 28nm at 1E18/cm^3 B levels respectively. The appropriate infusion energies were used to match these Xj values. Other process performance parameters such as uniformity of <0.5%, metal contamination of <5E9 for all metals, and up-time of >200hrs with wafer-to-wafer repeatability of <2% one sigma and low particle levels all appear production worthy and are described in detail elsewhere [11].

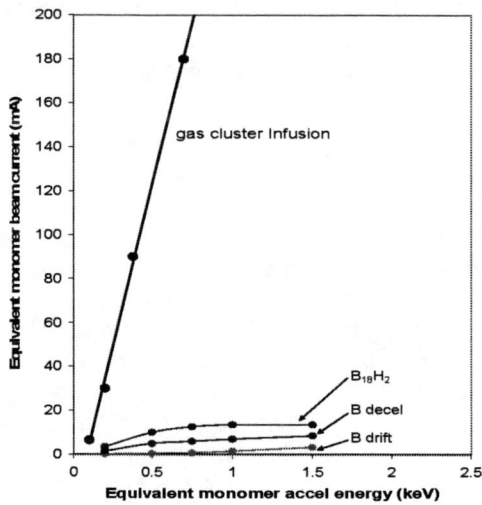

FIGURE 6. Productivity comparison of representative monomer drift and decel mode beamline currents [10] molecular doping equivalent current [8] and gas cluster infusion monomer equivalent current. The data is normalized to Xj values for B implantation at a given energy.

Device Performance

The advantages of infusion's ultra-abrupt profiles and lack of EOR are observed in device properties. Renesas has measured improved short channel effects (SCE) for 50nm gate length pMOSFETS using 5keV B infusion in comparison to 200eV implants [12,13]. Ho Lee and colleagues at Samsung [9] have also observed improved SCE using GCIB. They also measure a 4x reduction in STI-bounded p+/well junction leakage of infusion doped samples compared with the BF_2+Ge-PAI implantation for a standard 90nm logic process with 50nm physical gate lengths. Using a hybrid laser spike anneal (LSA) and a reduced temperature spike RTA, infusion doped ultra shallow junctions with Xj~18nm and Rs~800 Ohm/sq were produced exhibiting superior activation, lower DIBL (120mV/V) and a systematic inversion Tox reduction compared to the control implants. Specific halo and spacer schemes were needed for preservation of low parasitic series resistance, gate induced leakage and Vth variation in the MOSFETs designed for shallow and abrupt extension junctions.

CONCLUSIONS

The distinct physical differences between the doping mechanisms of infusion and the other implant technologies represent several advantages for USJ and DRAM manufacturing for GCIB technology.

REFERENCES

1. M.E. Mack, *Nucl. Inst. Meth.*B237 (2005) 235-239.
2. M.E. Mack, et al; IIT 2002 Proceedings, IEEE, Piscataway, 2003, p.665.
3. J. Bachand, et al; IIT 2002 Proceedings, IEEE, Piscataway, 2003, p. 669.
4. J.Borland, W. Skinner, J. Hautala; *Solid State Technology*, June 2004, p. 53.
5. J. Borland , *Nucl. Inst. Meth.*B237 (2004) 6-11.
6. J. Borland, *Solid State Technology*, June 2002. p. 83.
7. J. Borland,et al; *Semiconductor International*, January 2005.
8. D. Jacobson; Ext. Abs. IWJT 2005, S2-1
9. Ho Lee et al, IWJT 2006, Shanghai, China, IEEE Press
10. J. Borland et al;, 8th International Workshop on Ultra-Shallow Doping, June 5-8, 2005, p. 201-8.
11. W. Skinner, J. Hautala, M. Gwinn, T. Kuroi; Advanced Semiconductor Manufacturing Conference (2006) Boston, MA. IEEE Press
12. T.Yamashita, et al; Ext. Abs. IWJT 2005
13. T. Kuroi, and Y. Kawasaki, 8th International Workshop on Ultra-Shallow Doping, June 5-8, 2005,p.4-9.

A Vaporizer for Decaborane and Octadecaborane

Doug Adams*, Tom Horsky*, Glen Gilchrist*, Robert Milgate*,
Joe Sweeney** and Paul Marganski**

*SemEquip Inc., 34 Sullivan Road, North Billerica MA 01862, USA
**ATMI, 7 Commerce Drive, Danbury, Connecticut 06810, USA

Abstract. Decaborane ($B_{10}H_{14}$) and Octadecaborane ($B_{18}H_{22}$) are two promising new doping materials for performing very shallow boron implants at high implanter throughput. However, because these new materials are low-vapor pressure solids at room temperature, their delivery to the implanter's ion source requires specialized techniques to deliver the desired mass flow without condensation. Data are presented which describe several features of a vaporizer for producing Decaborane and Octadecaborane flows in a production environment. This paper will also focus on the critical design aspects of the vapor delivery system, including the effects of vaporizer geometry on vapor flow rate, the performance of various flow control systems, and the overall thermal design. In addition, data on physical and environmental, safety, and health properties of these materials are presented. The effectiveness of this system as a stable vapor source for an ion implanter will be described.

Keywords: Decaborane, Octadecaborane, ClusterBoron®, Vaporizer, Ion Source.

INTRODUCTION

Ion implantation utilizing borohydride molecules such as decaborane and octadecaborane allows for the extraction and transport of ion beams at much higher energies. These implant materials offer a solution to the problems of throughput and energy contamination.

The use of these feed materials for ion sources have introduced several technical challenges that must be practically overcome in order to safely and effectively deliver these materials into an implanter for the manufacturing of semiconductor devices.

In order to use decaborane and octadecaborane in implanters, the system must be able to deliver toxic material in vapor form to an ionization source located in a high vacuum system. These materials are solids and have low vapor pressure at room temperature. The vapor stream must be accurately controlled in order to achieve a stable and consistent flow of molecules to the ionization chamber. Reliably controlling these vapor streams requires heating the solid material in order to increase the rate of sublimation and therefore the vapor pressure.

Raising the temperature of borohydrides entails certain safety hazards. These hazards are toxicity, chemical decomposition and chemical reaction of the vapors if brought in contact with certain substances.

For a production implanter, the consumption of solid materials requires their periodic refill. This routine maintenance requires the correct sequence of breaking and making vacuum seals, system re-qualification, and special handling and service procedures to ensure personnel safety.

A comprehensive understanding of the characteristics of these materials, and the incorporation of this understanding into the equipment design, results in a safe and effective vapor delivery system.

VAPOR PRESSURE & THERMAL SENSITIVITY

Solid borohydrides have relatively low vapor pressures at room temperature. Figure 1 shows vapor pressure for decaborane and octadecaborane. These materials have relatively low vapor pressures when compared to conventional implant gases such as those delivered in high pressure gas bottles or SDS™ bottles. The low pressure prevents the application of conventional Mass Flow Controllers (MFC's) for vapor regulation. Raising the temperature in order to increase pressure is limited due to chemical decomposition of the borohydride compounds. Safety considerations also become a factor at elevated temperatures. It is possible that a sudden exposure to air at elevated temperatures can result in ignition of the material.

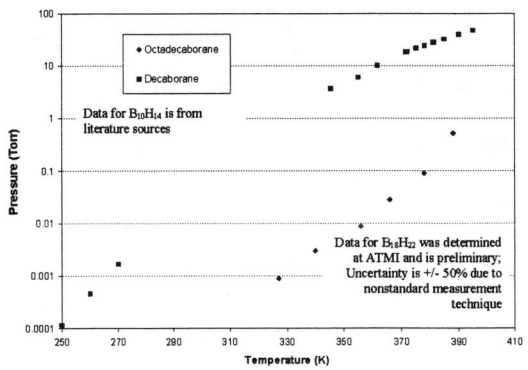

FIGURE 1. Vapor Pressure of decaborane [1,2] and octadecaborane.

Both decaborane and octadecaborane must be heated in order to achieve the pressure (>100mTorr) necessary for vapor transport to an ionization chamber. A pressure difference and a simple means of adjusting the conductance between the vapor source and the ionization chamber must exist in order to regulate the vapor mass flow rate. Heating of the borohydrides must be controlled with sufficient heat transfer surface area (conduction and radiation) to provide an adequate rate of sublimation. Providing excessive heating or heating too quickly may lead to disassociation of the molecules. Perel *et al* report [3] that rapid decomposition of decaborane [1] above 350°C can be readily observed, as indicated in Figure 2. Below 300°C, hydrogen partial pressure is much lower,

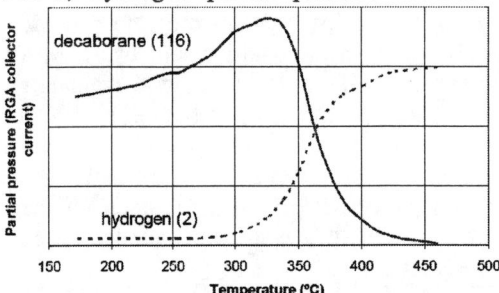

FIGURE 2. Partial pressure of Decaborane and Hydrogen while heating Decaboane until complete decomposition

indicating a very low rate of disassociation. Initial studies by ATMI into the decomposition of octadecaborane is represented in Figure 3. An experiment was conducted wherein the total vapor pressure of a heated sample was observed over time at five different temperatures. As temperature was increased, the rate of pressure rise appeared to increase. This is thought to be due to the buildup of hydrogen disassociated from octadecaborone, much as seen by Perel et al [3] for decaborane.

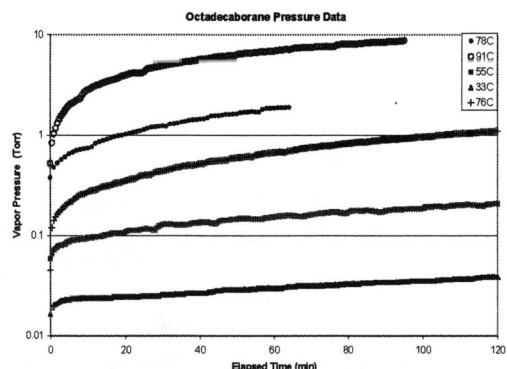

FIGURE 3. Measured vapor pressure of octadecaborane over time at various temperatures.

SUBLIMATION AND CONDENSATION

The solid borohydrides must be heated in order to create sufficient vapor pressure for transport to the source ionization chamber, where the pressure is maintained in the milliTorr range, and the region immediately outside of the ion source is at high vacuum. The gas vapors must be transported at temperatures below the melting point' however, sufficiently high temperature must be maintained to prevent condensation of the vapor. Condensation occurring between the vaporizer and the point of ionization can lead to transport failure. When the vapor path becomes partially occluded, this results in wasted source materials, poor vapor flow control, and beam instabilities that lead to degraded Implanter productivity. The vaporizer and vapor transport system were designed to thermally manage the temperature of the wetted surfaces, as shown in Figure 4.

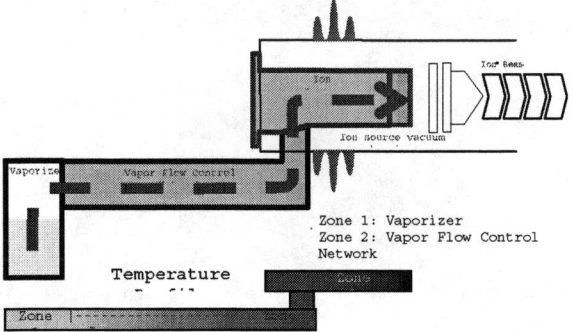

FIGURE 4. Vapor flow path from vaporizer to source ionization chamber

As the vapor travels from its point of sublimation, the walls along the vapor path are at increasingly higher temperature until reaching the ionization chamber. This is accomplished by managing three loosely coupled thermal zones and the effective distribution of heat sources. Aluminum is used for vapor passages and thermal insulation is used to maximize thermal uniformity.

THE VAPORIZER

The vaporizer consists of a two-piece factory refillable canister and isolation valve, shown in Figure 5. The bottom section provides the heat transfer surfaces necessary to sublimate the solid materials.

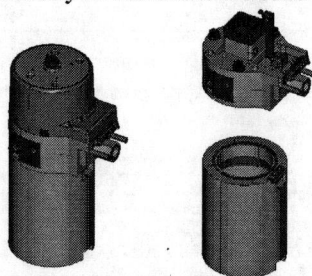

FIGURE 5. Commercial borohydride vaporizer shown with valve and heating head separated.

A temperature sensor is located in the lower section to provide temperature feedback to a closed loop PID controller. The top serves several functions: Closure to the canister, location for a vapor isolation valve, mounting position for heating elements, and thermal mass for distributing thermal energy to lower section.

This design achieves a positive thermal gradient from the bottom to the top, as indicated in Figure 6. The advantage is improved serviceability to the vaporizer isolation valve and control of condensing vapors. The upper section remains warmer than the

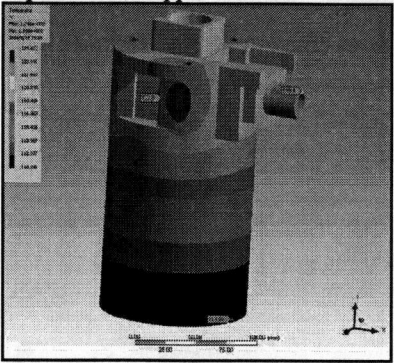

FIGURE 6. Thermal analysis of commercial borohydride vaporizer.

lower section. Gas vapors are most likely to condense back onto the lower section (coolest region) of the vaporizer thereby eliminating solid buildup on moving surfaces (*i.e.* valve). Additionally, a consumed vaporizer is unlikely to have any residual solid material remaining in the vapor path.

For production implanters, rapid recovery following a shutdown is important. The vaporizer must contain sufficient thermal heat transfer surfaces (primarily aluminum) and thermal energy to enable a thirty minute recovery from room temperature, see Figure 7. The surfaces in contact with the solid material being vaporized are located away from the heating elements. This heater location provides for a thermal gradient between the vaporizer exit and the bottom of the vaporizer. During heating, less than a 10°C variation exists between the vapor exit and bottom of the vaporizer.

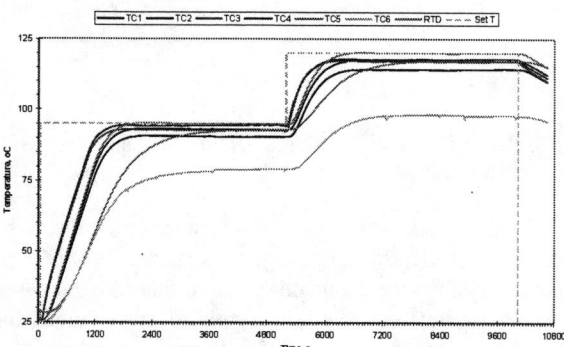

FIGURE 7. Thermal ramp of instrumented vaporizer, 25°C to 95°C hold one hour; 95°C to 120°C hold for one hour

This allows for sufficient thermal power to be added to the vaporizer for rapid heat up without the risk of disassociation of the vaporized material.

VAPOR FLOW CONTROL

An Implanter source must have a carefully regulated supply of feed gas in order to provide a stable ion beam. Conventional ion sources use MFC's for this function. MFC's are not able to regulate gas flows for octadeceborane and decaborane due to their requirement for a relatively high inlet pressure and pressure drop across the MFC. Figure 8 provides an example of a valve network that provided regulated molecular flow of gas vapor to an ion source.

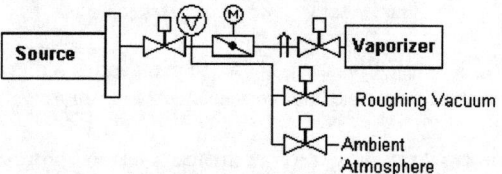

FIGURE 8. Vaporizer to source flow diagram.

The system consists of a vaporizer device capable of sublimating solids at a sufficient rate to provide a positive pressure across a conductance throttling device, and a vaporizer isolation valve to provide positive shut off of vapors from the vaporizer. A variable conductance is achieved using a commercial available servo-actuated vacuum butterfly valve controlled with a PID controller. Feedback control to the servo controller comes from a downstream heated pressure transducer. Other valves are shown that aid in vacuum pump down and venting for service.

The conductance between the throttle valve and the ionization source is fixed. The pressure of the vaporizer will change with vaporizer temperature setting, and is based on the amount of solid material available in the vaporizer canister. As the solid material depletes, the vapor pressure in the vaporizer drops due to a reduced mass rate of vaporization, since the volume and surface area of the solid material is reduced as the material is consumed. In addition, the heat transfer flux (in W/cm^2) is proportional to the temperature of the vaporizer. This results in reduced thermal energy being absorbed into the solid; thus, if the temperature is left constant, the rate of sublimation and also the inlet pressure to the servo valve diminishes as the solid material is consumed. The servo valve will open further to compensate for the lower inlet pressure. Over time, it is therefore necessary to raise the temperature of the vaporizer to compensate for the reduced sublimation rate. Since the mass flow of vapor into the ion source directly influences the ionization rate in the source, maintaining a stable flow necessary to produce a stable beam current.

Figure 9 shows the transient system response

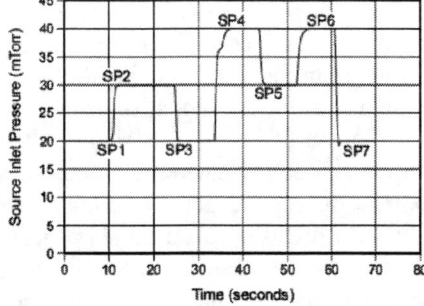

FIGURE 9. Flow control system response to step changes in control setpoint. SP1—SP7 refer to pressure set points.

of the valve controller when subjected to a series of 50% and a 100% changes in pressure (proportional to mass flow) set points. The valve controller was able to compensate for in flow setting. Since hanges to pressure set point typically occur during beam tuning events, rapid system response is critical in order to quickly achieved desired ion beam properties.

VAPOR SOURCE SERVICING

Conventional implant gases are distributed in gas bottles. The technology used may be conventional high pressure gas cylinders, or more advanced technology such as SDS or VAC gas cylinders. These devices are engineered to meet safety regulations assigned by industry and international commerce. Due to the characteristics of decaborane and octadecaborane, none of these existing distribution techniques are applicable. The concept of a replaceable bottle was chosen so as to remain consistent with current service techniques and periodicity. A distribution container that can also serve as a point of vaporization would have many advantages. The container is directly mounted to the vapor delivery system of Figure 8. In order to eliminate operator exposure and other potential hazards, a combination of hardware features were designed into the container and container mounting interface:

Potentially Hazardous Conditions:
- Toxic vapors emitted from the container
- Exposing oxygen to solid borohydride compounds at highly elevated temperatures
- Heated surface may cause burns to the skin

Safety Design Features:
- The vaporizer is mechanically interlocked from being removed unless the isolation valve is locked shut
- The vaporizer isolation valve is mechanically locked shut unless the vaporizer is properly secured to the vaporizer mount
- All readily accessible heated surfaces are thermally insulated to protected service personnel

CONCLUSION

We have developed a vaporizer system for decaborane and octadecaborane The vaporizer is capable of delivering the performance and reliability required by commercial ion source systems. The vaporizer has been tested on several SemEquip, Inc. ClusterIon® source systems and on a GSD/100 implanter that has been retrofitted with ClusterIon® source technology.

A means for safe delivery and integration of Borohydride materials into an implanter has been shown. This development effort further enables the use of decaborane and octadecaborane doping materials for performing very shallow boron implants at high implanter throughputs.

REFERENCES

1. Miller, G., "The Vapor Pressure of Solid Decaborane", Journal of Physical Chemistry, Vol. 67, pp. 1363-1364, 1963.
2. Furukawa, G., and R. Park, "Heat Capacity, Heats of Fusion and Vaporization, and Vapor Pressure of Decaborane ($B_{10}H_{14}$)," Journal of Research of the National Bureau of Standards, Vol. 55, No. 5, p. 255-260, November 1995.
3. Perel, A.S.., et al "Decaborane Ion Implantation" IIT(2000) pp304-307.

Simplifying the 45nm SDE Process with ClusterBoron® and ClusterCarbon™ Implantation

Wade Krull[1], Brian Haslam[1], Tom Horsky[1],
Kurt Verheyden[2] and Klaus Funk[2]

[1]*SemEquip, Inc., 34 Sullivan Road, North Billerica, Massachusetts 01862, USA*
[2]*ASM, Leuven Belgium and Munich Germany*

Abstract. The processes used to form an advanced PMOS SDE become increasing complex with each technology generation. Aggressive requirements for junction depth, sheet resistance and abruptness are difficult to satisfy with conventional technologies. The fundamental issues are the low productivity of implant systems at the required low boron implant energies and the diffusion rate of boron in silicon at the temperatures needed for activation. We propose an alternate process sequence with the goal of simplifying the process module by eliminating the Ge PAI implant and using a conventional spike anneal.

Our approach is to utilize a ClusterCarbon implant to control boron diffusion, a ClusterBoron® implant for high productivity at low process energy and a conventional spike anneal at moderate temperature. The ClusterCarbon™ implant uses the same ClusterIon® source and related system as the ClusterBoron implant. This process combination provides a solution to the technology requirements using conventional production tools, high productivity and a simple, direct path to optimizing the SDE conditions.

Keywords: cluster implantation; molecular implantation; boron diffusion control
PACS: 61.72Tt

INTRODUCTION

The requirements for the PMOS SDE are becoming increasingly difficult to achieve. The ITRS Roadmap for the 45nm node indicates that a junction depth of 7-10nm is required, consistent with the conventional scaling requirement that the SDE be around 1/3 of the physical gate length. In addition, there are requirements for low sheet resistance, extremely abrupt junctions and the necessity of managing leakage currents. Such requirements can only be met with a boron implant in the range of 200eV with pre-amorphization and a diffusionless anneal. However, the requirements listed in the ITRS have become increasingly abstract and divergent from typical processes used in the fabrication of advanced ICs. Many organizations are using processes less extremely scaled than the ITRS numbers. For these applications, SDE junction depths in the range of 12-20nm are acceptable, and this is the realistic target of the work described herein.

Most of the development activity directed at achieving the PMOS SDE is following a single path. First, a Ge PAI implant is used to create an amorphous surface layer and thereby avoid channeling issues, and achieve the shallowest possible as-implanted profile. A carbon implant is used to inhibit boron diffusion. Next, a BF2 boron implant is used to enhance the productivity of the low energy boron implant process. The fluorine inherent in the BF2 implant has both positive and negative aspects; it reduces the diffusion of boron, but also inhibits boron activation and complicates the elimination of EOR damage during the anneal. Finally, in order to achieve a diffusionless anneal with the maximum activation, a laser thermal or flash anneal is being developed but these processes require equipment which is new, expensive and not yet proven to be production worthy.

Our concept is to adopt an alternative approach with the goal of simplifying the process sequence while achieving very high productivity and using production proven equipment. The approach utilizes the demonstrated self-amorphization feature of cluster implants to eliminate the Ge PAI implant step [1,2]. Next, we make use of the very high productivity capability of ClusterBoron implant at very low process energies. This will allow for some diffusion to occur and still meet the junction depth requirements. To reduce the amount of diffusion, we will use a ClusterCarbon implant (describe herein for the first time) which has the same high productivity and self-amorphization advantages of the ClusterBoron implant technology. Finally, we will use a conventional, production proven spike anneal (at moderate temperatures) and the demonstrated high activation of ClusterBoron [3] to achieve the junction depth and sheet resistance requirements. A further goal of this work is to keep the amorphization depth low and below the active junction, so that the beneficial effect of the surface will enable complete elimination of the EOR damage. Since the amorphization depth is set by the ClusterCarbon energy in this process, we will restrict the carbon energy to 3keV per carbon atom and below. In all cases, the carbon is implanted first, then boron.

The goals of this work are to optimize the implant conditions of ClusterCarbon and ClusterBoron to achieve a low resistance, shallow and abrupt junction while using conventional spike annealing technology. Figure 1 below shows the benefit of using the carbon implant. The figure shows a 500eV boron equivalent implant, as-implanted and with 1050 spike anneal with and without carbon. It is seen that the carbon implant reduces the amount of diffusion, increases the active boron concentration, and improves the junction abruptness. The experiment was designed to explore the parametric implant space to achieve the best combination of Rs, Xj and junction abruptness.

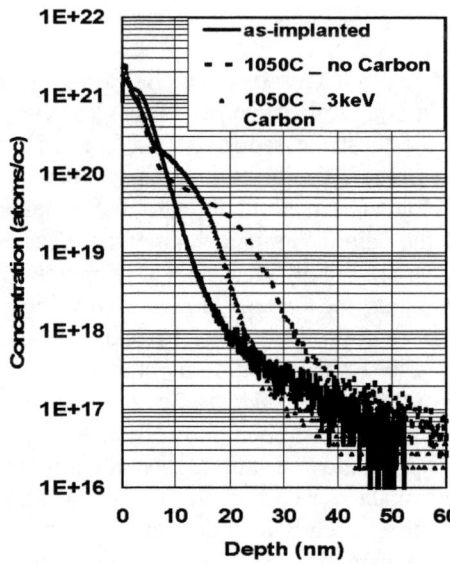

FIGURE 1: ClusterBoron implant of 500eV equivalent at 1E15 dose; as-implanted, and with 1050C spike anneal with and without carbon.

CLUSTERION TECHNOLOGY

The use of ClusterBoron implant has been established as an alternative approach to low energy boron implant. This technology utilizes a new source material (B18H22) and the ClusterIon ion source to provide a beam of B18 ions. The ClusterIon source uses soft ionization and temperature control to preserve the large borane molecule, in contrast to a conventional ion source which uses an intense plasma to disassociate molecules such as BF3. The advantages of using ClusterBoron implant are very high productivity down to very low process energies, freedom from energy contamination and self-amorphization. It has been shown that very high quality shallow junctions are formed with B18 for any annealing technology [3]. The ClusterBoron chemical is a solid at room temperature so the ClusterIon technology includes a low temperature vaporizer and vapor flow control to provide a stable flow of ClusterBoron vapor to the ion source [4]

We introduce a new cluster species for low energy carbon implant: ClusterCarbon. By appropriate choice of hydrocarbon molecule, we have a molecule whose physical properties are very similar to ClusterBoron. The ClusterCarbon chemical is also a solid at room temperature and vaporizes in the same temperature range as ClusterBoron. The soft ionization system developed for ClusterBoron also works very well for the ClusterCarbon vapor, which produces slightly higher electrical beam currents due to the narrower AMU spectrum of ClusterCarbon. In addition, the ClusterCarbon ion is in the same AMU range as ClusterBoron (~200AMU) so the remainder of the implant system works the same as with ClusterBoron. The implants in this experiment were performed on the SemEquip demo system; an Eaton GSD implanter upgraded with the ClusterIon system.

EXPERIMENT

The anneals in this work were all performed on the ASM Levitor demo system installed at Imec in Leuven, Belgium. The Levitor is a conventional, production proven rapid thermal annealing system capable of anneals over a wide temperature range with controllable ramp rates up to 1000C/s. For this work, we utilized spike anneals to a peak temperature in the range of 950-1075C. The anneals were done in a He environment with ramp rates in the range of 800C/s.

Since the application for this work is an advanced PMOS SDE, implant process conditions were chosen to be appropriate for such application. It is well established that a boron implant with energy of 500eV and dose of 1E15 is appropriate for the junction range of interest. This requires a ClusterBoron implant of 10kV. The carbon implant energy was limited to 3keV per atom and below and carbon doses were investigated between 1E14 and 2E15. For all implant combinations, a spike anneal sweep between 950C and 1075C was performed to evaluate annealing response of the implanted structure. The primary responses evaluated were sheet resistance (conventional 4pp), junction depth (SIMS at 5E18 concentration) and junction abruptness (1E18-1E19 concentration gradient). We also introduce a new response parameter, the product of Rs and Xj, to provide a single parameter corresponding to the net electrical concentration in the shallow junction.

One of the experimental designs for this study is related to leakage control. It is reported that carbon in silicon enhances the risk of leakage, so carbon energies were intentionally limited to values below the typical boron energy for the Source/Drain implant, or 3keV. This will keep the carbon out of the source/drain junction where the large area would produce high leakage current. This is consistent with recent work [5] where it was demonstrated that low energy carbon reduces leakage current. It remains to be evaluated with real structures what the leakage impact of such carbon implants would be.

RESULTS

The process conditions chosen produced shallow junctions in the range of 12-20nm for most of the experimental space investigated. Junction abruptness values in the range of 2-5nm/dec were achieved, with the shallower junctions showing improved abruptness. However, significant tradeoff with Rs is seen for all of the process conditions, as expected. The sheet resistances measured were in the range of 800-3000 ohms/square with the lowest values corresponding to the deepest junctions and the highest annealing temperatures.

The plots of Fig. 2 show the response values as a function of annealing temperature for the best case with carbon and without carbon. It is seen that the addition of the ClusterCarbon implant significantly reduces the junction depth and improves the junction abruptness. While the Rs values are higher than ideal, they are better than can be achieved by spike anneal without carbon for the same Xj.

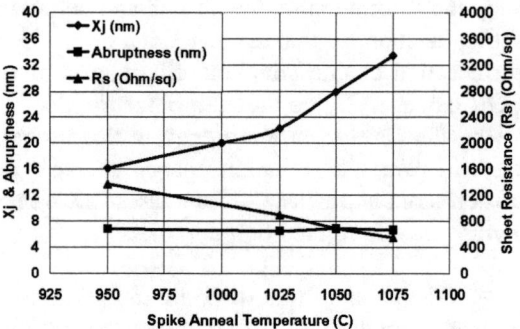

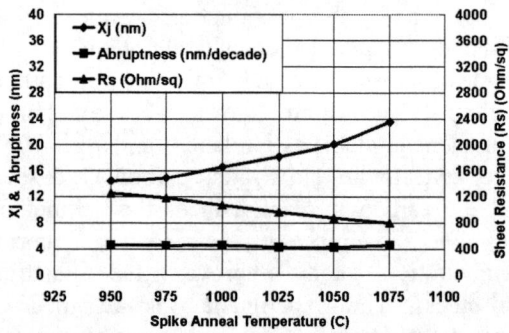

FIGURE 2. Results of Rs, Xj and abruptness for B18 implants of 500eV, 1E15 equivalent, without (above) and with 3keV 1E15 ClusterCarbon implant, as a function of anneal temp.

The response data is plotted in Rs/Xj space in Fig. 3. This figure includes data from several carbon implant recipes. It is clearly seen that the best carbon recipe is 3keV, 1E15 and that this process produces junctions with better Rs/Xj characteristics than can be obtained without the carbon implant. The datapoints on each curve are for different anneal temperatures. It is seen that the carbon implants with higher dose than 1E15 produce inferior results: shallow Xj but higher Rs than could be obtained without the carbon implant. Likewise, comparing 2keV carbon with 3keV carbon shows that the 2keV produces a shallower, but higher Rs junction. Since the shallower carbon also results in higher concentration of carbon in the active layer, it is likely that carbon acts as a scattering center, reducing mobility and thereby increasing Rs.

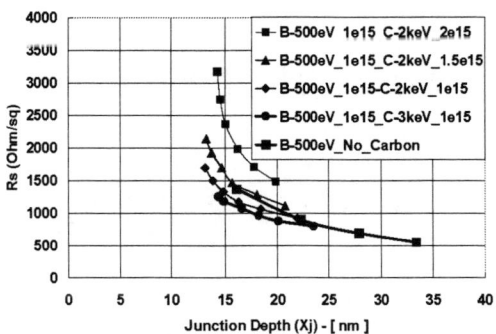

FIGURE 3. Data for B18 with various carbon implant process conditions plotted in Rs/Xj space.

The boron profiles in depth were evaluated by SIMS analysis. Figure 4 below shows the anneal sequence for the 3keV, 1E15 carbon case. As expected, lower anneal temps produced more shallow junctions and were the most abrupt. Solubility values in the range of 1-2E20cm-3 were observed for all but the highest anneal temperature

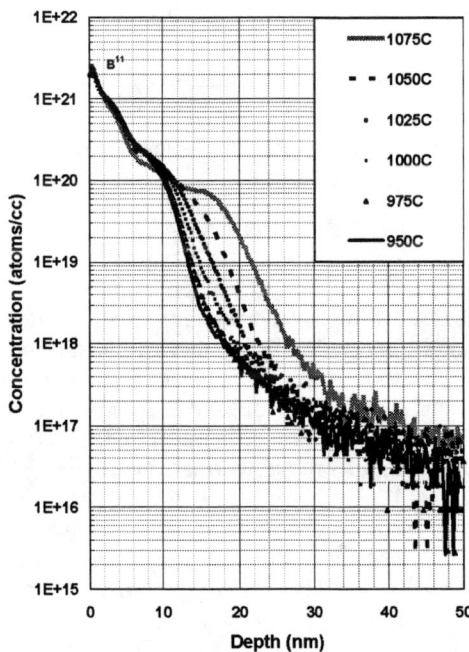

FIGURE 4 SIMS profiles for B18 500eV 1E15 with carbon 3keV 1E15 as a function of anneal.

This process is difficult to optimize due to the necessity of tradeoff between Rs and Xj, which naturally move in opposite directions as the conditions are modified. In an attempt to understand this tradeoff more explicitly, we examiine the product of Rs and Xj. It is noted that an ideal box profile would have Rs and Xj varying inversely and thus would produce a constant value of the product as the junction depth is reduced. The product Rs*Xj has units of inverse concentration, so a lower number indicates better activation. We show in Figure 5 the results for the Rs*Xj product for our experiment. The response without carbon anneal is a line which increases with reduced annealing temperature. The response with 1E15 carbon (either 2 or 3keV) shows a curve nearly flat, indicating near theoretical tradeoff between Rs and Xj. Higher doses of carbon produce higher values, indicating lower active concentration.

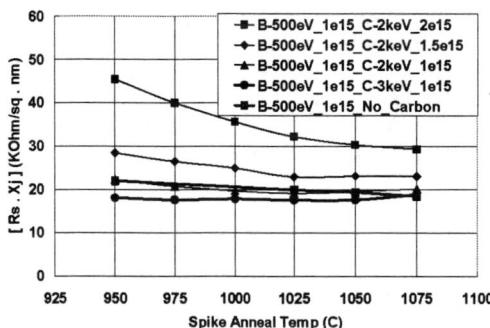

FIGURE 5: Product of Rs and Xj vs. anneal temp

SUMMARY

We introduce a new cluster species for low energy carbon implantation: ClusterCarbon. It is shown that ClusterCarbon inhibits boron diffusion during the anneal process, consistent with other developments using monomer carbon. Further, the combination of ClusterCarbon, ClusterBoron and conventional spike annealing technologies are shown to produce ultra-shallow junctions appropriate for the 45nm SDE. These junctions are shallower and more abrupt than can be achieved without the carbon implant. It is also shown that the optimized process has near ideal scaling of junction parameters as the anneal is varied. This process combination simplifies the process sequence for advanced PMOS SDE while using production proven tools with high productivity.

REFERENCES

1) J.O. Borland, IIT2004 Proceedings, Oct. 2004, p6.
2) N. Hamamoto, et al, IIT2004 Proceedings, Oct. 2004, p443.
3) J.O. Borland, et al, IWJT 2006, p4.
4) D. Adams, et al, IIT2006, June 2006
5) E.J.H. Collart IIT2006, June 2006

Implantation characteristics by boron cluster ion implantation

Tsutomu Nagayama*, Nariaki Hamamoto*, Sei Umisedo*, Masayasu Tanjyo*, and Takayuki Aoyama**

*Nissin Ion Equipment Co., Ltd.,
575 Kuze-Tonoshiro-cho, Minami-ku, Kyoto, 601-8205, Japan
Tel: +81-75-922-4611, Fax: +81-75-922-4615, e-mail: Nagayama_Tsutomu@nissin.co.jp
**Fujitsu Laboratories Ltd., 50 Fuchigami, Akiruno-Shi, Tokyo, 197-0833, Japan

Abstract. Recently, boron cluster implantation (i.e. decaborane: $B_{10}H_x^+$) is regarded as a promising technology for the formation of P-type Ultra Shallow Junction (USJ) because of the equivalent high beam current with less beam divergence compared to the conventional B^+ or BF_2^+ implantation. Also as-implanted and after-annealing characteristics are different due to the appearance of self-amorphized layer by the cluster ion bombardment, which suppresses the channeling and enhances the boron activation. However, it is anticipated that the properties caused by this amorphous layer will vary with different implantation conditions or a presence of Pre Amorphization Implantation (PAI) process, which should be understood well to maintain a good process control. From this point of view, we have measured the decaborane implantation characteristics by a couple of different related conditions, for instance, the beam energy and current. Sheet resistance vs junction depth (Rs-Xj) are also evaluated in different annealing methods with combination of PAI processes. In addition, a brief comparison is made by implanting the different boron cluster ions (i.e. $B_8H_x^+$) by mass selecting the ions extracted from decaborane ionization chamber. In this paper, these characteristics of boron cluster implantations are reviewed.

Keywords: Implanter, cluster, USJ
PACS: 85.40.Ry, 61.46.Bc, 61.72.Tt

INTRODUCTION

A formation of the ultra-shallow junction (USJ) is one of the key technology to fabricate the next generation semiconductor devices. Beyond the 45nm node, the p+ USJ for extension is requiring low energy implantation below 1 keV for shallower junction formation.

However, this low energy implantation has some problems such as decel-mode implantation (energy contamination) and spot beam blow-up effect (beam divergence). Cluster ion implantation is expected as one of the methods for solving these problems. Decaborane molecular ion implantation ($B_{10}H_x^+$) has been proposed for p-type dopant [1,2] and it was applied to p-type source drain extension (SDE) on sub-40nm-gate-devices [3,4].

Now we are developing a decaborane implantation system for medium current implanter, which can perform a precisely controlled implantation in terms of the implant dose, the uniformity, and the incident angle [5,6]. A decaborane ion source is installed in Nissin medium current implanter EXCEED2300 and the beam and implantation characteristics have been measured.

In the following, we report the advantages of $B_{10}H_x^+$ implantation such as the beam transport property, self-amorphization and device characteristic. These characteristics are also compared with B^+ implantation or B^+ implantation into pre-amorphized layer.

In addition, some results on the evaluated damages caused by molecular ion implantation comparing $B_8H_x^+$ and $B_{10}H_x^+$ are reported.

EXPERIMENTAL

Decaborane molecular ions were implanted using EXCEED2300H implanter which is a medium-current implanter with parallel beam and mechanical scan system. The tool is equipped with front and back faraday cups [5] which are utilized in the measurement of beam size and divergence. Implanting dopants were activated using the rapid thermal annealing (RTA) method.

The beam transport characterization for B^+ was done on the same machine, while low energy B^+ monomer implantations at 0.5 keV were performed using a conventional high-current implanter.

Germanium ions (Ge^+) were implanted for pre-amorphizaion implantation (PAI).

Doping profiles and activation profiles were analyzed using secondary ion mass spectrometry (SIMS) and the relationship of sheet resistance (Rs) versus junction depth (Xj) was determined using four-point probe (4pp). The surface amorphous layers were observed with XTEM and the surface roughness measurement was carried out by atomic force microscope (AFM).

P-channel devices were fabricated without channel

mechanical stress control in Fujitsu 40nm-gate CMOS logic lines. We used $B_{10}H_x^+$ implantation for source/drain extension (SDE) formation.

RESULTS AND DISCUSSION

Cluster beam characteristics

Figure 1 shows decaborane ($B_{10}H_{14}$) molecular structure. 10 bron atoms are implanted by one $B_{10}H_x^+$ ion. Mass is 10 times heavier than one of monomer boron atom and equivalent energy per monomer boron atom is 1/10. These properties give rise to the advantages for the beam transportation and USJ formation.

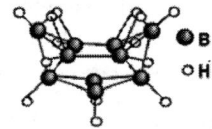

FIGURE 1. Decaborane ($B_{10}H_{14}$) molecular Structure.

As figure 2 shows, low energy implantation process meets some difficulties such as beam blow up and divergence driven by the space charge effect within the beam current. In $B_{10}H_x^+$ implantation, we can use 1/10 of the ion beam current, 10 times heavier mass with 10 times higher acceleration energy than those in B^+ monomer implantation.

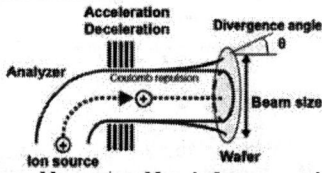

FIGURE 2. beam blow-up problem in low energy implantation.

Thus, space charge effect is suppressed by 1/100 of that of B^+ monomer implantation in the case of $B_{10}H_x^+$ implantation. We measured the divergence angle and the beam size of $B_{10}H_x^+$ beam, which were compared with B^+ monomer beam in the same tool. The divergence of $B_{10}H_x^+$ ion beam is around 0.3° and which is about one order of magnitude smaller than that of B^+ ion beam (Figure 3). The beam radius of $B_{10}H_x^+$ ion beam appear also to be in smaller size (Figure 4).

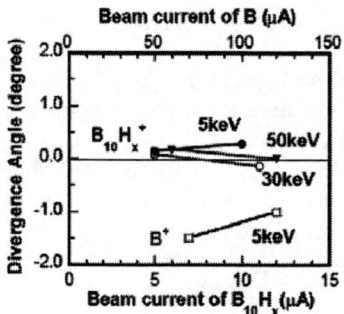

FIGURE 3. divergence angle of $B_{10}H_x^+$.

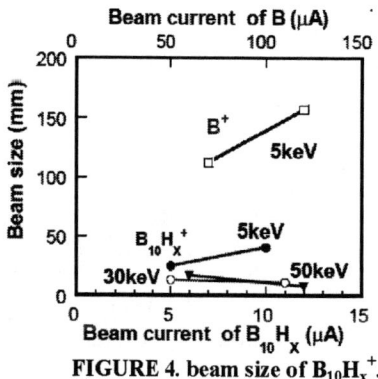

FIGURE 4. beam size of $B_{10}H_x^+$.

Implantation characteristics

Figure 5 shows as-implanted profiles at equivalent 0.5 keV. The SIMS profiles implanted using $B10H_x^+$ was shallower and steeper than that of B+ monomer ions. In addition, the as-implanted profile of $B_{10}H_x^+$ is less sensitive to its surface conditions, which are HF last or SC2 (HCl-H2O2-H2O mixture) surfaces before the implants, compared to that of B^+. It is conjectured that $B_{10}H_x^+$ ions are implanted with little channeling effect due to the creation of an amorphous layer as seen below in Figure 11, which is mentioned as self-amorphization effect.

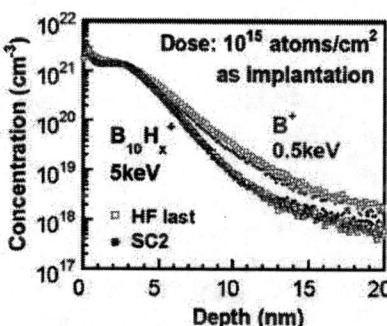

FIGURE 5. As-implanted boron profiles.

After RTA 1000°C, The SIMS profile of $B_{10}H_x^+$ is also shallower and steeper (Figure 6). Therefore, sheet resistance (Rs) of a $B_{10}H_x^+$ ions implanted layer is lower than that of B^+ monomer at same junction depth (Xj) as figure 7 shows. It is known that an implanted layer after RTA sometimes include defects such as end-of-range (EOR) defects near amorphous/crystalline interface which concerns with junction leakage and activation. High activation in $B_{10}H_x^+$ implanted layer, however, implies that a $B_{10}H_x^+$ ions implanted layer has less EOR defects than that of B^+ ions implanted layer. XTEM observation also indicates that there is no distinct defects in $B_{10}H_x^+$ implanted layer after RTA (Figure 8).

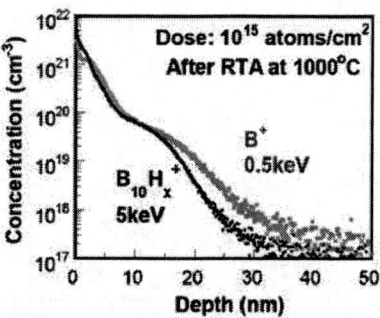

FIGURE 6. Implantation profiles after RTA.

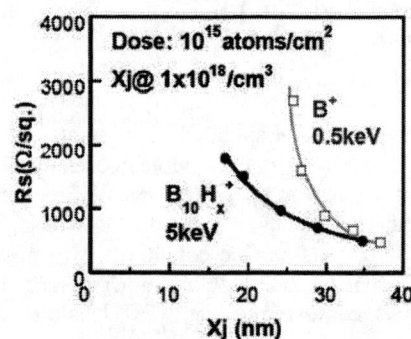

FIGURE 7. Xi vs Rs characteristic.

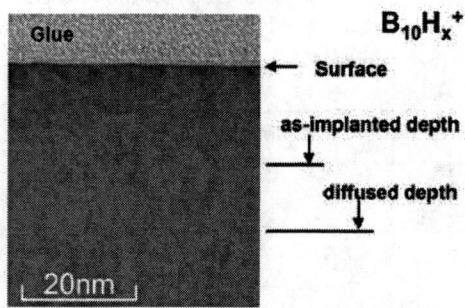

FIGURE 8. XTEM image after RTA.

In addition, we investigated junction leakage current comparing B^+, $B_{10}H_x^+$ and PAI+B^+ (Figure 9). In spite of forming a steep USJ, the junction leakage current of $B_{10}H_x^+$ implanted SD biased at -1 V was very small (0.47pA/μm) and was lower than that of B^+ with Ge PAI.

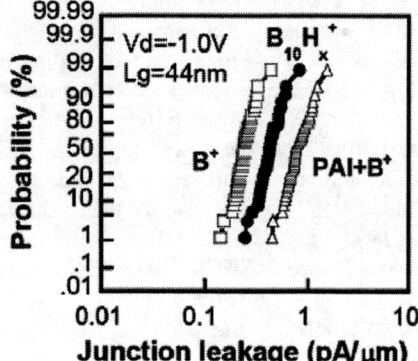

FIGURE 9. Junction leakage currents of source/drain junction.

Furthermore, we evaluated Rs versus Xj of $B_{10}H_x^+$ and B^+ monomer ion implantation with or without Ge^+ PAI (Figure 10). Rs implanted using $B_{10}H_x^+$ was the lowest at Xj below 30 nm. When the conditions of Ge^+ PAI are changed, the Rs become lower than that of layer implanted using only B^+ without Ge^+ PAI. However, the Rs was never lower than that implanted using $B_{10}H_x^+$. We believe that the low sheet resistance implanted using $B_{10}H_x^+$ means the high carrier activation, which is useful for USJ formation or fabrication of high performance device. Figure 11 shows XTEM images of as-implanted samples with $B_{10}H_x^+$ or Ge^+ PAI. A thin amorphous layer is formed by $B_{10}H_x^+$.

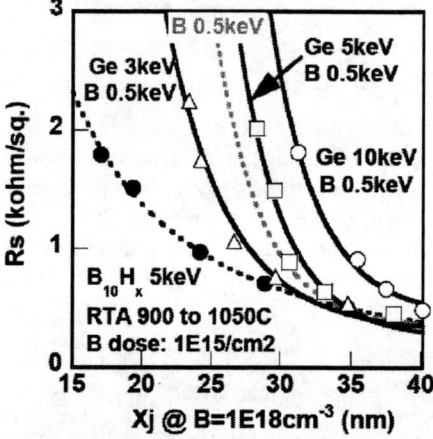

FIGURE 10. Comparison of Rs-Xj characteristics between $B_{10}H_x^+$ and B^+ monomer ion implantation with or without PAI.

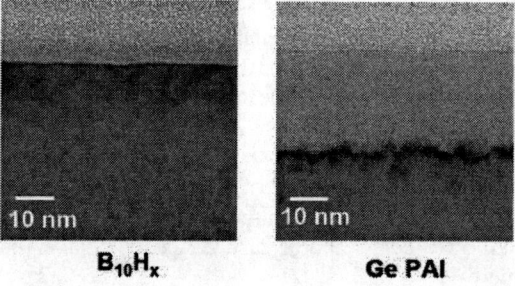

FIGURE 11. Cross-sectional TEM images of $B_{10}H_x^+$ and Ge PAI.

The amorphous/crystalline (a/c) interface is very smooth and any defect such as EOR-defect was not observed. On the contrary, in case of Ge^+ PAI, the a/c interface is very rough and some small defects were observed in the crystal region near the amorphous layer. We speculate that the high carrier activation created by $B_{10}H_x^+$ implantation results from a model as mentioned below (Figure 12).

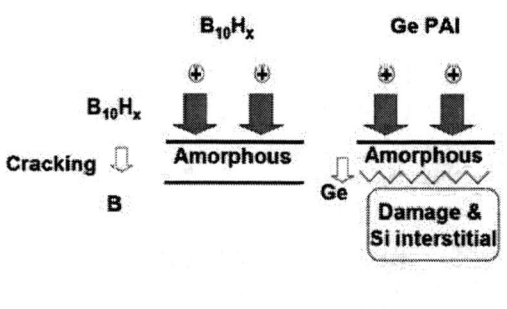

FIGURE 12. Schematic drawings of (a) $B_{10}H_x$ implantation and (b) Ge PAI.

In Ge^+ PAI, some Ge^+ ions implanted stop near the a/c interface, and Ge^+ ions may knock on some Si atoms in amorphous layer because of their heavy mass. Thus, Ge^+ PAI causes many interstitial Si atoms in crystal region. On the other hand, in $B_{10}H_x^+$ implantation, $B_{10}H_x^+$ cracks instantly into Boron atoms in amorphous layer. Because of their light mass, they rarely damage crystal region or knock on many Si atoms interstitial.

We investigated Ra versus dose on $B_8H_x^+$ and $B_{10}H_x^+$ implanted wafers by mass selecting from decaborane ion source extraction current (Figure 13). Ra of wafer implanted $B_8H_x^+$ and $B_{10}H_x^+$ didn't increase with implantation dose. While the increase of Ra can be regarded as the increase of the surface roughness due to the sputtering, Ra of wafer implanted $B_8H_x^+$ and $B_{10}H_x^+$ kept the same Ra of original Si wafer until $5E15/cm^2$ implantation dose.

We found out that $B_{10}H_x^+$ implantation has advantage of self-amrphization layer making possible few defects near a/c interface and few sputtering on Si surface.

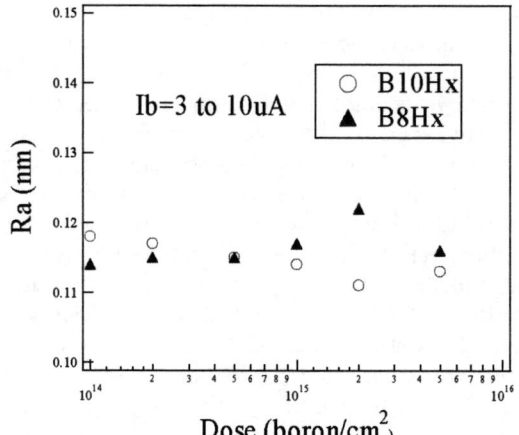

FIGURE 13 Comparison of implantation dose vs Ra between $B_8H_x^+$, and $B_{10}H_x^+$.

CONCLUSION

We report the advantages of $B_{10}H_x^+$ implantation as follows.

$B_{10}H_x^+$ implantation can provide precisely controlled beam without blow-up compared with the B^+ monomer implantation and can form a shallow and steep USJ with the low resistivity and the low leakage current (0.47 pA/μm).

$B_{10}H_x^+$ implantation can achieve a high carrier activation and smooth amorphous/crystal interface compared with conventional B^+ implantation into a pre-amorphization layer. $B_{10}H_x^+$ implantation can form self-amorphization layer with few sputtering on Si surface.

We clearly believe that $B_{10}H_x^+$ implantation will be useful for USJ formation in near future device technology. We are developing the special implanter for cluster implantation.

ACKNOWLEDGEMENT

This work is supported by Japan Science and Technology Agency (JST) for the equipment development and The New Energy and Industrial Technology Development Organization (NEDO) of Japan for the process development.

REFERENCES

[1] K. Goto, J. Matsuo, T. Sugii, H. Minakata, I. Yamada, and T.Hisatsugu, Extended Abstract of IEDM 1996, pp.435-438
[2] K. Goto, J. Matsuo, Y. Tada, T. Tanaka, Y. Momiyama,T.Sugii, and I. Yamada, Extended Abstract of IEDM 1997,pp. 471-474.
[3] K. Goto, Y. Tagawa, H. Ohta, H. Morioka, S. Pidin, Y. Momiyama, H. Kokura, S. Inagaki, N. Tamura, M. Hori, T.Mori, M. Kase, K. Hashimoto, M. Kojima and T. Sugii,Extended Abstract of IEDM 2003, pp. 623-626.
[4] T. Aoyama, M. Fukuda, Y. Nara, S. Umisedo, N.Hamamoto, M. Tanjo, T. Nagayama, Extended Abstracts of International Workshop on Junction Technology (IWJT)2005, pp. 27-30.
[5] S. Umisedo, N. Hamamoto, S. Sakai, M. Tanjyo, N. Nagai, and M. Naito, Extended Abstracts of International Workshop on Junction Technology (IWJT) 2004, pp.27-30.
[6] N. Hamamoto, S. Umisedo, T. Nagayama, M. Tanjyo, S. Sakai, N. Nagai, T. Aoyama, and Y. Nara, Final Program and Abstract Book of 15th International Conference on Ion Implantation Technology (IIT) 2004, p.95.

Sputtering and Chemical Modification of Solid Surfaces by Water Cluster Ion Beams

G.H. Takaoka, K. Nakayama, D. Takeda and M. Kawashita

*Ion Beam Engineering Experimental Laboratory, Kyoto University,
Nishikyo, Kyoto 615-8510, Japan*

Abstract. Water was introduced into a cluster source, and heated up to 150°C by a wire heater attached around the source. When the vapor pressure was larger than 1 atm, the vaporized water clusters were produced by an adiabatic expansion. The irradiation effects of water cluster ions on solid surfaces such as Si(100) and Ti substrates were investigated. The sputtered depth increased with increase of ion dose and acceleration voltage. The sputtering yield of the Si and Ti surfaces by the water cluster ion beams was approximately 10 times larger than that by Ar monomer ion beams at the same acceleration voltage. In addition, the XPS measurement showed that the sputtered surface had an oxide layer such as SiO_x and TiO_x. It was found from the depth profile of O_{1s} peak that the oxide layer thickness increased with increase of acceleration voltage, and it was about 10 nm at an acceleration voltage of 6 kV. Furthermore, the contact angles for the sputtered surfaces were measured, and they were about 80 degrees for the Ti surfaces and about 5 degrees for the Si surfaces, respectively. The contact angle for the unirradiated surface was about 45 degrees for Si surface and about 30 degrees for Ti surface, and the change of the contact angles was due to the oxide layer formation by the water cluster ion irradiation.

Keywords: cluster ion beam, water cluster, sputtering, chemical modification, oxidation
PACS: 36.40.Wa, 61.46.Bc, 61.80.Jh, 68.08.Bc, 68.37.Ps, 68.49.Fg, 81.65.Mq, 82.80.Pv

INTRODUCTION

The materials surface processes, in which dry and wet processes have been available, have been of much importance in the fabrication of LSI semiconductor devices.[1,2] It is well known that water has an important role in surface cleaning for solid surfaces in the wet process. Although water is not possible to be applied to etching the solid surface, it is used to remove acid or alkali molecules remained on the surface after etching. For the fabrication of advanced semiconductor devices, the demand for material processing technologies has recently been increasing,[3,4] and the surface treatment process also has specific requirements such as surface etching without damage, surface etching with chemical modification, high speed etching including flat surface formation at an atomic level. It is hard to achieve these requirements by conventional wet processes as well as conventional ion beam processes.

Liquid cluster ion beams have been demonstrated to be valuable for surface etching and chemical modification.[5-7] Equivalently low-energy and high-current ion beams can be realized using cluster ion beams. The impact process on solid surfaces by cluster ions has the capability of achieving extremely high temperature on the impact surface.[8] Therefore, cluster ions such as water cluster ions can enhance the oxidation of the solid surfaces by hydroxyl radicals, which are shallow-implanted near the surface.[9] As a result, the chemical modification of materials surface is performed by shallow implantation of water cluster ions. In addition, the acceleration energy for the water cluster ions can be used to sputter the solid surfaces effectively. Based on the above features of water cluster ions, we have investigated their irradiation effects on solid surfaces such as semiconductor and metal surfaces. In this article, sputtering effects of silicon (Si) and titanium (Ti) surfaces by water cluster ion beams are described. Furthermore, chemical modification of the surfaces irradiated by the water cluster ion beams is discussed.

EXPERIMENTS

We have developed a liquid cluster ion source,[5,10] which can provide water cluster ion beams for the surface treatment. Water was introduced into a cluster source, and heated up to 150 °C by a wire heater attached around the source. When the vapor pressure was larger than 1 atm, the vaporized water clusters were produced by an adiabatic expansion. The cluster size was measured by the time-of-flight (TOF) method.[9] It was distributed between a few hundreds and a few tens of thousands, and the peak size was about 2500 molecules-per-cluster. The neutral clusters were ionized by an electron bombardment method. The electron voltage for ionization (Ve) was adjusted between 0 V and 300 V, and the electron current for ionization (Ie) was adjusted between 0 mA and 250 mA. The water cluster ion beams were accelerated toward a substrate, which was set on a substrate holder. The acceleration voltage (Va) was adjusted between 0 kV and 10 kV. The substrates used were Si(100) and Ti, and the substrate temperature was at room temperature. The background pressure around the substrate was 6×10^{-7} Torr, which was attained using a diffusion pump.

RESULTS AND DISCUSSION

The sputtering process by irradiation of water cluster ions on Si(100) and Ti surfaces was investigated. The sputtered depth was measured by the step profiler (Veeco Instruments: DEKTAK-3173933). Figure 1 shows the dependence of sputtered depth for Si and Ti surfaces on acceleration voltage. The cluster size was larger than 100 molecules-per-cluster, and the ion dose was 1.0×10^{16} ions/cm^2. As shown in the figure, the sputtered depth increases with increase of the acceleration voltage. When the acceleration voltage is 9 kV, they are 35.7 nm for Si surface and 22.3 nm for Ti surface, respectively. Taking account of the sputtered depth and the ion dose, the sputtering yield was calculated by estimating the density of Si and Ti such as 2.42 g/cm^3 and 4.5 g/cm^3, respectively. The sputtering yield at an acceleration voltage of 9 kV for the ethanol cluster ions was 17.8 atoms-per-ion for Si and 9.1 atoms-per-ion for Ti, which are approximately 10 times larger than that by argon (Ar) ion beam sputtering.

The sputtered Si and Ti surfaces were measured by using an atomic force microscope (AFM). Figure 2 shows the surface morphologies (a) and (b) for the Si substrates and (c) and (d) for Ti substrates unirradiated and sputtered by water cluster ion beams, respectively. The acceleration voltage was 6 kV, and the ion dose was 1.0×10^{15} ions/cm^2. As shown in the figure, the surface roughness for the unirradiated substrates is 0.16 nm fro Si and 0.68 nm for Ti, respectively. On the other hand, the sputtered surfaces become rough, and the surface roughness is 0.55 nm for Si and 0.74 nm for Ti, respectively. Smooth surface with a roughness less than 1 nm is obtained even after sputtering. The lateral sputtering effect is responsible for the smooth surface formation by the water cluster ion irradiation. The water cluster ion beam process has unique characteristics suitable for surface treatment such as high sputtering yield and smooth surface at an atomic level, which are not achieved by the conventional wet process.

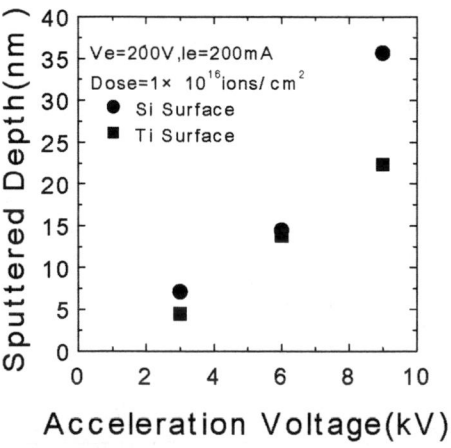

FIGURE 1. Dependence of sputtered depth for Si and Ti surfaces on acceleration voltage

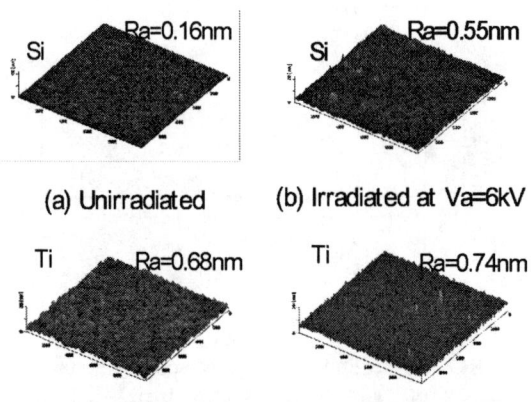

FIGURE 2. Surface morphologies for the Si substrates ((a) and (b)) and for Ti substrates ((c) and (d)) unirradiated and irradiated by water cluster ion beams.

The irradiated Si and Ti surfaces were investigated using X-ray photoelectron spectroscopy (XPS). Si_{2p} and Ti_{2p} spectra as well as O_{1s} spectra were observed from the irradiated surfaces. Figure 3 shows the dependence of the XPS peak intensities for (a) Si_{2p} spectra and (b) O_{1s} spectra at the etched depth. The Si_{2p} peak for the Si surface irradiated by the water cluster ion beams is shifted to the higher value of binding energy, which corresponds to the peak for SiO_2. The peak decreases with increase of the depth, and another peak corresponding to the Si appears. At a depth of 10.5 nm, the SiO_2 peak disappears, and only the Si peak is observed. On the other hand, O_{1s} peaks move to the lower value of binding energy at larger depths. The peak intensity decreases with increase of the depth, and is very weak at a depth of 10.5 nm. This indicates that the silicon oxide layer is formed by the water cluster ion irradiation, and the oxide layer thickness is about 10 nm.

peak is split, and two peaks such as $Ti2p_{1/2}$ and $Ti2p_{3/2}$ appear. These peaks are shfted to the higher values of binding energy on the irradiated surface, and they move to the lower energy with increase of the depth. In addition, the O_{1s} peak decreases with increase of the depth, and it is very weak at a depth of 8.8 nm. This indicates that the titanium oxide layer is formed by the water cluster ion irradiation, and the oxide layer thickness is about 9 nm.

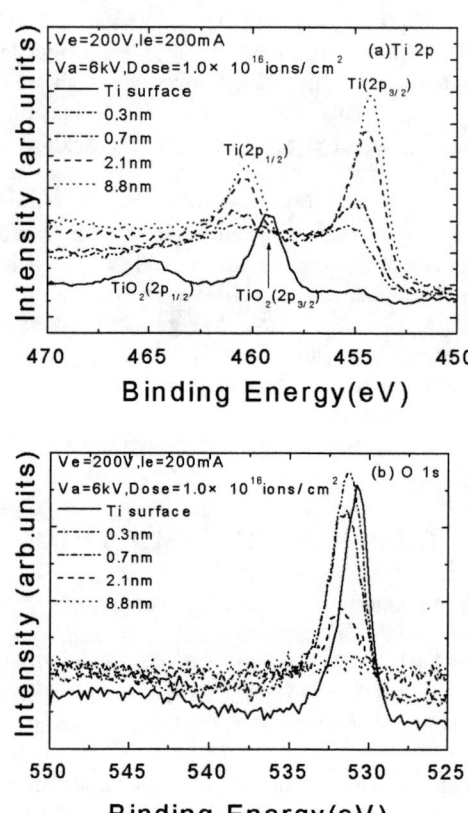

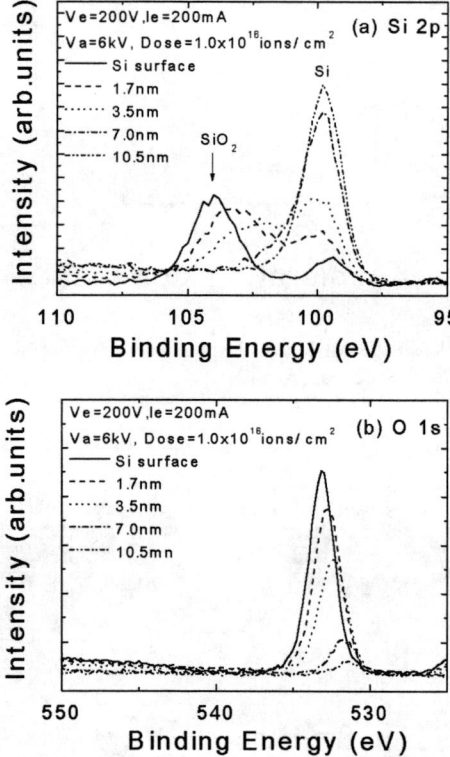

FIGURE 3. Dependence of XPS peak intensities for (a) Si_{2p} spectra and (b) O_{1s} spectra on etched depth.

Figure 4 shows the dependence of the XPS peak intensities for (a) Ti_{2p} spectra and (b) O_{1s} spectra on the etched depth. As shown in the figure, the Ti_{2p}

FIGURE 4. Dependence of XPS peak intensities for (a) Ti_{2p} spectra and (b) O_{1s} spectra on etched depth.

The wettability of the Si and Ti surfaces irradiated by water cluster ion beams were investigated by measuring the contact angles for water droplet, which was put on the Si and Ti surfaces immediately after their removal from the vacuum chamber. Figure 5 shows the dependence of the contact angle on the ion dose for the water cluster ion irradiation. The electron voltage for ionization (Ve) was 200 V, and the electron current for ionization (Ie) was 200 mA. The acceleration voltage (Va) was 6 kV. The cluster size was larger than 100 molecules-per-cluster. The contact angles for the unirradiated surfaces were also measured, and they were about 45 degrees for Si surface and about 30 degrees for Ti surface, respectively. As shown in the figure, the contact angle for the Si surface decreases with increase of the ion dose, and it is less than 5 degree at ion doses larger than 1×10^{15} ions/cm^2. The wettability of the Si surface is much enhanced by the irradiation of the water cluster ions. On the other hand, the contact angle for the Ti surface increases with increase of the ion dose, and it is about 80 degrees at ion doses larger than 1×10^{15} ions/cm^2. The Ti surface becomes hydrophobic by the irradiation of the water cluster ions. This is ascribed to the difference in the chemical modification of the Si and Ti surfaces by the hydroxyl radicals, which are produced after impact of the water cluster ions on the surfaces. The dangling bonds produced on the Si surfaces as well as Ti surfaces have a bond with the hydroxyl radicals, which result in the oxide layer formation on the surfaces.

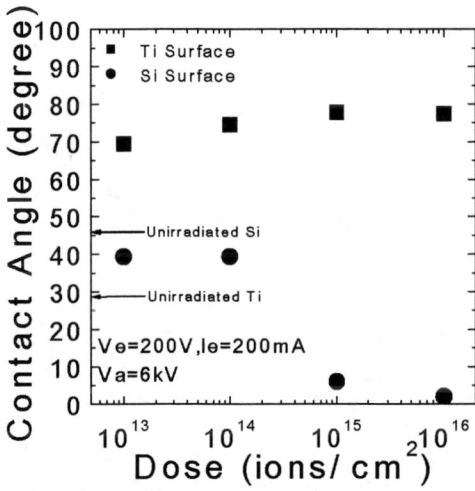

FIGURE 5. Dependence of contact angle for Si and Ti surfaces irradiated on ion dose.

CONCLUSION

Water was introduced into a cluster source, and heated up to 150°C by a wire heater attached around the source. When the vapor pressure was larger than 1 atm, the vaporized water clusters were produced by an adiabatic expansion. The irradiation effects of water cluster ions on solid surfaces such as Si(100) and Ti substrates were investigated. The acceleration voltage was changed between 0 kV and 10 kV, and the ion dose was between 1×10^{13} ions/cm^2 and 1×10^{16} ions/cm^2. The sputtered depth increased with increase of ion dose and acceleration voltage. The sputtering yield of the Si and Ti surfaces by the water cluster ion beams was approximately 10 times larger than that by Ar monomer ion beams at the same acceleration voltage. The XPS measurement showed that the sputtered surface had an oxide layer such as SiO_x and TiO_x. The oxide layer thickness increased with increase of acceleration voltage, and it was about 10 nm at an acceleration voltage of 6 kV. Furthermore, the contact angles for the sputtered surfaces were measured, and they were about 80 degrees for the Ti surfaces and about 5 degrees for the Si surfaces, respectively. The contact angle for the unirradiated surface was about 45 degrees for Si surface and about 30 degrees for Ti surface, and the change of the contact angles was due to the oxide layer formation by the water cluster ion irradiation.

ACKNOWLEDGMENTS

A part of this work was supported by "Nanotechnology Support Project" of the Ministry of Education, Culture, Sports, Science and Technology (MEXT), Japan.

REFERENCES

1. A. O'Neil and J.J. Watkins, Mater. Res. Bull. 30(12) (2005) 967.
2. J.O. Borland, Nucl. Instr. Meth B237 (2005) 6.
3. J.W. Faul and D. Henke, Nucl. Instr. Meth B237 (2005) 228.
4. O. Bakajin, E. Fountain, K. Morton, S.Y. Chou, J.C. Sturm and R.H. Austin, Mater. Res. Bull. 31(2) (2006) 108.
5. G.H. Takaoka, H. Noguchi, T. Yamamoto and T. Seki, Jpn. J. Appl. Phys. 42, (2003) L1032
6. G.H. Takaoka, H. Noguchi and Y. Hironaka, Nucl. Instr. Meth, B242, (2006) 100.
7. G.H. Takaoka, H. Noguchi and M. Kawashita, Nucl. Instr. Meth, B242, (2006) 417.
8. Z. Insepov and I. Yamada, Surf. Rev. Lett. 3 (1996) 1023.
9. G.H. Takaoka, K. Nakayama, H. Noguchi and M. Kawashita, Proc. 5th Int. Symp on Atomic Level Character. for New Mat. Dev. (ALC'05), Hawaii, USA, (2006)
10. G.H. Takaoka, H. Noguchi, K. Nakayama, Y. Hironaka and M. Kawashita, Nucl. Instr. Meth, B237, (2005) 402.

Cross-sectional TEM Observations of Si Wafers Irradiated With Gas Cluster Ion Beams

Hiromichi Isogai*, Eiji Toyoda*, Takeshi Senda*, Koji Izunome*,
Kazuhiko Kashima**, Noriaki Toyoda***, Isao Yamada***

*Processing Technology, Silicon Business Group, TOSHIBA CERAMICS CO., LTD. 6-861-5 Higashikou Seiroumachi Kitakanbaragun, Niigata, Japan.
**New Buisness Creation, TOSHIBA CERAMICS CO., LTD. 30 Soya, Hadano City, Kanagawa, Japan.
***Laboratory of Advanced Science and Technology for Industry, University of Hyogo, 3-1-2 Kouto, Kamigori, Hyogo, Japan.

Abstract. Irradiation by a Gas Cluster Ion Beam (GCIB) is a promising technique for precise surface etching and planarization of Si wafers. However, it is very important to understand the crystalline structure of Si wafers after GCIB irradiation. In this study, the near surface structure of a Si (100) wafer was analyzed after GCIB irradiation, using a cross-sectional transmission electron microscope (XTEM). Ar-GCIB, that physically sputters Si atoms, and SF_6-GCIB, that chemically etches the Si surface, were both used. After GCIB irradiation, high temperature annealing was performed in a hydrogen atmosphere. From XTEM observations, the surface of a virgin Si wafer exhibited completely crystalline structures, but the existence of an amorphous Si and a transition layer was confirmed after GCIB irradiation. The thickness of amorphous layer was about 30 nm after Ar-GCIB irradiation at 30 keV. However, a very thin (< 5 nm) layer was observed when 30 keV SF_6-GCIB was used. The thickness of the transition layer was the same both Ar and SF_6-GCIB irradiation. After annealing, the amorphous Si and transition layers had disappeared, and a complete crystalline structure with an atomically smooth surface was observed.

Keywords: Gas Cluster Ion Beam, GCIB, Si (100), planarization, flatness, high temperature anneal
PACS: 61.72.Ff, 81.40.-Z, 81.10.Jt, 68.37.Lp

INTRODUCTION

An extremely flat surface is required for the use of a Si wafer as a substrate for ULSI semiconductor devices because high resolution in the lithography process is necessary in order for microfabrication to progress. The flatness of a Si wafer has been improved by the development of mechanical and chemical processing technology such as lapping or polishing. However, conventional Si wafer manufacturing processing is considered inadequate to deal with the flatness at the several tens of nanometer levels demanded by next generation processes[1]. Gas cluster ion beam (GCIB) technology can etch the Si wafer at high accuracy, and is, therefore, one of the technologies which could be of use to resolve this issue.

In the case of GCIB etching, the etch rate varies with the species of source gas and the acceleration energy[2]. If a high etching rate is utilized to gain high productivity, the depth of the damage by GCIB irradiation may degrade.

The evaluation of damage to the Si wafer surface is important in order to include GCIB technology in the manufacturing process. When damage is created in the Si wafer by GCIB irradiation, a process will be required to remove it. For instance, if the damage is removed with chemical mechanical planarization (CMP), Si wafer flatness would be deteriorated by virtue of increased Si removal after having been flattened by GCIB.

In this study, the surface structure of Si (100) wafer after GCIB irradiation was analyzed with a cross-sectional transmission electron microscope (XTEM). Ar and SF_6-GCIB with 30 keV acceleration energy were used. In addition, the Si wafer irradiated with GCIB was annealed at 1200 °C for 1hour in a hydrogen atmosphere. After anneal, the near surface damage layer in the Si wafer, caused by GCIB irradiation, had disappeared, and a complete crystalline structure with an atomically smooth surface was obtained by the high temperature annealing.

TABLE 1. Conditions of GCIB irradiation.

	Ar-GCIB	SF$_6$-GCIB
Source gas	Ar	SF$_6$ (5 %) + He (95 %)
Gas flow [sccm]	300	13.4 (SF$_6$), 250 (He)
Acceleration energy [keV]	30	30
Beam current [μA]	200	200
Ionizing electron Voltage [V]	161	150
Ion dose [ions/cm^2]	1.0E+16, 5.0E+16, 1.0E+17	1.0E+15, 3.0E+15, 1.0E+16

EXPERIMENTAL

Si (100) wafers (1-3 Ωcm, Phos. doped) with a mechanically ground or polished surface were prepared. These Si wafers were irradiated with GCIB, under the conditions shown in Table 1. Ar and SF$_6$ were used as the source gases. The acceleration energy of cluster ions was 30 keV. The ion dose ranged from 1.0E+15 to 1.0E+17 ions / cm^2. GCIB was irradiated while covering a part of the Si wafer with a dummy Si wafer as a mask. The difference in the surface height of the GCIB irradiated part and the un-irradiation part was measured with a stylus, in order to estimate the amount of Si etched.

Near surface structure of the Si wafer after GCIB irradiation was analyzed with a high-resolution cross-sectional transmission electron microscope (XTEM). The surface roughness was measured by atomic force microscopy (AFM). The measurement area was 10 μm x 10 μm.

Next, following GCIB irradiation, an anneal was performed at 1200 ℃ for 1 hour in a hydrogen atmosphere. After annealing, the near surface structure was analyzed with XTEM.

RESULTS and DISCUSSION

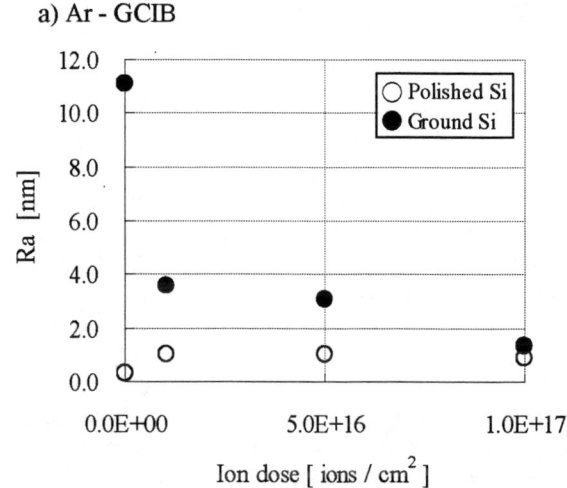

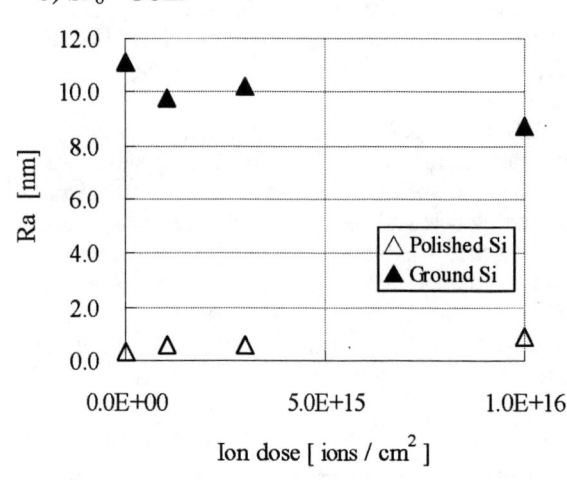

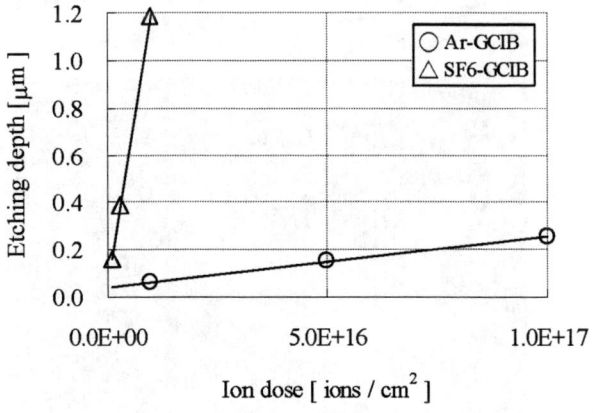

FIGURE 1. Ion dose dependence of the etching depth of the Si wafer with 30 keV Ar-GCIB and SF$_6$-GCIB.

FIGURE 2. Ion dose dependence of the surface roughness of polished Si and ground Si, irradiated with 30 keV Ar-GCIB (a) and SF$_6$-GCIB (b).

TABLE 2. XTEM images of the Si surface irradiated with 30 keV Ar-GCIB and SF_6-GCIB at an ion dose of 1.0E+16 ions/cm^2.

	Before irradiation	Ar-GCIB	SF_6-GCIB
Ground Si	Dislocations and cracks layer	Amorphous Si / Transition layer	Transition layer
Polished Si	(undamaged)	Amorphous Si / Transition layer	Transition layer

Figure 1 shows the etched depth of the Si wafer after Ar-GCIB and SF_6-GCIB. The relation between ion dose and etching depth is linear. The depth by SF_6-GCIB is about 20 times as faster than that by Ar-GCIB with the same ion doses. In Ar-GCIB, the sputtering is caused by the exchange of the kinetic energies between the gas cluster ions and the near surface Si atoms (so-called physical sputtering). On the other hand, reactive sputtering occurs with SF_6-GCIB[2]. Figure 2 shows the relationship between the average roughness (Ra) and the ion dose. Ra decreases with increasing ion dose for the ground wafer. The decrease in Ra is more effective with the Ar-GCIB than with SF_6-GCIB because of the lateral sputtering effect[2]. With the polished wafer, the Ra is slightly increased by GCIB irradiation, but the ion dose has little influence. In these experimental conditions, it is shown that Ra is saturated to this value.

XTEM images of the samples etched with GCIB are shown in Table 2. With the ground wafer, dislocations and cracks caused by the mechanical processing were observed and the depth of damage from surface was about 100 nm. The polished wafer surface exhibits undamaged crystalline structure. After GCIB irradiation, the surface damage layer was the same both the ground and the polished Si wafer. This means that grinding damage can be removed by GCIB irradiation.

An amorphous Si region was confirmed after each GCIB irradiation. The interface under the amorphous Si layer was rough. (This interface is called a transition layer in this paper.) The deeper edge of the transition layer was shown to be single crystalline by electron beam diffraction. The thickness of amorphous Si layer was about 30 nm after Ar-GCIB irradiation. However, a very thin (< 5 nm) amorphous layer was observed when SF_6-GCIB was used. This might be due to the difference between physical sputtering and reactive sputtering. With both beams, the thickness of the transition layer was about 15 nm.

Figure 3 shows the XTEM image of samples

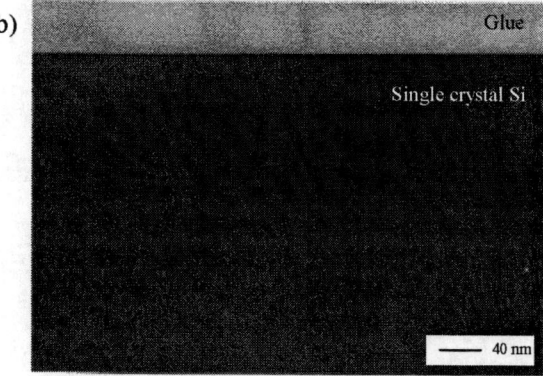

FIGURE 3. XTEM images of Si surface irradiated with 30 keV GCIB (a : Ar-GCIB, b : SF_6-GCIB), followed by annealing at 1200°C for 1 hour in a hydrogen atmosphere.

irradiated with GCIB, followed by a high temperature anneal. It can be observed that the surface of the sample irradiated with GCIB became atomically smooth. The amorphous Si layer and the transition layer on the Si surface disappeared completely. This is because the solid phase epitaxy was produced from the lower side in the amorphous Si layer.

CONCLUSION

The damage caused by GCIB irradiation on the surface of Si wafers was analyzed by XTEM observation. Ar-GCIB and SF_6-GCIB with different etching rates were compared. The thicknesses of the amorphous Si layer that had been formed by irradiating GCIB differed with the ion species. The transition layer thickness was the same. When Si wafers were annealed at 1200 °C for 1 hour in a hydrogen atmosphere after GCIB irradiation, the amorphous Si layer and the transition layer on the Si surface disappeared completely, and an atomically smooth surface was formed.

The results suggest that Si wafers could be manufactured with enhanced flatness by using GCIB followed by high temperature annealing.

ACKNOWLEDGMENTS

Special thanks are due to Professor J. Matsui, Hyogo Science and Technology Association, Hyogo, Japan and Dr. Y. Tsusaka, University of Hyogo, Hyogo, Japan for useful discussions and comments.

REFERENCES

1. http://www.itrs.net/Common/2005ITRS/FEP2005.pdf.
2. I. Yamada, J. Matsuo, N. Toyoda and A. Kirkpatrick, *Mater. Sci. Eng. R.*, **34**, (2001) 231-295.

Boron Beam Performance and in-situ Cleaning of the ClusterIon® Source

Thomas N. Horsky, Glen F. R. Gilchrist, and Robert W. Milgate III

SemEquip, Inc., 34 Sullivan Road, North Billerica, MA 01862 USA

Abstract: The implantation of borohydride ions has enabled very high dose rates at low implantation energies. Using the borohydride material $B_{18}H_{22}$, an on-wafer equivalent current of 18 mA at 1 keV are readily achieved using a ClusterIon® source on a conventional high current implanter. As is frequently the case when running condensables, when borohydrides are introduced into the source in the vapor phase, boron-containing deposits tend to condense and accumulate in and around the ion source over extended periods of operation. In order to achieve production-worthy source lifetimes, we have developed a means of controlling the temperature of those surfaces exposed to the vapor to reduce condensation, and an in-situ cleaning process to efficiently remove deposits from the system. We show beam recovery after periodic cleaning cycles which enable good beam stability and performance. The in-situ cleaning process is also beneficial in reducing potential exposure to toxic fumes during source removal.

Keywords: High current implantation; molecular implantation; ion sources

PACS: 61.72Tt

INTRODUCTION

The propagation of high currents of very low energy boron and arsenic beams has been an area of intense development for more than a decade. In particular, CMOS device scaling has driven boron implant energies below 1 keV, for example, to form source/drain extensions. The transport of such low energy boron beams is limited by space charge effects as embodied by the Child-Langmuir law. Deceleration techniques have proven successful as a means to increase beam currents in high current (HC) implanters by transporting them at higher energy and abruptly decelerating the beam prior to its reaching the wafer. However, upon deceleration, space charge effects tend to blow up the beam, increasing beam angular divergence and spatial extent. This creates unwanted non-uniformity and resist mask shadowing during the implant; these are negative effects for serial implantation systems, in which implant uniformity is more sensitive to beam quality than in batch systems. Energy purity is also affected by some deceleration techniques due to the partial neutralization of the beam prior to deceleration.

An attractive technique to increase transport energy and reduce space charge is through molecular implantation, using beams of the borohydride ions $B_{18}H_x^+$ or $B_{10}H_x^+$. The utility of implanting $B_{18}H_x^+$ ions can be appreciated through the observation that at a given accelerating voltage, the increased mass of the ion reduces the implant energy per boron atom by about twenty-fold, while at the same time producing an eighteen-fold increase in dose rate [1,2,3]. Thus, large conglomerates or "clusters" of like dopant atoms are useful for both ultra-low energy and ultra-high dose applications.

THE CLUSTERION® SOURCE

To efficiently produce molecular ions without dissociation, a source using a "soft" ionization method is preferred over a conventional arc discharge source. An electron impact source has been developed [4,5] for this purpose, the ClusterIon® source. By using ClusterBoron® feed material (a form of octadecaborane, $B_{18}H_{22}$), and SemEquip's integrated vapor delivery and reactive fluorine cleaning modules, molecular implantation can now be performed on implant tools having traditional beam line architectures.

Source Design

The source uses electron impact to provide the gentle ionization necessary to preserve the integrity of the molecules being ionized. While the source is designed to provide the best possible beam current performance using ClusterBoron® feed material, it also produces several mA of As and P beams (including As_4^+ and P_4^+) from arsine and phosphine gas, using a traditional gas box and gas feed to the system. The design of the source takes advantage of the remote electron emitter location made possible by the electron injection optics. By placing the emitter as shown in Figure 1, filament wear associated with ion erosion is minimized, helping to ensure long filament life.

The electron beam generated by the electron gun is deflected through 90 degrees by a magnetic dipole field. Once deflected, the beam is then injected into the ionization chamber where it is magnetically confined. The magnetic confinement is optimized to maximize the ionization efficiency of the injected electron beam.

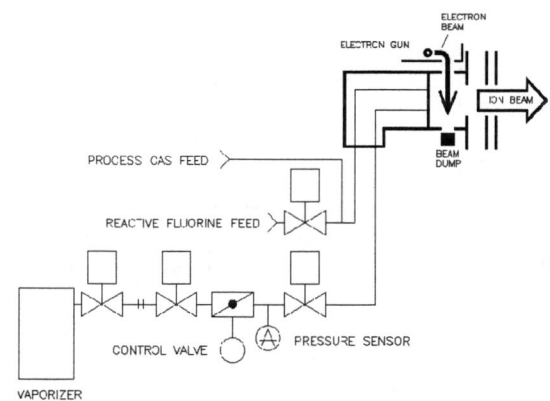

FIGURE 1. Schematic layout of the ClusterIon® source and associated vapor and cleaning gas delivery systems.

Temperature Control of the ClusterBoron Delivery System

The ClusterIon® system has been designed with the requirements of low temperature vaporization in mind. The vapor delivery system is designed to provide the thermal management necessary to avoid condensation and deposition by methods which include the creation of a positive temperature gradient along the vapor delivery path.

Thermal Management and Borohydride residues

In addition to controlling the wetted surface temperatures in the delivery system, it is desirable to control the temperature of the source exit aperture and the extraction electrode to minimize the condensation and deposition of borohydride residues. Experience suggests that while it is important to keep surfaces which come into contact with the material warm enough to avoid material deposition by cooling from the vapor phase, it is also necessary to avoid high temperature, which can lead to thermal disassociation of the vapor phase material. The disassociation yields lower order borohydride compounds, which then condense, creating deposits in the source region. Hence, active thermal management is provided for all the component surfaces in the ClusterIon® system.

SOURCE PERFORMANCE

Beam Current

The source produces high effective boron currents over a wide range of extraction energies. The data presented here are given in boron atomic ion equivalent beam currents which are eighteen fold greater than the $B_{18}H_x^+$ molecular-ion beam current, and boron atomic ion equivalent extraction energy which are twenty fold lower than the $B_{18}H_x^+$ molecular-ion extraction energy; thus, a 2 mA $B_{18}H_x^+$ extracted at 40 kV has dose rate and extraction energy equivalent to a 36 mA B^+ beam extracted at 2kV.

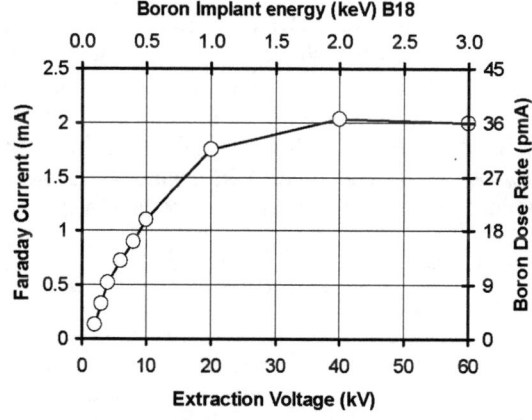

FIGURE 2. Octadecaborane boron equivalent dose rate versus equivalent implant energy.

Figure 2 plots octadecaborane electrical current and equivalent boron dose rate), versus extraction voltage (and equivalent boron implant energy) from 4 kV to 60 kV. These data were collected using a model 350 ClusterIon® source with a post-analysis Faraday, and operated in drift mode.

Decaborane ($B_{10}H_x^+$) also provides the advantages of molecular implantation, the difference being each $B_{10}H_x^+$ molecular-ion contains ten boron atoms and has mass approximately eleven fold greater than that of a single boron atom. Figure 3 shows decaborane beam current versus implant energy for extraction voltages from 2 kV to 60 kV. The equivalent boron current rises rapidly from 1.2 mA at 200 eV to 19 mA at 2 keV, rising slowly to 20.5 mA at 6 keV. These data were collected in the same system used to generate the data of Figure 2.

FIGURE 3. Decaborane boron equivalent beam current versus equivalent implant energy

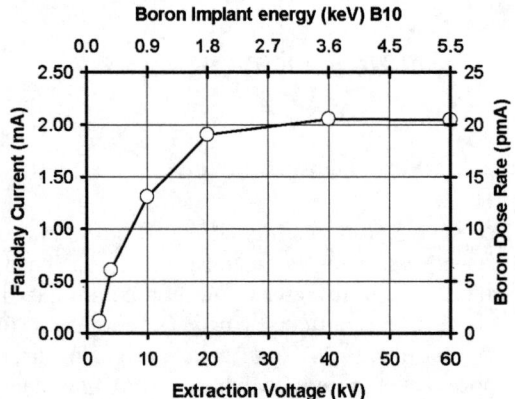

Beam Stability and Glitch Rate

Beam stability and glitch rate are important metrics for ion implantation. The ClusterIon® source model 350 has been operated for periods of over seven hours without adjustment of any source tuning parameters. Only the emission current of the electron gun was stabilized through closed-loop control; no other automatic tuning was used. With manual or automatic tuning, the beam current can be stabilized for longer periods of operation.

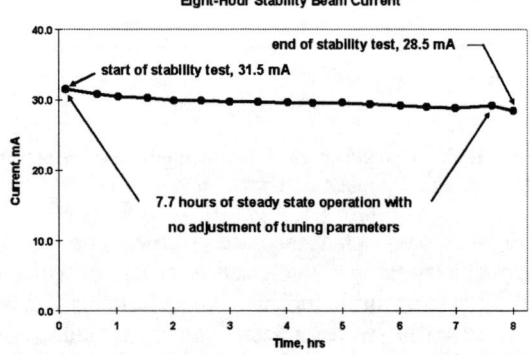

FIGURE 4. Eight hour stability run, without source tuning.

Figure 4 shows eight hours of continuous operation including a period of 7.7 hours during which no adjustment of source tuning was made. During this period the beam current fell from 31.5 mA to 28.5 mA or about 9.7% which equals a drift rate of about 1.3% per hour. Manual or automated beam tuning can be used to keep drift close to zero for a similar period.

Electric discharge across the extraction gap (*i.e.*, glitching) can affect implant dose uniformity and should be minimized. Typical glitch rates for ion implanters may be greater than five glitches per hour. During the entire eight hour period displayed in Figure 4, there were four glitches recorded.

Material Deposition & In Situ Cleaning

Cleaning Methodology

The ClusterIon® source, like other ion sources, tends to accumulate residual material in the ionization chamber and on other components, which will eventually cause a reduction in beam current. By introducing reactive fluorine gas into the ionization chamber these residues are converted to high vapor pressure compounds and pumped away. By adjusting fluorine flow, source housing pressure, and component temperatures, deposits are effectively removed from the ionization chamber, extraction electrode and source housing.

Cleaning recipe and schedules should be tuned to provide the highest degree of implanter availability, and generally accepts recovery time of the ion beam current to its spec level (including cleaning time, recovery, and beam restart). As of this writing, we routinely see one hour beam to beam performance with full beam current recovery.

The fluorine cleaning chemistry is highly reactive. While the focus of current development is on cleaning borohydride residues, preliminary testing indicates that the *in situ* clean is also effective on other typical implant chemistries.

In Situ Cleaning Simplifies Manual Cleaning

In addition to extending the time between mechanical source maintenance, *in situ* cleaning also simplifies maintenance. When it is necessary to remove the source for servicing, a cleaning recipe optimized for source housing cleaning reduces or eliminates the need for mechanical cleaning of the source housing and extraction electrode. This not only reduces the total time required for manual maintenance of the source but also improves the vacuum recovery of the system, reducing the pump down time associated with a source exchange. It is also beneficial in reducing personnel exposure to toxics.

Efficacy of the In Situ Clean

Figure 5 illustrates the efficacy of the *in situ* cleaning process. Photographs a, c, and e show typical borohydride deposition in a Model 320 ClusterIon® source and its associated extraction electrode. Photographs b, d, and f are of another Model 320 source with a similar operation history and a 30 minute *in situ* clean. These photographs illustrate that the cleaning process is very effective at removing deposited boron residues from the ionization chamber and extraction electrode. In fact, an optimized cleaning recipe maximizes source availability and beam current over the mechanical life of the source.

FIGURE 5. Model 320 ClusterIon® source, before and after *in situ* clean.

Material Evolution During in situ Cleaning

During the process of cleaning recipe development, a quadrupole residual gas analyzer was used to study material evolution during the cleaning process. Mass spectra were collected continuously, resulting in trend partial pressure plots (ion current over time) for multiple evolved gas phase species.

Figure 6 shows trend plots for: F*, F_2, BF_2, and N_2F_3 as well as the beginning and end of F* injection. The partial pressure of BF_2, P_{BF2}, is of particular interest because it can be used to evaluate the reaction rate and end point of the F*-B_xH_y reaction, and hence the effectiveness of the cleaning process. When the F* flow is switched on, P_{BF2} rises quickly and levels off, indicating a constant reaction rate, for about 12 minutes. The falling P_{BF2} indicates that most of the B_xH_y has been consumed, *i.e.*, the reaction rate is decreasing and the end point has been reached. While P_{BF2} is falling we see a rise in P_{F*}, P_{F2} and P_{N2F3} which is consistent with the absence of B_xH_y and the recombination of reactive species.

While RGA data is useful for cleaning process development it is not required for regular operation. For any given source operation regimen and cleaning recipe, the repeatability of the cleaning signature indicates that simple time-based control should work well.

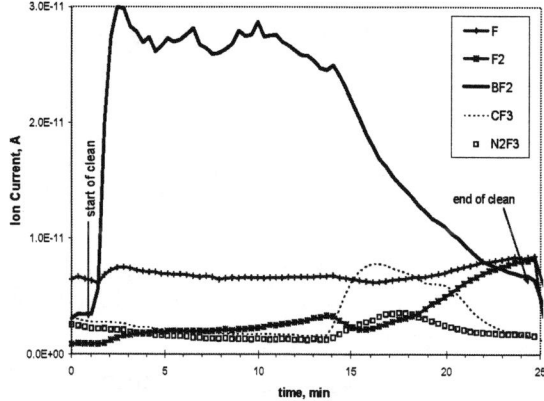

FIGURE 6. Typical RGA plot showing time varying ion currents for several species of interest.

CONCLUSIONS

The SemEquip ClusterIon® source is an effective tool for achieving high equivalent boron ion currents at low delivered implant energies in traditional beam line implanters. It provides for high throughput operation and eliminates the need for deceleration techniques. *In situ* cleaning ensures that a high level of source performance and availability can be maintained over the mechanical life of the source.

ACKNOWLEDGMENTS

The Authors wish to acknowledge the outstanding efforts made by Jeff Buda, Brian Haslam, Dennis Klesel, Neil Call, Scott Repucci, and Don Bistany Without their tireless efforts it would not have been possible to collect the data here presented in any reasonable amount of time.

REFERENCES

1. Y. Kawasaki, T. Kuroi, K Horita, Y. Ohno, M. Yoneda, T. Horsky, D. Jacobson, and W. Krull, *Ibid.*, pp. 25—29.
2. J. O. Borland, *Ibid*, pp. [ ].
3. J. O. Borland, M. Tanjyo, N. Nagai, T. Aoyama, and D. Jacobson, "Applying Equivalent Scaling to USJ Formation", *Semiconductor Int.*, Jan 2005, pp. 52—55.
4. T. Horsky, D. Jacobson, W. Krull, and H. Glavish, "Performance of the SemEquip Ion Source", 14th International Conference on Ion Implantation Technology, 22—27 September 2002, Taos, NM (unpublished).
5. T. N. Horsky, "ClusterIon™ Source for Cluster Implantation", 15th International Conference on Ion Implantation Technology, October 25—29, 2004, Taipei, Taiwan, ROC (unpublished).

P-type Gate Electrode Formation Using $B_{18}H_{22}$ Ion Implantation

Dietmar Henke, Frank Jakubowski, and Josef Deichler

Qimonda Dresden GmbH & Co. OHG
P.O. Box 10 09 40, D-01079 Dresden, Germany

Vincent C. Venezia, M. S. Ameen, and M. A. Harris
Axcelis Technologies
108 Cherry Hill Drive, Beverly, MA, 01915 USA

Abstract. We have investigated the use of octadecaborane ($B_{18}H_{22}$) cluster ion implantation to form highly active p-type gate electrodes in a 90 nm CMOS process. As device dimensions scale, the influence of poly-depletion and short channel effect control on device performance continues to become more significant. Increasing gate electrode doping via high dose ion implantation is a standard method for reducing poly-depletion. Poly-silicon gate doping with the molecular ion $B_{18}H_{22}$ offers throughput advantages over monatomic B ion implantation. For instance each molecular ion introduces 18-B atoms, thereby reducing the implant dose. In addition, each B constituent of the molecular ion is implanted with 1/20th the ion energy, making it possible to achieve low energy dopant distribution while taking advantage of higher beam energy currents. In this work, $B_{18}H_{22}$ implantation conditions (energy, dose) were matched to those of the standard B^+ process of record (POR) used for gate electrode doping. We show that the poly-depletion, threshold voltage, and yield of devices implanted with $B_{18}H_{22}$ are comparable to those implanted with the POR. We combine this device results with materials data to demonstrate that the high dose implants necessary to form p-type gate electrodes with minimum poly-depletion can be achieved with $B_{18}H_{22}$ ion implants without impacting the device performance.

Keywords: Cluster Ion Implantation, Octadecaborane ,, DRAM, CMOS, Poly Gate Doping
PACS:: 36.40.Wa, 61.72.–y, 61.72.Ss, 61.72.Tt, 73.30.+y 81.70.Jb, 85.40.Ry, 85.30.Tv

INTRODUCTION

The next performance boost throughout the DRAM industry will be provided by introduction of Dual Work Function (DWF), which is obtained by introducing n- and p-type polysilicon gates to enhance transistor performance. Additionally, high active dopant concentration in the poly of the gate stack is necessary to reduce the gate depletion effect. Accordingly, the device short channel properties will improve at a given physical gate oxide thickness [1].

Suppression of boron penetration through the gate oxide is the key to the implementation of DWF. The implant profile has to be well within the gate poly layer to avoid threshold voltage relevant parasitic doping. Low energy and high dose implants are normally done in deceleration mode of the implanter to ensure a production worthy throughput. The ratio of the deceleration than can practically be implemented is limited by the range of the high energy component, and is referred to as "energy contamination". The sum of implant range plus 4.5σ (corresponding to a transmission of 3ppm) needs to be below the thickness of the poly layer. The effect of energy contamination limits thus the maximum possible throughput of beam line implanters.

To reach the goal of highly doped gates new implant technologies using cluster implantation [2] could be considered as a highly productive solution.

In this experiment, a cluster ion implant of octadecaborane, $B_{18}H_{22}$ into the p-poly gate has been investigated. The goal was to find a matching of the cluster boron ion implant without changing of other process parameters such as the thermal budget, device implants, or gate oxide properties in the process flow. We compared electrical parameters and yield of devices implanted with $B_{18}H_{22}$ or Boron. In addition the boron profiles of implants in blanket poly layers were evaluated.

EXPERIMENTAL

Experimental assembly

The experiment was performed on a modified Axcelis HC3 ion implanter, equipped with a SemEquip ClusterIon® $B_{18}H_{22}$ ion source. The existing HC3 magnet can only bend an ion as large as $B_{18}H_{22}$ up to 1.4 keV equivalent B^+ energy. Therefore, the standard magnet was replaced with one that is capable of bending the large borane ion up to 9 kV B^+ equivalence, which is well past the extraction power supply limit of 4 kV. An electrostatic quadrupole was installed for beam size control and the resolving apertures were enlarged to increase beam current.

The reference B^+ implants were done on a standard Axcelis HC3 High Current Implanter in drift mode. The reader is referred to [3] for further information on the equipment parameters, source conditions, beam currents, and beam spectra that were used to generate the implants for this study.

Implant profiles and matching

In order to ensure proper device characteristics, the SIMS profiles and sheet resistance values were tested for B^+ and B_{18} implants into Si and poly-Si wafers.

Figure 1 shows the sheet resistance results after a 950°C 60 second anneal for two different doses. The sensitivity of R_s to dose at this anneal temperature is small. The small offset between B^+ and B_{18} could be due to slight differences observed in the SIMS profiles, shown in Figure 2.

Figure 2 show the as-implanted SIMS profiles at 3 kV and 8E15 cm^{-2} into poly-Si. Small differences in the deeper tail can be observed between B^+ and $B_{18}H_{22}^+$. The difference in channeling properties between the B_{18} and B^+ caused by the amorphizing properties of B_{18} has been reported previously [2]. Figure 3 indicates the profiles after anneal at 950°C for 60s of implants into a 1000 Å poly-Si on oxide. The B_{18} shows a higher peak concentration in the near surface than the Boron implants, but the activated profiles are identical throughout the thickness of the poly.

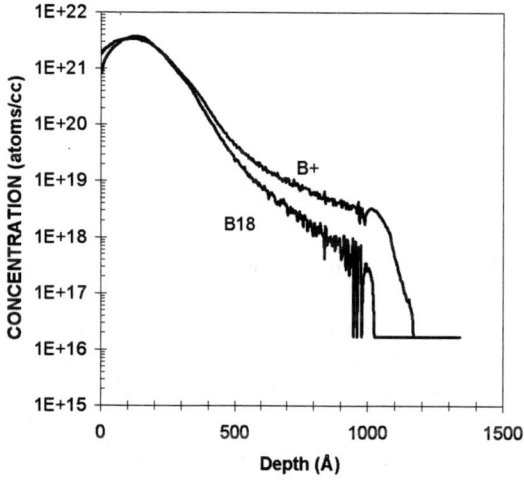

Figure 2 As-Implanted SIMS profiles of B+ and B_{18} at 3 kV and 8E15 cm^{-2} implanted into 1000 Å poly-Si on oxide.

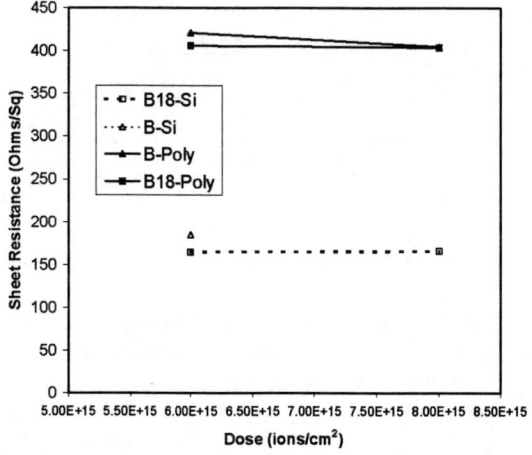

Figure 1: R_s Matching for B^+ and $B_{18}H_{22}$ implants into crystalline Si and poly-Si. Anneal was 950°C 60 s.

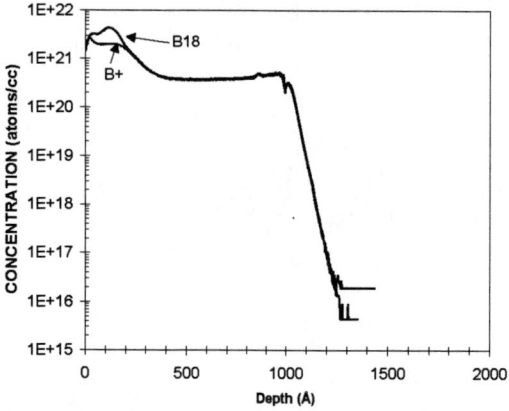

Figure 3. Profiles after 950°C 60 sec anneal for implants into 1000 Å poly-Si on oxide.

Process sequence and parameters

After the gate poly deposition, either $B_{18}H_{22}^+$ or B^+ was implanted with adjusted dose and energy. The reference implant of monomer B+ was at 4 keV and 6E15 cm^{-2}.

$B_{18}H_{22}^+$ was implanted at 3 and 4 keV equivalent B+ energy. The doses of the 4keV $B_{18}H_{22}^+$ energy was varied between 6E15 and 8E15 cm^{-2}.

In the process flow of the 90 nm CMOS technology all pre and post processes of the gate implant have not been changed including the gate poly boron activation thermal budget.

One of the most important device parameters for the characterization of the poly gate doping are the depletion in the inversion mode (see Figure 4) and the threshold voltage VTS. The depletion can be characterized from the capacitance-voltage (CV) measurement by extracting the capacity C (V_{bias_accu}) in accumulation and C (V_{bias_inv}) in inversion at fixed voltages V_{bias} (see Figure 4). Different splits can then be classified by the ratio r of the capacitances.

$$r = \frac{C(V_{bias\_inv})}{C(V_{bias\_accu})} \qquad (1)$$

For alternative technologies like cluster ion implant this ratio may not be lower than the reference to keep the device performance. That means that the amount of activated dopants has to be comparable or higher than that of the reference doping.

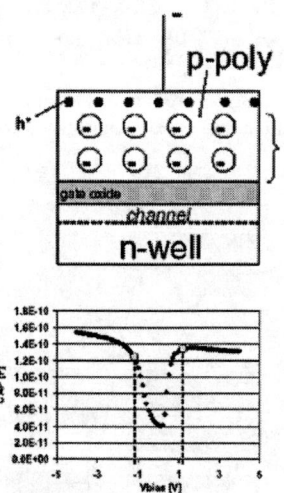

Figure 4: Examples for depletion layer in inversion mode and CV curve

It will be shown in this research, that there is neither an increase in depletion nor a deviation in VTS while using well adjusted $B_{18}H_{22}$ implants.

Results and Discussion

Electrical Results

In the first experiment the implant energy of $B_{18}H_{22}$ was kept constant at an equivalent Boron energy of 4 keV while varying the dose.

Figure 5 shows the difference of the ratio r in comparison to the reference implant. At a dose of 6E15 cm^{-2} the depletion of the $B_{18}H_{22}$ is identical to the reference. A higher activation and therefore a lower depletion are reached at the higher dose of 8e15 cm^{-2}.

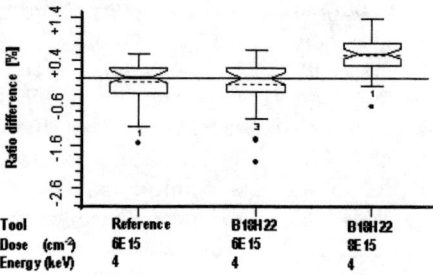

Figure 5 Ratio for different doses

The threshold value VTS of a pFET is given in negative values. The deviation of the $B_{18}H_{22}$ groups to the reference is shown in Figure 6. There are similar VTS when using $B_{18}H_{22}$ compared to the reference. The higher dose for the 3rd group leads to a minor increased tail of boron penetrating into the channel resulting in a little lower absolute VTS for the pFET.

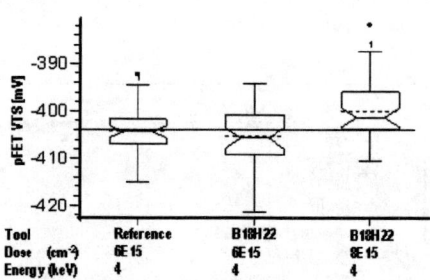

Figure 6 Threshold voltage of pFET for different doses

In a further experiment the depletion was compared at a constant dose of 6E15 cm^{-2} and different energies.

Figure 7 shows the same depletion for the $B_{18}H_{22}$ at 6E15 cm^{-2} and 4 keV compared to the reference. The reduction of the $B_{18}H_{22}$ energy to 3 keV equivalent Boron energy results in a lower value of the ratio r or higher depletion.

This observation matches with the experience that different depths of the as-implanted profiles in the poly need an adjustment of the thermal budget in the process flow to get identical results.

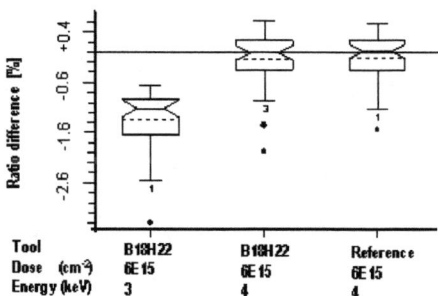

Figure 7 Ratio for different energies

It can be seen that for 3 keV the threshold value VTS follows the higher depletion to higher absolute numbers (see Figure 8). This results from a decreased boron penetration into the channel of the pFET due to the shallower as-implanted dopant distribution.

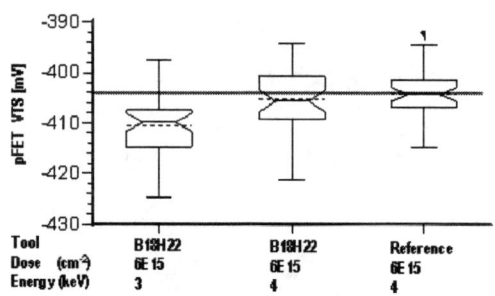

Figure 8 Threshold voltage of a pFET for different energies

The comparable within wafer uniformity of the ratio at 4kev 6E15 cm^{-2} for the reference and $B_{18}H_{22}$ is shown in the Figure 9. The range across the wafer is in order of 2% for both implant techniques. The within wafer non uniformity is mainly driven by the thermal processes.

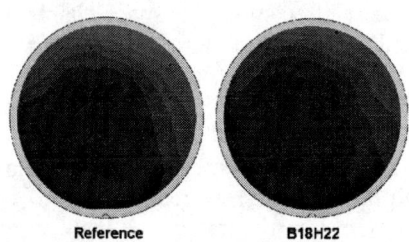

Figure 9 Within wafer uniformity of ratio at 4 keV and 6E15 cm^{-2}

Yield Results

Figure 10 shows the product yield which exhibits comparably for all groups with standard 90 nm CMOS product test program. The slight yield drops for two of the $B_{18}H_{22}$ groups were caused by other process steps than the implants.

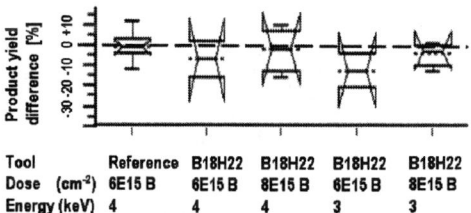

Figure 10 Relative product yield of all split groups

SUMMARY

It has been verified that the $B_{18}H_{22}+$ implantation can be used for the poly gate doping of pFET's in a 90 nm CMOS technology. A comparable device performance has been achieved. The product yield was not influenced by the cluster implant.

The cluster boron ion implant can be considered as highly productive solution for high dose low energy implants without the risk of energy contamination.

New implanters equipped with the cluster boron ion source need strong analyzing magnets to advance into the 4 keV equivalent B+ range.

The possibility of an accurate dose and energy matching of the implant profiles of B and $B_{18}H_{22}$ fits this new technology in existing process flows without additional adjustments in pre or post implant processes. From a production point of view it is important that all standard implanters can be used as backup tools in the case of special or dedicated cluster ion beam implanters.

REFERENCES

1. J. W. Faul and D. Henke, Nucl. Instr. and Meth. in Phys. Res. B 237 (2005) 228-234.
2. D.C. Jacobson, K. Bourdelle, H-J. Gossmann, M. Sosnowski, M. A. Albano, V. Babaram, J.M. Poate, A. Agarwal, A. Perel, T. Horsky, "Decaborane, an Alternative Approach to Ultra Low Energy Ion Implantation", Proceedings IIT 2000, p. 300.
3. D. Tieger, W. DiVergilio, E. C. Eisner, M. Harris, T.J. Hsieh, J. Miranda, W. Reynolds, V. Venezia, "ClusterBoron™ Implants on a High Current Implanter", this conferenc

ClusterBoron™ Implants on a High Current Implanter

Daniel R. Tieger, William DiVergilio, Edward C. Eisner, Mark Harris, T.J. Hsieh, John Miranda, William P. Reynolds

Axcelis Technologies Inc., 108 Cherry Hill Drive, Beverly, MA 01915 USA,

Tom Horsky

SemEquip, 34 Sullivan Rd. North Billerica, MA 01862 USA

Abstract. Advanced p-junction process tool throughput continues to be one of the principal drivers of the industry. First results from an octadecaborane ($B_{18}H_{22}$) ClusterIon® source integrated on an existing high current implant tool are presented. Beam current, throughput and process results are reported. The dose multiplication effect of the use of $B_{18}H_{22}$ means that an electrical current of 1mA produces a dopant flux equivalent to 18mA, while the energy equipartition means that a 20keV octadecaborane ion is process equivalent to a 1keV boron beam. Some modifications to a traditional high current beamline design were made in order to take advantage of the opportunities presented by this new ion source. A somewhat larger extraction slot was used and this, coupled with the fact that the ions have a large mass (210 amu) and therefore have high magnetic rigidity even at modest energies, drove the optics design toward a parallel-to-point configuration. Good mass resolution and control of beam size were demonstrated. Beam currents and throughput that are significantly higher than those available from traditional high current implanters were achieved, along with good process results.

Keywords: High current implantation; molecular implantation
PACS: 61.72Tt

INTRODUCTION

A key figure of merit in high current, or high dose as it is increasingly described, ion implantation is beam current. As devices shrink and the energies get lower, the ability to generate and transport high ion beam currents becomes increasingly more difficult. Traditional high current tools have thus resorted to a number of techniques such as shortening the beamlines and using deceleration to access the lowest beam energies. Molecular implantation offers another alternative since the effective energy of the implanted atom is lower while the transport energy is higher. Therefore, both deceleration mode operation and energy contamination are avoided. Well known examples of molecular implantation include BF_2 and the dimers of arsenic and phosphorus (As_2 and P_2)[1,2]. New source materials for molecular implantation include the boranes (B_xH_{x+4}) which have the additional feature that there are multiple B atoms per molecule, thus also multiplying the dose[3].

Octadecaborane ($B_{18}H_{22}$) has both of these advantages. Since there are 18 B atoms per molecule, the dopant flux for a given electrical beam current is increased by that factor and thus also the effective beam current: 1 mA of delivered beam becomes 18 mA of dopant flux. In addition its high mass (210 AMU at the peak of the spectrum) leads to essentially a factor of 20 decrease in the effective energy: 60 keV extraction and transport yield 3 keV effective energy. Both of these effects mitigate space charge blowup of traditional high current tools since high extraction energy and relatively low current are transported to the wafer while yielding high throughput beams for enhanced productivity.

The 20:1 effective energy reduction allows some key high dose process applications to fall within the capabilities of mostly traditional high current tools. For example, the Dual Poly Gate (DPG) application space was chosen for this first implementation and thus the effective energy of 1.5 keV to 4 keV B is the target energy range. The rest of this paper will discuss the equipment, show some resulting beam currents and throughput, typical mass spectra and resolution are discussed and beam size is shown, and finally include implant process results obtained to date.

EQUIPMENT

Ironically, the features previously mentioned, which allow for ease of transporting high beam currents at low effective energy, pose some new challenges for beamline design. The combination of high mass – up to 220 AMU and up to 80 keV extraction energy - implies very high magnetic rigidity in the analyzer magnet. Typical high current AMU magnets are designed for transporting 80 keV Arsenic and thus have a mass – energy product of about 6000 AMU – keV. Octadecaborane, on the other hand, requires roughly a factor of 3 higher rigidity of its analyzer magnet in the energy range of interest.

In addition, the ClusterIon® source allows for an increased length extraction slot to maximize extraction area and thus extraction current, further increasing the challenge of designing an analyzer magnet with sufficient capability. A novel approach of using a smaller pole-gap magnet in a horizontal configuration was used to take advantage of an existing magnet design as a drop-in replacement to a standard high current beamline. The existing magnet is essentially the same as is used in the Axcelis Paradigm high energy tools as the final energy magnet. Since the magnet's pole gap was not wide enough to allow full acceptance of the slot when mounted vertically, for this teststand the source was mounted horizontally, which changed the beamline optics from point-to-point to parallel-to-point. In addition a downstream electrostatic quadrupole was installed to control beam spot size. Shown in Fig. 1 are the elements of the beamline in this first teststand.

However, for less than the highest DPG energies, a standard high current beamline can be used for integration of the ClusterIon® source and a second teststand was, therefore, outfitted in this way. For that second teststand the maximum extraction energy was limited by the analyzer magnet to less than 30 keV (max effective B energy of less than 1.5 keV) but it served to also demonstrate beam current and throughput but in a more traditional optics setting.

In moving forward with productization of this technology a dedicated analyzer magnet was designed that is capable of handling the high magnetic rigidity demands of the DPG application, while allowing for preservation of the traditional optics and a vertical source orientation.

BEAM CURRENTS AND THROUGHPUT

Octadecaborane is transported with reasonable efficiency in a high current beamline. Shown in Table 1 are the parameters for 18 mA of effective beam current at 2 keV effective energy, which current is significantly higher than available from a traditional high current implanter. Yet, this first source had not been fully optimized for extraction area. A second iteration with a larger extraction slot yielded nearly 28 mA of effective beam current and this is shown in Fig. 2, where the beam current and transmission (electrical mass-resolved beam current to extraction current ratio) are shown plotted versus the source's emission current for 20 keV extraction energy. Future improvements are expected to yield 40 mA effective beam currents by the end of the year, driven largely by optics improvements to both the source and beamline. For a 1.0E16 dose at a beam current of 40 mA, more than 50 wafers per hour is expected to be the typical DPG throughput for 300 mm process.

TABLE 1. 40-keV B18, (2-keV B)		
Currents	Electrical	B Effective
Beam current	1.0 mA	18 mA
Extraction current	4.7 mA	
Suppression current	0.1 mA	

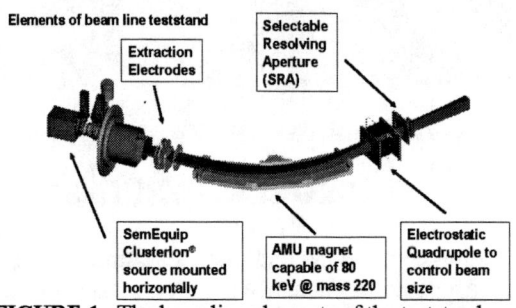

FIGURE 1. The beamline elements of the teststand.

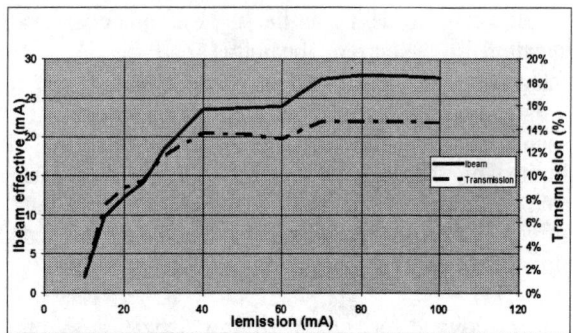

FIGURE 2. Beam current and transmission versus source emission current.

MASS SPECTRA, MASS RESOLUTION AND BEAM SIZE

The dominant feature of the mass spectrum of octadecaborane consists of a large, broad peak centered just beyond mass 200. While $B_{18}H_{22}$ is formally mass 220, the width of the spectrum peak (fine structure) can be understood to be made up of different numbers of hydrogen atoms in the molecule and the binomial distribution of the isotopes present in natural boron material[3]. Similarly, the next lowest mass borane, $B_{17}H_{21}$, has its endpoint mass at 208 but is peaked at approximately mass 196. Shown in Fig. 3a is a high resolution spectrum and in Fig. 3b is a comparison of two spectra obtained with different widths of the resolving aperture. With a narrow, 8 mm aperture, the feature of the $B_{17}H_x$ is visible while it is contained in the wider, 33 mm aperture. In order to maximize beam current a wider slit is desirable while the minimization of energy and dose contamination calls for a narrower slit. While these requirements go in opposite directions, it has been seen that accepting 15 AMU at the peak of the $B_{18}H_{22}$ spectrum (mass resolution of 14) is sufficient to include less than 1% of B_{17} in the accepted beam current which guarantees less than 1/18% dose error[4]. The next most prominent feature of the mass spectrum is the doubly charged octadecaborane peak located at approximately mass 100.

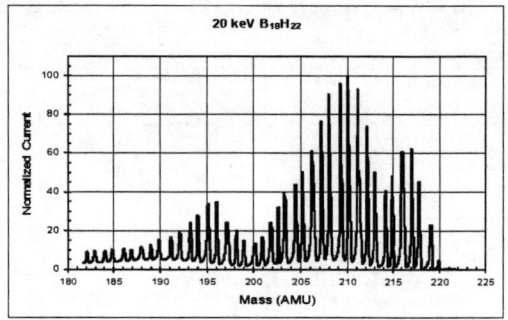

FIGURE 3a) Ultra high resolution spectrum showing individual hydrogen atom components

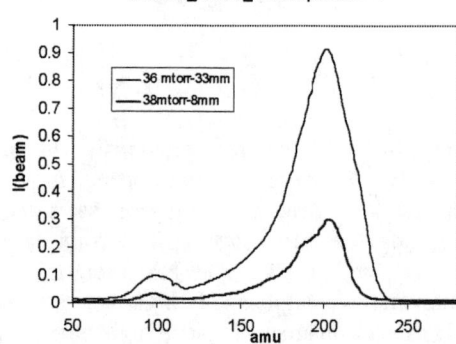

FIGURE 3b) Comparison of mass spectra obtained with two resolving slits.

The beam size as obtained in the high current implanter was found to be quite reasonable, even without utilization of the quadrupole lens downstream of the AMU magnet. Shown in Fig. 4 is a typical profile of the beam at the disk for a 40 keV extraction energy (2 keV equivalent) and an effective beam current of 12 mA.

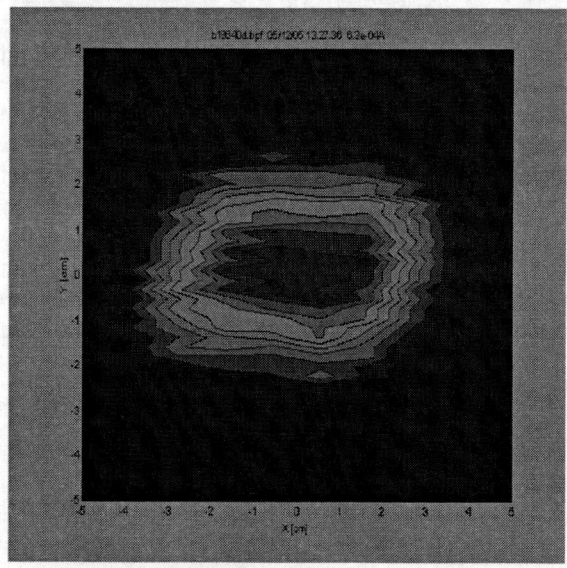

FIGURE 4. Typical beam size measured at disk.

PROCESS RESULTS

Above and beyond beam current, the next characterization challenge for octadecaborane occurs in the assessment of process results. Uniformity and Rs have both been found to be within specification for the equivalent B recipe and are indicated in Table 2.

The difference in R_s can be compensated with a dose trim factor. Particles have been seen to be within spec on the test stand and are shown in Fig. 5 indicating no fundamental difference for molecular implant. Metals contamination also is within spec and is shown in Table 3. One additional process note is that the profiles from B18 are always shallower than those of B when implanted into crystalline silicon as a result of suppression of channeling which results in a self-amorphization with B18 implants. This is shown in Fig. 6.

TABLE 2. Uniformity: 40-keV B18, 5.6e13 (2-keV B, 1e15)		
Condition	Rs (mean)	%SD (mean)
B18	136	0.43
B	126	0.65

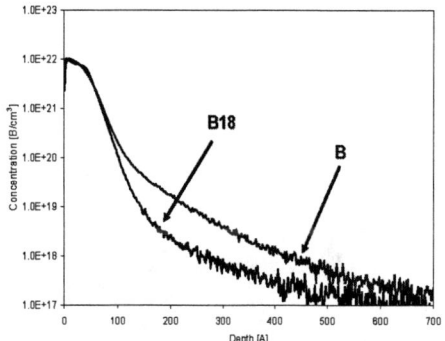

Channeling is suppressed in B18 implants – self amorphizing when implanted in crystalline silicon

Figure 6. Profiles of B18 and B.

SUMMARY

First results from the integration of a ClusterIon® source on a high current ion implanter have been shown. Beam currents that are significantly higher than available from traditional high current tools have been demonstrated with good process results.

ACKNOWLEDGMENTS

The authors wish to thank Vincent Venezia and Frank Sinclair for valuable discussions and early contributions to this effort.

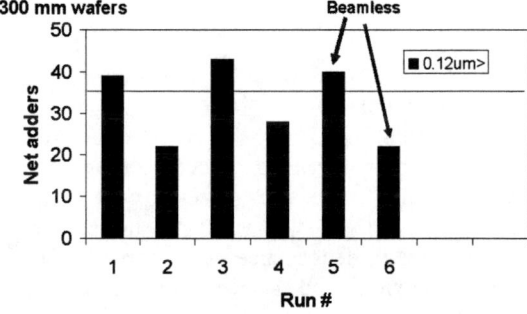

FIGURE 5. Sample particle data.

REFERENCES

1. M. Graf et al., "Low Energy Ion Beam Transport", Proc. 14th International Conference on Ion Implantation Technology, p. 359, 2002.
2. A. Agarwal et. al., "Molecular N-type Dopant Implants", Proc. 14th International Conference on Ion Implantation Technology, p. 119, 2002.
3. D. Jacobson et. al., "Ultra-high Resolution mass Spectroscopy of Boron Cluster Ions", Proc. 15th International Conference on Ion Implantation Technology, p. 406, 2004.
4. D. Jacobson, private communication

TABLE 3. Metals: 60-keV B18, 2.8e14 (3-keV B, 5e15)		
Metal	Results (10^{10}/cm^2)	Spec (10^{10}/cm^2)
Al	64	100
Cr	0.095	5
Cu	0.015	5
Fe	1.1	5
Ni	0.13	5
Ti	0.44	5

Cluster size effects of gas cluster ion beams on surface modification

N. Toyoda and I. Yamada

Laboratory of Advanced Science and Technology for Industry, University of Hyogo
3-1-2 Kouto, Kamigori, Hyogo, 678-1205, JAPAN

Abstract. In the gas cluster ion beam (GCIB) processes, number of atoms in a cluster ion or "cluster size" is one of the most important parameter on the surface interaction processes because it defines the equivalent bombarding energy of individual atom (eV/atom). In this study, cluster size effects on damage formations, sputtering yields were studied using size controlled GCIB systems. Damage formation in crystalline Si substrate showed significant drop at the cluster size where energy per atom was below 5 eV/atom. Sputtering yields of Au by one Ar atom in cluster showed linear relationship with the energy per atom of cluster ions. Threshold energy for sputtering was far below that with Ar monomer ions due to the near surface energy deposition by Ar cluster ion bombardments.

Keywords: Cluster ions, Low damage, Surface modification, Sputtering yield
PACS: 36.40.Wa, 81.65.Cf, 34.50.Dy

INTRODUCTION

Cluster ion bombardments have been received considerable attention due to their unique impact processes on solid surface. By utilizing characteristics of cluster ions, industrial applications using gas cluster ion beams (GCIB) has been started such as shallow ion implantation [1,2], surface smoothing [3,4,5], low-damage etching[6,7] and high quality thin film assisted depositions [8,9,10].

However, it is very difficult to understand the impact processes with conventional linear collision theory because of complex collisions between target and cluster atoms near surface. Therefore, molecular dynamics (MD) simulation have been widely used for large cluster ion impacts. From MD simulations, it was demonstrated that cluster size (number of atoms or molecules in a cluster) plays important roles in damage formations or surface modifications induced by energetic cluster impacts [11]. In irradiation experiments, however, it had been difficult to perform high ion-dose experiments due to low ion current. Also, difficulties of mass separation drastically increase with cluster size. As the average cluster size of the current gas cluster ion beam (GCIB) system is typically several thousands, it is difficult to perform experiments that require high ion-dose by size-selected cluster ion beams.

We have developed a high-current and size-selected GCIB system using a strong permanent magnet filter [12]. Using this GCIB system, the cluster size distribution can be reduced almost 1/20 of the original one. In this study, size-selected Ar-GCIB was used to irradiate a silicon and Au surface to study the cluster size dependence on damage formations, sputtering yields and surface modifications.

EXPERIMENT

Figure 1 shows a schematic diagram of the size-selected GCIB system. A permanent magnet which provides 1.2T of magnetic field (Field length: 450mm, Gap: 50 mm). Neutral cluster beams formed by supersonic nozzle was ionized and accelerated in a acceleration electrodes up to 30 keV. Then ionized cluster ions were bended in the magnetic field and

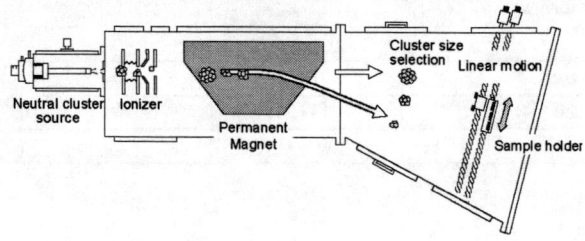

FIGURE 1. Schematic diagram of size-selected GCIB irradiation system.

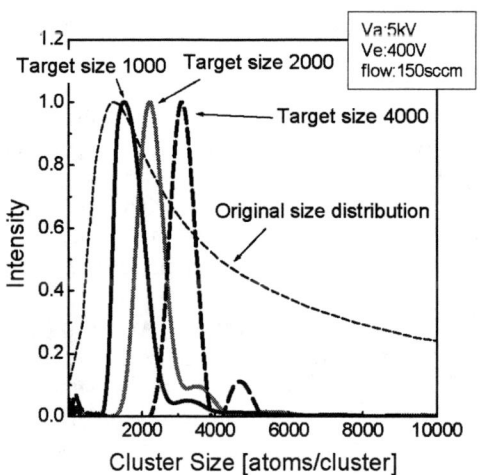

FIGURE 2. Cluster size distributions after magnetic mass separation for various target sizes.

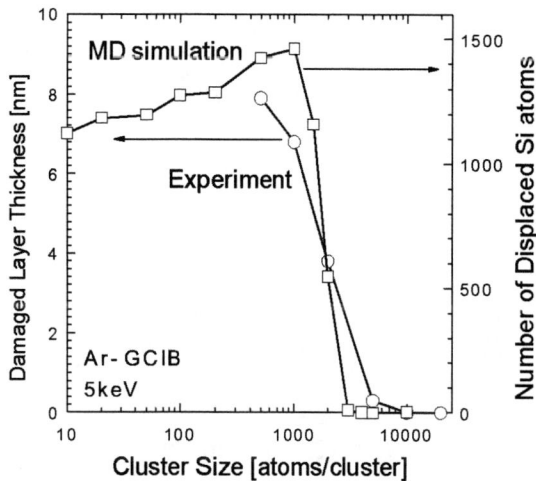

FIGURE 3. Cluster size dependence on damage layer thickness (experiment) and number of displaced Si atoms (MD) after Ar-GCIB irradiation [13].

ejection angle was defined by a total acceleration voltage and a cluster size. A Faraday cup and sample holder was mounted on the linear motion rail and desired size of cluster ions could be irradiated by locating them at a certain position calculated from the acceleration voltage and the desired cluster size.

Figure 2 shows time of flight mass spectra measured at down stream of the permanent magnet. As the deflection angle is determined by energy and mass of cluster ions, certain size of cluster ions can be captured by changing the position of a detector. Compared to the original cluster size distribution, the half width of full maximum is about 400 atoms, which was 1/20 of the original distribution. Also, the peak position of the cluster size distribution shifted with changing the detector positions. The mass resolution of this system is $M/dM = 3.5$ at $X = 250$ mm (Ar cluster size of 1250). Ion current obtained after mass separation was several hundreds of nA to several tens of nA depending on the target cluster sizes.

In this study, two types of experiments were performed. One is a cluster size dependence on damage formation in Si substrates. Another is that on sputtering yields of Au surfaces. For the first experiments, crystalline silicon substrates were irradiated with Ar cluster ions and subsequently damaged layer thickness was measured with spectroscopic ellipsometer. The total acceleration voltage of Ar-GCIB was between 5 and 30 kV and ion dose was 1×10^{14} ions/cm^2. Ar cluster size was changed from 500 to 20000.

For sputtering experiments, the total acceleration voltages of Ar-GCIB were 10 and 20 kV. Ion dose was fixed at 5×10^{15} ions/cm^2. Sputter deposited Au targets were irradiated with size-selected Ar-GCIB over a mask. After irradiation, the mask was removed and the sputtered depths were measured with a contact surface profiler (Veeco Dektak 3). Also, surface morphologies were observed with an atomic force microscope (Veeco Dimension 3000).

RESULTS AND DISCUSSIONS

Cluster size dependence on damage formations

Figure 3 shows both experimental and simulated results of cluster size dependence on damage formation in Si[13]. In both cases, total acceleration voltage was 5kV. In figure 3, number of dislocated Si atoms from lattice and the damaged layer thickness measured with ellipsometer were plotted together. MD simulations results show that the number of Si atoms displaced from their lattice sites increases with Ar cluster size, beginning from size 10 and showing a maximum around 1000 atoms/cluster. Then, a sudden decrease with increasing the cluster size is observed. When the cluster size was larger than 5000 atoms/cluster, no atom displaced was observed.

In the case of irradiation experiments, the damaged layer thickness also decreased around cluster size of 1000 and very thin damage layer was formed over this size. It shows very good agreement with MD simulations. These results indicate that the energy has to be controlled within several eV/atom in order to reduce irradiation damage. MD simulations also suggest that, even though no damage is caused by a large cluster ion impact, a large fraction of its acceleration energy can be deposited. For example, when the acceleration energy of Ar-GCIB and cluster size were 20 keV and 20000, approximately 6 keV was

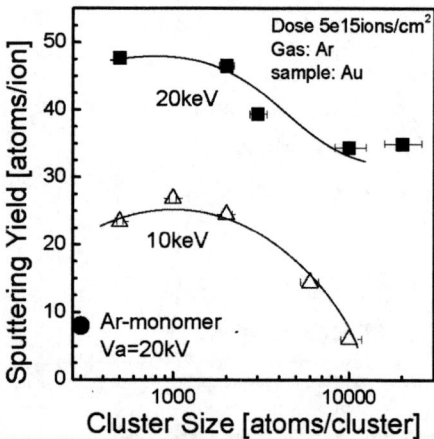

FIGURE 4. Cluster size dependence on sputtering yield (atoms/ion) by 10 and 20 keV Ar-GCIB.

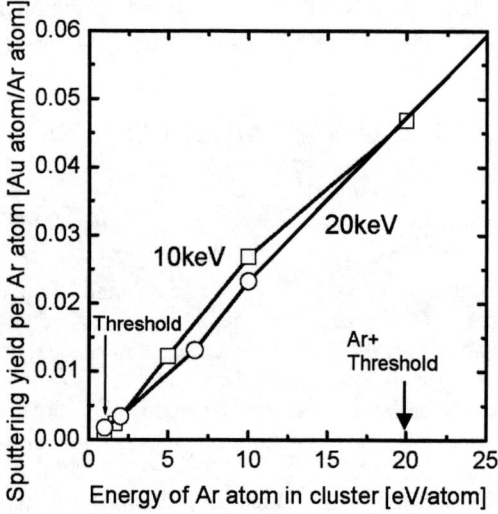

FIGURE 5. Sputtering yield per Ar atom [Au atoms/Ar atom] vs Energy per atom of cluster ion.

deposited on Si substrates without irradiation damage. This indicates that there is a possibility of surface processing with no damage formation.

Cluster size dependence on sputtering

It has been reported that the sputtering yields of various materials by gas cluster ions were one or two orders of magnitude higher than those of atomic or molecular (monomer) ions. However, the cluster size was not selected and it was discussed using average cluster size. As shown in the MD simulations, energy per atom plays very important role during the impact process on solid surfaces, the sputtering yield should be measured using narrow cluster size distribution. In this study, sputtering yield from Au surface and surface morphologies after size selected Ar-GCIB were studied using the system shown in the figure 1.

Figure 4 shows cluster size dependence of sputtering yield (atoms/ion) obtained from the sputtered depth and the ion doses. The sputtering yield was defined that number of Au atoms ejected by one Ar cluster ions. Total acceleration voltage was 10 and 20 kV.

In the case of acceleration voltage 10kV, the sputtering yield was 23.5 atoms/cluster at Ar cluster size 500. The yield was almost the same until Ar cluster size 2000, however, it decreased suddenly at Ar cluster size 5000. At Ar cluster size 10000, it was 6 atoms/cluster. In this case, the energy per atom was only 1.0 eV/atom. Considering that the threshold energy of sputtering of Au with Ar monomer ions is 20 eV [14], sputtering occurs at extremely low energy region in the case of cluster ion bombardments. When total acceleration voltage was increased to 20kV, the sputtering increased from 23.5 to 47.7 atoms/cluster at Ar cluster size 500. With increasing the cluster size, the sputtering yield decreased in the same manner as the case of 10 kV.

As the sputtering yield discussed in the figure 4 was number of ejected atoms by one cluster ion. By dividing the sputtering yield by cluster size, number of ejected atoms by one Ar atoms having the same energy per atom can be obtained.

Figure 5 shows relationship between energy of individual Ar atom in the cluster ion (energy/atom) and the sputtering yield per Ar atoms (Sputtered Au atoms/ Ar atoms in cluster ion). For both of the acceleration voltages, the plotted line showed almost linear relations. It indicates that the sputtering effects will be the same if the same number of Ar atoms having the same energy per atom is irradiated. For example, the number of ejected Au atoms would be the same for both ONE impact of 20keV Ar_{2000} (Energy: 10eV/atom, total number of Ar atom: 2000) and TWO impacts of 10keV Ar_{1000} (Energy: 10 eV/atom, total number of Ar atom: 2000).

In figure 4, the sputtering threshold by Ar monomer ion is plotted, which is much higher than those by Ar cluster ion (0.61 eV/atom). As the total acceleration energy was deposited near the surface, it is expected that surface atoms are effectively ejected by cluster ion bombardments.

SUMMARY

Size-selected GCIB irradiation system had been developed and cluster size dependence on damage formation and sputtering yields had been studied. By controlling the cluster size, damage formation in Si substrate can be controlled. MD simulations indicated that several keV of acceleration energy was deposited even in the no-damage conditions. Sputtering yields of Au by one Ar atom in cluster showed linear relationship with the energy per atom of cluster ions. Threshold energy for sputtering was far below that with Ar monomer ions due to the near surface energy deposition by Ar cluster ion bombardments.

ACKNOWLEDGMENTS

This work is supported by New Energy and Industrial Development Organization, Japan.

REFERENCES

1. K.Goto, J.Matsuo, T.Sugii, H.Minakata, I.Yamada and T.Hisatugu, IEDM Tech. Dig. IEEE (1996) 435.
2. epion infusion
3. N.Toyoda, H.Kitani, N.Hagiwara, J.Matsuo and I.Yamada, Mat. Chem. and Phys., 54 (1998) 106-110.
4. N.Toyoda, N.Hagiwara, J.Matsuo and I.Yamada, Nucl. Instr. and Meth. B, 148 (1999) 639-644.
5. J.A.Greer, D.B.Fenner, J.Hautala, L.P.Allen, V.DiFilippo, N.Toyoda, I.Yamada, J.Matsuo, E.Minami and H.Katsumata, Surface and Coating Technol., 133-134 (2000) 273-282.
6. N.Toyoda, H.Kitani, J.Matsuo and I.Yamada, Nucl. Instr. and Meth. B, 121 (1997) 484-488.
7. M.Nagano, S.Yamada, S.Akita, S.Houzumi, N.Toyoda and I.Yamda, Jpn. J. Appl. Phys., Vol.44, No.4 (2005) L164.
8. K.Shirai, Y.Fujiwara, R.Takahashi, N.Toyoda, S.Matsui, T.Mitamura, M.Terasawa and I.Yamda, Jpn. J. Appl. Phys., Vol.41, (2002) 4291-4294.
9. N.Toyoda and I.Yamda, Appl. Surf. Sci., 226 (2004) 231-236.
10. T.Kitagawa, I.Yamada, N.Toyoda, H.Tsubakino, J.Matsuo, G.H.Takaoka and A.Kirkpatrick, Nucl. Instr. and Meth. B, 201 (2003) 405-412.
11. T.Aoki, T.Seki, J.Matsuo, Z.Insepov and I.Yamada, Nucl. Instr. and Meth. B, 153 (1999) 264-269.
12. N.Toyoda, S.Houzumi and I.Yamada, Nucl. Instr. and Meth. B, 242 (2006) 466-468.
13. N.Toyoda, S.Houzumi, T.Aoki and I.Yamada, Mat. Res. Soc. Symp. Proc. 792, (2004) p.623.
14. L.Maissel and R.Grang, *Handbook of Thin Film Technology*, MaGraw-Hill (1970).

High-Speed Nano-Processing with Cluster Ion Beams

T. Seki and J. Matsuo

Quantum Science and Engineering Center, Kyoto University, Gokasyo, Uji 611-0011, Japan

Abstract. The gas cluster ion beam process has a high potential for material processing in nano-technology devices, such as photonic crystals, thin film transistors (TFTs) and micro-electromechanical systems (MEMS). In order to fabricate the devices, one needs to etch target materials with a high-speed, low-damage and ultra-smooth process. Extremely high rate sputtering was realized by high-energy cluster ion beam. We have been using this technique for poly-Si TFTs. There are many hillocks on poly-Si films formed by using a laser anneal technique, and they cause degradation of devices. When the laser crystallized poly-Si film was irradiated with cluster ion beam, the higher hillocks could be etched selectively and the surfaces of poly-Si films could be processed with low ion dose. High-speed nano-processing was realized by cluster ion beam.

Keywords: cluster, ion beam, smoothing, polycrystalline silicon, etching
PACS: 36.40.-c, 61.46.Bc, 52.77.Bn, 61.80.Jh, 79.20.Rf, 42.79.Kr

INTRODUCTION

Poly-Si TFTs (polycrystalline silicon thin-film transistors) are widely noticed as key devices of future flat panel display (FPD). In order to fabricate high performance Poly-Si TFTs, it is necessary to form high quality and smooth poly-Si films. A laser anneal technique is utilized as one of the methods for high quality poly-Si films formation. However, the technique produces a rough surface. Many hillocks are created on the poly-Si films during the lateral crystallization of the silicon crystal [1]. In order to use the films as active layers in the poly-Si TFTs, these hillocks must be removed. If a MOS (metal-oxide-semiconductor) structure is formed on this surface, the electronic field at the tip of the hillock is enlarged and degradation of the device occurs. Although the hillocks must be removed, the etching depth of the poly-Si films must be minimized because the poly-Si films are very thin. Moreover, a high-speed surface processing method is needed because FPDs have a large area.

The gas cluster ion beam process has a high potential for material processing in nano-technology devices, such as photonic crystals, TFT and MEMS (micro electro-mechanical system). In order to fabricate the devices, one needs to etch targets in a high-speed, low-damage, and ultra-smooth process. A cluster is an aggregate of a few to several thousands atoms. When the many atoms constituting a cluster ion bombard a local area, high-density energy deposition and multiple-collision processes occur simultaneously. Because of the unique interactions between cluster ions and surface atoms, new surface modification processes could be developed, and surface smoothing [2-5], shallow implantation [6,7], high rate sputtering [8] and low damage surface processing [9] have been demonstrated using this technique. For example, cluster ion beam could also smooth both the bottom surface and the sidewall of Si pillars, when a photonic crystal was etched with cluster ion beam [10]. Sidewall smoothing is essential for ultra-low-loss photonic crystals. Thus, high-speed and precise nano-processing can be realized with cluster ion beams.

In order to realize high-speed surface processing with gas cluster ion beams, a high-current and high-energy cluster ion beam is needed. In a previous work, a maximum beam current of 1 mA was achieved [11]. With this beam current, 12-inch wafers can be treated with 2×10^{15} ions/cm^2 in about 4 minutes, and this is sufficiently high for next generation processes. On the other hand, etching yields by cluster ion beams are expected to increase with acceleration energy [12]. In this work, a high-energy gas cluster ion beam irradiation system was developed, in which high-energy Ar and SF$_6$ cluster ion beams were generated. The effects of increasing accelerating energy on the sputtering yield with high-energy gas cluster ion beams were investigated. Additionally, smoothing of poly-Si films was demonstrated by using cluster ion beam.

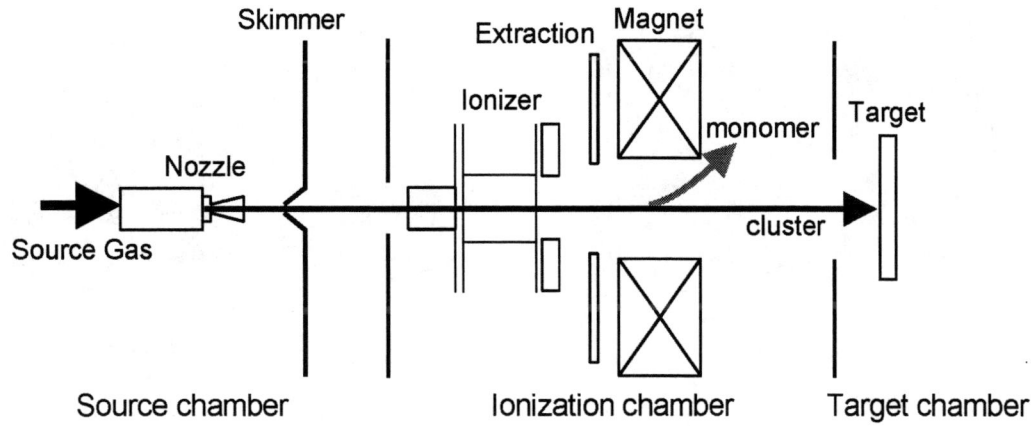

FIGURE 1. Schematic diagram of the high-energy gas cluster ion beam irradiation system.

EXPERIMENTAL

Figure 1 shows a schematic diagram of the high-energy gas cluster ion beam irradiation system. Adiabatic expansion of a high-pressure gas through a nozzle is utilized for the formation of the gas cluster beam. When a supersonic flow ejects from the nozzle, shockwaves are generated [13], which disturb the generation of neutral cluster beams. To avoid formation of such shockwaves, a skimmer is utilized to pass the core of the supersonic flow into high vacuum. The neutral clusters are then ionized by electron bombardment. The ionizer consists of a number of filaments and an anode. Electrons ejected from the hot filaments are accelerated towards the neutral cluster beam and ionize the clusters. The ionized clusters are extracted and accelerated to energies up to 80keV towards the target. Monomer ions are eliminated by a magnetic field. The mass distributions were measured with a compact time-of-flight (TOF) system. The system can be installed in cluster irradiation machines and used as a cluster size monitor. The mean Ar cluster size was about 1800 atoms when the source gas pressure was 0.53 MPa (4000 Torr), the ionization energy was 400 eV and the emission current was 200 mA. The mean SF_6 cluster size was about 650 molecules when the source gas pressure was 0.80 MPa (6000 Torr), the ionization energy was 400 eV and the emission current was 200 mA. In order to investigate the dependence of the sputtering yield on acceleration energy, Si substrates were irradiated with the Ar and SF_6 cluster ion beams at the above-mentioned conditions. Targets were irradiated at normal incident angle. The sputtering yields were measured with a contact profiler. Poly-Si films formed by the laser anneal technique were also irradiated with the Ar cluster ion beam. The irradiated surfaces of poly-Si films were investigated with an Atomic Force Microscope (AFM; Shimadzu SPM9500J2). For average roughness the data was derived from 10×10 μm^2 surface images measured by AFM.

RESULTS AND DISCUSSION

Figure 2 shows the sputtering yields of Si with Ar cluster, Ar monomer and SF_6 cluster as a function of acceleration energy. The sputtering yield of Si with Ar monomer, calculated using TRIM [14], did not increase with the acceleration energy. The sputtering yields of Si with cluster, however, have increased with acceleration energy. The sputtering yield of Si with Ar

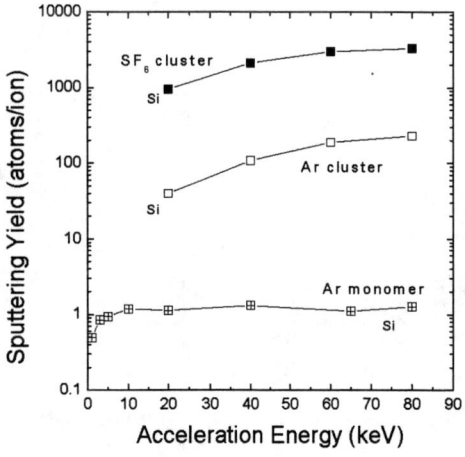

FIGURE 2. Dependence of sputtering yields of Si with Ar cluster, Ar monomer and SF_6 cluster on acceleration energy.

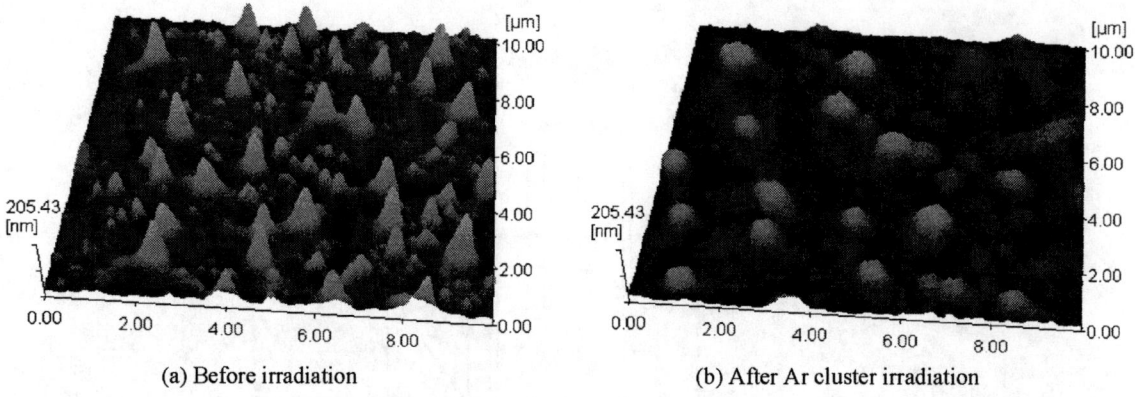

FIGURE 3. AFM images of the surfaces of poly-Si films (a) before and (b) after Ar cluster irradiation.

cluster reached about 230 atoms/ion at 80 keV, and this value is about 180 times higher than that obtained with Ar monomer ions, whereas the sputtering yield of Si with SF_6 cluster reached about 3300 atoms/ion at 80 keV, about 2600 times higher than with Ar monomer ions. With this beam, 12 inches wafers can be etched 0.3 μm per minute at a beam current of 1 mA. These results show that extremely high-speed etching can be realized with high-energy cluster ion irradiation.

Figure 3 shows the AFM images of the surfaces of poly-Si films before and after Ar cluster irradiation. The acceleration energy was 20 keV and the ion dose was 3×10^{15} ions/cm^2. Many hillocks were observed on the poly Si films before irradiation. Their height reached more than 200 nm and the average surface roughness (Ra) was 14.8 nm. On the other hand, the height of hillocks decreased and the average surface roughness (Ra) became 9.0 nm after Ar cluster irradiation. These results indicate that the surfaces of poly-Si films can be smoothed by Ar cluster irradiation. Figure 4 shows the histograms of hillock height before and after Ar cluster irradiation. These histograms were derived from the AFM images in figure 3. In this figure, the hillocks with a height lower than 50 nm were ignored because they could be confused with the undulating base plane. This figure shows that the hillocks higher than 100 nm on the initial surface were removed with Ar cluster irradiation. The etching speed of poly-Si after reduction of hillock height with ion dose of 5×10^{15} ions/cm^2 was similar to that of bulk Si surface. From figure 2 the sputtering yield of Si by irradiation with the 20 keV Ar cluster was about 40 atoms/ion. With the ion dose of 3×10^{15} ions/cm^2, the sputtered depth can be calculated at about 24 nm.

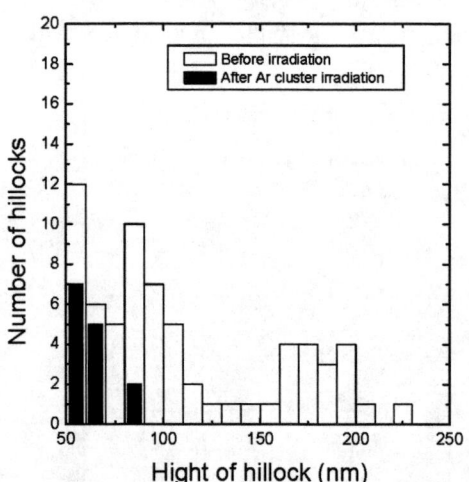

FIGURE 4. Histogram of hillock heights before and after Ar cluster irradiation.

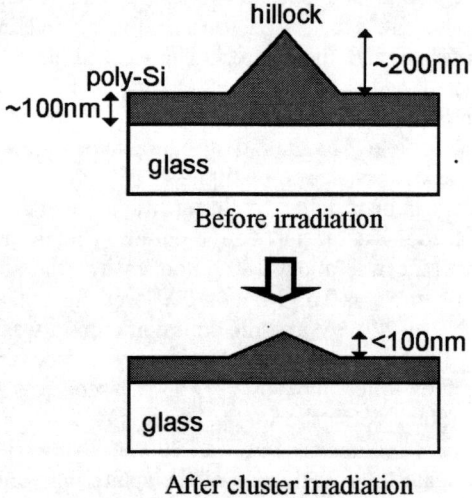

FIGURE 5. Schematic diagram of smoothing of poly-Si film with cluster ion beam.

Figure 5 shows a schematic diagram of smoothing of poly-Si film with cluster ion beam. The thickness of the initial poly-Si film was about 100 nm and the hillock height was about 200 nm. The height was larger than the film thickness. If the etching depth is similar to the height of the hillock, the film is removed completely. However, the hillocks could be removed with Ar cluster irradiation, although the etching depth was much lower than the initial height of the hillocks. These results indicate that the higher hillocks can be etched selectively and the surfaces of poly-Si films can be processed with low ion dose by cluster ion beam.

When the poly-Si film is irradiated with SF_6 cluster ion beam, it is expected to process the poly-Si surface with much higher speed than that of Ar cluster. Actually, the average surface roughness of a poly-Si surface became about 3 nm after SF_6 cluster ion beam irradiation with ion dose of 2×10^{14} ions/cm^2. The acceleration energy was 20 keV. Because the required ion dose for smoothing is very low, 12 inches wafers can be treated for about 0.4 minute at a beam current of 1 mA. A large number of surface smoothing processes by the cluster ion beam were demonstrated [2-5] and the smoothing effect of the cluster ion beam was preserved in the high-energy range up to 80keV [15]. It is expected that high-speed poly-Si surface processing can be demonstrated by using the high-energy cluster ion beam. Thus, the cluster ion beam processing is suitable for fabrication of thin film transistor.

CONCLUSION

When Si substrates are irradiated with Ar and SF_6 cluster ions, Si can be etched with much higher than that of Ar monomer ions. Because hillocks on poly-Si films can be etched selectively by cluster ion beam, the surfaces of poly-Si films can be processed with very low ion dose. Therefore, high-speed surface processing can be realized by cluster ion beam.

ACKNOWLEDGMENTS

This work is supported by the New Energy and Industrial Technology Development Organization (NEDO) of Japan.

REFERENCES

1. D.J.McCulloch and S.D.Brotherton, Appl. Phys. Letters 66, 2060 (1995).
2. H.Kitani, N.Toyoda, J.Matsuo and I.Yamada, Nucl. Instr. and Meth. B121 (1997) 489.
3. N.Toyoda, N.Hagiwara, J.Matsuo and I.Yamada, Nucl. Instr. and Meth. B148 (1999) 639.
4. A.Nishiyama, M.Adachi, N.Toyoda, N.Hagiwara, J.Matsuo and I.Yamada, AIP conference proceedings (15-th International Conference on Application of Accelerators in Research and Industry) 475 (1998) 421.
5. T.Seki and J.Matsuo, Nucl. Instr. and Meth. B216 (2004) 191.
6. D.Takeuchi, J.Matsuo, A.Kitai and I.Yamada, Mat. Sci. and Eng. A217/218 (1996) 74.
7. N.Shimada, T.Aoki, J.Matsuo, I.Yamada, K.Goto and T.Sugui, J. Mat. Chem. and Phys. 54 (1998) 80.
8. I.Yamada, J.Matsuo, N.Toyoda, T.Aoki, E.Jones and Z.Insepov, Mat. Sci. and Eng. A253 (1998) 249.
9. M.Akizuki, J.Matsuo, M.Harada, S.Ogasawara, A.Doi, K.Yoneda, T.Yamaguchi, G.H.Takaoka, C.E.Asheron and I.Yamada, Nucl. Instr. and Meth. B99 (1995) 229.
10. E.Bourelle, A.Suzuki, A.Sato, T.Seki and J.Matsuo, Jpn. J. Appl. Phys., Vol. 43, No. 10A (2004) pp. L 1253-L 1255.
11. T.Seki and J.Matsuo, Nucl. Instr. and Meth. B237 (2005) 455.
12. T.Seki, T.Murase and J.Matsuo, Nucl. Instr. and Meth. B242 (2006) 179-181.
13. H.W.Liepmann and A.Roshko, "Elements of Gas Dynamics" (John Wiley and Sons, Inc., New York, 1960).
14. J.P.Biersack and L.G.Haggmark, Nucl. Instr. and Meth., 174 (1980) 257.
15. T.Seki and J.Matsuo, Surf. and Coat. Tech., (2006) (In press).

Characterization of Molecular Clusters in the Supersonic Gas Jet

Takeshi Kagawa[1], Fuminobu Sato[1], Yushi Kato[1], Yoshifumi Ito[2], Toshiyuki Iida[1]

[1]*Division of Electrical, Electronic and Information Engineering, Graduate School of Engineering, Osaka University, 2-1 Yamada-oka, Suita, Osaka 565-0871 Japan*
[2]*Research and Development Department, Wakasa-wan Energy Research Center, 64-52-1 Nagatani, Tsuruga, Fukui 914-0192 Japan*

Abstract. Characterization of molecular clusters in a supersonic gas jet was performed using two probes of laser and ion beams. The two-dimensional distribution of average size of argon clusters was obtained by measurement of the Rayleigh scattering (RS) of the light from a scanning laser beam probe. A region of average cluster size of about 10^4 atoms/cluster was observed at a distance of more than 15 mm below the nozzle orifice of the cluster source in the condition that the backing pressure of the nozzle was 8 MPa, the initial temperature of argon gas was room temperature and the nozzle orifice diameter was 0.2 mm. In addition, the scintillation measurement of argon clusters impacted with energetic charged particles was performed by the photon counting method with the ion beam probe. It was found that the increase tendency of the scintillation photons with an increase in the backing pressure was similar to that of the cluster size obtained by the RS measurement. The scintillation efficiency of argon cluster depended on the cluster size.

Keywords: molecular cluster, Rayleigh scattering, scintillation
PACS: 36.40.-c; 39.10.+j

INTRODUCTION

Cluster ion beam technology has been found to offer solutions for a number of well-known problems associated with conventional monomer ion beam interactions [1,2]. For example, it enables ultra shallow implantation due to the very low-energy per atom, surface smoothing due to lateral sputtering, etc. These techniques have been expected to advance semiconductor device applications such as CMOS fabrication, smoothing of CVD diamond films, etc. [3]

Generally, a cluster ion beam is generated as follow: neutral clusters are produced in the supersonic gas jet by adiabatic expansion using a supersonic nozzle, which are subsequently collimated and ionized [4]. In these processes, the neutral cluster production is the most important process to generate a high-quality beam, because the spatial distribution of cluster size remarkably changes just below the nozzle orifice, and the intensity and size of cluster beam are almost determined in the gas jet profile. Therefore, it is significant to investigate the gas jet profile just below the nozzle for cluster production. In the present work, we have developed an analysis system for a supersonic gas jet profile using two probes of laser and ion beams, and have performed the characterization of molecular clusters, especially concerning cluster size and its scintillation efficiency.

EXPERIMETAL SETUP

Figure 1 shows a schematic diagram of an experimental setup for characterization of molecular clusters in the supersonic gas jet. This system consists of a vacuum chamber, a solenoid pulse nozzle (General Valve Series 99) for cluster production, a laser diode (LD), an ion gun and an optical measuring system. The vacuum pressure of 10^{-3} Pa in the chamber is kept by a diffusion pump (DP) under the operation of the pulse nozzle. The orifice diameter of the nozzle is 0.2 mm, and the maximum backing pressure of the nozzle is 8 MPa. The repetition rate of the pulse nozzle is 1 Hz, and the pulse duration is 0.2 ms. The high-pressure argon gas of over 1 MPa is injected through the nozzle into the high vacuum region, and large clusters over 10^3 atoms/cluster are formed in the gas jet by supersonic adiabatic expansion.

A fast response ionization gauge (IG) is set below the nozzle orifice to measure the molecular density. The molecular density is determined from the data obtained by the IG and calculated by the basis of

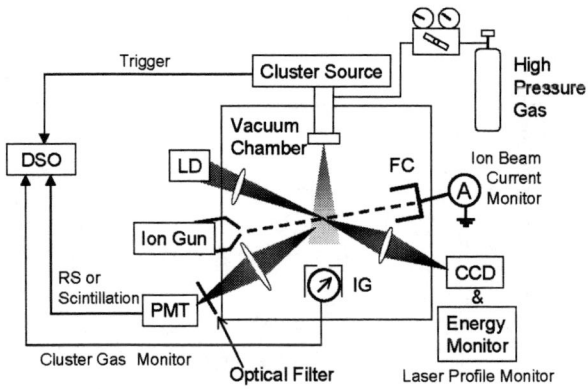

FIGURE 1. Schematic diagram of an experimental setup for characterization of molecular clusters in the supersonic gas jet.

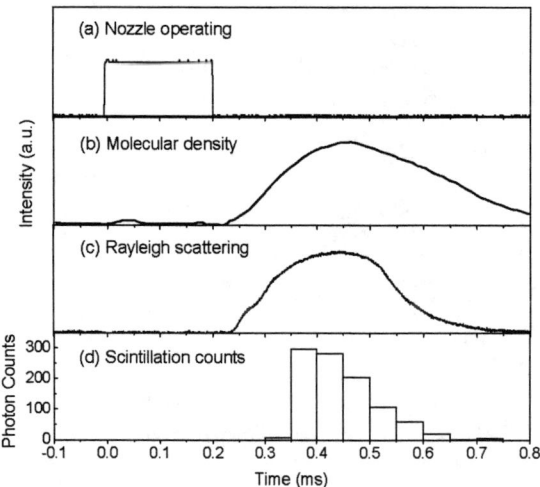

FIGURE 2. Typical example of time relationship of the nozzle operating (a), the molecular density (b), the RS photons (c) and the scintillation photons (d).

hydrodynamics [5,6].

The existence of clusters is confirmed with the measurement of the Rayleigh scattering (RS) light [7]. A laser beam probe at 635 nm is focused in the gas jet region. The focal spot image is obtained by a CCD camera, and its diameter was about 100 μm (FWHM). The RS light is detected by a photomultiplier tube (PMT, Hamamatsu Photonics R1635) through a laser line interference filter. The IG signals and the PMT signals triggered by the nozzle operating TTL signal are digitalized by a digital storage oscilloscope (DSO, LeCroy LT584), and the waveforms of the both signals are averaged one thousand times.

In the experiment with an ion beam probe, scintillation meaurement of argon clusters impacted with energetic charged particles was performed by the photon counting method using a PMT (Hamamatsu Photonics R955). The 3 keV Ar^+ beam is obtained utilizing an ion gun, and the maximum beam current is 1 nA. The ion beam probe is adjusted and focused in the cluster production region using a Faraday cup (FC) and ion optics. The PMT signals of scintillation photons triggered by the nozzle operating TTL signal are also digitalized by the DSO, and the scintillation photons are counted by analyzing each waveform.

RESULTS AND DISCUSSION

Figure 2 shows a typical example of time relationship of the nozzle operating, the molecular density, the RS photons and the scintillation photons. The molecular density, the RS photons and the scintillation photons were detected after a delay of 0.3 ms from the pulse nozzle operating. This delay was considered to be the transport time for the clusters to travel from the nozzle to the analysis point at a distance of 15 mm from the nozzle, and large clusters were formed at the peak time of RS signals. There was a correlation between the scintillation and the RS scattering. Therefore, collisions between large cluster and ion beam were mainly caused around 0.4 ms.

Determination of average cluster size with laser beam probe

The intensity of RS signal is related with the distribution of cluster size. The RS cross section is given by $\sigma \propto r^6/\lambda^4$, where r is the cluster radius and λ is the optical wavelength [8]. Average cluster size N_c is estimated from analysis of the RS intensity and the molecular density, which is roughly expressed by

$$N_c = N \frac{F_i \sigma}{E_{RS}} \qquad (1)$$

where N is the number of total molecules in the laser-cluster interaction region, F_i is the incident laser fluence, E_{RS} is the total energy of RS photons. In the RS measurement, the gathering rate is 12.7 % because the geometrical detection efficiency for the RS light is 1.6 sr, and the quantum efficiency of the PMT for the wavelength of 635 nm is 12.8 %, therefore the detection efficiency is 1.6 %. Figure 3 shows a two-dimensional distribution of the average size of the argon clusters obtained with a scanning laser probe

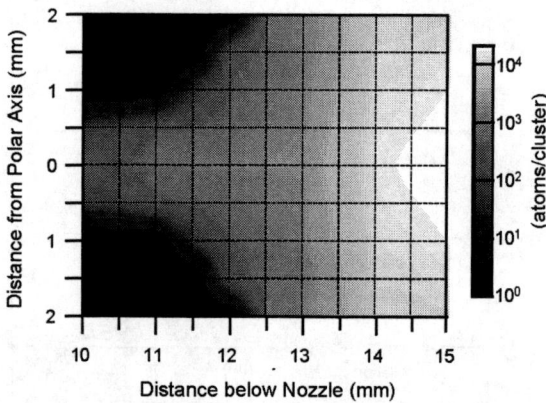

FIGURE 3. Two-dimensional map of average size of argon clusters obtained by scanning the laser beam probe. The backing pressure of cluster gas was 8 MPa, the initial temperature of gas was room temperature and the orifice diameter of the pulse nozzle was 0.2 mm.

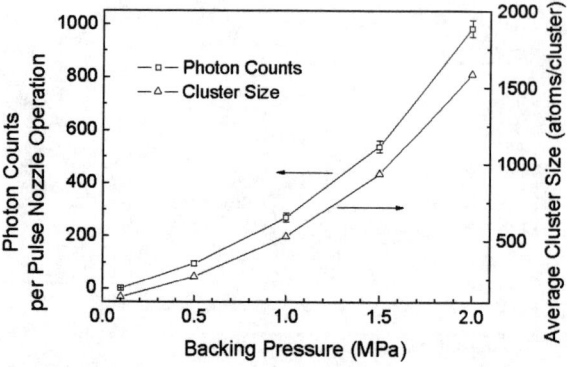

FIGURE 4. Photon counts per pulse nozzle operation and the average size of argon clusters as a function of the backing pressure of the nozzle.

beam [9]. The cluster production was observed along the polar axis, and no large clusters were observed in the distant region from the polar axis due to the barrel shock [5]. In the condition that the backing pressure, the orifice diameter and the initial temperature of the nozzle were 8 MPa, 0.2 mm and room temperature respectively, the region of average cluster size of 10^4 atoms/cluster was observed at a distance of over 15 mm from the orifice. According to hydrodynamics analysis, the mach number of the argon gas jet was about 58 at a distance of 15 mm from the orifice [5], and the relative temperature of argon atoms in the gas jet was estimated to be below 1 K. Therefore, a number of large clusters were formed with the van der Waals force of argon atoms. Also, it was found that the average cluster size increased with an increase in the backing pressure, the control of average size of gas clusters was able to be performed with adjustment of the backing pressure.

Scintillation measurement with ion beam probe

It has been reported that the major wavelength peak of argon gas scintillator is 250 nm [10]. Therefore, the scintillation measurement of argon clusters was performed assuming that the luminescent central wavelength of an argon cluster is near 250 nm. An ultraviolet transmitting filter was set in front of the PMT to cut off background light. The transmitting wavelength range of the filter is from 230 to 410 nm (transmittance > 10 %). In the scintillation measurement, the gathering rate of the scintillation light is 1.6 % because the geometrical detection efficiency is 0.2 sr, and the quantum efficiency of the PMT for the wavelength of 250 nm is 22.0 %; therefore the detection efficiency is 0.35 %. Figure 4 shows the photon counts per pulse nozzle operation and the average size of argon clusters as a function of the backing pressure of the nozzle. The focal spot of the ion beam probe and that of the laser beam probe was adjusted at a distance of 15 mm from the nozzle orifice. The number of scintillation photons per pulse nozzle operation was not proportional to the backing pressure, i.e. the molecular density in the interaction region. The increase of the scintillation photons per pulse nozzle operation for the backing pressure was similar to that of the average cluster size.

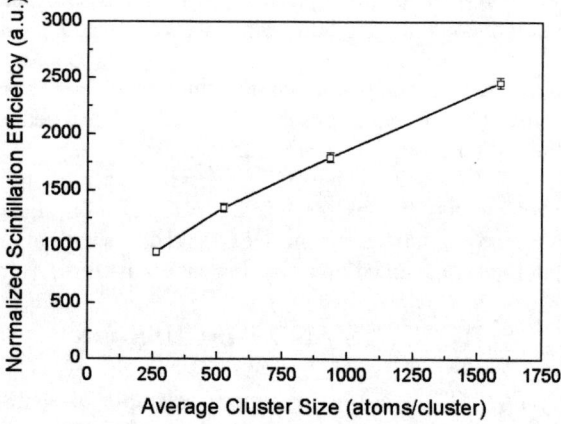

FIGURE 5. Normalized scintillation efficiency as a function of the cluster size.

Figure 5 shows the normalized scintillation efficiency as a function of the cluster size. The vertical axis in Fig.5 expresses the number of scintillation photons per the molecular density. The normalized scintillation efficiency increased with an increase in cluster size. This is considered to be involved in cluster formation in the gas jet. Compared with monoatom, the collision between an energetic ion and clustered atoms caused many-body collision and multiple scattering, consequently induced more scintillation photons. Therefore, the scintillation efficiency of clustered atoms is higher than that of monoatoms at the same density, that is, the scintillation efficiency depended on the cluster size.

SUMMARY

Characterization of argon clusters in the supersonic gas jet was carried out with two probes of laser and ion beams. By use of a typical cluster source with a pulse nozzle, argon clusters were formed in the gas jet by supersonic adiabatic expansion. The two-dimensional distribution of argon cluster size was obtained by the measurement of the Rayleigh scattering (RS) lights with the laser beam probe scanning. The cluster formation was observed along the polar axis, and no large clusters were observed at a distance from the polar axis. A region of cluster size of about 10^4 atoms/cluster was detected at a distance of more than 15 mm below the nozzle orifice in the condition that the backing pressure of the pulse nozzle was 8 MPa, the initial temperature of argon gas was room temperature and the nozzle orifice diameter was 0.2 mm.

In addition, scintillation measurement of the argon clusters was performed with the ion beam probe. It was found that the number of scintillation photons was not proportional to the backing pressure, *i.e.* the molecular density. The increase tendency of the scintillation photons with an increase in backing pressure was similar to that of the cluster size obtained by the RS measurement. This implied that the scintillation efficiency of argon cluster depended on cluster size.

REFERENCES

1. I. Yamada, N. Toyoda, *Nucl. Instr. and Meth.* B **232**, 195-199 (2005).
2. I. Yamada, J. Matsuo, Z. Insepov, T. Aoki, T. Seki, N. Toyoda, *Nucl. Instr. and Meth.* B **164-165**, 944-959 (2000).
3. I. Yamada, N. Toyoda, *Nucl. Instr. and Meth.* B **242**, 143-145 (2006).
4. I. Yamada, J. Matsuo, N. Toyoda, A. Kirkpatrick, *Mat. Sci. Eng. R* **34**, 231-295 (2001).
5. H. Ashkenhas, F.S. Sherman, *Proc. 4th Int. Symp. on Rarefied Gas Dynamics* **2**, 84-105 (1966).
6. F.P. Boynton, *AIAA J.* **5**, 1703-1704 (1967).
7. R.A. Smith, T. Ditmire, J.W.G. Tisch, *Rev. Sci. Instrum.* **69**, 3798-3804 (1998)
8. J. Larsson, A. Sjögren, *Rev. Sci. Instrum.* **70**, 2253-2256 (1999).
9. T. Kagawa, F. Sato, T. Iida, *J. Nucl. Sci. Technol.* **42**, 423-427 (2005).
10. J.B. Birks, *The Theory and Practice of Scintillation Counting*, Pergamon, Oxford, 1964.

PIII AND PLASMA DOPING

PLAsma Doping For P+ Junction Formation In 90 nm NOR Flash Memory Technology

Dario Bigarella [1*], Valter Soncini [1], Deven Raj [2], Vikram Singh [2], Steve Walther [2]

[1] *STMicroelectronics srl, Via C. Olivetti 2, 20041 Agrate Brianza, Italy*
[2] *VSEA Varian Semiconductor Equipment Associates, 35 Dory Road, Gloucester, MA 01930-2297, USA*

Abstract. For MOS devices belonging to 65 nm technology node and beyond, ultra-shallow LDD junctions are needed in order to match requirements in terms of sheet resistance and doping profile. PLAsma Doping has been proposed and developed as an effective and viable technology capable to produce such junctions, while keeping high productivity. Furthermore, as the equipment is simpler and smaller than a common implanter, PLAsma Doping can be considered, from the cost of ownership point of view, an attractive solution also for those applications whose requirements are not so demanding; an example can be the junctions of a Flash Memory. Aim of this study is to evaluate, electrically, the compatibility of PLAsma Doping with a NOR Flash Memory belonging to 90 nm technology node. Results of PLAsma Doping experiment concerning the matching of the device parameters will be here presented. Further investigation is needed in order to exclude any possible effect on device reliability.

Keywords: PLAD, Flash memory, S&D implant
PACS: 52.77.Dq, 61.72.Tt, 85.40.Ry

INTRODUCTION

As ULSI circuits are scaled down in order to increase their performance, vertical and lateral dimensions of the doped regions in Silicon must be reduced. Shallow doped Source/Drain extensions are required to control such issues as the "Short Channel Effects" that are detrimental to device functionality.

To produce such junctions, very low energy ion implantation must be performed, combined with proper thermal treatments.

Implanters with traditional architecture (Beam-line equipment), as the ion energy is reduced, may face a reduction of productivity: achievable beam currents are lower, so that process times are longer and the throughput is cut down to values that, industrially, are no more cost-effective.

PLAsma Doping (PLAD) [1, 2, 3] has been developed and already well characterized as a novel technology able to fulfil Ultra-Shallow Junctions requirements without losing productivity.

Ions from pulsed plasma are implanted into the wafer that is simultaneously pulse biased with a negative voltage. Ion energy is determined by the wafer bias itself and the trajectory of the ions is perpendicular to the wafer.

Different from traditional beam-line implanters, PLAD has no mass filter: it means that multiple species co-implantation and a wide energy distribution are present in the process, thus leading to a unique (typically surface-peaked) doping profile [4, 5].

Furthermore, as the whole wafer is implanted simultaneously, there is no need for a scanning system: the resulting hardware is simpler than a standard beam-line implanter, although aspects of the PLAD process and process control are certainly more complex. Consequently, PLAD offers the potential to make challenging high-dose low-energy processes more cost-effective.

In the field of NOR Flash memories, for the technology node actually in production (90 nm), requirements in terms of junction depth are not very severe; PLAD implementation for such a production can be attractive as a reduced cost alternative to standard implantation.

Aim of this study is a first evaluation of the compatibility of PLAD process with NOR Flash Memories: p+ S&D implant ($^{11}B^+$, energy 6 keV, total dose $3.5 \cdot 10^{15}$ atoms / cm^2) was chosen as test vehicle. It will be matched with PLAD process and transistor and memory cell parameters will be compared with the standard ones.

---

* Corresponding author. Tel.: +39 039 6032767, fax: +39 039 6035233, e-mail address: dario.bigarella@st.com

EXPERIMENTAL

Activities were divided into two steps: first, the process was matched on flat wafers with PLAD; then, some Flash memory device wafers were processed with the identified conditions.

Flat wafers (n-type (100) silicon, 2.5-8.5 Ω·cm resistivity) were prepared as follow, trying to emulate the process flow of device wafers: 16 nm thermal oxide growth, implant, RTP activation anneal.

Three wafers were implanted with the standard process in order to provide a reference for sheet resistance and junction depth; other wafers were processed with PLAD, as reported in Table 1, with two different equipment configurations. Note that PLAD doses are typically higher than the reference beam-line dose due to Faraday detection of lower energy dopant species as well as inert species such as Hydrogen.

Other parameters, such as chamber pressure and the spacing between anode and cathode, were set at values optimized for best uniformity within wafer.

After oxide removal with HF, sheet resistance of the wafers was measured with 4-point probe (49 points map).

A Dynamic Secondary Ions Mass Spectrometry (D-SIMS) was performed on one wafer for each bias/dose condition of the group processed with equipment configuration A, plus one wafer processed with standard beam-line implanter.

From the results measured on flat wafers, optimized processing conditions for the device wafers were identified. Those were split into three groups of four wafers each: a group for standard process as reference, a second one for PLAD configuration A and the last one for configuration B.

After completing their full process, the wafers were measured with the standard parametric testing.

RESULTS

Flat wafers

Measurements of sheet resistance are summarized in the following charts. With the broken lines the values measured with the beam-line process are reported.

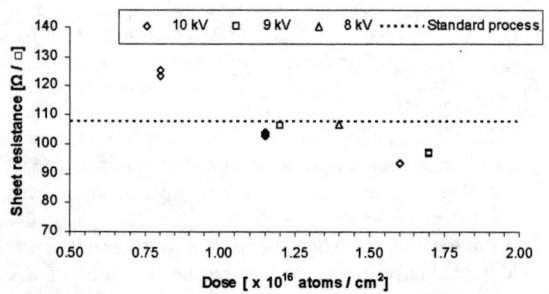

FIGURE 1. Configuration A sheet resistance average within wafer

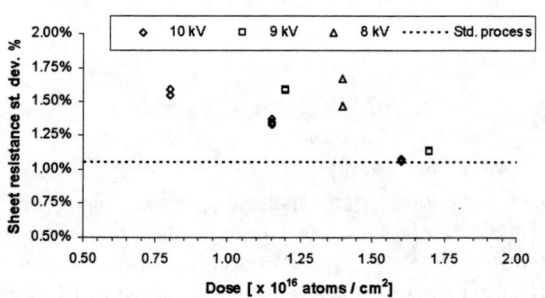

FIGURE 2. Configuration A sheet resistance uniformity within wafer

TABLE 1. Experimental matrix for flat wafers

PLAD equipment configuration	Wafer bias [kV]	Dose [x 10^{16} atoms/cm^2]	Gas used	Number of wafers
A	10	0.80	BF_3	2
A	10	1.15	BF_3	3
A	10	1.60	BF_3	2
A	9	1.20	BF_3	2
A	9	1.70	BF_3	2
A	8	1.40	BF_3	2
B	10	0.90	B_2H_6	2
B	10	1.20	B_2H_6	3
B	10	1.70	B_2H_6	2
B	9	1.30	B_2H_6	2
B	9	1.80	B_2H_6	2
B	8	1.60	B_2H_6	1

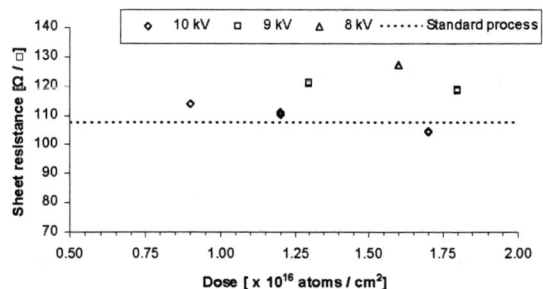

FIGURE 3. Configuration B sheet resistance average within wafer

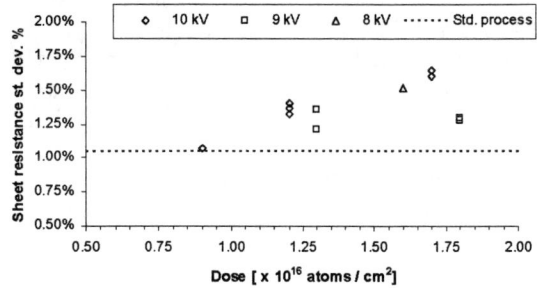

FIGURE 4. Configuration B sheet resistance uniformity within wafer

With configuration A, the standard value can be matched at each level of wafer bias voltage, while with configuration B, due to a lower sensitivity to dose, target can be achieved only at 10 kV. Uniformity within wafer is always worse than standard process, and it shows different trends with dose and bias voltage, depending on configuration used: with A, standard deviation is decreasing with dose and bias, while with B standard deviation is increasing with dose and not clearly affected by bias.

Figure 5 reports part of the D-SIMS profiles of A configuration processed wafers: after anneal all the profiles are similar; with PLAD, they are slightly shallower than standard.

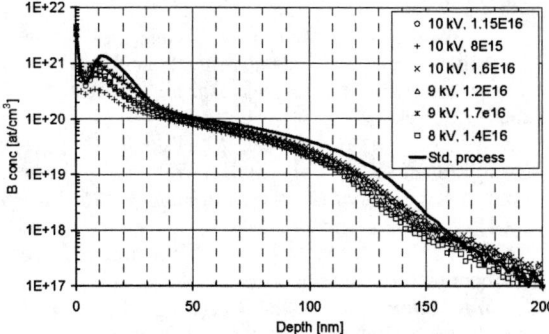

FIGURE 5. Boron D-SIMS profiles (wafers A only)

Junction depth is defined to be the depth at 10^{18} atoms/cm^3 Boron concentration: standard value is approached only at 10 kV.

Starting from these results, simple models for sheet resistance and junction depth were derived and optimized parameters were identified, considering also some constraints.

For A equipment, 10 kV is the best bias voltage: it can match the resistance with good uniformity and junction almost as deep as standard; but 10 kV is the maximum bias allowed with this configuration. For a safer process, as it would be chosen in a manufacturing environment, bias was limited to 8 kV, and dose fixed at $1.43 \cdot 10^{16}$ atoms / cm^2, thus matching sheet resistance. Uniformity is expected to be slightly worse and the junction shallower: these factors will be evaluated on devices' parameters.

For B equipment, the limit to the bias voltage is higher: this parameter was set to be 10 kV, with dose $1.45 \cdot 10^{16}$ atoms / cm^2.

Device wafers

Here follow some charts reporting the most important among the measured parameters for transistors. Nine different sites were measured for each wafer and all the values shown in the chart are normalized so that the average of the standard part is equal to 100. Error bars mean the 3σ of the average within each group of wafers. Bipolar transistors, resistances, p-channel MOS threshold voltages and saturation currents are aligned between the three groups. P-channel MOS breakdown voltages show a little but significant improvement for the equipment B group: this can be explained with a different doping profile, shallower and less abrupt. It is interesting that threshold voltage and saturation current are not affected: transistors are not becoming slower with the increase of their robustness.

Uniformity within wafer and repeatability within group are aligned between the splits for all the parameters.

Parameters related to memory cells, not reported here, are not affected in any way by the splits: this was expected, since the implant mask covers the memory array.

Effects on retention and reliability, if any exists, cannot be evaluated from this testing.

Figure 8 shows an example of the measurements on plasma damage monitoring structures: these are MOS transistors with an attached polysilicon antenna. Their threshold voltage and gate current are measured before and after an electrical stress is applied: the difference between the two is related to the damage induced by plasma environment; the higher is the variation, the higher is the damage.

As in PLAD technology wafers are immersed directly in plasma, this kind of damage can be a potential concern [6], but, for these wafers, no worsening can be seen between PLAD and standard process. The higher spread of PLAD splits is to be considered not significant.

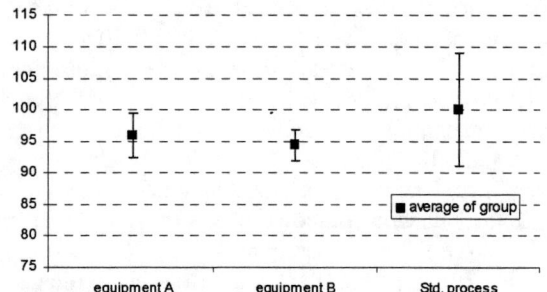

FIGURE 6. p+ resistances.

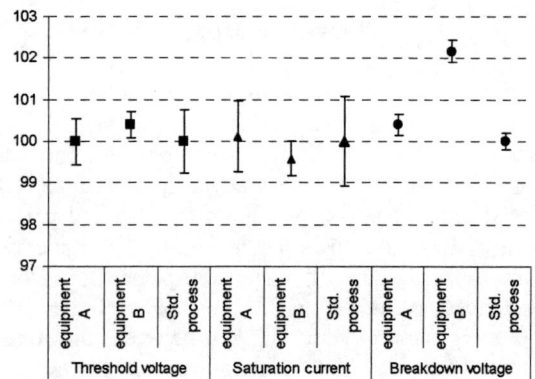

FIGURE 7. p-channel MOS transistor parameters.

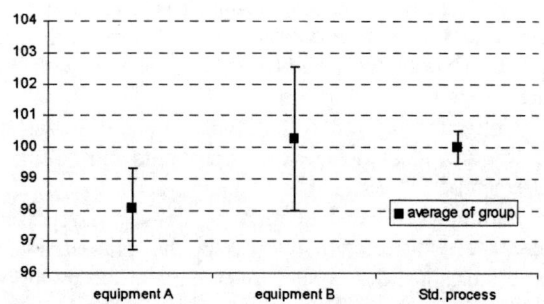

FIGURE 8. Plasma damage monitor; transistor threshold voltage variation.

CONCLUSIONS

PLAD technology was evaluated from the process point of view, with the aim of stating its compatibility with 90 nm NOR Flash Memory devices.

S&D p+ implant was chosen as test vehicle.

PLAD process was set-up on flat wafer through sheet resistance measurements and D-SIMS in order to match the values typical of standard beam-line implantation and then some device wafers were processed with the selected conditions. Structures tested on these wafers (resistances, bipolar and MOS transistors, plasma damage monitoring structures) resulted to be aligned with standard process. Breakdown voltages of p-channel MOS showed a little improvement for the wafers processed with equipment B, due likely to a different doping profile. No damage induced by plasma appears to be evident.

We can conclude that PLAD proved to be able to match a standard beam-line implantation process, without any major detrimental effect on device parameters. A deeper and finer analysis is required in order to evaluate if any negative effect on memory cell retention and reliability is induced by PLAD processing.

Then an economic evaluation must be performed, based on the flexibility that this equipment can provide, in terms of number of different processes (species, dose, energy) that can be run, in order to achieve full productivity.

ACKNOWLEDGMENTS

The authors would like to thank Camillo Bresolin, Alessandro Grossi and Anna Maria Conti of STMicroelectronics and Matt Blago, Phil Mohan, Louis Paiva and Justin Tocco of VSEA for their support and help in this work.

REFERENCES

1. E. C. Jones, N. W. Cheung, "Plasma doping optimization for ultra-shallow junctions", *Nuclear Instruments and Methods in Physics Research B* **121**, 216-220 (1997)
2. W. Skorupa, R. A. Yankov, W. Anwand, M. Voelskow, T. Gebel, D. F. Downey, E. A. Arevalo, "Ultra-shallow junctions produced by plasma doping and flash lamp annealing", *Materials Science and Engineering B* **114-115**, 358-361 (2004)
3. S.R. Walther, S. Mehta, U. Jeong, D. Lenoble, "Formation of extremely shallow junctions for sub-90 nm devices", *Surface & Coatings Technology* **186**, 68-72 (2004)
4. D. T.K. Kwok, P. K. Chu, M. Takase, B. Mizuno, "Energy distribution and depth profile in BF plasma doping", *Surface and Coatings Technology* **136**, 146-150 (2001)
5. B. P. Linder, N. W. Cheung, "Modeling of energy distributions for plasma implantation", *Surface and Coatings Technology* **136**, 132-137 (2001)
6. D. Henke, S. Walther, J. Weeman, T. Dirneker, A. Ruf, A. Beyer, K. Lee, "Characterization of Charging Damage in Plasma Doping", IIT2002 Conference Proceedings

Deep Trench Doping by Plasma Immersion Ion Implantation in Silicon

S. Nizou[1,2], V. Vervisch[3], H. Etienne[3], M. Ziti[1], F. Torregrosa[3], L. Roux[3], M. Roy[2] and D. Alquier[1,*]

[1]Université François Rabelais, Tours, L.M.P, 16, rue Pierre et Marie Curie, B.P. 7155, F37071 TOURS Cedex, France
[2]STMicroelectronics, 16, rue Pierre et Marie Curie, B.P. 7155, F37071 TOURS Cedex, France
[3]Ion Beam Services, Rue Gaston Imbert Prolongée, ZI Peynier-Rousset, 13790, PEYNIER, France

Abstract. The realization of three dimensional (3D) device structures remains a great challenge in microelectronics. One of the main technological breakthroughs for such devices is the ability to control dopant implantation along silicon trench sidewalls. Plasma Immersion Ion Implantation (PIII) has shown its wide efficiency for specific doping processing in semiconductor applications. In this work, we propose to study the capability of PIII method for large scale silicon trench doping. Ultra deep trenches with high aspect ratio were etched on 6" N type Si wafers. Wafers were then implanted with a PIII Pulsion system using BF_3 gas source at various pressures and energies. The obtained results evidence that PIII can be used and are of grateful help to define optimized processing conditions to uniformly dope silicon trench sidewalls through the wafers.

Keywords: Plasma Immersion Ion Implantation, deep trench, BF_3 doping, silicon
PACS: 52.65.–y, 52.77.Dq, 61.72.Tt,

INTRODUCTION

Plasma Immersion Ion Implantation (PIII) becomes key equipment for several microelectronic application developments [1]. Indeed, its benefits have been already demonstrated in Ultra Shallow Junction (USJ) formation [2], Silicon-On-Insulator (SOI) fabrication [3] or in other applications such as solar devices [4]. Among the various possible applications, trench doping using PIII technique has been intensively investigated during the last 15 years. Hence, the dimension reduction requested by the Semiconductor Industry Association (SIA) roadmap [5] has led to a large development of trench USJ based devices. In such a case, the trenches exhibit generally only submicron dimensions at the surface when the aspect ratio could reach up to 35:1. First research in this field led by B. Mizuno et al. [6] gave to PIII a great interest for micro trenches implantation. Later, other attractive results were reported on the sidewall doping for DRAM application and also confirmed the potentiality of PIII for efficient conformal trench doping [7, 8].

In the last few years, Deep Reactive Ion Etching (DRIE) has also demonstrated large progresses enabling to obtain very deep trenches in Silicon. The trenches may today cross an entire wafer with extremely sharp sidewalls. The combination of these two plasma techniques (PIII and DRIE) may lead to a new generation of applications going to real 3D devices. The first breakthrough requested to obtain such devices is the ability to dope trenches. Depending on the expected applications, the trench doping may be conformal or localized. Moreover, the scale of the designed trenches is here much larger than the one studied generally in the literature. X.Y. Qian et al. [9] already demonstrated that structures made of Si slices (0.5 mm wide and 12.5 mm deep), simulating a very large trench, can be doped by PIII. These interesting results enlighten that PIII is certainly usable for deep trench doping.

The goal of this work is to evaluate the capability of PIII method to dope large scale silicon trenches. After presenting the realization of ultra deep trenches with high aspect ratio on 6" Si wafers, we will shortly introduce the PIII Pulsion system developed by IBS as well as the implantations conditions that were performed. Through measurements such as cross-sectional observations, we will evidence the trench sidewall doping observed after annealing depending on plasma conditions.

EXPERIMENTS

Sample preparation

For these experiments, bulk N-type <100> 5-10 Ω.cm 300 µm thick Si wafers were used. A 1.2 µm thick oxide was first thermally grown (wet conditions) on the wafers. The oxide was then patterned by a plasma dry etching. The remaining photoresist was removed using O_2 plasma. Subsequently, a DRIE stage was performed to obtain deep trenches. The deep etching of Si was operated in an ADIXEN AMS 200 reactor with $SF_6/C_4F_8/O_2$ plasma gas. Trench morphology varied with apertures going from 30 µm to 56 µm wide and 200 µm to 250 µm deep. The targeted aspect ratio for the used plasma conditions was around 5:1. In figure 1, we presented the optical image of the trench morphology that is obtained after plasma etching of silicon in defined conditions. It can observed that sidewalls are extremely sharp and that the trench depth is perfectly regular.

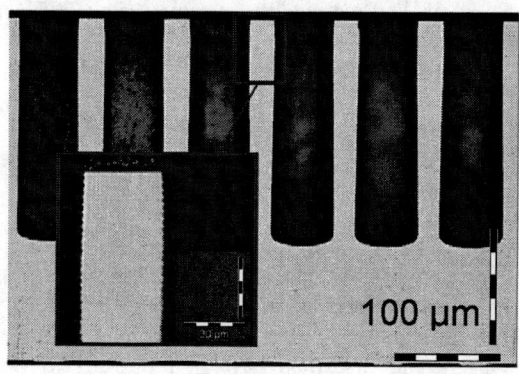

FIGURE 1. Trench morphology after plasma etching of silicon

To dope the so-formed trench sidewalls, PIII was performed using BF_3 as doping gas varying energy and process pressure conditions in the PULSION system developed by IBS. The system is presented in the next section. For all the conditions, Si plain wafers were implanted at the same time to serve as reference sample for the implanted dose. The reference samples were analyzed using spreading resistance profiling (SRP) measurements. Samples were then furnace annealed (FA) at high temperature (>1000°C) for several hours in nitrogen ambiance, leading to deep diffusion. Such annealing will allow sufficient dopant diffusion in order to obtain well-defined observations and measurements on trench sidewalls. Nitrogen environment was used during annealing in order to avoid silicon oxidation. Oxide thickness, measured by ellipsometry after annealing, had never exceeded 5nm.

PIII system and process description

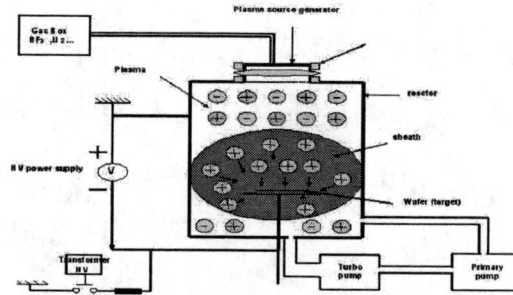

FIGURE 2. Schematic of the PIII PULSION system

In this section, plasma implantation, carried out in a PIII PULSION reactor developed by IBS [4], is described. The system has been designed to be versatile although it is generally used for USJ formation. Figure 2 presents a schematic of the PIII PULSION® system. In this work, we proposed to evaluate the capability of this technology to dope silicon trench sidewalls. For this purpose, first experiments were done using BF_3 gas at various pressures in the range $2.10^{-4} - 3.2.10^{-2}$ mbar. Silicon samples were negatively biased from 2 to 10 kV, plasma was pulsed. For all samples, the dose was adjusted at 10^{16} cm^{-2}. This dose was chosen in order to obtain a sufficient remaining dose along trench sidewalls assuming that dose at the wafer surface was probably higher that the one implanted in the trench.

Characterization methods

For this work, the measurements that enable to evaluate doping on these structures are extremely few and mainly related to staining techniques, allowing only qualitative information. Moreover, the quality of sample preparation for such measurements is crucial. Indeed, on the contrary to small trench analysis, a simple cleavage does not provide good surface topography for precise measurements. Only a meticulous sample preparation, comparable to SCM (Scanning Capacitance Measurement) one, allows to characterize the doping layer formed on the trench sidewalls. Samples were studied in both plan view and cross-section preparation. Figure 1 shows trench morphology after polishing where surface is extremely smooth. The preparation enables to visualize the roughness defaults created by the silicon plasma etching process, evidencing that the internal sidewall surface is not damaged by the polishing step. Scalloping phenomenon due to the consecutive etching and passivating steps is at the origin of the periodical little spikes (<1 µm) along the etched sidewalls.

The implantation of Boron in N-type Si indubitably leads to (p+/n) junction formation. The junction depth can then be determined using staining techniques. For this purpose two techniques were used. First, samples can be immersed in a Sirtl etch solution. By preferentially etching P type doped regions, junction depth measurement is possible. Moreover, copper sulfate solution is then used to confirm our measurements. Indeed, with an appropriate illumination when dipped in a $CuSO_4$ solution, copper ions are reduced in Cu metal and specifically plated on the N-type silicon while the P region remained without copper. A precise measurement of the metallurgical junction is then observed, done by simple optical microscopy. As mentioned in previous section, such annealing will allow sufficient dopant diffusion in order to obtain well-defined observations and measurements on trench sidewalls.

Other requested information concerns the measurement implanted dose during PIII process. An evaluation of this dose, using SRP measurements in the reference sample after an annealing step, is also presented.

RESULTS AND DISCUSSION

As mentioned in the introduction, the trench doping may be influenced by the dimension of the trenches that here are almost crossing the entire wafer. In these first experiments, we have focused on 56 μm wide and 200 μm deep trenches. In figure 3, we present typical micrographs obtained after trench junction staining for the following plasma conditions: BF_3 gas flow at 9 sccm for a chamber pressure of 2.10^{-4} mbar and negatively biased sample at 5 kV. The annealing condition in this particular case was 1100°C for 18h. A visible (p+/n) junction is clearly formed along the trench sidewalls after staining. The P doped region delineation has been obtained with copper sulfate dipping and a direct illumination of the junction for 30 secondes. At first order, the doping is conformal all over the trench sidewall. This result is consistent with the observations already done on micron and submicron trenches [1,7,8] and tended to prove that the PIII technique may be generalized in trench doping.

In this work, the shallow junction depths compared to the trench dimension require to perform deep dopant diffusion for accurate observations. For such annealing condition, optical microscopy was sufficient to measure junction depth. A 3 μm deep junction was measured along vertical sidewalls. The trench bottom exhibits a slightly deeper junction of 4.5 to 5 μm. All these observations and measurements were confirmed using Sirtl etch.

FIGURE 3. Junction staining on a conformaly doped 200 μm deep trenches

The observed junction variations between trench walls and trench bottom can be explained by the angular distribution of incoming ions as proposed by X.Y. Qian [9]. When negatively biased, silicon substrate becomes target for plasma positively charged ions. The formed plasma sheath in our conditions, despite on the large scale of trench, is supposed to be wider than the trench top opening. Consequently, ions are perpendicularly accelerated to the wafer surface, and anisotropic ion implantation happens. If ion angular distribution is probably the main doping source of vertical sidewall, the trench bottom is highly exposed to direct ion implantation and tends to receive the main dose of accelerated ions. This ratio between vertical and horizontal implanted dose is directly proportional to implant energy. In this condition, high energy implantation will certainly reduce angle distribution while low energy will increase it.

At this point of the study, it is important to notice that no specific variation of junction depth was observed depending of the pressure range studied [2.10^{-4} – $3.2.10^{-2}$ mbar] or gas flow (3 to 9sccm) when low negatively biased. This result is in agreement with our expectations. On the contrary, noticeable changes should be found when increasing DC bias, i.e. ion acceleration energy probably combined with high pressure conditions. In order to verify this point, an experiment was set up using these conditions, i.e. high energy at high pressure. With such plasma conditions, the ion density is expected to be important but the ion angular distribution due to scattering is largely reduced compared to the low energy case. After plasma implantation, different annealing conditions were held. Figure 4 shows micrograph obtained after Sirtl etch for a plasma implantation using chamber pressure of 3.10^{-2} mbar and a negatively biased sample at 10kV further annealed for 18h at 1100°C under nitrogen ambient. As expected, only the trench bottom appears to be doped.

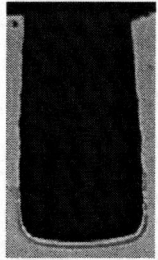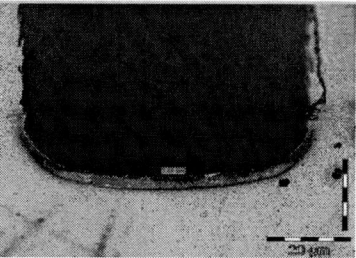

FIGURE 4. Sirtl etch sample for a plasma implantation (3.10^{-2} mbar, biased sample at 10kV) annealed for 18h at 1100°C under nitrogen ambient

The phenomenon, already observed for small trench features, is here increased as no doping is observed on vertical sidewalls. The whole ion dose seemed to be located at the bottom of the trench.

This observation is confirmed and even enhanced using higher temperature conditions. Indeed, annealing at 1280°C for 12h under nitrogen atmosphere has created an extremely deep diffusion at the trench bottom as presented in Figure 5. This result evidences that a local ion implantation can be obtained using specific conditions while vertical sidewalls remain visibly unaffected by the PIII process. Another important point to remind is that silicon trench profile may also greatly impact implantation efficiency.

FIGURE 5. Localized doping on trench bottom after long time high temperature annealing

Finally, it is important to evaluate the dopant dose obtained in such conditions. In figure 6, we present a typical SRP profile obtained for a reference sample after an annealing of 5h at 1000°C. The SRP measured dose is only $\sim 10^{15}$ cm^{-2}. Several assumptions can explain this observation such as dopant out diffusion during annealing, already observed in similar conditions or sputtering effects at low energy implantation, which tend to saturate the implanted dose. However, measured dose in the plasma system itself can drift by a factor up to 10 as the implantation current is composed of multiple charges and various elements. The encouraging result is that a high dose remains in the sample to form p$^+$/n junction in such annealing conditions.

CONCLUSION

Recent progresses in both DRIE and PIII processes has open a way to a new generation of applications going to real 3D devices. In this work, we enlighten that PIII treatment followed by furnace annealing can be suitable for conformal doping of vertical silicon trench sidewalls. Nevertheless, unsuited plasma conditions may lead to localize dopant at the trench bottom. By controlling plasma conditions and silicon trench morphology, PIII has a great potential efficiency as a future doping technology for complex shape doping even for large scale structures

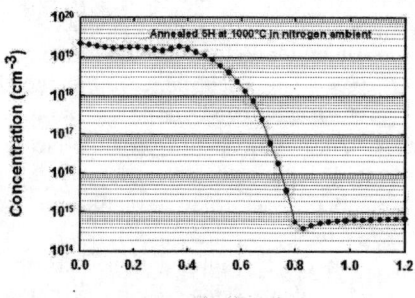

FIGURE 6. Typical SRP profile obtained after a 1000°C /5h annealing for a reference sample

REFERENCES

[1] P. K. Chu, "Semiconductor application of Plasma Immersion Ion Implantation", Plasma Phys. Control. Fusion 45 (2003), 555-570

[2] D. Lenoble and A. Grouillet, "The fabrication of advanced transistors with plasma doping", Surface and Coatings Technology Volume 156, (2002) 262-266

[3] P.K. Chu et al. "Instrumental and process considerations for fabrication of silicon-on-insulator (SOI) with Plasma Immersion Ion Implantation", IEEE Trans. Plasma Sci. 26 (1998) 79

[4] F. Torregrosa et al. " Realization of ultra shallow junctions by PIII, application to solar cells.", VIIth international workshop on Plasma Based Ion Implantation (PBII-2003), San Antonio, USA, 16-19 Sept. 2003.

[5] Semiconductor Industry Association, The International Technology Roadmap for Semiconductors, Front end processes, January 2005, 1-20

[6] B. Mizuno et al."New doping method for subhalf micron trench sidewalls by using an electron cyclotron resonance plasma", Appl. Phys. Lett. 53 (21), 21 November 1988.

[7] K. Lee, "Plasma Immersion Ion Implantation as an Alternative Doping Technology for ULSI" International workshop on Junction technology, 2001

[8] Crid Yu and Nathan W. Cheung, "Trench Doping Conformality by Plasma Immersion Ion Implantation (PIII)" IEEE Electron Device Letters, Vol. 15, No. 6, June 1994.

[9] X. Y. Qian et al., "Conformal implantation for trench doping with plasma immersion ion implantation", Nuclear Instruments and Methods in Physics Research B55 (1991) 898 - 901

Nitrogen Plasma Ion Implantation of Al and Ti alloys in the High Voltage Glow Discharge Mode

R.M.Oliveira[1], M.Ueda[1], J.O.Rossi[1], H.Reuther[2], C.M.Lepienski[3], A.F.Beloto[4]

[1]*Associated Laboratory of Plasma, National Institute for Space Research,*
Av. dos Astronautas 1758, São José dos Campos, São Paulo, Brazil
[2]*Institute of Ion Beam Physics and Materials Research, Rossendorf, Dresden, Germany*
[3]*Department of Physics, Federal University of Paraná, Curitiba, Paraná, Brazil*
[4]*Associated Laboratory of Materials and Sensors, National Institute for Space Research*

Abstract. Enhanced surface properties can be attained for aluminum and its alloys (mechanical and tribological) and Ti6Al4V (mainly tribological) by Plasma Immersion Ion Implantation (PIII) technique. The main problem here, more severe for Al case, is the rapid oxygen contamination even in low O partial pressure. High energy nitrogen ions during PIII are demanded for this situation, in order to enable the ions to pass through the formed oxide layer. We have developed a PIII system that can operate at energies in excess of 50keV, using a Stacked Blumlein (SB) pulser which can nominally provide up to 100 kV pulses. Initially, we are using this system in the High Voltage Glow Discharge (HVGD) mode, to implant nitrogen ions into Al5052 alloy with energies in the range of 30 to 50keV, with 1.5μs duration pulses at a repetition rate of 100Hz. AES, pin-on-disc, nanoindentation measurements are under way but x-ray diffraction results already indicated abundant formation of AlN in the surface for Al5052 treated with this HVGD mode. Our major aim in this PIII experiment is to achieve this difficult to produce stable and highly reliable AlN rich surface layer with high hardness, high corrosion resistance and very low wear rate.

Keywords: plasma immersion ion implantation, high voltage glow discharge mode, Al5052, Ti6Al4V.
PACS: 52.77Dq

INTRODUCTION

Aluminum and its alloys are highly demanded materials in the modern industries, especially in applications requiring light-weight, high corrosion resistance and reasonable toughness, as in aerospace components [1].

Because of its excellent combination of properties regarding mechanical and chemical stability, Ti6Al4V is one of the mostly used titanium alloys in aeronautical and biomedical applications [2, 3]. However, Ti6Al4V presents inadequate tribological properties. To improve them, ion implantation has been used previously [4, 5].

Possibilities of nitrogen ion implantation in the surface of Al alloys to produce AlN compound layer opened up new hopes to manufacture industrial components made of Al alloys with even more enhanced surface properties (both mechanical and tribological). In the case of the Ti alloy, compounds as TiN or Ti_2N are produced to favor improved surfaces.

Plasma immersion ion implantation (PIII) is a well suited technique for this transformation in reasonably complex shaped or large area components. However, it has been shown by many recent experiments that Al and its alloys are prone to rapid contamination by oxygen not only in air but also in relatively low O partial pressures (as residual gas during plasma treatments) [6, 7]. This results in a fast build-up of alumina layer which blocks the nitrogen ion penetration into the bulk, for low energy implantations. One way to circumvent this problem is to use sufficiently high energy nitrogen ion, of the order of 30keV or more, to pass through this oxide layer [8].

We have developed a PIII system that can operate at energies in excess of 50keV, using a Stacked Blumlein (SB) pulser which can nominally provide up to 100 kV pulses [9]. Initially, we are using this system in the High Voltage Glow Discharge (HVGD) mode, to implant nitrogen ions into Al5052 alloy with energies in the range of 30 to 50 keV, with 1.5 μs duration pulses at a repetition rate of 100 Hz. For Ti alloys, the problem of oxide layer is less severe.

In both cases of Al and Ti alloys, the thickness of the treated layer is usually small for implantation of nitrogen at such energies, unless hybrid process combining implantation with a diffusive process is used. Typically ~ 100 nm layers are useful for many applications. For Ti alloys, PIII at 800°C is highly effective to achieve treated layers as thick as 3 μm [10, 11].

EXPERIMENTAL

A schematic drawing of high energy PIII (HEPIII) system including all its components is shown in Fig.1. Using typical nitrogen gas pressures of 5×10^{-3} mbar, firstly, the filament is turned-on to start the glow discharge plasma with the electrode located at the upper part of the 180 liter volume stainless steel vacuum chamber. A base pressure of 5×10^{-6} mbar is achieved with a turbomolecular/mechanical pumping system. High voltage (HV) pulses are applied to the sample support through a high voltage feedthrough. Plasma parameters (T_e, n_e) were determined by using a double Langmuir probe. Typical parameters were $n_e \sim 5 \times 10^9$ cm^{-3} and $T_e \sim 5$ eV.

The HV pulses originates from a high voltage pulse generator with an expected performance of 100kV/200A, based on a Stacked Blumlein technology, which is usually used in applications such as x-ray generation, breakdown tests, etc [12]. According to our design, the output pulse should reach voltages of 150 kV. However, in practice, corona discharges (between metallic structure and connection) have limited the output voltages to up to 100 kV at most. Electrical details of the Blumlein pulser can be found elsewhere [13].

To characterize the Al5052 and Ti6Al4V samples submitted to the HEPIII system in the high voltage glow discharge mode (explained later), we used x-ray diffraction (Philips 3410 diffractometer in the Seeman-Bohlin 2θ mode) for microstructure information and Auger Electron Spectroscopy (FISONS Instruments Surface Science, model MICROLAB 310-F).

RESULTS

For the results shown here, the HV pulser was operated at 35 kV for Ti alloy and at 48 kV for Al alloy, with pulse duration of 1.5 μs and 100 Hz repetition frequency, for both cases, with the plasma load. When HEPIII system is activated in such a condition, with the working pressure above 5×10^{-3} mbar, the ion implantation occurs by means of a high voltage glow discharge (HVGD) mode (i.e., the plasma is produced by the HV pulse itself).

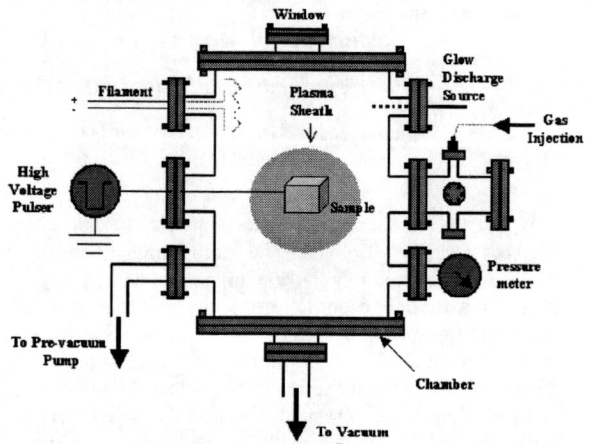

FIGURE 1. Schematic drawing of high energy PIII system

Typical voltage and current waveforms for the HEPIII system operated in the HVGD mode is shown in Fig.2(a) and (b), respectively. The voltage peak reaches 48 kV and the current peak, about 10A, at the implantation phase (excluding the first peak which is mainly due to displacement current). The inversion of the polarity in the voltage signal, after the main pulse, could be eliminated by using free-wheeling diodes that should also cut-off the negative current part. Voltage overshoot in this case is due to a mismatch between the characteristic impedance of the line and the plasma impedance.

In Fig. 3 we present the result of AES analysis performed in a sample of Al5052 treated with nitrogen PIII at 48 kV, 1.5 μs pulse duration, 100 Hz frequency, for a period of 1 h. Nitrogen implantation with ~ 10% atomic percent peak concentration and penetration of nitrogen of up to around 100 nm was achieved. Large concentration of oxygen was implanted (up to 50% atomic concentration) and its penetration was near 60 nm. It is clear from this result that insufficient amount of nitrogen was implanted because of two reasons: low delivered dose and very thick oxygen barrier to nitrogen influx formed as a native layer or during the

PIII processing. Increased total treatment time and cleaning of the surface by argon bombardment should allow better implantation of nitrogen in such condition of treatment. This is necessary for a successful formation of AlN layer with good mechanical and tribological properties. Preliminary measurements of corrosion resistance showed improved performance of the implanted surface compared to the unimplanted one.

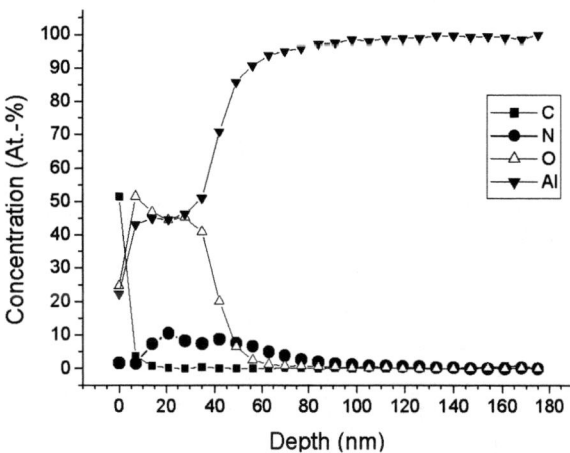

FIGURE 3. Atomic concentration profiles for the sample of Al5052 treated with nitrogen PIII, 48 kV/1.5μs/100Hz/1h.

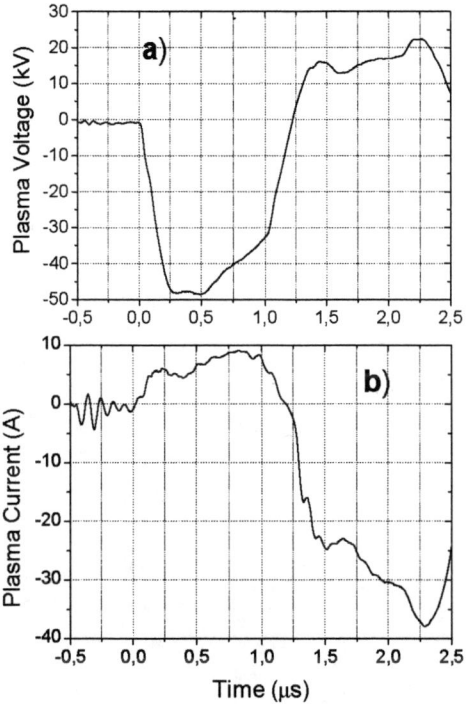

FIGURE 2. Typical voltage (a) and current (b) waveforms for the HEPIII system operated in the HVGD mode.

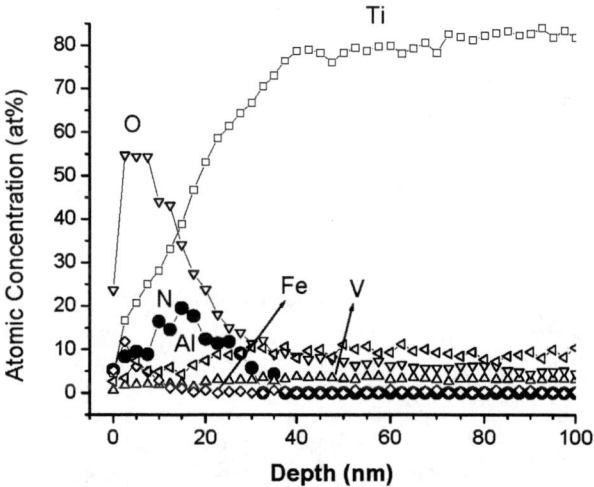

FIGURE 4. Atomic concentration profile for the Ti6Al4V treated at 35 kV/ 1.5 μs/ 100 Hz/ 1h.

In Fig.4 AES analysis of Ti6Al4V treated at 35 kV, 1.5 μs, 100 Hz, for 1 h is shown. Besides the profiles of atomic concentration of normal elements in the alloy, Ti, V, Al, we can also notice the presence of O and Fe that are impurities introduced during the treatment. Most importantly, we can see the profile of N which indicates near 20% atomic concentration at the peak and penetration depth of up to 40 nm. The oxygen contamination is very high, reaching 55% near the surface. Once again, the oxygen atoms come from the native oxide layer and the residual gas in the treatment chamber. This result is comparable to one obtained under typical PIII conditions [14] in Ti6Al4V samples without diffusion effects.

CONCLUSIONS

Al and Ti alloys were implanted by nitrogen PIII at energies of 48 and 35 keV, respectively, with frequencies of 100 Hz and very short pulses (1.5 μs), for up to 1 h. High voltage glow discharge mode was used for these treatments. High concentration of O was co-implanted as a result of native oxided layer and residual gas during the treatment, in both cases, being more critical in the case of Al alloy.

For better nitrogen implantation, attempts are underway to reduce the oxygen contamination by Ar cleaning before the PIII treatment of Al alloy, and also to increase the retained dose by longer treatments. For improved nitrogen implantation in Ti alloy, we are performing high temperature PIII with successful results [15].

ACKNOWLEDGEMENTS

This project is supported by FAPESP, MCT and CNPq from Brazil.

REFERENCES

1. A.K.Vasudeyan, R.D.Doherty (Eds.) Aluminum Alloys – Contemporary Research and Applications, Academic Press, Inc., London, 1989.

2. M.A.Khan, R.L.Williams, D.F.Williams, Biomaterials 20 (1999) 631.

3. Z.Cai et al, Biomaterials 20 (1999) 183.

4. D.Muster et al, MRS Bulletin (2000) 25.

5. M.Ueda, M.M.Silva, C.Otani et al, Surf. Coat. Technol. 169-170 (2003) 408.

6. W.Möller, S.Parascandola, O.Kruse et al, Surf. Coat. Technol. 116-119 (1999) 1.

7. E.Richter, R.Gunzel, S.Parascandola, T.Telbizova, O.Kruse, W.Möller, Surf. Coat. Technol. 128-129 (1995) 21.

8. K.C.Walter, R.A.Dodd, J.R.Conrad, Nucl. Instrum. Meth. B 106 (1995) 522.

9. J.O.Rossi, I.H.Tan, M.Ueda, Nucl. Instrum. Meth. B 1-2 (2006) 328.

10. M.Ueda, M.M.Silva, C.M.Lepienski, P.C.Soares Jr, J.A.N.Gonçalves, H.Reuther, to be published in Surf. Coat. Technol., 2006.

11. V.Fouquet, L.Pichon, A.Straboni, M.Drouet, Surf. Coat. Technol. 186 (2004) 34.

12. F.Davanloo, C.B.Collins, F.J.Agee, IEEE Trans. Plasma Sci. 26 (1998) 1463.

13. J.O.Rossi and M.Ueda, Proceed. Of Int. Pulsed Power Conf. 2005, paper 10097, CD-ROM.

14. K.Volz et al, Surf. Coat. Technol. 103-104 (1998) 257.

15. M.M.Silva, M.Ueda, L.Pichon eta al, to be submitted to Nucl. Instrum. Meth. B 2006.

Plasma Immersion Ion Implantation with a 4kV/10kHz Compact High Voltage Pulser

M.Ueda[1], R.M.Oliveira[1], J.O.Rossi[1], H.Reuther[2], G.Silva[1,3]

[1]*Associated Laboratory of Plasma, National Institute for Space Research,*
Av. dos Astronautas 1758, São José dos Campos, São Paulo, Brazil
[2]*Institute of Ion Beam Physics and Materials Research, Rossendorf, Dresden, Germany*
[3]*Department of Aeronautics and Mechanics, Technological Institute of Aeronautics, São José dos Campos, São*
Paulo, Brazil

Abstract. Development of a 4 kV/10 kHz Compact High Voltage Pulser and its application to nitrogen plasma immersion ion implantation (PIII) of different materials as Si, Al alloys, SS304 stainless steel and Ti alloys are discussed. Low voltage (1-5 kV) pulses at high frequencies (up to 20 kHz for 2 kV) were obtained with maximum power delivered at 5 kV, 7 kHz. These conditions were not sufficient to reach temperatures above 200°C in the samples because of short duration of the pulses. However, very shallow implantations of nitrogen in Si, Al5052, SS304 were observed by Auger electron spectroscopy and improved corrosion resistance was obtained for Al5052 when it was treated by nitrogen PIII at 2.5 kV, 5μs and 5 kHz pulses.

Keywords: plasma immersion ion implantation, compact high voltage pulser, Si, Al5052, SS304.
PACS: 52.77Dq

INTRODUCTION

Compact High Voltage Pulsers (CHVP) are important electronic devices that are being developed and improved with the aim to applying them in as diverse fields as aerospace, nano and microelectronics [1], low voltage plasma immersion ion implantation (PIII) nitriding [2], surface enhancement of polymers [3] and optoelectronic materials [4], medical [5], etc. Their light-weight and compactness in comparison to commonly used large and heavy high voltage pulsers are the reason of their possible success in the above mentioned fields that require somewhat modest voltages (1 to 5 kV) but high frequencies (over 5kHz). We have utilized semiconductor switching and step-up transformers instead of hard tube system for this pulser downsizing [6, 7, 8].

PIII is a relatively newly developed technique for surface modification of materials, specially suited for industrial components. This treatment consists of immersing the samples or workpieces in a plasma and applying a negative high voltage pulses to the target. Ions of interest are then extracted from the plasma, accelerated towards the surface (mainly at normal incidence) and implanted tridimensionally.

Normally, plasma implantation process imposes several requirements on the power supplies such as square wave high voltage pulses (in the range of 5kV to 40 kV), short rise-times (less than 1.0 μs) and pulse duration varying from 5 to 50μs at a certain repetition rate (typically between 300Hz and 1kHz). Compared with conventional ion beam implantation (IBI), PIII process has several advantages such as no need of beam guiding or target manipulation and ability of implanting objects with complex and irregular forms. In addition, by using this technique it is possible to obtain an improved conformal implantation if the ion sheath is completely conformal around the target. In practice, however, corners and edges of the real objects increase the non-uniformity of implant, especially if the plasma sheath is too thick. One way of circumventing this problem is to apply lower voltage pulses (1 kV to 5 kV) of shorter duration (1 μs to 5 μs) for producing thinner plasma sheath, which improves the uniformity of ion implantation of treated pieces (especially those with irregular shape) but at the cost of decreasing the thickness of the implanted layer.

However, in certain applications as ultra-shallow junctions, very thin treated layers are being pursued. Also in some applications in biomedical engineering, thin layers of modified materials with enhanced properties can be obtained through low energy PIII.

Therefore, we devised for plasma implantation systems a solid-state compact modulator, in which a small capacitor discharges through a forward converter composed of a low blocking voltage IGBT switch (of only 1 kV) and three step-up pulse transformers, rather than employing comparable hard-tube devices such as in conventional implant pulsers that are of large size, expensive and cumbersome [8].

Previously, we have reported mainly on the electrical performance of CHVP with 4kV/2A/5kHz nominal capability [8]. In the present paper, emphasis will be given to the surface modification results obtained by using an improved version of CHVP for plasma immersion ion implantation. With a maximum output of 4kV and operating at 10kHz, this pulser was applied to nitrogen PIII of Al5052, SS304, Si, polymers and Ti6Al4V, showing shallow implantation of N, as expected. In the Al alloy, a significant improvement of its corrosion resistance was measured even with such thin modified layer. Comparison of such material surface performances obtained at different operating conditions of CHVP are discussed in this paper, confirming the usefulness of this kind of low cost pulsers in PIII research.

EXPERIMENTAL METHODS AND APPARATUS

The details of the circuit and the electronic performance of the developed CHVP solid-state pulser are discussed elsewhere [8, 9].

For the implantation tests, the pulser was connected to a PIII system described in Fig. 1. The DC glow discharge is powered by simple power supply capable of providing 900V, 0.6 A maximum output. The nitrogen plasma is initiated between the cylindrical stainless steel rod and the neck connected to the PIII cylindrical chamber with 27 cm diameter and 47 cm length, as shown in the same figure. The stainless steel rod is connected to a cooper feed-through to apply the negative high voltage pulse. We typically run the discharges at powers of 180 W, with corresponding discharge voltages of 600 V and currents of approximately 0.3 A. Under these conditions, the plasma potential near the center of the PIII plasma chamber reaches 350 V, which in some cases can be deleterious for PIII processing due to the sputtering of the surface under treatment. To reduce plasma potential, an electron shower hot filament is used. By adjusting the power of this electron source, the plasma potential can be controlled from zero to 350V [10].

For the measurements of the microstructure of the nitrogen implanted materials, Philips PW 1830, x-ray diffractometer in the standard 2θ scan mode was used. For the Auger Electron Spectroscopy (AES) measurements, used for the determination of elemental concentration in the treated samples, a spectrometer from FISONS Instruments Surface Science, model MICROLAB 310-F was used.

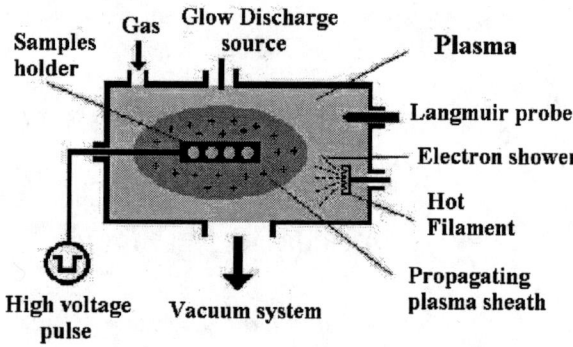

FIGURE 1. Schematics of the plasma immersion ion implantation experiment.

The corrosion resistance of the treated materials was measured by a potenciostat/galvanostat AUTOLAB, model PGSTAT30, with three electrode configuration inside an electrolytic cell. The cell contained a solution with 3.5% ppm of NaCl with pH equal to 6. The potentials were measured with respect to a reference electrode made of Ag/AgCl.

For the rocking curve measurements, a Philips X-PERT MRD, high resolution x-ray diffractometer was used.

RESULTS AND DISCUSSION

The experimental results concerning the electronic performance of CHVP in the present configuration are discussed shortly. Tests in the resistive load of 2 kΩ showed that pulses of 5 μs duration with rise and fall times of 1 μs each were obtained at frequencies of up to 10 kHz. Currents of up to 10 A were obtained during the pulses. With a nitrogen plasma load, we obtained pulsed voltages with square forms and currents with opposed peaks at the start and the end of the pulses with a plateau in between. For 2.5 kV, 5 μs pulses we achieved current plateau of 0.15 A, while pulsing at 5 KHz. We have performed many materials treatments under this condition.

We also performed parameter survey of the plasma and the pulser conditions. We were able to achieve conditions of very high frequency of 20 kHz but with lower voltages, below 2 kV. On the other hand, 5 kV pulses were maintained at 7 kHz, while pulse currents in plasma as high as 0.4 A were attained. We summarize the PIII operating conditions with CHVP in Table I.

In Fig.2 we show the profiles of implanted nitrogen and impurity oxygen, for the sample of Si treated under condition 1. The nitrogen penetration in this case was less than 10 nm, with peak concentration of about 15%, at around 5 nm. The contamination of oxygen was very high (up to 70% at the very surface) due to native oxide layer or oxygen residual gas in the chamber during PIII treatment. This situation could be improved by using better pumping systems and argon cleaning discharges.

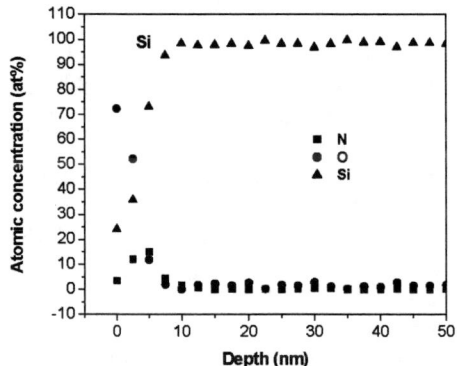

FIGURE 2. Atomic concentration profiles for the samples of Si treated by nitrogen PIII

In Fig. 3 we show the profiles of implanted nitrogen and impurities for the Al5052 sample treated under condition 1. Oxygen was again the predominant impurity but C and Fe were also present in this case. Nitrogen peak concentration was 15% and the reach was near 30 nm. Other elements shown are the usual constituents of the alloy.

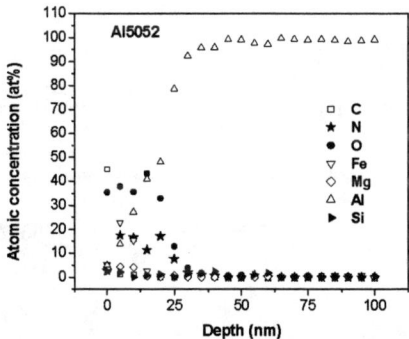

FIGURE 3. Atomic concentration profiles for the samples of Al5052 treated by nitrogen PIII

Samples of SS304 implanted with nitrogen treated under condition 1 showed AES profiles as in Fig.4. The peak concentration was near 25% while the maximum reach of nitrogen was less than 20 nm. Oxygen contamination was also high (near 40%) while some carbon was present near the surface. Chromium concentration was reduced at the surface but nickel concentration was increased (segregation effect?).

These implantations resulted in the following changes for the SS304: a) As was seen in the XRD results (not shown here), we could only observe the formation of the α-phase, indicated by a diffraction

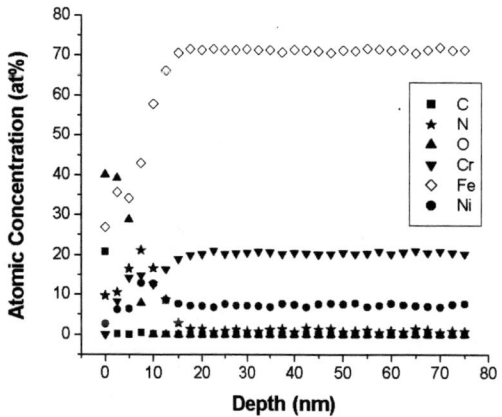

FIGURE 4. Atomic concentration profiles for the samples of SS304 treated by nitrogen PIII

peak to the right of the γ(111) peak at 2θ= 44°. No changes were seen for the corrosion resistance or microhardness of the surface.

Meanwhile, for the Al alloy treated under condition 1, there was a significant improvement in the corrosion behavior of the surface. As shown in Fig.5, the corrosion current density was reduced by an order of magnitude. Furthermore, the corrosion potential was improved from -680 mV to -250 mV and the alloy was passivated in the range of -250 mV to 1.5 V, after the implantation.

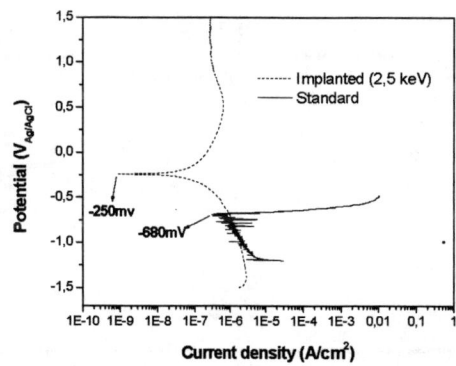

FIGURE 5. Polarization curves of Al5052 for untreated and nitrogen PIII treated samples, respectively

As for Si treated with nitrogen PIII under condition 1, it showed clear distortion in its rocking curve compared to the untreated one, as shown in Fig.6. This distortion to the left of the Si diffraction peak is indicative of strained layer due to the nitrogen implantation or defects resulting from the bombardment by those ions. Samples of materials treated under other conditions listed on Table I are being analyzed presently.

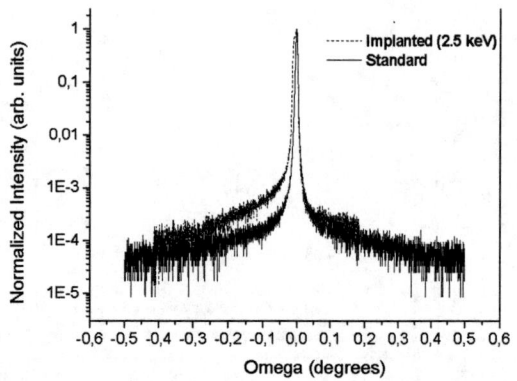

FIGURE 6. Rocking curves of untreated and nitrogen PIII treated samples of Si

CONCLUSIONS

A compact high voltage pulser (CHVP) nominally rated 4kV/5µs/10kHz was tested in terms of electronics and successfully applied to surface modification of different materials by plasma immersion ion implantation (PIII) at low energies (1- 5 kV) at high frequencies (up to 20 kHz at 2 kV). Our CHVP can stand continuous operation at such conditions for up to 1 to 3 hours, routinely. Plateau currents in plasma of up to 0.4 A were attained at 5 kV, when frequencies of 7 kHz were used. The CHVP combining IGBT switching with step up transformers (x3) allowed to reach such parameters in a very compact, lightweight, low cost configuration.

As a first application, this CHVP was applied to the nitrogen PIII treatment of different materials: Si, Al5052, SS304, Ti6Al4V. Polymers are also being tested. Because of short pulse length (5 µs), and relatively low voltages, very conformal PIII treatments of real workpieces are expected. Furthermore, very shallow implantations of nitrogen were achieved: less than 10 nm in Si, less than 20 nm in SS304 and 30 nm in Al5052.

For Si targets, this result is encouraging for shallow ion implantations needed for new frontiers in nano and microelectronics. For Al5052, despite of such thin treated layer, we observed substantial improvement of the surface against corrosion, when compared to the untreated specimens. In the case of SS304, only a transformation to alpha-phase was observed for the near surface layer, when 2.5 kV and 5 kHz (the same conditions of treatments for the Si and Al5052 also) was applied. We are presently pushing further the parameters of CHVP to achieve a thin layer of γ_N-phase in SS304 which presents nobler properties than its alpha-phase.

ACKNOWLEDGMENTS

This work was supported by FAPESP, CNPQ, CAPES and MCT, Brazil.

TABLE 1. Summary of operating conditions of CHVP for materials treatment by PIII processing

Treatment conditions	Implantation voltage (kV)	Pulse frequency (kHz)	Pulse duration (µs)	Implantation time (min)
1	2.5	5.0	5.0	60.0
2	4.0	10.0	5.0	60.0
3	2.5	10.0	5.0	180.0
4	2.0	20.0	5.0	60.0
5	5.0	7.0	5.0	60.0

REFERENCES

1. S. B. Felch, Z. Fang, B. -W. Koo, R. B. Liebert, S. R. Walter, D. Hacker, *Surf. Coat. Tecnol.* **156** (2002) 229.
2. X. Tian, X. Wang, B. Tang, P. K. Chu, *Rev. Sci. Instrum.* **70** (1999) 1824.
3. I. H. Tan, M. Ueda, R. S. Dallaqua, J. O. Rossi, A. F. Beloto, M. H. Tabacniks, N. R. Demarquette, Y. Inoue, *Surf. Coat. Technol.* **1-2** (2004) 234.
4. J. C. N. Reis, A. F. Beloto, M. Ueda, *Nucl. Instrum. Meth. B* **240** (2005) 219.
5. S. Mukherjee, M. F. Maitz, M. T. Pham, E. Richter, F. Prokert, W. Moeller, *Surf. Coat. Technol.* **196** (2005) 312.
6. L. M. Redondo, N. Pinhão, E. Margato, J. F. Silva, *Surf. Coat. Technol.* **156** (2002) 61.
7. X. Tian, Z. Zeng, X. Zeng, B. Tang, P. K. Chu, *J. Appl. Phys.* **88** (2000) 2221.
8. J. O. Rossi, M. Ueda, J. J. Barroso, Proceed. of the 15th IEEE Intern. Pulsed Power Conference (poster 10097), CD-ROM, 2006.
9. J. O. Rossi, M. Ueda, J. J. Barroso, G. Silva, H. Reuther, submitted to IEEE Trans. On Plasma Sci., 2006.
10. M. Ueda, G.F. Gomes, L. A. Berni, A.F.Beloto, E. Abramof, H. Reuther, *Nucl. Instrum. Meth. B* **161-163** (2000) 1064.

Effects Of Ion Energy On Nitrogen Plasma Immersion Ion Implantation In UHMWPE Polymer Through A Metal Grid

M. Ueda[1], R.M. Oliveira[1], J.O. Rossi[1], C.M. Lepienski[2], and W.A. Vilela[3]

[1]*Associated Laboratory of Plasma, National Institute for Space Research,*
Av. dos Astronautas 1758, São José dos Campos, São Paulo, Brazil
[2]*Department of Physics, Federal University of Paraná, Curitiba, Brazil*
[3]*Associated Laboratory of Materials and Sensors, National Institute for Space Research,*
São José dos Campos, SP, Brazil

Abstract. Herein, we consider the potential application of plasma immersion ion implantation (PIII) for treatment of polymer surfaces. This paper presents some experimental data for ultra-high molecular weight polyethylene (UHMWPE) implanted with nitrogen using PIII process. This polymer is widely used in medical prosthesis and PIII treatment has revealed to be an ease and cheap way to improve the lifetime of prosthesis made with UHMWPE. Here we show the latest results for UHMWPE surface treatment obtained with the use of a high voltage pulser of 100kV/200A based on coaxial Blumlein technology.

Keywords: plasma implantation, Blumlein line, high voltage, nitrided surfaces, ultra-high molecular weight polyethylene, and diamond-like carbon
PACS: 52.77.Dq

INTRODUCTION

To explore new operating regimes in surface treatment by plasma immersion ion implantation (PIII), high voltage pulses of the order of 100kV are needed. For instance, high energy is essential to obtain great depth of penetration of ion species into the surface of polymers, which increases the hardness factor and wear resistance of polymeric components under certain manufacturing processes [1]. On the other hand, working with short pulses (around 1µs) leads to the improvement of the conformal implantation of small pieces by decreasing the width of plasma sheath. For instance, using shorter pulse durations (≈1µs), higher currents (>100A), and high-density plasma can produce sheath thickness below 2.5cm as described elsewhere [2].

Recent studies, conducted by the plasma group at INPE, has shown a tremendous potential of the PIII treatment [3],[4] on improvement of mechanical properties of a ultra-high molecular weight polyethylene (UHMWPE), a kind of polymer widely used in traditional and modern industries. As an example of the traditional case, we can cite the protection layer in mine ore transporting dump trucks, valves or connecting tubes in food processing or chemical factories, etc. The most representative modern application of UHMPWE is probably the acetabular cap used in repairing component of femoral prosthesis. Despite its excellent properties regarding biocompatibility, machineability, and mechanical qualities, the extremely demanding conditions of real prosthesis operations in the human body require further improvements in surface properties, especially with respect to wear. Nitrogen plasma implantation (PIII) applied to UHMWPE material synthesizes a DLC layer on its surface with improved mechanical properties [3]. To avoid severe charging dielectric effects during PIII treatments of such non-conducting materials, we used a metal grid during ion implantation process. We found that the ion energy affects the homogeneity of the treated layer, being favorable for higher energies (for the range of 5 to 15keV with pulse duration of the order of 10µs). Much higher implantation energies up to 50keV with very short pulses of 1.2µs were tested also, and analyses based on Raman spectroscopy carried out for identifying diamond-like carbon (DLC) distribution over the surface.

In next section, we present two experimental set-ups used to carry out the tests while in section III we show the latest results obtained so far.

EXPERIMENTAL SET-UP

For the implantation tests, we used two different experimental set-ups. For the first tests carried out with lower voltages and long pulse duration (5kV/15kV&10μs), we used the system depicted in Fig. 1. In this case, a hard-tube (HT) pulser with maximum V/I capability of 30kV/30A (connected to the PIII system as shown in Fig. 1) was able to produce pulse duration in the range of 5-10μs at a repetition frequency of 100Hz [5], [6]. A simple power supply capable of providing maximum V/I parameters of 900V/0.6A gives output power for the DC glow discharge. The nitrogen plasma discharge initiated between the cylindrical stainless steel rod and the neck connected to the PIII cylindrical chamber with 27 cm diameter and 47 cm length as shown in Fig. 1. The stainless steel rod linked to a copper feed-through connector applied the negative high voltage pulse from the HT pulser. We typically run the discharges at powers of 180 W, with corresponding discharge voltages of 600 V and currents of approximately of 0.3 A. Under this conditions the plasma potential near the center of the PIII plasma chamber reached 350 V, which in some cases can be deleterious for PIII processing due to the sputtering of the surface under treatment. By adjusting the power of an electron source filament made of heated tungsten, we could control the plasma potential from zero to 350V. In the present experiment, we worked with reduced plasma potential of the order of 100V.

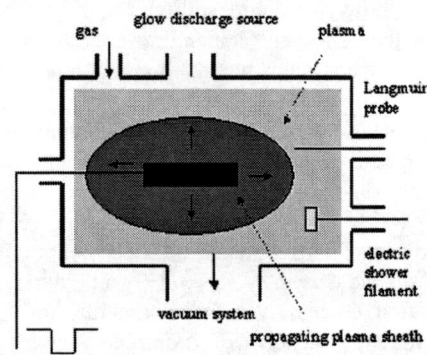

FIGURE 1. Experimental set-up for implantation tests with lower voltage and longer pulse duration.

Fig. 2 shows the experimental set-up used for the PIII treatment of UHMWPE polymer with shorter pulses and higher voltages (in the range of 1.2μs with applied voltages of 30-50kV). Because of the higher voltage applied, we use a stainless steel (SS) vacuum chamber with bigger dimensions (diameter of 58cm and length of 68cm) as well as a high voltage feed-through device rated at 60kV. To generate the high voltage pulse, a pulser of 100kV/200A/100Hz with fixed pulsed duration of the order of 1.2μs based on coaxial pulse forming line technology has being used, commonly known as Blumlein pulse generator [7], [8]. Similarly to the previous arrangement, we put the sample under treatment inside the chamber on an electrode support connected to the high voltage pulser and immersed in nitrogen plasma. The vacuum base pressure for the plasma production (provided by one turbo-molecular pump installed at the bottom of the vacuum chamber) was around 8.0×10^{-6} mbar with a typical working pressure of about 5.0×10^{-3} mbar. We produced the plasma by means of a pulsed glow discharge using the same high voltage pulse applied to the target. Again, to start the pulsed glow discharge we used an electric shower filament made of thin tungsten wire of 0.2mm diameter. To reduce plasma potential for avoiding surface sputtering (which could compromise the implantation process), we adjusted the AC voltage of the electric shower filament to approximately 11V. A pink luminescence emitted from the nitrogen glow discharge through a glass window at the chamber top indicated the plasma production.

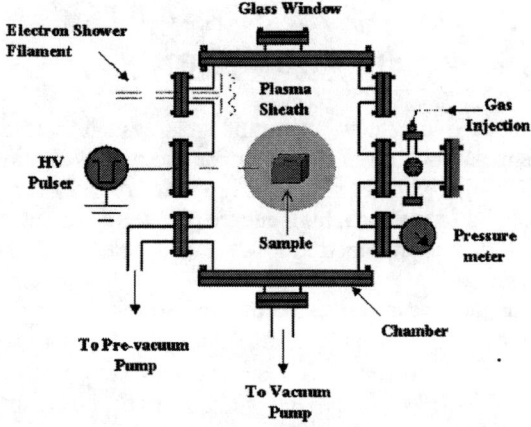

FIGURE 2. Experimental set-up for implantation tests with higher voltage and shorter pulse duration.

In both cases, we treated the polymer samples (25 mm in diameter × 5mm in length) by putting them on a metallic sample holder to a stainless steel rod, which was linked to the pulser via a high voltage copper feed-trough device. Because of the isolation properties of the polymer, a thin wire grid with 70% transparency (connected to the high voltage potential of the holder) was laid near the sample (2mm of distance) to attract the ions from the plasma. Since polymers are non-conducting materials and do no tolerate elevated temperature increase, thermal diffusion of the ions is not important and, hence, pulse repetition frequency was limited in the range of hundreds of Hz, more specifically 100Hz for both cases with higher and lower voltages.

EXPERIMENTAL RESULTS

The PIII results obtained so far with UHMWPE treatment have demonstrated that implantation process with higher energy ions can play an important role on getting implanted surfaces that are more homogenous. Experiments performed with UHMWPE, using applied lower voltages (5 & 15kV) with long pulse duration (of the order of 10µs), have shown that ion energy represents a key parameter on the dehydrogenation of polymer surface. Dehydrogenation correlates directly with cross-linked bond formation, which is related to DLC formation on polymer surface. Figs. 3 and 4 show the Raman shift for the UHMWPE treated by PIII. Each figure contains three Raman shifts for three different implanted locations in the sample. We chose the locations in order to cover the implanted sample surface, from one side to another. In this way, it was possible to realize that the DLC formation, which is correlated to the ratio of the areas under the two shoulders, one centered on $1330cm^{-1}$ and the other on $1560cm^{-1}$, is more homogenous when the pulse amplitude is higher. When the pulse amplitude is low, the discrepancies among the Raman shifts are visible, as we can see in the Fig 3. Theses discrepancies are a result of the shadowing effects of the grid placed in front of the polymer during the ion implantation. Since the metal grid is 2mm away from the polymer surface, some of the ions extracted from the plasma with a much thicker sheath are stopped at the grid lines and others go directly through the grid apertures. Hence, at places where correspond to the shadow of the grid lines, the Raman intensity is lower because of the lower implantation doses. However, if the voltage is increased the ions tends to spread across the shadow regions due to the charging effect, leading to an implanted surface with more homogeneous properties illustrated by the DLC formation identified in Raman spectroscopy as shown in Fig. 4. Beyond the pulse amplitude, the surface homogeneity has also a strong dependence on the length of the pulse. For instance, at higher energies (30kV & 45kV), but with very short pulses of about 1.2 µs, the same correlation, as shown in Figs. 5 and 6, can be observed for these results if we compare them to that obtained for lower energies (5kV & 15kV). Fig. 5 shows the Raman spectroscopy for ion energies of about 30kV obtained for eight location points varying along the sample surface during an exposure time of 120 min, while Fig. 6 gives the same diagnostic for higher energies up to 45kV measured in fourteen different location points during an exposure time of 60 min. In Fig. 5 with lower energy, we note less homogeneity on the surface when compared to that of Fig. 6. As before, for the previous case with longer pulse, this is explained by the fact that surface charging is induced by higher voltage application. Nevertheless, the same homogeneity is obtained at lower voltages around 15kV with application of a pulse with longer duration (of the order of 10µs as shown in Fig. 4) instead of a higher voltage with shorter pulse duration. This means that surface homogeneity is also dependent on pulse duration since the plasma sheath formed around the target has more time to induce the spreading of the ions over the surface shadow regions, increasing the surface charging and homogeneity. Again, for Figs. 5 and 6 we observe the same DLC formation with both characteristic shoulders in the graphics around Raman wavelength shifts of $1330cm^{-1}$ and $1560cm^{-1}$ corresponding respectively to the D and G peaks.

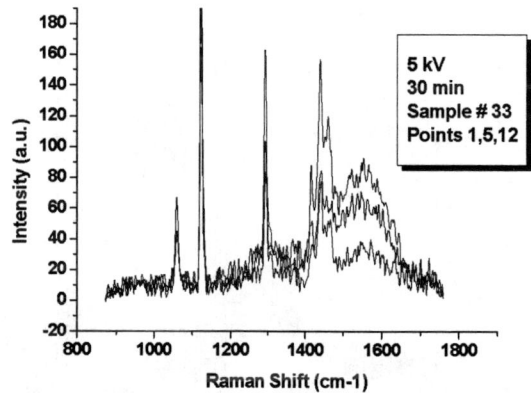

FIGURE 3. Raman shift for less energetic ions at 5kV with long pulse duration (10µs).

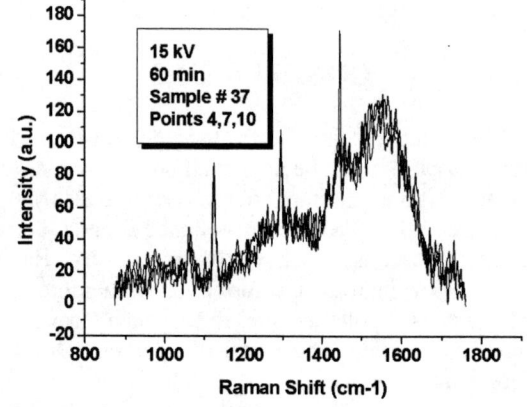

FIGURE 4. Raman shift for less energetic ions at 15kV with long pulse duration (10µs).

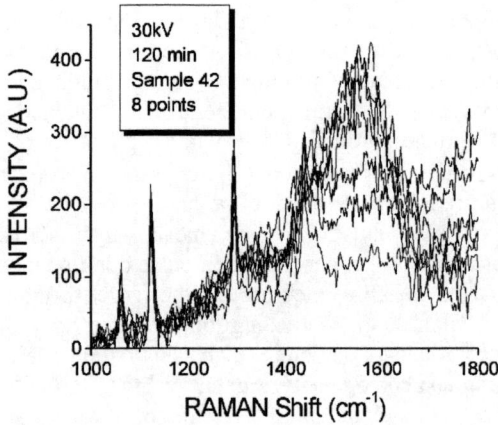

FIGURE 5. Raman shift for high-energy ions at 30kV with shorter pulse duration (1.2µs).

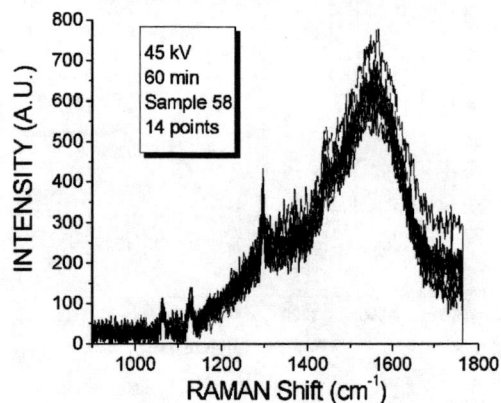

FIGURE 6. Raman shift for high-energy ions at 45kV with shorter pulse duration (1.2µs).

CONCLUSIONS

The results obtained so far have demonstrated that high-energy ions can be implanted on polymer surface structure without the formation of precipitates by self-clustering, and lead to cross-linked bonds among polymeric chains. Surface diagnostics by Raman spectroscopy indicate the formation of a more rigid structure on polymer surface, which could be identified as being Diamond-Like Carbon (DLC). A more homogeneous DLC formation on the polymer surface can alter the material properties, producing increased hardness factor with reduced friction coefficient, which would be ideal for applications of UHMWPE in prosthesis devices. To verify this statement, hardness and friction measurements should be performed. Moreover, the results indicate that the homogeneity and intensity of DLC formation on polymer surface is directly dependent on both duration and amplitude of the pulse provided by the high voltage pulser. In the case of pulse amplitude, this means that this dependence is directly related to the energy of the implanted ion species. Because of the need of the electrode grid structure near the polymer surface during the implantation process of non-conducting materials, non-homogeneous implantation can occur in the grid shadows over the sample surface. Obviously, this can be circumvented by applying higher voltages with short or long pulses as illustrated by the experimental results obtained. In view of that, our group has found interesting achievements when implanting UHMWPE samples with more energetic ions, especially in the range of 30-45kV. Moreover, the HV Blumlein pulser of the second experimental set-up (see section II) has shown excellent prospects for the future studies of implantation process with high-energy ions since it can theoretically provide pulses with amplitudes up to 100kV.

ACKNOWLEDGMENTS

The authors would like to thank Research Foundation of the Sao Paulo State (FAPESP) for its financial support. Other research funding agencies (MCT & CNPq) are also acknowledged.

REFERENCES

1. S. Han, Y. Lee, H. Kim, G. -H. Kim, J. Lee, J. -H. Ion, and G. Kim, *Surface & Coatings Technology* 93, 1997, pp. 261-264.
2. R.J. Adler, R.J. Richter-Sand, E.J. Clark, and C.W. Gregg, *J. Vac. Sci. Technol.* B 17(2), 1999, pp. 883-887.
3. A.R. Marcondes, M. Ueda, K.G. Kostov, A.F. Beloto, N.F. Leite, G.F. Gomes, and C.M. Lepienski, *Brazilian Journal of Physics* 34(4B), 2004, pp. 1667-1672.
4. I.H. Tan, M. Ueda, R.S. Dallaqua, J.O. Rossi, A.F. Beloto, M.H. Tabacnicks, N.R. Dermaquette, and Y. Inoue, *Surface & Coatings Technology* 186, 2004, pp. 234-238.
5. J.O. Rossi, M. Ueda, and J.J. Barroso, *Surface & Coatings Technology* 136, 2001, pp. 43-46.
6. J.O. Rossi, M. Ueda, J.J. Barroso, and V.A. Spassov, *IEEE Transactions on Plasma Science* 28(5), 2000, pp. 1392-1396.
7. J.O. Rossi, I.H. Tan, and M. Ueda, *Nuclear Instruments and Methods in Physics Research* B 242, 2006, pp. 328-331.
8. J.O. Rossi, M. Ueda, and J.J. Barroso, *IEEE Transactions on Plasma Science* 30(5), 2002, pp. 1622-1626.

Modified Phasor-Particle Model of Treating A Blocking Capacitor As A Phasor Element in Simulation of Plasma Coupling with An External Auto-Matching Network

Dixon T. K. Kwok [1] and Paul K. Chu [2]

[1] *APPG, School of Physics A28, University of Sydney, NSW 2006, Australia*
[2] *Dept. of Physics and Materials Science, City University of Hong Kong, Kowloon, Hong Kong*

Abstract. A phasor-particle model has been developed to numerical simulated the radio-frequency (RF) capacitive coupling between an external matching network and a low pressure electrical plasma [1]. Gauss's law was applied to the blocking capacitor in this model. Non-linearly current was observed flowing into the blocking capacitor [1]. A new method of treating the blocking capacitor as a phasor element is developed such that identical current is flowing through the blocking capacitor. A big impact on the circuit voltages and currents are observed when simulating the asymmetric system by the new method. The proposed method can be applied to quasi-matching condition.

Keywords: Phasor model, Auto-matching, plasma simulation, capacitive coupling.
PACS: 52.65.-y, 52.65.Rr, 52.65.Ww

INTRODUCTION

A phasor-particle model was developed to numerical simulated the radio-frequency (RF) capacitive coupling between an external matching network and a low pressure electrical plasma [1]. In this model, the external matching network or any complicated circuit is modeled by electrical phasor [1]. The chamber and electrodes, either RF powered or grounded, can be regarded as a good capacitor. The accumulated surface charges of the electrode and consequently the potential are modified by the plasma. The plasma is simulated by a one dimension hybrid particle-in-cell (PIC) ions and Boltzmann distribution of electrons [2,3]. The advantage of this model is that any complicated external network can be described by several linear equations that can be solved directly [1]. The amplitude and phase of the voltage/current of the external matching network were numerically estimated, an auto-matching network was simulated, the negative direct current (DC) biased voltage of an asymmetry system associated with a blocking capacitor was generated by this model [1,4].

The schematic of the model simulating an asymmetric system with a blocking capacitor connecting to a matching box was depicted in Fig. 1. An alternate voltage source Vs of absolute value 300 volts, radio frequency of 13.56 MHz with input resistance Ro equal to 50 ohms was connected to an asymmetric chamber represented by a cylindrical capacitor. The outside cylinder of radius r_b equal to 0.2 meter was grounded. The power was transferred to the inner electrode of radius r_a equal to 0.1 meter. The cylinders had a length of 0.2 meter given the chamber capacitance of 16.05 pF.

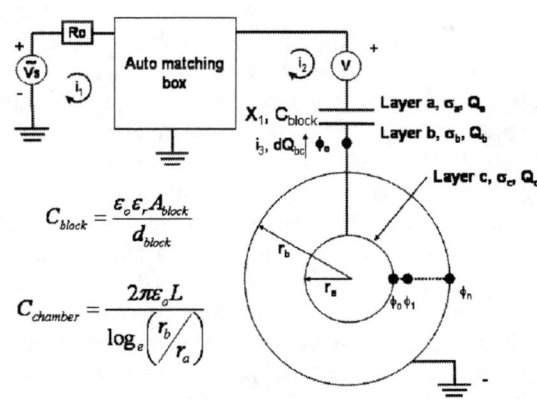

FIGURE 1. The schematic of the asymmetric system with the blocking capacitor is depicted. The chamber is presented by two cylinders of radius r_a and r_b with length L.

A capacitor C_{block} of 200 pF was used to block direct current from flowing into the circuit. An uniform electrical Ar$^+$ plasma of density equal to 1.0×10^{16} m^{-3} was initially filled up the chamber. The ions were at room temperature and given a zero initially velocity. The ions were mainly driven by electric field. The electrons were in thermal equilibrium and was described by Boltzmrium distribution [1-3]. The distance between the two electrodes was divided into uniform cells of length equal to 1 mm given a total of 100 cells. 1000 PIC particles were uniformly placed in each cell given a total of 100,000 PIC particles. The time step was equal to an RF cycle divided by 200, i.e., dt=3.68732×10^{-10} seconds. The simulation ended at 20 μsec. The blocking capacitor C_{block} was modeled by applying Guass's Laws to the three layers a, b, and c as depicted in Fig. 1 [1]. The current i_3 between the blocking capacitor and the top electrode was defined as dQ_{bc}/dt as depicted in Fig. 1. At 5 μsec, a negative dc biased voltage of -116 volts was observed at the inner electrode of the system as depicted in Fig.2a. The calculated loading resistance R_L was equal to 162 ohms [1]. However, the current i_2 flowing between the matching box and the blocking capacitor was difference from i_3, the current flowing between the inner electrode and the blocking capacitor as depicted in Fig. 2b. Talking Vs as the reference, it shown that i_1, the current flowing before the matching box, was 7.2° or 0.13 radians behind Vs with amplitude of 3.38 amps, i_2 was 50.4° or 0.88 radians behind with an amplitude of 1.71 amps, and i_3 was 27° or 0.47 radians behind with an amplitude of 1.98 amps. According to the circuit or phasor theories [5,6], it is abnormal to have a difference currents flowing through the blocking capacitor, since it is common treating the blocking capacitor as a circuit element. In circuit model approach [6], a difference current can flow into the ion sheath regions but not for the blocking capacitor. The aim of this work is to investigate the cause of the mismatched between i_2 and i_3 in the previous approach by applying three times Gauss's Law at the three layers. A difference approach of simulating the blocking capacitor will be used. In the new approach, the blocking capacitor is treated as a circuit element and combined into the phasor equations. In other words, we are forcing identical current flowing through the blocking capacitor. The proposed method will be described in the next section. It will be validated by simulating an asymmetric system with an empty chamber. Finally, the simulation results of modeling the asymmetric system coupling with a plasma will be presented and discussed.

SINGLE CURRENT LOOP FOR THE BLOCKING CAPACITOR

In the new approach, the blocking capacitor C_{block} is treated as a phasor element X_1 of value

$$X_1 = \frac{-j}{\omega C_{block}} \quad (1)$$

where $j = \sqrt{-1}$. We can derive a new phasor equation as;

$$V = \varphi_0 + I_2 X_1 \quad (2)$$

Solving with the previous complex number form of the phasor equation [1],

$$V = V_s B - I_2 A \quad (3)$$

gives

$$\varphi_0 = V_s B - I_2 A' \quad (4)$$

where $A' = A + X_1$. The phasor elements A and B are derived from the matching box [1]. It was shown that under matching condition, the phase of A θ_A is always equal to zero [1]. In real number form, it becomes

$$\phi_0 = |B||V_s|\cos(\omega t + \theta_B) - |I_2||A'|\cos(\omega t + \theta_2 + \theta_{A'}) \quad (5)$$

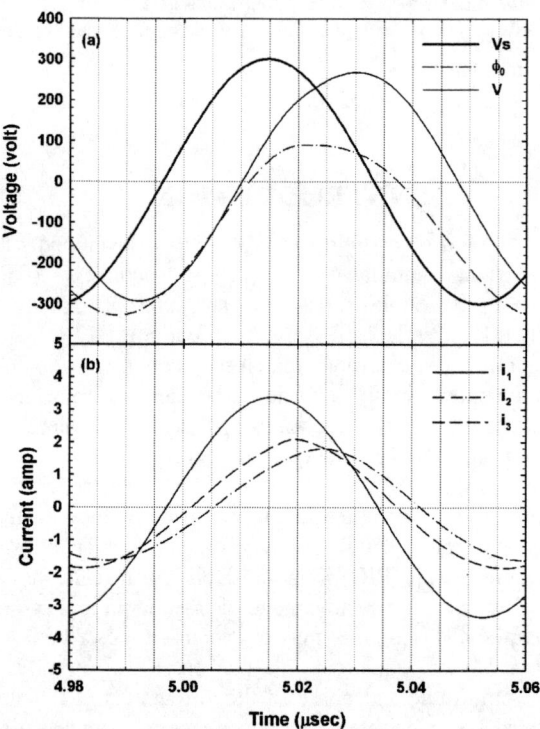

FIGURE 2. The voltages Vs, ϕ_0, and V at 5 μsec are plotted in (a). The currents i_1, i_2, and i_3 at 5 μsec are plotted in (b). The simulation was conducted by applying Gauss's law at the two layers of the blocking capacitor.

The phase of A' $\theta_{A'}$ is not equal to zero even under matching condition. The electrode potential $\phi_0(t)$ can be calculated by another formula [1],

$$\phi_0(t) = \phi_0(t - \Delta t) + \frac{i_2(t)\Delta t}{C_{chamber}} \quad (6)$$

With the presence of a plasma, it is modified as,

$$\phi_0(t) = \phi_0^p + \frac{i_2(t)\Delta t}{C_{chamber}} \quad (7)$$

where ϕ_0^p is calculated from the Gauss's law [1,7],

$$\phi_0^p = \phi_1 + \frac{\Delta r}{\varepsilon_o}\left(\sigma_c + \rho_0 \frac{\Delta r}{2}\right) \quad (8)$$

where ϕ_1 is the potential of the next node, σ_c is the surface charge density (Cm^{-2}), ρ_0 is the charge density of the plasma at the node (Cm^{-3}), Δr is the length of a cell, and ε_o is the dielectric constant. Combining Eqns. (5) and (7) gives,

$$\phi_0^p + \frac{i_2(t)\Delta t}{C_{chamber}} = |B||V_s|\cos(\omega t + \theta_B) - |I_2||A'|(\cos(\omega t + \theta_2)\cos(\theta_{A'}) - \sin(\omega t + \theta_2)\sin(\theta_{A'})) \quad (11a)$$

$$\phi_0^p + \frac{i_2(t)\Delta t}{C_{chamber}} = |B||V_s|\cos(\omega t + \theta_B) - |A'|\cos(\theta_{A'})i_2(t) + |I_2||A'|\sin(\omega t + \theta_2)\sin(\theta_{A'}) \quad (11b)$$

After rearrangement, it gives

$$i_2(t) = \frac{|V_s||B|\cos(\omega t + \theta_B) - \phi_0^p + |I_2||A'|\sin(\omega t + \theta_2)\sin(\theta_{A'})}{|A'|\cos(\theta_{A'}) + \frac{\Delta t}{C_{chamber}}} \quad (11c)$$

At the start of the simulation within 3 cycles, the $|I_2||A'|\sin(\omega t + \theta_2)\sin(\theta_{A'})$ term of Eqn. (11c) is ignored giving

$$i_2(t) = \frac{|V_s||B|\cos(\omega t + \theta_B) - \phi_0^p}{|A'|\cos(\theta_{A'}) + \frac{\Delta t}{C_{chamber}}} \quad (12)$$

The amplitude $|I_2|$ and phase θ_2 of the current i_2 are estimated from the previous cycle. After 3 cycles, Eqn. 14c is introduced in the simulation. To make the transformation less chaotic, $|I_2|$ and θ_2 are calculated from an average of two cycles, i.e., the previous cycle and the cycle before the previous cycle. Once i_2 is estimated, V, ϕ_0, and i_1 follow [1].

The method was validated by simulating the asymmetric system as depicted in Fig. 1 with an empty chamber, i.e., no plasma was presented inside the chamber. It was shown that an empty chamber was a good capacitor [1,7] and the asymmetric system with a blocking capacitor was modeled by pure phasor equations [1]. The currents at 5 μsec of a pure phasor model of the asymmetric system with a blocking capacitor of 200 pF were plot in Fig. 3a. The voltage source Vs was also plot as a reference. In the pure phasor model, it was assumed that only one current loop i_2 flowed in loading circuit. The currents at 5 μsec of the proposed method were plot in Fig. 3b. The simulation went into steady state in one μsec that $|I_2|$ and θ_2 did not vary at all. The currents at 5 μsec of

$$\phi_0^p + \frac{i_2(t)\Delta t}{C_{chamber}} = |B||V_s|\cos(\omega t + \theta_B) - |I_2||A'|\cos(\omega t + \theta_2 + \theta_{A'}) \quad (9)$$

If $\theta_{A'}$ is equal to zero, the current can be directly solved by,

$$i_2(t) = \frac{|V_s||B|\cos(\omega t + \theta_B) - \phi_0^p}{|A'| + \frac{\Delta t}{C_{chamber}}} \quad (10)$$

For finite $\theta_{A'}$, we rewrite Eqn. (9) as

the method by applying Gauss's laws to the blocking capacitor were also plot in Fig. 3c. Without the presence of the plasma, the current i_2 overlapped i_3. There was hardly a difference between Fig. 3a, 3b, and 3c. It was also founded that the proposed method had an error of 1% [1]. The error was calculated by dividing the average power deposited to the chamber, 4.70 watt (in principle it should be zero), with the average power input from the source, 649.2 watt. Therefore, the proposed method was validated with a 1% error and will be used to simulate the asymmetric system coupling with a plasma.

SIMULATION RESULTS WITH THE PRESENCE OF A PLASMA AND DISCUSSION

The validated proposed method was used to simulate the asymmetric system with the presence of a plasma. The cylindrical chamber was also filled up with an uniform electrical Ar$^+$ plasma of density equal 1.0×10^{16} m^{-3}. The temperature was chosen to be 3.0 eV. The simulation lasted for 20 μsec. Steady state was acquired after one μsec. The voltages at the electrode ϕ_0, V, the applied potential Vs, the currents i_1, and i_2 at 5 μsec were plotted in Fig. 4. A negative DC bias of -63.3 volt was observed at the electrode potential ϕ_0. A negative DC bias of -113.0 volt was also observed at the potential between the matching box and blocking capacitor V. It was different from the previous founding as depicted in Fig. 2a. A negative DC bias was never observed at the potential V. Moreover, a positive DC current bias of 0.47 amps was observed at the current i_2 as depicted in Fig. 4b. No bias in currents was observed as depicted in Fig. 2b although the currents i_2 and i_3 did not match each other. A big impact on the circuit voltages and currents was observed when a single current loop was assumed flowing through the blocking capacitor. In a non-linearly system, it is no guarantee that an uniform current shall flow into the system. We believe our previous method of applying Gauss's law to the layers of the blocking capacitor is a more appropriate approach. Experiment will be conducted to confirm the findings.

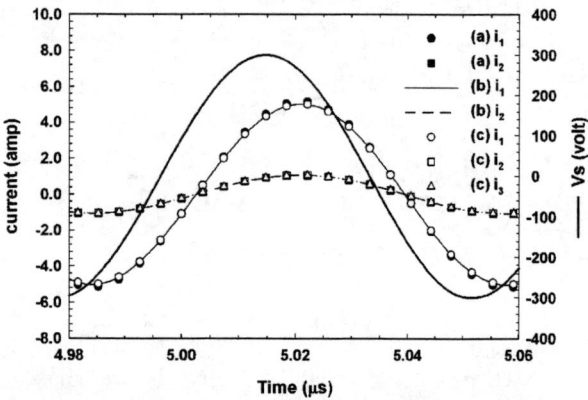

FIGURE 3. The currents i_1, i_2, and i_3 at 5 msec simulated from (a) pure phasor equations, (b) treating the blocking capacitor as a phasor element with an empty chamber, and (c) applying Gauss's law at the blocking capacitor layers, are plotted. The apply voltage Vs are plotted as reference.

CONCLUSION

Numerical simulations of treating the blocking capacitor as a phasor element such that a single current was flow through it in an asymmetric system of plasma coupling with an auto-matching network were carried out. To overcome the problem that the current $i_2(t)$ can not be solved directly with a finite phase $\theta_{A'}$ in Eqn. (9), a modified equation (11) was used. The amplitude $|I_2|$ and phase θ_2 were estimated from the previous cycle. The proposed method was validated by simulating the asymmetric system with an empty chamber. It was founded 1% error was associated with the proposed method. The modified Eqn. (11) can be used to handle quasi-matching condition when the phase θ_A of Eqn. (3) is finite. A negative bias was observed at V, the voltage between the matching box and blocking capacitor, and a positive bias was observed at i_2, the single current loop flowing through the blocking capacitor, by the proposed method. The findings are different from our previous results and experiment has to be conducted to confirm the results.

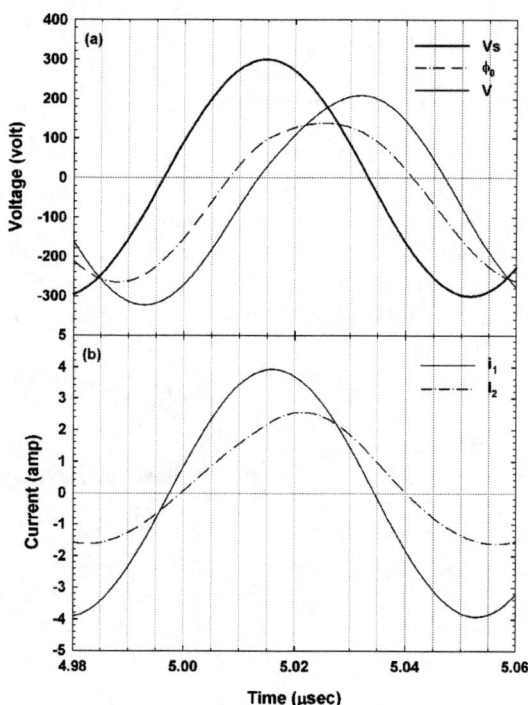

FIGURE 4. The voltages Vs, ϕ_0, and V at 5 μsec are plotted in (a). The currents i_1, and i_2 at 5 μsec are plotted in (b). The simulation was conducted by treating the blocking capacitor as a phasor element assuming identical current flowing through it.

ACKNOWLEDGMENTS

Dixon T. K. Kwok was supported by the Denison Fellowship of the School of Physics, University of Sydney. This work was supported by City University of Hong Kong Strategic Research Grant (SRG) #7001981. The authors appreciated the helpful discussion from Dr. Yongbai Yin.

REFERENCES

1. D. T. K. Kwok, Journal of Applied Physics.
2. D. T. K. Kwok, Y. Yongbai, M. M. M. Bilek, and D. McKenzie, Applied Physics Letters **87**, 181501-1-3 (2005).
3. R. C. Powles, D. T. K. Kwok, D. R. McKenzie, and M. M. M. Bilek, Physics of Plasmas **12**, 93507-1-6 (2005).
4. M. Chandhok and J. W. Grizzle, IEEE Transactions on Plasma Science **26**, 181-9 (1998).
5. L. P. Huelsman, *Basic Circuit Theory*, 2nd ed. (Prentice-Hall, Englewood Cliffs, N.J., 1984).
6. E. Kawamura, V. Vahedi, M. A. Lieberman, and C. K. Birdsall, Plasma Sources Science & Technology **8**, R45-64 (1999).
7. J. P. Verboncoeur, M. V. Alves, V. Vahedi, and C. K. Birdsall, Journal of Computational Physics **104**, 321-8 (1993).

B_2H_6 PLAD Doped PMOS Device Performance

Z. Fang, T. Miller, E. Winder, H. Persing, E. Arevalo,
A. Gupta, T. Parrill, and V. Singh

Varian Semiconductor Equipment Associates Inc., 35 Dory Road, Gloucester, MA 01930, USA

S. Qin and A. McTeer

Micron Technology Inc., 8000 S. Federal Way, Boise, ID 83707, USA

Abstract. Plasma doping (PLAD) achieves high wafer throughput by directly extracting ions across the plasma sheath. PLAD profiles are typically surface peaked instead of retrograde as obtained from beamline (BL) implant. It may require optimization of PLAD energy and dose in order to match BL doping results. From device optimization point of view, it is necessary to understand the impact of doping parameters to device characteristics. In this paper we present the PMOS device performance with the poly gate and source drain (SD) implants carried out using B_2H_6 PLAD. The BL control conditions are 2-5 keV $^{11}B^+$ 4-6×10^{15} cm^{-2}. Equivalent device performance for p$^+$ poly gate doping is obtained using PLAD with B_2H_6 / H_2. In SD doping using same gas mixture, nearly 50% reduction in SD contact resistance is observed in the PLAD splits. The reduction in SD contact resistance leads to 10-15% increase in device on-current, hence demonstrating the process advantages of using PLAD in addition to having a high wafer throughput.

Keywords: Plasma doping, PMOS, process optimization.
PACS: 85.40.Ry, 52.77.Dq, 85.30.De

INTRODUCTION

The fabrication of advanced CMOS device calls for production worthy solutions for low-energy high-dose applications. Doping of p$^+$ poly gate and source drain (SD) faces significant throughput challenge. It has been demonstrated [1] that the productivity of a beamline (BL) tool drops quickly at low energies due to beam transportation difficulties. Efforts to improve BL productivity include the use of molecular ion sources ("cluster beams") [2-3], although these do not ultimately offer the high throughput and species flexibility of plasma doping (PLAD) [4]. PLAD completely eliminates beam transportation limitations so it is able to achieve high throughput regardless of wafer size [4-9]. As a result, it is viewed as an attractive doping solution for device manufacturing.

Process matching between different BL tools is a common practice in a manufacturing environment where the goal is simply to achieve a good match in beam species, energy, dose, and incident angle distribution. With PLAD, however, differences in co-implanted species, energy distribution, and damage [7] must be accounted for in the device matching process. Both post-implant cleaning and thermal budgets must usually be included, followed by evaluations at the device level to establish an optimum matching condition. This approach ensures that the resultant PLAD recipe is a complete production solution, producing the desired device characteristics and compatible with the integrated process flow. At the same time, working at the device level also allows one to take advantage of PLAD's throughput flexibility and unique doping features to significantly improve device performance.

In this paper we present the steps taken to match PLAD with BL for an advanced PMOS process. The utility of this procedure is proven by both dopant profiles and PMOS device characteristics.

EXPERIMENT

The experimental procedure to match a doping process starts with tests carried out on blanket wafers. For p$^+$ poly gate doping, the gate stack comprises of 700 Å undoped poly-Si / 60 Å nitride / Si substrate.

For p+ SD doping, bare n-type wafers are used. PLAD implant parameters are selected from the bare wafer tests to determine suitable conditions for device wafer tests. The device structure is a standard self-aligned poly gate PMOS transistor with a channel W/L of 11.7μm/ 0.078μm, and gate oxide thickness is 50Å. MOS capacitor structures are used for CV test. Standard post-implant strip, clean, and rapid thermal annealing process (RTP) are used to activate the dopants. The RTP condition is 965°C for 20s in N_2 ambient. The BL control for poly gate doping is 4.5keV 6×10^{15}/cm^2 $^{11}B^+$, and for SD doping it is 2keV 4.5×10^{15}/cm^2 $^{11}B^+$, at 0° incident angle for both. All PLAD implants used 15% B_2H_6 / 85% H_2 gas mixture to generate ions that are accelerated into substrates biased at negative potential using a pulsed DC power supply. The SD contact is formed by a Ti/W-base deposition. When the poly gate is doped by PLAD, the SD is doped by BL, and vice versa, to facilitate PLAD doping characterization in each case.

There are more parameters other than energy and dose in PLAD that require careful consideration. For example, the DC wafer bias pulse width can affect throughput as well as gate oxide integrity [8]. Dilution ratio of B_2H_6 in H_2 can affect surface deposition as well as implanted boron profile characteristics. A mixture of hydride and fluoride such as B_2H_6 and BF_3 can be used to reduce or even eliminate deposition. Detailed discussion on these parameters is outside the scope of this paper. PLAD parameter optimization was completed prior to blanket wafer test.

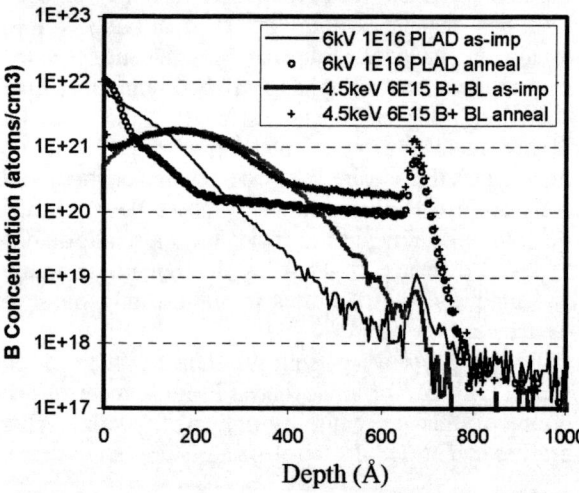

FIGURE 1. Comparison of SIMS profiles in p+ poly gate doping. Peaks around 700 Å mark the poly-oxide interface.

RESULT AND DISCUSSION

In Fig. 1 we show the as-implanted and annealed SIMS profiles in poly Si for both BL control and PLAD. It is clear that the 6kV B_2H_6 PLAD profile is shallower and surface peaked. The profile tail cannot be moved deeper by increasing PLAD energy because boron penetration to the gate oxide would cause undesired V_T shift. For the same reason, no or very little energy contamination is allowed in the BL implant for this application.

One of the goals for poly gate doping is to provide adequate dopant activation at the poly-oxide interface. Under the same thermal budget, the shallower PLAD profile means less B atoms diffused to the poly-oxide interface, causing lower dopant activation or higher poly depletion at the interface. Dopant activation is improved at higher PLAD dose. Electrical tests such as Rs, SRP, and Hall effect (not shown) on blanket wafers suggest that beyond 1×10^{16}/cm^2 dose, there is little difference between PLAD and BL doped samples. However these tests are not sufficient to ensure that equivalent poly depletion level is reached.

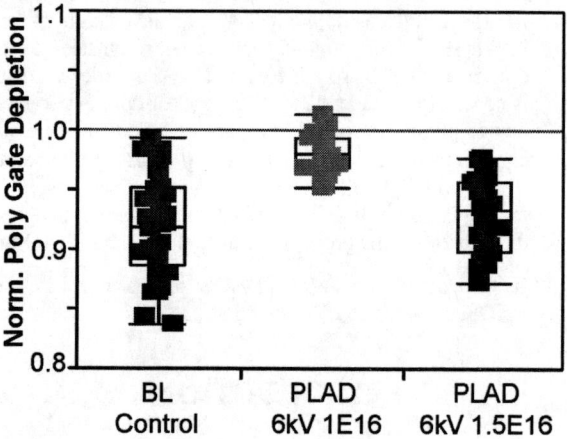

FIGURE 2. Poly gate depletion effect as measured by CV test for two PLAD doses.

In Fig. 2, we compare poly depletion for two PLAD splits with different doses. Even though Rs is saturated above 1×10^{16}/cm^2, it can be seen that PLAD dose increase to 1.5×10^{16}/cm^2 reduces poly gate depletion to within 2% of BL control. It demonstrates that device wafer test is the best approach to determine how much PLAD dose is needed to achieve same device performance.

Despite the small increase in poly depletion, PLAD reduces the device poly line resistance by 2-3%. It is likely that the PLAD surface-peaked profile accounts for these observations. In Fig. 3 we compare V_T and I_{DS}. The overlap and linear fitting of the data indicates that there is no significant performance difference between PLAD and BL poly-doped devices. The selected PLAD conditions also did not induce gate oxide damage due to charging, which agrees with previously published Q_{BD} data [7]. Finally, the tested device parameters are summarized in Table 1 with the corresponding PLAD vs. BL comparison results.

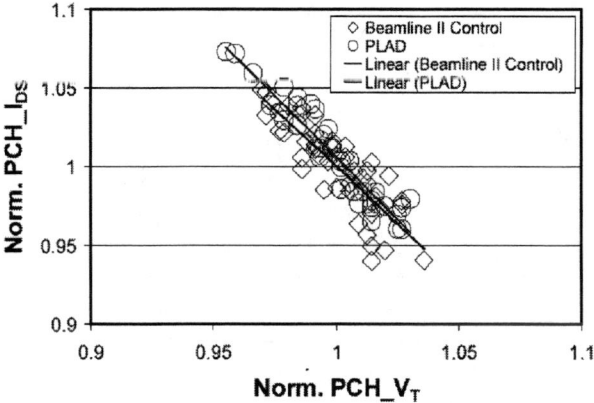

FIGURE 3. PMOS drive current I_{DS} against threshold voltage V_T after p+ poly gate doping.

TABLE 1. Device Performance Comparison.

Doping Step	Device Parameter	Compared to BL PLAD Is
p+ poly gate doping	Poly gate depletion	less than 2% higher
	Poly line resistance	2-3% lower
	Threshold voltage V_T	statistically equal
	Sub-threshold slope	statistically equal
	Drive current I_{DS}	less than 2% higher
	SD breakdown voltage BV_{SD}	statistically equal
	Gate oxide V_{BD}	statistically equal
	Gate oxide Q_{BD}	statistically equal
PMOS SD doping	SD contact resistance	~50% lower
	SD series resistance	2-3% higher
	Drive current I_{DS}	15-20% higher
	Threshold voltage V_T	statistically equal
	Sub-threshold slope	statistically equal
	SD breakdown voltage BV_{SD}	statistically equal

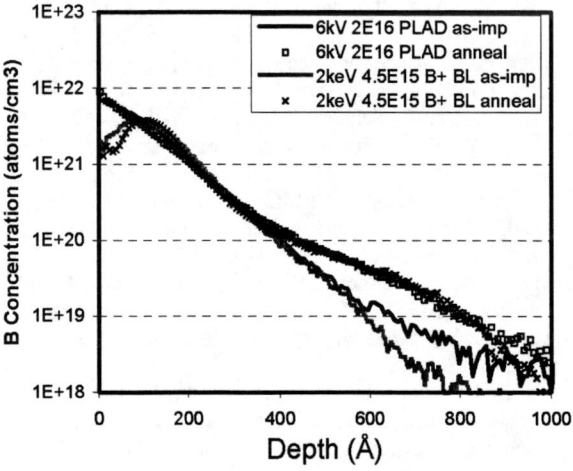

FIGURE 4. Comparison of SIMS profiles in c-Si for SD doping conditions by PLAD and BL control.

The device parameters tested for PMOS SD doping are also summarized in Table 1. In Fig. 4 we show the as-implanted and annealed SD profiles in c-Si for BL and PLAD doping conditions. Although the as-implanted PLAD profile is slightly deeper than BL control, a very good match is achieved after anneal. This close match is a result of optimization within PLAD bias voltage and dose range 4-8kV and 0.5-$2 \times 10^{16}/cm^2$ respectively. At first glance, it is surprising that a PLAD voltage of 6kV matched both the SD and poly processes, since the BL control process was 2keV vs. 4.5keV. We suggest that this result is associated with differences in channeling into crystalline Si.

In B_2H_6 PLAD, there are more than one ion species being implanted. Under the same PLAD bias, light ions go deeper but their contribution to profile tail is limited due to their relatively low concentration. The majority ions in plasma are $B_2H_x^+$ and they made much less contribution to profile tail even with channeling. Meanwhile, the molecular species $B_2H_x^+$ tends to cause more crystal damage [7] and therefore reduces channeling for all ion species. In addition, deposition of surface neutrals during PLAD creates a thin layer that tends to de-channel accelerated dopant species. It is possible that one or more such mechanisms leads to less channeling of the PLAD process that must be compensated by increasing the bias voltage.

The origin of the surface peaked profile in PLAD has been explained elsewhere [9]. We shall focus on its impact on device performance when the SD region is doped with such a profile. As indicated in Fig. 4, the close match in annealed profile would suggest similar junction characteristics, which is confirmed by Rs, SRP, and Hall measurements. According to Table 1, the SD series resistance measured on device wafers has only 2-3% increase for PLAD. However, the SD contact resistance has dropped ~50% for PLAD as indicated in Fig. 5. This significant improvement in

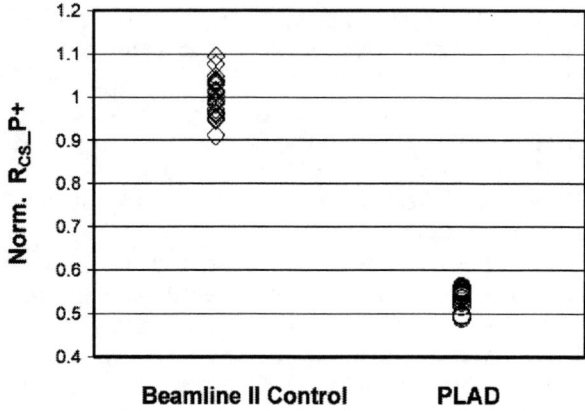

FIGURE 5. Comparison of SD contact resistance after SD region doped by PLAD and BL control.

contact resistance is attributed to the surface peaked profile. It is known that contact resistance is sensitive to contact material type as well as to the carrier properties near the contact region. Higher surface B concentration leads to higher carrier concentration for the same anneal, therefore a surface peaked profile is advantageous in contact resistance reduction.

The lower contact resistance is translated into more drive current as indicated in Table 1. The SD drive current I_{DS} is increased by 15-20% despite a 2-3% increase in SD series resistance. Although the PLAD bias is 6kV compared to BL control of 2keV, there is no detectable V_T and sub-V_T slope change. This is a result of successful device parametric tuning with a careful control of lateral dopant profile.

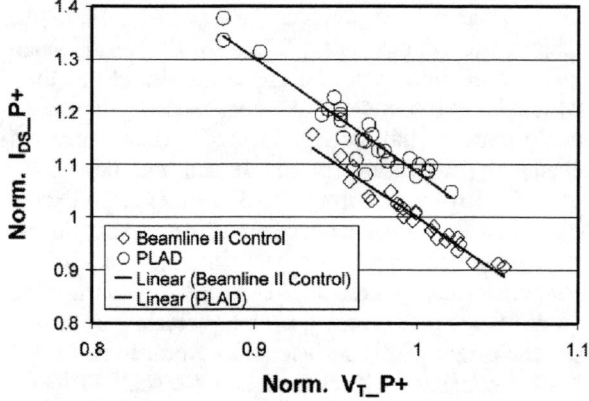

FIGURE 6. PMOS drive current I_{DS} against threshold voltage V_T for SD doping.

The device performance boost with V_T aligned can be seen clearly in Fig. 6. The linear fitting to I_{DS} - V_T plots suggests that there is 10-15% increase in I_{DS} for PLAD doped devices, primary due to reduction in SD contact resistance. This serves as a good example of taking advantage of new features from alternative doping technology. As pointed earlier, there are other PLAD parameters that can be further optimized for better results.

The SD breakdown voltage BV_{SD} is a test of junction quality after forming the SD region. In Fig. 7 we compare BV_{SD} for PLAD and BL doped devices. There is no difference found in BV_{SD} as indicated in Table 1 as well as in Fig. 7.

In conclusion, we have demonstrated that PMOS devices with its poly gate doped by PLAD perform equally well as those doped by BL implantation. For devices with SD region doped by PLAD, significant device performance enhancement is observed, which is attributed to the surface peaked profile from PLAD.

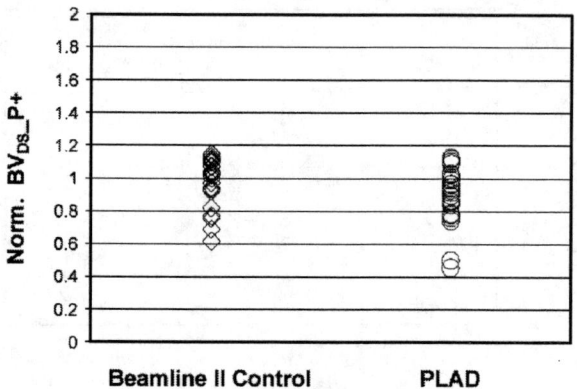

FIGURE 7. Comparison of SD breakdown voltage after SD region doped by PLAD and BL control.

ACKNOWLEDGMENTS

The authors would like to take this opportunity to thank the PLAD technical team, Steve Grimmett and Steve Romero for their great support.

REFERENCES

1. A. Renau, D. Downey, "Review of low energy doping and activation technologies", Semi Forum Japan, June 2003.
2. D. Jacobson, T. Horsky, W. Krull and K. Cook, "Cluster Boron™: A New Doping Material for P-Type Ultra-Shallow Junctions," Int. Conf. on Ion Implant. Tech., 2004, p126.
3. J. Borland et al., Solid State Technology, May 2004, p.53.
4. R. Liebert, S. Walther, S. Felch, Z. Fang, B. Pedersen, O. Pedersen, D. Hacker, "Plasma Doping System for 200 and 300mm Wafers", Int. Conf. on Ion Implant. Tech., 2000, p.472.
5. S.R. Walther, S. Mehta, U. Jeong, and D. Lenoble, "Formation of extremely shallow junctions for sub-90nm devices," Surface and Coatings Technology, Vol. 186, No. 1-2, 2 August 2004, p. 68.
6. A. Renau and J. Scheuer, "Comparison of Plasma Doping and Beamline Technologies for Low Energy Ion Implantation", Int. Conf. on Ion Implant. Tech., 2002.
7. Z. Fang, E. Arevalo, T. Miller, H. Persing, E. Winder, and V. Singh, Ext. Abstr. 5th International Workshop on Junction Technology, June 2005, p. 67.
8. D. Henke, S. Walther, J. Weeman, T. Dirnecker, A. Ruf, A. Beyer, and K. Lee, "Characterization of Charging Damage in Plasma Doping", Int. Conf. on Ion Implant. Tech., 2002.
9. L. Godet, Z. Fang, S. Radovanov, S. Walther, E. Arevalo, F. Lallement, J.T. Scheuer, T. Miller, D. Lenoble, G. Cartry, and C. Cardinaud, JVST B, to be published.

Plasma Immersion Ion Implantation applied to P+N junction solar cells

Vanessa Vervisch [1,2], D. Barakel [1], F. Torregrosa [2], L. Ottaviani [1] and M. Pasquinelli [1]

(1) UMR TECSEN, Université Paul Cezanne Aix Marseille III, 13397 Marseille cedex 20, France
(2) Ion Beam Services, Av Gaston Imbert prolongée, 13790, ZI Rousset, France

Abstract: Plasma immersion ion implantation is an alternative doping technique for the formation of Ultra Shallow Junctions in semiconductor. In this study, we present the PIII technology developed by the company Ion Beam Services and called PULSION®. We explain the advantages of PIII for the conception of thin emitter solar cells and the use of N type silicon in the fabrication of photodiode. Electrical characterisations of solar cells prepared by immersion of silicon wafer in BF_3 plasma are presented, showing a satisfying photovoltaic behaviour and more specially an increase of internal quantum efficiency in the short wavelength range, due to the thickness of the emitter.

Keyword: Plasma immersion ion implantation, shallow junction, N-type Silicon, Solar cells.
PACS: 52.77.Dq ; 85.40.-e

INTRODUCTION

Among different existing doping techniques, beam line implanter has a predominant place in semiconductor's industry. But due to the decrease of transistor size and so the decrease of source/drain extension, new ways of doping techniques are being studied, among them Plasma immersion ion implantation offers possibility to reach ultra low acceleration energies [2-7]. A PIII prototype called PULSION®, developed by Ion Beam Services Company, is described in Fig.1. [8-9]

The PIII technique is constituted by a vacuum chamber containing a chuck, a plasma source and a power supply.
A specific voltage permits the formation of a sheath around the wafer. Then, doping ions are projected on the sample and implanted in the material at depth varying from some nm to several hundreds of nm.
This process allows a high dose implantation of doping species during a short time, which promises to be interesting for low cost solar cells. Moreover, this technique gives the advantage of processing a large area and ensuring a great productivity.

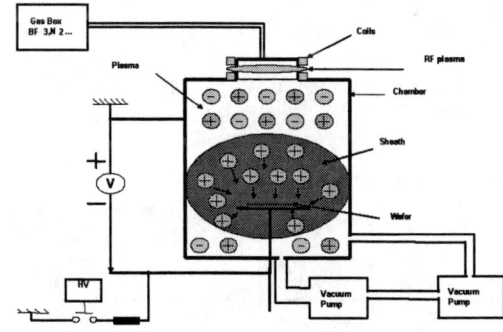

FIGURE.1: Schema of Plasma immersion ion implanter

Time-of-Flight Secondary Ion Mass Spectrometry (ToFSIMS) measurements have been realised in order to analyse the metallic contaminations. Fig.2 presents the level of contamination for Plasma immersion ion implantation by PULSION® in comparison with the limits given by ITRS2005. Even if Ni and Cu concentrations are slightly high for PULSION®, the global contamination level remains acceptable for semiconductor applications like solar cells.

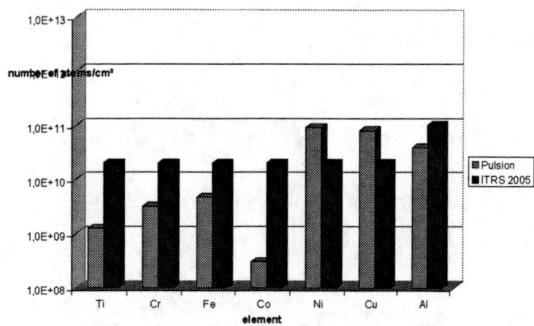

Fig.2: ToFSIMS Measurements: Metallic contaminations

Why N-type for Photovoltaic industry?

Former studies of ultra shallow junctions in N-type silicon with BF_3 PIII convinced us to study the fabrication of P-type shallow emitter photodiodes.
In a previous work [10], a comparison between N^+PP^+ and P^+NN^+ solar cell structures was elaborated by means of PC1D4 software simulations.

The simulated internal quantum efficiency of solar cells revealed to be 1% higher with P^+NN^+ structures.
Besides, the simulations pointed out the reduction of emitter depth results in an increase of sensitivity in the blue range.

From a material point of view, in P-type (boron-doped silicon) the boron-oxygen complexes formed after a prolonged exposure of solar cells become minority carrier recombination centers.
This problem is avoided with the use of N-Type phosphorous-doped silicon.
S.Martinuzzi and al. have demonstrated that in N-type silicon the capture cross sections of metallic impurities are neatly smaller than in P-type [11]. Table 1 illustrates this matter.

Impurity	E_t (eV)	S_n (cm^{-2}) P-TYPE	S_p (cm^{-2}) N-TYPE
Fe_i	+ 0.38	$5\ 10^{-14}$	$7\ 10^{-17}$
Ti_i	+ 0.27	10^{-14}	$3\ 10^{-17}$
Mo_i	+ 0.28	$1.6\ 10^{-14}$	$6\ 10^{-16}$
Cr_i	- 0.22	$2\ 10^{-13}$	$8\ 10^{-14}$
Mn_i	+ 0.27	$3\ 10^{-15}$	$2\ 10^{-18}$
V_i	+ 0.2	10^{-16}	$2\ 10^{-18}$
Co_s	+ 0.41	$2\ 10^{-15}$	$5\ 10^{-18}$

Table 1: Capture cross sections of electrons and holes for some metallic impurities

Besides, there is about $2\ 10^6$ kg/year of N-type wastes which could be used for low cost solar cells.
The goal of this work is to demonstrate the interest of solar cells based on thin and highly B-doped emitter.

Their sensitivity to short wavelengths prove that such cells can be used for space applications.

EXPERIMENTAL

All the wafers used in this study are N-type Cz (100) and N-type poly-crystalline silicon substrates with a resistivity of 5-10 ohm.cm. First, a LYDOP diffusion of $POCl_3$ during 20 minutes at 850°C on the back side of the wafers is realised in order to optimise the ohmic contact. The emitter is created by BF_3 PULSION® process: three different acceleration voltage are compared (1, 5 and 10 kV) as well as two doses (5e15 and 1e16 at/cm^2). Boron is activated by an annealing at 900°C during 30 minutes under N_2H_2. A chemical evaporation of aluminium is then processed on both sides of the wafer: for the collected grid on the P^+ emitter and for the ohmic contact on the rear side (followed by an annealing).
The complete structure is shown in Fig.3

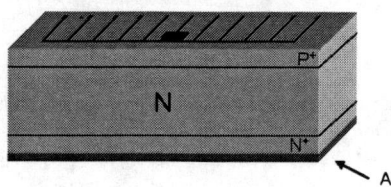

Fig.3 Schema of a solar cell

RESULTS AND DISCUSSION

After the BF_3 plasma implantation and the activation annealing, wafer sheet resistance is measured to define the doping activation and the conductivity type of the layer. A Cascade four points probe coupled with Keithley precision multi-meter were used.
Table 2 gives the corresponding results.

Acceleration voltage (kV)	1	5	10
Sheet Resistance (Ω/sq)	456	294	262

Table 2: Sheet resistance measurements

The Spreading Resistance Profile (SRP) obtained with an N-Type Cz (100) sample implanted by PULSION® at an acceleration voltage of 10 kV and a dose of 5×10^{15} at/cm^2 is shown in Fig.4.
At such energy, junction depth seems to be located at around 150 nm from the surface. Moreover, we observe that the maximum concentration measured is about 2E19 cm-3. This result is surprising as it is lower than expected. There is only partial diffusion of boron atoms probably due to the annealing, with a target temperature superior at the real temperature. Others analyses are in hand.
Secondary Ion Mass Spectrometry (SIMS) profiles of PULSION® BF_3 5 kV 1e15 at.cm^{-2} are presented in Fig.5. They compare the as-implanted boron profile, with the one obtained after a classical annealing at 950°C during 30 minutes. Few diffusion is observed compared with the theoretically calculated diffusion length:

$$L_{diff} = \sqrt{Dt} = 0.02\ \mu m$$

As we can expect, the junction depth is lower (75 nm@1E18 cm^{-3}) at 5 kV than at 10 kV.

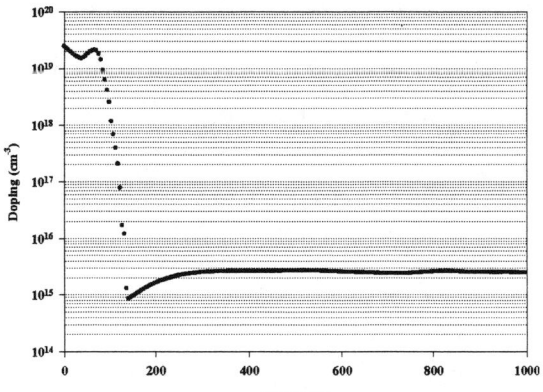

FIGURE.4: SRP profile: PULSION® BF$_3$ 10kV and annealed sample

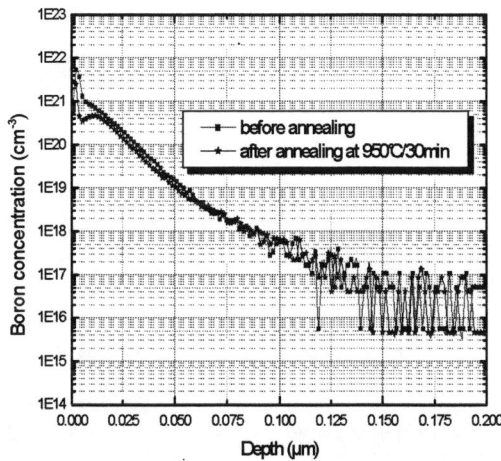

FIGURE 5: SIMS profiles PULSION® BF$_3$ 5kV

Electrical characterisations like I-V measurements under sunlight and spectral sensitivity inform us about photovoltaic properties of our different photodiodes.

Fig. 6 compares the internal quantum efficiency of different photodiodes versus the incident wavelength.

Solar cells realised by PIII show better internal quantum efficiency than a conventional Fz solar cell in the short wavelength i.e. in the range 400-700 nm. This result is due to a shallower emitter obtained by PIII (150 nm at 10 kV) than in a conventional one which is around 300 nm. Table 3 presents some electrical properties under sunlight of PIII solar cells.

We can see that minority carrier diffusion length value is about 250 µm, which indicates a rather low metallic contamination.

Electrical Properties	1 kV	5 kV	10 kV
Voc(mV)	477	590	568
Jsc (mA/cm^2)	16.07	23.4	29
Length Diffusion(µm)	261	232	210
Sheet Resistance(Ω)	1.02	0.53	0.4
Shunt Resistance(Ω)	20.58	564	1454

Table 3: Electrical Properties under sunlight of PIII solar cells.

In the case of shunt resistance, 5 kV and 10 kV solar cells have acceptable values. Only 1 kV solar cell has a low shunt resistance but this is probably due to process contamination of the sample edge.

Voc and Jsc values reveal to be slightly lower than those expected. However we can easily optimise the process by passivating the surface and improving the front side metal grid, with an adjunction of a proper anti reflective layer. This should also result in a decrease of the sheet resistance and an increase of the shunt resistance, leading to a better behaviour of the solar cells.

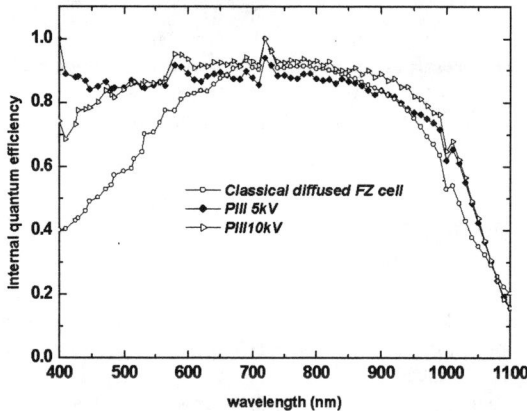

FIGURE 6: Internal quantum efficiency results obtained on PULSION® solar cells

Light Beam Induced Current (LBIC) scan map is presented in Fig.7a. Fig 7b gives a map of minority carrier diffusion lengths L$_{diff}$, extracted from LBIC results. We can see that even if the global value is around 250 µm, we obtain more than 300 µm (corresponding to the thickness of the samples) in many of grains.

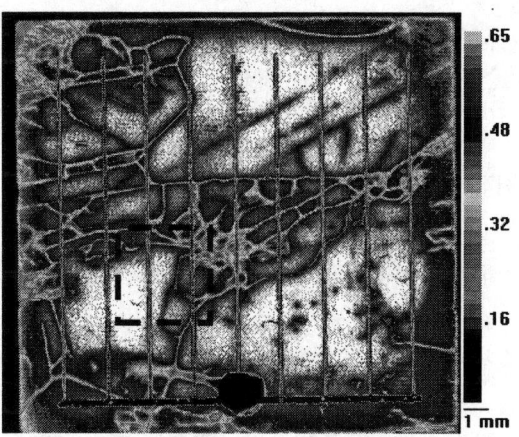

Fig. 7a PULSION® solar cell LBIC scan map (white light)

255

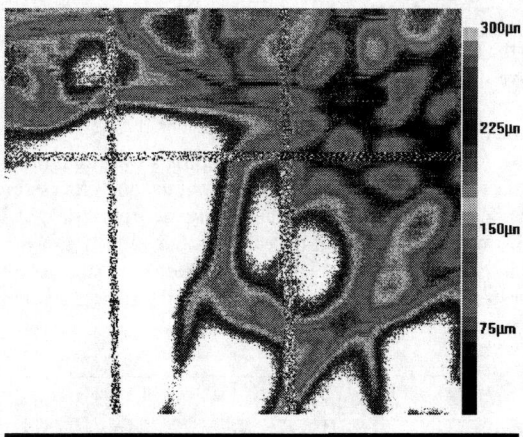

Fig 7b PULSION® solar cell map of L_{diff} (zoom)

SUMMARY

Solar cells have been realised by plasma immersion ion implantation at different acceleration voltage on N-type poly crystalline silicon. SIMS profiles and SRP measurements have shown that PIII prototype, PULSION®, was able to realise shallow junctions and more specifically thin emitters for a photovoltaic application. The electrical properties obtained show a satisfying photovoltaic behaviour.

The spectral variation of the internal quantum efficiency is practically constant in the short wavelength, due to the thickness of the emitter. LBIC scan maps have shown that grains present very good properties.

In conclusion, thanks to the lower cost of the equipment, and thanks the fact that implant time is independent from the surface, PIII technique seems to be a good candidate for the realisation of low cost solar cells.

ACKNOWLEDGMENTS

The authors would like to thank Dr W. Vervisch of LMP-CNRS-Tours laboratory for performing SRP profiles.

REFERENCES

1. The international Technology Roadmap for Semiconductors. 2005.
2. P.K. Chu et al, Mater.Sci. Eng. R 17 (1996) 207
3. P.K. Chu et al, Solid State technol (1999), october.
4. F. Le Coeur et al, Surface and coating Technology 93 (1997) 265-268
5. D. Lenoble and A. Grouillet, Surface and Coating Technology 156 (2002) 262-266
6. S.B .Felch et al, Surface and coating Technology 156 (2002) 229-236
7. D Lenoble Thesis, Etude, realisation et integration de jonctions P+N ultra fines pour les technologies CMOS inferieures à 0,18μm. Soutenue en 2000
8. F. Torregrosa et al. Surface and coating Technology 186 (2004) 93-98
9. F. Torregrosa et al. Nuclear Instruments and Methods in Physics research B 237(2005) 18-24
10. D. Barakel thesis, Implantations d'ions H^+ et BF_2^+ dans du Silicium par faisceaux d'ion et immersion plasma. Soutenue en 2004
11. S. Martinuzzi et al, IEEE 0-7803-8707-4/05 (2005)

Ion behaviour in pulsed plasma regime by means of Time-resolved energy mass spectroscopy (TREMS) applied to an industrial radiofrequency Plasma Immersion Ion Implanter PULSION®

M. Carrere*, F. Torregrosa**, V. Kaeppelin*

*Equipe Plasma Surface, Laboratoire PIIM, UMR CNRS 6633, Université de Provence,
case 241,13397 Marseille cedex 20, France
**ION BEAM SERVICES, ZI Peynier-Rousset, rue Gaston Imbert Prolongée, 13790 Peynier, France

Abstract. In order to face the requirements for P+/N junctions requested for < 45 nm ITRS nodes, new doping techniques are studied. Among them Plasma Immersion Ion Implantation (PIII) has been largely studied. IBS has designed and developed its own PIII machine named PULSION®. This machine is using a pulsed plasma. As other modern technological applications of low pressure plasma, PULSION® needs a precise control over plasma parameters in order to optimise process characteristics. In order to improve pulsed plasma discharge devoted to PIII, a nitrogen pulsed plasma has been studied in the inductively coupled plasma (ICP) of PULSION® and an argon pulsed plasma has been studied in the helicon discharge of the laboratory reactor of LPIIM (PHYSIS).Measurements of the Ion Energy Distribution Function (IEDF) with EQP300 (Hidden) have been performed in both pulsed plasma. This study has been done for different energies which allow to reconstruct the IEDF resolved in time (TREMS). By comparing these results, we found that the beginning of the plasma pulse, named ignition, exhaust at least three phases, or more. All these results allowed us to explain plasma dynamics during the pulse while observing transitions between capacitive and inductive coupling. This study leads in a better understanding of changes in discharge parameters as plasma potential, electron temperature, ion density.

Keywords: Plasma Immersion Ion Implantation (PIII), Plasma diagnostics, ICP and helicon discharge..
PACS: 52.77.Dq ; 85.40.-e

INTRODUCTION

High density, low pressures plasmas are widely used, everywhere in research laboratories or in the industrial world for material processing applications as Plasma Enhanced Chemical Vapour Deposition (PECVD), Reactive Ion Etching (RIE), Plasma Implantation Ion Immersion (PIII) or sputtering. All these processes are based on the good control of the plasma properties. By plasma properties we mean the plasma potential (V_p), the ion density (n_i) and the electronic temperature (Te) but also the homogeneity (profiles). The size of the wafer increases in the same range as the size of the components decreases. Therefore the engineer has to increase the size of the active plasma. Moreover the improvement of the plasma quality in terms of density and temperature but also homogeneity should lead to a faster industrial treatment over a large wafer. Furthermore the size of the patterns on the wafer decrease and led to the use of high density plasma. These high densities could be obtained by pulsing the injected power on the antenna. One has to control the process during a plasma pulse, which is divided in three phases: the ignition or the breakdown, the stationary state and the post-discharge. In this article we focus on the breakdown phase.

In order to study high density plasma we design a helicon source baptised: PHISIS [2]. Then the methodoly was applied to the industrial reactor [3] named Pulsion® designed by Ion Beam Services (IBS). Firstly we present briefly the two reactors. Secondly we present the ignition behaviour on "PHISIS" and then measurements on Pulsion® and thirdly our conclusion.

EXPERIMENTAL APPARATUS

PHISIS is a classical helicon reactor (cf fig. 1), which consists of an upper source chamber and a lower diffusion chamber. The source has been designed by ANU; it is a 150 mm diameter and 200 mm long Pyrex cylinder, surrounded by a double loops antenna (Boswell antenna). The matching box is a (L type) with a 0-2000 pF tune capacitor and a 0-1000 pF load capacitor and is controlled manually. Our radio-frequency generator (13.56 MHz) can deliver up to 3 kW power and is able to operate in continuous or pulsed regimes. The main pulsed regime that we used is 10% duty cycle with the plasma ON during 2 ms. The total volume of the reactor is around 10 litres. Three coils allow to obtain an axial magnetic field of 0-150 G in front of the antenna. The magnetic field is applied to improve the plasma density in the chamber and to allow the propagation of whistler waves (helicon). A base pressure better than 10^{-7} mBar (Penning gauge limit) is achieved in the chamber (10 l) by a 150 l/s turbomolecular pump. The stainless steel chamber (spherical vessel 200 mm diameter) could be baked over 250°C. During plasma operation the pressure is controlled by a Baratron gauge. We use two diagnostics : Langmuir's probes (scientific systems) and mass spectroscopy (EQP 300).

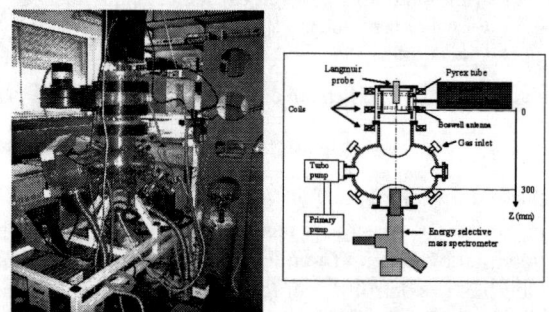

FIGURE 1. Picture and structure of PHISIS Helicon reactor

The structure PULSION® is described in figure 2

The pulsed plasma is created thanks to a self designed ICP source located above a vacuum chamber. Wafer is located on a chuck at the bottom of the chamber. The chuck is polarized through a capacitor and thanks to a current source (High voltage capacitor charging system). To improve uniformity on large substrates, the substrate holder rotates under the non centered plasma source.

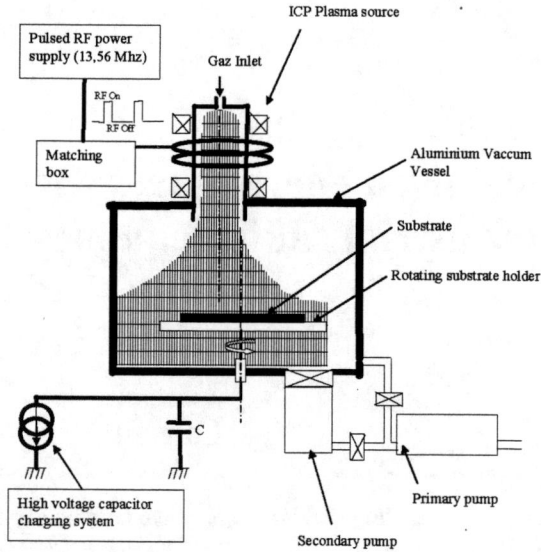

FIGURE 2. Machine structure of PULSION®

MEASUREMENTS ON PHISIS

We measure different ion mass signals for given energy show on figure 3. Roughly speaking at a first glance we could distinguished the classical three phases: breakdown, stationary and post-discharge. But very narrow peaks are clearly present during the breakdown phase. As far as we known we are inspired by the article of Boswell and Vender 1995 [1] in order to explain this peak. We also perform a reconstruction of the Ion Energy Distribution Function (IEDF) versus time but we will not show these results in this paper.

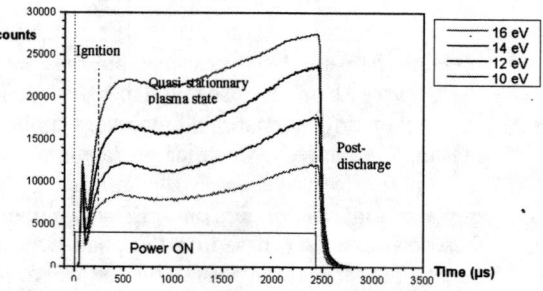

FIGURE 3. four signals of Ar+ at different energies versus time

The initial step is defined by the small quantity of free electrons. These electrons are created continuously by the cosmic and gamma rays. The multipactor effect at low pressure consists in accelerate electrons with the RF electric field and give them enough energy to generate secondary electron emission by striking inner surfaces. In fact a cloud of electrons oscillates with the RF electric field. Hatch and William have shown that after several half periods (37 ns each), an electron cloud average energy increases up to 5 KV for a 100 VRF applied to a capacitive antenna. Boswell [1] suggests that in his helicon the initial average energy of electron cloud is around 1000 eV. In our case the initial step and the multipactor effect lasts 40 µs. In fact as we measure ion signal, ions arrived in the channeltron of the detector 40 µs after the beginning of the injected power.

The second step as it is shown on the figure 4, is called high energy ions. The ionisation rate is very high at the beginning of the ion creation due to the very high electron temperature. Moreover the high energy of the electron gas is slowly reduced by multiplying the number of ions. We measure ions up to 100eV but they could have much more [1]. This second phase ends 55 to 60 ms after the beginning of the injected power. Then capacitive coupling is activated, which means the coupling (E as electrical) of the helicon antenna is given by the acceleration of the electron in the RF sheath [3]. Generally this phase is followed by a transition between capacitive to inductive regime (H for helical) and sometimes at high injected power (more than 1.5 KW) to helicon regime also called wave sustained mode (W).

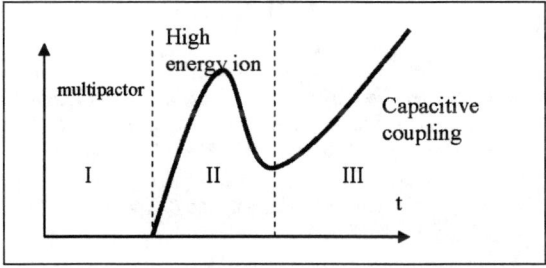

FIGURE 4. ion signal for a given energy exhibits three phases during the breakdown of a pulsed plasma

MEASUREMENTS WITH THE INDUSTRIAL PIII REACTOR (PULSION®)

The measurements have been made with the same mass spectrometer (EQP 300) in the centre of the expanding chamber. As the figure 5 shows the breakdown does not exhibit an electron phase and a high energy ion phase. This is mainly due to the fact that a small quantity of injected power is releases between two pulses. This small quantity (around few Watts) of injected power maintains a very low density plasma during the off phase and then suppress the two primary phases. Without High energy electrons the high energy ions could not be created. So the coupling between antenna and plasma is directly a capacitive coupling. This injection of a small power between pulses could be an advantage because it allows to create a pulsed plasma at a very low pressure.

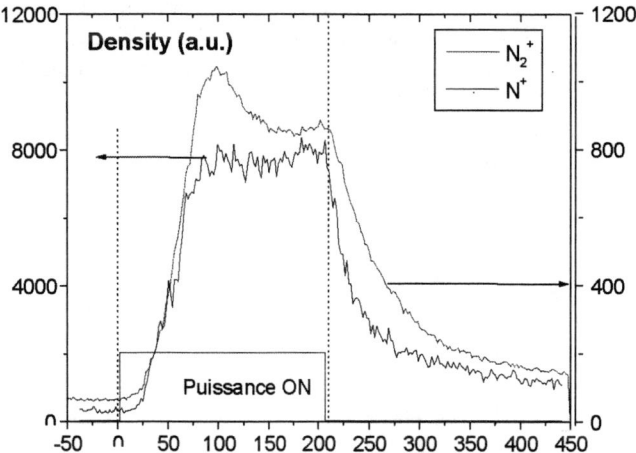

FIGURE 5. Ion signal for two mass ratio during one plasma pulse in Pulsion®.

We have characterized in a previous paper [5] the capacitive, inductive and helicon coupling for continuous plasma. In the expanding chamber the capacitive coupling exhibits a high plasma potential and a lower ion density than the inductive. In fact a capacitive coupling gives a flat profile for plasma potential from the source through the expanding chamber and a decrease of the ion density. In the opposite inductive coupling induces a flat ion density profile and a decrease in plasma potential profile. Figure 6 shows transition between a capacitive coupling (E) and an inductive coupling (H).

The peak on the ion signal is not the signature of high density ions but the balance between density and plasma potential in the expanding chamber. Plasma potential goes from 55 Volts to 28 volts and the ion density goes from 10^8 to few 10^9. During this transition the balance in the ion density is also changed. In nitrogen, two kinds of ions are produced N_2^+ and N^+. The N_2^+ production is easier than N^+. The dissociation energy of the molecule is needed to produce N^+ add to ionisation energy. We present on figure 7 the ratio between N^+/N_2^+ which oscillate from 5 to 9 percents. Let focus on the black curve which represent percentage of N_2^+, then two peaks are clearly visible. The first one is induced by the electron temperature during low density and capacitive regime, and the second is induced by the high density in inductive regime. This second peak ends by switching off of the power supply. We explain this relaxation effect in a previous paper [2]. The plasma potential, electron and ion density change drastically. We demonstrate in our argon plasma that the competition of these three parameters induces a high electron current peak on the Langmuir's probe. Here, the number of N_2^+ decreases faster than the number of N^+ when the power supply is switch off. Of course the global ion density decreases, but a small capacitive coupling should lead to an increase of the plasma potential and of the electron temperature.

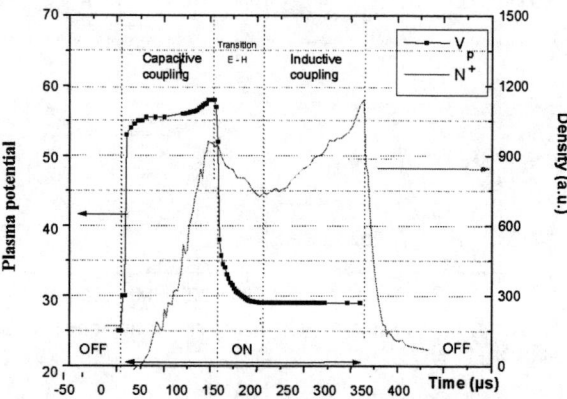

FIGURE 6. ion signal for and plasma potential one plasma pulse in Pulsion®, which exhibits the transition between capacitive coupling and inductive coupling..

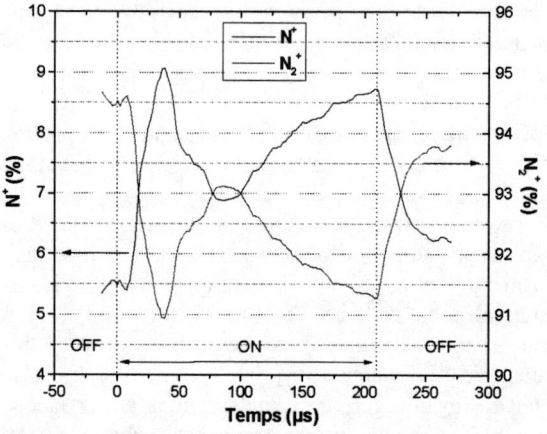

FIGURE 7. ion signal N^+ and N_2^+ during one plasma pulse in pulsion, the % of N_2^+ increase in the inductive coupling with the ion density, but decrease with the transition due mainly by the decrease of the electron temperature.

CONCLUSION

We present a study by TREMS (time energy resolved energy ion spectroscopy) of two different plasma discharges in pulsed regime. Our helicon (PHISIS) allows to clearly describe the different phases during the breakdown, thus we extend the classical three phases of the pulsed regime (ignition or breakdown, stationary and post-discharge) to at least six phases. Our physical point of view of the breakdown is composed by an electron phase, a high energy ion phase a capacitive coupling phase and depending to the injected power an inductive coupling phase and sometimes a helicon coupling phase.

In the case of an industrial discharge (Pulsion®) we demonstrate the creation of a small plasma during the plasma off phase, this effect suppress primary phases. The peak in the ion density is induced by the transition between the capacitive to the inductive coupling. Moreover the balance of ion percentage N^+/N_2^+ in the pulse is explained by the coupling of the antenna and by the plasma parameter. These measurements lead to a better understanding of pulsed plasma discharge, which allows to perform in the case of Pulsion® a well controlled ion implantation.

REFERENCES

1. R.W. BOSWELL and D VENDER, *"An experimental study of breakdown in pulsed helicon plasma"*, Plasma Sources Science & Technology, vol.4, p.534-540 (1995)
2. V. Kaeppelin, M. Carrère, J.M. Layet, *"Langmuir probe study in the post-discharge of a pulsed helicon discharge"*, Plasma Sources Science & Technology, vol.11, p.53-56 (01/2002)
3. V. Kaeppelin, M. Carrère, J.M. Layet, *"Ion energy distribution functions and Langmuir probe measurements in low pressure argon discharges"*, Journal of Vacuum Science & Technology A, vol.20 (2), p.526-529 (03/2002)
4. V. Kaeppelin, M. Carrère, F. Torregrosa, G. Mathieu, *"Characterisation of an industrial plasma immersion ion implantation reactor with a Langmuir probe and an energy selective mass spectrometer"*, Surface and Coatings Technology 156 (2002) 119-124

Boron Ion Implantation into Silicon by Use of the Boron Vacuum-Arc Plasma Generator[1]

J. M. Williams*, C. C. Klepper*[†], D. J. Chivers[¶], R. C. Hazelton[†], J. J. Moschella[†] and M. D. Keitz[†]

*Brontek Delta Corporation, 6580 Valley Center Drive, Radford, VA 24141, USA
[†]HY-Tech Research Corporation, 105 Centre Court, Radford, VA 24141, USA
[¶]Ion Links Int. Ltd., 32 St. Mary's Place, Bathgate, Scotland.

Abstract. This paper continues with presentation of experimental work pertaining to use of the boron vacuum arc (a.k.a. cathodic arc) plasma generator for boron doping in semiconductor silicon, particularly with a view to the problems associated with shallow junction doping. Progress includes development of an excellent and novel macroparticle filter and subsequent ion implantations. An important perceived issue for vacuum arc generators is the production of copious macroparticles from cathode material. This issue is more important for cathodes of materials such as carbon or boron, for which the particles are not molten or plastic, but instead are elastic, and tend to recoil from baffles used in particle filters. The present design starts with two vanes of special orientation, so as to back reflect the particles, while steering the plasma between the vanes by use of high countercurrents in the vanes. Secondly, behind and surrounding the vanes is a complex system of baffles that has been designed by a computer-based strategy to ultimately trap the particles for multiple bounces. The statistical transmittance of particles is less than 5 per coulomb of boron ions transmitted at a position just a few centimeters outside the filter. This value appears adequate for the silicon wafer application, but improvement is easily visualized as wafers will be situated much further away when they are treated in systems. A total of 11 silicon samples, comprising an area of 250 cm^2, have been implanted. Particles were not detected. Sample biases ranged from 60 to 500 V. Boron doses ranged from 5×10^{14} to 5×10^{15}/cm^2. Exposure times ranged from 20 to 200 ms for average transmitted boron current values of about 125 mA. SIMS concentration profiles from crystalline material are presented. The results appear broadly favorable in relation to competitive techniques and will be discussed. It is concluded that doubly charged boron ions are not present in the plume.

Keywords: boron, cathodic arc, vacuum arc, macroparticles, ion implantation, shallow junction doping
PACS: 61.72Tt

INTRODUCTION

For some ten years the semiconductor industry has been seeking better methods of boron ion generation and delivery in order to cope with the demands for increasingly shallower ion implants in solid state circuits. This need is a result of the continuing imperative for miniaturization of integrated circuits, as dealt with in the International Technology Roadmap for Semiconductors (ITRS). For boron, the problem relates to circuit architecture, together with the low atomic mass of B. This property means that B requires low energy for a given range in the solid and is also more difficult to transport at low energy by traditional beam line practice in ion implanters of the prevailing design. Quite a few approaches have been explored to ameliorate this problem. However, thus far, only two have gained much traction, and they appear to be nearing commercial maturity.

The first technique is the plasma immersion ion implantation process (PIII) [2]. In this process a plasma of large volume is generated from BF$_3$ gas. The entire wafer is immersed in the volume. Cracking of the gas produces several ion constituents. These are drawn into the wafer by polarization of the wafer. This process obviates the need for much transport of ions. The design is for use of the most prominent cracked constituent, BF_2. Other constituents are accepted, needed or not.

The other approach is the cluster ion technique [3], where a cluster in this case refers to a large molecule containing several atoms of B, such as decaborane (contains 10 boron atoms) or octadecaborane (18 atoms of B). Single ionization of such molecules means that the cluster can be transported with rather large energy, but the fraction of energy apportioned to each B atom is low, and the velocity and, therefore the range in the solid, is also small.

The present approach arrives as a late entry, but in most ways would appear to be the simplest. The idea is to generate a robust plasma of pure boron ions by use of the cathodic arc, a.k.a., the vacuum arc, technique. This technique is well established as an industrial plasma generation and coatings technique. The idea of using it for boron was conceived by Richter [4], who made several fundamental contributions. For this ion generation process, it is visualized that the ions might be delivered to the wafer either by a beam technique or by an immersion approach.

ISSUES, STATUS, AND PROGRESS

From the work of Richter, other literature [5], and the known industrial practice of vacuum arcs, it was judged that there were five fundamental issues pertaining to use of the vacuum arc source for boron doping of semiconductors at the outset of development. These were:
1. *Cathode integrity and purity*
2. *Possible energy contamination due to doubly charged ions*
3. *Macroparticles*
4. *Acceleration techniques*
5. *Dosimetry and uniform distribution over wafers*

Actually, as to category 4, techniques that would work for the other two might work for this case, but there might also be both new opportunities and new challenges for the cathodic arc. Opportunities could spring from the fact that the plasma in cathodic arc is born with useful energy and directionality (see below). The present paper presents new results directed primarily to issues 2 and 3, which were two of the least well addressed before the present research.

As to item 1, however, Richter had problems with cathodes, because the combination of mechanical strength, elasticity, purity, and heat conductivity needed to form a successful cathode presents difficulties. The present authors experienced many of the same issues as those of Richter in trying to procure suitable consolidated B materials. The solution has previously been addressed [6]. Aside from operational problems, purity is the only one of interest to the present application. The present paper adds one new datum to that issue, a further measurement of purity. Item 2 will appear as obviated by the implantation results to be presented. The same experiments involve successful acceleration over some practical range of acceleration voltage.

A well-known aspect of cathodic arc sources (item 3) is that much of the cathode comes off as solid (or liquid) particles with a range of sizes, but approximately 10 μm is a typical average. Despite the sensitivity of the semiconductor industry to particle contamination, it has never appeared that the macroparticle issue was totally intractable, as far as application of the cathodic arc to ion implantation was concerned. There are perhaps thirty or more macroparticle filters in the literature. Boxman et.al. cited 24 designs in 1996 [7]. There are commercially vended models, which make claims of high performance, but are expensive. In addition, filtering B macroparticles poses different issues than filtering metal macroparticles, for which most of the existing designs apply. Metal macroparticles tend to be either molten or rather soft and plastic. As a result, the metal particles can be deposited on baffle surfaces and be captured much more readily than can B particles. The latter particles, along with those of C, are refractory, solid, hard, and elastic. They tend to bounce around in the system and not be captured. In view of these considerations, it was decided to develop a custom macroparticle filter, for which the performance could be fully quantified when married to the B arc source.

EXPERIMENTS AND RESULTS

Purity

Much of the research reported here was supported by the National Institutes of Health. The boron ion source is prolific enough to make coatings. It has turned out that, as a coating, B is biocompatible. For reasons that will not be detailed, a program has existed to form B coatings on the orthopedic alloy, CoCrMo. A coating thickness that was favorable for the program goal was 150 nm, or in atomic density, 2 X 10^{18}/cm^2. Such a coating was analyzed by combined Rutherford backscattering (RBS) and non-Rutherford elastic scattering (NRES). The He-ion incident energy was 4.27 MeV. This energy was chosen because of the huge NRES resonance cross section there for carbon, which is 130 times the RBS cross section. The carbon concentration was 3 at. % for the coating. This would presumably be the maximum possible C concentration in the cathode, given that there was some carbon on the surface before coating, and perhaps some possibility of further contamination from the system. From the synthesis method of the cathode, it would seem likely that C is the most important impurity. This amount of C is putatively acceptable in the doping process, since C is 4-valent, as with Si, and fits nicely in to the lattice. A similar experiment was performed for the NRES oxygen resonance energy of 3.03 MeV for the incident He. No O was detected. Previous analyses [8] have not detected metallic impurities. In particular, the anode is of tungsten, which has a very high RBS cross section. Tungsten has never been detected.

Macroparticle Filter

The filter started with the idea of Ryabchikov [9] as a nucleus. This design uses vanes or blades, which are parallel in the long direction, but which may be rotationally tilted as to flat plane orientation, if desired. Large counter currents in the vanes produce complementary magnetic field contributions threading the channels between the vanes. Plasma is channeled between the vanes, but the vanes present an optically dense pathway for particles, so that particles are supposed to be deposited on the vanes. The idea presumably works well for soft or liquid particles, as suggested above. For the elastic B particles it was necessary to use vanes that were rotationally oriented at near 90 degrees to each other, and with overlap at the downstream edge, so as to provide a relatively narrow escape channel for the filtered plasma. With the vane orientation chosen, the first bounce of particles is then backwards or at no worse than 90 degrees. Secondly, these blades were then surrounded by a complex system of baffles. The configuration of the baffle system was determined by a computer study that took into account the kinematics of particle recoils for 10 bounces of particles that had originally made the first bounce backward from the blades. Thus, the present filter has a more objective design basis than the other filters. The others use magnetic steerage of some sort to steer the plasma away from the particle plume, and then may use some baffling that is intuitively designed.

Figure 1 is a photograph showing a view of the filter blades from the top with the plasma passing through.

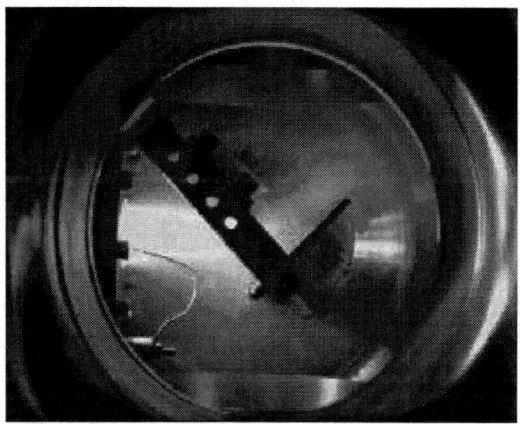

FIGURE 1. Passage of plasma through the filter. The unfiltered plume of about 2 A boron enters at about 9 o'clock. The filtered plume of about 400 mA exits at about 2 o'clock. Then the filtered plume is re-channeled in the original direction by magnetic ducting (not shown). The baffle system is also not shown.

For the coating of the CoCrMo orthopedic coupons, as described above, some had the thickness stated above, which was approximately optimal for the purpose. Others had much thicker coatings and many had thinner areas. The coupons were positioned just a very few centimeters down stream from the filter, so as to receive the maximum possible dose rate and also, so as to have a high probability of intercepting any particles that might come through. Many of these coupons had no particles at all. The entire collection was coated with enough boron to have ion implanted many 30-cm silicon wafers. The statistical estimate was that a maximum of 5 particles/coulomb of boron ions were transmitted. That value would represent about 1 particle/wafer if wafers were at that position for implantation. Further transport of plasma and/or effective expansion for wafers will provide opportunity for still further filtering if needed.

It is concluded that the macroparticle issue is not an intransigent obstacle for application of the present technology to ion implantation of silicon wafers.

Implantations and Results

A total of eleven silicon samples were ion implanted. Sample dimensions for most of the samples were 5 cm X 5 cm for a total area of about

250 cm². The sample position was about 20 cm downstream from the filter, much further away than the orthopedic coupons. Beam diameter at the position was about 10 cm, so that the sample dimension was only about half the beam diameter. The samples were positioned somewhat off to the side so as to keep much of the area out of the most intense part of the beam; thus there was a gradient in dose across the samples. There were apparently no particles on the samples. Particles, when present, are usually very easily detected, especially on Si.

Three of the samples, all crystalline, have been analyzed for boron concentration versus depth by SIMS (secondary ion spectroscopy). Bias voltages were 60, 100 and 500 V. Bias was essentially d.c. with down spikes at a frequency of 20 kHz for the spark arrested power supply. It is presumed that the incident plume energy before acceleration through the sheath was 20 to 30 V, where the lower value would be inferred from the careful work of Anders and Yushkov [10] on carbon for a similar situation, but the value of 29 eV of Richter [3] cannot be ruled out. These incident energy values should be added to the energies imparted by the biases to obtain true energies. Our analysis is only in terms of bias voltage. Exposure times were as high as 200 ms. Average currents were approximately 125 mA over the sample areas. Average doses were estimated to be about $4 \times 10^{15}/cm^2$ for the sample at 500 V and about $1 \times 10^{15}/cm^2$ for the other two.

The SIMS profiles are shown in Figure 2 for the respective biases. The junction depths at $1 \times 10^{18}/cm^3$ and the inverse logarithmic decrements (angstroms/decade) for the main drop in concentration are plotted versus bias voltage in Figure 3. The junction depths were 59, 67 and 155 angstroms, respectively, for the bias voltages of 60, 100, and 500 V.

DISCUSSION

The implications of the purity measurement and the filter performance have been mentioned as part of the presentation of results. It appears likely that neither of these issues is a fundamental obstacle to the further consideration of the process for wide use.

The most important part of the discussion relates to the assessment of the concentration depth profiles. The limited data available seem to be indicative of shallower profiles for the given boron ion energy than most of the other techniques produce (155 angstroms for the 500 V bias or 520 eV actual). However, we

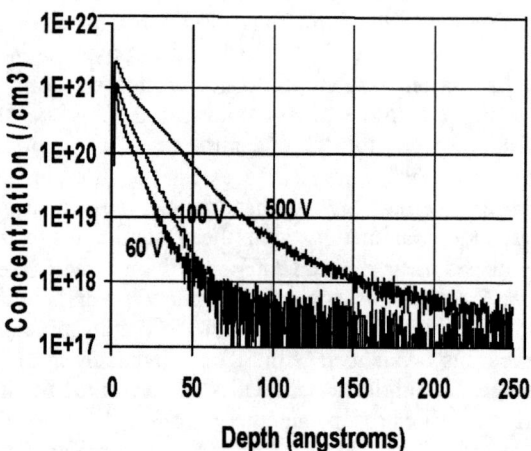

FIGURE 2. Concentration versus depth for boron implanted silicon for three bias voltages by SIMS. Crystalline silicon.

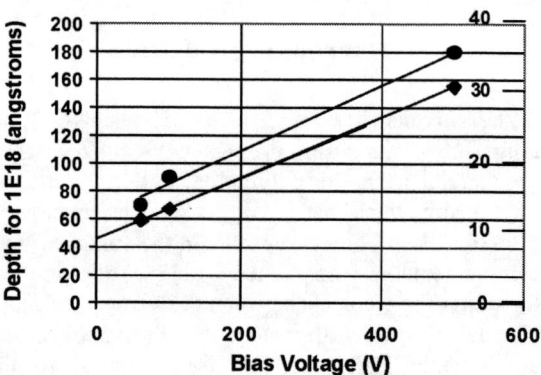

FIGURE 3. Junction depth at $1 \times 10^{18}/cm^3$ vs bias voltage (lower curve referred to left hand scale) and inverse logarithmic decrement (angstroms/decade) for the main decline (Fig. 2) referred to the right hand tick marks (angstroms/decade).

have few statistics. As far as the state of development of this technique at present is concerned, the data appear to be at least comparable to other similar results for the same ion energies. In particular, there is no evidence of energy contamination due to doubly charged boron ions.

As to the prospect that this technique might produce shallower junctions for the given B ion energy than some other techniques, that idea could be over interpreted on the basis of the present data. The present technique does produce the minimum possible total radiation damage for the given ion injection, but also a very high damage rate. It might be argued that the high rate enhances interstitial-vacancy recombination (second order reaction), resulting in less dynamic radiation enhanced diffusion during the implant and leaving less residual damage than occurs at ordinary implant rates. On expansion of the plasma to production level, however, the damage rate may be less than for the present experiments.

One other point worthy of note might be that the differences between the data for the 100 V and the 60 V biases remain clearly perceptible both in the junction depth and in the inverse decrement, and these two plots are clearly self consistent. This result means that this technique has not yet reached a limit as to lower values of junction depth.

ACKNOWLEDGEMENTS

It is a pleasure to thank J. H. Freeman for technical advice and encouragement pertaining to this program. Funding has been from the National Science Foundation, the National Institutes of Health, and Virginia's Center for Innovative Technology.

REFERENCES

1. Patent pending
2. P. K. Chu, S. Qin, N. W. Cheung, and L. A. Larson, *Materials Science and Engineering,* **R17,** 207-280 (1996).
3. Marek Sosnowski, *Applications of Accelerators in Science and Industry,* ed. by J. L. Duggan and I. L. Morgan, AIP Conference Proceedings 576, American Institute of Physics, Melville, NY, 2001, pp904-907
4. Frank Richter, Siegfried Peter, Volodymyr, Gert Flemming and Michael Kuhn, *IEEE Transactions on Plasma Science,* **27,** 1079-1083 (1999).
5. I. G. Brown, A. Anders, S. Anders, M. R. Dickinson, R. A. MacGill, and E. M. Oks, *Surface and Coatings Technology,* **84,** 550-556 (1996).
6. C. C. Klepper, R. C. Hazelton, E. J. Yadlowsky, E. P. Carlsen, M. D. Keitz, J. M. Williams, R. A. Zuhr, and D. B. Poker, *J. Vac. Sci. Technol,* **A20,** 725-732 (2002)
7. R. L. Boxman, V. Zhitomirsky, B. Alterkop, E. Gidaleveich, I. Beilis, M. Keidar, and S. Goldsmith, *Surface and Coatings Technol.,* **86-87,** 243-253 (1996).
8. J. M. Williams, C. C. Klepper, and R. C. Hazelton, *Nucl. Inst. Methods in Phys.Res.,* **B,** 278-283 (2005)
9. A.L. Ryabchikov, *Surface and Coatings Technology* **96,** 9-15 (1997).
10. Andre Anders and George Yu Yushkov, *J. Appl. Phys.,* **96,** 970-974 (2004).

MATERIALS—NOVEL TECHNIQUES AND APPLICATIONS

Micro-patterned porous silicon using proton beam writing

M. B. H. Breese, D. Mangaiyarkarasi, E. J. Teo*,
A. A. Bettiol and D. Blackwood*

*Centre for Ion Beam Applications, Department of Physics,
National University of Singapore, Singapore 117542*
Materials Science and Engineering Department, National University of Singapore, Singapore 117542

Abstract. A high-energy beam of hydrogen or helium ions focused to a small spot in a nuclear microprobe selectively damages a silicon lattice. This damage acts as an electrical barrier during subsequent formation of porous silicon by electrochemical etching of p-type wafers, so the un-irradiated regions are preferentially anodized. This process has opened up new research directions for the fabrication of a variety of high-aspect ratio, multi-level microstructures in silicon, such as gratings, photonic lattices and waveguides. The same process enables local modifications and control of the properties of the porous silicon, leading to new luminescent micro-structures in which both the spatial location and the wavelength and intensity of emission can be carefully controlled. The reflectivity and transmission of Bragg reflectors and microcavities fabricated in porous silicon can also be controlled using this same process and examples are given of each of these applications.

INTRODUCTION

Porous silicon (PSi) is of interest due to its photoluminescence and electroluminescence properties [1,2], raising the possibility of producing light emitting devices made of PSi and microelectronics compatibility. A potentially important application of PSi is the production of combined optical/electronic devices incorporating patterned porous material directly onto a single-crystal Si substrate with a high spatial resolution. The photoluminescence from PSi can be greatly improved by placing it in a Fabry-Perot microcavity with dimensions comparable to the optical emission wavelength. One dimensional photonic structures based on alternating high and low porosity PSi layers have found applications as dielectric mirrors in the form of Distributed Bragg Reflectors [3], microcavities [4,5] and waveguides [6]. PSi exhibits optical properties consistent with a single effective refractive index despite its nanoscale structural inhomogeneity. The refractive index of PSi is lower than for bulk silicon; it is inversely proportional to the etch current density used for anodisation. A Distributed Bragg Reflector selectively reflects a band of incident wavelengths and is formed in PSi by periodically lowering and raising the etch current density, resulting in a sequence of porous layers with alternating high and low refractive index. Highly-doped p-type silicon (~0.01 Ω.cm) is commonly used to fabricate such PSi-based photonic structures because a wide range of refractive index can be achieved.

Many new technologies require the fabrication of precise three-dimensional structures in silicon [7,8]. One major limitation of conventional lithography and silicon etching technologies for these applications is the multiple processing steps involved in fabricating free-standing multilevel structures. Electrochemical etching of silicon in hydrofluoric acid is emerging as an alternative technique for micromachining due to its low cost, fast etching process and easy implementation. The porous silicon produced can be easily removed by immersion in a potassium hydroxide (KOH) solution.

ION IRRADIATED POROUS SILICON

Figure 1 shows the basis of producing micro-patterned silicon surfaces, controlled photoluminescence from bulk PSi and controlled reflectivity from PSi Bragg reflectors. A finely-focused beam of MeV ions is scanned over the wafer surface using a nuclear microprobe [9], in this case at the National University of Singapore, which can focus MeV ion beams to spot sizes of less than 50 nm [10]. The ion beam loses energy as it penetrates the semiconductor and comes to rest at a well-defined range, equal to ~8 µm for 2 MeV helium ions, and ~50 µm for 2 MeV protons in silicon. The stopping process causes the silicon crystal to be damaged, by producing additional vacancies in the semiconductor. A higher beam fluence at any region produces a higher vacancy concentration, so by pausing the focused beam for different amounts of time at different locations, any pattern of localized damage can be built up.

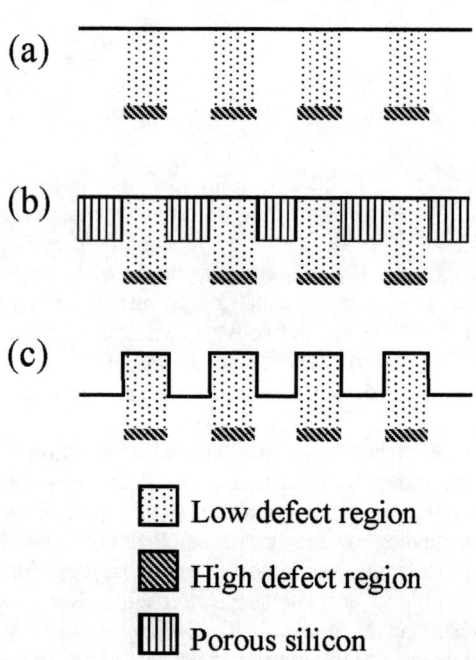

FIGURE 1. (a) Patterning of *p*-type silicon with proton beam writing, (b) electrochemical etching to selectively form patterned porous silicon and (c) removal of porous silicon with dilute KOH solution to form a micromachined structure

The irradiated wafer is then electrochemically etched in a dilute solution of hydrogen fluoride. An electrical current is passed through the wafer which causes the formation of porous silicon at the surface. The buried regions of high vacancy concentration inhibit this formation process, so a thinner layer of porous silicon is produced at the irradiated regions. A very large beam dose may reduce the etch rate to zero at the irradiated regions. The etched sample comprises patterned areas of porous silicon which may emit light with greater or lesser intensity, or different wavelength, compared with the surrounding unirradiated regions. If the anodisation is repeated with alternating low and high current densities then the range of wavelengths of incident light which are reflected from the resultant Bragg reflector depends on the ion irradiation dose. If the etched sample is immersed in KOH, the porous silicon is removed, leaving the final patterned structure on the wafer surface as a three-dimensional representation of the scanned pattern area and fluence.

MICROMACHINING RESULTS

Recent results on fabricating high-aspect ratio, multilevel structures in silicon have been published in Refs. [11-13]. Fig. 2a,b shows Scanning Electron Microscope (SEM) images of a linear waveguide with a grating at the top surface with a 1 µm period. This is designed to selectively transmit at certain infra-red wavelengths along the waveguide while the grating modifies which wavelengths are transmitted, depending on the periodicity. This waveguide was formed by first fabricating the waveguide using the above micromachining process, then spinning on a layer of polymer resist and creating the grating in the polymer using a second irradiation with a beam spot focused to 100 nm or less [14]. The pattern was then transferred to the silicon waveguide top surface using reactive ion etching.

After etching beyond the end of range, the isotropic electrochemical etching process starts to undercut the irradiated structure, so multi-level micromachined structures can be created by irradiating with two different proton energies. The regions irradiated with lower energy become undercut at a shallower etch depth while regions irradiated with higher energy protons continue to increase in height. In this way, multi-level free-standing microstructures can be fabricated in a single etch step. To demonstrate this capability, bridge structures were irradiated with 0.5 MeV protons (range ~6 µm), and supporting pillars with 2 MeV protons (range ~48 μm). Fig. 2c shows the resultant 'StoneHenge' structure after prolonged etching. The bridges are fully undercut and separated from the substrate. They remain supported by the pillars irradiated by higher energy.

FIGURE 2. SEM images of (a) linear waveguide with a grating at the top surface with a 1 μm period, (b) higher magnification. (c) micromachined 'StoneHenge' structure, 100 microns in diameter.

PATTERNED POROUS SILICON RESULTS

We have undertaken a comprehensive study of the effects of ion irradiation on a wide range of different resistivity p-type wafers, and recorded the resultant photoluminescence intensity and wavelength [15-18]. Three different resistivity p-type silicon wafers were patterned with a focused beam of 2 MeV helium ions. After irradiation the wafers were anodized, producing a porous layer several microns in thickness. Fig. 3 plots the measured photoluminescence intensity and peak wavelength emission as a function of ion dose. There are two resistivity regimes in p-type silicon where photoluminescence is affected in different ways by ion irradiation: for low resistivity (~0.01 Ω.cm) wafers irradiation primarily results in a large PL increase whereas for moderate resistivity (0.1-10 Ω.cm) wafers irradiation primarily results in a large PL wavelength red-shift.

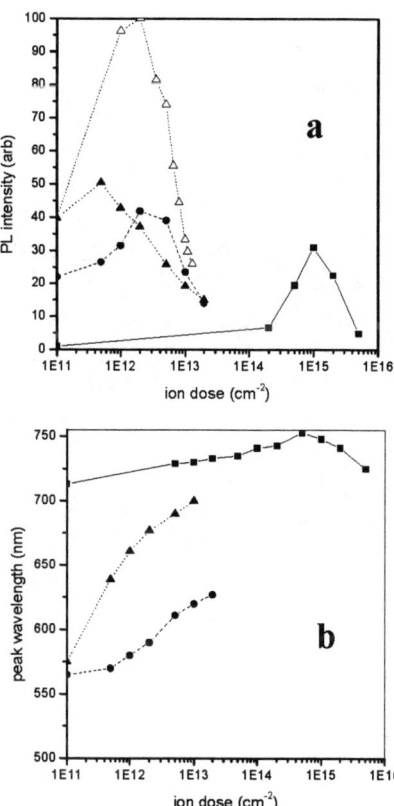

FIGURE 3. (a) Measured PL intensity and (b) peak wavelength emission as a function of dose, for low (0.02 Ω.cm, solid line) and moderate resistivity (0.3 Ω.cm, dotted line) and (3 Ω.cm, dashed line) p-type silicon wafers.

A demonstration of this different behaviour is shown in Fig. 4a,b. In low resistivity p-type silicon the irradiated dragon produces bright, red PL compared with the faint unirradiated background. In moderate resistivity silicon bright, orange/red PL is produced from the dragon whereas the background produces green PL of a similar intensity. In Fig. 4c,d a range of doses have been irradiated in a continuous distribution in moderate resistivity silicon, producing a gradual change in PL emission wavelength. In Fig. 4c a circular region was irradiated with a dose which increased linearly towards the centre, resulting in a gradual PL red-shift. The tapered region in Fig. 4d was produced by a gradually increased irradiation dose from right to left, again resulting in a gradual PL red-shift.

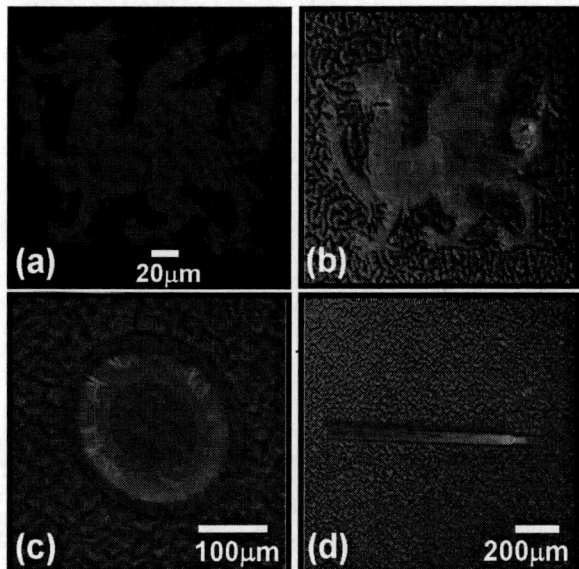

FIGURE 4. Photoluminescence images of dragons formed by irradiating a p-type (a) 0.02 Ω.cm wafer, (b) 3 Ω.cm wafer. (c) Concentric ring pattern in a 3 Ω.cm wafer formed by linearly increasing the dose from the outer edge to the centre. (d) Colour bar showing a gradual wavelength red-shift as the dose increases from the right to the left in a 3 Ω.cm wafer.

resistivity p-type wafer [19]. The wafer was then etched with an alternating high/low current density for 4 seconds per layer, with a total of 15 bilayers formed, then cleaved perpendicular to the line direction for cross-section imaging. The etched layers appear with light/dark contrast, with high porosity (low refractive) regions appearing darker. The etch rate is progressively slowed by a larger irradiation dose, resulting in thinner porous layers which reflect shorter incident wavelengths. The reflectivity spectra in Fig. 5c were recorded from similar Bragg reflectors produced by irradiating 3×3 mm² areas with different proton doses. The unirradiated wafer reflects a central wavelength of 740 nm, which is continuously blue-shifted to 490 nm for a proton dose of $2\times10^{15}/cm^2$.

Fig. 6a shows an optical micrograph of a 500×500 μm² region which was irradiated with different overlaid scan patterns, with different doses. The wafer was etched as in Fig. 5 except the etching period for each layer was chosen to give the central wavelength of 850 nm in the unirradiated regions. Each dose produces a different reflected colour when illuminated with white light, with red/orange colours corresponding to areas irradiated with the lowest dose. The potential of this approach to form patterned array of colour pixel and

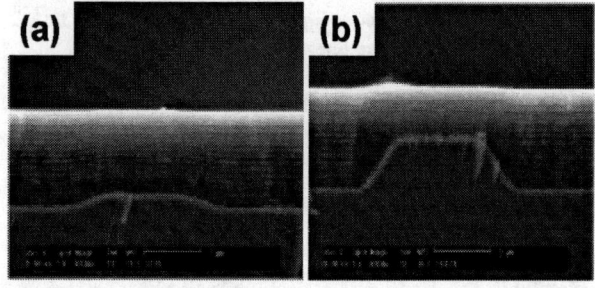

FIGURE 5. SEMs of lines irradiated in a 0.02 Ω.cm wafer with doses of (a) $1\times10^{15}/cm^2$, (b) $7.5\times10^{15}/cm^2$. The irradiated lines (locations shown by the arrow) were 2 mm long and 2 μm wide. (c) Reflectivity spectra of MCs fabricated with different proton doses in the same material. The height of each spectra is normalized to a maximum of 1.0.

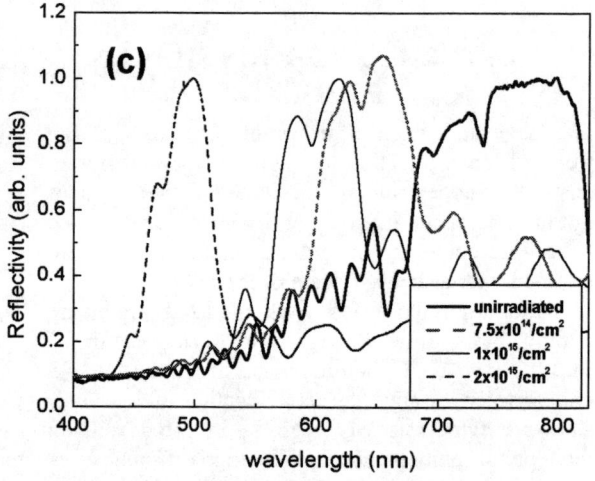

DISTRIBUTED BRAGG REFLECTORS

Fig. 5a,b show SEMs of a sample in which lines were irradiated with different doses in a 1×1 cm² low

lines for display applications is shown in Fig. 6b, where vertical lines, each 10 μm wide, were irradiated to form alternating red-green-blue stripes. The lateral resolution in the image is about 1 micron.

 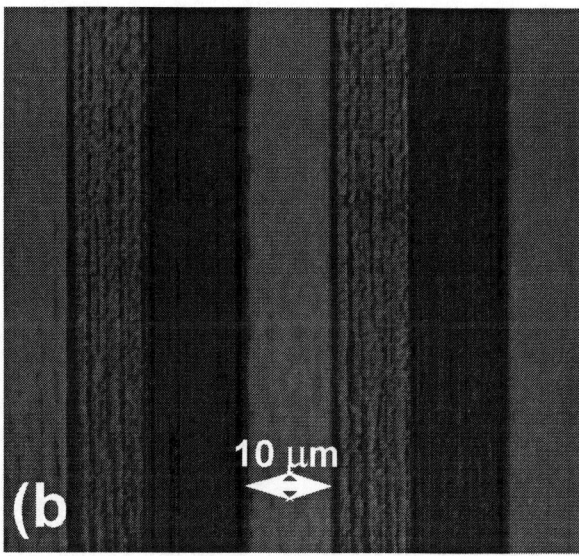

FIGURE 6. Optical reflection images of (a) the painting "La Musique", by Henri Matisse (1939), created by irradiating a 500×500 µm^2 area. (b) vertical lines, each 10 µm wide, irradiated to form alternating red-green-blue stripes. In each recorded image the sample was illuminated with white light and the reflected light was recorded for 30 seconds using a Nikkon Eclipse ME600 microscope with a ×10 objective.

CONCLUSIONS

We have developed the use of high-energy, focused ion beam irradiation to produce patterned porous silicon for a range of different areas of application. In high resistivity silicon, ion irradiation tends to produce patterned PSi with reduced PL intensity but controllably red-shifted. In low resistivity silicon, focused proton beam irradiation can be used to selectively enhance the light emission in the irradiated regions. Tuning of the PL intensity has been obtained by controlling the local resistivity as a function of fluence. If the irradiated wafers are etched with alternating high/low current density, then the reflectivity of the resultant Bragg reflectors is controllably blue-shifted. Focused MeV ion beams in a nuclear microprobe may be used to produce high-aspect ratio, multilevel microstructures in silicon. The feature height of the irradiated structure can be accurately controlled with the fluence of the incident beam, enabling the production of a multilevel structure by multiple fluence exposures in a single irradiation.

REFERENCES

1. A. G. Cullis, L. T. Canham and P. D. J. Calcott, J. Appl. Phys. **82**, 909 (1997)
2. S. Ossicini, L. Pavesi and F. Priolo, Light Emitting Silicon for Microphotonics, SMTP 194 75 (2003), Springer-Verlag,
3. M. G. Berger, R. Arens-Fischer, M. Kruger, S. Billat, H. Luth, S. Hillbrich, W. Theiß, P. Grosse, Thin Solid Films, **297**, 237 (1997)
4. M. Araki, H. Koyama, N. Koshida, Appl. Phys. Lett. **69**, 2956 (1996)
5. L. Pavesi, *Riv. Nuovo Cimento* **20**, 1 (1997)
6. S. Nagata, C. Domoto, T. Nishimura, K. Iwameji, Appl. Phys. Lett. **72**, 2945 (1998)
7. H. G. Craighead, Science **290**, 1532 (2000)
8. S. Y. Lin, J. G. Fleming, D. L. Hetherington, B. K. Smith, R. Biswas, K. M. Ho, M. M. Sigalas, W. Zubrzycki, S. R. Kurtz and Jim Bur, Nature **394**, 251 (1998)
9. M. B. H. Breese, D. N. Jamieson, P. J. C. King, *Materials Analysis using a Nuclear Microprobe*, Wiley, New York, **1996**.
10. J. A. van Kan, A. A. Bettiol and F. Watt, Appl. Phys. Lett. **83**, 1629 (2003)

11. E. J. Teo, M. B. H Breese, E. P. Tavernier, A. A. Bettiol and F. Watt, M. H. Liu and D. J. Blackwood, Appl. Phys. Lett. **84,** 3202 (2004)
12. E. J. Teo, E. P. Tavernier, M. B. H. Breese, A. A. Bettiol, F. Watt Nucl. Instrum. Methods Phys. Res. B **222,** 513 (2004)
13. E. J. Teo, M. H. Liu, M. B. H. Breese, E. P. Tavernier, A. A. Bettiol, D. J. Blackwood and F.Watt, Proceedings of SPIE **5347,** 264 (2004)
14. J. A. van Kan, A. A. Bettiol and F. Watt, Nano Letters **6,** 579 (2006)
15. E. J. Teo, D. Mangaiyarkarasi, M. B. H. Breese, A. A. Bettiol, D. J. Blackwood, Appl. Phys. Lett. **85,** 4370 (2004)
16. D. Mangaiyarkarasi, E. J. Teo, M. B. H. Breese, A. A. Bettiol, D. J. Blackwood, J. Electrochem. Soc. **152,** D173 (2005)
17. E. J. Teo, M. B. H. Breese, A. A. Bettiol, D. Mangaiyarkarasi, F. J. T. Champeaux, F. Watt, D. J. Blackwood, Adv. Mater. **18,** 51 (2006)
18. M. B. H. Breese, F. J. T. Champeaux, E. J. Teo, A. A. Bettiol, D. J. Blackwood, Phys. Rev. B **73,** 035428 (2006)
19. D. Mangaiyarkarasi, M. B. H. Breese, Y. S. Ow, C. Vijila, Applied Physics Letters **89,** 021910 (2006)

Grazing Incidence angle X-ray Diffraction of implanted stainless steel: comparison between simulated data and experimental data.

J. Dudognon, M. Vayer, A. Pineau, and R. Erre

Centre de Recherche sur la Matière Divisée, 1b rue de la Férollerie, 45071 Orléans Cedex 2, France

Abstract. Austenitic stainless steel was implanted with specific elements using specific conditions. The goal of this studies was to compare the predicted structural modifications within the implanted layer with the experimental ones observed by Grazing Incidence angle X-ray Diffraction (GIXD).
During ion implantation implanted austenite steel layer undergoes modifications such as austenite lattice expansion, ferrite apparition and structure destruction. The X-ray diffraction austenite peak shape was predicted using the concentration depth profile of implanted element, the information depth profile of diffracted intensity and a linear relationship between implanted element concentration and lattice parameter. Experimental and predicted austenite X-ray diffraction peaks are in good accordance as long as the implanted layer contents mainly austenite. Whatever the nature of implanted element, ferrite appears above a given threshold of incoming energy amount. Moreover, the structure of the implanted layer is destroyed above a given amount of incoming energy.

Keywords: ion implantation, stainless steel, X-ray diffraction
PACS: 61.82.Bg, 61.72.Ww, 68.55.Ln, 68.55.Nq, 61.10.Nz, 61. 80. Jh

INTRODUCTION

Ion implantation induces modifications such as structural and chemical ones which are not completely elucidated.

Within the implanted layer, the implanted element concentration varies with the depth. Concentration depth profiles have been investigated by many authors [1] and have been predicted using simulation such as TRIM (transport range of ions in matter) code [2]. In numerous cases the simulated implanted element concentration depth profile is in good agreement with the experimental ones.

As the implanted element concentration varies with the depth, local characteristics of the implanted layer such as the structure could be expected to also vary with depth. Investigations of the structural modifications induced by ion implantation are the object of numerous studies via Transmission Electron Microscopy or X-ray diffraction (XRD). However, these techniques are not depth selective. Few tentative measurements of structural depth profiling have been performed via ionic abrasion and X-ray analysis or by using Grazing incidence X-ray diffraction (GIXD) with different incidence angles [3,4]. Consequently, structural modifications induced by implantation are still under study. As observed by numerous authors, and especially in case of nitrogen, implantation of austenitic stainless steel leads in a first time to both shifted and broadened X-ray austenite peaks. These spectral modifications have been attributed to the formation of a new phase named expanded austenite (γ_N) but the exact nature and characteristics of this new phase are subjects of controversies and investigations [5].

The goal of this paper is to establish links between ion implantation conditions (element and energy) and the induced structural and chemical modifications.

Austenitic stainless steel was implanted with - inert elements (Ar, Xe),- N which inserts in the lattice, alloy elements which substitute steel elements and stabilize ferrite (Mo, Cr) and alloy elements which also substitute steel elements but stabilize austenite (Ag, Pb). The nature of implanted ion was also chosen to have a wide range of covalent radius of element, smaller or larger than the steel elements ones. The relationship between austenite structure expansion and the size of implanted element will be thus addressed.

The energy of implanted ion was such that the implanted layer has a thickness of 70 nm with a maximum at 20 nm and the fluences were chosen to have an implanted element concentration enough to be detected by Rutherford Backscattering Spectrometry (RBS) and to enable observable modifications in GIXD (2×10^{16} ions/cm^2 and an averaged concentration of 4 %).

The concentration depth profiles were established using TRIM code and the sputtering coefficient then experimentally confirmed by RBS.

The structure was explored using GIXD varying the incidence angle in order to study the structure variation with depth: as the X-ray beam penetration depth varies, the information depth varies.

The austenite X-ray diffraction peaks were also predicted for different incidence angles and compared with experimental ones.

MATERIALS & METHODS

Material

The material used was AISI 316LVM cold-rolled stainless steel (18.71 Cr, 13.08 Ni, 1.71 Mn, 1.61 Mo, 1.03 Si, 0.29 N, 0.07 C, 0.04 P, balance Fe (atomic %)). Steel density was 8.57×10^{22} at.cm^{-3}. Samples cut into a bar were mechanically polished up to obtain a mirror-like surface with a 2 nm (RMS) roughness.

Implantation

Ion implantations were performed under the conditions defined in Table 1 at room temperature in the IRMA ion accelerator with the SIDONIE isotopic separator at CSNSM (Orsay, France). The fluence was always the same 2×10^{16} ions/cm^2.

TABLE 1. Implantation conditions.

ion	N	Ar	Cr	Mo	Ag	Xe	Pb
E^a(keV)	28	75	90	150	170	190	280
P^b	1.1	3.8	5.0	7.3	8.0	8.9	12.2

[a] E: ion energy
[b] P: sputtering coefficient (number of sputtered atoms by incident ion)

Grazing Incidence X-ray Diffraction

Asymmetric in plane Grazing Incidence X-ray Diffraction was performed with a Phillips parallel beam horizontal X-pert diffractometer. Cu Kα was used as incident beam.

The peak positions were evaluated after corrections of refraction effect due to grazing incidence. Angles were measured with a 0.03° precision that means 5×10^{-4} nm precision for the lattice parameter. Incidence angles were ranged between 0.5 and 1.5°.

Prediction of Austenite XRD Spectra

The austenite XRD spectra [4] were simulated taking into account some considerations:
• The concentration c of the implanted element as a function of depth z is known (simulated by TRIM).
• The implanted layer and the bulk have the same structure austenitic structure with a constant lattice parameter a_0 for the bulk and a variable lattice parameter a for the implanted layer. We suppose that a linear law $a = a_0 + K \times c$ exists between the concentration c of implanted element and the lattice parameter a with a proportional parameter K.
• The XRD experimental observed signal is the convolution of the instrumental transfer function and the XRD theoretical signal. The instrumental transfer function was evaluated by recording the XRD spectra of a polycrystalline silicon for the different used incidence angles.
• To calculate the theoretical XRD signal coming out from the sample, this one is divided in n superimposed i layers of a thickness e and a concentration c_i. The number n of layers is chosen so that the contribution of nth layer is less than 0.1% of the total theoretical signal.
• Each i layer at a depth $[z_{i-1}, z_i]$ has an averaged lattice parameter a_i which corresponds to a $d_{i,hkl}$ interplanar distance for (hkl) planes and contributes to the signal at a diffracting angle 2θ given by the Bragg law for $I_0 * \{exp(-z_{i+1}*\tau) - exp(-z_i*\tau)\}$. I_0 is the diffracted signal of a layer of thickness e without taking into account the absorption coefficient τ. This coefficient, which describes the attenuations of the incident and diffracted signals, is supposed to be constant within the depth but depends on the incident angle. $1/\tau$ is called the information depth and 63 % of the intensity comes from a layer of thickness $1/\tau$. Table 2 displays the information depth as a function of the incidence X-ray angle.
• Finally the theoretical spectra is obtained by summing the signals coming from all i layers.

TABLE 2. $1/\tau$ as a function of α.

α (°)	0.5	0.6	0.7	1.0	1.5
$1/\tau$ (nm)	26.5	39.4	50.6	80.7	127.3

RESULTS

Concentration Depth Profiles

TRIM 2000 code provides the presence probability of implanted ion as a function of depth [2]. Although sputtering is not accounted in our version of TRIM code, it provides the ion sputtering coefficients (Table 1) and allows us to predict the concentration c of the implanted element as a function of depth z including sputtering [4]. Figure 1 displays the TRIM predicted concentration depth profiles for the implantation conditions given in Table 1. The simulated concentration depth profiles were confirmed by RBS. Except for N and Ar where the sputtered coefficients are very low (displayed in table 1), the depth profiles for the other ions are similar.

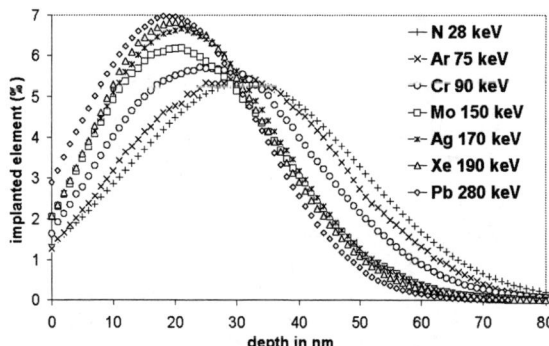

FIGURE 1. Element concentration depth profile as calculated using the TRIM code for ion and energy displayed in Table 1 and a fluence of 2×10^{16} ion/cm^2.

Structure Analysis

The austenite structure undergoes several modifications or transformations with the ion implantation as shown in Fig. 2.

Implantation induces expansion of austenite (γ) with a shift of the austenite peak towards lower diffraction angle and in some cases the apparition of new austenite peaks. For Cr, N and Ar, only expansion of austenite is observed. For Mo, Ag, Xe, and Pb implantation went through deeper transformation with moreover ferrite (α) formation and for Pb amorphization of the surface layer.

Austenite

For all the samples, austenite X-ray diffraction peaks for different incidence angles were simulated on the base of concentration depth profile evaluated using TRIM data.

Even with a continuous repartition of element into the implanted layer, the X-ray peak could present two or three components (Fig. 2 and 3). This fact has not to be attributed to the presence of two different phases but is only due to the convolution of the pseudo-Gaussian concentration depth profile analyzed with an exponential information depth profile.

The proportional coefficient between implanted element concentration and lattice expansion K was adjusted to have the best correlation between experimental and simulated spectra for all incidence angles.

For samples where implantation induces only austenite expansion, an excellent correlation exists between simulated and experimental (220) austenite peak as, for example, in the case of implantation of N at 28 keV shown in Fig. 3. The correlation is valid as long as the ferrite is not the main component of the implanted layer which is the case for Ag and Xe.

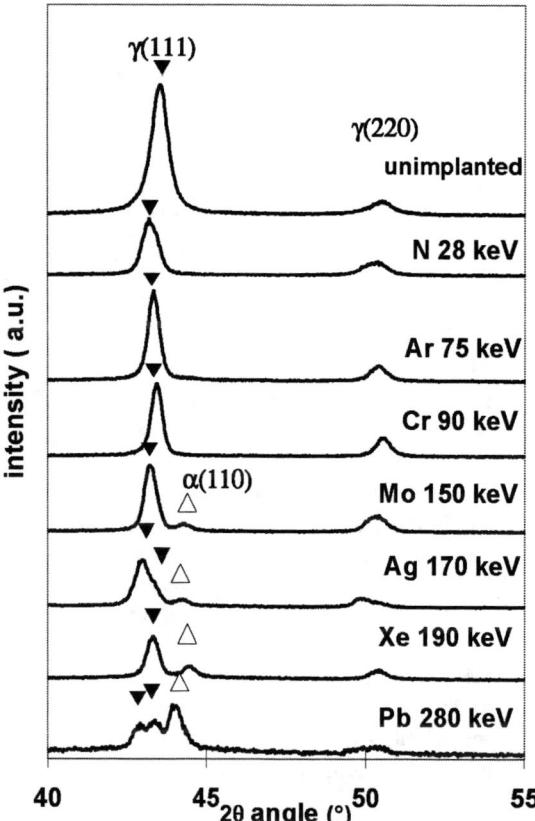

FIGURE 2. X-ray diffraction spectra of the 2 θ region for unimplanted and implanted samples between 42 and 50°. γ(111), α(110) and γ(200) are shown in this region for 0.5° incidence angle (fluence of 2×10^{16} ion/cm^2).

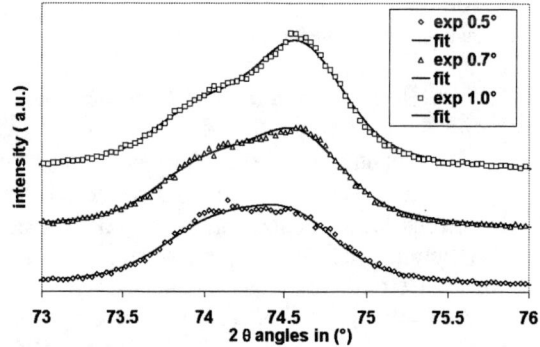

FIGURE 3. X-ray diffraction γ(220) peaks at 0.5°, 0.7°, 1.0° for N 28 keV implanted sample (fluence 2×10^{16} ion/cm^2).

For Pb 280 keV at 2×10^{16} ions/cm^2, shown in Fig. 4, it is impossible to superpose the simulated and the experimental X-ray spectra. The experimental X-ray spectra is very noisy and moreover the signal coming from the unimplanted bulk layer is overcounted due to a strong amorphization of the implanted layer.

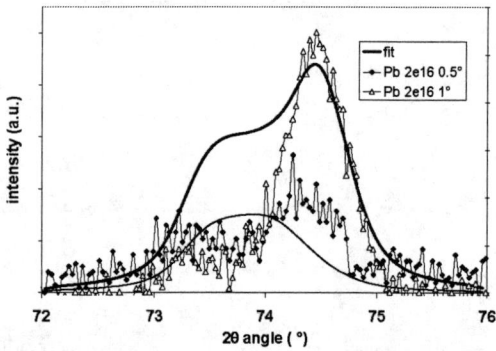

FIGURE 4. X-ray diffraction γ(220) spectra at 0.5°, 1.0° for Pb 280 keV implanted (fluence 2×10^{16} ion/cm^2).

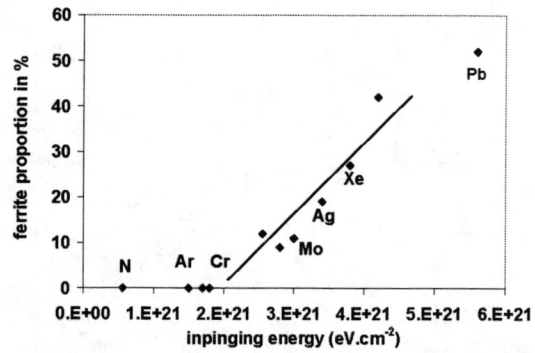

FIGURE 5. Ferrite proportion R (area under α(110) at 0.5°) as a function of the total impinging energy (product of fluence by energy).

The K values have to be considered taking into account the nature and the covalent radius of the elements. For alloy elements (Cr, Mo, Ag) which substitute steel element, the evolution of K values follows the one of covalent radius. Although N is a small size element, as it goes into interstitial position, the K value is larger than for Cr which only can substitute the steel elements. When important amount of ferrite is formed or when amorphization such as for Pb takes place, K value is strongly modified and becomes meaningless.

TABLE 3. K values

Ion, E[a]	N, 28	Ar, 75	Cr, 90	Mo, 150	Ag, 170	Xe, 190	Pb, 280
K[b]	6.1	2.5	1.2	5.3	8.2	2.3	9.0
R[c]	0.75	0.98	1.18	1.30	1.34	1.31	1.47

[a] energy in keV
[b] ($\times 10^3$)
[c] covalent radius (in Å), r(Fe) = 1.17 Å.

Ferrite Formation And Amorphization

The quantity of ferrite formed during the ion implantation was estimated from the α and γ X-ray intensities for an incident angle of 0.5° making the assumption that the implanted layer is only constituted of diffracting austenite and ferrite.

As shown in Fig. 5, there is an energy threshold (2×10^{21} eV/cm^2) under which no ferrite is formed in the implanted layer. Above this threshold, the energy amount brought by the implanted ions is sufficient to form ferrite. The amount of ferrite is proportional to this energy amount. Even for elements which stabilize the austenite structure (Ag, Pb), ferrite is formed. It is useless to recall that austenite is a metastable phase whereas ferrite is a stable one. Bringing energy leads to the formation of the most stable phase.

There is also an energy amount above which the proportionality between energy amount and ferrite is no longer valid since implantation give rise to amorphization. We estimate the corresponding value to 5×10^{21} eV/cm^2.

CONCLUSIONS

In this paper, we show that the concentration depth profile of implanted element could be calculated using TRIM code and sputtering coefficient.

Under implantation, austenite undergoes first expansion, then ferrite formation and in some cases amorphization. Grazing incidence angle X-ray diffraction performed at different incidence angle allows to investigate the structure depth profile.

For "soft implantation", austenite XRD peaks could be totally simulated by taking into account only the repartition of implanted ion in samples.

Energy amount thresholds for ferrite formation and amorphization have been shown up.

ACKNOWLEDGMENTS

The authors would like to thank T. Cam for GIXD experiments, T. Sauvage of CERI (Orléans, France) for RBS experiments and O. Kaitasov and S. Jacob of CNSM (Orsay, France) for ion implantations.

REFERENCES

1. S.H. Gazestani, M.G. Shahraki, M.R. Hantehzadeh, M. Ghoranneviss and A. Shokouhy, Appl. Surf. Sci. **237**, 2004, 332.
2. J.F. Ziegler, J.P. Biersack and V. Littmark, The Stopping and Range of Ions in Solids 1, Pergamon, New York, 1996.
3. H. Pelletier, D. Müller, P. Mille, A. Cornet and J.J. Grob, Surf. Coat. Tech. **151-152**, 2002, 377.
4. J. Dudognon, M. Vayer, A. Pineau and R.Erre, Surf. Coat. Techn. **200**, 2006, 5058.
5. X. Xiaolei, W. Liang, Y. Zhiwei and H. Zukun, Surf. Coat. Tech. **192**, 2005, 220.

Investigation Of The Impact On Device Parameters of Fluorine Enhanced Oxide In A Power Trench MOSFET

Jeffrey H. Rice and Chun-Tai Wu

Fairchild Semiconductor, 3333 W. 9000 S., West Jordan, UT, USA
jeff.rice@fairchildsemi.com, chun-tai.wu@fairchildsemi.com

Abstract. Oxide growth can be enhanced at the bottom of a trench MOSFET by implanting the MOSFET structure with fluorine ions at a zero-degree angle. The impact to the MOSFET device caused by the fluorine implant is reported. Trench MOSFETs were implanted with a range of fluorine ion doses and energies, with and without a sacrificial oxide. Gate oxide was subsequently grown. The gate oxide integrity (GOI) is evaluated and is shown to improve with higher dose and energy of fluorine ions. Gate-to-drain charge (Q_{gd}) is shown to be reduced with increasing fluorine dose.

Keywords: Fluorine implant, gate oxide, trench MOSFET, gate-to-drain charge
PACS: 85.40.Ry + 85.30.Tv + 61.82.Fk + 68.55.Ln + 76.30.Da

INTRODUCTION

An advantageous feature in a trench-based power MOSFET is a thicker gate oxide in the bottom of the trench relative to the sidewall gate oxide thickness. The characteristics of the trench bottom have an impact on the voltage breakdown and Miller capacitance (or gate-to-drain charge, Q_{gd}) [1]. A thicker trench bottom oxide is expected to improve the breakdown characteristics as well as reduce the Miller capacitance. Various methods, such as a LOCOS and selective deposition, have been used to grow the thicker trench bottom oxide [2].

In this work, a novel method for growing a thicker bottom gate oxide using fluorine implantation is evaluated. It has been shown that fluorine will enhance the oxidation rate. One possible reason is that the fluorine increases the oxidation rate at the Si/SiO$_2$ interface. The Si-F bonds formed at the interface are positively charged and attract the negatively charged oxygen ion to the silicon, acting as a catalyst for the SiO$_2$ reaction. Additionally, the lower packing density of the fluorinated oxide might enhance the oxygen diffusion [1, 3].

Gate oxide reliability is a concern when using fluorine to enhance the oxidation. Prior studies have shown that excess fluorine incorporation reduces the quality of the oxide, most likely due to Si/SiO$_2$ interface degradation and charge trapping. However, there is a range of fluorine concentrations that have been shown to improve the quality of the oxide. [4].

The optimum fluorine ion implant dose and energy, as well as when in the process fluorine is introduced, are critical in creating a MOSFET with the beneficial characteristics of a thick bottom oxide. A zero-degree fluorine ion beam can be used to preferentially dope the trench bottom (and mesa) as compared to the trench sidewall. The heavier doped areas will then grow more gate oxide. The results indicate that fluorine at an energy of 45 keV and a dose of 2E+16 ions/cm^2, implanted through a sacrificial oxide prior to gate oxide growth, can create a MOSFET device with improved gate oxide integrity and reduced Q_{gd}.

EXPERIMENTAL

Fluorine ions were implanted into silicon trench MOSFET devices fabricated on 0.28 ohm-cm <100> n-type silicon wafers. The fluorine was implanted at 10 keV, 20 keV and 30 keV directly into the silicon prior to gate oxidation for the first sample and at 25 keV, 35 keV and 45 keV through a 400 Å sacrificial oxide for the second sample. The energies were chosen so that the fluorine range was equivalent in the silicon regardless of whether sacrificial oxide was used or not. Fluorine doses ranged from 3E+15 to 2E+16 ions/cm^2. Implantation was performed at a zero-degree angle so that the trench bottom received more implant dose than the trench sidewall. The sacrificial oxide was removed (on the second sample) and the wafers were subjected to an 1175 °C oxidation to grow 400 Å of gate oxide. Oxide thickness was measured on device wafer test structures by ellipsometry. Figure 1 details the process flow.

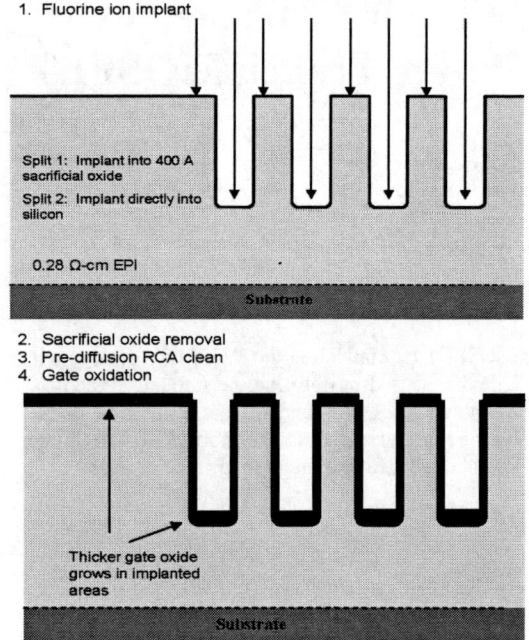

Figure 1. Process flow.

The trench MOSFET devices were used to investigate the impact of the fluorine implant on gate oxide integrity (GOI) and gate-to-drain charge (Q_{gd}).

GOI is evaluated by Charge-to-breakdown (Q_{bd}) measurement with a voltage ramp algorithm, where gate voltage is ramped from 10V up to the point where oxide breakdown occurs. Q_{bd} is calculated by:

$$Q_{bd} = \frac{\sum I_g \cdot \Delta t}{A}$$

Where I_g is the gate current, A is the gate active area (2.82E-2 cm^2), and Δt is the time interval for measurement. Fifty samples are stressed positively. Scanning electron microscope (SEM) photographs were used to verify the increased oxide thickness at the trench bottom on fluorine implanted samples.

For gate charge measurements, 15V of drain voltage and 20nA of gate current are supplied while V_g is monitored. Thus, gate-to-drain charge (Q_{gd}), gate-to-source charge (Q_{gs}) and total gate charge at $V_g = 5V$ (Q_{g5V}) can be calculated.

RESULTS AND DISCUSSION

Oxide Growth

The fluorine implant was shown to enhance the oxidation rate of the gate oxide. There was a 60% increase in oxide thickness, from 400 Å to 640 Å as measured on a test structure, for the device implanted with a 2E+16 ions/cm^2 dose at 45 keV. No difference was noted between the gate oxide thickness for devices with fluorine implanted through a 400 Å sacrificial oxide at 45 keV and those with fluorine implanted directly into silicon at 30 keV (with subsequent gate oxide grown on damaged silicon). SEM cross-sections for representative implant doses at 45 keV were taken to verify that thicker trench bottom oxide was actually grown (Fig. 3).

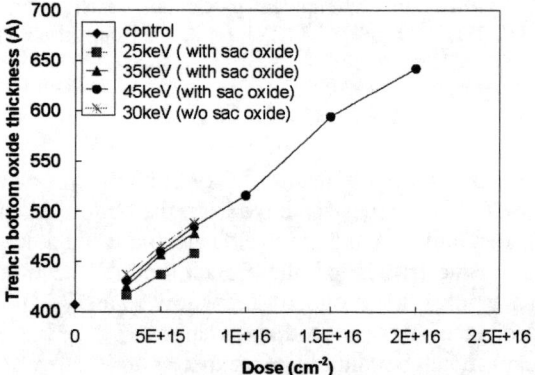

Figure 2. Gate oxide thicknesses as a function of fluorine ion implant dose and energy.

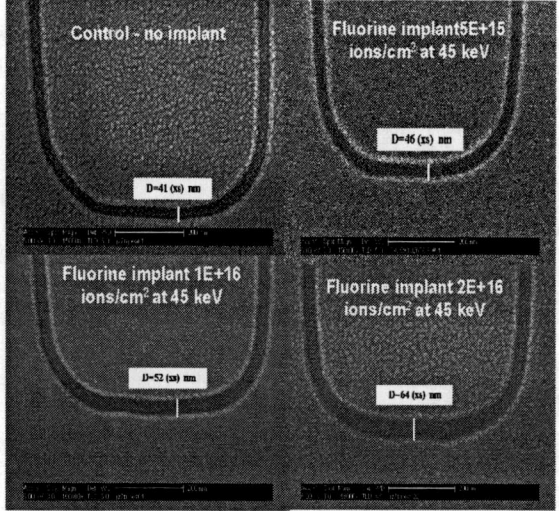

Figure 3. Trench bottom oxide thickness SEM photos.

Gate Oxide Integrity (GOI)

Improved GOI is obtained by implanting fluorine through a sacrificial oxide as compared to implanting directly into the silicon surface to be oxidized, as shown in Fig. 4. The sacrificial oxide is removed prior to gate oxide growth, providing an undamaged silicon surface for the gate oxide growth. Thus, compared to implant without sacrificial oxide and grow gate oxide on damaged Si surface, the quality of gate oxide grown on this surface is better and results in fewer weak devices with shorts or lower breakdown values. Unless specified otherwise, all results discussed below are from samples that were implanted through sacrificial oxide before gate oxidation.

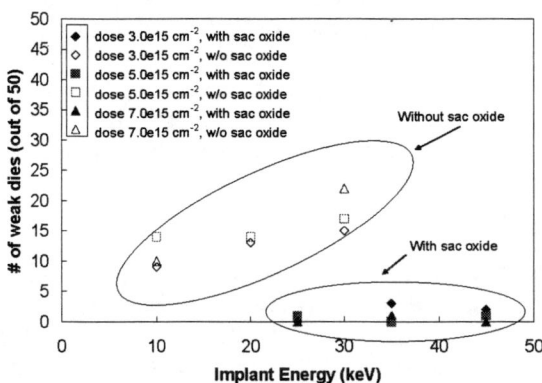

Figure 4. The influence of sacrificial oxide on number of weak dice.

The Q_{bd} Weibull distributions of samples which had a 7E+15 ions/cm^2 implant dose through sacrificial oxide are shown in Fig. 5. First, it can be observed that both control (no implant) and fluorine implanted samples have very uniform and high Q_{bd} values. This indicates excellent GOI on these wafers. Second, it is found that Q_{bd} is improved when the fluorine implant energy is 35 keV or greater.

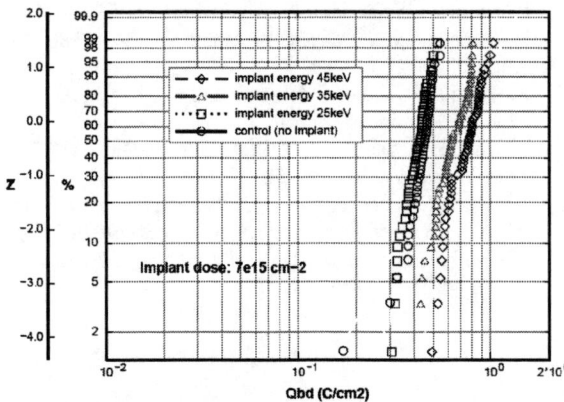

Figure 5. Q_{bd} Weibull distributions of control vs. fluorine implant energy at 7E+15 ions/cm2 dose.

The Q_{bd} Weibull distributions of samples implanted with fluorine doses up to 2E+16 ions/cm^2 at 45 keV through sacrificial oxide are shown in Fig. 6. All samples exhibited very uniform and high Q_{bd} values. It is observed that increasing fluorine dose results in additional Q_{bd} improvement.

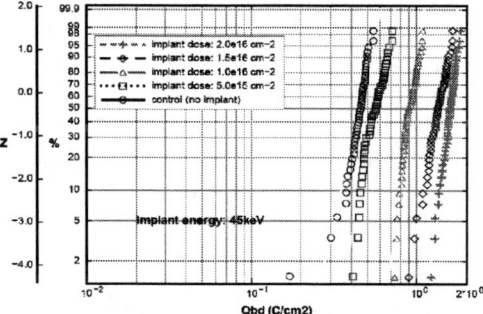

Figure 6. Q_{bd} Weibull distributions of control vs. fluorine implant dose at 45 keV.

The effect of fluorine implant energy and dose on the characteristic Q_{bd} value is summarized in Fig. 7. It is clear that when the implant energy is below 25 keV, practically no effect on Q_{bd} is observed. However, when implant energy is higher than 35 keV, Q_{bd} increases with implant dose and energy. This can be explained as follows: When fluorine atoms are implanted into silicon, there are two mechanisms that are occurring simultaneously. First, the oxidation rate is enhanced so thicker oxide is grown. For trench MOSFETs, the weakest part of the gate oxide is usually located at the bottom of trench, where finite curvature and facets are presented. Therefore, thicker gate oxide at the trench bottom improves the overall GOI of the device. Second, the interface density is increased due to additional implanted fluorine atoms; therefore, GOI is degraded. As shown at Fig. 2, the oxidation rate is increased with higher implant energy and dose. When the implant energy is higher than 35 keV, the benefits from the first mechanism outweigh the negative impact from the second mechanism. Thus, GOI is improved with higher implant energy and dose.

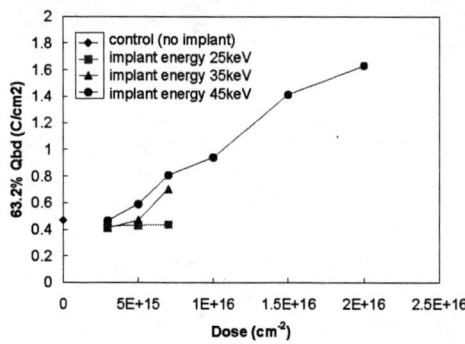

Figure 7. Characteristic Q_{bd} by fluorine dose and energy.

Gate-to-drain Charge (Q_{gd})

Gate-to-drain charge (Q_{gd}), or Miller capacitance, is an important parameter in power MOSFETs. A lower gate charge is desired because it results in reduced switching losses in DC-DC converters. A thicker oxide in the bottom of the trench will reduce the capacitance, thus reducing the Q_{gd} [5].

Figure 8 shows the plots of the gate voltage as a function of time for the MOSFETs implanted with fluorine at 45 keV through a sacrificial oxide. The first portion of the plot, where the gate voltage is rising, is when the gate-to-source capacitance (C_{gs}) dominates. At this stage, the device is still off and the charge injected is called Q_{gs}. The next section of the plot, where the gate voltage flattens out, is where the gate-to-drain capacitance (C_{gd}) is dominant. Here, the channel is starting to turn on and the injected charge, called Q_{gd}, reduces the depletion regions near the drain. The third section is where both capacitances are charged enough to switch the required current and voltage. Table 1 summarizes the results of each dose. Charge is determined by I_g (20 nA) multiplied by time.

Only the devices implanted at 45 keV with sacrificial oxide were considered, as this was shown to provide the best GOI. It can be seen that the Q_{gd} was reduced by 30% on the device implanted with 2E+16 ions/cm^2. This is because the thicker bottom trench oxide results in smaller C_{gd}, therefore less Q_{gd} is needed at the second stage. On the other hand, Q_{gs} is increased with the additional fluorine implantation. This can be explained by the thicker oxide that is also grown on the trench mesa region, inhibiting the source implant. This creates a shallower source-body junction and results higher body region peak concentration at the source-body junction and thus higher threshold voltage is presented.

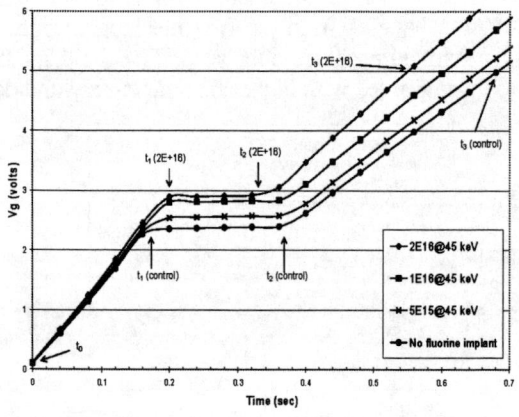

Figure 8. Gate charge plot. Vd = 15 V, Ig = 20 nA.

TABLE 1. Q_{gd}, Q_{gs}, and Q_{g5V} (fluorine implant at 45 keV through sacrificial oxide)

Fluorine Dose	Q_{gd} (nC)	Q_{gs} (nC)	Q_{g5V} (nC)
No implant	4.0	3.2	13.2
3.0E+15	3.9	3.4	13.1
5.0E+15	3.7	3.6	13.0
7.0E+15	3.6	3.8	12.6
1.0E+16	3.2	4.0	12.2
1.5E+16	3.0	4.0	11.2
2.0E+16	2.8	4.0	11.0

CONCLUSIONS

A thick bottom oxide was grown in a trench MOSFET by implanting the trench with fluorine ions at a zero-degree angle through a sacrificial oxide. It is observed that gate oxide thickness is increased with higher implant dose and energy. Additionally, GOI is improved when the implant energy is higher than 35keV. Since the weakest part of the gate oxide is typically located at the bottom of the trench where finite curvature and higher electric field are present, GOI is improved with a thicker trench bottom oxide. A 30% reduction in Q_{gd} was observed as the result of thicker bottom gate oxide. By incorporating a thicker bottom gate oxide, C_{gd} is reduced and thus Q_{gd} is also reduced. Conversely, Q_{gs} is increased due to higher body peak concentration at the source-body junction due to the source implant blocking effect of additional oxide growth on the mesa region caused by the fluorine implantation.

ACKNOWLEDGMENTS

Thanks to Fairchild Semiconductor for the support of this work.

REFERENCES

1. D. Woolsey, *Solid State Technology* **45**, 73-78 (2002).
2. H. Hurst and J. Murphy, U.S. Patent No. 6,437,386 (16 August 2000).
3. M. Morita, T. Kubo, T. Ishihara and M. Hirose, *Appl. Phys. Letters* **45**, 1312-1314 (1984).
4. Y. Mitani, H. Satake, Y. Nakasaki, A. Toriumi, *IEEE Trans. Electron. Devices*, **50**, 2221-2226 (2003).
5. J. Brown, *Proceedings of the 19th Annual Applied Power Electronics Conference and Exposition, Anaheim, CA, 22-26 February, 2004*, (IEEE, Piscataway, NJ, 2004), vol. 3, pp. 1864-1869.

Rare Gas Ion Implanted-Silicon Template for the Growth of Relaxed $Si_{1-x}Ge_x$/Si (100)

G. Regula[1], M. Raissi[2], J.-L. Lazzari[3], F. Chevrier[4], N. Burle[1], E. Ntsoenzok[5,6]

[1] *TECSEN, UMR-CNRS 6122, Case 262, Université Paul Cézanne, F-13397 Marseille cedex 20*
[2] *FSM, Faculté des Sciences de Monastir, Avenue de l'Environnement, T-5019 Monastir*
[3] *CRMCN, UPR-CNRS 7251 associée aux Universités de la Méditerranée et Paul Cézanne, Campus de Luminy, Case 913, F-13288 Marseille cedex 9*
[4] *IES-CEM2, UMR CNRS 5507, Case 067, Université de Montpellier 2, F-34095 Montpellier Cedex 05*
[5] *LESI, Université d'Orléans, 21 rue Loigny La Bataille, F-28000 Chartres;* [6]*CERI-CNRS, 3A rue de la Férollerie, F-45071 Orléans*

Abstract. P-type (100) float zone Si wafers were implanted at room temperature by Ne or He at high dose ($5 \times 10^{16} cm^{-2}$) and low energy (50keV or 10keV respectively). Some of them were annealed at low (700°C) and high (1050°C) temperatures to form a close surface buried layer of tiny cavities and to test their thermal stability. Microscopic studies of these samples revealed that He-implanted wafers were adequate substrates for epitaxial growth. Hence, as-implanted He samples were carefully chemically cleaned to grow by low pressure chemical vapor deposition thin $Si_{0.80}Ge_{0.20}$ layers (about 170nm thick) at 600°C with a growth rate of (11.6 ± 0.5) nm.mn^{-1}. X-ray diffraction measurements demonstrated that the final layer was fully relaxed. Meanwhile atomic force microscopic scans revealed that the surface roughness was low enough (0.3nm) to stand further silicon deposition. Etch pit counts yield a threading dislocation density of $(2.3 \pm 0.7) \times 10^4$ cm^{-2}. Since this density value is about two orders of magnitude lower than what was obtained so far with thin SiGe buffer layers obtained by assistance of gas ion implantation, it is concluded that a new mechanism of threading dislocation annihilation had occurred. Meanwhile, the implanted ductile area was probably very efficient in bending down the threading segments.

Keywords: Implantation, He, Ne, SiGe, thin buffer layer, LP-CVD, relaxation, threading segments, etch pits, AFM, XTEM, XRD
PACS: 81.05.-t, 81.05.Hd, 81.10.-h, 81.15.Jj, 81.20.-n

INTRODUCTION

Formation of voids in silicon-based materials by rare gas ion implantation has been intensively studied since it generates a bundle of possible technical applications for semiconductor industry such as exfoliation, live time control, metal chemisorption, self interstitial or impurity diffusion barrier, low dielectric constant material manufacturing, luminescence and so on. Since year 2000, a new appliance was stepped forward in the field of Si/SiGe heterostructures for strain engineering. Indeed, the latter have potential applications in both optoelectronic and high speed devices, since the band gap value of SiGe as well as SiGe/Si band offsets are very sensitive to both Ge level and strain induced by the rather large lattice mismatch (4.17%). For instance, the effective electron mobility of a 12.5nm thin Si on relaxed $Si_{0.80}Ge_{0.20}$ can double due to about one GPa tensile strain[1]. Many recipes were developed using He implantation, assisted[2] or not by ultrasonic treatments[3], performed during or after the SiGe growth on Si substrate. Some process consisted in strain transfer[4] and involved an implantation through a thin Si/strained-SiGe/Si system followed by an annealing step in the 650°C-1000°C temperature range.

The role of the implantation in the SiGe growth is expected to be double: i) it can swell the point defect concentration, which can boost the number of dislocation nucleation sites; ii) it can help in confining the threading segments (TSs) in the ductile-implanted region[5].On one hand, the above methods did never achieve the best results obtained by growing graded SiGe layer on Si[6]. The relaxation rate R never exceeded 82% and the threading dislocation density N_{TD} was hardly below or equal to $1-2 \times 10^6 cm^{-2}$. On the other hand, the thick graded SiGe layer process resulting in the best crystalline quality had other drawbacks: first, the used method was very time consuming, second, the resulting films had a low SiGe conductivity and are therefore expected to induce self heat during device operation, and third it results in excessive surface roughness.

The current challenge is to grow thin SiGe layer with low surface roughness and N_{TD}. In this respect, this paper presents a new approach in coupling gas ion implantation and growth of thin SiGe layer by low pressure chemical vapor deposition (LP-CVD). It is based on the simple idea of preserving the SiGe layer from implantation damage since the implantation step was scheduled before the growth. Hence, possible annihilation of TSs with cavity-punched dislocations[3,7] can occur throughout the duration of the whole layer deposition. Note that the SiGe layer growth plays the role of an annealing step, thus the cavity ripening can take place concomitantly.

To increase the Ge content in SiGe, Ne implantation seems to be more suitable since it has a higher atomic number than He and hence it should induce more damage. Moreover, the literature data concerning Ne implantation in Si is very scarce. Therefore this paper also reports results concerning upstream studies by cross section electron transmission microscopy (XTEM), optical micrographs and atomic force microscopy (AFM) on both He and Ne implantation carried out on the Si substrate only, in the aim to define the most appropriate candidate for SiGe epitaxial growth.

EXPERIMENTAL PROCEDURE

P-type (100) float zone (FZ) Si wafers were implanted at room temperature by Ne or He at 50 or 10 keV respectively, with a dose of $5 \times 10^{16} cm^{-2}$ and a flux of about $(9.0 \pm 0.3) \times 10^{12}\ cm^{-2}s^{-1}$. During implantation, the samples were 7° tilted to minimize any channeling effect. The implantation parameters were chosen to obtain a layer of tiny bubbles close to the surface without reaching it, as advised in the literature[3].

Part of the as-implanted samples was thermally heated under argon atmosphere at 700°C during 1 hour or 1050°C during 2 hours using a conventional induction furnace, without any quenching. The roughness of the 700°C annealed samples was tested by AFM. All the samples were observed by XTEM. The thin foils were prepared by focused ion beam (FIB) using a gallium source and a protective Pt coating before ion cutting. Nevertheless, this cannot avoid a surface amorphization of about 25nm.

Adequate samples were chosen for deposition of $Si_{0.80}Ge_{0.20}$ at 600°C by LP-CVD with a growth rate of (11.6 ± 0.5) nm.mn^{-1}. The cleaning process prior to the SiGe growth as well as the growth procedure itself is described elsewere[6]. Note however that hydrogen terminated surface obtained by HF dip were used to avoid the oxide removing by annealing at high temperature. With a lattice parameter $a[Si_{0.80}Ge_{0.20}]=0.5472nm$[8], the misfit m is 0.75% between the layer and the substrate. The critical thickness h_c is expected to be in the 10nm-200nm range depending on the chosen model or the growth experimental technique and thermodynamics conditions. Thus, the $Si_{0.80}Ge_{0.20}$ layer thickness was chosen to be close to the upper limit of that range.

The layer surface was studied by AFM and the relaxation rate and homogeneity as well as the exact Ge composition was determined by X-ray diffraction (XRD), using the well known two configurations for both a symmetric (004) and an asymmetric (115) reflections. The N_{TD} was revealed chemically using CrO_3 (0.2M) and HF (40%) 5:4 diluted SECCO like-solution for 3s up to 20s with an etch rate of about 6nm.s^{-1}. Several optical micrograph assemblies were performed for pits count statistics.

RESULTS AND DISCUSSION

Thermal stability of the He/Ne bubbles in Si (100) substrate

The implanted-He sample (Fig.1a, c, e) was found to have a damaged area or a bubble layer at a depth of 120±10nm, from the surface to half width of the damaged region. This is compatible with the projection range R_p=110±50nm obtained by Monte Carlo simulation. After an annealing at 700°C or 1050°C temperatures, neither platelet can be observed by post-mortem XTEM observations or any exfoliation, as expected from the literature. Moreover, AFM studies showed that, even after an annealing of 700°C for one hour, the root mean square roughness (RMS) of the surface is as low as 0.3 nm. Thus, this sample can be a perfect substrate for SiGe alloy deposition at temperature below or equal to 700°C.

Ne-as implanted sample (Fig.1b) exhibited a damaged half width of 125±5nm which is in fair agreement with the simulated R_p=110±40nm. The XTEM micrographs of the 700°C and 1050°C annealed samples reveal that they still contain platelets as displayed in Fig1.d and Fig1.f, respectively. As observed by AFM, a beginning of a partial exfoliation at the exact depth of the platelet chain eliminates the possibility of a following deposition step. However Ne implanted Si could be an appropriate substrate candidate if the dose was lowered by an order of magnitude and the Si was heated during implantation[9]. Indeed, according to the model proposed by Holländer et al[7], the punch-out regime of squared {100} dislocation loops did not come yet to an end. The platelets are still expanding even at 1050°C and thus the TD annihilation process could occur during all the SiGe layer growth.

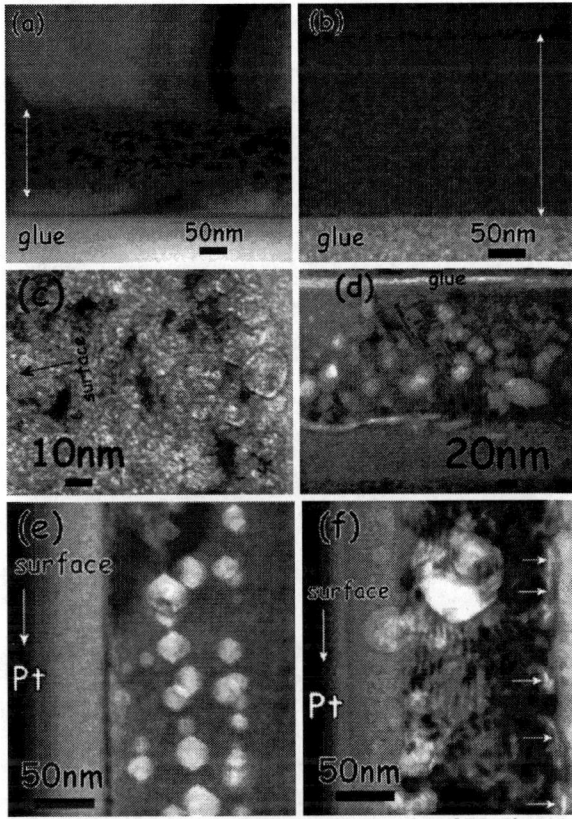

FIGURE 1. Bright field XTEM micrographs of FZ Si (100) substrates implanted with the same dose (5×10^{16} cm^{-2}) with of He$^+$ ions at 10keV (left hand side) or with Ne$^+$ at 50keV (right hand side). a) and b) are as-implanted samples. The white double arrows show the width of the implanted zones; c) and d): after 1h annealing at 700°C; e) and f) after 2h annealing at 1050°C. The platelets still present in f) are pointed out by white arrows.

Characterization of the Si$_{0.80}$Ge$_{0.20}$ template

Figure 2a displays a typical AFM micrograph where the wakes due to few misfit dislocation paths are visible (grey lines). This is the fingerprint of at least a partial relaxation of the SiGe layer. The RMS is as low as 0.8nm that allows growing subsequent strained layers. X-ray topography (Figure 2b) demonstrates that this relaxation is homogeneous on a few squared millimeters.
XRD measurements allowed determining precisely the relaxation rate of the layer as depicted in Figure 3. Under the uncertainties due to experimental angular error when the peaks are fitted by Gaussian functions ($2 \times 0.005°$), the relaxation rate R is found to be almost complete (97%<R<100%). Note that the peak intensities are weak and their width are large due to both the low thickness of the film (170nm) and the loss of coherence due to strain relaxation. The final Ge composition of the layer was calculated taking the Poisson's coefficient given by Wortman et al[10] and found to be 0.206 and 0.196 for (115) and (004) configurations, respectively.

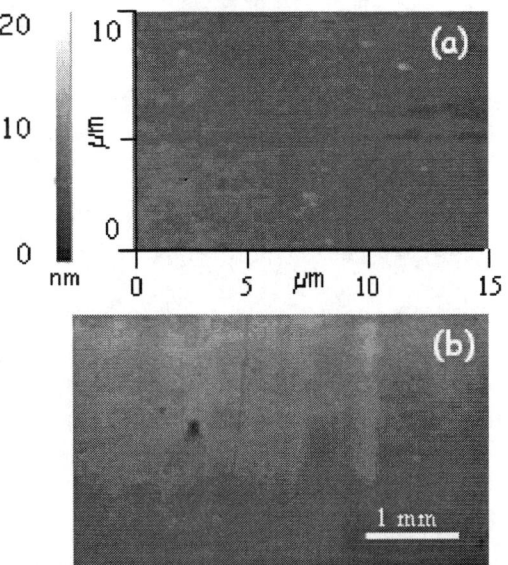

FIGURE 2. Micrographs revealing the relaxation of the Si$_{0.80}$Ge$_{0.20}$ template layer grown on He-implanted Si(100). a) 10×15 μm^2 AFM scan. Note the horizontal fingerprints due to the paths of threading dislocation segments. b) X-ray imaging in transmission mode (g=220). No residual curvature of the sample or dislocation cell structure is noticed.

The threading dislocation density N_{TD} was determined via optical microscopy as displayed in Figure 4. N_{TD} values of $(2.3 \pm 0.7) \times 10^4$ cm^{-2} are about two orders of magnitude of what was reported so far for a process combining implantation and growth. Since the probability of the annihilation of TD segments suggested by the model of Holländer et al[7] is rather low, some other multiplication/interaction mechanisms are probably involved or many threading segments are bent towards the substrate. Despite the low threading segment density, it seems interesting to perform XTEM to identify the bubbles / dislocations interaction and to determine which mechanism occurs among those reported by Vdovin et al[11]. It should actually be possible to clarify the character of the dislocation loops punched out during the bubble/platelet growth.

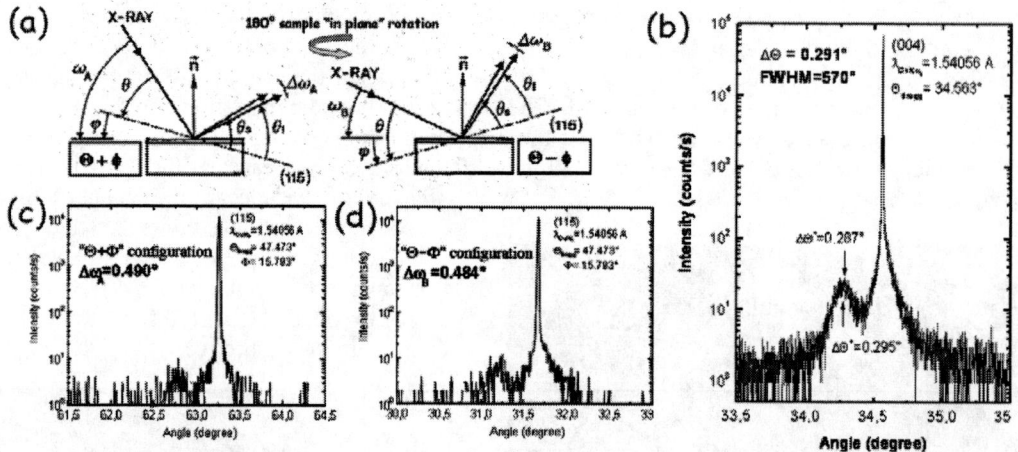

FIGURE 3. X-ray diffraction study of a $Si_{0.80}Ge_{0.20}$ template layer grown by LP-CVD at 600°C on (100) He implanted/annealed Si(100) wafers. a) Experimental configurations labeled "$\Theta \pm \Phi$" shown for (115) asymmetric reflection; b) rocking curves on (004) in both configurations; c) and d) rocking curves on (115) in each configuration.

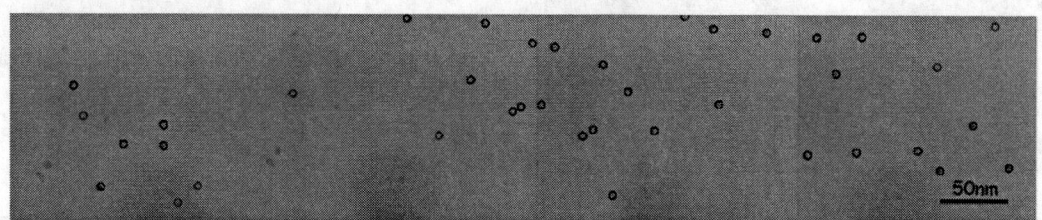

FIGURE 4. Optical micrographs of the chemically etched (CrO_3 (0.2M) and HF (40%) 5:4 combination for 3s, 6nm.s^{-1}) $Si_{0.80}Ge_{0.20}$ layer. The fine dots are underlined by open circles.

Nonetheless, the injection of point defects (and the bubble formation during implantation and annealing) induces a huge benefit by lowering drastically the number of threading segments, as it was reported for SiGe alloys deposited on low temperature grown layers[12-13].

ACKNOWLEDGMENTS

G. Regula is grateful to O. Kaitasov for providing large surface implanted samples, M. Gailhanou and M. Regula for their availability for scientific discussions and technical support.

REFERENCES

1. T. A. Langdo, M. T. Currie, Z.-Y. Cheng, J.G. Fiorenza, M. Erdtmann, G. Braithwaite, C. W. Leitz, C. J. Vineis, J. A. Carlin, A. Lochtefeld, M. T. Bulsara, I. Lauer, D. A. Antoniadis and M. Somerville, Solid State Electronics 48 (2004) 1357-1367.
2. B. Romanjuk V. Kladko, V. Melnik, V. Popov, V. Yukhymchuk, A. Gudymenko, Ya. Olikh, G. Weidner and D. Krüger, Materials Science in Semiconductor Processing **8** (2005) 171-175.
3. M. Luysberg, D. Kirch, H. Trinkaus, B. Holländer, St. Lenk, S. Mantl, H.-J. Herzog, T. Hackbarth and P. F. P. Fichtner, J. Appl. Phys., **92**, 8 (2002).
4. D. Buca, S. F. Feste, B. Holländer, S. Mantl, R. Loo, M. Caymax, R. Carius and H. Schaefer, Solid State Electronics (2006) 32-37.
5. A. R. Powell, S. S. Iyer and F. K. LeGoues, Appl. Phys. Lett. **64**, 1856 (1994).
6. S. Bozzo, J.-L. Lazzari, C. Coudreau, A. Ronda, F. Arnaud d'Avitaya, J. Derrien, S. Mesters, B. Holländer, P. Gergaud, O. Thomas, J. Crystal Growth 216 (2000) 171-184; S. Bozzo, J.-L. Lazzari, B. Holländer, C. Coudreau, A. Ronda, S. Mantl, F. Arnaud d'Avitaya, J. Derrien. Appl. Surf. Sci. 164 (2000) 35-41.
7. B. Holländer, St Lenk, S. Mantl, H. Trinkaus, D. Kirch, M. Luysberg, T. Hackbarth, H.-J. Herzog and P. F. P. Fichtner, Nuc. Instr. and Meth. in Phys. Res. **B** 175-177 (2001) 357-367.
8. D.J. Lockwood and J.M. Baribeau, Phys. Rev. B **45** (1992) 8565-857.
9. S. Peripolli, E. Oliviero, and P. F. P. Fichtner, M. A. Z. Vasconcelos and L. Amaral, Nucl. Instr. and Meth. in Phys. Res. B 242 (2006) 494-497.
10. J. J. Wortman and R.A. Evans, J. Appl. Phys. **36** (1965) 153.
11. V. I. Vdovin, Phys. Stat. sol. (a) **171**, 239 (1999)
12. K. Lyutovich, J. Werner, M. Oehme, E. Kasper and T. Perova, Materials Science in Semiconductor Processing **8** (2005) 149-153.
13. M. Rzaev, F. Schäffler, V. Vdovin and T. Yugova, Mat. Sc. in Semiconductor Processing **8** (2005) 137-141.

High Temperature Implantation of Aluminum in 4H Silicon Carbide

M. Rambach[1], A.J. Bauer[2], and H. Ryssel[1],[2]

[1]*Universität Erlangen-Nürnberg, Lehrstuhl für Elektronische Bauelemente,
Cauerstrasse 6, 91058 Erlangen, Germany*
[2]*Fraunhofer Institut für Integrierte Systeme und Bauelementetechnologie (IISB),
Schottkystrasse 10, 91058 Erlangen, Germany*

Abstract. The influence of implantation temperature on the resistivity of aluminum implanted 4H-silicon carbide was determined. A dose of $1.2 \cdot 10^{15} cm^{-2}$ aluminum ions was implanted at temperatures between room temperature and 1000°C. A decrease of resistivity down to $0.35 \Omega cm$ was found with increasing implantation temperature. The influence of implantation temperature on resistivity was identified by modeling temperature dependent resistivity data. The results showed an increase of mobility due to the lowering of compensating centers.

Keywords: Aluminum, High Temperature Implantation, Mobility, Silicon Carbide.
PACS: 72.80.Jc, 73.61.Lc, 81.40.Rs

INTRODUCTION

Silicon carbide (SiC) is a promising material for high power, high frequency and high temperature electronic devices due to its wide band gap (3.2 eV for 4H-SiC) and its high electrical breakdown field of 2.4 MV/cm compared to Si (band gap of 1.1 eV and electric breakdown field of 0.3 MV/cm). Laterally structured doped layers are necessary for the fabrication of electronic devices. But, doping by diffusion is not possible due to the low diffusion coefficients of dopants in SiC. Therefore, ion implantation is presently the only method to manufacture laterally structured doped layers in SiC.

The implanted ions occupy mainly interstitial lattice sites where they are not electrically active. Thus, a post implantation high temperature annealing step is mandatory to bring the implanted atoms on lattice sites and to reduce implantation induced defects. The temperatures used for the activation of implanted aluminum, the acceptor with the lowest ionization energy in SiC, are 1700°C or even more [1,2]. The achieved resistivity, however, is rather high. A promising way for further reduction of resistivity lies in the use of implantations at elevated temperatures.

EXPERIMENTAL

The substrates used for the implantation and annealing experiments were (0001)-oriented n-type 4H-SiC wafers with an epitaxial layer thickness of 10μm and a net donor concentration of $9 \cdot 10^{15} cm^{-3}$.

For electrical characterization, the samples were implanted with aluminum ions through a scattering oxide at four different energies between 30keV and 180keV in order to obtain a box-shaped profile. The total aluminum dose used was $1.2 \cdot 10^{15} cm^{-2}$. This corresponds to an implantation density of $5 \cdot 10^{19} cm^{-3}$ at the plateau of the profile. The implantation temperatures used were between room temperature (RT) and 1000°C. Annealing of the samples was performed in a resistive heated vertical furnace at a temperature of 1700°C for 30min in argon atmosphere. The resistivity was measured by the 4-point method. The measurement temperature was varied between 10°C and 210°C.

Investigations of crystal structure depending on implantation temperature can be found in the literature [3].

RESULTS AND DISCUSSION

Fig. 1 shows the temperature dependent resistivity measured for implantation temperatures between RT and 1000°C. A decrease of resistivity with increasing measurement temperature is found. The decrease is about 80% for all implantation temperatures. An increase of implantation temperature results in a further reduction of resistivity for a given measurement temperature between 0°C and 250°C. E.g., the decrease of resistivity for a RT implantaton and a 1000°C implantation at 25°C is about 40% with absolute value of 0.35Ωcm. An even more pronounced dependence on implantation temperature is found for 200°C measurement temperature. Resistivity is reduced nearly by a factor of 2 for an implantation temperature of 1000°C compared to a RT implantation.

To get a better understanding of the influence of high temperature implantation on resistivity, the measured data are fitted with a resistivity model.

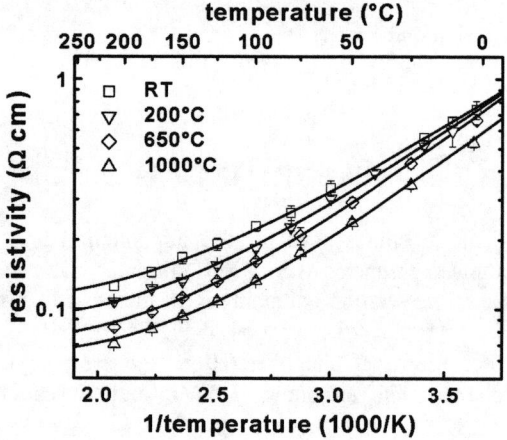

Figure 1. Resistivity vs. measurement temperature for different implantation temperatures. Symbols represent experimental results and lines are fits with the resistivity model.

The model is based on the description of resistivity by

$$\rho(T) = \frac{1}{p(T) \cdot q \cdot \mu(T)} \quad (1)$$

$\rho(T)$ is the resistivity, $p(T)$ the hole concentration, q the elementary charge and $\mu(T)$ the hole mobility. Expressions for $p(T)$ and $\mu(T)$ have to be found in order to model the measured resistivity data.

$p(T)$ is described by the neutrality equation shown in Eq. (2) [4].

$$p(N_a, N_{comp}, T) = 0.5 \cdot \left(-N_{comp} - x + \sqrt{(N_{comp} - x)^2 + 4 \cdot N_a x} \right)$$

$$\text{with} \quad x = \frac{N_V}{g_a} \cdot \exp\left(-\frac{E_{ion}}{kT}\right) \quad (2)$$

The parameters used are the acceptor concentration N_a, the concentration of compensation centers N_{comp}, the effective density of states of the valence band N_V, the degeneracy factor g_a, the ionization energy of aluminum acceptors E_{ion} and the Boltzmann constant k.

Mobility $\mu(T)$ is given by a Thomas-Caughey model [5]. An adapted version of this model for SiC is described in detail elsewhere [6]. In this adapted model, mobility depends on temperature T, concentration of acceptors N_a, and concentration of compensation centers N_{comp}.

The free parameters used for modeling the resistivity using Eq. (1) are therefore the ionization energy E_{ion} and the concentration of compensating centers

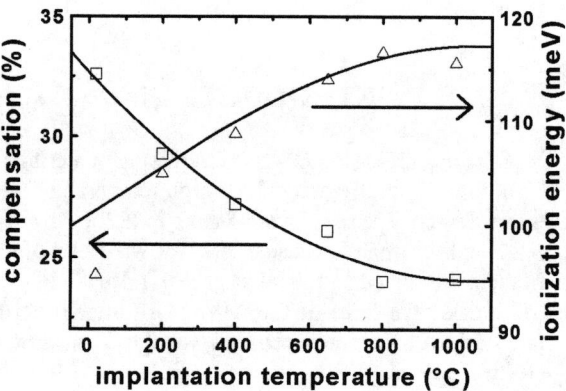

Figure 2. Ionization energy and concentration of compensation centers vs. implantation temperature.

N_{comp}. Both parameters are independent of measurement temperature. An activation of 100% was achieved for the applied aluminum doping and used annealing step, corresponding to an acceptor concentration N_a of $5 \cdot 10^{19}$cm^{-3}, as already presented in [6]. The results of the modeling are shown in Fig. 1 as lines.

The modeled data coincide with the measured temperature dependent resistivity for all analyzed implantation temperatures.

Furthermore, the ionization energy and the concentration of compensation centers is determined by the model. These results are shown in Fig. 2.

The compensation is the ratio of compensation centers N_{comp} to the concentration of acceptors N_a. A decrease of compensation is found with increasing implantation temperature. A room temperature implantation results in a compensation of about 33%, whereas for an implantation temperature above 800°C a compensation below 25% is achieved.

The ionization energy is increasing with implantation temperature. Values of about 100meV are found for room temperature implantation and 115meV for implantation at 1000°C. This behavior can be explained by Eq. (3) [7,8].

$$E_{ion} = E_0 - f(comp) \cdot \frac{q^2 N_a^{1/3}}{4\pi\varepsilon_{SiC}\varepsilon_0} \quad (3)$$

E_{ion} is the ionization energy, E_0 the ionization energy for low doping, $f(comp)$ the correcting factor for compensation, ε_{SiC} and ε_0 are the dielectric constants for SiC and the vacuum, respectively. The function $f(comp)$ is a monotone increasing function for compensation ratios between 0% and 50% [7]. The compensation values found for the investigated samples of Fig. 2 lie between theses values. Therefore, the ionization energy is increasing with decreasing compensation. Due to the decreasing compensation with increasing implantation temperature, the increase of ionization energy with implantation temperature can be explained. The line in Fig. 2 is a calculation of the ionization energy using Eq. (3). The value of E_0 used is 210meV. This value was obtained by fitting Eq. (3) to experimental data of ionization energy for different doping concentrations and compensation ratios (not shown). Published values lie between 190meV [9] and 240meV [10].

The calculated and measured values of ionization energy coincide well. Only for the room temperature implantation, calculated ionization energy is about 5meV higher than the measured one.

The free carrier concentration can be calculated using Eq. (2) by taking into account the extracted values of compensation ratio and ionization energy. The calculated data are shown in Fig. 3 for implantation temperatures between RT and 1000°C.

At lower temperatures down to 0°C, a small difference in carrier concentration is seen between different implantation temperatures. Lower implantation temperatures result in higher free carrier concentrations. This is due to the lower ionization energy of theses samples. But, the difference in free carrier concentration between the implantation temperatures is quite small. Furthermore, the free carrier concentration is comparable for measurement temperatures higher than 200°C for all implantation temperatures. No significant difference can be seen. The maximum deviation of free carrier concentration between different implantation temperatures is only about 10%.

Therefore, the difference in resistivity between different implantation temperatures can not be explained by the free carrier concentration.

Thus, mobility was calculated using a modified Thomas-Caughey model, as describe above. The results for the different implantation temperatures are shown in Fig. 4.

The mobility decreases below 5cm²/Vs for high measurement temperatures above 500°C. This is caused by the increasing scattering at phonons with increasing temperature [11]. This effect is independent of the implantation temperature. Whereas for lower measurement temperatures down to 0°C, the mobility depends on implantation temperature (e.g., a very low mobility of about 15cm²/Vs for a implantation at RT).

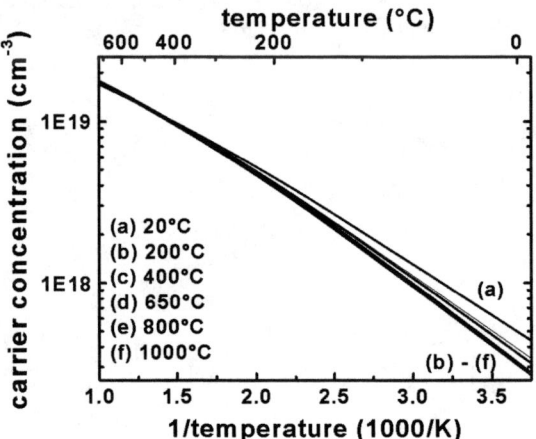

Figure 3. Calculated free carrier concentration vs. temperature for implantation temperatures between RT and 1000°C.

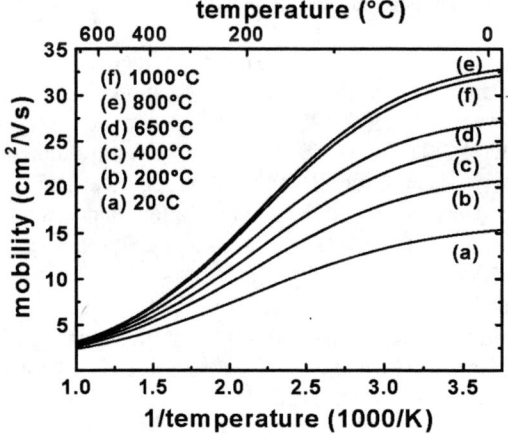

Figure 4. Calculated mobility vs. temperature for different implantation temperatures.

Mobility can be significantly enhanced by increasing implantation temperature. Only for implantation temperatures above 800°C no clear difference in mobility is visible anymore. The highest mobility achieved is about 32cm²/Vs for the measurement at RT. This value is more than twice the value of the RT implantation. An explanation for this mobility behavior is the decreasing concentration of compensation centers with increasing implantation temperature. Therefore, the concentration of ionized scattering centers is decreasing with implantation temperature, resulting in enhanced mobility. The compensation ratio for 800°C and 1000°C implantation temperature is identical, resulting in comparable mobilities, as shown in Fig. 4.

But, even at 200°C measurement temperature a difference in mobility between the tested implantation temperatures is present. Therefore, the differences in resistivity, shown in Fig. 1, are mainly caused by different mobilities rather than different free carrier concentrations.

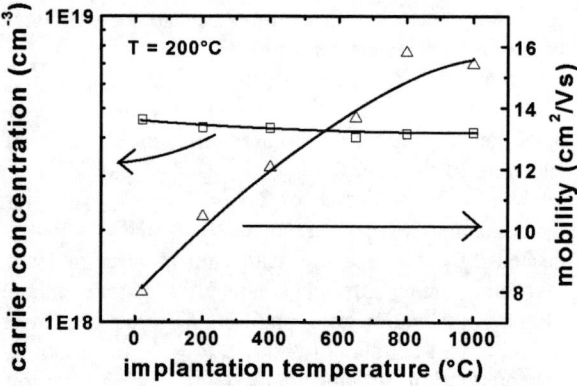

Figure 5. Calculated mobility and free carrier concentration vs. implantation temperature at 200°C.

Hence, in Fig. 5 the free carrier concentration and the mobility are shown for implantation temperatures between RT and 1000°C for a measurement temperature of 200°C. The free carrier concentration stays nearly constant for all implantation temperatures and is $4.5 \cdot 10^{18} cm^{-3}$ for a RT implantation and decreases down to $4.2 \cdot 10^{18} cm^{-3}$ for an implantation temperature of 1000°C. The difference between these values is only about 7%.

In contrast, mobility shows a strong dependence on implantation temperature for a measurement temperature of 200°C. The value for a RT implantation is 8cm²/Vs and increases up to 16cm²/Vs for an implantation temperature of 800°C. A further to 1000°C causes no significant change in mobility.

CONCLUSION

Implantation of aluminum into 4H silicon carbide was performed at implantation temperatures between RT and 1000°C. The temperature dependent resistivity was measured. Due to the simulation of resistivity, an increase of mobility with implantation temperature was found, whereas free carrier concentration is nearly unaffected by the implantation temperature. Therefore, the main effect of implantation at high temperature is the increase of mobility.

REFERENCES

1. T. Troffer, M. Schadt, T. Frank, H. Itoh, G. Pensl, J. Heindl, H.P. Strunk and M. Maier, Phys. Stat. Sol. (a) **162**, 277 (1997).
2. Y. Negoro, T. Kimoto and H. Matsunami, Mat. Sci. Forum **483-485**, 599 (2005).
3. T. Ohno, H: Onose, Y. Sugawara, K Asano, T. Hayashi and T. Yatsuo, J. of Elec. Mat. **38 No.3**, 180 (1999)
4. F. Schmid, M. Krieger, M. Laube, G. Pensl, G. Wagner, Silicon Carbide - Recent Major Advances, edited by W.J. Choyke et al., Springer, Berlin, 517 (2004).
5. D.M. Caughey and R.E. Thomas, Proc. IEEE **55**, 2192 (1967).
6. M. Rambach, L. Frey, A.J. Bauer, H. Ryssel, Mat. Sci. Forum, accepted for publication (2006).
7. A.L. Éfros, N. Van Lien, B.I. Shklovskii, Solid State Physics **12**, 1869 (1979).
8. B.I. Shklovskii, A.L. Éfros, Sov. Phys. Semicond. **14 (5)**, 487 (1980).
9. Y. Negoro, T. Kimoto, H. Matsunami, F. Schmid and G. Pensl, J. Appl. Phys. **96 No. 9**, 4916 (2004).
10. A. Schöner, "Electrical Characterization of shallow and deep impurities in 4H-, 6H- and 15R-Silicon Carbide", Ph.D. Thesis, University of Erlangen, 1994.
11. J. Pernot, S. Contreras, J. Camassel and J.L. Robert, Mat. Sci. Forum **483-485**, 401 (2005).

Annealing of TiO_2 Films Deposited on Si by Irradiating Nitrogen Ion Beams

Katsuhiro Yokota*, Yoshinori Yano*, and Fumiyoshi Miyashita**

*Faculty of Engineering and HRC, Kansai University, Suita, Osaka, 564-8680, Japan
**Faculty of Informatics, Kansai University, Takatsuki, Osaka, 569-1095, Japan

Abstract. Thin TiO_2 films were deposited on Si at a temperature of 600 °C by an ion beam assisted deposition (IBAD) method. The TiO_2 films were annealed for 30 min in Ar at temperatures below 700 °C. The as-deposited TiO_2 films had high permittivities such 200 ε_o and consisted of crystallites that were not preferentially oriented to the c-axis but had an expanded c-axis. On the annealed TiO_2 films, permittivities became lower with increasing annealing temperature, and crystallites were oriented preferentially to the (110) plane.

Keywords: dielectric, IBAD, TiO_2, Si
PACS: 77.55.+f, 81.15.Ji, 81.05Je

INTRODUCTION

Downsizing of dynamic random-access memory (DRAM) for future ultra-large scale integration (ULSI) requires a reduction in the thickness of gate-oxide as an important material of metal-oxide-semiconductor field-effect transistors (MOSFET). The conventional silicon oxide and silicon oxide-nitride-oxide triple layers will reach their applied physical limits in terms of reduction of thickness due to a marked increase in the direct tunnel current [1]. Many people have participated in the development of high dielectric materials showing sufficiently low leakage current densities. Insulating oxides, such as tantalum oxide (TaO_x) [2, 3] and titanium dioxide (TiO_2) [4, 5] have permittivity much higher than silicon oxide, silicon nitride, and oxide-nitride-oxide triplet. Titanium oxide has permittivity higher than TaO_x [3]. Single-crystalline TiO_2 has a permittivity of 89 ε_o for planes normal to the c-axis and 173 ε_o for planes parallel to the c-axis [6], where ε_o is the permittivity of free-space. However, the permittivity of conventionally deposited TiO_2 films is small, at most 30 ε_o [4, 5]. The textured O-deficient TiO_2 films deposited by an ion-beam assisted deposition (IBAD) method [7] had high permittivities of ~ 160 ε_o, approximately as the same as that for planes parallel to the c-axis of the bulk TiO_2 single-crystals [8]. Permittivities degraded at high temperatures.

EXPERIMENTAL PROCEDURE

N-type phosphor-doped (100) Si wafers of 1 - 2 Ωcm were used in this experiment. The Si wafers were ultrasonically cleaned in an acetone bath for 5 min and subsequently in a methanol bath for 5 min. The cleaned Si wafers were charged into the deposition chamber immediately after removed the native oxide layer. Titanium oxide was deposited onto the Si wafers by an IBAD technique constructed from a Ti electron beam evaporation and an ECR ion source to ionize O_2 gas [9]. Ti atoms were impinged onto the Si substrates a rate of about 1.5×10^{14} atoms/cm^2/s. Furthermore, O ions with an energy of 0.5 keV were irradiated on the Si substrates at an intensity of 0.16 mA/cm^2 by flowing O_2 gas at rates of 20 - 30 sccm. TiO_2 films with thicknesses of 50 – 200 nm were deposited on the Si wafers at a temperature of 600°C for 60 min in a vacuum of $\sim 10^{-2}$ Pa. The TiO_2 films were annealed at temperatures below 700 °C for 30 min in flowing Ar gas at a rate of 80 sccm.

The composition and thickness of the TiO_2 films deposited on the Si substrates was measured by the Rutherford backscattering spectrometry (RBS) technique with He ions at the energy of 2.0 MeV. The detection angle was 165° and the angle of incidence was 60°. The spectra were simulated using the computer code RUMP [10]. The thicknesses of the TiO_x films were also measured using a laser probe roughness tester. The crystalline structure of the TiO_2 films was studied using an X-ray diffraction (XRD) meter with a Cu Kα radiation source, collimeters, and a nickel filter. The diffraction angle was accurately measured within 0.016 deg. The permittivities of the TiO_2 films were measured by capacitance-voltage (C-V) characteristic measurements with a time constant of 0.1 s at a frequency of 1 MHz on the MIS structures.

RESULTS and DISCUSSION

Figure 1 shows XRD spectra on as-deposited and annealed TiO_2 films. The TiO_2 films consisted of many crystallites belonged to a rutile-phase and were oriented preferentially to the (110) plane [14]. RBS measurements revealed that the as-deposited and annealed TiO_2 films had an atomic ratio of Ti : O $\approx$ 1 : 2. The lattice constants of the TiO_2 crystallites are shown in Fig. 2. TiO_2 crystallites in the as-deposited films were expanded to the c-axis direction. The volume of the unit cell in the as-deposited films is larger than that on the rutile-TiO_2 seen in the AMTS card [11] by about 3.3 %. Thus, on the as-deposited TiO_2 films, a Ti atom in the unit cell can gain a high freedom of movement compared with other annealed TiO_2. The average sizes of the crystallites in the TiO_2 film can be calculated from the full width at half maximum (FWHM) of the XRD peaks from the (110)

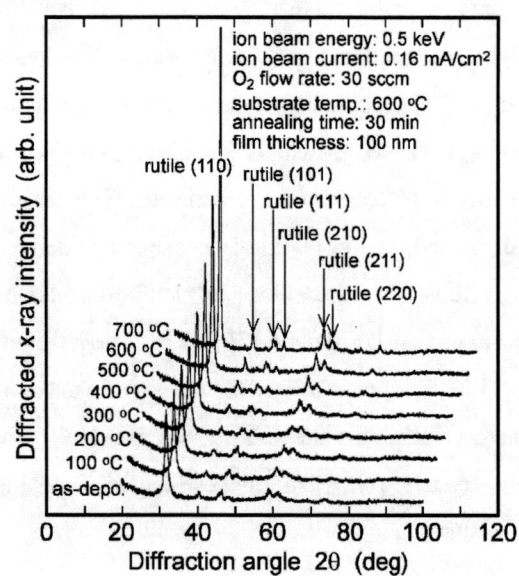

FIGURE 1. XRD spectra on as-deposited and annealed TiO_2 films.

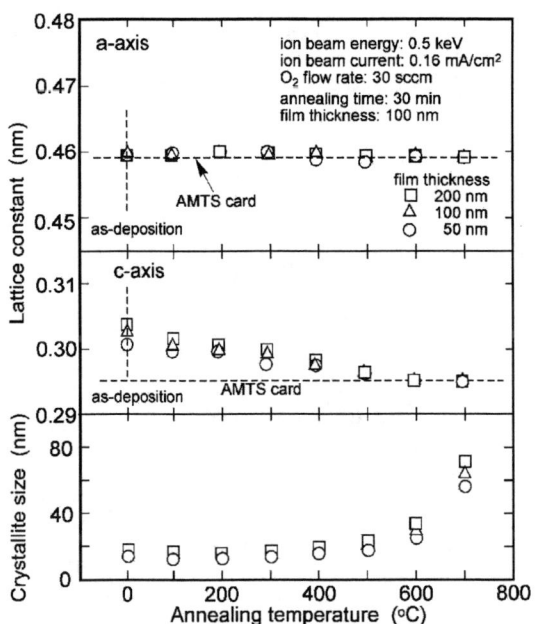

FIGURE 2. Lattice constants and crystallite sizes on as-deposited and annealed TiO$_2$ films.

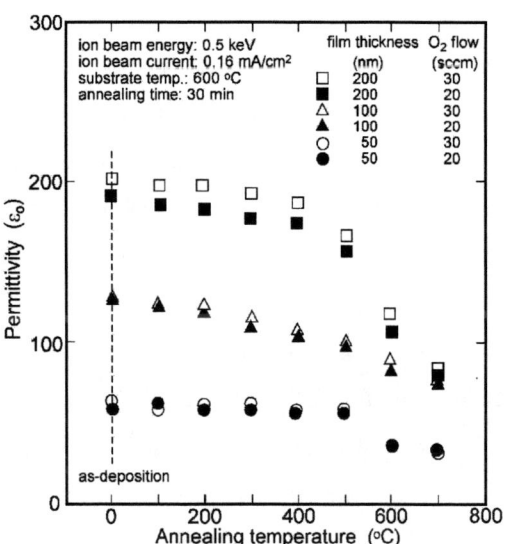

FIGURE 3. Permittivities as functions of film thickness and annealing temperature on TiO$_2$ films.

c-TiN crystallites and the XRD peak angle by using Scherrer's equation [12]. The as-deposited TiO$_2$ films had crystallites with average sizes of about 14 nm.

XRD peaks from various rutile TiO$_2$ crystalline planes grew with increasing annealing temperature. The intensity of the (110) XRD peak increased exponentially with increasing annealing temperature, and the values of FWHM for the (110) XRD peaks were larger at temperatures above 500 °C: the (110) XRD peak started to grow strikingly. The crystallite's sizes increased from about 14 nm on the as-deposited TiO$_2$ films to about 70 nm on the 700 °C-annealed TiO$_2$ films. A decrease in the length of the c-axis on the rutile TiO$_2$ crystalline lattice occurred in conjunction with the growth of the strikingly growth of the (110) XRD peak and the crystallite's size. The long c-axis of the as-deposited TiO$_2$ films had the same length as that given in the AMTS card [11] on the 600 °C-annealed TiO$_2$ films. That is, the deformation in the unit cell by collision of energetic oxygen ions was diminished by annealing the TiO$_2$ films at high temperatures. Small crystallites merged with each other and became large crystallites.

Figure 3 shows permittivities on annealed TiO$_2$ films. The permittivities decreased with increasing annealing temperature and film thickness. Larger compressive stress is generated in the thin rutile TiO$_2$ film and tensile stress is generated in Si wafers. The compressive stress is a maximum value near the TiO$_2$ film surface at the interface between the TiO$_2$ film and the Si wafer, and decreases with increasing distance from the interface. The averaged compressive stress over all the film becomes smaller with increasing film thickness. The compressive stress in materials restricts the movement of atoms by electric field. The permittivity on materials reflects polarization caused by the displacement of ions in the unit cell by application of an electric field. Thus, the permittivities are small on thin TiO$_2$ films, and large on thick TiO$_2$

films.

A striking decrease in permittivity started on the TiO$_2$ films annealed at temperatures around 500 °C. This is caused by a prompt increase of the rutile (110) XRD peak corresponded on the annealing temperature dependence of the crystalline structure: the permittivities decreased with shortening the length of the c-axis. When an electric field is applied to the TiO$_2$ film, O$^-$ ions in the unit cell displace toward the positive electrode and Ti$^+$ ions displace toward the negative electrode. The large displacements of ions in the unit cell resulted in the high permittivities on the TiO$_2$ films. On the as-deposited TiO$_2$ films, Ti$^+$ ions can be displaced largely in the unit cell expanding by the oxygen ion bombardment. The as-deposited TiO$_2$ films had approximately the same permittivity as bulk TiO$_2$ [6]. The c-axis length of the unit cell became shorter with increasing annealing temperature as shown in Fig.2. This shortened c-axis restricted the displacement of Ti$^+$ ions in the unit cell, and the TiO$_2$ films had low permittivities.

CONCLUSION

50 – 200 nm TiO$_2$ films were deposited on Si at a temperature of 600 °C by an IBAD method. The TiO$_2$ films had permittivities proportional to film thickness because they consisted of crystallites with an expanded c-axis. The 200 nm as-deposited TiO$_2$ films had high permittivities near 200 ε_o. On the annealed TiO$_2$ films, many crystallites were oriented preferentially to the (110) plane and the expansion of the c-axis diminished. Thus, the permittivities of the annealed TiO$_2$ films were lower with increasing annealing temperature.

REFERENCES

1. F. Hamzaogle and M. R. Stan, Circuit-Level Technique to Control Gate Leakage for Sub-100 nm CMOS, in proc. of ACM/IEEE ISLPED, Aug. 2002, pp.60-63.
2. D. M. Hausmann, P. de Rouffugnac, A. Smith, R. Gordon, and D. Monsma, Thin Solid Films, **443**, 1-4(2003).
3. G. B. Alers, D. J. Werder, and Y. Chabal, Appl. Phys. Lett, **71**, 1517-1529(1998).
4. T. Fuyuki and H. Matsunami: Jpn. J. Appl. Phys. **25**, 1288(1986).
5. T. Nakayama, K. Onisawa, M. Fuyama and M. Hanazono: J. Electrochem.Soc. **139**, 1204(1992).
6. D. R.Lide: editor-in-Chief, Handbook of Chemistry and Physics, 80th edition, CRC, New York, 1999, pp. 12-60
7. S. M. Sze, Semiconductor Devices Physics and Technology, Wiley New York, 1985, p. 335.
8. K. Yokota, K. Nakamura, T. Sasagawa, T. Kamatani, Jpn J. Appl. Phys. **40**, 718(2001).
9. E. Bergmann: Surf. Coat. Technol. **57**, 133(1991).
10. L. R. Doolittle: Nucl. Instrum. Methods Phys. Res. **B9**, 344(1985).
11. Powder Diffraction Files, JCPOS, Pennsylvania, 1991, Card No. 21-1276.
12. Cullity, B. D., 1978, Elements of X-Ray Diffraction, Werley, New York, Chap.3.

Thermal Diffusion Barrier for Ag Atoms Implanted in Silicon Dioxide Layer on Silicon Substrate and Monolayer Formation of Nanoparticles

Hiroshi Tsuji, Nobutoshi Arai*, Naoyuki Gotoh, Takashi Minotani, Toyotsugu Ishibashi, Tetsuya Okumine*, Kouichiro Adachi*, Hiroshi Kotaki*, Yasuhito Gotoh and Junzo Ishikawa

Department of Electronic Science and Engineering, Kyoto University (Nishikyo-ku, Kyoto 615-8510, Japan)
*Advanced Technology Research Laboratories, SHARP Corporation (Ichinomoto-cho, Tenri, Japan)

Abstract. We have investigated thermal diffusion behavior of implanted Ag atoms in SiO_2 by using a high-resolution RBS method in the formation process of monolayered Ag nanoparticles. Ag atoms were implanted by negative ion implantation at 10 keV with 5×10^{15} ions/cm^2 into the 25 nm-thick SiO_2/Si. Samples were annealed at 500 - 800°C for 1 h under Ar gas flow. At annealing temperature of 500°C, implanted Ag atoms distributed at the surface and at a depth corresponded to the calculated profile. It is expected that the surface accumulation of Ag atoms resulted from thermal diffusion of implanted atoms during implantation. At 500°C, the very small peak in concentration was observed at a depth of 22 nm. This means that a diffusion barrier for Ag atoms exits in this depth. The diffused atoms accumulated at this depth. At 700°C, the main peak of concentration was appeared at 20 nm in depth, where FWHM was 7 nm. These results well corresponded to the mono-layered Ag nanocrystals observed by HR-TEM.

Keywords: Negative-ion, Ion implantation, Thermal diffusion, Silver nanocrystal, RBS, Diffusion barrier.
PACS: 41.75.Cn, 68.55.Ln, 81.07, 66.30.-h, 68.35.Fx, 61.43.Hv, 82.80.Yc, 81.70.Jb

INTRODUCTION

Single electron devices by using nanoparticles (NPs) are expected as multi-value distinctive and low-power consumption devices, because that insulator film including NPs show Coulomb blockade phenomenon at room temperature [1,2]. As formation method of such NPs in a thin gate oxide layer in MOSFET, there were reported several methods by using ion implantation technique for monolayered NPs of Sn by A. Nakajima [3], Cu by Y, Takeda [4], Si by K.H. Heinig [5] and Ag [6, 7] and Au [7, 8] by H. Tsuji. In the above formation methods, the both Sn and Ag NPs were formed at a deep depth and just near the interface between SiO_2 and Si by appropriate thermal annealing after ion implantation. This position was deeper than the projected range at 10 keV. Heinig also formed Si NPs near the interface, but he implanted Si ions at a relatively high energy to result an excess of Si atoms there by recoils from Si substrate. Therefore, redistribution of implanted atoms in SiO_2 is important for Sn and Ag NPs, in which the depth position of NPs was both at a vicinity distance of 2 nm from the interface of SiO_2/Si. Although a diffusion barrier can be expected in the vicinity region at the interface, there is no report for such diffusion barrier for Sn and Ag atoms. The formation mechanism of NPs aligned along the interface of SiO_2/Si substrate is still unknown.

In this paper, we have investigated the thermal diffusion of implanted Ag in a thermally grown thin silicon dioxide layer on silicon substrate by using a high-resolution Rutherford backscattering spectrumetry (HRBS). After annealing, the reaggregation of Ag atoms was found in the SiO_2 layer. This was well corresponded to the monolayered Ag NPs observed by TEM.

EXPERIMENTAL

We implanted ^{107}Ag negative ions from an RF plasma-sputtered type heavy negative ion source and extracted at 10 keV [9, 10] after mass-separation into a thermally grown silicon dioxide layer with thickness of 25 nm on silicon substrate (15 mm x 15 mm) at ion energy of 10 keV with dose of 5×10^{15} ions/cm^2 at a

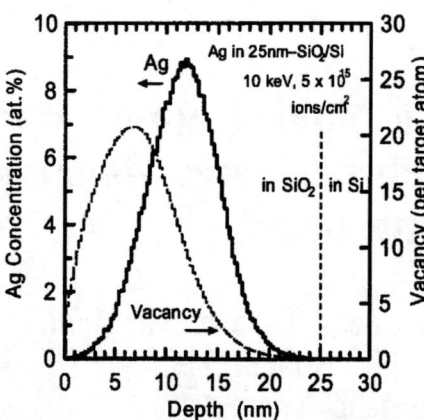

FIGURE 1. Estimated concentration of Ag atoms implanted into SiO_2 medium at 10 keV with 5×10^{15} ions/cm^2 calculated by TRIM-DYN program.

room temperature. The implantation was performed in a Faraday cup in an implantation chamber of the negative ion implanter (Nissin Electric Corp., Japan). The projected range of Ag atoms is calculated to be about 12 nm in amorphous SiO_2 with 2.20 g/cm^3 by the transport of ion in matter (TRIM-DYN) program [11] and shown in Fig.1. This range corresponds to about a half depth of the SiO_2 thickness. Negative-ion implantation was applied this study since it has almost "charge-up free" feature [12] for insulators. Therefore it is expected that almost same depth profile as the calculated is obtained without any charge compensator in the collector cup. After implantation, these implanted samples were annealed at various temperatures of 500, 700 and 800°C for 1 h under Ar flow (50 ml/min) condition in an evacuated quartz tube of an electric oven.

The depth distribution of implanted Ag atoms in the thin SiO_2 layer were measured by HRBS in an ultra high vacuum condition of 10^{-6} Pa by using relatively low-energy He ion beam as a probe primary. The beam energy and current intensity of the probe beam (2×2 mm^2) in the HRBS were 400 keV and 30 nA, respectively. The reflected ions at 100 degrees were analyzed of their energy by a magnetic energy analyzer in a range from 230 to 410 keV with an energy step of 0.388 keV. The depth resolution under the above total conditions is considered to be 0.5 nm at the surface region in a silicon dioxide. Random spectrum was searched by rotating the sample surface and recorded. The depth profile of Ag atoms in the SiO_2 layer was obtained by backscattering cross-section and scattering yield ratio of Ag and Si atoms in amorphous SiO_2. The typical error in RBS measurement is considered to be ±5%.

The direct observation of nanoparticles in the samples was investigated by cross-sectional transmission electron microscope (XTEM) by 200 keV electron beam. The XTEM specimens were prepared by FIB (Ga$^+$, 30 keV).

RESULTS

Ag Depth Profiles by HRBS

In the calculation by TRIM-DYN, the implanted Ag atoms at 10 keV with 5×10^{15} ions/cm^2 distribute as like a Gaussian profile with a peak at a depth of 12 nm with Ag concentration of about 9 at.%. The total sputtering yield was 1.8. But, the surface recession due to sputtering was about 0.65 nm as assuming expansion by implanted Ag atoms. The Ag ions passed through the layer were only 0.018%.

Figure 2 shows obtained HRBS spectra near the surface region, where black arrows at 364 and 284 keV indicate the Ag and Si (in SiO_2) edges. The spectra of as-implanted and after annealing at 500°C samples show two Ag peaks and many small ones. The peaks near 364 eV corresponded to the surface accumulation of Ag atoms. The Other peak near 350 keV corresponded to the implantation peak of Ag near the 12-nm depth. The yields below 255 keV were contributed by Si atoms in Si substrate. From these yields, we can calculate Ag depth profiles in SiO_2.

The obtained Ag depth profiles are shown in Fig. 3. The SiO_2 layer is amorphous and the Si-yields allows us to use it as a standard even when channeling in Si substrate might happened. Then, in the calculation of Ag concentration, we used Si yield around 280 keV from the SiO_2 as the standard density, after fluence correction of He ions. The accumulation of Ag atoms at the surface for two samples of as-implanted and after 500°C-annealing was observed. These two samples also showed concentration peak of Ag atoms at the depth of 12 nm (black dawn allow). In the sample annealed at 500°C, there are other two small peaks around 6 nm and 22 nm (white dawn allows), indicating small Ag-segregation due to the annealing.

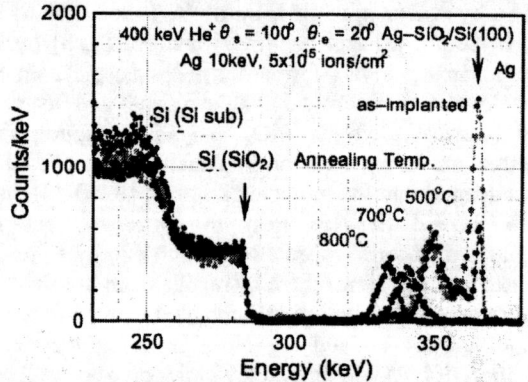

FIGURE 2. HRBS yields of Ag-implanted SiO_2/Si samples before and after annealing at various temperatures of 500°C, 700°C and 800°C.

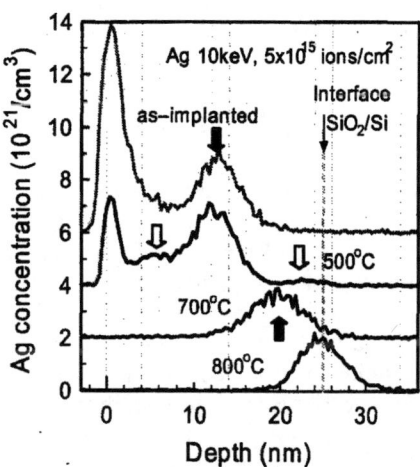

FIGURE 3. Depth profiles of Ag atoms implanted SiO$_2$ layer before and after annealing at various temperature of 500°C, 700°C and 800°C by HRBS.

The remained Ag atoms in samples were 5×10^{15} and 3.3×10^{15} atoms/cm^2 for the as-implanted and after annealing at 500°C.

After annealing at 700°C, Ag atoms moved to the deeper depth and segregated at 20-nm depth. In the surface side from the peak, there is no any other Ag segregation. The Ag segregations at 22 nm for 500°C-annealing and 20 nm for 700°C suggested an existence of a certain diffusion barrier of Ag atoms in the SiO$_2$ layer near the interface of SiO$_2$/Si. At 800°C-annealing, Ag atoms reached the Si substrate. The Ag atoms remained in the samples after annealing at 700°C and 800°C are both 1.4×10^{15} atoms/cm^2, which is about 28 % of implanted Ag atoms.

Ag Nanoparticles and Position

Cross-sectional TEM images are shown in Fig. 4, although the images of Ag nanoparticles formed in the SiO$_2$ layer were already reported [6-8]. As shown in Fig. 4a, Ag NPs with 3 - 4 nm in diameter were formed in the SiO$_2$ layer at 500°C-annealing. After annealing at 700°C in Fig. 4b, all the Ag NPs were obtained at the same depth of 20 nm, with the minimum distance of 2 nm from the interface. The particle size was almost 7 nm in diameter. The position of formed Ag NPs well agreed with the peak depth of remained Ag atoms in the HRBS at 700°C. After annealing at 800°C, no particle was observed inside the SiO$_2$ layer, but some wedge-like things were observed in the interface region. This was considered to be reactants of Ag and Si.

As for Ag NPs at 500oC-annealing, the depth distribution of 47 Ag NPs obtained from counting Ag NPs every 2-nm depth of Fig. 4a is shown in Fig. 5. This histogram shows two main peaks in Ag NP number at depths of 13 nm and 21 nm. No Ag NP at the surface was obtained.

DISCUSSION

Ag Thermal Diffusion

The surface accumulation of Ag for the as-implanted sample and after annealing at 500°C was considered to be due to thermal diffusion during implantation procedure, beam heating by nuclear collisions of ion and recoils with solid atoms.

After annealing at 500°C, three segregations of Ag atoms, excepting the Ag layer at the surface, obtained by HRBS are at depths of 6, 12 and 22 nm. In the energy loss process in ion implantation at 10 keV in SiO$_2$, the depth distribution of displacement energy of target atoms is calculated to have a peak at a depth of 6.5 nm by TRIM-DYN as shown in Fig.1. Such displacement energy might result many damages such as oxygen deficit or voids even in the amorphous SiO$_2$. Therefore, the segregation of Ag at 6 nm is considered to be contributed by ion-induced effect. The main segregation of Ag atoms at 12-nm depth in HRBS well agreed with the estimated projected range of Rp, this depth was also corresponded to the peak depth of 13 nm of observed Ag NPs and in the histogram of Fig. 5. The small Ag segregation at deep depth of 22 nm

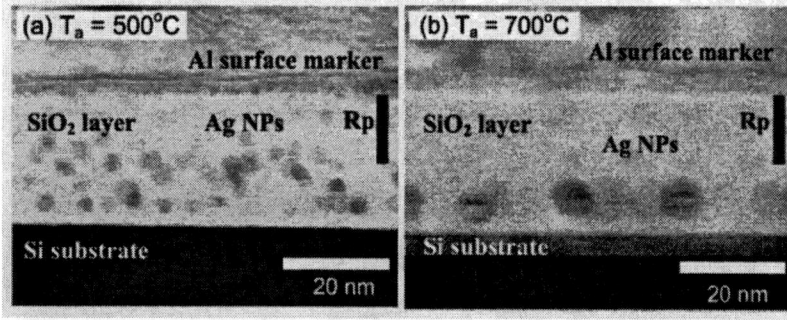

FIGURE 4. Cross-sectional TEM images of Ag- implanted 25-nm-SiO$_2$ layers on Si samples at 10 keV with 5×10^{15} ion/cm^2 after annealing at various temperatures of (a) 500°C and (b) 700°C.

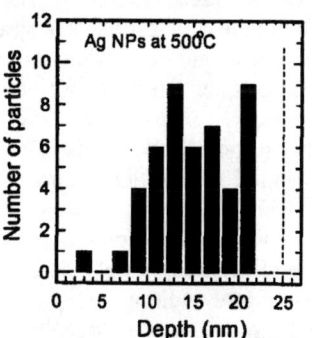

FIGURE 5. Histogram of Ag NPs formed in 25-nm-SiO$_2$ layers after annealing at 500°C.

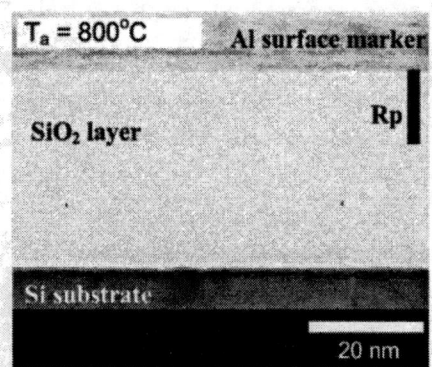

FIGURE 6. Cross-sectional TEM images of Ag-implanted 25-nm-SiO₂ layers on Si after annealing at 800°C.

obtained by HRBS analysis is fairly corresponded to the second peak at 21 nm in the Ag NP histogram in Fig. 5. These deepest Ag NPs are considered to sharply align. This alignment suggests a certain diffusion barrier near the depth of 22 nm. After annealing at 700°C, the segregation of Ag atoms in HRBS at the depth of 20 nm well agreed with the position of monolayered of Ag NPs in TEM shown in Fig. 4b. The implanted Ag atoms are considered to diffuse toward the deeper side from the implanted site and to be trapped at the depth of 22 nm and then accumulated here. As a result, the diffusion barrier exists at a vicinity distance from the interface of SiO_2/Si. However, the barrier lost its effect to the thermal diffusion of Ag atoms at 800°C.

Although the reason for the barrier is not understood, the followings are considered for resulting density change in SiO_2. The thermally grown silicon dioxide on Si substrate is generally amorphous. On the other hand, Si substrate is a single crystal with a periodic regularity. In the oxidation process, oxygen atoms at first bind the surface Si atoms. Then, O atoms gradually invade between Si-Si bonding to grow the SiO_2. Therefore, the regularity of Si crystal is expected to affect SiO_2 formation in configuration in a vicinity region at the interface SiO_2/Si. The transition region in SiO_2 facing to Si substrate was stressed to increase the density a little in compassion to the normal amorphous SiO_2. This is considered to act as a diffusion barrier of Ag atoms.

The image of the sample after annealing at 800°C is shown in Fig.6. In the Si surface region of the image, some wedge-like things were observed. This was reactant of after diffusion of Ag atoms to the interface. In general, Ag and Si make eutectic alloy, which has a relatively low melting point. The TEM-image was taken from the <110> direction of Si substrate. The ridge lines of both sides in the Si are almost parallel to the planes of Si (111). So, Ag atoms were considered to penetrate along the plane of Si (111) to make some alloy.

Diffusion Barrier

The followings are considered for resulting density change in SiO_2, although the reason for the barrier is not fully understood, The thermally grown silicon dioxide on Si substrate is generally amorphous. On the other hand, Si substrate is a single crystal with a periodic regularity. In the oxidation process, oxygen atoms at first bind the surface Si atoms. Then, O atoms gradually invade between Si-Si bonding to grow the SiO_2. Therefore, the regularity of Si crystal is expected to affect SiO_2 formation in configuration in a vicinity region at the interface SiO_2/Si. The transition region in SiO_2 facing to Si substrate was stressed to increase the density a little in compassion to the normal amorphous SiO_2. This is considered to act as a diffusion barrier of Ag atoms. Owing to this diffusion barrier, monolayered Ag NPs were formed at a bottom region of the oxide layer.

CONCLUSION

The depth profiles of Ag atoms implanted at 10 keV in a 25-nm-thick thermally grown SiO_2 layer on silicon substrate after annealing at various temperatures have measured by using the HRBS method. The implantation was performed with negative Ag ions at 10 keV. The calculated projected range is about 12 nm, which corresponded to a half of the thickness of the oxide layer. Even this very shallow depth, the HRBS by He ions at 400 keV with a magnetic energy analyzer was able to detect the depth profile with an accuracy of about 0.5 nm.

The surface Ag accumulation for samples as-implanted and after annealing at 500°C is considered to be due to local heating by ion beam during implantation. After annealing at 500°C, one large and two small segregations of Ag were detected inside the layer. The main peak is placed at the depth corresponding to the calculated projected range of 12 nm. Others were at 6 nm in depth near the surface and at 22 nm in the oxide layer near the interface of SiO_2/Si. The deep segregation is expected to be due to density change of SiO_2 in the transition layer in the vicinity of the interface. This transition layer worked at a diffusion barrier up to a temperature of 700°C. Owing to this barrier, the monolayered Ag NPs were obtained after annealing at 700°C, aligning along the interface, from which the minimum distance of 2 nm. The aligned Ag NPs observed by cross-sectional TEM image were well corresponded to the distribution of Ag atoms obtained by HRBS.

ACKNOWLEDGMENTS

The authors are grateful to Prof. K. Kimura and Dr. K. Nakajima for assistance of the HRBS measurement in Nanotechnology Support Project in Kyoto University.

REFERENCES

1. K. Yano, T. Ishii, T. Hashimoto, T. Kobayashi, F. Murai, and K. Seki, IEEE Transactions **ED 41,** IEEE, 1994, pp. 1628-1638
2. H. Tsuji, N. Arai, T. Matsumoto, K. Ueno, Y. Gotoh, K. Adachi, H. Kotaki, and J. Ishikawa, *Appl. Surf. Sci.,* **238,** 132-137 (2004).
3. A. Nakajima, T. Futatsugi, N. Horiguchi, and N. Yokoyama: *Apple. Phys. Lett.,* **71,** 3652-3654 (1997).
4. Y. Takeda, C.G. Lee, N. Kishimoto, *Nucl. Instr. and Meth. in Phys. Res.* **B 191,** 422-427 (2002).
5. K. H. Heinig, T. Mueller, B. Schmidt, M. Strobel, and W. Moeller, *Appl. Phys.,* **A 77,** 17-25 (2003).
6. H. Tsuji, N. Arai, T. Matsumoto, K. Ueno, K. Adachi, H. Kotaki, Y. Gotoh, and J. Ishikawa, *Surf. Coat. Tech.,* **196,** 39-43 (2005).
7. J. Ishikawa, H. Tsuji, N. Arai, T. Matsumoto, K. Ueno, K. Adachi, H. Kotaki, Y. Gotoh, *Nucl. Instr. and Meth. in phys. Res.* **B 237,** 422-427 (2005).
8. H. Tsuji N. Arai, K. Ueno, T. Matsumoto, N. Gotoh, K. Adachi, H. Kotaki, Y. Gotoh and J. Ishikawa, *Nucl. Instr. and Meth. In Phys. Res.* **B 242,** 125-128 (2006).
9. H. Tsuji and J. Ishikawa, *Rev. Sci. Instrum.* **63,** 2488-2490 (1992).
10. H. Tsuji, J. Ishikawa, Y. Gotoh, and Y. Okada, AIP Conf. Proceeding 287, American Institute of Physics, Melville, NY, 2002, 1994, pp. 530-536.
11. J.P. Biersack, *Nucl. Instr. and Meth. In Phys. Res.* **B 27,** 21-36 (1987).
12. H. Tsuji, Y. Gotoh, J. Ishikawa, *Nucl. Instr. and Meth. in Phys. Res.* **B 141,** 645-651 (1998).

Water Splitting and Hydrogen Emitting Catalytic Function of Hydrogen-Implanted Oxide Ceramics Studied Using Ion Beam Technology

K. Morita, B. Tsuchiya[1], S. Nagata[1], K. Katahira[2], M. Yoshino[3], J. Yuhara[3], Y. Arita[5], T. Ishijima[4] and H. Sugai[4]

Department of General Education, Faculty of Science and Technology, Meijo University, 1-501 Shigamaguchi, Tenpaku-ku, Nagoya 468-8502, Japan
[1] *Institute for Materials Research, Tohoku University, 2-1-1 Katahira, Aoba-ku, Sendai 980-8577, Japan*
[2] *Advanced Research Institute for Materials, TYK Co. Ltd., Ohata-cho, Tajimi 507-8607, Japan*
[3] *Department of Quantum Science and Energy Engineering and* [4] *Department of Electronic Engineering, Graduate School of Engineering,* [5] *Institute for EcoTopia Science, Nagoya University, Furo-cho Chikusa-ku, 464-8603, Japan*

Abstract. Temperature dependence of the D-H replacement speed in D-implanted oxide ceramics exposed to H_2O vapor has been studied using the ERD technique. It is found that the D-H replacement speed increases when the exposed specimen is heated above the implantation temperature and the amount of H uptaken by the D-H replacement is almost the same as the implantation amount of D, while both D as-implanted and H uptaken in the replacement are released by heating the specimen in the vacuum. No reduction in the amount of H uptaken by exposure to H_2O vapor at elevated temperatures is ascribed to enough supply of H from the surface due to splitting of H_2O faster than their thermal release. Based on the one way diffusion model of both dipole-induced water splitting at the surface and hydrogen gas emission from the bulk, it is shown that thermal detrapping of D implanted and H replaced in the specimen heated enhances the D-H replacement speed higher than the one extrapolated from the temperature dependence of the D-H replacement speed for the specimens cooled below the implantation temperature, in which the thermal detrapping hardly takes place.

Keywords: Hydrogen isotopes replacement, Oxide ceramics, Water splitting, Hydrogen gas emission, Ion beam analysis
PACS: 61.72.Ww, 68.43.Mn, 68.47.Gh   **Corresponding Author:** K. Morita, E-mail: kmorita@ccmfs.meiju-u.ac.jp

INTRODUCTION

The dynamics of hydrogen isotopes in materials has received intensive interest from fundamental and applied points of view, in relation to the hydrogen gas production, the hydrogen storage and the hydrogen fuel cell in electro-chemical devices [1]. Combined use of hydrogen ion implantation and elastic recoil detection analysis techniques is very suitable to study the hydrogen dynamics in materials, especially the isotope effects [2~4]. Recently, we have found a big isotope effect that D, of $1 \times 10^{17}/cm^2$ implanted into the oxide ceramics is almost completely replaced by H, when the specimen is exposed to air vapor at room temperature, while in the vice versa case H is hardly replaced by D and a few monatomic layers of D is adsorbed at the surface and in grain boundaries [2]. The experimental finding has been explained in terms of the one way diffusion model: adsorption of OH^- and absorption of H^+ due to dipole-induced splitting of H_2O at the surface, diffusion of H^+ in the bulk and emission of D in traps as DH gas due to molecular recombination with H^+ and subsequent trapping of H^+ in the vacant trap [5]. According to the model, hydrogen-implanted oxide ceramics provides with a water splitting and hydrogen gas emission catalytic function, which is applicable to produce hydrogen gas from water vapor at room temperature. In order to produce hydrogen gas at a rate, 1 Nm^3/hr, the surface area of the specimen is estimated to be 500 m^2, using the present champion data on $BaCe_{0.95}Y_{0.05}O_{3-\delta}$. For practical use of the catalytic function to hydrogen gas production, it is required to clarify the mechanism of water splitting and hydrogen gas emission and to find higher specimens in the hydrogen gas emission rate.

Based on the atomistic model to explain that the $1 \times 10^{17}/cm^2$ of D implanted is completely replaced by H, it is proposed that water splitting should always bring about due to attractive Coulomb interaction of dipole-charge of H_2O adsorbed with charge of lattice defects such as oxygen vacancy, Y^{3+} ion and Ba^{2+} ion at Ce^{4+} site at the surface. According to a cluster calculation of the band structure, the doping of trivalent ion and implantation of D (or H) [6] into the perovskite-type oxide ceramics introduces two kinds of donor levels and acceptor levels in the forbidden band near the bottom of the conduction band and the top of the valence band, respectively. The previous experimental results on the D-H replacement[2~5] give clear evidence that even at room temperature, the trivalent ion is negatively charged by electron capture (charge transfer) and the oxygen vacancy is positively double-charged by ionization of donor electrons due to thermal activation (or ionization). Moreover, it confirms that the charge of the lattice defects recovers rapidly just after the attractive Coulomb interaction has brought, because the $1 \times 10^{17}/cm^2$ of D atoms are completely replaced by H for a short time by exposure to air vapor. The recovery time is directly connected to the D-H replacement speed, namely the hydrogen gas emission rate.

Recently, we have found in the D-H replacement experiments with D-implanted $BaCe_{1-X}Y_XO_{3-\delta}$ and $SrCe_{1-X}Yb_XO_{3-\delta}$ that the decay rate of D increases and after a maximum at X=0.05~0.10 decreases as X increases and the absolute values for $BaCe_{1-X}Y_XO_{3-\delta}$ are larger by an order of magnitude than those for $SrCe_{1-X}Yb_XO_{3-\delta}$. The latter result is primarily ascribed to difference between the energy positions of donor levels from the bottom of the conduction band. The energy position has been regarded to reflect the activation energy in the temperature dependence of the decay rate of D in the D-H replacement speed [7].

In this paper, we report the experimental results on the temperature dependence of the decay rate of D in the D-H replacement for both specimens. It is found that the D-H replacement speed increases as the specimen temperature is elevated above the implantation temperature and the uptake amount of H is almost comparable to the implantation amount of D, while both D as-implanted and H uptaken in the replacement are released by heating the specimens in the vacuum. It is shown that thermal detrapping of D implanted and H uptaken at elevated temperatures enhances the D-H replacement speed higher than the ones extrapolated from the temperature dependence of the D-H replacement speed below the implantation temperature, based on the one way diffusion model.

EXPERIMENTS

The specimen of $BaCe_{0.9}Y_{0.1}O_{3-\delta}$ used was a disc of 15 mm in diameter and 2 mm in thickness, which was prepared by a solid state reaction and sintering processes [8]. The specimen was placed on a manipulator in contact with a ceramic heater in a conventional UHV chamber, which was connected to a beam line of the Tandem Accelerator at Tohoku University. The details of the experimental arrangement were described elsewhere [9]. Prior to deuterium implantation, the specimen was heated at 973 K for several tens min in order to remove residual hydrogen (H). The specimen was implanted with 10 keV D_2^+ ions at room temperature up to saturation concentration (one D atom/ unit cell of $BaCeO_3$). The specimen was exposed to water vapor carried by Ar gas of 47~54 % humidity at the dew point -60°C passing through a water bubbler in order to reduce the dose rate of water vapor. The dose rate of water vapor was calibrated by use of the experimental data obtained by exposure to atmospheric air. After the exposure, the concentrations of D and H in the specimen were measured by means of the ERD technique. The exposure of the specimen to water vapor and ERD measurement were repeated at several times, and the decay curve of D and the uptake curve of H were obtained as a function of the exposure time.

EXPERIMENTAL RESULTS

Typical ERD spectra for the specimen, measured at several stages of the exposure to water vapor at 313 K, are shown in Fig.1, where the as-implanted and the total exposure time of 5 min, 10 min, 40 min and 70 min are shown and the peak at around 450 channel number originates from D, the one at around 280 channel number does from H uptaken during the D_2^+ implantation and by exposure to water vapor. It is clearly seen from Fig.1 that as the exposure time increases, the peak height of D implanted becomes lower and that of H uptaken does higher and then saturates. The saturation concentration of H uptaken was almost the same as that of D as-implanted.

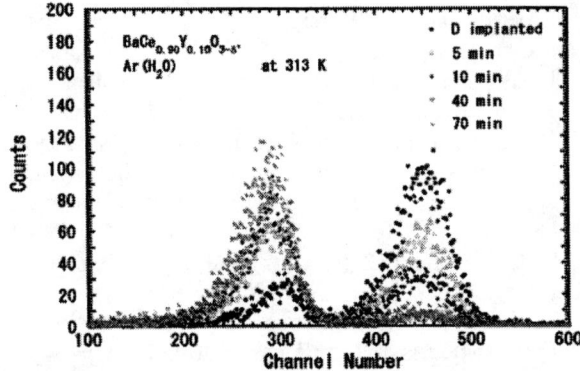

FIGURE 1 ERD spectra measured for the specimen as-implanted with D_2^+ ions at room temperature and subsequently exposed to water vapor for the total time of 5 min, 10 min, 40 min and 70 min at 313 K.

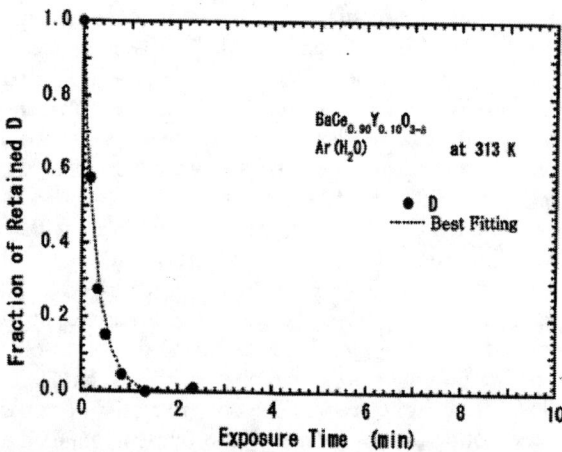

FIGURE 2 Decay curve of D implanted in $BaCe_{0.9}Y_{0.1}O_{3-\delta}$ exposed to water vapor at 313 K. The dotted line represents the best fitting curve calculated according to the one way diffusion model.

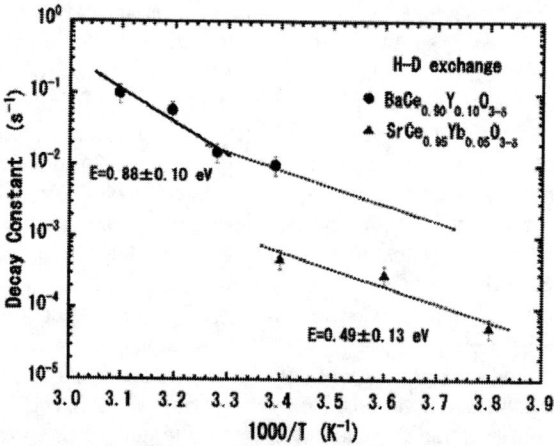

FIGURE 3 Temperature dependence of decay constant of D in $BaCe_{0.9}Y_{0.1}O_{3-\delta}$ exposed to water vapor (●) at elevated temperatures above the implantation temperature and in $SrCe_{0.95}Yb_{0.05}O_{3-\delta}$ (▲) at cooling temperatures below the one.

The decay curve of D in the specimen exposed to water vapor at 313 K, which was obtained from the area of D peaks in Fig.1, is shown as a function of the exposure time in Fig.2, where the horizontal axis represents the exposure time for atmospheric air, corrected from the water vapor carried by Ar gas through the bubbler in the present experiment and the vertical axis does the retained fraction of D normalized by the initial implantation one. It is seen from Fig.2 that the dotted line of the best fitting curve calculated theoretically on the one way diffusion model, reproduces excellently well the experimental decay curve. The decay constant of D, namely the D-H replacement speed was determined to be 6×10^{-2}/s. Similar decay curves of D were obtained and the corresponding decay constants were determined in the D-H replacement experiments at different exposure temperatures.

The result on the temperature dependence of the decay constants obtained from the D-H replacement experiments at different exposure temperatures is shown as a function of 1000/T (K) by symbol (●) in Fig.3, where the results on the D-H replacement speed for $SrCe_{0.95}Yb_{0.05}O_{3-\delta}$ [10] obtained at temperatures lower than the implantation temperature is also shown by symbol (▲) for comparison. The apparent activation energy of the decay constant of D in the D-H replacement for $BaCe_{0.9}Y_{0.1}O_{3-\delta}$ is estimated to be 0.70 eV from Fig.3, which is rather higher than that for $SrCe_{0.95}Yb_{0.05}O_{3-\delta}$. This difference between both activation energies is ascribed to thermal detrapping of D implanted and H uptaken due to D-H replacement, as shown in the experimental results below.

Thermal release curves of D as-implanted and H uptaken in the specimen by subsequent exposure to water vapor at room temperature which were measured on isochronal annealing for 10 min, are shown in Fig.4. It is seen from Fig.4 that both D and H are released by elevating the specimen temperature in the vacuum by 30 K. It is also seen from Fig.4 that the release rate of H is larger than that of D, which is so-called isotope effect, although the retained fraction of H is higher that of D. The reason is ascribed to the absorption effect of H at the specimen surface due to dissociation of H_2O, which is main component in the residual gas of the vacuum chamber. The present experimental result for $BaCe_{0.9}Y_{0.1}O_{3-\delta}$ shown in Fig.4 is very similar to those for $SrCe_{0.95}Yb_{0.05}O_{3-\delta}$, from which the activation energy for thermal detrapping of D implanted was determined to be 0.56 eV[11]. The thermal release of hydrogen isotopes from these oxide ceramics is reasonably explained in terms of their thermal detrapping from the traps and local molecular recombination and rapid emission of the molecule [11].

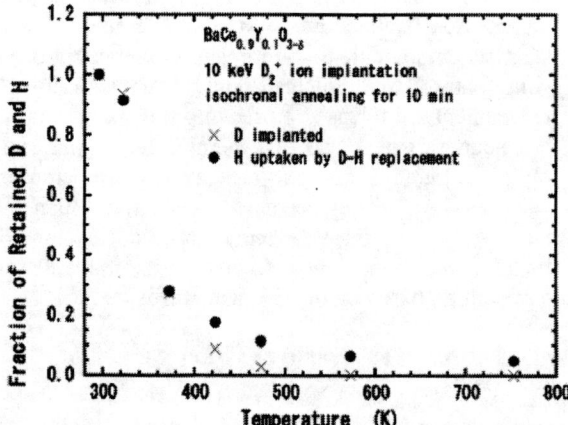

FIGURE 4 Thermal release of D implanted and H uptaken in $BaCe_{0.9}Y_{0.1}O_{3-\delta}$ specimen by exposure to water vapor at room temperature. The vertical axis represents the retained fraction normalized by the initial one, respectively.

DISCUSSION

In the preceding section, it has been shown in Fig.3 that the apparent activation energy of the decay constants of D in $BaCe_{0.9}Y_{0.1}O_{3-\delta}$ exposed to water vapor at elevated temperatures is larger than that in $SrCe_{0.95}Yb_{0.05}O_{3-\delta}$ exposed to water vapor at cooling temperatures. The difference between both activation energies is ascribed to thermal detrapping of D implanted and H uptaken in the D-H replacement, because thermal detrapping of D and H almost hardly takes place in $SrCe_{0.95}Yb_{0.05}O_{3-\delta}$ cooled below the implantation temperature. Experimental observation that the maximum concentration of H uptaken after the complete decay of D at elevated temperatures is almost the same as that at room temperature, namely at implantation temperature, indicates that the release of D and H by thermal detrapping is compensated by fast enough supply of H from the surface due to splitting of H_2O. The thermal detrapping of D and H increases the concentration of freely mobile species of D and H and results in enhancement of the hydrogen gas emission rate due to local molecular recombination.

The apparent activation energy of the decay constants of D in $BaCe_{0.9}Y_{0.1}O_{3-\delta}$ is regarded to be the sum of the activation energy of thermal detrapping and that of the decay constant of D at cooling temperatures. The dotted line passing through the data point for $BaCe_{0.9}Y_{0.1}O_{3-\delta}$ at 295 K represents the activation energy of 0.49 eV. Since the activation energy of D-decay constant for $BaCe_{0.9}Y_{0.1}O_{3-\delta}$ at cooling temperatures is expected to be lower than that for $SrCe_{0.95}Yb_{0.05}O_{3-\delta}$ from fact that the absolute values for the former is by an order of magnitude larger than those for the latter, the dotted line is considered to correspond to the lowest values for the former.

It has been found above that the decay constants of D, namely the D-H replacement rate increases as the exposure temperature rises. When it continuously rises, however, the concentration of H uptaken in the D-H replacement is expected to decrease, because the activation energy of D-decay constant is lower than the activation energy for thermal detrapping of D and H, as described for $SrCe_{0.95}Yb_{0.05}O_{3-\delta}$ in the preceding section. Therefore, the hydrogen gas emission rate is expected to have a maximum at optimum temperature. In order to find the optimum temperature for practical use, further systematic study is required.

CONCLUSIONS

The temperature dependence of the D-H replacement speed in D-implanted oxide ceramics exposed to H_2O vapor has been studied using the ERD technique. It has been found that the decay constant of D, namely D-H replacement rate increases as the exposure temperature is elevated above the implantation temperature and the uptake amount of H is almost the same as the implantation amount of D, while both D as-implanted and H uptaken in the D-H replacement are released by heating the specimens in the vacuum. No reduction of H uptaken in the D-H replacement by exposure to H_2O vapor is ascribed to e fast enough supply of H due to splitting of H_2O at the surface, compared with its thermal release. Based on the one way diffusion model of both the water splitting at the surface and the hydrogen gas emission from the bulk, it has been shown that thermal detrapping of D implanted and H uptaken at elevated temperatures enhances the D-H replacement rates higher than the ones extrapolated from the temperature dependence of the D decay constants at cooling temperatures, at which little thermal detrapping takes place,

ACKNOWLEDGEMENTS

This was supported by Industrial Technology Research Grant Program in 05A50002c from New Energy Development Organization (NEDO) of Japan and also supported by Feasibility Study of Japan Science and Technology (JST) Tokai Plaza. The authors are grateful to the members of the Tandem accelerator group in Institute of Materials Research, Tohoku University for valuable co-operation.

REFERNCES

1) H. Iwahara: Solid State Ionics 77 (1995) 289 and 86-88 (1996) 9.
2) B. Tsuchiya, E. Iizuka, K. Soda, K. Morita and H. Iwahara: J. Nucl. Mater. 258-262 (1998) 555.
3) E. Iizuka, T. Horikawa, B. Tsuchiya, K. Soda, K. Morita, H. Iwahara: Jpn. J. Appl. Phys. 40 (2001) 3349.
4) K. Soda, E. Iizuka, B. Tsuchiya, K. Morita and H. Iwahara:J.Nucl.Sci. and Technol.39 (2002) 359.
5) K. Morita, H. Suzuki and K. Soda: Nucl. Instru. Meth. B206 (2003) 228.
6) H. Yukawa, K. Nakamura and M. Morinaga: Solid State Ionics 116 (1999) 89.
7) K. Morita, B. Tsuchiya, S. Nagata and K. Katahira: Nucl. Instru. Meth. B (2006) in print.
8) H. Uchida et al: Solid State Ionics 36 (1989)89.
9) B. Tsuchiya, S. Nagata, N. Ohstu, K. Toh and T. Sikama: Materials Transactions 46 (2005) 196.
10) K. Morita et al: J. Nucl. Sci. and Technol. 38 (2001) 930.
11) B. Tsuchiya, K. Soda, J. Yuhara, K. Morita and H. Iwahara: Solid State Ionics 117 (1998) 311.

STRUCTURAL CHANGES IN POLYMER FILMS BY FAST ION IMPLANTATION

M. A. Parada[1], R. A. Minamisawa[1,2], C. Muntele[2], I. Muntele[2], A. De Almeida[1] and D. ILA[2]

[1] Departamento de Física e Matemática – FFCLRP, Universidade de São Paulo, Brazil
[2] Center for Irradiation of Materials, Alabama A&M University, Normal, AL, 35762-1447, USA

Abstract. In applications from food wrapping to solar sails, polymers films can be subjected to intense charged particle bombardment and implantation. ETFE (ethylenetetrafluoroethylene) with high impact resistance is used for pumps, valves, tie wraps, and electrical components. PFA (tetrafluoroethylene-per-fluoromethoxyethylene) and FEP (tetrafluoroethylene-hexa-fluoropropylene) are sufficiently biocompatible to be used as transcutaneous implants since they resist damage from the ionizing space radiation, they can be used in aerospace engineering applications. PVDC (polyvinyllidene-chloride) is used for food packaging, and combined with others plastics, improves the oxygen barrier responsible for the food preservation. Fluoropolymers are also known for their radiation dosimetry applications, dependent on the type and energy of the radiation, as well as of the beam intensity.

In this work ETFE, PFA, FEP and PVDC were irradiated with ions of keV and MeV energies at several fluences and were analyzed through techniques as RGA, OAP, FTIR, ATR and Raman spectrophotometry. CF3 is the main specie emitted from PFA and FEP when irradiated with MeV protons. H and HF are released from ETFE due to the broken C-F and C-H bonds when the polymer is irradiated with keV Nitrogen ions and protons. At high fluence, especially for keV Si and N, damage due to carbonization is observed with the formation of hydroperoxide and polymer dehydroflorination. The main broken bonds in PVDC are C-O and C-Cl, with the release of Cl and the formation of double carbon bonds. The ion fluence that causes damage, which could compromise fluoropolymer film applications, has been determined.

Keywords: Polymers, Ion bombardment, Damage.
PACS: 72.80.Le

INTRODUCTION

Polymers are chemical compounds formed of small monomers and their chemical and physical properties differ from those of the original monomers [1]. Generally, they do not react with acids or alkalis and present desirable mechanical, thermal and electrical properties. When a polymer is exposed to ionizing radiation, it can suffer damage according to the type, energy and intensity of radiation. The radiation can break the polymeric chains releasing species such as fluorine, carbon and hydrogen for fluorocarbon polymers and chlorine release from PVDC film. Radiation damage in fluoropolymers from low and high energy of photons and ions had been described [2,3,4,5] and these results are important for application such as food preserving packaging, aerospace materials coating and dosimetry [6,7]. We have determined the damages from ETFE, PFA, FEP and PVDC through RGA, OAP, FTIR techniques.

MATERIAL AND METHODS

To generate the ion beams a 2 MV tandem Pelletron accelerator (AAMU/CIM) was used to provide MeV and keV protons and keV N and Si. Polymer film samples with 20x20 mm^2 and 125 μm, were submitted to several ions fluences (one film for each fluence): ETFE with fluences between 1×10^{11} and 1×10^{16} ions/cm^2 for 1 MeV p, N and Si keV. PVDC was bombarded at fluences between 1×10^{11} and 3×10^{14} ions/cm^2 for 1 MeV protons. PFA and FEP were bombarded at fluences between 1×10^{11} and 1×10^{16} protons/cm^2 for 1 MeV proton. The current was kept below 1 μA to avoid excessive sample heating.

In order to infer the damage mechanisms from bombardment, the following techniques were used: residual gas analyses (Stanford Research System/200) in real time during protons, Optical absorption photospectrometry for p, Si and N (Cary/5000 spectrophotometer for ultraviolet-visible region), Fourier Transform Infrared equipped for ATR measurements. Raman spectra (LabRam spectrophotometer) were also acquired.

RESULTS AND DISCUSSION

The RGA results, shown in Figure 1, presents the gas emitted due to the ions bombardment. In figure 1(a) there are indications that bonds were broken, been HF the main specie formed released besides *F, *F$_2$CCH$_2$CH$_3$, F$_3$CCH$_3$ and FCCCCH. Figures 1(b) and

(c) show that, CF3 radicals account for the greater part of partial pressure detected during the PFA and FEP films bombardment. PVDC films, in figure 1(d) show that the species preferentially emitted are H_2, Cl_{35} and Cl_{37}. These emissions during the bombardments, indicates that bounds were broken in the polymeric chains, resulting in modifications in the polymers.

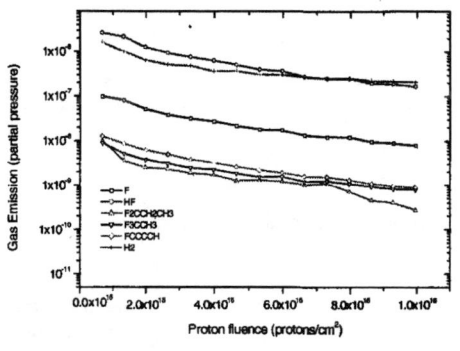

(a)

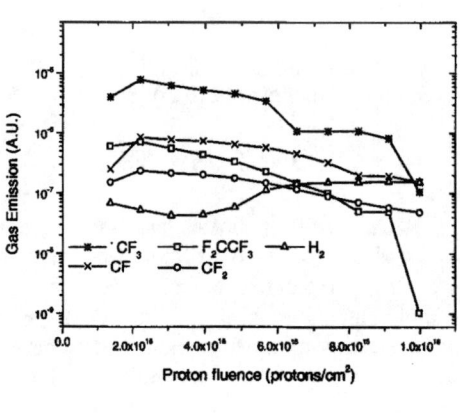

(b)

(c)

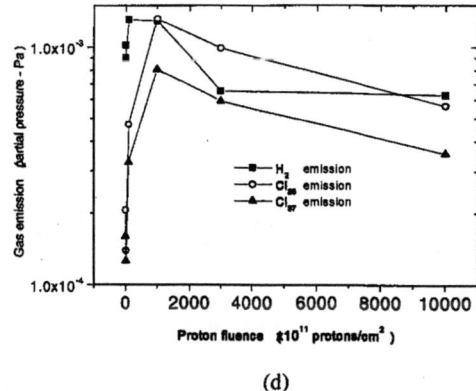

(d)

Figure 1. RGA analysis of the polymer bombarded with 1 MeV protons with different fluences: (a) ETFE, (b) PFA, (c) FEP and (d) PVDC.

The OAP results of ETFE bombarded with proton, Si an N, presented in figure 2, show a shift to higher wavelengths of the absorption edge, indicating increasing carbonization.

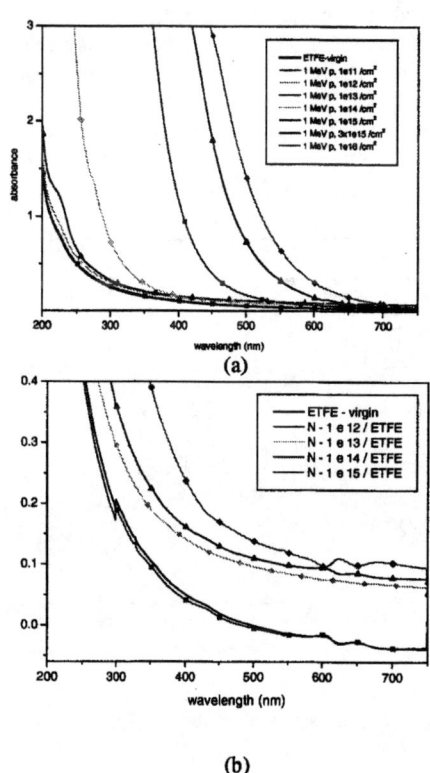

(a)

(b)

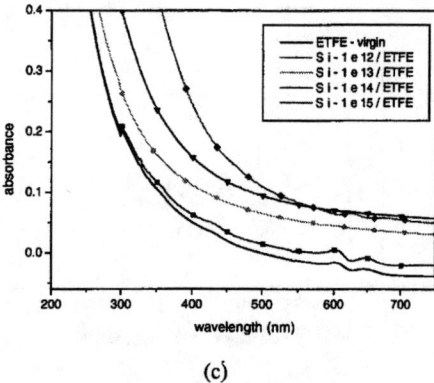

(c)

Figure 2. OAP analysis of the ETFE polymer bombarded with: (a) 1MeV proton, (b) 170 keV N and (c) 210 keV Si.

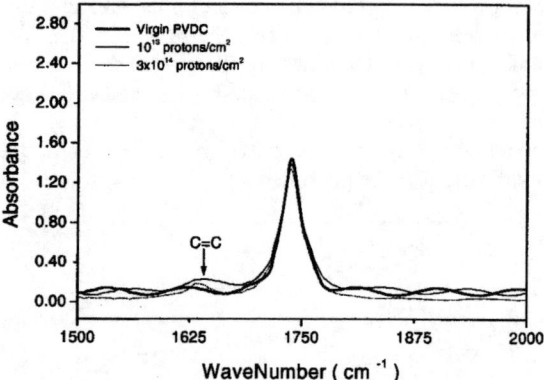

Figure 4. ATR analysis of the PVDC polymer bombarded with 1 MeV protons.

FTIR spectra for 1 MeV proton bombardment showed a carbon double bond formation for ETFE film (Figure 3) and C-O and C-Cl single bonds destruction for PVDC films (Figure 4). The spectra for non-bombarded PFA and FEP films showed similarity between the two films, only differing by the oxygen presence in the PFA. PFA and FEP films spectra presented the carbon double bond formation appearing for the highest fluence (for 1×10^{16} p/cm^2 could not be obtained, due to damage produced in the film). ATR spectra for ETFE films bombarded with keV N and Si indicated an additional peak at 1100 cm^{-1} due to C-O, related to the hydrogen and fluorine leaving the sample. Also the C=C and C-O evolution were observed with the fluence.

Raman spectra of ETFE (Figure 5) bombarded with keV Si and N ions, show the peaks with the same appearance (even for the highest fluence), but a broad peak due to graphitic formation, at 1580 cm^{-1}.

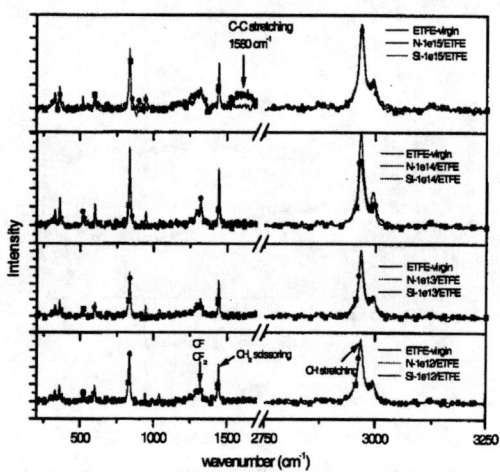

Figure 5. Raman measurement of the ETFE polymer bombarded with 170 keV N and 210 keV Si.

CONCLUSIONS

RGA results indicate that several chemical species are ejected out of the film surface; one also can see that $^\bullet$CF$_3$, a radical in the polymeric chain, is the specie preferentially emitted. In the OAP and FTIR spectra, one can note the formation of new bands after proton bombardment, indicating that these techniques are adequate to evaluate the film damages due the proton bombardment. The C-F bonds are easily broken when compared with C-C bounds in the PFA and ETFE

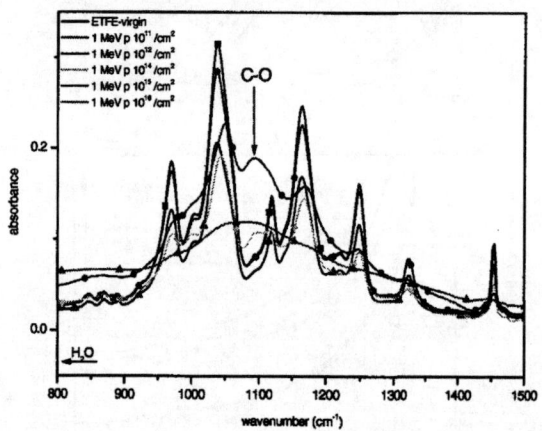

Figure 3. FTIR measurement of the ETFE polymer bombarded with 1 MeV proton.

polymers structures, possibly due to the fluorine electro negativity.

RGA, OAP and ATR-FTIR techniques are appropriate to evaluate damage from MeV proton bombardment. The C-F and C-H bonds are easily broken by MeV proton bombardment compared with C-C bonds in ETFE. For bombardment to 1×10^{16} protons/cm^2 conjugated bonds were formed in ETFE rendering the electrical conductivity too high for use as an electret dosimeter.

ETFE films RGA during bombardment by protons and nitrogen show that HF and H_2 are the dominant emitted entities. FTIR and Raman spectroscopic analysis of the films after bombardment do not detect the formation of CC double bonds, but transmission measurements of UV-visible-near IR light indicate absorption of blue light perhaps by these bonds. The RGA analyses indicate that C-F and C-H bounds are broken allowing hydrogen and hydrogen fluoride to escape. Beyond the penetration depth of keV protons and nitrogen ions, there is no detectable damage or structural change. We expect no significant loss of electrical and mechanical properties in ETFE for accumulated fluence of keV nitrogen or protons of 10^{15} ions/cm^{-2}.

RGA and FTIR techniques are appropriate to evaluate the damage from proton bombardment. C-O and C-Cl are the main broken bonds during the proton bombardment accompanied by the chlorine release and the carbon bonds creation. The optical measurements performed on the implanted ETFE with Si and N, reveal low damage levels and little or no evidence of any chemical reaction between the implanted species and the host material. There is a certain amount of carbonization following higher dose implantation (10^{15} cm^{-2}), equal for both types of ions used, as well as formation of hydroperoxides following dehydroflorination of the polymer.

ACKNOWLEDGMENTS

The present work was supported by grants from CAPES/CNPq (Brazil) and AAMU (USA)

REFERENCES

1. C. R. Brundle, C. A. Evans Jr., S. Wilson, *Encyclopedia of Materials Characterization: surfaces, interfaces, thin films*, Butterworth-Heinemann, Boston, 1992.
2. M. A. Parada, R. A. Minamisawa, A. de Almeida, I. Muntel and D. Ila, *Damage Effects of Gamma and X-rays Polymer film electrets* (Accepted).
3. M. A. Parada, P. C. D. Petchevist, A de Almeida, N. C. Silva, J. S. C. Campos and D. Ila, ISE12 proceedings of the *IEEE dielectrics and Electrical Insulation Society*, 428-430 (2005).
4. M. A. Parada, A. de Almeida, C. Muntele, I. Muntele, N. Delalez and D. Ila, *Nuclear Instruments and Methods in Physics Research Section* B **241**(1-4), 521-525 (2005).

Hydrogen ion implantation mechanism in GaAs-on-insulator wafer formation by ion-cut process

H.J. Woo, H.W. Choi, G.D. Kim, J.K. Kim, W. Hong and H.R. Lee

Ion Beam Application Group, Korea Institute of Geoscience & Mineral Resources, Daejeon, 305-350 Korea

Abstract. The results of the basic study on depth distribution of hydrogen atoms and corresponding damage profiles produced by 40 keV hydrogen ion implantation in (100) GaAs are reported. Depth distribution of hydrogen was measured by SIMS, and the lattice disorder in the samples was studied by RBS/channeling analysis. The influence of the ion fluence and the implantation temperature, and subsequent annealing on blistering and/or flaking was studied, and the optimum conditions for achieving blistering/splitting only after post-implantation annealing were determined. In addition, the microscopic evolution in the damaged layer was also studied by XTEM analysis. Our results suggest that the ion-cut process is sensitive to both the implant temperature and the fluence. At implant temperatures below 100□, hydrogen is unable to form into the defect structure which is responsible for blistering, and if the implant temperature is too high, the platelets are not able to evolve and blistering is less prolific. It was found that the optimum implant temperature window lie in 120~150°C, which is markedly lower than the previously reported implant temperature window probably due to the inaccuracy in temperature measurement in other laboratories. Optimum post-implantation annealing temperature was less than 300°C, and low temperature splitting is of importance for layer transfer between dissimilar materials with very different thermal expansion coefficients as well as for processed wafers containing temperature-sensitive devices.

Keywords: GaAs-on-insulator, ion-cut, hydrogen implantation
PACS: 61.80.Jh; 68.55.Jk

INTRODUCTION

Much scientific interest has been focused on the combination of III-V semiconductors with mature silicon technology for many years. Especially, monolithic integration of GaAs into silicon technology presents a huge potential of interest as it combines the superior electrical and optical properties of GaAs with the mechanical and economical advantages, and density of integration of silicon. To obtain this structure, heteroepitaxial growth has been investigated extensively, but due to the notable lattice mismatch of about 4%, an unacceptable high density (typically > $10^7/cm^2$) of threading dislocations could not be avoided [1,2].

From technical and economical points of view, one of the most promising techniques for joining thin GaAs single crystal layer with various substrates is an ion-cut technology proposed by Bruel in 1995 [3] in the name of 'smart-cut' as a method for the fabrication of high-quality silicon-on-insulator (SOI) wafers. The ion-cut technology consists in hydrogen ion implantation into the semiconductor wafer, bonding implanted wafer with an unimplanted one, and final exfoliation by thermal treatment leading to the transfer of the surface layer of hydrogen-implanted wafer. However, layer splitting can be achieved only if appropriate implantation conditions (fluence, flux and actual implantation temperature due to beam heating), specific to each material, are employed. As opposed to Si, the optimum implantation temperature range for GaAs (160~250°C) is known to be rather narrow, and such condition is somewhat difficult to control when standard ion implanters are used [4,5]. Several attempts to transfer GaAs layers onto Si by layer splitting have been reported [6-8]. However, until now, there is no consensus as to the implantation temperatures required to split GaAs [9] most probably due to the inaccuracy in the measurement of wafer surface temperature.

Recently, Radu et al. proposed He+H co-implantation technique which enables room temperature implantation, and in addition, the minimum fluence necessary to induce blistering and exfoliation of silicon can be decreased by a factor of 2~3 by this technique [10]. Originally, it was proposed as a low-temperature (<250°C) splitting and layer transfer of GaAs. When dissimilar materials are used, it is desirable that the splitting temperature is low enough to allow bonded wafer pairs to withstand stresses associated with a difference in thermal expansion coefficients.

We have tried to develop ion-cut process to allow a large dimension GaAs thin film to be transferred onto a full silicon wafer as shown in Fig. 1, based on the already developed ion-cut technology for the formation of SOI wafers with various SOI thicknesses [11]. The aim of this paper is to investigate the role of implantation temperature and ion fluence on blistering of the surface after a subsequent annealing, which is a prerequisite of ion-cut splitting, in detail and to further investigate hydrogen implantation mechanism.

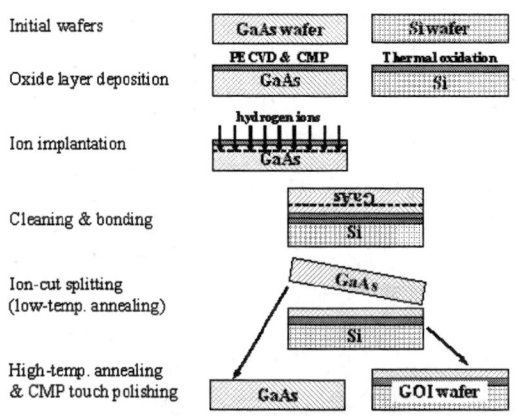

FIGURE 1. Schematic of the ion-cut process for GaAs-on-insulator wafers formation.

EXPERIMENTS

In order to determine the optimum conditions (implantation temperature and fluence, and annealing parameters) for the layer transfer of GaAs thin film, hydrogen ions were implanted into semi-insulating (100) oriented, 2" GaAs wafers at the temperature ranging from 40°C to 300°C. The actual surface temperature of the implanted wafer was directly measured to increase the accuracy in temperature measurement during implantation. Implantation energy of hydrogen ions was 40 keV, and fluences were varied from 4×10^{16} H^+/cm^2 to 2×10^{17} H^+/cm^2. To minimize channeling effects, implantation was performed under 7° sample tilt. In order to ignore the dependence of hydrogen flux on the ion-cut process [9], it was kept almost constant at about 1×10^{13} $H^+/cm^2/s$.

The implantation profile of hydrogen ions in GaAs was simulated by SRIM2003 (The Stopping and Range of Ions in Matter) code. A normal hydrogen depth distribution with a single peak at 320 nm depth corresponding to the projection range of 40 keV hydrogen ion beam was confirmed by secondary ion mass spectrometry (SIMS) measurements as shown in Fig. 2, and the profile of hydrogen atoms didn't alter remarkably either upon hot implantation and or upon annealing. It means that only short-range rearrangement of hydrogen atoms in the matrix can occur at the temperature less than 350°C [12].

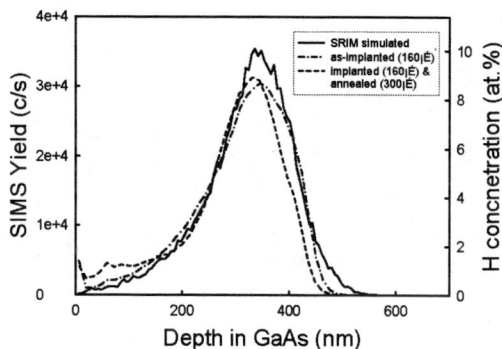

FIGURE 2. Hydrogen concentration profiles in GaAs implanted with 8×10^{16} H^+/cm^2 at 160°C and annealed at 300°C as determined by the SIMS analysis.

Surface morphology after hydrogen bombardment and thermal treatment finally leading to the surface layer exfoliation was studied by means of optical microscopy and field emission scanning electron microscopy (FE-SEM). Structural defect formations were studied by means of Rutherford backscattering spectrometry (RBS)/channeling, and formation of platelets in the as-implanted GaAs and their evolution with annealing were analyzed by cross section transmission electron microscopy (XTEM).

RESULTS and DISCUSSION

Blistering is directly representative of the formation of hydrogen-induced microcracks, a condition required to obtain layer splitting, and blistering in GaAs is strongly dependent on both the substrate temperature and implantation fluence. For limited implantation fluences, no morphological change effect is observed on the surface of as-implanted wafers, and a bonding step with other wafer is possible. In the samples implanted at an optimum temperature and fluence, the mean size of the microcavities increases during annealing at an elevated temperature (>200°C) inducing deformations of the surface as blistering and flaking as shown in Fig. 3.

For the samples implanted at 120°C ~160°C with a fluence in the range 1x10^{17} to 2.0x10^{17}/cm^2, denoted with gray color in Fig. 4, the surface blistering phenomenon was observed after annealing at 200~300°C for about 30 min. The surface blisters have about 2~3 μm lateral size as shown in Fig. 4(a), and the areal density of generated blisters (~10^6 blisters/cm^2) is found to be sufficient for ion-cut splitting. The minimum implantation temperature for obtaining blisters is a direct function of the hydrogen implanted fluence.

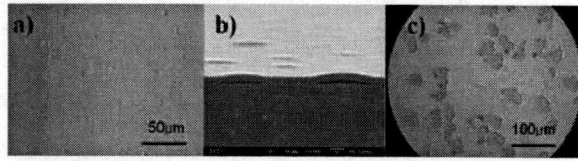

FIGURE 3. Microscopic images on GaAs surfaces after implantation (40 keV, 1.6x10^{17} H$^+$/cm^2) and annealing for 30 min at 300°C (a: optical, b: FE-SEM) and at 400°C (c: optical).

that are known to be responsible for blistering. The common crystal damage curing effect at elevated temperatures doesn't appear in this case against expectation. It is worth to point out that at this temperature only defects in the Ga sublattice are mobile, as defect mobility threshold for As sublattice lies above 300°C [8]. If the implant temperature is too low, the hydrogen does not appear to be mobile enough in the lattice to rearrange into the micro-defect structures. This is in agreement with the observation that no detectable surface deformation were observed after low temperature hydrogen implantation up to the fluence of 2.0x10^{17}/cm^2. At temperatures above 200°C, RBS/ channeling data has not been measured because blistering begins to appear at the high temperature. However, it is reported that at the high temperature the lattice disorder is lower, suggesting that the evolution of these defects is less prolific because of a pronounced out-diffusion of hydrogen [9]. These phenomena are responsible for the existence of a implantation temperature window for the layer splitting of GaAs wafer by an ion-cut process.

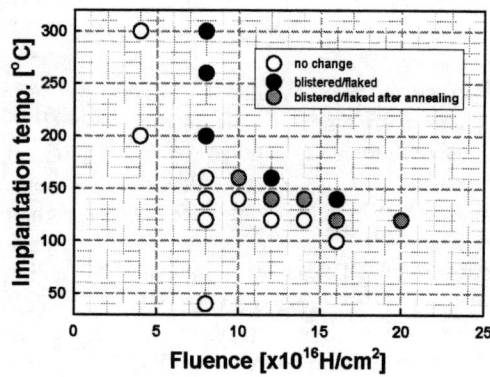

FIGURE 4. Fluence and temperature boundaries of blister formation in hydrogen-implanted GaAs.

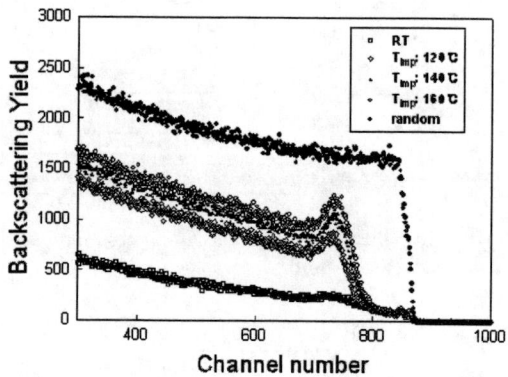

FIGURE 5. Random and aligned RBS spectra for GaAs single crystals implanted with 8x10^{16} H$^+$/cm^2 at different temperatures.

Actually the optimum temperature window for GaAs ion-cut process is not yet clearly known probably because the actual temperature on a wafer surface in implantation is not directly measured for most of the implanters. Tong et al. [4] reported that the appropriate window for the implantation temperature for GaAs ion-cut process is 160~250°C. However, this range is markedly different from our new result which indicate that the temperature window should be in the range between 120°C and 160°C for fluences ranging from 1.0 to 2.0x10^{17} H$^+$/cm^2.

As shown in Fig. 5, it is clear from the RBS/channeling measurement that the lattice disorder substantially increases with increasing temperature due to rearrangement of hydrogen into the microcavities

Fig. 6(a) shows a XTEM image of the damaged layer in a GaAs wafer implanted at a fluence of 1.2x10^{17} H$^+$/cm^2 at 140°C. It revealed that the microcavities are preferentially oriented along {111} and {100} planes and are localized at a depth of 320 nm from the surface close to the maximum concentration of hydrogen implanted into GaAs at 40 keV calculated by SRIM2003. These microcracks are hydrogen-filled platelets responsible for blistering and/or exfoliation, and have a relatively large (100~300 nm) lateral size in comparison to those in case of silicon. Those microcavities are also found in Si, SiC and Ge, and these results show that hydrogen-related splitting mechanism in GaAs is similar to that observed in other semiconductors.

During annealing, hydrogen is released from the trapping sites and accumulates on the platelets formed during implantation. Due to an increase of the inner pressure, platelets grow and overlap, eventually leading to formation of large cracks following a zigzag path but parallel to the wafer surface on the whole as shown in Fig. 6(b). It is known that during thermal treatment a vertical rearrangement of the platelets occurs, following an Ostwald ripening mechanism, leading to formation of microcracks in a narrow layer, where cracking would occur.

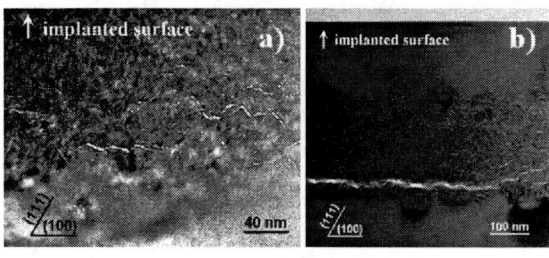

FIGURE 6. Cross section TEM images of GaAs wafer after a) hydrogen ion implantation (40 keV, 1.2×10^{17} cm^{-2}) at 140°C and b) annealing at 300°C.

When a handle wafer is bonded to the surface of implanted donor wafer as a stiffener, the blistering and/or flaking phenomena do not take place and a propagation of the cavities is observed. During annealing, gas-containing microcavities gradually grow and an interaction between neighboring microcavities results in the propagation of a crack along the cavity plane and the complete separation parallel to the bonded surface. The originality of the ion-cut process is to use this propagation of cavities as a way to induce in-depth splitting over the whole wafer. Transferred film thickness can be correlated to projected range of implanted hydrogen ions, and as cavities are located in a layer with thickness corresponding to the simulated projected range straggling, a somewhat rough surface is obtained after splitting.

CONCLUSIONS

Our results suggest that the ion-cut process for the GaAs layer transfer is sensitive to both the implant temperature and the hydrogen ion fluence. At implant temperatures below 100°C, hydrogen is unable to rearrange into the defect structure which is responsible for blistering, and if the implant temperature is too high, the platelets are not able to evolve and blistering is less prolific because of a pronounced out-diffusion of hydrogen atoms due to increased mobility. It was found that the optimum implant temperature window lie in 120~160°C for fluences ranging from 1.0 to 2.0×10^{17} H$^+$/cm^2, which is relatively lower than the previously reported implant temperature window probably due to the inaccuracy in temperature measurement in other laboratories.

Optimum post-implantation annealing temperature for the ion-cut layer splitting was found to be less than 250°C, and the low temperature splitting is of importance for layer transfer between dissimilar materials with very different thermal expansion coefficients as well as for processed wafers containing temperature-sensitive devices.

ACKNOWLEDGEMENTS

This work was supported by the Proton Engineering Frontier Project under the 21st Century Frontier Research Program of the Korean Ministry of Science and Technology.

REFERENCES

1. J. A. Carlin, S. A. Ringel, E. A. Fitzgerald, M. Bulsara and B. M. Keyes, *Appl. Phys. Letters* **76**, 1884-1886 (2000).
2. I. Radu, "Layer transfer of semiconductors and complex oxides by helium and/or hydrogen implantation and wafer bonding", Ph.D. Thesis, Martin-Luther University, Halle-Wittenberg, 2003.
3. M. Bruel, *Nucl. Instr. Meth. B* **108**, 313-319 (1996).
4. Q. Y. Tong, L. J. Huang, and U. M. Goesele, *J. Electron. Mater.* **29**, 928-932 (2000).
5. M. Alexe and U. Goesele, *Wafer bonding; Application and Technology*, Berlin, Springer-Verlag, 2004, p.297.
6. E. Jalaguier, B. Aspar, S. Pocas, J. F. Michaud, M. Zussy, A. M. Papon, and M. Bruel, *Electron. Lett.* **34**, 408-409 (1998).
7. I. Radu, I. Szafraniak, R. Scholz, M. Alexe and U. Goesele, *J. Appl. Phys.* **94**, 7820-7825 (2003).
8. G. Gawlik, J. Jagielski and B. Piatkowski, *Vacuum* **70**, 103-107 (2003).
9. M. Webb, C. Jeybes, R. M. Gwillian, Z. Tabatabaian, A. Royle and B. J. Sealy, *Nucl. Instr. Meth. B* **237**, 193-196 (2005).
10. I. Radu, I. Szafraniak, R. Scholz, M. Alexe and U. Goesele, *App. Phys. Letters* **82**, 2413-2415 (2003).

11. H. J. Woo, H. W. Choi, J. K. Kim, G. D. Kim, W. Hong, W. B. Choi and Y. H. Bae, *Nucl. Instr. Meth. B* **241**, 531-535 (2005).
12. G. Gawlik, R. Ratajczak, A. Turos, J. Jagielski, S. Bedell and W. L. Lanford, *Vacuum* **63**, 697-700 (2001).

High Dose Hydrogen Implant Blistering Effects As a Function of Selected Implanter And Substrate Conditions

Ronald Eddy [a], Chuck Hudak [a], Pamla Bettincurt [b], Sandra Delgado [b]

[a] *Core – Div of Implant Sciences, 1050 Kifer Rd. Sunnyvale, CA USA 94086*
[b] *Accurel Systems Int'l - Subsidiary of Implant Sciences. 785 Lucerne Dr. Sunnyvale, CA USA 94085*

Abstract. Buried dielectrics using high dose Hydrogen with subsequent wafer splitting and bonding (Smart-Cut™ for example) is a well-known process [1]. The ion implantation portion of this popular process in silicon is a key step and it is not fully characterized in silicon as to dose rate, dose duty cycle or wafer platen differences. Failures or problems in wafer splitting are not uncommon due to improper or non-uniform heat sinking or other seemingly common implanter problems such as parameter holds or changes in beam spot size. This paper will focus on ion implanter variations and will use the generation of surface blisters and sub surface damage formation under various conditions. One common anneal temperature and time is proposed. Various microscopy and metrology tools will be correlated with respect to H dose, the dose variations and various wafer platen conditions. There will be a brief discussion of implanter scan types related to high dose H^+.

INTRODUCTION

The authors' workplace, an implant foundry, performs a wide range of H+ implants as well as H+ with a co-implant species for various thin film exfoliation applications. There are occasional problems with total exfoliation or total uniformity of "damage' due to a number of implanter issues – some not usually thought of as major inhibitors of good performance. Some of the root causes have included: improper heat sinking or wafer temperature. Problems in a given application are often solved in one run or after a short series of experiments. This study is an overview of some of the parameters studied but in a full test under controlled pre and post implant conditions as compared to former studies done at our lab under conditions that had a variety of different pre and post implant processing. This test was designed to determine sensitivities to various implant conditions. As such, anneal conditions were not optimized for each implant and the blister data is quoted to indicate differences between implants rather than as an indicator of overall wafer quality. In most cases, deviations in blistering or exfoliation may easily be accommodated with slight changes to post processing.

EXPERIMENTAL

A number of wafers were run on two different types of implanters. On some wafers, small implant areas were used so as to reduce implant time on a serial implanter. Wafer types Czochralski (CZ) Si <100>, p-type were used.

High dose Hydrogen implanted bare wafers that are annealed without bonding will manifest surface blisters [2]. This provides a simple method for determining the effects of implant or anneal temperature, dose and energy [3,4]. For investigating Hydrogen agglomeration or growth as a function of common/possible implanter problems we kept implant energy, dose and the anneal conditions the same since most H^+ implant papers involve various changes to the implant recipe as a function of blister density. [3]. We wanted to see what differences might be discernible on silicon with the following: Three different substrate temperatures: +250C, RT and –120C, two dose duty cycles – full/100% as well as ¼ x 4 duty cycle to observe any damage change to due to implant interruption whether planned or not and we ran wafers with two different wafer "clamping" or heat sinking problems. The test also included three different beam densities of 0.5, 1.0 and 3.8 $\mu A\ cm^{-2}$, which bracket the beam densities of most high current implanters. 3.5 $\mu A\ cm^{-2}$ is equivalent to ~30 mA on a 200 mm wafer implant disk/wheel. In addition, a H/He implant combination was also run with the He at three different energies. Table 1 shows the test set.

Table 1. Variables for H (and H/He) Test			
Parameter	Number	Values	Notes
Energy	1	40 keV	
Dose (H/cm^2)	2	3.6 & 7.2 E16 cm-2	
Beam Density	2	1 & 3.8 uA cm-2	
Substrate Temperature	3	250C, RT, -120C	
Dose Duty Cycle	2	100%/Full, 1/4 x 2	1/4 x 4 simulates Implant HOLDS and Interrupts
Heat Transfer Integrity	2	Normal & Damaged, Diminished	
He Implant Supplement	3	Shallower, Same as or Deeper than H	On one standard H implant (H, 40 keV, 1E16)

For the He implants, the energies were tailored to have one run at the same depth of the H, one profile 500Å shallower and one 500Å deeper than the H profile. One anneal (550C, 25 min) was used for most off the wafers while a few wafers with H and He were annealed at the same temperature but a slightly longer time. (33 min). The characterization involved the use of standard optical microscopes SEM, AFM – done using a NanoScope™ III, Dimension 5000. We also used an optical 3D profiler – a Zygo NewView in order to look at blister depth and cross sections as a supplement to the AFM. A series of optical microscope fields were used, generally 5 or 9 fields on each wafer, in order to see the uniformity of blister density across the wafer.

RESULTS

The data collection began with optical microscopy using at ~1000X with views ranging from ~16,000 - 64,000 μm^2. SEM pictures were taken on selected wafers. On the optical microscope, the wafer was scanned in several axes to see the general uniformity of blisters and a minimum of five views per wafer was noted to determine relative blister uniformity SD% using a counting template on the views. The blister density and blister evolution (onsets and full blistering and % broken blisters) were generally consistent within our data and in one significant case, the low temperature implants, were different from results seen on other papers [5]. Some of these differences were related to % broken blister density and may be a result of the high beam current density used here in these cases. Fig 1 shows a very general comparison of blister density. We have included the lower dose H & He for reference purposes. The density and the % of broken blisters on H$^+$, high dose only, was highest on the wafers implanted at full dose, full duty cycle with low temperature substrate (-120C). See Fig. 2. This is inconsistent with other reports but two different tests were done with similar results.

A substrate with full dose and dose duty cycle that was heated (250C) showed significantly less exploded blisters (30% less blisters and 90% less exploded blisters) than wafers with the same dose and DDC and at RT. When comparing all implant cases or substrate conditions but where the doses were different, we see where the dose is high, the blister density is high and the % of exploded blisters is higher than with lower doses.

In all cases where the dose was interrupted three times (25,50,75% complete) during the implant for high dose and regardless of platen temperature, the number of blisters per unit area was lower by 1/3 to 1/2 than with a full, dose duty cycle. This is consistent with dose measurements on "crystal damage measurement" tools. This needs further study since the H$^+$ implants are a relatively high dose and there is a chance of beam interruptions or even a quad mode implant in the case of some serial high current implanters might manifest sensitivity.

Some 10-15% of our implants were done on a batch implanter using a beam density of 0.5 – 1 mA H$^+$. Wafers in this group that used typical clampless holders were evaluated at the top, bottom and center along the axis of slow scan. Although the top, center and bottom have similar blister density and sizes, there is a noticeable difference the top or bottom versus the center with regards to "blister lids" – the tops of broken blisters that are of similar topography and size as surrounding broken blisters (or open craters). These lids or flakes are large 3 – 5 μm flakes which when in the AFM for assessment of crater information, tend to be dragged along by the probe. The top/bottom versus center "unbroken flake-lid" difference is likely due to difference in damage and/or slight temperature differences from the turn-around on the wafer annulus.

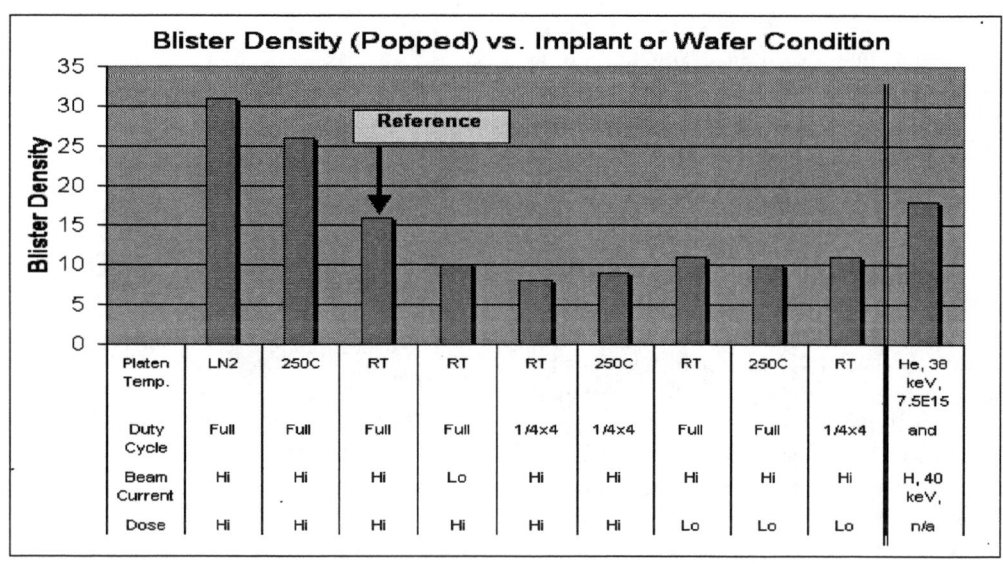

FIGURE 1. Blister Density (counting template area ~ 0.09mm2) comparison. Note the Reference Implant of full dose, high beam current, full DC and RT.

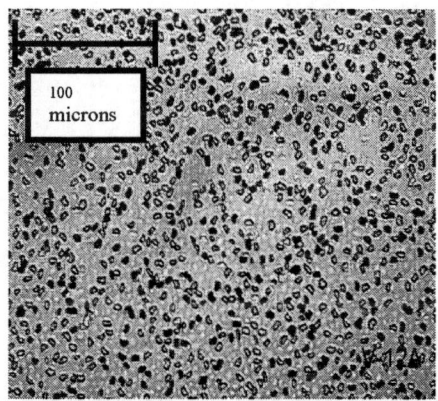

FIGURE 2. Optical micrograph picture of a high density of broken blisters and flakes (blister lids?). High dose, high beam current density, low substrate temperature.

See Figure 3 for a comparison on wafer center vs. wafer top, for example, on a full dose, full beam current implant on a batch implanter. Note that "top of wafer" denotes top of slow scan. This might also occur on serial high current tools running high dose H^+ although the beam density is much lower. In cases where broken blister density was high, they also manifested, as one might expect, the highest % of larger unbroken blisters – blisters that had coalesced almost to the point of breaking. Test samples run with lower duty cycle, i.e., planned, interrupted implant, showed lower blister density than tests run with the same implant conditions except for full duty cycle. This might infer that

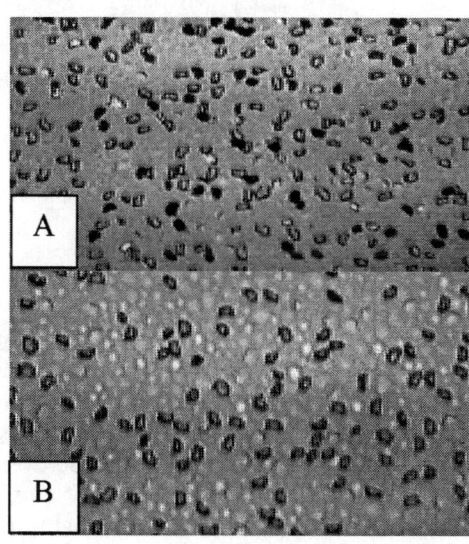

FIGURE 3. Sample S-13 done on a batch implanter shows the difference between A - Center of the Wafer and B - Bottom of the Wafer. Implant was high dose, high beam density, RT.

interruptions in some implants especially for long, high dose implants, may need remedial post implant processing for proper exfoliation. Wafers that had the normal heat sinking compromised with 2 layers of Kapton™ tape or with a dummy wafer beneath the test wafer showed very close approximation of blister density (broken as well as larger unbroken blisters). So called "blister lids" were more abundant on wafers with compromised heat sinking. It is interesting to note in Fig. 4 the AFM broken blister depth data was consistently in agreement with range data from Profile Code™ while the Zygo depth data (Fig. 5) was consistently in agreement with SRIM 2003. Both are within ~10% of each other for this H$^+$ implant recipe. In our study, and our conditions, with the H$^+$ and He$^+$ implant we observed that the implants with H$^+$ at 40 keV with the accompanying deeper and shallower (500 Å each) He$^+$ implant, the shallower implant had a 4X higher blister density. Although all three He$^+$ implants had scattered clusters of blisters, the deeper energy He$^+$ (55 keV) substantially less clusters and the average density within a cluster was very low whereas the shallower He$^+$ with its H$^+$ was about equal to the baseline H$^+$ that had > 4X the H$^+$ dose.

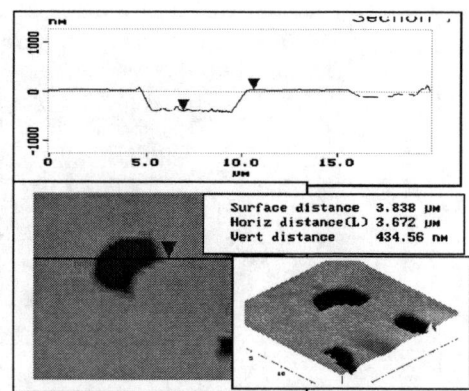

FIGURE 4. AFM pictures of typical blister showing the depth as 434.56 nm.

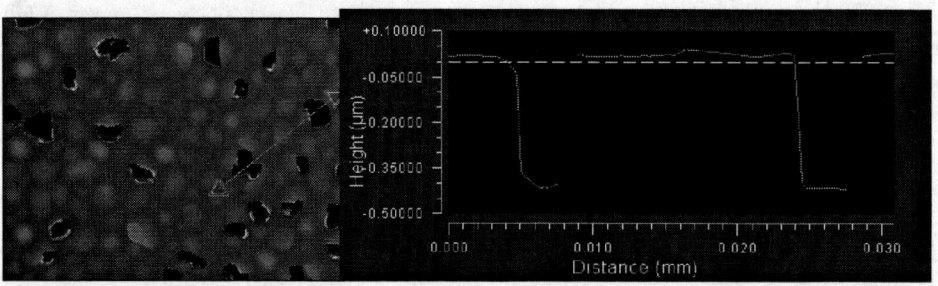

FIGURE 5. Zygo 3D Profiler showing blister depth = 0.400 µm. The depth of >25 measured blisters over 3 different wafers is the same within the precision of the measurement

SUMMARY

We have observed a number of various implanter and substrate heat sink differences. Improper clamping shows some blister density differences but the blister uniformity is still consistent across the wafer and any exfoliation issues may be overcome with post implant process adjustments. Some H/He combinations show blister densities equal to that of full dose full DC high dose H$^+$ only. Since the H/He combination has a much lower total dose than H$^+$, this could alleviate any subtle problems due to dose rate or wafer condition. Dose interruptions during a high dose H$^+$ also show a blister density and size difference. This needs further investigation for various scan techniques.

REFERENCES

1. M.Bruel. "Silicon On Insulator Material Technology". Electronic Letters. July 6, 1995. Vol 31 #14.
2. O.W.Holland, D.K.Thomas, R.B.Gregory. "Optimization of the Ion-Cut Process in Si and SiC". MRS Symposium Proceedings 2001. Vol 647. PO6.1
3. L.-J.Huang, Q.-Y.Tong, J.-L.Chao, T.-H.Lee, T.Martini, U.Gösele. "Onset of Bistering in Hydrogen-Implanted Silicon". Appl. Phys. Lett. 74, No.7. 15 Feb. 1999.
4. U.Gosele, M.Alexe. "Wafer Bonding – A Flexible Approach to Materials Integration". Electrochemical Society – Interface. Summer 2000. 20-25
5. V.C.Venezia, T.E.Haynes, A.Agarwal, D.J.Eaglesham, O.W.Holland, M.K.Weldon, Y.J.Chabal. "The Role of Implantation Damage in the Production of of Silicon-on-Insulator Films by Co-implantation of He+ and H+ ". Proc. of the Eighth International Symposium on Silicon Materials Science and Technology. Ed. H.Huff, U.Gösele, H.Tsuya. Electrochemical Society. NJ. 1998
6. S.Falk, R.Callahan, P.Lunquist, "Accurate Dose Matching Measurements Between Different Implanters". Proc. of 11th International Conference on Ion Implantation Technology. IEEE, NJ. 1997
7. A.Giguere, J.Beerens, B.Terreault. "Creating Nanostructures on Silicon Using Ion Blistering and Electron Beam Lithography". IOP Nanotechnolgy. Vol 17. 20

Germanium Nanoparticle Formation into Thin SiO₂ Films by Negative Ion Implantation and Their Electric Characteristics

Nobutoshi Arai*, Hiroshi Tsuji, Naoyuki Gotoh, Tetsuya Okumine*, Toshio Yanagitani*, Masatomi Harada*, Takeshi Satoh*, Hitoshi Ohnishi*, Takashi Minotani, Kouichirou Adachi*, Hiroshi Kotaki*, Toyotsugu Ishibashi, Yasuhito Gotoh and Junzo Ishikawa

Department of Electronic Science and Engineering, Kyoto University (Nishikyo-ku, Kyoto 615-8510, Japan)
*Advanced Technology Research Laboratories, SHARP Corporation (Ichinomoto-cho, Tenri 632-8567, Japan)

Abstract. Germanium nanoparticles in a thin SiO₂ film on Si have been formed by negative ion implantation for the development of very low power consumption electron devices using nanoparticles. Their electrical properties of 25-nm-SiO₂/Si films including Ge nanoparticles were investigated with CV method after subsequent annealing at various temperatures. Ge atoms were implanted at 10 keV with fluencies of 1×10^{15} and 5×10^{15} ions/cm². Samples were annealed at 300, 500, 700 and 900°C for 1 h. Depth profiles of implanted Ge atoms in the SiO₂ films were measured by using a high-resolution RBS technique. The formed Ge nanoparticles were studied by cross-sectional TEM observation. After annealing at less than 700°C, Ge nanoparticles were confirmed in the film. After 300°C-annealing, a CV curve had so small hysteresis that could not be applied to memory devices. After 500°C-annealing, both samples with 1×10^{15} ions/cm² and with 5×10^{15} ions/cm² had obvious hysteresis curves. Calculations of charge and nanoparticle intensity from flat band shift and implanted Ge dose lead about one electron in one nanoparticle with 3nm diameter. These results suggest that thin SiO₂ films including Ge nanoparticles formed with negative ion implantation can applied with memory devices.

Keywords: Negative-ion implantation, Heat treatment, Germanium nanoparticle, CV method.
PACS: 61.72.Tt; 41.75.Cn; 61.72.Ww; 81.07.-b; 81.40.Gh; 84.37.+q

INTRODUCTION

Electrical conductor nanoparticles embedded in insulator have so small capacitance that they can show Coulomb blockade phenomena at room temperature [1, 2]. Therefore, metal nanoparticles in SiO₂ are attractive materials for the development of single electron memories with nanoparticles embedded in gate insulators of MOSFETs. As MOSFET devices shrink, nanoparticles have to be embedded in a very thin SiO₂ film on Si substrate. Formations of nanoparticles in thin SiO₂ films with implantation technique have been already reported [3-7]. These nanoparticles should be chargeable and dischargeable to be used as memory devices. Si nanoparticles with CVD method were applied to floating gate type memory with about 0.25 V of threshold voltage shift [8]. We think that spherical nanoparticles in high-quality thermally oxidized silicon created by negative ion implantation are better to store electrical charge.

So we used negative ion of Ge, which fit semiconductor processing. Thermally grown SiO₂ was implanted by Ge ion implantation and annealed to form nanoparticles. Electrical characteristics of the Ge implanted SiO₂ were investigated with Capacitance-Voltage (CV) method. The results showed that the Ge implanted SiO₂ charged and discharged electrical charge.

EXPERIMENTAL

Germanium negative ions were generated in an RF (radio frequency) plasma-sputtering type heavy negative ion source and extracted at 10 keV [9, 10]. After mass separation by a sector magnet, the Ge ion beam was introduced into a collector cup with a limiting aperture of diameter 8 mm in an implantation chamber of the negative ion implanter (Nissin Electric Corp., Japan) [11]. In the collector cup, Ge negative ions were implanted at room temperature into a silicon dioxide film with thickness of 25 nm thermally grown on a silicon substrate (15 mm x 15 mm), at ion energy of 10 keV with doses of 1×10^{15} and 5×10^{15} ions/cm^2.

The depth profiles of implanted Ge atoms calculated by using the transport of ion in matter (TRIM-DYN) program [12] are shown in Fig.1. The Ge profiles show almost Gaussian distributions with a peak concentration of 1.5 and 7.0 at.% at the depth of 12 nm, this is approximately a half of the thickness of the SiO$_2$ film.

The implanted samples were then annealed for 1 h in an evacuated quartz tube of an electrical oven, at temperatures of 300, 500, 700, and 900°C respectively.

After the annealing, Al electrodes were formed on the samples with vacuum evaporation.

Capacitance-Voltage (CV) characteristics of the samples were measured by sweeping voltage of the Al electrode with ground state of Si substrate. The voltage was swept from positive to negative and return from negative to positive voltage.

The real depth profiles of Ge atoms implanted into the 25-nm thick SiO$_2$ on Si before and after the heat treatment were measured by the high-resolution Rutherford backscattering spectrometry (HR-RBS) by using an apparatus of HRBS500 (KOBELCO, Japan).

Cross-sectional transmission electron microscope (XTEM) images of the samples were obtained using a scanning-type TEM of EM-002B (Topcon Techno, Japan), operating at 200 keV, after cutting the samples with focus ion beam (FIB) of Ga$^+$ with 30 keV.

RESULTS AND DISCUSSION

The Ge concentration in the SiO$_2$ obtained from the HR-RBS measurements are shown in Fig. 2. The implanted depth-profiles of Ge atoms showed considerably agreement with the prediction by TRIM-DYN calculation for annealing below 700°C although Ge atoms were depleted by about 10% of the initial

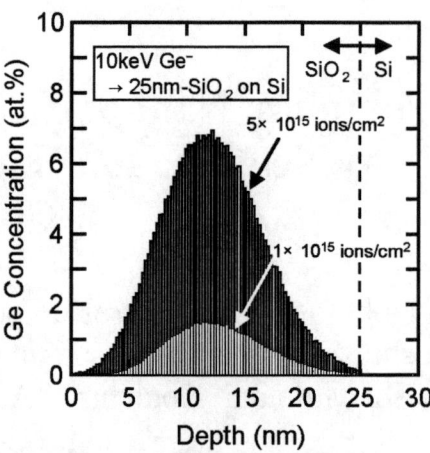

Figure 1. Depth profile of Ge atoms implanted into SiO$_2$ medium calculated by using TRIM-DYN program under conditions at 10 keV with 1×10^{15} and 5×10^{15} ions/cm^2

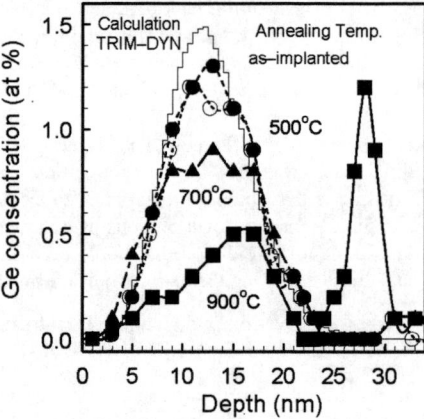

Figure 2. Depth profiles of Ge atoms obtained from HR-RBS for Ge-implanted 25-nm-SiO$_2$/Si samples at 10 keV with 1×10^{15} ions/cm^2 after annealing at various temperatures.

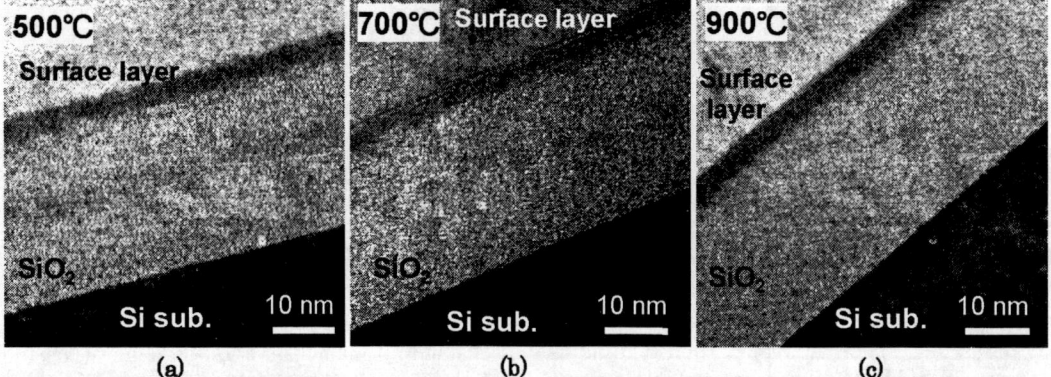

Figure 3. Cross-sectional TEM images of Ge-implanted 25-nm-SiO$_2$/Si samples at 10 keV with 1×10^{15} ion/cm^2 after annealing at various temperatures of : (a) 500°C; (b) 700°C and (c) 900°C.

dose. After annealing at 900°C, the other Ge concentration peaks of 1.2 at.% that appeared around the interface of SiO$_2$/Si is higher than the peak is in the SiO$_2$ film. Besides, a peak position in the SiO$_2$ film shifted about 4 nm to the interface side from the peak position for the as-implanted sample.

The XTEM images are shown in Fig. 3 for each sample implanted with 1 x 10^{15} ions/cm^2, following annealing at (a) 500°C, (b) 700°C and (c) 900°C, respectively, and in Fig. 4 for a sample implanted with 5 x 10^{15} ions/cm^2 and annealed at 700°C. Results show thin black shadows that are we think Ge nanoparticles with about several-nm diameter were appeared around the middle of the SiO$_2$ film for the samples annealed at temperatures less than 700°C. The Ge concentration of about 1 at.% in the conditions and the size of Ge nanoparticles of about several-nm diameter agree with the relations between the results of nanoparicle size and atomic concentration from our previous studies of Ag nanoparticles formation [13].

Although, we could not identify any nanoparticles after annealing at 900°C in the XTEM image of Fig. 3(c), the RBS results suggest that 900°C annealing makes many Ge atoms diffuse and disappear from the SiO$_2$ film. After annealing at high temperature of 900°C, remaining of Ge atoms was not enough to form nanoparticles, or the remained Ge concentration of 0.5 at.% might form so small nanoparticles that can not be observed by our XTEM apparatus. In Fig. 4 with 5 x 10^{15} ions/cm^2, nanoparticles with a diameter of around 3-nm are observed.

The CV characteristics of samples implanted by Ge negative ions and annealed under various conditions are shown in Fig. 5. Hysteresis curves with a range of about 0.6 V appeared for the sample with 1 x 10^{15} ions/cm^2 and 500°C annealing, as shown in Fig. 5(a). At larger dose of 5 x 10^{15} ions/cm^2 and 500°C annealing, the range of hysteresis was about 6.6 V, as shown in Fig. 5(b). The reason for the large hysteresis is considered to be formation of numerous large nanoparticles to store electrical charges. Size of the large nanoparticles is about 3-nm on a basis of a XTEM image of Fig. 4 with 700°C-annealing.

Figure 5(c) shows a result of CV measurement for the sample with 1 x 10^{15} ions/cm^2 and 900°C annealing. Figure 5(c) clearly indicates that the sample cannot be evaluated with CV method. The reason for this abnormal CV curves is considered as the Ge accumulation around the interface of SiO$_2$/Si due to thermal diffusion obtained by the RBS. At 300°C, only small hysteresis appeared although the implanted doses were as high as 5 x 10^{15} ions/cm^2. This result was considered to be caused by too low annealing temperature to form enough large nanoparticles to store electrical charges.

An electrical hysteresis characteristic is an important requirement of memory film for application for semiconductor memory device. Memory film with larger hysteresis in CV characteristics is better because the range of the hysteresis, flat band shift (ΔV_{FB}) corresponds to threshold voltage shift (ΔV_{th}) of MOS memory transistor which includes the film as the gate oxide.

The ΔV_{FB} is obtained by

$$\Delta V_{FB} = -(x_c / T_{ox})(Q_s / C_{ox}),$$

where x_c is the average distance of the stored charge from interface of the gate electrode, Q_s is the surface

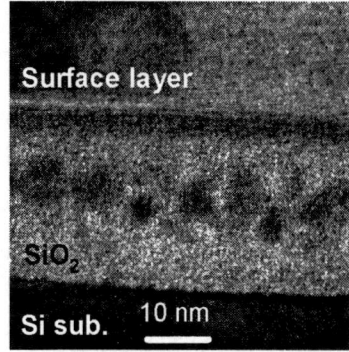

Figure 4. Cross-sectional TEM images of Ge-implanted 25-nm-SiO$_2$/Si samples at 10 keV with 5 x 10^{15} ion/cm^2 after annealing at 700°C.

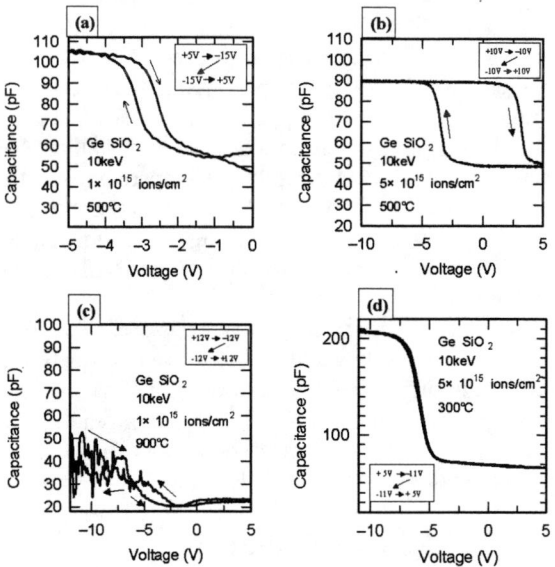

Figure 5. CV characteristics of Ge-implanted 25-nm-SiO$_2$/Si samples at 10 keV with (a) 1 x 10^{15} ion/cm^2 after annealing at 500°C, (b) 5 x 10^{15} ion/cm^2 after annealing at 500°C, (c) 1 x 10^{15} ion/cm^2 after annealing at 900°C and (d) 5 x 10^{15} ion/cm^2 after annealing at 300°C.

density of the stored charge, C_{ox} is the capacitance per unit area of the gate insulator, and T_{ox} is the thickness of the gate insulator.

The stored charge surface density, Q_s, can be calculated with the equation and the ΔV_{FB} of 6.6 V. The x_c is 12.5 nm on the assumption that Ge nanoparticles are at middle depth of the 25nm-SiO_2 film by consideration of the XTEM images. If relative dielectric constant of SiO_2 is 3.9, Q_s calculated is 1.8×10^{-2} C/m^2, so, changing to numbers of electron is 1.1×10^{17} e/m^2, or 1.1×10^{13} e/cm^2.

Next, surface density of Ge nanoparticles is calculated. The implanted Ge atoms of 5×10^{15} $ions/cm^2$ form Ge nanoparticles of 7×10^{12} $/cm^2$, if all implanted Ge atoms form particles of 3-nm diameter and Ge density is 5.4 g/cm^3. The above calculations lead to 1.4 electrons per particle on average (e/p).

For the sample implanted with 1×10^{15} $ions/cm^2$ and annealed at 500°C, the ΔV_{FB} of 0.6 V leads to 0.8 electrons per particle on average.

These results of about one stored electron per particle suggest that Ge nanopaeticles formed by negative ion implantation can charge and discharge and be used as memory devices.

Although we assume that the size of all nanoparticles is roughly 3-nm diameter, and that all implanted Ge atoms are used to form the nanoparticles with 3-nm diameter, some implanted Ge atoms actually are not used to form nanoparticles and still keep atomic state, or re-sputtered or thermally diffused to out of the SiO_2 film. The Ge atoms with atomic state do not contribute to charging electrons.

In case of low dose implantation, the Ge atom should require more long distance than that in case of high dose implantation to combine with other Ge atoms to form a nanoparticle. Comparison of the nanoparticles formation of Ge atoms at the same annealing conditions, the percentage of formation at high dose is much more than that of the formation at low dose.

Therefore it is possible for actual stored electrons per particle to be larger than that of our estimations, and for the difference of stored electrons per particle between in the sample of 1×10^{15} $ions/cm^2$ and in the sample of 5×10^{15} $ions/cm^2$ to be smaller than that in our estimations.

CONCLUSION

Electrical characteristics of the samples were measured with CV method. The samples of thermally grown SiO_2 on Si substrate were implanted with Ge negative ions and annealed at various temperatures. The implanted Ge atoms were evaluated by RBS and XTEM. For the sample annealed at 900°C, Ge atoms diffused and accumulated around the SiO_2/Si interface. For 300°C, the hysteresis was too small to be used for memory devices. For 500°C, the hystereses of the samples implanted with both 1×10^{15} and 5×10^{15} $ions/cm^2$ are 0.6 and 6.6 V, respectively. The rough evaluations of stored charge densities and Ge particle densities from the hysteresis width and the implanted doses indicated one or several stored electrons per particle on average. These results suggest that the thermally grown SiO_2 including Ge nanoparticles formed with negative ion implantation can be used to memory devices.

ACKNOWLEDGMENTS

The authors are grateful to Mr. C. Ichihara and H. Tamagaki of Kobe steel., Ltd. for the support of the HR-RBS measurement.

REFERENCES

1. K. Yano, T. Ishii, T. Hashimoto, T. Kobayashi, F. Murai, and K. Seki, IEEE Transactions **ED 41,** IEEE, 1994, pp. 1628-1638
2. H. Tsuji, N. Arai, T. Matsumoto, K. Ueno, Y. Gotoh, K. Adachi, H. Kotaki, and J. Ishikawa, *Appl. Surf. Sci.,* **238,** 132-137 (2004).
3. A. Nakajima, T. Futatsugi, N. Horiguchi, and N. Yokoyama: *Apple. Phys. Lett.,* 71, 3652-3654 (1997).
4. Y. Takeda, C.G. Lee, N. Kishimoto, *Nucl. Instr. and Meth. in Phys. Res.* **B 191,** 422-427 (2002).
5. K. H. Heinig, T. Mueller, B. Schmidt, M. Strobel, and W. Moeller, *Appl. Phys.,* **A 77,** 17-25 (2003).
6. H. Tsuji, N. Arai, T. Matsumoto, K. Ueno, K. Adachi, H. Kotaki, Y. Gotoh, and J. Ishikawa, *Surf. Coat. Tech.,* **196,** 39-43 (2005).
7. J. Ishikawa, H. Tsuji, N. Arai, T. Matsumoto, K. Ueno, K. Adachi, H. Kotaki, Y. Gotoh, *Nucl. Instr. and Meth. in phys. Res.* **B 237,** 422-427 (2005).
8. S. Tiwari, F. Rana, H. Hanafi, A. Hartstein, E.F. Crabbe, and K. Chan: Appl. Phys. Lett., 68 (10) (1996) 1377
9. H. Tsuji and J. Ishikawa, *Rev. Sci. Instrum.* **63,** 2488-2490 (1992).
10. H. Tsuji, J. Ishikawa, Y. Gotoh, and Y. Okada, AIP Conf. Proceeding 287, American Institute of Physics, Melville, NY, 2002, 1994, pp. 530-536.
11. H. Tsuji, Y. Gotoh, J. Ishikawa, *Nucl. Instr. and Meth. in Phys. Res.* **B 141,** 645-651 (1998).
12. J.P. Biersack, *Nucl. Instr. and Meth. In Phys. Res.* **B 27,** 21-36 (1987).
13. N. Arai, H. Tsuji, T. Matsumoto, Y. Gotoh, K. Adachi, H. Kotaki, J. Ishikawa, *Nucl. Instr. and Meth. in Phys. Res.* **B 206,** 629-633 (2003).

Size Analysis of Ethanol Cluster Ions and Their Sputtering Effects on Solid Surfaces

G. H. Takaoka, K. Nakayama, T. Okada and M. Kawashita

Ion Beam Engineering Experimental Laboratory, Kyoto University,
Nishikyo, Kyoto 615-8510, Japan

Abstract. When vapors of liquid materials such as ethanol were ejected through a nozzle into a vacuum region, ethanol clusters were produced at the vapor pressures larger than 1 atm. The cluster size measured by the TOF method was distributed between a few hundreds and a few thousands, which was different depending on the vapor pressure. Several kinds of solid surfaces such as Si(100) and metal surfaces were irradiated at different acceleration voltages with ethanol cluster ion beams, and chemical and physical sputtering with a high sputtering yield was achieved. In addition, the sputtered surfaces were very flat at an atomic level. The dependence of the sputtered depth on the incident angle of the ethanol cluster ion beams was investigated for the Si(100) surfaces. The sputtered depth had a peak value at an incident angle less than 60 degrees, and the incident angle corresponding to the peak value decreased with decreasing acceleration voltage.

Keywords: cluster ion beam, ethanol cluster, sputtering, TOF method.
PACS: 36.40.Wa, 61.46.Bc, 61.80.Jh, 68.47.Fg, 68.49.Fg, 81.65.Cf, 82.80.Rt, 82.65.+r

INTRODUCTION

The physical and chemical properties of clusters, which consists of a few tens to several thousands atoms, are not the same as those of bulk state matter. Clusters represent a new phase of matter, and they have attracted much interest.[1,2] The investigation of clusters opens up a new field of materials science, which bridges the gap between individual atoms and bulk state matter. Furthermore, cluster ion beams can be applied as useful tools for wafer processing in the LSI device fabrication as well as the investigation of the fundamentals of solid-state physics, chemistry and related materials science. The impact of accelerated cluster ion beams on the solid surfaces exhibits unique characteristics such as multiple collision and low-energy irradiation effect, which are not obtained by monomer ion beams. As a consequence, the properties of the bombarded surface are significantly altered through shallow implantation, sputtering and deposition by the cluster ion beams.[3-6]

A liquid-source cluster ion beam system, which has several advantages for surface treatment, has been developed.[7,8] With regard to liquid materials including organic materials, one of their features is the presence of various kinds of structures and chemical properties as well as the weakly bound state. The deposition of liquid cluster ions on the solid surface does not normally lead to the thick film, but the cluster material itself is deposited on the surface by forming extremely thin layer. This feature is useful for the chemical modification of solid surface using various kinds of liquid materials such as water, alcohol and paraffin. Furthermore, the inherent fluid property of liquid materials as well as gas materials is effective for their continuous supply to the source, which is useful feature in engineering applications of the cluster ion beam process. In addition, the interactions of cluster ions with solid surface atoms are occurred in small area and short time at nano-level, and liquid cluster ion beams offer many possibilities for new and unique applications in surface processing.[9,10] For example, they can be used in physical sputtering and chemical etching of the substrate material, for making surfaces smooth, and for cluster-activated chemical modification of the substrate surfaces. In this article, the size distribution of liquid cluster ions such as ethanol cluster ions are investigated, and the fragmentation of the ethanol cluster ions after impact on the solid surfaces is described. Furthermore, in order to clarify various applications of the ethanol cluster ion beam process, sputtering effects of the ethanol cluster ions are described, and the surface state after sputtering is investigated based on AFM and XPS measurements.

EXPERIMENTS

The liquid cluster source has been described elsewhere [7,8]. Briefly, it composed of a liquid container and a nozzle. The nozzle was made of glass, and it was a converging-diverging supersonic nozzle.

The diameter of the nozzle at the throat was 0.1 mm. Liquid materials such as ethanol was introduced into the container and heated up to 150°C. The vapors of ethanol were ejected through the nozzle into a vacuum region. When the vapor pressure was larger than 1 atm, the vaporized ethanol clusters were produced by an adiabatic expansion. The clusters were ionized by an electron bombardment method. The electron voltage for ionization (Ve) was adjusted between 0 V and 300 V, and the electron current for ionization (Ie) was adjusted between 0 mA and 250 mA. The cluster ions were extracted by applying an extraction voltage to the extraction electrode. The extraction voltage (Vext) was adjusted between 0 kV and 2 kV. The extracted cluster ions were size-separated by applying a retarding potential such as 27 V, and the cluster ions with the size larger than approximately 100 molecules per cluster were separated. These cluster ions were accelerated toward the substrates by applying an acceleration voltage. The acceleration voltage (Va) was adjusted between 0 kV and 10 kV. The substrates used were Si(100) substrates and metal films, and the substrate temperature was room temperature. The background pressure around the substrate was 1×10^{-7} Torr, which was attained using a turbo-molecular pump.

RESULTS AND DISCUSSION

The cluster size of ethanol cluster ions was measured by a time-of-flight (TOF) method. In the cluster size measurement, it was assumed that the ethanol cluster ion was a singly charged ion. If the multiply charged cluster ions were produced, the dissociation of cluster ions were assumed to be occurred due to the Coulomb repulsion forces between the constituent atoms of a cluster ion.[11] Figure 1 shows a size distribution for ethanol cluster ions as a parameter of vapor pressure. The electron voltage for ionization (Ve) was 200 V, and the electron current for ionization (Ie) was 200 mA. The acceleration voltage (Va) was 6 kV. As show in the figure, the cluster size is distributed between a few hundreds and a few thousands, and the peak size is about 500 to 1000 molecules per cluster. In addition, the beam intensity as well as the peak size of the cluster ions increases with increase of the vapor pressure. This is ascribed to the enhancement of condensation and nucleation of vaporized ethanol molecules by increasing the vapor pressure, which resulted in increasing growth of clusters.

When accelerated cluster ions impact on the solid surface, the cluster ions are broken up, and the constituent atoms (or molecules) of a cluster migrate on the surface.[12] For the ethanol cluster ions, the

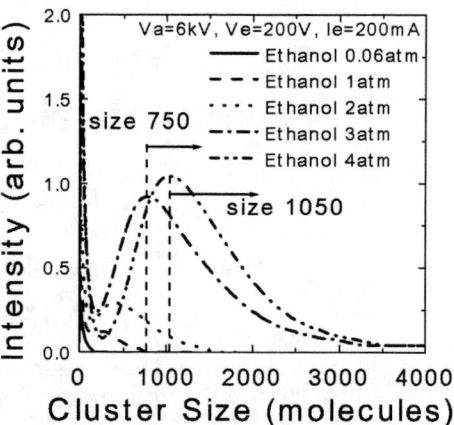

FIGURE 1. Size distribution of ethanol cluster ions as a parameter of vapor pressure.

ethanol molecules are present on the surface through the migration, and the surface is covered with the ethanol molecules. When the impact of other ethanol cluster ions are followed, some ethanol molecules remained on the surface might be dissociated. Figure 2 shows the Q-mass spectra (a) after impact of the ethanol cluster ions and (b) at the atmosphere of the ethanol vapors. The cluster size used was larger than 100 molecules per cluster, and the vacuum pressure at

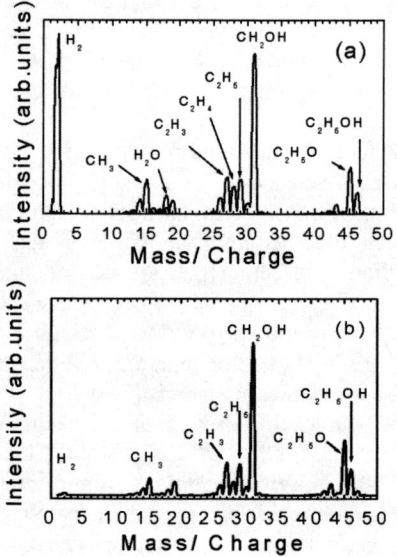

FIGURE 2. Q-mass spectra (a) after impact of the ethanol cluster ions and (b) at the atmosphere of the ethanol vapors.

the atmosphere was 10^{-5} Torr. As shown in the figure, the fragmentation of the ethanol molecule is occurred due to the ionization by an electron bombardment method in the Q-mass system. To be compared with Fig. 2(b), H_2 and C_2H_4 peaks for the ethanol cluster irradiation become large. The dissociation of the ethanol molecules after impact is enhanced. In addition, the conversion efficiency from ethanol molecule to hydrogen molecule is high, which is not achieved at room temperature in the conventional wet process.

The sputtering of Si(100) and metal surfaces by the ethanol cluster ion beams was performed at room temperature. Taking account of the sputtered depth and the ion dose, the sputtering yield was calculated by estimating the density of the Si and metal films. Figure 3 shows the sputtering yield for Si, Ti, Ni, Cu, Ag and Au surfaces at an acceleration voltage of 9 kV. In the figure, the sputtering yield by irradiation of argon (Ar) ion beam at an acceleration voltage of 9 kV is also shown. In the case of the ethanol cluster ion irradiation, the sputtering yield for several metal films is about ten times larger than that by Ar monomer ion irradiation, even if the incident energy of an ethanol molecule as a constituent particle of a cluster ion is less than 90 eV. In particular, Si and Ti surfaces are sputtered more effectively, and the sputtering yield is approximately 100 times larger than that of Ar monomer ions. This is considered to be due to the enhancement of chemical sputtering, in which alkyl compound of Si or Ti is formed as a volatile product. Alkyl radicals consisting of the ethanol molecule, which are produced after impact of the ethanol cluster ions on the solid surfaces, have an important role in chemical sputtering for Si and Ti surfaces.

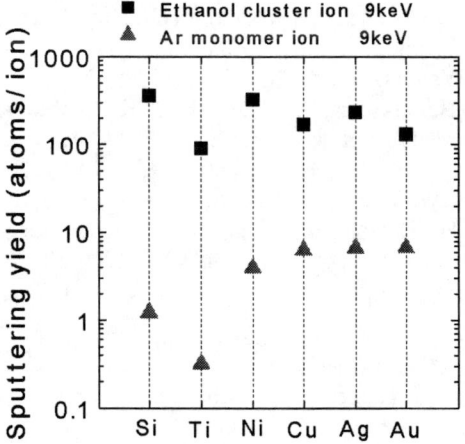

FIGURE 3. Sputtering yield for Si, Ti, Ni, Cu, Ag and Au surfaces by irradiation of ethanol cluster and Ar monomer ion beams at an acceleration voltage of 9 kV.

Figure 4 shows the dependence of the sputtering yield for Si surfaces on the incident angle. The sputtering yield is normalized by the value at an incident angle of zero, which corresponds to the normal direction to the surface. The acceleration voltage was adjusted at 3 kV and 6 kV, and the ion dose was 1.0×10^{15} ions/cm^2. The cluster size used was larger than 100 molecules per cluster. As shown in the figure, the sputtering yield of Si surfaces increases with increase of incident angle, and it has a maximum at an incident angle. Furthermore, the angle at the maximum value increases with increase of the acceleration voltage. For the ethanol cluster ion irradiation on the Si surfaces, the chemical sputtering occurs. It is enhanced at an optimum acceleration voltage, which corresponds to the optimum energy of chemical reaction between ethanol and Si. It is estimated to be a few keV. Therefore, the sputtering yield increases with increase of the incident angle, when the cosine component of the incident energy is larger than a few keV. The incident angle at the maximum value of the sputtering yield also increases with increase of the acceleration voltage, when the acceleration voltage is larger than a few kV. This is much different from the physical sputtering by Ar cluster ion irradiation, in which the sputtering yield

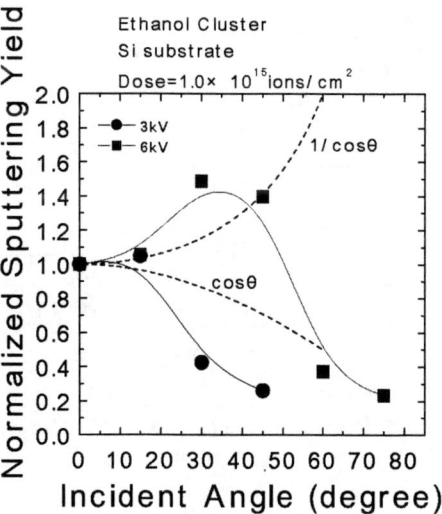

decreases with increase of the incident angle according to cosine law, as shown in the figure.

FIGURE 4. Dependence of the sputtering yield for Si surfaces on the incident angle for the ethanol cluster ion irradiation.

The surface state of the Si substrates after irradiation by the ethanol cluster ion beams was investigated by X-ray photoelectron spectroscopy (XPS). Figure 5 shows the depth profile of (a) O1s and (b) C1s peaks. After etching the irradiated surface by 0.5 nm, both the O1s and C1s peaks disappear. This indicates that ethanol molecules or its fragments are present on the Si surfaces irradiated, and they are not implanted into the Si surface. Furthermore, an atomic force microscope (AFM) observation showed that the Si surfaces sputtered by the ethanol cluster ions were very flat, and the average surface roughness was less than 1 nm.

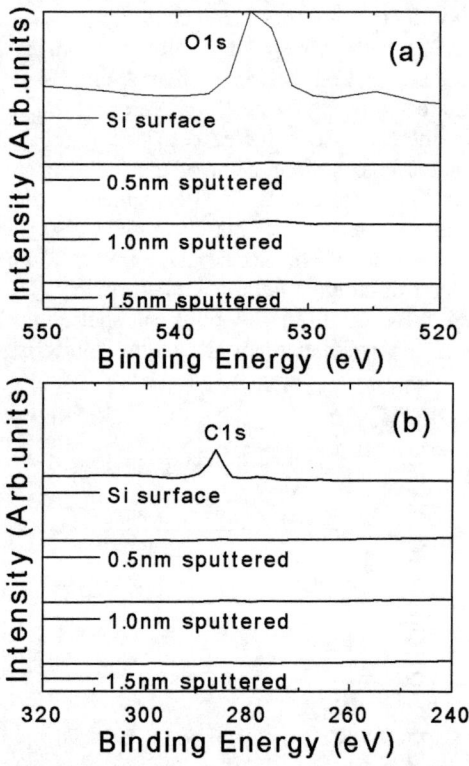

FIGURE 5. Depth profile of (a) O1s and (b) C1s spectra for the Si surfaces irradiated by the ethanol cluster ions.

CONCLUSION

Liquid cluster ion beams have several advantages for surface modification, and we have investigated the cluster size distribution as well as their irradiation effects on solid surfaces. When vapors of ethanol were ejected through a nozzle into a vacuum region, ethanol clusters were produced at the vapor pressures larger than 1 atm. The TOF measurement showed that the cluster size was distributed between a few hundreds and a few thousands, and the peak size increased with increase of the vapor pressure. Several kinds of solid surfaces such as Si(100), Ti, Ni, Cu, Ag and Au surfaces were irradiated at different acceleration voltages with ethanol cluster ion beams, and chemical and physical sputtering with a high sputtering yield was achieved. In addition, the sputtered surfaces were very flat at an atomic level. For the chemical sputtering of Si(100) surfaces, the sputtered depth had a peak value at an incident angle less than 60 degrees. Furthermore, the incident angle corresponding to the peak value of the sputtered depth decreased with decreasing acceleration voltage. Thus, the sputtering process by the ethanol cluster ion beams was much different from that by conventional ion beams.

ACKNOWLEDGEMENTS

A part of this work was supported by "Nanotechnology Support Project" of the Ministry of Education, Culture, Sports, Science and Technology (MEXT), Japan.

REFERENCES

1. M.A. Duncan and D.H. Rouvray, Scientific America, December 1989, p.110.
2. M.E. Mack, Nucl. Instr. Meth B237 (2005) 235.
3. K. Goto, J. Matsuo, T. Sugii, H. Minakata, I. Yamada and T. Hisatsugu, Tech. Dig. IEEE IEDM, Washinbgton, DC, 1996, p.435.
4. I. Yamada, Euro. Phys. J. D9 (1999) 55.
5. G.H. Takaoka, M. Kawashita, K. Omoto and T. Terada, Nucl. Instr. Meth B232 (2005) 200.
6. G.H. Takaoka, H. Shimatani, H. Noguchi and M. Kawashita, Nucl. Instr. Meth B232 (2005) 206.
7. G.H. Takaoka, H. Noguchi, T. Yamamoto and T. Seki, Jpn. J. Appl. Phys. 42 (2003) L1032.
8. G.H. Takaoka, H. Noguchi, K. Nakayama, Y. Hironaka and M. Kawashita, Nucl. Instr. Meth, B237 (2005) 402.
9. G.H. Takaoka, H. Noguchi and Y. Hironaka, Nucl. Instr. Meth, B242, (2006) 100.
10. G.H. Takaoka, H. Noguchi and M. Kawashita, Nucl. Instr. Meth, B242, (2006) 417.
11. J.G. Gay and B.J. Berne, Phys. Rev. Lett. 49 (1982) 194.
12. I.Yamada and G.H.Takaoka, Jpn. J. Appl. Phys. 32, (1993) 2121.

Tuning of Etching Rate by Implantation: Silicon, Polysilicon and Oxide

Rémy Charavel and Jean-Pierre Raskin

Université catholique de Louvain,
Research Center in Micro and Nanoscopic Materials and Electronic Devices,
Place du Levant, 3, B-1348 Louvain-la-Neuve, Belgium

Abstract. Ion implantation is the mostly used method for semiconductor doping but can also be of interest to change locally the etch rate of silicon and silicon dioxide. Indeed damages induced by ion implantation can increase the etch rate of silicon dioxide by a factor of 200 when etched in vapor HF. Doping species of n or p type increases drastically the etching rate of silicon in HNA. Substitutional boron implanted in silicon can reduce the etch rate of silicon in TMAH from a factor of 200. Etch rate modification by ion implantation can be of interest for the fabrication of buried mask, sacrificial layers and etch stop layers.

Keywords: Ion implantation, damage, HNA, TMAH, VHF, selective etching, buried mask.
PACS: 61.72.-y, 61.72.Tt, 68.55.Ln

INTRODUCTION

The main application of ion implantation in semiconductor processing is doping of semiconductors but can also be used to modify the composition of the target material as for example in SIMOX processing[1]. In this paper we propose new fabrication process and applications of ion implantation based on the etch rate modification of silicon and silicon dioxide. Etch rate modification is related to the effects of implantation such as damage creation in the target material or on the presence in the target material of impurities. Ion implantation damages the target material by breaking the covalent bonds and amorphization. Annealing can reconstruct the lattice target and include the implanted atoms at substitional sites, which changes the electrical properties of the semiconductor but can also stress the lattice if there is a mismatch in atomic radii of target and implanted atoms. These three effects of ion implantation, bond breaking, doping, and internal stress can cause etch rate modification of oxide and silicon.

EFFECTS OF ION IMPLANTATION

Ion implantation consists in accelerating ionized species to an energy of a few tens of keV toward a target to introduce ions in this target at a given depth. The dissipation of incoming ion energy in the host target damages the target.

Amorphization and Damages

Each incoming ion penetrating a crystalline target loses its energy through collision with the atoms of the target. The energy is lost through collision with the nuclei of the target atoms and through constant slowdown in the electron sea. The energy dissipated through the collision with the nuclei, which is responsible for the target damage such as amorphization and bond breaking is called the nuclear energy loss denominated S_n. For ion energy lower than 1 keV/amu, the nuclear energy loss dominates whether it becomes negligible in front of the electronic energy loss at higher energy[2]. This implies that for a given energy, heavy ions will induce more damages to the target than lighter ions but their depth of penetration will be shallower than the one of light ions. The total damage induced to the target increases with the implantation dose ϕ according to the following equation[3]:

$$E_n = \phi \times S_n \qquad (1)$$

However, the number of broken bonds tends to saturate for high doses as reconstruction occurs thanks to thermal energy dissipated in the target during implantation. Garrido et al.[4] have shown that at saturation only 15.5% of the bonds are broken.

Annealing and Lattice Reconstruction

After implantation the target crystal has been amorphized. High temperature annealing allows lattice reconstruction and permits to include the implanted species at substitutional sites in the lattice. Rapid Thermal Annealing (RTA) is preferred to long time annealing for keeping the diffusion of the species as low as possible.

Once the implanted atoms sit at substitutional sites in the host lattice they form covalent bonds with the neighboring atoms. In the case of a semiconductor like silicon, if the implanted atoms are from the third or fifth column from the Mendeleïev chart, the impurities at substitutional sites have one default or respectively one extra electron to share with the neighbor atoms. They thus become negatively or respectively positively ionized by getting or releasing an electron into the lattice. This lack of electron or default of electron is then free to move in the lattice and is called a free carrier.

If the implanted atom has a smaller atomic radius than the host atoms, it induces tensile stress[5] in the lattice. It is the case for boron in silicon which has an atomic radius of 0.88 Å versus 1.17 Å for silicon. The stress induced is proportional to the boron concentration[6] but dislocation can relax the lattice if the stress is too high.

ETCH RATE MODIFICATION

These three effects induced by ion implantation:
- bond breaking
- doping
- internal stress

can be used to modify the etching properties of common microelectronics materials.

Etch Rate Reduction of Boron Doped Silicon and Polysilicon

Tetra Methyl Ammonium Hydroxide (TMAH) is a popular base used for silicon etching. The etch rate is decreased for highly boron doped samples[7]. Fig. 1 plots the etch rate versus the doping concentration for etching of silicon and polysilicon in TMAH concentrated at 20% and at 80 °C. The etch rate decreases when increasing the boron concentration. The origin for this etch rate decrease is the induced stress[8] due to silicon doping. The etch rate of polysilicon is higher than the one of silicon, due to the presence of joint of grains. The etch rate decrease in function of the doping concentration is higher for silicon than for polysilicon. In crystalline silicon, all the implanted boron atoms sit at substitutional sites after anneal and contribute to lattice stress.

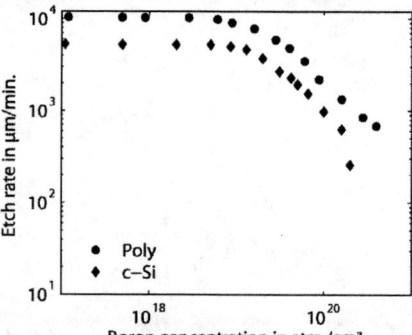

Figure 1: Etch rate of polysilicon and silicon in TMAH 20% at 80 °C.

In polysilicon some of the boron is trapped into the grains boundaries and cannot contribute to the tensile stress of the atomic lattice.

Etch Rate Enhancement of Silicon Dioxide

Bonds breaking of silicon dioxide due to ion implantation makes it more reactive to hydrofluoric acid (HF)[9]. The etch rate of silicon dioxide in HF 3% has been measured on oxide implanted with boron, phosphorous, argon and arsenic. Fig. 2 plots the etch rate in function of the nuclear deposited energy as it is the parameter describing the damages made to the target.

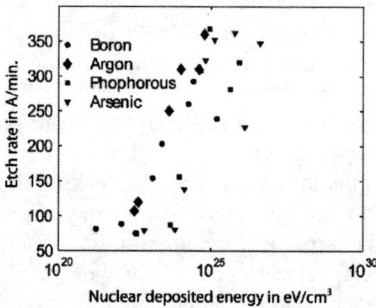

Figure 2: Etch rate in HF 3% of implanted oxide in function of the nuclear deposited energy.

For nuclear deposited energy lower than 1.10^{23} eV/cm³, the etch rate of oxide is similar to the one of unimplanted oxide, i.e. 80 Å/min. Above 3.10^{24} eV/cm³ the etch rate saturates to a value of 320 Å/min, which corresponds to the nuclear deposited energy at which the amount of broken bonds saturates at 15.5%. The etch rate does not depends on the type of implanted species but only on the nuclear deposited energy[10]. Annealing of the implanted samples at 1000 °C during 30 min. under nitrogen flow leads to a drastic reduction of the oxide etching rate, in fact similar to the unimplanted oxide.

Etch Rate Enhancement of Silicon and Polysilicon

The etching solution composed of hydrofluoric acid, nitric acid and acetic acid (HNA) is well known to be a strong silicon etchant[11]. HNA of composition [1:1:2] is known to etch faster the doped silicon. Fig. 3 plots the etched thickness from uniformly boron and arsenic doped polysilicon films etched in HNA of composition [1:1:2].

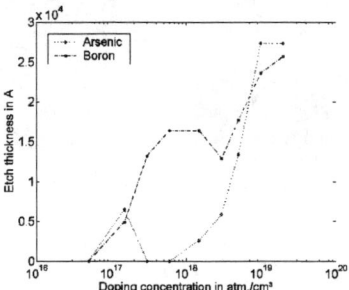

Figure 3: Etched thickness of Boron and Arsenic doped sample in HNA [1:1:2] without stirring.

The etching mechanism can be decomposed in two steps. Oxidation of silicon by the nitric acid followed by dissolution of the oxide in hydrofluoric acid. The oxidation process in nitric acid, described by Robbins et al[12]., is rather complex and involves excited states on the silicon surface and reaction by products issue from nitric acid dissociation. Therefore, presence of free carriers enhances the etch rate of silicon in HNA.

Improved Etching Selectivity of Autocatalytic Reactions

Etching of implanted oxide and doped polysilicon shows etching selectivity but without any possible application due to the too poor etching selectivity for oxide etching and due to the erratic behavior of the polysilicon etching in function of the doping.

HF etching of oxide as whole as HNA etching of silicon are autocatalytic reaction, which means that they are using reaction byproducts as catalyst, and that any start of the reaction sustains further etching.

Vapor HF etching

The etching of silicon dioxide in HF is decomposed in the following steps:

Dissociation of HF thanks to water:
$$HF + H_2O \rightarrow F^- + H^+ \sim H_2O \quad (2)$$

Binding of the fluorine ion to HF:
$$HF + F^- \rightarrow HF_2^- \quad (3)$$

And finally dissolution of oxide which produces water:
$$SiO_2 + 3HF_2^- + H^+ \rightarrow SiF_6^{--} + 2H_2O \quad (4)$$

Water can thus be seen as an autocatalyst as the reaction does not start without water, but produces water once it started.

When etching in aqueous HF, this autocatalytic phenomenon does not impact the reaction rate as the produced water is negligible compared to the water present in the solution.

In vapor HF (VHF) etching, the amount of water present in the ambient is limited due to evaporation therefore the water produced by the reaction impacts the etch rate of oxide[13]. Highly selective etching is obtained between unimplanted and implanted oxide.

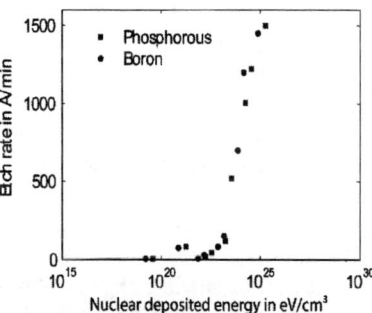

Figure 4: VHF etching of oxide implanted with boron and phosphorous in function of the nuclear deposited energy.

Indeed, implanted oxide is more reactive to HF and will produce water which will enhance the reaction rate and produce even more water than evaporation can remove. Thanks to this avalanche effect very high etch rate are obtained in implanted oxide. For unimplanted oxide, reaction occurs but the produced water evaporates before contributing to a new reaction cycle. Fig. 4 plots etch rate of boron and phosphorous implanted samples in function of the nuclear deposited energy for an etching of 2 min. in VHF at 38 °C. The selectivity which was in the order of 4 for etching in aqueous solution reaches a value of 200 for etching in VHF. Etching in VHF rather than in aqueous HF emphasizes the autocatalytic character of the reaction and therefore increases the etching selectivity.

HNA in ultrasonic bath

The overall reaction of silicon etching in HNA is the following:

$$Si + 6HF + HNO_3 \rightarrow H_2SiF_6 + HNO_2 + H_2O + H_2$$

HNO_2 is produced by the reaction but it is also needed for the oxidation of silicon. Which means that like in the case of oxide etching in VHF, the reaction sustains itself once it has started. Fig. 3 shows that the start of the reaction was quite erratic and that undoped samples could be etched faster than doped samples.

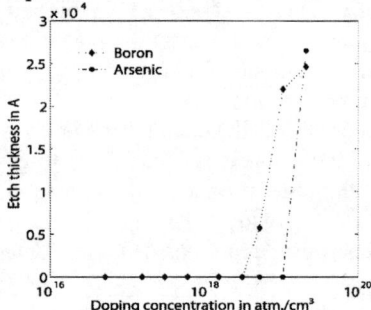

Figure 5: Etched thickness of Boron and Arsenic doped sample in HNA [1:1:2] with stirring and ultrasounds.

The idea is, like in the case of VHF, to remove the reaction byproducts from the reaction site in order to have an etching only where the intrinsic properties of the material could sustain the etching reaction although the lack of catalyst. Etching in ultrasonic bath and stirring of the solution are very effective to remove the reaction by products from the reaction sites. Only highly doped areas produce more catalyst than the agitation of the solution can remove and show very high etch rate. Fig. 5 plots the etched thickness from uniformly boron and arsenic doped polysilicon films etched in HNA of composition [1:1:2] in ultrasonic bath and stirred solution. The etching at low doping concentration vanishes.

APPLICATIONS AND CONCLUSION

Based on these highly selective techniques of silicon and oxide using implantation, several test structures have been analyzed to evaluate the interest of these techniques in CMOS and MEMS technologies. Implantation allows the definition of etch rate modified zones deeply into the material. It is thus a good technique for the creation of buried mask. Fig. 6 shows examples of buried masks created thanks to ion implantation and HNA, TMAH and VHF revelation. After a brief review of the ion implantation mechanisms and their effects on the target material we proposed selective etching based on bonds destruction, silicon doping and stress induced in the lattice. These three selective etching methods being promising solutions for the fabrication of buried masks and the creation of buried etch stop or sacrificial layers which are of interest in the case of the fabrication of self-aligned double gate MOSFETs or the definition of MEMS anchors.

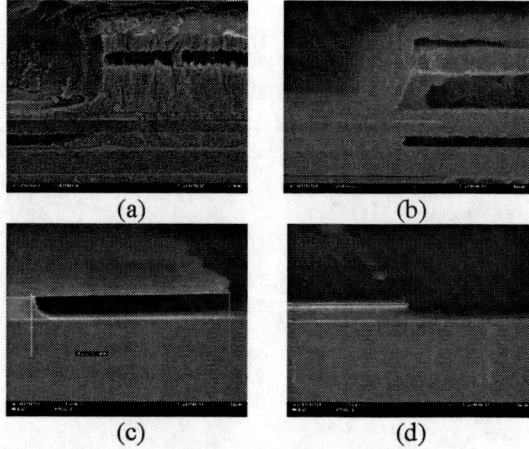

Figure 6: Examples of buried mask. (a) HNA etching, (b) TMAH etching, (c) implanted, (d) unimplanted oxide after VHF etching.

REFERENCES

[1] V. Yakovlev, et al., *2001 IEEE Int. SOI Conf. Proc.*, p. 39-40, 2001.

[2] J. Ziegler, *The stopping and range of ions in solids*, 1995.

[3] A. Hiraiwa et al., *App. Phys. Let.*, vol. 54, n° 12, p. 1106-1109, 1989.

[4] B. Garrido et al., *J. of Non Cryst. Solids*, vol. 187, p. 101-105, 1995.

[5] W. Chu et al., *IEEE Trans. on Elec. Dev.*, vol. 40, n° 7, p. 1245-1250, 1993.

[6] X. Ning, *J. Elec. Soc.*, vol. 143, n° 10, p. 3389-3393, 1996.

[7] R. Charavel, *Proc. SPIE Int. Soc. Opt. Eng.*, vol. 5116, p. 699-709, 2003

[8] M. Decarli, *Int. Conf. On Micro. Test Struc.*, p. 25-28, 2003.

[9] L. Liu, *IEEE Elec. Dev. Meet.*, p. 17-20, 1996.

[10] R. Charavel, *Elec. and Solid-State Let.*, vol. 9, n° 7, 2006

[11] H. Robbins et al., *J. of the Elec. Soc.*, vol. 106, p. 505-508, 1959.

[12] H. Robbins et al., *J. of the Elec. Soc.*, vol. 107, n° 2, p. 108-111, 1960.

[13] A. Witvrouw, *Proc. SPIE Int. Soc. Opt. Eng.*, vol. 4174, p. 130-141, 2000.

The Influence of Ion Implantation on cell Attachment to Glassy Polymeric Carbon

R. Zimmerman[1], I. Gurhan[2], F. Ozdal-Kurt[3], B. H. Sen[4], M. Rodrigues[5], D. Ila[1o]

1. Center for Irradiation of Materials, Alabama A&M University, Normal, AL USA
2. Ege University Faculty of Engineering, Izmir, Turkey
3. CBU Faculty of Science, Manisa, Turkey
4. EU Faculty of Dentistry, Izmir, Turkey
5. University of São Paulo, Ribeirão Preto SP, Brazil

Abstract. *In vitro* biocompatibility tests have been carried out with model cell lines to demonstrate that near surface implantation of silver in Glassy Polymeric Carbon (GPC) can completely inhibit cell attachment on implanted areas while leaving adjacent areas unaffected. Patterned ion implantation permits precise control of tissue growth on medical applications of GPC. We have shown that silver ion implantation or argon ion assisted surface deposition of silver inhibits cell growth on GPC, a desirable improvement of current cardiac implants.

Keywords: Carbon biocompatibility, Cell growth, Ion implantation
PACS: 87.68.+z

INTRODUCTION

Greater cell adhesion to Glassy Polymeric Carbon (GPC) can be achieved by making the surface rough. The texture dimensions and chemical affinity of the surface should be adequate to allow strong adherence of tissue on the material surface. Alternatively, surface treatment by an appropriate ion implantation may be expected to inhibit cell growth or give the cells no adherence at all. Both enhancement and inhibition of cell attachment to GPC are desirable modifications of GPC for particular medical applications.

The electrical, mechanical and chemical properties of Glassy Polymeric Carbon (GPC) make it an exceptionally versatile biomaterial [1-5]. Like other forms of pure carbon, GPC objects implanted in living tissue remain inert and have no biological influence on neighboring organs. Cells of connective tissue eventually encapsulate the GPC object. The formation of a completely biocompatible interface with biological tissue is an important attribute for carbon trans coetaneous electrodes or semi permanent fluid delivery tubes.

The strength, durability and low density make GPC a favored material for the manufacture of replacement heart valves. However, the low adhesion at the interface with biological tissue [6] has the potential of creating an embolism when endothelial tissue that naturally encapsulates the implant is released into the blood stream.

We have previously reported the use of oxygen ion bombardment to increase the surface roughness, to enhance cell adhesion, and here we show that implanted silver ions near the surface of GPC inhibit cell attachment and adhesion.

EXPERIMENTAL METHODS

The GPC samples used in the study being reported were molded with dimensions 20x10x1 mm, then pyrolized to 700°C. Available porosity is known [1] to vary greatly with heat treatment temperature. For these cell adhesion studies, we chose GPC samples heat treated to 700°C that produces the maximum available porosity.

Samples were implanted with silver atoms, either by 6 MeV silver ions delivered by a Pelletron electrostatic accelerator, or by physical vapor deposition of silver atoms while simultaneously bombarding the surface with 700 eV argon atoms. The first process leaves the silver atoms in a layer about 4 microns beneath the surface of the GPC. The second process, Ion Beam Assisted Deposition (IBAD), mixes the carbon and silver atoms in the first 5 nm of the GPC surface.

Rutherford Backscattering Spectrometry (RBS) was employed to determine the silver concentration in samples implanted by one or the other process and expressed as a fluence in silver atoms/cm^2

Cell adhesion to GPC was tested with L929 cells mouse connective tissue fibroblast cell-line, from the Animal Cell Culture Collection HUKUK, Ankara, Turkey). L929 cells were cultured in Dullbecco's modification of Eagle's medium (DMEM) with 10% fetal bovine serum (FBS) in 75 cm^2 tissue culture flasks. They were harvested from monolayers by 0.25% trypsin in 0.53M EDTA solution at 80-90% confluence. Cells were centrifuged and resuspended in the culture media. After staining with the trypan blue (0.4%), the cell

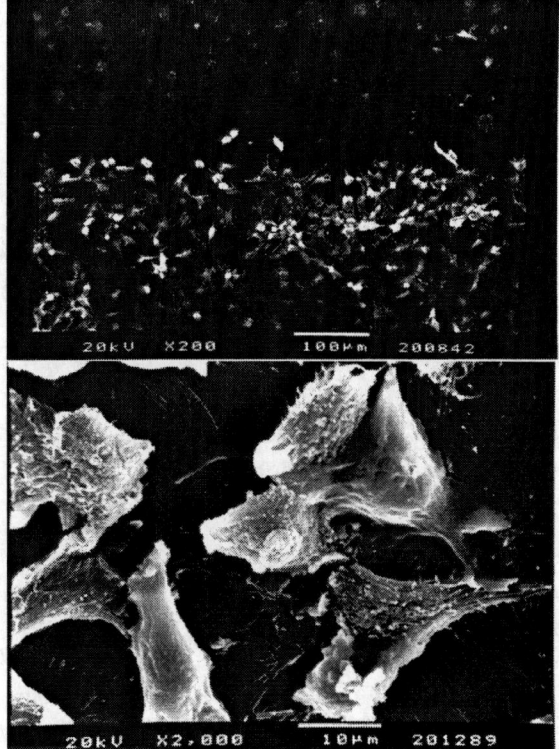

Figure 1. SEM images of cell attachment and adhesion on the sample with and without argon ion beam assisted deposition of silver. The upper image shows prolific cell attachment on areas with no silver (lower half of the image) and none where 9×10^{15} silver atoms per cm^2 have been implanted (upper half). The higher magnification image at the bottom shows the long filapodial extensions that indicate strong adhesion of the cells to the untreated GPC substrate.

density N_S was counted with haemocytometer (Bürker) and adjusted to 3.5×10^5 cells/ml.

Each GPC sample was placed in a separate microbiologic petri dish and 1 ml cell suspension was seeded into each well. After 3 hours in the incubator (Hereaus, Germany) with a humidified 5%CO$_2$

Table 1, Cell adhesion to the surface of GPC implanted with 6 MeV silver ions [7]

Ag fluence (cm^{-2})	Adhesion %
3.9×10^{16}	0
3.9×10^{15}	8
3.9×10^{14}	28
3.9×10^{13}	75
Zero	85

atmosphere, the density of non adherent cells N_N was counted. The density of attached cells N_A was determined by subtraction $N_S - N_N$. The percentage adherence A_D is given by

$$A_D = \frac{N_S - N_N}{N_S} \qquad (1)$$

Scanning electron microscopy (SEM) was employed to obtain images of the cell attachment, growth and adherence.

RESULTS and DISCUSSION

After incubating the samples in a solution seeded with mouse connective tissue fibroblasts, table I shows the inhibition of cell attachment of MeV silver ion implantation, more so when the silver fluence is increased. Figure 1 shows scanning electron micrographs of cells attached and adhering to GPC in areas free from the silver implantation. The formation of a flattened morphology of cell growth with long filapodial extensions is evidence of strong adhesion [7,8]. A desirable surface treatment of GPC causes complete inhibition of cell growth. Silver implantation appears to accomplish this alternative. A previously reported result [9] has shown that cell attachment is inhibited also by IBAD deposition of silver atoms near the surface of GPC.

The available porosity of GPC is increased by the MeV silver ion bombardment. The enhancement of porosity may be inferred from lithium percolation into the GPC surface and a measurement of the lithium concentration by a nuclear reaction analytic method [10,11]. The lithium atoms have been shown to sequester more at the end of range of the 700 eV argon ions, about five nanometers [13]. The silver atoms are shown by RBS also to concentrate at the end of the argon ion range. Owing to the increased porosity at the GPC surface, both cell attachment and adherence at the surface are expected to be influenced chemically by the guest atoms of silver beneath the carbon surface.

Figure 1 compares cell attachment to a GPC surface implanted with and without silver near the surface of GPC. Strong adhesion is indicated by the long filapodial extensions from each cell that has attached to the surface. Examination of the SEM micrographs in figure 1 shows strong adherence of cells to untreated GPC, and no adhesion and few attached cells to a GPC surface that has implanted silver atoms. Table 1 shows that a silver ion fluence more than 10^{15} cm^{-2} inhibits cell adhesion.

We believe that the cells were inhibited from attaching by a mechanism other than toxicity. The cell adhesion experiments were performed on single samples with sharp boundaries between areas with different silver atom fluence and with no silver atoms. Cell adhesion on the silver free areas was uniform within 20 microns of a silver treated area where no cells attached.

The risk of silver depletion from the carbon surface by blood serum or plasma such that cell attachment occurs has been studied. No such diminution of the inhibitory effect of silver has been observed after six months in physiological solution. Longer term leaching of the silver, which would reduce the benefits of treatment and might possibly produce undesirable collateral effects in vivo, have been accomplished on IBAD silver GPC after 11 months in physiological solution. No diminution of the inhibitive effect has been observed. These studies are continuing.

CONCLUSIONS

Silver atoms/cm^2 implanted 4 microns beneath the surface of Glassy Polymeric Carbon by MeV ion bombardment significantly inhibits cell attachment, consistent with previous studies [7] with a concentration of 10^{15} cm^{-2} or more silver deposited close to the surface by argon Ion Beam Assisted Deposition. A concentration of 10^{15} cm^{-2} or more silver atoms close to the surface either by MeV ion implantation or by argon IBAD proves entirely to prevent cell attachment, a desirable property for GPC cardiac implants.

ACKNOWLEDGMENTS

We have received partial support from the Alabama A&M University Research Institute (AAMURI) and the AAMU Center for Irradiation of Materials. We are grateful to the Ege University Bioengineering Department where the cell adhesion and SEM studies were performed.

REFERENCES

1. G. M. Jenkins and K. Kawamura, *Polymeric Carbons-Carbons Fiber* (Cambridge University Press 1976).
2. H. Maleki, L.R. Holland, G.M. Jenkins, R.L. Zimmerman, *Journal of Material Research* **11-9**, 2368 (1996).
3. H. Maleki, D. Ila, G.M. Jenkins, R.L. Zimmerman, A.L. Evelyn, *Material Research Society Symposium Proceeding* **371**, 443 (1995).
4. G.M. Jenkins, C.J. Grigson, *J. Biomedical Materials Research* **13**, 371 (1979).
5. G.M. Jenkins, D. Ila, H. Maleki, *Mat. Res. Soc. Symp. Proc.* **394**, 181 (1995).
6. N.S. Braunwald, L.I. Bonchek, *J. Thoracic & Cardiovasc. Surg.* **54-5**, 127 (1967).
7. R. Zimmerman, I. Gurhan, S. Sarkisov, C. Muntele, D. Ila and M. Rodrigues, submitted *Research Society Symposium Proceedings*, 2005Jockusch BM, Bubeck P, Giehl K, Kroemker M, Moschner J, Rothkegel M, Rudiger M, Schluter K, Stanke G, Winkler J., *Annual Review of Cell and Development Biology* **11**, 379-416 (1995).
9. R. Zimmerman, I. Gürhan, C. Muntele, D. Ila, M. Rodrigues, F. Özdal-Kurt and B. H. Sen, *Surface Modification of Materials by Ion Bombardment*, Izmir, Turkey, (2005).
10. R.L. Zimmerman, D. Ila, G.M. Jenkins, H. Maleki, D.B. Poker, *Nuclear Instruments and Methods in Physics Research* **B106**, 550 (1995).
11. R.L. Zimmerman, D. Ila, D.B. Poker, S.P. Withrow, *Application of Accelerators in Research and Industry*, Duggan & Morgan (Eds), New York, 957 (1996).
12. H. Maleki, D. Ila, R. L. Zimmerman, G. M. Jenkins and D. B. Poker, *Materials Research Society Symposium Proceedings* **414**, 107 (1996).
13. J. F. Ziegler, J. P. Biersack and U. Littmark, *The Stopping and Range of Ions in Solids* (Pergamon Press Inc., New York, 1985).

IMPLANT TECHNOLOGY

Chicane Deceleration –
An Innovative Energy Contamination Control Technique in Low Energy Ion Implantation

N. White [a], J. Chen [a], C. Mulcahy [b], S. Biswas [b], R. Gwilliam [c].

[a] *Advanced Ion Beam Technology Inc., 33 Cherry Hill Drive, Danvers, MA 01923, USA*
[b] *Cascade Scientific Ltd., ETC Building, Brunel Science Park, Uxbridge, Middlesex UB8 3PH, UK*
[c] *Surrey Ion Beam Center, Advanced Technology Institute, University of Surrey, Guildford GU2 7XH, UK*

Abstract. High current beams suitable for USJ implantation were generated by 'Chicane Deceleration' involving an s-bend to block contaminants. Implanted wafers were analyzed with 200eV O_2^+ beams at 45° to resolve the sources of dopant profile variation in fine detail. Energy contamination is essentially eliminated, but for B^+ channeling remains important. Unannealed X_j values from 5 to 7 nm are reported for different implant species.

Keywords: Ion implantation; SIMS; Boron; USJ; Energy Contamination; Deceleration.
PACS: 61.72.Tt; 61.72.-y; 66.30 Jt; 66.30.Pa; 85.40.Ry; 85.40.Ss.

INTRODUCTION

The ITRS identifies the need for sub-40nm S/D junction depths. In the case of P-MOS devices, implant process requires very shallow implants, probably optimized to control damage mechanisms, to minimize phenomena such as transient-enhanced and boron-enhanced diffusion during the anneal. Ultra-shallow dopant placement (<< 10 nm) is desirable.[1] Alternatives to conventional boron implants have been developed, including implantation of decaborane[2] and octadecaborane[3] molecular ions, true gas cluster doping[4], and plasma doping.

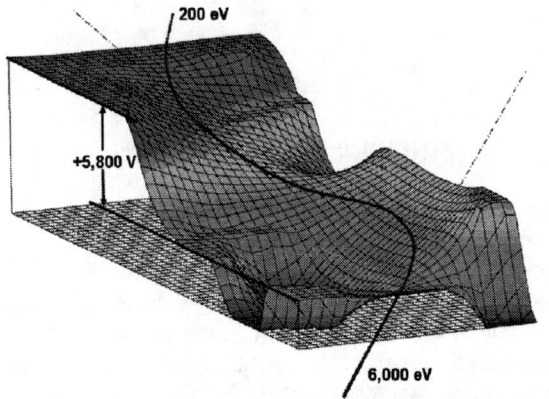

Figure 1. Chicane Deceleration. Ions are retarded by a potential field which deflects left and right around obstacles.

This paper describes a) a deceleration technique to deliver high current beams of boron or BF_2 at pure ultra-low energies, and b) SIMS data taken under conditions which optimize resolution of the ultra-shallow dopant profiles produced with the new technique. We explore a number of effects on the resulting measured dopant profile, including the effects of SIMS parameters. We quantify energy contamination and channeling (which could lead to deep tails in the dopant profile), as well as dose and species-related effects which can modify the as-implanted profile. At this stage, annealing was not investigated.

EXPERIMENTAL DETAILS

Deceleration a short distance before the implantation station of an implanter is desirable to increase the available beam current, but in most commercially available ion implanters, results in energy contamination.[5,6]

We have developed an apparatus and technique for decelerating ion beams to small fractions of their original energy (from 1/3 to 1/30) while preserving a high percentage of the beam current and eliminating these high-energy contaminants. Fig. 1 shows how this technique works. Because of the obvious resemblance, it is named a 'Deceleration Chicane' after chicanes used in motor racing. The beam is deflected through an s-bend by an electric field which contains a strong component of deceleration.[7] Boron ion beams were decelerated from energies in the range of 4 to 10 keV down to final energies of 200eV to 1 keV.

The beam of a relatively conventional beamline is first steered off axis by 5° to direct it into the chicane. In the chicane electric fields bend it 45° left, then 40° right. Ions which do not at all times have the correct energy-to-charge ratio cannot be transmitted around this chicane, and they leave the beam path to the left or right, where they are carefully intercepted by beam stops angled to block forward transmission of any unwanted emitted particles, neutral or charged. Neutral atoms formed within the beam leave the beam at the first bend. Additionally, ions which are neutralized within the chicane are also blocked.

Although space-charge forces are strong, they are controlled well, because the electrostatic bends are strongly focusing[9], and the focusing and defocusing forces can be balanced. Data is presented in the next section to confirm this balance.

Beam currents were measured in a magnetically suppressed Faraday cup, whose field had been modeled and calculated to permit 100% transmission of 200eV boron. Dosimetry was largely manual; the desired mechanical velocity was hand-calculated and programmed. Since we are considering channeling, we also measured the angular divergence of the beam under several conditions.

SIMS analyses were initially performed on pilot samples under a variety of implant and analysis conditions. The effect of using different O_2^+ energies was explored, and was compared with data obtained by others. For the pilot implants we used argon preamorphization, and saw variability in the dopant background between about 1e17 and 1e18 per cc. However, RBS analysis confirmed that the material had been fully amorphous, eliminating channeling as a potential cause.

The present implant matrix was designed to remove uncertainty about the background, and to be sure to resolve any deep tails present in the dopant profiles. Prime n-type wafers were used. Most wafers were given a pre-amorphization implant (PAI) of 100keV of Ge at 2E15. The high energy and dose were designed to ensure that any higher energy beam components remained in amorphous silicon, to aid interpretation. It also ensured that any end-of-range damage existing beyond the fully amorphous layer was beyond the region of interest in these measurements. Some wafers were left crystalline and were implanted at 0° to permit axial channeling. Some implants used photoresist over 50% of the implanted surface. The effect upon the dopant profile of implanting to different doses was explored, since there was interest in observing saturation effects.

RESULTS

Metrology

Figure 2 shows two overlaid measurements of the same sample implanted with BF_2^+ at 981eV (equivalent to 200eV B), using different probe beams. Both use O_2^+ at 45°, but the energies are 200 and 750 eV. The normal velocity of the oxygen atoms is lowered by the use of molecular ions and by the 45 degree incidence, giving them a normal velocity equivalent to energies *per atom* of 70.7 and 265 eV respectively. Figure 2 shows that the result of analyzing these profiles with 750eV O_2 is an overestimate of depth of ~ 2 nm. We have based our conclusions on the 200eV O_2 45° results, estimating the uncertainty in depth at 0.6nm. The data has an internal consistency of better than 0.1 nm.

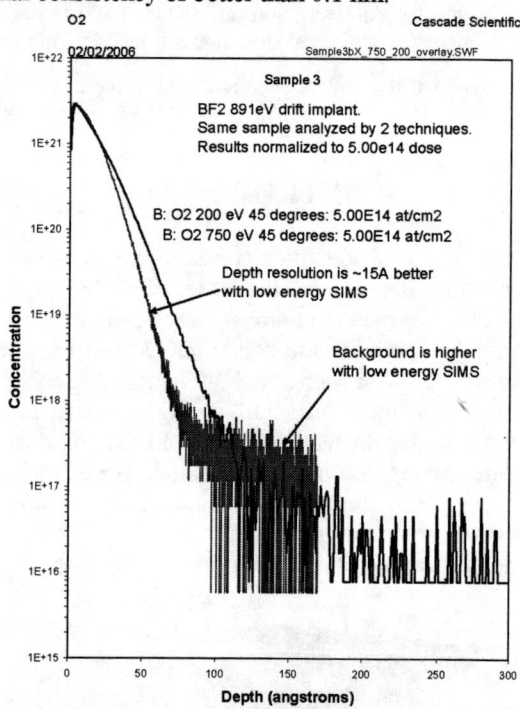

FIGURE 2. Effect of SIMS energy.

The background level is a source of concern and uncertainty. It can be seen from Fig. 2 that the background is higher in the case of 200eV analysis than of 750eV analysis. BF_2^+ produces a profile differing from the equivalent boron profile both in a substantially lower background (at about 1/3) and in a deeper peak concentration; further the amount of self-sputtering dose loss is greater. (This can be seen below in the data presented in Fig. 4.) The reason for the lower background is not clear, and is the subject of further study. The difference is not caused by the presence of energy contaminants or by channeling; these mechanisms have been eliminated by careful use of controls, and will be further discussed below. The use of BF_2^+ implants at 891 eV as a reference standard for 200eV boron implants is therefore invalid.

The angular distribution of the boron ion beam was measured by passing the beam through a set of holes in a plate, and allowing the resulting beam to mark a target 150mm behind the holes. In this manner we confirmed that the centroid of the incident ion beam was normal to the wafer surface within about 0.3°. The beam contained a distribution of angles about this centroid, 90% of the ions lying within about +/-2.5° of the centroid. There was no discernible variation of incident angle across the surface of the wafer using this method.

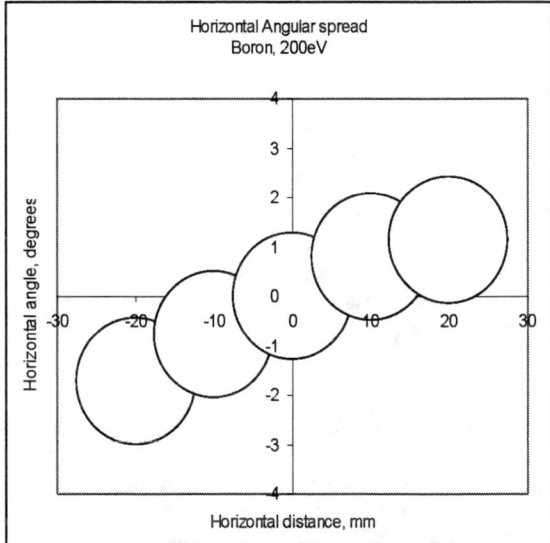

FIGURE 3. The angular divergence of the 200eV beams measured by passing the beam through five 3mm dia. holes, drifting 150mm, and recording the beam burns on paper.

Analysis

Key observations

The profile of each peak is well resolved. No measurement artifacts are apparent at the surface. Surface oxide was approx 1.3nm based on elapsed time since pre-amorphization, except for the channeled wafers, where oxide thickness was unknown.

We can see differences in Fig. 4 in the position of the maximum concentration as a function of dose, presumably because of knock-on during implantation. For 200eV B at 8.0e14 per cm^2, the peak concentration occurs at 0.11nm deep. At 1.9e15 this increases to 0.21 nm. The peak position of a 5.8e14 BF_2 891eV implant is deeper than that of the same dose of 200eV B, although the ion velocities are the same, but matches that of the 1.9e15 boron profile. This can be accounted for by assuming that one effect of the co-implanted fluorine atoms in the lower dose BF_2^+ implant is to knock on previously implanted boron at a similar rate to primary boron ions.

Above 3nm the effect of photoresist-induced pressure changes can be resolved. Above 4nm the effects of channeling can be resolved. Whereas the channeling leads to a very noticeable tail about 4nm deeper than the PAI implants, the effect of pressure dependence cannot be resolved above 10nm deep.

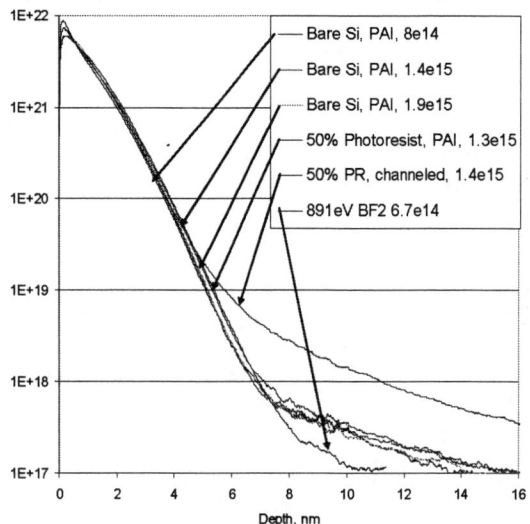

FIGURE 4. All profiles 200eV B^+ using chicane deceleration from 6 keV, except 891eV BF_2 from 8 keV for reference. *All normalized to 8.0e14 for comparison.* Data is smoothed by a rolling average to reduce noise, width varying from 0.01 to 0.1 nm with depth. SIMS O_2^+ 200eV 45°. Boron currents 2.1 to 2.5 mA. BF_2 3.6 mA. Preamorphization Ge 100 keV 2e15 except where noted.

Energy contamination.

It was deemed impractical to obtain a valid reference sample of silicon implanted with drift-mode Boron at 200eV. The least ambiguous way to determine the level of energy contamination is to compare results from a bare wafer and from one with a 50% covering of photoresist (PR). Measurement of the actual pressure close to the wafer has little meaning since the outgassing hydrogen is a directed stream, and effective pressure in the beam is far higher than gauges will record. Dose control was based on measured beam current before the implant; the dose received by the wafer with photoresist was found to be 5% *lower* than the bare wafer. This can be explained if the pressure rise in the beamline upstream of the chicane structure was high enough to increase beam neutralization by 5%. It is thus estimated that the flux of 6keV neutral boron atoms was at least 5% of the delivered 200eV boron. The projected range of 6 keV boron is 27 nm with 13 nm straggle, *and the SIMS data shows no sign of such a contaminant.*

But Fig. 4 shows a slight shift in the profile with the photoresist present. By normalizing the two profiles to the same total dose, and subtracting the bare wafer's profile from the photoresist profile, we can look at this difference in detail. This difference is plotted in Fig. 5. We have a high quality signal extending to 6.5 nm. Beyond this depth, the noise-swamped signal changes sign several times, and no clear evidence can be extracted from it. The data indicates that the profile depth at 6.5 nm is increased by 0.25nm (i.e. to 6.75nm), by the presence of photoresist. At this depth the concentration is around 1E18 per cm^3.

mode energy by about 98eV would give a profile falling within the chicane decel profile.

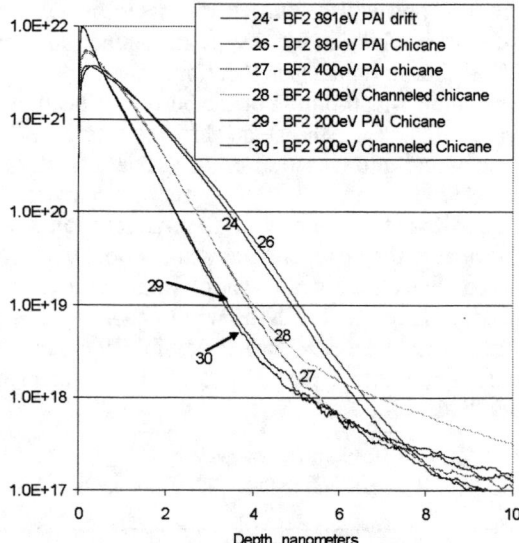

FIGURE 6. BF2 implant profiles, all doses normalized to 5e14 for comparison and smoothed by a rolling average increasing from 0.01 to 0.1nm with depth

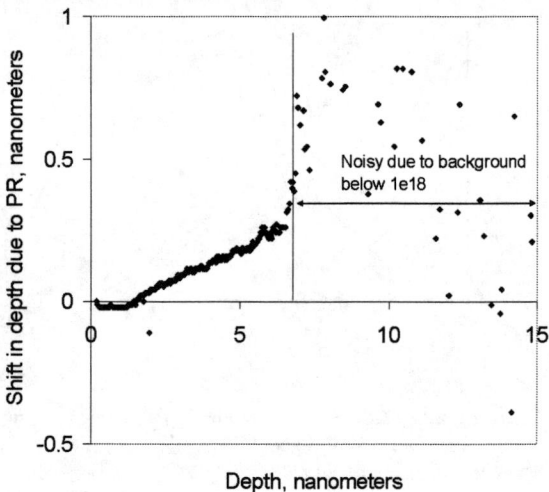

FIGURE 5. Effect of photoresist is to increase profile depth by up to 0.25nm. This is the limit of any energy contamination.

Fig. 6 shows two 891eV BF_2 implants, number 24 being a drift-mode beam and number 26 using chicane deceleration from 8 keV. There is a shift in profile depth of about 0.25 nm. The background in both BF_2^+ implants is 1/3 that of the 200eV B^+ implants.

The highest contaminant energy present is clearly very small, being only tens of volts greater than the 200eV present. The mechanism by which it is formed is shown in Fig. 7. Any high-energy contaminant which can reach the wafer must be formed right at the exit of the chicane in the final region of electric field. We can estimate the magnitude. For 200eV boron, we can estimate that raising the energy by 22 eV on a bare wafer would give a profile which would fall within the observed PR profile. For 891eV BF2, raising the drift-

The chicane is highly effective in eliminating energy contamination. The excess energy is much lower than that produced by Einzel lenses and electron suppression schemes in some systems claiming to deliver 'drift' beams. Evaluation of a percentage energy contamination is almost impossible when the increase in energy is very slight, and has no practical value. Stating the effective maximum energy is more useful.

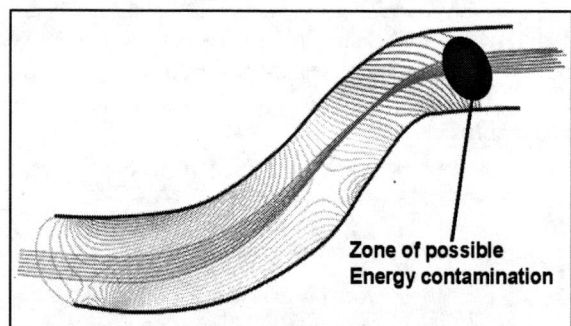

FIGURE 7. Source of residual contamination.

Profile Shape.

We implanted 200eV boron at several doses under otherwise identical conditions, and analyzed them in one batch under identical conditions.

For bare wafers, with PAI and 1.3nm oxide, we observed that the peak concentration occurred at a depth of 0.11 nm for a SIMS dose of 8.5e14, but increased to 0.22 at a dose of 1.9e15. Comparison of the implanted doses and SIMS doses indicates loss of dopant at higher doses, reaching a relative 25% loss of dose by self-sputtering at 1.9e15.

It appears that as the dose is increased, the boron is redistributed by the collision cascades of the implanted ions. Dose retention would be seriously compromised at doses above 4e15, but up to 2e15 is acceptable. The same mechanism that removes part of the dose pushes the peak concentration deeper.

891 eV BF_2^+ implantation produces a profile similar to a 200eV B^+ implant *at three times the dose*, as can be seen in the figures. Presumably fluorine atoms in the molecule also displace previously implanted atoms at a similar rate to boron ions.

Channeling.

For 200 keV B^+ implants at 0°, Fig. 4 shows that failure to pre-amorphize the substrate caused the 'as-implanted X_j' depth of dopant at a concentration of 1E18 to increase by 4 nm. More than 1% of the implant channeled to a small extent, about 0.1% contributed to this deep tail. It has been reported elsewhere[8] that tilting the substrate is ineffective at low energies in reducing channeling.

For BF_2^+ implants, the situation was different. We compared channeled and PAI implants at 400 and 200eV. In analyzing the results shown in Fig. 6, one should bear in mind that the effective boron energy is 90 and 45 eV respectively, while the oxygen effective energy in the SIMS is 71 eV. Therefore the resolution of the depth profile of these implants is compromised because the SIMS energy is not low enough.

At 400eV the implant doses were 5e14. Channeling did produce a significant effect at this energy, probably disqualifying this energy for use without PAI.

At 200 eV the implant doses were 3e14. It is striking that there is no discernible difference between the PAI and the crystalline substrates. The depth at which the concentration reaches 1E18 is about 4.6 nm. The beam current for these implants was 1.0mA, which is sufficient for 33 300mm wafers per hour using reasonable generic handling times. With no requirement for PAI this is a throughput worth consideration for production use.

CONCLUSIONS

We presented the Chicane deceleration technique for generating high-current ion beams (several mA) of various ions at low energies, and we present data in the range from 200eV to 1 keV. The technique produces beams with very low angular spread, dominated by the thermal physics of ion production. It virtually eliminates high-energy contaminants from the ion beam.

We showed the need to use very low energy O_2 beams in the SIMS analysis, in order to avoid significant distortion of the results. We further discussed issues of control and background. We performed SIMS analyses of a number of implants using this technique, to quantify the mono-chromaticity of the ion beam, and elucidate the effects of channeling and saturation.

We demonstrated that USJ implants using B^+ ions can be performed *with only ~22eV energy contamination* at 200eV and 2.5 mA, sufficient for >60 wph. However PAI is required. We further showed that 200eV BF_2^+ implantation can give still shallower junctions, and does not require PAI implants.

ACKNOWLEDGMENTS

We thank Ed Petersen and Yap Han Chang for beamline operation and performing the implants, Neil Montgomery and Paul Ebblewhite of Cascade Scientific for the SIMS analyses, Chris Jeynes of Surrey University for RBS measurements.

REFERENCES

1. J Chen et al, The Effects of non-Chromaticity of ^{11}B ion beams on ^{11}B diffusion, *Nucl. Instr. and Meth.* B 237 (2005) p. 155.
2. N. Hamamoto et al., Decaborane Implantation with the Medium Current Implanter, *Nucl. Instr. And Meth.* B 237 (2005) p. 25.
3. Y. Kawasaki et al., Ultra-shallow junction formation by $B_{18}H_{22}$ ion implantation, *Nucl. Instr. And Meth.* B 237 (2005) p. 443.
4. M.E. Mack, Gas Cluster Ion beams for Wafer Processing, *Nucl. Instr. And Meth.* B 237 (2005) p. 235.
5. US Patent 6,489,622 B1
6. US patent 6,710,358 B1
7. US patent application 2006/0113494 A1.
8. M. Foad and D. Jennings, Formation of ultra-shallow junctions by ion implantation and RTA, in *Solid State Technology*, December 1998.
9. O.A. Anderson, D.S.A. Goldberg, W.S. Cooper, L. Soroka, A Transverse Field Focusing (TFF) accelerator for intense ribbon beams, *IEEE Trans NS-30 No. 4* (1983) p. 3215.

Characterisation Of The Beam Plasma In High Current, Low Energy Ion Beams For Implanters

J. Fiala[1], D. G. Armour[1], J. A. van den Berg[1], A. J. T. Holmes[1], R. D. Goldberg[2], and E. H. J. Collart[2].

[1] *Institute of Materials Research, University of Salford, Salford M4 5WT, UK*
[2] *Applied Materials, Foundry Lane, Horsham, RH13 5PY, UK*

Abstract. The effective transport of high current, positive ion beams at low energies in ion implanters requires the a high level of space charge compensation. The self-induced or forced introduction of electrons is known to result in the creation of a so-called beam plasma through which the beam propagates. Despite the ability of beams at energies above about 3-5 keV to create their own neutralising plasmas and the development of highly effective, plasma based neutralising systems for low energy beams, very little is known about the nature of beam plasmas and how their characteristics and capabilities depend on beam current, beam energy and beamline pressure. These issues have been addressed in a detailed scanning Langmuir probe study of the plasmas created in beams passing through the post-analysis section of a commercial, high current ion implanter. Combined with Faraday cup measurements of the rate of loss of beam current in the same region due to charge exchange and scattering collisions, the probe data have provided a valuable insight into the nature of the slow ion and electron production and loss processes. Two distinct electron energy distribution functions are observed with electron temperatures ≥ 25 V and around 1 eV. The fast electrons observed must be produced in their energetic state. By studying the properties of the beam plasma as a function of the beam and beamline parameters, information on the ways in which the plasma and the beam interact to reduce beam blow-up and retain a stable plasma has been obtained.

Keywords: ion implanter, Langmuir probe, low energy beam transport.
PACS: 52.40.Mj

INTRODUCTION

The work described in this paper is concerned with an investigation of the detailed physics of high current ion beams aimed at improving the performance of ion implantation machines at low energy when using non-decelerated beams. It is commonly believed that improvements in forced space-charge neutralisation are responsible for the maintenance of high currents down to energies of about 3 keV. However, despite major improvements in plasma flood neutraliser design, drift mode currents below this energy are still too low. Neither the nature of the neutralisation process nor the role of the "so called" beam plasma that is created in the beam environment is fully understood. There is also evidence, based on the observation of beam profiles, that some form of magnetic confinement of the beam can occur even when there is no externally applied magnetic field in the beam environment (1).

This paper describes the early results obtained in an on-going, detailed study of the plasma created in the environment of a high current ion beam. Langmuir probe measurements have been carried out on Ar^+ beams at the wafer position in a specially adapted, commercial ion implanter (Applied Materials xR LEAP). The main parameters of the beam plasmas, electron and ion densities, electron temperatures, and plasma and floating potentials have been measured as functions of beam energy, beam current, beamline pressure and beam tuning settings.

It has been found that in common with the plasmas in ion source and plasma flood plasmas, the electrons in the beam plasma are characterised by two temperatures. The populations of the two distributions involved, together with the ion density, the transverse profile and the plasma and floating potentials, all depend critically on the details of the tuning parameters and the beamline pressure.

EXPERIMENTAL SYSTEM

The measurements were carried out in a specially designed target chamber attached to a standard beamline as shown schematically in Figure 1. By using a high speed cryopump, base pressures below 1×10^{-7}

mbar and operating of 3×10^{-6} mbar were achieved. This allowed a comprehensive study of the effects of pressure on the beam plasma to be carried out.

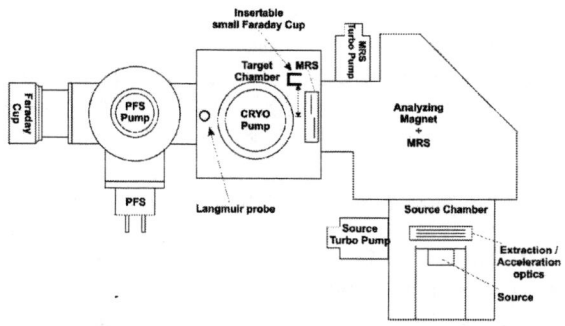

FIGURE 1. Schematic diagram of the implanter showing the probe position.

In view of the need to collect very small currents (down to <μA) of plasma particles on the probe within the volume of the multi-mA ion beam, the probe geometry was required to present a large collecting area to the plasma while minimising beam collection. Consequently, a planar probe with dimensions of $3 \times 3 \times 0.3$ mm was applied with the narrow end facing the beam and the 3×3 mm face of the probe accurately aligned parallel to the beam. Its size was chosen to avoid a rapid change in collection area with probe potential, as occurs in low pressure plasmas. The probe was positoned in the target chamber as shown in Figure 1 and could be scanned across the ion beam.

RESULTS AND DISCUSSION

The Langmuir probe characteristics obtained for all pressures and positions in the beam were very similar to those obtained from magnetically confined discharges such as those found in ion sources and plasma flood systems. The electron saturation current was always very poorly defined and, where it could be evaluated, was significantly smaller than that expected on the basis of the ratio of the ion and electron masses. This is intriguing since in the beamline region under study, there was no externally applied magnetic field.

On the basis of the assumption that the electron and ion densities in the plasma are equal, the ratio of the electron and ion currents on the probe in the saturation regions of the characteristic is given by:

$$R = \frac{I_e}{I_+} = \frac{1}{2.4}\sqrt{\frac{8 \cdot M}{m \cdot \pi}} \qquad (1)$$

where m and M are the masses of the electron and ion, respectively. For an argon plasma, the ratio is approximately 180. All the ratios observed so far have been below about 50, indicating that if the beam plasma is truly quasi-neutral, the ability of the plasma electrons to reach the probe is being significantly impeded.

If the electron energy distributions in the plasma are Maxwellian, important parameters such as the electron temperatures, electron and ion densities and the plasma potential can be evaluated by plotting the Langmuir probe characteristics on a semi-logarithmic graph, as shown in Figure 3. In this figure, two linear regions, associated with two groups of electrons characterised by different temperatures, are clearly observed.

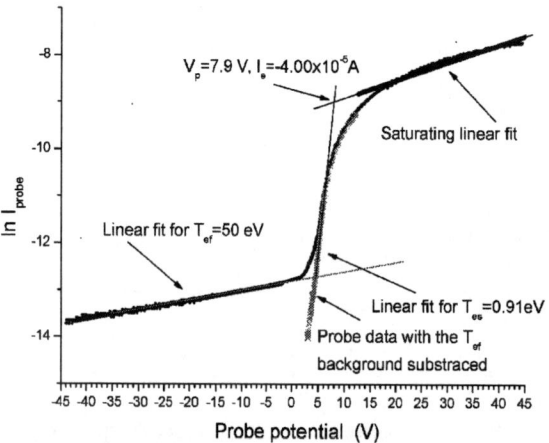

FIGURE 2. Langmuir probe characteristic for 10 keV Ar$^+$ beam in logarithmic scale (p = 5×10^{-5} mBar, I_{arc} = 4 A, V_{arc} = 80 V, I_{beam} = 4.5 mA).

A summary of the information that can be obtained from such probe characteristics is provided in Table 1, where the parameter values relating to the plasma created in a 10 keV, 5 mA, Ar$^+$ beam at a beamline pressure of 5×10^{-5} mBar are shown. They are typical of those observed for a wide range of beam conditions. The temperatures of the slow electrons are almost always in the range from 0.5 to 2 eV while those associated with fast electrons cover the range from about 25 to 55 eV. The density of the fast electrons is only a few percent of the total electron density.

The presence of a significant fast electron content is, at first sight, inconsistent with the observation of a positive floating potential. The potential in the ion beam environment is known to be a complex function of the slow ion and electron production and loss processes and its value, with respect the earth reference used for the probe measurements, is a function of the precise transport conditions. If the

TABLE 1. Parameters obtained from the Langmuir probe characteristics.

Parameter	Symb	Formula	Typ. Value	Unit
Floating potential	V_f	from graph	3.22	V
Plasma potential	V_p	from graph	7.90	V
Temperature slow electrons	T_{es}	from graph	0.91	eV
Temperature fast electrons	T_{ef}	from graph	50.00	eV
Density of plasma ions	n_+	$\dfrac{I}{A_p e\, c_S}$	3.04×10^{15}	m^{-3}
Density of plasma slow electrons	n_e	$\dfrac{4 I_e}{A_p e \sqrt{\dfrac{8 e T}{\pi m}}}$	1.74×10^{14}	m^{-3}
Density of plasma fast electrons	n_f	$\dfrac{4 I_f}{A_p e \sqrt{\dfrac{8 e T_f}{\pi m}}}$	1.18×10^{12}	m^{-3}

coupling between the beam plasma and the beam is not capable of pinning the beam potential to the plasma potential, the beam plasma and floating potentials may, actually, be referenced to a potential that is positive with respect to earth. Slow ion spectroscopy measurements, in which the energy of the slow ions leaving the beam are taken as a measure of the potential at which they were created, have confirmed that in beams that are not fully neutralised, positive potentials between about two and twenty volts may exist (2). Consequently, in the present studies in which no forced neutralisation systems were in operation, the observation of a positive floating potential using an earth-referenced probe could result from the fact that the beam environment is at a positive potential with respect to ground.

A second feature of interest concerning the fast electrons is that their density as a fraction of the total electron density is similar to that observed in the plasmas in IHC sources operating at low magnetic field settings (3). However, the IHC plasmas contain a source of fast primary electrons and the pressure conditions are such that some electrons are able to retain a significant fraction of their energy as they pass through the plasma volume. The source of the fast electrons in the beam plasma is more difficult to identify and, in fact, the details of the production and loss processes associated with these particles and the more numerous slow ions and electrons in the beam plasma are still poorly understood. The processes that may be involved are charge exchange, ionisation in the gas phase in ion-atom or electron-atom collisions and beam scattering leading to secondary electron emission from bombarded surfaces. Since all these processes depend on pressure, beam loss and probe measurements were carried out over the pressure range from about 3×10^{-6} to 4×10^{-4} mbar in order to assess their contributions to the formation of the beam plasma.

The loss of beam in the target chamber in which the probe was located was measured by monitoring the beam current at the entrance to the chamber using a moveable Faraday cup and on the beam dump at the end of the beamline. Since the beam loss over a distance, l, is given by:

$$I = I_0 \exp(-N \sigma l) \qquad (2)$$

where N is the gas density and σ the cross-section for charge exchange. The results, for a 10 keV, 5mA Ar$^+$ beam, are shown on the semi-logarithmic graph shown in Figure 3.

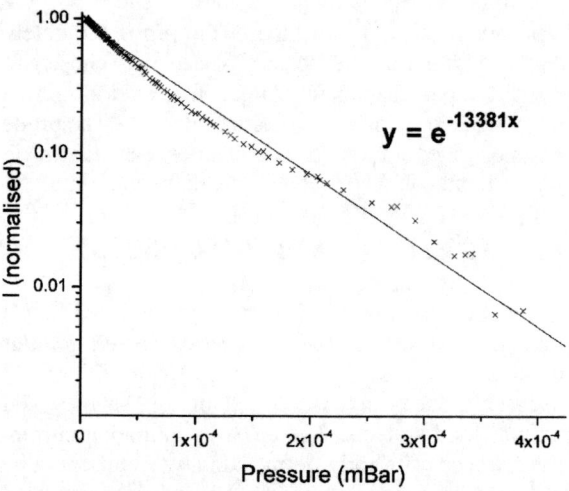

FIGURE 3. Beam loss in the probe chamber as a function of pressure for a 10 keV, Ar$^+$ beam in Ar.

The slope of the line indicates a loss cross-section of 4.5×10^{-15} cm^2. Ionising collisions in the gas phase, leading to the production of an electron and slow ion, leave the fast primary ion in its charged state and do not contribute to the beam loss. Consequently, the loss cross-section would be expected to represent the sum of the charge exchange and scattering events. The charge exchange cross-section for 10 keV, Ar$^+$ on Ar is 2×10^{-15} cm^2 (4) and, because of the resonant nature of the process and the large impact parameters involved, is expected to be considerably higher than the scattering cross-section. This implies that beam expansion due to incomplete space charge neutralisation may also be contributing to the loss at the lower end of the pressure range studied and that only quite small scattering angles are required.

At each pressure at which the beam loss was measured, the Langmuir probe was scanned across the beam at fixed potentials of +25 V to measure the electron current in the saturation region of the characteristic and –42 V to measure the saturation ion current. The full scan was necessary to ensure that the measured probe currents were not being distorted by variations in the transverse density distribution of the plasma as the pressure was changed. The results of these measurements for a 10 keV, 9mA, Ar$^+$ beam are shown in Figure 4. The normalised electron and ion current curves represent the change per unit beam at the input of the target chamber. For comparison, the inverted, normalised beam loss curve has been included.

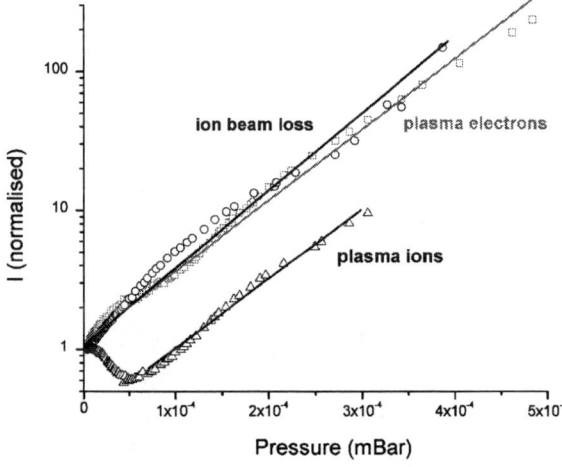

FIGURE 4. Comparison of plasma ion, plasma electron and beam loss changes with beam line pressure for 10 keV, 9mA, Ar$^+$ beam.

The behaviour shown is very similar to that seen in other beams studied. The electron currents to the probe follow the beam loss quite closely over the full range of pressures studied while the slow ion currents fall initially before ultimately rising at a rate commensurate with the beam loss rate. The initial fall in slow ion current appears to reflect the reduction in the ion beam current at the probe position. If this is the case, it implies that at low pressures, the slow ion density at any point along the beam is dictated by the local production and loss rates and the probe ion current falls as the beam current and hence the slow ion production rate falls. It also implies that the slow ions are not highly mobile along the beam direction and that they are lost mainly through a combination of diffusion and drift (if there is a significant space charge potential in the beam) out of the sides of the beam.

In contrast to the slow ion currents, the electron currents rise almost exponentially over the entire pressure range studied, despite the reduction in beam current at the probe position. From the point of view of electron production, the fall in beam current is not as severe as for slow ion production since the fast neutrals formed in charge exchange collisions upstream of the probe scatter and generate secondary electrons in the same way as the beam ions. However, there is still a loss of beam due to scattering or space charge expansion prior to reaching the probe. Consequently, the observation of an increase in the electron current implies that the electron density at the probe position is not determined by very local effects and that the electrons trapped in the beam must be mobile along the beam direction. The generation rate of secondary electrons within the target chamber is directly related to the amount of scattering if the fast neutrals and ions behave in the same way. Therefore, the increase in the electron current to the probe as the pressure rises and the beam current to the beam dump falls is an indication that the electrons must be effectively trapped within the beam.

As the electron density within the beam volume rises, it begins to affect the transport behaviour of the slow ions. At a pressure of approximately 5×10^{-5} mbar, the ion losses have been reduced to the extent that the ion current to the probe starts to increase at a similar rate to the electron current. At this stage, it appears that a genuine beam plasma has been created. The fact that the rate of increase in plasma density corresponds so closely to the beam loss rate implies that the plasma confinement is highly effective even when the potential in the beam has been reduced to a value that is probably close to the plasma potential.

The effectiveness of this confinement, combined with the nature of the Langmuir probe characteristics, suggests that a small, azimuthal, magnetic field may be present in the beam environment once a quasi-neutral beam plasma has been created. Effects that could result from the presence of such fields have been observed in other, medium energy, high current beam systems but the field strengths have never been measured (1). If it is assumed that a magnetic field is responsible for the electron confinement, an indication of the field strength can be obtained from the expression derived by Bickerton (5) to explain the diffusion of electrons across magnetic field lines due to scattering events. The normal diffusion is reduced in a magnetic field by the factor:

$$G = (1+\omega_c^2\tau^2)^{-1} \qquad (3)$$

where ω_c is the electron cyclotron frequency {eB/m} and τ is the mean time between collisions {1/Nσv}. In the present plasmas, the electron saturation currents are reduced by factors of about 4. If it is assumed that the electron scattering collisions are mainly with the background gas atoms, a magnetic flux density of 3.5 µTesla would be required. This field may be produced by a plasma current flowing along the beam direction.

SUMMARY

Langmuir probe studies of the behaviour of electrons and slow ions created within high current, 10 keV, Ar$^+$ ion beams being transported through an Ar gas environment have been described. The data have provided a valuable insight into the properties of the beam plasma and the way in which it is formed and have highlighted the role of effective electron confinement. At low pressures, when there are insufficient electrons present, the slow ions produced in charge exchange collisions are rapidly lost from the beam and do not contribute to the space charge neutralisation process. At higher pressures, the increased production of electrons, largely via scattering and secondary emission rather than gas phase ionisation, and their effective trapping in the beam reduces the ion loss and facilitates the development of a quasi-neutral plasma in the beam environment. Typical plasmas contain fast and slow electrons characterised by temperatures in the ranges 0.5 to 2 and 25 to 55 eV. The effectiveness of the electron confinement, combined with the shape of the Langmuir probe characteristics, suggests that a small, azimuthal magnetic field may account for the reduced electron mobilities perpendicular to the beam direction.

REFERENCES

1. A. J. T. Holmes (private communication).
2. J England, N Bryan, H Ito, D G Armour, J A van den Berg, I Fotheringham and P Kindersley, *Proc. Ion Implantation Technonology* (IIT-92), edited by D F Downey et al., 1993, Elsevier Sci Publ, pp. 613-616.
3. J. E. Mefo, *PhD Dissertation*, University of Surrey (2005).
4. J A van den Berg, G Wostenholm, D G Armour, C E A Cook and A Murrell, *Proc Int. Conf. on Ion Implantation Technology* (ITT 2000), edited by. D. F. Downey et al., IEEE Operations Services Piscataway, NJ 08855-1331, 2001, pp. 627-10.
5. M I Swift and Schwar, *Plasma Diagnostics*, London, Iliffe Books 1970, p. 267

Next Generation Medium Current Product: VIISta 900XP

A. Renau

Varian Semiconductor Equipment Associates, Inc., 35 Dory Road, Gloucester, MA 01930, USA

Abstract. Device fabrication requirements for 65nm and 45nm technologies further increase the demands on medium current implanters for higher process quality as well as for increased productivity and flexibility. Application space continues to grow beyond the traditional Vt adjustment implants to include other well and halo implants, creating the need for a broader energy range. Pressure for reduced CoO drives the need for higher productivity and this, in turn, drives requirements for tuning time, wafer handling speeds, beam currents and reliability that are substantially more aggressive than for older generation tools. Additionally, dose control, angle control and defect control standards have become more stringent. We discuss these requirements in more detail and describe the improvements that have been made to Varian's next generation medium current tool in order to meet them.

Keywords: Ion implantation; Energy contamination; Angle control.
PACS: 41.75.Ak; 61.85.+p; 85.40.Ry; 61.72.Tt

INTRODUCTION

Medium current tools are used for lower dose implants, where they have high productivity and low cost per wafer, and for implants that require the high angle and dose accuracy that can be achieved with the sophisticated control capabilities designed into this type of tool. For CMOS transistor fabrication, medium current implants can be classified as either well or channel engineering (figure 1).

Loosely speaking, channel engineering is used to control the drive characteristics of the transistor and well engineering to control its isolation. For medium current tools, the three channel engineering steps are the halo, threshold voltage adjustment and super-steep retrograde implants. The three well engineering implants are for the shallow well, the deep well and the channel stop.

The dopant ions that are generally used for well and channel engineering are B, BF_2, As, P, In and Sb. The typical dose and energy requirements for these implants are shown in Table 1. Many, if not all, of these implants have precise placement requirements and need 1.5% (3σ) dose control or better, and 0.2° (3σ) angle control.

TABLE 1. Mid Current Implants

Implant	Dose Range (/cm2)	Energy Range (keV)	Tilt
Channel Eng.			
Halo	1e13-6e13	2-40	Yes
Vt	5e11-5e12	5-40	No
SSR	5e11-8e12	100-200	No
Well Eng.			No
Shallow well	8e12-3e13	100-600	No
Deep well	6e12-1e13	800-1200	No
Channel stop	5e12-1e13	80-200	No

Varian has developed the VIISta 900XP generation of medium current implanters to provide a highly productive solution for addressing a wider range of these applications. The VIISta 900XP is over 25% more productive than its predecessor, the VIISta 810XE, and has improved process capability and a wider process window. These performance improvements are discussed below, along with a brief description of the system changes.

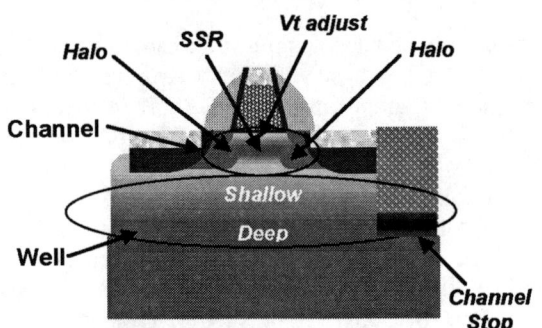

FIGURE 1. Medium current implant applications on a CMOS transistor

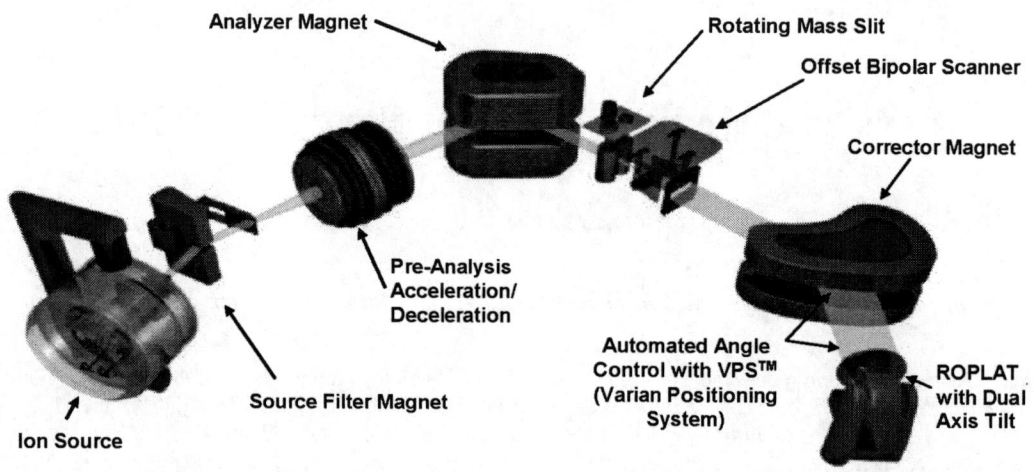

FIGURE 2. VIISta 900XP beam line

SYSTEM DESCRIPTION

The VIISta 900XP beam line is shown in figure 2. The overall architecture is similar to the original VIISta 810 [1]. Output from the source is filtered before entering the beam line to control contamination. The beam is at final energy before mass analysis to prevent any energy contamination. The dose uniformity and angle uniformity are close loop controlled to ensure precise dopant placement [2].

The changes that have been made to the system for the VIISta 900XP are: to the accel/decel column of the beam line, to increase lower energy beam currents; to the wafer handling and wafer processing systems, to increase the maximum throughput; to the operating voltage of the tool, to increase the application space; to the dose control system, to eliminate the detrimental effects of photo-resist outgassing; and to the software and hardware, to reduce beam tuning time and increase reliability. The effect of each of these improvements is described in more detail below.

PRODUCTIVITY

An implanter's productivity is controlled by the wafer handling time, the implant time, the tune time and the availability.

For most of the implants shown in Table 1, the dose and energy are such that the steady-state throughput is not limited by beam current. So, for the majority of mid-current implants, the implant time is controlled by the mechanical throughput. For the VIISta 900XP mid-current tool the maximum wafer scanning velocity has been increased, and significant improvements have been made to the timing sequence

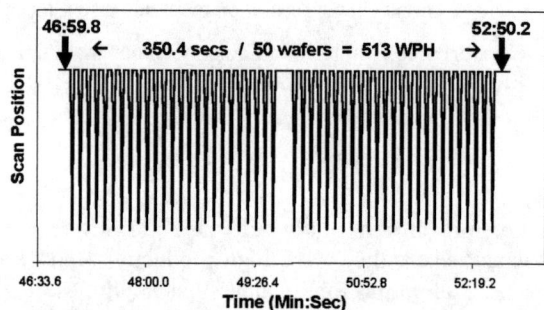

FIGURE 3a. Throughput calculated from scan position

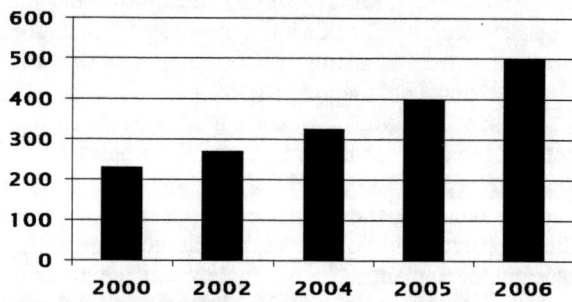

FIGURE 3b. VIISta medium current throughput increase since introduction

for atmospheric and vacuum wafer handling and to the vent and rough cycles. These have resulted in a maximum throughput for the tool in excess of 500WPH (figure 3a). Since its introduction, the mechanical throughput of the VIISta 810 has more than doubled (figure 3b).

When throughputs are as high as they are on the VIISta 900XP, tuning time becomes a critical aspect of productivity. For example, with a steady state throughput of 500WPH, a lot of 25 wafers will be processed in 3 minutes. If the average tune time is 3 minutes then a fully utilized tool running these types of lots will spend 50% of its time tuning recipes. Significant effort has therefore been dedicated to tune

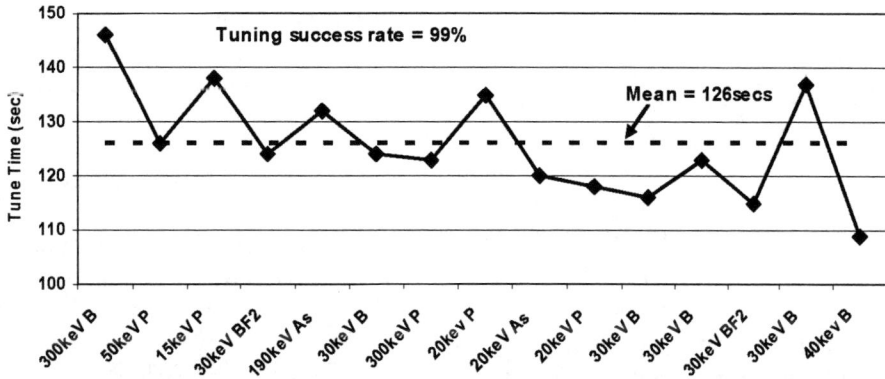

FIGURE 4. VIISta 900XP auto-tune performance for the VSEA 'standard memory and logic recipe set'. Data is from a one week marathon. Each point represents the average of many set ups. The spec tune time for this recipe set is 2.5 min.

time reduction using a 'VSEA standard memory and logic recipe set' as representative of the traditional (excluding deep well) application space. The average tune time specification for this recipe set is 2.5 minutes. The tool's performance is shown in figure 4.

For implants that are lower energy or higher dose, such as the halo implants, beam current may limit productivity. For this reason, some modifications have been made to the accel/decel column of the beam line, to substantially increase the beam currents in the low to mid energy range. These are illustrated in Figure 5. For beams at or below 40keV the spec currents have been increased by between 25% and 100%.

The final component of productivity is tool reliability. Figure 6 shows the performance of some of the over 200 VIISta medium current tools that are in production worldwide.

PROCESS CONTROL

The important components of process control for low dose applications are dose control, angle control, energy contamination control, metals contamination control and particle control. The angle control capabilities of the VIISta 810 have been discussed elsewhere [2], as have the aspects of the beam line design that ensures zero energy contamination [3].

This architecture, along with the simple end station design, also provides excellent control of metals and particle contamination. An example of the particle performance is shown in figure 7, which shows that

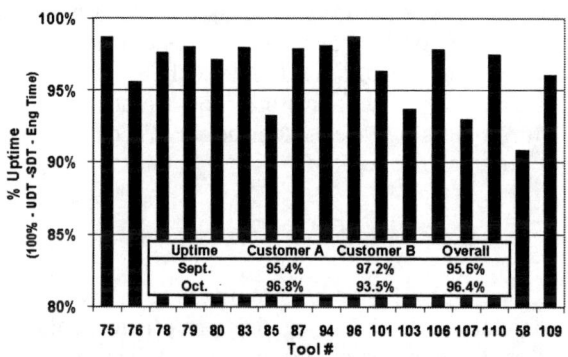

FIGURE 6. Reliability of 17 VIISta 810 implanters measured over 2 months at 2 customer sites.

the tool's performance exceeds its spec of <21 adders ≥0.12μm. The metal contamination of the tool is also well within its spec of 5ppm for heavy metals and 30ppm for aluminum. Moreover, a significant improvement beyond these levels has been made possible by new technologies that have been developed for the tool, such as the new plasma flood gun [4].

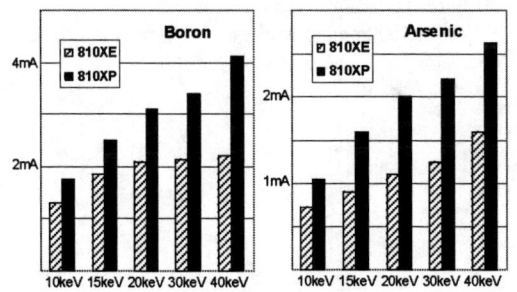

FIGURE 5. Low and mid energy beam currents. The 810XP referred to here is a VIISta 900XP without the extended terminal voltage.

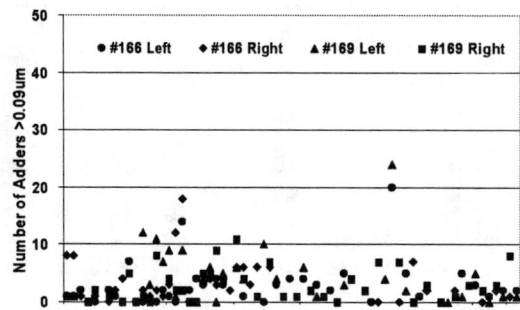

FIGURE 7. Particle adders (>0.09μm) as measured over 3 months on two VIISta 810 tools in production at a memory fab.

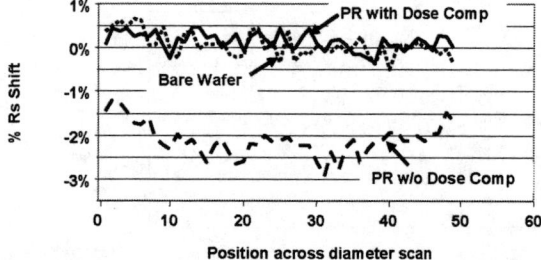

FIGURE 8. Control of photo-resist out-gassing effects. The dashed line shows that out-gassing effects can cause significant dose shift from the bare wafer results (dotted line). The VIISta 900XP's automatic dose compensation (solid line) prevents this.

The dose control system has been improved significantly for the VIISta 900XP by incorporating a patented [5] means for controlling the effects of photo-resist outgassing. This operates automatically and, unlike other schemes for dealing with neutralization, it is not based on pressure compensation. The effectiveness of the automatic dose compensation feature is illustrated by the data in figure 8.

APPLICATION RANGE

Traditional high energy tools use complex and less efficient LINAC [6] or tandem [7] technologies. These tools are much less productive than one based on DC technology with a simple high throughput end station like that used by the VIISta tools. Moreover, CMOS scaling continues to reduce the required energies of many of the low dose implants. So, in order to provide a more productive solution, the upper energy range of the VIISta 900XP has been increased to 300keV (singly charged) and 600keV (doubly charged). This extended application window allows the VIISta 900XP to perform well implants at more than double the productivity of a high energy tool and, in some cases, enables it to provide all the required low dose implants for a semiconductor fab. This benefit is illustrated in figure 9.

CONCLUSION

The VIISta 900XP offers a dramatic improvement in productivity for low dose implants. All of the key components for productivity have been improved significantly. Substantial improvements have also been made to the process integrity of the tool, particularly in the area of photo-resist gas load handling. The application window of the tool has also been extended by increasing its energy range.

ACKNOWLEDGMENTS

I would like to thank Sean Cloherty and Dennis Rodier for providing some of the data discussed here. I would also like to thank Jay Scheuer and Joe Olson for their comments and discussions.

REFERENCES

1. A. Renau and D. Hacker, "The VIISta 810 medium current ion implanter", *Proc. 12th Int. Conf. on Ion implantation Tech.*, Kyoto, Japan (1998)
2. J.C. Olson and A. Renau, "Control of channeling uniformity for advanced applications", *Proc. 13th Int. Conf. on Ion implantation Tech.*, Alpbach, Austria (2000)
3. A. Renau, "Beam line architecture of the VIISta 810 medium current ion implanter", *Proc. 12th Int. Conf. on Ion implantation Tech.*, Kyoto, Japan (1998)
4. P. Kurunczi et al, "Advanced charge control for single wafer implanters", *Proc. 16th Int. Conf. on Ion implantation Tech.*, Marseilles, France (2006)
5. S. Walther, "Method and apparatus for controlling implantation during vacuum fluctuations", *U.S. Patent no. 6,323,497*
6. S. Wilson and T. McIntyre, "Introducing the Eaton NV-GSD/HE high energy implanter", *Proc. 10th Int. Conf. on Ion implantation Tech.*, Catania, Italy (1994)
7. J. Tokoro, D. Holbrook and D. Hacker, "Introduction of the VIISta 3000 single wafer high energy ion implanter", *Proc. 13th Int. Conf. on Ion implantation Tech.*, Alpbach, Austria (2000)

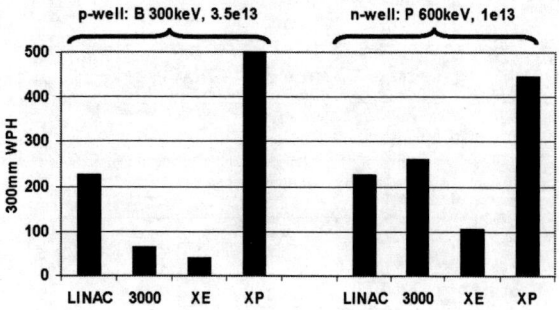

FIGURE 9. Well implant throughputs on a) LINAC with 225WPH mech. limit, b) VIISta 3000, c) VIISta 810XE which had to use B^{++} and P^{+++}, d) VIISta 900XP with 3.8mA B^+ and 0.75pmA P^{++}

Implant Angle Control on Optima MD

R. D. Rathmell, B. Vanderberg, A. M. Ray, D. E. Kamenitsa, M. Harris, and K. Wu

Axcelis Technologies Inc, 108 Cherry Hill Drive, Beverly, MA 01915

Abstract. Implant angle control is increasingly important with each new device node. Some devices have demonstrated a sensitivity of threshold voltage of about 100 mV/deg for implant angle and require implant angles to be held within +/- 0.2° for process control. There are many sources of angle variation in single wafer implanters. Mechanical orientation can usually be controlled to high precision, but an accurate control of the implant angle requires knowledge of the actual beam angle relative to the surface or crystal planes of the wafer. In-situ methods to measure beam angles in both the horizontal and vertical planes are required and it is necessary that these methods be calibrated to the surface or crystal planes of the wafer to achieve the required angle control. Optima MD has incorporated methods to automatically measure beam angles prior to implant in both planes, and correct for any deviation from the desired implant angle. The symmetric parallelizing lens that corrects angles without bending the beam enables a method of calibrating the horizontal angle mask to crystal planes with one or two wafers. This paper discusses the methods of measurement, calibration, and accuracy of the Optima MD angle control system.

Keywords: ion implant, angle control
PACS: 61.72.Tt, 85.40.Ry, 06.30.Bp

INTRODUCTION

Each succeeding device node has placed more stringent requirements on angle control, and implant angle control has become a necessary component of implanter process control. Early medium current implanters used raster scanned beams to span wafers up to 150 mm diameter with scan angles of up to +/- 2° across the wafer. Channeling effects on profile depth were tolerated or managed by using non-channeling crystal orientations. While this was adequate angle control for devices at that time and these tools are still in use today, it became clear that 200 mm wafers would need better angle control. Parallel beam implanters were introduced to avoid the scan angle variation [1-3]. These offered implant angle control of <1°, which was satisfactory for devices until recently. Device structures, which can lead to asymmetry in source/drain extensions of transistors due to shadowing of gates or narrowly spaced photoresist patterns are placing even tighter limits on angle control. Devices in which Vt varies by up to 100 mV/degree have been reported requiring angle control of 0.2° [4]. Such tight control requires in-situ measurement of the implant angle in horizontal and vertical planes and is smaller than the present tolerance on crystal cut error. However, it motivates the implanter designers to minimize their contribution to the total variation.

SOURCES OF VARIABILITY

All ribbon beam or ribbon-like beam implanters today use a magnetic or electrostatic lens to image ions to infinity after they diverge in one plane from a waist near the focal point of the lens [1-3,5]. The focal lengths of these lenses are in the range of 600 to 1000 mm. Magnetic lenses achieve parallelism by bending the beam in a wedge shaped dipole magnet through an angle ranging from 40°-70° for the central ray. Electrostatic lenses allow the central ray to continue in a straight line, while the scanned rays have the scan angle cancelled by the vector sum of velocities in the acceleration or deceleration gap between the shaped electrodes which are symmetric about the central ray.

For some simple effects, the focal properties of these lenses can be approximated by the thin lens equation. If the object point of the lens or position of the scan vertex moves to one side of the focal point of the lens, the image is still focused to infinity (parallel rays), but at an angle Θ, where Δx is the lateral displacement and F is the focal length,

$$\Theta = \arctan(\Delta x/F), \quad (1)$$

Also if the implanter focusing elements make the apparent vertex move backward or forward along the beam by Δz, the beam will be converging or diverging at the edge of a 150 mm radius wafer by

$$\varphi = \arctan(150*\Delta z/(F^2+F*\Delta z)). \quad (2)$$

For small errors of 5 mm in position and F = 800 mm, these amount to 0.36° and 0.07° angles, respectively.

Electrostatic lenses are designed for a fixed ratio of lens voltage to extraction voltage and for the Optima MD lens, a 1% error in the ratio results in a 0.05° error at the edge of the 300 mm wafer. A magnetic lens is designed for a fixed bend radius or ratio of $(ME/Q^2)^{1/2}/B$. For a 50° bend angle, a 1% error in the field B due to hysteresis, for example, can result in an error of just under 0.5° in the average angle. The field must be set to achieve parallelism across the wafer, however, and any residual error in the average angle must be corrected by tilting the wafer about a vertical axis [6]. Charge exchange of ions that have partially completed the bend on the corrector magnet may strike the wafer at the wrong angle, which can also result in dose non-uniformity [7].

Deceleration/acceleration of a beam with an angle error magnifies/demagnifies the angle by a factor M,

$$M = (E_{in}/E_{out})^{1/2}, \quad (3)$$

where E_{in} is the energy before decel/accel and E_{out} is the energy after. Other effects such as beam blowup and portions of divergent beams interaction with apertures also affect the final effective angle at the wafer.

Vertical angle errors may occur due to source/extraction alignment or by vertical deflectors or focusing aberrations [8].

HORIZONTAL ANGLE MEASUREMENT AND CALIBRATION

This paper will review the methods of angle measurement and calibration of those methods for the Optima MD. The beamline layout can be seen in a separate paper presenting an overview of the Optima MD [9].

If a recipe has a limit on maximum implant angle, Axcelis beam tuning software, Autotune™, automatically measures and corrects angles to satisfy the recipe limit. If angles are diverging or converging as one moves from the center toward the left/right edges of the wafer, this is corrected by adjusting the voltage of the Parallelizng lens, P-lens, to make the lens stronger or weaker. The P-lens voltage can be adjusted independently of the beam energy, it is just a lens. Average angle errors are corrected by steering the pencil beam before the scanner and P-lens horizontally by two electric elements of the middle quadrupole. For a given angle error, the required change in the P-lens voltage and steerer is computed in a model based closed-loop control algorithm including the effects of acceleration or deceleration, and angles are usually corrected with one iteration.

Angle measurement devices can be calibrated to the crystal planes of the wafer to provide a proper reference for channeled implants and to confirm that a zero degree implant angle is actually normal to the surface of the wafer. The axially symmetric parallelizing P-lens of the Optima MD enables a unique method for horizontal angle calibration. During normal operation the P-lens is set to a voltage that cancels the scan angle to produce parallel rays. The horizontal beam angles of the scanned beam are measured using a moving profiler behind a mask with vertical slots at seven points across the wafer [10]. If the P-lens voltage is turned off, and with zero voltage on the accel column, the scanned beam travels at the extraction energy of 45 keV in straight lines from the scanner to the wafer, as shown in Figure 1. In that case, the implant angle ranges linearly approximately +/- 4° across the wafer, and the beam angle is easily predicted from the geometry. When a wafer is implanted in this mode, at some point on the wafer the ions will have the angle that is parallel to the crystal planes and a measurement of Therma Wave or sheet resistance will reach a minimum.

As shown in Figure 2 for a portion of the wafer, beam angles measured across the wafer form a straight line plot of angle vs. position while the TW values form an approximately parabolic shape with a minimum at the position of maximum channeling.

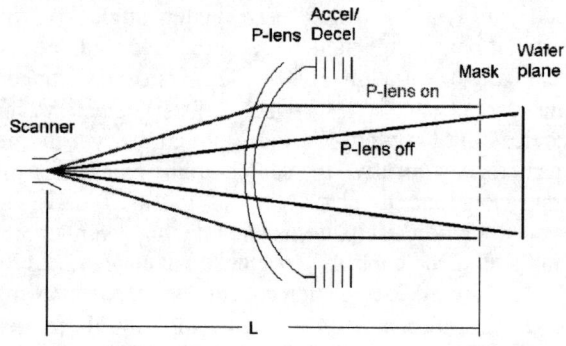

FIGURE 1. Illustration of ion rays from the scanner to the wafer with the P-lens on or off.

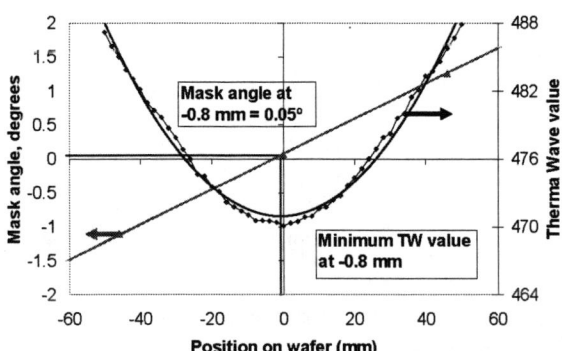

Figure 2. A horizontal diameter scan of Therma Wave values compared to horizontal mask angles for an implant of 45 keV, 100 uA, B+ at a dose of 1E12 and tilt/twist of 0°/0°.with the P-lens off.

The corresponding position on the angle plot is +0.05° for this wafer. If the wafer had no crystal-cut error, one would calibrate the mask to read 0° at that position. However, to compensate for crystal-cut error one typically uses two wafers implanted with 180° difference in twist.

An example of a horizontal angle measurement is shown in Figure 3.

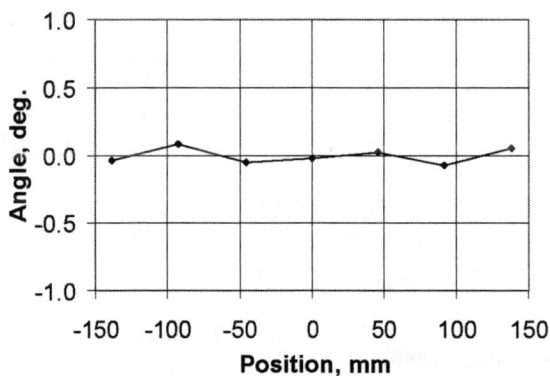

Figure 3. Horizontal angles for 2 keV, 750 uA B+.

VERTICAL ANGLE MEASUREMENT

The vertical beam angle monitor, VBA, is rigidly mounted on the arm that supports the electrostatic chuck such that there is a fixed offset between the angle of the surface of the VBA and the surface of the wafer of about 30°. The VBA is a faraday with a thick slotted mask as shown in Figure 4 that limits the current transmitted to the collector as the slots in the mask are tilted relative to the beam angle. This will produce a current as a function of the tilt angle of the VBA that peaks when the slots are aligned with the

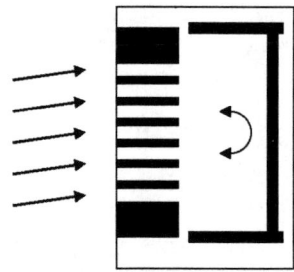

Figure 4. An illustration of the vertical beam angle faraday

beam and that allows accurate determination the "center of mass vertical angle" of the beam.

The fixed offset between the VBA and the wafer surface allows one to set the chuck tilt relative to the actual beam angle to the value specified in the recipe.

The vertical beam angle is measured at the start of a batch by tilting the chuck so that the VBA is in line with the ion beam, as shown in Figure 5a. The chuck and VBA are tilted +/- 5° about this point to collect the data. At these tilt angles, the wafer is safely above the ion beam. Then the chuck moves to the implant position, as shown in Figure 5b, and scans through the

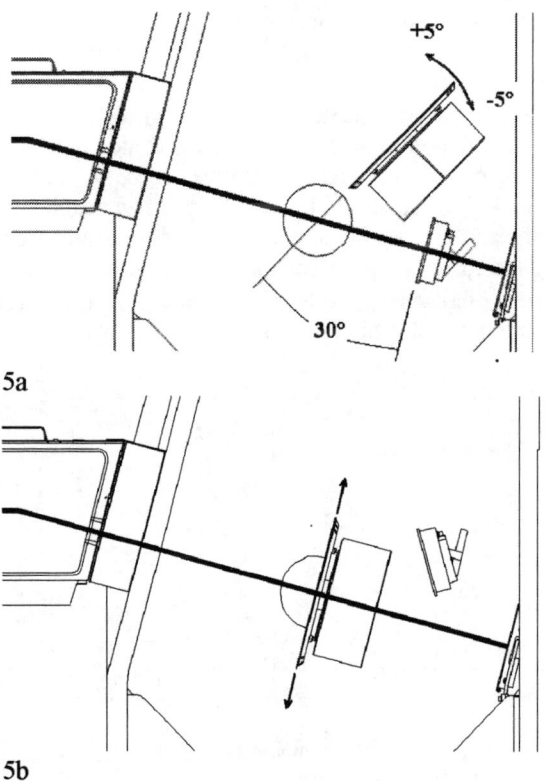

Figure 5. a) Scan arm at tilt angle to measure vertical beam angle. b) Scan arm at an implant position with VBA out of the beam.

beam at the desired tilt angle relative to the measured beam angle, while the VBA remains fixed to the tilt mechanism and is safely out of the beam.

VERTICAL BEAM ANGLE CALIBRATION

As with horizontal angles, it is preferred to calibrate the vertical beam angle measurement to the crystal planes of a wafer as the best way to verify that a zero degree tilt is indeed perpendicular to the wafer. In this case implants are done at a range of tilt angles with wafers from a single boule. The VBA is used to measure the beam angle, then implants are performed at tilts ranging +/- 1.5° and the average TW data are fit with a second order trendline to determine the minimum accurately. The data in Figure 6 show an offset of 0.54° between the average angle of the uncalibrated VBA and the crystal planes. Results are not shown, but these implants are also run with 180° twist to compensate for crystal-cut error. The angle offset is then zeroed so that subsequent angle measurements are referenced to the ideal crystal plane orientation, normal to the surface of the wafer. Implants are then performed with constant focal length scanning ensuring the wafer has the same beam focus and angle content at all positions regardless of tilt.

Using these angle control schemes, an implant of 450 keV P++ was made at a tilt/twist of 0°/0° and the implanted profile was measured on center and at a radius of 140 mm at four points, 3, 6, 9, and 12 o'clock. As shown in Figure 7, the profiles match well with each other indicating all points of the wafer were implanted with ions at the same angle. Profiles from wafers implanted at tilts of 0.2° and 0.5° show the sensitivity of this implant to small angle errors.

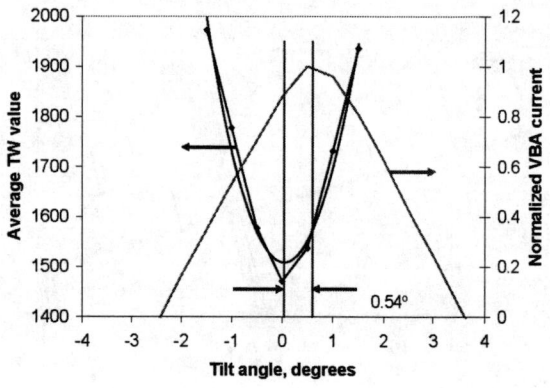

Figure 6. Calibration of VBA (or beam tilt angle) measurement to crystal planes using 370 keV P++, 5E13 at near 0°/0° orientation.

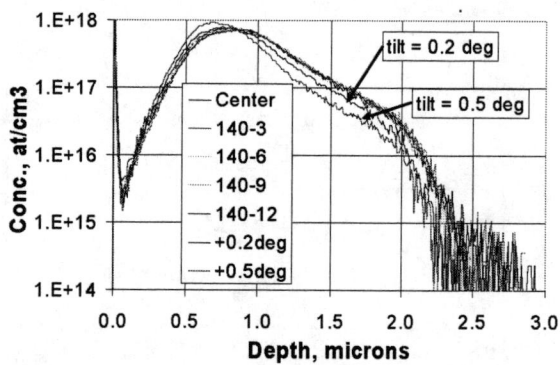

Figure 7. SIMS profiles of channeled implants of 450 keV P++, 5E13 at 0°/0° in {100} wafers at 5 points compared to implants at 0.2° and 0.5° tilts.

SUMMARY

Optima MD has complete angle control in horizontal and vertical planes of the scanned beam. Measurements are made at the start of a batch and are automatically corrected in the horizontal plane by adjusting the voltage of the parallelizing lens or horizontal steerer and in the vertical plane by tilting the wafer to the specified recipe tilt as referenced to the actual vertical beam angle. This method allows angles to be controlled within 0.2 degrees at most conditions. The angle measurement devices can be calibrated to crystal planes to ensure beam orientation is actually referenced to the wafer. This calibration is needed only once as long as the wafer chuck or measurement devices are not moved from the calibrated positions by a disassembly of components.

REFERENCES

1. D. W. Berrian, et al, Nucl. Instr. and Meth. B37/38 500-503 (1989)
2. A. M. Ray, J. P. Dykstra, and R. B. Simonton, Nucl. Instr. and Meth. B74 401-404 (1993)
3. T. Kawai, et al, Nucl. Instr. and Meth. B96 470-473 (1995)
4. A. Agarwal, private communication (2006)
5. N. R. White, M. Sieradzki, and A. Renau, IEEE Proc. of 11th Intl. Conf. on Ion Implantation Tech., Austin, TX 396-399 (1996)
6. J. C. Olson and A. Renau, IEEE Proc. of Intl. Conf. on Ion Implantation Tech., Alpbach, Austria 670-673 (2000)
7. M. R. LaFontaine, et al, IEEE Proc. of Intl. Conf. on Ion Implantation Tech., Alpbach, Austria 246-250 (2000)
8. D. C. Sing and M. J. Rendon, Nucl. Instr. and Meth. B237 318-323 (2005)
9. K. W. Wenzel, et al, "Optima MD: Single-Wafer, Mid-Dose, Hybrid-Scan Ion Implanter", These proceedings, Marseille (2006)
10. R. D. Rathmell, et al, IEEE Proc. of Intl. Conf. on Ion Implantation Tech., Kyoto, Japan 392-395 (1998)

Process Transferability from a Spot Beam to a Ribbon Beam Implanter: CMOS Device Matching

Vincent Kaeppelin[1], Zdenek Chalupa[2], Laurent Frioulaud[3], Sandeep Mehta[4], Baonian Guo[4], Kyu-Ha Shim[4], Horst Lendzian[4] and Yuri Erokhin[4]

[1] *Philips Semiconductors, 850, rue Jean Monnet, 38920 Crolles, France*
[2] *Freescale Semiconductor, 870, rue Jean Monnet, 38920 Crolles, France*
[3] *ST Microelectronics, 850, rue Jean Monnet, 38920 Crolles, France*
[4] *Varian Semiconductor Equipment Associates, Inc., 35 Dory Road, Gloucester, MA 01930, United States*

Abstract. The exercise of dose and energy matching is the standard way to integrate a new implanter into a manufacturing fab. Sheet resistance and secondary-ion mass spectroscopy (SIMS) measurements on bare silicon wafers have been the conventional metrologies to establish dose/energy equivalence between implanters. Invariably, matched performance on bare silicon wafers translated into matched device performance between implanters of the same kind. However, as devices scale down to 90 nm and beyond, the implanter design can become a significant factor in terms of process matching. In this paper we discuss the dynamics of transferring 120-90nm logic processes from a traditional batch, spot beam implanter to a single wafer (SW), parallel ribbon beam implanter. The results show that the traditional approach to dose matching involving the basic parameters of specie, dose and energy, although necessary, is inadequate to provide matched device performance between the two implanter types. 3-dimensional effects which cannot be represented by bare silicon wafers necessitate the use of device wafers to meet the target requirements.

Keywords: Ion Implantation; Batch implanter; Single wafer implanter; Spot beam; Ribbon beam
PACS: 61.72.Tt, 85.40.Ry

INTRODUCTION

Spot beam batch tools have been the traditional workhorse for high current implanters for more than 2 decades. In the past several years, due to the increased sensitivity of devices to critical factors such as defects, beam angle, contamination and productivity etc, the need for single wafer (SW) high current implanters has become more and more important. Parallel ribbon beam, single wafer implanter is one of the approaches to meet these needs. This tool delivers a parallel and uniform beam across the wafer and requires only vertical mechanical scanning of the wafer. Since many companies had developed their devices on batch implanters, the switch to SW tool requires meticulous understanding of the device effects especially as dimensions scale down to 90 nm and beyond.

In this paper, we present process matching results from a traditional batch implanter to a SW ribbon beam implanter. The paper will focus on 120 and 90nm logic devices, on which interesting results have been observed concerning angle control, energy purity and dose rate effects.

EFFECTS OF ANGLE CONTROL RIBBON VERSUS SPOT BEAM

Beam blow up due to space charge effects is a well known phenomenon in ion beam transport. These effects are more significant at lower energies. Consequently, in a self aligned implant such as source drain extension (SDE), a blown up beam leads to dose loss at the edges due to shadowing of the beam caused by the gate electrode. Devices developed on a tool with spot beam architecture inherently exhibited this behavior although such effects went unnoticed until recently when the parallel ribbon beam ion implanter was developed. It is in the transfer of processes between these two tool types, when such effects become very evident. The phenomenon of dose loss due to shadowing by the gate is best illustrated in Figure 1. A controlled, parallel beam delivers the desired dose at the edges of the gate-offset spacer whereas a dispersed spot beam leads to loss of dose and encroachment under the spacer. Transferring a SDE process from spot beam to the parallel beam case would intuitively imply that the latter would have to lower the dose to match transistor characteristics.

Previous investigations [1] with 130nm NMOS devices have demonstrated that the As^+ SDE dose on the parallel ribbon implanter had to be lowered by ~ 22% to achieve matched transistor performance between the two tool types.

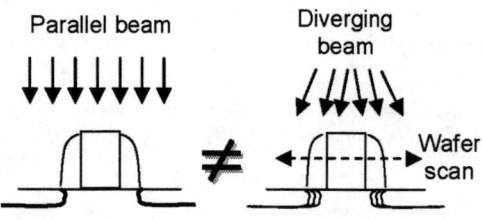

FIGURE 1. Different dopant interactions around spacer edge on parallel ribbon beam versus spot beam

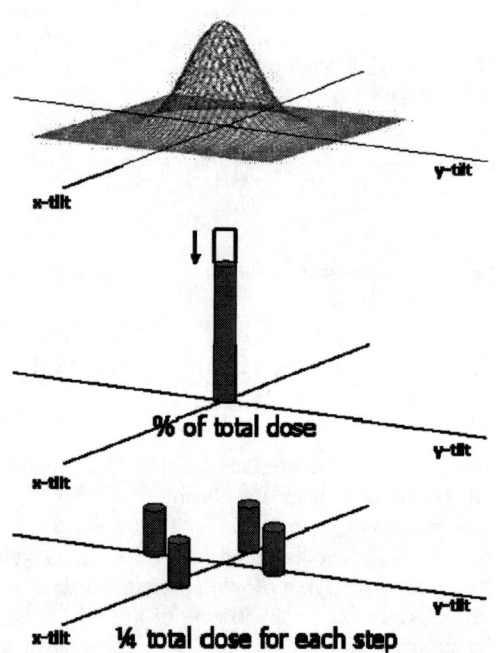

FIGURE 2. (a) Illustration of a spot beam, (b) parallel ribbon beam mimicking a spot divergent beam by dose reduction for 0° and (c) by low tilt quad implant

This paper discusses the transfer of a self-aligned SDE process for a 90nm logic device from the spot beam batch implanter to the single wafer, parallel ribbon beam implanter. Two approaches to match the transistor performance were investigated – (i) lowering the dose on the ribbon beam tool with 0° tilt to mimic previous investigations [1] and (ii) low tilt angle quad mode to mimic the dispersion effects of the spot beam Both approaches are illustrated in Figure 2.

90nm p-SDE matching

The SDE implant for 90nm PMOS is a BF_2, 2keV, mid E14 cm^{-2} implant. The actual dose used is proprietary. As usual, traditional R_s and SIMS analysis was done to match the tools as closely in dose and depth as possible. TCAD simulation using the device parameters was also done to provide guidance on the dose skew that should be employed to define the correct dose scalar which would lead to matched device performance between the tools. Based on the results of this pre-work, the dose was skewed from 90% to 75% of the dose matched with Rs and SIMS. Drive current (I_{dsat}), threshold voltage (V_t) and R_s (Poly) were compared. The I_{dsat} results from the various dose skews and tilt tests are depicted in Figure 3. Due to space limitations of the paper, only the I_{dsat} results have been included. Mean value is represented by the diamond box while the other symbols are standard: data distribution is represented by a box plot that contains 25 to 75% of the distribution and other bars show the following limits: 0%, 10%, 90% and 100%.

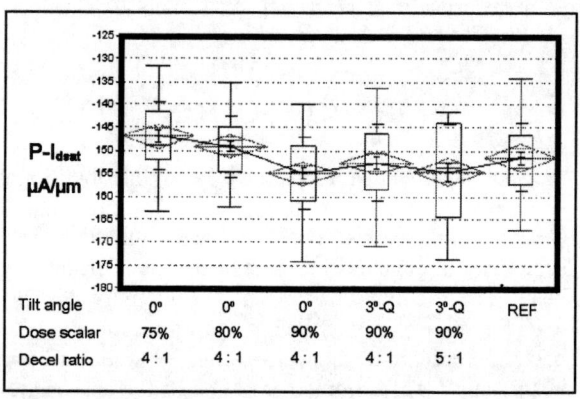

FIGURE 3. I_{dsat} as a function of various SDE implant conditions including dose skew, tilt angle and decel ratio

All dose skew tests were done with 0° tilt. The effect of changing dose on the ribbon beam tool is strongly evident in the first 3 data sets from the left where the dose was varied from 75% to 90%. This data shows that near perfect matching of the I_{dsat} is achieved with a dose scalar setting of ~ 85%. This result is consistent with previous observations although the exact dose required for matching is different depending upon the device design and other factors.

The 4th data set from the left shows that a 3° tilt quad mode implant, with almost equivalent dose to the reference tool also provides adequate matching. The choice of which one to adopt is up to the user, but these results demonstrate that there are multiple options with a controlled, parallel ribbon beam to mimic the performance of a spot beam tool. Such effects would not be evident if only bare silicon wafers were employed in process matching of tools. This is further evident in the P-SD tests described later.

Effects of Energy Contamination

It is to be noted that the aforementioned tests for the p-SDE were done in decel mode on the ribbon beam tool, whereas the reference tool runs the same implant in a drift mode. Preliminary work, although not included here due to space limitations showed that a decel ratio of 5:1 and higher, skewed the I_{dsat} results. This is reflected in the 5th data set from the left in Figure 3. This is due to the increased energy contamination (EC) stemming from high decel ratios. Previous work [2] has shown that skew in drive current and V_t resulting from EC in SDE implants can be recovered by adjusting the dose of the halo implant. However, we did not choose to take this approach. Instead, maintaining a decel ratio of 4:1 was adequate for device and production applications.

90nm p-SD matching

The 90nm PMOS source drain (p-SD) implant is a 0°, 2 keV Boron implant with mid E15cm^{-2} dose. Both tools employ decel mode operation for this implant.

As was the case with p-SDE, traditional R_s and SIMS with bare silicon wafers was used to achieve the first order matching between tools. SIMS profiles from the annealed samples on both tools are presented in Figure 4 and show that despite the higher decel ratio on the ribbon beam tool than the reference tool, no difference in the depth profiles is observed.

Based on R_s and SIMS results, the first split lot was run with a dose scalar of 114%. Results showed that p-I_{dsat} was 7% higher than the reference tool. Correspondingly, active and polysilicon resistances measured on devices were lower by 7%. These results would indicate that the ribbon beam implanter was overdosing relative to the reference spot beam tool.

Subsequently, another device split was performed, whereby the dose was reduced by 5% and 10% on the ribbon beam tool. In addition, to mimic the spot beam characteristics, a 3° tilt test was also performed in quad mode as was done in the p-SDE case described earlier.

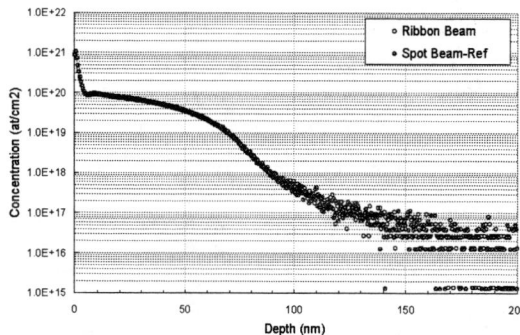

FIGURE 4. Matched dopant depth profiles for 90nm p-SD implant (B$^+$, 2keV) after spike anneal

Table I summarizes the key device parameters from these tests.

TABLE 1. Main PT results of 90 nm pSD split lot

Split conditions	PI$_{dsat}$ (µA/µm)	V$_t$ (mV)	Poly Rs (Ω/sq)
Batch tool (reference)	-173.8	-499.0	445.2
SW 109% trim – 0° tilt	-172.8	-501.4	436.3
SW 104% trim – 0° tilt	-163.8	-509.0	449.6
SW 104% trim – 3° tilt	-167.8	-506.6	452.2

Results show that dose adjustment alone is not adequate to match all device parameters. Indeed, the 0° tilt approach would require a dose scalar ~ 109%, but polysilicon resistances are not matched for devices of different sizes. The combination of lowering the dose scalar to 106% with a 3° tilt provides the best matching conditions. In essence, the effect of tilting the wafer by 3° on the ribbon beam implanter is to increase I_{dsat} by 4µA/µm, which is equivalent to a 2% increase in dose. This is attributed to having a higher concentration of the dopant under the spacer.

DOSE RATE EFFECTS

120nm P-SD matching

This section discusses the preliminary findings from tests conducted to match the p-SD process between the tools for a 120nm logic device. The implant in question is a 15 keV, BF$_2$ beam at 7° tilt with a dose in the mid E15cm^{-2} range. Again, the exact dose is not mentioned due to proprietary considerations.

Same methodology as for 90nm was used for the process matching of this recipe on bare silicon. A dose trim of 100% was found to match both R_s and as-implanted depth profiles.

Based on these results, a first device split was performed with a dose trim at 100% on the ribbon beam tool. Results are summarized in Table II.

Table II. Mean values of device parameters (120nm)

Split conditions	PI_{dsat} (μA/μm)	P+ Active Rs (Ω/sq)	P+ Poly Rs (Ω/sq)
Spot beam tool (ref)	-284.7	129.0	330.8
Ribbon tool 100% Dose scalar	-280.0	121.4	325.0

First results show that the 120nm devices processed on the ribbon beam tool exhibit lower I_{dsat} by ~ 5 μA/μm compared to reference devices. Associated P+ poly resistance and active resistance values were lower by ~ 6 and ~ 8 Ω/sq respectively. Relative to the reference devices, shift in Poly resistance can be explained by a 2~3% overdosing on ribbon beam implanter, however the 8 Ω/sq shift in active resistance can not be explained by a simple dose change. Indeed, based on our experience, a 10% change in dose changed this parameter only by ~3 Ω/sq.

These apparently contradictory results (lower resistances, but lower I_{dsat}) led us to run another device split with different implant conditions on the ribbon beam tool. Dose scalar was reduced by 3% to match P+ polysilicon resistance. To investigate the I_{dsat} behavior, different conditions were tested – (i) lowering the dose rate by reducing beam current from 13mA to 3mA, (ii) reduce the decel ratio for the implant and (iii) increase tilt angle from 7° to 10° (same concept as used on 90nm p-SDE). Results are summarized in Table III.

Table III. Results of second 120nm pSD device split

Split conditions	p-I_{dsat} (μA/μm)	P+ active Rs (Ω/sq)	P+ Poly Rs (Ω/sq)
Ref tool	-286.3	127.3	325.2
7° tilt - 13 mA	-280.0	120.5	324.4
10° tilt - 13 mA	-278.0	122.1	325.0
7° tilt - 3mA	-284.9	122.0	322.8
7° tilt - Lower decel ratio	-285.2	120.6	324.1

It is observed that with 3% dose reduction, poly resistance is better matched to the reference regardless of the implant conditions. As expected, I_{dsat} is still lower by ~ 6μA/μm for the standard implant conditions (7° tilt, 13mA). In addition, increasing the tilt angle from 7° to 10° did not have any impact on the results. This is expected as the SD implant is isolated from the gate edges by the spacer. Interestingly enough, lowering the beam current and reducing the decel ratio both resulted in I_{dsat} values closely matched to the reference value. These results indicate that dose rate is probably the key factor for this implant level. However, the P+ active resistance still shows a mismatch. The behavior of defects and activation as a function of dose rate has been more rigorously investigated by our co-workers and will be published in these proceedings [3]. Indeed, the results of this work show that the density of defects in the spot beam case is higher. Consequently, after anneal, a higher fraction of the Boron is trapped in the defect region and is not substitutional. We hope that tuning the beam with a lower density will enable the matching of both parameters. This is the focus of our future investigations.

CONCLUSIONS

With devices becoming increasingly sensitive beyond the 90nm node, traditional dose matching methods involving TW, R_s and SIMS measurements on bare silicon wafers are no longer adequate. This paper demonstrates that, to fully consider the impact of angle, energy purity and others effects such as dose rate, findings on bare silicon wafers must be validated through device measurements and appropriate adjustments must be made before releasing the process to production.

REFERENCES

1. U. Jeong, S. Mehta, C. Campbell, Z. Zhao et.al, "Effects of Beam Incident Angle Control on NMOS S/D Extension Applications," 14th International Conference on Ion Implantation Technology, 2002, pp. 64-68
2. U. Jeong, S. Mehta, G. Li and J. Liu, "Energy Contamination in Low Energy Implantation," Semiconductor International, October 2003
3. N. Cagnat, C. Laviron, N. Auriac, J. Liu, S. Mehta, L. Frioulaud & D. Mathiot, "Defect Behavior in BF_2 Implants for S/D Applications as a Function of Ion beam Characteristics," These proceedings.

Application of Stencil Mask Ion Implantation Technology to Power Semiconductors

T.Nishiwaki[1], H.Saito[1], K.Hamada[1], K.Tonari[2], T.Nishihashi[2]

[1] *Toyota Motor Corporation, 543 Kirigahora, Nishihirose-cho, Toyota, Aichi, 470-0309, Japan*
tsuyoshi@nishiwaki.tec.toyota.co.jp

[2] *ULVAC, Inc., 1220-1 Suyama, Susono, Shizuoka, 410-1231, Japan*

Abstract. We have been studying the application of ion implantation technology to power semiconductors. The ion implantation technology uses a stencil mask instead of a photo-resist mask in order to reduce the number of processes and achieve high-energy mask ion implantation. Deterioration in implantation pattern accuracy, which is caused by thermal deformation of the stencil mask due to ion beam incidence, has become a critical issue because of the heavy use of high doses of ion implantation in the power semiconductor. In this research, we developed a stencil mask structure that drastically reduces thermal deformation, and technology capable of realizing the ring pattern needed in power semiconductors.

Keywords: Stencil Mask Ion Implantation, Power Semiconductor, Thermal Deformation, Ring Pattern
PACS: 85.40.Ry

I. INTRODUCTION

Our efforts to tackle the global issue of lowering CO_2 emissions in order to prevent global warming have included the mass production of the world's first hybrid vehicle starting in 1997. The hybrid combines a gasoline engine with a motor and achieves twice the fuel economy of a conventional gasoline vehicle. The production volume of hybrid vehicles is growing fast, but still only makes up approximately 1% of all vehicles now manufactured. To increase production further, the price of the hybrid vehicle must become competitive with conventional vehicles. The inverter module controlling the motor of the hybrid vehicle uses many power semiconductors such as IGBTs and diodes, which account for approximately 45% of the surface area of the semiconductor chips used in one hybrid [1,2]. However, the processes that selectively implant impurities (photo-resist coating, exposure, photo-developing, ion implantation, ashing, and wet treatment) for this 45% account for approximately 27% of all the wafer engineering processes in the case of IGBTs. We focused on stencil mask ion implantation technology [3,4,5,6] that is capable of performing all these processes with one machine.

Application to power semiconductors is hampered by the following two issues. (1) It has been confirmed in simulations and actual measurements that in a stencil mask using a thin silicon film with a 20μm thickness, the mask temperature increased to 400°C and above when the ion implantation exceeded a beam power of 100W. Furthermore, the pattern openings were shifted on the level of several μm due to thermal expansion of the membrane. As a result, only implantation with a beam power up to 2W can be achieved for a power semiconductor with an acceptable accuracy of 0.5μm. (2) A ring pattern must be formed in the power semiconductor to maintain the breakdown voltage in the area surrounding the chip, but this is difficult to achieve as the stencil mask must have openings.

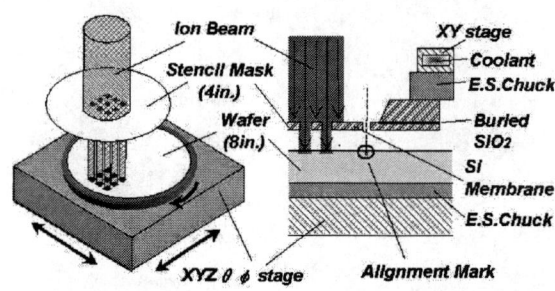

FIG. 1 End-station chamber of the stencil mask ion implantation system (ULVAC, Inc.; SLIM-I)

II. NEW STENCIL MASK STRUCTURE & SIMULATION

< Thermal Deformation Suppression>

A mask structure capable of suppressing thermal deformation is proposed (FIG.2). In order to suppress temperature increases caused by the increased heat of the mask and improve mechanical rigidity, struts with a thickness of several hundred μm were disposed in areas other than the openings (proposed mask 1). Simulations and actual measurements confirmed that thermal conductivity deteriorates when silicon is used in the membrane, and therefore heat accumulates as a result of repeated implantation. To reduce heat accumulation, material with high thermal conductivity is disposed in the membrane of proposed mask 1 (proposed mask 2). Using the model shown in FIG.3, simulations verified the heatproof effect of the proposed masks. Note that CVD diamond is used in proposed mask 2 for its high thermal conductivity. The thermal conductivity of CVD diamond is approximately five times greater than silicon and Young's modulus is also approximately four times greater. However, its heat capacity is approximately 1/17 that of silicon. Hence, the membrane structure uses CVD diamond while the strut that accounts for most of the surface area uses silicon.

<Physical Properties of Material>

	Thermal conductivity W/(m·K)	Young's modulus GPa	Density kg/m³	Specific heat J/(kg·K)
Silicon(a)	148	130	2330	710
CVD (b) diamond	700	516	3300	29
(b)/(a) ratio	4.7	4.0	0.06 (1/17)	

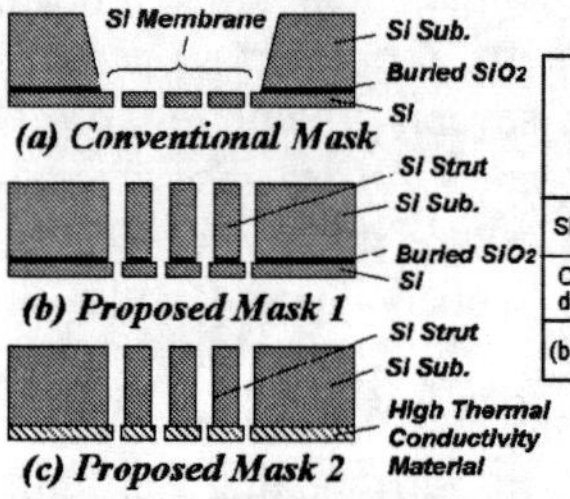

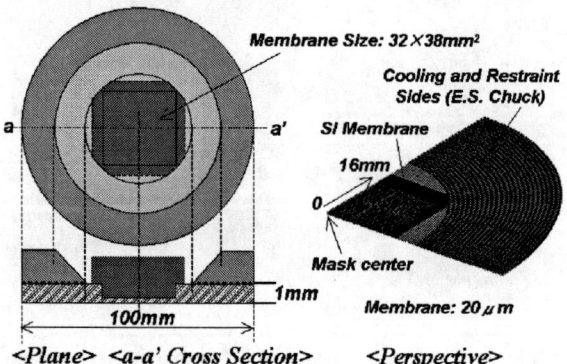

FIG. 2 Heatproof mask structure

FIG. 3 Thermal deformation simulation model (struts area ratio of membrane: approximately 8%)

TAB. 1 Verification results of proposed masks

< Simulation (1) >

Heat input condition I: 120W, On/Off=0.2/1 sec, 1cycle

Structure \ Item	Max. temperature (℃)	Max. displacement (μm)	Max. pattern shift (μm)
(a) Conventional Mask	466	15	2.0
(b) Proposed Mask 1	121	2	0.2

< Simulation (2) >

Heat input condition II: 120W, On/Off=2/1 sec, 10cycles

Structure \ Item	Max. temperature (℃)	Max. displacement (μm)	Max. pattern shift (μm)
(b) Proposed Mask 1	804	25	3.2
(c) Proposed Mask 2	373	9	1.0

TAB.1 shows the conditions and results for simulations (1) and (2). In simulation (1), the conventional mask and proposed mask 1 were compared under heat input condition I. The results confirmed that in proposed mask 1, the temperature was reduced to approximately 1/4 the temperature of the conventional mask. The pattern shift was also approximately 1/10 that of the conventional mask. In simulation (2), proposed mask 1 and proposed mask 2 were compared under heat input condition II. The results confirmed that in proposed mask 2, the temperature was reduced to approximately 1/2 the temperature of proposed mask 1. The pattern shift was also approximately 1/3 that of proposed mask 1, respectively. Based on the above, proposed mask 2 can be assumed to have 1/30 the pattern shift of the conventional mask, in consideration of the fact that pattern shift is almost proportional to heat input. This means that in the case of a semiconductor with a 0.5μm pattern shift tolerance, the applicable range can be expanded up to 60 W by using proposed mask 2 (FIG.4). Prototypes and verification of the effect of proposed masks 1 and 2 are planned.

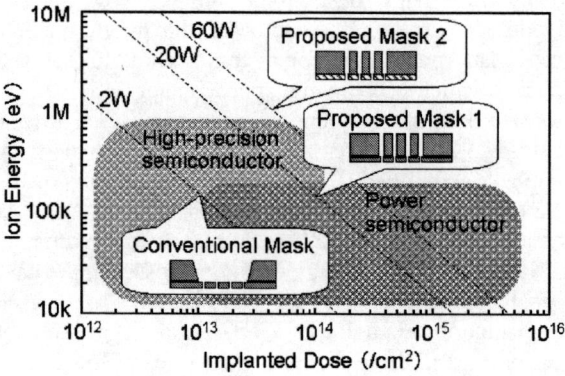

FIG. 4 Expansion of applicable range using proposed masks (pattern shift tolerance: 0.5μm, implanted area: 32x36 cm^2, implantation time: 2 sec)

<Ring Pattern Formation>

A mask structure and implantation method capable of realizing a ring pattern formation is proposed. A mesh pattern was adopted in the ring pattern. Through the use of such a ring pattern, it was confirmed that a consecutive ring diffusion layer can be formed by adjusting characteristics including the opening width, opening pitch, ion beam divergence, and gap between the mask and wafer (FIG.5). The results of the implantation simulation are shown in FIG.6. It was confirmed that the implant concentration range can be adjusted up to approximately 0.2%, while varying the gap from 20μm to 100μm.

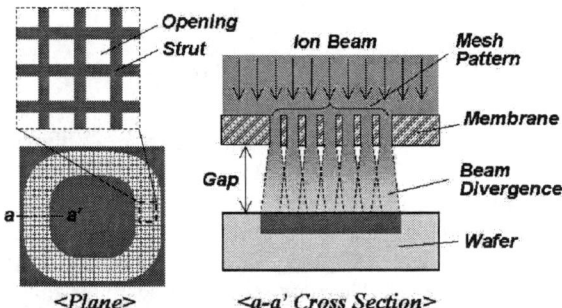

FIG. 5 Concept of mesh pattern implantation

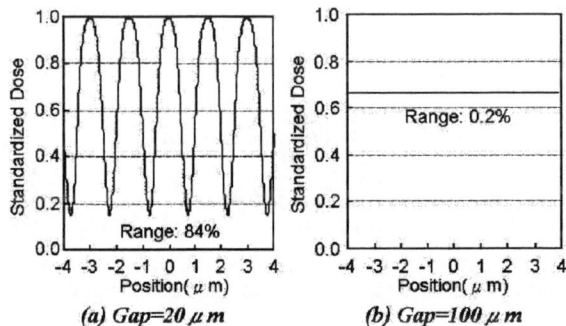

FIG. 6 Results of mesh ion implantation simulation (beam divergence: 0.5°, strut width: 1μm)

III. EXPERIMENTAL RESULTS & DISCUSSION

<Mesh Pattern Implantation>

A prototype was made of a mesh pattern mask with a 20μm membrane thickness, and the distribution was measured subsequent to boron implantation (FIG.7). A mesh-like contrast was observed in the case of a 100μm gap, although this became less conspicuous in the case of a 1000μm gap. Thus, we confirmed that adjusting the gap can achieve a more uniform

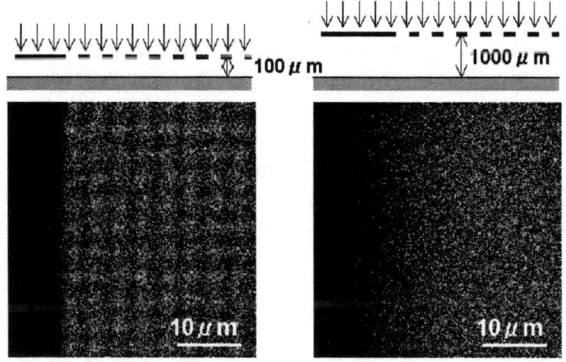

FIG. 7 Image of surface boron secondary ion immediately after mesh pattern implantation (mesh pitch: strut width =3.75/0.75μm, $^{11}B^+$, 50keV, 3E15/cm^2)

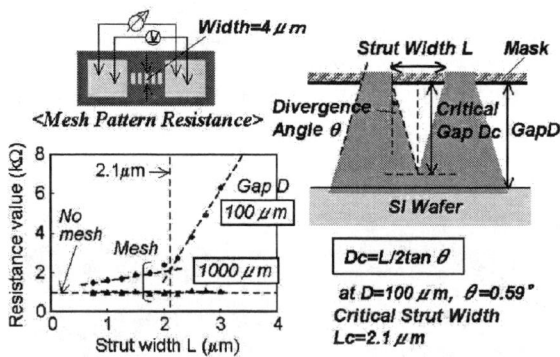

FIG. 8 Strut width dependency of diffused resistor formed by mesh pattern implantation and mesh pattern implantation model (pitch adjusted to achieve constant 44% open area ratio independent of strut width, $^{11}B^+$, 50keV, 3E15/cm^2, 950°C, N$_2$, 30min annealing)

implantation. Furthermore, the mesh pattern was used to form a diffused resistor in order to study the conditions that can achieve mesh pattern implantation. The relationship between the resistance value, strut width and gap was examined. The evaluation results are shown in FIG.8. In the case of a 100μm gap, the resistance value wildly fluctuates up to the vicinity of a 2.1μm strut width, but only gradually changes at widths smaller than 2.1μm. Conversely, in the case of a 1000μm gap, the resistance value is constant with almost no dependency. The illustrations in FIG.8 are simple models of mesh pattern implantation. An ion beam passing through the mask opening at a divergence angle θ overlaps with the ion beam passing through an adjacent opening as shown in the figure. The critical gap at the exact point where the beams radiating from the two adjacent openings overlap can be expressed as $D_c=L/2\tan\theta$ through geometric calculation. If the actual measured gap and divergence angle are 100μm and 0.59°, respectively, then the critical strut width is 2.1μm and coincides with the experimental results. Also, the critical strut width is 21μm in the case of a 1000μm gap, and sufficiently uniform in the range of the actual measured strut width. Therefore, the resistance value was found to have no dependency on strut width.

<Ring Pattern Formation>

Shown in FIG.9 is the boron concentration distribution immediately after implantation in the ring pattern. There are 8 rings and the pitch is 80 μm for a line/space ratio of 1:1. It was verified that the ring pattern can be formed with one mask. In addition, FIG.10 shows a mask where the open area ratio of the mesh pattern is varied over 4 stages in the range of 20% to 80%. The boron concentration distribution is also shown in FIG.10. It was verified that around 4 gradations in concentration can be obtained after one implantation. A scanning capacitance microscope image of 100μm and 1000μm gaps is shown in FIG.11. More uniform implantation is achieved with a 1000μm gap.

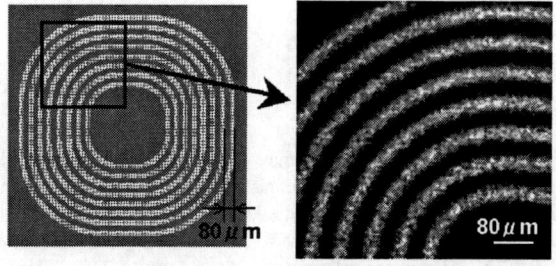

<Ring Pattern Diagram>   <^{11}B Concentration Distribution>

FIG. 9 Boron concentration on silicon surface implanted with mesh pattern (mesh pitch: strut width=3/1μm, ^{11}B$^+$, 50keV, 3E15/cm^2)

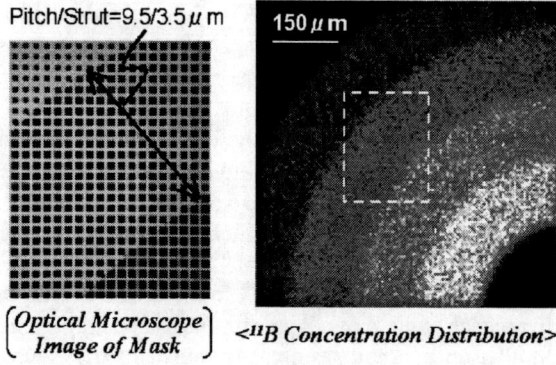

(Optical Microscope Image of Mask)   <^{11}B Concentration Distribution>

FIG. 10 Boron concentration with mesh pattern open area ratio varied over 4 stages (open area ratio: 20/40/60/80% from outer periphery, gap: 1000μm, beam divergence angle: 0.59°, ^{11}B$^+$, 50keV, 3E15/cm^2)

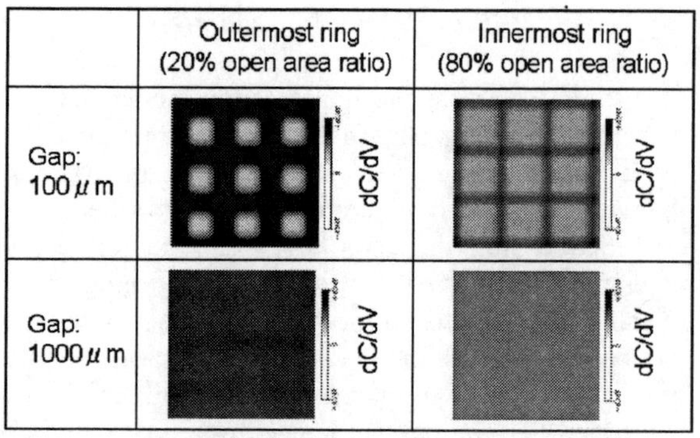

FIG. 11 Scanning capacitance microscope image of different gaps (^{11}B$^+$, 50keV, 3E14/cm^2, 950°C, N$_2$, 30min annealing, analyzed area: 30x30μm^2)

IV. CONCLUSIONS

It was confirmed in simulations that mask thermal deformation, a critical issue in the application of stencil mask ion implantation technology to power semiconductors, can be suppressed to approximately 1/30 that of a conventional mask, thus expanding the applicable range. The proposed mask achieved this by using struts in the mask and a material with high thermal conductivity in the membrane. In addition, tests confirmed that the ring pattern required by power semiconductors can be formed with one mask using the proposed mesh pattern implantation technology.

ACKNOWLEDGMENTS

The authors would like to thank Toppan Printing Co., Ltd. for its cooperation in the manufacture of the masks. This work was supported by the New Energy and Industrial Technology Development Organization (NEDO).

REFERENCES

[1] A.Kawahashi, "A New-Generation Hybrid Electric Vehicle and Its Supporting Power Semiconductor Devices", Proceedings of ISPSD2004, pp.23-29
[2] T.Fujikawa, "Semiconductor Technologies Support New Generation Hybrid Car", Proceedings of Symposium on VLSI Circuits, 2004, pp.6-9
[3] T.Shibata et al., "Stencil mask ion implantation technology for high performance MOSFETs", Proceedings of IEDM2000, pp.11-13
[4] T.Nishihashi et al., Abstracts of EIPBN2001, pp.85
[5] T.Nishihashi et al., "Lithographyless Ion Implantation Technology for Agile Fab", IEEE Trans. Semiconduct. Manufact., vol.15, No.4, pp.464-469, 2002
[6] K.Tonari et al., "Stencil mask Ion Implantation Technology for realistic approach to wafer process", Proceedings of IIT2006, P220

Ion Implanter Cross Contamination And Maintenance Safety Considerations With High Dose Phosphorus

Ron Eddy*, Brett Ostrowski*, Ming Hong Yang¶ and Darryl Huntington†

*Core- A Division of Implant Sciences Corporation, 1050 Kifer Rd, Sunnyvale, CA 94086, USA
¶Evans Analytical Group, 810 Kifer Rd. Sunnyvale, CA, 94086, USA
†Implant Sciences Corporation, 107 Audubon Rd. #5, Wakefield, MA 01880, USA

Abstract. The contamination on previously implanted Phosphorus into other species, known as "implant memory" has been previously reported with emphasis on diffusivity of the P in As (or Sb) [1,2,3]. This study continues some of the investigations done earlier but with some additional focus on some safety considerations. The residual Phosphorus or Arsenic in beamlines, target chambers and on implant disks has resulted in emissions of PH_3 or AsH_3 exceeding the TLV locally for extended periods during maintenance. Measurements taken during the cleaning of P contaminated ion source chambers in situ or on a bench top show that the maintenance technician can be exposed to PH_3 levels in excess of several hundred ppb or more. Freshly created PH_3 is available during maintenance due to the reaction of water vapor and AlP and AlAs in quantities that need attention. Certainly self-contained breathing apparatus is used for gas bottle changes but that use is not always dictated for cleaning of major subassemblies whether in place (on the implanter) or under a designated, remote work area – and it should be in these circumstances.

Keywords: Cross Contamination, Autodoping, Toxic Gas, Phosphorus, Hydrides, Aluminum Phosphide
PACS: 61.72.Tt, 06.60.Wa, 66.30.Jt

INTRODUCTION

The safety protocols for proper handling of Phosphine and Arsine gas bottles are well documented [4,5]. These protocols have remained essentially unchanged with the introduction in late 1992 of sub-atmospheric gas delivery systems [5,6]. Many fabs use solid materials (elemental Arsenic, elemental Antimony and other solids) in a vaporizer to make a gaseous state available in the ion source, due often to company or local regulations. This study is a result of data and information collected at many locations and from results at an implant foundry where a higher than normal % of Phosphorus is run. A reluctance to running Phosphorus in some fabs, especially high dose Phosphorus is due to the cost of implanter dedication in order to avoid cross contamination [3] at their own location or to simply reduce the long, high dose implants.

Hydride emissions from two types of high current implanters were assessed – Axcelis GSD and Varian VIISion. Both of these implanters use full wheel disk/platens. Care was taken to assure that a minimum of exposure to the local atmosphere (with a general relative humidity/RH = 50%) was maintained during measurements.

CONTAMINATED IMPLANTER ASSEMBLIES

There have been many discussions in the implant industry of the so-called "implant odor" when opening a beamline or a target chamber after prolonged use with As or P operation. This odor is present even in cases where no hydride precursors have been used [8]. Maintenance routines that expose Phosphorus or Arsenic coated implanter assemblies to the water-laden atmosphere can create potentially hazardous levels of PH_3 and AsH_3 if proper precautions are not taken. Two very enlightening reports [7,8] show how the hydrides of Arsenic and Phosphorus can be created from deposited P and As - additional mechanisms for the creation of PH_3 and AsH_3 are shown later. Given a further understanding of the various mechanisms for hydride formation, proper safety guidelines can be incorporated into maintenance procedures. One product group at the authors' Sunnyvale facility refurbishes implanter disk/wheels for fabs worldwide. When the disks are cleaned, using a wet-slurry cleaner, there are often noticeable odors during the early stages of the cleaning. Extensive tests have been done using a portable gas detection system (MST-Satellite) bearing in mind the LDL of 0.05 ppm for PH_3 and a LDL of 0.005 ppm of AsH_3. During these times when heavily implanted (with P and/or As) disks were cleaned different reports of odor were acknowledged, and measurements were taken. There has been no detectable PH_3 or AsH_3 using the appropriate gas detector sensor (for MST-Satellite). The odor threshold of PH_3 is often reported to be 0.14 ppm and the odor threshold of AsH_3 is often reported as 0.5 ppm. No detectable PH_3 or AsH_3 during disk decontamination is consistent with a previous study [8]. Several people have been questioned during times when odors are present and they odors are characterized as "fishy", "garlic", "metallic" or "carbide". The fishy and carbide are indicated as characteristic odors in selected AlP MSDS (11). Selected agencies have indicated the odor threshold of Phosphine in a range from as low as 0.01 PPM up to 5.0 ppm depending on various conditions, i.e., relative humidity. See Table 1 for selected agency or organization reports.

Table 1. PH_3 Odor Threshold Limit

Agency	Odor Threshold (ppm)
NIOSH (Traditional)	0.14
CHRIS; Chemical Hazard Response Information System (US Coast Guard)	0.14
AAR; American Association of Railroads. Bureau of Explosives.	0.02
AIHA; American Institute of Industrial Hygienists	0.01 - 5.0

The strength of the odor gives no indication of the PH_3 concentration [13].

HYDRIDE PRODUCTION MODELS

One report [7] related to MBE chambers showed that relatively high amounts of AsH_3 were detected when no AsH_3 was used – only solid Arsenic. Two tentative reactions (1,2) were subsequently proposed:

$$As_4 + 3 H_2O \rightarrow 2 AsH_3 + As_2O_3 \quad (1)$$
$$2 As_2 + 3 H_2O \rightarrow 2 AsH_3 + As_2O_3 \quad (2)$$

The same types of reactions are expected with P_4 or $2P_2$ respectively [9]. Another well documented hydride release is from the formation of Aluminum Arsenide (AlAs) or Aluminum Phosphide (AlP) [10]. The majority of the implanter vacuum system (including the disk in the case of many high current, batch implanters) is made of 6061 T6 Aluminum. The AlP (or AlAs) in the implanter may react with water vapor - generally a fixed, repeatable % in fabs, and create measurable amounts of PH_3 (or AsH_3). These reactions can be represented (3,4) as:

$$AlP + 3H_2O \rightarrow Al(OH)_3 + PH_3 \quad (3)$$
$$2AlP + 3H_2O \rightarrow Al_2O_3 + 2PH_3 \quad (4)$$

Table 2 shows emissions from two disks (GSD) in separate implanters in the same fab. One was a silicon coated disk and the other a non-coated disk that were run with a high dose of P (~1.2E17 cm-2) just before opening the target chamber. There is a measurable difference in PH_3 levels between the two disks immediately after door opening whereas the empty bare Aluminum chambers show little difference. The silicon-coated disk does not have the AlP levels that the uncoated disk has. Note also that most users of the silicon-coated disks notice a reduction in overall Phosphorus cross contamination compared to bare Aluminum disks [3]. There are many reports of AlP releasing PH_3 in agricultural applications where AlP pellets are used for fumigation of certain food commodities for human or livestock consumption as well as tobacco [12]. These and other agricultural reports provide a very large world-wide human toxicology database.

Table 2. PH_3 Emissions from Disks and Chamber (ppb)

Disk Type	Disks Tested (#)	With Disk in Place	Chamber Walls (Disk Removed)	Notes
GSD	2	a) 30 - 50 (coated) b) 80 - 100 (uncoated)	a) 20 - 30 b) 30	Measurement < 10 sec after door open (High P dose run just prior to test)
VIISion	1	40	20 - 40	Disk pulled out of chamber. Chamber with N2 bleed. Phosphorus was ~ 40% of use.

Several series of tests were done on implanters or implanter assemblies at several locations in the US and Asia. Some of the maintenance activities showed some serious shortcomings on personnel protection. Some of these were a result of poor training, even today, with 3rd party or even 4th party personnel. Many fabs may employ a 3rd party "full maintenance contractor" (not employed by the fab or by the OEM). In many instances, that 3rd party, after establishing maintenance protocols with the fab personnel – production and maintenance management, hires a sub-contractor ("4th party") or temporary employees to do many of the lower skilled tasks, i.e, pump changes, maintenance activities and so forth.

Figure 1 shows PH_3 measurements from a freshly opened source housing of a high current, batch implanter that had seen a total of > $1.2E18$ P cm^{-2} over a 3 – 4 day period. The implanter is used almost exclusively for high dose P (60keV, 1.2E16) Maintenance activities consisting of scraping and periodic vacuuming began within minutes of source removal. PH_3 levels were recorded using an Analytical Technology PortaSense II at several intervals over 2.5 hours.

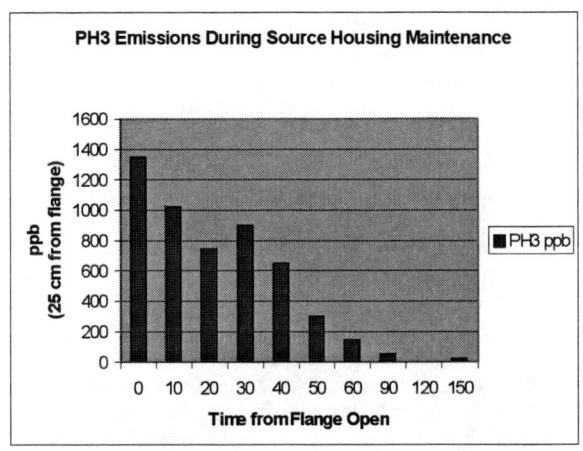

FIGURE 1. PH_3 Levels Measured 25 cm from an Open Source Chamber/Housing During Maintenance (High Current Implanter with dedicated P)

There were unexpected bursts up to 1200 ppb in the measurement possibly due to PH_3 in the newly generated, air-borne flakes during the scraping exercise. It is suggested that since the implanter vacuum system had been purged with clean, dry N_2 it is unlikely to see any AlP from water reaction until some of the high volume of AlP or P particles became airborne as a result of local high volume exhausting and aggressive scraping. The graph shows the level of PH_3 over time with the probe fixed at a point 25 cm from the flange face. This shows the need for appropriate (per local, regional, national codes) particle masks and aggressive air exhaust or SCBA during the entire cleaning process. This becomes especially important if the source housing is removed for a modular PM and is cleaned offline at a later time

Note that the levels stay high for a long time period during aggressive cleaning. There are burst levels during the early part of the cleaning and these bursts tended to remain >800 ppb for up to 10 to 20 seconds, approximately - even after scraping was stopped. The possibility of delays in measurement response was evaluated three times by removing the probe away from the source flange and the decay in the reading was almost immediate (< 3 sec). At the 50-minute interval, a continuous DI water mist was used on the deposited residue as an attempt to reduce any PH_3 from entering the general area beyond the source flange.

Contamination – P in P

Another aspect of Phosphorus is the insidious nature of even low levels (500ppm and lower) of P cross contaminating a high concentration of As and Sb and the disproportionate change in Rs due to a compromise of the junction. [1,2,3]. Implant disks contribute heavily (up to 80%) to the overall sputter of previous dopants [2]. The trend of Rs for a B implants following runs of high dose P are well known where the P simply counter dopes the B for up to 30 – 60 minutes of implants following the P operation – but without any diffusion anomaly. Beyond the As, Sb and B – could P "self-contaminate" or autodope?

A "P in P" test was conducted with a number of wafers (p-type, <100>, 10 – 50 Ω-cm). Half of the wafers were bare and the other half coated with a screen oxide of 200Å. These wafers were all implanted together with P, 120 keV, 5E15 cm^{-2} in one run on a high current, batch implanter which is dedicated to all high dose P implants (and selected B implants) and which also had a series of P implants run (accumulation 8E16) so there was likely to be a saturation [2] of Phosphorus present. These wafers were evaluated for discernible surface contamination on the bare and oxide coated wafers. SIMS using a high precision protocol with better than 2% precision against a NIST standard was performed. The SIMS was done on annealed (oxide-on and oxide-stripped) and on unannealed wafers to determine if any added surface concentration affected the junction due to any possible high dose damage enhancement. We also wanted to see if there was any profile effect. In Table 2, notice that there is a small difference shown in the near surface (<200 and <300 Å) whereas the total SIMS dose is within 2.4% (1σ) for all wafers. No discernible different in P SIMS dose was seen on the profile or the surface/near surface using HPIC SIMS (See Table 2). See Fig 2 for an overlay of annealed P wafers profiled with SIMS on wafers with and without the implanted oxide layer. There is also no profile anomaly observed.

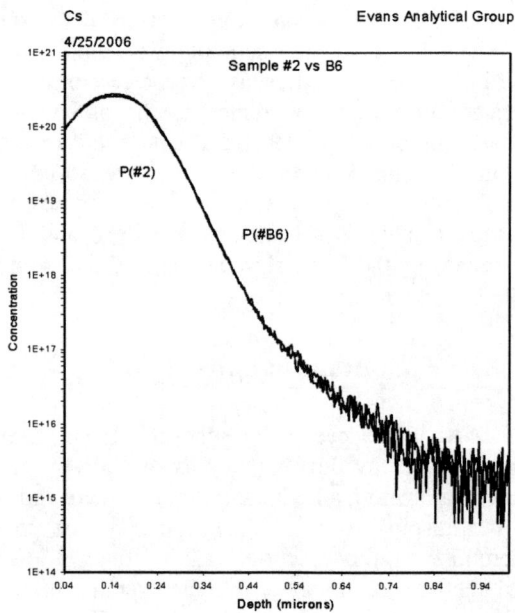

FIGURE 2. SIMS overlay of two wafers run with P, 120 keV, 5E15. Wafers were coated with 200 Å oxide through anneal. Sample B6 had oxide stripped prior to SIMS, Sample 2 did not.

Table 2. P SIMS Dose Measurements			
Depth Measurement	Bare (Not Annealed)	Oxide & Stripped (Annealed)	Oxide (Annealed)
<300 Ang	3.53E+13	1.34E+14	1.93E+14
>200 Ang	4.95E+15	4.71E+15	4.67E+15
<300 Ang	3.5E+15	1.34E+15	1.93E+14
Total SIMS Dose	4.95E+15	4.77E+15	4.73E+15

SUMMARY

Hydrides of Phosphorus and Arsenic (and we expect Antimony although that was not studied here) can be generated long after the parts were exposed to Phosphorus or Arsenic beams. Although the odor threshold of these emissions is often well below the TLV proper monitoring and safety guidelines are required. Heavily implanted Aluminum parts are especially prone to releases of PH_3 or AsH_3. Fabs should ensure that all levels of contractors – and their subcontactors follow the fab or local/regional safety guidelines for implanter and parts cleaning and that they have full access to training and support equipment. Local/direct monitoring at or near the part(s) is strongly recommended. Phosphorus self-contamination was investigated using a high dose "P dirty-up implant". More investigation is needed using lower energy P with a corresponding higher concentration but at this point, the SIMS results show very little or no effects of P self contamination in high dose applications.

ACKNOWLEDGMENTS

Thanks are due to Rob Wipprecht of Dominion Technologies for his valuable input.

REFERENCES

1. Fair, R.B., Meyer, W.G., "Modeling Anomalous Junction Formation in Silicon by the Co-diffusion of Implanted Arsenic with Phosphorus". Silicon Processing. ASTM-STP 804. D.C.Gupta, Ed. American Society for Testing and Materials, 1983, pp 290-305
2. Current, M., Larsen, L., "Ultra-Pure Processing: A Key Challenge for Ion Implantation Processing for Fabrication of ULSI Devices". MRS Proceedings. N.Cheung, N., Marwick, A.D., Roberto, J. Spring Meeting (Vol 147) 1989
3. "Advances in Cross Contamination Control Using Single-Wafer High Current Implantation". Todorov, S., Bertuch, A., Piscitello, W., Eddy, R., Robertson, T. Proceedings of the 13th International Conference on Ion Implantation Technology. Alpbach, Austria, 2000.
4. "Guide to Safe Handling of Compressed Gases". Matheson Gas Products (3rd Ed). 1983.
5. Roberge, S., Ryssel, H., Frey, L. "Safety Considerations for Ion Implanters". Ion Implantation Technology. J. Zeigler Ed. Chapter 14 pp 642-680. Ion Implantation Technology Co. Edgewater, MD 2000
6. Eddy, R., Plotnik, I., Walther, S., Higashi, H. "Ion Implanter Testing and Process Results Using Low Pressure, Zeolite-based Gas Bottles in Medium and High Current Ion Implanters". Nikkei BP Technology Conference, Tokyo, Japan, November 1995.
7. Asom, M.T., Mosovsky, J., Leibenguth, R.E., Zilgo, J.L. "Transient Arsine Generation During Opening of Solid Source MBE Chambers". Journal of Crystal Growth, 112, 1991, pp 597-599.
8. Filipp J., Hunsaker, H., Herring, P. "Investigation of Hydride Emissions During the Maintenance of Ion Implantation Equipment". Presented at the AIHA Conference and Exposition, Boston, MA. 6/92
9. Private Communication with J.Mosovsky - Jan. 23, 2006
10. Private Communications with M.T.Asom – Jan – May 1993
11. Various MSDS; for example - MSDS – Spent Aluminum Phosphide. Degesch-America
12. U.S. Environmental Protection Agency; Fact Sheet 69.1; Oct 1986. Aluminum Phosphide/Phostoxin–applications.
13. "Three Fatalities Involving Phosphine Gas". Willer-Russo, J.L. J. Forensic Science. Vol 44 –3. May 1999

Source Life Improvement For Germanium Implant

Andrew Allen, Peter Banks
Applied Materials UK Ltd., Horsham, UK.

Sukanta Biswas, Christopher Mulcahy
Cascade Scientific Ltd., Uxbridge, UK

Abstract. In recent years germanium has been increasingly used in ion implant applications, for example for pre-amorphization and materials-properties modification. Almost without exception the pre-cursor feed material used is GeF_4 gas, introduced into the ion source from pressurized or sub-atmospheric storage. However, a common limitation of this arrangement is significantly reduced ion-source life due to tungsten redeposition within the arc chamber.

This paper reports the evaluation of an alternative feed-gas, specifically GeH_4. Previously avoided in the implant community, at least in part because of undesirable physical and chemical properties, recent advances in sub-atmospheric delivery have prompted review of this material. The effect of this gas on source life is studied, as is the ionization fragmentation pattern as revealed by beam characterization and SIMS isotope profiling of implanted samples. Comparisons are drawn with GeF_4 in highlighting future potential for this feed gas in ion implant applications.

Keywords: ion implant; GeF_4; GeH_4; Germanium; amorphization.
PACS: 41.75.Ak ; 61.72.Tt ; 07.77.Ka

INTRODUCTION

Germanium has long been used as a means to pre-amorphise crystalline silicon (prior to low energy dopant implantation) to reduce the effects of ion channelling [1].

Previous investigations have shown that a satisfactory feed material from which to generate Germanium ion (Ge^+) beams is Germanium Tetrafluoride (GeF_4) [2],[3]. This has become almost universally adopted on ion implant systems even though it has a detrimental effect on the ion source lifetime when run as a dedicated species or for prolonger periods without species change [3].

The typical source life performance (for an Applied Materials Ultra Life Source™) running a mix of species is ~200 hours, which reduces to less than 30 hours for dedicated Ge^+ use.

Reported in this paper are the results of a study into using the gas Germanium Tetrahydride (GeH_4) as an alternative feed material for the generation of Germanium beams.

EXPERIMENTAL PROCESS

Much data exists for the Ge^+ processes using GeF_4 source feed material [2] [3].

Germanium beams were generated from both GeF_4 and GeH_4 using the same implant system. Data acquired from running tests on GeF_4 are used as a source of reference when judging the performance of GeH_4.

One of the key considerations when qualifying a new material is process performance and transfer. The tests performed have been aimed at determining whether GeH_4 can generate ion beams which will have similar or better properties than those of GeF_4.

Germanium ion beams generated from both GeF_4 and GeH_4 source feeds were analysed for isotope distribution chararacteristics using the magnet spectrum functionality of the implant system. Bare silicon wafers have also been implanted for secondary ion mass spectrmetry (SIMS) analysis.

Equipment Used

The equipment utilised was a prototype tool used to develop beam line and wafer scanning hardware. It was not a standard, commercially available ion implanter. This further required careful baseline tests to allow the data obtained to be representative of the effects that would be observed on an implanter in a production environment.

The test tool consisted of an Applied Materials Quantum™ beam line together with a prototype X-Y wafer scan system: wafers were held on an electrostatic chuck (e-chuck) during implant. Wafers were transferred from a 200mm SMIF cassette to the e-chuck via a vacuum load lock by a robot wafer handling system.

Arc Chamber Configuration

The first tests were carried out using a standard Ultra Life Source™ shown in Figure 1.

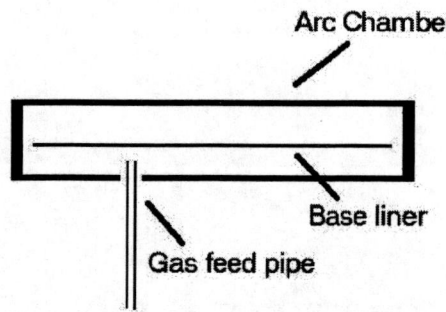

FIGURE 1: Standard arc chamber configuration

This configuration was baselined using GeF_4, and QuantumX™ specification performance was achieved to demonstrate correct system setup. Testing of GeH_4 feed material on the same configuration yielded significantly lower beam current.

Investigation of the underside of the baseliner revealed evidence of deposits and it is believed that the GeH_4 gas was being decomposed upon contact with this hot liner and unable to reach the active region of the arc chamber.

Modifications were made to the source configuration as shown in Figure 2. In this case the feed pipe extends through a small aperture in the base liner to direct gas flow straight to the plasma region.

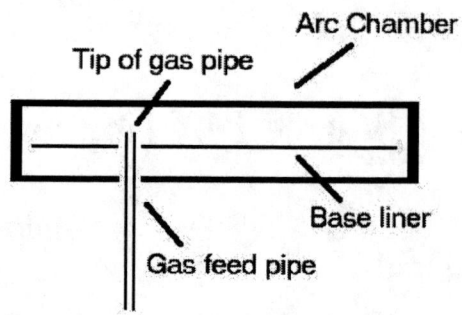

FIGURE 2: First modification of arc chamber configuration

This configuration gave good initial beam current when running GeH_4, however after 1 hour the beam current reduced until the beam disappeared and the arc plasma failed. It was not possible to run other species and the source chamber pressure did not change when gas was injected. The source was removed and it was found that the gas feed pipe (close to the tip) was blocked. The heat of the source plasma is suspected to have decomposed the GeH_4 leaving a solid Germanium metal residue blocking the end of the pipe.

A second modification was made and tested. This used an extended graphite feed pipe with a large bore to conduct the gas to the arc chamber as shown in Figure 3.

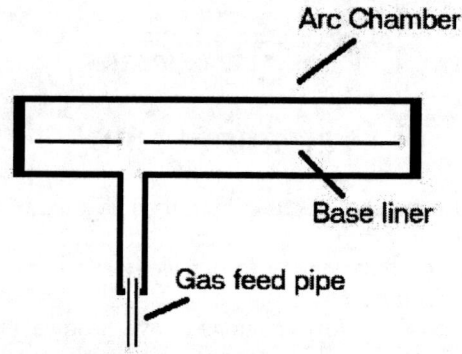

FIGURE 3: Second modification of arc chamber configuration

This gave comparable Ge^+ beam currents to those achieveable running GeF_4 and continued to do so for over 50 hours of beam generation. At this point in the source life the GeH_4 feed material is already demonstrating close to a 100% improvement over GeF_4 feed material and with no signs of performance degradation. At this point the system was vented and the hardware inspected. There were no visible signs of tungsten build up (associated with running GeF_4) or any unusual signs of deterioration for a source of that age.

IMPLANTS

Ion beams of 30keV energy at maximum achievable currents were run.

Tests wafers were implanted at masses 72AMU and 74AMU; at two different mass resolution settings using each of the feed materials (GeF$_4$ and GeH$_4$), giving a matrix of 8 implants - these wafers were analyzed using secondary ion mass spectrometry (SIMS).

The beamline control software allows the analyzing magnet current to be automatically scanned across a pre defined range, while recording beam current. This is used to construct a spectrum of magnet current versus beam current. The software converts the magnet current to atomic mass units (AMU) by an internal calculation before outputting final data.

RESULTS

Magnet spectra were performed to determine the relative abundance of Germanium isotopes and other components in Ge$^+$ beams generated from GeF$_4$ and GeH$_4$.

The results of a magnet spectrum taken at 30keV using the highest mass resolving power, comparing GeF$_4$ and GeH$_4$ are presented in Figure 4, below. This shows no significant difference in ion beam mass ratios between GeF$_4$ and GeH$_4$ source feeds.

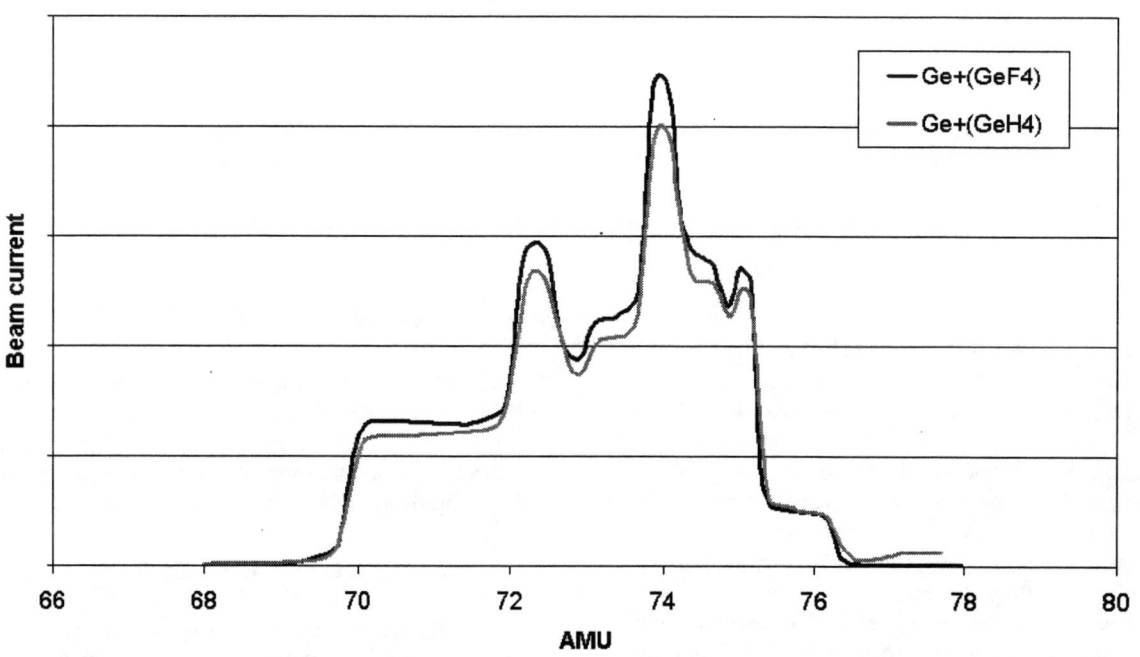

FIGURE 4: Spectral comparison of Ge+ beam generated from GeF4 and GeH4.

Germanium has 5 stable isotopes with relative abundances as shown in Table 1 [4].

TABLE 1. Stable Isotopes of Germanium.

Atomic Mass (AMU)	Relative Abundance
70	20.84%
72	27.54%
73	7.73%
74	36.28%
76	7.61%

The SIMS analysis for Ge$^+$ 30keV 1e15 is shown in Figure 5, with the beamline optimised for transmission of ions centred around mass 72 as shown in Figure 4. The profiles demonstrate a good match between GeF$_4$ and GeH$_4$ feed materials. The projected range of germanium ions in silicon is 269Å (from SRIM simulation [5]) which aligns well with the SIMS data shown.

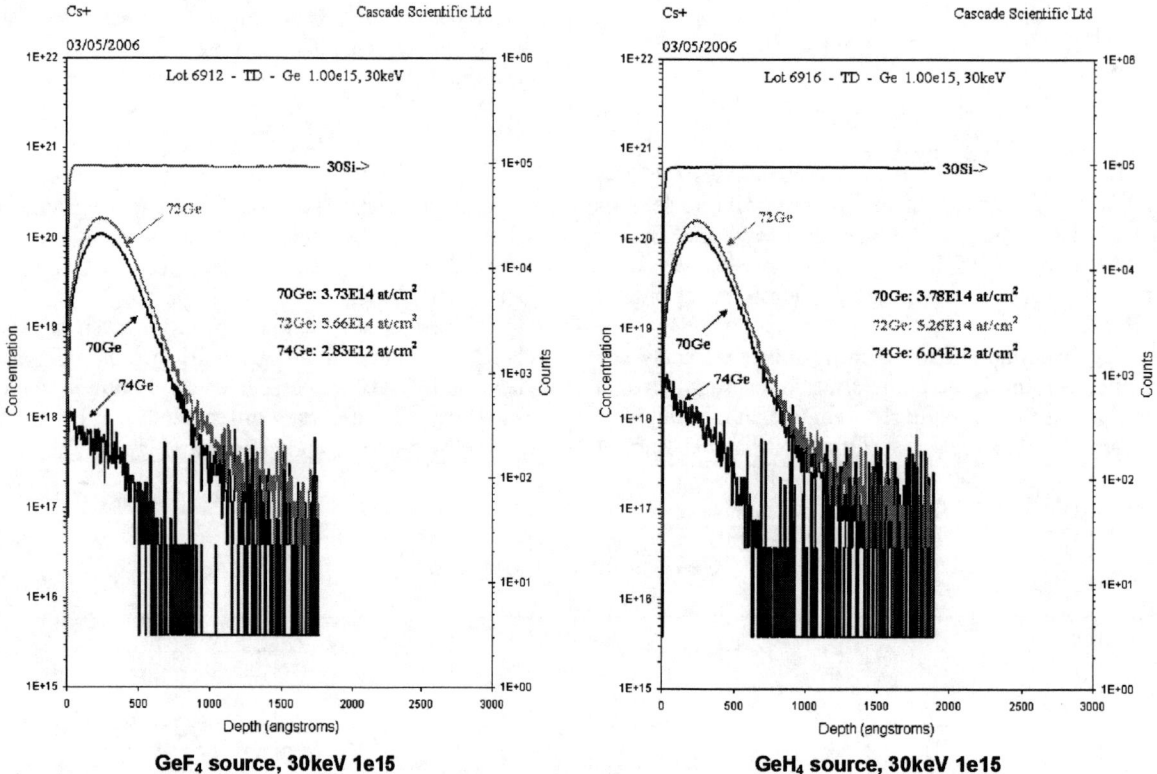

FIGURE 5: SIMS analysis of implants done from GeF_4 and GeH_4 source feed material.

CONCLUSIONS

The results of the SIMS analysis show that the implanted doses and relative abundances of Germanium isotopes is the same for both GeF_4 and GeH_4 feed materials. The magnet spectra recorded for the two feeds also match well and support this conclusion. In this respect GeH_4 is shown to be a suitable substitute for GeF_4.

The source performance at 50 hours of GeH_4 continuous running demonstrates significant improvement over GeF_4 and at this point in the source life had not shown any signs of degradation of performance. An examination of the internal arc chamber parts showed no damage or tungsten build up as would be expected if the source had run GeF_4 (for only 20-30 hours).

ACKNOWLEDGMENTS

Support from Air Products by way of supplying a cylinder of GeH_4 for this test is gratefully acknowledged. Many thanks go to the other members of the application laboratory for their help and support with the hardware modifications and wafer processing.

REFERENCES

1. A. Al-Bayati, et. al., "Exploring the limits of pre-amorphization implants...", Ion Implantation Technology Conference, 2000.
2. P. Banks, et al., "Novel Species Implantation using Applied Materials 9500xR™ and xR LEAP™ implanters", Ion Implantation Technology Conference, 1998.
3. B. Freer, et al., "Germanium Operation on the GSDIII/LED and Ultra High Current Ion Implanters", Ion Implantation Technology Conference, 2004.
4. www.webelements.com
5. J. F. Ziegler, J. P. Biersack and D. V. Littmark, "The Stopping and Ranges of Ion in Solids", http://www.srim.org

Ion Sources for High and Low Energy Extremes of Ion Implantation

A. Hershcovitch[1], V. A. Batalin[2] A. S. Bugaev[3], V. I. Gushenets[3], B. M. Johnson[1], A. A. Kolomiets[2], G.N. Kropachev[2], R. P. Kuibeda[2], T. V. Kulevoy[2], I. V. Litovko[3], E.S.Masunov[4], E. M. Oks[3], V. I. Pershin[2], S. V. Petrenko[2], S.M.Polozov[4], H. J. Poole[5], I. Rudskoy[2], D. N. Seleznev[2], P. A. Storozhenko[6], A. Ya. Svarovski[7], G. Yu. Yushkov[3]

1-Brookhaven National Laboratory, Upton, New York 11973, USA
2-Institute for Theoretical and Experimental Physics, Moscow, Russia
3-High Current Electronics Institute Russian Academy of Sciences, Tomsk, 634055 Russia
4-Moscow Engineering Physics Institute, Kashirskoe sh. 31, Moscow, 115409, Russia
5-PVI, Oxnard, California 93031-5023, USA
6-State Research Institute for Chemistry and Technology of Organoelement Compounds 38, sh. Entuziastov, Moscow, 111123, Russia
7-Siberian Divisions of Russian National Research Center "A.A. Bochvara Scientific Research Institute for Inorganic Materials," Seversk, 636070 Russia

Abstract. A joint research and development effort focusing on the design of steady state, intense ion sources has been in progress for the past two and a half years. Our ultimate goal is to meet the two, energy extreme range needs of mega-electron-volt and 100's of electron-volt ion implanters. This endeavor has already resulted in record steady state output currents of higher charge state Antimony and Phosphorous ions: P^{2+} (8.6 pmA), P^{3+} (1.9 pmA), and P^{4+} (0.12 pmA) and 16.2, 7.6, 3.3, and 2.2 pmA of Sb^{3+} Sb^{4+}, Sb^{5+}, and Sb^{6+} respectively. For low energy ion implantation our efforts involve molecular ions and a novel plasmaless/gasless deceleration method. To date, 1 emA of positive Decaborane ions were extracted at 10 keV and a somewhat smaller current of negative Decaborane ions were also extracted. Initial results also indicate that a Boron fraction of over 70% was extracted from a Bernas-Calutron ion source.

Keywords: ion sources for ion implantation
PACS: 41.75Ak, 41.75Cc, 81.15Jj, 52.77Dq,

INTRODUCTION

Various types of ions are implanted, over a wide range of energies into some of the materials used in the construction of semiconductors. These energies range from as low as approximately 100 eV for shallow surface implantations, to as high as multi-MeV for deep implantation into the substrate. State of the art ion sources meet industry needs for the energy range of about 10 keV to about 300 keV. But at the two extremes (100's of eV and at multi-MeV) of the energy range, there is a lot of room for improvement due to space charge limitations at the low energy range and due to inefficiency in acceleration at the higher energy range. This paper is a synopsis of an extensive ion source R&D program designed to address industry needs.

Originally, the collaboration started to develop pulsed metal vapor ion sources with enhanced charge states. We utilized an external electron beam in two ion sources provisionally dubbed E-MEVVA. Lead and Bismuth, which previously achieved doubly charged ions, were ionized to ion charge states of Pb^{+7} & Bi^{+8} with ion currents exceeding 200 mA [1,2].

The natural next step was to adapt these charge enhancement characteristics to ion sources that generate steady state multi-charged B, P, As, and Sb ions. These technical enhancements can be adapted to DC ion implanters [3] in order to improve upon present day high-energy ion implanters that use rf accelerators. Progress in generating higher charge state B, P, and Sb ion beams is reported in section I.

However, we soon realized the semiconductor industry has greater needs in the area of low energy (100's of eV) ion implantation, where space charge problems associated with lower energy ion beams limit implanter ion currents, thus leading to low production rates. To tackle the space charge problem, two approaches were followed: using molecular ions and ion beam deceleration with space charge compensation. Recent results from a Decaborane ion source are briefly described in section II; while a novel gasless/plasmaless ion beam deceleration method is also mentioned in this section.

Finally, a spin-off result of an ion source, from which over 70% of the extracted ion beam consists of singly charged boron, is described in section III.

I. HIGH CHARGE STATE ION SOURCES

Ion beams containing record high charge states of Phosphorous and Antimony have been extracted from ion sources located at HCEI and at ITEP respectively. For some of the higher charge states, the improvement was greater than an order of magnitude over existing technologies.

At HCEI the ion source is a modified Bernas-Calutron ion source with 1mm x 40mm aperture. The source employs a design similar to that of the Russian ion implanter "Vesuvius" [4] which can generate record high charge states of Phosphorous ion beams. This kind of ion beam generator could be considered as a combination of Bernas ion source [5] and Calutron ion source [6]. A standard Calutron ion source has a filament cathode outside the arc chamber with a collimating slot. The Bernas ion source has a filament inside the arc chamber. In our modification of the ion sources we have removed the second filament cathode and placed a Ta plate outside the discharge chamber. The anticathode can be electrically coupled to the anode, allowed to float, or connected to the cathode (filament). When the anticathode is electrically floating or connected to the cathode, an electron oscillation discharge occurs. This mode is characterized by high efficiency of multiply-charged ion production. This is due to most of the electron energy being expended during ionization.

In this ion source a conventional gas delivery system was replaced by an oven. After optimizing all ion source operating parameters: power, magnetic field and oven temperature, record yields of P^{2+} (8.6 pmA), P^{3+} (1.9 pmA), and P^{4+} (0.12 pmA) were extracted from the modified Bernas-Calutron ion source [7] (spectrum is displayed in figure 1). It is significant to observe that the previous best results [8,9] were P^{2+} (3 pmA), P^{3+} (0.2 pmA), and only a miniscule P^{4+} output. Further details and experimental results can be found in reference 7. Since the ion source contains a magnetic field perpendicular to ion beam extraction, charge state distribution measurements are being repeated with magnetic separation for confirmation.

Additionally, from this ion source (when operating with Boron), close to 1emA of B^{+2} ions were extracted.

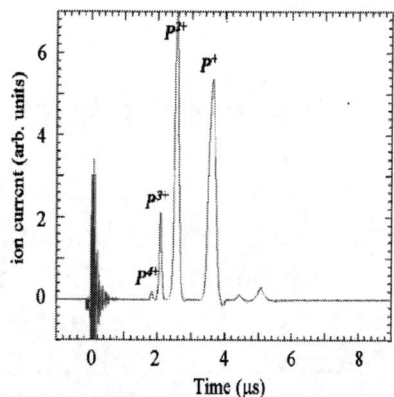

FIGURE 1. Phosphorous TOF ion beam spectrum.

Record enhancement of Antimony charge states were obtained in an ITEP Bernas ion source in which a staggered, oscillating electron beam was generated [10]. Figure 2 shows the spectrum of Antimony extracted from the ITEP Bernas ion source. Current levels reaching a Faraday cup after magnetic separation are 16.2, 7.6, 3.3, and 2.2 pmA of Sb^{3+}, Sb^{4+}, Sb^{5+}, and Sb^{6+} respectively. Additional results as well as a detailed investigation can be found in reference 10. Ion source extraction area is 20 mm^2.

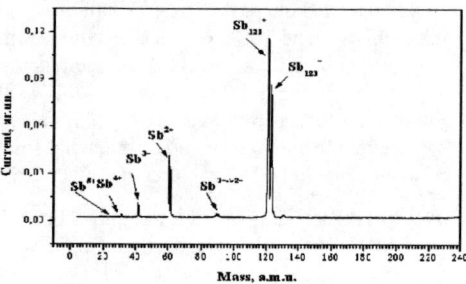

Figure 2. Antimony ion beam spectrum.

II. IONS FOR SHALLOW IMPLANTATIONS

Since the invention of the transistor, the trend has been to miniaturize semiconductor devices. This has resulted in the need to decrease ion implantation energy, since shallow profile implantation is desired. But, due to space charge (intra-ion repulsion) effects, low energy ion beams are characterized by low current. Neutralizing plasmas, utilized in today's implanters, to reduce space charge offer only a partial solution and often result in implanting undesirable impurities. Therefore, low energy ion implanters have low production rates. Consequently, increasing the current of pure, low energy ion beams is of paramount importance to the semiconductor industry.

To mitigate the contamination problem, our collaboration is involved in two projects: molecular ions and beam decelerator that compensates for space charge effects without gas or plasma. The latter is a highly proprietary novel technique for a low energy high current ion beam propagator. This technology produces ion implantations that are contaminant free! A record of invention has been filed by the collaborators.

Decaborane ($B_{10}H_{14}$) was introduced into the ITEP Bernas source and the spectrum shown in figure 3 was obtained [11]. A Decaborane current of 1 emA was extracted. Very recently, a somewhat smaller current of negative Decaborane was also obtained [12]. This significance result opens the possibility of merging negative and positive Decaborane beams, while slowing them down to further reducing the space charge problem.

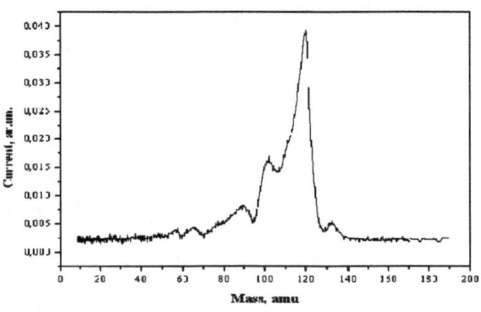

Figure 3. Spectrum of Decaborane ion beam.

Aperture size of the ITEP Bernas ion source is only 1mm x 20mm due to small bending magnet acceptance, which is located 1m from source extractor. There is an inherent difficulty in transporting various species from Bernas type ion sources (for even less than 1m), since they contain magnetic fields perpendicular to the extraction direction that bend and separate ion species during extraction. The challenge is to find a transport system where electro-static forces compensate for this effect of the magnetic field. Such a system was successfully set up by adding a focusing element followed by a deflection element (to compensate for source magnetic field bending) and another focusing element (before the bending magnet). Simulation of extraction and transport of a ribbon beam over a wide range of masses from Boron to Decaborane was performed [13]. The latest results [12] are displayed in figure 4.

Simulations were performed with a modified (ITEP) version of KOBRA. Initial results (very preliminary) indicate that this transport system functions experimentally as well (at least for Decaborane).

Presently, we think that Decaborane current is limited by bending magnet acceptance. Enlarging source aperture size by a factor of 6 might result in a proportional increase in Decaborane ion beam current.

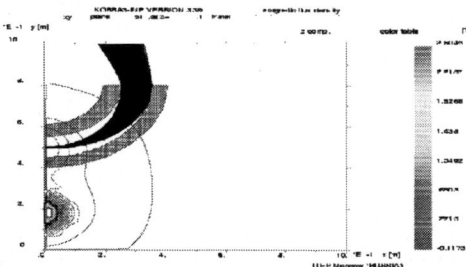

FIGURE 4. Simulation results of sector cylindrical electrostatic deflector for Decaborane beam indicating that the whole beam is being transported.

III. HIGH FRACTION BORON YIELD

Final results to report are that intense beams of boron ions were extracted from the HCEI modified Bernas-Calutron ion source. The anticathode was placed inside a discharge chamber and instead of using the conventional boron-trifluoride (BF_3) gas, a solid lithium-boron-tetrafluoride ($LiBF_4$) compound was heated to release boron. For optimal ion source parameters beams of up to 41 mA were extracted. Singly charged boron made up over 70% of the total ion beam [14]. "Optimal" extracted ion spectrum is shown in figure 5.

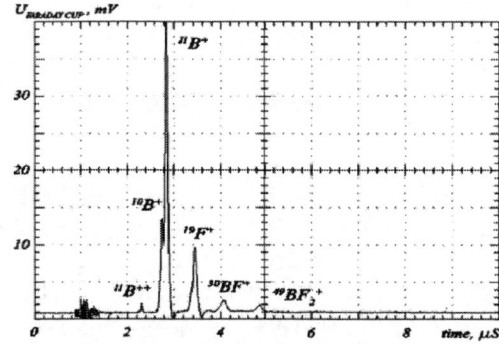

FIGURE 5. Ion beam charge state distribution for beam extracted from the Bernas-Calutron ion source for ion source discharge current of 4 A and a corresponding discharge voltage of 110 V.

By comparison, no more than 25% of the extracted beam from conventional ion sources is Boron. Additionally, BF_3 is extremely toxic, while $LiBF_4$ is a safe compound.

Ion source performance is optimized by adjusting the temperature of the oven containing the solid $LiBF_4$ compound. Once steady optimal boron vapor flow into the ion source is achieved, discharge current and voltage is adjusted by increasing hot cathode emission current. Table 1 shows source performance (from low top row, to high, bottom row). More details can be found in reference 14.

TABLE 1. Boron charge-state fractions for some operating parameters of the Bernas-Calutron ion source.

Source Current & Voltage	Total Extracted Ion Beam	$^{11}B^+$ Fraction	$^{11}B^{++}$ Fraction
2 Amp, 250 Volt	25 mA	72 %	-
3.6 Amp, 190 Volt	34 mA	71 %	0.6 %
4 Amp, 110 Volt	41 mA	68 %	1.1%

DISCUSSION

The main objective of our program has been to develop commercial ion implantation sources for the semiconductor industry. We started to develop high charge state ion sources for high energy implantation. After achieving record results with P & Sb ions and trying to interest potential clients, we learned that the real interest is in B ions. While shifting emphasis to Boron ions, we realized that low energy ion implantation is what the industry needs, hence, diversion of our efforts to molecular ions.

Over the past two and a half years substantial results were obtain by our collaboration in spite of the fact that many changes in the research program have been introduced during this relatively short period. A major obstacle to development of a commercial source has been reluctance of potential clients in the ion implantation industry to reveal their true needs out of fear that the competition may find out anything about their R&D plans. Therefore, in addition to tackling technical and scientific challenges, one needs to guess what a secretive market wants.

ACKNOWLEDGMENTS

One of the authors (A.H) is grateful to V. Benveniste, K. Saadatmand, B. Vanderberg, M. Graf, J. Poate W. DiVeriglio, R. Rathmell, and A. Perel for many stimulating and very useful discussions. Work is supported by research contracts between BNL and HCEI with ITEP under the IPP Trust-2 program and by the U.S. Department of Energy.

Notice: This manuscript has been authored by Brookhaven Science Associates, LLC under Contract No. DE-AC02-98CH1-886 with the US Department of Energy. The Untied States Government retains, and the publisher, by accepting the article for publication, acknowledges, a world-wide license to publish or reproduce the published form of this manuscript, or others to do so, for the United States Government purposes.

REFERENCES

1. A.S. Bugaev, V.I Gushenets, E.M. Oks, G.Yu Yushkov, T.V. Kulevoy, A. Hershcovitch and B.M Johnson, Appl. Phys. Lett., **79**, 919 (2001).
2. V.A. Batalin, A.S. Bugaev, V.I. Gushenets, A. Hershcovitch, B.M. Johnson, A.A. Kolomiets, R.P. Kuibeda, T.V. Kulevoy, E.M. Oks, V.I. Pershin, S.V. Petrenko, D.N. Seleznev, and G.Yu. Yushkov, Journal of Applied Physics, **92**, 2884 (2002).
3. V.A. Batalin, A.S. Bugaev, V.I. Gushenets, A. Hershcovitch, B.M. Johnson, A.A. Kolomiets, R.P. Kuibeda, B.K. Kondratiev, T.V. Kulevoy, I.V. Litovko, E.M. Oks, V.I. Pershin, H.J. Poole, S.V. Petrenko, D.N. Seleznev, A. Ya. Svarovski, V.I. Turchin, and G.Yu. Yushkov, Rev. Sci. Instrum. **75**, 1900 (2004).
4. V.V. Simonov, L.A. Kornilov, A.V. Shashelev, E.V. Shokin, *Ion Implantation Equipment* (Radio i Svjaz', Moscow, 1988, in Russian).
5. I. Chavet and R. Bernas, Nucl. Instrum. Methods 51, 7 (1967).
6. H. Rassel and H. Glawischnig, Eds., *Ion Implantation Techniques* (Springer, New York, 1982), p.39.
7. V.I. Gushenets, A.S. Bugaev, E.M. Oks, A. Hershcovitch, B.M. Johnson, T.V. Kulevoy, H.J. Poole, and A.Ya. Svarovski, Rev. Sci. Instrum. **76**, 083301 (2005).
8. T.N. Horsky, Rev. Sci. Instrum. 69, 840 (1998).
9. K. Saadatmand, Rev. Sci. Instrum. 69, 859 (1998).
10. T.V. Kulevoy, R.P. Kuibeda, S.V. Petrenko, V.A. Batalin, V.I. Pershin, N. Kropachev, A. Hershcovitch and B.M. Johnson, V.I. Gushenets and E.M. Oks, and H.J. Poole, Rev. Sci. Instrum. **77**, 03C110 (2006).
11. T.V. Kulevoy, S.V. Petrenko, R.P. Kuibeda, V.A. Batalin, V.I. Pershin, A.V. Koslov, and Yu. B. Stasevich, A. Hershcovitch and B.M. Johnson, E.M. Oks and V.I. Gushenets, H.J. Poole, P.A. Storozhenko, E.L. Gurkova, and O.V. Alexeyenko, Rev. Sci. Instrum. **77**, 03C102 (2006).
12. A. Hershcovitch, V. A. Batalin, A. S. Bugaev, V. I. Gushenets, B. M. Johnson, A. A. Kolomiets, G.N. Kropachev, R. P. Kuibeda, T. V. Kulevoy, I. V. Litovko, E.S.Masunov, E. M. Oks, V. I. Pershin, S. V. Petrenko, S.M.Polozov, H. J. Poole, I. Rudskoy, D. N. Seleznev, P. A. Storozhenko, A. Ya. Svarovski, G. Yu. Yushkov, Rev. Sci. Instrum. **77**, 03B510 (2006).
13. S.V. Petrenko, G.N. Kropachev, R.P. Kuibeda, T.V. Kulevoy, and V.I. Pershin, E.S. Masunov and S.M. Polozov, A. Hershcovitch and B.M. Johnson, H.J. Poole, Rev. Sci. Instrum. **77**, 03C112 (2006).
14. V.I. Gushenets, A.S. Bugaev and E.M. Oks, Ady Hershcovitch and B.M. Johnson, T.V. Kulevoy, H.J. Poole, A. Ya. Svarovski, Rev. Sci. Instrum. **77**, 03C109 (2006).

An Electron Cyclotron Resonance Ion Source with Cylindrically Comb-Shaped Magnetic Field Configuration

Yushi Kato[*], Hiroshi Sasaki[*], Toyohisa Asaji[*,†], Takashi Kubo[*], Fuminobu Sato[*] and Toshiyuki Iida[*]

[*]*Devision of Electrical, Electronic and Information Engineering, Graduate School of Engineering, Osaka Univ. 2-1 Yamada-oka, Suita-shi, Osaka 565-0871, Japan.*

[†] *Tateyama Machine Co., Ltd., 30 Shimonoban, Toyama 930-1305, Japan.*

Abstract. A new concept on magnetic field of plasma production and confinement has been proposed to enhance efficiency of an electron cyclotron resonance (ECR) plasma for broad and dense ion beam source under the low pressure. The magnetic field configuration is constructed by a pair of comb-shaped magnet which has opposite polarity each other, and which cylindrically surrounds the plasma chamber. This magnetic configuration suppresses the loss due to ExB drift, and then plasma confinement is enhanced. The resonance zones of the fundamental and the second harmonics for 2.45GHz microwaves detach from the wall of the chamber. The connection length of the magnetic field lines through the resonance zone is elongated, and the confinement is better than that of the simple multipole magnetic field. The 2.45 GHz microwaves are fed from the side wall by the rod antenna. The electron density attained to about four times cutoff density for the 2.45GHz microwave at the low Ar pressure below 0.08Pa and also the low microwave power below 300W. We compare profiles of the electron density and temperature in the comb-shaped magnetic field configuration with those in the simple multipole magnetic field.

Keywords: ECR, Ion source, Comb-shaped magnet, Board ion beam process, Second harmonics
PACS: 29.25.Ni, 52.75.-d

INTRODUCTION

We have already proposed a concept to enhance efficiency of the electron cyclotron resonance (ECR) plasma for production of multicharged ions [1,2] and large-area ion source [3,4] by constructing microwave cavity, and then making the maximum electric field correspond to the ECR zone. We can position the peak of the electric field of the standing waves in the ECR zone of the cavity resonator, *i.e.*, the vacuum chamber. The experiment of the large-area ion source has been conducted in the simple multipole magnetic field configuration. [4]

Now, a noble magnetic field configuration for plasma production and confinement has been proposed in order to enhance efficiency of the ECR plasma as for broad and dense ion beam source. Tentatively the typical diameter and the length of the plasma production chamber are about 200mm and 400mm, respectively.

The magnetic field configuration is constructed by a pair of comb-shaped magnet which has opposite polarity each other, and which cylindrically surrounds the plasma chamber. In addition to multipole magnets, each end of the magnet is connected, and forms a pair of ring-like magnet with opposite polarity each other. This magnetic configuration suppresses the loss due to ExB drift along the longitudinal direction to the inner surfaces of magnets, and then plasma confinement is enhanced.

We compare profiles of the electron density n_e, the electron temperature T_e, the space potential, and the floating potentials in the comb-shaped magnetic field configuration with those in the simple multipole magnetic field configuration, experimentally.

DESIGN ASPECTS AND EXPERIMENTAL APPARATUS

We choose the number of poles on multipole field from point of view of constructing the ECR surfaces (zones) of the fundamental and the second harmonics; the ECR zones form closed surface and are good enough to be detached from the chamber wall. It is suitable that connection length of the line force through the ECR zone from the wall to the wall is long as possible as it is. For the present we apply 2.45GHz frequency microwave and we are planning to increase frequency of microwave. [5] We calculate magnetic field and estimate the line of force on the basis of dipole modeling. Therefore we choose the octupole magnetic field as shown in Fig.1, because the ECR zone of the fundamental and the second harmonics for 2.45GHz, and also the later for 11-13GHz, form the detached closed surface.

Figure 1 (a) and (b) depicts setting the octupole magnets to the chamber and contour plots of dotted lines of equi-strength surfaces at the cross section of the center of chamber, respectively. The resonance zones of the fundamental (0.0875T) and the second harmonics (0.0438T) of the electron cyclotron frequency corresponding to 2.45GHz microwaves in the simple multipole are formed around about 120mm and 100mm diameters, respectively. These surfaces construct cylindrical, and detach more than 40-50mm apart from the wall of the chamber. Figure 1 (c) shows radial profiles of magnetic field strength at cross sections A and B in Fig.1(b). Figure 1 (d) shows strength of the magnetic field along to the magnetic field lines labeled by (1) ~ (3). Because these resonance zones are assigned around the bottom of the local mirror configuration by the magnets, the stability

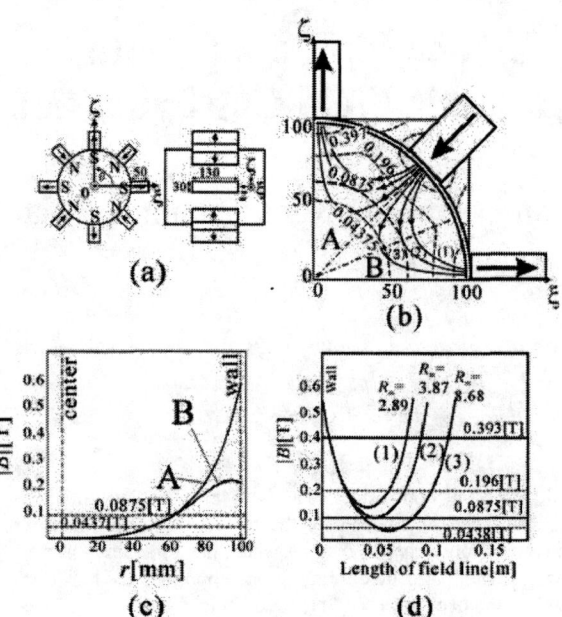

FIGURE 1. Magnetic configuration of multipole (octupole) magnetic field.

is considered to be good as well as confinement. The local mirror ratios, i.e. $R_m = B_{max}/B_{min}$ are about 2.9 ~

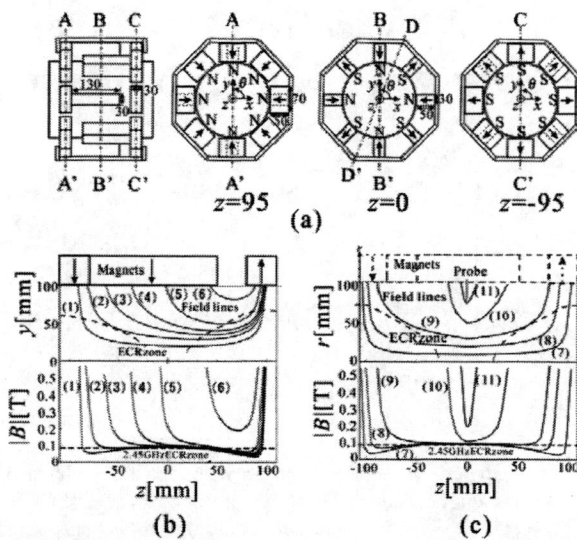

FIGURE 2. Magnetic configuration of comb-shaped magnetic field.

8.7. Because the mirror ratio of the second harmonics is larger, the confinement of the trapped particles is better than that of the fundamental ECR.

Figure 2 indicates field configurations in the comb-shaped magnets. In the comb-shaped case, the relevant ECR surfaces change to a pair of cone-shaped surface. These resonance zones are also assigned around the bottom of the mirror configuration, and then the stability and confinement are also considered to be good. Furthermore the connection length of the magnetic field lines through the resonance zone is elongated along to axial direction by a pair of ring-like magnets, and the confinement is considered to be better than that of the simple multipole magnetic field. This magnetic configuration constructed by permanent magnets is similar to that of the multipole magnetic field, i.e. octupole magnetic field, superimposed by magnetic mirror field.

Schematic drawings of setting the comb-shaped magnets to the chamber are shown at several cut surfaces in Fig.2 (a). The configuration of the magnetic lines on the y-z cross section ($\theta=\pi/2$) of Fig.2 (a) and the strength along them are depicted in the upper and the lower figures in Fig.2 (b), respectively. The projection of the field lines through r-axis at z=0 to the r-z plane at $\theta=3\pi/8$ in the cross section D-D' of Fig.2 (a) is indicated in the upper figure of Fig.2 (c). The probe measurements are conducted along the r-axis ($\theta=3\pi/8$) in this cross section. The lower figure of Fig.2 (c) is the strength along magnetic field lines. The positions of the ECR zones are indicated by the dotted lines in both of the upper figures in Fig.2 (b) and (c).

However there is not exist surface of ECR zone at the cut surface of the center of the multipole-

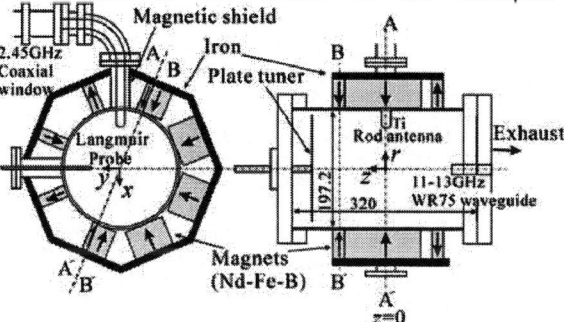

FIGURE 3. Schematic drawing of experimental device of 2.45GHz with comb-shaped magnets.

magnets part (B-B'). The field lines through the ECR zone position lie on at $z=0$ within $r=30$mm along the r-axis on this r-z plane.

Figure 3 shows the schematic drawing of the experimental device. The multipole field magnets and comb-shaped magnets are set on four parts of iron yorks. First we set simple only multipole and conducted the experiments, and then we modified to the comb-shaped magntes. The vacuum vessel is made from 2.8mm thick stainless steel. The x- and y-axis's are the horizontal and vertical coordinates, respectively. The z-axis was chosen as the geometrical axis directed from the upper stream of the waveguide for 11-13 GHz microwave inlet to the down stream. The origin is the center of the multipole magnets on the geometrical axis.

We used mainly 2.45GHz microwaves in this study. The rod antenna for the 2.45 GHz microwave is set from the side wall with several magnetic shields. The fixed and mobile plates are set at the end of the magnets for tuning microwave. The microwave powers are usually 50-300W and 1.3kW at the maximum. The Ar operating pressure is about 10^{-3} ~10^{-1} Pa. The electron density n_e and the temperature T_e are measured by the Langmuir probe which is installed along y-axis at the center ($z=0$). At the first we conducted plasma production with the simple multipole field, and measured the n_e and the T_e. And then we constructed the comb-shaped magnets, conducted plasma production, and measured the n_e and the T_e. We compare the characteristics and the features of them by using 2.45GHz microwave.

EXPERIMENTAL RESULTS AND DISCUSSIONS

We measured profiles of the n_e and the T_e in multipole and comb-shaped magnetic field configurations at the center of the chamber (see Fig.3).

We found that plasma density attained to the cutoff density of 2.45GHz microwaves (7.45×10^{16}m^{-3}) under

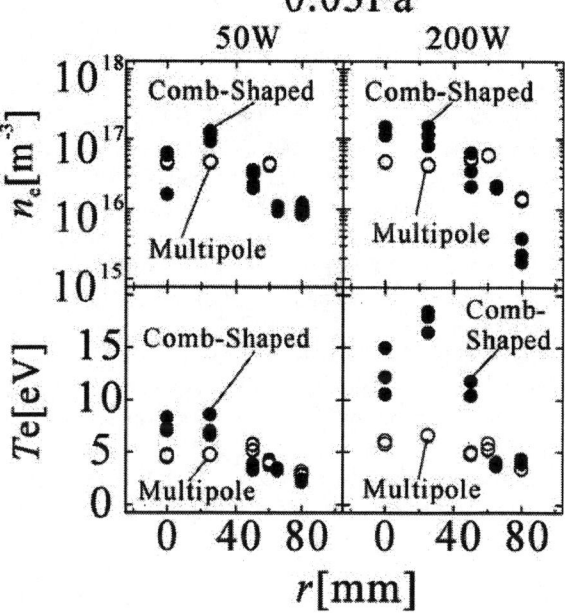

FIGURE 4. Typical dependence of the n_e and T_e profiles on the power of the microwaves at the low pressure (0.03Pa).

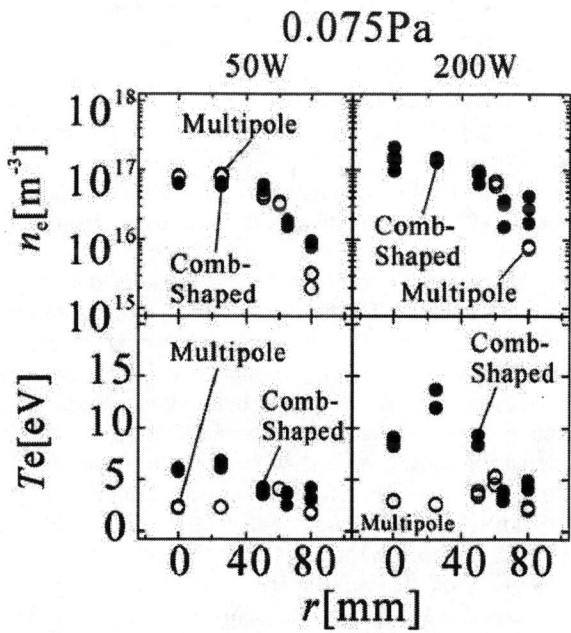

FIGURE 5. Typical dependence of the n_e and T_e profiles on the power of the microwaves at the high pressure (0.075Pa).

rather lower microwave powers (~50W) in both of magnetic field configuration.

Figure 4 and 5 indicate the microwave power dependence of the n_e and T_e profiles at the operating pressure 0.03Pa and 0.075Pa, respectively. At the lower pressure (Fig.4), the n_e at the center on the microwave power of 50W has the similar value around the cutoff density of 2.45GHz microwave in both of the multipole and comb-shaped magnetic field. However the n_e of high microwave power of 200W in the multipole field keeps the same level, that in the comb-shaped magnets increases drastically around the center. The T_e profiles shows drastic increase around the center in the comb-shaped magnetic field as the microwave power increasing, while those in the multipole field hardly change. On the contrary, in the higher pressure, difference of the n_e between both of magnetic configurations become small except of increase in the peripheral region of the comb-shaped case as show in Fig.5. But the maximum n_e value of $2.1 \times 10^{17} m^{-3}$ is obtained at the center in the comb-shaped magnetic field at the microwave power of 200W. The dependence of the T_e behaves similar manners as the lower pressure, but the increments become smaller than the case of 0.03Pa. (Fig.4) General speaking about plasma production, the n_e and the T_e have conflict behavior each other, i.e. tendency which the n_e increase and the T_e decrease as the pressure increases. While our results have slightly the same tendency, the n_e and the T_e increase coincidently as the microwave power increases in the comb-shaped magnets. The manner like this is prominent in the lower pressure region. It is considered that these results indicate the direct evidence of improvements of plasma particle confinements of the ECR plasma in the comb-shaped magnetic field and moreover energy confinements.

In the multipole magnetic field, the local peaks of the T_e at the second harmonic ECR position around r=50mm are observed at low microwave power (50W) as shown in both of Fig. 4 and 5. These peaks sift to the fundamental ECR position around r=60mm at high microwave power (200W) as shown in both of them. According to detail measurements, it tends that the fundamental ECR zone contributes direct production and ionization, and the second harmonics does increasing the T_e in the low microwave power in the multipole field. Therefore the second harmonics is so effective for confinement and heating as to increase the T_e, in the low pressure and the low microwave power region. The similar effects are also observed in the comb-shaped magnetic configuration at the 11-13GHz microwave frequency. [6] While we have not yet investigated the effect of the second harmonics in the comb-shaped magnetic field configuration in the case of 2.45 GHz, we will get use of advantage of the second harmonics for ECR in order to production of another plasma/ion sources for processing, and also that of multicharged ion.

Figure 6 shows the dependence of the maximum value of the ion saturation currents against the microwave power in the comb-shaped magnetic

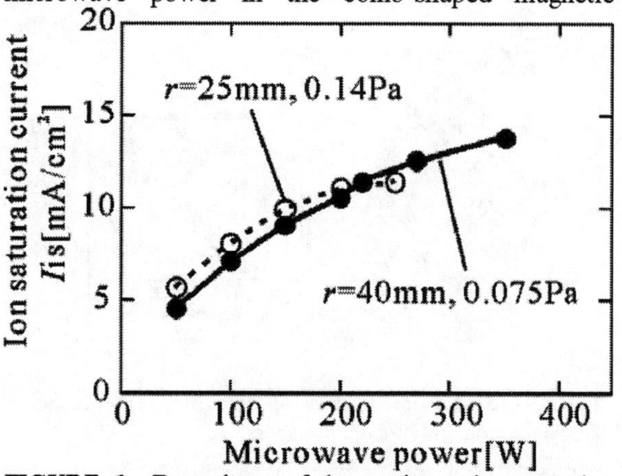

FIGURE 6. Dependence of the maximum ion saturation current against the power of microwaves.

field. Within the microwave power of 360W, the values at 0.075Pa dose not indicate the saturation, and the estimated maximum n_e we obtained is about $3 \times 10^{17} m^{-3}$; the values is about four times larger than the cutoff density of the 2.45GHz microwaves.

The comb-shaped magnetic configuration, which has detached and elongated magnetic field lines of the fundamental and the second harmonic ECR, has feasebility of use for large-area ion beam and multicharged ion sources, because of high efficiency of the ECR, particle and energy confinement under the low pressure.

ACKNOWLEDGMENTS

The authors wish to thank an emeritus professor S. Ishii, Toyama Pref. Univ. for continuous support and suggestions. The authors also wish to thank a professor Z. Yoshida, Toyo Univ., and Dr. A. Kitagawa and Mr. M. Muramatsu, National Institute of Radiological Sciences for useful discussions and comments.

REFERENCES

1. Y. Kato, H. Furuki, T. Asaji, H. Sasaki and S. Ishii, *Neucl. Instrum.& Methods*, B **237**, 256-261(2005).
2. Y. Kato, H. Furuki, T. Asaji, F. Sato, and T. Iida, *Rev. Sci. Instrum.*, **77**, 03A336 1-4(2006).
3. T. Asaji, H. Sasaki, H. Furuki, Y. Kato, S. Ishii, M. Kanazawa, and J. Saito, *Neucl. Instrum.& Methods*, B **237**, 262-266(2005).
4. T. Asaji, H. Sasaki, Y. Kato, F. Sato, T. Iida, and J. Saito, *Rev. Sci. Instrum.*, **77**, 03C104 1-3(2006).
5. T. Asaji, Y. Kato, H. Sasaki, T. Kubo, F.Sato, T. Iida and J. Saito, contribution paper in this conference.
6. T. Asaji, Y. Kato, F.Sato, T. Iida and J. Saito, *Rev. Sci. Instrum.*, submitted and under review.

Advantages of Dual Magnet Ribbon Beam Architecture for Particle Control in Single Wafer High Current Implant

C. Campbell, G. Redinbo, J. Blake, P. Kellerman, E. Moore and N. Variam

Varian Semiconductor Equipment Associates, Inc., 35 Dory Road, Gloucester, MA 01930, USA

Abstract. The issue of particle contamination in ion implant has received renewed interest recently due to the confirmation of killer ballistic particles in batch-style high current implanters. Single wafer high current is now the preferred method of high current implant and device limited yield is now being driven by particles that mask the implant and kill devices. Managing particles will become even more critical to device yields as devices continue to scale to tighter line widths and overall die sizes increase. Just as the batch-style implanters were found to limit yield at smaller device dimensions, the specifics of single wafer implant architectures can affect particle performance. Here we demonstrate the particle performance capability of the Varian VIISta HC dual magnet ribbon beam architecture, analyze the sources of particles in the system and offer an explanation of the physical mechanism that enables on-wafer particle performance.

Keywords: High current implanter, single wafer, particle.

INTRODUCTION

Single wafer high current implant has been adopted as the implant platform of choice for advanced device manufacturing because of the limitations of batch system ballistic particle damage and resulting yield loss [1]. The impact of particle density on device limited yield has been widely understood for some time. In single wafer high current implant the device limited yield is now being driven by the "masking" particles that block the implant and act as the new killer defect. The minimum size of these killer defects has decreased as devices feature sizes have continued to scale. The simultaneous increase of die size has put significant pressure on advanced device manufacturers who need to complete a larger number of high current implant steps while still maintaining acceptable yield. The smaller particles that now limit yield are harder to remove in typical post-implant cleaning steps. This has led to extremely tight particle performance requirements for single wafer high current implant. It is preferable to keep the particles off the wafer than to try to clean them off later.

While implant itself is generally acknowledged to be one of the cleaner processes in the fab, high current implant brings special challenges. The higher currents, power densities, photo-resist out gassing and beam strike make the high current system design requirements more rigorous, both in terms of mechanical wafer handling and beamline design. Between these two major particle sources, mechanical adders can usually be managed through simplified end-station design, reduced mechanical contact and forces, vacuum sequencing and other best practice wafer handling approaches. In terms of beam-borne particles, remedial steps can be taken to reduce sources of beamline particles in the implanter (materials selection, process residue reduction, optics beamstrike, etc) but once these steps are implemented it is clearly best to prevent any particles from being transported to the wafer once they are generated.

It has been established that particles are subject to electrostatic forces in the ion beam itself which can trap and transport the charged particles around an analysis magnet, down the beamline and onto the wafer surface [2]. In this paper we demonstrate the particle performance capability of the dual magnet ribbon beam architecture of VIISta HC, analyze the sources of particles in the system and offer an explanation of the physical mechanism that enables on-wafer performance.

SYSTEM DESCRIPTION

The VIISta HC system is shown in Figure 1 is based on a dual magnet ribbon beam architecture where the first magnet serves to analyze the beam and the second magnet makes the beam uniform and parallel prior to the wafer. The end-station is comprised of a simple one-dimensional wafer scan that moves the wafer vertically through the parallel ribbon beam at low velocity and acceleration.

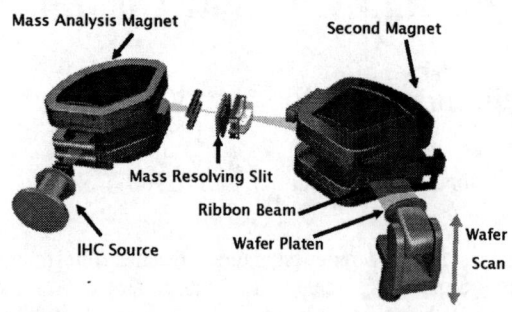

FIGURE 1. VIISta HC dual magnet ribbon beam architecture single wafer high current implanter.

PARTICLE PERFORMANCE

Typical particle performance of VIISta high current systems is shown in Figure 2 where beam borne adders in a logic production process flow are shown. The monitor recipe is a mid-range energy (< 15keV) and high dose (> 1E15 cm^{-2}) with cumulative particles > 0.1μm.

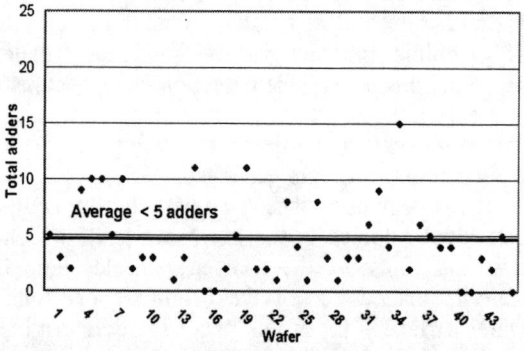

FIGURE 2. Particle adders on a VIISta high current system in logic production. Particle size ≥0.1um over approximately 6 weeks of data collection..

Beam line cleanliness is one critical element of particle performance in high current implanters. Figure 3 shows the inside beam-guide of the mass analyzer and the second magnet of a VIISta HC system that had been run through various recipes for > 700 hours. As is typical in high current beam line implanters, the beam guide of the analysis magnet (Figure 3a) shows evidence of flaking and debris. The mass analysis function of the analyzer magnet results in mass filtering and ion beam strike on beam guide surfaces, which form surface coating, flaking and surface sputtered particulate. In contrast, the inside of the VIISta HC second magnet (Figure 3b) does not show evidence of flaking or particulate, due to the ion beam properties, the magnet properties and the beam-guide design features within the second magnet of VIISta HC. The relative cleanliness of this second magnet beam guide just prior to the wafer, may partially explain the on-wafer particle results achieved in this architecture.

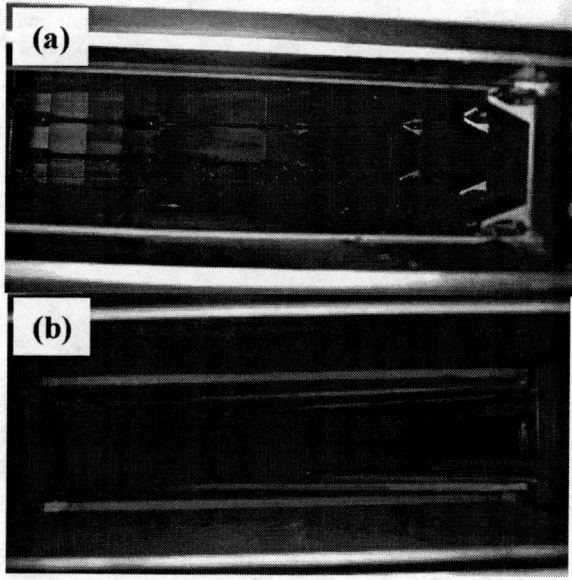

FIGURE 3. Photographs of inside of mass analysis magnet (a) and beam precision magnet (b) after ~ 700 hours of implanter operation.

EXPERIMENT

In order to further investigate the source of particle performance of the VIISta HC system, particle measurements were taken at the entrance of the second magnet with a special wafer edge grip holding fixture. Comparison particle measurements were also taken from wafers in the end-station, after the second magnet sampled with a wafer on the platen. Particle control base-line measurements were collected during a controlled no-beam rough/vent sequence to accurately partition the particle adders to the mechanical transport and rough vent sequence. Beam transport particle measurements were collected both at the wafer plane with a P+ 20keV, 10mA ion beam normally tuned with a wafer positioned at the entrance of the second magnet as in Figure 4b. Specifically, wafer exposure time was recorded and the wafer was re-positioning to the implant position as shown in Figure 4a. The wafer was implanted with the recorded exposure time with the same beam implant conditions, equivalent to a 5E14 atoms/cm^2 dose. Comparison particle measurements were taken with steady-state ion beam conditions and with induced particle excursion scenarios in the extraction and analyzer magnet regions. Generated particle excursions were induced from analyzer beam guide hammer impacts and intentional mass resolving slit (MRS) beam strike.

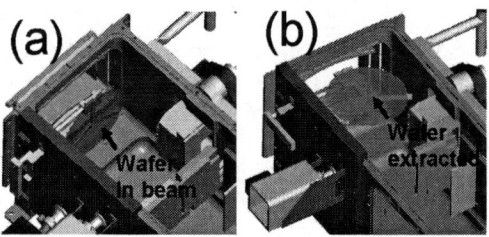

FIGURE 4. Experimental fixture for holding a particle wafer between the magnets. Figure 4a shows a wafer in the ion beam and Figure 4b shows the wafer extracted.

RESULTS

Pre and post-second magnet results from the steady-state P+ 20keV, 10mA, 5E14 atoms/cm^2 implant are shown in Figure 5. Wafer particle measurements taken with KLA-Tencor Surfscan SP1 TBI. Measurements indicate that the pre-second magnet particles are a factor of 6 times higher than the post-second magnet measurements for the .09 - .5µm particles with larger pre vs. post magnet particle reduction factors for the >.5u particle sizes.

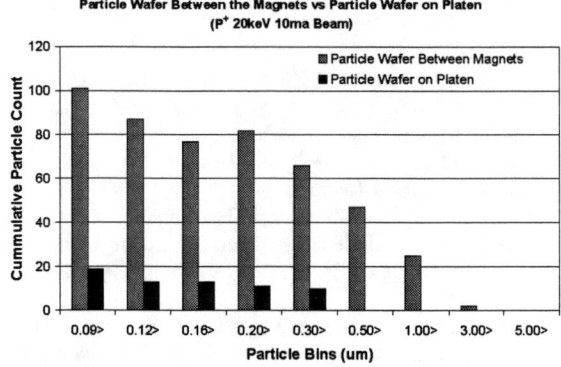

FIGURE 5. On-wafer particle data comparing a wafer placed between the magnets vs a wafer on the platen.

Experimental measurements taken with induced particle excursions are shown in Figure 6. Controlled hammer impact tests were used to introduce particles into the ion beam path within the analyzer magnet region. High speed video cameras positioned at the entrance and exit of the second magnet verified the occurrence of particle "fire-fly's" trapped within the beam entering the second magnet during this impact test, though no "fire-fly" particles were visible at the magnet exit. Figure 6 particle measurement results from the impact test show significant particle excursions levels at the pre-second magnet wafer position in comparison to negligible particle increase levels at the post-second magnet wafer position. Particle measurements collected with induced ion beam contact with the mass resolving slit (MRS) also show substantially elevated particle levels at the pre second magnet wafer position compared to zero particle levels at the post second magnet position. This type of beam strike condition simulates poorly optimized analysis AMU tuning, for instance.

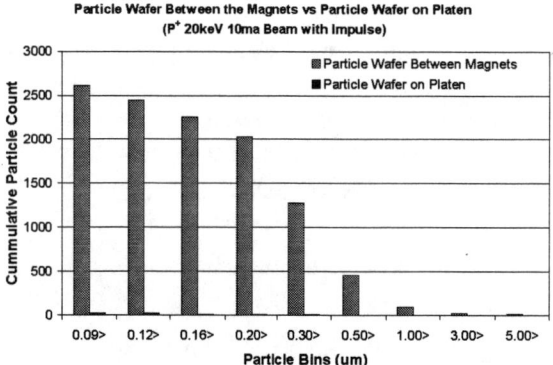

FIGURE 6. On-wafer particle data comparing a wafer placed between the magnets vs a wafer on the platen. An impulse was provided on the analyzer magnet to initiate beam-borne particles.

DISCUSSION

Reference [2] provides strong evidence that particles can be trapped electrostatically within an intense ion beam and transported down the beam-line by momentum transfer from the ions, even through bending magnets. This evidence appears at first to contradict the experimental data presented here. However, differences between the spot beam used by Sferlazzo et al and the nature of the ribbon beam examined here can modify the beam's ability to trap particles. To extend this previous work [2] we have constructed a simple model to compare spot and ribbon beams. Consider the electric field in a beam with uniform charge density over an elliptical cross section as shown in Figure 7. The uniform charge density required for dose uniformity in a ribbon beam tool makes this a reasonable approximation. While this is not necessarily true for a spot beam tool, the uniform charge density is consistent with minimizing the beam potential.

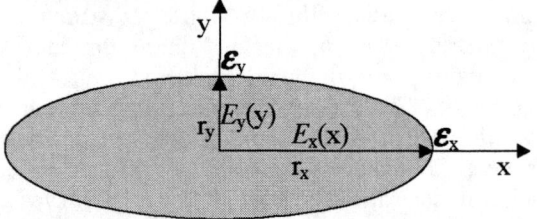

FIGURE 7. Elliptical cross section ion beam with uniform charge density and electric fields E_x and E_y.

A solution to Poisson's Equation requires that

$$E_x(x) = (\mathcal{E}_x/r_x)x, \quad E_y(y) = (\mathcal{E}_y/r_y)y$$

where $\mathcal{E}_x$ and $\mathcal{E}_y$ are the values of the field components at the boundary (i.e., at the points $x = r_x$ and $y = r_y$). Following Kellerman [3], $\mathcal{E}_x$ and $\mathcal{E}_y$ are determined by the charge density, $\rho = \dfrac{I}{Area}\sqrt{\dfrac{m}{2E}}$, and boundary conditions. For the case where the beam is surrounded by a conducting wall,

$$\frac{\mathcal{E}_x}{\mathcal{E}_y} = \frac{1}{k},$$

$$E_x = \frac{\rho}{\varepsilon_0}\frac{x\, r_y^2}{(r_x^2 + r_y^2)} \quad E_y = \frac{\rho}{\varepsilon_0}\frac{y\, r_x^2}{(r_x^2 + r_y^2)}$$

The question is whether this field can provide sufficient force on a particle to cause it to curve along with the beam as the beam is bent by a magnetic field. The magnitude of this required centripetal force depends on the particle's diameter d and magnet radius R.

$$F_{cent} = \frac{\pi d^3 \delta v^2}{R},$$

where δ is the density of the particle, and v is the velocity of the particle due to momentum transfer from the beam ions (v typically ranges from ~10m/s – 50m/s [2]).

To find the electrostatic force on a particle, we need its charge in the beam plasma. This depends on the beam's electron temperature, and is difficult to calculate [2]. However, it is reasonable to assume that the particle will charge (negatively) such that its "floating" potential V_f will be within ~10V. From this assumption we can obtain a reasonable upper limit on the particle's charge Q, assuming it is spherical:

$$Q_{particle} = 2\pi\varepsilon_0 V_f d$$

As a figure of merit, we consider the following situation: A graphite particle of diameter .1µm is traveling within a 20kV P 10mA ribbon beam that is assumed to be 99% neutralized. The beam is then magnetically deflected horizontally with a radius of .6m. If the particle has attained a velocity of 10m/s, then the minimum electrostatic force needed from the space charge of the beam to transport the particle through the magnet is 1.75×10^{-16} N. If the beam's half width and height are r_x=15cm and r_y=2cm, the maximum horizontal electrostatic field at the edge of the beam is 1.2N/C, and since $Q_{particle}$~5.6×10^{-17} C, the maximum beam confinement force is 6.5×10^{-17} N, which is less than the needed centripetal force to trap the particle. However, for a spot beam of diameter 6cm, the electrostatic confinement for the same conditions is 4.7×10^{-15} N, which more than an order of magnitude greater than the required confinement force.

SUMMARY

The Varian VIISta HC single wafer ion implanter architecture incorporates a dual magnet design. Demonstrated world-class particle level performance is the standard for this architecture. Experimental results show that this second magnet in conjunction with the ribbon beam optics properties serves as a critical particulate transport barrier. The derived model provides the basis for the particle performance of this architecture.

ACKNOWLEDGMENTS

We would like to thank Alex Perel, Joe Olson and Frank Sinclair for their contributions to this paper.

REFERENCES

1. L. Pipes, M. Taylor, G. Zietz, A. Al-Bayati, M Castle, T. Marin and J. Simmons, "Characterization and reduction of a new particle defect mode in sub-0.25um semiconductor process flows," *Proc. 15th Int. Conf. on Ion Implantation Tech.* (2004).
2. P. Sferlazzo, D.A. Brown, S.E. Beck and J.F. O'Hanlon, "Experimental Evidence of Beam Particulate Transport in Ion Implanters," *Proc. Conf. on Ion Implantation Tech.* (1993) p. 565.
3. P. Kellerman, "Advanced Modeling Techniques for Analysis of High Current Ribbon Beam Transport and Control," *Proc. 16th Int. Conf. on Ion Implantation Tech* (2006)

Optimized Autotuning for Single Wafer High-Current and Medium-Current Implanters

J.T. Scheuer, A. Cucchetti, M. Welsch, W. Callahan, K. Luey, and J.C. Olson

Varian Semiconductor Equipment Associates, 35 Dory Road, Gloucester, MA 01930, USA

Abstract. As semiconductor lot sizes decrease the impact of autotuning performance on productivity is increased. The dual magnet architecture of the VIISta series of single wafer ion implanters provides unparalleled defect prevention and beam control while presenting unique challenges for autotuning. We present data from field installations of high-current and medium-current VIISta implanters demonstrating excellent autotuning performance in terms of tuning speed and success rate.

Keywords: Ion Implantation, Beam Tuning.
PACS: 41.75.Ak; 61.85.+p; 85.40.Ry; 61.72.Tt

INTRODUCTION

Over the past several years semiconductor fabrication facilities have been transitioning from batch ion implanters to single wafer systems. In batch tools the beam enters the process chamber where wafers are rotated at high speed through the beam, while the rotation mechanism is translated back and forth through the beam. An advantage of the batch system is the ability to provide a uniform dopant dose by mechanical means. This eliminates the need to adjust the beam shape to provide a uniform implant. However, several disadvantages of this technique, including mechanical stress and ballistic particle damage of semiconductor device features, and cone angle effects have led to the increasing use of single wafer systems. [1]

Single wafer implanters can be divided into categories. Architectural categories include single magnet and dual magnet designs. Two-dimensional wafer scanning with a spot beam and one-dimensional wafer scanning with a ribbon beam are both employed.

In a two-dimensional wafer scan implanter, a stationary spot beam is incident on a wafer that is scanned both vertically and horizontally. The beam line design is simple, but the end station hardware is complex. Other disadvantages include an inherent compromise between throughput and dose uniformity for low dose implants as well as issues with wafer temperature control for high power beams [2].

Systems utilizing one-dimensional wafer scan (such as the VIISta platform) rely on a uniform beam in the horizontal direction and a wafer scan in the vertical direction to achieve uniform implants. The VIISta 810 medium-current implanter achieves horizontal beam uniformity by electrostatically scanning a spot beam while the VIISta HC high-current implanter uses a parallel ribbon beam. Advantages of this technique include mechanical simplicity without the throughput and thermal constraints of the two-dimensional scan system. The beam tuning approach required to produce a uniform beam is technically more challenging, however.

The dual magnet architecture of the VIISta platform has several advantages. First, the second magnet isolates the wafer from particles generated by beamline elements such as the mass resolving slit, resulting in two to five times lower particles on the wafer [3]. Second, the closed loop Varian Positioning System (VPS™) delivers accurate, repeatable and interlocked incident angle control over the full range of angles needed for high current applications. The result is precise wafer-to-wafer and lot-to-lot beam steering. Finally, in the case of decel operation on the high-current VIISta HC, the second magnet also isolates neutrals generated by beam impact at the mass resolving slit which has line of sight to the wafer on single magnet systems. A consequence of the dual magnet architecture is the added beam tuning effort to optimize the second magnet to achieve angle control across the wafer.

In this paper we discuss the approach to beam tuning in the single-wafer, one-dimensional wafer scan, two-magnet architecture of the VIISta 810 and VIISta HC. We present recent tuning algorithm improvements and production field data demonstrating state of the art tuning performance.

CP866, *Ion Implantation Technology*,
edited by K. J. Kirkby, R. Gwilliam, A. Smith, and D. Chivers
© 2006 American Institute of Physics 978-0-7354-0365-9/06/$23.00

IMPACT OF TUNE TIME ON IMPLANT PRODUCTIVITY

Semiconductor manufacturers demand the highest possible productivity from ion implantation equipment in order to decrease their costs of manufacturing. In many cases the implant equipment customer requests an increase in ion beam current for his recipe set in order to improve run rate and therefore productivity. In some cases, however, decreasing the time to tune the beam can have an equal or larger impact on productivity.

Tables I and II show the effective productivity as a function of implant run rate and tune time for implant lot sizes of 200 and 20 wafers, respectively. Lot size is defined as the average number of wafers implanted before a new recipe setup is required, while run rate is defined as the number of wafers implanted per hour after the beam is setup. Base run rates of 100 and 400 wafers per hour were chosen to highlight typical applications for high current and medium current implanters, respectively.

In Table I the lot size is 200 wafers, typical for a logic manufacturer focusing on relatively few types of products. From this table, it is clear that tune time has little effect on productivity for large lot sizes. For each case the effective productivity only changes by a few percent as the tune time is decreased from 5 min to 2 min.

In Table II the lot size is 20 wafers, typical for a semiconductor foundry which runs many different products for a wide array of customers. From this table, it is clear that tune time has a large effect on productivity for small lot sizes. In each case, a reduction of tune time by 1 minute has an effect equal to or greater than a 10% increase in implant run rate.

TABLE 1. Effective implant productivity (wafers per hour) as a function of implant run rate and tune time for a lot size of 200 wafers (typical for logic application).

Run rate (wph)	Tune time (min)			
	2	3	4	5
100	98	98	97	96
110	108	107	106	105
400	375	364	353	343
440	410	396	384	372

TABLE 2. Effective implant productivity (wafers per hour) as a function of implant run rate and tune time for a lot size of 20 wafers (typical for foundry application).

Run rate (wph)	Tune time (min)			
	2	3	4	5
100	86	80	75	71
110	93	86	80	75
400	240	200	171	150
440	254	210	178	155

BEAM TUNING IMPROVEMENTS

In order to produce superior implant beam control the VIISta family of single wafer dual magnet implanters uses more beam line elements compared to traditional single-magnet ion implant systems. In addition to adjusting the mass analysis magnet to achieve the highest beam current, the VIISta family leverages the second (beam precision) magnet to produce a uniform and parallel beam. Since the VIISta single wafer implanters do not rely on complicated two-dimensional mechanical wafer scan to achieve implant uniformity, the beam itself is instead made uniform across the wafer diameter. The methods of achieving parallel, uniform beams on the medium- and high-current VIISta implanters are described elsewhere [4-7].

Advanced tuning algorithms have been developed to simultaneously achieve beam angle and uniformity performance, fast tune times and high tuning success rate.

Several improvements have been made to the VIISta HC hardware to produce a beam which is more easily tuned [7]. These improvements include a triply indexed analyzer magnet with neutralizing gas bleed, two quadrapole lenses and an indirectly heated cathode ion source.

In addition to these hardware upgrades, numerous improvements have been made to the VIISta HC tuning algorithms to improve the tune time and success rate. Certain components require software control over electro-mechanical systems (e.g. the extraction optics manipulator). Improvements were made to speed up the interface between the software and the electro-mechanical components. A unique feature of the VIISta HC is the ability to measure and correct the steering angle of the beam relative to the wafer. The typical time to make this measurement was reduced from 28 seconds to 8 seconds with no loss in measurement accuracy. Many of these algorithm improvements have a favorable impact on both tune time and tuning success rate.

The common strategy in designing the beam tuning algorithm on the VIISta tools relies on developing an intelligent sequence that executes only those tuning steps required to achieve the desired beam current and beam uniformity. In this way, the highest success rate with the fastest tune time can be achieved [8].

During the initial transition from one recipe to another, the settings of gas feed type, gas flow rate, extraction voltage, analyzer magnet setting, etc. must be changed from those of the previous recipe to those of the new recipe. The software was improved to maximize the number of these transitions occurring in parallel with one another to reduce tune time. However, several built-in recovery routines have also been added to prevent failures and instabilities during arc strike as well as ramping of power supplies.

The time spent optimizing source and beam-line parameters is minimized. Parameters are tuned only if the target beam current has not already been met or if their optimization affects beam uniformity. Furthermore, if the current parameter optimization does not yield a significant beam current improvement, the tuning sequence will quickly skip ahead and focus on the next parameter.

Uniformity is achieved by electrostatic scanning on the medium current tool and by magnetic steering of beamlets on the high current tool. In both cases, the fine tuning of a large number of electrostatic waveform or magnetic elements is required (an electrostatic waveform with up to 100 elements for the medium current and up to 35 magnetic dipoles for the high current tool). The uniformity routine relies on learned recipe values as a starting point. Furthermore, an approach similar to downhill simplex is employed to optimize all the parameters in parallel and converge to a solution in the shortest amount of time [9].

Finally, given the commonality of the Varian control system [10] and the significant overlap in hardware components between product-lines, improvements in accuracy and speed of controlling, measuring, and tuning common hardware can be propagated across all products. In particular, the lessons learned in tuning components such as the IHC source, the extraction manipulators, the analyzer and corrector magnets [11], are utilized wherever possible on both medium and high current. Furthermore, the common end station allows the benefits of faster and better beam metrology to be realized on all VIISta products.

BEAM TUNING PERFORMANCE

Figs. 1 and 2 show beam tuning performance for field installations of VIISta HC and VIISta 810HP. This data demonstrates excellent tuning performance for single-wafer, high-current and medium-current ion implanters. The data shown in Fig. 1 for the VIISta HC is a compilation of 5181 setups from 12 tools over a 30 day period at one customer site. The median tune time is 3:36 while the overall success rate is 98%. The data shown in Fig. 2 for the VIISta 810HP is a compilation of 19728 setups from 7 tools over a 30 day period at one customer site. The median tune time is 2:45 while the overall success rate is 99%.

The average number of setups per tool per month is 2818 for the VIISta 810HP and 431 for the VIISta HC. The VIISta 810HP averages more than six times the number of setups per machine than the VIISta HC. This large discrepancy is due to the fact that the medium-current tools are typically running low dose ($<1e14$ atoms/cm^2) implants at or near the mechanical throughput limit (500wph) while the high-current tools run mostly high dose implants ($>1e14$ atoms/cm^2) at lower average throughput. This difference leads to nearly 4 setups per hour on the medium-current tools and roughly 1 setup every 2 hours on the high-current tools.

While achieving the optimum tune time is important for high-current tools, it is absolutely essential for medium current tools. A savings of 1 minute of setup time over each of the 2181 medium current setups results in a savings of 47 hours (nearly 2 days) of productive implant time per machine per month.

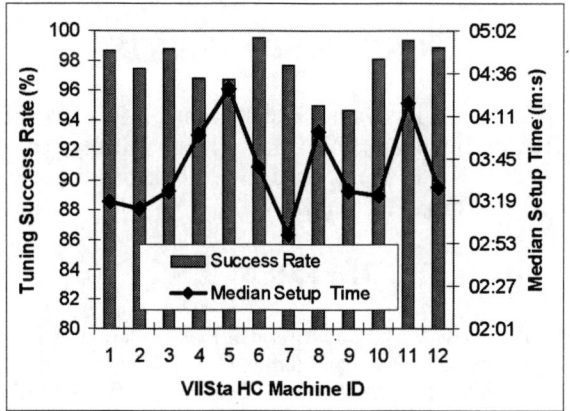

FIGURE 1. VIISta HC beam setup performance. Tuning success rate (%) and median beam setup time is shown for a total of 5181 setups from 12 tools in production for a 30 day period. The number of setups per tool ranges from 240 to 631 which results in varied weighting of each tool's performance in the overall results. Overall success rate is 98% and median tune time for the entire population is 3:36.

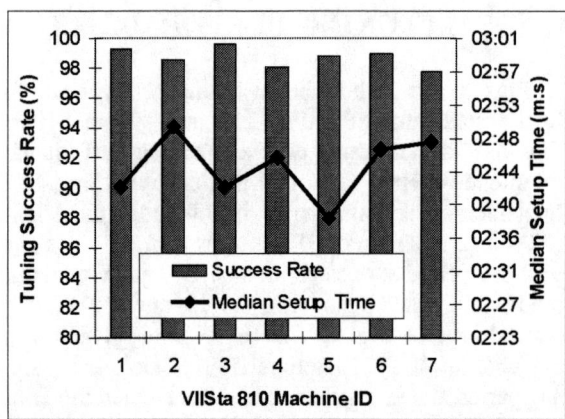

FIGURE 2. VIISta 810HP beam setup performance. Tuning success rate (%) and median beam setup time is shown for a total of 19,728 setups from 7 tools in production for a 30 day period. The number of setups per tool ranges from 2255 to 3068 which results in varied weighting of each tool's performance in the overall results. Overall success rate is 99% and median tune time for the entire population is 2:45.

CONCLUSIONS

Single wafer, dual magnet ion implantation systems with one dimensional wafer scan have demonstrated excellent performance for dose control, angle control, and prevention of device particle defects. Improvements in beam line hardware and tuning algorithms have resulted in excellent tuning performance. Field data from production implanters show 99% success rate and median tune time of 2 min. 45 sec. for the VIISta 810 and 98% success rate with a median tune time of 3 min. 36 sec. for the VIISta HC.

ACKNOWLEDGMENTS

The authors would like to thank Chris Larson, Stewart Hitelman, Rob Doody, Norm Hussey, Doug Fielder, Keith Pierce and Greg Redinbo for their contributions to this work.

REFERENCES

1. A. Renau, et al., "Comparison of Plasma Doping and Beamline Technologies for Low Energy Ion Implantation, *Proc. 14th Int. Conf. on Ion implantation Tech.*, Taos, NM (2002)
2. A. Renau, *Nucl. Instr .and Meth. In Phys. Res. B* **237**, 2005, pp 284-289.
3. C. Campbell et al., "Advantages of Dual Magnet Beamline Architecture for Particle Control in Single Wafer High Current Implant", *Proc. 16th Int. Conf. on Ion implantation Tech.*, Marseilles, France (2006)
4. A. Renau and D. Hacker, "The VIISta 810 medium current ion implanter", *Proc. 12th Int. Conf. on Ion implantation Tech.*, Kyoto, Japan (1998)
5. J.C. Olson and A. Renau, "Control of channeling uniformity for advanced applications", *Proc. 13th Int. Conf. on Ion implantation Tech.*, Alpbach, Austria (2000)
6. A. Renau, "Beam line architecture of the VIISta 810 medium current ion implanter", *Proc. 12th Int. Conf. on Ion implantation Tech.*, Kyoto, Japan (1998)
7. G. Redinbo et al, "Advanced Single Wafer High Current Beamline Architecture for Sub 65nm", Proc. 16th Int. Conf. on Ion implantation Tech., Marseilles, France (2006)
8. A. Cucchetti, D Distaso, R. Mollica, J.C. Olson, A. Renau, J.T. Scheuer, D.L. Smatlak, "Beam Autotuning on the VIISta 810 Ion Implanter," *Proc. 13th Int. Conf. on Ion implantation Tech.*, Alpbach, Austria (2000)
9. J.C. Olson, A. Renau, J. Buff, "Scanned beam uniformity control in the VIISta 810 ion implanter", *Proc. 12th Int. Conf. on Ion implantation Tech.*, (1998
10. G. Viviani, N.A. Parisi, W.G. Callahan, L.M. Zimmermann, "Achieving Next Generation Performance of Ion Implanters with the Varian Control System (VCS)," *Proc. 13th Int. Conf. on Ion implantation Tech.*, Alpbach, Austria (2000)
11. A. Cucchetti, et al., "Methods and Apparatus for Alignment of the Ion Beam System using Beam Current Sensors" U.S Patent No 6,430,972, Issued June 11th 2002.

Backing up Medium Current Implanters using Single Wafer High Energy Implanter for Manufacturing Efficiency

H.L. Sun**, Woojin Lee**, and Knight Xu**
HY Tsun*, KT Peng*, LS Juang*, and HP Tseng*,
* Taiwan Semiconductor Manufacturing Company (Shanghai), Fab10
4000, Wen Xiang Road, Songjiang, Shanghai, China,

**Varian Semiconductor Equipment Associates
35 Dory Road, Gloucester, MA 01930 U.S.A

Abstract — Today most high volume manufacturing fabs are experiencing an inefficient distribution of resources between medium current (MC) and high energy (HE) implanters. The utilization levels of medium current implanters and high energy implanters differ greatly due to increasing amounts of medium current implant recipes. Medium current implanters are very highly utilized, while high energy implanters are not used as much. Thus, the high energy implant tools can be used to optimize manufacturing efficiency and production cost reduction by backing up the medium current tools. Traditionally, there were only well implants that can be processed on both single wafer medium current (SWMC) and batch HE implanters. However, by using the Varian VIISta 3000HP Single Wafer High Energy implanter (SWHE) with high tilt and true zero degree implantation capability, Vt, anti-punch-through and pocket/halo implants can also be used as backups. The data of the electrical function and yield comparison between SWHE VIISta 3000HP and SWMC EHP500 was shown to be compatible. The throughput was also evaluated to match the productivity of medium current implanters which are higher than the batch tools. This paper shows that the SWHE can maximize the manufacturing efficiency and minimize the production cost.

Keywords: Single Wafer High Energy implanter, SWHE
PACS: 85.40.Ry

I. INTRODUCTION

As the market moves into advanced technology nodes, the number of medium current applications (i.e. halo, pocket, anti-punch-through, and Vt implants) keep increasing. In addition, as the energy requirements for traditional well applications decrease, it becomes a medium current application rather than a high energy application. For instance, by running a B 300keV 3.5e13 P-Well implant on the medium current implanter, the productivity will be much higher (> 200%) than that of a traditional batch high energy implanter [1]. However, this creates an un-balanced distribution of tool-to-tool utilization between medium current and high energy implanters. Therefore, backing up medium current implants using high energy implanters, thus optimizing the production utilization and reducing production costs, becomes more important. The traditional batch HE implanter has fundamental limitations due to its limited backup performance (no high tilt, no zero degree, and novel species). Thus the high energy tools remain unused while medium current tools face an overload of work. Utilizing a Single Wafer High Energy (SWHE) tool will significantly improve the in-efficiency between MC and HE tools. However, in order to utilize the high energy tools for the medium current implants without compromising process and device performance, a thorough process and device evaluation will be required [2].

In this paper, the entire performance of SWHE was tested to demonstrate its MC backup capability in the areas of productivity, process matching, and device performance.

II. VIISTA 3000HP – SINGLE WAFER HIGH ENERGY IMPLANTER

The Varian Semiconductor VIISta 3000 is a single wafer ion implant system designed to meet high-energy and medium current implant process needs in 200mm and 300mm [3,4] semiconductor wafer manufacturing. An IHC (Indirected Heated Cathode) ion source was developed in order to improve ion source lifetime of the VIISta 3000 [5] with better beam stability. The VIISta 3000HP ion implanter provides accurate setting of the final ion energy through use of a nuclear resonance technique for energy calibration of the generating voltmeter (GVM) in the feedback loop on the acceleration power supply driver.

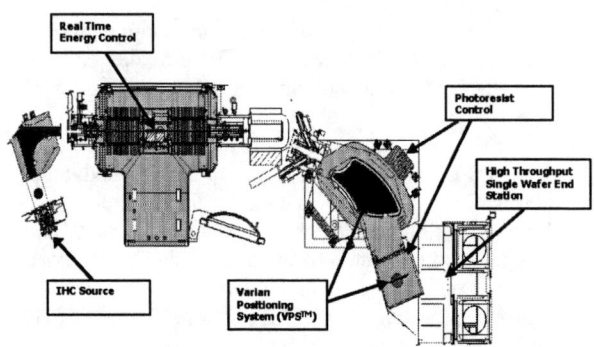

Fig. 1. Schematic layout of the Varian VIISta 3000HP high-energy ion implanter.

A high throughput (340 WPH @ 200mm) single wafer end station based on patented parallel path wafer handling architecture provides reliable, low particle processing. The VIISta 3000HP's high vacuum conductance end station and beam filter eliminates energy contamination and dose shift while sustaining the maximum usable beam current. Angle control becomes an important consideration as devices get smaller. The VIISta 3000HP's closed loop Varian Positioning System (VPS^{TM}) provides an unmatched uniform, zero degree implantation across the wafer which enables device shrinkage to meet sub 130nm device requirements [6]. The VIISta 3000HP also ensures high utilization in a production environment by providing true medium-current backup capability [7].

III. VIISTA 3000HP MEDIUM CURRENT BACKUP CAPABILITY

As shown in Fig. 2, due to high degree of commonality between the two tools, VIISta 3000HP's single wafer parallel beam technology provides identical process performance for medium current applications. Both zero degree and high tilt applications can easily be transferred between two products, making the VIISta 3000HP the perfect backup for medium current implanters. Also, based on the patented parallel wafer handling architecture, the VIISta 3000HP offers even higher throughput for medium current applications than the EHPi-500 tools.

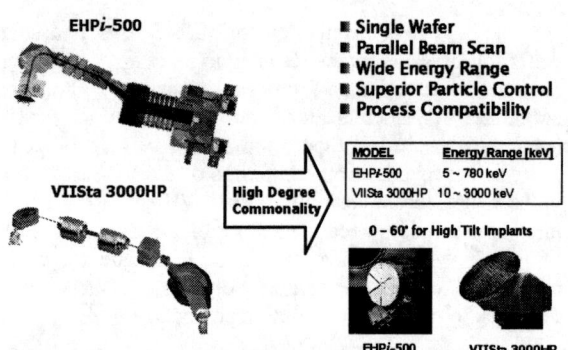

Fig. 2. High degree of single wafer commonality between the EHPi-500 and VIISta 3000HP provides easy process transfer, highest productivity, and reliability.

Based on single-to-single backup capability, the VIISta 3000HP can cover a broad range of medium current applications as shown in Fig. 3. Traditionally, the batch high energy implanter can only cover very limited and non-critical medium current applications due to its systematic 'Cone Angle Effect' and limitation of tilt angle (0~11°). However, the single wafer high energy VIISta 3000HP with true zero degree and high tilt (0~60°) implantation capability can cover most medium current applications including critical layers such as Vt, anti-punch-through, LDD, and pocket/ halo implants.

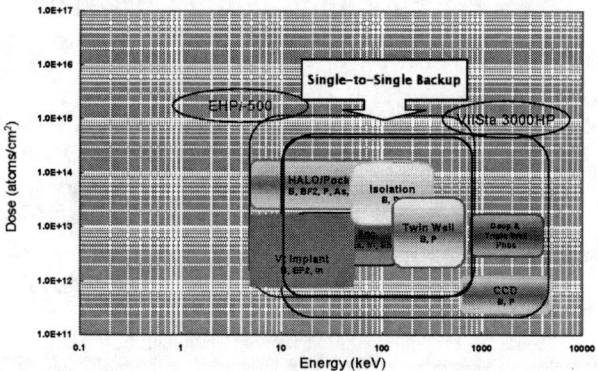

Fig. 3. The VIISta 3000HP Application covers a broad range of medium current applications as a single-to-single backup.

IV. Experimental

The high energy implanter used in this study was the Varian VIISta 3000HP high energy implanter featuring a single wafer end station. Varian EHP-500 series medium current implanters were used for comparison as a baseline. Several key medium current applications such as Vt, LDD, and Halo implants were performed on both tools. The productivity performance (run-rate including tune time) in those medium current applications was measured on each tool. After successful tool matching on monitor wafers, the split lots were performed to verify and prove the yield results on product in 0.25-0.35 μm logic technology.

V. Results and Discussion

A. Throughput Comparison

Even if a high energy implanter is capable of running medium current applications, it will not be useful as a backup to medium current implanters unless the productivity of the high energy tool is comparable to medium current. Fig. 4. shows the average run rate of the VIISta 3000HP and the EHP-500 series tools during the month of Nov. 2005. The VIISta 3000HP's average run rate was 20~85% higher than the EHP-500 series tools.

Fig. 4. Actual production throughput data comparing the VIISta 3000HP tool and the EHP-500 series tools.

B. Device Yield Comparison

The tool matching becomes more and more important for smaller devices. Different tool architectures and differences in tool design may challenge the tool matching. The process results on monitor wafers may look similar, but differences in electrical parameters can appear by processing structured wafers.

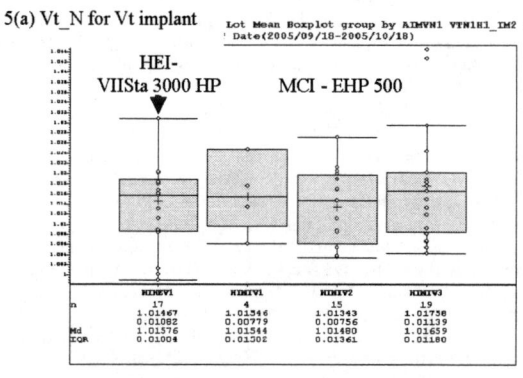

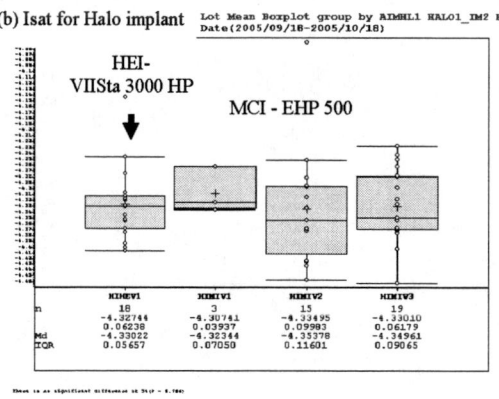

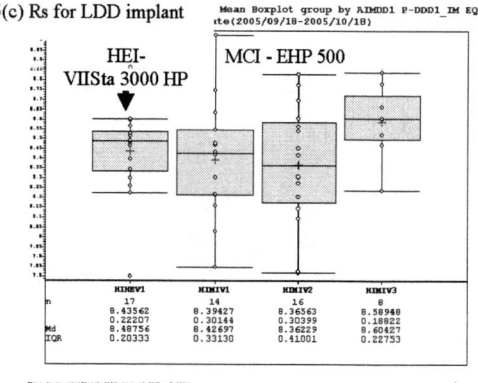

Fig. 5 (a), (b), and (c): Comparison of device electrical feature performance between the VIISta 3000HP and EHP-500 series tools.

Therefore, before using the single wafer high energy implanter as a medium current backup, the device matching must be performed. As shown in Fig. 5, the VIISta 3000HP can match with the EHP-500 series tools without compromising the device performance of Vt, Isat, and Rs.

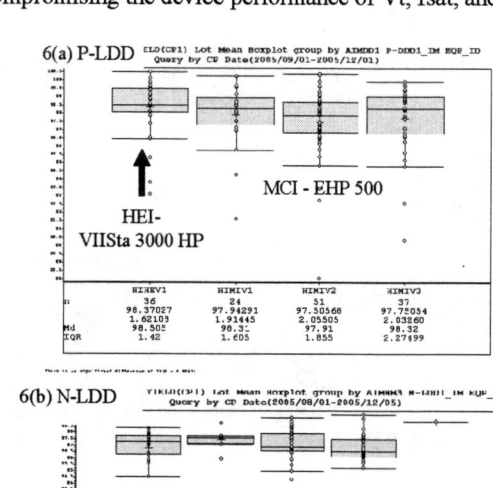

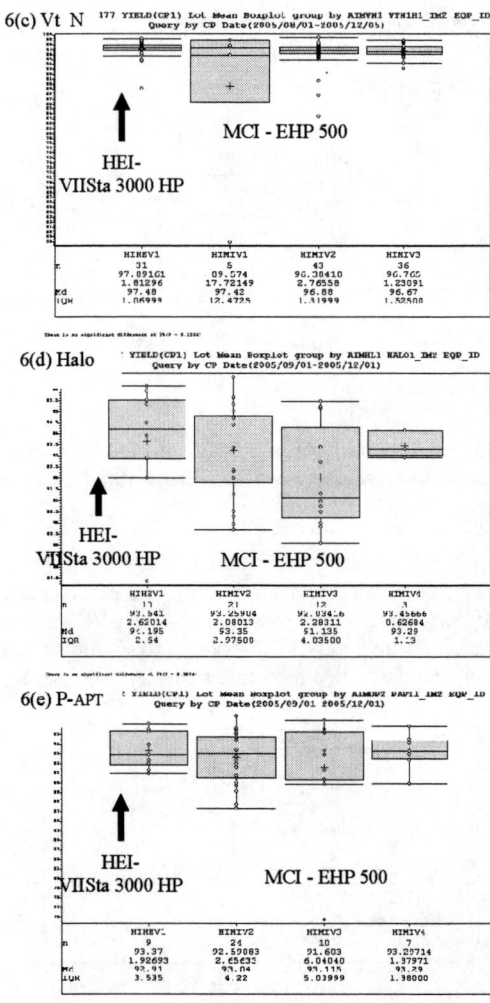

Fig. 6 (a), (b), (c), (d), and (e): Comparison of device yield performance between VIISta 3000HP and EHP-500 series tools for LDD, Vt, Halo and Anti-punch-through implant process.

In Fig. 6, the standard deviation of device yield across the wafers for the implant of P_LDD, N_LDD, Vt_N, P_Halo, and P-anti-punch-through (APT) on the VIISta 3000HP are much tighter than the EHP-500 series tools, mainly due to the unmatched VIISta 3000HP's single wafer parallel beam technology.

VI. CONCLUSION

In order to increase manufacturing efficiency and reduce production cost, optimizing tool-to-tool utilization between medium current implanters and high energy implanters becomes more and more important. In this paper, the VIISta 3000HP single wafer high energy implanter clearly demonstrated its ability to perform as a full medium current backup without compromising any productivity. Also, while the device performance on the VIISta 3000HP was comparable with the EHP-500 series tools, the standard deviation across the wafer on a VIISta 3000HP was much tighter than the EHP-500 series tools, mainly due to the unmatched VIISta 3000HP's single wafer parallel beam technology with the closed loop Varian Positioning System (VPS^{TM}). In the current status quo, there is an inefficient distribution of resources between medium current and high energy implanters, thus leading to a decrease in manufacturing efficiency. This work proves that the VIISta 3000HP can be used as a backup for EHP-500 series medium current tools, and have comparable or better productivity performance. The VIISta 3000HP can clearly optimize production costs and manufacturing efficiency.

REFERENCES

[1] P. Layne, G. Gammel, J. Dzengeleski, G. Gibilaro, J. McLane, "Increased Medium Current Mechanical Limited Throughput", 2004 International Conference on Ion Implantation Technology Proceedings, Taipei, Taiwan, pp. 44-47.

[2] N. Variam, S. Mehta, S. Norasetthekul, and B.N. Guo, "Seamless Transferability of Doping Processes Between the VIISta Platform of Ion Implanters", 2002 14th Int. Conf. on Ion Implantation Technology Proceedings, Taos, NM, Sept. 2002, pp. 283-286.

[3] N. Tokoro, D. Holbrook, D. Hacker, "Introduction of the Varian VIISta 3000 Single Wafer High-Energy Ion Implanter", 2000 International Conference on Ion Implantation Technology Proceedings, Alpbach, Austria, pp. 368-371.

[4] M. R. Lafontaine, P. J. Murphy, E. Bell, D. Holbrook, "Beam Optics of the VIISta 3000 Ion Implanter", 2000 International Conference on Ion Implantation Technology Proceedings, Alpbach, Austria, pp. 403-406.

[5] J.C. Olson, P. E. Maciejowski, S. Chang, L. Klos, "Varian Indirectly Heated Cathode Ion Sources", 2002 14th Int. Conf. on Ion Implantation Technology Proceedings, Taos, NM, Sept. 2002, pp. 283-286.

[6] A. Thornton, S. Chang, and D. Hacker, "Performance Characteristics of the Varian Semiconductor VIISta 3000 Single Wafer High Energy Ion Implanter", 2002 14th Int. Conf. on Ion Implantation Technology Proceedings, Taos, NM, Sept. 2002, pp. 501-504.

[7] J.C. Olson, A. Renau, J. Buff, "Scanned Beam Uniformity Control in the VIISta 810 Ion Implanter", 1998 International Conference on Ion Implantation Technology Proceedings, Kyoto, Japan, pp. 169-172.

Rising Microwave Frequency of a Broad-Ion-Beam ECR Source with Cylindrically Comb-Shaped Magnetic Field Configuration

Toyohisa Asaji[*,†], Yushi Kato[*], Hiroshi Sasaki[*], Takashi Kubo[*], Fuminobu Sato[*], Toshiyuki Iida[*] and Junji Saito[†]

[*]*Division of Electrical, Electronic and Information Engineering, Graduate School of Engineering, Osaka University, 2-1 Yamadaoka, Suita, Osaka 565-0871, Japan*
[†]*Development Center of Advanced Technology, Tateyama Machine Co., Ltd., 30 Shimonoban, Toyama, 930-1305, Japan*

Abstract. An 11-13 GHz electron cyclotron resonance (ECR) plasma source with a cylindrically comb-shaped magnetic field configuration has been examined in order to apply to ion beam processing. The ion saturation current density has been measured using a Langmuir probe. It was found that the ion density linearly increases as gas pressure and microwave power increases. The maximum ion density at 13 GHz microwaves is 37.4 mA/cm^2 under low microwave power. The ion beam extractor which has multihole apertures has been constructed at the end of the magnet. The ion beam current has reached 20 mA at the microwave power of only 300 W. The ion beam current has clearly increased by rising microwave frequency as well as the tendency of the plasma density.

Keywords: Ion source, ECR, Microwave frequency, Permanent magnet
PACS: 29. 25. Ni, 29. 27. Ac

INTRODUCTION

Ion beams have several advantages for material processing. Those are the controllability of ion energy and directivity, and low-damage processes due to materials are not exposed to plasma directly. We have developed a high-density electron cyclotron resonance (ECR) plasma source for producing broad ion beams [1,2]. We have improved the plasma density by applying an 11-13 GHz traveling-wave tube (TWT) amplifier and a cylindrically comb-shaped magnetic-field configuration. The ECR plasma reached the plasma density of about 10^{18} m^{-3} at the microwave power of only 350 W [2]. In those works, we clarified the increase of plasma density by rising microwave frequency and improving plasma confinement.

As the next step, we have constructed an ion beam extractor with multiple apertures. The diameter of the ion beam is about 60 mm. We have investigated the extracted ion beam as functions of electrode parameters, i.e. the gap distance between two extraction electrodes, the position of the electrodes and extraction voltage. And then, we have confirmed that the rising microwave frequency is effective for the increase of ion beam current. In this article, we present the measurement of ion density in the plasma source and extracted ion beam.

We will apply the ion source to broad ion beam processing such as reactive ion beam etching and surface modification.

EXPERIMENT

The 11-13 GHz ECR plasma source with the cylindrically comb-shaped magnetic field was described [2]. We have constructed the broad-ion-beam extractor which has multihole apertures. A schematic diagram of the plasma/ion source and the extractor system is shown in Fig. 1. The cylindrical chamber is 197.2 mm diameter and 320 mm length. We use the cylindrical coordinates (r, φ, z) with an origin at the center of the chamber. The direction of the Langmuir probe is assigned to $\varphi = 0°$. The microwave frequency and the maximum power of the TWT amplifier are 11-13 GHz and 350 W, respectively. The microwaves are launched from WR75 waveguide through a quartz window for vacuum sealing. The magnetic field for ECR is formed using Nd-Fe-B permanent magnets. The ECR zones for 11 and 13 GHz microwaves, i.e. 0.393 and 0.464 T, are formed near each magnet within 6.5 and 3 mm from the inner surface of the chamber, respectively. The ion saturation current density is measured by using a Langmuir probe. Plasma density is almost proportional

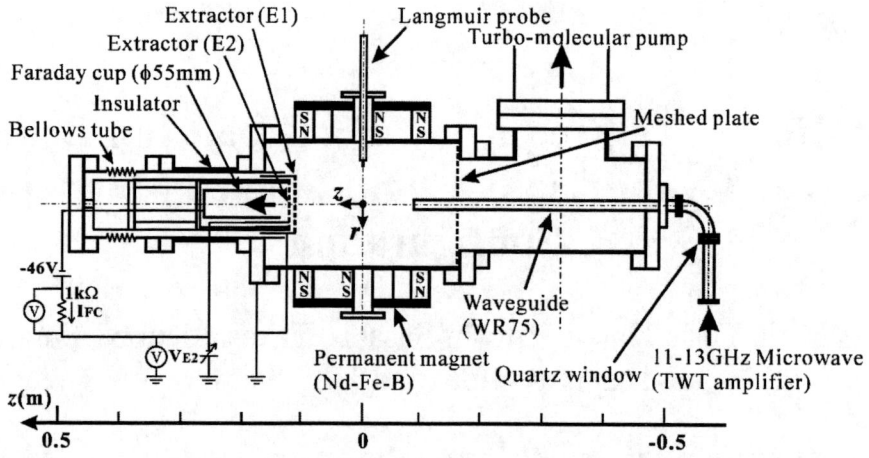

FIGURE 1. A schematic diagram of the ECR plasma/ion source and the setup of the extractor system.

to ion saturation current density.

The extractor consists of an electrode facing the plasma (E1) and an extraction electrode (E2). The electrodes are 80 mm diameter and have 21 apertures of 10 mm diameter. The gap distance d between E1 and E2 electrodes can be set at 10, 15 and 20 mm. The position of the E1 electrode can be varied from $z = 100$ mm to 150 mm by using a bellows tube. The E1 electrode is the same electric potential as the discharge chamber, i.e. ground potential. The E2 electrode is insulated from the chamber; the maximum extraction voltage is 3 kV. The extracted ion-beam is measured by using a Faraday cup. The Faraday cup which is 55 mm diameter and 130 mm length is set at 15 mm apart from the E2 electrode. A part of experiment which has been used with the Faraday cup of 25mm diameter is shown by an arbitrary unit. The measurements with the Langmuir probe and the Faraday cup have been conducted by using argon for operation gas. The ion saturation current density has been measured at the pressure of 0.03-1 Pa. The experiment of ion beam extraction is kept at the pressure of 0.1 Pa

RESULTS AND DISCUSSION

Measurement of Ion Saturation Current Density

We have measured the ion saturation current density in the ECR plasma source. The Langmuir probe is set at $r = 60$ mm and $z = 0$ mm, i.e. the position where we obtained the maximum plasma density [2]. Figure 2 shows the ion saturation current density as a function of microwave power. The input power of 11 and 13 GHz microwaves is varied between 100 and 350 W. The gas pressure is kept at 0.3 Pa. The ion density lineally increases with the microwave power within this power range. Figure 3 shows the ion saturation current density as a function of gas pressure. The pressure is varied between 0.03 and 1 Pa and the microwave power is set at 350 W.

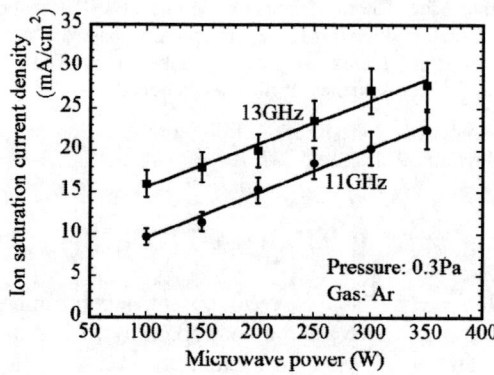

FIGURE 2. Ion saturation current density as a function of power of 11 and 13 GHz microwaves. The probe is set at $r = 60$ mm and $z = 0$ mm.

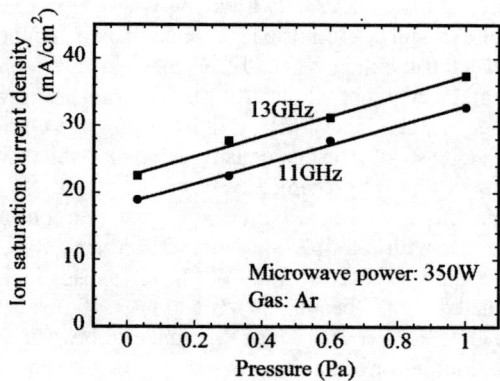

FIGURE 3. Ion saturation current density as a function of Ar gas pressure. The 11 and 13 GHz microwaves are 350 W. The probe is set at $r = 60$ mm and $z = 0$ mm.

The ion density also increases lineally. At this pressure range, the ion saturation current density using 13 GHz microwaves is about 1.15 times larger than that using 11 GHz. The ion density has typically reached 37.4 mA/cm^2. This plasma source can produce the same ion density as the previous plasma source which consumed the microwave power of several kW [3]. The electron density is about 2-4 eV.

We have examined a 2.45 GHz ECR plasma generated in this source. The maximum ion saturation current density was below 15 mA at the microwave power of 350 W although the experimental conditions were different from these conditions [4]. Therefore, the ion density in this source is obviously influenced by rising microwave frequency. We consider that the main effect of rising microwave frequency is the increase of the energy gain of electrons which are accelerated by ECR. The details were described in previous article [2].

Optimization of Ion Extraction

Figure 4(a) indicates the axial distribution of the magnetic-field intensity on z-axis. The distribution has the minimum value at z =100 mm; i.e. the vicinity of the end position of the permanent magnet. The maximum magnetic-field intensity is approximately 0.1 T at z = 0 mm. Figure 4(b) shows the axial distribution of the ion saturation current density on z-axis. The pressure was kept at 0.1 Pa. The microwave frequency and power are 11 GHz and 180 W, respectively. The peak position of the ion density coincides with the lowest position of the magnetic-field intensity. It is thought that the ions are trapped at the position by the gradient of the magnetic field. The dependence of the ion beam current I_{FC} through the Faraday cup on the extractor position is shown in Fig. 4(c). The extraction voltage V_{E2} is set at -1 kV. The peak of the beam current is observed at z = 140 mm; the position is slightly apart from the ion density peak. The extractor probably disturbs the plasma confinement by the magnetic field where that nears the plasma. The ion saturation current density is about 2 mA/cm^2 around z = 140 mm. Assuming that the ions of the density can be extracted through the all apertures, the beam current is 33 mA. The efficiency of the beam extraction can be estimated at about 12 % because the maximum beam current was 3.8 mA at the extraction voltage of -1 kV.

We have carried out the measurement of ion beam current at the higher voltages at the gap distance of 10 mm and the E1 electrode position of z = 140 mm. Figure 5 shows the ion beam current as a function of extraction voltage. The experimental conditions were the microwave frequency of 11 and 13 GHz, the microwave power of 180 and 300 W, and gas pressure of 0.1 Pa. In the case of 11 GHz and 180 W, the beam current was 12.3 mA at the extraction voltage of -2.6 kV. The efficiency of the beam extraction improves to 37 %. However, the beam current tends to saturate. In the case of 300 W, the beam current increases as the extraction voltage increases in accordance with the Child-Langmuir low [5]. The broken line in Fig. 5 indicates the proportion of $V_{E2}^{3/2}$. The results indicate

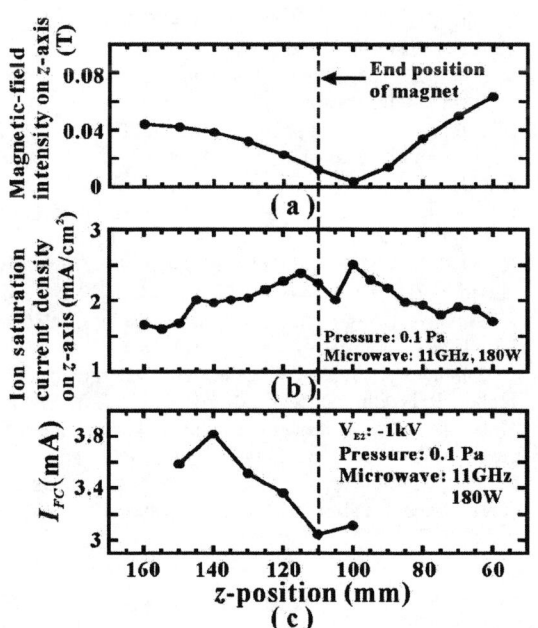

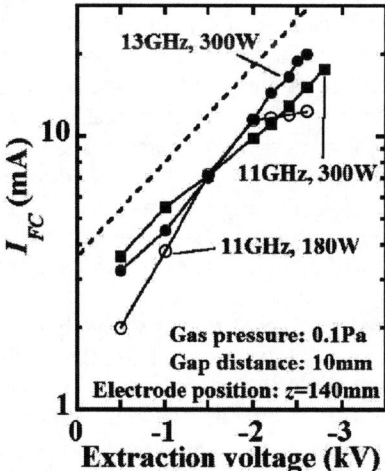

FIGURE 4. Comparison of (a) axial distribution of magnetic-field intensity on z-axis, (b) axial distribution of ion saturation current density on z-axis and (c) measurement of the ion beam current as a function of E1 electrode position.

FIGURE 5. Ion beam current as a function of extraction voltage at the Ar gas pressure of 0.1 Pa. The microwave conditions are the power of 180 and 300 W, and the frequency of 11 and 13 GHz.

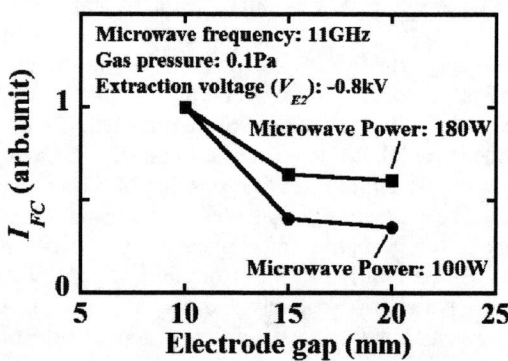

FIGURE 6. Ion beam current through the Faraday cup as a function of the gap distance between the electrodes at the Ar gas pressure of 0.1 Pa. The power of 11GHz microwaves is 100 and 180W.

that the ion beam is effectively extracted. The maximum ion beam current is 20 mA. At above -1.5kV, the beam current at 13 GHz is higher than that at 11 GHz.

Figure 6 shows the dependence of the ion beam current through the Faraday cup against the gap distance at the pressure of 0.1 Pa. The 11GHz microwave power was 100 and 180 W. The extraction voltage is -800 V. The beam current increases as the gap distance decreases. The efficiency of beam extraction is higher when electrode gap distance is nearly equal to the ion sheath [6]. Assuming that experimental conditions are the plasma density of 10^{17}-10^{18} m^{-3}, the electron temperatures of 2-4 eV and the negative extraction voltages of 1-3 kV, the ion sheath is estimated at 1.0-8.2 mm. However, we cannot set at $d < 10$ mm due to the experimental limitation. We will change the structure and furthermore investigate the efficiency of beam extraction less than $d = 10$ mm in detail for producing high-current broad ion beams.

The feature of our plasma source is that the comb-shaped magnetic field has good plasma confinement. However, the ion density around z-axis is lower than that of surrounding part ($r \geq 40$ mm) because the ECR zones are formed near the sidewall of the chamber and the plasma confinement is better at the surrounding part. We will optimize the extractor in order to extract ions from the surrounding part.

We have attempted to polymer etching by the ion beam. The processing is suitable for the applications of the ion source because polymers are weak against heat. Although the charge built-up on polymer surfaces probably becomes an issue in the process, the neutral beam etching is effective as the solution. We are also interested in fullerene beams as an application of other than polymer etching. Some groups recently have reported producing fullerene beams using ECR ion sources [7,8]. We expect that our plasma source is more suitable for trapping fullerene ions because that have larger diameter than the ECR ion sources.

CONCLUSION

We have examined our high-density ECR plasma source in order to apply to ion beam processing. The ion saturation current density at 13 GHz microwaves has reached 37.4 mA/cm^2. The density is about 1.15 times that at 11 GHz. As a next step, we have constructed the ion beam extractor which has multiple round apertures. The maximum ion beam current is 20 mA at the microwave power of only 300 W. The beam current depends on the ion saturation current density, at the extraction voltage above 1.5 kV. The effect of the rising microwave frequency is clear.

We will present the applications such as reactive ion etching of polymers and producing fullerene beams in near future.

ACKNOWLEDGMENTS

Authors would like to thank Mr. Kagawa and Mr. Satani of Osaka University for technical assistance.

REFERENCES

1. T. Asaji, H. Sasaki, Y. Kato, F. Sato, T. Iida and J. Saito, Rev. Sci. Insturm. 77, 03C104 1-3 (2006).
2. T. Asaji, Y. Kato, F. Sato, T. Iida and J. Saito, Rev. Sci. Insturm. (under review).
3. R. Hidaka, T. Yamaguchi, N. Hirotsu, T. Ohshima, M. Tanaka and Y. Kawai, Jpn. J. Appl. Phys. 32, 174-178 (1993).
4. Y. Kato, H. Sasaki, T. Asaji, T. Kubo, F. Sato and T. Iida, this conference.
5. R. Keller, *The Physics and Technology of Ion Sources*, edited by I. G. Brown, Wiley, New York, 1989, pp. 25.
6. J. Ishikawa, *Ion-gen Kougaku*, Ionics, Tokyo, 1986, pp. 178-179 (in Japanese).
7. L. Maunoury, J. U. Andersen, H. Cederquist, B. A. Huber, P. Hvelplund, R. Leroy, B. Manil, J. Y. Pacquet, U. V. Pedersen, J. Rangamma and S. Tomita, Rev. Sci. Instrum. 75, 1884-1887 (2004)
8. S. Biri, É. Fekete, A. Kitagawa, M. Muramatsu, A. Jánossy and J. Pálinkás, Rev. Sci. Insturm. 77, 03A314-1-3 (2006)

Profile and Angle Measurement System of SHX

Yuji Kikuchi, Mitsuaki Kabasawa, Mitsukuni Tsukihara and Michiro Sugitani

SEN Corporation
1501 Imazaike, Saijo, Ehime 799-1362 Japan

Abstract. To cope with the manufacturing processes for shrunk semiconductor devices, a precise implant angle control is required for the latest generation of ion implanters. Various ideas are incorporated into the SHX, a single wafer type high current ion implanter developed by SEN Corporation, to meet the requirements not only with a newly designed beam line but also with an accurate angle monitoring system.

In the SHX, an ion beam is transported to the electrostatic beam scanning system after a mass analysis. The scanned beam passes through Parallel Lens to be arranged in the parallel direction. Next, the beam is bent vertically by the energy filter and reaches the wafer platen, finally. The beam profile measurement system, Beam Profiler, is positioned on the same plane as the wafer.

The Beam Profiler can measure horizontal uniformity of the scanned beam current. Using the Divergence Mask, information about the horizontal beam parallelism at the wafer position also can be acquired. In addition, 2-dimensional profile measurement mechanism is equipped on the same Beam Profiler. A tilt angle can be corrected with vertical angle information from the 2-D profile.

Data from the SHX about a relation between a horizontal beam profile and a sheet resistance distribution within a wafer, which proves the reliability of the Beam Profiler will be described. Well-controlled parallelism even for low energy ion beams is also presented.

Keywords: Angle Control, Ion Implantation
PACS: 61.72.Tt, 85.40.Ry

INTRODUCTION

Multi-wafer ion implanters with a spinning disk have been a main stream for almost 30 years as high-dose application tool because of their high throughput capability. Recently, advanced devices are revealing the system limitation in two points of view. The first one is a particle attack, which means a collision between a particle generated in the system and rotating wafers at a high speed. This collision causes pattern damages[1]. This interaction can be reduced significantly or eliminated with a control of the disk rotation speed[2].

The other point is angle variation. The zero-degree angle implant with a multi-wafer type implanter inevitably has different implant angles between the center and edge of a wafer[3]. This angle variation causes asymmetric device characteristics in source-drain behavior[4].

As a countermeasure to overcome the discrepancies, several types of single-wafer high-current implanters were developed[5-7]. For the particle attack, this single-wafer approach could be a very effective solution because the wafer motion is much slower than the multi-wafer type. On the other hand, the angle control capability on the single-wafer type is not so superior to the multi-wafer type because it strongly depends on a beam line design, and then an angle measurement becomes very important to ensure the exact implant angle.

The SHX, a single-wafer high-current implanter developed by SEN Corporation, has a carefully and precisely designed beam line, but still has errors in the implant angle, especially for low energy beams such as B+ sub-keV, because of limitation of space-charge controllability. In order to ensure accuracy of the implant angle as well as dose uniformity, the SHX is equipped with a newly developed beam monitor system.

In this paper, capability of the measurement system is described and the angle controllability and accuracy of the SHX is also reported.

MEASUREMENT SYSTEM

The SHX has an electrostatic scanning system after mass analysis. Using Parallel Lens, the scanned beam is arranged in the parallel direction. And then, the beam flux is transferred to a wafer platen through Angular Energy Filter (AEF) as described in figure 1.

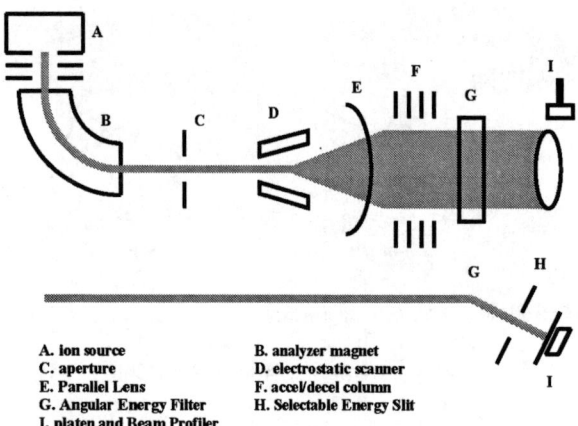

A. ion source B. analyzer magnet
C. aperture D. electrostatic scanner
E. Parallel Lens F. accel/decel column
G. Angular Energy Filter H. Selectable Energy Slit
I. platen and Beam Profiler

FIGURE 1. Beam line of the SHX.

In such a beam system, it is important to ensure the beam parallelism as well as the beam current uniformity in the horizonal direction. The newly developed beam measurement system for the SHX was designed to measure both the beam current uniformity and implant angle in sufficient accuracy.

The system is mainly composed of two parts; Selectable Energy Slit (SES) and Beam Profiler. SES has two sets of rotators with 4 fins as shown in figure 3. It is located just after AEF and can control the beam height for 3 kinds of ions, such as Boron, Phosphorus and Arsenic. This is called "Slit Mode". A pair of the 4th fins form 3 vertical slits when it is fully closed. It is called Divergence Mask and positioned in 260 mm upstream of the wafer platen, and the 3 slits are of 50 mm height and 5 mm width at intervals of 100 mm.

Beam Profiler has two monitors; one is a simple Faraday cup (Profiler Cup) and the other is multiple Faraday cup (Vertical Cup). Profiler Cup consists of a 100 mm height and 10 mm width slit on a front cover and a Faraday cup in the unit. Vertical Cup consists of 20 small holes and 20 small Faraday cups. They are aligned in 2 lines in a 180 degree different phase and can measure the charges individually. The combination is shown in figure 3.

Beam Profiler is positioned to trace the same plane of a wafer to be implanted in order to eliminate an error coming from the imperfect parallelism. Running the area where a wafer is exposed to the beam flux, the Faraday cups count ion charges. The unit travels horizontally at a constant speed during measurement, awaiting at a side of the platen during implantation.

A. BEAM PROFILING IN THE HORIZONTAL DIRECTION

In the SHX, the beam current uniformity in the horizontal direction is strictly required because the uniformity of implantation is strongly and directly reflected by the current uniformity. The beam flux is sampled 256 times while Beam Profiler is running 310 mm path. This means digital data is sampled by a 1.2mm interval.

When SES is in Slit Mode, this measurement can obtain the horizontal uniformity.

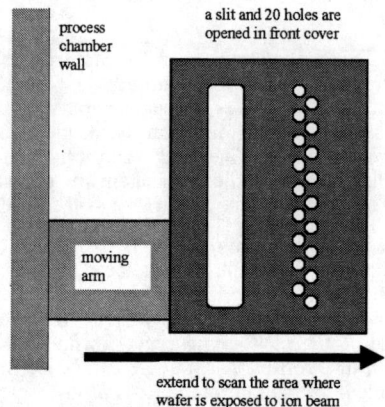

FIGURE 3. The front cover of Beam Profiler has one long slit and 20 small holes.

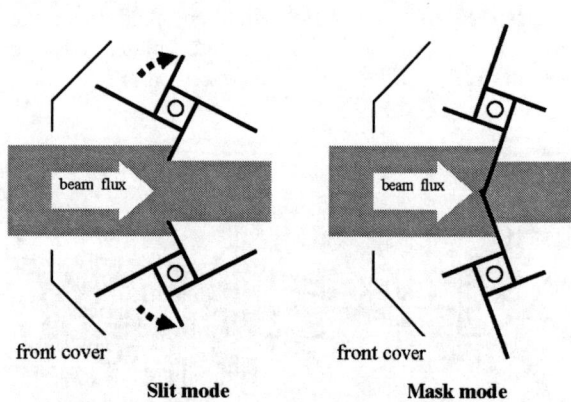

FIGURE 2. Selectable Energy Slit consists of 2 rotators with 4 fins each. Each pair of 2 fins form a horizontal slit. The 4th fins form Divergence Mask when it is fully closed.

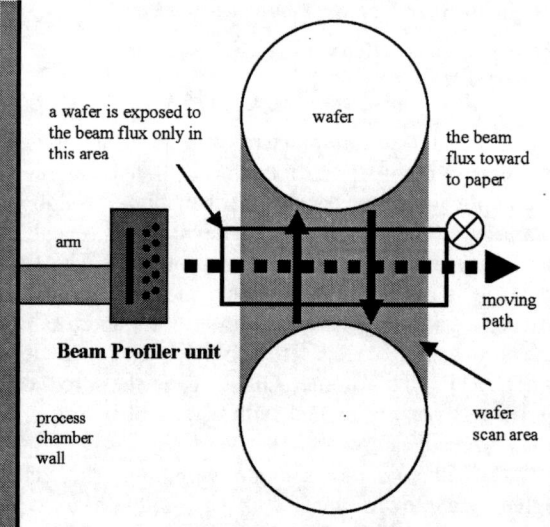

FIGURE 4. Location of Beam Profiler. It is just on the same plane where a wafer is to be implanted.

When SES is in Divergence Mask Mode, this Profile measurement can acquire information on the beam parallelism and beam divergence. The beam direction is estimated from the following equation,

$$\text{beam angle} = \arctan\left(\frac{d}{L}\right), \quad (1)$$

where d is the difference between the peak position and the corresponding slit position and L is a distance between Divergence Mask and a wafer (260 mm). The total measurement error can be derived from
(1) error of measurement position
   = ±0.6mm (half of measurement interval),
(2) accuracy of vertical slit width = ±0.1mm
(3) effects of slit width of SES: negligible
(4) effects of slit width of Beam Profiler: negligible
The estimation of the error of the peak position is

$$\sigma^2 = \sigma_{pos}^2 + \sigma_{slitwidth}^2$$
$$= (0.6)^2 + (0.1)^2 \quad (2)$$
$$\sigma = \pm 0.6 (mm).$$

It corresponds to ±0.1 degree implant angle in the horizon direction from equation (1). The parallelism in the SHX is calculated with this accuracy.

B. 2-DIMENSIONAL BEAM PROFILING

The parallelism discussed above only means an accuracy of the implant angle in the horizontal direction. Vertical charge distribution of the unscanned beam flux is measured by Vertical Cup continuously for 64 times for the same range as the horizontal direction with Slit Mode of SES. After the measurement, a 2-dimensional beam profile image is obtained as well as information of the center position of the beam.

Similar to the horizontal direction, the vertical implant angle can be described as

$$\text{vertical angle} = \arctan\left(\frac{d'}{L_{AEFcenter}}\right), \quad (3)$$

where d' is the peak position on the vertical axis and $L_{AEFcenter}$ is a distance between the center of Angular Energy Filter unit and a wafer (720 mm).

Size of each hole in front of Faraday cup has a radius of 5mm, and an angle error of measurement can be estimated to ±0.20 degree through equation (3).

EXPERIMENTS

Some examples obtained from the SHX using the monitoring system are presented below.

A. RELIABILITY OF BEAM PROFILER

Figure 5 shows a beam profile of the B+ 2keV beam. A standard deviation of the profile is 0.44 %. A sample wafer was implanted at a dose of 1E15 in this condition and annealed at 1,150 degree for 30 seconds. Then, a standard deviation of sheet resistance was 0.42%. This good correspondence means the measured profile well reflects the beam current distribution.

B. BEAM PARALLELISM

Figure 6 shows the beam profile of the B+ 0.5keV, 1.2mA through Divergence Mask. The 3 peaks passing through the 3 slits are observed. The result of calculated parallelism is shown in table 1 and figure 7. It can be said that the beam angles in the horizon direction are well controlled.

And as you see at Table.1, the beam divergence angle is big in low energy because of the space-charge. Where usual beam divergence at the medium energy from 10 to 50 keV is < 0.5 degree. These values are computed as FWHM of each peak.

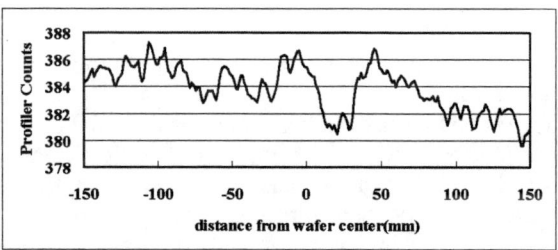

FIGURE 5. Beam Profile of the B+ 2keV beam of 4mA. A standard deviation of the profile is 0.44%.

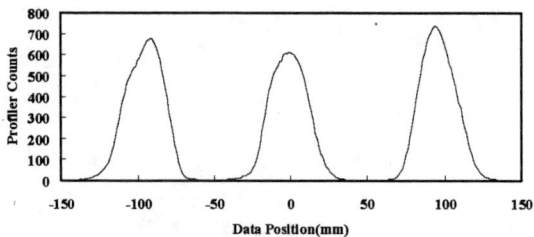

FIGURE 6. Profile data of B+0.5keV 1.2mA beam through Divergence Mask.

TABLE 1. Estimated parallelism of B+ 0.5keV beam. A positive value means the beam is turning left from the beam advancing direction.

position in wafer	difference from the slit pos.(mm)	parallelism (degree)	beam divergence (degree)
left	0.58 ± 0.6	0.13 ± 0.1	2.7
center	-0.40 ± 0.6	-0.09 ± 0.1	2.7
right	0.59 ± 0.6	0.13 ± 0.1	2.5

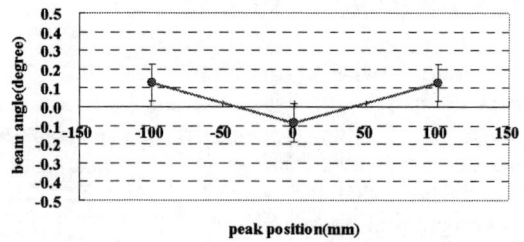

FIGURE 7. The result of calculated parallelism. The beam angles in the horizontal direction are well controlled.

C. CENTER OF BEAM IN VERTICAL DIRECTION

Figure 8 shows a example of the 2-D profile. This is a result for the B+ beam 15keV and 4mA. The longer axis means the horizontal axis and the shorter one means vertical one. A center of the beam is calculated after measurement. In this example, the center in the vertical direction is 4.1 ± 2.5mm from beam line center. The beam angle differs 0.33 degree from the designed beam plane. This information can be used for correction of the tilt angle.

If the correction works effectively, implant angle in vertical direction is estimated with an accuracy of ± 0.2 degree.

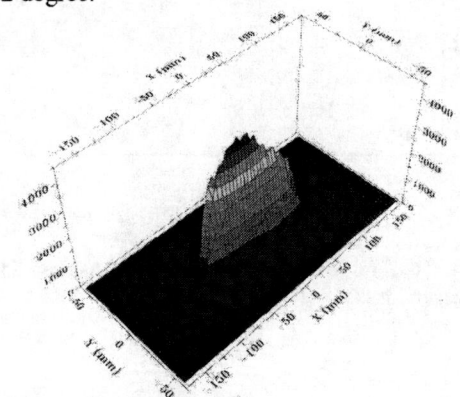

FIGURE 8. An example of the 2-D Profile in case of B+ 15keV beam.

SUMMARY

The new beam monitoring system has been developed for the single-wafer high-current implanter, the SHX. The system showed good capability to measure both the beam uniformity and angle integrity in the horizontal and vertical direction.

Using this system, the beam quality of the SHX has been proven to be suitable for use of 65nm and 45nm device processes.

For more advanced devices the system should be improved, having more precision about a half, maybe.

ACKNOWLEDGEMENTS

The authors would like to thank Mr. Kazumori Nishibara and Mr. Takashi Murakami for their vigorous work on data taking. They also would like to thank Mr. Yoshito Fujii, Mr. Yoshitaka Amano, Mr. Takashi Kuroda, and Mr. Masaki Ishikawa for excellent design of mechanism and software.

Finally, they would like to express a deep appreciation to Dr. Genshu Fuse and Dr. Shiro Ninomiya for their helpful discussion.

REFERENCES

1) Y. Kawasaki, K. Tokunaga, K.Horita, K.Mitsuda, A. Yamaguchi, A. Ueno, A. Teratani, T. Katayama, K. Hayami, M. Togawa, Y. Ohno and M. Yoneda, IWJT2004, A 1.6pp. 39-41
2) E.Ohga, H.Izutani, G.Fuse and M.Sugitani, to be presented in IIT 2006
3) A.M.Ray, J.P.Dykstra and R.B.Simonton, IIT Conf.Proc., 1992,pp.401-404
4) K. Yoneda, and M. Niwayama, IWJT2002, S2-3 pp.
5) G.Mezack, T.Callahan, S.Mehta, U.jeong, IIT Conf.Proc.,2000,pp431-434
6) A.Murrell, D.Hacker, P.Edwards, R.Mitchell, P.Banks, M.Foad, B.Harrison, G.Ryding, T.Smick, M.Farley, T.Sakase, Proc.IIT2004,PartII,pp 20-24
7) S. Patrick, to be presented in IIT2006

Implant Angle Deviation Reduction in Batch-Type High Energy Implanter

Noriyuki Suetsugu, Tatsuya Yamada, Mitsukuni Tsukihara and Michiro Sugitani

SEN Corporation
Saijo, Ehime, 799-1362 Japan

Abstract. Angle deviations at zero-degree implants are very large with an ordinary disk, which has a cone angle of 5 degrees. Dopant depth profiles are different within a wafer due to the angle deviations since the channeling at zero-degree is very steep. To achieve more uniform dopant depth profiles at zero-degree implants, the NV-GSD-HE3/RD (RD), which has a process disk with a small cone angle of 1.5 degrees, was developed. Using the reduced cone angle disk, the angle deviations at zero-degree implants are reduced to 1/3 of those with the ordinary disk, improving drastically the uniformity of dopant depth profiles within a wafer.

Keywords: Implant angle deviation
PACS: 61.72 Tt

INTRODUCTION

Shrinkage of device sizes in advanced LSI's makes a large tilt angle in ion implants an issue, especially in well formation with high energy implantations, because it becomes difficult to shrink the isolation width owing to skew of the well boundary[1]. In addition, since the shadowing becomes more serious as the device size shrinks and aspect ratio of the device pattern increases, the low tilt angle down to 0 degree is frequently used in ion implantations. In such low tilt angle implantations channeling effect becomes a problem, especially in high energy implantations. The channeling degrades dopant profile uniformity within a wafer.

A batch type implanter having a spinning disk with a cone angle is usually used for high energy implantations. It is pointed out that the optimization of the disk inclination relative to the beam is important for the reduction of angle deviations in a 200mm wafer and that the usage of 2 degree cone angle disk becomes more effective for the angle control at low angle implants[2]. As the angle deviation becomes more serious in implantations into 300mm wafers, the precise analysis and finding of effective methods are very important to decrease the deviations in the wafers.

In this paper, the methods of minimizing these angle deviations in a batch type high energy implanter for 300mm wafer; NV-GSD-HE3 (HE3)[3] are introduced with precise analysis and confirmed to be effective by measuring SIMS impurity profiles. An adjusting function which is installed in HE3 is confirmed to be useful to correct the angle error of an incident beam.

I. ANALYSIS OF THE IMPLANT ANGLE DEVIATION

Angle deviations occur more or less in all batch type implanters and in all implant angle conditions[4]. It is thought to be important to clarify the mechanism of these angle deviations and to find a measure to minimize them. In this section, the angle deviation will be explained for HE3.

I-1. Mechanism of the Implant Angle Deviation

HE3 has a two-axis gyro disk system for the implant angle control[5]. Fig. 1 shows the definitions of tilt and twist angle relative to the notch position of a wafer and the direction of rotation angles of α and β in the gyro system. HE3 has a disk with 5 degrees cone angle ordinarily. Tilt angle can be set arbitrarily from zero-degree to 7 degrees by controlling the angles α and β. The relation between the angles α and β and the implant angles tilt (γ), twist (θ) is expressed by the equations (1) for the cone angle ψ.

$$\gamma = \cos^{-1}(\cos(\alpha-\psi)\cos\beta\cos\psi + \sin(\alpha-\psi)\sin\psi)$$
$$\theta = \tan^{-1}\left[\frac{\cos(\alpha-\psi)\sin\beta}{\cos(\alpha-\psi)\cos\beta\sin\psi + \sin(\alpha-\psi)\cos\psi}\right] \quad (1)$$

When the disk rotates by an angle φ, the relation is changed to the equations (2).

$$\gamma = \cos^{-1}\left[\cos(\alpha-\psi)\cos\beta\cos\psi + \cos(\alpha-\psi)\sin\beta\sin\psi\sin\varphi - \sin(\alpha-\psi)\sin\psi\cos\varphi\right]$$
$$\theta = \tan^{-1}\left[\frac{\cos(\alpha-\psi)\sin\beta\cos\varphi + \sin(\alpha-\psi)\sin\varphi}{\cos(\alpha-\psi)\cos\beta\sin\psi - \cos(\alpha-\psi)\sin\beta\cos\psi\sin\varphi + \sin(\alpha-\psi)\cos\psi\cos\varphi}\right] \quad (2)$$

The differences of γ and θ between equations (1) and (2) correspond to the angle deviations Δγ and Δθ when the disk rotates by an angle φ. Maximum angle deviation within a wafer occurs through the disk rotation by the angle of wafer diameter.

※Twist angle deviation for zero-degree implant cannot be defined.

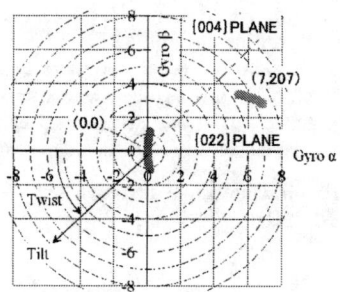

Fig.2 The angle deviations; γ and θ in GSD system with the 5 degrees cone angle disk. Radial and azimuthal directions indicate tilt and twist, respectively.

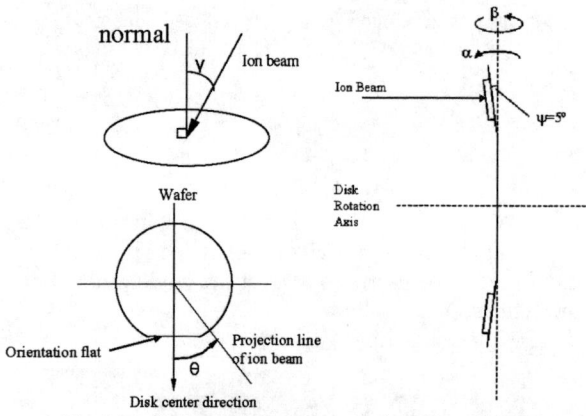

Fig.1 Definition of implant angles tilt γ, twist θ and inclinations α, β in GSD gyro system.

Table 1 shows the calculated angle deviations for the two settings of α and β which correspond to the tilt angles of 0 degree and 7 degrees. In the both settings, tilt deviations are more than 1 degree. Figure 2 shows two dimensional expression of the angle deviations in the implant at tilt 0 and tilt 7 degrees in HE3.

The magnitude of channeling effect is very sensitive to an angle deviation, especially, in the case of zero-degree implant, because this implant angle is involved in {022} and {004} channel planes. Suppression of the angle deviation is very important to achieve uniform dopant profiles in low tilt angle implants.

Table 1. Angle deviations at tilt 0/7 deg implant

Tilt	0	7
Twist	-	207
gyro α	0	6.3
gyro β	0	3.2
Δ tilt	1.1	1.02
Δ twist	※180	9.5

I-2. The Optimization of the Wafer Load Angle

The implant with the tilt angle of 7 degrees and twist angle of 27 degrees are generally used as an effective setting to suppress the channeling effect in high tilt implants. In the case of HE3, a gyro setting at α=-6.3deg and β=-3.2deg corresponds to the above implant angles. But this setting causes very large twist deviation of more than 40 degrees as shown in Fig.3, and the deviation will result in certain channeling effects within a wafer.

Table 2. Angle deviations for the implant at tilt angle of 0, 7 degrees and optimized notch angle.

Tilt	0	7	7	7	7
Twist	-	207	207	27	27
gyro α	0	6.3	7	-6.3	7
gyro β	0	3.2	0	-3.2	0
Load Angle	0	0	207	0	207
Δ tilt	1.1	1.02	0.04	0.97	0.04
Δ twist	※180	9.5		41.6	7.37

Even in such a condition, the twist angle deviation can be suppressed by optimizing the wafer notch angle in wafer loading. As shown in the Table 2 and Fig. 3, the optimization of notch angle reduces the tilt angle deviations and twist angle deviation effectively to less than 1 degree and less than 10 degree, respectively, for the high tilt implant at 7 degrees.

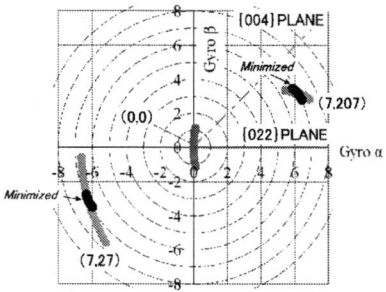

Fig.3 The angle deviations; γ and θ for minimized angle deviations at the settings γ=7, θ=207 and γ=7, θ=27 by optimization of the wafer notch angle on the 5 degrees cone angle disk.

The effect of reducing the angle deviation is remarkable for the optimization of the wafer–loading angle in high tilt angle implant, but not in low tilt implant.

I-3. Reduction of the Cone Angle

When the two axis gyro disk system is used, generally the implant angle deviation becomes small under the gyro conditions below;
Gyro α : close to disk cone angle,
Gyro β : close to zero-degree.
As the setting of angles α and β must be low for the implant with low tilt angle, the employment of a disk of small cone angle is able to suppress the angle deviations even for low tilt implants. Table 3 and Fig. 4 show the results of the calculations adopting the low cone angle disk of 1.5 degrees. The deviation in the low tilt angle implant is shown by A in Fig 4. The adoption of a low cone angle (1.5 degree) disk achieves the small angle deviation at the setting less than 3 degrees within a wafer in the implant with low tilt angle. Tilt angle deviation can be suppressed, especially, to 1/3 compared to that of the 5 degree cone angle disk as shown in Table 3.

Table3. Angle deviations for the implant at tilt angle of 0, 7 degrees and 7, 207 degrees with the 1.5 degree cone angle disk

Cone Angle	5		1.5	
Tilt	0	7	0	7
Twist	-	207	-	207
gyro α	0	6.3	0	6.3
gyro β	0	3.2	0	3.2
Load Angle	0	0	0	0
Δ tilt	1.1	1.02	0.34	0.3
Δ twist	※180	9.5	※180	20.95

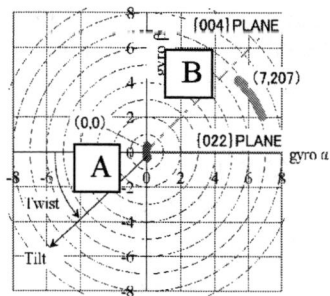

Fig.4 The angle deviations; γ and θ for the implant at the settings γ=0, θ=0 and γ=7, θ=207 on the 1.5 degree cone angle disk.

II. CHANNELING RESULTS AND DISCUSSION

II-1. Tilt 0 Degree Implant with Standard HE3

Dopant profiles implanted with the 5 degrees cone angle disk at zero degree tilt angle vary widely with positions in a wafer as shown in Fig. 5. This measured results verify the calculated result that the angle deviation can not be reduced less than to 1 degree in the implant with 5 degrees cone angle disk at low tilt angles.

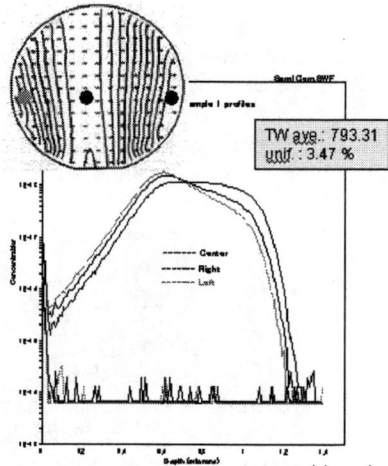

Fig.5 Dose uniformity in a wafer measured by Therma-Wave after zero-degree implant with 5 degrees cone angle disk and SIMS profiles at the points indicated in the map.

II-2. Tilt 0 Degree Implant with "RD"

Effect of the reduced cone angle disk shown in table 3 is evaluated in zero-degree tilt angle implants using NV-GSD-HE3/RD which is equipped with a 1.5 degrees cone angle disk. The dopant profiles measured with SIMS of very high uniformity are obtained as shown in Fig. 6. This result agrees well with that of the analysis.

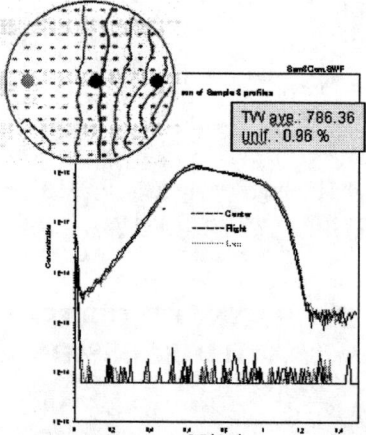

Fig.6 Dose uniformity in a wafer measured by Therma-Wave after zero-degree implant with RD and SIMS profiles at the points indicated in the map.

II-3. Channeling Risk of the HE3/RD

In the case of the HE3/RD, twist angle deviations for the high tilt implant become very large as can be seen by B in Fig.4. This results in very narrow angle margins to {004} and {022} channeling plane. When the disk with cone angle of 5 degree is used, the margin to the channeling for high tilt implants is large enough with twist setting 207 degrees as shown in Fig.3. But in the case of RD, the channeling occurs occasionally on the left or right edge of the wafer due to the errors in beam incident angles and the surface orientation of the wafer as shown in Fig.7.

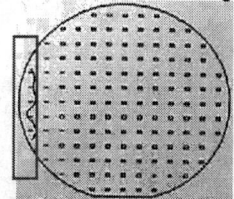

Fig.7 {004}channeling in phosphorous implant at 150 keV to a dose of 5×10^{12} cm^{-2} with the beam current of 100 µA at the settings $\gamma=7$, $\theta=207$.

The angle deviation by optimization of the angle of notch position is suppressed even in the implant using RD, but suppression effect is not enough to avoid perfectly the channeling due to the angle errors caused by aforementioned beam incident angle and surface orientation.

HE3 system is equipped with a unique function that can adjust implant angles. The beam incident angle is derived from the measured beam position, and the correct angle is set automatically. Figure 8 shows an example of the effect of this adjusting function. A wafer setting in which correct beam generates a channeling in the central area of the wafer without the adjusting function was used. Without the angle adjusting function, channeling occurred on the right edge because of the beam angle error. The adjusting function corrected the channeling position to nearly the center of the wafer. Beam angle error converted into twist angle is suppressed from 4 to 1 degree.

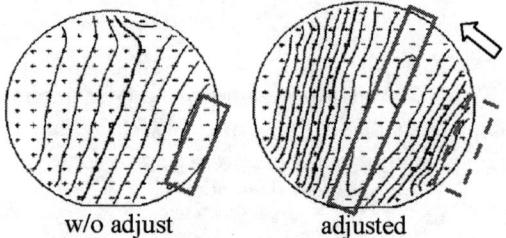

w/o adjust   adjusted

Fig.8 Effect of implant angle adjusting function.

However, the accuracy of the surface orientation of the wafer is also accountable for the angle mismatch between beam and surface orientation. It is important to use wafers having precisely controlled surface orientations for channeling control in advanced LSI's.

III. CONCLUSION

Origins and mechanisms of the angle deviations in a batch type high energy implanter: NV-GSD-HE3 were addressed precisely. It is clarified that the following measures are very effective for suppression of the angle deviation.

- Optimization of the wafer loading angle.
  Effective in high tilt implant
- Reduction of the cone angle.
  Effective, especially, in low tilt implant including zero-degree.

Using the RD, the angle deviation at zero-degree implant is reduced to 1/3 of that with the ordinary disk. Consequently, uniformity of the dopant depth profiles within a wafer is improved successfully.

REFERENCES

1. T. Yamashita et al., Jpn. J. Appl. Phys. 41, pp.2399 (2002).
2. Y. Kawasaki et al., IWJT 2002, S2-2, pp.15 (2002).
3. M. Sugitani et al., Proc. of the 12th IIT, pp.192-195, June 1998.
4. Andy M. Ray and Jerald P. Dykstra, Nucl. Instr. and Meth. B55 (1991) 488.
5. T. Tamai et al., Nucl. Instr. and Meth. B55 (1991) 408.

Stencil mask Ion Implantation Technology for realistic approach to wafer process

Kazuhiko Tonari, Tsutomu Nishihashi, Michio Ishikawa, Junki Fujiyama

Ulvac,Inc.,
1220-1 Suyama, Susono, Shizuoka, 210-1231, Japan
kazuhiko_tonari@ulvac.com

Abstract. We are researching and developing the stencil mask ion implanter (SLIM), which enables direct implantation on wafer through a stencil mask. This technology eliminates photo resist lithography processes for ion implantation. The basic performances of SLIM as an ion implanter were the same level as those of conventional ion implanters. "SLIM process" has an effect of shortening the process time to 1/4. We present the solution to magnification alignment error, and also introduce three examples of applying this process: reduction of the threshold voltage errors, formation of ring ion implantation area and formation of isolated diffusion layer.

Keywords: Stencil mask, ion implanter, stencil mask ion implanter, agile-fab, mini-fab
PACS: 80.40.Ry

INTRODUCTION

Applications of semiconductor devices are increasing in recent years. These include, for example, DVD player, digital camera and flat panel television. The hybrid vehicle and the electric vehicle also need a lot of semiconductor devices. These demands necessitate quick and flexible manufacturing method of semiconductor devices, and also reduction of cost. One of the answers to these demands is mini-fab or agile-fab [1]. This requires semiconductor-fabrication equipments to be multitask and multifunction. To meet the requirement, we are studying and developing the Stencil mask Lithographic Ion Implanter (SLIM), which enables direct implantation on wafer through a stencil mask and, thus eliminating photo resist lithography processes for ion implantation [2].

In a few years, we are planning to apply this new ion implantation technology to power semiconductor devices.

Two technologies are key points: 1) keeping the error of overlay less than 500nm (the allowable error depends on an applied ion implantation processes), 2) formation of the ring ion implantation area with a single stencil mask.

In this paper, we will review the basic performance of SLIM: ion implantation uniformity and repeatability, performance of ion beam angle controls, comparison of raw process time between the SLIM process and conventional ion implantation process, metal contamination and particle contamination. In addition, we will present the result of study on alignment between wafer pattern and mask pattern. In the applications section, we will give three examples of applying this process: reduction of the threshold voltage errors, formation of ring ion implantation area and formation of isolated diffusion layer

I .BASIC PERFORMANCE

A. Implant Uniformity and Repeatability

SLIM has the in-situ ion beam intensity monitor, which controls the uniformity of ion implantation. Fig1 shows typical sheet resistance map in the area of $22 \times 24 mm^2$. The area was implanted for B^+, 40keV, $3.8E14 ions/cm^2$ and the map was measured after annealing 950°C 30minutes. Implant uniformity was 1sigma 0.29%, and repeatability was 1sigma 0.10%. Implant uniformity varies in the axis of scanning ion beam when using conventional ion implanters because the incident ion beam angle and the beam spot size varies in this axis. However, the implant uniformity of SLIM in a whole wafer is 0.10%, which is the implant repeatability between implantation areas of SLIM, because repeating the implantation in a square of 20-30mm forms the whole implanted wafer. This value has an advantage over the performance of conventional ion implanters.

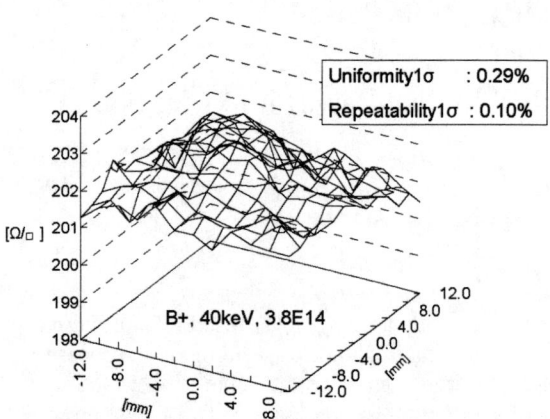

FIGURE 1. Typical sheet resistance map in the area of 22x24mm².

B. Beam Angle Control

SLIM has high beam parallelism using parallel beam monitor (PAM) because this process requires a highly accurate implantation pattern formation.

We measured a beam parallelism, which is calculated from two overlay errors obtained at different mask-wafer distances. In the implanted region, ion incident angle ranged from 0.01° to 0.07°, and the average was 0.03°. The average overlay error was 30nm calculating with a typical mask-wafer distance of 50um.

Figure2 shows the examples of SIMS profiles for B^+, 100keV, 5E14ions/cm² measured by changing the incident beam angle from –0.2° to +0.2°. Incident beam angles were measured by PAM. Depth profiles vary as the incident angle because of channeling phenomenon [3]. According to this result, it is clear that SLIM can control the beam incident angle up to the order of 0.1°, and also reduce the depth profile error due to the channeling phenomenon, which cause sheet resistance error.

C. Particle Contamination

In SLIM process, not only wafer but also stencil mask undergoes collisions with ions. Table1 shows one of result to ensure the effects of these collisions on particle contamination for Ar^+, 100keV, 3E11(ions/cm²)/shot. Stepping and repeating 44shots ion implantation through stencil mask, a whole wafer area was implanted. Condition (A) is the result of the normal SLIM process. Condition (B) is the result repeated 8times of SLIM process to a same wafer. The result proves that the ion collisions with stencil mask and the movement of wafer-mask stage have dangerous effect on particle contamination. In addition, it proves that there were few particles falling from the beam line.

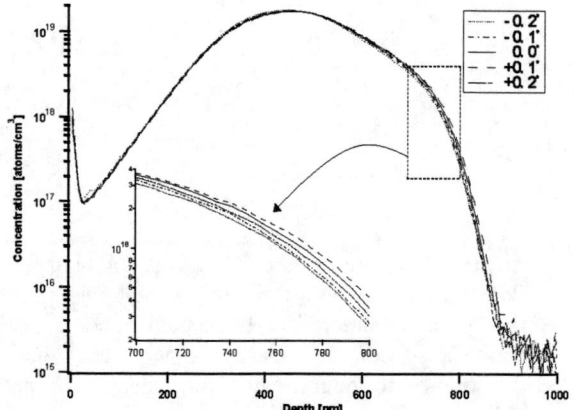

FIGURE 2. SIMS profiles for B+, 100keV, 5E14ions/cm², changing the incident beam angle from –0.2° to +0.2°.

TABLE 1. Particle Adders

Particle size um	Condition (A) 44shots	Condition (B) 44shots ×8times
0.12-0.13	1	1
0.13-0.14	0	0
0.14-0.15	1	1
0.15-0.16	1	0
0.16-0.17	0	1
0.17-0.20	0	0
0.20-0.30	1	1
0.30-0.50	0	0
0.50-0.70	0	2
0.70-1.00	1	1
SUM	5	7

D. Metal Contamination

Table2 shows metal contamination evaluated on 200mm wafers (edge erased 3mm) utilizing VPD-ICP-MS. The implants measured were BF_2^+, 40keV, and 5E14 ions/cm². Condition (A) is the result of metal contamination on wafer surface, and Condition (B) is the result on wafer backside after ion implantation.

These results show that the stage parts and stencil mask have no severe influence on metal contamination.

E. Raw Process Time

The reduction of raw process time is one of the advantages of SLIM process.

Figure3 shows estimations of typical raw process time. The estimation condition of SLIM process was implant time 0.2sec/shot, 50shots/wafer and 25wafers/lot. It is clear from this result that SLIM process does actually reduce the process time by 1/4 over conventional ion implantation and lithographic processes.

TABLE 2. Metal Contamination

	Control Wafer	Con. (A) Surface	Con. (B) Back Side
Al	9.5E+09	1.7E+10	5.2E+10
Fe	1.5E+10	1.2E+09	1.2E+10
Cr	-	-	5.1E+09
Pb	3.2E+07	3.3E+08	8.6E+07
Cu	1.2E+09	6.7E+08	1.7E+09
K	5.4E+09	-	2.0E+10
Mo	-	-	-
Ni	3.4E+08	-	8.5E+08

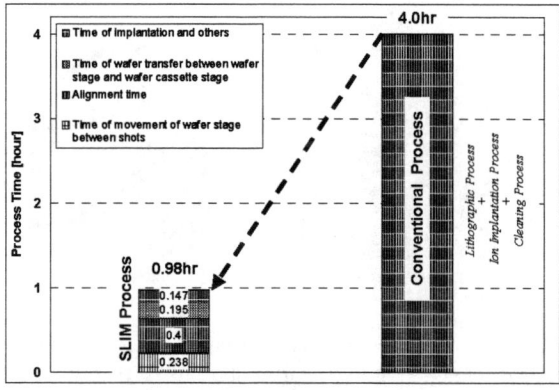

FIGURE 3. Comparison of raw process time between SLIM process and conventional process. The SLIM process time was calculated on the condition: implant time 0.2sec/shot, 50shots / wafer and 25wafers/lot.

II. ALIGNMENT BETWEEN WAFER PATTER AND MASK PATTERN

Alignment is one of the keys for this technology. The origin of alignment errors may be classified as follows:

1) Mechanical alignment error is caused by misregistration of pattern centers and rotation angle between mask pattern and wafer pattern. The value of this error is 170nm (3sigma) at the corner of the implantation area in a square of 22x22mm^2.

2) Manufacturing process of stencil mask causes magnification pattern error because of stress adjustment of membrane.

3) Thermal deformation of mask membrane causes alignment error. To reduce the deformation, it is important to increase heat capacitance, strength of membrane and heat conductivity [4]. According to this reference, the heatproof mask will reduce this error to 0.5um at the ion beam power 60W, and will be able to apply SLIM process to power semiconductor devices.

Figure4 is results of an experiment to reduce the magnification pattern error by controlling temperature difference between mask and wafer. A magnification pattern error of 1ppm causes alignment misregistration 1nm at a position 1mm away from the pattern center. This result proves that both X-axis and Y-axis magnification pattern errors increase in proportion to the temperature difference between mask and wafer. This result also shows that the total of the alignment errors in the X-axis and the Y-axis can be minimized when the temperatures difference of the mask and the wafer is 2.7 degrees, and that the pattern error in 22mm implantation area is calculated to 62nm.

It has been proved that this technology can be applied to the semiconductor devices that pattern shift tolerance is above the total alignment error of SLIM, and this error is calculated to 0.23um-0.73um.

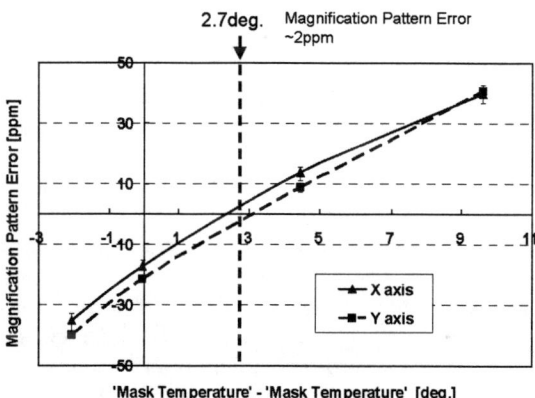

FIGURE 4. Magnification pattern error by controlling temperatures fluctuates between mask and wafer.

III. APPLICATIONS

A. Reduction of Threshold Voltage Errors

Some of semiconductor device manufacturing processes cause threshold voltage error of the concentric circle in a wafer. In many cases, chips with large threshold error are located in the outer part of wafer and it is possible to increase yield by correcting those errors. One of methods of correcting this error is to change dosage of channel ion implantation in outer part of wafer and in the other part. Figure5 shows the threshold voltage errors of MOSFET obtained by using this method. It turned out these methods produces approximately a half of the errors compared with conventional process. It means that this method is especially effective for the improvement of yield of the devices of which the errors of threshold voltage is severely managed.

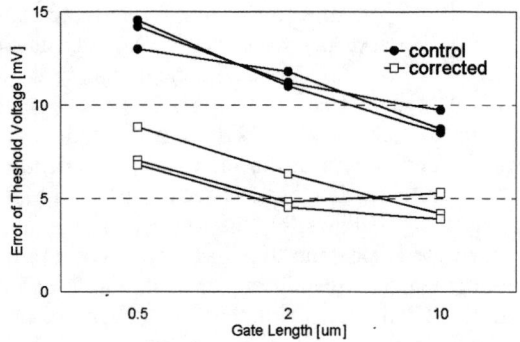

FIGURE 5. The threshold voltage errors of MOSFET corrected by SLIM process were approximately a half of the errors of control devices.

B. Formation of Ring Ion Implantation Area

The formation of ring ion implantation area is important to apply this process to power semiconductor devices. Using the divergence of ion beams, we tried to form the area without two complementary patterned masks [4]. Figure6 is a part of ring ion implantation area, which has 4 variations of open area ratio (20/40/60/80%). "Mask image" is the original formation of stencil mask. "SIMS image" is the image of ion implantation area, which was measured by surface boron secondary ion. The ion implantation condition was B^+, 50keV, $3E15/cm^2$, and the gap of 1mm between stencil mask and wafer.

C. Ion Implantation Profile

Figure7 shows the measurements by spread resistance after ion implantations. At the edge part of the mask (inside white dot rings), these two processes show a clear difference about the implantation areas. And also it is clear that ion implantation through stencil mask can make isolated diffusion layer, which is expected to improve the performance of semiconductor devices.

IV. CONCLUSION

It has been confirmed that the basic performance of SLIM as an ion implanter is in the same level as those of conventional ion implanters, and also has been shown that SLIM process has an advantage that it reduce the process time up to 1/4. We have shown solution on the magnification pattern error of alignment. In addition, as application of this process, reduction of threshold voltage errors, formation of ring ion implantation area and formation of isolated diffusion layer have been described.

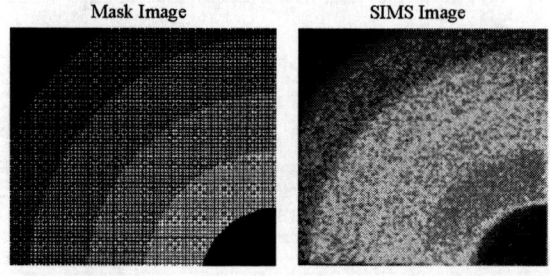

FIGURE 6. Part of ring ion implantation area, which has 4 variations of open area ratio (20/40/60/80%).

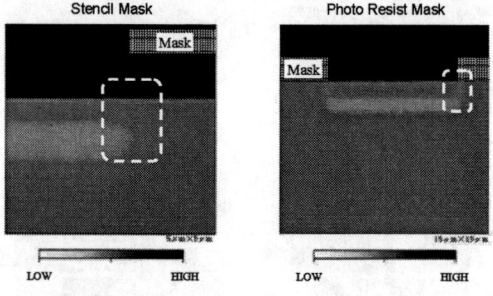

FIGURE 7. Ion implantation profile implanted through stencil mask was different from the profile of photo resist mask.

ACKNOWLEDGMENTS

The New Energy and Industrial Technology Development Organization (NEDO) supported this work. The authors would like to thank Toppan Printing Co., Ltd. for its cooperation in the manufacture of the masks, and Toyota Motor Corporation for valuable data.

REFERENCES

[1]. Y.Mikata et al, "Improvement of Mini-Fab Uptime by Using Multitask and Multifuncton Tools," *IEEE Trans. Semiconduct. Manufact.*, vol.17, NO.3, August 2004, pp.273-280
[2]. T.Nishihashi et al, "Lithography-less ion implantation technology for agile fab," *IEEE Trans. Semiconduct. Manufact.*, vol.15, NO.3, Nov. 2003, pp.464-469
[3]. T.Shibata et al, "The Control of Channeling Phenomenon," *IEEE Trans. Semiconduct. Manufact* vol.17, NO.3, August 2004, pp.299-304
[4]. T.Nishiwaki,et al, "Application of Stencil Mask Implantation Technology to Power Semiconductors" IIT 2006, Proceeding iit2006_0179_o1203

Enhanced Dosimetry For Single Wafer High-Current Implanters

J.T. Scheuer, J. Dzengeleski, D. Distaso, D. Timberlake, J. Cummings, and J.C. Olson

Semiconductor Equipment Associates, 35 Dory Road, Gloucester, MA 01930, USA

Abstract. Decreased energy of halo implants with tighter requirements on beam angle control have driven many lower dose (<1e15 cm^{-2}) implants from medium current tools to the VIISta high-current, single-wafer line of ion implanters. These implants demand enhanced dosimetry performance relative to that of previous high current implanters. We will present uniformity and repeatability data demonstrating improved dosimetry performance throughout the operating envelope of doses on the VIISta HC implanter.

Keywords: Ion Implantation, Dose Control.
PACS: 41.75.Ak; 61.85.+p; 85.40.Ry; 61.72.Tt

INTRODUCTION

Accurate and repeatable dose control is becoming increasingly important for high current ion implantation. The decrease in the energy of halo implants, and the relatively low throughput of medium current implanters at these energies, has resulted in the migration of halo implants to high-current implanters at some semiconductor manufacturers. Halo implant doses are below that of typical high-current applications (<1e15 cm^{-2}). This has resulted in an increased emphasis on dosimetry control for high-current implanters.

The VIISta HC implanter employs a ribbon beam and vertical wafer scan to achieve uniform dosing [1]. The beam current is monitored during the implant by a series of Faraday cups and the velocity of the wafer scan is adjusted on each wafer pass to account for any change in beam current.

Alternative high current implanter architectures rely on a stationary spot beam with a two dimensional wafer scan to achieve dose uniformity. In addition to the added mechanical complexity of the end station, this concept suffers from an inherent trade-off between wafer throughput and dose uniformity particularly at low doses.

In this paper we describe several improvements to the VIISta HC dosimetry system and present process wafer results for Rs uniformity and repeatability.

DOSIMETRY IMPROVEMENTS

Several VIISta HC dosimetry system improvements have been released to production tools while others are tested and nearing full release. In this section we will describe these improvements and their effects.

A key component of the VIISta HC dosimetry system is the Faraday system used to measure the beam current during implant. Several improvements have been made to the Faraday hardware as well as the beam current measurement software algorithm in order to improve the signal to noise ratio of the beam current measurement. Table 1 shows Rs repeatability and uniformity results for 200 and 300mm implants on the VIISta HC for a number of recipes. The 200mm data testing was performed using 2 monitors per run over 10 runs, while the 300mm data was obtained with 2 wafers per run over 5 runs. This data demonstrates excellent dosimetry performance following the improvements to the dose system.

TABLE 1. Rs repeatability and uniformity data for 200 and 300 mm implants on VIISta HC with improved Faraday system and dosimetry algorithm.

Recipe	200mm wafers		300mm wafers	
	Rs Rep. (%)	Rs unif. (%)	Rs Rep. (%)	Rs unif. (%)
B 40keV 1e15	0.27	0.71	0.22	0.55
B 35keV 1.5e14	0.17	0.52	0.12	0.65
P 20keV 1e14	0.22	0.42	0.39	0.68
As 2keV 3e15			0.22	0.75
B 500eV 1e15			0.43	0.70

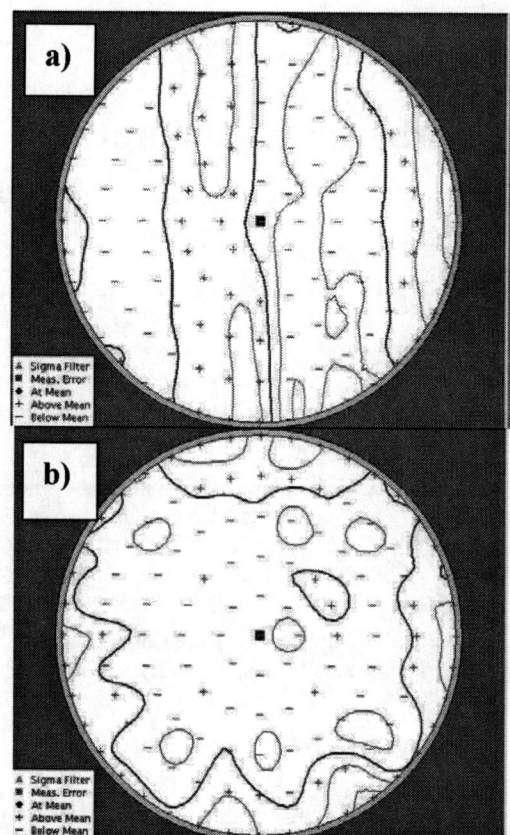

FIGURE 1. Rs wafer maps for boron 4keV, 1e15 cm^{-2} implant for a) single step implant and b) 4 step implant with 90 degree rotation between steps. Beam was intentionally mis-tuned to 1.75% to emphasize effect of rotation.

One component of the VIISta HC dosimetry system is the rotating platen, or roplat [2]. The roplat has several features. It is used to electrostatically clamp the wafer during implant and provide backside wafer cooling. The roplat also allows horizontal and vertical wafer tilt to align the wafer to the measured beam incident angle. The feature most critical to dosimetry is its ability to rotate the wafer between passes through the beam during implant. This feature allows the effect of minor beam non-uniformities to be reduced.

Fig. 1 shows the effect of wafer rotation between steps of a 4 keV, 1e15 cm^{-2} boron implant on the VIISta HC. For this test the beam uniformity was intentionally mis-tuned to 1.75%. Fig. 1a shows the Rs wafer map for a single step implant following anneal. The Rs uniformity for this wafer was 1.77%. Fig. 1b shows the Rs wafer map for a 4 step implant with 90 degree rotation between steps. The dose for each step was 0.25e15 cm^{-2}. As a result of the 4 step implant with rotation, the Rs uniformity was improved to 0.87%.

Fig. 2 shows the effect of the roplat for daily monitor implants run on a VIISta 80HP high current implanter in production at a customer site. For the first 90 implants, single step implants were performed with an average Rs uniformity of 1.7%. After installation of the roplat and use of 4 step implants with 90 degree rotation between steps, the average Rs uniformity improved to 0.7%.

An important aspect of accurate dose control is precise detection and compensation for beam current changes due to photoresist (PR) outgassing [3]. During high power implants into patterned photoresist wafers, the hydrogen released raises the pressure in the endstation and beam line. This elevated pressure increases the rate of charge exchange collisions which convert ions into fast neutrals.

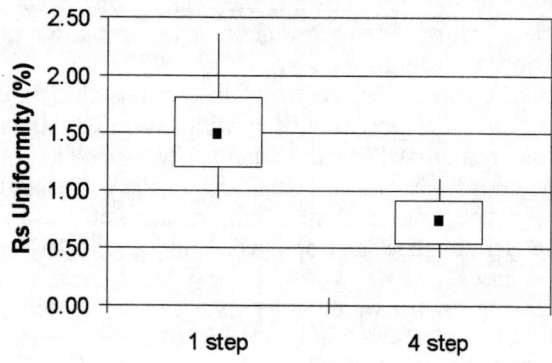

FIGURE 2. Rs uniformity for boron 10keV, 8.8e14 cm^{-2} single step (90 monitor wafers) and four step (14 monitor wafers) implants with 90 degree rotation between steps.

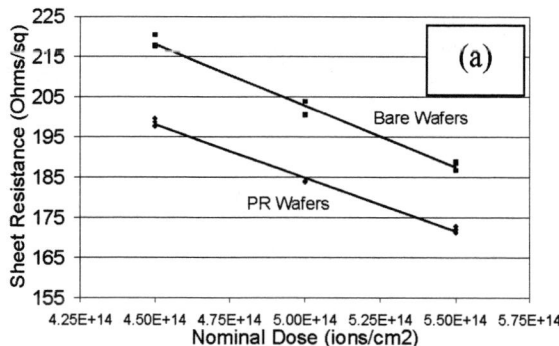

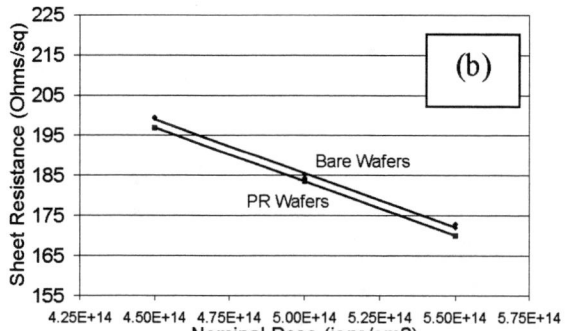

FIGURE 3. Average Rs for boron 80keV, 4.5-5.5e14 cm^{-2} implants into bare and photoresist wafers: (a) with dose compensation improvements disabled and (b) with dose compensation improvements enabled.

These neutrals contribute to the implanted dose, but are not detected by the Faraday system that measures beam current. If this effect is not properly compensated a dose shift can arise in comparing bare to PR wafers. This dose shift is an indication of the worst case for dose repeatability for varying PR coverage or varying PR condition.

Recent improvements to the VIISta HC Faraday system and dose control algorithm have greatly improved the response to PR outgassing. Fig. 2 compares average Rs results for a B 80keV implant into bare and 75% coverage, patterned PR wafers. Wafers of each type were implanted to target doses of 4.5, 5.0 and 5.5e14 atoms/cm^2. For the baseline case (improvements disabled) there is an average Rs offset of 38.3 ohms/square from bare to PR wafers which represents a 2.5% dose shift. With the improvements in dose hardware and software enabled, the Rs offset is improved to 2.1 ohms/square which equates to a 0.14% dose shift, well within the specification for dose repeatability.

Occasionally during implant a momentary interruption in beam current (glitch) will occur. This drop in beam current results in a dose non-uniformity which must be filled in. This fill in process presents a challenge when using a dc ribbon beam such as the VIISta HC when a subsequent glitch occurs during the fill in process [4]. This challenge has been overcome by employing a new, proprietary glitch recovery algorithm. Data obtained using this technique is shown in Table 2. Wafers were implanted with a B 35keV 1.5e14 cm^{-2} recipe. This implant is typically completed in six passes of the wafer through the beam. For the glitched wafers, the beam current was interrupted during the first wafer pass, and while the recovery algorithm occurred a second, third or fourth glitch was triggered. The implant was then allowed to complete normally.

The data in Table 2 shows excellent glitch recovery performance. The average Rs of 513.2 ohms/sq. for the three glitched wafers are within experimental error compared to the un-glitched case. The Rs standard deviations from the glitched wafers compare well with that of the un-glitched case and all are within the 0.8% specification for the VIISta HC.

The algorithm and hardware changes described above deliver improved wafer dose uniformity and repeatability. We have also designed a system for the VSEA HC implanter to perform a dosimetry self check to improve machine-to-machine dose matching and process integrity over time.

TABLE 2. Glitch test results. Rs results for un-glitched reference wafer and wafers with multiple glitches for a B 35keV 1.5e14 cm^{-2} 6 pass implant.

Wafer #	Notes	Mean	% Std Dev	Min	Max
1	Reference (not glitched)	512.8	0.622	505	520
2	2 glitches (1 fill-in)	513.5	0.688	505	523
3	3 glitches (2 fill-ins)	513	0.723	501	521
4	4 glitches (3 fill-ins)	513.1	0.742	501	520

The beam current measurement electronics has been redesigned to include precision current sources to exercise the I/V converter and ranging circuitry throughout its operating range. This 'self-test' will be executed automatically at opportue times, and maintain a running log of test results. These results will be used to verify the initial accuracy of the system as compared to a standard calibrated current source and against all other HC dose systems. This type of automated test has been operating on the VSEA MC tools for more than 2 years [5]. Results have shown long term stability of individual units of < 0.25%, and matching between units (machines) of <0.5%.

In addition to automatic checking of the electronics, the machines will periodically check the integrity of the dose Faradays. This is accomplished by re-directing the dose controller internal current source out through the Faraday cabling harness to the cups, and through a separate connection back to the dose controller I/V converter. The results of this current measurement will be compared against the I/V results to insure that there are no current leakage paths that may corrupt a critical dose control measurement. This system will also check the integrity of the cabling, electrical connections and Faraday selection relays on every cycle. The recorded data can be monitor and used for SPC control, and as a troubleshooting diagnostic tool to help eliminate the possibility that a Faraday cup has caused a problem.

CONCLUSIONS

The migration of halo implants to high current ion implanters has increased the emphasis on dose control for these tools. We have presented a number of improvements to the VIISta HC including, Faraday and dose algorithm improvements, rotated implants, compensation for photoresist outgassing, improved glitch recovery and dosimetry system self check. Results from testing of these improvements demonstrate superior dose control across the full range of operation.

ACKNOWLEDGMENTS

The authors would like to thank Greg Redinbo, Morgan Evans, Keith Pierce, and Mark Saunders for their contributions to this work.

REFERENCES

1. A. Renau, *Nucl. Instr .and Meth. In Phys. Res. B* **237**, 2005, pp 284-289.
2. J.C. Olson, A. Renau, J Lu, D.L. Smatlak, "Uniformity control using multiple tilt axes, rotating wafer and variable scan velocity," US Patent.Publication Number: US20050263721 (2005)
3. G.Gammel, et. al, Further Performance Improvements for High Power Implants into Photo-Resist on the V810EHP, IIT 2004, pp 48-51
4. G.C. Angel,. R.J. Low,. US Patent Publication Number: US7005657
5. J. Dzengeleski, et.al. Advances in Dose Control Calibration and Verification. IIT 2004, pp 52-54

Using Multiple Implant Regions To Reduce Development Wafer Usage

S. R. Walther, S. Falk, S. Mehta, Y. Erokhin, and P. Nunan

Varian Semiconductor Equipment Associates, Inc., 35 Dory Road, Gloucester, MA 01930, United States

Abstract The cost of new process development has risen significantly with larger wafer sizes and the increased number of fabrication steps needed to create advanced devices. The high value of each 300 mm development wafer has spurred efforts to find a way to explore more than a single process setting with each wafer. Traditional methods of defining multiple spatially distinct implant regions on a single wafer achieve poor utilization of device die. The need for efficient utilization of the die and wide process latitude for defining multiple implant regions per wafer has led to the development of an implant proximity mask (vMask™), which permits sharply defined borders between implant regions that may have different species, energy, angle, or dose. The capability of this system to achieve multiple spatially resolved implant conditions per wafer with high die utilization and using the same process parameters as production implants will be described. Specifically, results for measurement of the uniform process area, process repeatability, and cleanliness will illustrate the potential of this technique to dramatically reduce implant process development costs.

Keywords: ion implantation, VLSI technology, silicon
PACS: 85.40.Ry, 61.72.Tt, 81.05.Cy

INTRODUCTION

The rapidly increasing cost of process development is driven in large part by the implant split lots needed to characterize the new process [1]. An entire wafer must be used to characterize any given implant condition. The cost of this poor utilization of silicon drives the need to perform multiple split conditions on a single wafer. In addition, a single wafer split can isolate implant effects from the process variation of other tools. Two ways to achieving this capability are physical masking of the wafer or altering the two dimensional scanning of the beam/wafer to implant a defined area. The latter method is quite limited, as the actual usable uniform implant region is small and the change to the dose rate causes different crystal damage and wafer charging when compared to the full wafer production implant. The physical mask permits production recipe parameters to be used with a very large proportion of uniform implant area. The feasibility of a one-quarter (90° pie section opening) physical mask, referred to here as a 'vMask™', is demonstrated with process and device test results. The cost benefit of four split conditions per wafer, used to develop a process for a new node, can be very large [2], in addition to accelerating process development.

Description and Operation

The vMask™ is used in 300 mm VIISta implanters [3] with a rotating platen to allow one-quarter of a wafer to be implanted by masking the other three-quarters of the wafer from the beam. The wafer can then be implanted four times, each time exposing a different quadrant to a unique set of implant conditions, such as dose, implant angle, energy or even dopant species. Each quadrant can be implanted with multistep or chained recipes in the same fashion as a whole wafer and using the exact same recipe conditions. This system is compatible with the VIISta series common end station and thus can be used for medium current, high current, or even MeV implants.

The vMask™ system has two main components, the mask itself, and a loader. The physical mask is composed of a proprietary high strength carbon material with a pure carbon CVD surface as shown in figure 1. This permits a robust lightweight design that requires no modification to the vertical scanning system. When not in use, the mask is attached to the loader in a safe position in the end station, out of the path of the ion beam. If the implant operator instructs the implanter to use the mask for a split lot, the loader automatically places the mask over the platen, locks it

FIGURE 1. A picture of the physical mask component of the vMask™ system, showing the 90° cutout for implant exposure and the three precision attachment points that align, position, and fasten the mask to the rotating wafer platen.

into place, and it then moves with the rotating platen assembly. The mask is displaced sufficiently from the platen so that wafers may be exchanged with the mask in place. After the wafer is clamped, only one quarter, or quadrant, of the wafer is exposed for implant.

Proper exposure of each quadrant is handled automatically by unloading, reorienting and reloading the wafer, where each wafer quadrant has a unique implant recipe associated with it. After completing masked split lots, the mask can be unloaded, moved to its safe location and the implanter returned to normal production mode with standard implants. Loading or unloading the mask takes less than a minute without affecting the end station vacuum. Since the same implant recipe conditions can be used with or without the mask the recipe software is interlocked to prevent mask use for production whole wafer recipes (production mode) and to require the mask for development recipes that specify it. Interlocks also ensure that the mask is properly positioned on the platen after loading, with independent mask position sensors.

PROCESS AND DEVICE TESTING

The vMask™ provides a sharply defined implant region with a minimal exclusion boundary between quadrants. Methods that define an implant region using only scanning, without a physical mask, suffer from excessively large boundaries between regions as illustrated by figure 2, which shows that the boundary is a factor of 10 larger than the physical mask case. In figure 3, a sheet resistance map illustrates how each

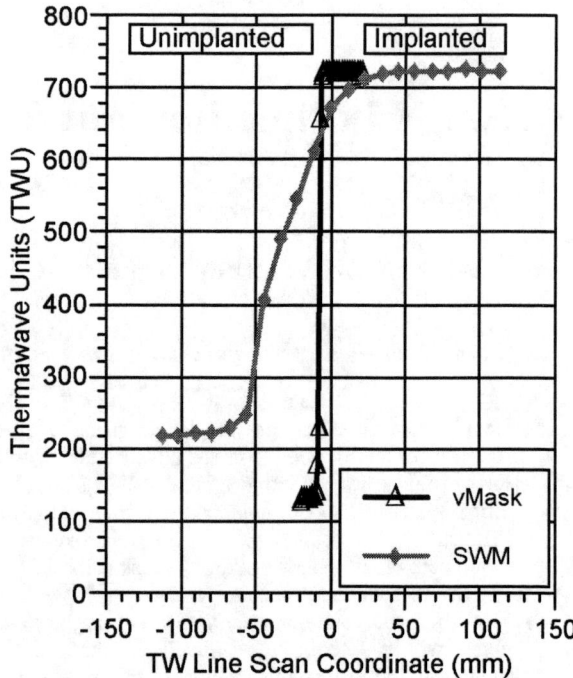

FIGURE 2. Thermawave (TW) line scans showing the transition from unimplanted to fully implanted using a physical mask (<10 mm transition width) and a beam scan alternative (SWM, >100 mm transition width) for a HALO implant (7 keV B^+ 7E13, 45° tilt).

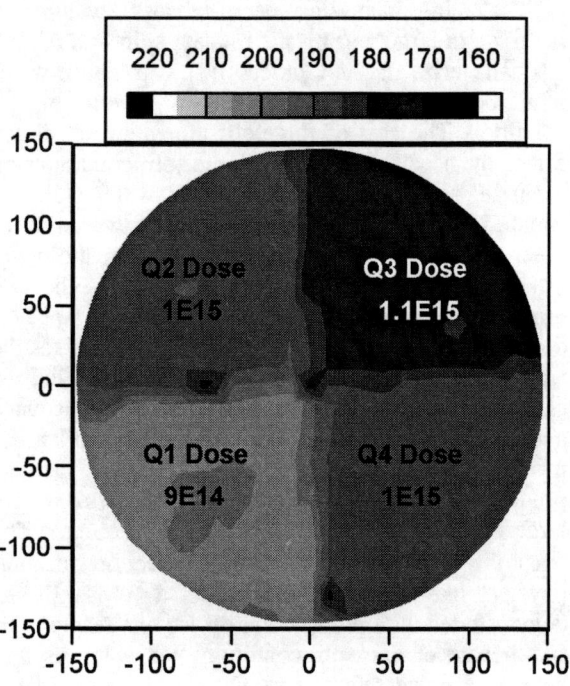

FIGURE 3. A sheet resistance contour fill map for a dose split completed on a single wafer for a 5 keV B^+ implant with zero degree tilt/twist angles.

quadrant implanted with the mask provides a large usable area. Each distinct implant region is implanted using the same recipe parameters as a production implant and with comparable dose uniformity.

Contamination levels for implants utilizing a mask have been compared to identical reference implants. Particle levels are evaluated in Table 1 and show the masked implant adders are somewhat higher than baseline, although still quite acceptable for split lot testing. The particle level returns to baseline once the mask is unloaded from the platen. Measured metals levels, detailed in Table 2, are very similar to baseline values, which is not surprising as the mask is entirely of carbon composition. Despite the proximity of the carbon surface to the wafer, the level of added carbon appears to be negligible, as illustrated by a SIMS depth profile for carbon in figure 4. Likewise, the level of cross contamination when switching dopant species with the mask installed is similar to baseline values as detailed in Table 3. Thus the contamination levels with the mask in place are well controlled and generally similar to baseline values.

TABLE 1. Particle Test (15 keV As$^+$ 1E15 cm^{-2})

Particle Bin (μm)	Pre-map	Post-map	Net Adders	Areal Density (cm^{-2})
0.09 to 0.12	4	32	28	
0.12 to 0.16	3	14	11	
0.16 to 0.20	0	5	5	
0.20 to 0.30	1	14	13	
0.30 to 0.50	1	11	10	
0.50 to 1.0	1	3	2	
1.0 to 3.0	1	0	-1	
3.0 to 5.0	0	0	0	
5.0 to 20.0	0	1	1	
All ≥ 0.09			69	0.102
All ≥ 0.12			41	0.060
Baseline Spec ≥ 0.12			34	0.050
vMask™ Spec ≥ 0.12			102	0.150

TABLE 2. Metals Test (60 keV As$^+$ 1E16 cm^{-2}) TXRF Results in units of E10^{-2} (equivalent to ppm)

X (mm)	Y (mm)	W	K	Ca	Ti	V	Cr	Mn	Fe	Co	Ni	Cu	Zn
0.1	0.1	0*	0	0	0	0	0	0	0	0	0	0	0
-75	75	0	0	0	0	0.08	0	0	0	0.01	0	0	0
75	-75	0	0	0	0	0	0	0	0	0	0	0	0
75	75	0	0	0	0.02	0	0	0	0.11	0	0	0	0
-75	-75	0	0	0	0	0	0	0	0.15	0	0	0	0
-37.5	-37.5	0	0	0	0	0	0	0	0	0	0	0	0
-37.5	37.5	0	0	0	0	0	0	0	0	0	0	0.01	0
37.5	-37.5	0	0	0	0	0	0	0.01	0	0	0	0	0
37.5	37.5	0	0	0	0	0	0	0	0	0	0	0	0
* Zero reading is equivalent to 'not detected'													

TABLE 3. Cross Contamination Test (35 keV As$^+$ 4E15 cm^{-2}, contamination implant: 40 keV P$^+$ 5E15 for one hour, result measured by SIMS)

Condition	Pre-test P level (cm^{-2})	Post-test P level (cm^{-2})	Net % P dose in As implant
Baseline	1.9E12	1.0E13	0.25%
Baseline Spec			1.00%
vMask™ (center)	3E11	5.6E12	0.14%
vMask™ (edge)	6.2E11	5.8E12	0.14%

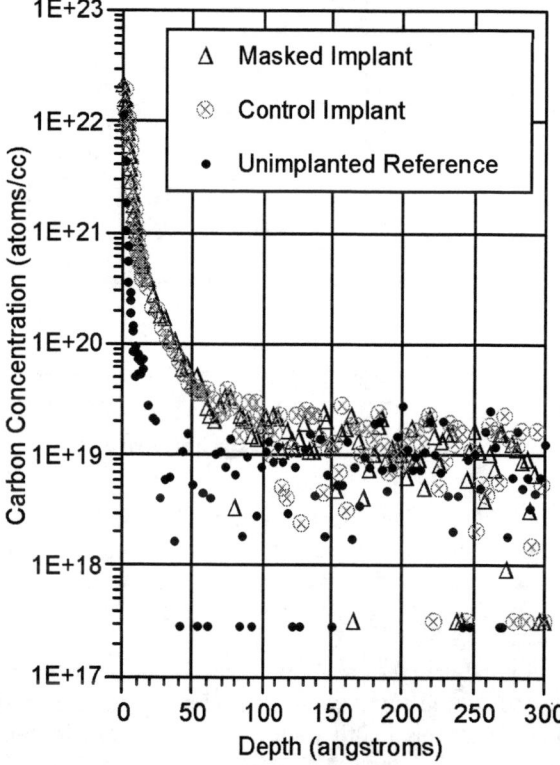

FIGURE 4. SIMS carbon depth profile data from patterned photoresist wafers implanted with 35 keV 4E15 As$^+$, with and without a mask along with an unimplanted reference wafer.

The dopant depth profiles for the vMask™ are identical to whole wafer implants. Figure 5 compares the depth profiles for three cases; the center of the masked quadrant, the edge of the quadrant, and a reference whole wafer implant, all using the same 35 keV 4E15 As$^+$ recipe. All three profiles overlay as expected since the ion energies are unaffected by the presence of the mask. The mask itself is electrically grounded through the wafer platen and has no effect on ion trajectories. Similarly, implants with the grounded mask show no degradation of charging performance, as measured using antenna structure devices and compared with standard implants.

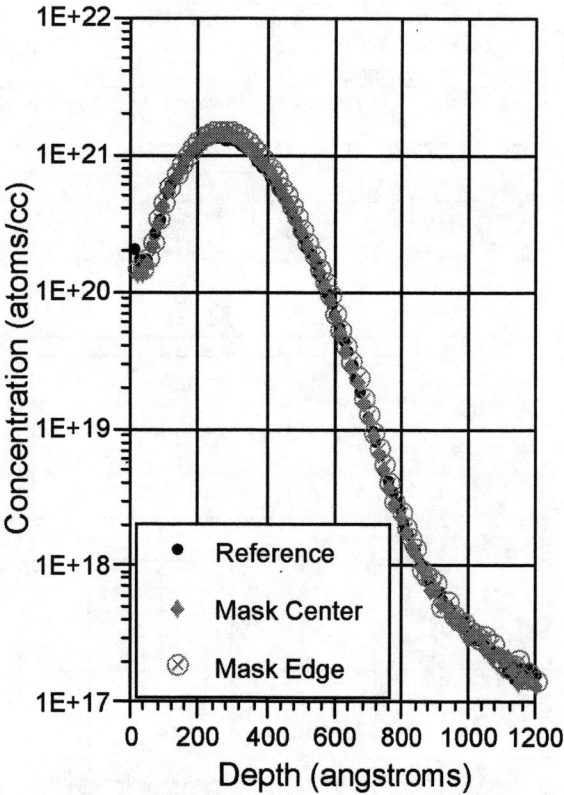

FIGURE 5. SIMS depth profiles of a reference non-masked implant compared with a masked implant measured at both the center and edge of the open quadrant for a 35 keV 4E15 As$^+$ implant.

To validate the capability of the vMask™, an initial device split test was completed. The implant used was a 2 keV arsenic lightly doped drain (n-LDD) with four doses – 4, 5, 6, and 8E14 cm^{-2} completed on a single wafer. The device structures were 45 nm CMOS transistors that were tested at the second level metallization stage. The split on wafer quadrants behaved as if completed on whole wafers, as indicated in Figure 6, which shows the expected correlation of the I_{on} current with the n-LDD dose.

CONCLUSION

The accelerating costs of semiconductor process development are highly correlated to the requirements for implant wafer splits. The ability of the vMask™ to produce sharply defined, reliable four region implant splits on a single wafer offers a dramatic cost savings over conventional split lot processing. Masked implants demonstrate no significant degradation in cleanliness, have the capability to run single wafer splits of dose, angle, energy, or even species, with minimal edge exclusion at quadrant boundaries.

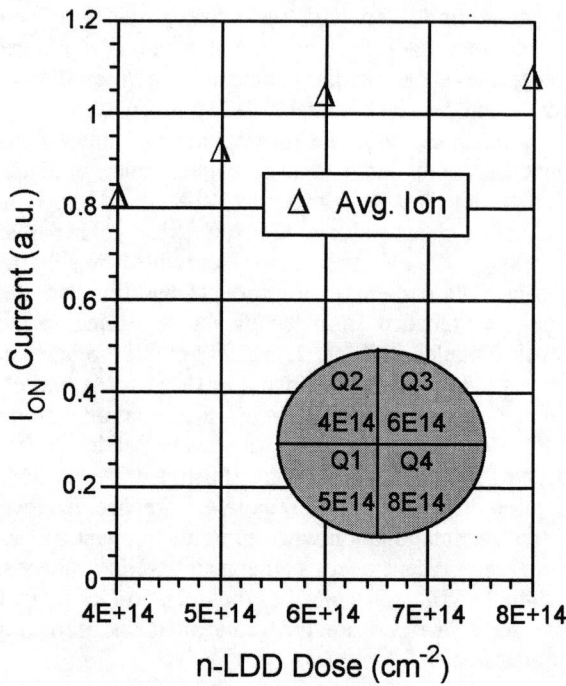

FIGURE 6. I_{ON} current (four die average) for 45 nm CMOS transistors as a function of the dose applied to each quadrant of a single wafer n-LDD dose split completed using the vMask™.

Virtually any implant is compatible with this system, as it works across the suite of 300 mm VIISta single wafer implanters. Device results illustrate the capability to reproduce full wafer implant results using the same implant parameters, including dose rate. The single wafer split will provide a powerful means of isolating implant parameter effects from the variability introduced during the rest of the process flow. This may lead to use beyond development split lots to become a standard tool for yield enhancement work and foundry implant qualification.

ACKNOWLEDGMENTS

We gratefully acknowledge the creativity and dedication of the engineering team and especially Tim Kirk, Chuck Teodorczyk, Andrew Poitras, Jeff Krampert, and Larry Ficarra.

REFERENCES

1. Ben McKee, VLSI Research, Doc. 600201, 1/28/05.
2. Dan Hutcheson, 'The Chip Insider', VLSI Research, 4/27/06.
3. N. Variam, S. Mehta, S. Norasetthekul, and B. Guo, Proc. 14th Int. Conf. Ion Implantation Tech., IEEE Cat. No. 02EX505, (2003) p.283.

ADVANCED MODELING TECHNIQUES FOR ANALYSIS OF HIGH CURRENT RIBBON BEAM TRANSPORT AND CONTROL

Peter Kellerman and Frank Sinclair

Varian Semiconductor Equipment Associates, Inc., 35 Dory Road, Gloucester, MA 01930, United States

Abstract. A computer model capable of analyzing electrostatic lenses, including the effects of space charge, yet simple enough to run an entire beamline system calculation in seconds, has been developed and applied to VSEA's latest high current beamline. Such a code is a useful tool in exploring tuning possibilities when running in various modes of drift and decel, and can be used to find solutions that optimize transport, as well as providing a basis for beam tuning. This paper details the basis of the model and compares results with more rigorous models.

Keywords: ion beam, first order model, space-charge, electrostatic lenses
PACS: 41.75.Ak; 61.85.+p; 85.40.Ry; 61.72.Tt

INTRODUCTION

High current ion implanters in today's market are expected to provide high productivity (i.e. high beam currents down to sub-keV energies) while maintaining the highest control of process purity, implant uniformity, angles, beam shape, and charging. These requirements have driven beamline developers to more complex beamlines, which, in turn, create new challenges in beam optics analysis and modeling. In particular, high beam currents at energies below 1 keV have required deceleration, so that modeling codes must handle both magnetic as well as electrostatic elements and include the effects of space charge in beam propagation. Although there are several codes available that perform such calculations (even in 3 dimensions), such as Opera, Lorenz, and Cobra, these codes are very time consuming, both in specifying as well as running problems. They are therefore generally used to perform the detailed analysis of individual optics elements, but not entire beamlines. In order to explore the advantages of different architectural approaches to entire systems, as well as explore different tuning approaches used in such systems, it is useful to have a first order model which can be modified and run in minutes and seconds, rather than weeks and hours. However, there are currently no available first order codes that include electrostatic slit lenses for ribbon beams or include the first order effects of space charge. This paper describes an approach for adding such capabilities to a first order optics model.

Model

In order for a first order model to be useful in describing a high current, low energy, ribbon beam which undergoes deceleration, it must provide a model that is sufficiently detailed to include the following effects:

1. Vertical (y) focusing due to fringe fields in the slit lens, including dependence on electrode voltages and geometry (gaps, widths),
2. Horizontal (x) focusing due to curvature in slit lens electrodes,
3. Space charge defocus within the stripped region of the slit lens
4. Partial space charge defocus within the drift regions and within magnetic elements
5. Focus effects of magnetic elements, such as sector and quadrupole magnets.

The treatment of magnetic elements is quite standard, and so will not be discussed here. It is the first order treatment of slit lenses with space charge that will be detailed.

The treatment of slit lenses is based on the paraxial ray equation [1] [2], which relates the transverse component of the electric field, $E_y(y)$, which causes vertical focusing, to the second derivative of the potential with respect to z on axis, $\nabla^2 V$, while the first derivative of the potential is simply related to the longitudinal component of E (producing refraction).

$$\frac{d\theta}{dz} = -\frac{\nabla V}{2\Xi}\theta - \frac{\nabla^2 V}{4\Xi} y ,$$

where θ is the angle the particle trajectory makes with respect to the z axis and Ξ is the particle energy. The trick is how to obtain a good approximation to V(z) within the slit lens, without explicitly solving Poisson's equation. It is found that the following heuristic algorithm works quite well. Referring to figure 1, we first extend the potential into end electrodes by $\alpha \cdot a$ (where a is half the electrode gap), reflecting the real situation of the location of the beam plasma boundary. The interior electrode voltages are then averaged over a window that depends on the gaps of those electrodes (half width = $\beta \cdot a$), reflecting the shielding that occurs from neighboring electrodes. Finally, the dots are smoothly connected by clamped cubic spline interpolation.

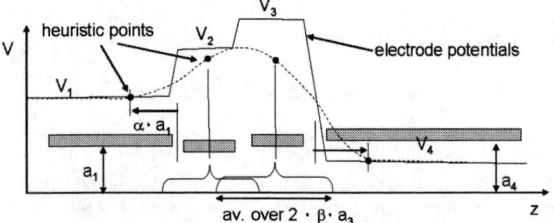

Figure 1. Heuristic assignment of V(z).

It has been found (by comparison with exact Poisson solvers) that using $\alpha=1.1$ and $\beta=1.4$ works well for a wide range of situations. Figure 2 shows the agreement of resulting beam focusing, (as calculated the Poisson solver). Note that the beam current is set to zero, so that there is no space charge considered at this point.

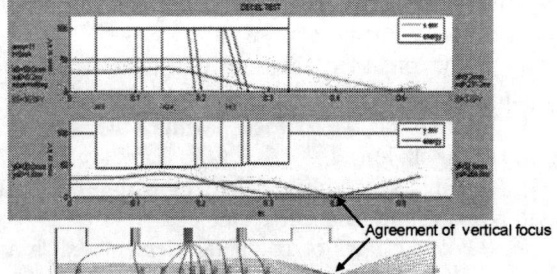

Figure 2. Comparison with Poisson Solver- no space charge

In order add the affect of space charge, we consider the electric field in a beam with uniform charge density over an elliptical beam cross-section

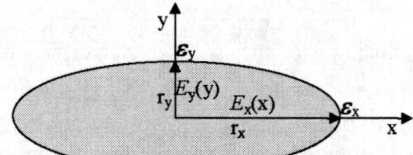

Figure 3. Elliptical beam with uniform current density

A solution to Poisson's Eq.

$$\nabla^2 V = \rho/\varepsilon_0$$

within the beam is:

$$E_x(x) = \frac{\mathcal{E}_x}{r_x} x, \quad E_y(y) = \frac{\mathcal{E}_y}{r_y} y$$

where $\mathcal{E}_x$ and $\mathcal{E}_y$ are the values of the field at the boundary [3]. We thus see that the defocus is linear, therefore producing a first order lens, so that the beam will retain an elliptical shape with a uniform ρ. In order to determine $\mathcal{E}_x$, $\mathcal{E}_y$, we need to determine the boundary conditions. There are two cases that can be solved explicitly: "bounded", where the beam is surrounded by a conducting wall, and "unbounded", where the potential is taken to be zero at infinity. These boundary conditions lead to the following solutions:

"Bounded":

$$\frac{\mathcal{E}_x}{\mathcal{E}_y} = \frac{1}{k},$$

$$E_x = \frac{\rho}{\varepsilon_0} \frac{x \, r_y^2}{(r_x^2 + r_y^2)} \quad E_y = \frac{\rho}{\varepsilon_0} \frac{y \, r_x^2}{(r_x^2 + r_y^2)}$$

"Unbounded":

$$\mathcal{E}_x = \mathcal{E}_y,$$

$$E_x = \frac{\rho}{\varepsilon_0} \frac{x \, r_y}{(r_x + r_y)} \quad E_y = \frac{\rho}{\varepsilon_0} \frac{y \, r_x}{(r_x + r_y)}$$

By considering the deflection of charged particles within the beam, one can then obtain the beam deflection and focal length f in the path length L:, in terms of the beam perveance P

Bounded

$$\frac{f_x}{f_y} = k^2$$

$$f_x = \frac{1}{P} \frac{r_x^2 (r_x/r_y + r_y/r_x)}{L} \quad f_y = \frac{1}{P} \frac{r_y^2 (r_x/r_y + r_y/r_x)}{L},$$

where $P = \dfrac{I}{2\pi\varepsilon_0 \left(2e/m\right)^{1/2} E^{3/2}}$

Free space

$$\frac{f_x}{f_y} = k$$

$$f_x = \frac{1}{P} \frac{r_x(r_x + r_y)}{L} \quad f_y = \frac{1}{P} \frac{r_y(r_x + r_y)}{L}$$

Figure 4 shows the same example as in figures 2, but with beam current added (16mA Boron). Good agreement can be seen with the Poisson solver model

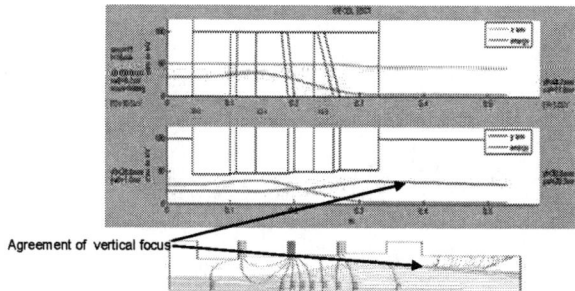

Figure 4. Comparison with Poisson Solver- space charge (16mA)

Note that the affect of conducting walls is to reduce the space-charge defocus in x and increases the defocus in y. This can be understood in terms of image charges in walls, which effectively shield the beam from itself along the ribbon in x, yet add attraction to the walls (thus adding to the deflection) in y.

The expressions for the space charge focal lengths explain the inherent advantage of ribbon beams over spot beams for the transport of high current at low energy. Consider the situation of transporting a 1keV 10mA Boron beam 0.5m, assuming 99% neutralization. For spot beam with r=2.5cm, the focal

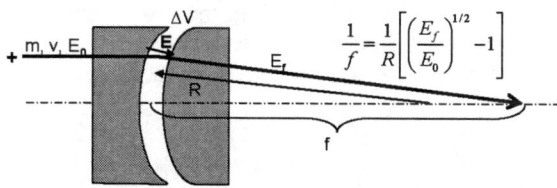

$$\frac{1}{f} = \frac{1}{R}\left[\left(\frac{E_f}{E_0}\right)^{1/2} - 1\right]$$

Figure 5. Curved lens focusing

length is .18m, yielding a diverging angle of +- 7.8°. For a ribbon beam with half width 15cm and half height 2cm, the resulting focal lengths are f_x=25.3m and f_y=.45m, corresponding to divergence angles of only .34° in x and 2.5° in y.

One more thing needs to be added into the model, namely the focus in x that can be achieved by curving the electrodes. Curved gaps produces transverse components of the field, proportional to x (for small x/R), resulting in first order focusing, as shown in figure 5.

These electrostatic elements can then be inserted into the usual first order matrix formulation [4], where an elliptical beam is described by the Twiss parameters α, β, γ, and ε. The advantage of this approach is that beam distributions with elliptical trace-space boundaries transform to other elliptical distributions (by linear optical elements).

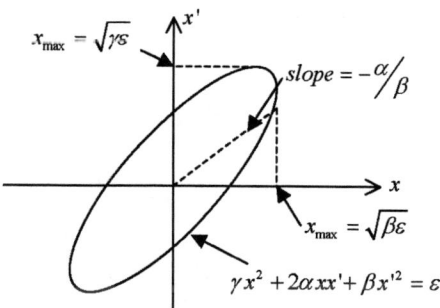

Figure 6. Twiss parameters for elliptical beams

Twiss or Courant-Snyder representation of ellipse: α specifies beam focus, β specifies maximum x, and γ specifies maximum x'. If there is no acceleration, α β γ are not independent, and are related by the "Courant-Snyder invariant": $\gamma\beta - \alpha^2 = 1$. Individual rays transform according to

$$\begin{pmatrix} x \\ x' \end{pmatrix} = \begin{pmatrix} m_{11} & m_{12} \\ m_{21} & m_{22} \end{pmatrix} \begin{pmatrix} x_0 \\ x'_0 \end{pmatrix}$$

where
$m_{12} \sim$ drift distance
$m_{21} \sim -1/f$
$m_{22} \sim n$ ('index of refraction' = v_0/v)

If there is no acceleration, the Twiss parameters transform according to

$$\begin{pmatrix} \gamma \\ \alpha \\ \beta \end{pmatrix} = \begin{pmatrix} m_{22}^2 & -m_{21}m_{22} & m_{21}^2 \\ m_{12}m_{22} & 1+2m_{12}m_{21} & -m_{11}m_{21} \\ m_{12}^2 & -2m_{11}m_{12} & m_{11}^2 \end{pmatrix} \begin{pmatrix} \gamma_0 \\ \alpha_0 \\ \beta_0 \end{pmatrix}$$

and emittance is conserved, i.e. $\varepsilon = \varepsilon_0$.

However, if there is acceleration, Twiss parameters transform differently

$$\begin{pmatrix} \gamma \\ \alpha \\ \beta \end{pmatrix} = \left(\begin{pmatrix} m_{22}^2 & -m_{21}m_{22} & m_{21}^2 \\ m_{12}m_{22} & m_{11}m_{22}+m_{12}m_{21} & -m_{11}m_{21} \\ m_{12}^2 & -2m_{11}m_{12} & m_{11}^2 \end{pmatrix} / \det M \right) \begin{pmatrix} \gamma_0 \\ \alpha_0 \\ \beta_0 \end{pmatrix}$$

Emittance is not conserved:

$$\frac{\varepsilon}{\varepsilon_0} = \frac{v_0}{v} = \sqrt{\frac{E_0}{E}} \equiv n \text{ (index of refraction)} = \det M$$

In addition, the expression for x_{max} is more complicated

$$x_{max} = \frac{-\alpha^2 \varepsilon^{1/2} + \varepsilon^{1/2}(\alpha^4 - \beta^2\gamma^2)^{1/2}}{\gamma\beta^{1/2}(\gamma\beta - \alpha^2)^{1/2}} \quad \Rightarrow \quad \sqrt{\beta\varepsilon}$$

(where $\Rightarrow$ holds if there is no acceleration)

Figure 7. Example of ribbon beamline with double deceleration

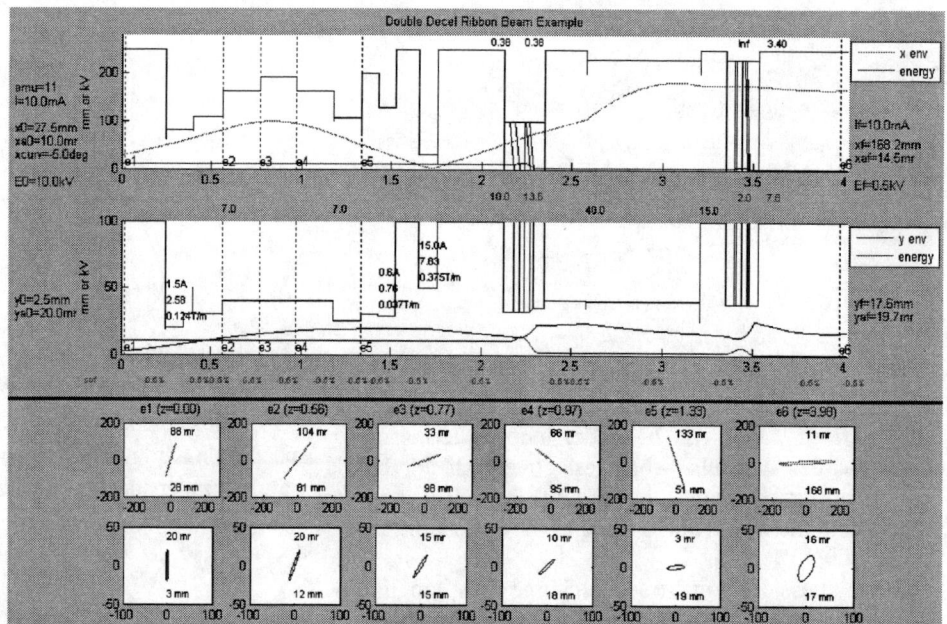

For the electrostatic elements, we have that from the complete paraxial equation,

$$\frac{d\theta}{dz} = -\frac{\nabla V}{2E}\theta - \frac{\nabla^2 V}{4E}X - \frac{1}{f_{sc}} - \frac{1}{f_{curv}}$$

So that for each increment dz

$$1/f = -\frac{\nabla^2 V}{4E} - \frac{1}{f_{sc}}, \quad n = -\frac{\nabla V}{2E}.$$

By setting $r_x = x_{max}$ and $r_y = y_{max}$ in the expressions for space-charge focal lengths for each dz, we can then evaluate the m_{ij}, form the transform matrix for Twiss parameters α β γ, and perform the transformation to get the elliptical distribution for the next dz. All this is coded in MATLAB, with the different beamline elements specified in a table, providing easy editing, while still allowing beamline relationships and constraints to be maintained. The display shows the x and y beam envelopes, along with the beam energy along z. Emittance plots can be specified anywhere along the beamline.

Example

VSEA's latest high current ribbon beamline, the VIIsta HCP, includes many components (quadrupoles, AMU magnet, collimating magnet, decel lenses D1 and D2, and multipole correction elements) to routinely deliver the high current ribbon beam with controlled uniformity and angles. The first order model presented here aids our understanding of how to best utilize these components. Figure 7 shows a boron .5kV, 10mA beam, achieved in double decel mode, where the beam is extracted at 10kV, then decelerated to 2keV in D1, and finally to .5keV in D2. The model shows how the beam size is carefully controlled within the decel lenses to balance the beam's space charge and strong lens focusing. With the low beam density of a ribbon beam, space charge expansion is not substantial, even with the .5% un-neutralized drift regions.

CONCLUSIONS

The first-order model presented here is a useful supplement to the arsenal of modeling tools available today. It enables rapid comparison of different ribbon beamline architectures over wide ranges of beam conditions (including the affects of space charge, and including decel elements), allowing the optimization the sophisticated beamlines needed to meet the simultaneous market requirements of high productivity and high beam quality, as well as providing guidance in development of tuning procedures for such beamlines.

REFERENCES

1. S. Humphreys, Charged Particle Beams (John Wiley & Sons, New York, 1990)
2. J. Orloff, ed., Handbook of Charged Particle Optics (CRC Press, New York, 1997)
3. G. Gillespie and T. A. Brown, Optics Elements for Modeling Electrostatic Lenses and Accelerator Components, Nuclear Instruments and Methods in Physics Research B (1998)
4. S. Humphreys, Principles of Charged Particle Acceleration (© Stanley Humphreys, 1997)

Increase of Beam Current Mass-Separated by Long Gap Dipole Sector Magnet for S/D Process in FPD manufacturing

Shojiro DOHI, Yasunori ANDO, Yutaka INOUCHI, Yasuhiro MATSUDA, Masashi KONISHI, Junichi TATEMICHI, Masaaki NUKAYAMA, Kazuhiro NAKAO, Kohichi ORIHIRA and Masao NAITO

NISSIN ION EQUIPMENT CO., LTD.
(SHIGA PLANT) 29, HINOKIGAOKA, MINAKUCHI-CHO, KOKA, SHIGA 528-0068, JAPAN
PHONE 81. (748) 65-0801   FACSIMILE 81.(748) 65-0802
E-MAIL: Dohi-Syojiro@nissin.co.jp
URL http://www.nissin-ion.co.jp/

Abstract. A mass analyzing ion implantation system (called Ion Doping iG4) was developed for FPD manufacturing. One of most important concept of iG4 is to transport a sheet ion beam maintaining its current density profile from the ion source to the target, which leads good mass resolution and simple control of the beam profile. The system has a bucket type ion source which provides a sheet ion beam whose longer dimension of the cross section is 800 mm the 4th generation FPD glass substrate generally sized 730mm x 920mm. The sheet ion beam is mass-analyzed with a dipole sector magnet with a long pole gap. In order to enhance through-put for Source Drain implantation processes, we modified the ion source to increase high beam currents and obtained 300μA/cm for Boron ion beams and 500μA/cm for Phosphorus ion beams. Better uniformity and higher mass resolution were achieved by optimizing shape of the analyzing magnet pole faces.

Keywords: Ion implanter, Mass separation, Bucket type I/S, Sheet ion beam, Long pole gap sector magnet, Thin-film-transistor, S/D process, Flat-panel-display
PACS: 61. 72. Tt; 68. 55. Ln; 85. 40. Ry

I. INTRODUCTION

Recently, as advanced application of Low Temperature Poly-crystalline Silicon Thin Film Transistors (LTPS-TFT), system-on-glass devices which mean the integration with a display unit, memories and CPU on one glass panel are going to be developed by FPD manufacturer. To make such higher integrated circuits on a glass, it is required for ion implantation equipments to have good beam current uniformity, wide dose range (to meet channel implant, LDD and S/D implants), and mass-separated beam to avoid contaminations.

The developed Ion Doping iG4 fulfills these requirements (ref.1). One of the most important concepts of the iG4 is to transport a long length sheet ion beam maintaining its current density profile from the ion source to the target. The sheet ion beams were extracted from the bucket ion source and their current and current density profile were easily controllable.

We developed an analyzing magnet with a long pole gap which could transport longer sheet ion beams.

Ion Doping iG4 is recently used for not only channel implant and LDD but also S/D implants. But Ion Doping iG4 had not so good performance as non-mass analyzing systems for fabrications of S/D implants such as processes that needed a high beam current. Thus, we continue the developments of the main items to increase ion beam currents and to achieve good uniformity of a sheet ion beam. In this paper, we especially describe about the performance of beam currents and uniformity.

II. FEATURE OF ION SOURCE AND MASS-ANALYZING MAGNET

Figure 1 shows the schematic diagram of the Ion source of Ion Doping iG4. This ion source is a bucket type and can produce a sheet ion beam with about 30 mm width and 800 mm length which is corresponding to narrow side of 4th generation FPD glass substrate (730 mm x 920 mm). In order to control the plasma density profile, six filaments are placed into the arc chamber. Uniformity of the sheet ion beam is obtained by independently controlling of the six filaments with a feedback signal from multiple faraday cups placed at the back of the glass substrates.

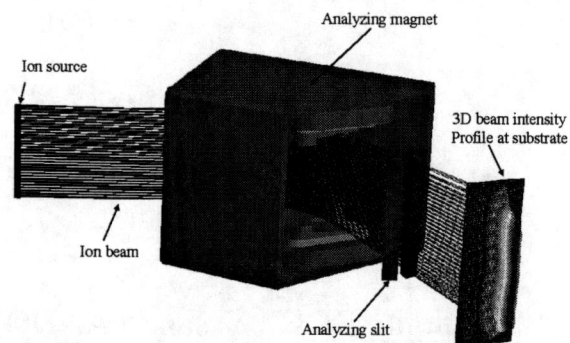

Fig. 2 The schematic diagram of Ion Doping iG4 beam line

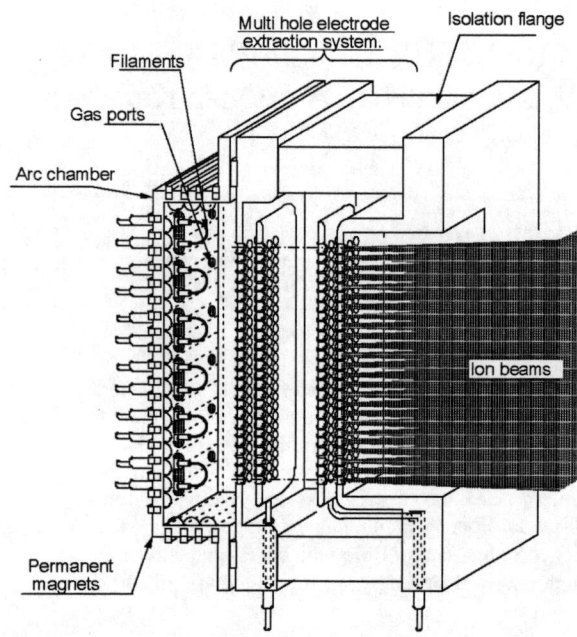

Fig. 1 The schematic diagram of the ion source of Ion Doping iG4

Figure 2 shows the schematic diagram of Ion Doping iG4 beam line. Ion Doping iG4 beam line is very simple and mainly consists of the mass-analyzing magnet and the analyzing slit stood at downstream of the mass analyzing magnet.

This figure shows that the sheet ion beams were bended and transported horizontally through a long pole gap in the analyzing magnet with almost the same length. The reason we use this long pole gap magnet is as follows; a good controllability of beam profile by ion source parameters, and a high mass resolution. In order to obtain the enough mass resolution of $M/\Delta M = 10$ for LTPS-TFT manufacturing, the bending angle of this mass-analyzing magnet is designed to be 70 degrees. Furthermore the analyzing slit can remove dispensable ions at downstream of the analyzing magnet. By changing the analyzing slit width, Ion Doping iG4 can control the mass resolution.

III. IMPROVEMENT FOR ION BEAM CURRENT AND MASS ANALYZING MAGNET

Ion source for higher ion beam current

In Ion Doping iG4, BF_3 gas is used for a production of B^+ ions as the source gas. In the plasma of BF_3, there are several ions; they are B^+, F^+, BF^+ and BF_2^+. The ratio of BF_2^+ was several times of others. By the way, B^+ ion is needed for implantations into LTPS-TFT. Thus the increasing of the ratio of B^+ ion is important. Other ions hit to the wall of the vacuum chamber. This results in outgassing and contaminations.

In order to increase an input power density for a plasma production, we reduced to about 90% the volume of the plasma production region in the arc chamber. The plasma electrode was positively biased by several volts against the arc chamber, which reduced F^+, BF^+ and BF_2^+ ions in extracted beams.

Figure 3 shows a mass spectrum of an extracted ion beam from BF_3 plasma in the improved ion source. The ratio of B^+ ion was larger than others. The beam current density of B^+ ion beam was 300 µA/cm.

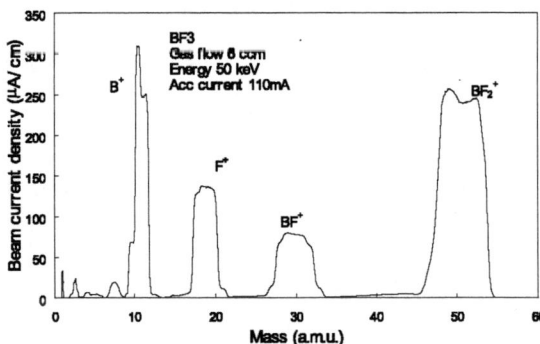

Fig. 3 Mass spectrum of an extracted ion beam from BF₃ plasma in improved ion source

Figure 4 shows the typical profiles of the high current densities; for 50keV B⁺ 100, 200, 300μA/cm. We obtained the sheet ion beams which uniformity of beam current density was about 2%.

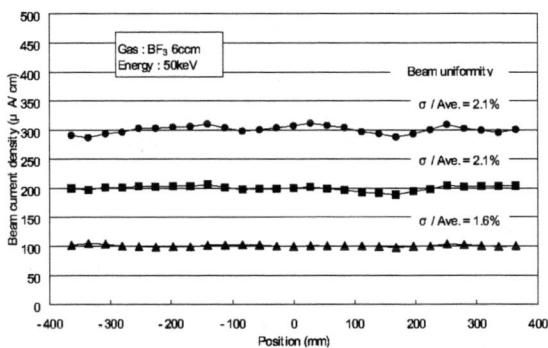

Fig. 4 B⁺ ion beam profiles at 50keV

Next we present the results of PH_x^+ ion beam. As well as BF₃ plasma, phosphine plasma also contains several ions, H_x^+, PH_x^+ and $P_2H_x^+$. The ratio of PH_x^+ ion is usually higher than others. And PH_x^+ beam currents can be easily increased by a high arc power input. Figure 5 shows PH_x^+ ion beam profiles. The beam current density was about 7 μA/cm to 500 μA/cm.

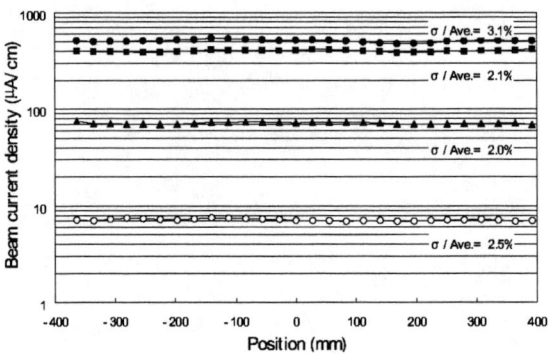

Fig. 5 PH_x^+ ion beam profiles at 65keV

Improvement of mass analyzing magnet performance

As the ion source is a long size bucket type, extracted ion beams are expected to be uniform along the long side of sheet ion beams. We continue to develop the mass analyzing sector magnet as to maintain the beam current density profile and uniformity. Recently, we adjusted a pole face of each sector to optimize the magnetic fields between poles.

Figure 6 shows the dependence of the analyzing slit width on the average beam current density, the beam uniformity and the mass resolution for the previous type of magnet. Ar⁺ ion beam of 30keV and 200 μA/cm was obtained at the analyzing slit width of 160 mm. We changed the slit width from 160 mm to 40 mm. In the Fig.6, the open circle, the square and the triangle show the average beam current density, the beam uniformity and the mass resolution respectively. Narrowing the slit width, the beam current density became smaller and the beam uniformity grew worse. These imply the beam profile was not straight and a part of beams was cut off by the analyzing slit. Figure 7 shows each beam profile of Fig.6.

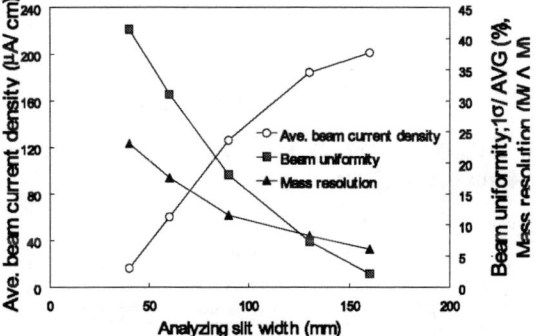

Fig. 6 Analyzing slit width dependence of sheet ion beam (previous magnet)

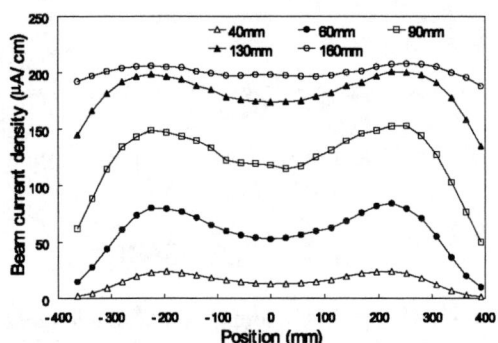

Fig. 7 Beam profiles of the previous magnet

For the new type of magnet, figure 8 shows the data that the analyzing slit width was only changed a method similar to the previous magnet. The symbols in Fig.8 are the same as Fig.6. The slit width was changed from 153 mm to 38 mm. Ar$^+$ ion beam of 30keV and 200 µA/cm was at the slit width of 153 mm. For the new type magnet, it was understood that the reduction of the beam current density and the deterioration of beam uniformity was improved. Fig.9 shows each beam profile of Fig.8. At the new type magnet, a current density of an ion beam was almost evenly reduced at each position. At the high mass resolution, uniformity of the ion beam was kept. This implies that the sheet ion beam became straight in a perpendicular. As a result, the beam profile at the ion source is simply reflected in the beam profile at the target.

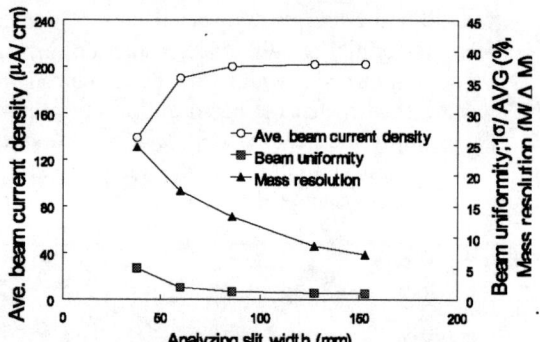

Fig. 8 Analyzing slit width dependence of sheet ion beam (New magnet)

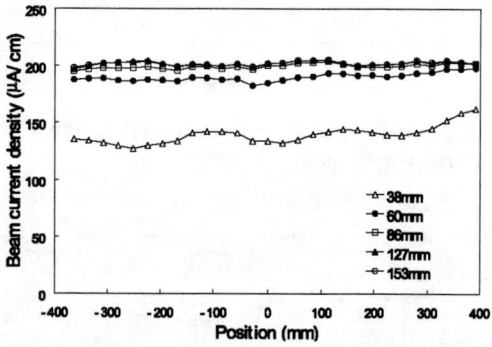

Fig. 9 Beam profiles of the new magnet

IV. CONCLUSION

We developed Ion Doping iG4 with a long size ion source and a long gap sector magnet for FPD manufacturing.

To increase an input power density into plasma and to optimize the pressure in the arc chamber and to be positively biased the plasma electrode against the arc chamber, the ratio of B$^+$ ion in extracted ion beams was increased. As a result, B$^+$ ion beam with 300 µA/cm was obtained. And PH$_x^+$ ion beam with 500 µA/cm was obtained.

By optimizing shape of the pole faces of analyzing magnet, sheet ion beams became almost straight and uniformity of sheet ion beams was maintained at the high mass resolution. As a result, ion implantations by Ion Doping iG4 became more uniform.

REFERENCES

1. S. Maeno et al. "iG4 High Current Ion Implanter for FPD Manufacturing Installed with Remarkably Long Pole Gap Sector Magnet", Proc. 2004 Int. Conf. on Ion Implantation Technology, Taipei, Taiwan, Oct. 25-29.
2. Y. Inouchi et al. "High Current Ion Implanter iG4 with Long Pole Gap Sector Magnet for FPD Manufacturing", IDW/AD'05 p.297-300

Nissin Ion Equipment Indirectly Heated Cathode Ion

Kohei Tanaka, Sei Umisedo, Kenji Miyabayashi,
Hideki Fujita, Toshiaki Kinoyama, Nariaki Hamamoto, Takatoshi Yamashita and
Masayasu Tanjyo

Nissin Ion Equipment Co., Ltd.,
575, Kuze-Tonoshiro-Cho, Minami-Ku, Kyoto, 601-0825, Japan

Abstract. To apply a multiply-charged ion beam to device fabrication, it is indispensable to improve the beam current while also increasing the short lifetime of ion source caused by the wear-out of the filament. The advantage of the indirectly heated cathode (IHC) configuration is well known as a means of extending the lifetime in a severe plasma environment. From this point of view, the IHC ion source, which is designed for the EXCEED2300 & 3000 series, has been developed to expand the capability of EXCEED series as the single wafer high energy ion implantation tool. We report the performance of IHC ion source with higher multiply-charged ion beam current and longer lifetime than a Bernas type ion source.

Keywords: indirectly heated cathode, multiply-charged ion beam, medium current, high energy.
PACS: 52.77.Dq, 52.59.Tb, 52.50.Dg

INTRODUCTION

In general, the medium current implanter is regarded as one which can cover the energy range up to 200keV or so with precise implantation control, but today it is also expected to cover a higher energy range (without degrading its controllability) previously reserved for a conventional high energy implanter (as shown in Figure 1).

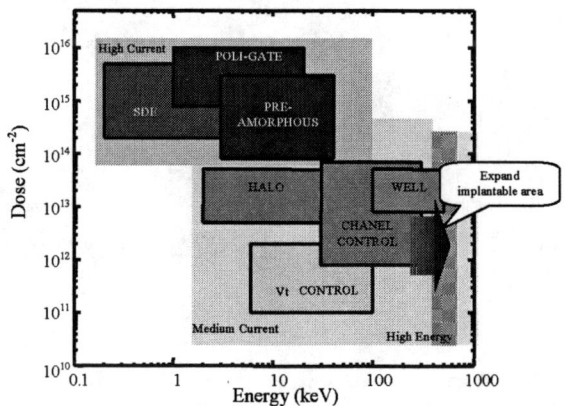

FIGURE 1. Dose-Energy Map. Yellow area is covered by current medium implanter.

To achieve productivity in such an energy range, sufficient multiply-charged ion beam current is required.

However, when the conventional Bernas type ion source is operated to obtain such ion beam current, the main problem from a production point of view is the short lifetime. Since the higher density plasma with high arc voltage is required to achieve sufficient multiply-charged ion, the filament exposed to plasma directly is worn easily. Therefore this kind of source is supposed to be unsuitable for this purpose.

It is well known that the IHC type ion source has a longer lifetime than Bernas type ion source [1]. Since the cathode which emits the primary electron for maintaining the plasma has a large area compared to the conventional filament type source, it is possible to generate the multiply-charged ion beam with a longer lifetime. Therefore, we have developed an IHC type ion source which can apply to our present medium current implanter. By installing the IHC type ion source, it can cover a higher energy range with the precise implantation control facility of the EXCEED series.

In this paper, we present the details of the Nissin IHC type ion source and the availability for EXCEED3000AH by presenting the beam current, mass spectra, effective throughput and implantation results.

EXCEED3000AH WITH THE IHC TYPE ION SOURCE

Nissin IHC Ion Source

A Nissin IHC type ion source is shown in Figure.2 and also the source control system diagram is shown in Figure.3.

The cathode is heated up by the thermal radiation of the filament and the electron bombardment from the filament with the aid of the heat voltage as shown in Figure.3. Plasma is established by the electrons which are emitted from the cathode. Since the cathode is designed to have enough thickness to bear sputtering, it is possible to set the arc voltage higher and generate sufficient multiply-charged ion beam current for a long time.

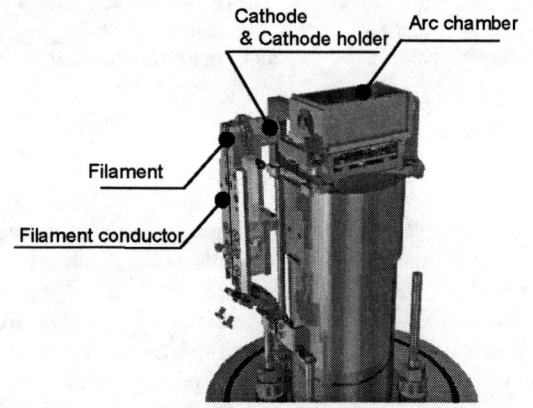

FIGURE 2. NISSIN IHC ion source.

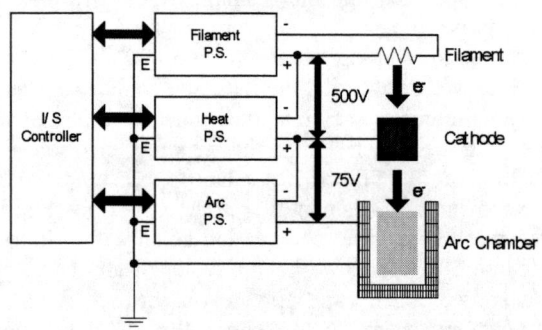

FIGURE 3. The source control system diagram.

Beam Current

The comparisons of the beam currents of ^{31}P and ^{75}As between the IHC type and Bernas type sources are shown in Figure.4 and 5. As shown in these figures, the IHC type ion source makes it possible to obtain multiply-charged ion beam currents more than 2 - 20 times of that of the Bernas type ion source. Suppose we implant a wafer in the dosage 1×10^{13} (dose/cm^2) with the beam current obtained in these results. Then we find that IHC type source shortens the processing time from 332 seconds (Bernas type) to only 16 seconds for $^{75}As^{+++}$ and 10 seconds to 5 seconds for $^{31}P^{++}$ & $^{75}As^{++}$ respectively, which is still half of that of the Bernas type source.

Hence, it is expected that a machine equipped with the IHC type source improve the productivity in the high energy range suggested above.

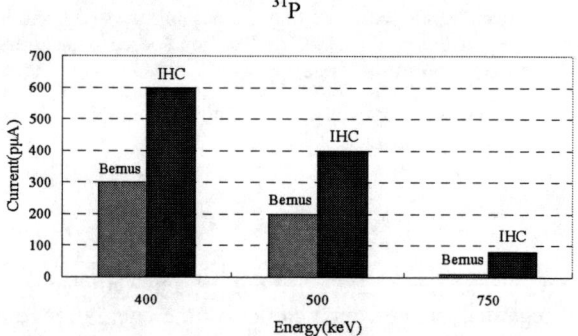

FIGURE 4. Comparison of the beam current for ^{31}P between Bernas type & IHC source.

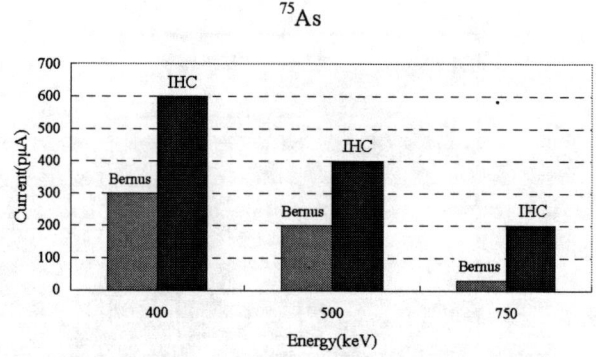

FIGURE 5. Comparison of the beam current for ^{75}As between Bernas type & IHC source.

TABLE 1.

Multiply-charged Ion	Bernas	IHC	Improvement
$^{31}P^{++}$	96wfs/hr	108wfs/hr	+12.5% UP↑
$^{31}P^{+++}$	23wfs/hr	47wfs/hr	+104% UP↑
$^{75}As^{++}$	96wfs/hr	108wfs/hr	+12.5% UP↑
$^{75}As^{+++}$	8wfs/hr	82wfs/hr	+925% UP↑

Lifetime

As seen in the previous section, the beam current of the multiply-charged ions significantly improved. However, if the lifetime of the source becomes comparably short in order to achieve such current, the productivity will not increase as much. Therefore, the lifetime of each source under continuous operation with a maximum multiply-charged ion beam current was compared. The result is shown in Figure.6. Running condition is $^{31}P^{+++}$ 300keV 80pμA for IHC source and 30pμA for Bernas type source. As is clearly seen in Figure 6, the lifetime of the IHC source is more than 50 hours which proves that it is more than 3 times longer compared to the Bernas type source.

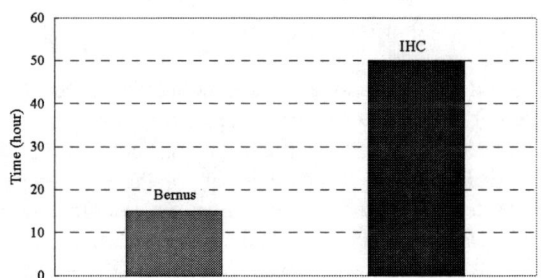

FIGURE 6. Comparison of the lifetime of Bernas type & IHC source.

Effective Throughput

Since it was revealed that both in beam current and lifetime the IHC type source exceeded the Bernas type source, it is expected that the effective throughput is also improved significantly. In this section, therefore, we provide some calculations to estimate the improvement of the effective throughput are shown. We considered the case of implanting a wafer in the dosage 1×10^{13} (dose/cm^2) with the maximum beam current of each ion species ^{31}P and ^{75}As. In this calculation, we assumed that the auto setup time 6 minutes for the IHC type source, while 3.6 minutes is used for the Bernas type source as described in EXCEED3000AH system performance [2]. The result is shown in Table.1. As is shown here, in all conditions, the throughput with the IHC type source exceeded the one with the Bernas type source. The effective throughput increases by 74 wafers/hour (+925%) in the best case, although the auto setup time of the IHC type is set a little longer than the Bernas type source.

Cross Contamination

With all these promising results on the productivity shown above, we further examined the reliability of the application for high energy implantation. First, it was studied that whether EXCEED3000AH equipped with the IHC type source can analyze the desired multiply-charged ions to be implanted. In order to clarify the problem, possible contaminant species generated around the ion source during the implantation of the multiply-charged ions of ^{31}P and ^{75}As are listed in Table.2. Typical mass spectrums of ^{31}P are also shown in Figure.7.

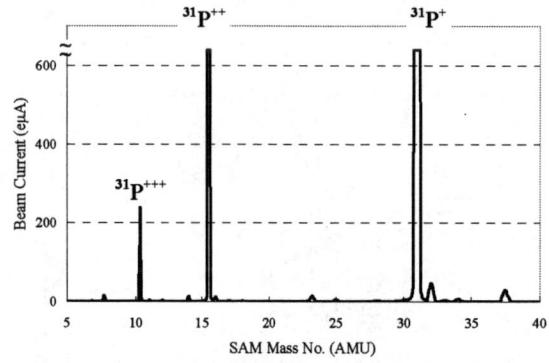

FIGURE 7. The mass spectrum of ^{31}P with arc voltage of 75V, arc current of 3,600mA

The mass resolution required to analyze the desired ion species from contamination species having a close trajectory is listed in the second column of Table.2. Since all the required resolutions in Table.2 are less than that which is achieved with EXCEED3000AH

TABLE 2.

Species to be implanted	Contaminant species having close trajectory	Resolution (m/Δ m) required
$^{31}P^{++}$	$^{14}N^+, ^{15}N^+, ^{16}O^+, ^{17}O^+, ^{29}BF^{++}, ^{30}BF^{++}$	31
$^{31}P^{+++}$	$^{10}B^+, ^{11}B^+, ^{12}C^+, ^{13}C^+, ^{29}BF^{+++}, ^{30}BF^{+++}$	34
$^{75}As^{++}$	$^{36}Ar^+, ^{38}Ar^+$	75
$^{75}As^{+++}$	$^{23}Na^+, ^{27}Al^+, ^{49}BF_2^{++}$	50

(more than 80), it is obvious that any suspected contaminations are readily eliminated.

Implantation Results

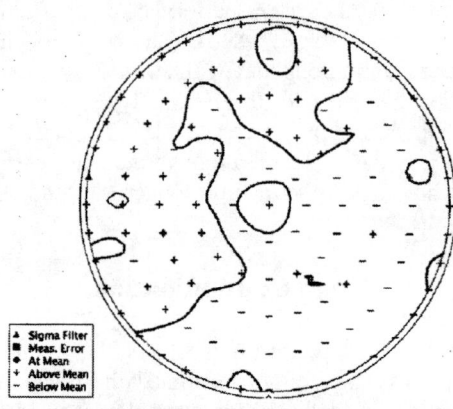

FIGURE 8. The Rs map of $^{31}P^{+++}$
Std Dev/Mean=0.384%.

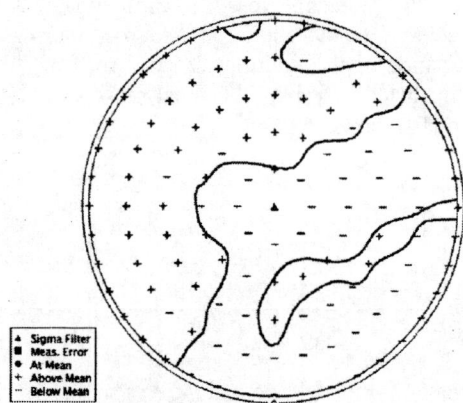

FIGURE 9. The Rs map of $^{75}As^{+++}$
Std Dev/Mean=0.303%.

Finally, Rs maps as implantation result are shown in Figure.8 and 9. We have implanted 2×10^{13} (dose/cm^2) of $^{31}P^{+++}$ and $^{75}As^{+++}$ to the wafers rotated by 7 and 23 degrees for tilt and twist at an energy 700keV. It was confirmed that the implantation is performed with good uniformity with σ less than 0.5% which verifies the applicability of these high energy implantation recipes with multiply-charged ions using the EXCEED3000AH.

CONCLUSION

In this paper, we have demonstrated that by replacing the Bernas type ion source with the IHC type ion source, the Nissin EXCEED3000AH can run twice the beam current for the doubly-charged ion beam, and a 20 times higher triply-charged ion beam of ^{75}As, which results in an the effective throughput increase of about 10 times. Lifetime of the source has increased more than 3 times compared to the Bernas type source in spite of the severe condition for ion source operation. The mass resolution (m/dm) of the tool of more than 80, was concluded to be high enough to prevent implantation of any cross contamination. Since a good implantation uniformity was confirmed with σ less than 0.5% in the implantation results, it was proved that the IHC type source is superior as a tool for production processing.

As described above, it is concluded that the IHC type ion source clearly promotes the system performance of the medium current implanter EXCEED3000AH to cover a part of the high energy range as substitute for the conventional high energy implanter, with sufficient productivity and reliability.

REFERENCES

1. R. Kirchner and E. Roeckl, "A cathode with long lifetime for operation of ion sources with chemically aggressive vapors," Nucl. Inst. & Meth., vol.127,pp.307-309,1975.
2. S. Sakai, "High Performance Medium Current Ion Implanter EXCEED3000AH,", Proceedings of the 2004 Int'l Conf. on Ion Implantation Tech, pp.70, 2004.

Improved Ion Beam Incident Angle Control for Varian E220 and E500 Implanters

David Hendrix, Zhiyong Zhao*, Reuel Liebert, Ken Gifford[†] and Pierre Mitchell[¶]

*Spansion LLC, Austin, TX, USA
[†] Varian Semiconductor Equipment, 35 Dory Road, Gloucester, MA, USA.
[¶] Therma-Wave, Fremont, CA. USA.

Abstract. This paper discusses the need and a solution for improved implant angle control required to support advanced device production at 90nm on Varian E220/E500 series ion implanters. The paper characterizes the software and hardware improvements made to the implanter in order to achieve improved control over implant angle. In-situ Therma-wave metrology checks along with split-lot test methodology and device electrical performance results are also characterized.

Keywords: Ion Implantation, Angle Control, Semiconductors
PACS: 61.72, 29.27, 41.75Ak, 41.85Ja, 41.85Si

INTRODUCTION

The challenges presented by ever-advancing nodes of semiconductor production have led to rapid obsolescence of older generation ion implantation systems. Upgrades in wafer size and angle control have been major drivers of this trend. While wafer size increase is primarily a productivity and cost of ownership issue, the degree of angle control affects the basic capability of the implanter to control dopant placement to the level needed for advanced device fabrication. By the 130nm node, it became evident that the accuracy of implant angle is critical to control the lateral dopant profile of pocket implants and the substrate channeling effect that can dominate the PMOS device sensitivity [1]. Device makers had to very carefully select their device layout and architecture to minimize these effects. Angle requirements become even more stringent with more advanced nodes [2].

This paper evaluates an upgrade to the E220/E500 implanter series that substantially improves its level of implant angle control. First introduced in 1989, the E220 effectively operated with an implant angle accuracy of ±1.2° (combining in quadrature the effects of tilt angle and beam entry angle uncertainty). The upgrade brings the implant angle ("beam incident angle" or BIA) accuracy from ±1.2° to a specification of ±0.5°. The corresponding dose error arising solely from the cosine projection at an implant angle of 45° should improve from 2.5% to 1.25% at the extreme of the tolerance.

DESCRIPTION

The Beam Incident Angle (BIA) upgrade hardware consists of an improved accuracy high-resolution dual-encoded platen tilter assembly and control electronics for positioning the wafer in the process chamber, a new "focus cup" faraday (used conventionally during scan setup), a new positive position optical sensor for tilter initialization, and a new traveling faraday centerline alignment sensor. The new focus faraday and the traveling faraday assembly, which is normally used for setting up the scan uniformity, are used with new software packages to measure the beam entry angle (BEA) with respect to the wafer plane at zero tilter angle. The implant angle (BIA) is calculated from the set tilter angle and the measured BEA.

The architecture of the E220/E500 implanter has been described elsewhere [3]. It features an electrostatic scanner and a dipole angle correction magnet before final acceleration. The angle correction magnet converts the angled scan that emerges from the electrostatic deflection plates to a parallel-scanned beam. The alignment of the implanter and the setting of the field in this magnet will determine the BEA for a given ion beam. The implanter alignment was verified before the upgrade was installed. The system computer sets the field in the magnet according to a table lookup ordered by beam stiffness. The software provided with the upgrade provides for automated generation of table entries using a machine BEA routine known as the "Refraction Method". The Refraction Method measures the difference in position of the ion beam's centerline location in the process chamber when voltage is applied to the acceleration tube. If the angle of attack is perfect, i.e. the BEA is zero degrees, then there should be no shift when the acceleration tube voltage is turned on. The traveling faraday is moved through the beam in 0.1mm steps and the software calculates the position of the beam center from the resulting plot of beam current vs. profiler position. An initialization routine verifies the calibration of faraday position readout by calibrating to the new zero-position sensor. The resolution of the method is ~0.004° based on geometry and faraday position accuracy. . By using this method, the magnet constant table used in beam setup was filled with constants that were adjusted using an automated routine to iteratively minimize the measured shift.

Once the table was constructed, a second method was used to measure the BEA at the time of each beam setup. The second method (the Shadow Method) is faster, does not require changes in the beam position, and measures the angle of the beam at its final energy in the process chamber. The tradeoff for using this fast method is a small degradation in the resolution to ~0.015°. The Shadow Method employs an ion beam profiler faraday (new focus cup) located along the back wall of the implant chamber and the traveling faraday

The Shadow Method uses the motion of the traveling faraday across the process chamber to block the ion beam from entering the narrow slit in the focus faraday mounted in the back of the implant chamber. By measuring the drop in beam and correlating it to the traveling faraday encoded position, the software can measure and calculate the BEA. During beam setup, this method is used to assure the BEA is within the ±0.2° interlock that guarantees the BIA accuracy to within ±0.5°. If the measurement is outside the allowed error, then the angle correction magnet current is adjusted, the beam is repositioned in the focus faraday cup, and the measurement is repeated. In this manner, closed loop control of the BEA is achieved.

SYSTEM TESTING

The upgraded tilter assembly uses encoders for the motor drive and also for the driven shaft of the tilter. The software requires that the two encoders agree when the tilt angle is set. Figure 1 shows the position variation of the driven shaft encoder during a test that cycled the system 7669 times through various angles from the implant to load position. The specification of the design is that the tilter be held to within ±0.35° to assure the overall goal and the data in the figure (with 22.76 counts/degree) show variation well within that target.

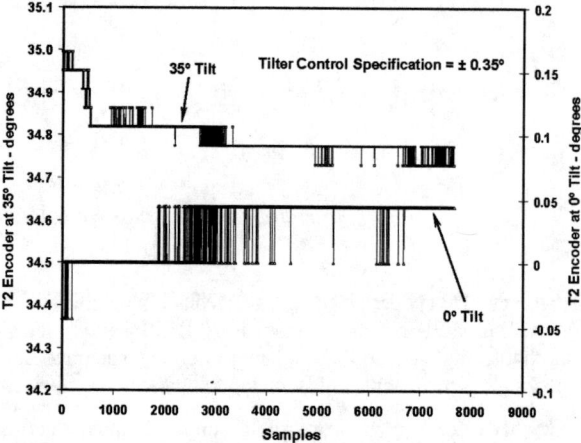

Figure 1. Position Repeatability of the Upgraded Tilter Measured with New Driven-Side Encoder

The Refraction and Shadow methods were compared to verify that the angles measured by the two methods are equivalent. Testing of the prototype unit at Varian and the Beta unit at Spansion showed the upgraded implanter could routinely achieve a BEA specification of ±0.20° on a variety of implant recipes.

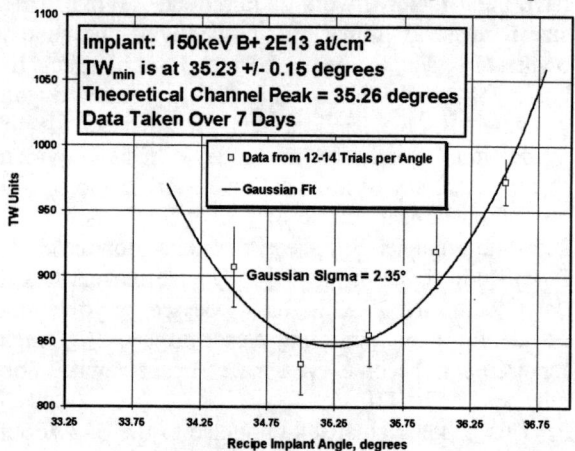

Figure 2. Alignment of the BEA system and implanter is shown using a <112> channeling test

Alignment of the implanter and BEA system at the Beta site was verified by a channeling test using the <112> axis that should align at a tilt angle of 35.26°. Fig. 2 shows the results of a seven day run in which five angles spanning the expected channeling minimum were measured by Thermawave. The data for each angle (12-14 trials) were used to determine an average and an error-weighted Gaussian fit was used to determine that the alignment was good to 0.03° within the fit error of 0.15°.

90NM PROCESS QUALIFICATION

To monitor beam incident angle at high tilt, the production B+ beam is implanted into test wafers at 34, 35 and 36 degrees over the period of months. Because of axial channeling, a minimum Thermawave measurement is expected at near 35 degrees. A Thermawave Model TP-500 was used for all the testing. A parabolic fit through the three Thermawave measurements serves to estimate the programmed tilt that would give in the lowest measurements. At this tilt, the channeling is greatest. Typically, good agreement between the values calculated from Thermawave and sheet resistance measurements demonstrated the accuracy of this method.

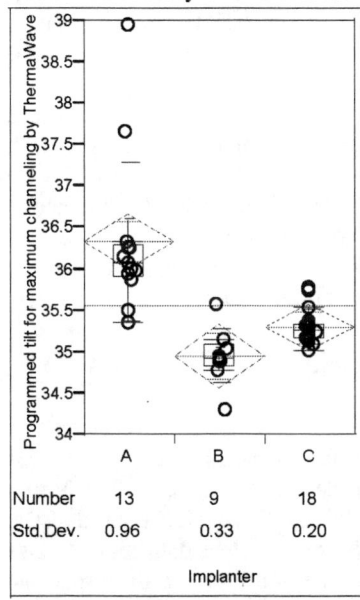

Figure 3. The best achieved channeling angle for the <112> axis based on Thermawave measurement. Three groups are given: an E500EHP implanter before and after the BIA upgrade as well as an E500EHP without the BIA upgrade. (B,C and A). The theoretical tilt angle at the channel is 35.26°

These calculated values are plotted in Figure 3. Implanter A is a virtually identical E500HP without the upgrade, B is the E500EHP before the upgrade, and C is the same E500EHP after the upgrade The programmed implant tilt angle determined to give the minimum measurements is plotted for the E500EHP for making comparisons before and after the BIA upgrade. Comparing the data from E500HP without the upgrade to data from the E500EHP with the upgrade illustrates that the variability of beam incident angle is reduced by at least 50%. Comparison of (a weighted average of) the standard deviations of Thermawave data from A and B to the standard deviation of C suggests that the upgrade reduced this variability by 71%.

Split-Lot Tests

A Flash Memory device lot was run to test the BIA upgrade. The splits were arranged to have three wafers run on each of three different implanters for several consecutive days. The splits are: six wafers for two days on an E500HP without the BIA upgrade, six wafers for two days on a VIISta810EHP, and thirteen wafers for four days on the E500EHP with the BIA upgrade. VIISta810EHP was used as a preferred performance standard since it has an implant angle correction package that has been proven to perform better than the standard E500 implanters at Spansion fab 25. The E500HP without the BIA upgrade is utilized as a reference of a nominal E500 implanter. The 90nm feature size device was utilized for this test. A high tilt angle channel engineering implant into the memory cell that is angle-sensitive was used to enhance the response.

Two more device lots of 110nm feature size were also tested. The results were similar to the 90nm lot. This paper will present the data of the 90nm device lot only for the purpose of clarity.

The following three figures represent the electrical evaluation of the split lot. The data has been normalized for proprietary reasons. The parameters related to this implant did show responses in favor of the BIA upgrade.

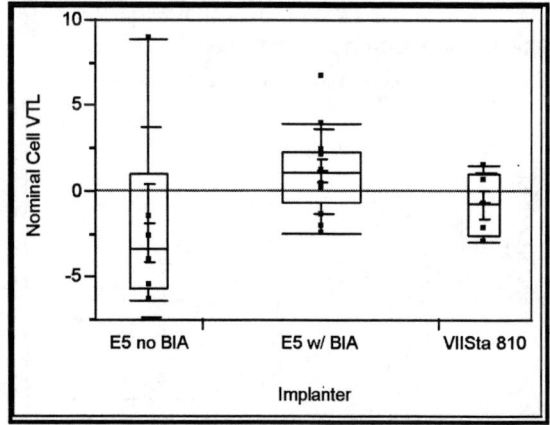

Figure 4. The linear Vt variation in percentage as measured on the tested device wafers.

Figure 4 shows the percent variation of the nominal memory cell's linear Vt. From the left, the first data group is from an E500HP implanter without the BIA upgrade. The second data group is from an E500EHP with the BIA upgrade. Finally, the third data group is from a VIISta 810 implanter serving as a preferred performance standard. The data shows that the E500EHP with BIA achieves Vt variation 56% smaller than the E500HP without BIA. Further, the BIA E500EHP performance is closer to that of a VIISta 810EHP, which has a reduced Vt variation of 66% over the non-BIA E500HP.

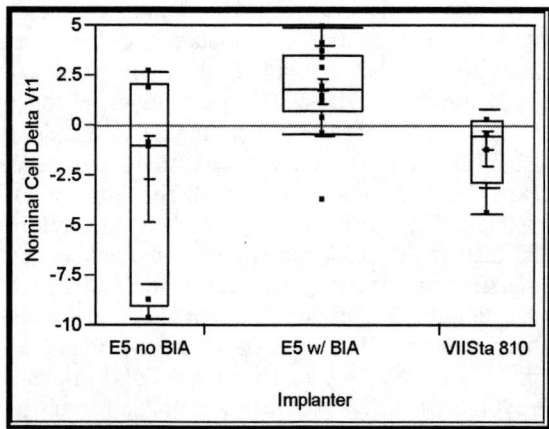

Figure 5. The Delta Vt variation in percentage as measured on the tested device wafers.

Figure 5 shows the corresponding delta Vt of Figure 4. The E500EHP with the BIA upgrade has an improved delta Vt over the non-BIA E500HP and it behaves closer to a VIISta 810EHP. The magnitude of improvement is 58% for the E500 with the BIA and 63% for the VIISta 810EHP. This correlates to the Vt variation improvement observed in Figure 4. Figure 4 illustrates that the Delta Vt for the BIA E500EHP tends to have higher values than either of the other implanters while its data distribution is the tightest. This phenomenon is not understood at this time.

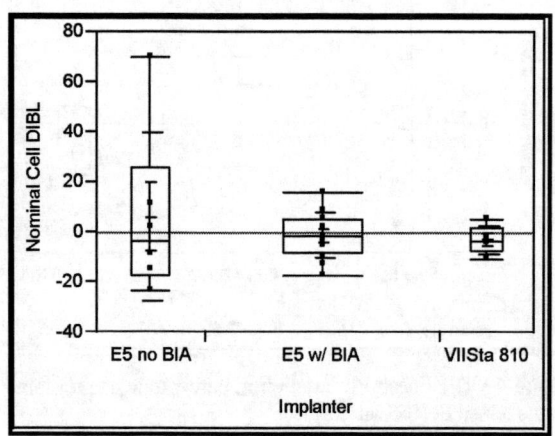

Figure 6. The Drain induced barrier lowering variation in percentage as measured on the tested device wafers.

The smaller variations of Vt and delta Vt as observed in Figure 4 and Figure 5 can be attributed to a smaller implant angle variation. As evidence, Figures 6 presents DIBL (drain induced barrier lowering) variation of the same memory cell. With a better defined drain or equivalently, a better defined channel, variation in DIBL is reduced by 73% for the E500EHP with the BIA and 83% for the VIISta 810EHP over the non-BIA E500HP. If this channel engineering implant (pocket implant) is the only contributor to the difference, then this would imply that the smaller Vt and delta Vt variations are due to a better controlled implant angle during the implantation process.

CONCLUSIONS

The BIA Upgrade on E500EHP, based on the limited data on test wafers and device wafers, has a measurable improvement over the angle control capability of standard E500 series implanters. The design goal for the BIA on E500-series implanters is to improve BIA from a capability of $\pm 1.2°$ to a specification of $\pm 0.5°$. This is an improvement of 58% over its current performance. The data on the Vt and delta Vt suggests that the E500 with BIA can control the beam incident angle to its design specification of $\pm 0.5°$ and is compatible with 90nm processes. In fact, the data shows it behaves closer to the VIISta 810 implanter that has an angle controllability of $\pm 0.2°$.

ACKNOWLEDGEMENTS

The Spansion authors would like to thank Linda Wang of Fab25, Spansion LLC, for her helpful discussion on device parameters. The Varian authors would like to thank Dave McClellan and Dave Olson for their help with the test data and Moussa Haddad for his help and guidance with respect to all the software tasks.

REFERENCES

1. Kenn S.Y. Yeh, M.C. Chiang, C.J. Tsai, Y.L. Wang and J.K. Wang, "Optimization of High Tilt Pocket Implant Process for Improving Deep Sub-micro PMOS Device Sensitivity" in Proc. of 14th Int. Conf. on Ion Implantation Technology, 2002, p.13-16
2. U. Jeong, Z. Zhao, B.N. Guo, G. Li, and S. Mehta, "Requirements and Challenges in Ion Implanters for Sub-100nm CMOS Device Fabrication," CP680, Application of Accelerators in Research and Industry: 17th Int'l. Conference, edited by J.L.Duggan and I.L.Morgan, AIP press, New York, (2003), 697
3. D.W. Berrian, R.E. Kaim, J.W. Vanderpot, and J.F.M. Westendorp, Nucl. Instr. And Meth. B37/38 (1989) 500

Development of an Ion Beam Aligner for Liquid Crystal Displays

Takeshi Matsumoto, Yasuhiro Matsuda, Masahiro Tanii, Masashi Konishi, and Yasunori Andoh

Nissin Ion Equipment Co., Ltd., 29 Hinokigaoka, Minakuchi-cho, Koka, Shiga 528-0068, Japan

Abstract. We have developed an ion beam irradiation system for an alignment process of liquid crystal displays. This system is applicable to glass substrates of up to 550 x 650 mm^2 in size. The ion beam exhibited an excellent uniformity of ion beam current density, incident angle and divergence angle. Twisted nematic cells with ion-irradiated polyimide films showed a well-aligned liquid crystal texture. This result indicates that ion beam alignment is a promising method alternative to the conventional rubbing method which has problems such as dust particles and generation of electrostatic charges.

Keywords: Ion beam, Liquid crystal display, Alignment, Incident angle, Divergence angle
PACS: 41.75.Ak, 42.70.Df

INTRODUCTION

In the liquid crystal display (LCD) industry, rubbed polyimide is generally used as alignment films. However, the rubbing method has drawbacks such as the generation of dust particles and electrostatic charges, and an alternative method is strongly demanded. A new alignment method using an ion beam is a non-contact alignment method and does not have these problems. Several research groups have reported on the ion beam alignment method applied to polyimide films [1, 2] and inorganic films [3-6]. They indicated that the method had a high potential as an alternative way of producing alignment films instead of the conventional rubbing method.

In order for the ion beam alignment method to be employed in the LCD industry, the development of an ion source with a large beam size is indispensable because the display panel size has become larger. In addition, a high current ion beam is essential for the high throughput of LCD panels, and good current uniformity of the ion beam is necessary for the production of high quality alignment films.

We have recently developed an ion beam irradiation system called "Beam Aligner" which is designed for alignment film production and equipped with a wide ion source applicable to glass substrates of up to 550 x 650 mm^2 in size. In this paper, performance of the Beam Aligner is detailed.

SPECIFICATION OF THE BEAM ALIGNER

Figure 1 shows a schematic diagram of the Beam Aligner. A substrate is placed on the sample setting table at the loading stage and is transferred into the load lock chamber, then the chamber is evacuated by a vacuum pump. After the load lock chamber is evacuated, the gate valve opens and the ion source runs with a low-energy beam. A hot filament

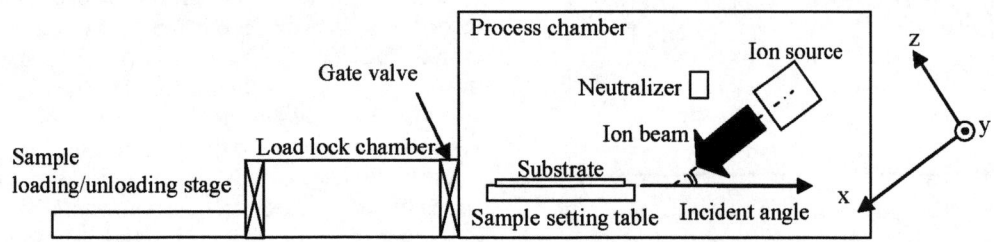

FIGURE 1. Schematic of the Beam Aligner.

TABLE 1. Specification of the Beam Aligner.

Items	Comments
Maximum substrate size	550 x 650 mm
Ion incident angle (Ion source tilt angle)	10 - 90 deg
Ion species	Argon
Ion energy	200 - 1000 eV
Maximum ion beam current	500 uA/cm @ 500 eV
Beam uniformity	(Max-Min)/(Max+Min)< ±5%
Beam monitoring system	Beam uniformity monitoring system
	Beam incident & divergence angles analyzing system
Operation	Semiautomatic

neutralizer is included to neutralize ion beam charges accumulated on the substrate surface. The ion source is placed with its longitudinal dimension parallel to the y axis. The substrate is moved into the process chamber, and transferred to be irradiated with Ar ions at a constant speed. The substrate is moved back to the load lock chamber after the ion beam irradiation.

Ion incident angle to the substrate surface is controlled by tilting the ion source inside the chamber. This motion is mechanically performed using a motor without breaking the vacuum environment of the chamber. This is helpful for research that requires frequent changes in ion incident angle.

Table 1 shows the specification of the Beam Aligner. This system is compatible with a substrate size of up to 550 x 650 mm^2. The sample setting table is rotatable to set the substrate at any angle. The ion incident angle to the substrate surface is variable from 10 to 90 deg. The acceleration energy ranges from 200 to 1000 eV. The maximum ion beam current density per unit length along the y axis is 500 uA/cm at 500 eV. Uniformity of the ion beam density along the 640 mm width is better than 5%.

In the process chamber, a multiple Faraday cup array is placed to measure the distribution of ion beam current density. We can check the distribution before the ion beam irradiation, and adjust parameters for the ion source operation if necessary. In addition, the Beam Aligner has a beam monitor which enables us to measure the distribution of the incident and divergence angles of the ion beam along the 640 mm width.

The ion source is controlled through the computer. Since there are quite a few parameters for ion source operation, it is troublesome to input all the parameters each time. To avoid this, it is possible to register a set of those parameters as one recipe, and use it to operate the ion source instead of inputting the parameters each time.

RESULTS AND DISCUSSION

Figure 2 shows a typical distribution of ion beam current density along the y axis. The excellent beam uniformity of 2.8% and 3.2% over 670 mm was observed at the ion acceleration energy of 500 and 250 eV, respectively. This large and uniform ion beam ensures the application to the alignment process of 550 x 650 mm^2 substrates. Since the ion source has multiple filaments whose currents are adjusted individually, the beam current distribution can be made uniform even if it becomes worse for any reason.

We measured incident and divergence angles of the ion beam by the monitor described above. It is capable of measuring these angles in the x-y plane. Before measuring incident angle distribution of the ion beam used for alignment film irradiation, we examined the path of the ion beam which was curved deliberately by permanent magnets. As shown in Figure 3 (a), three pairs of magnets were fixed at the edge of the ion beam outlet. The ion beam was expected to curve under the influence of the magnetic field shown in the figure. The ions passing around the pairs of the magnets at ±160 mm should curve to the –y direction, while the ions passing around the pair of the center magnets should curve to the +y direction. The monitor successfully detected this change in ion incident angle caused by the magnetic field, as shown in Fig. 3 (b). In this graph, zero degree means the ion trajectory is parallel to the x axis.

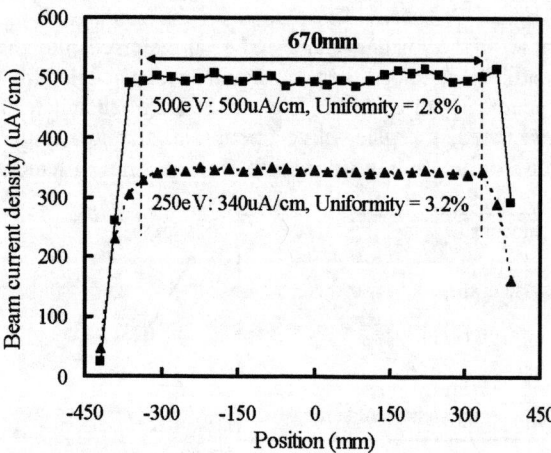

FIGURE 2. Typical distribution of ion beam current density along the y axis.

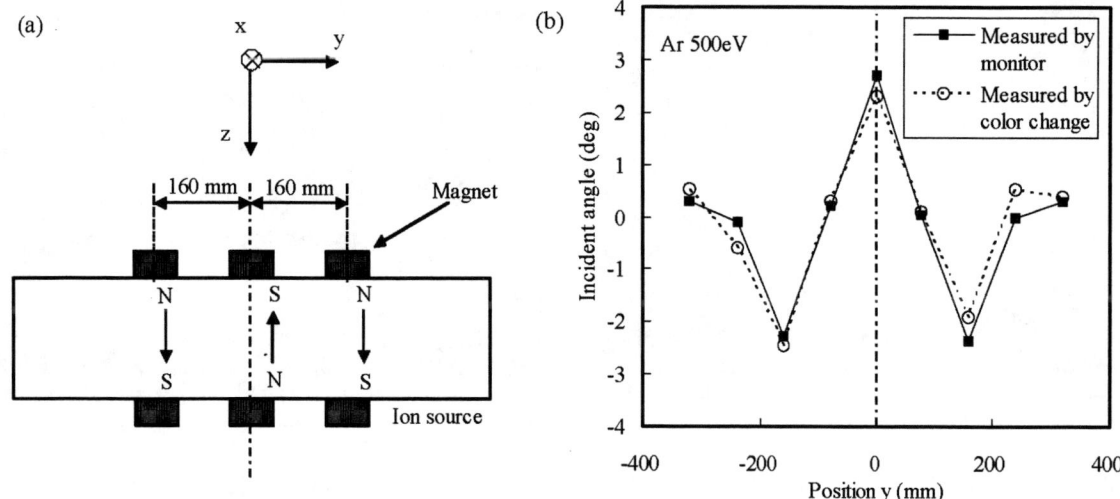

FIGURE 3. (a) Arrangement of magnets in the ion source outlet (Top view). (b) Distribution of incident angle measured by the beam monitor and change of plastic film in color in the case of the ion source with three pairs of magnets.

To briefly investigate performance of the monitor, we measured the incident angle in another way, and compared the results. A plastic film was placed under the ion source and irradiated with ions. A mask with multiple slits was set between the ion source and the film so that a part of the ion beam could pass through the slits and reach the film. The irradiated part of it changed its color. Incident angle was obtained by measuring the positions of the color. Comparison of incident angle distribution between the two methods is shown in Fig 3 (b). Both methods showed very similar distributions. The small discrepancy between them is most likely to be caused by measurement uncertainty of the positions where the color changed. This result indicates the usefulness of our beam monitor for incident angle observation.

Next, we removed the magnets and measured the incident angle distribution of the ion beam used for alignment film irradiation using the monitor. Figure 4 shows a typical distribution of incident and divergence angles of the ion beam at the ion energy of 500 eV. These two angles are very uniform over a width of 640 mm. Incident angle in the x-y plane is believed to have a significant influence on the alignment direction of liquid crystal in the plane of the film, whereas divergence angle is considered to be linked to the quality of the alignment texture. From this point of view, a nonuniform incident angle could result in nonuniform brightness of a LCD panel, while a smaller divergence angle could improve brightness and darkness. We obtained a divergence angle of about 1.4 degrees on average, and the LCD panel exhibited good black and white images.

In the Beam Aligner, the distribution of the beam density, divergence angle and incident angle are automatically measured and recorded by the computer before the irradiation of a substrate with ions. This logged data can be used for the purpose of quality management and process development.

In order to investigate the performance of alignment film irradiation, twisted nematic mode cells were assembled using a pair of polyimide films on indium-tin-oxide (ITO) coated glass substrates. The ITO electrodes were coated in a square at the center of the substrates. The alignment texture of the liquid crystal showed a uniformly bright image under a crossed polarizer configuration without voltage between the electrodes, whereas the cells showed uniformly dark image with a voltage of 5 V, as shown in Figure 5. This result indicated that the ion-irradiated polyimide surface can be a useful alignment layer. We

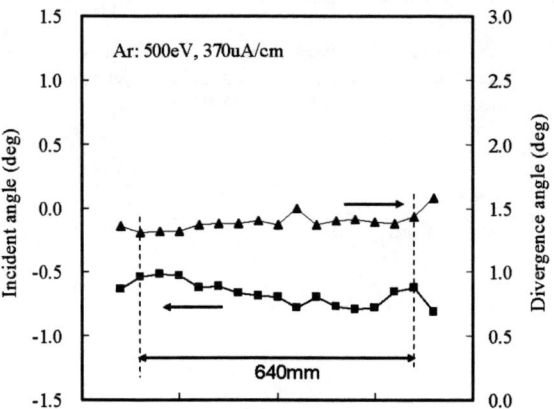

FIGURE 4. Typical distribution of incident and divergence angles.

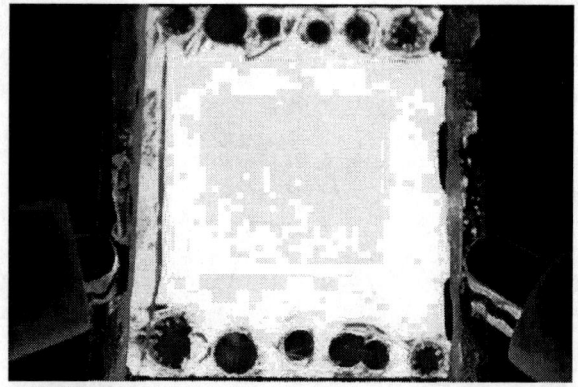 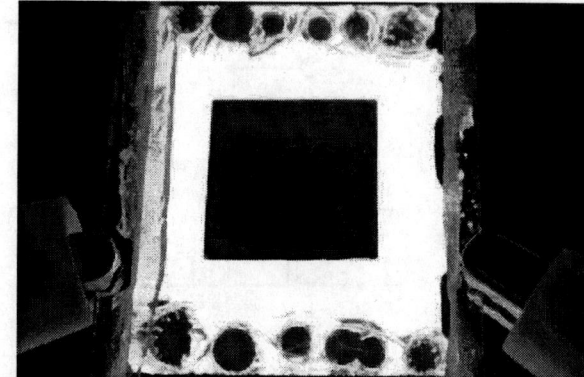

FIGURE 5. Photographs of twisted nematic cells assembled using a pair of ion beam-treated polyimide films. At the center of the glass substrates, indium-tin-oxide electrodes were coated in a square. On-state (left) and Off-state (right).

assembled LCD panels of up to 550 x 650 mm^2 in size using the ion beam alignment technique, and all of them showed a well-aligned liquid crystal texture.

We are developing a future Beam Aligner aiming at larger glass substrates and mass production. This next-generation Beam Aligner will process substrates of up to 730 x 920 mm^2 in size, have an upright substrate transfer which is free of dust particle accumulation on a substrate, and have fully automated operation. Once a glass cassette is set, substrates are sequentially taken in the vacuum chamber, irradiated with ions, and brought back to the cassette. The only thing the operator has to do is to choose a recipe of ion source operation parameters. Maximum ion beam current density is more than doubled in order to increase the productivity.

CONCLUSION

We have developed an ion beam irradiation system for an alignment process of liquid crystal displays. This system is applicable to glass substrates of up to 550 x 650 mm^2 in size and has systems to monitor beam current density, incident angle and divergence angle. As to these beam properties, an excellent uniformity across the substrate was exhibited. Twisted nematic cells with ion-irradiated polyimide films showed a well-aligned liquid crystal texture. This result indicates that ion beam alignment is a promising method alternative to the conventional rubbing method which has problems such as dust particles and electrostatic charges.

ACKNOWLEDGMENTS

We are grateful to Dr. Y. Iimura of Tokyo University of Agriculture and Technology for many useful discussions.

REFERENCES

1. O. Yaroshchuk, R. Kravchuk, A. Dobrovoskyy, L. Qiu and O. D. Lavrentovich, *Liquid Crystals* **31**, 859-869 (2004).
2. P. Chaudhari, J. A. Lacey, S.-C. A. Lien and J. L. Speidell, *Jpn. J. Appl. Phys.* **37**, L55-56 (1998).
3. P. Chaudhari et al, *Nature* **411**, 56-59 (2001).
4. D. K. Lee, S. J. Rho, H. K. Baik, J.-Y. Hwang, Y.-M. Jo, D.-S. Seo, S. J. Lee and K. M. Song, *Jpn. J. Appl. Phys.* **41**, L1399-1401 (2002).
5. S. J. Rho, D.-K. Lee, H. K. Baik, J.-Y. Hwang, Y.-M. Jo and D.-S. Seo, *Thin Solid Films* **420-421**, 259-262 (2002).
6. A. Asahara, A. Horibe, H. Kimura, J. Nakagaki, T. Nishiwaki, H. Tokushige, T. Yamada, H. Kitahara and Y. Shiota, *Proceedings of IDW/AD'05*, 309-312 (2005).

Real-time Optimization Method for Optical Parameters of Ion Implanters

Seiji Ogata[1], Tsutomu Nishihashi[2], Kazuhiko Tonari[2], Hidekazu Yokoo[3], Hideo Suzuki[3], Takeshi Hisamune[3] and Masasumi Araki[4]

1) ULVAC, Inc., Research and Development Div., 2500 Hagizono Chigasaki Kanagawa 253-8543, JAPAN
2) ULVAC, Inc., Inst. Semiconductor Technology, 1220-1 Suyama Susono Shizuoka 410-1231, JAPAN
3) ULVAC, Inc., Semiconductor Equipment Div. 2, 1220-14 Suyama Susono Shizuoka 410-1231, JAPAN
4) ULVAC, Inc., Software Development Control Solution Div., Hagizono Chigasaki Kanagawa 253-8543, JAPAN

Abstract. Real-time optimization for optical parameters, such as applied voltage to the electrostatic quadrupole lens, has been realized by using newly developed high-speed computation algorithm for charged particle beams. The virtual optimization code has been incorporated in the control system of SOPHI-200, which is the ULVAC's new medium current ion implanter. Automatic setup within 5minutes is achieved for any recipe of implantation.

Keywords: ion implanter; real time; optimization; beam optics
PACS: 85.40.Ry; 41.85.-p

I. INTRODUCTION

Aiming at the high performance, shorter set-up time is required for the industrial application of semiconductor manufacturing equipments. At ion implanters, the set-up time is mainly restricted by three tunings. They are the tuning of the ion source, the tuning of the scan waveform and the tuning of optical lens to attain the maximum beam current with the minimum beam distribution on the target. We have developed the real-time optimization algorithm for optical parameters and incorporated the algorithm into the control system of the ULVAC SOPHI-200. The overview and the performance of SOPHI-200 is discussed elsewhere [1]. In this paper, framework of the real-time optimization and performance of the optimization are discussed.

II. FRAMEWORK OF THE REAL-TIME OPTIMIZATION

The bird's eye view of SOPHI-200 is shown in figure 1. Optical system of SOPHI-200 has been designed by utilizing both a high-speed envelope computation and a precise ray tracing. The optical design of SOPHI-200 is characterized as low aberration with high performance. Low aberration optics has achieved the high dose uniformity as low as 0.5% without invoking the modulation of the scan waveform. Two pairs of electrostatic quadrupole doublet lenses are mounted at the both sides of the acceleration column to achieve a wide range focusing of the ion beam.

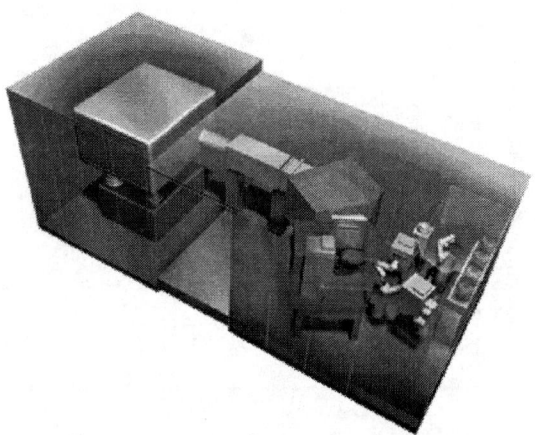

FIGURE 1. Bird's eye view of SOPHI-200

1. Virtual Optimization

Since the space charge effect is substantially non-linear, optical lens parameters act non-linearly on the beam performance such as the transmission to the target or the beam distribution on the target. Non-linear parameters are usually optimized through monitoring the variation of each parameter.

In conventional implanters, non-linear parameters such as applied voltage to electrostatic lenses are tuned by monitoring the variation of the beam performance with varying the actual applied voltage. The lens parameters of SOPHI-200, on the contrary, are optimized in the virtual space. Since the variation of parameters is executed without varying the actual applied voltage, any electrical discharge or any unwanted irradiation to the chamber does never occur. The parameters obtained by this virtual optimization have been confirmed to be superior to those obtained manually.

2. Classification of the Computational Algorithms

There are two different approaches to the beam optical computation considering the space charge effect. One approach is the macro particle computation, which solve simultaneously the kinetic equation of charged particles in the electric and/or magnetic field and the Poisson equation. The spatial distribution and the aberration are obtained through this approach with somewhat long, several tens of seconds or more, computational time. Another approach is the fluid model such as the Kapchinskij and Vladimirskij equation [2,3]. We have extended the K-V equation to be able to describe the non-uniform distribution of the beam cross section.

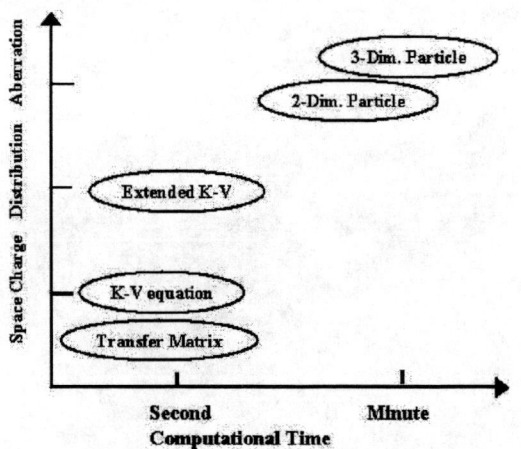

FIGURE 2. Classification of the computational algorithms for the charged-particle beams. The abscissa and the ordinate show the general idea of the computational time and the obtained information

The computational algorithms for the charged-particle beams are classified in the viewpoints of the computational time and the obtained information as shown in figure 2. Though the computation with the transfer matrixes is the fastest algorithm, non-linear effects such as the space charge or the aberration can't be considered. On the other hand, the macro particle computation, which is capable of full consideration of non-linear effects, requires iterative converging computation. The computation with extended K-V equations is the fastest algorithm that can consider the non-uniform distribution and the space charge effect.

The computational time with the extended K-V equation is nearly proportional to the number of the ellipsoidal shells and thus the computation has been executed almost within one second.

3. Extended Kapchinskij and Vladimirskij Equation

The K-V distribution is a hyper-ellipsoidal shell in four-dimensional phase space as follows,

$$f(x, y, x', y') = \delta(\frac{x^2}{a^2} + \frac{y^2}{b^2} + \frac{a^2 x'^2}{\varepsilon_x^2} + \frac{b^2 y'^2}{\varepsilon_y^2}), \quad (1)$$

where ε_x and ε_y are the emittance in the x and y-plane.

Since the two-dimensional projections on the x-y plane of each hyper-ellipsoidal shell is the ellipse with uniform density, the two-dimensional projection of overlaps of K-V distributions is an overlap of ellipses as illustrated in figure 3.

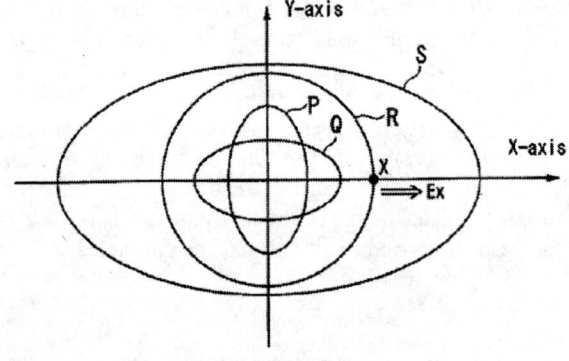

FIGURE 3. Two-dimensional projection of overlapping of the K-V distributions

The exterior electric field along the x-axis due to the charge of ellipsoidal distribution is given semi-analytically as

$$E_x(x) = \frac{2neab}{\pi\varepsilon_0} \frac{1}{\sqrt{x^2 + (b^2 - a^2)}}$$
$$\times \int_{-1}^{1} \frac{(1-t^2)^2}{(1+t^2)(1+\alpha t^2 + t^4)} dt, \quad (2)$$
$$\alpha = 2\frac{x^2 - (b^2 - a^2)}{x^2 + (b^2 - a^2)},$$

where n is the number density of the charged-particle, e the elementary charge, ε_0 is the permittivity of the vacuum. The interior electric field along x-axis is given as

$$E_x(x) = \frac{nebx}{\varepsilon_0(a+b)}. \quad (3)$$

In figure 3, the electric field at point X due to the ellipse P or Q is represented by equation (2) and the electric field due to the ellipse R or S is represented by equation (3).

The original K-V equation is represented as

$$\frac{d^2x}{dz^2} + F_x(x)x - \frac{E_x(x)}{2\phi} - \frac{\varepsilon_x^3}{x^3} = 0,$$
$$\frac{d^2y}{dz^2} + F_y(y)y - \frac{E_y(y)}{2\phi} - \frac{\varepsilon_y^3}{y^3} = 0, \quad (4)$$

where $F_x(x)$ and $F_y(y)$ are the external focusing, $E_x(x)$ and $E_y(y)$ are the electric field along the x-axis and y-axis at the boundary of the ellipse, ϕ is the electrostatic potential. When the beam is represented by a overlap of the K-V distributions, the electric field in the third term of the K-V equation is replaced by summation of the electric fields of each ellipse.

4. Emittance

As the initial condition, any algorithm for charged-particle beams requires the emittance at the ion source. The emittance of the ion beam extracted from rectangular shaped slit is well approximated as the product of the emittance in the longitudinal direction of the slit (y-plane) and that in the latitudinal direction (x-plane). The ellipse, which fits the experimentally observed emittance [4], is used as the emittance in the longitudinal direction of the slit. On the other hand, the emittance in the latitudinal direction of the slit changes as the operational condition varies [5]. The emittance at the ion source is estimated inversely from the spatial distribution measured at the scan profile monitor by solving the extended K-V equation [6]. The scan profile monitor, whose aperture width is 30mm, is incorporated in the main Faraday-cup of SOPHI-200. Several sets of emittance are prepared to cope with the various operations.

The beam profile measured at the scan profile monitor is compared with the simulated profile in figure 4. Sufficient agreement is confirmed.

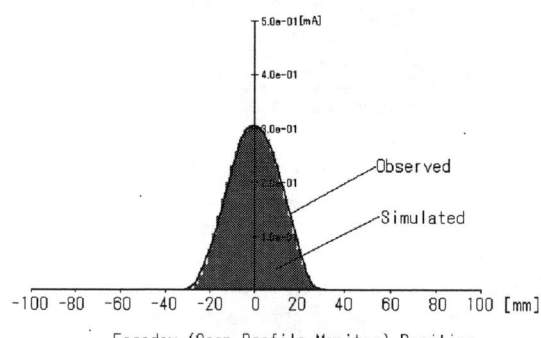

FIGURE 4. Beam profile measured at the scan profile monitor of SOPHI-200. Solid line shows the experimentally observed beam profile and the histogram shows the computer simulated beam profile

III. OPTIMZATION FOR OPTICAL PARAMETERS

When the implantation recipes such as ion species, implantation energy and beam current are specified, the operational condition of four quadrupole lenses must be tuned. For the recipes not stored in the memory of the control system, the operational condition is determined through the virtual optimization code based on the extended K-V equation. The highest transmission of the ion beam to the target with minimum spatial distribution on the target is searched

1. Optimization Algorithm

Since the transmission and the spatial distribution are the complex functions of the applied voltages of lenses and the maximum voltage applied to the lenses are restricted within finite values, the searching algorithm is categorized as the constrained non-linear optimization [7]. The multiplier method has been coded as the searching algorithm.

2. Performance of the Optimization

A part of the screen of the control system of SOPHI-200 is shown in figure 5. Though the screen is written in Japanese, the picture shows the screen when the virtual optimization code is executed. The parameters are optimized within from 5 to 30 seconds.

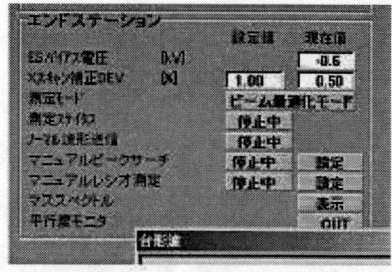

FIGURE 5. The screen of the control system of SOPHI-200.

The beam envelopes from the ion source to the target, which are obtained by solving the extended K-V equation, are shown in figure 6. Upper half of the optical axis in each figure is the envelope in the horizontal plane and the lower half is that in the vertical plane. The diverging beam under non-optimized operational condition as figure 6(A) is optimized as figure 6(B). The spatial distributions on the target are shown in figure 7. Figures 7(A) and 7(B) correspond to the operation of figures 6(A) and 6(B) respectively. The spatial distribution on the target is enhanced by one order of magnitude.

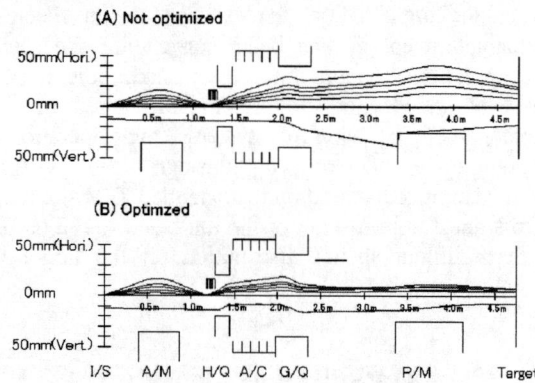

FIGURE 6. Beam envelopes of SOPHI-200. The ion is B+ of 0.35mA at the extraction voltage of 33kV and the implantation energy of 50keV. I/S is the ion source, A/M is the mass analyzing magnet, H/V is the quadrupole double on the high voltage terminal, A/C is the acceleration column, G/Q is the quadrupole double on the ground potential and P/M is the parallelizing magnet.

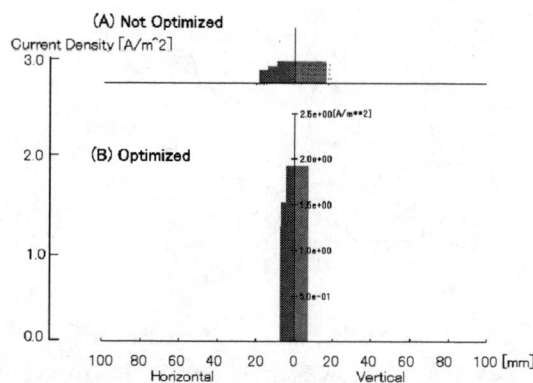

FIGURE 7. Spatial distribution on the target. The left half of the center position in each figure is the spatial distribution along the horizontal axis and the right half is that along the vertical axis. Ordinate of figure (A) and (B) are current density at the same scale.

IV. CONCLUSIONS

The virtual optimization code, which utilizes the extended K-V equation, has been developed and incorporated into the control system of SOPHI-200. The optical parameters for the new recipes not stored in the memory are optimized within from 5 to 30 seconds. Automatic setup within 5minutes has been realized by using the virtual optimization.

REFERENCES

1. H.Yokoo et al., "New medium current ion implanter SOPHI-200", in these proceedings.
2. I.M.Kapchinskij and V.V.Vladimirskij, Proceedings of the Int. Conf. On High Energy Accelerators, CERN, Geneva 274-288 (1959)
3. J.D.Lawson "The Physics of Charged-Particle Beams", Oxford : Clarendon Press, 1988, pp171-175.
4. S. Ogata et al., Nuclear Instruments and Methods **A363** 468-472 (1995).
5. T.Kunibe et al., Proceedings of Int. Conf. On Ion Implantation Technology-98, Ed. I.Yamada IEEE98EX144 424-427 (1999).
6. ULVAC, Patent Pending
7. R.Fletcher, "Practical Methods of Optimization", Chichester : John Wiley & Sons, 1987, pp195-228.

Implant Angle Monitoring – A Comparison of Channeling Features

Mark A. Rathmell

National Semiconductor, 5 Foden Rd., Mail Stop 04-05, S. Portland, ME 04106 USA

Abstract. High tilt halo implants are widely used to reduce short channel effects in modern CMOS devices. As device sizes decrease, the control of the tilt angle of the implant becomes more critical and can become one of the largest sources of variability in electrical results between implanters if not properly monitored. For implanters without hardware based in-situ angle monitoring systems the tools can be calibrated to each other by monitoring implants oriented intentionally into channeling features. In this paper we compare resistivity monitors based on channeling features at the tilt/twist angles of 0/0, 35/0, and 45/45 to evaluate each for the sensitivity of the monitor to implant angle and the implications to manufacturing.

Keywords: Ion implant, angle control
PACS: 85.40.-e, 85.40.Ry

INTRODUCTION

Tilt angle control at high tilt angles (>30°) is important in controlling the electrical repeatability of devices that use halo or pocket implants. Small variations in the tilt angle will significantly impact the device threshold voltage, drive current and leakage as the dopant moves further under the poly and into the channel region with increasing tilt. Lenoble et al. [1] noted a Vtp shift of 11 mV/° of pocket tilt angle on a 0.11 μm technology which accounted for 30% of the total of all process variation of the Vtp. In order to control the electrical variation caused by changes in the tilt angle a monitor is needed. Current generation implanters have in-situ hardware based tilt monitoring systems to ensure tilt angle repeatability, but many older implanters lack such a feature. In the absence of a hardware monitor a process monitor can be used to check that the tilt is on target.

One method of measuring the tilt angle independently of the implanter is to implant parallel to a channeling feature in the silicon. Small changes in tilt angle around a channel will strongly impact the distribution and depth of the dopant as the number of ions that scatter deep into the lattice will increase as the implant angle becomes coincident with the channel. These distribution differences can be measured with a variety of methods including SIMS, Therma-Wave (TW), and sheet resistance (Rs). For the purpose of an in-line factory monitor quick turnaround time is important after maintenance events, so TW and Rs are logical choices for a measurement method. In the literature, both Rs and TW have been widely used to measure implant channeling [2]. Even though Rs has been shown to have a lower sensitivity to tilt variation than TW for some implant and channel combinations [2] and using Rs adds the additional variation of the anneal process into the results, Rs was chosen as the method to explore in this study due to the relative abundance of Rs tools in National Semiconductor Corporation's (NSC) factories and the existence of an implant dose monitor at similar implant conditions.

EXPERIMENT

Three separate channeling features were evaluated. The channel tilt/twist angles and corresponding crystal orientations tested were; 0/0 <001>, 35/0 <112>, and 45/45 <011>. Each feature was evaluated for sensitivity to tilt and twist angle. The most tilt sensitive channel was further tested for sensitivities to implant dose, energy, beam current, filter angle, anneal temperature, and wafer oxide thickness to establish good repeatability and easy implementation into a manufacturing environment.

Wafers were N-type, non-epi, bare silicon with a <100> cut and the notch parallel to {011}. The wafers had a native oxide only. The implant was done on a serial, mid-current implanter with boron, 100 keV,

1E14 dose with varying tilt and twist angles. The anneal used was 1060° C for 30 seconds in an N_2 ambient. 49 site Rs maps were measured with a 4 point probe. Boron was chosen for the dopant to maximize the potential for channeling of the ions into the lattice, and the dose and energy were picked to match another Rs monitor where NSC had extensive repeatability experience.

RESULTS

Sheet Resistance

Plots of % Rs change vs. tilt angle are shown for the 0/0, 45/45 and 35/0 channels in Figure 1. For this figure 0° tilt from the ideal channel angle signifies a tilt of 0.0° for the 0/0 channel, a tilt of 45.0° for the 45/45 channel and a tilt of 35.25° for the 35/0 channel as dialed into the recipe on the implanter. The implanter's tilt and twist angles were calibrated using the 35/0 channel with part of one lot of wafers. A second lot was then used to evaluate all three channels with wafers from a common stockpile in an attempt to minimize oxide and crystal cut variation.

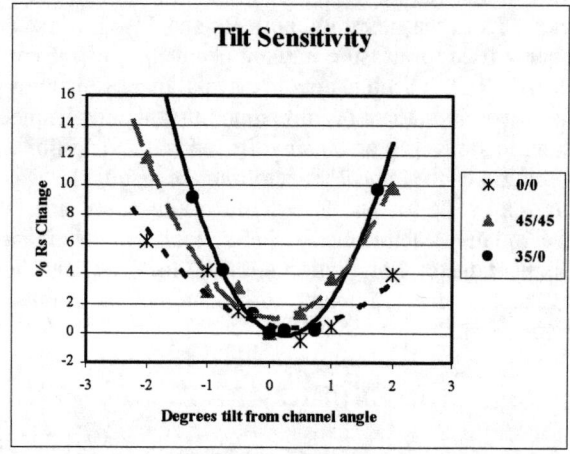

FIGURE 1. Sensitivity of Rs to changes in tilt angle.

Note how the 35/0 channel has a steeper Rs response for a 1° change in tilt indicating that the 35/0 channel has a higher sensitivity to tilt than the 0/0 or 45/45 channels. A second order polynomial was fit to each set of data to determine the tilt value for the minimum of each curve which should indicate the true channel position. The 0/0 channel was at 0.4°, the 45/45 channel was at 45.1° and the 35/0 channel was at 35.5°. All three channels are simultaneously within 0.5° of their respective target values. Basing the calibration off of the more sensitive 35/0 channel the data indicates the tilt is ~0.3° lower than ideal, and the other two channels agree with that estimate to within 0.2°. Possible sources of the offset from target include the tilt repeatability of the implanter, the repeatability of the monitor method and the crystal cut of the wafers.

Other sensitivities were explored to test the robustness of the 35/0 monitor. Table 1 shows the relative amount of Rs change compared to the 35/0 channel Rs minimum for a number of variables likely to be encountered in a fab environment.

TABLE 1. % Rs Change for 35/0 Channel

Variable	% Rs Change
1° Tilt	4.1
1° Twist	2.4
+/-5% Energy Filter V	1.0
+ 10° Anneal	-0.8
+ 10% Beam current	0.7
+ 1% Dose	-0.6
+ 1Å Oxide	0.2
+ 1 keV Energy	-0.1

SIMS

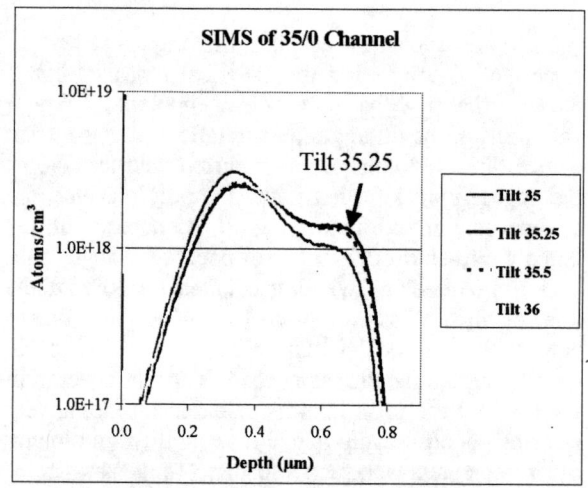

FIGURE 2. SIMS close to 35/0 channel.

SIMS profiles in Figure 2 were generated from the same lot that was used for the Rs curve data in Figure 1. The profiles for 35.25° and 35.5° tilt show the most channeling and coincide closely with the <112> pole of silicon, which theoretically lies at 35.26° tilt and 0° twist. Both SIMS and Rs data indicate the tilt is 0.2° - 0.3° lower than the theoretical position of the channel, and the agreement between the two methods shows that Rs can be used to calibrate the tilt angle.

DISCUSSION

Tilt Angle

All three channels show a minimum in Rs near the center of the channeling feature in Figure 1, but differ in the slope of the curve leading down to the minimum. The 35/0 channel has a steeper slope when measured as a percentage of Rs change. The higher slope enhances the tilt angle resolution of the 35/0 orientation and makes it a better choice for monitoring tilt than the 0/0 or 45/45 channels.

All three channels had a flat region near the minimum of the curve. In this region a change in tilt of less than 0.5° from the minimum would shift the Rs by only 1% of the total value. This limits the monitor's ability to detect small tilt variations at the minimum Rs. The tilt angle could be off of the channel by up to 0.5° in either direction without producing a significant response in the monitor. This is acceptable if the devices being manufactured can tolerate 1° tilt variation or more and still function well within the specified electrical parameters. Increasing the energy of the implant can increase the sensitivity of the monitor to tilt if higher tilt angle precision is required [2].

Twist Angle

The 0/0 angle is insensitive to twist due to the rotational symmetry of the <100> orientation of silicon. Both the 35/0 (Table 1.) and 45/45 (not shown) angles are approximately half as sensitive to twist as tilt. In general, tilt angle has a higher impact on device electrical results than twist. Tilt directly influences the ultimate penetration of a halo implant under a poly gate due the shadowing of the transistor channel by the poly. Twist angle can have a significant impact to the projected range of the implant if the implant lines up with a channeling feature or if the change in twist angle is large, but small changes in twist may be unnoticeable in the electrical results. One example of this is a 0.18 μm process at NSC that had a Vtp sensitivity of 18.0 mV/° of tilt, but no significant response for a +/-5° change in twist.

If the processes run in the factory are electrically insensitive to small twist angle changes then the 35/0 channel should provide protection against significant twist deviations.

Other Manufacturing Variables

Table 1 shows the percent change in Rs for a number of common manufacturing variables. A 10° C change in RTP anneal temperature only caused a 0.8% shift in Rs, less than 1/5 of the impact of a 1° change in tilt. A more typical amount of RTP variation might be on the order of <5° C (depending on the RTP tool and monitoring methods), or less than 1/10 of a 1° tilt change. The amount of RTP induced variation in this case would be small enough to not significantly impact the performance of the tilt monitor.

As noted in Sing & Rendon [3] energy filters that deflect the beam can have an impact on the tilt angle. A 5% voltage shift on the energy filter should shift the tilt by 0.75°. The data in Table 1 shows a shift of only 1% for a 5% energy filter change, but shows a larger shift of 4.1% for a 1° change in tilt. The flat region at the minimum of the Rs curve in Figure 1 reduced the response to the smaller angular deviation of the energy filter and resulted in the smaller than expected shift. Tool SPC showed the standard deviation of the energy filter voltage to be approximately 1% of the total voltage, so the expected tilt angle and Rs variation would be on the order of 0.2° tilt and 0.2 % Rs.

Ten percent changes in beam current and small changes in dose and energy produced shifts smaller than 1/5 of a 1° tilt change. If dose was mismatched between tools by 2 % it could be mistaken for a ¼ of a degree of tilt difference. Therefore, it's important to verify that dose is matched at a non-channeling angle using SIMS or Rs prior to using the tilt monitor to ensure the best angle matching between implanters.

The thickness of the native oxide on the wafer surface had a small impact on the monitor (0.2% Rs/Å) and normal variation of up to 4Å would still be less than 1/5 of a 1° change in tilt. Wafers that had not been cleaned for an extended time period, on the order of a month or more, may have native oxides or airborne molecular contamination (AMC) measuring close to 20Å. In those cases it may be preferable to reclean the wafers with HF followed by SC1 to grow a uniform, thin chemical oxide prior to implanting them.

Monitor Results

At NSC the tilt monitor is run after maintenance events to ensure that tilt is returned to the proper angle and it is also run periodically to monitor the drift of mechanical systems.

Typical performance of the monitor for one implanter is shown in Figure 3. Mechanical maintenance that disturbed the tilt angle was caught by the monitor and corrected prior to running production. The standard deviation of the data after the last maintenance event was 4.5 ohm/sq, or 0.9% of the Rs mean target. This is equivalent to a standard deviation of <0.2° tilt, showing repeatable performance for both the implanter and the monitor. However, due to the reduced sensitivity to angle at the minimum Rs the actual tilt angle variation could be up to 0.5° and produce the same results.

Wafer crystal cut variation can be additional source of noise in the tilt monitor that could show up as an Rs shift from lot to lot [4, 5]. Over a period of three months and more than ten different lots of stockpiles NSC observed no systematic lot to lot shifts (Figure 3.). Crystal cut specifications may be different within a factory for production grade wafers vs. test grade wafers. It is important to use test wafers with a crystal cut accuracy better than the desired tilt angle repeatability to prevent miscalibrating the implanter based on an off angle set of test wafers.

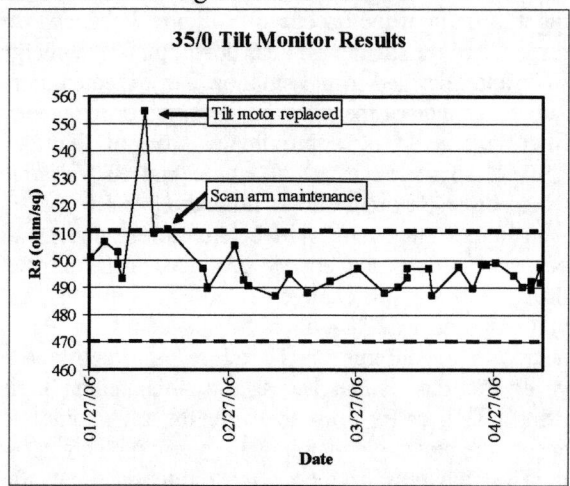

FIGURE 3. 35/0 tilt monitor results for one implanter over a 3 month time period.

The implanter's wafer aligner may be an additional source of monitor variation. If the aligner has more than one notch detector then either both should be calibrated independently or only one should be left enabled. If both detectors are active a bimodal distribution may occur in the monitor due to the different twist angles resulting from each detector.

CONCLUSION

Control of implanter tilt angle for halo and pocket implants can be critical in reducing electrical variability. The tilt angle can be successfully controlled by monitoring the Rs of wafers that have been implanted into channeling features in silicon. The 35/0 channel of silicon is more sensitive to tilt angle variation than the 0/0 or 45/45 channels, and it can be used to monitor the tilt angle of an implanter with an accuracy and repeatability of approximately +/-0.5° when using a 100 keV, 1E14 boron implant.

ACKNOWLEDGMENTS

Many thanks to Rita White for her support in generating much of the data used in this study, and thanks to Thanas Budri for the excellent (as usual) SIMS analysis.

REFERENCES

1. D.Lenoble, F. Wacquant, E. Josse, F. Arnaud, "Impact of angle variation on device performance", IEEE Proc. of the XX Intl. Conf. on Ion Implantation Technology, Taos, NM, USA, Sept. 2002, pp. 44-47.
2. R. Simonton and A. Tasch, "Channeling effects in ion implantation into silicon" in Ion Implantation - Science and Technology 1996 Edition, ed. By Ziegler, J.F., Ion Implantation Technology Co., pp. 329-390.
3. D.C. Sing, M.J. Rendon, "Implant process control: Going beyond particles and RS", Nuclear Instruments and Methods in Physics Research B 237 (2005) pp. 318-323.
4. R.D. Elzer, P. Oakey, J. Chen, D. Sing, "Two methods for improved accuracy calibration and control of ion implanter incidence angle", IEEE/SEMI Proc. of the Advanced Semiconductor Manufacturing Conference, 1999, pp. 343-347.
5. B.N. Guo (private communication)

Defectivity Reduction and Control In Ion Implant Systems Through Hardware and Process Optimization

Dirk-Wito Franke, Falk Hundt, Tobias Guenther, Matthias Schmeide*, Ronald N. Reece**, Christopher E. Ferrell**, Bernhard Krimbacher***, Falk Haerting***, John Grant**

Qimonda Dresden, Königsbrücker Str. 180, D-01099 Dresden, Germany
*Infineon Technologies Dresden GmbH & Co. OHG, Königsbrücker Str. 180, D-01099 Dresden, Germany
**Axcelis Technologies Inc., 108 Cherry Hill Drive, Beverl,y MA01915
***Axcelis Technologie GmbH, Otto-Hahn-Strasse 20, D-65609 Aschheim-Dornach

Abstact: Defect monitoring becomes increasingly important for smaller device structures. Previous publications have shown the damage of particles impacting device structures. The fast movement of the wafers in batch ion implant systems, or a fast moving particle in any implanter, can generate damage on patterned wafers. Defects generated are visible after the implant on an in-line defect inspection tool as well. Yield loss can be observed at final test if poly-silicon lines or photo resist patterns become permanently damaged. Reducing the speed of the wafers reduces the force of the particle-wafer interaction, changing the defect type and severity. Adjusted maintenance procedures are shown to reduce the particle levels of an Axcelis HC3 implanter and lower defectivity levels. Reduced spin speed and specific hardware modifications in the source and beam line areas will be discussed. The influence of these changes on defect density and type is shown. The use of an HYT in-situ particle monitor helps to detect the occurrence of these events.

Keywords: High current implanter, Axcelis HC3, defectivity reduction
PACS Codes: 68.55.LN, 85.40.Ry

INTRODUCTION

A mechanism of device damage occurring on wafers implanted in batch ion implanters has been published since the late 1990's. [1,2,3,4] Most of the publications describe damaged polysilicon lines as the source for the yield loss in circuits, although other pattern level damage and other materials may be affected as well. The authors of the reference publications have discussed several methodologies for reducing the pattern damage on the wafer. Lower spin speeds, change of the gate orientation on the wafer, use of photo resist as a protective layer, and hardware modifications were discussed as opportunities for reducing defectivity.

It is recognized that higher particle levels can lead to higher defectivity on the wafer. As stated, defectivity can be enhanced due to the energetic interaction of a particle and a fast moving wafer. Strategies for reducing the particle related defectivity focused on two areas, reducing particle levels, and reducing the energy dissipated in a particle interaction with the moving wafer. The reduction of wafer energy through reduced rpm is well known and reported previously [4], so we focused on the reduction of particles through incremental hardware improvements and maintenance practices.

Thin films are generated in the implanter primarily through two mechanisms, condensation of vapor materials, and re-deposition of sputtered materials. Both of these mechanisms may result in the creation of particles. As these deposited films build up in thickness, stress can fracture and peel the film microscopically releasing small sections of the film as particles. On rare occasions, larger sections of the film may fail, causing a larger particle burst. The proximity of these films to the wafer is directly proportional to the probability of the particles causing wafer level defectivity, as transport can be short line-of-sight, or longer if the particle is transported by the beam. [5, 6]

At Qimonda, damage was first observed in 140 nm DRAM technology on 300 mm wafers. The yield loss on the wafers, was related to structural damage at the pattern level. Often a signature of just 12 or 13 wafers affected out of one lot implicated the batch ion implanter. The event was called "half lot signature", when only one part of the lot was affected. Initially the defectivity was not discovered during in-line electrical testing because of the small test structures and the limited probability of the test structure being affected. As a result, the "half lot" lot damage was not observed until the first full wafer circuit test. Up to two half lots per week were observed across 3 systems running full production, with yield loss caused by physical damage to small structures. To reduce the affect of energetic particle interactions, several corrective actions were implemented.

EXPERIMENTAL

In-situ and in-line particle and defectivity monitoring was introduced to increase the correlation of particle bursts to "half lot" events and improve feedback regarding maintenance activity.

An in-line defect density measurement was implemented after one particularly sensitive implant step (22 keV As+ 3E15 cm-2) in 110 nm DRAM technology. About 10 percent of all implanted lots for this process were tested. Rules stated that in case of a violation of certain defectivity level the wafer with the highest defect counts was reviewed by a scanning electron microscope (SEM) to categorize the defects. After detection of defects on product wafers a particle check on monitor wafers was performed. In case of violation of specification the ion implanter was given to the maintenance group for troubleshooting.

To improve the in-situ process control an HYT sensor was installed to observe particle behavior and excursions. The HYT in-situ particle sensor is positioned to have the greatest sensitivity to detecting a particle. Later the HYT data were correlated to defect measurements on product wafers.

A dedicated team of people from Axcelis and Qimonda was formed into a joint task force with the goal of lowering the probability of, and the damage caused, by a "half lot" event. The strategy involved 3 areas:

1. Improving maintenance techniques and practices. Optimize maintenance intervals.
2. Implement reduced spin speed, (Optimized Fast Scan or OFS) to minimize the kinetic energy.
3. Implement incremental hardware improvements, (Block 3-5) to reduce the generation of particles, and minimize the transport of particles to the process chamber.

1. The joint team performed an evaluation of maintenance activities and started the optimization of the maintenance procedures. Improvements in maintenance practices centered on a methodical approach to performing preventative maintenance in all areas of the ion implanter. Each region of the implanter was examined for the presence of deposited films, and the ideal methodology for cleaning these areas was developed.

The maintenance activities were performed every ten days proactively. This procedure included the source change and the change of the selectable resolving aperture as well as cleaning of the following parts and areas: extraction head, extraction manipulator, source defining aperture, gas baffle, disk, and process chamber.

2. To reduce the force of the particle-wafer interaction Axcelis offered the option of reducing the spin speed from 815 rpm down to 293 rpm on the HC3 300 mm high current ion implanter. Tests on monitor wafers and on product lots were performed to evaluate the influence of the lower spin speed on uniformity, repeatability and wafer.

3. Several hardware modifications (Block implementations) were performed to reduce the defectivity level during the implant. The improved hardware focused on reducing the sputter yield of any surface struck by the beam, improving the mechanical adhesion of evaporated or sputtered films to internal surfaces, and by limiting the mechanisms and probability of a particle being transported into the process chamber region.

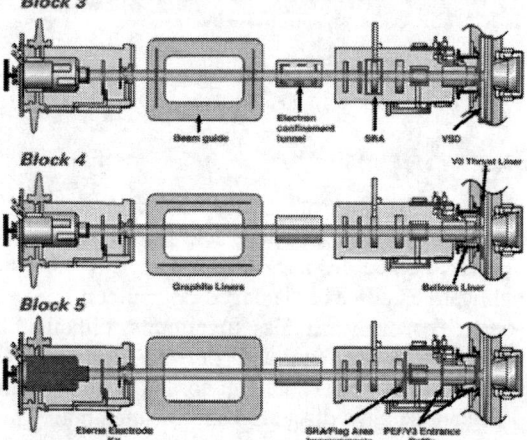

FIGURE 1. Block implementations for the defectivity reduction program.

Modifications in the source area and the beam line were provided by Axcelis (Fig. 1). Modifications should prevent excessive coating and material flaking, catastrophic delaminating of films, and transportation of particles. Changes in the source and extraction electrode area should reduce beam instabilities, glitches and film built up.

RESULTS AND DISCUSSION

Up to 60% yield loss between good and bad wafers within one lot were observed. In average 10% overall yield loss across two lots per week has been seen prior to the start of our corrective actions.

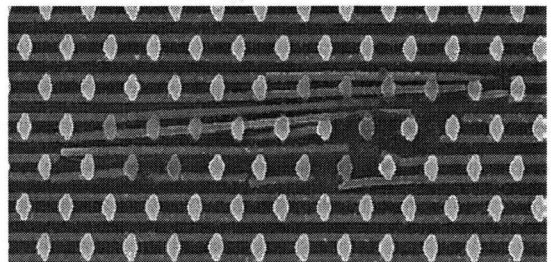

FIGURE 2. Broken poly lines on wafer in 110 nm DRAM technology, yield loss up to 60% compared to the good run, analysis performed after the final test

The first action to mitigate yield impact was limiting the implantation of structured wafers without photo resist, as the photo resist film added structural stability and protection to steps with high defectivity.

After one specific implant step product wafers were inspected on a KLA AIT Fusion tool to detect the number of defects after the implant and to make the distribution across the wafers visible. The defects of the reviewed wafers by SEM were classified into different categories like surface particles, scratches, and holes (Fig. 3).

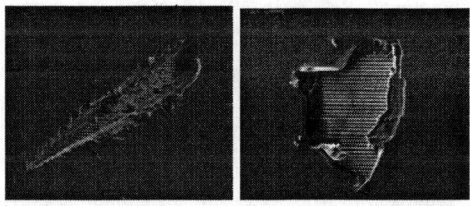

FIGURE 3. Damage of particle strike into photo resist covered wafer during the implant, defect classification: scratch (left, hole (right)

Focused maintenance activities and more frequent interventions reduced the particle excursions. Parts and regions of the ion implanter which were conspicuous to create particles on the wafers were inspected and if needed cleaned or changed. The number of defects generated by particle bursts during the half-lot excursions became lower. Particle high flyers detected on bare test wafers were reduced as well.

The reduced implant spin speed evaluated on monitor wafers by verifying the particle behavior, uniformity, repeatability and wafer cooling did not show differences compared to implanted wafers at 815 rpm. The lower spin speed applied on product wafers matched the POR (process of record) and was released for production.

The implantation of wafers at 293 rpm did not change the total number of defects as much as reducing the severity of the defectivity, e.g., the number of scratches and holes was reduced while the number of surface particles increased (Figure 4, 5).

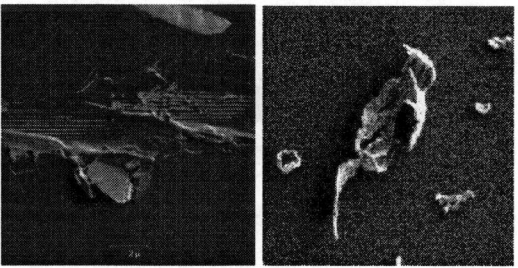

FIGURE 4. Damage of particle strike during the implant at 815 rpm (left) and surface particles generated during implant at 293 rpm (right)

The Figure 5 shows the defect level measured on product wafers and a comparison between HC3 implanters with 815 rpm, 293 rpm and graphite liners (2_graphite) in the analyzer magnet. The test implanter (golden tool) received in addition optimized maintenance procedures. The installed graphite liners (Block 4 improvement) did not change the particle base line level (median) but lead to a reduced excursion level (mean value).

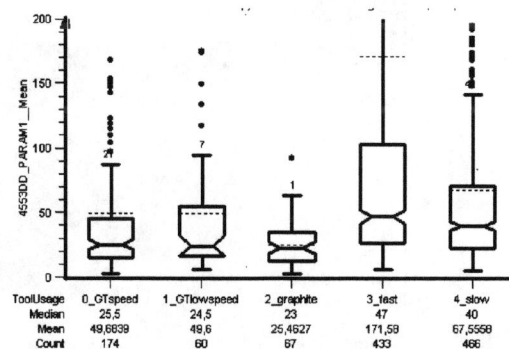

FIGURE 5. Defect adders measured on product wafer during the implant, 0: golden tool 815 rpm, 1: golden tool 293 rpm, 2: golden tool 293 rpm and graphite liners, 3: production tool with 815 rpm, 4: production tool with 293 rpm

Optimization of the surface of the introduced beamline material changed the particle behavior of the ion implanter too. The improved adhesion of films especially in the analyzer magnet area (Fig.

6) reduced the number of flakes in the beam line and reduced the number of defects detected on product wafers.

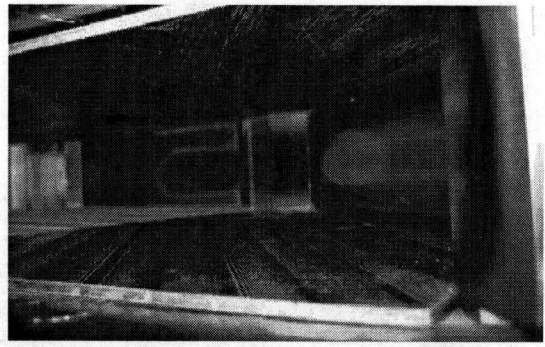

FIGURE 6. Beam guide area after 900 hours of operation, textured graphite surfaces.

The correlation of HYT to low particle counts is weak, but a strong correlation of HYT counts and defects detected at in-line measurement was possible. As a result, the HYT could be used to halt implantation on a system exhibiting a HYT count. This prevented continuous processing under questionable conditions.

The HYT sensor registers the time during implant, when particle events happen. It was recognized, that most particle excursions occur within a few seconds, and these events are visible on wafers as striping or accumulation of defects. (Fig. 7)

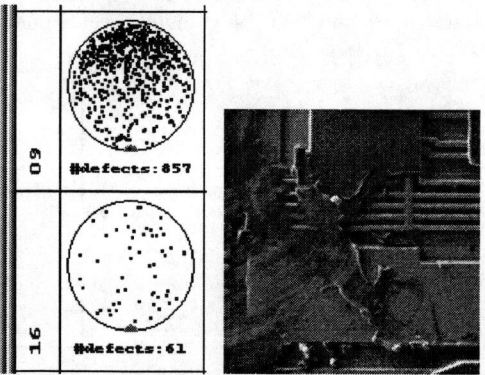

FIGURE7. Image from defect distribution after implant on product wafers, wafer 9 belongs to first, wafer 16 to second run. Indication for half lot signature and reviewed defect by SEM, Run 1: 337 HYT counts, Run 2: 20 HYT counts

CONCLUSION

Device structures may be damaged by particles interacting or colliding with fast moving wafers. The defects become more critical for smaller devices. The amount of particles generated during the implant can be reduced by focusing maintenance procedures and optimizing the maintenance intervals, while the affect of the damage from a particle-wafer interaction can be reduced by lowering the spin speed and kinetic energy of the wafers. Improvements of the design in some dedicated areas of the ion implanter decrease the particle generation and transport to the wafers. A balance between maintenance cost and tolerable particle level needs to be found. In-situ particle detection with HYT sensor allows fast response on particle events and prevents further product wafers from damage if the production is stopped after critical particle detection.

ACKNOWLEDGMENTS

The authors thank Uwe Vater and Dietmar Henke from Qimonda Dresden for helpful discussions and advice.

REFERENCES

1. L. Pipes, M. Taylor, G. Zietz, A. Al-Bayati, M. Castle, T. Marin, J. Simmons: "Characterization and reduction of a new particle defect mode in sub-0.25 μm semiconductor process flows", 15th International Conference on Ion Implantation Technology, IIT 2004
2. O. Diop, S. Blain: "Reducing yield impact at ion implant process due to particles using in situ particle measurement at sub 0.13μ geometries", 2005 IEEE 0-7803-9144-6/05
3. B. Dunham, R. Anundson, Z.Y. Zhao: Characterization of missing poly defects in ion implantation in ULSI manufacturing, Characterization and Metrology for ULSI Technology: 2003 International Conference
4. Y. Kawasaki et. Al " The Collapse of Gate Electrode in High-Current Implanter Batch Type" 2004 IEEE 0-7803-8191-2/04
5. T.C. Smith, K. Larsen "Photoresist and particulate problems". 14th International Conference on Ion Implantation Technology, IIT 2002
6. D. Brown, P. Sferlazzo, J. O'Hanlon "On the transport and heating of particulate contamination entrapped in an intense cylindrical ion beam" Journal of Vacuum Science & Technology A: Vacuum, Surfaces, and Films -- September 1991 -- Volume 9, Issue 5, pp. 2808-2812

Advanced Charge Control for Single Wafer Implanters

P. F. Kurunczi, A. S. Perel, E. Wright, S. Kikuchi, and J. T. Scheuer

Varian Semiconductor Equipment Associates
35 Dory Rd., Gloucester, MA 01930, USA

Abstract. Supplying low energy electrons to insure neutralization of implanted wafers requires an advanced charge control system. Some devices are extremely sensitive to low levels of metal contamination. We have designed a radio frequency (rf) plasma flood gun that eliminates the use of a hot filament. This makes this PFG more reliable and not susceptible to filament failures. We will present data showing that this advanced PFG has excellent charge control with an electron energy distribution that floods the wafer with an abundant supply of low energy electrons.

Keywords: Ion implantation, charge neutralization, plasma flood gun, electron energy distribution.
PACS: 41.75.Ak, 52.70.-m, 61.72.Tt, 85.40.Ry

INTRODUCTION

During ion implantation positively charged ions from the beam are incident on various parts of the semiconductor device. For conducting materials, these ions can be neutralized within the device by electrons from the bulk material. However, when these positive ions strike insulating materials (e.g. gate oxide) they create a positive charge on the insulator, which can result in a high electric field across the thin oxide. Above a certain threshold this electric field can damage the gate oxide and therefore the device. As semiconductor devices shrink, the need for neutralization of the positive ion beam is increased.

Most ion implanters have plasma or electron flood guns to provide negative electrons which neutralize the positive ions in the beam. Historically these devices have employed hot tungsten filaments to provide

FIGURE 1. Varian's Advanced Plasma Flood Gun. Ribbon beam for high current single wafer implanter is neutralized along the full width of the beam.

thermionic electrons which, in turn, ionize an inert gas within a small arc chamber. This ionized gas then diffuses through an aperture to the beam providing the neutralizing electrons.

The presence of the hot filament near the wafer carries a risk of tungsten contamination depositing on the wafer. Also, the inert gas plasma sputters the filament which eventually breaks. This requires production to stop in order to replace the failed components.

Varian Semiconductor has developed an Advanced Plasma Flood Gun (PFG) that meets the requirements of next generation semiconductor devices.

ADVANCED FLOOD GUN

Varian's single wafer beam line implanters produce either a ribbon beam (the high-current VIISta HC) or a scanned spot beam (the medium-current VIISta 810 and high-energy VIISta 3000). These beams span the width of the wafer and the wafer is scanned vertically through the uniform beam. Proper charge control requires introduction of electrons along the entire width of the wafer. Fig.1 shows Varian's Advanced PFG that floods both the ion beam and wafer along their entire widths by creating an rf plasma in a large volume chamber.

The plasma is created by injecting rf energy and inert gas into an elongated discharge chamber thereby providing a uniform and distributed source of electrons across the full beam and wafer widths. A hot filament discharge would require numerous arc chambers with filaments running in parallel, compounding the risk of tungsten contamination and filament failure. Since the

Advanced PFG does not contain any tungsten the risk of heavy metals contamination is eliminated.

Wafer Neutralization with Low Energy Electrons

Low energy electrons are preferred for charge neutralization for two reasons. The first is related to "negative charging." Just as positive ions can damage devices, a surplus of negatively charged electrons striking thin gate oxide can create high electric fields that can damage devices. However, these electrons cannot charge a wafer to a potential that exceeds their energy. If the energy of the electrons is low enough, the probability of negative charge damage is eliminated.

The second reason low energy electrons are preferred pertains to the mechanism by which electrons are transported to the wafer to counteract the positive charge flux due to the ion beam. The positive charge of the ion beam creates a positive potential well. This generates electric fields that can significantly perturb the path of a PFG electron that enters the beam. If the initial electron energy is comparable to, or smaller than, the beam potential, then these fields may contain the electrons. However, high energy electrons can pass through the beam and will not serve to neutralize it. Therefore, if electron energies are small compared to the beam potential, then the beam will help transport electrons to the wafer. Although the space-charge potential of the beam can, unabated, reach many hundreds of volts [1], on most implanters this is reduced to tens of volts (or lower) by neutralizing electrons that come from beam collisions or from the PFG itself [1,2]. The beam is therefore an important conduit for charge neutralization, provided the electron energy is low enough to allow containment by the beam potential.

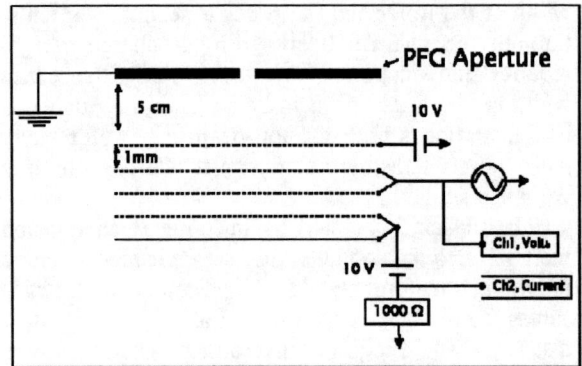

FIGURE 2. Schematic of the energy analyzer used to measure the electron energy distribution.

A retarding field analyzer (Fig. 2) was used to measure the energy distribution of the PFG electrons at the output of the Varian Advanced PFG. The analyzer consists of a positively biased grid to repel the PFG ions and accelerate the electrons toward the detector, a second grid with a variable applied retarding potential, and a collector plate that measures the electron current. The electron energy distribution is then obtained by differentiating the collector current with respect to the retarding voltage and subtracting out the shift caused by the initial applied accelerating potential. The data was acquired by using a storage oscilloscope to record the signals from a periodic potential applied to the retarding grid (Ch1). The collector plate current was determined by measuring the voltage across a resistor at the collector plate (Ch 2).

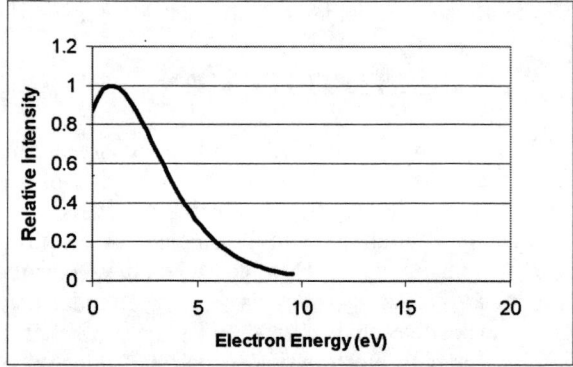

FIGURE 3. Electron energy distribution from Varian's Advanced PFG. Peak in electron energy distribution is at 1eV.

Fig. 3 shows the normalized electron energy distribution as measured by the retarding field energy analyzer. Note that the majority of the PFG electrons are below 5 eV with the peak of the distribution at 1eV. These particularly low energy electrons are suitable for efficient coupling to the ion beam resulting in improved wafer charge control. Moreover, the coupling spans the entire width of the ion beam due to the broad spatial geometry of the PFG.

Charge Control

With gate oxides as thin as 10 angstroms, and oxide breakdown voltages in the range of 10 MV/cm, charging potentials ~10V can cause device damage. Assessing charge neutralization systems, such as the Varian Advanced PFG, requires implantation of charge-sensitive test devices with high dose implants. In this study we implanted charge-sensitive test wafers with the high dose of $5e15/cm^2$ As at 20 keV on the VIISta HC and compared it with identical implants using a filament-based plasma flood system at a gas flow and arc current appropriate for a high current implant.

Charge-sensitive Test Element Group (TEG) wafers, made for evaluating charging damage, were used for these tests. The devices on the TEG wafers are referred to as antenna MOS devices and are manufactured by Philtech, Inc. [3]. An antenna MOS device is a MOS transistor with a large conducting area connected to the gate. The antenna ratio refers to the ratio between the large area antenna and the area of the gate. The antenna and the gate are electrically connected and thus the charge accumulated on the antenna builds up on the gate.

The oxide thickness for these devices was 40 angstroms. If charge on the gate results in an electric field that exceeds the dielectric strength of the oxide, then the device will fail [4]. For these tests data are shown with antenna ratios of 2,500:1, 10,000:1, and 100,000:1. A gate with a 100,000:1 antenna ratio accumulates a charge 100,000 times larger than the gate alone would accumulate. With an oxide of 40 angstroms thick, the high antenna ratio devices are extremely sensitive to charge accumulation.

The devices are tested by measuring the breakdown voltage between the gate and the substrate ground. Devices that break down during the implant as a result of too much charge will show a very low breakdown voltage when the devices on the wafer are tested. Those that survive the implant will exhibit a large breakdown voltage, in the vicinity of 5-10 Volts.

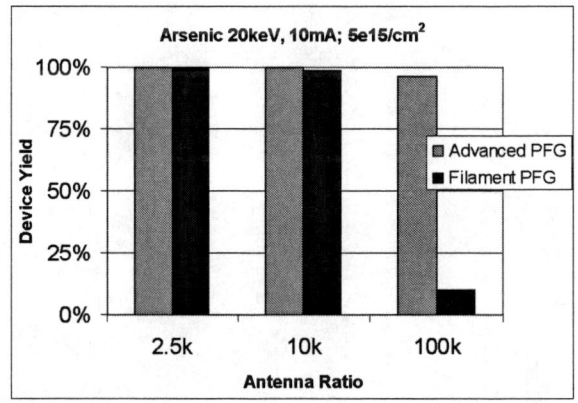

FIGURE 4. Yield of antenna ratio devices for advanced PFG and filament type PFG. Advanced PFG shows superior yield for 100k ratio devices.

Fig. 4 summarizes data from the charging tests. Both PFG systems neutralized the wafers sufficiently for 2.5k and 10k antenna ratio devices. Very nearly 100% of those devices survived the As 20keV 10mA 5e15/cm² implant. The very sensitive 100k antenna ratio devices indicated differences in the charging capabilities of the two PFG types. The yield for the 100k devices was more than 96% for the Advanced PFG; for the filament-based PFG the yield was about 10%. The low yield of the high antenna ratio devices indicate that the wafer is not fully neutralized and may result in damage to sensitive devices.

Metals Contamination

Tungsten contamination during implantation at levels above 1e10/cm² can result in degradation of device performance. The 2005 version of the International Technology Roadmap for Semiconductors (ITRS) [5] sets the limit of transition metal contamination at 1e10/cm². Device manufacturers for many applications require less than 1e10/cm² of transition metal contamination for doses as high as 1e16/cm².

Most plasma or electron flood systems neutralize wafers with the help of a hot tungsten filament. Because these neutralizing systems are located near the wafer, contamination from the hot filament may be deposited on the wafer.

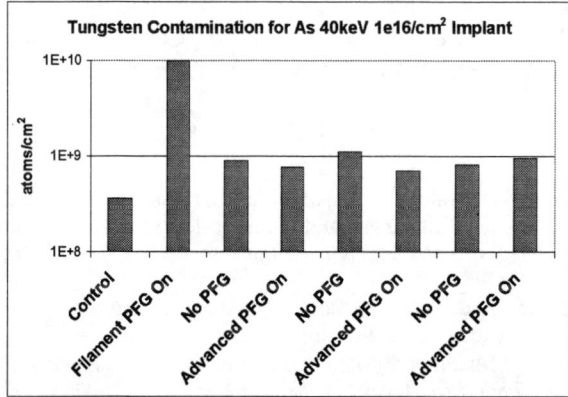

FIGURE 5. Concentration of tungsten as measured by SIMS for an arsenic 40 keV 1e16/cm² implant. Control wafer did not enter the implanter. Data set consists of measured tungsten on a control wafer, a wafer implanted with a filament PFG in operation, and 3 sets of implants with the advanced PFG off and on.

The non-filament based technology of the Advanced PFG makes it an excellent charge neutralization system without heavy metal contamination of the wafer. Fig. 5 shows the SIMS tungsten levels for a control wafer that did not enter the implanter, a wafer implanted with a filament-based PFG in operation, three wafers implanted with PFG-on, and three with PFG-off. All implants were with a dose of 1e16/cm² of As at an energy of 40 keV. The filament-based PFG deposited 1e10/cm² of tungsten. In comparison, there was no increase in the tungsten levels with the Advanced PFG on. This is not surprising since there are no tungsten components in the design. Both PFG-on and PFG-off conditions indicated tungsten levels in the range of 1e9/cm^2 for the Advanced PFG. This number is near the SIMS detection limit and below the level capable of affecting device performance as stated in the ITRS [5].

CONCLUSION

Gate oxides have become thinner with device scaling, leading to a greater sensitivity to charging damage and heavy metal contamination. Varian's Advanced PFG neutralizes the beam and reduces the risk of charging damage during implant by flooding the width of the beam with low energy electrons. This design has been shown to provide superior wafer neutralization compared to filament-based plasma flood guns. In addition, the risk of tungsten contamination on the wafer from the plasma flood system has been eliminated.

ACKNOWLEDGEMENTS

We would like to thank Mary Lou Pascucci, Alex Soskov, Leo Klos, and Eric Cobb for their contributions to the design of the Advanced PFG.

REFERENCES

1. M.S. Ameen and M.E. Mack, "Wafer Cooling and Wafer Charging," Ed. J.F. Ziegler, Ion Implantation Science and Technology, Ion Implantation Technology Co 2004.
2. S. Radovanov, P. Corey, G. Angel, D. Brown,"Wafer Floating Potential for a High Current Serial Ion Implantation System", *1998 Int. Conf on Ion Implant. Tech. Proc.* Kyoto, Japan, p.482.
3. Philtech, Inc., Tokyo, Japan, http://www.philtech.co.jp/ (June 2006).
4. Z. Fang, S. Felch, J. Weeman, S. Mehta, "Evaluation of Different Wafer Charging Metrology Protocols for Thin Dielectrics," *2000 Int. Conf on Ion Implant. Tech. Proc.* Alpbach, Austria, p..565.
5. *Int. Tech. Roadmap for Semi; 2005 Edition;Front End Processes*, Semiconductor Industry Association, p.17.

APC Implementation on VIISta Ion Implanters

Youn Ki Kim, Bret Adams, Nick Parisi, Sandeep Mehta, Jim Hamilton

Varian Semiconductor Equipment Associates, Inc., 35 Dory Road, Gloucester, MA 01930, United States

Abstract. As transistor features shrink, small variations in process steps have a significant impact on device performance and yield. The industry is adopting Advanced Process Control (APC) to control wafer-to-wafer variations in several process steps, but to-date APC has had limited application for implant. For example, lot-to-lot adjustments are being used today by chip manufacturers to adjust halo implants. The controls now possible due to the use of single wafer implant systems to adjust for blanket film variations has generated increased interest in implementing APC in the implant bay. Emerging implant APC applications are being implemented using Within wafer (WiW) control and feed-forward methods to compensate for transistor threshold voltage non-uniformities introduced during gate etch process steps. In the past, APC applications in implant have lagged other processing steps because of the inability to adjust parameters on batch systems and the lack of on-board metrology. New in-situ metrology technologies being developed on the VIISta platform for monitoring critical yield parameters will enable APC on implanters for meeting emerging challenges in advanced transistor designs.

Keywords: Ion implantation; APC, FDC

INTRODUCTION

As transistor features shrink, small variations in processing steps have a significant impact on device performance and yield. For example, inherent variations in critical dimension (CD) which did not have a measurable impact on device performance in >90nm technology nodes are becoming intolerable at sub 65nm nodes. Fabs have embraced APC as a major tool to enable tighter process windows and reducing variability leading to improved device performance and time to yield. [1]

The VIISta platform family features new in-situ metrology options which are being used to monitor parameters that are critical to yield. The parameters most often being considered in APC implementation are beam angles, beam shapes, energy contamination, and particles. [2] Data synchronization, fidelity, and sampling rate, are designed to support very short process times, often less than 5 seconds. This information is well suited to existing fab APC infrastructure as it is available in real time to the host.

As Fabs consider the return on investment of APC implementation, ion implant applications have lagged behind lithography, CMP, CVD, sputter, diffusion, and etch processing. [3,4] Several factors have made implant less suitable for APC, including insufficient data resolution, data synchronization, and extensive data sets.

Fabs have implemented Run-to-Run (R2R) APC methods to compensate for variations in some implant process process steps. A common implant example is compensating for etch via variation by modifying the implant dose.

The most significant difficulty in the implementation of implant APC, the lack of individual wafer control on batch system, has now been eliminated, creating new opportunities. With single wafer systems equipped with on-board metrology systems[2] for implant angle and energy contamination, it is now possible to have effective individual wafer control coupled needed in APC. Already, Within Wafer (WiW) APC techniques are being exploited on single wafer implant systems to control the impact of CD variation introduced during furnace or etch processing steps. Tailoring the across wafer profile of the dose during implant compensates for previous processing step non-uniformities, improving device C_{pk}.

Another opportunity in APC that is gaining ground in the implant bay is Fault Detection and Classification (FDC) [6]. Kyek [3] discussed methods to measure the health of the tool by counting the number of alarms and the duration of the alarm and FDC provides systematic approach to monitor the health of the tool. By employing more advanced detection and classification algorithms such as multi-variate analysis, more than one sensor signal is combined to improve the detection rate while reducing the false alarm rate. This overcomes limitations of Statistical Process Control (SPC) to control as the number of signals to monitor increase.

This paper presents examples of implementation of APC techniques and investigations of new technologies available in the implant bay.

I. CURRENT IMPLANT APC

The Varian Control System (VCS) is one component of the common VIISta platform. VCS provides fast, high resolution data collection for all the key parameters in VIISta tools. For example, on the VIISta HC implanter, VCS samples and stores over 8000 individual values at a 20 Hz rate. These values include in-situ sensor readings, actuator positions, and calculations from internal control models. Data fidelity is achieved through an I/O system that is operated synchronously and deterministically. Each value is collected at equally spaced time periods, and at the same time relative to all of the other values. This level of control over the timing of the data collection is a prerequisite for valid statistical calculations and for APC.

Other key VCS features that aid APC implementation are:
- Closed loop control of wafer handling, the beamline, and general systems (vacuum, gas flow, cooling, power, etc.).
- In-situ APC -- Process monitoring and control is achieved through in-situ sensors, such as multipixel beam profilers, the plasma flood gun emission current monitor, and power supply feedback such as suppression current.
- SPC control
- Event Pareto and query tools

The above features can help detect secondary effects. For example, a small, intermittent oscillation of beamline magnet power supplies can cause a small oscillation in the ion beam. These variations are visualized and analyzed using the VCS Statistical Process Control (SPC) Viewer, and tight controls can be put into place using the VCS SPC Monitor to detect and interlock this fault condition.

II. NEW DEVELOPMENT IN IMPLANTER APC

A. R2R

Wafer to wafer compensation

Vt variation is a more significant issue in advanced scaled devices than it has been in the past. Long channel devices were unaffected by the small, yet inherent variations in the other processes such as gate CD etch. Variations in Vt stemming from variability in the gate CD definition step are critical. Device manufacturers constantly adjust the Vt, and/or halo implants to compensate for such variations. Depending upon the consistency of the etch step, the variations could be lot-to-lot or even within a wafer.

APC Variable Recipe Emulator

A barrier in implementing R2R is that unless the tool is fully integrated into the fab APC infrastructure, the total number of recipes needing to be created and maintained becomes unmanageable. One method to circumvent this problem is "Variable recipe parameter" specified in the SEMI standard E42, recipe management standard [7]

The Variable Recipe Parameter specification allows a user to maintain one recipe while only the APC parameters supplied by the MES at run time per incoming job are changed without affecting the rest of the recipe parameters. This method provides a secure way of managing the recipes and keeping the beam characteristics controlled in a systematic manner. The APC parameters supported on VIISta tools are the dose, energy, tilt angle, and twist angle.

VCS also provides an on-tool functionality that emulates host driven "Variable Recipe Parameter" capability. This emulator provides immediate R2R simulation so that Design of Experiments (DOE) can be run to assess the impact of potential APC implementations prior to investing in the host modifications necessary.

Within Wafer(WiW) Compensation

Typically, variations in the gate CD etch step produce a radial pattern on the wafer. This etch variation results in Vt variations. The implant recipe can be set to avoid Vt variation by deliberately implanting a uniform radial pattern, countering the impact of the etch non-uniformities. An example is depicted in Figure 1 where multiple radial zones on the wafer have been implanted during a single implant recipe. Some fab has demonstrated significant improvement on device characteristics by employing this method and the Vt uniformity was improved up to 47%.[8]

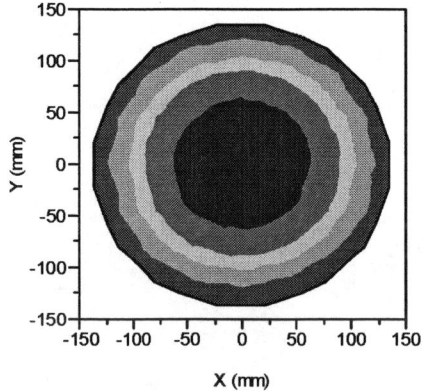

Figure 1. Implant with multiple radial zones

B. FDC

Fault Detection and Classification (FDC) is intended to identify systematic offsets and drifts between tools and ensure tool-to-tool matching.

In alarm analysis, Kyek [3] discussed measuring the health of the tool by counting the number of alarms and the duration of the alarms. The event analysis tool on VCS performs similar functions by manipulating sophisticated filtering and querying capability.

Enhanced Equipment Quality Assurance (EEQA) [9] a concept proposed by Japan Electronics and Information Technology Industries Association and Semiconductor Leading Edge Technologies, Inc. (SELETE/JEITA) provides extended FDC capability. Varian Semiconductor has developed an EEQA application to analyze data collected during system testing and then statistically compare the results. Figure 2 shows the graphical user interface which provides a summary screen of all the calculated measurements. Figure 3 shows the applications capability to drill down feature for viewing the actual measurements for the various system tests.

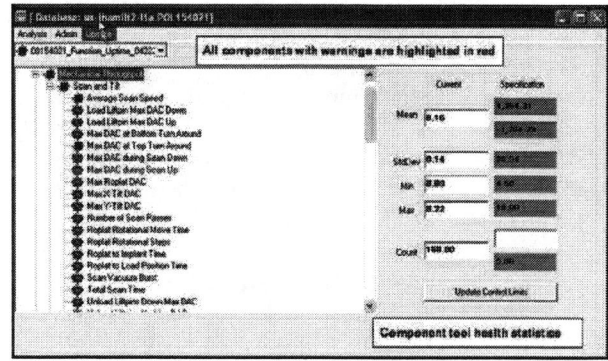

Figure 2 EEQA Main User Interface

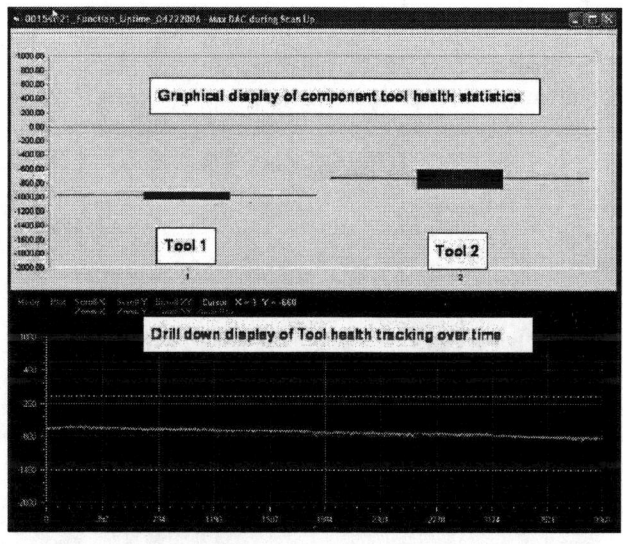

Figure 3 EEQA Measurement Drill-Down

Multi Variate Analysis

Inherent limitations of SPC, univariate analysis is that it lacks a single view of the health trends of the tool. A user must select and monitor several relevant parameters. Multivariate analysis combines multiple variables to provide a single output indicating the health of the tool. Research [10] has shown that it improves the detection rate and reduces false alarms.

C. NEW IN-SITU METROLOGY ON IMPLANTERS

Olson [2] discusses the sensitivity of device performance to the various beam properties and how the VIISta platform measures such beam properties. The beam properties include: verification of the angle integrity of each setup and implant, monitoring of energy contamination, and real time uniformity monitoring. Work is underway to assess the benefit of using multivariate analysis to track these variables to predict beam health; initial results are promising.

CONCLUSION

Fabs are now taking advantage of the migration to single wafer systems, new implant metrology, and new data analysis capability to implement APC in the implant bay. In-situ metrology, WiW APC control, high data rate, and multivariate analysis are creating opportunities for improving device performance, reducing C_{pk}, and maintaining tool performance. We have presented examples of APC applications in production and under investigation that are taking advantage of these opportunities.

REFERENCES

[1] Sonderman, Tom, "APC at AMD" in Proc. APC Conf. (2005)
[2] Olson, J., et al., "Metrology Requirements For Single Wafer Ion Implanters" in Proc. 16th Int. Conf. on Ion implantation Tech., Marseilles, France (2006)
[3] Kyek, Andreas, in Proc. 15th Int. Conf. on Ion implantation Tech., Taipei, Taiwan (2004)
[4] Passow, M., et al., "Implementing FDC with Minimal Pain and Maximum Return on Investment", SEMATECH AEC/APC XVII, (2005)
[5] Wang, Yiping, "FDC Technique Application in Implant Process" SEMATECH AEC/APC XVII, (2005)
[6] SEMI Electronic Control Standards [ECS] Task Force 10/97
[7] SEMI E42 Recipe Management Standards
[8] Lee, M., et al., "The Concept of LDSI (Locally-Differentiated-Scanning Ion Implantation) for the Fine Threshold Voltage Control in Nano-Scale FETs" in Proc. 16th Int. Conf. on Ion implantation Tech., Marseilles, France (2006)
[9] Equipment Engineering Capabilities (EEC) Guidebook (Phase 2.5). ISMI & JEITA/Selete Collaboration.
[10] Hendler, L and Ambrozic, C. "A Comparison between Univariate Data Analysis and Multivariate Data Analysis for FDC", SEMATECH AEC/APC XVII, (2005)

Optimization of High-Energy Implanter Beamline Pumping

Marvin LaFontaine, Michel Pharand, <u>Yongzhang Huang</u>, Ilya Pokidov, and Joseph Ferrara

Axcelis Technologies, Incorporated (ACLS)
108 Cherry Hill Drive
Beverly, Massachusetts 01915

Abstract. A high-energy implanter process chamber and its pumping configuration were designed to minimize the residual gas density in the endstation. A modified Nastran™ finite-element analysis (FEA) code was used to calculate the pressure distribution and gas flow within the process chamber. The modified FE method was readily applied to the internal geometry of the scan chamber, the corrector magnet waveguide, and the process chamber, which included the scan arm assembly, 300mm wafer, and plasma electron flood gun (PEF). Using the modified Nastran code, the gas flow and pressure distribution within the beamline geometry were calculated. The gas load consisted of H_2, which is generated by photoresist (PR) outgassing from the 300mm wafer, and Xe from the plasma electron flood gun. Several pumping configurations were assessed, with each consisting of various locations and pumping capacities of vacuum pumps. The pressure distribution results for each configuration are presented, along with pumping efficiency results which are helpful in selecting the optimum pump configuration. The analysis results were compared to measured data, indicating a good correlation between the two.

Keywords: Molecular Flow, Residual Gas, Cryo Pump, FEA.
PACS: 51.10

INTRODUCTION

PR outgassing from beam-implanted wafers can cause dose shift in the process chamber of an ion implanter. When photoresist outgas enters the beam line, the resulting interaction of the energized beam ions and the residual gases can cause charge exchange and charge neutralization of the implant ions. This can result in dose and angle error during the implant.

The PR outgassing typically generates gas molecules of H_2, H_2O, CO_2, CO, as well as other hydrocarbons [1]. Thermalization and chemical decomposition of the photoresist, as well as sputtering, are the mechanisms which cause the outgassing. These gas molecules then enter the implanter chamber and increase the background residual gas pressure. The gases can also migrate up the beam line into other regions of the implanter, such as the corrector magnet and electrostatic scanner. However, in these regions, the gas usually has less effect on the wafer dose uniformity [2].

A more serious effect of PR outgassing is the direct streaming which occurs directly in front of the wafer. Streaming directs the gases outward from the wafer, where the direction of the gases follows a cosine, or "Lambertian," distribution into the beamline. This is particularly detrimental when the wafer is oriented at low tilt angles.

Another source of residual gas is the PEF, which typically uses Xe gas to reduce charge up of the wafer and platen. The Xe gas stream is beneficial when it is directed into the region in front of, and parallel, to the wafer. However, the Xe must be pumped out of the process chamber in order to avoid a buildup of gas pressure.

Calculating the pressure distribution and gas flow in the molecular flow regime is often accomplished using Monte-Carlo methods of gas molecule ray tracing. However these methods usually require over simplification of vacuum chamber geometries. Then the results are less applicable to the physical design of vacuum systems.

Another method used to calculate molecular flow is the application of classical textbook relationships, which also suffer from the requirement of simplified geometries [3], [4], and [5]. The method used in the present study relies on FE analysis, which uses the actual geometry of the beam line chamber construction.

TABLE 1. Cryo Pump Configuration Table And resulting Calculated H₂ Pressure At The Entrance Of The Process Chamber

Pump Config-uration	Chamber Front Upper (C)	Chamber Front Lower (D)	Chamber Left Side (E)	Chamber Back Right (B)	Chamber Back Left (A)	Wave guide (G)	Scan Housing (F)	Total Pumping Capacity (klit/sec)	Average Pressure at PC Entrance (torr)
1	X	Y	Y	Y	Y	X	X	71	3.615E-05
2	X	Y	0	Y	Y	X	X	58.5	4.334E-05
3	X	0	Y	Y	Y	X	X	58.5	4.341E-05
4	X	0	0	Y	Y	X	X	46	5.447E-05
5	0	Y	Y	Y	Y	X	X	64	4.17E-05
6	0	Y	0	Y	Y	X	X	51.5	5.18E-05
7	0	0	Y	Y	Y	X	X	51.5	5.186E-05
8	0	0	0	Y	Y	X	X	39	6.887E-05

TABLE 2. Cryo Pump Speeds Used For Analysis.

Pump Designation	Pump Type	H₂ Speed (lit/sec)	Xe Speed (lit/sec)
X	250FE	7000	870
Y	320FE	12500	1550

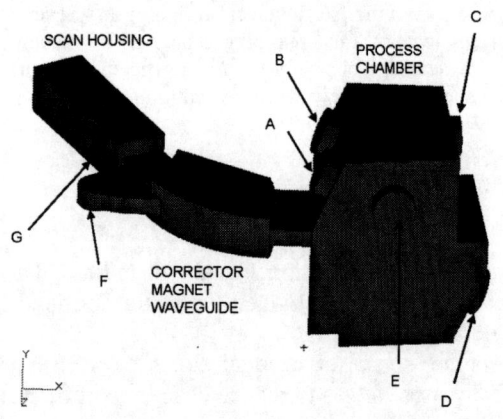

FIGURE 1. Layout of implanter beam line and cryo pump locations. The solid represents the vacuum space.

CALCULATION METHODOLOGY

The method used in this study to calculate the pressure distributions using FEA was first described in [6], which provides the theoretical basis. A large benefit of using FEA is that the actual geometry of the beamline is utilized. By contrast, the Monte-Carlo method of ray tracing, which is the most common method of calculating molecular flow, can only typically be used for simple geometries. The geometry of the high-energy implanter beam line was originally created on ProEngineer™ and Solidworks™ CAD workstations, and then imported into Nastran™ for FE analysis.

In setting up the FE problem, the molecular-flow boundary conditions have to be determined; namely, the outgassing rate from the wafer and PEF, as well as the associated gas pressure which is required to generate the flows on those surfaces. The relationship between pressure **P**, in torr, and flow rate **Q**, in molecules/second, is [7]

$$\frac{Q}{A} = 3.513 \times 10^{22} \frac{P}{\sqrt{MT}} \quad (1)$$

where **A** is the area of the outgassing surface in cm², **M** is the molecular weight of the gas which is being considered, and **T** is the gas temperature in °K. For this study, the H₂ outgassing rate of photoresist was extrapolated from Horsky [8]. For a 1500 watt implant, the H₂ outgases at about 170 sccm. Using the relation (1), the associated 300mm wafer outgassing pressure is then approximately P_{H2}=6.84E-5 torr. Likewise, for Xenon, which is pumped out of the PEF at ≤ 1.0 sccm, the associated pressure is about P_{Xe}=5.11E-6 torr. When multiple gases are analyzed, then the pressure distribution due to each gas must be solved for individually. This is due in part to the fact that the pumping efficiencies are generally different for different gases. The total gas pressure is then simply the sum of each component pressure distribution.

PRESSURE DISTRIBUTION IN THE IMPLANTER BEAM LINE

The physical layout of the implanter beam line is shown in figure 1. The solid model here represents the vacuum space within the chamber walls, including the scan housing, the corrector magnet waveguide, and the process chamber. The pump locations have been designated with letters which correspond to the descriptions in Table 1. Only two types of cryo pumps (Table 2) have been considered in the analysis: the "250FE" and the "320FE." Also, the wafer platen, scan arm assembly, and PEF have been included in the model.

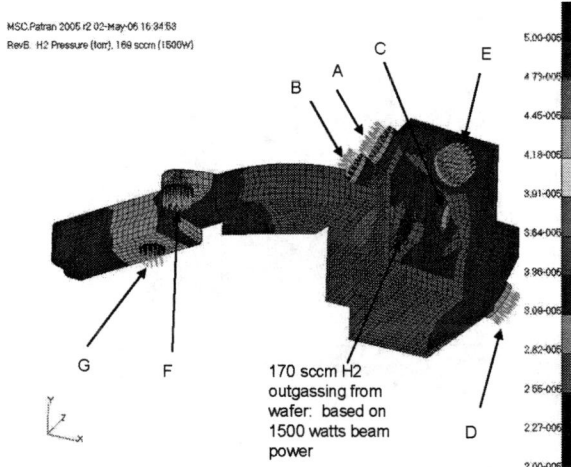

FIGURE 2. Simulated H_2 pressure distribution, in torr, inside the implanter beam line. Results are for the following conditions: implant angle = 7°; H_2 outgassing flow rate = 170 sccm from the 300mm wafer, and with all pumps operating (configuration 1).

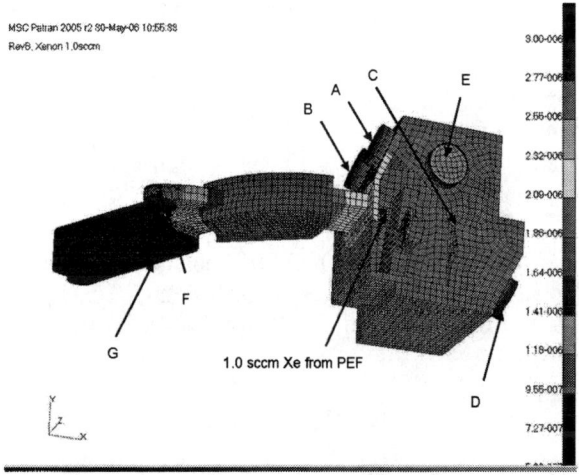

FIGURE 3. Simulated Xe pressure distribution, in torr, inside the implanter endstation. Results are for the following conditions: implant angle = 7°; Xe gas flow rate = 1.0 sccm from PEF, and with all pumps operating (configuration 1).

The simulated H_2 pressure distribution, based on an outgassing rate of 170 sccm and with all pumps operating, is presented as contour bands in figure 2. The direct streaming effect of the outgassing can be seen on the wall directly in front of the wafer, similarly to previous work [8]. The gas pressure is of course highest on the wafer itself, and the pressure is a minimum in the scan housing.

The simulated Xe pressure distribution, figure 3, is based on the same pumping configuration as in figure 2. The Xe pressure levels are much lower than those for H_2, due to the lower outgassing rate as well as the

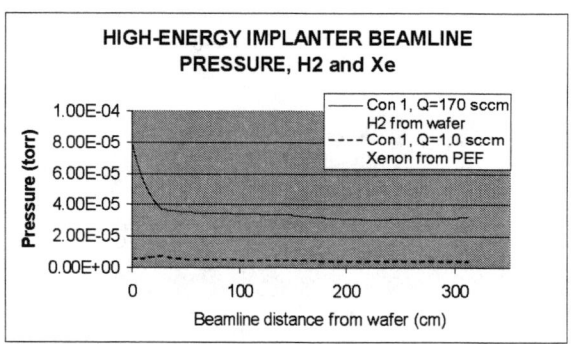

FIGURE 4. H_2 and Xe pressure, in torr, as a function of distance from the 300mm wafer, with all pumps operating. The Xe pressure exhibits a maximum in the region which is just above the plasma electron flood gun.

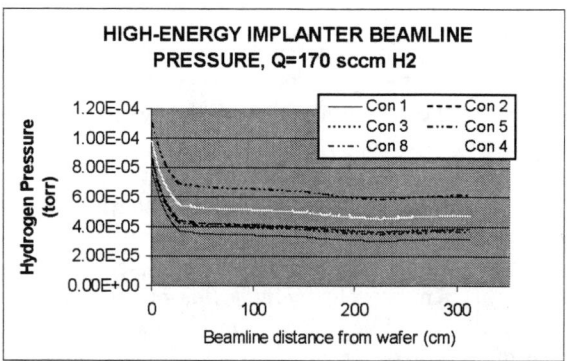

FIGURE 5. H_2 pressure, in torr, as a function of distance from the 300mm wafer, for several pump configurations.

lower pumping capacities for Xe. The pressure distribution along the beam line for both gases is graphed in figure 4. The Xe exhibits a maximum, 7.2E-6 torr, in the region directly above the PEF by design.

In this study, it was desirable to calculate the benefit of placing more cryos on the process chamber, in addition to the two 320FE's which are located on the back of the process chamber. Therefore we set up a '2^3' factorial matrix for pumps in the locations 'C,' 'D,' and 'E.' The pressure distributions along the beamline for several pump configurations are graphed in figure 5. The best case is 'configuration 1' where all the pumps are operating. The worst case is 'configuration 8' where the C, D, and E pumps are not operating. In comparing the performance of each pump configuration, it is convenient to examine the resulting pressure at the entrance of the beamline into the process chamber, which is a good metric for the pressure.

Figure 6 demonstrates the effect of factoring in the pumping capacity of each configuration. Here the "configuration throughput" Q_T is defined as

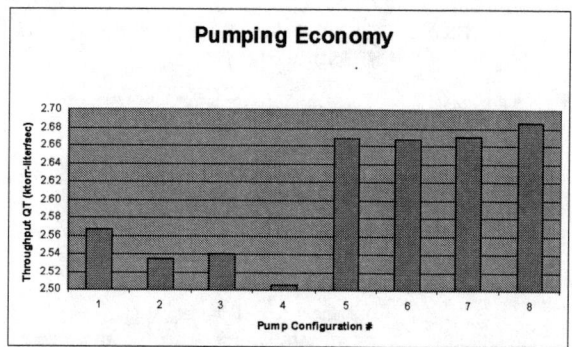

FIGURE 6. Comparison of the effect of factoring in the pumping capacity of each configuration. The most efficient pumping strategy is #4.

$$Q_T = S_{TOTAL} P_M. \quad (2)$$

where S_{TOTAL} is the total pumping capacity and P_M is the reference pressure which is used as the metric, in this case the entrance pressure. From this relation, Q_T has the units of mass flow rate, or torr-liter/sec. A lower Q_T indicates better configuration performance. A rather surprising result of this is that configuration 4 is the best economical solution for the pumping strategy, meaning that it is the most efficient pumping strategy. An even better indicator of configuration efficiency would be to factor in the actual monetary cost of the pumps and their associated hardware, such as compressors and gate valves. However, the actual cost of each configuration can change on a daily basis, so that it is more general to use Q_T for comparing different pump configurations.

COMPARISON OF ANALYSIS RESULTS WITH DATA

Figure 7 shows the differences between the analysis results and the measured data for both gases, H_2 and Xe. The differences are 28% and 27% for each gas, respectively. All of the data is for the case of the beam power of 650 watts. For both gases, the analysis result is higher than the measured result. The sources of these differences might come from both the calculated and the measured results. The differences are actually very reasonable.

CONCLUSIONS

The method of using FEA to calculate steady-state vacuum-pressure distributions has been successfully applied to the implanter beamline. A surprising result was found when the pressure results were combined with the total pumping capacity of the pump

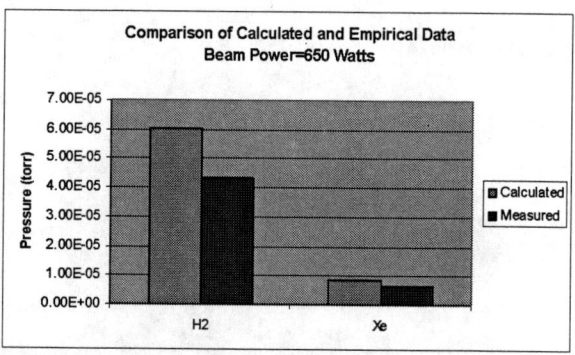

FIGURE 7. Comparison of analysis results and measured data. The difference is about 28% and 27%, respectively, for H_2 and Xe.

configurations, giving a "pump throughput" number for each configuration. The throughput number gave an indication of the efficiencies of the configurations. In comparing the throughput numbers, it turned out that the most efficient pump configuration consisted of 2-320FE pumps on the back wall and 1-250FE pump on the front wall of the process chamber. Thus the method of FE could be applied to determining the optimum pump configuration of the implanter process chamber.

ACKNOWLEDGMENTS

The authors are thankful to Leonard Rubin for his input on implanter process requirements. Thanks also go to Robert Rathmell and Tariq Fasheh for supplying the test data.

REFERENCES

1. A. Renau, M.E Mack, J.P. O'Connor, and N. Tokoro, Nucl. Instr. And Methods B55(1991) 61-65.
2. N. Tokoro, "Viista 3000 Ion Implanter System Paper," IIT Conf. Proc., 2000.
3. Venema, A., 1973/1974, The Flow of Highly Rarefied Gases, Philips Tech. Rev. 33-43.
4. Levenson, L.L., N. Milleron, and D.H. Davis, 1963, Le Vide 18, 42.
5. P. Clausing, The Flow of Highly Rarified Gases Through Tubes of Arbitrary Length, J. Vac. Sci. & Technol. 8, 636 (1971) [originally published: 1932 Ann. Physik 12, 961].
6. M. LaFontaine, "Calculation of molecular flow using the heat-transfer analogy," ANSYS Conf. Proc., 1998,.
7. A. Roth, Vacuum Technology, 3rd ed., Amsterdam: Elsevier Science, 1990, pp. 36.
8. T. Horksy, T. and A. Perel, Eaton Internal Memorandum, 2003R.
9. M. LaFontaine, N. Tokoro, "Distribution of residual gas in an ion-implant chamber," IIT Conf. Proc., 1998.

Reduction of the Wafer Pattern Damage on the Batch-type High Current Ion Implanters

Emi Oga, Hisaki Izutani, Genshu Fuse and Michiro Sugitani

SEN Corporation
1501, Imazaike, Saijo, Ehime, 799-1362, Japan

Abstract

The more shrinkage of device patterns becomes, the more device patterns, themselves, can be damaged by collisions between particles and wafers which moves very fast on batch-type implanters. The measures against this problem are to decrease the particles source, not to reach at the wafers, and to prevent the damage to device patterns by reducing the disk rotation speed. As a result, this method is extremely effective to minimize the yield loss without any faults.

Keywords: Wafer pattern damage, particle and disk rotation speed

PACS: 61.72Tt

Introduction

Recently a yield loss by the particle attack on the batch-type implanters becomes a severe issue for devices with short channel gates. (1, 2) Therefore, single-wafer implanters have been much expected. But, there are also some other issues on single-wafer implanters. Batch-type implanters are simple and high-quality finished form. So the following solutions are taken measures to prolong the batch-type implanter's life. The first concept is to decrease the particle sources. The second is to prevent particles to reach at the wafers. The thirds is to prevent particle damages to device patterns on the wafers. The easiest method is to reduce the disk rotation speed. (2) A collision energy between particles and wafers is proportional to a square of the disk rotation speed. Therefore, reduction of the rotation speed can lead to prevent effectively the damage to the devices by the particle attack. Table 1 shows energies calculated at a 200 mm batch-type high-current ion implanter; the LEX.

In this case, by slowering the disk rotation speed from 1215 rpm to 415 rpm, the collision energy decreases to 12%. Surprisingly, the energy decreases to 3% by the reduction to 215 rpm. However, by the reduction of the disk rotation speed would include concerns about the wafer temperature, implant uniformity and wafer charge-up. It is also known that the slower rotation speed leads to higher damage formation and sheet resistance changes often. (3)

On the other hand, at the single-wafer implanter using a two-dimensional mechanical scan system the wafer speed corresponds to about 10 rpm of the batch system. This fact indicates that such a type of a single-wafer implanter can hardly help having some very difficult issues. In this paper discussions are focused on the reduction of the disk rotation speed.

Experimental Procedure

The LEX manufactured by SEN for 200 mm wafers is used for all experiments. The unique method is used for an observation of a correlation between the disk rotation speed and wafer pattern damages. Some kinds of thermal labels are used for the wafer temperature evaluation. The implant uniformity is checked by the traditional sheet resistance (Rs) and the implant uniformity (σ) after annealing at temperatures of 1150 and 1000 degree centigrade for 20 seconds. The implantations at a dose of 1×10^{16} of As^+ ions cm^{-2} at an energy of 40 keV are performed to investigate the wafer pattern damages. Silicon powder is set in the beam line as shown in Fig. 1. Particles can be generated deliberately in the beam line using this method, and the scratches checked as shown in Fig. 2.

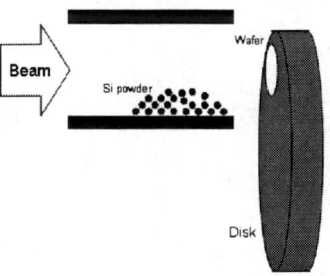

Fig.1 Diagrammatic illustration of experiment.

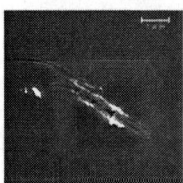

Fig.2 Wafer scratch

Table.1 Energy (1 μm³ size of Si) gained from the disk rotation.

No.	Disk rotation (rpm)	Energy of 1 μm³ of Si piece gained from disk rotation (ev)	Ratio
1	1215	3.5E+07	1.00(criterion)
2	915	2.0E+07	0.57
3	765	1.4E+07	0.40
4	415	4.2E+06	0.12
5	215	1.1E+06	0.03

The disk rotation speeds are 1215 rpm, 765 rpm, 615 rpm, 415 rpm and 215 rpm. After implantation the defects on the wafer are counted by a particle counter; Hitachi LS6800, and details of the defects are observed by a wafer review SEM; Hitachi RS-3000.

Result and Discussion

A. Wafer Pattern Damage

Table 2 shows counts of the defects, scratches and rate of these two numbers as an occurrence rate with respect to each disk rotation speed. These are cumulative counts of the 5-time implantations. The scratch generation can be reduced easily by slowering the disk rotation speed. It characterizes that the slower disk rotation speed is an extremely effective approach to reduce the pattern damage.

Table.2 Counts of the defects and scratches within a wafer

Disk rotation (rpm)	Total Defect count	Total Scratch	Rate of occurrence
1215	1911	76	4.0%
765	309	8	2.6%
615	1909	12	0.6%
415	406	3	0.7%
215	702	1	0.1%

These particles are deliberately for this experiment.

B. Implant Uniformity

In general, the slower disk rotation speed leads to a fewer number of painting counts at a wafer point. It could cause uniformity deterioration within a wafer. Table 3 shows a dependence of the sheet resistance (Rs) and implant uniformity (σ) on the disk rotation speed. This result indicates that it does affect neither the shift of sheet resistance nor increase non-uniformity. The disk rotation speed of 215 rpm is found to be able to be used for manufacturing from a uniformity point of view.

Table.3 Sheet resistance and uniformity was inspected at the various disk rotations decreased to 215rpm.

Implant condition	Disk rotation (rpm)	Rs ($\Omega/\square$)	σ (%)	Amount of shift
B+ 50keV 4mA 1E15	1215	93.962	0.530	criterion
	215	93.987	0.521	0.03%
B+ 2keV 1mA 1E15	1215	167.34	0.494	criterion
	215	167.21	0.534	-0.08%
As+ 30keV 8mA 1E15	1215	127.83	0.242	criterion
	215	128.36	0.260	0.41%

Annealing system:Axcelis Reliance850
Measuring equipment:KLA-Tencor OmniMap RS-100

C. Wafer Temperature Control

The second concern is wafer temperature rising due to a weaker centrifugal force. Figure 3 shows a correlation between the wafer temperature and implant power with respect to each disk rotation speed. Even when the disk rotation speed is 215 rpm, the wafer temperature is kept under 80 degrees centigrade at the 300-W implant beam condition, such as 10 mA and 30 keV.

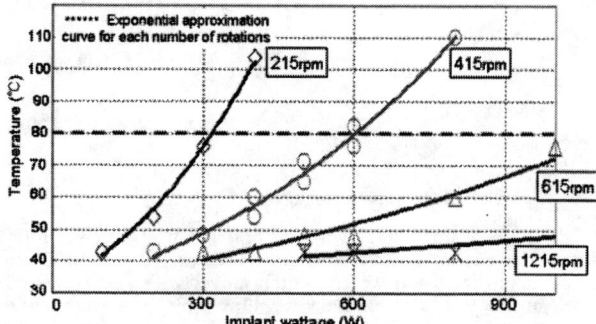

Fig.3 Dependency of wafer surface on rotation speed

D. Field Data

Implantation to actual devices with shrunk device patterns can be performed without any drastically change on characteristics of the fabricated devices by the reduction of the disk rotation speed.

Some users already have carried out the reduction of the disk rotation speed, and have succeeded in significant decrease of the wafer pattern damages. One of them obtains the almost same result as ours as shown in Fig. 4. It shows that pattern damages decrease responsibly to the problem-free level by the reduction of the disk rotation speed at a certain customer's site. This data seems to be corresponding to the particle attack result in this work. One of the user introduces 315 rpm as a solution of this issue.

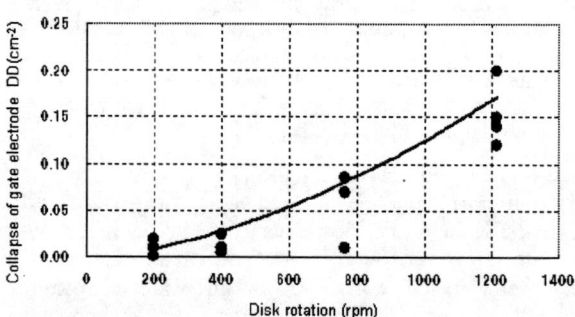

Fig.4 Dependence of defect density of gate electrodes on disk rotation.

Figure 5 shows changes of the yield loss before and after taking the three actions wihich were discussed in the introduction. The result indicates that the severe yield losses drastically decrease or disapper. By taking SEN's recommended measures, especially, the reduction of the disk rotation speed, even the batch-type ion implanter can keep higher production yields, clearing up the wafer pattern damage.

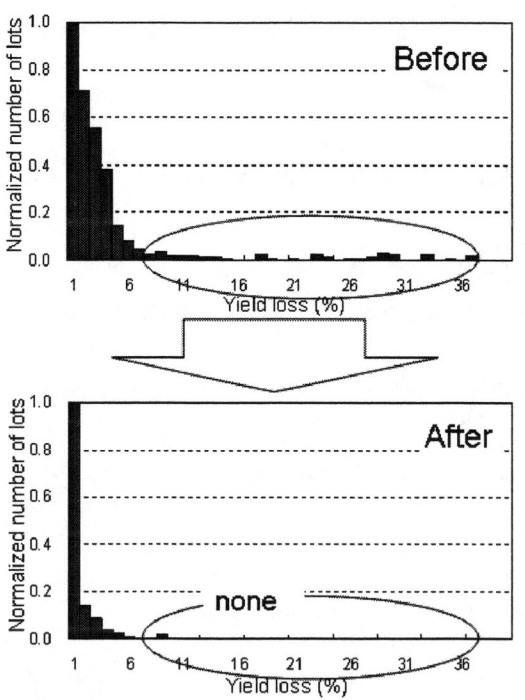

Fig.5 Reduction of yield loss before and after the actions.

SUMMARY

The solutions of the particle attack problem at the batch-type high-current ion implanter have been investigated. Especially, the reduction of the disk rotation speed has been studied. The following effects were confirmed.

1. The gate pattern damage issue can be reduced extremely by the slow disk rotation speed.
2. The uniformity within a wafer is kept at a sufficient level for the practical use even at the slow disk rotation speed.
3. The wafer temperature is lower than 80 degree centigrade even for a 300-W ion beam power even at a slow disk rotation speed of 215 rpm..
4. By taking SEN's recommended measures, lower yield loss or higher production yield can be obtained even with the batch-type ion implanter.

It promises that the batch-type implanter still, can be used for the advanced device fabrication.

ACKNOWLEDGMENT

The authors would like to thank Mr. T. Nakamura and Mr. D. Hamada of SEN Corporation for generous support, and to thank Dr. M. Kabasawa, Mr. T. Morita, Mr. M. Tsukihara, Dr. T. Yagita of SEN Corporation for fruitful discussions.

REFERENCES

1) Y. Kawasaki, K. Tokunaga, K.Horita, K.Mitsuda, A. Yamaguchi, A. Ueno, A. Teratani, T. Katayama, K. Hayami, M. Togawa, Y. Ohno and M. Yoneda, IWJT2004, pp. 39-41
2) L. Pipes, M. Taylor, G. Zietz, A. Al-Bayati, M. Castle, T. Marin, J. Simmons, Nwuclear Instruments and Methods in Physics Research B 237 (2005) 330-335
3) G. Fuse, M. Sano, H. Murooka, T. Yagita, M. Kabasawa, T. Siraishi, Y. Fujino, N. Suetsugu, H. Kariya, H. Izutani and M. Sugitani, Nuclear Instruments and Methods in Physics Research B 237 (2005) 77-82

Charging mechanism during ion implantation without charge compensation

[1]Shigeki SAKAI, [2]Hung-chi FAN, [2]Emily CHEN and [1]Masayasu TANJYO

[1]*Nissin Ion Equipment co.,ltd, 575 Kuze Tonoshiro cho Minami-ku 601-8205 Kyoto Japan*
[2]*Nissin Allis Union Ion Equipment co.,ltd,4 Fl.6 No.371 Sec.1, Guangfu Rd., Hsinchu Taiwan 300 R.O.C.*

Abstract. The charge accumulated on an electrode electrically floated during implantation is calculated using a simple model. The model includes the ion beam current, the secondary electron emission, the neutralizing electron current and the leak current through resistance between the electrode and the bulk of wafer. Expressing the leak current with a parameter of relaxation time, it is found that amount of the accumulated charge at the completion of implantation reached at a constant irrespective of the beam current in the condition that the relaxation time is long enough. The experimental result showed that the measured potential of the floated electrode depends on a beam potential as well as the beam current. This result implies that the neutralizing electrons are more effectively transported to the isolated electrode in the high beam current condition compared with in the low beam current condition. We discuss possibility that such charging phenomena may occur in the implantation for BiCMOS or SOI device fabrication.

Keywords: Ion implantation, beam potential, charging
PACS: 34.50.Dy, 41.75.-I

INTRODUCTION

Charging phenomena is one of the big problems in ion implantation. In the past 10 years people say that it is necessary to use charge compensation technique such as plasma flood gun even in medium current implanter for reducing charging damage to device structures on the wafer[1]. However, in the conventional device production line, ion implantation with medium current implanter is still used without any active charge compensation technique.

Now it is thought that without charge compensation the charging damage is much likely to happen because the device size has been shrinking and new types of wafer such as SOI and BiCMOS are being used instead of silicon bulk wafer.

In the implantation for conventional device fabrication one can say the charging voltage increases as beam current becomes higher. Can we apply this simple idea to the implantation for fabricating SOI or BiCMOS device? We think that the higher electric resistance between an electrode and bulk wafer in these new device structures compared with the conventional device structures may cause different charging phenomena. Using a simple model we calculate the charging voltage of electrode as a function of a resistance-related parameter.

We also study experimentally the charging voltage of electrically isolated electrode and discuss what the key parameter is.

CHARGE ACCUMULATION SIMULATION

When charged particles are implanted to an electrically isolated electrode with a resistance to

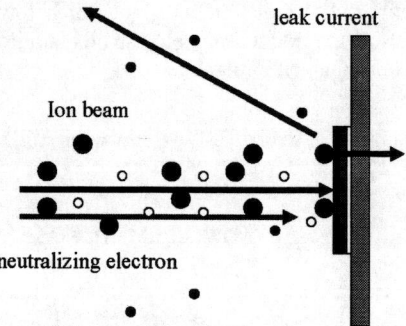

Figure1. A charge accumulation model on isolated electrode by ion implantation

wafer, electric charges are accumulated in time. Figure 1 shows a simple model of this charge accumulation. In the following discussion, the ion charge current is denoted as I_0, the secondary electron emission factor as γ, the leak current through the resistance as I_L, the neutralizing electron current transported by beam potential as I_n. The accumulated charge dQ in Fig.1 can be expressed as follows,

$$dQ = (I_0 + \gamma I_0 - I_L - I_n)dt. \quad (1)$$

First we discuss the effect of leak current to charge accumulation.

Influence of leak current

The leak current I_L is related to the resistance and capacitance between the electrode and wafer. Using the charge relaxation time τ of the leak current is, the leak current can be expressed by

$$I_L = \frac{Q}{\tau}, \tau = CR, \quad (2)$$

where τ is given by the product of the electrical resistance R and the capacitance C. For easy estimation of the effect of leak current upon the accumulated charge, we assume the secondary electron emission factor γ is zero and also the neutralizing current I_L is zero. So the Eq.(1) becomes

$$dQ = \left(I_0 - \frac{Q}{\tau}\right)dt. \quad (3)$$

Figure 2 shows the accumulated charge on the isolated electrode centered on the wafer, which is

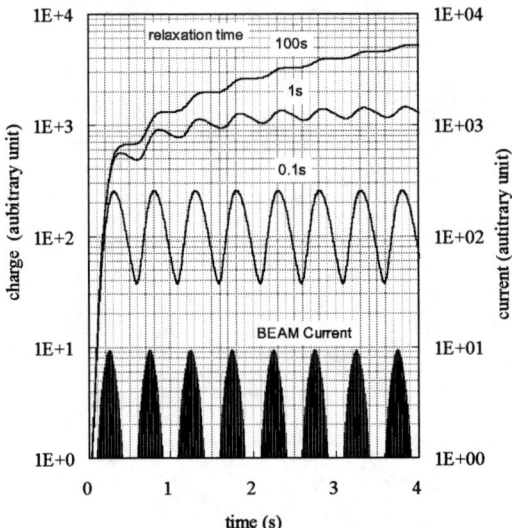

FIGURE 2. Calculated charge accumulation on an electrode on wafer with a parameter of relaxation times

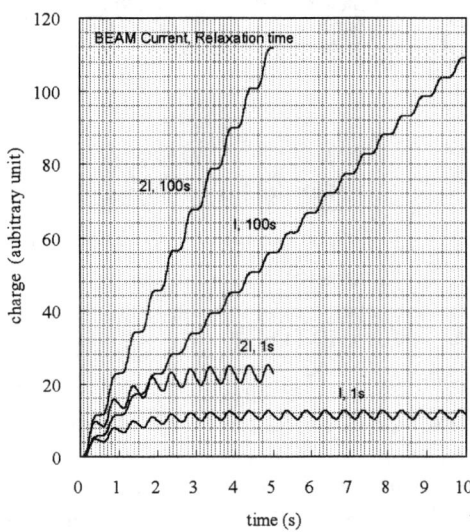

FIGURE 3. Calculated charge accumulation on an electrode for two cases of t relaxation time, in each case the beam current being changed doubly.

calculated using Eq.(3) with a parameter of charge relaxation time. The calculation is made for the implantation in EXCEED series implanters, which uses hybrid scan of 173 Hz magnetic beam scan and 2 Hz mechanical scan. Also we presume that the beam size follows 20mm Gaussian circular distribution. Note that also in Fig.2 the ion beam current to the electrode centered on the wafer is illustrated by solid line, and is seen as an array of cone-shaped trees; the frequency of 173 Hz makes the picture painted. It is seen that the maximum amount of the accumulated charge becomes higher as the relaxation time becomes longer. Also seen is that in a condition that the relaxation time is shorter than the mechanical scanning time, the maximum charge becomes constant irrespective of dosage.

Figure 3 shows the accumulated charges with changing beam current doubly for two cases of the relaxation time. Of course, in same dosage, doubling the beam current makes the implantation time half for the same recipe of dosage. In the condition of the relaxation time of 100sec, the accumulated charges at the completion of the implantation are same for the two cases of the beam current. On the other hand in the condition of the relaxation time of 1sec, the accumulated charge becomes higher by doubling the beam current.

This result can be summarized that when the relaxation time is much longer than the mechanical scanning time the charging doesn't change with beam current, and that on the other hand when the relaxation time is almost the same or less than the scanning time the charging decreases as the beam current becomes small.

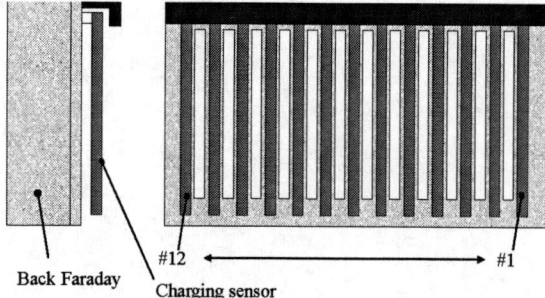

Figure4. The charging sensor array in front of the Back Faraday system in EXCEED2000A

Influence of neutralized electrons

The charging in BiCMOS or SOI devices sometimes shows complicated results and is likely to happen even with lower beam current. Regarding Eq. (1) we have to consider the charge neutralization. An array of charging sensors has been equipped in front of the Back Faraday System in EXCCED2000A in order to measure the charging voltage. Figure 4 shows a schematic diagram of the charging sensor array. The charging voltage is picked up by the high impedance voltmeter. Figure 5 shows the charging voltage measured by the twelve charging sensors. One can see in Fig.5 that t the charging voltage is high when the beam hits the charging sensors, while the voltage becomes when the beam is away from the sensors. The highest voltages are seen in both sides of charging sensor array, that is, at the sensors #1, #2, #11 and #12. As the relaxation time becomes longer than the period of the beam sweep, charges accumulate before they

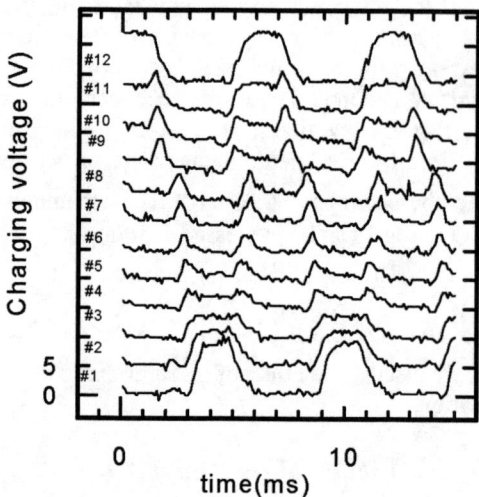

FIGURE 5. Charging voltage measured by the charging sensor array.

leak. In Fig.6 the highest charging voltages measured by the charging sensor array are plotted for various beam conditions. One can see in Fig.6 that the highest charging voltages decrease as the beam energy goes higher.

In the case of the circular cross-section beam the potential at the center of the beam with respect to the grounded beam line wall is represented [2] by

$$\varphi = \frac{I}{4\pi\varepsilon_0}\sqrt{\frac{m}{2eV}}\left[1+2\ln\left(\frac{R}{r}\right)\right]. \qquad (4)$$

where I is the beam current, m is the ion mass, eV is the ion energy, r is the beam radius and R is the radius of the cylindrical beam line wall. To roughly estimate

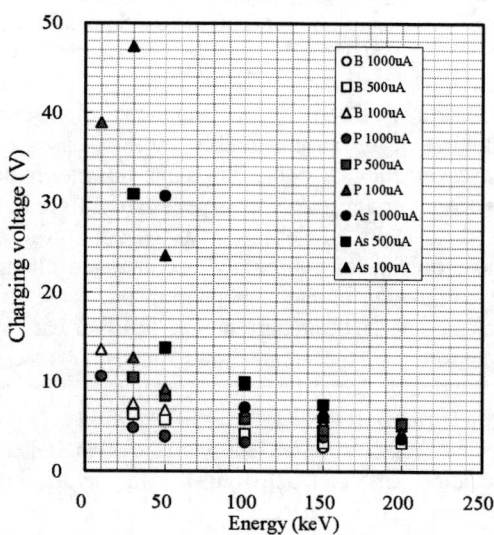

FIGURE 6. Maximum charging voltage as a function of ion energy changing ion species and beam current

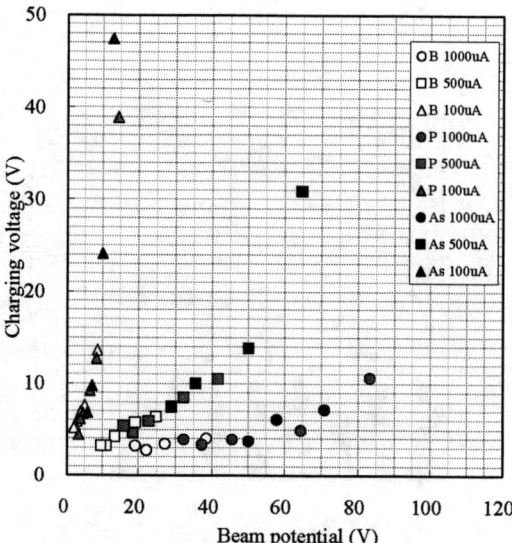

FIGURE 7. Highest charging voltages as a function of beam potential for various ion species and three different

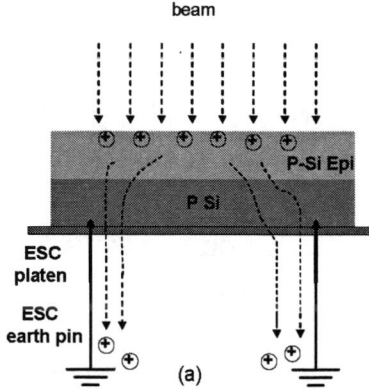
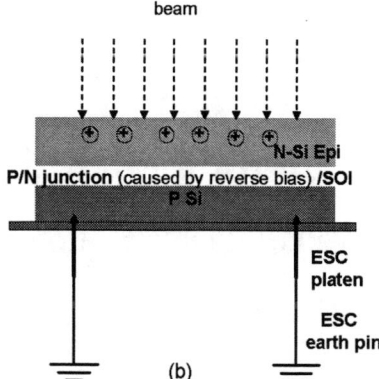

FIGURE 8. Model of charging comparing (a) conventional CMOS technology and (b) BICMOS or SOI technology

the effect of the potential we use the following 'estimate beam potential',

$$\varphi_{estimate} = \frac{I}{4\pi\varepsilon_0}\sqrt{\frac{m}{2eV}}. \quad (5)$$

This potential neglects the effect of the term R/r, which can be treated as a constant because of the beam property at the Faraday system in the EXCEED2000A. In Fig 7, the highest charging voltages are re-plotted as a function of the estimate beam potential defined by Eq.(5). It is seen in Fig.7 that the re-plotted points lies on three lines which correspond to three different conditions of beam current. More interesting in Fig.7 is that the charging voltage decreases as the beam current increases.

We construe this data as follows. The beam potential as expressed by Eq.(4) or (5) develops in front of the isolated electrode and the isolated electrode potential follows the beam potential. The fact that the measured charging voltage decreases as the beam current increases at the same beam potential may be due to the effect of the neutralizing electrons. The neutralizing electrons are generated as secondary electrons by ion beam hitting a target beside the charge sensors or the beam line wall. Electrons are gathered by positive beam potential. Gathered electrons are transported to the isolated electrode. Therefore the charging voltage decreases as the beam potential. There should be a mechanism of gathering neutralizing electrons much more in the higher beam current condition than in the lower beam current condition.

Charging model comparing CMOS and BiCMOS or SOI technology

The previous calculation and experimental results shows that if the relaxation time of isolated electrode is shorter than scanning speed, charging voltage decreases as the beam current on the same implanted dosage. If the isolated electrode is perfectly insulated, the charging voltage is the same as that without neutralization. With neutralization due to electrons in the beam line, the charging voltage decrease as the transported electrons increase by beam potential and the beam current.

In conventional CMOS technology the accumulated charge easily flows out because the relaxation time of the materials and device structure is small. Therefore the charging voltage increases as the beam current. On the other hand in the BiCMOS or SOI technology, the accumulated charges does not easily flows out because the relaxation time of such structure is substantially higher, and thus the charging voltage decreases as the beam current.

Figure 8 shows the structure of conventional CMOS technology and BiCMOS or SOI technology.

CONCLUSION

In order to consider the charging phenomena in ion implantation, we used the simple charge accumulation model. Studying the leak current and neutralizing electrons how they influence charging phenomena in no active charge compensation system, we found that the amount of charge at the completion of implantation reaches at a constant irrespective of the beam current in the condition that the relaxation time is long enough. The experiment showed the highest measured charging voltage is dependent upon the beam potential as well as the beam current. We think this result indicates that the neutralizing electrons are very much easily transported to the isolated electrode in higher beam current at the same beam potential.

REFERENCES

1. M. Tanjyo ct. al., Proceedings *of the 2000 Int'l Conf. on Ion Implantation Tech,* pp.588-591
2. D. M. Jamba, Nucl. Inst. Meth. 189, 253 (1981)

The Impact of Mass Resolution on Molybdenum Contamination for B, P, BF$_2$, and As Implantations

Volker Häublein*, Lothar Frey,[†] and Heiner Ryssel*,**

*Fraunhofer Institut für Integrierte Systeme und Bauelementetechnologie, Schottkystrasse 10, 91058 Erlangen, Germany
[†]Qimonda Dresden GmbH & Co. OHG, Königsbrücker Straße 180, 01099 Dresden, Germany
**Lehrstuhl für Elektronische Bauelemente, Cauerstrasse 6, 91058 Erlangen, Germany

Abstract. This paper discusses the impact of the mass resolution of an implanter on molybdenum contamination for B^+, P^+, BF_2^+, and As^+ implantations by means of simulated mass spectra. In all cases, the simulations are based on experimentally identified transport mechanisms of molybdenum through the analyzer magnet. The mass spectra were simulated using the ENCOTION (ENergetic COntamination simulaTION) software, which is a powerful tool for the simulation of mass interferences and mass spectra.

Keywords: ion implantation, contamination, mass interferences, encotion
PACS: 41.85.Qg, 33.15.Ta, 85.40.Ry

INTRODUCTION

The mass resolving power of an implanter is expressed by the mass resolution $M/\Delta M$, where M represents the mass of an ion and ΔM is the full width at half maximum of a peak in a mass spectrum [1]. Typical values lie between 50 and 150.

For a constant mass resolution, the peaks in a mass spectrum get broader with increasing mass which makes mass separation more effective for smaller masses. This is illustrated in the simulated mass spectrum in Fig. 1, where the peaks of B^+, BF_2^+, and Mo^+ ions are shown for mass resolution values of 50, 100, and 150. As far as the separation of the boron isotopes is concerned, mass resolution appears to have a very low impact on the efficiency of mass separation. This is, however, entirely different as regards the heavier BF$_2$ molecules, where a mass resolution of 50 is completely inadequate in order to separate the two BF$_2$ peaks.

In case of molybdenum, which has seven isotopes in the range of 92–100 amu, the separation of the molybdenum isotopes is even worse. For a mass resolution of 50, the Mo spectrum appears to be a block rather than a spectrum of single isotopes. For a mass resolution of 100, the Mo isotopes are noticeable, for a mass resolution of 150, they are evident. At first glance, this might appear not to be a big issue for standard implantations, such as B, P, and As, since the masses of these elements and the mass of Mo differ considerably. There are, however, several mechanisms which "project" the Mo spectrum to other positions in the mass spectrum [2]:

- multiply charged Mo ions
- Mo ions that undergo a charge exchange prior to mass analysis
- Mo containing molecules that dissociate prior to mass analysis.

The ENCOTION software, which is described in the next section, simulates a variety of potential Mo transport mechanisms for B, P, As, and BF$_2$ implantations. In order to evaluate the impact of mass resolution on molybdenum contamination, the respective Mo transport mechanisms have to be known. Experiments showed that there are dominant mechanisms for each of the mentioned implantations. In Tab. 1, these transport mechanisms are listed together with the favored Mo isotopes. In case of BF$_2$ implantations, Mo contamination by ^{98}Mo^{++} ions is a classical problem which has already been discussed in several papers (e.g. [3, 4]). A similar molybdenum transport mechanism occurs in P implantations, where ^{92}Mo^{+++} and ^{94}Mo^{+++} ions interfere with ^{31}P$^+$ ions [5]. The Mo transport mechanisms for ^{11}B$^+$ and ^{75}As$^+$ implantations are more complicated, since charge exchange reactions are involved. In comprehensive studies involving mass spectra analysis and SIMS measurements, the electron stripping mechanism Mo$^+$ →Mo^{+++} and the electron capture mechanism Mo^{+++} →Mo^{++} have been found to be responsible for Mo contamination in boron and arsenic implantations, respectively [6].

SOFTWARE ENCOTION

The simulation tool ENCOTION has been developed in order to check the potential of any element of the peri-

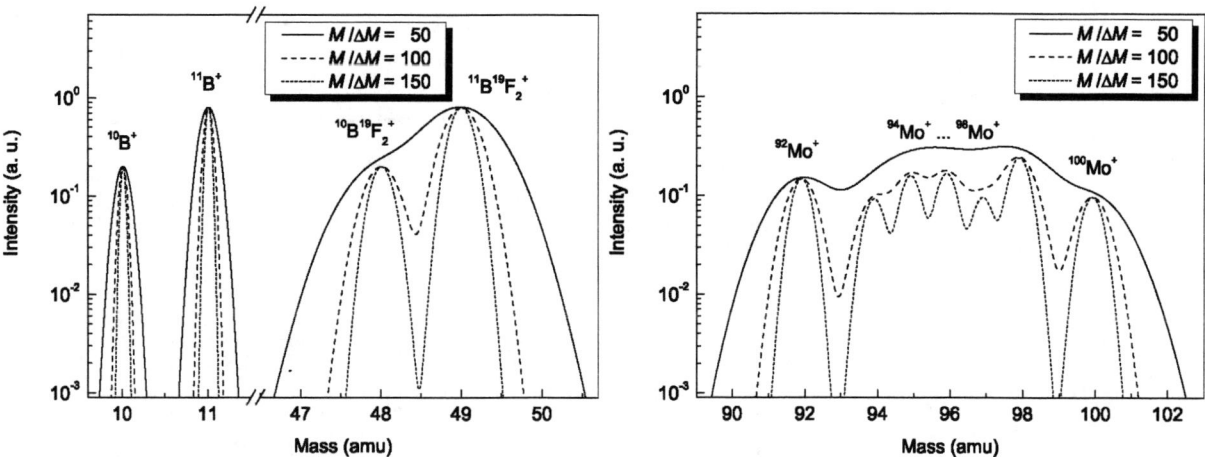

FIGURE 1. Simulated B, BF$_2$, and Mo signals for mass resolution values of 50, 100, and 150.

TABLE 1. Mechanisms of Mo contamination for B, P, BF$_2$, and As implantations [6]

Implanted ion species	Mo contamination mechanism	Favored Mo isotope
^{11}B$^+$	Mo$^+$ →Mo^{+++}	^{100}Mo
^{31}P$^+$	Mo^{+++}	^{92}Mo
^{11}B^{19}F$_2^+$	Mo^{++}	^{98}Mo
^{75}As$^+$	Mo^{+++} →Mo^{++}	^{100}Mo

odic table to contaminate a wafer during ion implantation [7]. Since the analyzing magnet of an implanter does not separate the real mass of an ion but the so-called apparent mass, mass interferences and energetic contaminations may occur. ENCOTION enables the user to simulate expeditiously transport mechanisms of contaminants through the analyzer magnet in order to evaluate the contamination risk in ion implantation.

Since the introduction of ENCOTION at the IIT 2002 conference [7], the functionality of the software was extended to simulate mass spectra. Therefore, ENCOTION is, on the one hand, an optimum tool for identifying peaks in measured mass spectra, on the other hand, it allows an examination of the correlation between contamination and the mass resolution of the implanter.

The mass resolution not only depends on the physical dimensions of the implanter, such as the radius of the magnet or the width of the beam slit, but also on the focus of the ion beam. The smaller the beam diameter, the steeper the rising and the flatter the maximum of the corresponding signal in a mass spectrum and vice versa. ENCOTION, therefore, offers two modi to simulate mass spectra. The first mode simulates the peaks by Gaussian curves and is sufficiently accurate in most cases. The second mode allows the user to set up separately the magnet radius, the width of the beam slit and the standard deviation of the ion beam. This mode is especially useful for an implanter set up where the resolving slit is much larger than the deviation of the ion beam.

The following considerations about Mo contamination are based on Gaussian shaped signals.

SIMULATION OF MO CONTAMINATION FOR STANDARD IMPLANTATIONS

B implantation

For B implantations, Mo contaminations are mainly caused by the stripping of singly charged to triply charged Mo ions prior to mass separation (Mo$^+$ →Mo^{+++}). Figure 2 shows the corresponding simulated Mo spectra. The shapes of the Mo spectra have not changed compared to the original Mo spectra in Fig. 1, i. e. the original Mo spectrum is projected in the mass range from about 9.8–11.5 amu, showing the same inadequate separation of the Mo isotopes.

The Mo contamination levels for B implantations can be extracted at the interception point of the B peak with the Mo spectrum. Starting from a mass resolution of 50, the Mo level decreases drastically with increasing mass resolution. An increase from 50 to 100 leads to a decrease of the Mo level by approximately an order of magnitude. When the mass resolution is increased from 50 to 150, the Mo level is decreased by even more than two orders of magnitude. Therefore, mass resolution has a high impact on the Mo contamination levels. In addition, the simulations verify that ^{100}Mo is the favored isotope being implanted together with ^{11}B.

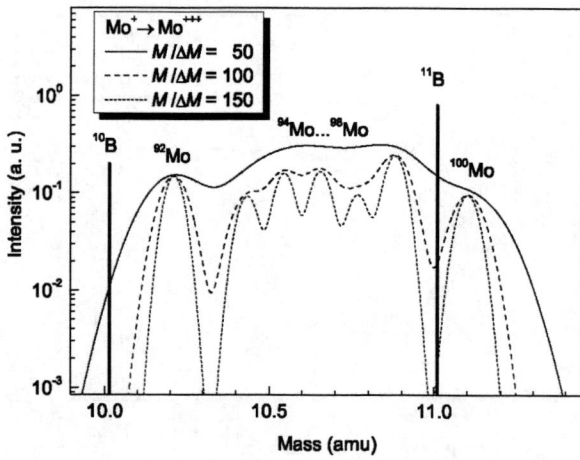

FIGURE 2. Using ENCOTION simulated Mo signals resulting from the mechanism $Mo^+ \rightarrow Mo^{+++}$. The peak of $^{11}B^+$ is located between the two peaks which are caused by the ^{98}Mo and ^{100}Mo isotopes.

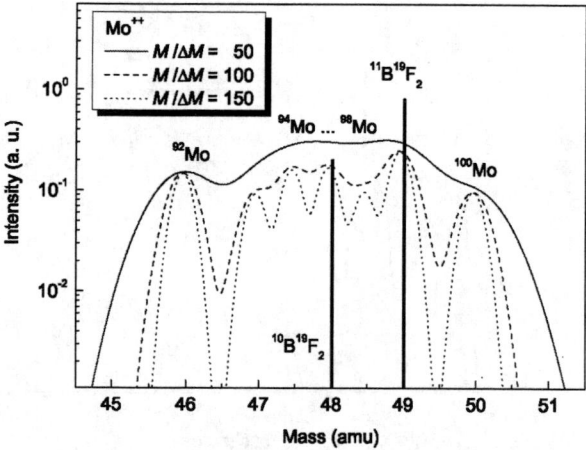

FIGURE 4. Using ENCOTION simulated Mo signals of the Mo^{++} isotopes. The signal of the isotope $^{98}Mo^{++}$ is practically at the same position as the $^{11}B^{19}F_2$ signal, the $^{96}Mo^{++}$ nearly overlaps with $^{11}B^{19}F_2$.

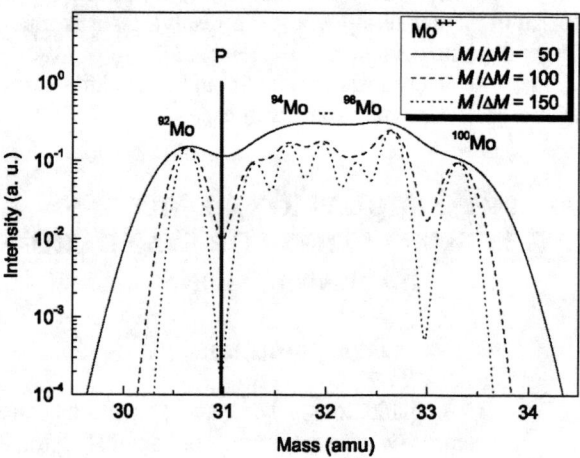

FIGURE 3. Using ENCOTION simulated Mo signals of the Mo^{+++} isotopes. The peak of $^{31}P^+$ is located between the peaks of $^{92}Mo^{+++}$ and $^{94}Mo^{+++}$.

P implantation

For P implantations, Mo contamination is mainly caused by Mo^{+++} ions. In Fig. 3, the respective simulated Mo spectra are illustrated. In this case, the original Mo spectrum from Fig. 1 is projected in the mass range from approximately 29.5–34.5 amu. Since the P peak is located approximately in the middle between two Mo peaks, the role of mass resolution is similar as in the case of B implantations. An increase from 50 to 100 reduces the Mo level by an order of magnitude, an increase from 50 to 150 by almost three orders of magnitude. Again, mass resolution is a very effective parameter to reduce Mo contamination.

BF₂ implantation

In BF₂ implantations, Mo^{++} ions pass the analyzer magnet. The corresponding Mo spectra are shown in Fig. 4. As can be seen by the interception point of $^{11}B^{19}F_2$ with the Mo spectra, the mass resolution value has only a very low impact on the Mo contamination levels, since the $^{11}B^{19}F_2$ and the Mo^{++} peaks nearly overlap. An increase of the mass resolution from 50 to 100 leads to a decrease of the Mo level by approximately 20 %. An increase from 50 to 150 reduces the Mo level by approximately 22 %.

As implantation

For As implantations, Mo contaminations are mainly caused by the electron capture mechanism $Mo^{+++} \rightarrow Mo^{++}$. Figure 5 shows the corresponding simulated Mo spectra. There is, analog to the case of BF₂ implantations, an overlap of the peak which is caused by the ^{100}Mo isotope and the As peak. According to the following figures, the impact of mass resolution is even smaller than for BF₂ implantations: An increase of the mass resolution from 50 to 100 leads to a decrease of the Mo level by approximately 12 %, a further increase of the mass resolution is practically without effect.

Figure 6 shows the Mo contamination for the discussed implantations against mass resolution. The curves are normalized on the total amount of Mo ions of the respective mechanism (including all isotopes). As far as BF₂ and As implantations are concerned, Mo contamination levels can only be reduced up to a mass resolution of approximately 50, larger values keep the Mo contami-

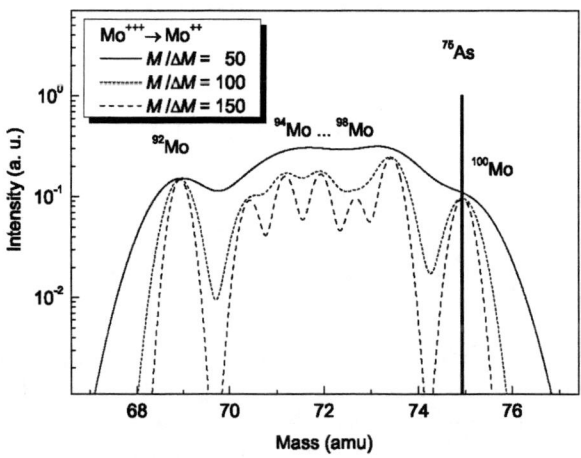

FIGURE 5. Using ENCOTION simulated Mo signals resulting from the mechanism $Mo^{+++} \to Mo^{++}$. The signal of the isotope ^{100}Mo is practically at the same position as the As signal.

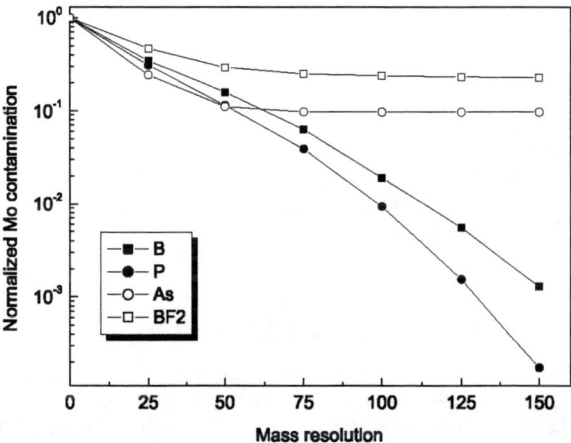

FIGURE 6. Mo contamination as a function of mass resolution for B, P, As, and BF$_2$ implantations. For each implantation, the Mo contamination is normalized on the total number of Mo ions belonging to the respective transport mechanism as per Tab. 1.

nation on a constant level. This level is for BF$_2$ implantations higher by a factor of 2.3 than for As implantations, which corresponds to the ratio of the natural abundances of the favored isotopes ^{98}Mo and ^{100}Mo.

For B and P implantations, however, mass resolution has to be considered a very effective parameter for reducing Mo contamination: for a mass resolution of 150, the Mo contamination is reduced by four orders of magnitude and three orders of magnitude for P implantations and B implantations, respectively.

CONCLUSION

The impact of mass resolution on mass interferences was studied for the case of Mo contamination for standard implantations of B, P, BF$_2$, and As.

The studies were performed using the ENCOTION simulation software. The simulations of mass spectra show that the mass resolution of an implanter could have a large impact on the contamination level, as it is the case for B and P implantations, where Mo levels can be reduced by several orders of magnitude. In the case of BF$_2$ and As implantations, however, the ^{98}Mo and the ^{100}Mo isotopes, respectively cannot be screened out by the analyzer magnet. Therefore, mass resolution has hardly any impact for values in excess of 50.

REFERENCES

1. M. Farley, "Ion Source Operation and Maintenance / Ion Source Problems," in [8], pp. 457–486.
2. H. Ryssel, M. I. Current, and L. Frey, "Contamination Control for Ion Implantation," in [8], pp. 564–601.
3. A. Cubina, *Nuclear Instruments and Methods in Physics Research* **B55**, 160–165 (1991).
4. G. Curello, D. Carroll, J. Marley, X.-M. Zhang, N. Brooks, and P. Mason, "Effect of BF$_2$ induced Mo^{++} on long retention time in 0.25 μm DRAM technology," in *1998 International Conference on Ion Implantation Technology Proceedings*, edited by J. Matsuo, G. Takaoka, and I. Yamada, IEEE, Piscataway, NJ, USA, 1999, pp. 550–553.
5. K. Funk, V. Häublein, H. Chakor, M. Ameen, L. Frey, H. Ryssel, and A. Ramirez, "Investigation of Molybdenum Contamination in ^{11}B$^+$ and ^{31}P$^+$ Implants," in *2000 International Conference on Ion Implantation Technology Proceedings*, edited by H. Ryssel, L. Frey, J. Gyulai, and H. Glawischnig, IEEE, Piscataway, NJ, USA, 2000, pp. 711–714.
6. V. Häublein, *Kontaminationsprozesse in der Ionenimplantation*, Ph.D. thesis, University Erlangen-Nuremberg, Cauerstrasse 6, 91058 Erlangen, Germany (2006).
7. V. Häublein, L. Frey, and H. Ryssel, "ENCOTION – A new Contamination Analysis Software," in *2002 14th International Conference on Ion Implantation Technology Proceedings*, edited by B. Brown, T. L. Alford, M. Nastasi, and M. C. Vella, IEEE, Piscataway, NJ, USA, 2003, pp. 101–104.
8. J. F. Ziegler, editor, *Ion Implantation – Science and Technology*, Ion Implantation Technology Co., Edgewater, 2000.

Maximizing Productivity for Well Implantation

Qing Zhai, Youn Ki Kim, Dennis Rodier, and Naushad Variam

Varian Semiconductor Equipment Associates, Inc., 35 Dory Road, Gloucester, MA 01930, United States

Abstract. Until recently most semiconductor manufacturers have used high-energy ion implanters for deep well formation. This was partially driven by the fact that these well implants were beyond the productive operating range of conventional medium current implanters to achieve in a productive and cost effective fashion. Recently, VSEA has increased the energy range of the VIISta medium current implanter to 300keV, 600keV and 900keV for single charge, double charge and triple charge implantation, respectively. A combination of this new capability with a reduction in well implant energies due to device scaling, has brought most well implants in the regime of the VIISta 900XP system. This paper describes the superior process performance that is achieved and how this capability results in increased productivity and reduced cost of ownership.

Keywords: Ion Implantation, Productivity, Well Doping
PACS:

I. INTRODUCTION

Tight control of well implant parameters is key in manufacturing leading edge semiconductor devices [1, 2], due to increasing device packing density. Historically, sub-MeV well doping has been conducted on high energy implanters. With the device geometry scaling down, most of the well implant energies have been reduced to lower than 600keV. In addition, true zero degree well implants are required to avoid shadowing from the photo resist mask and encroachment. The effectiveness of zero well implants to control intra-well isolation has been reported [1, 2]. The new generation VIISta medium current implanters from Varian Semiconductor Equipment Associates (VSEA) has the capability of processing 300keV, 600keV and 900keV implants using single charge, double charge and triple charge ions, respectively. The VSEA single wafer architecture guarantees excellent implant angle accuracy. The VIISta 900XP medium current implanter satisfies all the process requirements of well doping, including dose uniformity and repeatability within wafer, wafer-to-wafer, and lot-to-lot, and beam incident angle accuracy. The system can also provide high productivity, comparing to high-energy implanters which significantly reduces the cost of ownership (CoO). In this paper, we will demonstrate the process performance of the VIISta 900XP, and its capability of maximizing productivity using a representative set of recipes with energy ranging from 30keV to 2.5MeV.

II. EXPERIMENTAL

In this paper, a B^+, 300keV, $3E13cm^{-2}$ recipe was chosen for process performance characterization. This implant is sensitive to small variations in ion channeling due to small incident angle differences and can easily be quantified by standard bare wafer techniques such as Secondary Ion Mass Spectroscopy (SIMS), thermawave, and sheet resistance. All wafers used were 300mm in diameter and were implanted on a VIISta 900XP ion implanter. After implant, wafers were annealed for 10 seconds at 1000°C in 10% oxygen atmosphere for dopant activation. Sheet resistance was measured by four-point probe with an edge exclusion of 3mm. As-implanted dopant profiles were characterized using SIMS. Thermawave was used for as-implanted dose uniformity characterization.

III. PROCESS PERFORMANCE RESULTS

Dopant profiles obtained with SIMS from 5 points (center, left, right, top, and bottom) representative of three points each in the slow scan and fast scan direction for a 300keV, B^+, $3E13cm^{-2}$ at 0 degree tilt are shown in Figure 1. The top, bottom, left, and right points are obtained 5mm from the wafer edge. These dopant profiles consist of two peaks; the deeper peak at 1.15 micron consists of channeled ions while the peak at the lower depth of 0.75 micron consists of de-channeled ions. SIMS profiles obtained from different wafer locations overlay at both de-channeled and channeled regions. This behavior reflects the beam parallelism across the wafer. A $0.1°$ un-parallelism in the beam across the wafer will show clearly on the overlay, since the Rp sensitivity per $0.1°$ at $0°$ is about 43.2nm. With $0.1°$ angle variation, the non-channeling portion will increase noticeably at the expense of channeling portion. In Figure 2, four dopant profiles taken from the center of four wafers implanted over a 20 day time period (including the center profile from Figure 1) are shown. The complete overlay of the four profiles indicates a well under $0.1°$ variation dopant profile can be achieved repeatedly, demonstrating outstanding run-to-run repeatability of zero degree well implants on the VIISta 900XP implanter. The sheet resistance and Thermawave data from the four runs, as shown in Table I, confirms both within wafer dose uniformity and run-to-run dose repeatability of this channeled implant. Figure 3 contains the typical Thermawave and sheet resistance maps for this implant.

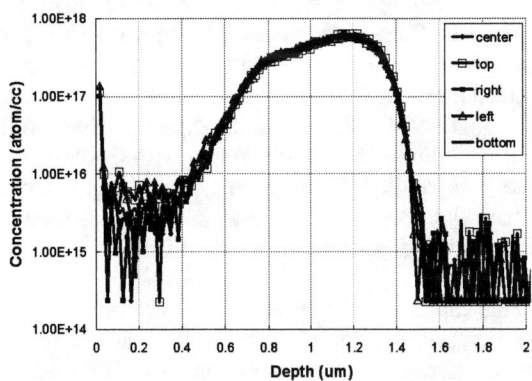

FIGURE 1. As-implanted dopant profiles from 5 locations of a wafer implanted with B^+, 300keV, $3E13cm^{-2}$, $0°/0°$, with an edge exclusion of 5mm.

This beam angle performance on the VIISta 900XP medium current ion implanter is controlled by the use of Varian Postioning System (VPS^{TM}), a closed loop angle control system integrated into all VIISta implanters [5]. With this system, both beam parallelism and beam steering angles are measured and corrected with each beam setup and at user selectable wafer intervals. This ensures accurate angle control, which is critical for $0°$ tilt well implants.

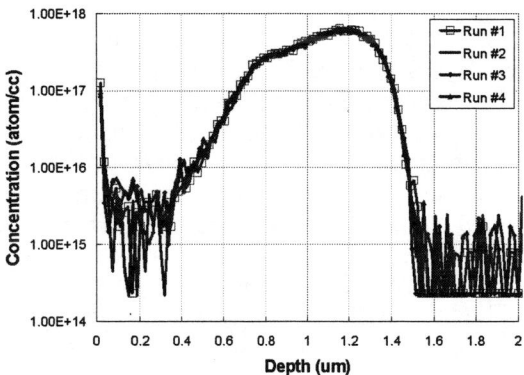

FIGURE 2. As-implanted dopant profiles from center of four wafers implanted with B^+, 300keV, $3E13cm^{-2}$, $0°/0°$ over 20 days.

Table I. Sheet resistance and Thermawave data of B^+, 300keV, $3E13cm^{-2}$, $0°/0°$ over 20 days

Implant Date	TW Unit	TW Unif. (%)	Rs (Ohm/sq.)	Rs Unif. (%)
Day 1	654.33	0.26	971.31	0.194
Day 4	654.11	0.33	974.89	0.266
Day 13	656.68	0.42	973.57	0.254
Day 20	655.92	0.39	978.44	0.261
Average	655.26	0.35	974.55	0.24
Repeat. (%)	0.19		0.31	

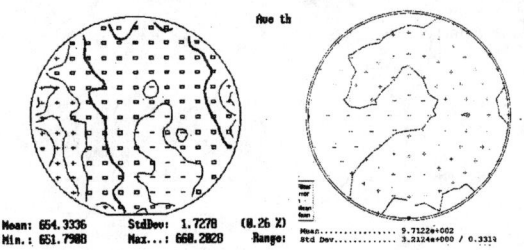

FIGURE 3. Typical Thermawave and sheet resistance maps of B^+, 300keV, $3E13cm^{-2}$, $0°/0°$ implant.

A true $0°$ tilt implant has channeled doping profile, as shown in Figures 1 and 2, although the degree of channeling varies, depending on factors like substrate crystal orientation, beam energy, ion species, doping dose and implantation induced crystal damage, etc. [6, 7]. Figure 4 shows three doping profiles of B^+, 300keV, $3E13cm^{-2}$ implants with implant tilt angles at $0°$, $0.5°$ and $1°$. It can be clearly seen that with the tilt angle changing from $0°$ to $0.5°$ to $1.0°$, the extent of channeling becomes less while the de-channeling portion becomes more dominant. In other words, the mean projected range of the ions (R_p) becomes smaller with implant tilt angle. The Thermawave data from these wafers, as summarized in Table II, are consistent with the conclusions drawn from the SIMS dopant profiles. The increase in thermawave unit with implant tilt angle shows that the 0° tilt implant has the most channeling. From these results it can be discerned that a variation 0.5 degree in implant angle will result in large variations in dopant profiles. This proves again that the VIISta 900XP can provide excellent beam parallelism and accurate beam incident angle control consistently, as shown in Figures 1 and 2. Surface concentrations of well implants are affected and have implications in definition of halo/well junctions, critical for transistor performance. Thus even a +/- 1 degree ion beam incident angle variation across the wafer, like that observed when using a relatively flat disc on a batch implanter, is not sufficient to enable small incident angle well implants.

Table II. Thermawave data from B^+, 300keV, $3E13 cm^{-2}$, with various implant angles

Implant Angle (deg)	TW Unit
0	654.33
0.5	674.43
1.0	717.63

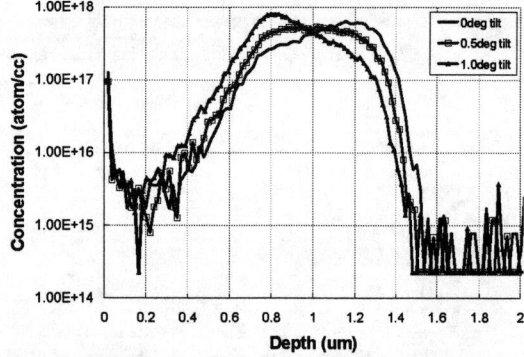

FIGURE 4 As-implanted dopant profiles of B^+, 300keV, $3E13 cm^{-2}$, implant with various implant angles.

III. PRODUCTIVITY IMPROVEMENT

To illustrate the productivity improvement by transferring sub-MeV well implants from high energy tools to medium tools with 300keV and 600keV capability, a representative recipe set, including implant energy from 30keV to 2.5MeV, as summarized in Table III, is used for modeling. The productivity simulation that follows has the following assumptions:
1. 90% of specification beam current
2. 70% of tool utilization rate
3. 300mm wafers, 100 wafer lot

The following two scenarios are simulated, based on the above assumptions.
1. High energy implanter and a medium current implanter without increased beam energy capability, where implants 3, 4, 5, 12, 13, 26 and 27 are dedicated to high energy implanter, and the rest to a medium current implanter
2. High energy implanter and medium current implanter with increased beam energy capability combination, where implants 4, 5, 26 and 27 are assigned to high energy implanter, and the rest to medium current tool with increased beam energy capability.

The simulation results are shown in Figure 5. All data are presented in percentage, after being normalized to scenario 1.

When a medium current tool without increased beam energy capability is used, seven out of the twenty seven recipes are dedicated to the high-energy tool. With the increased beam energy capability, only four of the twenty seven recipes are dedicated to the high energy implanter for maximizing productivity. As a results, the number of total tools required decreases by 25% to 40% for wafer start per month (WSPM) ranging from 10k to 50k as compared to scenario 1. These advantages will continue to occur with the addition of other medium current recipes. For customers who do not have recipes with energy higher than 900keV, all the implants can be carried out on medium current implanter with 300keV and 600keV capability. With fewer types of implanters in the bay, the operation overhead, such as maintenance will be reduced, resulting in reduced CoO.

This capability of optimizing the tool requirements as a function of wafer utilization is enabled by the seamless process transfer between the VIISta 900XP medium current and VIISta 3000HP high energy ion implanters [1]. More over, the commonality between these tools allow transparency of their use as well as optimized utilization in a dynamic environment with varying product mixes

Table III, recipe set for bay capacity simulation

Implant Name	Species	Energy (keV)	Dose
1	B	200	3.50E+13
2	B	250	9.00E+12
3	B	300	5.00E+13
4	B	400	2.00E+13
5	B	600	2.00E+12
6	P	200	1.00E+13
7	P	250	4.00E+12
8	P	400	1.00E+13
9	P	400	2.00E+13
10	P	400	3.00E+13
11	P	500	3.00E+13
12	P	600	3.00E+13
13	P	600	8.00E+13
14	As	180	8.00E+12
15	B	70	9.00E+12
16	B	70	1.00E+13
17	B	30	6.00E+12
18	As	30	1.00E+13
19	P	50	1.00E+13
20	B	30	2.00E+13
21	P	30	1.00E+13
22	P	40	1.20E+14
23	B	250	1.00E+13
24	B	120	8.00E+12
25	B	120	6.00E+12
26	P	1000	3.00E+13
27	P	2500	2.00E+13

The expanded energy range of the VIISta 900XP enables the transfer of most high-energy well implant applications to the VIISta 900XP medium current implanter. If manufacturers do not have any implants with energy higher than 900keV, they can select either VIISta 900XP only or combining it with VIISta high energy tool, depending on the recipe set. All of these can be accomplished without any compromise to process performance, due to the tight process control capabilities on VIISta 900XP and VIISta 3000HP implanters.

ACKNOWLEDGMENTS

The authors would like to thank Paul Rizk and Marcus Monell at VSEA for their support.

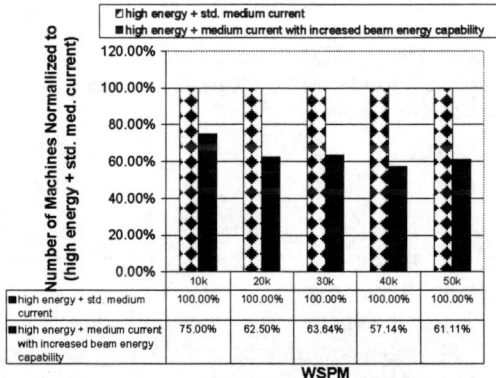

FIGURE 5 Calculated implanter requirement for various wafer start per month, all data are normalized to scenario 1 and are presented in percentage

This capability of optimizing the tool requirements as a function of wafer utilization is enabled by the seamless process transfer between the VIISta 900XP medium current and VIISta 3000HP high energy ion implanters [1]. More over, the commonality between these tools allow transparency of their use as well as optimized utilization in a dynamic environment with varying product mixes.

IV. CONCLUSIONS

REFERENCES

1. S. Norasetthekul, B. N. Guo, J. Flanagan, N. Variam, and S. Mehta, "Meeting the Well Doping Requirement of sub 100nm Devices- Process Performance Characteristics of the VIISta 3000 Implanter, IEEE 2002 International Conference on Ion Implantation Technology Proceedings, Taos, New Mexico, USA, pp530 - 533
2. T. Yamashita, M. Kitazawa, Y. Kawasaki, H. Takashino, T. Kuroi, Y. Inoue, and M. Inuishi, "Advanced Retrograde Well Technology for 90-nm-node Embedded Static Random Access Memory Using High Energy Parallel Beam", Jpn. J. Appl. Phys., Vol. 41, Part 1, No. 4B, pp2399-2403
3. A. Stuber, S. Jang and W. Kim, "Ion Implant Requirements for High Volume DRAM Manufacturing", IEEE 2002 International Conference on Ion Implantation Technology Proceedings, Taos, New Mexico, USA, pp157-160.
4. E. McIntyre, J. P. O'Connar, S. Mehta, U. Jeong, S. Norasetthekul, "Gains from Angular Integrity", Eropean Semiconductor, March 2002
5. J.C. Olson and A. Renau, "Control of Channeling Uniformity for Advanced Applications", IEEE 2000 International Conference on Ion Implantation Technology Proceedings, Alpback, Austria, pp670-673.
6. J. F. Ziegler, "Ion Implantation Physics", Ion Implantation Science and Technology, pp 4-1 to pp 4-50
7. B. N. Guo, N. Variam, U. Jeong, S. Mehta, "Experimental and Simulation Studies of the Channeling Phenomena for High Energy Implantation", IEEE 2002 International Conference on Ion Implantation Technology Proceedings, New Mexico, USA, pp131-134.

Efficient High Current Process Transfer and Device Matching Strategies for sub-90nm Manufacturing

Dennis Lee[1], NyenSiong Loh[1], Miow Chin Tan[1], Kyuha Shim[2], Scott Falk[2], Baonian Guo[2], Scott Jillson[2], Bryan Wong[3], Kherchong Loh[3], Sandeep Mehta[2], Yuri Erokhin[2]

[1]Chartered Semiconductor Manufacturing Ltd., 60 Woodlands Industrial Park D, Street 2, Singapore
[2]Varian Semiconductor Equipment Associates, Inc., 35 Dory Road, Gloucester, MA 01930, United States
[3]Varian Semiconductor Equipment Associates Asia, Ltd 10 Upper Aljunied Link #04-09, York International Building, Singapore.

Abstract. As devices scale down to 90nm and beyond, process transfer between different high current (HC) implanter platforms requires precise matching of multiple characteristics of implant process environment. Matching traditional recipe parameters such as dose, energy and tilt/twist no longer secures achieving target device characteristics with tolerances needed for high volume manufacturing ($\Delta<1\%$ for I_{dsat}, V_t). Thus, beam characteristics, previously considered to be second order, such as beam angular properties (divergence and steering), beam current density, and energy contamination (EC) levels must be taken into account to achieve accurate device matching between different high current implanter platforms. While device sensitivity to beam angular properties increases as devices shrink, at reduced ion energies traditional implanter matching procedures relying on SIMS and Rs measurements on bare wafers lose sensitivity to variations of beam angular characteristics. Thus, to achieve efficient qualification of new high current implant tools new approaches delivering precise process and device matching using minimum number of device split lot runs are needed. In this paper, using 90nm CMOS device data, we demonstrate that in order to transfer existing HC processes to new high current tool successfully, key parameters determining ion beam characteristics need to be quantified and matched between the tools with considerations for unique responses of 3D PR patterned device structures to differences of beam angular and EC components of implant process environment.

Keywords: Device matching, high current, 90nm CMOS
PACS: 61.72.Tt, 85.30.De

INTRODUCTION

Traditional implant process matching between different ion implanters has been done on bare wafer which satisfied device electrical parameters matching for >0.18um device. As CMOS devices are shrunk for circuit level improvements, however, dose matching on bare wafer no longer secures achieving target device characteristics.

Typical high current applications, Source/Drain (SD) and Source/Drain Extension(SDE) doping , are self-aligned process defined by gate or gate spacer patterns. Depending on the incident angle of the ions relative to the gate, two competing phenomena, gate shadowing effect and lateral encroachment effect will govern key device parameters. Since the device sensitivity to beam angular properties increases as devices shrink, traditional methods, Rs measurement and SIMS on bare wafers, can not capture beam angular effect properly.

Thus, to achieve efficient qualification of new high current implanter, device split lot runs after well-known dose matching on bare wafer are needed.

In this paper, we demonstrate the importance of beam incident angle and dose control to match device performance to Production Tool of Record(PTOR).

EXPERIMENTAL

As 2keV 1E15 tilt 7 degree quad rotation for NMOS SDE formation and BF2 5keV 1E15 tilt 7 degree quad rotation for PMOS SDE implants were carried out on a parallel ribbon beam single wafer high current implanter, VIISta HC. In order to understand device sensitivity to angle or dial-in dose, implant recipe was modified with +/- 3 degree tilt split or reduced dose by 10%. Table 1 shows the detailed test matrix.

Key transistor parameters such as Overlap capacitance(Cov), Saturation Vt (Vtsat) and Saturation Drive Current (Idsat) were measured.

TABLE 1. Test Conditions

POR Implant Conditions	Angle Sensitivity		Dose Sensitivity
As 2keV 1E15 tilt 7_quad rotation for NMOS	Tilt 4° quad rotation	Tilt 10° quad rotation	9E14 at/cm^{-2}
BF$_2$ 5keV 1E15 tilt 7_quad rotation for PMOS	Tilt 4° quad rotation	Tilt 10° quad rotation	9E14 at/cm^{-2}

RESULTS

Figure 1 shows the effect of implant angle on NMOS transistor Vtsat and overall capacitance (Cov). Any change (positive or negative) in the implant tilt angle from 7 degrees resulted in Vt lowering and increased Cov.

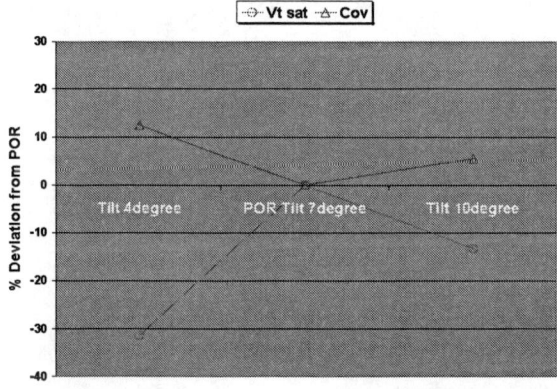

FIGURE 1. NMOS Vt, Cov vs. Tilt angle of quad mode.

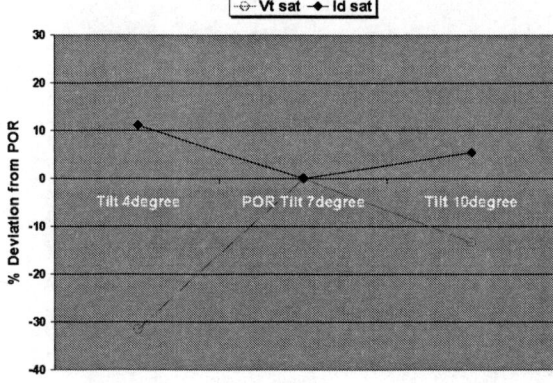

FIGURE 2. NMOS Vt, Idsat vs. Tilt angle of quad mode.

The reduction in Vtsat results in a higher drive current as observed in Figure 2. An interesting feature to note is the asymmetry in the response for these parameters. The decrease in implant tilt angle results in a more dramatic decrease in Vt (10% Vt shift per 1 degree) while the decrease in Vt is less dramatic when the tilt angle was increased beyond 7° (5% Vt shift per degree).. It seems that dopant loss by gate shadowing effect from 4° through 7° tilt is significantly larger than effective channel length shorten by lateral encroachment effect.

Figure 3 shows the device sensitivity to dose variations. A 10% decrease in dose for the SDE implant POR resulted in a <4% decrease in Vtsat and an even smaller (~2%) change in Idsat. The implant angle thus plays a far more important role in the device performance and incident angle matching was recognized as a critical knob for NMOS SDE process transfer.

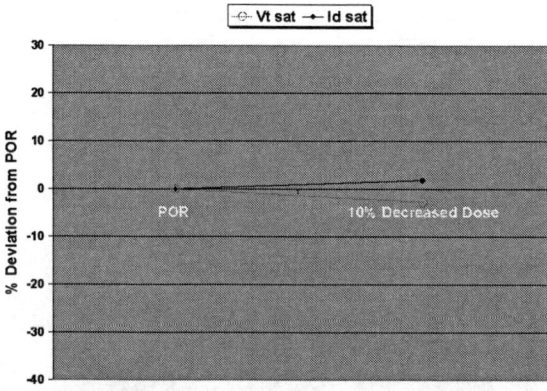

FIGURE 3. NMOS Vt, Cov vs. Implant dose

A similar tilt angle split was used to evaluate the SDE implant for the PMOS devices. Unlike the NMOS case, key device parameters of PMOS transistor remained relatively unaffected by the +/-3 degree change in tilt angle from the 7 degree POR tilt as shown in Figure 4. This is attributed to the fast Boron diffusion process during the post implant anneal that dominates over any doping profile differences caused by gate shadowing or encroachment effect due to tilt angle errors. This mechanism is not as dominant in the NMOS case where the As dopant atoms have a much lower diffusivity than the B atoms.

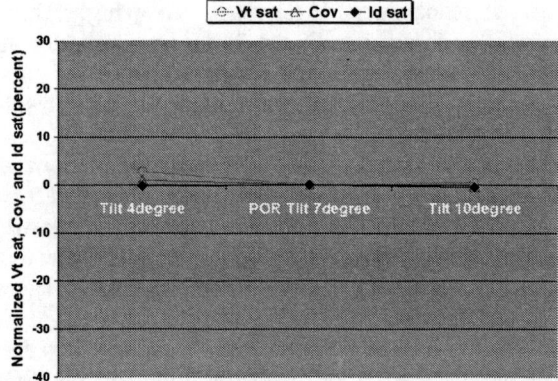

FIGURE 4. PMOS Vt, Cov Idsat vs. Tilt angle of quad mode

Figure 5 shows PMOS Vt/Idsat sensitivity to dose. We observe that a 10% decrease in the dose resulted in a 10% increase in Vt and an 8% decrease in Idsat. This is again explained by diffusion dominating the B profiles in the SDE region since the concentration of diffused B under the gate is directly proportional to the implanted dose. Thus, the SDE implant dose is the more important process control knob for the PMOS devices.

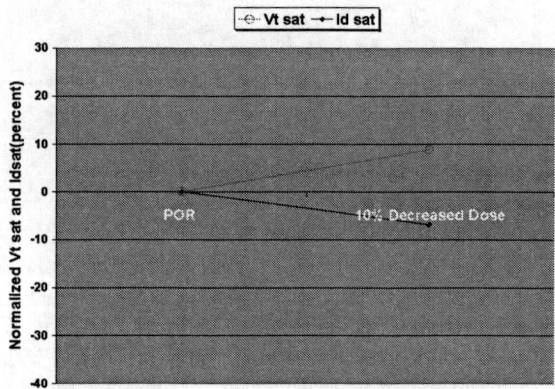

FIGURE 5. PMOS Vt, Idsat vs. Implant dose

CONCLUSIONS

In this paper, we have shown the 90nm CMOS device sensitivity to angle and dose for both the NMOS and PMOS devices. The NMOS devices are more sensitive to beam incident angles and hence beam angle control is a critical parameter for matching NMOS devices. On the other hand, accurate dose control is the more critical parameter to get repeatable PMOS transistor performance and process matching.

ACKNOWLEDGMENTS

The authors would like to thank Suresh Kumar and Peter Benyon for their contributions to this paper.

REFERENCES

1. U. Jeong et al. "Devices dictate control of implant-beam incident angle", Solid State Technology, Oct. 2001
2. U. Jeong et al, "Effects of beam incident angle control on NMOS source/drain extension applications", IIT 2002
3. S.Falk, "Accurate dose matching measurements between different ion implanters", IIT 1996
4. Y.Erokhin, "Precision Implant Requirements for SDE junction Formation in sub-65nm CMOS Devices", IWJT2006

IMPLANT TECHNOLOGY (MATERIALS)

The Development of In-Situ Ion Implant Cleaning Processes

Steve Bishop, Robert Kaim, Sharad Yedave, Josep Arnó, Frank DiMeo Jr.[*], Mike Wodjenski

ATMI Inc., 7 Commerce Drive, Danbury, CT 06810, USA

Abstract. Considerable gains in implanter utilization efficiency can be attained with in-situ cleaning of deposited material, particularly in and around the ion source. Different methods of in-situ cleaning are described, and we discuss the relative merits of several chemical reagents. We introduce XeF_2, a new and promising reagent for in-situ cleaning and present some preliminary experiments showing its ability to etch dopant materials. We also show that in some cases etching by XeF_2 can be selective with respect to ion source construction materials such as tungsten.

Keywords: Process Efficiency, In-situ cleaning, Implant Efficiency, Ion-implanter Maintenance, Chemical Cleaning

PACS: 85.40.Ry; 81.65.Cf; 61.72.Vv; 52.77.Dq; 52.77.Bn

INTRODUCTION

Since the introduction of ion implantation tools into high volume semiconductor manufacturing more than 20 years ago, implantation equipment has become increasingly automated, efficient and productive. However there is one area in which there has been little improvement: maintenance procedures to remove deposited material from interior surfaces of the implanters remain manual and potentially hazardous. Build-up of deposited material inside the tool is known to cause voltage instability and dopant cross-contamination, and yet implanter engineers are often forced to continue to operate with significant interior deposits because their removal is labor intensive and requires lengthy downtime for maintenance.

In this paper we will discuss the use of in-situ chemical means to remove deposited material, and the potential of such methods to reduce manual cleaning and improve tool performance.

CURRENT IMPLANTER MAINTENANCE PROCEDURES

Ion implanter maintenance in the fab is typically organized into periodic categories. The first and most frequent category is the ion source change. Ion sources are preferably changed at the end of cathode life; however, premature source failures due to deposits on source insulators, high voltage electrodes or bushings are relatively common, depending on the dopant species and usage history of the source. At the end of its life, the source is removed, disassembled and parts which can be re-used are usually bead/slurry blasted. The mechanical action of the beads removes the deposits, but often damages the part, particularly if it is made of a soft material such as graphite or boron nitride. Once used, the beads are contaminated with toxic material and must be disposed of appropriately.

The second maintenance category is for the region in the immediate vicinity of the ion source: the extraction electrodes, the source high voltage bushing and the source vacuum chamber must be cleaned periodically. Some fabs will schedule separate maintenance for these activities, others will combine some or all of them with the source change. The ground and suppression electrodes are removed and cleaned, and the interior surfaces of the bushing and vacuum chamber are manually scrubbed with low particulate wipes, usually wetted with alcohol to avoid generating hazardous airborne particulates. Wiping down these interior surfaces is unpleasant and potentially hazardous work, and it generates large quantities of toxic waste in the form of discarded wipes, gloves, body suits and face masks. A large

---

[*] Current address: SAIC, 1710 SAIC Drive, McLean, VA 22102, USA

production fab will generally have to dispose of up to 100 barrels of solid toxic waste and 20 barrels of beadblast/slurry generated by implanter maintenance procedures.

Less frequent maintenance is required for components of the implanter which are more remote from the ion source. These include magnet waveguides and other vacuum chambers, interior electrodes, and wafer handling components such as clamp rings, wheels and discs. These components are removed at major maintenance intervals, typically every 6 – 12 months, and cleaned with manual or liquid chemical means, depending on the nature of the deposits. For the components located closest to the wafer, the deposits are composed mainly of photoresist products. In this paper, we will only discuss chemical removal of deposits which are directly related to the dopant species. Generally, these occur on the components closest to the ion source.

CHEMICAL CLEANING

The concept of cleaning a reactor *in situ* is widely and successfully used in chemical vapor deposition (CVD) processes[1] Expanding the concept by introducing a vapor into an ion implanter in order to chemically remove deposited residues is attractive from the point of view of safety and convenience because solid toxic waste is converted into a gaseous waste stream which is easily and safely handled by existing fab abatement systems. Chemical cleaning may also significantly improve implanter operational efficiency by allowing recovery of ion sources which are failing prematurely due to deposition on internal electrodes and insulators, by reducing the time and labor for maintenance schedules, and by avoiding damage to sensitive parts during manual cleaning.

Cleaning Methods

There are several ways in which a reactive cleaning vapor can be introduced into an ion implanter. For cleaning the ion source, the vapor can be introduced while the source is operating, so that cleaning is effected by means of a plasma. The cleaning vapor may be introduced simultaneously with the dopant species, or a separate cleaning plasma may be used to clean the source during species changes. The ion source can also be chemically cleaned while it is not operating – in this case the cleaning vapor must be capable of effectively cleaning a cold source without plasma. If the vapor cannot clean without plasma activation, a separate source of plasma radicals must be supplied external to the ion implanter.

For the cleaning methods discussed in the previous paragraph, the vapor is flowing continuously from its point of entry and out through the pump exhaust manifolds. This might be referred to as "dynamic" cleaning, and removal of deposits would be expected to occur mostly along a flow path of the vapor. An alternative method, which could be used to clean all surfaces within a vacuum chamber, might be called "static" cleaning. In this case, the vacuum chamber is sealed, vacuum pumps are isolated or turned off, and cleaning vapor is introduced into the chamber from any convenient location until a suitable pressure is attained. Static cleaning methods have been demonstrated in the thermal cleaning of tungsten using F_2 gas.[2] The vapor, which must react with deposited material at room temperature and without plasma activation, is allowed to remain in the chamber until it is consumed or an endpoint in the reaction is reached, and the gaseous byproducts are then pumped away. Multiple cycles of filling, reacting and pumping away byproducts may be required to effectively remove all the deposited material.

Chemical Cleaning Agents

Effective in-situ cleaning is primarily dependent on three factors: the reactive nature of the cleaning precursor, the volatility of the by-products, and the reaction conditions. Optimum reactive cleaning materials are required to selectively remove unwanted residue while minimizing damage to the materials of construction of the implanter. To a certain extent, the activity of the cleaning material can be tuned by adjusting its flow rate and the pressure and temperature of the reaction. The byproducts generated by the reaction between the cleaning agent and the residues have to be volatile enough to facilitate their removal by the vacuum system. Fluorinated cleaning agents are ideal to remove arsenic, phosphorous, germanium, antimony, silicon and boron residues as they readily form volatile fluoride species. However, indium fluorides have very low volatility and chlorine-containing reactants are more suitable for cleaning purposes.

As mentioned in the previous section, external energy sources can also be added to enhance the reactivity and speed of the cleaning process. Typical gases used to etch or clean reactors using plasma processes rely on electron collisions to generate highly reactive atomic fluorine from perfluorinated gases such as NF_3, CF_4, C_2F_6, SF_6 and C_3F_8. In most cases, plasma dissociation is induced upstream of the chamber using expensive remote plasma reactors.

While the cleaning precursors are relatively inert (except NF_3), perfluorinated gases used in these processes are among the strongest greenhouse gases with global warming potentials (GWPs) 3 and 4 orders of magnitude higher than CO_2. Goals to reduce their usage in 160 countries were negotiated in the Kyoto Climate Protection Protocol.

Thermal cleaning agents are highly active materials that react with residue in the reactor via thermodynamic dissociation. Examples used in CVD include ClF_3 and fluorine gas. Handling these reactive materials generates significant health, environmental, and safety concerns.

Xenon difluoride (XeF_2) offers several advantages for cleaning most ion implant residues *in-situ*. The material is a sublimable white solid with a vapor pressure of 4 Torr at room temperature. While XeF_2 is stable, this strong oxidizer reacts readily and selectively with other metals to generate fluorides and xenon gas. Its efficacy as an etching material has been demonstrated in MEMS and focused ion beam (FIB) applications[3]. Additional data is included in the next section. Effective and reproducible vaporization of solid materials requires high surface areas and optimum heat transfer properties of the delivery vessel or vaporizer. A gas box compatible XeF_2 cylinder was developed that combines an internal macropore core to enhance the surface area and thermal transfer to provide accurate and sustainable XeF_2 vapor delivery. Xenon difluoride is a toxic and corrosive material (TWA= 0.3 ppm). Appropriate abatement methods are recommended to minimize health and environmental hazards associated with a release. Chemisorbtion-based scrubbers are most suitable for abating both hydrides and fluorinated implanter species including XeF_2.

ETCHING RESULTS WITH XENON DIFLUORIDE

Initial studies of etching with XeF_2 were carried out in a small test reactor using test samples prepared by e-beam deposition of Aluminum, Boron, Tungsten, and Silicon on glass microscope slides. The Aluminum was used as a bottom barrier layer on the glass slide. Some samples were given a protective silicon cap layer, and others were left uncapped and allowed to oxidize in room air. The test samples were sequentially placed into the test reactor and etched for 16 pulse-etch cycles. In each cycle the chamber was filled with XeF_2 to a pressure of 300 – 400 mTorr, allowed to react for one minute and then evacuated.

The pulse-etch cycles are shown by peaks in the RGA pressure plots in Figure 1. In Figure 1A the BF_3 partial pressure shows etching of a boron film until it is completely removed after about 4 or 5 cycles. In this case there was no silicon film protecting the boron during sample transfer, but boron etching was found to be independent of whether or not a silicon film was used. Etching properties of As, P, Si and Ge are expected to be quite similar to boron.

Figure 1B shows a similar measurement with a tungsten film. In this case the tungsten was not significantly etched, showing the potential selectivity of XeF_2 etching.

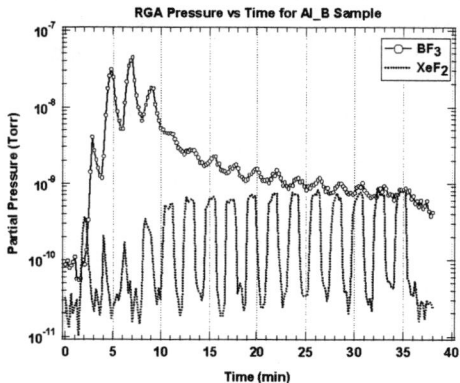

FIGURE 1A. Etching of Boron sample with XeF_2.

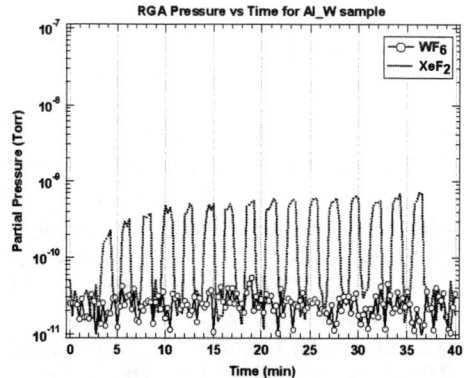

FIGURE 1B. Tungsten sample exposed to XeF_2 but not etched

Figures 2 and 3 show data taken on an ion source test stand located in ATMI's Ion Implant Process Efficiency Research Laboratory[4]. The test stand features a single filament Bernas Source with molybdenum arc chamber, and an 80° analyzing magnet with faraday cup measurement of the analyzed beam. A cylinder of XeF_2 operating at ambient temperature was installed in the gas box and connected to the ion source with a conventional high conductance gas line designed for use with SDS® cylinders.

Figure 3 shows an example of use of XeF_2 plasma to improve stability of the ion source. The source was run with BF_3 plasma for about 12 hours at its beam current specification limit, which corresponds to an analyzed beam of about 10 mA of $^{11}B+$. When the beam current measurements shown in Figure 3 began, the source was showing severe instability at 50kV extraction voltage, and this was not much improved by reducing the voltage to 40kV. After running XeF_2 plasma for 20 minutes, the source ran much more stably for about two more hours. No definite conclusions could be reached about the mechanism for the stability improvement, even after the source was opened and examined, but it was assumed that the XeF_2 plasma was able to remove deposits either from a source insulator, or from the high voltage gap.

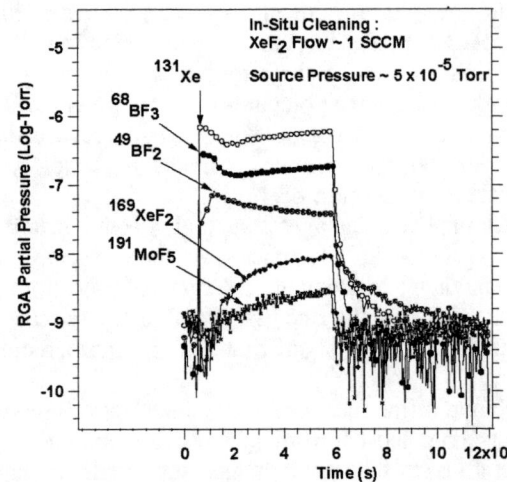

FIGURE 3. Time dependence of RGA peaks while XeF_2 is flowed in a cold ion source which had previously run boron plasma.

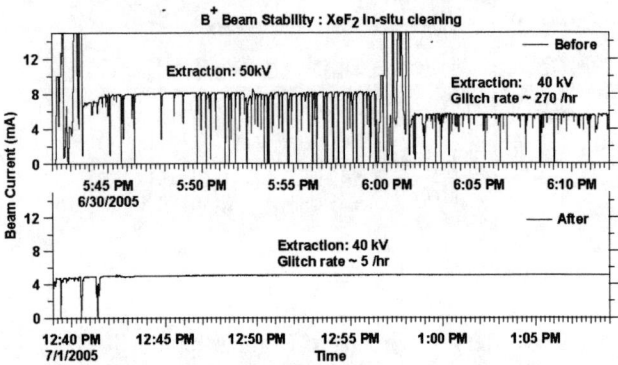

FIGURE 2. Improvement of stability of B+ from BF3 plasma by running 20 minutes of XeF_2 plasma.

Figure 3 shows measurements from an RGA located on the source vacuum chamber immediately above the ion source. In this case, XeF_2 vapor was flowed for about two hours through a cold ion source which had previously been running high current B+ beam from BF_3. The BF_3 and BF_2 RGA peaks from etching of boron can be seen clearly. The intensity of MoF_5 from etching of the molybdenum arc chamber is about 100 times smaller than the BF_3 intensity. The time dependence of the signals is not well understood, although it can be seen that XeF_2 is strongly consumed at the beginning of the measurement, and the signal subsequently rises. Apparently boron was still being etched when the XeF_2 flow was shut off after 6,000 seconds.

CONCLUSIONS AND FURTHER WORK

We have discussed the significant contribution to implanter operational efficiency that can be made by in-situ cleaning of deposited materials, particularly in the ion source region. Different methods of in-situ cleaning have been presented, and the relative merits of suitable fluorinating cleaning agents have been described.

XeF_2 has properties which make it a promising candidate for an in-situ cleaning reagent, and we have shown preliminary data on cleaning of boron deposits.

Our future work will concentrate on tests and measurements required to qualify this promising material for use in a production implant environment.

REFERENCES

1. J. Langan, J. Maroulis, and R. Ridgeway, "Strategies for Greenhouse Gas Reduction" *Solid State Technology*, 39(7), 115-122, 1996.
2. J. Arnó, W.K. Olander, and E. Sturm "Fluorine Gas for Thermal Reactor Clean Applications: Safe Packaging and Cost Considerations" ISESH 9th Annual Conference, San Diego, CA. June 9-13 (2002)
3. H.F. Winters and J. W. Coburn, *Appl. Phys. Letters* **34**, 70-73 (1979).
4. S, Yedave, F. DiMeo, R. Kaim, S. Bishop, J. Arnó and L. Wang, *These Proceedings*

Qualification of the GASGUARD® SAS GGT Arsine Sub-Atmospheric Gas Delivery System for Ion Implantation

James P. Dunn*, James L. Rolland*, James S. Grim*, Reinaldo M. Machado¶, and Christopher L. Hartz¶

Atmel Corporation, Colorado Springs, Colorado, 80906, USA
¶Air Products and Chemicals, Allentown, Pennsylvania, 18195, USA

Abstract. A beta level evaluation of the GASGUARD® SAS GGT Arsine ion implant dopant supply developed by Air Products and Chemicals, Inc. was conducted by Atmel Corporation. The evaluation included characterization of the normalized wafer yield, mass spectra, ionization efficiency, flow rate, beam current, extraction of usable material and cylinder lifetime.

This new and novel sub-atmospheric dopant gas delivery system utilizes a unique electrochemical process, which can generate, on demand, high flows of arsine at a constant 400 torr pressure while limiting net inventory of arsine to only 1 gram. This paper illustrates how Atmel Corporation evaluated and released this new arsine dopant delivery system for commercial production and verified high delivery capacity, resulting in reduced gas costs and increased cylinder life compared to the traditional adsorbent based technology.

Keywords: ion implant, implanter, source, dopant, arsine, generator, Air Products
PACS: 41.75.Ak; 61.72.Tt; 07.77.Ka

INTRODUCTION

A beta level evaluation of the new GASGUARD® SAS GGT Arsine delivery system was conducted by Atmel Corporation, Colorado Springs, Colorado. The evaluation included spectrum analysis, ionization efficiency, flow rate, beam current, extraction of usable material and cylinder lifetime.

The purpose of the beta level evaluation was to determine the purity of the deliverable dopant, confirm the quality of the ionized dopant available for a given gas flow, demonstrate sustained flow capabilities of the extracted gas, establish the extraction effectiveness as the delivery system approaches depletion and to quantify the overall capacity of deliverable product. The results of a successful evaluation are evident in reduced operating costs, demonstrated consistent performance of the implant tools and the achievement of dependable operation.

Sub-atmospheric delivery of dopant gases for ion implantation has been established as a standard in the IC fabrication industry. In the traditional adsorbent based sub-atmospheric delivery systems, the capacity is both pressure and flow rate dependent and the ultimate capacity is dependent upon the ability of the ion implanter to draw gas at high vacuum. A new sub-atmospheric gas delivery system for arsine, GASGUARD® SAS GGT, uses a proprietary electrochemical generator to manufacture arsine in situ; delivering only what is required by the ion implanter [1]. Among the advantages of this technology are operation at constant pressure (insuring stable and consistent gas flows); ability to generate high beam currents during the life of the generator; zero arsine inventory in the generator until installation and operation; accurate measurement of the gas delivery amounts; and similar product performance compared to traditional adsorbent based systems. This approach to gas delivery mitigates the reduction in gas flows often experienced with adsorbent based delivery systems at lower pressures.

THE GAS GENERATION PROCESS

The GASGUARD® SAS GGT arsine generator uses an electrochemical process to convert a purified arsenic cathode into a pure arsine gas (92-95%) with hydrogen (5-8%). The current generator design is a "drop in" vertical installation that includes all necessary hardware for installation and to interface with the existing tool configuration and power source. The generator system consists of the gas generator in a nominal 2.2-liter cylinder of approximate JY cylinder dimensions, a power supply and the controller, which are mounted in the electrical panel of the ion implanter gas box.

The generator interfaces to the controller with a gas pressure sensor to control generation pressures and a pneumatic pressure sensor to interlock operation to the implant tool pneumatic control system. Installation was essentially "plug-and-play" with the tool specific controller panel and specially designed cable for accessing electrical power. The controller readout, in generation-hours, is a visual indication of generator life. Each generation-hour equates to 2.3 grams of arsine [2], with a guaranteed run time of 313 hours or 720 grams of arsine continuously generated at 10 sccm. The controller cycles the low voltage electrical current to the generator to maintain the pressure at 400 torr +/- 5 torr. The net inventory of arsine in the generator during operation is approximately 1 gram in the generator headspace.

The GASGUARD® SAS GGT arsine generator was evaluated by third party safety reviewers, Lewis Bass International, Inc. [3] and found to meet the relevant safety requirements including SEMI S2-0703A, SEMI S10 Safety Guidelines for Risk Assessment and European Union CE marking directives.

EXPERIMENT

Before any new, different or modified product or process is used in the wafer fabrication environment, it must first be evaluated and qualified to insure conformity to all relevant manufacturing and tool specifications. The qualification process is the responsibility of the designated process and equipment engineers, and is specific to the needs of each individual manufacturer.

The experimental parameters were selected based on their significance in proving no material change in the composition or concentration of the extracted gas, the ability to derive the same or better ionized product from the available gas, comparable or better flow capabilities and comparable or better deliverable product mass.

Behavior of the generator at depletion of the cathode was another concern to address since the material composition of the available gas headspace could be expected to change significantly.

Qualification Parameters

Gas Purity

Gas purity was evaluated by use of tool spectrum analysis. The criterion was an analysis of historic specific ion peak data for comparable or reduced ratios of the undesired energetics. This was an easy test for assessment of the energetic component of material contaminants. The novel approach to generation, as opposed to de-adsorption, warranted a close look at any potential byproducts of the electrolyte. Figure 1 shows an Arsine spectrum from the test Varian VIISion PLUS implanter while running the GASGUARD® arsine generator and Figure 2 shows an arsine spectrum from another VIISion implanter while running a sub-atmospheric adsorbent based arsine gas source.

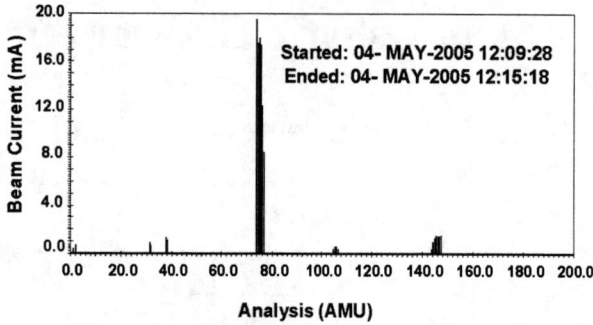

FIGURE 1. Full scale Arsine spectrum using GASGUARD® SAS GGT arsine generator.

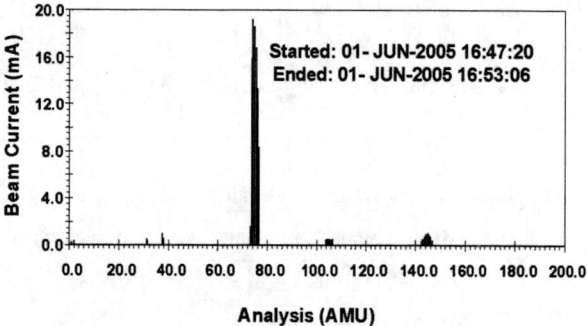

FIGURE 2. Full scale Arsine spectrum using industry standard sub-atmospheric adsorbent based arsine gas source.

Both implanters had been previously running phosphorus implants and the phosphorus contaminant peaks are notable in each spectrum. It can be seen from these full-scale spectrums that the dopant sources are similar (at full scale). Spectrums were also performed during the life of the arsine generator on the 100 micro amp scale and no unusual energetic contaminant peaks were noted as compared to the standard sub-atmospheric gas source. In addition to spectrum data, a periodic chemical test on implanted pilot material is performed at Atmel, which tests for certain metallic contaminants. No abnormal levels of contaminants were noted.

Product wafer die electrical test and yield data can also be used to infer contaminant levels of the dopant species. Analysis of a variety of product devices that were implanted before and after installation of the GASGUARD® arsine generator indicated that there were no statistically significant changes in either e-test or yield while running the arsine generator for any of the product devices analyzed as illustrated in Figure 3.

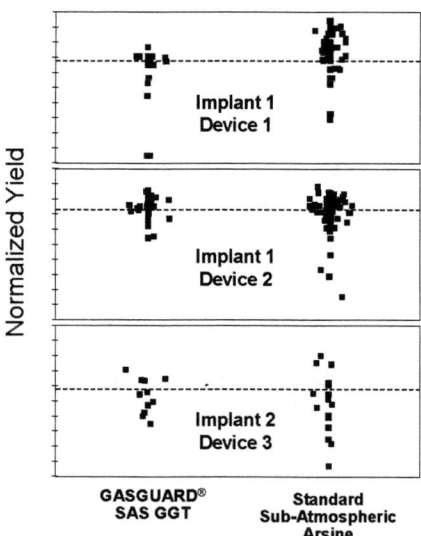

FIGURE 3. Normalized yields for two implant recipes and three devices comparing GGT and the traditional subatmospheric arsine supply.

The arsine generator was run past the recommended capacity limit of 720 grams to characterize the ultimate depletion limits of the generator, which was 816 grams. Once the arsenic is exhausted the generator is no longer able to maintain a constant pressure at 400 torr and water vapor is depleted from the in situ purifier as the pressure in the generator drops. The spectrum in Figure 4 shows the presence of water when the generator is exhausted beyond 800 grams of arsine.

Also noteworthy is that the product lot running at the time the generator could no longer produce arsenic beam currents was completed on a different implanter (using the standard sub-atmospheric adsorbent based arsine gas). There was no difference in wafer die yields or electrical test parameters between the wafers implanted with the two different gas delivery systems.

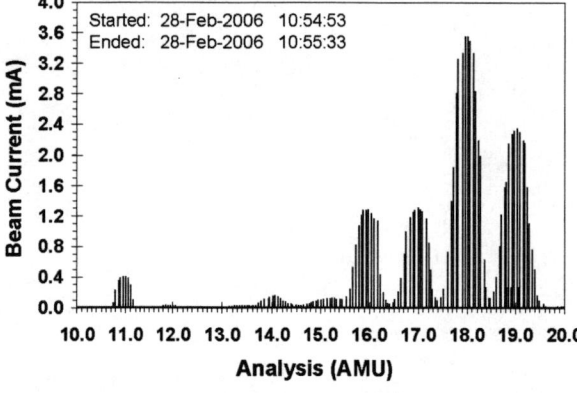

FIGURE 4. Spectrum for the arsine generator after 816 grams of arsine delivered showing water vapor. Note that this level of depletion is past the recommended delivery capacity of 720 grams at < 2ppm water.

Ionization Efficiency

Ionization efficiency was tested by a comparison of available beam currents versus ion source arc current and gas flow settings for the GASGUARD® SAS GGT delivery system and the standard sub-atmospheric adsorbent based arsine gas source. The criterion was beam currents equivalent to or better to those obtained with the standard sub-atmospheric arsine gas source. No changes in recipe tuning were needed for any recipes. Implant statistics were compared between 2673 implants using the GASGUARD® SAS GGT arsine source and 3209 implants using the standard sub-atmospheric arsine gas source.

Six production recipes were selected for high volume and compared using a created value labeled "transmission". The value of "transmission" equals the beam current divided by the extraction current x 100 (Figure 5). The "transmission" value is indicative of the ratio of desired dopant to the total of the extracted ionizable species and is an indicator of the purity of the dopant gas. Source arc voltage is fixed in the production recipes and other tunable parameters are optimized by the implanter software, therefore the derived parameter is a good indicator of the performance of the GASGUARD® SAS GGT dopant gas. The data indicates the two dopant sources have similar performance levels.

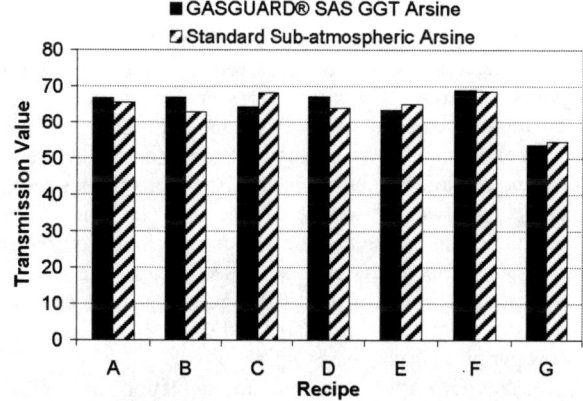

FIGURE 5. GGT vs. standard sub-atmospheric gas source using "transmission value = beam current/extraction current x 100" as a created value.

Flow Characteristics

A comparison of maximum continuous flow and flow at depletion was done to test the flow characteristics of the GASGUARD® SAS GGT generator. Maximum continuous flow criterion was the ability to flow at maximum demand for the MFC rating of 10 sccm for the VIISion tool. As the cylinder approached depletion, periodic flow testing at 4 sccm was done to watch for possible rate of flow changes.

The generator was able to meet the maximum flow demand at turn on and end of life. Throughout the lifespan of the generator, flow was consistently between 3.98 and 4.0 sccm for a demand of 4.0 sccm. No changes in these flow characteristics were seen until actual depletion.

Cylinder Lifetime

Cylinder lifetime was evaluated by running the GASGUARD® SAS GGT generator to depletion. One generator was run in a VIISion PLUS ion implanter for initial lifetime testing. A second generator was installed in an Applied 9500XR ion Implanter for continued testing.

There were some issues with an extended down time of the VIISion PLUS test tool that impacted actual tool run time with the arsine generator. Nevertheless, total run time was thirty-eight weeks, of which eight weeks the tool was unavailable. This includes four weeks of wait time for the initial test lots to clear e-test. No significant change in performance was observed due to sit time. The second evaluation continues to operate with the installed generator, which has been in use for more than 39 weeks. This tool has experienced no significant down time during the test.

Deliverable Product

Deliverable product was determined by evaluation of guaranteed run time as measured by actual generator hours of use in a production environment. The guaranteed capacity of 720 grams (313 hours on the controller timer) was compared against actual delivered product. Note that each on-time hour corresponds to 2.3 grams of arsine delivered[1]. The criterion was establishment of the specific amount of product extracted for comparison against advertised capacity. This information was also used for a comparison of contract cost per gram based on guaranteed capacity versus actual cost per gram used.

The generator was run to depletion at 357.45 generator on-time hours. This corresponds to 822 grams of product actually delivered against a guarantee of 720 grams. This is a 14.2% increase in actual deliverable against advertised capacity. At the point the tool would no longer set up a beam, pressure had dropped to 300 torr. A mass analysis measurement of the product showed 815 grams used which is within 1% of the estimated use.

Conclusions

The GASGUARD® SAS GGT generator performed as specified. The increased deliverable capacity, reduced risk of product loss from cylinder backfill, and comparable ionization efficiency compared to the current adsorption technology delivery system, results in more production time and a reduced risk of intervention.

No appreciable change in wafer product electrical test or yield was seen from start to finish of the evaluation period. This includes product running on the tool during the time the generator was finally depleted.

Installation was simple and straightforward. In addition, the exposure risks during shipment and operation are minimized by the low inventory of arsine in the system, adding to the product safety and meeting all the goals of a sub-atmospheric gas delivery system.

ACKNOWLEDGMENTS

Stephanie Campbell for regularly collecting flow rate data on the ion implanters.

Atmel ion implant EES technicians for their support in the gas installation and maintaining the test tools in a consistent production worthy state for the evaluation period.

REFERENCES

1. R. M. Machado, et. al, " Electrochemical Generator for Sub-Atmospheric Supply of Arsine to the Ion Implant Market," Ion Implant Technology, Proceedings of the 15th International Conference on Ion Implantation Technology Part II, Taipei, Taiwan, Republic of China, 25-27 October, 2004.
2. R. M. Machado, "Determination of Arsine Generator Yield in the GASGUARD® SAS Generated Gas Technology from the Controller Timer Reading", Air Products and Chemicals, Inc., Applications Note, 2006.
3. Lewis Bass International, Inc., 621 E. Campbell Avenue, Suite 11A, Campbell, CA 95008

Manufacturing Assessment of SDS3® Gas Upgrade

James P. Dunn*, James Rolland*, James Grim*, Bob Brown¶,

Atmel Corporation, Colorado Springs, Colorado, 80906, USA
¶ATMI, Danbury Connecticut, 06810, USA

Abstract. The deliverable capacity of adsorbent-based sub-atmospheric storage and delivery systems for dopant gas is proportional to the storage capacity and low-pressure flow capability of the adsorbent material. SDS3 is the next generation of adsorbent-based sub-atmospheric storage and delivery system for ion implant. A unique higher density adsorbent material is used to provide an increased deliverable capacity.

This paper shows how Atmel Corporation, Colorado Springs, Colorado evaluated this new dopant delivery system and verified the increased delivery capacity. The increased deliverable capacity combined with a lower price per gram provides lower operating costs.

Keywords: ion implant, implanter, source, dopant, arsine, boron, phosphine, SDS, ATMI
PACS: 41.75.Ak; 61.72.Tt; 07.77.Ka

INTRODUCTION

Prior to introduction, manufacturing assessment of new or improved products is required. An evaluation of the new SDS3 delivery system was conducted at Atmel Corporation, Colorado Springs, Colorado.

The tests were based on an evaluation of material purity, beam current and sustained gas flow, with a focus on gas flow at the end of cylinder life. A successful evaluation is determined by reduced operating costs, consistent performance and compatibility with Atmel safety standards.

Atmel is committed to sub atmospheric gas delivery as a responsible approach to toxic gas handling. All gas delivery alternatives are currently measured against SDS as the standard.

SDS3 ADSORBENT

A method for mass-producing high density microporous carbon blocks was developed in order to offer the increased gas storage capacity required for this new adsorbent. Although these materials are typically described as microporous, the dimensions of the actual pores are measured in nanometers. These unique adsorbent blocks or monoliths are the heart of this newest generation of SDS. The carbon monolith combines zero void space with uniform pore size to create measurable capacity improvements over the previous product generation (Tab. 1).

TABLE 1. Physical Properties of SDS Adsorbents

Properties	SDS3	SDS2
Shape	Monolith	Sphere
Material	Carbon	Carbon
Density (g/ml)	1.1	0.56
Dimensions (mm dia)	100 (block)	0.7 (bead)
Surface area (m^2/g)	1013	1150
BF$_3$ Capacity at 650 torr (g/L)	295	170
AsH$_3$ Capacity at 650 torr (g/l)	645	440
PH$_3$ Capacity at 650 torr (g/l)	310	130

The proprietary polymer precursor material is pyrolized in special furnaces to produce high-density carbon blocks. During this thermal process, the molecular conformity of the polymer chains is slowly transformed into the desired microporous structure. At every step of the multiple stage reaction, precise temperature controls and tight thermal uniformities maintain the structural integrity of the carbon blocks. The resulting monoliths are of extreme purity and uniformity.

SDS3 CYLINDER

In order to package the new monolithic adsorbent it was necessary to come up with a specially designed cylinder. The cylinder is welded instead of extruded in order to accommodate the carbon blocks. While the cylinder was designed to meet the necessary requirements of physical dimensions and durability it also was necessary for it to conform to the regulations and restrictions of state, national and international regulatory agencies.

The minimum burst pressure specification is 1500 psig for cylinders that have an actual burst pressure of about 1900 psig. This specification complies with the requirements of the major implanter manufacturers. More recent data shows that the typical burst pressure for this welded JY size cylinder is actually around 2530 psig.

The cylinder Maximum Allowable Working Pressure (MAWP) is rated at 75 psig. As per industry Standard Operating Procedure (SOP), anytime the MAWP is exceeded, the cylinder must be pulled and tagged for return. The supplier will then evaluate the cylinder for compliance to specification before it is released back into the usable population. The implanters used for this evaluation have a purge pressure of 85 psi, which is hard coded in the software and not subject to change. The cylinder MAWP value can and should be increased to a value above that of any purge gas pressures. This will insure the ability to recover from accidental back filling of a cylinder and lessen the economic impact of premature removal of the cylinder. Interestingly, the US DOT working and test pressure specifications for the SDS3 cylinders are less stringent than what is required for the implant application, where high pressure cylinders may share the same manifold as sub-atmospheric cylinders.

EXPERIMENT

The tests were selected to prove no material change in the composition of the gas or adsorbent material, the ability to derive same or better ionized product from the available gas and comparable or better deliverable gram product.

Applied Materials 9500XR high current implanters were selected for use in the test because of their higher gas flows and related gas usage. All flow and pressure data was extracted from implanter logs. To insure maximum cylinder evacuation, gas stick conductance was verified prior to testing. This was to ensure the tool specific configurations were consistent with SDS requirements.

Qualification Parameters

Gas Purity

Spectrum analysis was used to assess the gas purity. The criterion was a comparative analysis of expected ion peaks for comparable or reduced ratios of abundance of any undesirable inclusions and detection of volatile contaminants or by-products from reaction between the dopant and adsorbent materials. Any effects on the dopant material from the changes in the composition or manufacturing process of the adsorbent or dopant used would show up as a deviation from historical spectrum data. This allowed us to readily evaluate the energetic component of the dopant material and detect any undesirable byproducts.

Spectrums were run at each ion source change and a periodic Atmel specific chemical test, to evaluate metallic contaminants, was also run. No abnormalities were noted.

The boron spectrum (Fig. 1) and the arsenic spectrum (Fig. 2) were consistent with historical spectrum data. The phosphorus spectrum (Fig. 3) was run with an included argon carrier gas, which is evident in the spectrums. This is also consistent with historical spectrum data.

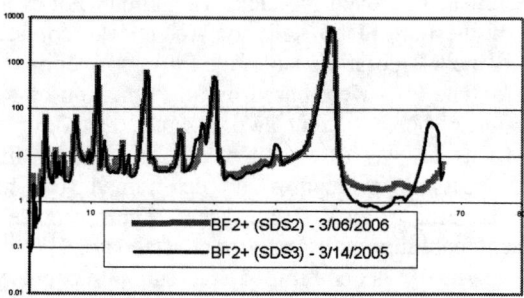

FIGURE 1. SDS2 and SDS3 spectrums for boron

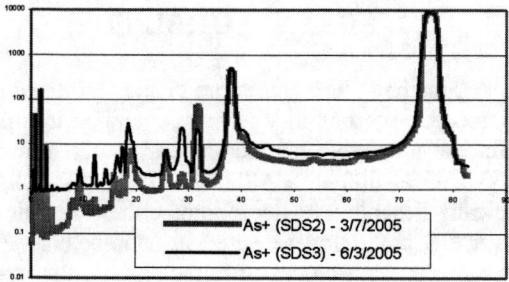

FIGURE 2. SDS2 and SDS3 spectrums for arsenic

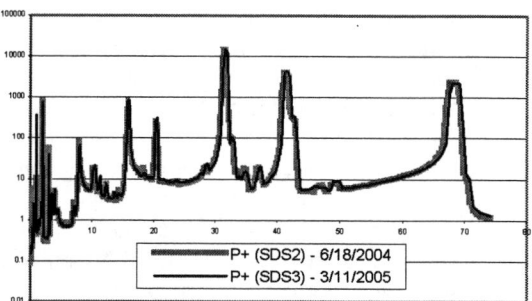

FIGURE 3. SDS2 and SDS3 spectrums for phosphorus with an argon carrier gas.

Beam Current

Beam current at given flow rates were compared. The criterion was a beam current equivalent to or better than that obtained with the same gas flow using the SDS2.

No changes were necessary in any of the recipes to achieve nominal beam current. SDS3 end of implant data was compared to historical end of implant data. Beam current was also used as a measure of possible non-energetic contaminants.

Most recipe parameters are locked down, but arc current and beam current have some latitude. There are still numerous variations to consider, such as source life, pump life, source base pressure, etc., so a setup value given as beam current divided by arc current was created. This allows us to make a generalized comparison of the setups with some confidence. Data from 40 recipes was evaluated, and the results were similar. One recipe was selected as representative of the data (Fig. 4). Comparable gas flows gave identical beam currents to those seen with the previous generation of SDS.

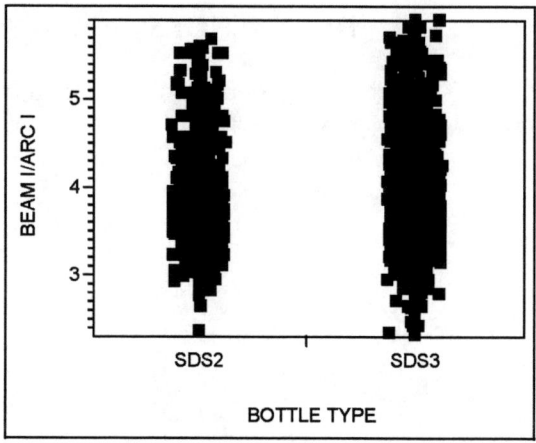

FIGURE 4. SDS2 and SDS3 at nominal recipe setup.

Flow Characteristics

Comparing flow at maximum, nominal and depletion established the characteristics. Maximum flow was the ability to flow with the MFC set to fully open. Nominal flow was the ability to flow at normal recipe demands. As the cylinder approached depletion, periodic flow measurements were done to check for any change in flow capability. Depletion is defined as failure to set up to the recipe beam current requirement. Depletion occurred when the SDS gas source could no longer meet the gas flow required for set up.

The 9500XR pressure transducers are inaccurate at low pressure. Atmel does not rely exclusively on cylinder pressure, as read at the implanter, as an indicator of gas depletion. When cylinder change pressure is reached, the gas is first checked for adequate flow. As long as the gas flow is sufficient to run production, the cylinder is shut off, the gas line is evacuated and the transducer zero is calibrated before the gas is turned back on. This insures all gas cylinders are emptied before being changed.

No flow errors occurred until cylinder pressure as read at the tool reached approximately –14.7 psig (Fig. 5), which is equivalent to 4 torr. The pressure drop across the MFC is typically between 4 and 6 torr. Optimum gas extraction occurs when the cylinder pressure at end of life is equal to the pressure drop across the MFC ($P_{EOL} = MFC_{\otimes P}$). This test cylinder was essentially emptied of product and no loss of flow was experienced. The arsine and phosphine test cylinders were similarly emptied.

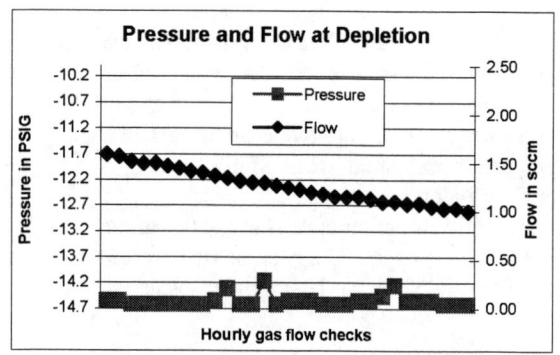

FIGURE 5. SDS3 boron flow at depletion

The behavior at depletion was of particular interest since most of the gas availability occurs at the lowest pressures. Once the tool was unable to establish the required beam current, the gas flow was evaluated to see how quickly the remaining gas was being depleted and to see if the loss of flow was precipitous or gradual.

Flow dropped from 1.6 sccm to 1.0 sccm over a period of 31 hours while the MFC was held open. The initial flow fault occurred when the cylinder could not achieve 1.54 sccm of flow during a recipe setup. The subsequent testing began after a short rest period. This allowed cylinder pressure to recover enough to flow at 1.6 sccm (Fig. 5). Continuous demand did not again drop below the 1.5 sccm flow capability for just over 5 hours.

Cylinder Lifetime

Cylinder lifetime was evaluated by running the SDS3 cylinders to depletion. Depletion is defined as the failure to set up to the nominal recipe beam current requirements. Three SDS3 cylinders, one of each species, were run in parallel on 3 different 9500XR ion implanters for lifetime testing.

Typically, the SDS2 cylinder will last 3 to 4 months in a high current tool. The three test cylinders lasted approximately 8 months.

Deliverable Product

Precision weighing and pressure measurements, before and after use, determined deliverable product. The actual gas used was then compared to historical data for SDS2 and the advertised capacity for SDS3.

Weight measurements show 373 grams of SDS3 PH_3 was extracted as compared to 310 grams specified. The 373 grams is more than twice the historical capacity of 175 grams extracted from SDS2, averaged over the last 7 cylinders.

Weight measurements show 318.4 grams of SDS3 BF_3 was extracted as compared to 350 grams originally specified. The advertised fill capacity has since been revised down to 295 grams to more accurately reflect the actual extraction capability. The 318 grams is more than twice the historical capacity of 150 grams extracted from SDS2, averaged over the last 3 cylinders.

Weight measurements show 700.4 grams of SDS3 AsH_3 was extracted as compared to 645 grams specified. The AsH_3 cylinder was accidentally backfilled with argon a week after installation and some capacity was bled off to allow for continued use in production and continued evaluation. Time in use was not accurately determined as a result, though the cylinder remained in use for 10 months and actual gas extracted was readily determined by precision weighing. The 700.4 grams of extracted product is a 60% increase in capacity compared to the 50% advertised increase. This also represents a 30% improvement over the actual gas extracted from SDS2, averaged over the last 9 cylinders.

CONCLUSIONS

SDS3 is comparable to the prior SDS generation in all aspects tested, with the notable exception of increased capacity of usable product for the same cylinder. This next generation product was fully compatible with existing hardware and presented no risks to product or tools. The monolithic adsorbent block raised some questions regarding flow at low pressure, but no flow problems were encountered until the source gas was fully depleted.

ACKNOWLEDGMENTS

Stephanie Campbell for regularly collecting flow rate data on the ion implanters.

Atmel ion implant EES technicians for their support in the gas installation and maintaining the test tools in a consistent production worthy state for the evaluation period.

All cylinder pre and final weights supplied by ATMI

REFERENCES

1. Cylinder performance specification information was supplied by ATMI

ATMI's Ion Implant Process Efficiency Research Laboratory (IIPERL)

Sharad Yedave, Josep Arnó, Steve Bishop, Frank DiMeo Jr.[*], Robert Kaim, Luping Wang

ATMI Inc., 7 Commerce Drive, Danbury, CT 06810, USA

Abstract. We describe an ion source test stand recently installed by ATMI, and show data illustrative of the research being carried out at the new facility.

Keywords: IIPERL, Process Efficiency Research, Implant Efficiency Solutions, Chemical Cleaning

PACS: 85.40.Ry; 81.65.Cf; 61.72.Vv; 52.77.Dq; 52.77.Bn

INTRODUCTION

Process efficiency is crucial in implant technology and as the number of ion implantation steps increases in leading edge technologies, process efficiency gains in the implant process can have substantial impact in the fab. ATMI is continually working towards providing process efficiency solutions to the semiconductor industry. ATMI's recent research efforts have been directed at continuing to improve both the safety and efficiency of ion implantation processes and tools. As part of these efforts we recently installed a state of the art ion source test facility at our Danbury CT facility. In this paper, we describe the test stand, and show data to illustrate its capabilities and the type of research in progress.

SOURCE TEST STAND

The ion source test stand (Figure 1) features a single filament Bernas source, an 80 degree analyzing magnet with a mass resolution $M/\Delta M > 50$, and a magnetically suppressed Faraday located immediately downstream of the analysis slit. The gas box is currently configured for three low pressure gases and one high pressure cylinder, with a fifth gas slot available for future expansion. Beam energy is variable from 10 to 60 keV with analyzed beam currents of about 10 mA for B^+, P^+ and As^+.

The primary purpose of the test stand is to facilitate ATMI's research into innovative materials which can optimize operation of ion implant sources and beamlines. In the remainder of this paper we give examples of some of the research that has been carried out since the test stand installation was completed in September 2005.

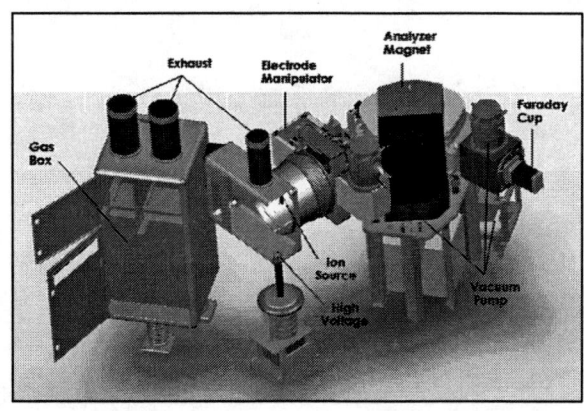

FIGURE 1. Schematic of ATMI's Ion Source Test Stand

---

[*] Current address: SAIC, 1710 SAIC Drive, McLean, VA 22102

RECENT DATA FROM TEST STAND

Relative Performance of SDS2 and SDS3 Dopant Cylinders

At the time of installation of the ion source test stand, ATMI was introducing its new SDS3 dopant cylinders, which are based on use of a very high density graphite adsorbent and offer higher storage capacity than standard SDS2 cylinders. As part of ATMI's effort to demonstrate the operational equivalence of SDS2 and SDS3, the first measurements carried out on the test stand were comparisons of ion source operational characteristics and mass spectra when using SDS2 or SDS3 dopant material. Figure 2 shows comparison of BF_3, PH_3 and AsH_3 spectra: in all cases there was no observable difference between spectra from SDS2 and SDS3 dopants. Source stability measurements (not shown) were also identical for the two dopant source types.

Contamination from Alumina Insulators in Bernas Source

While characterizing the mass spectra from a Bernas source running about 6mA of $^{11}B^+$ beam from BF_3 dopant, we noticed an unexplained peak with intensity of almost 1mA at mass 27 (see Figure 3). Suspecting that the peak was caused by ^{27}Al derived from alumina (Al_2O_3) insulators, we replaced all alumina internal source insulators with boron nitride (BN). The intensity of the mass peak at 27AMU was reduced by about a factor of 10. While in the case of a BF_3 plasma, the Al^+ and AlF_n^+ peaks are well resolved from boron peaks of interest, contamination by aluminum could clearly be a problem when using other fluorinated dopants, particularly $^{28}SiF_4$.

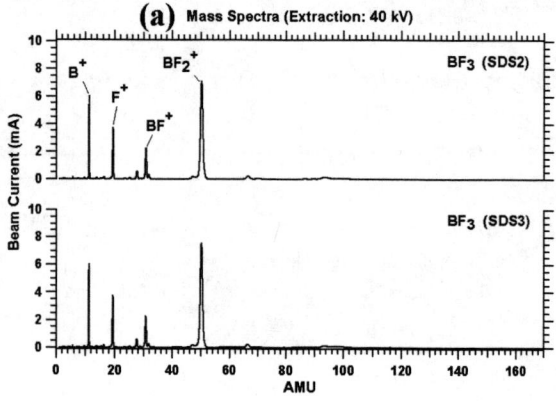
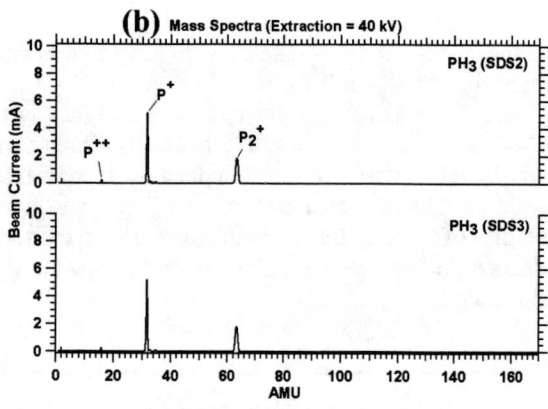
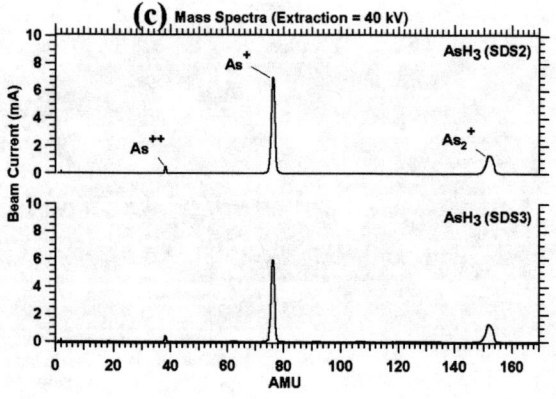

FIGURE 2. Comparison of mass spectra from SDS2 and SDS3 dopant sources: (a) Boron Trifluoride (BF_3), (b) Phosphine (PH_3) and (c) Arsine (AsH_3).

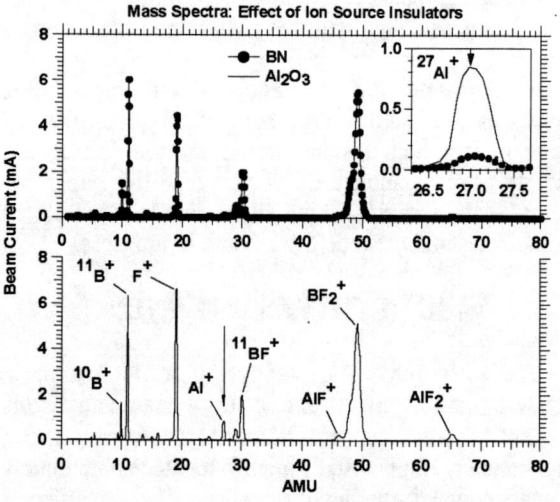

FIGURE 3. Mass spectra from Bernas source running BF_3, showing presence of ^{27}Al mass peak from alumina source insulators.

Erosion of Ion Source Consumable Parts by Fluorine Radicals

As part of our work on in-situ chemical cleaning[1], we have been studying effects of fluorinated reagents on ion source arc chamber materials (usually molybdenum or tungsten). In Figure 4, we show that even for a conventional fluorinated dopant such as BF_3, there is significant reaction between fluorine radicals in the plasma and the arc chamber materials. The figure compares mass spectra with argon plasma and BF_3 plasma in a Bernas source with tungsten filament and molybdenum arc chamber. In the spectrum from BF_3 plasma, there is clear evidence of ions of molybdenum and its fluorides. In contrast, the Mo^+ peaks are absent from the spectrum with argon plasma, indicating that insignificant amounts of Mo^+ ions are created by sputtering alone. Evidence for ions of tungsten and its fluorides is inconclusive in the accessible mass range. Future experiments are planned to investigate chemical effects with tungsten.

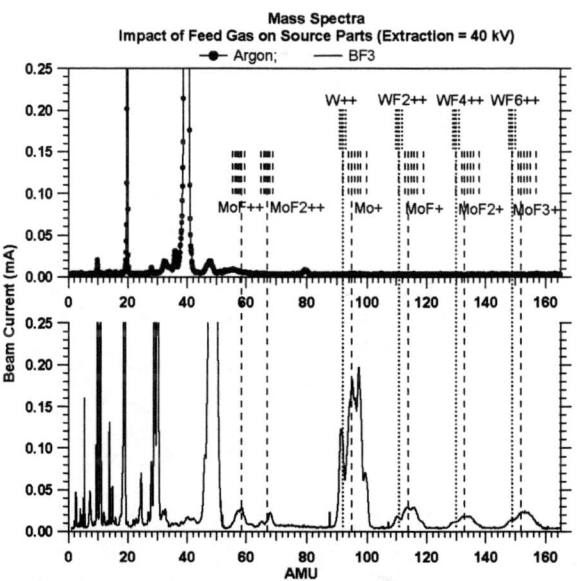

FIGURE 4. Comparison of mass spectra from argon and BF_3 plasma, showing molybdenum peaks created by the chemical effect of fluorine radicals in the BF_3 plasma.

Temperature Dependence of Conductance Limited XeF_2 Vapor Flow

We recently installed an SDS GasGauge Monitor[2] in the IIPERL ion source test stand. This device comprises two electronic modules, a data acquisition unit (DAQ) and a display unit, communicating via fiber optic cable. A DAQ unit installed in the gas box of the IIPERL test stand is used to measure the pressure of any combination of up to four SDS®, VAC® or high pressure cylinders. The display unit is located near the test stand control console.

The SDS GasGauge DAQ module can also measure temperature at up to four locations inside the gas box. This capability proved useful while monitoring the flow characteristics of XeF_2 vapor for in-situ chemical cleaning[1]. The vapor is derived from the equilibrium vapor pressure in a container of solid XeF_2 powder, which is approximately 4 Torr at room temperature. Figure 5 shows measurements for a XeF_2 container which was installed in the gas box and connected to the ion source arc chamber via conventional SDS-type gas line with mass flow controller fully opened. The lower part of the figure shows the XeF_2 gas flow and the source vacuum gauge measurements. The upper part displays temperature measurements from two thermistors, one for the gas box ambient and the other attached to the XeF_2 container.

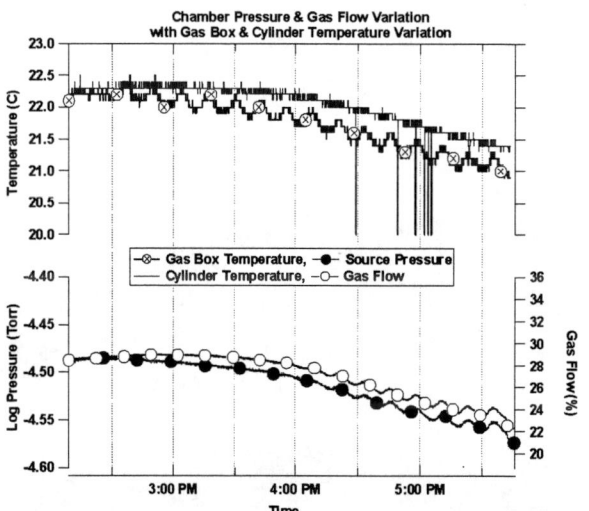

FIGURE 5. Temperature effect of conductance limited flow of XeF_2 from a container into the ion source chamber.

The data in Figure 5 clearly show the dependence of XeF_2 vapor flow on temperature, which is characteristic of conductance limited flow from the vapor source into the ion source chamber.

CONCLUSIONS

In this paper, we have given some examples of work done at ATMI's IIPERL laboratory since its establishment less than 12 months ago. The laboratory is focused on providing solutions to improve the efficiency and safety of implant processes and tools. The ion source test stand will provide support services for existing ATMI products and serve as a platform for development of new products, including innovative dopant precursors and in-situ cleaning chemistries.

REFERENCES

1. S. Bishop, R. Kaim, S. Yedave, J. Arno, F. DiMeo and M. Wodjenski, *These proceedings*
2. S. Bishop, M. Wodjenski, R. Kaim, S. Lurcott and J. McManus, *These proceedings*

A Safe Solution to Dopant Gas Desorption from Metal Surfaces

Tsutomu Nakanoya, Maki Egami

Takachiho Chemical Industrial Co., Ltd.
4-6, Hiroo 1-chome, Shibuya-Ku, Tokyo, Japan

Abstract. TOXICAPTURE[TM] is used to further minimize trace toxic dopant gas inside cylinder valve outlets, which, over time, may desorb from metal surfaces. When outlet caps or connections to ion source gas cylinders are disconnected in order to perform installations or bottle changes, there always is some risk that toxic fumes resulting from desorption of the metal surface in contact with dopant gas are released in air and inhaled by the operator. TOXICAPTURE[TM] is a simple and easy solution to reduce this risk that may damage human health or may pollute clean room environment. TOXICAPTURE[TM] will react with the poison gas vapor to form nontoxic and solid material through irreversible chemical reactions. TOXICAPTURE[TM] prevents contamination and corrosion on gas contact surfaces of gas pipings, pressure regulators, pneumatic valves, mass flow controllers, and other parts in a gas box. TOXICAPTURE[TM] is highly effective in shortening the time to achieve high vacuum and in extending the lifetime of devices in the gas box. In this paper, we introduce the structure, functions, reactivity, applications, and effectivity of TOXICAPTURE[TM].

Keywords: TOXICAPTURE[TM], safety device, arsine, phosphine, boron trifluoride
PACS: 85.40.Ry; 28.41.Te; 0.60 Wa; 89.60.Ec

I. INTRODUCTION

Ion implantation is a critical step in semiconductor manufacturing process. Ion implanters are used to form p-n junctions at the exact point and the depth where aimed by ions ionized from dopant gases, accelerated by high voltage, and driven into the silicon substrate. Ion implanters typically use dopant materials such as solid arsenic, phosphorus, and boron -- or more commonly use dopant gases such as arsine, phosphine, and boron trifluoride. Such gases are delivered in small high-pressure gas cylinders or sub-atmospheric containers. These dopant materials are highly toxic and should be safely and environmentally controlled to protect both equipment and operators. As shown in Table 1, ACGIH (American Conference of Governmental Industrial Hygienists) prescribes TLV of these dopant gases – AsH_3 0.05ppm, PH_3 0.3ppm, BF_3 1ppm, and solid sources – arsenic 0.01mg/m3, phosphorus 0.1mg/m3, boron oxide 10 mg/m3 [1]. These TLVs show that these elements themselves are toxic, and not only their gaseous compounds. It is obvious that we need to consider such strong toxicity with whatever compound they are rendered in – oxides, metallic compounds, or halogenides.

This paper describes an efficient safety device, TOXICAPTURE[TM], developed by Takachiho Chemical Industrial Co., Ltd., which can be attached to cylinder valves of dopant gases in place of regular valve caps.

TABLE 1. ACGIH TLVS 2005

Dopant	TWA ppm	Element	TWA mg/m^3
arsine	0.05	arsenic	0.01
phosphine	0.3	phosphorus (yellow)	0.1
boron trifluoride	1	boron oxide	10

II. BACKGROUND

TOXICAPTURE[TM] was developed to ensure the safety in clean rooms for people working with ion implanters. Ion implant dopant gases are typically delivered in cylinders. Since these gases are highly toxic, rigid leak checks are conducted on all source cylinders using highly sensitive leak detectors before taking them inside a clean room for cylinder change.

FIGURE 1 shows one actual method of a rigid leak check for a cylinder of arsine or phosphine. The picture shows a hand sticking a monitor probe into the connection port of a valve. This method effectively detects any slow leak, but due to its high sensitivity, it also detects trace desorption of the dopant gas from the inner metal surface of valves. Even if there is no actual leakage, the detector often reacts to trace fumes rising up from the valve's metal surface.

Also, when an outlet cap is loosened to detach the cap from the valve of a boron trifluoride cylinder, occasionally, small amount of white smoke may be observed.

FIGURE 1. Dopant gas cylinder receiving valve leak check before cylinder is brought inside a clean room
Detector Model: RIKEN KEIKI SC-90

The possible causes of diffusion of toxic gases from valve when detaching the outlet cap are as shown below:
1. Leakage due to insufficient seal of valve
2. Loosening of valve handle due to vibration during transportation
3. Mishandling of valve handle before use
4. Desorption from inner metal surfaces of valve that comes in contact with the gas (inner structure of a valve actually has a lot of surface area that comes in contact with the gas due to its complicated design)
5. Dopant gas emission due to chemical reaction of moisture in air and solid compound build up on metal surface (Contact of metal material with dopant gas such as hydrides over time can result in solid compound build up on metal surfaces [2]. Especially boron trifluoride corrodes metals markedly and produces white smoke when the corroded metal contact moisture in the air.)

To better service the ion implantation market, Takachiho needed to develop solutions for the above-mentioned circumstances. Conventionally, as preventive measures against causes No. 1 and No. 2, valve-tightening torque is severely controlled and valve handle is securely fixed to avoid rotation during transportation. Cause No.3, where possible, is addressed through increased or improved training in gas cylinder handling. To eliminate causes No. 4 and No. 5 (which is typically called 'outgas'), the following measures are implemented:

A) Continuous blowing of valve outlet with inert gas (e.g.- N_2)
B) Intermittent pressurizing and releasing of inert gas (repetition of this process increases cleanliness)
C) Physical polishing of metal surface of valve outlet by rotating wire brush
D) Swab cleaning of valve outlet with isopropyl alcohol (IPA)

However, the above conventional measures cannot eliminate 'outgas' completely. Figure 2 shows the occurrence of 'outgas' from valve outlet before shipment at Takachiho Chemical Industrial Co., Ltd.

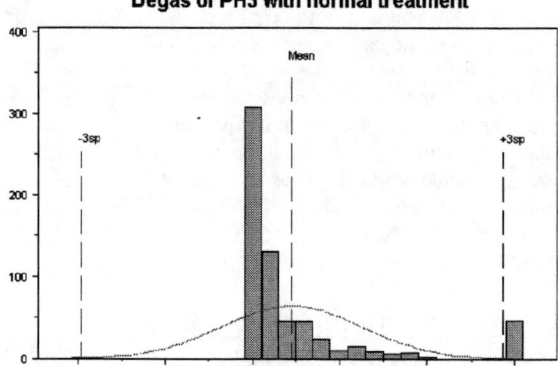

FIGURE 2. Outgas occurrence after conventional cleaning

About 10% of high pressure phosphine cylinders were detected to have more than 300ppb of 'outgas' and were required to be re-cleaned. If those detected had not been re-cleaned, they may have emitted trace amount of phosphine inside the clean room when the outlet cap was detached.

Thus to provide a better solution toward the 'outgas' phenomenon, we developed TOXICAPTURE™ to help maintain safe work environment and to prevent health impairment.

III. STRUCTURE AND USAGE

During the process of filling a high pressure gas cylinder with dopant gas, metal around the valve outlet comes in contact with the filled dopant gas for a considerable amount of time. Dopant gases possess properties where they are readily adsorbed by metals. When the cylinder filling process is completed, all dopant gas within the valve outlet is completely purged, but the dopant molecules absorbed strongly on the metal surface cannot be completely removed.

The absorbed gas desorbs gradually, and slowly diffuses from the metal surface. This effect where the gas that come in contact leaves its trace is called 'memory effect' of gases. The parts of the valve where this 'memory effect' may take place consists of stainless steel base, diaphragm, and spring. The typical volume inside a valve outlet is 1.0 - 1.5ml. Thus the gradually built up 'outgas' stored inside the outlet cap is suddenly released into the atmosphere when the outlet cap is detached.

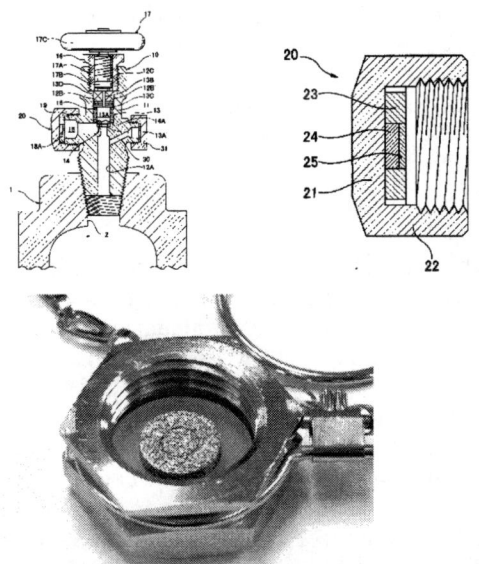

FIGURE 3. TOXICAPTURE™ structure diagram and photograph

Figure 3 shows the design of TOXICAPTURE™ and TOXICAPTURE™ implanted inside a typical valve outlet cap [3].

The upper left diagram shows a valve with TOXICAPTURE™ and the upper-right diagram shows an enlarged view of TOXICAPTURE™ (No. 20), that consists of a gasket (No. 23), reagent (No. 24) which reacts with poison gases to form nontoxic and solid material through irreversible reactions, and metal barrier (No. 25) which helps to avoid contact with operator's skin without disturbing penetration of toxic fumes nor disturbing reaction of detoxifying reagent. The photograph shows the JIS outlet cap type TOXICAPTURE™. Various types of TOXICAPTURE™ can be made to fit different types of valve outlets, and no modifications to valves nor equipment is necessary in using TOXICAPTURE™.

IV. EXPERIMENTAL

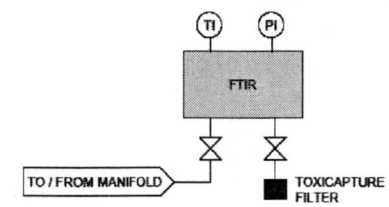

FIGURE 4. Experiment apparatus

The capability of TOXICAPTURE™ was quantified using the experiment apparatus shown above in Figure 4.

FTIR gas cell (370ml) was filled with the target gas, and when the concentration stabilized within the cell, the cell valve was opened to fill the TOXICAPTURE™ filter cell. The concentration was recorded up to 20 hours using FTIR. This test was performed for both phosphine and boron trifluoride.

After each test, TOXICAPTURE™ samples were tested for re-desorption of toxic fumes using a chemical detector. They were then placed in water to monitor for any signs of reaction.

The setup of testing apparatus, the data taking and evaluation were performed with the cooperation of Advanced Technology Materials, Inc. [4].

V. RESULTS AND DISCUSSIONS

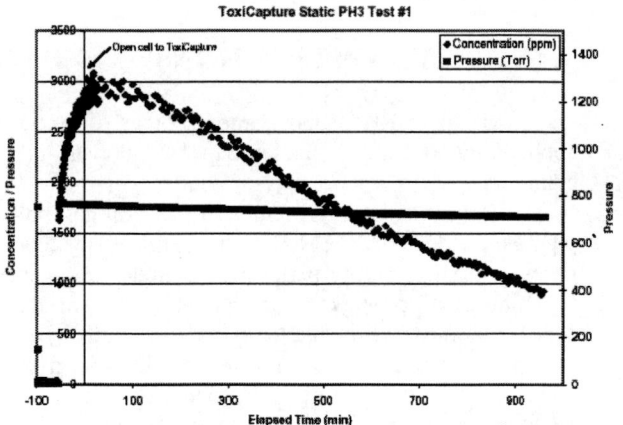

FIGURE 5. FTIR curve showing PH_3 concentration

Figure 5 shows phosphine concentration at the time of introduction of gas into the FTIR cell, when the cell is opened to TOXICAPTURE™, and after the gas is exposed to TOXICAPTURE™. Significant decrease of concentration was observed after the gas was exposed to TOXICAPTURE™.

The reaction capacity and reaction rate of TOXICAPTURE™ are calculated using this result. The capacity is calculated using the difference of the concentration at the beginning and at the end of the experiment. Reaction rate corresponds to the rate of concentration decrease, and it is calculated using the gradient of the graph.

The reaction capacity with phosphine is 970μL (43μmol), and reaction rate is 2.3μ mol/hr (1.4xE-5 cc/sec). This capacity suffices the total volume of a valve outlet even if a 100% concentration of phosphine is filled (versus a phosphine mixture) into the valve outlet. The reaction capacity with boron trifluoride is 300μL (14μmol).

No re-desorption of phosphine from TOXICAPTURE™ sample after use was observed (detection limit is less than 0.5ppm PH_3). Though very minimal boron trifluoride (less than 0.5ppm HF) was detected, it is within allowable limit, since it is below the TLV. For both gases, no re-desorption was detected from the samples placed in water after exposure to the target gases.

By these test results, it has been verified that TOXICAPTURE™ can be used repeatedly several times toward 'outgas' since 'outgas' is a trace amount of gas and TOXICAPTURE™ has enough capacity to treat substantial amount of toxic gas.

We can also assess that TOXICAPTURE™ is stable and safe for handling even after use, although proper disposals are recommended in compliance with local regulations.

VI. APPLICATIONS

TOXICAPTURE™ has various other potential applications outside of outlet caps for dopant gas cylinders.

One example is to prevent corrosion of gas box pipings. TOXICAPTURE™ can remove residual gases in pipings before performing cylinder changes. Cleanliness of pipings can remarkably shorten the necessary time of vacuum after cylinder changes.

Another example is to prevent gas diffusion from maintenance parts. During maintenance of parts, such as when disassembling cryopumps, strong odor is often noticed due to the emission of toxic gas from hydride build up [2]. Hydrides can build up due to reaction of dopant gas with metal parts (aluminum) of the ion implant machinery during ion implantation, and during maintenance, moisture in the air may react with these hydride compounds built up on the metal parts and emit toxic gas [2]. TOXICAPTURE™ can remove these emitted gases during transportation of the maintenance parts and prevent diffusion from occurring when the parts are opened for maintenance.

Besides ion implantation dopant gases, it has been reported that TOXICAPTURE™ is also effective on other toxic industrial gases such as carbon monoxide, nitric oxide, sulfur oxide, hydrogen chloride, chlorine, hydrogen fluoride, hydrogen sulfide, and ammonia, etc..

VII. CONCLUSION

Although minimized, use of toxic gas in ion implantation still contains risk of toxic fume exposure. TOXICAPTURE™ can further minimize this risk thru a miniaturized gas abatement system placed in the outlet cap of a cylinder. TOXICAPTURE™'s effectivity was quantified—it has enough capability to hold 'outgas' from valves. TOXICAPTURE™ will help to maintain safe cylinder changes in ion implantation. TOXICAPTURE™ in various forms may help to reduce: corrosion of gas supply system, hazard in work environment, and vacuum operation time.

We trust that TOXICAPTURE™ can be used in various applications as a simple, convenient, and highly effective safety device.

ACKNOWLEDGMENTS

The authors would like to thank Luping Wang, Ph.D., and Paul J. Marganski at Advanced Technology Materials, Inc., for their valuable time and effort assembling the experimental apparatus and collecting data to quantify the system's capabilities. The authors would also like to thank Ron Eddy of Core Systems for helpful discussions and references.

REFERENCES

[1]. Threshold Limit Values for Chemical Substances and Physical Agents Biological Exposure Indices, ACGIH 2005

[2]. Filipp, N., Hunsaker, H., Herring, P., "Investigation of Hydride Emissions during the Maintenance of Ion Implantation Equipment"

[3]. T. Nakanoya, M. Egami, R. Tada, Japanese Patent Application No. 2004-186189

[4]. Paul J. Marganski, "Evaluation of TOXICAPTURE™ Filter for BF_3 and PH_3 Removal", March 6th, 2006

Dopant Cylinder Lifetime Monitor

Steve Bishop, Michael Wodjenski, Robert Kaim, Steve Lurcott, Jim McManus

ATMI Inc, 7 Commerce Drive, Danbury CT USA

Gordon Smith

Freescale Semiconductor, 6501 William Cannon West, Austin TX. USA

Abstract: The cost of consumable materials is a significant component in the cost of implanter operation. With the higher cost of sub-atmospheric gas alternatives it is increasingly important to accurately monitor its usage. The ATMI® SDS® GasGauge™ monitoring system accurately monitors gas level in four cylinders simultaneously, throughout their lifetime, in order to optimize usage of gas and related implanter productivity.

This paper displays how the GasGauge monitoring system accurately monitors the cylinder contents in SDS®, VAC® and high pressure gas cylinders. Internal and customer test data is also presented to verify these claims.

Introduction

ATMI and partner company Matheson Tri-Gas verify the return pressure of all cylinders arriving at our manufacturing sites. This data has been collected and logged for several years. Each year this cylinder return pressure is reviewed and provided to the customer whose cylinders were monitored. After 12 years of mainstream SDS use, we continue to see a substantial number of returned cylinders above the 20 Torr limit for specified delivered dopant volume. In discussions with our global customers on this issue, we have found that much of the installed ion implanter base simply lacks accurate dopant cylinder pressure monitoring systems. In response to this shortcoming, ATMI has developed the SDS GasGauge monitoring system.

Pressure Monitoring System Criteria

In designing an accurate monitoring system, it was concluded that 4 primary needs must be met.
- Accurate to 10 Torr to provide 20 Torr accuracy resolution
- Programmable alert for impending cylinder expiration with pressure set points designated by the user
- Log data at a pre-described rate designated by the user
- No interference with the operating system of the implanter.

The last criterion will be explained here in greater detail. The ATMI system will not use signals from the implanters pressure transducers, therefore an additional transducer is required on the pigtail of each dopant cylinder monitored. Our system will not have an electrical interface with the implanter it is monitoring.

Theory of Operation

The SDS GasGauge monitoring system is designed to help optimize the use of SDS 2 and SDS 3 gas sources. This system allows users to monitor up to four dopant cylinders, of all pressure ranges, from outside the implanter on a remote display. The SDS GasGauge monitoring system acquires signals from pressure transducers and thermistors mounted inside the gas box of an ion implanter via a data acquisition unit (DAQ) mounted inside the gas box. The signals are transmitted from the DAQ to a user interface (Display Module) via fiber optic cables. The data is then transformed using proprietary algorithms by the Display Module into units of dopant volume or grams remaining for the specific SDS gas and cylinder size being used. Pressure, volume, or grams remaining can be displayed numerically on the Display Module screen. A visual indication of the amount of gas remaining in the cylinder is displayed on the screen in the form of a bar graph so the user can tell at a glance when a cylinder is nearing expiration. In addition to monitoring SDS 2 and SDS 3 gas sources, the SDS GasGauge system can also monitor the output pressure of a VAC or any high pressure dopant cylinder in the gas box.

The SDS GasGauge system offers definable cylinder low alarms to alert the user when a cylinder is nearing expiration. The alarms have a visual, audible, and remote alert capability so the user has a variety of ways to communicate a low condition. Up to 3000 pressure and temperature readings are stored in the memory of the Display Module and can be uploaded via an RS232 port. Pressure and temperature data can be continuously acquired from the Display to a remote device via the RS232 port as well.

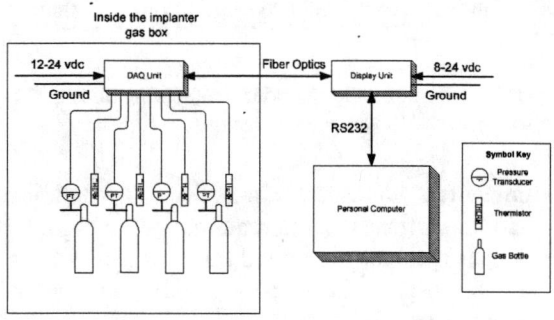

FIGURE 1. Graphical representation of the SDS GasGauge monitoring system architecture.

Accuracy

Total system accuracy of the SDS GasGauge system is defined as the accumulation of the pressure monitoring device (transducer) error, plus the error of the electronic modules at the low end of the pressure measurement scale. Our chosen pressure transducers, have a total voltage output accuracy of +/- 7.7 Torr from 0-50°C and +/- 3.0 Torr from 17-27°C. The SDS GasGauge system has a design accuracy of .010VDC from 0-50°C which equals +/- 1.5 Torr for 770 Torr range system. With this in mind, tests were undertaken on 10 sets of electronic modules, inputting precision voltages from a calibrated power supply, on each of the transducer input channels. These tests were conducted at 0.100, 1.500, 3.000, and 5.000VDC. The voltage inputs were compared with the output on the display module while it was in voltage test mode. Of these 40 voltage measurements, the largest offset voltage measured was .003 VDC, which equates to +/- .5 Torr. From this data, we can expect a design specified total system error of (SDS GasGauge + transducer) less than or equal to +/- 10 Torr for the entire 0-50°C range and a total system error of +/- 3.5 Torr for the 17-27°C range.

Cylinder Expiration Detection

A critical feature of the SDS GasGauge system is user alerting of impending dopant cylinder expiration. While the system senses pressure for each SDS, high pressure, or mechanically regulated (VAC)$_1$ dopant cylinder, it continuously monitors and compares the sensed pressure to the expiration set point programmed by the user. This capability allows the user to maximize dopant gas usage, while minimizing the risk of wafer scrap due to cylinder expiration during product implants.

In recent years, dopant material suppliers have developed mechanically regulated, sub atmospheric dopant delivery vessels. These vessels present a unique challenge in monitoring and predicting expiration. The dopant liquid or gas is stored internally at high pressure, but delivered at sub atmospheric pressure. In normal operation, pre regulated vapor pressures may range from 100 to over 1000 PSIG, while the post regulated pressures range from 200 to 500 Torr. The cylinder expiration detection system must be sensitive enough to detect post regulated pressure, while the cylinder is robust enough to maintain gas flow as pre and post regulated pressures equalize. In today's ever demanding implant processes, implant times are increasing due to decreasing beam energies and currents. It is not uncommon to see 15 or 20 minute implants on batch implanters. The cylinder and detection system must provide a level of robustness and resolution to give enough linear gas flow to complete the implant and mitigate the risk of scrap.

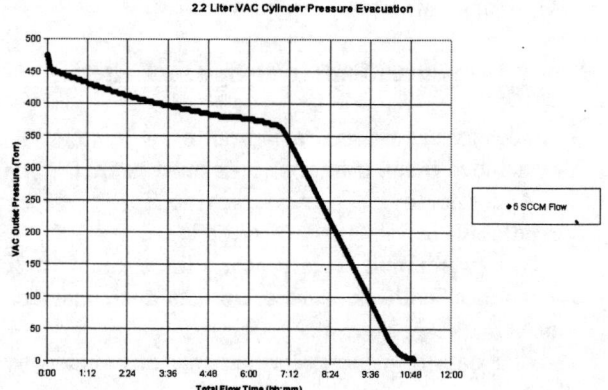

FIGURE 2. A VAC cylinder evacuation plot while monitored by an SDS GasGauge monitoring system, alarming at an expiration set point of 350 Torr, yet delivering over 3 hours of linear gas flow at 5 SCCM.

Reliability and Beta site Evaluations

In order to establish and estimate device reliability, 2 SDS GasGauge systems underwent a projected field failures (zero fails) approach life cycle test. This testing was designed to estimate the likely number of failures to occur in a 365 day period per 100 systems in operation. The testing consisted of 2 units being power cycled while exposed to a 3.000 VDC input voltage to simulate an output from a transducer and a resistor to simulate a thermistor output for temperature reading. Power cycling was chosen as the method of life cycle testing for these devices as it is the most stressful exposure the device will likely see during its practical application. The tests were conducted at ambient temperature (19-24°C). It takes 7 seconds for the system to power on, boot up, readings to stabilize, and 4 seconds for the system to completely power down. These parameters were the ones chosen as a complete power cycle for this test. At this rate of cycling, each unit under goes 7854 cycles per 24 hour period so, 15708 cycles per day were conducted. This testing was conducted for 28 days which entailed a total of 450,000 cycles without a failure. From this data we can calculate the expected failures per 100 modules in operation as 0.2 failures per year with a 90% confidence value.

Four beta site systems were installed globally, two in the Eastern United States, one in the central United States, and the last system in Europe. These installations exposed our systems to the harsh rigors of a production implanter, most notably, the DAQ module functioning in an electrostatic, RF noisy, high voltage gas box. The beta site in the central US had its stored data downloaded after 2 months. All data were intact and properly formatted. Our European beta site recorded an SDS2 BF_3 cylinder end pressure at 12 Torr. This cylinder was measured on the pressure standard at our manufacturing site. The returned cylinder measured 11 Torr. Additionally, our beta site partners were encouraged to overview the software operating system and provide feedback on suggested revisions that would be of benefit to them.

Beta Site Performance
- Zero Electronic Module failures during 17 months of on tool performance.
- One non-failure reliability issue- "Extra digit is displayed on pressure readout during ion source arcing". This was corrected with a software revision on 1/30/2006.

Customer Requested Software Revisions
- Dopant species displayed in volumetric measurement bar, installed 2/18/2006.
- Option for Grams remaining display with SDS cylinders, installed 2/18/2006.

Cost Reduction Evaluation

To perform an accurate cost comparison, we use actual customer annual cylinder pressure return data. The data is averaged for each of the dopant types used by this customer. A reduction of the customer's average cylinder pressure to 10 Torr is projected for the cost analysis, and only gas cost reductions are noted as percentages in this evaluation. Isotherms stored in the software of the SDS GasGauge monitoring system, provide dopant loading in grams at a given pressure for SDS 2 and SDS 3 $_2$ cylinders. Cost savings are projected by evacuating more gas from each cylinder when monitored by the SDS GasGauge system. Deriving more gas from each cylinder, fewer cylinders are used annually per implanter.

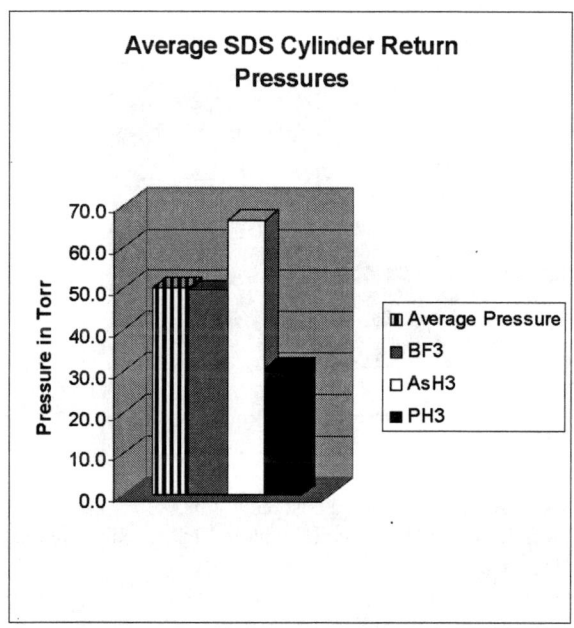

FIGURE 3. The average of cylinder return pressures by dopant species.

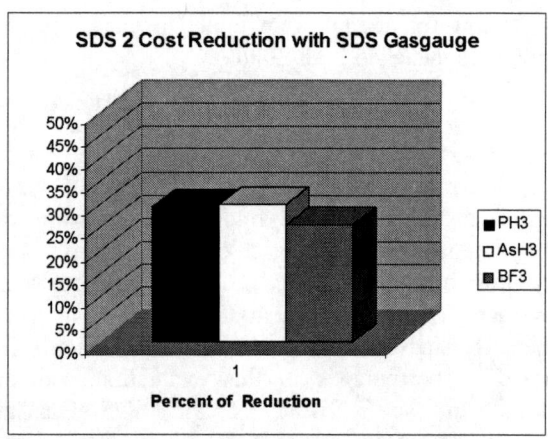

FIGURE 4. Annualized cost reduction, per implanter on SDS 2 cylinders when monitored by the SDS GasGauge system.

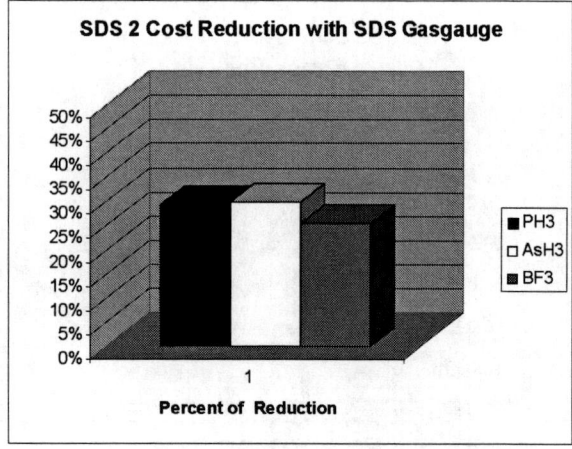

FIGURE 5. Annualized cost reduction, per implanter on SDS 3 cylinders when monitored by the SDS GasGauge system.

Summary

The SDS GasGauge monitoring system has been rigorously evaluated using beta sites, testing standards, and cycle testing. This system has proven its reliability in the harshness of production environments. Through the system's accuracy, the cost benefit has been demonstrated. In using the data logged by the SDS GasGauge system, end users can improve predictability in use of dopant materials, and potentially streamline the materials acquisition process.

References

1. W. K. Olander, J. Mayer, M. Donatucci, and L. Wang "Vacuum Actuated Gas Delivery" 13th IIT2000 proceedings, pg. 722
2. Josep Arnó, Donald Carruthers, Ed Sturm, Luping Wang, Jim Dietz "Development of Next Generation Safe Delivery System (SDS): SDS3" 15th International Conference on Ion Implant Technology, IIT2004 Taipei, Taiwan October 2004.

PROCESS CONTROL & YIELD

Real World Experience With Ion Implant Fault Detection at Freescale Semiconductor

David C. Sing, Terry Breeden, Hassan Fakhreddine, Steven Gladwin, Jason Locke, Jim McHugh, Michael Rendon

Freescale Semiconductor, Inc. 3501 Ed Bluestein Boulevard, Austin TX 78721, USA.

Abstract. The Freescale automatic fault detection and classification (FDC) system has logged data from over 3.5 million implants in the past two years. The Freescale FDC system is a low cost system which collects summary implant statistics at the conclusion of each implant run. The data is collected by either downloading implant data log files from the implant tool workstation, or by exporting summary implant statistics through the tool's automation interface. Compared to the traditional FDC systems which gather trace data from sensors on the tool as the implant proceeds, the Freescale FDC system cannot prevent scrap when a fault initially occurs, since the data is collected after the implant concludes. However, the system can prevent catastrophic scrap events due to faults which are not detected for days or weeks, leading to the loss of hundreds or thousands of wafers. At the Freescale ATMC facility, the practical applications of the FD system fall into two categories: PM trigger rules which monitor tool signals such as ion gauges and charge control signals, and scrap prevention rules which are designed to detect specific failure modes that have been correlated to yield loss and scrap. PM trigger rules are designed to detect shifts in tool signals which indicate normal aging of tool systems. For example, charging parameters gradually shift as flood gun assemblies age, and when charge control rules start to fail a flood gun PM is performed. Scrap prevention rules are deployed to detect events such as particle bursts and excessive beam noise, events which have been correlated to yield loss. The FDC system does have tool log-down capability, and scrap prevention rules often use this capability to automatically log the tool into a maintenance state while simultaneously paging the sustaining technician for data review and disposition of the affected product.

Keywords: Ion Implantation, Fault Detection, Automatic Process Control, Charge Control
PACS: 61.72 Tt

I. INTRODUCTION

Freescale Semiconductor has been utilizing an internally developed fault detection and classification (FDC) system since 2002 [1,2]. The FDC system downloads implant summary statistics to a central server which accepts data from Freescale facilities around the world. Compared to systems which store trace run-to-run data, the data storage requirements are relatively modest: as of January 2006 data from over 3.5 million implants were stored on a single server. Control rules are applied to the data when it is transferred from the tools to the database. If a control rule failure is detected, then several different actions are possible depending upon the pre-assigned severity of the failure: for low severity failures an email is sent to an list of engineers, for higher severity faults a page is also sent to a process technician to for further investigation, and in the most severe cases a command is sent to log the tool down to prevent the track-in of additional material into that tool.

With any FDC system, there is a delicate balance to be maintained when constructing control rules. Rules must be constructed with appropriate limits such that faults which are known to cause scrap are detected, yet do not result in an excessive number of false negatives which cause un-necessary downtime. The problem is compounded in a facility like Freescale's ATMC facility, which runs many lines of technology from quarter micron to 45 nm, with both high volume production and low volume R&D material. In a typical 30 day period over 900 different implant recipes are used. The shear number of different recipes combined with the hundreds of process variables could quickly result in thousands of control rules being active at a given time.

The initial set of control rules deployed when the FDC system came online were control limits on the fundamental recipe parameters such as the species, dose, and energy of an implant. Past catastrophic scrap events which have occurred within the industry, such as cross species contamination and charging, weighed heavily in the types of rules created. Statistically calculated control limits were applied to fundamental parameters such as AMU magnetic current, extraction voltage, total implant energy, and flood gun emission current for the high running implant recipes. However, these recipes already had recipe limits applied to these very same parameters. The application of both FDC control rules and recipe limits created redundant systems for detecting out of control conditions, and since recipe limit violations immediately prevent the processing of material while the FDC control rules violations provide after the fact notification of an out of control situation, recipe limits have become the control methodology of choice for fundamental implant parameters.

II. CLASSIFICATION OF IMPLANT FDC CONTROL RULES AND FAULTS

As more experience has been gained with the FDC system, the key application which was developed was the ability to detect and flag changes in tool signals which often indicate the need for routine maintenance, and if left undetected, could eventually lead to product scrap. An analysis of the type and distribution of the active control rules at the ATMC fab illustrates the strategy of using FDC signals to flag the need for PM's and to detect tool signal trends that monitor wafer based SPC methods cannot detect.

Table I lists the distribution of the ATMC fab's control rules broken down into general categories. Rules which monitor the pressure in either the process chamber or the beamline of the implanter account for more than half the active rules. These rules are tailored for specific process recipes, since the pressure in the tool depends upon the specifics of the recipe (species, energy, dose, etc) and the properties of the material (photoresist coverage and composition). The next most

Table I. Summary of ATMC Implant FDC Control Rule Classifications

Classification	Count
Pressure	265
Charging	123
Magnet Current	63
Implant_Time	58
Interrupts	18
Particles	16
Ref_Pwr	8
Beam Shape	4
Energy	4

populous category is charging, which accounts for over 20% of the active rules. These rules are a mixture of recipe specific rules and more generalized rules

Table II. 30 Day Summary of ATMC Fab Implant FDC Faults by Classification and Severity

ATMC Fab Implant fault Classification (30 days)	
Comment_Notification	Number
Charging_Email	228
Pressure_Email	101
Ref_Pwr_Email	100
Disk_Space_Email	33
Implant_Time_Email	13
Network_Down_Email	6
Interrupts_Pager	202
Pressure_Pager	127
Particles_Pager	81
Implant_Time_Pager	42
Magnet_Current_Pager	37
Charging_Pager	23
Energy_Pager	4
Beam_Shape_Pager	2
Ref_Pwr_Pager	2
Particles_Shutdown	20
Implant_Time_Shutdown	13
Sum Email Notification Faults	481
Sum Pager Notification Faults	520
Sum Shutdown Faults	33
Alarm Rate (Email Notification)	1.43%
Alarm Rate (Pager Notification)	1.54%
Alarm Rate (Shutdown)	0.10%

which are applied to categories of recipes based on filters on such parameters such as beam current, species, and energy. The next two most popular categories, each accounting for over 10% of the active rules, are rules to monitor magnet current and implant time. Rules are also in place to monitor the number of interrupts during implants and detect particle excursions. The interrupt rules are applied to all recipes on all tools, and the particle rules are applied to all recipes on the high energy and high current tools, which are equipped with in-situ particle monitors.

Table II lists number and type of faults flagged by the FDC system at the ATMC fab in a typical 30 day period. The fault category as well as the notification level are listed in the table. Over 1000 faults messages were generated by the FDC system in that time period, which is approximately 3% of all implants. About half of the faults have email only notification, which alerts process engineers of a possible trend in a tool signal. Half of the faults also generate a page to a process technician in the factory, who then follows a specific action plan for each type of fault, usually involving the review of additional data to determine if the signal is an isolated fault or part of a trend, and also to gauge the severity based upon the magnitude of the rule violation. Only 0.1% of all implants result in an automatic tool shutdown, which disables the ability of operators to track additional lots into a tool as well as alerting sustainers to a potential scrap causing situation.

III. APPLICATIONS OF FDC RULES FOR PM TRIGGERING AND SCRAP PREVENTION

Since the vast majority of FDC rules were written to provide either email or pager notification to engineers and technicians, the occurrence of these faults is not considered an immediate scrap risk, but are often used to trigger preventive maintenance (PM) events.

Examples of PM Trigger Applications

Charge Control System Monitoring

The high current tools at the ATMC fab employ secondary electron flood (SEF) guns for charge control. In these tools low energy electrons created by secondary emission from a graphite target provide charge neutralization to the wafer surface. Over a maintenance cycle, the level of secondary emission can vary as components age and surfaces are coated with photoresist and other materials. Figure 1 shows the typical response of the in-situ charging sensors and disk current sensor with changes in secondary flood emission. Positive charge signals decrease, negative charge signals increase, and disk current decreases as secondary emission increases.

FDC control rules are in place which monitor positive charging, negative charging, and disk current for all recipes in which the flood gun is active. Typically flood emission increase during the SEF lifecycle, as surfaces coat and become more likely to emit secondary electrons. Occasionally an SEF flood guns runs "hot" with flood gun emission increasing by several times. Hot flood gun conditions have developed spontaneously in some circumstances, in other cases an SEF runs hot immediately upon installation. An example of a hot flood gun signal is shown in Figure 2, in which the secondary current, normalized by the relatively stable primary current, has increased to greater than 20% of the primary current. In that case, disc current rules failed (negative) and the flood assembly was replaced. As a consequence of this incident, FDC rules to detect excessive secondary current levels relative to primary current were created to detect this specific failure mode.

Pressure Variations

Examination of Table I shows that the most common category of control rule are those to detect changes in pressure, either chamber or beamline pressure, in the tools. Rules for pressure are almost always customized for specific recipes. Figure 3 shows the pressure trend for a high power high energy implant recipe. The sharp spike occurred when the cryo-pump became loaded and pumping speed was severely degraded. FDC rule faults for process chamber pressure and for normalized implant time caused the tool to be logged down for a cryo-pump regeneration. High power implant pressure rules generally flag the need for pump maintenance, while low power implant pressure rules are the most common detection method for small vacuum leaks which cause an increase in the base pressure.

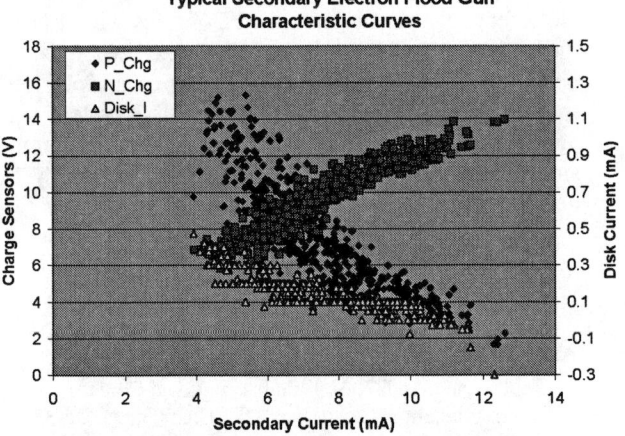

Figure 1. Response of charge sensors and disk current to changes in secondary emission current.

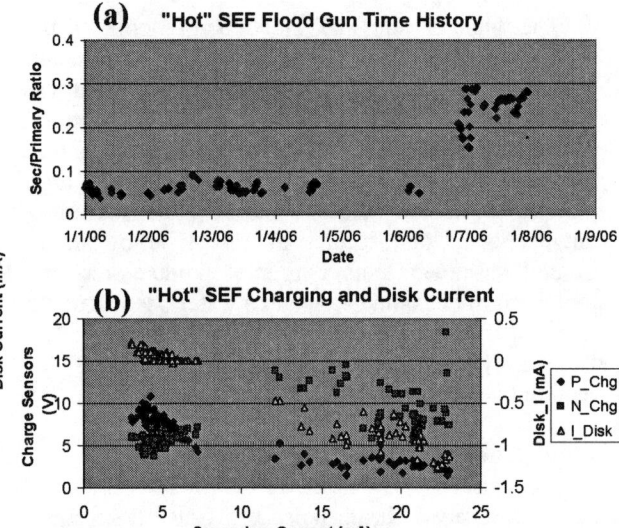

Figure 2. (a) "Hot" secondary flood emission failure and (b) response of disk current and charging sensors.

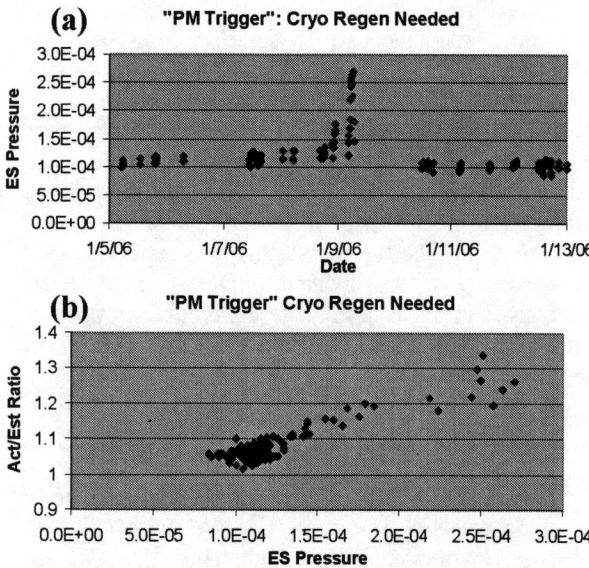

Figure 3. (a) End Station pressure (b) Normalized implant time response to end station pressure.

Scrap Prevention Rule Applications

Scrap prevention rules are written for specific failure modes and usually implement control by automatically logging the tools into maintenance upon failure. Some examples are discussed below:

Particle Excursions

The high current and high energy tools at the ATMC fab have in-situ particle monitors. Analysis by device and process engineering established a particle level at which yield loss was likely if the lots were not correctly dispositioned after such events. An FDC rule was implemented to log a tool into maintenance if high particle levels are detected and to page sustaining engineering. Upon receipt of this type of alarm, process sustainers remove affected product from the tool for further examination and a wafer based particle test is performed to determine if the event was an isolated "bursting" event or was severe enough to required maintenance activity to clean up the tool.

High Current Beam Current Fluctuations

High levels of beam noise can cause dose shifts and have been correlated to yield loss and scrap at Freescale's OHT fab. The detection of excessive beam noise is an example of a parameter which is monitored by the FDC system which the tool itself has either limited or no ability to detect and hence has no ability to control through recipe limits. In this application, the beam current mean, range, and standard deviation are combined into a mathematical formula to create a noise figure of merit. A FDC rule calculates this noise figure and shuts the tool down if a preset upper control limit is violated.

High Current Beam Steering

Beam profiling is performed by some of the high current tools used at Freescale factories. The tool software calculates a horizontal center of gravity which is used to insure uniform dosing. At the OHT fab yield fallout due to device asymmetry was correlated to vertical beam displacement for a specific product family. The FDC system combines beam profile data signals to derive a vertical center of gravity signal and a control rule was written to log the tool into maintenance if the center of gravity signal indicated the beam was significantly off center. Like the beam noise application, a composite signal that the tool itself does not calculate is used to detect a failure mode which can only be controlled using an FDC rule

IV. SUMMARY AND CONCLUSIONS

The Freescale FDC system has proven an effective supplement to traditional wafer based tool monitoring protocols. At the ATMC fab the FDC system is the primary system for monitoring the health of critical tool systems such as charge control systems and cryo pumps. Monitoring of these critical systems is performed on virtually every process run for virtually every recipe. Routine PM's for these systems are regularly triggered by the FDC system, and charging system failures and vacuum leaks are routinely detected by the system. Specific scrap causing events are effectively monitored with FDC rules which have the ability to log tools into maintenance when faults occur. Composite signals composed of multiple tool signals are easily calculated by the FDC system and provide detection and intervention of out of control situations that the implant tool itself has no ability to detect or control.

REFERENCES

1. M. Rendon, D.C. Sing, M. Beard, M. Hartig, J. Arnold, "Ion Implant Data Log Analysis for Process Control and Fault Detection," 2002 14th International Conference on Ion Implantation Technology, IEEE, 2003, pp 331-334.
2. D. Sing, M. Rendon. "Implant Process Control: Going Beyond Particles and RS," Nuclear Instruments and Methods in Physics Research Section B: Beam Interactions with Materials and Atoms, Volume 237, August 2005, Pages 318-323.

Angle Measurement and Control in High Current Ion Implantation

B.S. Freer, D.E. Hoglund, and M.A. Graf

Axcelis Technologies, Inc.
108 Cherry Hill Dr., Beverly, MA 02155, USA

Abstract. Precise beam-to-wafer angle control continues to become more important for critical high current implants where gate stack shadowing variations lead to parametric shifts. This paper examines contributors to variations in beam-to-wafer angle from a beamline, process chamber, and control system standpoint. Wafer positioning, beam steering, and beam angular spread are discussed, along with techniques to precisely measure and control them.

Direct, in-situ techniques are used to measure beam angular offset and spread in a 'spot beam' high current implanter. The relationships between position and angle, both on average and within the beam, are detailed. In a 2-D mechanical scan system, angle uniformity (the local angle variation at different points on the wafer) is influenced by how each point on the wafer samples the angles within the beam. Angle uniformity is shown to be less sensitive than dose uniformity to scan spacing. Furthermore, conditions which result in better dose uniformity also tend to give better angle uniformity.

Keywords: Angle Control, Ion Implantation, CMOS Devices
PACS: 61.72.Tt, 85.40.Ry

INTRODUCTION

Angle control in high current ion implantation is increasingly important as device scaling continues. The most critical high current implant is the source/drain extension (SDE). The lateral extent of the SDE is shrinking through the use of lower energy implants, which have less lateral straggle, and smaller thermal budgets, which reduce lateral diffusion. Gate stack heights are not being reduced at the same rate, resulting in higher gate stack aspect ratios. Gate overlap therefore becomes more sensitive to shadowing. Minimizing shadowing is important to control device parameters such as drive current [1-4].

In any ion implantation system, every point on the wafer sees a distribution of ion beam angles. At each point, one can describe this distribution of angles as a local average angle in two orthogonal directions, and as an angular spread in the same two directions. In most high current systems, the dispersive angle is in the horizontal direction, and the non-dispersive angle is in the vertical direction.

For a specific device geometry, at a given point on the wafer, the primary influence on shadowing is the component of local average angle that is perpendicular to the gate stack top edge. This is a combination of wafer positioning and beam angle. If different parts of the beam have different angles, then the local average angle also depends on the portions of the beam to which the point on the wafer is exposed.

Local angular spread is the distribution of beam angles around the local average angle at a given point on the wafer. Control of angular spread is believed to be beneficial for process control [5]. Local angular spread is dominated by beam angular spread. Modern high current implanters employ plasma flood systems, electron confinement, and optical design to control space charge [6] and angular spread. A beam with moderate angular spread can even mitigate the effects of shadowing due to local average angle, in much the same way bi- and quad-mode implants can [4].

Angle uniformity is the variation in local average angle across the wafer. The origins of angle uniformity are quite different in 'ribbon beam' and 'spot beam', 2-D mechanical scan systems.

In 'ribbon beam' systems[5], each point on the wafer sees only one stripe of the ribbon. The two contributors to angle uniformity are beam parallelism, which is the variation in angle at different stripes on the ribbon, in the direction along the ribbon; and variation in angle at different stripes on the ribbon, in the direction along the stripe. Angles in both directions must be carefully controlled so that no part of the wafer sees a stripe that has too large a local average angle. Beam current density uniformity must also be maintained for dose uniformity. Bi- and quad-mode implants can be used to provide some averaging in angle and dose. In the case, however, of a beam that is diverging or converging in the ribbon direction, a bi-mode (180°) implant does not help parallelism.

'Spot beam' systems with 2-D mechanical scan endstations (spinning disk, linear scan, or pendulum scan) do not make a uniform beam in density or angle. By exposing each point on the wafer to a representative sample of the beam, the variation in beam current density can be averaged out. In this paper we will show how internal angles of the beam depend on location within the beam. To achieve good angle uniformity each point on the wafer must be exposed to representative angles within the beam for good local averaging of angle.

Implants on spot beam systems with spinning-disk endstations typically have many 'passes per point', or number of times a point on the wafer fast-scans through the beam on a unique path. For a 90 second implant on 300mm wafers with a 50 mm tall beam, there can be over 100 passes per point. The averaging of angle and dose is therefore very good.

The 'cone angle', a wafer positioning effect due to the pedestal angle, is a significant contributor to angle non-uniformity, ~0.4° 1-sigma at 0° tilt. Because this is fixed by the geometry of the disk, the variation is limited to ~1.1° center-to-edge. It can still give good performance if cone angle is tolerable and the beam average angles are controlled [7].

Spot beam systems with single wafer, 2-D mechanical scan endstations have several advantages over spinning disks. The pedestal does not need to be angled to hold and cool the wafer, and there is no cone angle variation. Twist errors in a pendulum scan are eliminated by counter-rotation of the wafer. The number of passes per point is, however, lower than on a spinning disk system, possibly ten passes or fewer. In this paper, we show that this is sufficient for excellent angle averaging.

GLOBAL ANGLE OFFSET

One requirement of achieving good angle control is to minimize the global angle offset, which is the difference between the desired angle and the across-wafer average of the local average angles. Since it is a global value, it can be corrected for by tilting the wafer, steering the beam, or a combination of the two.

Wafer angle offsets caused by wafer positioning can be minimized by proper mechanical alignment of

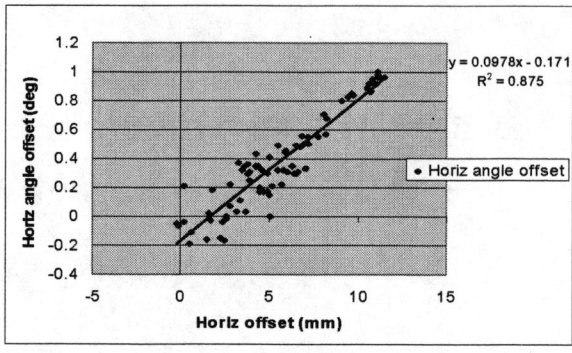

FIGURE 1. Horizontal angle and position offsets

the endstation. This is typically done once at setup, and may be checked periodically afterwards.

Beam angle offsets cause more day-to-day variation than positioning does. In spot-beam systems, horizontal angle offsets can be caused by analyzer and electrode tuning and by source and electrode alignment. A beam with a horizontal angle offset follows a non-ideal path through the beamline, and its position at the wafer is expected to be offset from the ideal location [8]. Fig. 1 shows the relationship between position and angle for one high current beam. It is therefore possible to reduce beam angle variation by measuring and controlling the beam offset.

On Axcelis HC3 Ultra high current implanters, BHO (Beam Height Offset) control allows automatic horizontal centering of the beam. First electrode steering variations are reduced by using the beam to calibrate the side axis after each source PM. Then side axis tuning is limited in the process recipe. Before each implant, the BHO is measured using the 2-D profiler [9]. If the BHO is outside of the limits set in the recipe, the analyzer magnet current is changed by a percentage proportional to the offset. This process is repeated if necessary until the BHO is within the limits, as shown in Fig. 2. This feature is also used on Axcelis Optima HD systems.

Direct beam average angle measurement can further improve global angle control. Ribbon beam implanters employ methods such as traveling beam shadow monitors to measure horizontal angle along the ribbon. Average angle is then corrected for by tilting the wafer.

In HC3 Ultra, an on-disk angle sensitive mask, described in more detail elsewhere [8,9], can be used to measure and control horizontal angle. In this system, the beam is first centered, then the current through an array of high aspect ratio slots is measured as a function of angle. From this the horizontal angle centroid is calculated, and the disk is repositioned to correct for the offset angle. This has been shown to reduce variation in device parameters beyond beam centering alone [9].

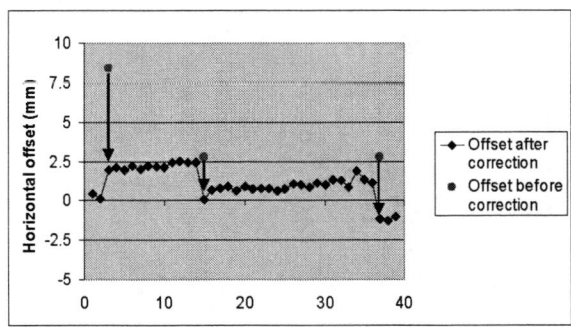

FIGURE 2. Horizontal beam height offset (BHO) correction

In spot beam systems, the variation in vertical position and angle is smaller than in the horizontal direction. Fig. 3 shows the vertical angle vs. vertical position offsets for B, BF2, and As beams from 2 keV to 50 keV, on the same scale as Fig. 1. The slope of the line is 0.0313 deg/mm or 32 mm/deg. Vertical steering can be easily detected and interlocked against, due to the high sensitivity of vertical position to angle.

The corresponding length over which the angle occurs is $l = 1/\tan(0.0313) = 1.8$ m, approximately the beamline length. This suggests that the origin of the vertical angle is in the extraction region in this case.

The tight angle distribution is due to the lack of tuning parameters that strongly affect vertical steering, and also to the collimating effect of the beamline length-to-height ratio. Optimizing for beam current favors reduction in the amount of beam lost to vertical steering of the beam into the top of bottom of the beamguide or apertures. This collimating effect is stronger for low energy beams, which tend to use more of the vertical acceptance of the beamline.

ANGLE UNIFORMITY

The second requirement of achieving good angle control is minimizing the variation in angle across the wafer. Angle uniformity is not a global value, so it cannot easily be corrected for by tilting the wafer.

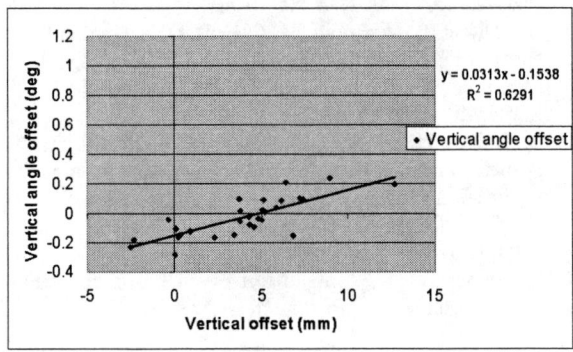

FIGURE 3. Vertical angle and position offsets

For ribbon beam systems, the angle uniformity is most strongly affected by the beam itself. Since each point on the wafer sees only one stripe of the ribbon, great care must be taken to minimize the variation in angle between various stripes in the beam. The angles at a few locations in the beam are measured by the traveling shadow profiler. Based on the profiler data, the beam divergence or convergence in the ribbon direction is determined and corrected by adjusting the corrector magnet. Higher order angle variation (i.e., non-linear divergence or convergence) cannot be trimmed by using a corrector, and variations in angle with spatial extents smaller than the size of the shadow faraday spacing (~40mm) cannot be seen.

For 2-D mechanical scan spot beam systems, dose averaging depends on the number of passes per point, which in turn depends on the beam intensity distribution in the vertical direction. To estimate angle uniformity, the vertical angular distribution must also be known, if different parts of the beam have different angles. We have measured the vertical angle and angular spread at different vertical locations of an As^+ 5 keV beam generated by a spot-beam beamline. As can be seen in Fig. 4, the angular spread or 'fuzziness' at all points in the beam is uniform and low, about 0.7°. The average angle of points within the beam did depend approximately linearly with the position of the point within the beam. The angular spread was only 1.26° 1-sigma, since edges of the beam with angles >1° had little intensity.

Dose and Angle Simulation

Dose/angle simulations for pendulum-scan, spot-beam systems were performed using a finite element model. The dose simulation examined how points on the wafer sampled the beam's intensity and angle distributions. Beam intensity profiles in the vertical and horizontal dimensions were measured using the 2-D profiler and entered into the model. For computation purposes, the profiles were fit by three gaussians, which properly accounts for over 97% of beam intensity for this vertical profile. The angle simulation used the same intensity distributions plus the measured data (Fig. 4) relating local angle to position in the beam. All of the angular variation was assumed to be related to position in the beam. Finally, scan paths in the pattern of Optima HD (interlaced pendulum scan with counter-rotation of the wafer) were used.

Angle uniformity is not straightforward to determine using wafer methods, while measurement of dose uniformity is common and well-characterized. Since similar sampling mechanisms are at work for

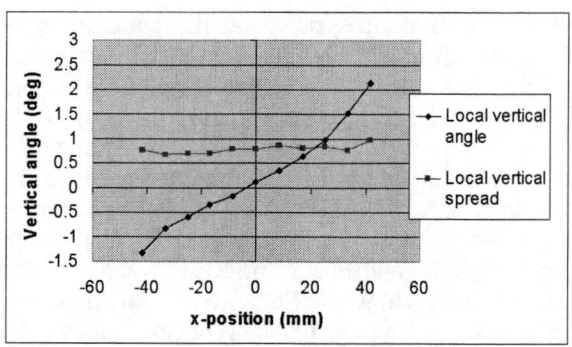

FIGURE 4. Internal vertical angular distribution vs. vertical position of an As 5 keV beam

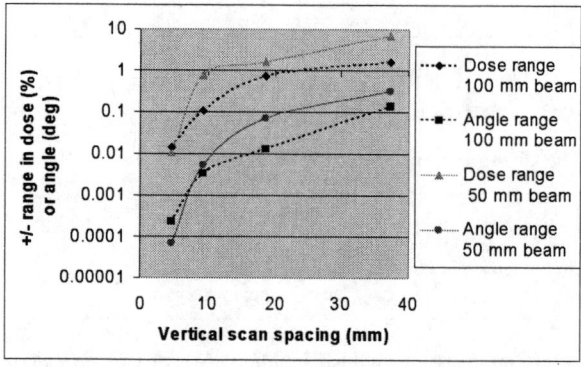

FIGURE 5. Modeled R_s and angle ranges, As 5 keV

both dose and angle averaging, we used dose uniformity to validate the model. Then angle uniformity can be simulated without relying on wafer-based angle measurements.

To estimate the sensitivity of angle averaging on beam size and scan spacing, the simulation was run with scan spacings from five to 37 mm and with beam heights of 100 and 50 mm. The 100 mm beam profile is measured data, and the 50 mm profile was derived by scaling the height axis of the measured profile by a factor of 0.5.

The results in Fig. 5 show the uniformity of angle and dose due to averaging only (contributors to dose non-uniformity that would dominate at small scan spacing were not modeled). It can be seen that the angle uniformity and dose uniformity curves have similar shapes, but that the angle uniformity is always much better. For example the 100 mm tall beam and a 20 mm scan spacing gives ±1% range in dose, which is probably tolerable but measurable, while the angle range would not be detectable. This difference is because the intensity of the beam varies from 0% to 100%, while the angle is only about ±2°.

CONCLUSIONS

The different types of angles that are important to high current ion implantation were described. Angle control consists of controlling global angle offsets and angle uniformity. Global offsets can be reduced by controlling beam position and by measuring angle directly and adjusting the wafer position. Angle uniformity in ribbon beam systems depends on the angle uniformity of the beam. In 2-D spot beam systems, averaging of intensity and angles within the beam gives good dose and angle uniformity, with angle uniformity being the easier one to achieve.

ACKNOWLEDGMENTS

The authors thank D. Polner, A. Wells, and A. Moulton of Axcelis for their contributions.

REFERENCES

1. H.-J. L. Gossmann, "Ion Implantation in Advanced Planar and Vertical Devices", *Proceedings of the 15th Int'l Conf. on Ion Implantation Technology*, Taipei, Taiwan, p. 1, 2004.
2. H.-J. L. Gossmann, L. Rubin, T. Parrill, and A. Agarwal, "Impact of Extension Implant Energy Purity and Angle on the Electrical Characteristics of a 65 nm Device Technology", to be published in *J. Vac. Sci. Tech. B*.
3. U. Jeong, S. Mehta, C. Campbell, and R. Lindberg, "Effects of Beam Incident Angle Control on NMOS Source/Drain Extension Applications," *IIT Proceedings*, Taos, NM, 2002.
4. J.D. Bernstein, J.J. McComb, A.W. Alvarez, J. Chow, and L.M. Rubin, "Quad-mode Source/Drain Extension Implants for Reduced Sensitivity to Angle Variation," *IEEE Trans. Semi. Manuf.*, submitted for publication.
5. C. Campbell, et. al., "Beam Angle Control on the Viista80 Ion Implanter", *IIT Proceedings*, Taos, NM, 2002.
6. M. Graf, B. Vanderberg, V. Benveniste, D. Tieger, and J. Ye, "Low Energy Ion Beam Transport", *IIT Proceedings*, Taos, NM, 2002.
7. B. S. Freer, L. M. Rubin, M. A. Graf, D. E. Hoglund, D. Newman, K. Ditzler, K. Elshot, and T. Romig, "Direct Measurement of Beam Angle in a High Current Ion Implanter", this conference.
8. B. S. Freer, R. N. Reece, M. A. Graf, T. Parrill, and D. Polner, "*In Situ* Beam Angle Measurements in a Multi-wafer High Current Ion Implanter", *Proc. of the XV Int'l Conf. On Ion Implantation Technology*, Taipei, Taiwan, 2004, p. 378.
9. P. Splinter, et. al., "In-situ Ion Beam Profiling by Fast Scan Sampling", *IIT Proceedings*, Austin, TX, 1996.

Process Control in Production-Worthy Plasma Doping Technology

Edmund J. Winder, Ziwei Fang, Edwin Arevalo, Tim Miller, Harold Persing, Vikram Singh, and T.M. Parrill

Varian Semiconductor Equipment Associates, Inc., 35 Dory Road, Gloucester, MA 01930, USA

Abstract. As the semiconductor industry continues to scale devices of smaller dimensions and improved performance, many ion implantation processes require lower energy and higher doses. Achieving these high doses (in some cases ~1×10^{16} ions/cm^2) at low energies (<3 keV) while maintaining throughput is increasingly challenging for traditional beamline implant tools because of space-charge effects that limit achievable beam density at low energies. Plasma doping is recognized as a technology which can overcome this problem. In this paper, we highlight the technology available to achieve process control for all implant parameters associated with modern semiconductor manufacturing.

Keywords: Plasma doping, ion implantation technology, optical emission spectroscopy.
PACS: 52.77.Dq, 85.40.Ry 52.70.Kz.

INTRODUCTION

The fabrication of advanced CMOS devices calls for production worthy solutions to meet the low-energy high-dose doping challenge. With a radically different architecture that eliminates beamline (BL) transport limitations altogether, plasma doping (PLAD) [1-3] allows challenging high-dose low-energy processes to run at production throughput and potentially opens up new applications not possible with BL technology. The challenge is particularly acute in advanced DRAM poly gate electrode doping applications, with doses >4×10^{15}/cm^2 and B energy (for p-type) <3keV driven by the need to achieve low poly resistivity and poly depletion in a film <80nm. In addition, it is advantageous for some DRAM processes to implement CMOS in a scheme along with in situ doped (ISD) poly [4]. This is most readily accomplished by counter-doping the ISD poly, a process requiring >1×10^{16}/cm^2 implanted dopant and one for which BL throughput may be completely inadequate. A similar challenge looms on the horizon for sub-45nm logic processes in which source drain extension (SDE) energy is decreasing for junction scaling and doses are increasing to reduce series resistance. In any event, the significant throughput advantage afforded by PLAD must be accompanied by process performance well established in ion implantation. First, the parameters that fully describe an implant process—dose, energy, and angle—must be understood and controlled. Second, potentially deleterious parameters, especially particles and metal contamination, must be reduced and maintained. State-of-the-art process control capabilities have now been developed to enable PLAD for production applications. In this paper we describe these technologies and demonstrate results relevant for the applications of interest, particularly gate electrode doping.

PROCESS CHARACTERISTICS

A PLAD chamber schematic is included in Fig. 1 for reference. Generally, after the substrate is loaded onto an electrostatic clamp within the process chamber, the chamber is backfilled with a suitable gas containing the dopant atoms of interest, such as B_2H_6, BF_3, AsH_3, or mixtures such as BF_3 plus B_2H_6 with or without diluents such as H_2 and He. An energy source creates plasma while fresh gas continually flows into the chamber. A bias applied to the wafer (and clamp) accelerates ions from the plasma into the wafer, implanting them at the desired energy corresponding to the peak wafer bias potential. It is useful to note that the bias itself can provide sufficient energy to generate plasma during the transient [5]. In this case, it is not necessary to strike plasma with an independent energy source before commencing the implant.

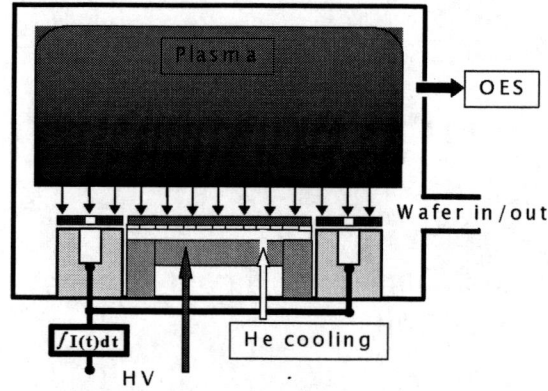

FIGURE 1. Plasma doping chamber schematic.

Implant Parameter Control

A variety of technological innovations enable process control of the primary parameters that describe all implant processes: dose, energy, and angle.

Dosimetry

It is possible to integrate a Faraday cup, the canonical approach to ion implant dose measurement and control, into a PLAD system as a sensitive in situ monitor. One possible implementation is included in Fig. 1. Just as in a conventional BL implant, the in situ Faraday counts ions as the implant proceeds, providing a signal to the dose controller which stops the implant automatically at a preset dose. Naturally such a capability improves repeatability control.

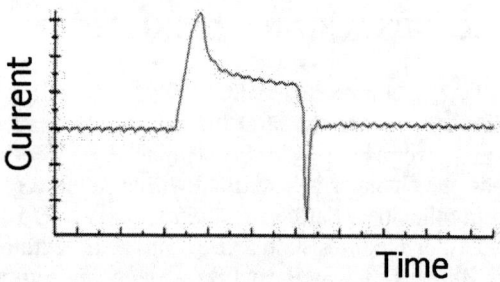

FIGURE 2. Single pulse dose pulse signal from an in situ Faraday system. An individual pulse dose out of limits can trigger an immediate action. The peaks observed at the leading and trailing edges of the pulse are related to plasma dynamics.

In addition, the PLAD Faraday can monitor ion current in real time, providing an assessment of the health of the implant process. This action is analogous to a BL Faraday detecting beam current drift during an implant. With PLAD, however, the Faraday can detect the dose-per-pulse (DPP), the number of ions associated with each wafer bias pulse, as shown in Fig. 2. In this case, rather than simply counting for the purpose of integrating total dose, the DPP signal immediately indicates any out-of-control conditions associated with the plasma. The PLAD system can then be directed to take any appropriate action, such as halting the implant process.

Energy Control

PLAD ion implant energy control is achieved in a manner equivalent to BL systems by using a DC wafer bias, so that the high voltage power supply controlling the ultimate ion energy can be calibrated and measured with an independent feedback monitor. When a pulsed DC supply is set at 6kV, as shown in Fig. 3 for example, the majority of ions are accelerated to 6 keV, and profiles scale in accordance with voltage changes.

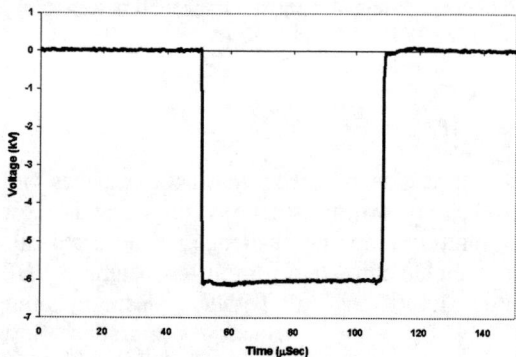

FIGURE 3. Wafer bias voltage for a single pulse applied using a pulsed DC system with ~1µs rise and fall time. Ion energy is well-controlled and controllable via calibration of the power supply.

It is important to note that a pulsed supply also allows substrate charge neutralization. Electrons readily available from the plasma neutralize the wafer during the off period between pulses without the need to include special apparatus (e.g. electron flood). Q_{BD} data from antenna devices show no dielectric charging damage for a variety of PLAD implants typical for gate electrode doping [6-7].

Angle Control

One often-overlooked and beneficial aspect of PLAD is its self-alignment at zero tilt. Since ions must follow the electric field generated during the DC pulse, the average trajectory of the ions is by definition normal to the wafer surface. No alignment of either the wafer or the plasma source to an external reference point is required to achieve this characteristic.

Furthermore, by incorporating a shield ring, biased to the same potential as the wafer, the angular integrity is completely extended to the wafer edge. The shield ring also provides a convenient location for Faraday aperture(s). Fig. 4 shows an ion trajectory model across a wafer and shield ring assembly.

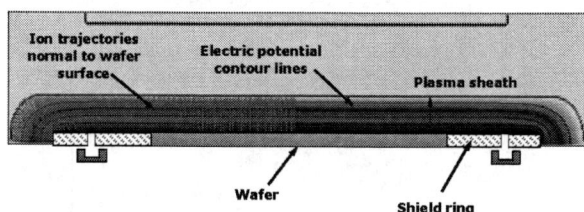

FIGURE 4. Ion trajectory is self-aligned at zero since ions follow the electric field lines. A shield ring extends the trajectory alignment beyond the wafer edge.

Contamination Control

Achieving low defect levels is just as important for the integrity of the implantation process. Particles and surface metal contamination, in particular, are critical areas that must be controlled for PLAD to be viable in manufacturing.

Particle Control

In general, hydride chemistries such as B_2H_6 tend to produce a net depositing environment in a plasma system, increasing deposition on the process chamber walls with every implant. Eventually, the deposited film will flake or otherwise separate from the chamber surfaces and emit particles. This phenomenon, known as a particle excursion, is quite well known in plasma CVD tools. Various reactor design considerations are effective at extending the number of wafers processed before an intervention is required. For PLAD, implementing an automated in situ clean is most effective. Important considerations for the in-situ clean process include the chemistry, plasma power, pressure, gas flow, wafers between cleans, and endpoint methodology.

The necessary cleaning interval may be determined empirically. Fig. 5 shows particle data collected periodically during an extended B_2H_6 PLAD run without in situ cleaning. It is observed that soon after the first particle excursion, additional excursions occur. Repeated tests establish the required interval to eliminate excursions to chamber film buildup.

Closed-loop control of the plasma clean is desirable to ensure that all the deposited film is completely removed and to eliminate excessive cleaning that costs time and wears chamber components. NF_3 is a particularly useful cleaning agent for PLAD technology since it reacts with deposited dopant elements, does not aggressively attack common chamber materials, and provides a strong endpoint detection signal.

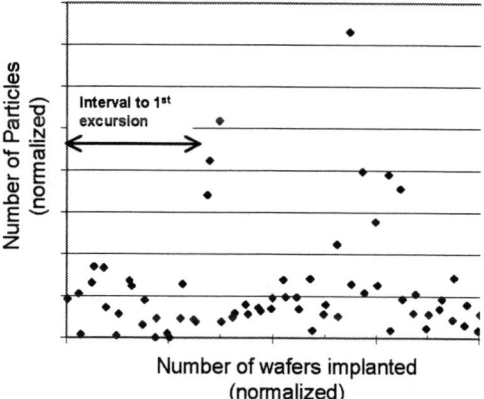

FIGURE 5. Wafer particle counts for an extended run of B_2H_6 PLAD implants. The interval to first excursion helps determine the required in situ cleaning frequency.

In plasma, NF_3 breaks down as follows:

$$2NF_3 (g) + e^- \rightarrow 6F + N_2 (g) \quad (1)$$

The F species is highly reactive and efficiently removes B deposits via:

$$B (s) + 3F \rightarrow BF_3 (g) \quad (2)$$

Thermal and ionic bombardment by the plasma provides the energy to drive this reaction. BF_3, a gas under these conditions, is then readily pumped out of the reactor.

F-based plasmas readily allow for closed-loop control using optical emission spectroscopy (OES). One implementation monitors the change in F optical emission intensity as cleaning progresses, as illustrated in Fig. 6. At the beginning of the clean, little F is present since it reacts with B according to Eqn. 2. The free F intensity increases with time as B in the chamber is consumed. By monitoring the derivative of the F intensity, OES easily detects the cleaning endpoint. The cleaning plasma then continues for a brief overetch period to ensure cleaning is complete in all regions of the chamber.

Surface Metal Contamination

After NF$_3$ cleaning, all deposited films within the process chamber are eliminated and the bare chamber, usually Al, is exposed.

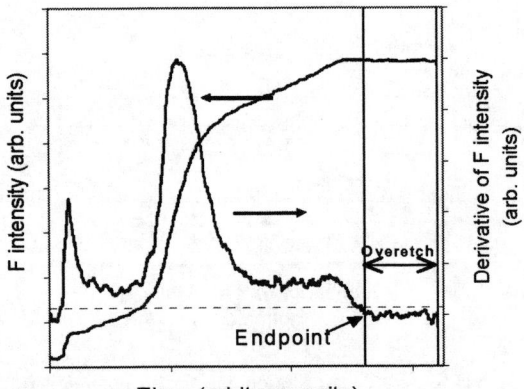

FIGURE 6. Optical emission spectroscopy using F signal, both absolute intensity and derivative signal from an NF$_3$ plasma in a PLAD chamber after multiple implantations.

The depositing nature of dopant hydride chemistries can be used to season the chamber and cover exposed Al prior to implanting product wafers. For example, before implanting with B, B$_2$H$_6$ plasma conditions the chamber immediately after NF$_3$ cleaning, coating all chamber components and virtually eliminating exposure to any metal components during the subsequent implant. The chamber environment, in effect, is composed of the same material as the implant species.

PRODUCTION SIMULATION RESULTS

A production simulation indicates whether all process control parameters are achieved simultaneously for real processes at production throughput. Results from such an extended test for an advanced DRAM electrode gate doping process are included in Fig. 7. Process conditions were: 6kV B$_2$H$_6$ (15% in H$_2$, 6 mTorr) at 1.0×10^{16}/cm^2. The activation anneal for sheet resistance was 965°C, 20s after a dry/wet post-implant clean. Dose uniformity, repeatability, particles, and metals are similar to BL tool performance. Implant time for this process is <20s per wafer.

ACKNOWLEDGMENTS

We gratefully acknowledge S. Qin and A. McTeer for many helpful technical discussions.

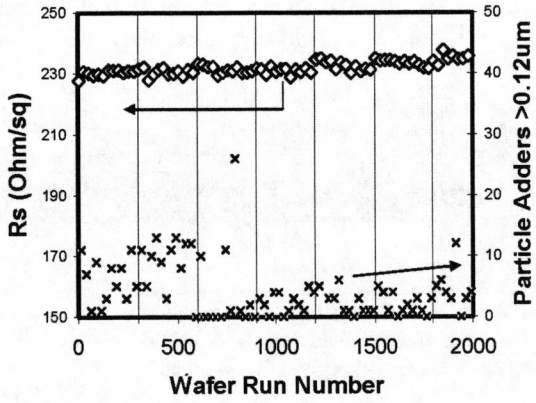

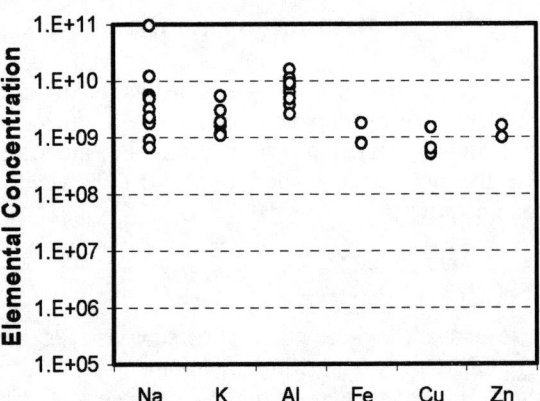

FIGURE 7A,B. Sheet resistance (Rs), added particles, and metal contamination results for a PLAD p-poly gate doping extended run. The vertical blue lines in Figure 7A, indicate in situ cleaning cycles.

REFERENCES

1. A. Renau and J. Scheuer, "Comparison of Plasma Doping and Beamline Technologies for Low Energy Ion Implantation", Proc. 2002 Int. Conf. Ion Implant. Tech., Taos, NM, Sept., 2002.
2. S.R. Walther, S. Mehta, U. Jeong, and D. Lenoble, "Formation of extremely shallow junctions for sub-90nm devices," Surface and Coatings Technology **186**, 68. (2004).
3. Z. Fang, E. Arevalo, T. Miller, H. Persing, E. Winder, and V. Singh, Ext. Abstr. 5th International Workshop on Junction Technology, June 2005, p. 67.
4. Y.S. Kim, et al, "Local-Damascene-FinFET DRAM Integration with p+ Doped Poly-Silicon Gate Technology for sub-60nm Device Generations," IEDM 2005, Washington, DC, Dec., 2005.

5. R. Liebert, S. Walther, S. Felch, Z. Fang, B. Pedersen, O. Pedersen, D. Hacker, "Plasma Doping System for 200 and 300mm Wafers", Int. Conf. on Ion Implant. Tech., 2000, 472.
6. Z. Fang, T. Miller, E. Winder, H. Persing, E. Arevalo, A. Gupta, T. Parrill, and V. Singh, "B_2H_6 PLAD Doped PMOS Device Performance," Int. Conf. on Ion Implant Tech. 2006, Marseilles, France, June, 2006.
7. D. Bigarella, V. Soncini, D. Raj, V. Singh, and S. Walther, "PLAsma Doping for P+ Junction Formation in 90nm NOR Flash Memory Technology," Int. Conf. on Ion Implant Tech. 2006, Marseilles, France, June, 2006.

Gate Dielectric Damage Due To High-Tilt Implant

S. B. Felch, R. Hung, B. Ninan, M. Smayling, N. Toshiyuki, H. Chen, and C-P. Chang

Applied Materials, 974 E. Arques Ave., M/S 81280, Sunnyvale, CA 94085 USA

Abstract. This paper reports an assessment of the gate dielectric damage caused by high-tilt implants using MOS capacitors fabricated with 50 Å SiO_2 and doped polysilicon gates. Capacitor area, structure, and perimeter-to-area ratio were varied to enable identification of the implant damage contributions. The implants studied were typical PMOS source/drain extension conditions that would be used with a diffusion-less anneal. Tilt angles up to 40 degrees were evaluated. Current-voltage sweeps from 0V to -8V were performed to characterize the dielectric quality of the capacitors. Some results have confirmed expectations, as BF_2 implants showed more damage than B implants of an equivalent energy. Additional data shows the dependence of the damage on tilt angle and polysilicon re-oxidation thickness.

Keywords: Dielectric damage, high tilt implant.
PACS: 61.72.Tt, 85.40.Ry

INTRODUCTION

Implantation of source/drain extensions at high tilt angles is being considered as a possible solution to obtain the necessary gate/extension overlap with the diffusion-less anneals required for the ultra-shallow junctions of sub-45 nm technology nodes [1,2]. However, the implant trajectories with high tilt angles penetrate further into the gate stack and might cause unacceptably high levels of dielectric damage [3]. In order to understand the impact of high-tilt implants, metal-oxide-semiconductor capacitors have been fabricated and characterized after implant. The effects of implant species, implant tilt angle, post-implant annealing, and polysilicon re-oxidation thickness have been evaluated.

EXPERIMENT

The gate dielectric damage caused by high-tilt implants was assessed using metal-oxide-semiconductor (MOS) capacitors fabricated with 50 Å SiO_2, 1500 Å doped polysilicon, and 450 Å high-temperature oxide. A thicker gate oxide was chosen to minimize the leakage current without implant so that the sensitivity to implant damage would be enhanced. Most of the capacitors had a 60 Å polysilicon reoxidation, although two wafers studied only had 20 Å polysilicon reoxidation. Capacitor area, structure, and perimeter-to-area ratio were varied to enable identification of the implant damage contributions. Checkerboard photoresist patterning was also used to generate in-wafer control on half of the dies in each wafer and expose the other half of the dies to implant.

The implants studied were typical PMOS source/drain extension conditions that would be used with a diffusion-less anneal. These included 1 keV B and 4.45 keV BF_2 at doses of 1×10^{15} at/cm^2, so both had equivalent boron implant energies. Tilt angles up to 40 degrees were evaluated using an Applied Quantum® X Implant system. Half of the dose was implanted at each of two twist angles (90 and 270 degrees) to ensure that the ions were being directed into the longest sidewalls of the capacitors. A few wafers were also annealed after implant to see the residual damage after typical thermal processing. Both a 1050C spike anneal performed on an Applied Radiance*Plus*™ RTP system and a furnace forming gas anneal (FGA) at 425C for two hours were examined.

A current-voltage sweep from 0V to -8V was performed to characterize the dielectric quality of the capacitors. The current flow across the capacitor was measured as a ramped voltage was applied. This measurement was then conducted on every die for the selected test structures on each wafer.

RESULTS

The starting quality of the capacitors was assessed by measuring the current-voltage characteristics for unimplanted plate capacitors, which are large rectangular capacitor structures. The data for control plate capacitors at the center of each wafer studied are presented in Fig. 1. All of the curves are nearly identical, demonstrating good repeatability for the control devices. The non-implanted controls also show good uniformity (data not shown here). So, the starting capacitors have sufficient quality to enable leakage current increases due to implant damage to be detected.

The effect of implant on the current-voltage characteristics of corner-intensive capacitors (CIS) is illustrated in Fig. 2. These capacitors are long, narrow stripes with many crosses to provide long perimeters with many corners. The upper three curves are for devices implanted with BF_2^+ at a tilt angle of 0°, while the lower three curves are for unimplanted controls. Data for three different perimeter lengths (18887, 37587, and 94010 μm) are given. The implanted devices clearly show about three orders of magnitude higher leakage current at low voltage, which demonstrates that the implant is damaging the dielectric. In addition, the leakage current increases as the perimeter of the device increases, suggesting that the damage is caused by the implantation of the ions into the sides of the capacitors. This increase is roughly proportional to the increases in both the perimeter and number of corners. However, Fig. 3 shows that stripes and CIS have similar leakage currents, demonstrating that implantation into the corners does not contribute significantly to the leakage current.

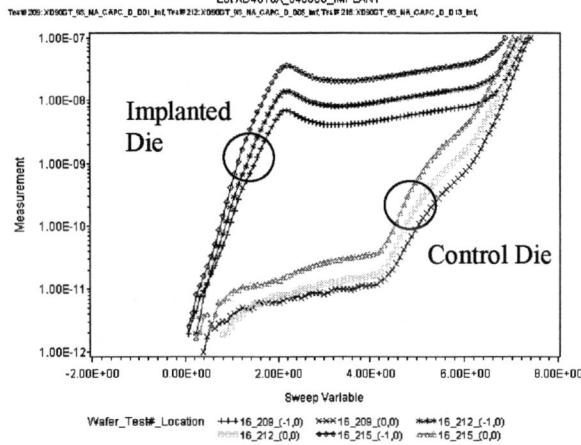

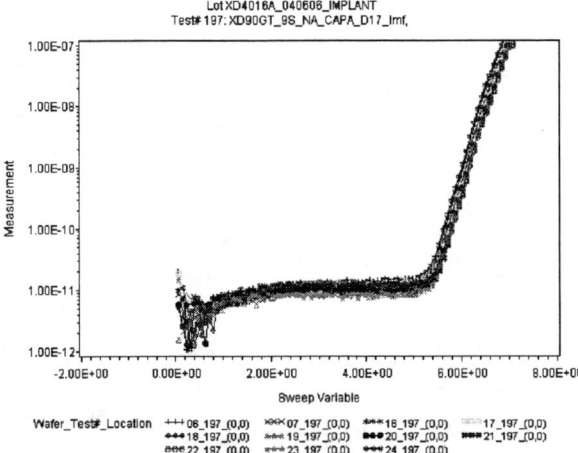

FIGURE 1. Current-voltage characteristics for the unimplanted, center plate capacitors on all wafers, showing good repeatability of the control devices.

FIGURE 2. Current-voltage characteristics for corner-intensive capacitors with three different perimeter lengths (18887, 37587, and 94010 μm) that have been implanted with BF_2^+ at a tilt angle of 0° and for the unimplanted controls.

Figure 3 shows the effect of implant species on the dielectric damage. Current-voltage characteristics for BF_2^+ and B^+ implants at 40° tilt are plotted for plate, stripe, and corner-intensive capacitors. The active perimeter length for the stripe and corner-intensive structures is 94010 μm. BF_2^+ implant caused higher leakage current than B^+ implant for both the stripe and corner-intensive capacitors, which can be explained by the greater physical damage to the dielectric produced by implant of the higher-mass species. However, the I-V curves for the plate capacitors, which have minimal perimeter, show little difference between the two species and are actually quite close to those for unimplanted control devices. This observation confirms that the dielectric damage is caused by implant into the sides of the devices and that traditional charging damage is not detectable for these implants.

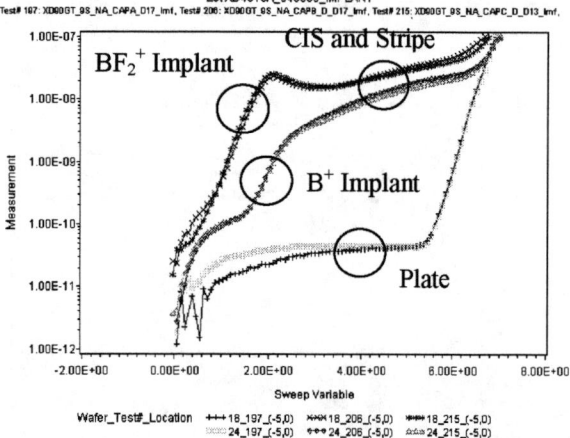

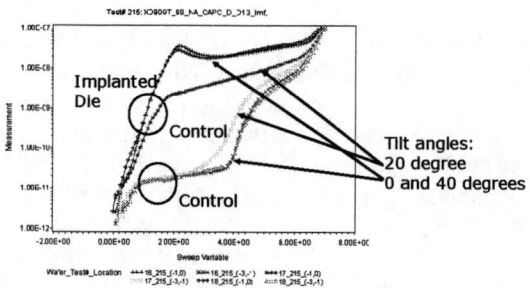

FIGURE 3. Current-voltage characteristics for BF_2^+ and B^+ implants at 40° tilt for the three capacitor structures. The active perimeter length for the stripe and corner-intensive structures is 94010 μm.

FIGURE 4. Current-voltage characteristics for corner-intensive capacitors that have been implanted with BF_2^+ at tilt angles of 0, 20, and 40°.

The tilt angle effect on the leakage current for corner-intensive structures that have been implanted with BF_2^+ is shown in Fig. 4. Implants at 0° and 40° tilt result in similar I-V characteristics, which represent significantly higher leakage currents than the control devices. However, implant at 20° tilt produces about one order of magnitude less leakage current and an I-V curve with a different shape, closer to that seen for B^+ implant. This observation is quite surprising, as increasing tilt angle increases the direct lateral penetration of the implanted ions into the sides of the dielectric which should create more damage. Nevertheless, ions that have been directly implanted into the surface silicon below the gate and then scattered up into the overlying dielectric also enter the dielectric and produce damage. The trend between 20° and 40° is expected, as the deeper penetration for the latter case would lead to deeper ion scattering and a larger amount of damaged dielectric. However, 0° tilt does produce ion channeling along the horizontal <110> axis that goes along the transistor channel, which could also lead to deeper ion scattering, more damage and higher leakage current [4]. So, the results of Fig. 4 suggest that the dielectric damage is dominated by scattered ions originally implanted into the silicon, rather than ions directly implanted into the dielectric, and that 20° tilt minimizes the damage. One should note that these implants were performed into crystalline silicon, so different results might be obtained if pre-amorphized silicon is used.

Figures 5 and 6 display the current-voltage characteristics for the three types of capacitors after implant with BF_2^+ at a tilt angle of 40° and after spike anneal and forming gas anneal, respectively. The pairs of curves for each capacitor type are nearly identical. Thus, neither the activation spike anneal nor the forming gas anneal was able to reduce the low field leakage due to tilted implant damage.

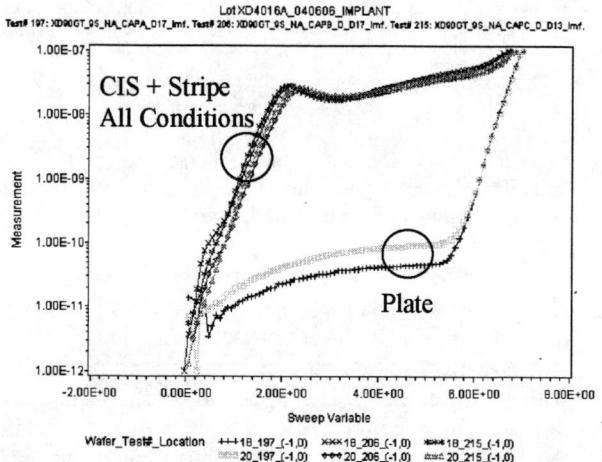

FIGURE 5. Current-voltage characteristics for the various capacitors that have been implanted with BF_2^+ at a tilt angle of 40° after implant and after spike anneal. The active perimeter length for the stripe and corner-intensive structures is 94010 μm.

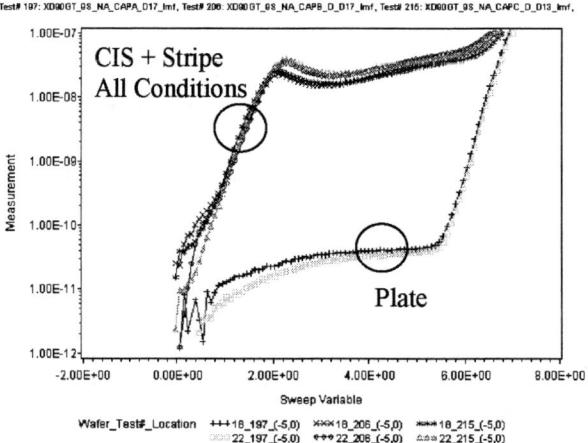
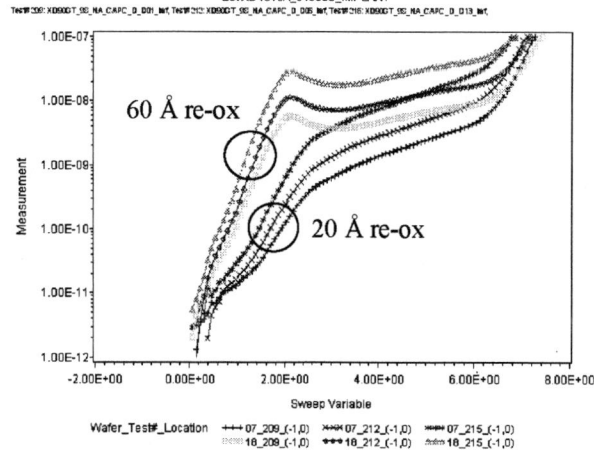

FIGURE 6. Current-voltage characteristics for the various capacitors that have been implanted with BF_2^+ at a tilt angle of 40° after implant and after forming gas anneal. The active perimeter length for the stripe and corner-intensive structures is 94010 µm.

FIGURE 7. Current-voltage characteristics for the corner-intensive capacitors with 20 and 60 Å re-oxidation and three different perimeter lengths (18887, 37587, and 94010 µm) that have been implanted with BF_2^+ at a tilt angle of 40°.

The current-voltage characteristics for the corner-intensive capacitors with 20 and 60 Å re-oxidation and three different perimeter lengths that have been implanted with BF_2^+ at a tilt angle of 40° are presented in Fig. 7. The most striking observation is that the leakage current is about one order of magnitude higher for thicker re-oxidation. In addition, the 20 Å re-oxidation I-V curve is remarkably similar to that for 20° tilt with thicker re-oxidation. This counter-intuitive result can be explained by realizing that thicker re-oxidation also means that the silicon surface has a thicker oxide layer. This will reduce the channeling of the implanted ions, forming a shallower implant with more ions located closer to the surface. In turn, more ions are then available for scattering into the overlying dielectric, causing greater damage and higher leakage.

CONCLUSIONS

An MOS capacitor study of dielectric damage due to high tilt implant showed more damage from BF_2^+ implant than from B^+ of the same equivalent energy and that neither 1050C spike or forming gas anneal removes the damage. In addition, 20° tilt produces much less damage than either 0° or 40° tilt, which implies that the damage is dominated by surface scattering rather than direct trajectories. Finally, higher leakage produced with thicker polysilicon re-oxidation is consistent with the hypothesis that surface scattering dominates the dielectric damage.

ACKNOWLEDGMENTS

The authors acknowledge Sunder Thirupapuliyur of Applied Materials and Victor Moroz of Synopsys for valuable technical discussions.

REFERENCES

1. J. Hwang, H. Kennel, P. Packan, M. Taylor, M. Liu, R. James and M. Kuhn, in *Advanced Short-Time Thermal Processing for Si-Based CMOS Devices*, edited by F. Roozeboom, E. P. Gusev, L. J. Chen, M. C. Ozturk, D.-L. Kwong and P. J. Timans (The Electrochemical Society, Pennington, 2003), pp. 35-42.
2. J. O. Borland et al., *Solid State Technology*, 52-58 (June 2003).
3. Y-H. Lee, R. Nachman, S. Hu, N. Mielke, and J. Liu, *IEDM 2004 Tech. Dig.*, IEEE, 2004, pp. 481-484.
4. S. Tian, V. Moroz, and N. Strecker, in *Silicon Front-End Junction Formation – Physics and Technology*, Vol. 810, edited by P. Pichler, A. Claverie, R. Lindsay, M. Orlowski, and W. Windl (Materials Research Society, Warrendale, PA, 2004), pp. 287-292.

High Current Implant Precision Requirements for Sub-65 nm Logic Devices

Yuri Erokhin[1], Terry Romig[2], Elshot Kim[2], JieJie Xu[3], Baonian Guo[1], Jinnig Liu[1], Kyu-ha Shim[1] and Peter Nunan[1]

[1]*Varian Semiconductor Equipment Associates, Inc., 35 Dory Road, Gloucester, MA 01930, United States*
[2]*TI DMOS6, 213011 TI Blvd., North Expressway, Dallas, TX 75243*
[3]*TI KFAB, 313575 TI Blvd., North Expressway, Dallas, TX 75243*

Abstract. As CMOS devices shrink they become increasingly sensitive to variations of ion beam angular properties and beam current density. In sub-65 nm devices beam divergence and beam steering variations at levels commonly seen in high current implanters for Source/Drain Extension (SDE) implants could significantly shift device characteristics compromising yield and robustness of manufacturing process. In this paper we review the implant precision requirements for Source/Drain Extension (SDE) formation for sub-65nm node devices. TCAD simulation was used to analyze the effects of beam divergence and steering errors for an on-axis (0°) SDE implant on sub-65 nm NMOS HP devices. Effects of energy contamination introduced along with decelerated low energy ions in p-type SDE implants in PMOS devices is also discussed. Response of device electrical characteristics to variation of beam angle properties is quantified and beam angle control requirements for state-of-the-art ultra-low energy implanters formulated.

Keywords: ion implantation, VLSI technology, silicon, TCAD modeling
PACS: 61.72.Tt, 85.40.Ry

INTRODUCTION

Traditionally, an implant process is defined by the implant species' atomic mass, energy and dose and angular wafer positioning with respect to the projected beam direction (beamline axis). For older device generations (>0.18 μm) meeting X_j and R_s targets for SDE junctions using bare wafer measurements would virtually guarantee meeting target device characteristics. As device dimensions shrink, this is no longer the case. Precise control of ion dose, wafer positioning and nominal implant energy alone is no longer adequate for fabricating devices with highest performance possible [1-3], while simultaneously achieving stable manufacturing process. In addition to traditional implant parameters, now one must also focus on process variables previously considered insignificant second tier beam properties. Of those, most important are beam divergence, variations of implant angle with respect to the wafer surface due to changes of beam trajectory in implanter beamline (beam steering) and beam current density (dose rate). Thus, novel design requirements for advanced high current implanters dictate development of capabilities to set, control and interlock these beam characteristics.

Another challenge unique to beam transport in ultra-low energy implanters is overcoming beam transmission loss due to space charge induced beam expansion and resulting loss of implanter productivity. It became a common approach to extract and transport ions through beamline at energy higher than final recipe value and to decelerate ion beam near the implanted wafer. In this approach, a small fraction of ions could become neutralized due to interaction with beamline residual gas, not affected by deceleration potential and implanted into the wafer with energy exceeding target value, resulting in "energy contamination". This effect leads to increased depth of as-implanted SDE junctions and compensation of a fraction of halo implant. On a device level it results in deterioration of device short channel effect (SCE) characteristics, manifested in V_{th} shift and increased I_{off} current. In this paper, we present results of TCAD modeling of the effects of beam divergence, beam steering and energy contamination on key electrical characteristics (V_t, I_{dsat}, and I_{off}) of sub-65 nm high performance (HP) NMOS and low power (LP) PMOS devices. We use these modeling data to formulate top level beam angle control requirements in advanced high current ultra-low energy implanters.

SIMULATION

In this section we describe methods and assumptions of our modeling of the device effects of beam divergence, beam steering and energy contamination. Beam divergence is defined as angular dispersion of ion trajectory within the beam. Assuming a Gaussian distribution of beam intensity across each beamlet, a series of twenty-one beamlets with varying doses and angles are introduced at the SDE implant energy to compose a Gaussian-like beam with designated divergence. Beam steering represents the offset between beam centroid and the wafer normal direction. In every modeled case the total dose that the wafer receives remains constant and the variation of device characteristics is the result of divergence change and/or steering angle offset. A Monte Carlo implant was used for the SDE implant simulation and implant tables were used for other implants.

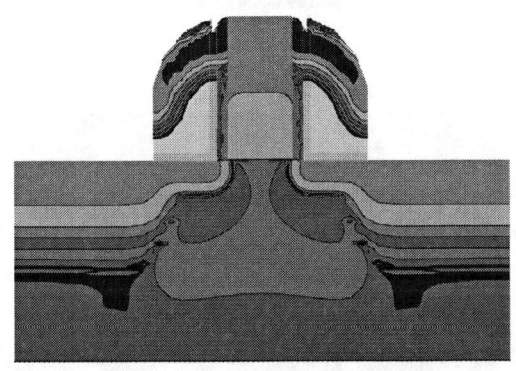

Device Dimensions	
L_{Poly} [nm]	26
T_{Poly} [nm]	50
T_{Gox} [nm]	0.9
SDE off-set spacer thickness [nm]	4.5
S/D Sidewall spacing [nm]	26

FIGURE 1. Cross section of simulated NMOS device with net doping contours and device dimensions.

A high performance 65nm NMOS with a physical gate length of 26nm, gate oxide thickness and operating voltage specified by the ITRS Roadmap was used in this simulation. Optimization of simulated process characteristics was made to closely match device to V_t, I_{dsat} and I_{off} values listed in ITRS. A V_t roll-off curve was produced to validate the simulation conditions. Figure 1 shows the cross-section of the device simulated and its dimensions. The contour areas represent variation in the net doping concentration.

A SDE implant with As 1keV $2x10^{15}cm^{-2}$ was used in this simulation. Beam divergence was varied between one sigma of 0.6° to 10°. Beam incident angle was varied between 0° and 3°. NMOS device was chosen due to its the higher sensitivity of variations of 2D As profile caused by beam angle variations.

The effect of energy contamination was simulated by intentionally introducing energy contaminants along with the monoenergetic SDE implant. A 65nm Low Power PMOS device based on ITRS Roadmap was used in the simulation. The SDE implant condition is B 500eV $1x10^{15}cm^{-2}$. The added contaminants have an energy of 3keV or 5keV and varied from 0.1% to 0.5% of the nominal implant dose. The physical poly gate length is 38nm. In a separate simulation, to compensate the increased I_{dsat} and I_{off} due to the presence of energy contamination, halo dose was increased proportionally. It is shown that such halo dose adjustment is sufficient in suppressing the device parametric shift from moderate levels of energy contamination.

RESULTS AND DISCUSSION

A. Effect of Beam Divergence

Figure 2 shows the net channel doping concentration as a function of the position along the channel. Two cases with divergence at one sigma of 0.6° and of 5° were compared. Not only the effective channel length is different in the two cases, but the lateral abruptness is also different, especially near the junction regions. Lower divergence presents higher junction abruptness which can translate into lower overlap capacitance, reduced SDE parasitic resistance and, ultimately higher device speed [4,5].

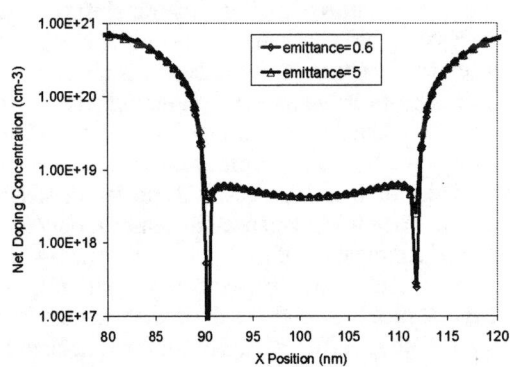

FIGURE 2. Net doping profiles across the channel beneath the gate for two beam divergences: 0.6°, 1σ, and 5°, 1σ.

Saturation threshold voltage (V_t) and leakage current (I_{off}) were shown to have a non-monotonic dependence on beam divergence, as shown in Figure 3 (a) and (b). A larger divergence is believed to contribute to both a dose loss into the gate edge and a dopant lateral encroachment underneath the gate.

While dose loss at the edge of the gate due to gate shadowing results in higher V_t and lower I_{dsat} and I_{off}, increased lateral encroachment yields the opposite trend. The net result of the two effects produces non-monotonic V_t dependence versus beam divergence. For the particular device structure modeled, at a divergence less than ~3°, dose loss dominates the trend and V_t increases with increasing divergence. At divergence greater than ~3°, lateral encroachment reverses the trend leading to V_t reduction with divergence.

Results have also shown similar trend in I_{dsat} and C_{ov}. Transistor delay from small signal AC analysis indicates that higher divergence increases device delay.

implant total implant dose is split in two (bi-mode) to four (quad mode) equal implants performed with wafer rotated by 180° (bi-mode) or 90° (quad-mode) between each step. Figure 5 shows I_{dsat} variation versus beam incident angle for a single mode implant and for a bimode implant, where a small tilt angle ranging of 0° to 3° was employed from device source and drain sides. Beam divergence was set at 1°, 1σ. SDE implant in bi-mode implant shows significantly reduced device sensitivity to beam steering errors and suppressed device skew.

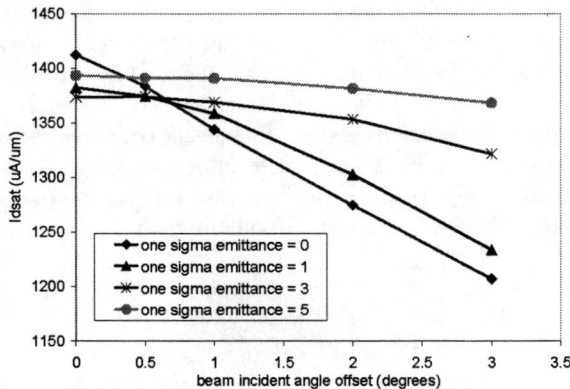

FIGURE 4. I_{dsat} as a function of beam incident angle offset for various beam divergence in a single mode (non-quad) implant.

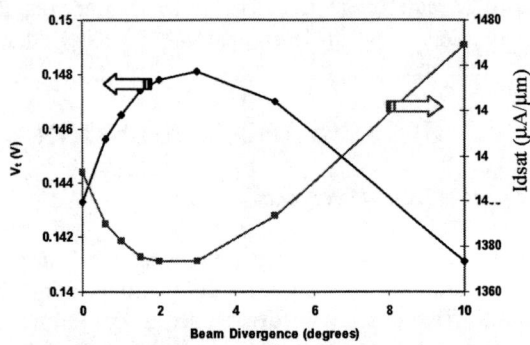

FIGURE 3. Effect of beam divergence on V_t and I_{off}.

B. Effect of Beam Angle Steering

Besides the beam divergence effect, the offset angle of beam centroid from the desired 0° tilt has a strong effect on device performance and introduces both device asymmetry and parametric shift. Figure 4 (a) shows I_{dsat} as a function of beam steering angle offset for various beam divergences. It is observed that I_{dsat} decreases as beam steering angle increases due to both effective channel length increase and insufficient gate/drain overlap. I_{off} performance (not shown) has trend similar to I_{dsat}. I_{dsat} sensitivity to beam steering reduces as beam divergence increases, but at an expense of degradation of I_{on}-I_{off} device merit figure.

For a single mode implant at a small tilt angle error, the source side and the drain side of the device receive the beam differently, i.e. an encroachment of one side versus an undercut on the other. This causes undesirable device asymmetry (skew). Figure 4 (b) shows I_{dsat} skew as a function of beam incident angle offset. Again, larger divergence angle displays smaller sensitivity, but degrades device performance.

The device sensitivity to beam incident angle offset in a case of small beam divergence is significantly reduced by utilizing a multimode implant. This implant mode can be easily implemented in single wafer high current implanters. In multimode

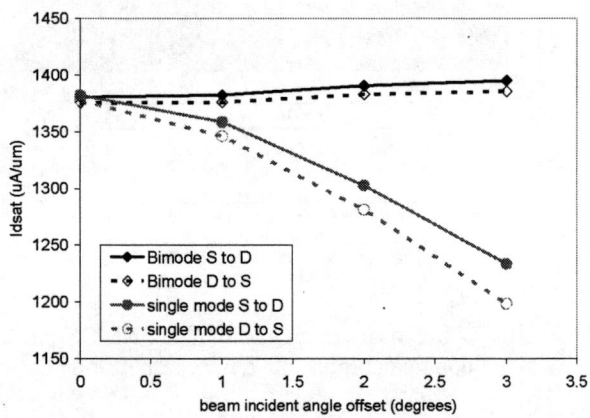

FIGURE 5. I_{dsat} as a function of beam incident angle offset for single mode and bimode implant.

While I_{dsat} was used in this section to demonstrate the effect of beam angle steering, other parameters such as I_{off} and V_t show similar trend. Quadmode implant is the recommended mode for SDE considering both the vertical and horizontal alignment of devices on a wafer. While a tightly distributed beam with accurate angle control is desired, quadmode implant ensures superior skew control in real life IC manufacturing.

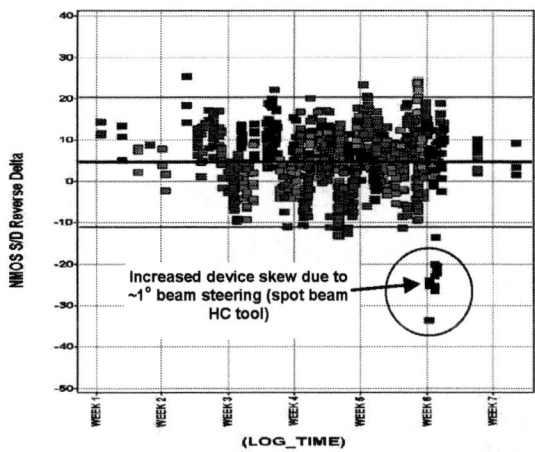

FIGURE 6. Production SPC 90 nm NMOS device skew chart with As SDE implant performed using four different high current tools.

Effect of beam steering during nSDE implant in single mode on device skew in an environment of high volume manufacturing illustrated ion Figure 6. Device lots produced using one of the tools (blue squares) show skew values outside of the allowed process control limits. Vertical beam steering ranging from 0.75 to 1° has been identified as a root cause of the increased device skew and corresponding yield loss.

C. Effects of Energy Contamination

The effect of energy contamination (EC) was simulated for SDE implant for a PMOS device. Figure 7 shows I_{dsat} dependence on amount of energy contamination for 3keV or 5keV implant added to 500eV 1×10^{15}cm^{-2} B$^+$ SDE implant. At fixed halo implant conditions I_{dsat} (and I_{off}) increase linearly with the EC dose. A 0.5% and 3.7% shift of I_{dsat} and I_{off}, respectively were observed per 0.1% EC dose increment for 5keV contaminant energy. When halo compensation is applied, the shift in I_{dsat} can be

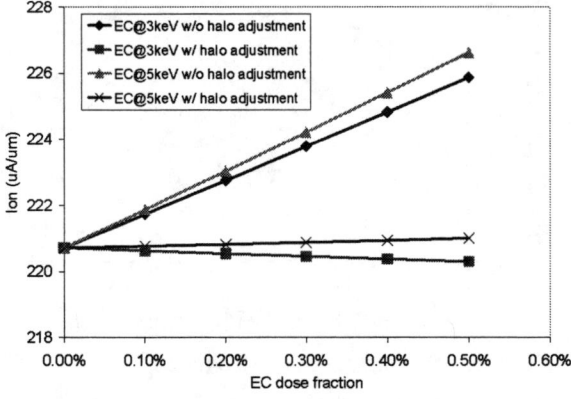

FIGURE 7. Effect of energy contamination on I_{dsat} with and without halo dose compensation.

brought back to its original value. For the PMOS device modeled, 0.36% halo dose adjustment per 0.1% EC dose increase was required to bring I_{dsat} to its original value. Thus, for SDE implant process tuned with 0.3% EC varying from 0.2% to 0.4% (+/-30% change) variation of I_{dsat} is estimated to be less than 0.5%, 3s, which is well within UCL/LCL margins of mature manufacturing process.

CONCLUSIONS

TCAD simulation study of the effects of beam divergence, beam steering and energy contamination during SDE implant on characteristics of sub-65nm CMOS devices is presented. Results demonstrate that SDE implants in sub-65 nm devices require that beam divergence must be controlled with accuracy better than 1°. SDE implants performed with low divergence ion beams form junctions with superior lateral abruptness yielding improvement of device I_{dsat}/I_{off} and C_{ov}, and, ultimately increased device speed. To maintain the required level of device symmetry, beam steering must be controlled to <0.25° for single step SDE implants, but is relaxed to ~1° for quad mode SDE implants with 90° wafer rotation between each step. Thus, design of state-of-the-art high current ultra-low energy ion implanters must integrate recipe driven tuning, monitoring and interlocking beam angle characteristics. To minimize device sensitivity to beam steering implanter, must enable quad implants with 90° wafer rotation fast enough to prevent tool throughput degradation. Moderate (<0.5%) levels of energy contamination in p-SDE implants in PMOS device could be compensated by minor adjustment of halo implant dose resulting in negligible variation of device I_{on} and I_{off} (not shown) characteristics, as long as EC level is controlled to <+/- 0.1% of its absolute value.

ACKNOWLEDGMENTS

The authors would like to thank Greg Redinbo for valuable discussions.

REFERENCES

1. U. Jeong et al, "Effects of Beam Incident Angle Control on NMOS Source/Drain Extension Applications", Proceedings of IIT-2002, p.64-68
2. H.-J.L. Gossmann, "Ion Implantation in Advanced Planar and Vertical devices", Proceedings of IIT-2004, p.1-5.
3. K.-H. Shim et al, "Impact of Dose Rate Effects and Damage Engineering on Device Performance", This Proceedings.

Production-Worthy USJ Formation by Self-Regulatory Plasma Doping Method

Y. Sasaki[a], H. Ito[a], K. Okashita[a], H. Tamura[a], C. G. Jin[a], B. Mizuno[a], T. Okumura[b],
I. Aiba[d], Y. Fukagawa[c], H. Sauddin[d], K. Tsutsui[c] and H. Iwai[d]

[a] Ultimate Junction Technologies Inc., 3-1-1 Yagumo-Nakamachi, Moriguchi, Osaka, 570-8501 Japan
[b] Matsushita Electric Industrial Co., Ltd., Kadoma, Osaka, Japan
[c] Interdisciplinary Graduate School of Sci, and Eng., Tokyo Institute of Technology, 4259 Nagatsuta, Midori-ku, Yokohama, 226-8502 Japan
[d] Frontier Collaborative Research Center, Tokyo Institute of Technology, 4259 Nagatsuta, Midori-ku, Yokohama, 226-8502 Japan
Tel/Fax: +81-6-6906-6208, Email: sasaki@ujtlab.com

Abstract. A new method of plasma doping that achieves tight control on dosimetry and uniformity has been developed. It uses a self-regulatory behavior of plasma processes that brings high accuracy on dose control and uniformity within 1.5%. The largest advantage of this self-regulatory plasma doping (SRPD) is that the accuracy of the process control is much less dependent on the uniformity of the plasma, which makes a revolutionary difference to the plasma process as it becomes free from the primary hardware constraint. A typical doping of boron using B_2H_6/He gas mixture at dose of 1×10^{15} ions/cm^2 can achieve a uniformity of less than 1.5% across a 300mm silicon wafer when the plasma uniformity above the wafer plane is as poor as 10%. The SRPD process also forms very abrupt junctions such as less than 2nm/decade at the junction depth of 10nm due to an instantaneous amorphization of the wafer surface within the first 5 seconds of the process duration. Combined with the throughput advantage at low energy against the conventional ion implantation, the SRPD offers an ideal performance for USJ formation for 45nm technology node and beyond.

Keywords: plasma doping, self-regulation, dose control, uniformity, ultra-shallow junction (USJ), amorphization
PACS: 52.40.Hf

INTRODUCTION

Plasma doping (PD) has been proposed as one of the most promising methods to replace low energy ion implantation (LII) for the formation of USJ for Xj<15nm. PD has several advantages over LII such as shallowness, steep abruptness, lower Rs and lower cost of ownership due to high throughput and small foot-print. We have demonstrated the actual results of the USJ formation in our previous paper that achieved Xj of 10-15nm and Rs of 870-330Ω/sq. [1-3] (Fig.1). However, there have been several concerns on PD as a production method in terms of metal contamination, uniformity, repeatability, dosimetry and junction leakage. Those concerns were the main focus of this study and the results proved that PD is readily applicable as a production process. A new concept of PD, namely, SRPD has enabled the breakthrough to achieve accurate control on uniformity and dosage.

EXPERIMENTS

Two different types of the plasma equipment were used; Tool-A (Helicon wave) and B (ICP) [1]. The typical range of operation for both tools includes source power of 1 to 2kW and total gas pressure of 0.9 to 2.0 Pa. The gas mixture of B_2H_6/He was used as a source gas. Typical process time ranges between 7 and 60 seconds.

After the doping processes were carried out, the wafers were annealed for the evaluation of Rs and uniformity. The Rs was measured on 121 points on a 300mm wafer using four-point probe with 3mm edge exclusion. The wafers used were <100> n-type bulk Si substrate with a resistance of 10 ohm-cm.

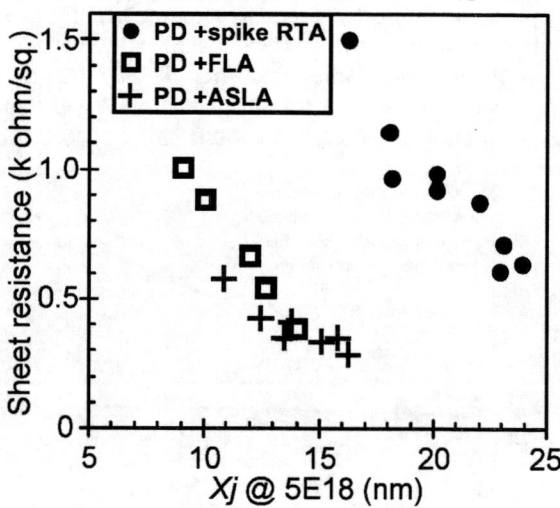

FIGURE 1. Relationship between Rs and Xj on PD using FLA, LA and spike RTA[1, 2].

RESULTS AND DISCUSSION

Determining the dosage has been one of the major difficulties associated with the use of PD. This problem was overcome by introducing the SRPD process. The basic behavior of the SRPD is schematically described in Fig.2. Under a given plasma condition, the boron dose increases rapidly as a function of time in the initial stage. The increase in dose begins to slow down and finally peaks at a unique value in the dose saturation stage, as long as the low gas ratio of B_2H_6/He is used such as below 1×10^{-2}. During this stage, the dose stays almost constant for a period of time (typically 5-15 seconds) within 1.5%, which makes it possible to control the dose with remarkably high accuracy. The value of the saturating dose can be controlled over a wide range typically between 10^{14} and 10^{16} cm^{-2} by changing the gas ratio (Figs.2 and 3). The SRPD process eventually moves to the end stage where the dose starts to decrease due to the self-sputtering effect.

The relatively long saturation time also helps improve the dose uniformity across the wafer. Fig.4 shows the dose and the uniformity plots using Tool B. The dose saturation stage is seen at around time 10. The uniformity was about 2% in the initial stage and was improved steadily as a function of time to 1.0-1.1% towards the dose saturation stage (Figs.4 and 5). Fig.6 shows the typical example of the plasma uniformity measured at approximately 10mm above the wafer plane in Tool B. The plasma density dropped significantly near the edge of the wafer and the overall uniformity at one sigma was 8.8%. Rs values over a 300mm wafer processed by Tool B using the same plasma condition are also plotted in Fig.6 showing the uniformity of 1.0%. This would mean that, even if the dose uniformity in the initial stage is poor, the entire wafer surface reaches the same dose during the saturation interval (Fig.2).

The depth of the dopant profiles is predominantly controlled by the bias potential applied to the wafer as shown in Fig.7. The abruptness of the as-doped profiles is as steep as 2.0nm/decade at 10nm using the SRPD process because of the simultaneous amorphization of the top surface layer. This abruptness is significantly steeper than those by LII and the conventional PD methods [4, 5] (Fig.8). The formation of the well-defined amorphous layer on the surface is seen after the PD process (Fig.9(a)), however, no remaining defects were observed after spike RTA (Fig.9(b)).

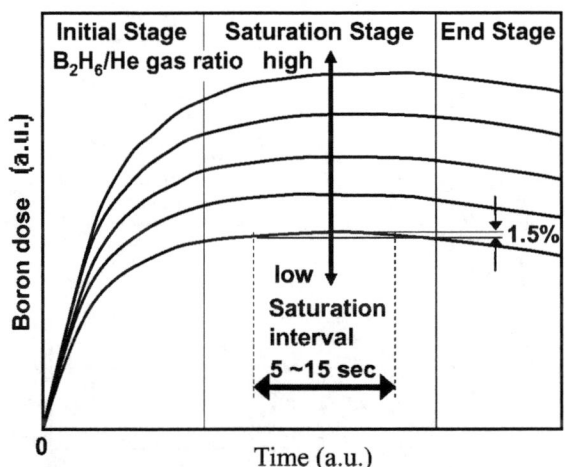

FIGURE 2. The basic characteristics of the SRPD process.

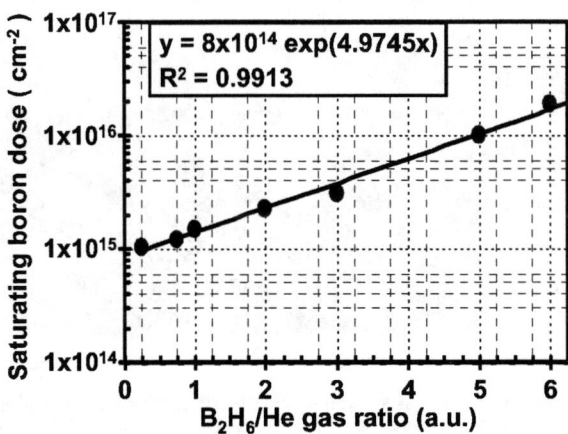

FIGURE 3. Relationship between B_2H_6/He gas ratio and the saturating boron dose.

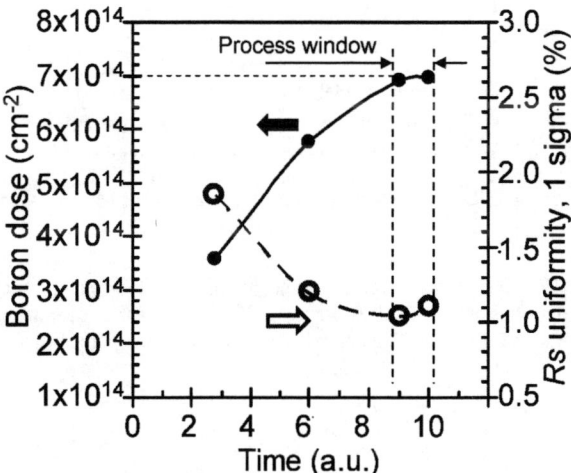

FIGURE 4. The dosimetry and the within-wafer uniformity on Rs of SRPD process performed by Tool B. The anneal condition is 1075°C for 20sec.

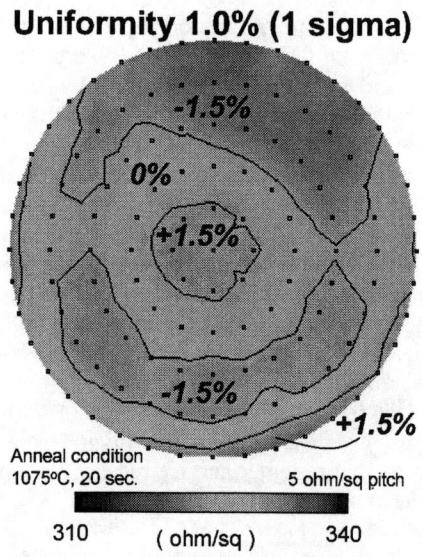

FIGURE 5. The distribution map on Rs of SRPD process performed by Tool B.

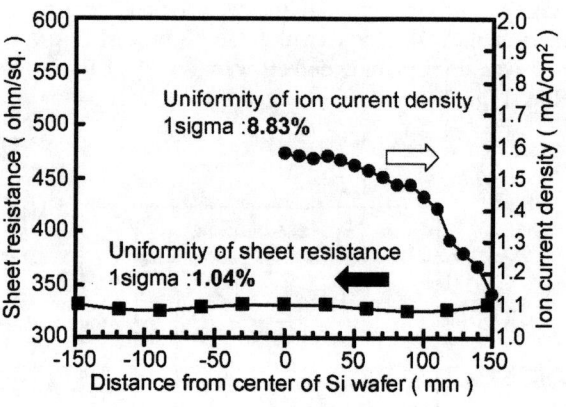

FIGURE 6. The comparison of the ion current density distribution of Tool B using B_2H_6/He gas plasma and the uniformity of Rs. The anneal condition was 1075°C for 20sec.

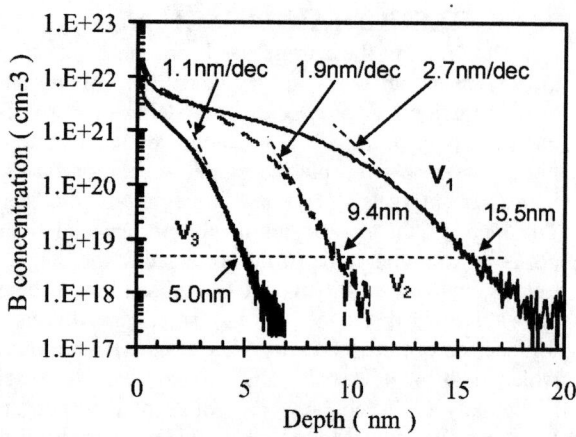

FIGURE 7. SIMS profiles of boron using SRPD at various bias voltages at V_1, V_2 and V_3 where $V_1>V_2>V_3$.

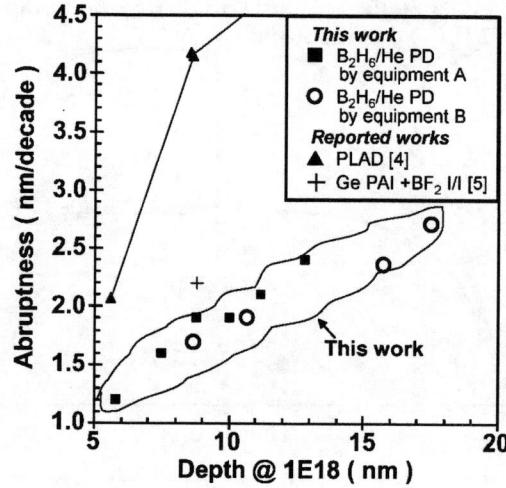

FIGURE 8. The comparison in abruptness between SRPD and the other doping methods [4, 5].

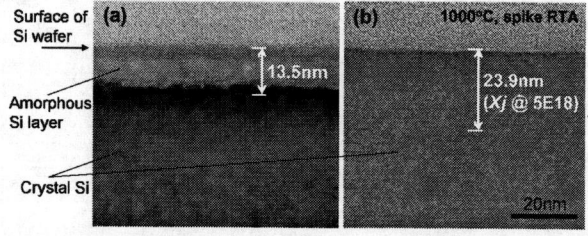

FIGURE 9. Cross sectional TEM images; (a) after PD process and (b) after spike RTA process.

Metal contamination was evaluated by ICP-mass analysis as shown in Table 1. The results of the PD show successful suppression of the contamination down to the same level as LII by conducting an appropriate coating on the PD chamber. The leakage current on pn diode processed by PD was in the order of 10^{-9} A/cm^2 at room temperature at most when the dopant (N_D) concentration in substrate is 4×10^{14} cm^{-3} (Figs.10(a) and (b)). The leakage increases to the order of 10^{-5} A/cm^2 when N_D exceeds 10^{18} cm^{-3} due to the tunneling diode effect (Fig.10(c)). Those values at both low and high N_D conditions are equivalent to those of LII [6].

TABLE 1. Metal contamination on bare Si wafer after PD process at 1×10^{15} cm^{-2} of boron dose and 10nm of the as-doped depth.

(unit : 1×10^{10} cm^{-2})

Fe	Ni	Zn	Cr	Na	K
0.68	< 0.081	0.18	< 0.091	< 0.21	< 0.12

Ca	Al	Mg	Cu	Co
0.3	1.8	0.2	< 0.075	< 0.08

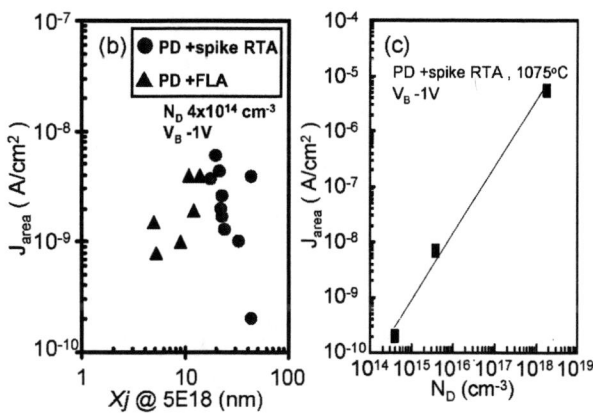

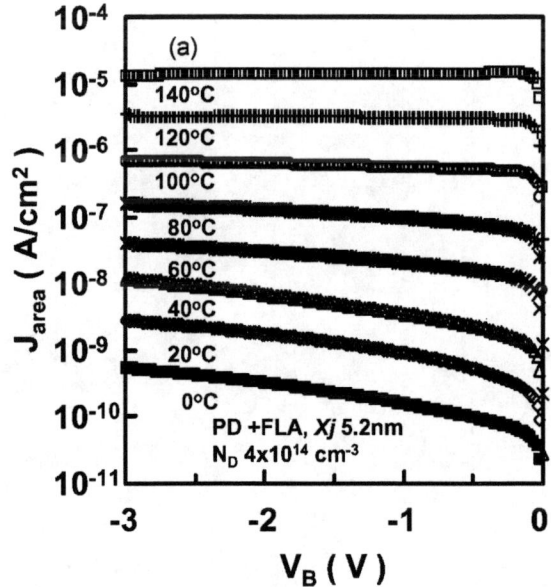

FIGURE 10(a). Temperature dependence of the leakage current on pn diode fabricated by plasma doping process (PD +FLA).

FIGURE 10(b) and (c). Leakage current on pn diode fabricated by plasma doping process; (b) relationship between Xj and the leakage current (PD +FLA, PD +spike RTA) and (c) relationship between N_D and the leakage current (PD +spike RTA).

CONCLUSIONS

Remaining concerns associated with PD as a production process were investigated. It was demonstrated that the uniformity and the dosage were tightly controlled with accuracy of less than 1.5% by introducing a new concept of plasma doping named SRPD. The metal contamination was successfully suppressed to the level of LII by applying appropriate coating on the PD chamber. The junction leakage on pn diode processed by PD was equal to that of LII. The SRPD method can provide an enabling technology for the manufacturing of advanced USJ devices as it achieves the high control accuracy on the process performance while maintaining the overwhelming advantage on throughput at ultra low energy against ion implantation.

ACKNOWLEDGEMENT

This work was partially supported by Special Coordination Funds for Promoting Science and Technology by Ministry of Education, Culture, Sports, Science and Technology, Japan.

REFERENCES

[1] Y. Sasaki, et al., Symp. on VLSI Tech. (2004) 180.
[2] Y. Sasaki, et al., NIM-B 237 (2005) 41–45.
[3] R. Higaki, et al., ESSDERC, p. 569, 2003.
[4] D. Lenoble, et al., IIT 2002, p. 36.
[5] T. Ito, et al., Symp. on VLSI Tech. (2003) 53.
[6] S. Severi, et al., Tech. Dig. of IEDM, p.99, 2004.

PROCESS CONTROL & YIELD (METROLOGY)

Surface Charge Profiling
– An advancement in Ion Implant Monitoring

Christian Krueger, Fab36, AMD, Dresden, Germany
Che-Hoo Ng, SDC, Spansion LLC, Sunnyvale, California
Zhiyong Zhao, Fab25, Spansion LLC, Austin, Texas
Gerard Krytsch, QC Solutions, Dresden, Germany

ABSTRACT *As the industry gets on the new technology nodes of 65nm and 45nm devices, implant monitor becomes even more crucial for consistent device performance. Common practice has been the use of 4-point probe with sheet resistance and thermal wave technique with the implant damage. However, both techniques have limitations on sensitivity. With the need of monitoring smaller variations in the ion implantation process, there is a need for a new and better approach on implant monitoring.*
A new non contact method using SCP (Surface Charge Profiling) is gradually gaining ground in the industry as a control technique for ion implantation. In this work, the authors compare the responses on implanted wafers with thermal wave, sheet resistance and SCP. Comparisons are made to implants of low doses, high doses and low energies.

1. Introduction and theory of operation

The QCS-ICT 300™ is manufactured by QC-Solutions™ and uses a simple non-contact technique to measure the Surface Photo Voltage (SPV) or the surface recombination lifetime of a normal Silicon bulk wafer. Therefore the tool is equipped with two light sources with different wave lengths (Blue-Light (BL) and Ultra Light(UL)).
In principal the silicon surface is illuminated by a collimated beam of chopped light of photon energy greater than the silicon bandgap and the light is absorbed close to the surface. The acquisition of the resulting SPV (Surface Photo Voltage) signal is achieved by means of capacitive coupling though an air gap. The implanted dose and energy can be calculated by measuring the width of the surface depletion region and the surface recombination lifetime. It is possible to support the measurement of low dose implants with a positive or negative charged Corona. [1-2]
The resolution of the tool is adjustable between
high: 7671 (Diameter 150) points per Wafer
medium: 4418 (Diameter 75) points per Wafer
low: 1134 (Diameter 38) points per Wafer

2. Ion Implant Metrology

The measured surface photo-voltage is a result of the ion implant crystal damage and depends from several implant conditions like dose, energy, beam density and indecent angle. This effect is the difference to normal techniques in the common implant control philosophy.
Therefore the QCS-ICT 300™ can be used to measure "as implanted" and "annealed" implant profiles.
The conditions of the measurement recipes are depending from these implant conditions and have several different settings. The authors used the experience from QC Solutions™ and created the following basic recipe groups divided by the implanted dose as the dominating factor[3]:

Boron_Less_5E13   (UL/BL, pos. Corona)
Phosph-Less_1E13   (UL/BL, neg. Corona)
Arsenic_Less_5E12   (UL/BL, neg. Corona)

Boron_5E13_and_more   (UL/BL, no Corona)
Phosph_1E13-and_more   (UL/BL, no Corona)
Arsenic_5E12_and_more   (UL/BL, no Corona)

LE-Boron_HD   (UL/BL, no Corona)
LE-Phosph_HD   (UL/BL, no Corona)
LE-Arsenic_HD   (UL/BL, no Corona)

Other recipe modifications were used as well.

Sensitivity
The sensitivity of the each recipe was measured on common implant qualification and production recipes. A nominal dose variation of ± 10% was used for each recipe. *Figure 1* is showing the very good sensitivity of the QCS-ICT 300™ on a 5E13 Boron recipe with an energy < 10keV. With all the tested matrices, SCP showed up to 10 times more sensitive than the other techniques.

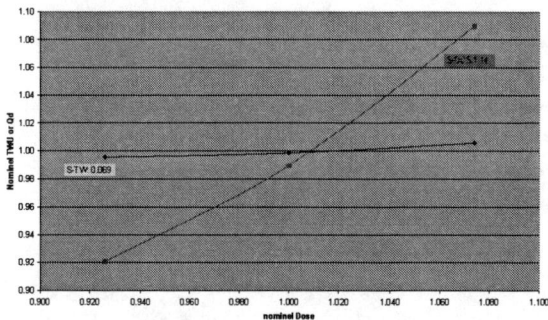

Figure 1: Sensitivity of QCS at ±10% Dose Variation

In addition to the sensitivity, SCP also shows advantages in its high resolution mapping capability with good throughput. While other techniques give good wafer uniformity values, SCP reveals implant micro non-uniformity issues such as striping in the slow and fast scan direction of hybrid scanning, glitch recovery success rate and different test wafer surface conditions.

Figure 2 is showing a QCS map of a normal E13 Arsenic implant on a medium current tool with a common mechanical scan system in the slow scan direction (S) and an electrostatic beam scan in the fast scan direction (F).

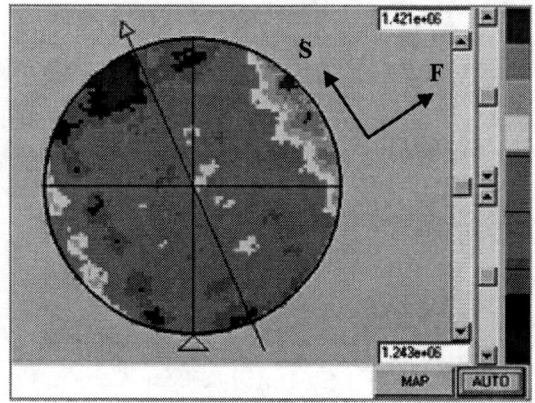

Figure 2: medium resolution map of a medium dose implant from an implanter with an electrostatic beam scan system

Figure 3a and *3b* are two cross section analyses in the slow and fast scan direction of the same wafer. The beam profile in the fast scan direction and the dose failure during surface outgasing in the slow scan direction is clearly visible in both cross sections.

Both analysis can be used to optimize the current implant regarding better beam profile, reduced edge effect, etc. and for failure analysis in general.

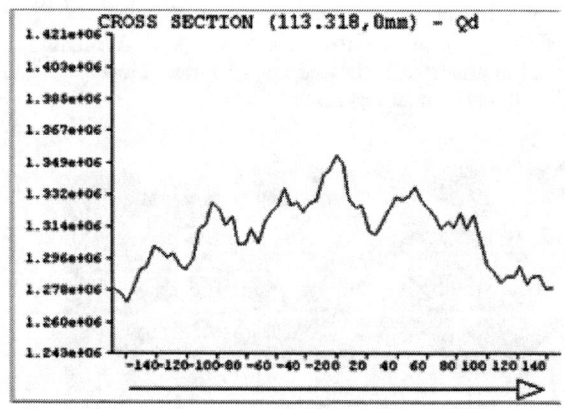

Figure 3a: Cross section in the slow scan direction

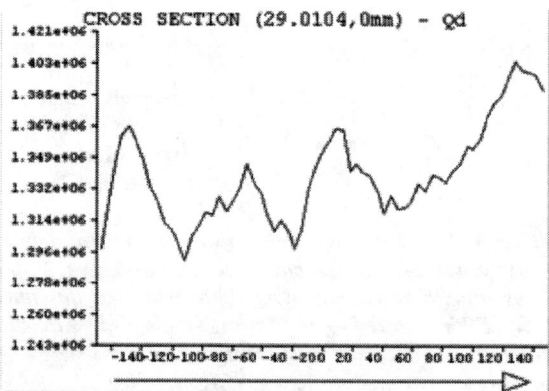

Figure 3b: Cross section in fast beam scan direction

Other typical implant tool finger prints were measured with the tool as well. *Figure 4a* is a typical map of an implanter with two mechanical beam scan systems. The implanted beam profile and the dose failure during outgasing are clearly detectable.

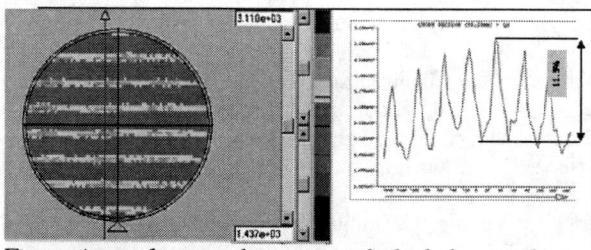

Figure 4a: medium resolution map of a high dose implant from an implanter with two mechanical scans

The micro-striping of the same wafer got reduced after a 1000°C/10sec anneal (*Figure 4b*) and is not detectable during the electrical transistor measurement of the wafer. This striping could become an important issue in Sub-45nm-Technologies again.

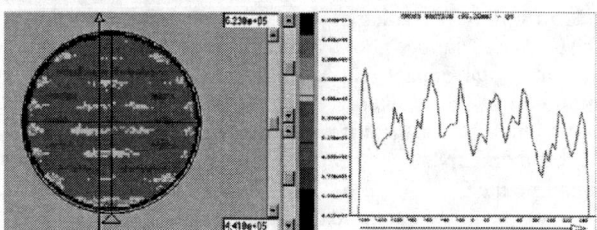

Figure 4b: same implant conditions as in 4a after annealing

Other typical implant effects become visible with this technique also.

Figure 5 clearly identifies two implant interruptions (Glitches) during a normal implant. Those non-uniformities got washed out during an annealing step and a high resolution sheet resistance map (>600points) is not able to detect these interruptions anymore.

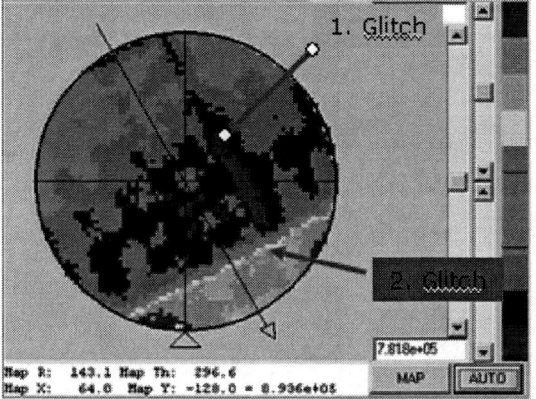

Figure 5: Two implant interruptions during a multi-twist implant

Statistical Process Control (SPC) and Stability

The QCS-ICT 300 ™ is easy to implement in the implant control loop. At AMD Fab36 it is used to measure the as-implanted conditions on daily implant qualification wafers. Several SPC-Charts were generated and indicating nearly every failure during the implant process now.

Some difficulties with Boron and with the substrate bulk material are discovered during this qualification process. The surface conditions and the silicon manufacturing process of each wafer can cause some different lifetime and SPV values. An implant qualification SPC-chart is shown in *Figure 6*. The usage of test-wafer material from an other silicon supplier generates Out-Of-Control (OOC) values in the QCS-SPC-chart suddenly while the sheet resistance chart seems to be more insensitive. It is not clear which surface conditions are generating these OOC-values, but the authors identified a much higher Oxygen level in these silicon wafers from that supplier.

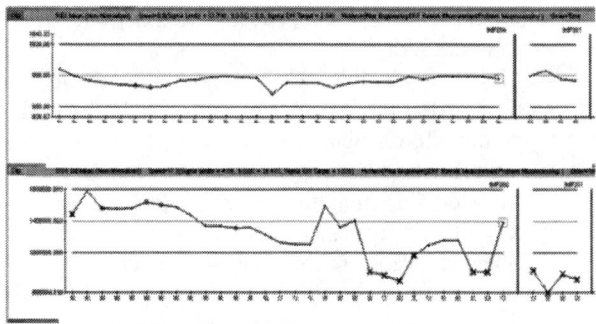

Figure 6: OOC QCS-signal change by silicon supplier change

Other usages

It is more and more important to control the implant tool prior the implant process and calibrate for instance the implant angle. Common techniques like the V-Curve-Method are used to adjust the 0°-Tilt-Position of each ion implanter. Several test wafer and implant processes at different tilt angles are necessary for this method. Normally the V-Curve got generated by 0°, ±1° and ±2° tilted implants at a certain dose and energy.

This is not necessary by using the QCS-ICT 300 ™. The authors used one test wafer and switched the electrostatic scan on the ion implanter off. The center of the beam should hit exactly the middle of the wafer at 0° tilt and the normal beam divergence will generate different implant angles on the edge of the wafer. The horizontal diameter scan / cross section should show the exact centered V-Curve caused by different implant conditions in the center versus the edge of the wafer.

The results of this test with and tilt angle failure of < 0.5° are shown in *Figure 7*.

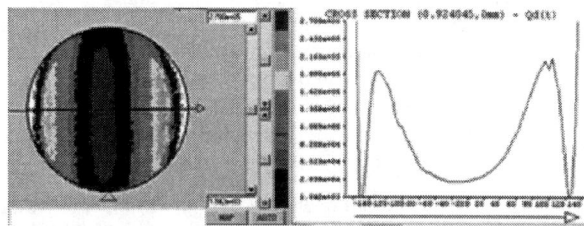

Figure 7: V-Curve measured with QCS on one test wafer

3. Summary

- QCS-ICT 300 can be used as an implant metrology tool
- The tool is able to measure „as implanted" and annealed profiles with a high sensitivity
- Highest sensitivity is in the E12 – E14 dose region
- Beam glitches, beam non-uniformities, changes in the dose measuring system can be identified
- The tool is very sensitive regarding the conditions in the build up process
- Good repeatability results are observed

The authors believe SCP can be a good technique for ion implant statistical process control and implant trouble shooting.

References
(1) QC Solutions User Manual
(2) Edward Tsidilkovski, ed.al.: Ion Implant Process Monitoring with a Dynamic Surface Photo-Charge Technique
(3) QCS as-Implanted Conditions/Recipe Guide

Metrology and High Resolution Mapping of Shallow Junctions Formed by Low Energy Implant Processes

Dr. Eric Don[+], Mr. Aron Pap[+], Mr. Peter Tutto[+], Dr. Tibor Pavelka[+]
Dr. Christophe Wyon[*], Mr. Cyrille Laviron[*], Mr. David Sotta[*],
Dr Richard Oechsner[#] and Mr. Marcus Pfeffer[#]

+ Semilab R.T. Budapest, HUNGARY.
* CEA-LET, STMicroelectronics & Philips Semiconductors, Crolles FRANCE.
 # Fraunhofer IISB, Erlangen, GERMANY

Abstract. This paper presents results obtained using a Junction Photo-Voltage (JPV) method optimized for characterization of the combined implant-annealing process. The tool was found to be particularly suited to measurement of ultra-shallow junction sheet resistivity and leakage. In this work the authors also evaluated the benefits of improved spatial resolution compared to previous equipment designs. Current technology USJ monitor wafers were made using a BF_2 or Arsenic implant followed by a spike anneal and also R&D USJ wafers were made by Plasma Immersion followed by laser annealing. All the wafers were measured using the non-contact JPV measurement tool. Results obtained from the JPV measurements were correlated to destructive off-line analytical measurement tools.

Keywords: JPV, Ion Implant Dose, Wafer Map, photovoltage.
PACS: 02.60.Cb, 06.20.-f, 61.72.Tt, 61.72.Vv, 73.50.Pz, 85.40.Ry

INTRODUCTION

The need for improved electrical methods for implant and anneal process control has been well documented (1). The theory and conditions for linear SPV are well known (2) and also commonly applied to JPV for diffusion length measurement (3). Verkuil et al extended this known methodology by measuring the lateral spread of JPV thus inventing a non-contact method to measure sheet resistance (4), which has already found wide application in solar cell process development and in-line control (5).

A wafer map of sheet resistance with mm spatial resolution is required when qualifying a new process or for process tool qualification after maintenance. The main requirement for process control applications is not wafer mapping but high repeatability (<0.1%) without need for "golden" reference wafers.

Principles of the JPV Measurement.

An amplitude modulated beam of monochromatic light, of wavelength short enough to be absorbed close to the junction and with a beam diameter of millimeter dimensions, illuminates the surface of the wafer at normal incidence.

The modulated light generates minority carriers near the p-n junction and those carriers generated within a diffusion length of the junction are swept across it creating a local time varying excess of majority carriers in the illuminated area. The majority carrier excess drives radial diffusion parallel to the junction away from the illuminated area. The majority carrier distribution creates a modulated junction Photovoltage that decays radially outside the illuminated area.

Two conductive electrodes held above the surface of the wafer and positioned radially in line from the illuminated area wafer pick up the time varying Photovoltage by capacitive coupling. The difference (in magnitude and phase) of the Photovoltage between the two electrodes is related to the sheet resistance in the implanted layer.

In previous publications (4,6) mathematical models of this measurement have been developed and explicit solutions in the form of Bessel functions have been found from which values for sheet resistance and junction leakage of an epitaxial or implanted layer can be extracted.

Equipment Design of JPV Tool.

We have optimized Verkuil's original sensor design (4) and found that we could obtain a higher spatial resolution and better signal-to-noise ratio than was possible when using our SPV transparent electrode sensor (5). With the coaxial symmetry dual metal ring electrode sensor we can achieve spatial resolution in the low mm range with excellent signal to noise ratio even though the sensor electrodes are larger in diameter (See fig 1) than the effective resolution. This is explained by the fact that the sensitivity of the dual metal ring sensor to deviations in resistivity is highest at the central area of the ring where the diffusion current is generated.

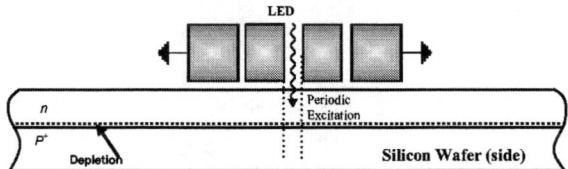

FIGURE 1. Semilab JPV sensor configuration (section).

All the measurement data published in this paper were taken using this type of dual metal ring sensor.

In order to analyze our data we adopted a different approach to others (4, 6). We model the total complex impedance $|Z|$ that the whole wafer presents to a photovoltage V_{jpv} generated near the center of the wafer. We developed a general numerical simulation-fitting algorithm to fit the measured photovoltage as a function light intensity and modulation frequency using the following differential equations derived from our model;

$$\left(\frac{\partial V_{jpv}}{\partial r}\right) = -I_{rad}\left(\frac{R_s}{2r\pi}\right) \quad (1)$$

$$\left(\frac{\partial I_{rad}}{\partial r}\right) = 2r\pi\left(\Phi - (G_d + (i\omega)C_d)V_{jpv}\right) \quad (2)$$

Where:
V_{jpv} is the Junction photovoltage;
I_{rad} is the lateral current density in the top layer;
R_s is the sheet resistivity of the top layer;
Φ is the photocurrent density (electron-hole).
C_d is the capacitance of the p-n junction;
G_d is the conductance of the p-n junction;
r is the distance from the axis of the probe;
ω is the modulation frequency of the excitation.

This allows us to set optimum test conditions to each wafer prior to wafer mapping by first moving the sensor to the center of the wafer and sweeping the modulation frequency while measuring the JPV at several illumination levels. Using our proprietary algorithm we can find the optimum region in the illumination frequency-intensity-JPV response that exhibits best linearity of JPV versus sheet resistance. We then can set to these conditions and map the wafer. Our procedure can remove the need for a calibration wafer. This method has been proven by correlation to four point probe measurements (see Fig 2).

Results from USJ Measurements

The following results were obtained on 300mm wafers supplied from the Crolles Alliance. In these wafers substrate carrier concentrations were in the range 1E15-1E16 cm^{-3}. All implants were given a dose < $5E14 cm^{-2}$ at energy < 1.5KeV and were spike annealed resulting in junction depths in the range 10-60nm. Two sets were supplied; one P+/N the other N+/P with Boron and Arsenic being the doping atoms. The wafers are characteristics of 65nm technology.

Sheet Resistance Correlation and Repeatability for Boron USJ implants

Our auto setting algorithms have been proven over a range of different implants. We have successfully correlated data from our measurements on Boron implanted USJ wafers against a mercury four point probe (see Fig 2).

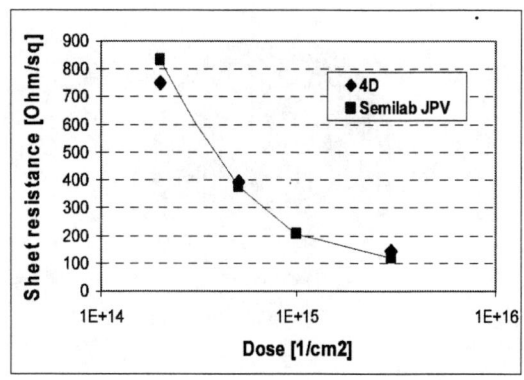

FIGURE 2. Correlation of sheet resistance for JPV average of 16,000 points and Hg4PP average of 5 points to Boron implants at 1KeV.

Arsenic implants proved more difficult to correlate due to a surface oxide which impaired the Hg4PP measurement. The JPV system provides the repeatability required for advanced implanter monitoring. The system also does not suffer from probe wear out. Measurement sensitivity is close to unity therefore standard deviation in resistivity equals standard deviation in dose.

TABLE 1. Typical Repeatability Boron USJ sheet res.

Site	Points	Mean	%Dev
1	10x16	882.1	0.064%
2	10x16	882.6	0.062%
3	10x16	882.1	0.034%
4	10x16	885.7	0.074%
5	10x16	885.1	0.043%

Sheet Resistance Mapping of Arsenic USJ implants in batch implanter.

A series of arsenic implants was made on a batch implanter holding 13 wafers on a wheel with rotation speed 900 r.p.m. and scanning 10-15 cm/sec.

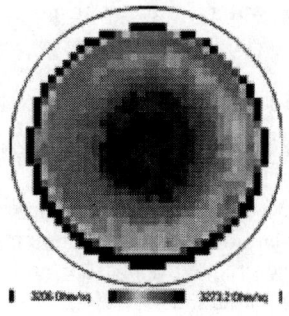

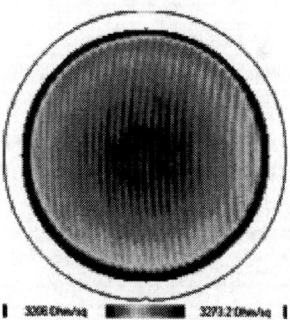

FIGURE 3. Top: Wafer map of sheet resistance with 8mm raster producing over 1,000 measurement points in 2.0 minutes. Bottom: Wafer map of sheet resistance with 2mm raster producing over 17,000 measurement points in 28 minutes.

Fig. 3 illustrates the importance of scanning resolution. The maps are both of the same wafer Arsenic implanted 1.5KeV dose $4E13 cm^{-2}$. Only the higher resolution map clearly shows a striping problem. The curvature of 120 cm and spacing of 8.5-9.0 mm fits the geometry and rotation speed of the implanter wheel.

The Fig 4 shows a sheet resistance map of an Arsenic implanted wafer at 0.5KeV with dose $4E13 cm^{-2}$ also showing striping but the additional rings in the lower energy implant pattern are due to non-uniform heating in the spike RTP tool. The line-scan shows that the peak-peak variation in the striping is about 1.5% of the mean sheet resistance value.

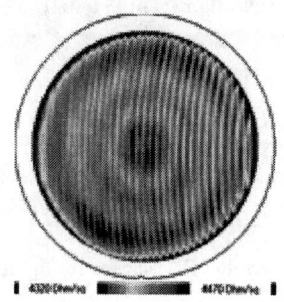

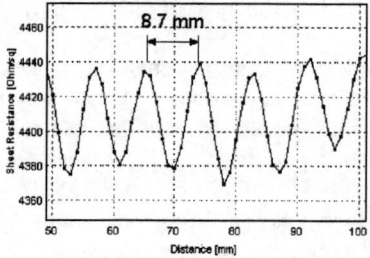

FIGURE 4. Top: Wafer map with 1mm raster with over 70,000 points in 4 hours. Bottom: Partial line scan showing excellent signal to noise and spatial resolution of the sensor.

Sheet Resistance Mapping of Boron USJ implants in single wafer implanter.

The wafer was quad implanted in a single wafer ribbon beam implanter. The wafer was given a 22 degree twist and a 7 degree tilt to eliminate channeling. The wafer was not given a pre-amorphisation implant.

The 22 degree twist is easily seen in the measured pattern. In a scanned beam system, if no scan angular correction is used one may expect the incident angle (and channeling) to be increased on one side of the wafer and reduced on the other side. The quad implant with four 90 degree rotations would introduce the observed four fold symmetry.

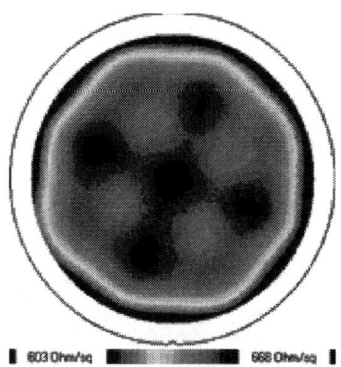

FIGURE 5. 300mm Si wafer Boron implanted 1KeV dose 4E14 cm^{-2} showing four fold symmetry pattern. Map has a 2mm raster giving 17,000 points in 28 minutes.

Resolution and Edge Exclusion.

Plasma immersion and laser annealing are two new processes under development as potential alternatives in manufacture of USJ. We have measured a series of such samples made in an IBS PIII system and then laser annealed by CNET. The results will be reported in future publications and are included here only for illustration of JPV resolution.

The test wafers were silicon 100mm diameter and following PIII of the entire wafers 2cm^2 areas were laser annealed each area with different laser energy and different number of scans. The laser beam was 1.5mm wide. These dimensions are reproduced correctly in the image confirming that the JPV system does indeed have mm scale resolution as seen by the minimal edge effect between the amorphous "as implanted" surface and the annealed surface. This can be explained if the "as implanted" areas form a highly conducting film isolated from the substrate so that the junction photovoltage voltage is sustained and spreads in the "as implanted" area.

This contrasts to the large edge effects seen at the edge of the wafers in Figs 3, 4 & 5. In this case movement of the electrodes over the edge of the wafer where there is no photovoltage signal amounts to an effective reduction in the capacitive coupling and therefore the measured Junction Photovoltage is reduced in proportion to the "reduction" in electrode area indicating that a simple correction algorithm should be applicable.

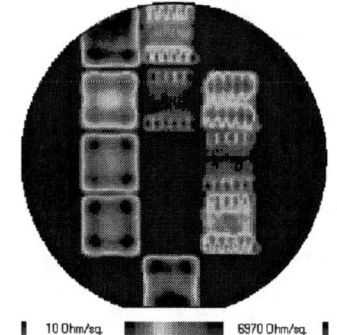

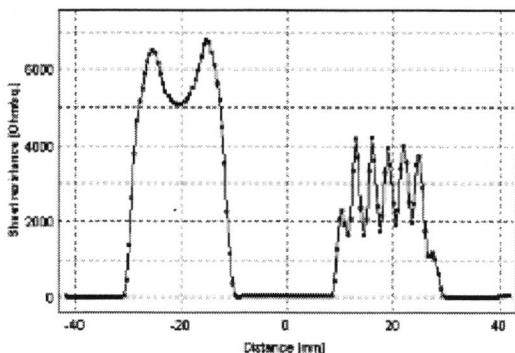

FIGURE 6. Top: Sheet Resistance Map of 100mm wafer after plasma immersion implant and laser annealed areas. Bottom: Line scan through two annealed squares.

ACKNOWLEDGMENTS

This work is part of the SEA-NET project with financial support of the European Community Framework 6 Program. We also thank IBS SA, Four Dimensions Inc. and Dr. R. Gwilliam for their contributions.

REFERENCES

1. Borland, J.O. The Fourth International Workshop on Junction Technology, 2004. IWJT apos; Volume 4, Issue 15-16 March 2004 Page(s): 8 - 11M.
2. A.M. Goodman J. Appl. Phys 32 p.2550 1961
3. L. Tarricone, E.R. Don, N.M. Pearsall and T.J. Coutts Solar Cells Vol. 7 p.281 1982.
4. US Patent 5,442,297 Verkuil, Roger. L. Filed 30 June 1994, Granted 15th August 1995.
5. E. Ruland, P. Fath, T. Pavelka, A. Pap, K. Peter & J. Mizsei in Proc. 3rd World Conf. on PV Energy Conversion May 12-16 2003
6. V. N. Faifer et al in MRS Spring 2004 Symp. C Vol. 810 p.475
7. US Patent 7,019,513 Faifer V. et al Filed 19th January 2005, Granted 28th March 2006

Metrology Requirements For Single Wafer Ion Implanters

Joseph C. Olson, Gordon Angel, Atul Gupta, Rosario Mollica, Daniel Distaso, Jinning Liu

Varian Semiconductor Equipment Associates, Inc., 35 Dory Road, Gloucester, MA 01930, United States

Abstract. As production CMOS devices shrink to 90nm and below, the requirements for high quality ion beams increase in medium and high current implant systems. Critical variations in device performance can be related to changes in ion beam properties, such as ion beam size, shape and density as well as ion beam angle steering and distribution. Transistor properties are known to be sensitive not just to mean beam angle but also to details of the distribution of angles contained within the beam. As a result, modern single wafer implanters are required to monitor more than just the energy and dose of implanted ions. The process security provided by single wafer ion implantation can be ensured by measurements which verify the angle integrity of each setup and implant. Monitoring of energy contamination, another element of beam quality, allows end users to utilize the improved productivity of decel operation on high current implanters while maintaining minimal and, more importantly, repeatable energy contamination. Finally, real time uniformity monitoring provides assurance that the beam uniformity established during setup is preserved during wafer processing. In this paper, example metrology systems from the VIISta platform will illustrate measurement of these beam quantities. Representative performance data for monitored beam quality and control will be presented.

Keywords: Ion implantation; Energy contamination; Angle control; Angle measurement.
PACS: 41.75.Ak; 61.85.+p; 85.40.Ry; 61.72.Tt

INTRODUCTION

As semiconductor manufacturing becomes more advanced, the need for process information increases. In a production environment the desire for more information is driven by process assurance, while in a research and development setting the information may be most useful as input to device modeling. In ion implantation this translates into requirements for more accurate and more complete measurements of any ion beam properties that can affect device performance.

Some examples include beam current, current density and uniformity, beam angles and beam angle distributions, and energy contamination. In this paper we will briefly discuss the sensitivity of device performance to the various beam properties, then more completely discuss systems on the VIISta platform of implanters to measure these beam properties and provide representative data.

I. PROPERTIES AFFECTING DEVICE MANUFACTURE

The role of ion implantation in semiconductor manufacturing is to deliver to a wafer a precise amount of dopant, implanted to a specific depth at a specific direction as uniformly as possible across the wafer. The measurement of beam current and uniformity are clear requirements to provide a precise uniform dose. Moreover, the current density of the beam can have an effect on damage to the crystalline structure of the wafer[1] and to the electrical conductivity achieved after activation of the dopant[2]. Measurement of the beam current density is needed in order to predict or plan for these effects.

The energy of the ion beam determines the implant depth. The use of decelerated ion beams to improve implanter productivity at low energy creates the possibility of energy contamination (ions or dopant atoms at pre-deceleration energy) reaching the wafer. Small amounts of energy contamination (EC) are often acceptable[3] particularly if included in the design phase of the transistor. Monitoring of the level of EC can therefore speed the development of recipes and provide process security for the most aggressive use of deceleration.

Beam incident angle with respect to the wafer can affect device performance[3]. The aspect ratio and shadowing of structures on the wafer and channeling effects contribute to device sensitivity to angle. For the discussion here we will divide the angle of ions as they impinge on the wafer surface into three categories: (1)

the average angle of all ions arriving at the wafer during an implant is the wafer-to-wafer steering angle; (2) at any point on the wafer the average angle of all ions arriving is the local steering angle, and the distribution of local steering angles over the entire wafer has a standard deviation that is the within wafer angle spread; and (3) the distribution of ions arriving at a single point on the wafer has a standard deviation that is the within die angle spread. The three categories of angle variation are illustrated in Fig. 1.

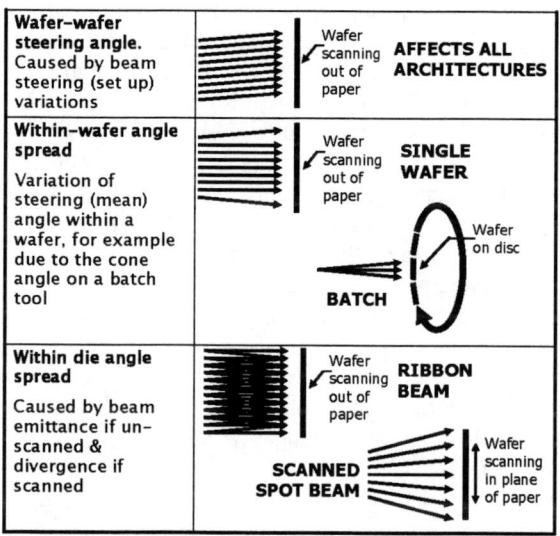

FIGURE 1. Schematic depicting the three categories of angle variation.

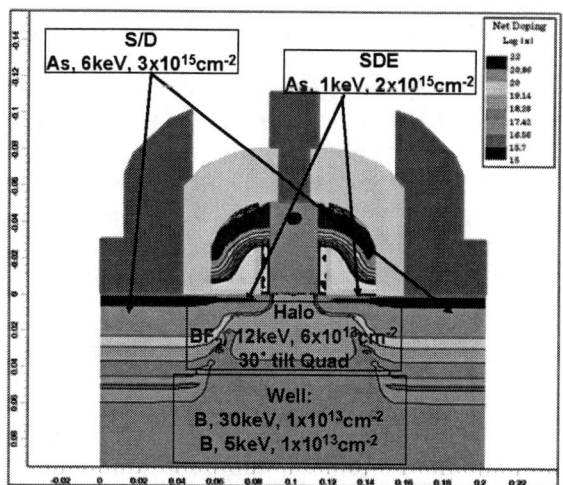

FIGURE 2. Cross section of a 65 nm NMOS device modeled to gauge angle sensitivity using TCAD.

Figure 2 shows a TCAD model of high performance logic transistor consistent with the ITRS roadmap for 65 nm. The source drain extension (SDE) implant in the model was varied to study the effect of changes in beam steering and within die angle spread on device performance. In Fig. 3a the calculated sensitivity of on state current to changes in wafer-to-wafer and within wafer steering angle is shown. Fig. 3b displays variations in on state current (I_{dsat}) as the within die angle spread is changed. Note that in Fig. 3b the sensitivity of on state current to within die angle spread exhibits a minimum at just above 2°, and therefore device repeatability could be enhanced by operating near that point. Both graphs highlight the benefit to having detailed beam angle information when determining how to optimize the desired transistor output.

One of the main benefits of single wafer ion implantation is the ability to monitor and reduce process risk for every wafer. Process risk can be further reduced by monitoring the beam uniformity during implant through the use of a real time uniformity monitor.

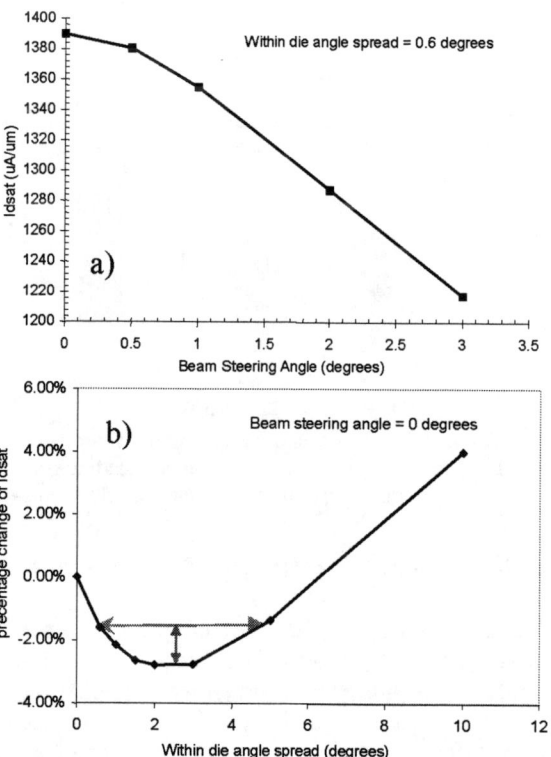

FIGURE 3. Sensitivity of on state current for 65 nm NMOS device due to change in a) beam steering angle and b) within die angle spread on the SDE implant.

II. VIISTA METROLOGY SYSTEMS

The VIISta family of ion implanters have been described in detail elsewhere[4,5,6]. For the VIISta family, VSEA has developed systems to measure beam current, uniformity and density, beam angles, and energy contamination, as well as to monitor the beam uniformity during ion implantation. In this section we describe the measurement systems.

II.A Beam Angle Measurements

VIISta metrology systems can provide up to 45 measurements of horizontal and vertical beam angles as part of automated beam setup. The wafer-to-wafer steering angle error is less than 0.1° on all products due to the Varian Positioning System (VPS™) which has been described elsewhere[7]. Within wafer angle spread is measured and interlocked by recipe. Figure 4 shows the measured within wafer angle spread from one month of running on a high current implanter at a leading-edge logic customer. Nearly 500 setups are included.

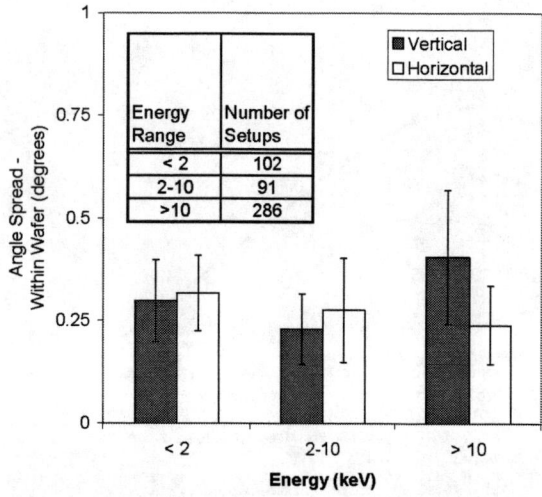

FIGURE 4. Average measured within wafer beam angle variation for 1 month's running at a leading edge logic manufacturer. Error bars are 1 standard deviation.

II.B Beam Size and Shape Measurements

The VIISta tools can be configured to provide measurements of beam size, shape, and current density. The measurements are typically displayed as a two dimensional (2D) profile plot. Fig. 6 shows such a 2D plot for a 3 keV arsenic beam, obtained from a VIISta HC high current tool equipped with a multi-pixel profiler to obtain the spatially resolved information.

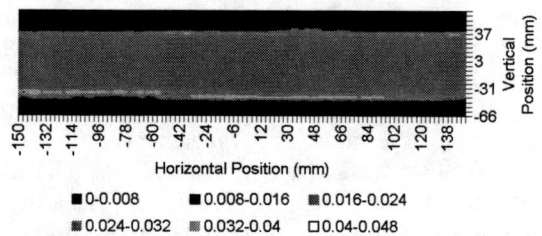

FIGURE 5. Two dimensional profile of 3 keV As. Units are A/cm².

Beam uniformity is established during automated setup and on VIISta tools is measured directly. Single wafer implanters in general rely on beam stability to maintain uniformity during wafer processing. Process assurance can be significantly enhanced if this stability is measured and interlocked by a monitor specifically geared to tracking uniformity changes in the beam. Figure 6 illustrates the output of such a monitor on a VIISta HC where the assumption of stable beam is being deliberately violated. Comparison of the change in monitor signal to the change in profiler uniformity, which is measured in the wafer plane, shows the monitor signal is about four times more sensitive to changes in the beam uniformity, which allows the monitor to respond to changes as early as possible.

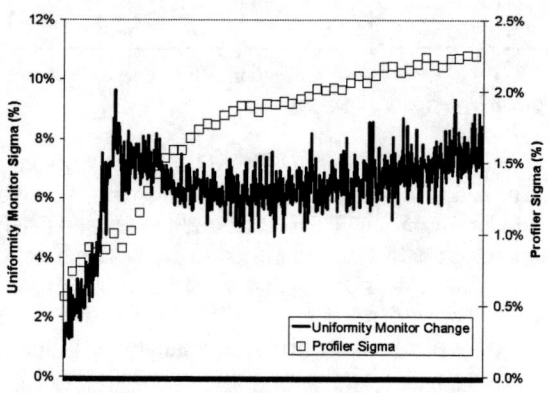

FIGURE 6. Uniformity monitoring data over 60 minutes. Note the 4x improvement in sensitivity between the uniformity monitor and the wafer plane

II.C Energy Contamination Measurements

Beamline architecture and machine operating modes preclude the possibility of energy contamination on the VIISta 3000 and VIISta 810 series implanters. In high current, low energy implantation, however, the challenge of productivity requires the use of deceleration and raises the possibility of energy contamination.

In order to mitigate any process risk of decel operation, VSEA has developed a monitoring system that measures the high energy contaminant flux within the beam. Fig. 7 shows the output of this monitor compared to SIMS measurements of energy contamination.

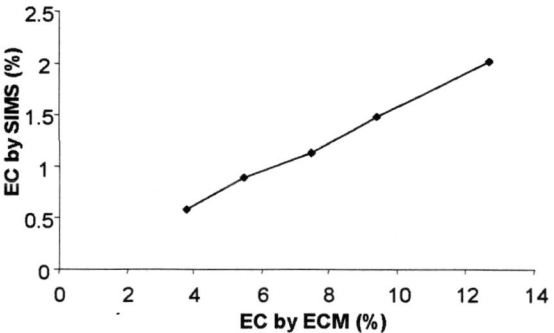

FIGURE 7. Energy contamination monitor output plotted against energy contamination as measured by SIMS for 3 keV B.

CONCLUSIONS

The VIISta platform of implants provides metrology solutions needed to meet the demands of device design, development and manufacturing. Measurements of ion beam current, uniformity, current density and beam angle are available and systems to characterize uniformity in real time and energy contamination have been developed. These systems enable the VIISta platform to enable higher productivity while monitoring to protect high yield.

ACKNOWLEDGMENTS

The authors gratefully acknowledge helpful discussions with Tony Renau, Greg Redinbo, and Jay Scheuer.

REFERENCES

1. Otto, G. *et al.*, *Nuclear Instruments and Methods in Physics Research B* **242**, 667-669 (2006).
2. Lai, Y. *et al.*, *Nuclear Instruments and Methods in Physics Research B* **244**, 338-342 (2006).
3. Mehta, S., *et al.*, *Materials Science and Engineering B* **114-115**, 72-76 (2004).
4. Renau, A. and Hacker, D., "The VIISta 810 medium current ion implanter" in *Proc. 12th Int. Conf. on Ion implantation Tech.*, Kyoto, Japan (1998).
5. Redinbo, G, Campbell, C., and Mullin, J., "Advanced Single Wafer High Current Beamline Architecture for Sub-65nm" in *Proc. 16th Int. Conf. on Ion implantation Tech.*, Marseilles, France (2006).
6. Tokoro, J., Holbrook, D. and Hacker, D., "Introduction of the VIISta 3000 single wafer high energy ion implanter" in *Proc. 13th Int. Conf. on Ion implantation Tech.*, Alpbach, Austria (2000).
7. Todorov, S., Lee, W., and Lacey, K., "VIISta 3000HP High Energy Ion Implater" in *Proc. 15th Int. Conf. on Ion implantation Tech.*, Taipei, Taiwan (2004), edited by L. J. Chen et al., pp. 39-43.

Local resistance measurement on polycrystalline silicon layer in low-temperature poly-Si thin film transistor using scanning spreading resistance microscopy

S. Abo[1], H. Yamagiwa[1], K. Tanaka[1], F. Wakaya[1], T. Sakamoto[2], H. Tokioka[2], N. Nakagawa[2] and M. Takai[1]

[1] *Center for Quantum Science and Technology under Extreme Conditions, Osaka University, 1-3, Machikaneyama, Toyonaka, Osaka, 560-8531 Japan*
[2] *Advanced Technology R&D Center, Mitsubishi Electric Corporation, 8-1-1, Tsukaguchi-Honmachi, Amagasaki, Hyogo, 661-8661 Japan*

Abstract. A local resistance of a low-temperature polycrystalline silicon in a thin film transistor (TFT) for a liquid crystal display was measured using scanning spreading resistance microscopy for investigating an activation of dopant atoms. The observed TFTs had lightly doped drain (LDD) structures with lengths of 0.6 μm and 1.0 μm. The measured local resistance ranges by 4 - 6 orders of magnitude for measuring points. The resistance data were separated into two groups: the higher resistance data were inside the grains and at the center of the grain boundaries, while the lower resistance data were near-by the grain boundaries. The lower resistance data were evaluated, since the TFT operation current usually flows through the lower resistance area. From the lower resistance data, the drastic resistance change from the channel to the drain regions, which is LDD, were observed in both samples. The LDD lengths of two samples are found to be approximately 0.6 and 1.0 μm, respectively.

Keywords: low-temperature poly-Si, thin film transistor (TFT), scanning spreading resistance microscopy (SSRM)
PACS: 61.72.Tt, 68.37.Ps, 68.55.Ln

INTRODUCTION

A thin film transistor (TFT) built in a polycrystalline silicon (poly-Si) layer on a glass is a key device for a peripheral circuit of a liquid crystal display. TFTs deteriorate during an operation by an abnormal current due to hot carriers generated at a boundary between a channel and a drain [1-3]. This issue is also reported to be related to grain boundaries in poly-Si TFT [4,5] and well known as a short channel effect in the bulk silicon devices. A solution for this issue is using a lightly doped drain (LDD) structure, placed between the channel and the drain regions, for suppressing the high electric field. The drain and LDD layers are made by a plasma doping process. Dopant atoms are activated by a low temperature annealing process. Thus, the distribution and the activation of dopant atoms are not accurately controlled in the poly-Si TFT.

In this study, the local resistances across the grain boundaries in the poly-Si TFTs with LDD structures with lengths of 0.6 μm and 1.0 μm have been investigated using scanning spreading resistance microscopy (SSRM) with a resolution of 50 nm.

EXPERIMENTAL

Fig.1 shows the schematic diagram of the SSRM. The SSRM system is based on a conventional contact mode atomic force microscopy (AFM). The local resistance data are measured by the current from a sample holder to a B-doped diamond coated conductive probe. The height is measured by the conventional AFM. The sample current was sensed by a logarithmic current amplifier with ranges of 10 pA to 0.1 mA. The local resistance is calculated from Ohm's law.

The fabrication process of a low-temperature poly-Si TFT is as follows. A Si-layer was deposited on a glass substrate as an amorphous Si and laser annealed for making a poly-Si. The n-type source/drain and LDD regions were fabricated with PH_3 plasma doping, followed by post annealing at 400 °C and hydrogenation process. Two samples with different

LDD mask lengths (0.6 μm and 1.0 μm) were measured for clarifying the LDD lengths. Fig.2 shows the schematic diagram of SSRM measurement of the poly-Si TFT. The top gate metal and the gate oxide were removed before the SSRM measurement. The poly-Si layer was scanned by the diamond coated probe from the drain to the channel region. The V_{DC} was applied not at the sample backside but at the drain contact because the substrate is an insulator. The radius of the probe used in this study was approximately 50 nm. The radius of the contact area, which was estimated from the probe radius, the probe force and the elastic parameters of the probe and the sample, was 8 nm [6]. The probe was scanned at 5 - 10 μm/s with a force of 2 μN (5GPa).

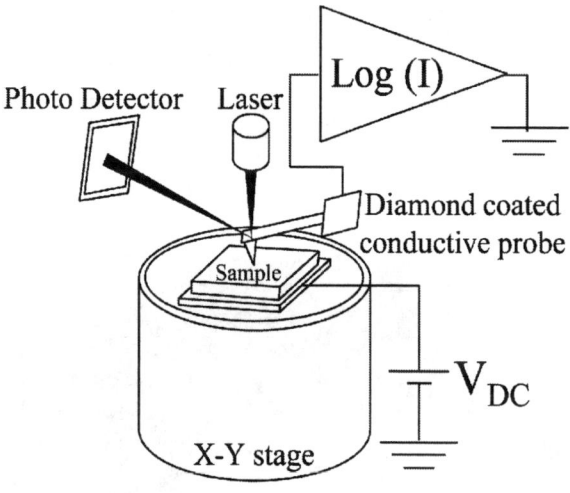

Fig.1 Schematic diagram of SSRM system.

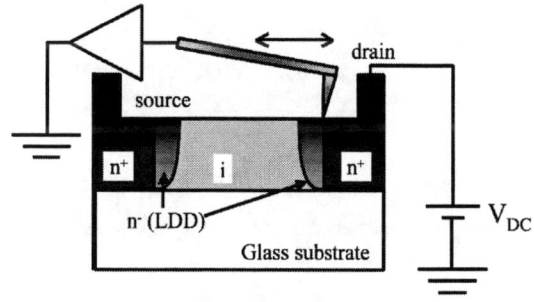

Fig.2 Schematic diagram of SSRM measurement of poly-Si TFT.

RESULTS AND DISCUSSION

Fig.3 shows the AFM and the SSRM images of the poly-Si TFT with 0.6 μm LDD structure for the same area of 10.0 x 5.0 μm² from the channel to the drain regions with V_{DC} of 500 mV. From the AFM image, the grain boundaries were 5 – 10 nm higher than surroundings. The grain size is approximately 300 nm. From the SSRM image, the local resistance gradually decreases from the channel to drain region. However, the local resistance reflects the morphology of the grain boundaries and does not homogeneously decrease. Fig.4 shows the zoomed AFM and SSRM images from Fig.3 for an area of 1.5 x 1.2 μm² with some grains. Some grain boundaries had a higher resistance area (dark contrast) sandwiched by lower resistance areas (light contrast). Fig.5 shows the cross

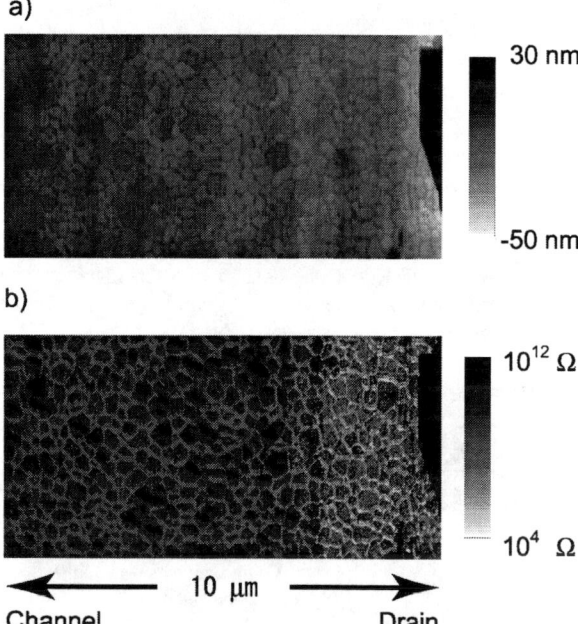

Fig.3 AFM (a) and SSRM (b) images of TFT with 0.6 μm LDD structure.

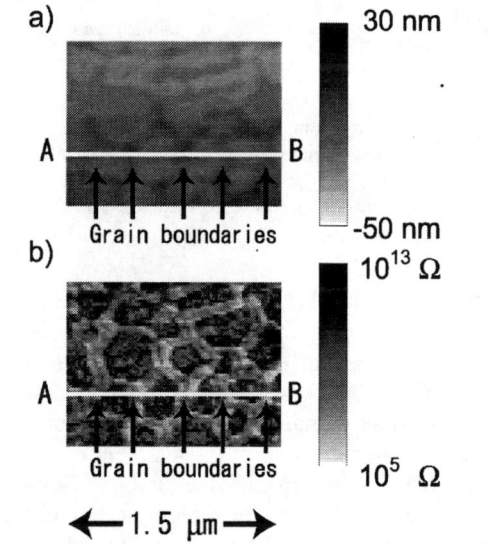

Fig.4 Zoomed AFM (a) and SSRM (b) images of TFT with 0.6 μm LDD structure.

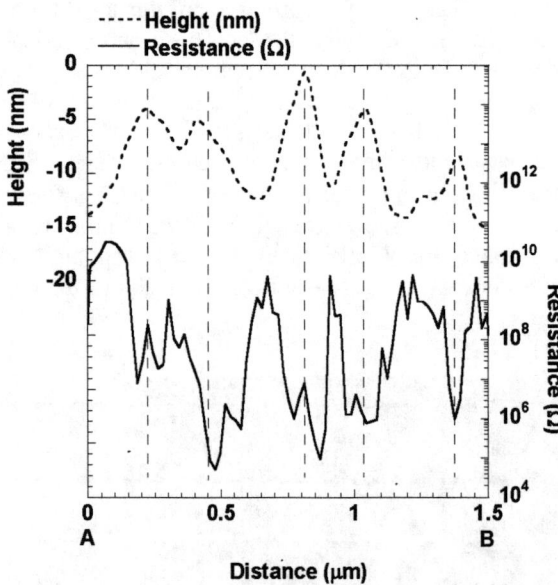

Fig.5 Height and Resistance as a function of distance with some grains (dashed lines indicate grain boundaries, which are the same points with arrows in Fig.4).

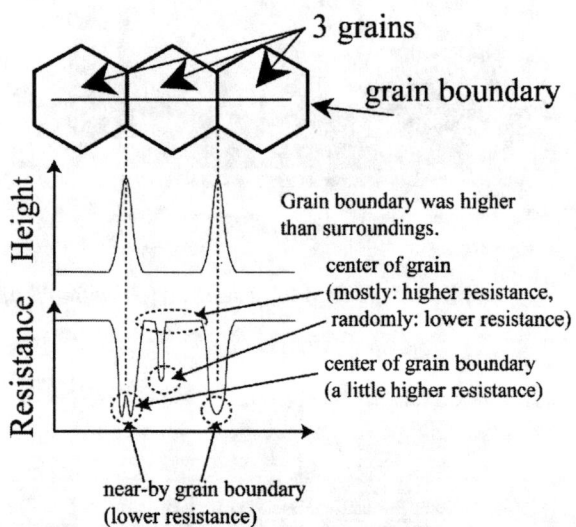

Fig.6 Schematic of height and resistance profiles across grain boundaries.

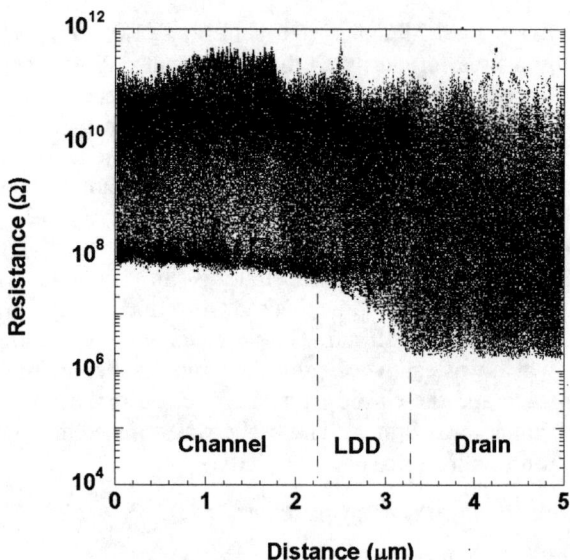

Fig.7 The local resistance as a function of distance in poly-Si TFT with 1.0 μm LDD length.

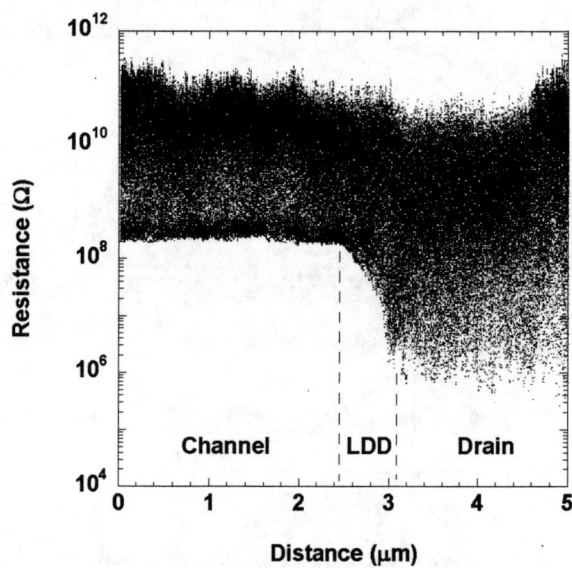

Fig.8 The local resistance as a function of distance in poly-Si TFT with 0.6 μm LDD length.

section of the height and the local resistance of Fig.4 from A to B. The points with arrows, which are the grain boundaries in the Fig.4, were shown as dashed line in Fig.5. The grain boundaries were 5 – 10 nm higher than surroundings. The local resistances near-by grain boundaries were 3 – 4 orders of magnitude lower than those inside the grains. Some grain boundaries had 4 – 10 times higher resistance area at the center of the grain boundaries. The low resistance points were randomly observed in the grains. Such characteristics are explained with Fig.6 as follows.

Fig.6 shows the schematic of the height and the resistance profiles across grain boundaries. The poly-Si TFT was fabricated with plasma doping process. Therefore, the dopant atoms were trapped at the grain boundaries. The concentration of dopant atoms near-by the grain was higher than that inside the grain. After annealing, the dopant atoms were activated near-by the grain boundaries. Because of this reason, the local resistances near-by the grain boundaries were lower than that inside the grains. At the center of grain boundaries, the local resistances were little higher than surroundings, which might be due to a low activation of dopant atoms on the grain boundaries. Inside the grains, most places had higher resistance. At some places inside the grains, an inhomogeneous activation occurred and resulted in lower resistance. The TFT operation current is through the lower resistance area, which is near-by the grain boundaries.

Figs.7 and 8 show the local resistance as a function of distance in poly-Si TFTs with the LDD lengths of 1.0 μm and 0.6 μm, respectively. The local resistance had ranges of $10^8 - 10^{12}$ Ω at the channel region and ranges of $10^6 - 10^{12}$ Ω at the drain region. It should be noted that the lowest resistance profiles clearly show LDD structures between the channel and the drain with a length of approximately 1.0 μm in Fig. 7 and 0.6 μm in Fig.8.

CONCLUSION

A local resistance of the low-temperature poly-Si in a TFT was measured for investigating an activation of dopant atoms. The TFTs were designed to have LDD structures with mask lengths of 0.6 μm and 1.0 μm. The measured local resistance ranges by 4 - 6 orders of magnitude for measuring points. The measured resistances were separated into lower and higher resistances. The higher resistances were inside the grains and at the grain boundaries. The lower resistances were near-by the grain boundaries. The lower resistance data were evaluated since the TFT operation current usually flows through the lower resistance area. From the lower resistance data, the drastic resistance change from the channel to the drain regions, which is LDD, is observed in both samples. The LDD lengths of two samples are found to be approximately 0.6 and 1.0 μm, respectively.

REFERENCES

1. Y. Uraoka, T. Hatanaka, T. Fuyuki, T. Kawamura and Y. Tsuchihashi, Jpn. J. Appl. Phys. **39**, 1209 (2000)
2. Y.Uraoka, Y. Morita, H. Yano, T. Hatanaka and T. Fuyuki, Jpn. J. Appl. Phys. **41**, 5894 (2002)
3. T. Shiba, T. Itoga and Y. Toyota, SID 02 Digest, 220
4. T. Yoshida.Y. Ebiko, M. Takei, N. Sakaki and T. Tsuchiya, Jpn. J. Appl. Phys. **42**, 1999 (2002)
5. Y. Uraoka, K. Kitajima, H. Kumiura, H. Yano, T. Hatanaka and T. Fuyuki, Jpn. J. Appl. Phys. **44**, 5A 2895 (2005)
6. C. Shafai, D. J. Thomson, M. Simard-Normandin, G. Mattiussi and P. J. Scanlon, Appl. Phys. Lett. **64** (3), 342 (1994)

Study on chemical binding states of silicon in conjunction with ultra-shallow plasma doping by using Hard X-ray Photoelectron spectroscopy (HX-PES)

C.G. Jin[1], Y. Sasaki[1], K. Okashita[1], H. Tamura[1], H. Ito[1], B. Mizuno[1], T. Okumura[2], M. Kobata[3], J.J. Kim[3], E. Ikenaga[3], K. Kobayashi[3]

[1] *Ultimate Junction Technologies Inc.,*
3-1-1, Yagumonakamachi, Moriguchi, Osaka, 570-8501, Japan
[2] *Matsushita Electric (Panasonic), Kadoma, Osaka, Japan*
[3] *Japan Synchrotron Radiation Research Institute (JASRI/SPring-8), Hyogo, Japan*
Tel: +81-6-6906-6208, Fax: +81-6-6908-8707, Email: jin@ujtlab.com

Abstract. We took HX-PES measurement (Si 1s) on ultra shallow plasma doped silicon samples before and after spike RTA, flash lamp anneal (FLA) and all solid-state laser anneal (ASLA) in SPring-8 for the first time. After PD, the carrier density of n-Si substrate decreased to intrinsic Si level due to defect induced carrier traps. After annealing by either spike RTA or FLA, the PD samples showed excellent chemical binding states with high impurity activation and recrystallization. After annealing by ASLA, PD samples showed ultimate high impurity activation at surface several nanometer layer.

Keywords: Plasma doping, Hard X-ray Photoelectron spectroscopy, chemical binding.
PACS: 52.77.-j , 78 , 81 , 85.

INTRODUCTION

Plasma doping (PD) has been proposed as one of the most promising alternatives to replace the conventional low energy ion implantation for the fabrication of advanced devices of 45nm technology node and beyond as it presents high productivity and unique advantages on junction performance due to low sheet resistance and steep junction abruptness[1-3]. Although the conventional electrical measurement such as four-point-probe can prove the benefits of the PD process, the fundamental reason that explains why PD can result in better activation of silicon is yet to be understood. HX-PES in SPring-8, Japan, offers an ideal capability to study the chemical bonding of silicon in atomic level as it provides the highest X-ray flux with photon energy of 6-10 keV from synchrotron light source which enables the measurement of 1s orbit of silicon up to 10nm deep from the wafer surface [4]. Unlike the ordinary XPS, HX-PES can give the intrinsic information of silicon atoms across the whole depth of ultra-shallow junction (USJ). In this study, we measured Si 1s spectra of ultra shallow plasma doped Si wafers combined with various types of thermal annealing including spike RTA, flash lamp annealing (FLA) and all solid-state laser annealing (ASLA), and evaluated the effects of the PD process and annealing process on chemical binding state.

EXPERIMENTS

An n-type 8-inch Si wafer (10 Ωcm) was introduced into the PD chamber in which a helicon wave plasma of B_2H_6 (diluted by He) was irradiated onto the Si wafer with an external DC bias potential. The PD conditions are summarized in Table 1. The doped wafers were annealed by the spike RTA, FLA and ASLA. For the spike RTA, the peak temperature was 950-1075°C. For the FLA, the intermediate temperature (pre-heating) was 750°C, the front side peak temperature was 1163-1348°C and the flash lamp irradiation time was 1 ms. Laser annealing was performed by ASLA with a green frequency-doubled diode pumped solid-state laser system (λ=0.53μm, pulse duration 100 ns), which has advantage of good stability of energy density (standard deviation = 0.3 %) , infrequent maintenance and small footprint

Table 1. The PD conditions of the processed samples

Sample No.	PD condition				
	B2H6/He concentration	Vdc (V)	time (s)	depth (nm)	Dose (cm−2)
PD1	0.025%	60	14	7.0	5.0E+14
PD2	0.025%	60	30	7.0	1.0E+15
PD3	0.20%	60	30	7.0	2.0E+15

Table 2. The annealing conditions of samples

Sample No.	Annealing condition		Rs (Ω/sq.)
	method	peak temperature or energy density	
PD3-RTA1	spike RTA	975 °C	1345
PD3-RTA2		1025 °C	686
PD3-RTA3		1075 °C	369
PD3-FLA1	FLA	1163 °C	1300
PD3-FLA2		1295 °C	540
PD3-FLA3		1348 °C	340
PD3-LA1	LA	1300 mJ/cm2	3620
PD3-LA2		1400 mJ/cm2	830
PD3-LA3		1500 mJ/cm2	430

(1.5m x 1.8m). The annealing conditions are summarized in Table 2.

HX-PES measurements were performed at an undulator beam line, BL47XU, of Spring-8. X rays monochromatized at 7936.7eV with a Si 111 double-crystal monochromator and a Si 333 channel cut monochromator.

RESULTS AND DISCUSSION

Fig. 1 shows Si 1s spectra of as-doped PD samples compared with that of n-Si substrate: (a) area normalized; (b) peak position offset with normalized peak height. The binding energy of PD samples was about 0.3eV lower than that of n-Si and it varied slightly depending on PD conditions. As the change in binding energy of Si 1s means the shift of Fermi level in band gap caused by carrier density variation, the energy shift by the PD process may be attributed to the change in carrier density caused by defect induced carrier traps. Knowing that the Fermi level of the intrinsic Si is about 0.3eV lower than that of n-Si substrate (10Ωcm), the carrier density of the substrate after PD appeared to drop to the intrinsic Si level. As shown in Fig.1 (b), the full width at half max (FWHM) was increased after PD, and the peak shape became asymmetric in lower binding energy region. The peak broadening associated with the PD process appears to be caused by the amorphization of the silicon substrate [2, 3]. When the silicon crystal is amorphized, it would

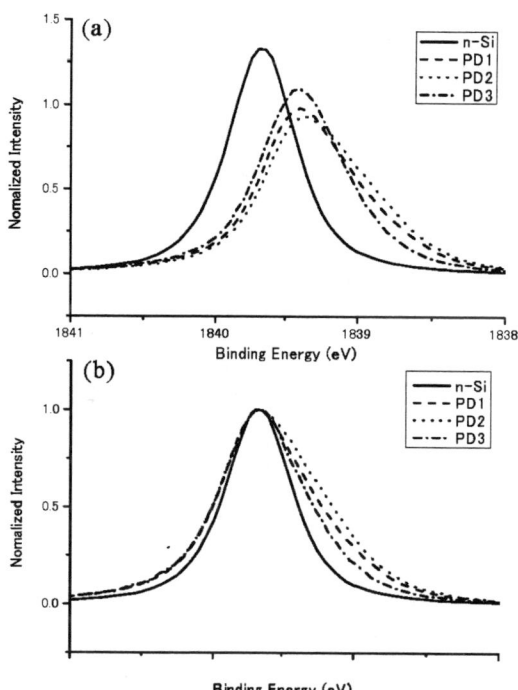

Fig. 1. Si 1s spectra of as-doped PD samples compared with that of n-Si substrate: (a) area normalized ; (b) Peak position offset with normalized peak height.

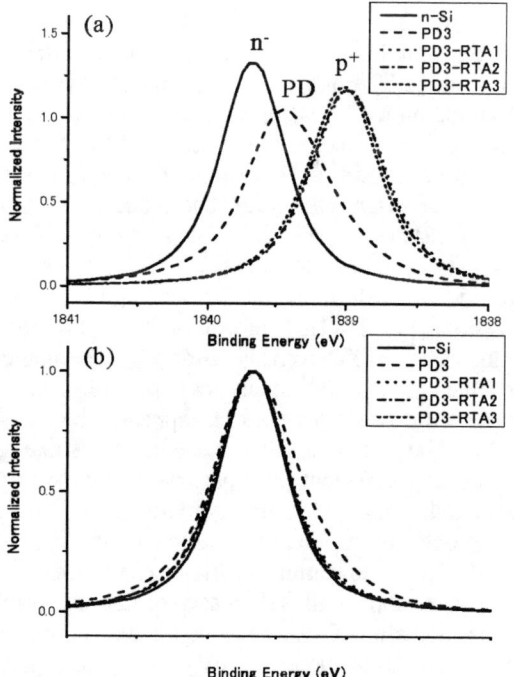

Fig. 2. Si 1s spectra of PD samples before and after spike RTA: (a) area normalized ; (b) peak position offset with normalized peak height.

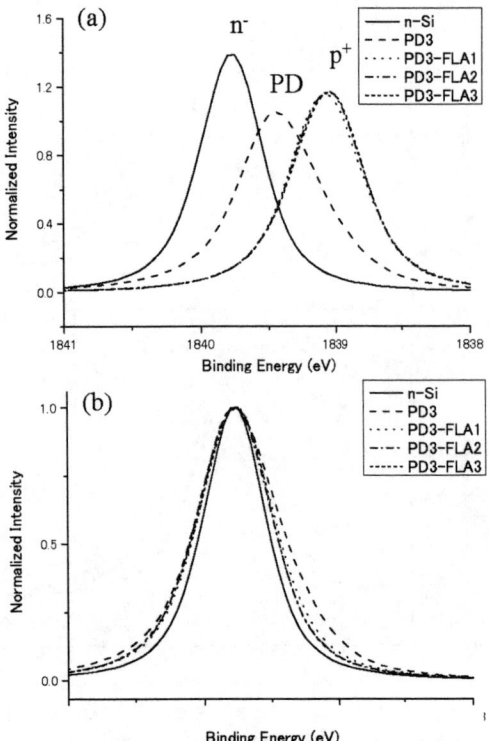

Fig. 3. Si 1s spectra of PD samples before and after FLA: (a) area normalized ; (b) peak position offset with normalized peak height.

result in a range of variation in length and angle of atomic bonding, which can lead to the peak broadening. The origin of the asymmetric peak shape observed on the PD samples, was thought to be a type of B-Si clusters and/or a presence of H-Si bonding.

Fig. 2 shows Si 1s spectra of the PD samples before and after spike RTA. The reduction of binding energy due to spike RTA was about 0.45eV. It suggests high hole density after spike RTA, considering that the half of Si band gap is 0.55eV. In addition, the FWHM of the spike RTA samples reduced by 0.1eV from the as-doped PD sample and became almost the same as that of n-Si. The asymmetric peak shape also disappeared by the spike RTA. Those results indicate that the spike RTA achieves excellent impurity activation and recrystallization. Variation in binding energy and peak shape between different peak temperatures was rather small in those annealing conditions examined.

Fig. 3 shows Si 1s spectra of the PD samples before and after FLA. The reduction of binding energy due to FLA was about 0.43 eV, which was almost the same as that of the spike RTA. Though the asymmetric peak shoulder did disappear by FLA, the FWHM after FLA was slightly larger by 0.06eV compared with the spike RTA. This would indicate a difference in re-crystallization of Si between FLA and spike RTA

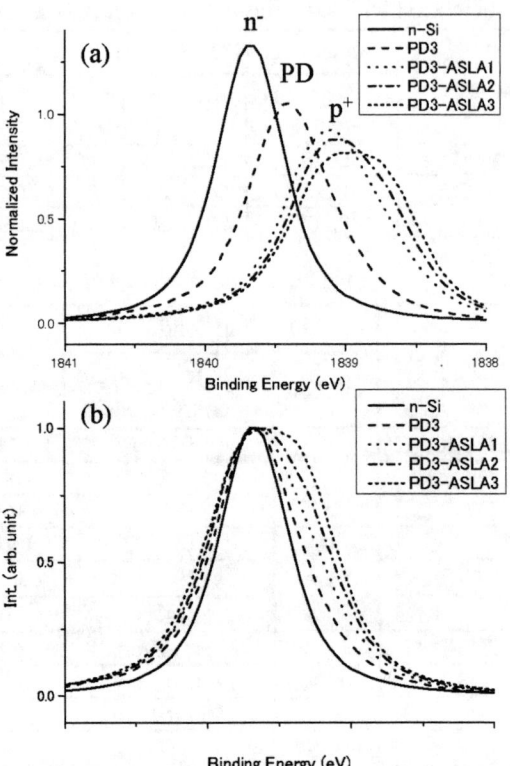

Fig. 4. Si 1s spectra of PD samples before and after ASLA with different laser energy density: (a) area normalized ; (b) peak position offset with normalized peak height.

cases as their thermal exposure is largely different (the ratio of annealing time: FLA/spikeRTA = 1/1000). Binding energy and peak shape did not change significantly over the different peak temperatures examined.

Fig. 4 shows Si 1s spectra of the PD samples before and after ASLA with different laser energy density. The binding energy of ASLA samples was shifted to lower binding energy region, and higher laser energy density resulted in lower binding energy. Unlike spike RTA and FLA, the FWHM was increased after ASLA. In addition, second component in lower binding energy region was appeared in ASLA samples.

Peak fitting using Voigt function was performed to separate the second component, as shown in Fig. 5. With the increasing of laser energy density, the area of second component is increased. It is correspond to the lower sheet resistance at larger laser energy density. All above Si 1s spectra from Fig. 1 to Fig. 5 were taken at take-off angle of 80° which had detection depth of 10 nm from the Si wafer surface. In order to investigate the origin of second component, we also measured the Si 1s spectra of sample PD3-ASLA3 for various take-off angles from 80° to 15°, which

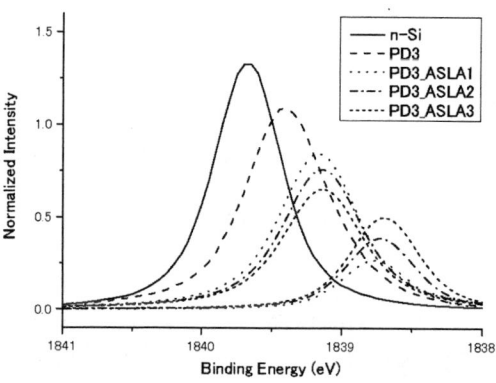

Fig. 5. Si 1s spectra of ASLA samples with different laser energy density after peak fitting using Voigt function.

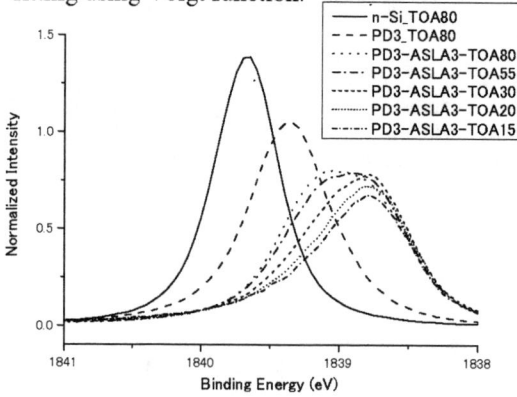

Fig. 6. Si 1s spectra of sample PD3-ASLA3 for various take-off angle (TOA) from 80° to 15°.

correspond to detection depth from 10 nm to 2.6 nm, as shown in Fig. 6. The binding energy of sample PD3-ASLA3 with low take-off angle below 30°, which was 0.2 eV lower than that of high take-off angle above 55°, was about 0.6 eV lower than that of sample PD3 with take-off angle 80°. Those results indicate that ASLA achieves ultimate high hole density in surface several nanometer layer caused by laser melt process.

CONCLUSIONS

HX-PES measurement was taken for the first time on ultra shallow plasma doped silicon substrate with Si 1s spectra before and after thermal activation. After PD, the carrier density of n-Si substrate decreased to the intrinsic Si level due to defect induced carrier trap. After annealing by either spike RTA or FLA, the PD samples showed excellent chemical binding states with high impurity activation and recrystallization. After annealing by ASLA, PD samples showed ultimate high impurity activation at surface several nanometer layer. HX-PES was able to measure the change of silicon properties before and after the doping and the annealing processes with high enough accuracy in terms of carrier concentration affecting the Fermi level and the chemical bonding. This measurement technique can be very useful for optimizing the semiconductor doping and thermal activation processes for advanced micro-devices that have the junction depth shallower than 10nm.

ACKNOWLEDGMENTS

The authors are grateful to Dr. Jeff Gelpey and Dr. Steve McCoy in Vortek/Mattson for their help in performing the FLA processes. The authors are grateful to Dr. Toshio Kudo of Sumitomo Heavy Industries Ltd. for his help in performing the ASLA processes. They are also thankful to Drs. K. Tamasaku, M. Yabashi, D. Miwa and T. Ishikawa of SPring-8 for their help in the X-ray optics instrumentation at BL47XU. This work was partially supported by a Nanotechnology Support Project of The Ministry of Education, Culture, Sports, Science and Technology.

REFERENCES

1. A. Hori and B. Mizuno, *Tech. Dig. of IEDM*, (1999) p.641.
2. C.G. Jin, Y. Sasaki, K. Tsutsui, H. Tamura, B.Mizuno, R.Higaki, T.Satoh, K. Majima, H. Sauddin, K. Takagi, S. Ohmi and H.Iwai, *International Workshop on Junction Technology* (2004), 102.
3. Y. Sasaki, C. G. Jin, H. Tamura, B. Mizuno, R. Higaki, T. Satoh, K. Majima, H. Sauddin, K. Takagi, S. Ohmi, K. Tsutsui and H. Iwai, *Symp. On VLSI Tech.* (2004), 180.
4. K. Kobayashi, M. Yabashi, Y. Takata, T. Tokushima, S. Shin, K. Tamasaku, D. Miwa, T. Ishikawa, H. Nohira, T. Hattori, Y. Sugita, O. Nakatsuka, A. Sakai and S. Zaima, *Appl. Phys. Lett.* 83 (2003) 1005

Characterization of Parasitic Transistor Phenomenon in Nano-scale NAND Flash Device by Blanket Tilt Implantation and Scanning Capacitance Microscopy

Dong-Ho Lee, Seung-Woo Shin, Choon-Kun Ryu, Moon-Keun Lee, Noh-Yeal Kwak, Hyun-Soo Shon, Byung-Seok Lee, Sung-Ki Park and Kae-Dal Kwack[1]

R&D Division Hynix Semiconductor Inc., San 136-1, Ami-ri, Bubal-eub, Ichon-si, kyoungki-do 467-701, Korea.
[1]*Department of Electronics and Computer Engineering, Hanyang University, Seoungdong-gu, Seoul, Korea*

Abstract. We present the application of scanning capacitance microscopy (SCM) in the failure analysis of 70nm NAND flash memory device. The SCM results are compared with chemical staining data and the feasibility of using the SCM are discussed. In order to suppress these anomalous hump characteristic, we perform numerical simulation study of trench sidewall implantation process and proposed the larger tilt angle implantation which can reduce dose factor of STI implant process for the DJBV (Drain Junction Breakdown Voltage) characteristic.

Keywords: hump, shallow trench isolation (STI), parasitic transistor
PACS: R79.20.R

INTRODUCTION

As the design rule of NAND-type memory decreases down to sub 100nm tech regime, one of important problems is the control of the parasitic transistor phenomenon. The parasitic transistor which causes subthreshold kink at high substrate bias is a common phenomenon for STI(shallow trench isolation) technology, especially for isolation whose pitch needs to be shrunk. Therefore, the subthreshold hump in the current-voltage(I-V) characteristics is significantly enhanced at n-channel MOSFET's (Cell and High Voltage NMOS). To resolve the degradation of device performance by the subthreshold hump, Blanket Tilt Implanted shallow trench isolation(BTI-STI) process has been widely used for deep submicron device[1]. Furthermore, due to the continuous shrinkage of semiconductor device, the use of a good 2-D profile analysis technique is essential as these structures are entirely two-dimensional.

In this paper, we present the numerical simulation study of BTI process factor to suppress hump effect and investigate feasibility of the application of scanning capacitance microscopy (SCM) in the failure analysis of 70nm NAND flash device.

EXPERIMENTAL

The devices were fabricated by a NAND Flash process with a 70nm design rule using retrograde triple well, SA-STI (Self-Aligned STI) and WSix gate. A 250nm deep trench was formed with target angle of ~82 degree. After lining oxidation, boron was implanted into the trench sidewall at 15~25 degree and with rotation of four times to compensation shadowing. The trench gap was filled by high-density plasma (HDP) chemical vapor deposition(CVD), and planarization was performed using chemical mechanical polishing(CMP) after densification annealing at 900 degree for 60min in N2 gas. Next, IPD (Inter Poly Dielectric) and control gate was composed of WSix/Poly-Si stacked structure. After that, the process flow of NDNF type Flash is followed.

The SCM and chemical staining were performed for failure analysis of hump characteristic. The dC/dV image and the corresponding surface morphology were obtained using a Digital Instrument/Dimension 3100 system. The SCM probes were used commercially available Pt-Ir coated silicon tips with tip radius of 20nm. Dynamic range of detection level of SCM system is from 10^{15} to 10^{20}(cm^{-3}). All SCM images were obtained using the constant voltage mode (ac bias of 0.5V at 89kHz)

The chemical stain was carried out using a mixture of 1:100:24 for 49 percent HF, HNO3 and CH3COOH. The samples were prepared either by cleaving or focused ion beam(FIB) cross-sectioning and was imaged using a scanning electron microscope(SEM).

Results and Discussion

The measured Id-Vg curve in NMOSFETs (W×L=10×1.3um) is shown in Fig.1. The anomalous parasitic transistor phenomenon was not monitored in donut gate transistor with the STI edge-less between source to drain. And the parasitic transistor causes the part of the transistor near the STI edge to turn on before the central part of transistor dose. So, the results imply that the parasitic transistor characteristics are very related with channel length and STI edge. The hump effect becomes more severe as substrate bias Vbs increases due to the Vt difference between the main transistor and STI edge parasitic transistor. Therefore, the hump characteristics reveal the parasitic transistor dependence and is observed in the Id-Vgs curve.

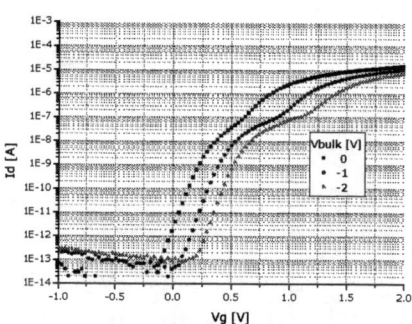

Fig.1 (a)

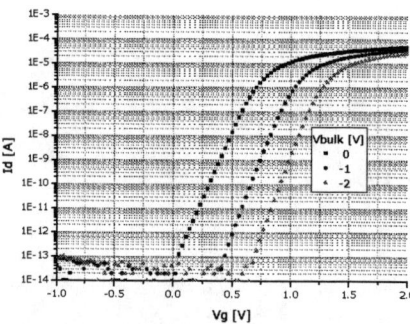

Fig.1(b)

Fig.1 Id-Vg characteristics of as a function of substrate voltage for (a) the normal type peripheral NMOSFET and (b) the donut type NMOSFET without STI edge pattern.

For the analysis of these parasitic transistor phenomenon, the feasibility of using chemical stain and SCM is demonstrated. Figure 2 shows the SEM image of the sample prepared through chemical staining in NMOSFETs regions. The result clearly reveals a uneven etch profile at the trench sidewall only in the case of the hump sample. It means a different concentration region exits in trench sidewall between the hump and no hump sample. However, the actual presence of a uneven etch profile remains uncertain as this could be an artifact of the chemical stain.

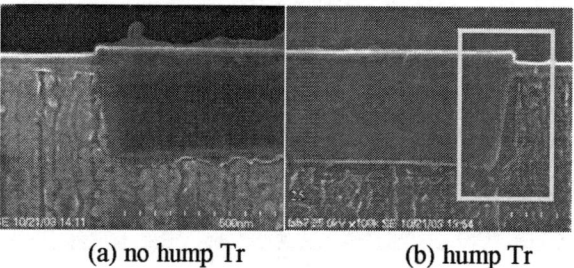

(a) no hump Tr   (b) hump Tr

Fig.2 Artifacts observed on the NMOSFETS after chemical staining

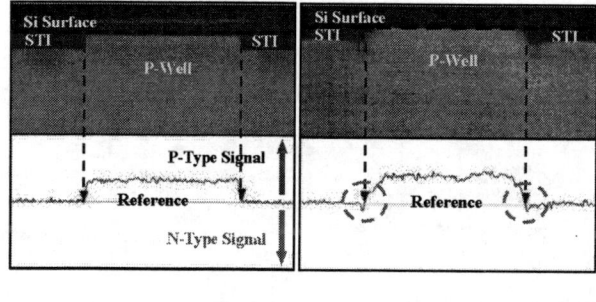

(a) no hump Tr   (b) hump Tr

Fig.3 SCM images of NMOSFETs at no hump(a) and hump(b)

In order to investigate these delineation SEM phenomena clearly, the SCM analysis was performed. Figure 3 shows the cross-sectional SCM images of the NMOSFETs regions. It clearly reveals the difference of dc/dv plot between the two samples in the trench sidewall region, which is well correlated with the anomalous hump transistor characteristic, repeatedly. Also, the SCM image (dc/dv) reveals the additional information about the trench sidewall region where abnormal n-type dc/dv signal in spite of p-type doping region (P-well region) was observed.

In the SCM dv/dc signal, the n-type signal is related with in two cases : 1)n-type doped region, 2)depletion region by induced low carrier concentration[2]. Considering that the measurement region was p-type well, it is assumed that the abnormal signal was caused by minor carrier-induced-depletion, not by n-type doping.

Although the chemical staining is a quick, efficient way to reveal the doping profiles, the etch rate varies greatly with different device structure and complexities, as dose the shelf life of the solution [3]. The SCM technique, on the other hand, eradicates these uncertainties and is quite sensitive to the effect of parasitic transistor which was caused by charge depletion around STI side wall.

To analyze anomalous hump effect, we performed a simulation of using TSUPREM4/DAVINC which is a 2D process and a 3D device simulator as a various trench sidewall doping conditions as shown in Table.1. According to swing curve fitting method, we can obtain the transistor hump factor (ΔVg).

Dopant	Energy [keV]	Dose [atoms/cm³]	Tilt [°]
Boron	10	3E11 ~ 7E11 (Rotation = 4)	15
			25
	20	3E11 ~ 7E11 (Rotation = 4)	15
			25
	30	3E11 ~ 7E11 (Rotation = 4)	15
			25

Table.1. Simulation split Table

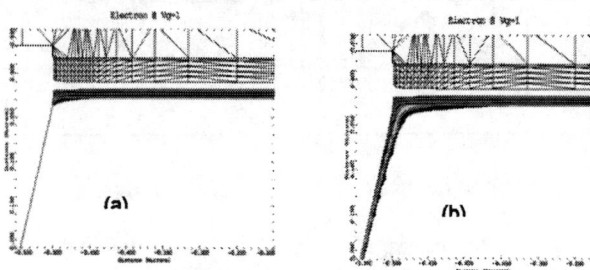

Fig.4. Simulation results of electron distribution at (a) no hump Tr and (b) hump Tr

Figure 4 shows the simulation analysis results for the parasitic transistor characteristics, which reveals electron distribution difference in the trench sidewall region only with a hump Tr, not without hump Tr. Also, the simulation results can explain the reason for the abnormal dc/dv SCM signal on the hump Tr as stated above. It can be predicted that the dc/dv SCM signal depends on depletion layer caused by electron distribution of the trench sidewall region.

Figure 5 shows the hump characteristics simulation results of the NMOSFETs as a function of STI implant conditions, energy, dose and tilt angle. Figure 5(a) shows the contour map of the hump factor with respect to the energy and the tilt angle at the dose of 5×10^{11} (ions/cm^2). The gray region indicates the hump-safe region where the hump factor (ΔVg) is less than 0.02. The wide hump-safe region was observed with the variation of tilt angle.

Figure 5(b) and 5(c) show the contour maps of the hump factor with respect to the energy and the dose at the tilt angle of 15 and 25 degrees, respectively.

As the tilt angle increases, the hump-safe area expands to the lower dose. At the larger tilt angle of 25 degree, the strength of the hump was effectively reduced, and the hump eventually disappeared above the dose of 4×10^{11} (ions/cm^2), as shown in Fig 5(c).

The hump characteristic can improve with increasing the tilt angle in spite of decreasing implant dose condition. In addition, the lower dose of implantation into trench sidewall is the more helpful for improving DJBV (Drain Junction Breakdown Voltage).

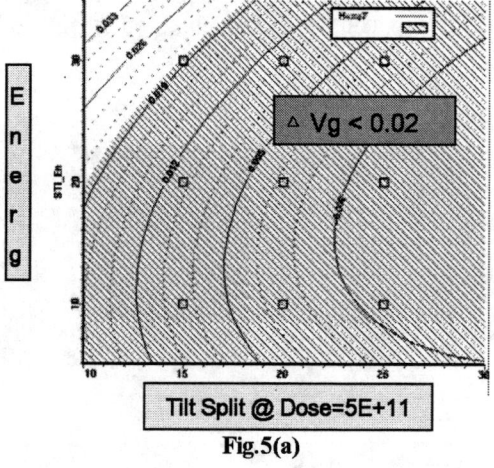

Fig.5(a)

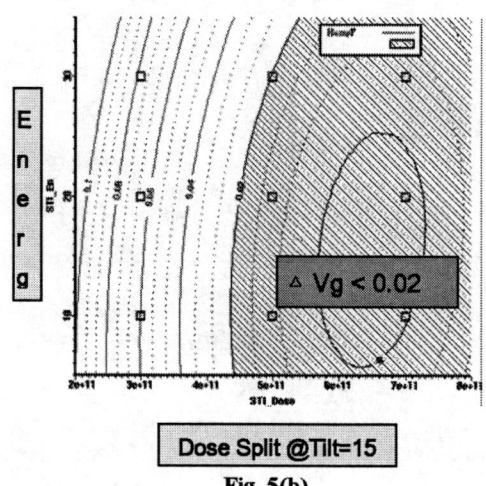

Fig. 5(b)

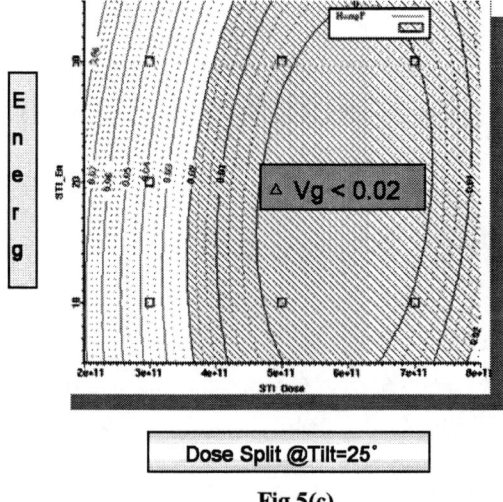

Fig.5(c)

Fig.5. The Simulated hump factor as function of (a) energy, (b) dose at 15° degree tilt angle and (c) dose at 25° tilt angle

CONCLUSION

We have found that the SCM technique is quite effective for investigating anomalous parasitic transistor phenomenon which was caused by depletion around STI side wall. In order to improve these anomalous device characteristic, the implantation onto trench sidewall by adjusting doping concentration was successfully applied. The larger tilt angle implant process was useful for the better DJBV characteristic in sub 50nm MLC (multi-level cell) NAND Flash memory.

REFERENCES

1. Jeong-Hwan Son et. al. " Blanket Tilt Implanted shallow trench isolation(BTI-STI) process for enhanced DRAM retention time characteristics" VLSI and CAD, 1999. ICVC '99. 6th International Conference, pp122 – 124.
2. H. Edwards et al, "An Analytical Technique for PN-Junction Delineation in Si Devices" APL Vol. 72, No 6, 1998, pp 698-670
3. C.H Tung et. al. " ULSI Semiconductor Technology Atlas", John Wiley & Sons Inc, ch.1, 2003, pp. 22-23.

Direct Measurement of Beam Angle in a High Current Ion Implanter

B. S. Freer[1], L. M. Rubin[1], M. A. Graf[1], and D. E. Hoglund[1]
D. Newman[2], K. Ditzler[2], K. Elshot[2], and T. Romig[2]

[1]*Axcelis Technologies, 108 Cherry Hill Dr., Beverly, MA 01915, USA*
[2]*Texas Instruments, 13011 TI Blvd., Dallas, TX 75243 USA*

Abstract: We report the first device results from a new method of direct measurement and real-time control of the average angle of an ion beam in a high current ion implanter. The angle detector consists of an array of high aspect ratio slots that are mounted directly on the same process disk containing the wafers. Beam profiling is achieved by measuring the ion current through the slots versus angle as the disk is rotated perpendicular to the slots. From this profile we determine an angle offset relative to the nominal implant angle. This offset may be a result of beam steering, mechanical positioning uncertainty, or both. The disk is then reoriented if necessary to ensure that the desired beam angle with respect to the wafer is achieved. We implanted the NMOS and PMOS source/drain extension implants for several dozen lots of 90nm and 120nm NMOS and PMOS devices. We showed tightened distributions of both transistor drive currents and asymmetry of drive currents under reverse biasing for 90nm and 120nm devices manufactured on 300mm wafers after the installation of the angle detection hardware. We also observed a tightening of the yield distribution for the 120nm devices.

Keywords: Ion Implantation; Angle Control; Ion Beam Steering; CMOS Devices
PACS: 61.72.Tt, 85.40.Ry

INTRODUCTION

Precise placement of all dopants is essential for optimum operation of advanced devices. For doping introduced by ion implantation, proper control of the dose, energy, and angle of the incoming ion beam is required. While the importance of dose and energy control have been known for a long time, it is only recently that the ion beam angle has been shown to significantly impact device performance[1-9]. The lightly doped drain (LDD) implant is particularly sensitive; ion beam angle errors in this implant adversely affect drive current (I_{ON}), and off state leakage current (I_{OFF})[1-4]. We report the first device data from a new method of *in situ* measurement and control of ion beam angles in a high current implanter.

DESCRIPTION OF APPARATUS

Direct angle measurement hardware and software were installed on an Axcelis HC3 Ultra high current ion implantation system. The angle detector hardware, described in more detail elsewhere[10], consists of an array of high aspect ratio slots mounted directly on the process disk, which also holds the wafers during implantation. The beam current that passes through the array and gets measured by the faraday behind the disk depends strongly upon the relative angle between the ion beam and the normal of the slots.

The angle correction algorithm used in this work has two main steps: beam height offset (BHO) correction and angle trimming. Before each implant the beam profile is taken, and the beam height offset is calculated. The BHO is the difference between the measured beam centroid (center-of-mass) and the center of the beamline in the dispersive (AMU) direction. Since beam location and beam angle are related in spot beam ion beam systems[10], beam position adjustments correspond to angle adjustments. If the BHO is outside of recipe-specified limits, the system iteratively changes the analyzer magnet current until it is within these limits. This centering of the beam serves as a coarse angle correction. All HC3 Ultras in this work used BHO correction.

After beam centering, a more direct beam angle measurement is made by sweeping the gyro in the dispersive plane of the beam and measuring the beam current that passes through the angle detector slots. Several hundred points are taken, one every disk rotation during the sweep, thus the angle measurement is very insensitive to noise. An angle centroid (average beam angle) in the dispersive plane is then calculated. The gyro is then repositioned as appropriate to remove any angle offset remaining after centering the beam.

EXPERIMENT

For a one month period, we implanted a 90nm n-LDD recipe and a 90nm and a 120nm p-LDD recipe with the *in situ* angle measurement and correction system enabled. These implants were selected because variation in device drive current (I_{ON}) are known to be related to angle errors in the beam centroid. Furthermore, for these implants, repeatability of these parameters had been improved by implementing centroid correction alone.

Drive current, transistor asymmetry (% change in drive current upon reversal of source and drain biasing), and yield of n- and p-channel devices were measured on devices implanted using both BHO and *in situ* angle correction. We measured a total of over 100 lots of 90nm and 120nm devices with an average of 22 devices per wafer. These measurements were compared to those on devices implanted using BHO correction alone.

RESULTS AND DISCUSSION
Measured Beam Angles

Table 1 shows the average beam angle error measured (and corrected for) after the BHO correction for the n-LDD and p-LDD over a two week period. The BHO alone corrected the beam to an average angle error of ~0.5° or better, with a maximum deviation from zero of ~1.0°. The gyro was then used to correct (up to 0.8°) for this residual angle error.

TABLE 1. Average correction required after *in situ* angle measurement using the slotted detector

	Beam angle (degrees)		
	90 nm N-LDD (n=28)	120 nm P-LDD-1 (n=26)	90 nm P-LDD-2 (n=45)
Average	0.21	-0.46	-0.51
Std. Deviation	0.22	0.24	0.19

Device Performance

Figure 1 shows measured I_{ON} currents for 120nm PMOS devices. While the median on current is not noticeably different, the devices fabricated after both the BHO correction and the *in situ* angle measurement correction (right side) had a significantly tighter distribution than those implanted after only the BHO correction (left side). The elimination of the small fraction of poorly performing devices is attributed to the improved beam angle accuracy and repeatability provided by the *in situ* angle measurement method. Similar results are observed for transistor asymmetry (Fig. 2) and overall yield (Fig. 3). Transistor asymmetry is defined as:

$$(\text{FWD } I_{ON} - \text{Reverse } I_{ON})/\text{FWD } I_{ON} * 100\% \quad (1)$$

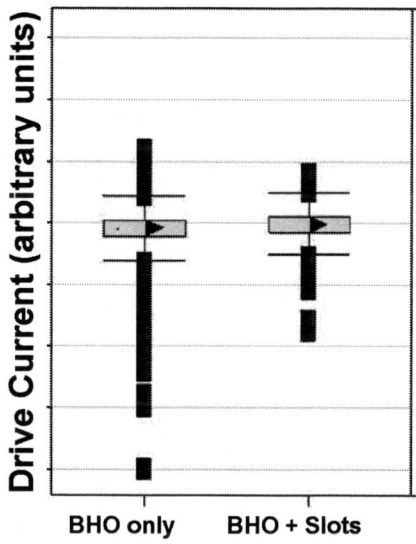

FIGURE 1. Normalized I_{ON} currents for 120nm PMOS transistors with (BHO + Slots) and without (BHO only) *in situ* angle measurement.

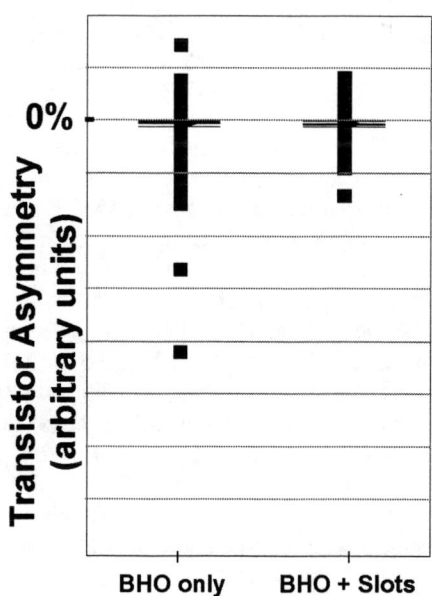

FIGURE 2. Normalized distributions of transistor asymmetry for the 120nm PMOS transistors

Fig. 3 also compares the yield distribution of the HC3 Ultra with and without the *in situ* measurement hardware to other implanters in the fab. The *in situ* measurement hardware slightly increases the overall yield, and sharply reduces the standard deviation by eliminating the tails of the yield distribution. For the duration of this test, there was nothing unique about implanter #2 (e.g. more frequent preventative maintenance) other than the use of *in situ* angle detection.

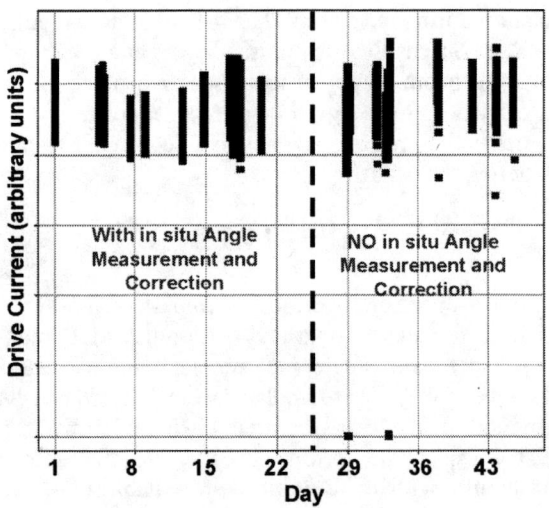

FIGURE 4. I_{ON} currents for 90nm NMOS transistors with (left side) and without (right side) *in situ* angle measurement.

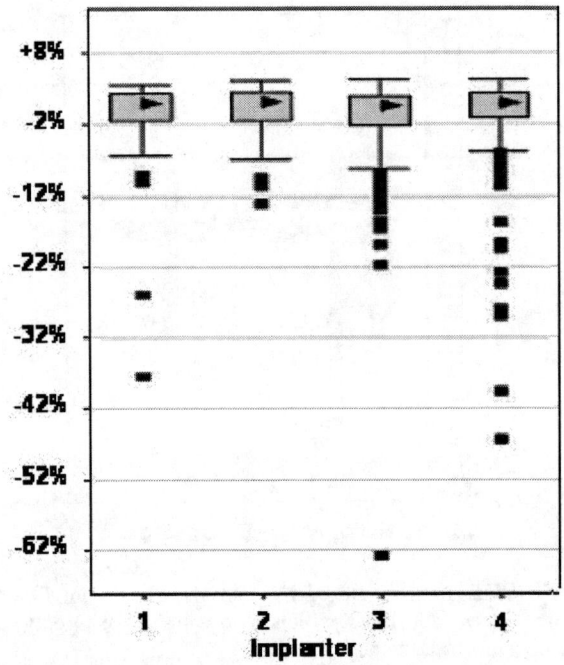

FIGURE 3. Normalized 120nm yield vs. implanter used for the p-LDD implant. Only implanter #2 was outfitted with slots.

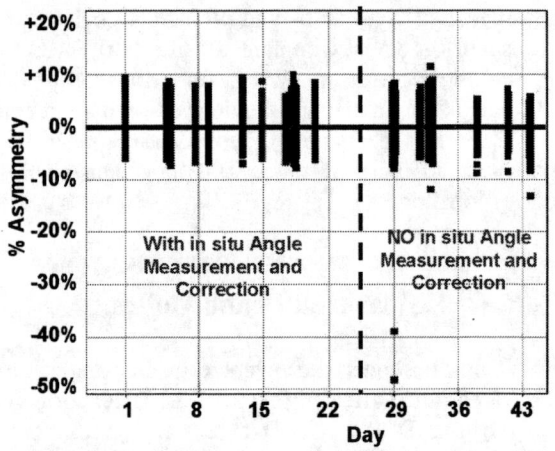

FIGURE 5. Normalized transistor asymmetry for the 90nm NMOS transistors with (left side) and without (right side) *in situ* angle measurement.

SUMMARY

We present the first electrical and yield data from a direct, *in situ* measurement of multi-wafer beam angle error in the dispersive plane followed by re-orientation of the wafer to correct for the error. The standard deviation of drive current, transistor asymmetry, and yield have been significantly reduced, and overall yield has been slightly improved. The reduced standard deviations in device parameters are attributed to reduced day-to-day variations in beam angles.

Drive current and transistor asymmetry data for 90nm NMOS and PMOS devices are shown in Figs. 4-7. For both parameters on both device types, use of the *in situ* angle detection and correction hardware showed a statistically significant reduction in the count of extreme outlying points (device failures) over the use of the beam height offset adjustment alone. It is the elimination of this tail of the distribution that is of most interest to fab operations. The 90 nm results are not as dramatic as the 120nm results because the 90nm data without *in situ* angle detection had a smaller distribution tail to begin with. This is due to two factors: greater care taken in general for all 90nm pilot processing; and a 90nm sample size only about ¼ as large as the 120nm dataset.

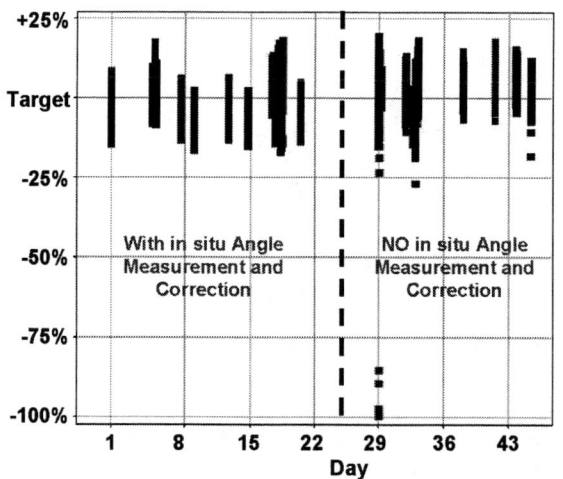

FIGURE 6. Normalized I_{ON} currents for 90nm PMOS transistors with (left side) and without (right side) *in situ* angle measurement.

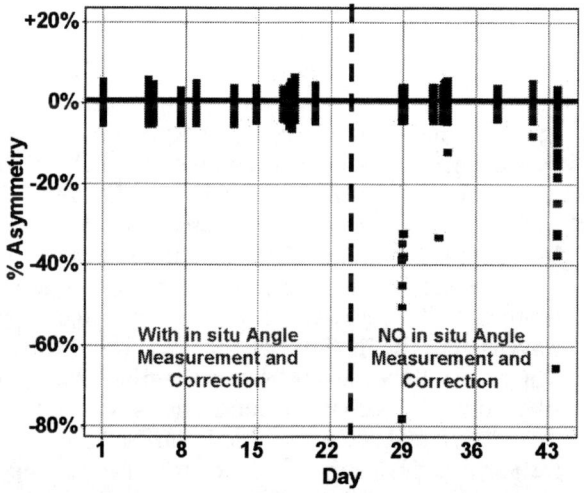

FIGURE 7. Transistor asymmetry for the 90nm PMOS transistors with (left side) and without (right side) *in situ* angle measurement.

REFERENCES

1. H.-J. L. Gossmann, L. Rubin, T. Parrill, and A. Agarwal, "Impact of extension implant energy purity and angle on the electrical characteristics of a 65 nm device technology", *to be published in J. Vac. Sci. Tech. B*.
2. J.D. Bernstein, J.J. McComb, A.W. Alvarez, J. Chow, and L.M. Rubin, "Quad-mode source/drain extension implants for reduced sensitivity to angle variation," *IEEE Trans. Semi. Manuf.*, submitted for publication.
3. T. Ohzone, M. Yamamoto, H. Iwata, and S. Odanaka, "Electrical characteristics of scaled CMOSFET's with source/drain regions fabricated by 7° and 0° tilt-angle implantations", *IEEE Trans. Electron Dev.*, vol. 42, Jan. 1995 pp. 70–77.
4. C. Jasper, P. Dahl, and Y. Y. Yang, "Electrical Comparison of a Parallel Beam and Batch Implanters", *Proc. of the XIII Int'l Conf. On Ion Implantation Technology*, Alpbach, Austria, Sept 18-22, 2000, pp. 376-379.
5. R.S. Santiesteban, G. C. Abeln, T. E. Beatty, O. Rodriguez, "Effect of tilt angle variations in a halo implant on V_T values for 0.14-μm CMOS devices", *IEEE Trans. Semi. Manuf.*, vol. 16, Nov. 2003 pp. 653–655.
6. K. Yoneda and M. Niwayama, "The drain current asymmetry of 130nm MOSFETs due to extension implant shadowing originated by mechanical angle error in high current implanter", *Extended Abstracts of the Third International Workshop on Junction Technology*, Dec. 2-3, 2002, pp. 19–22.
7. E. Lampin, E. Dubois, X. Hui, S. Bardy, F. Murray, "Accurate modeling of large angle tilt and pure vertical implantations: application to the simulation of n- and p-LDMOS backgates", *IEEE Trans. Electron Dev.*, vol. 50, May 2003, pp.1401–1404.
8. D.G. Borse, K.N. Rani, N.K. Jha, A.N. Chandorkar, J. Vasi, V. Ramgopal Rao, B. Cheng, J.C.S. Woo, "Optimization and realization of sub-100-nm channel length single halo p-MOSFETs", *IEEE Trans. Electron Dev.*, vol. 49, June 2002 pp. 1077–1079.
9. L. M. Rubin, W. Morris, and C. Jasper, "Process Control Issues for Retrograde Well Implants for Narrow n+/p+ Isolation in CMOS", *Proc. of the XIV Int'l Conf. On Ion Implantation Technology*, Taos, NM, 2002, pp. 17-20.
10. B. S. Freer, R. N. Reece, M. A. Graf, T. Parrill, & D. Polner, "*In situ* beam angle measurement in a multi-wafer high current ion implanter", *Proc. of the XV Int'l Conf. On Ion Implantation Technology*, Taipei, Taiwan, 2004, pp. 378-383.

Photoelectric Measurement Method For Implanted Silicon: A Phenomenological Approach

K. Steeples and E. Tsidilkovski

QC Solutions Inc, Billerica, MA, USA

Abstract. A photoelectric method based on surface photovoltage effect (SPV) has been developed for monitoring of implanted silicon. A phenomenological model explaining major implant correlations has been proposed. The mechanisms for implant dose and energy sensitivities of the method have been identified.

Keywords: Metrology, photovoltage, defects, lifetime.
PACS: 61.72.Tt, 61.80.-x, 85.40.Ry, 73.50.Pz, 73.50.Gr

INTRODUCTION

The real-time fab metrology of implanted silicon is traditionally associated with the two techniques: optical reflectance measurement of crystalline damage for as-implanted silicon and electrical resistivity measurement (4pp) of annealed silicon. We have developed a non-contact photoelectric method [1] based on a surface photovoltage technique that allows measurement of both as-implanted and implant-annealed silicon wafers.

The surface photovoltage (SPV) effect is a well-established method that is used for measuring various semiconductor parameters, such as minority carrier lifetime, doping density, interface defect density, as well as for processing equipment characterization, such as furnace oxide quality, level of metal contamination, etc. Practically, all the SPV parameters are not measured directly, but are calculated from the measured characteristics using appropriate theoretical models and comparing the results to some reference values. Therefore, it is important to establish the proper theory and data analysis technique for this method. In general, the SPV phenomenon is based on two effects: 1) generation of electron hole pairs by light, 2) separation of the pairs by the surface electric field. In the depletion state, the minority carriers are collected at the surface changing surface potential and thus creating the SPV. We consider a small signal ac-SPV – a method where frequency modulated band-gap light only slightly changes the surface potential $V_{SPV} \ll kT$. The first theoretical description of this method for a uniformly doped semiconductor in inversion was done by Nakhmanson [2]. We present here a model for the SPV effect in ion-implanted semiconductor.

SPV IN IMPLANTED SILICON: THEORY

The frequency-dependent small signal SPV measurement is well understood for the single crystal semiconductor with the uniform doping. Charge on the wafer surface induces a space charge of the width W_d. Incident light with the energy higher than a band gap generates electron hole pairs in the space charge and neutral bulk regions, depending on the light wavelength. The lifetimes of generated photo-carriers depend on the recombination mechanisms. In a pure single crystal silicon photo-carriers recombine through the mid gap defects, pre-dominantly at the wafer surface. In high frequency regime, the measured photo-voltage is proportional to W_d that allows calculating surface charge in the depletion or doping concentration in the inversion state.

The implantation process introduces a large number of crystal defects that act as recombination centers for photo-generated electron hole pairs (see Fig.1). This high defect density affects the surface photo-voltage through reducing the photo-carrier lifetime: the larger the defect density the shorter the lifetime, therefore less photo-carriers reach the surface and the smaller is the SPV. Another effect of implantation process on the SPV phenomenon is a local modification of the energy bands due to the charge re-distribution in the implanted areas. This effect may be caused by the net charge of introduced defects or through trapping of majority photo-carriers by the implanted trap centers. In general, this effect is less important for the SPV analysis, with the exception of a few special cases, described later in the paper

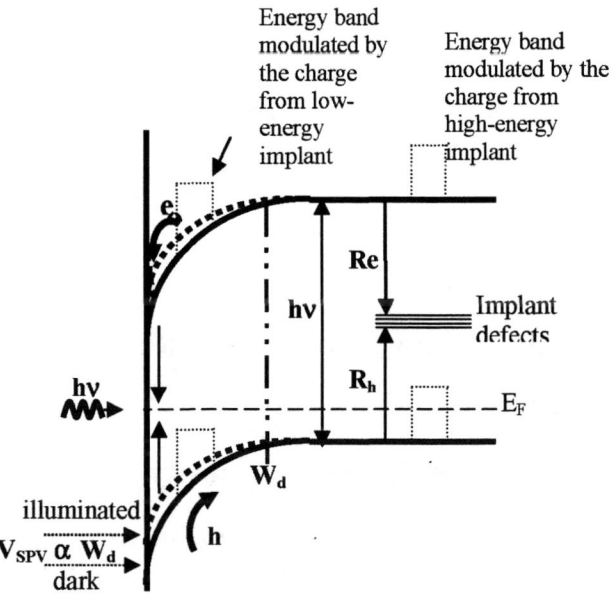

FIGURE 1. SPV band diagram of implanted p-type semiconductor in depletion; shows recombination through surface/space-charge states and through bulk implant defects

Equivalent Circuit

The standard set of differential equations describing the re-distribution of charge in semiconductor due to photo-excitation [2] can readily be analyzed with an equivalent electrical circuit.

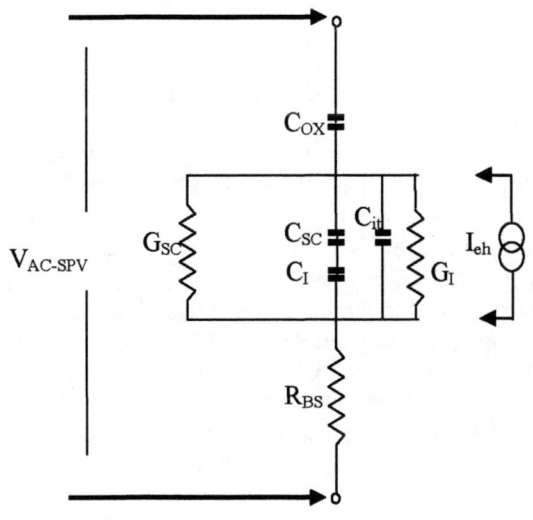

FIGURE 2. Equivalent electrical circuit of the SPV effect in semiconductor in full depletion with a contribution from implanted defects.

Conductance of the space charge region G_{SC}
Capacitance of the space charge region C_{SC}
Alternating photo current source I_{eh}
Oxide capacitance C_{OX}
Back contact series resistance R_{BC}
Implant area conductance G_I
Implant area capacitance C_I
Interface charge capacitance C_{it}

Figure 2, can be used to relate the capacitance and conductance to physical properties of the semiconductor, such as; doping density, carrier lifetimes, and defect density

From Figure 2 surface photo-voltage, V_{SPV} can be related to a familiar relationship [2]

$$V_{SPV}(\omega) = I_{eh}(\omega) Z_{eff}(\omega) \quad (1)$$

$$Z_{eff} = \frac{1}{G_{tot} + i\omega C_{tot}} \quad (2)$$

Where ω is a light modulation frequency, Z_{eff} is the effective impedance of the equivalent circuit, G_{tot} and C_{tot} are total conductance and capacitance of all contributing effects. At high frequency the oxide and interface traps capacitances can be neglected. The series resistance R_{BS} represent the majority carrier fast response and can be replaced by a short circuit, unless we deal with an intrinsic semiconductor. Further, in most practical cases the surface depletion layer width of the silicon substrates W_d used in IC processes (~1um) is much greater than implanted layer width (0.01-0.2um), that makes $C_I \gg C_{SC}$, and therefore we have $C_{tot} \sim C_{SC}$. Also the recombination time of photo-carriers in the implant induced damage area is normally several orders of magnitude shorter than the lifetime in un-damaged space-charge region. As a result, the implant area conductance is much smaller than the space charge conductance, leading to further simplification $G_{tot} \sim G_I$.

Model

We present below a model of a small signal ac-SPV in implanted silicon for a case when the probing light absorption depth $1/\alpha$ is greater than implant layer depth R_p. Other cases will be discussed qualitatively in the section on implant energy sensitivity.

It can be shown, that for low surface photovoltage $V_{SPV} < kT$, a relationship for the bulk (in our case implant area) conductance [3] reduces to:

$$G_I \approx \frac{q^2 n_i \Delta R}{\tau_{rec} kT} \quad (3)$$

with a characteristic recombination time

$$\tau_{rec} = \gamma^{-1} N_d^{-1} (kT/m)^{-1/2} \quad (4)$$

Here q is the elementary charge, n_i is an intrinsic carrier concentration, ΔR is an effective width of an implanted area, γ and m are a charge carrier capture cross section and an effective mass respectively, N_d is a concentration of defects-recombination centers.

Combining equations (1), (2) and (3) with a standard expression for space charge capacitance $C_{SC} \sim \varepsilon/W_d$, we find V_{SPV} to be:

$$V_{SPV} = \frac{I_{eh} W_d \tau_I}{(1 + i\omega\tau_I)\varepsilon} \quad (5)$$

With the implant area dominated lifetime given by:

$$\tau_I = \frac{kT}{q^2 n_i W_d \Delta R} \tau_{rec} \quad (6)$$

Where we assume the photocurrent density $I_{eh} \approx q\Phi$; Φ is the light flux density.

In the low frequency regime ($\omega\tau < 1$) SPV measurement gives:

$$|V_{SPV}| \approx \frac{q\Phi W_d \tau_I}{\varepsilon} \quad (7)$$

which is a direct measure of photo-carrier lifetime τ_I influenced by implant defects.

To have a convenient measure of implant characteristics we introduce a new parameter -- dynamic charge Q_D that represents a charge in a depletion layer modulated by the photo-carrier response $\widetilde{W}_d = W_d \times (\omega\tau_I)$. From a standard Poisson equation, at full depletion condition a charge in the near surface region can be related to a depletion layer width:

$$Q_D \propto \frac{kT}{q\varepsilon \widetilde{W}_d^2} = \frac{q^3 n_i^2}{\varepsilon m \omega^2} (\gamma \Delta R N_d)^2 \quad (8)$$

Leading to a dependence of the introduced parameter Q_D to depend on the square of implant induced defect density and implant area width.

Ion Implant Measurement

To illustrate the SPV effect in implanted material the following experiment was performed: silicon samples were implanted with the ions of different mass and energy set to obtain similar depth profiles at constant dose. Figure 3 shows that the SPV measurement of as-implanted silicon is highly sensitive to the implant atomic mass or the amount of induced damage: approximately 7x variation of the AMU corresponds to about five orders of magnitude change in the dynamic charge Q_D value.

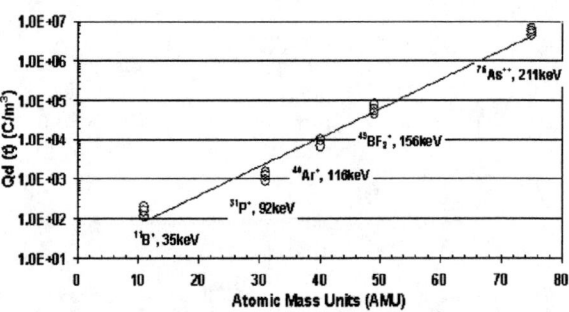

FIGURE 3. Sensitivity to implant damage/defect density at a constant dose of 6E13, with the same projected range of ~130nm [1].

Implant Dose Sensitivity

The mechanism of the SPV dependence on the implant dose is based on the straightforward relationship between the photo-carrier lifetime and the density of defects/recombination centers (6). Higher dose produces more defects that shortens the lifetime and reduces the photo-voltage signal. The SPV signal has monotonic behavior with respect to the dose up to the level of amorphization. Above the amorphization level the relevant lifetimes and the SPV measurement saturate, limiting the use of the method. Depending on other implant parameters affecting defect distribution – species, energy, tilt and implant temperature – the typical dose sensitivity factor is determined to be in the range from 1 to 3.

Implant Energy Sensitivity

The effect of the implant energy on the SPV measurement depends on the relationship of the damage profile and the probing light absorption. In the case described in a theory section, when the implant depth R_p is smaller than the absorption length $1/\alpha$, we find out from equations (5) and (6) that the photo-voltage is inversely proportional to the implanted area width ΔR, which is directly proportional to the implant energy. Thus, for all heavy implants or low energy light implants, the energy dependence of the calculated SPV parameter Q_D (8) has a monotonic behavior with a positive sign. (As^{75} implants with the energies < 200keV and B^{11} implants with the energy < 20keV are the examples of such measurement behavior, when the probing light wavelength is less than 0.5um.)

In case of the implant depth larger than $1/\alpha$, the energy dependence of the parameter Q_D has essentially non-monotonic behavior demonstrating first positive and later negative sign. To explain this energy dependence the non-uniform charge distribution, associated with implant has to be taken into account. The analysis of the non-uniform charge profiles requires solution of differential equations for multi-layer model that is beyond the scope of this paper; here we give a qualitative explanation of the negative energy dependence phenomenon. With the increase of implant energy/depth beyond the light absorption length, all the electron-hole pairs are generated in the area between the surface and the implanted region. After the majority carriers reach the implanted region, part of them get trapped and change the potential barrier associated with the implant area (Fig.1). This charge layer screens the remaining portion of the depletion area (in case of $R_p<W_d$) or creates an additional charge area at a distance R from the surface (when $R_p>W_d$), thus effectively modulating a measured capacitance/depletion width. Now the photo-voltage becomes proportional to the implant depth R_p, which replaces W_d in equation (5). So, for higher energy implants this mechanism dominates and is mainly responsible for the reverse behavior of $Q_D(E)$.

SPV Measurement of Annealed Silicon

The SPV measurement of the implanted and annealed silicon is essentially similar to the photo-voltage measurement of thin crystalline films [5]. The sequence of implantation and annealing processes creates a non-uniform doping profile in the near surface region that allows measurement of only an average value of the doping concentration. Due to the presence of several closely spaced and, in many cases, only partially depleted zones, the SPV measurement yields the net photo-voltage effect from one or more potential barriers. In case of the opposite doping types of implant and substrate (with the exception of ULE implants), the measurement result is highly inaccurate due to the compensating effects of the photocurrents oppositely directed at different barriers. So, normally SPV monitoring of annealed silicon requires same doping type of the implant and substrate that does not create additional p-n junctions. As it was mentioned earlier, the only exception is the ultra-low energy implants located within less than few nm from the surface. When implanted into the opposite doping type substrate, ULE implants make a very thin layer separated from the bulk by a potential barrier, resulting in a photovoltage signal dominated by the pn junction. One phenomenon characteristic to SPV measurement of ULE implants requires separate discussion. From Figure 4 one can see that a very distinct ring is visible at the edge of ULE wafer map.

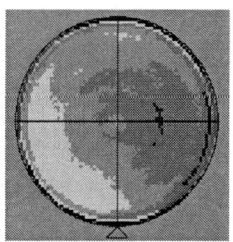

FIGURE 4. Typical wafer map of annealed ULE implant: B^{11} 0.5keV 1e15 implanted in n-type silicon substrate. The characteristic ring is due to the charge pile-up at the wafer edge.

Usually, for adequate explanation of major SPV related effects (strong inversion condition is one known exception) one-dimensional model is sufficient. This ring is an example of the case when a 3d modeling is required. Similar to the inversion layer, photo carriers in ULE area confined between the surface and p-n junction create lateral currents. Though, unlike the case of inversion, these lateral currents are formed by *majority* carriers, which have no other ways to move to the bulk because of the p-n junction presence. The majority carriers flow to the edges where the junction barrier can be lower due to the edge defects. The observed ring is clearly associated with build-up charge due to the major carriers piling up along the wafer edge.

REFERENCES

1. E. Tsidilkovski, K. Crocker, K. Steeples, "Ion implant process monitoring with a dynamic surface photo-charge technique", 15th Annual IEEE/SEMI ASMC Conference Proceedings, Boston MA, 2004, pp. 181-186.
2. R. S. Nakmanson,, *Solid State Electronics* **18,** 617-626 (1975)
3. D.K. Schroder, "Trends in Lifetime Measurements," in *High Purity Silicon* VI (C.L. Claeys, P. Rai-Choudhury, M. Watanabe, P. Stallhofer, and H.J. Dawson, eds.), Electrochem. Soc., Pennington NJ, ECS PV2000-17, 2000, pp. 365-382.
4. W. W. Gartner, *Phys. Rev* **116,** 84-87 *(1959)*
5. M. Leibovitch, L. Kronik,, E. Fefer, L. Burstein, V. Korobov, and Y. Shapira, J. Appl. Phys. **79,** 8549-8556 (1996)

Detection and Reduction of the Yield Impact of Particle Induced Structure Defects at Batch Ion Implanters

Matthias Schmeide, Michael Kokot, Dirk-Wito Franke*, Bernd Sauter

Infineon Technologies Dresden GmbH & Co. OHG, Königsbrücker Str. 150, D-01099 Dresden, Germany
*Qimonda Dresden GmbH & Co. OHG, Königsbrücker Str. 150, D-01099 Dresden, Germany

Abstract. This paper focuses on the introduction and qualification of the in situ particle monitor so called High Yield Technology (HYT) sensor on an Axcelis NV-GSD/E 200mm high current batch implanter to detect particles in real time during the implantation process. The particles on the wafer surface were measured with Surfscan and their composition was determined by means of Energy Dispersive X-ray (EDX) analysis. A good correlation between the HYT particle counts and surface particles on dummy wafers as well as defect densities measured on wafers structured with photo resist was found. Moreover, there is a well defined linear correlation of the HYT particle counts to the yield loss. To reduce the level of particle contamination, preventive maintenance procedures were optimized and the hardware in the beam line was modified. In order to minimize structural damage from high velocity particles, the disk drive was upgraded from a belt drive to a direct drive which offers the possibility to decrease the spin speed, thus to reduce the kinetic energy of the particles. Furthermore, measures to repair the already broken structures and to reduce the impact of particles on the products were taken. Depending on the values of the HYT in situ particle monitor, rework steps on various products were introduced. The results of these actions are discussed in combination with the costs.

Keywords: High current batch implanter, preventive maintenance, particle, structure defects.
PACS: 85.40.Ry

I. INTRODUCTION

In the last 10 years batch implanters were the most common tools for high current and high energy applications, due to their straightforward construction, good cooling performance and above all the good cost of ownership. Due to the short tuning time, batch implanters are still attractive for factories running lower batch sizes, such as logic factories.

The first in situ particle monitor, called the High Yield Technology (HYT) sensor, was installed on one of our Axcelis NV-GSD high current implanter at 1995. At this time, the design rules were about 0.5µm. Poly line, photo resist and structural damage due to large particles from particle excursion colliding with the spinning disk was not detected and reported. During the evaluation period of this sensor no correlation to the standard particle test, preventive maintenance (PM) intervals, defect density measurements or yield was found.

However, with reduction of the geometries under 200nm the structures on the wafer surface are very fragile and vulnerable to ballistic forces caused by particles emanating from the beam line in combination with the high spin speed of the disk. The destruction of structures parallel to the rotation direction is called avalanche effect and was described in the last years (e.g. [1]). Since in such a case the whole batch of wafers on the spinning disk is affected the yield impact is high. Thus, for factories equipped with batch implantation tools it is mandatory to solve this problem.

The software and hardware of the Axcelis batch implanters were prepared to use an in situ particle monitor, however, the reduction of the spin speed of the disk was not supported before 2005 and after this time, only possible on high current batch implanters with the newest hardware configuration. Therefore, it was necessary to quantify the risks and the yield loss due to damaged structures in combination to the costs of a hardware upgrade.

This paper focuses on the quantification of the particle problem, the introduction and qualification of the in situ particle monitor and the correlation between the HYT particle counts and surface particles on dummy wafers as well as defect densities measured on wafers structured with photo resist on an Axcelis NV-GSD/E 200mm high current batch implanter to detect particles in real time during the implantation process.

II. PARTICLE PERFORMANCE

The particles from implant tools are typically monitored by implanting monitor wafers. Typical frequencies are once a day, every 8 hours or more frequently. In most cases the particle monitor wafers are not implanted together with the production wafers. Thus, the tests are not sufficient to monitor the real particle performance of the implanter. Particle problems of the tool will be captured only later. Due to the nature of the particle events it is necessary to use a higher number of particle tests to get statistically significant information. However, this is not possible in the production environment.

Axcelis batch implanters have a batch size of 13 wafers on the disk. This means, for a batch size of maximum 25 production wafers after every second implant sequence the tool has to use at least one dummy wafer from the dummy wafer buffer.

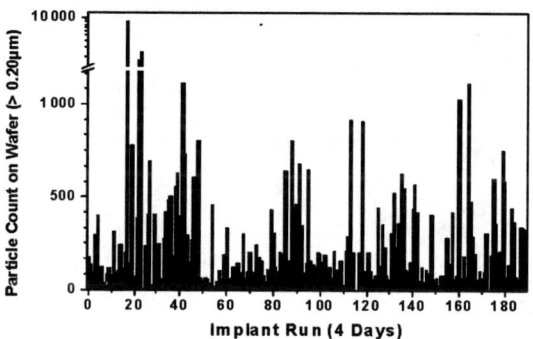

Fig. 1. Particle performance of an arsenic dedicated high current implanter.

To measure the real particle performance the particles on the dummy wafers from the dummy wafer cassette were measured using a KLA-Tencor Surfscan 6200. All particles greater than 0.20µm were counted. The particle performance obtained for an arsenic dedicated high current implanter over 4 days is shown in Fig.1. Compared to the standard particle test, the measurements on dummy wafers have shown a much greater number of particles. Within the 4 days of operation, 6 times the particle values were higher than 1000. At the highest value of 8500 the beam was very unstable and at least 9 interrupts were counted. At three other implants, the particle values were between 1000 and 2231 and the high chamber pressure resulted in 2 to 3 interrupts of the implant, but for two implants there were no difficulties observed.

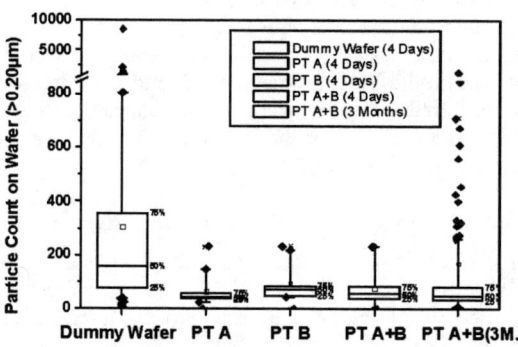

Fig. 2. Comparison of the particles on dummy wafers to the standard monitoring results.

Compared to the results of the dummy wafer measurement, the results of the standard particle monitoring tests are shown on Fig.2. Two standard particle tests are defined to control the defect density of the arsenic dedicated implanter. PT A is a test implant with an energy lower than 60keV, extraction voltage only, and PT B is a test implant using the maximum applied energy of the implant tool, 80keV arsenic, 60kV on the extraction electrode and 20kV on the accel electrode. As expected the median of the particle values of the PT B test is with 72 particles about 70% higher than the PT A test results of 43 particles. The median values from both tests over a period of 3 months are 50 particles compared to the median of the dummy wafer test running within 4 days with a median of 157 particles.

The results have shown that the used standard particle tests are not sufficient to control the particle performance of an implanter and are not suitable for product correlation. Due to the high costs of using the information captured by the dummy wafers this can be used only for short term analysis. Consequently, an in situ particle monitor has to be used.

III. INTRODUCTION OF AN IN SITU PARTICLE MONITOR

A. Description of the Particle Sensor

The in situ real time particle detection system consists of a High Yield Technology (HYT) 20SX/SXG particle sensor and a PM-250 electronic controller integrated into the implanter's system software. The main part of the system is the in situ particle detector shown in Fig.3. The sensor is mounted directly in the process chamber in the path of the particles accelerated from the spinning disk.

The sensor can detect particles as small as 0.19µm and the size of the electrical pulse depends on the particle size and the speed of the particles passing the laser beam.

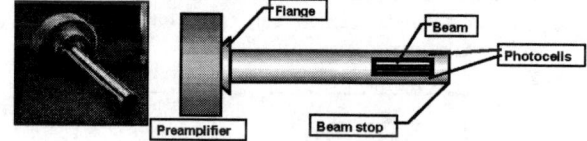

Fig. 3. Particle sensor 20SX/SXG.[2]

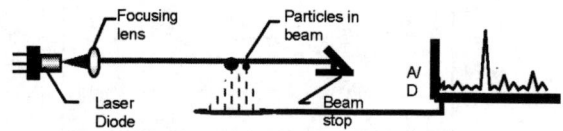

Fig. 4. Schematic of the particle sensor 20SX/SXG.[2]

The working principle of the sensor is shown in Fig.4 [2]. The particles pass through the laser beam and scatter light to the photocell. The photocell converts the light into an electrical signal; which will be amplified and processed to provide particle counts as a function of time. Particle counting will be done only when the disk is rotating. The implanter software is able to collect the results in real time and to compare those with recipe limits.

B. Correlation with Particles on Bare Dummy Wafers

The correlation was done using the same dummy wafer measurement described in section II and structured wafers with a 100% measurement. Over a period of two weeks the particles on the dummy wafer on an Axcelis NV-GSD/E high current implanter were measured with the Surfscan SP1. The measurement was performed periodically every 6 hours to be sure that each dummy wafer was implanted only once. For the final analysis SEM and EDX were used. As expected, the analysis results have shown that the particles are composed of the implant elements used (e.g. As, P, B), carbon from beamline components and resist deposition, silicon and SiO2.

The results show that there is linear correlation around 1 between particles on the dummy wafer and particles detected by the HYT sensor. An offset of around 50 particles is caused mainly by the handling system.

C. Correlation with Defect Density Measurement on Product Wafers and Yield Results

Due to the good correlation of the HYT counts to particle counts, affected production runs were used to measure the defect density on the structured wafers directly after the implant process step. In addition, the wafers were processed completely to measure the yield loss before they were scrapped.

In Fig.5, the correlation of the HYT counts to the defect density measured on structured wafers and to the yield loss is shown for one of the critical implants of the 110nm NROM technology. The difference between the HYT counts and the defect density measurement can partly be explained by the review area of the product wafer.

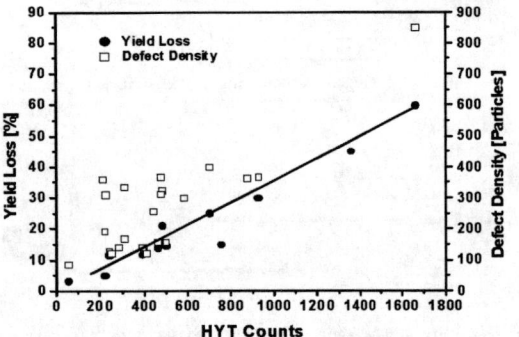

Fig. 5. Correlation between HYT counts, defect density and yield loss for a 110nm NROM technology.

Using the HYT monitor the final yield can be estimated directly after finishing the implant process. Compared to the 110nm NROM technology (cf. Fig.5) the sensitivity of the older 170nm NROM technology against particle related yield loss was only 50%. With decreasing structures the impact of the particles on the yield will increase dramatically. That means, particle excursion has to be minimized. For high volume products using high dose arsenic angle implants an excursion rate of to 0.5% was found.

D. Correlation with Implant Parameter

One of the most important parameters to describe the stability of the beam current during the implant process is the beam ratio, the relation between the beam current standard deviation and the mean value of the beam current, as described in [3].

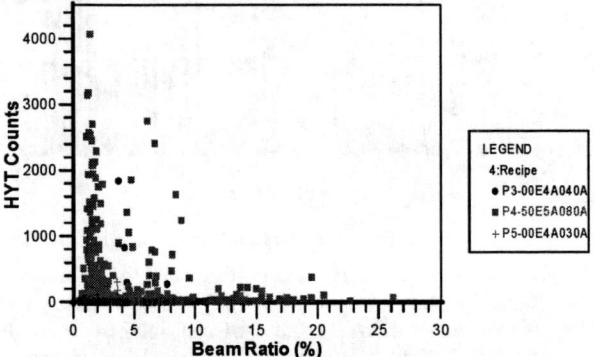

Fig. 6. Correlation between HYT counts and the important implant parameter beam ratio.

Typical values for the beam ratio for stable beam currents for high current implanters are 1% to 3%. If the beam current is more unstable, noisy, or if the implant is interrupted due to any reasons, the ratio will be higher. Due to this fact, the value is suitable to correlate to the HYT counts.

As an example the correlation is shown for 3 high dose arsenic implants in Fig.6. The expectation is that the HYT values are lower if the beam current is stable, but this is not true. There is no correlation between the beam ratio and the HYT counts in the data shown in Fig.6. It shows that very often the particles are from beamline components, from the disk or from the V3 bellow and not due to beam instabilities and thus, maintenance of these components is required. Without the HYT sensor the reason for the yield loss of the product could not be detected.

E. Example of the Effect of Rework Results

The software of the Axcelis batch implanters is able to set recipe selected limits for the HYT counter. This functionality was used to interrupt the implant process after reaching a defined HYT value. 50% of the affected wafers were reworked by removing the resist, cleaning the wafers, re-patterning and finally implanting using the abort implant recipe. This process was evaluated using products with a quarter micron design rule. The results have shown that even for older technologies the main yield detractors are damaged structures. For some products, no difference between affected wafers with and without rework is seen. Their yield loss compared to the clean implant run was nearly comparable. This shows that reworking the wafers without reduction of the spin speed is not sufficient.

F. Supervising Preventive Maintenance Activities

In addition to monitoring and controlling the product and the particle tool performance, the values from the HYT counter can be used to supervise maintenance and preventive maintenance (PM) activities. To reduce the level of particle contamination, PM procedures were optimized and divided into short and advanced PM's.

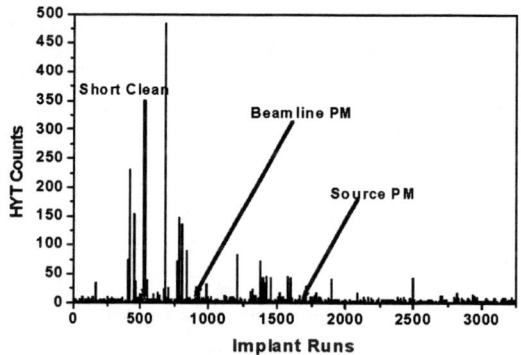

Fig. 7. Maintenance activities compared to the HYT counts for a time period of 23 days on an Axcelis NV-GSD/E² high current implanter.

In the example shown in Fig.7, the short PM activities including cleaning and exchange of many beamline components in combination with vacuum cleaning of the beam guide (analyzer magnet area) were carried out without reaching a better particle performance. Due to the still higher particle counts, an advanced beamline PM was done by exchanging all important parts of the beamline and the extraction area, including the V3 bellow and cleaning the beam guide. After 4 days of operation the tool again became a little more unstable. After performing a source PM the particle performance was stable over more than 10 days. The example has shown that supervising the PM activities can result in a better particle performance. However, the maintenance costs will increase and additional actions have to be taken to hold the costs and uptime stable, like using a counter for higher HYT counts in combination with our advanced process control system (APC), described in [3].

IV. REDUCTION OF THE SPIN SPEED

To be able to reduce the spin speed of the disk, the belt driven Axcelis GSD high current implanters have to be upgraded with a direct disk drive, a new dose controller and a new segmented disk. In addition, there are some process issues like wafer cooling, beam noise impact on uniformity, change of the diffusion, etc., which need to be carefully investigated before a product qualification.

Due to the high costs and the down time of nearly one week to do the upgrade, a cost of ownership calculation has to be done before, and to do this, the Axcelis ULTRA class implanter was used to evaluate the slow spin speed, called optimized fast spin speed (OFS) at Axcelis, due the availability of the hardware.

The results of the evaluation for the 130nm flash logic are shown in Fig.8. During the evaluation period no excursion was detected. A comparison of broken structures to baseline shows a yield advantage of nearly 0.6%.

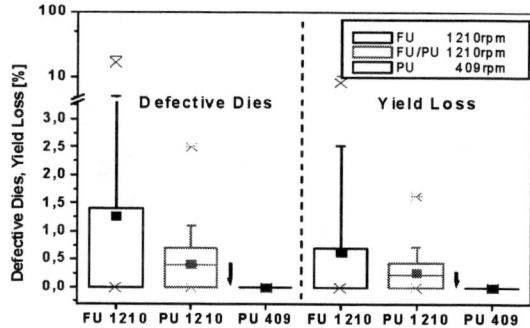

Fig. 8. Defective dies and yield loss depending on the disk spin speed.

V. CONCLUSION

The results have shown that in situ particle monitor data are an important addition to the standard implant parameters typically used to enhance product yield and monitor particle performance. Regular particle tests are not sufficient to meet the requirements of technologies lower than 170nm.

A yield advantage can only be reached if all activities fit together. This means, particle control by using the in situ particle monitor, optimization of the maintenance activities, reduction of the spin speed of the disk and rework of the affected wafers. Finally, the trend will go to single wafer implanters also for 200mm factories due to their better particle performance.

REFERENCES

[1] L. Pipes, M. Taylor, G. Zietz, "Characterization and reduction of a new particle defect mode in sub-0.25 µm semiconductor process flows", Proceedings of the XV International Conference on Ion Implantation Technology, pp. 330-335 (2004).

[2] "HYT Axcelis GSD Application Training", Hach Ultra Analytics, Grants Pass, OR, 2004.

[3] M. Schmeide, "Improved process control using statistical methods on Axcelis High Energy Implanters", Proceedings of the XV International Conference on Ion Implantation Technology, Part II, pp. 96-101 (2004).

[4] "Best methods and practices", Axcelis Technologies Inc., Beverly, MA, 2005.

Non-Contact, Image-Based Photoluminescence Metrology for Ion Implantation and Annealing Process Inspection

Andrzej Buczkowski[1], Zhiqiang Li[1], Tom Walker[1], Steven G. Hummel[1] and John O. Borland[2]

Accent Optical Technologies, 1320 SE Armour Rd., Ste. B2, Bend, OR 97702, USA
J.O.B Technologies, 98-1204 Kuawa St., Aiea, HI 96701, USA

Abstract. In this paper, we report results from a systematic evaluation of sensitivity, resolution and intrinsic capability of an RTPL system across a range of different implant conditions, including doses from 10^{11} to 10^{16} cm^{-2} and energy from sub-keV to the MeV level. Comparisons are made to existing non-contact and physical methods across this broad range of implant conditions. The RTPL system is shown to correlate well with all techniques investigated while offering sensitivity improvements for resolving critical parameters, such as energy, which is a reflection of the depth of the as-implanted junction. In addition, rapid, high resolution, full-wafer map imaging and smaller micron-scale scans enable visual characterization of uniformity under both as-implanted and annealed conditions. Visual inspection capability is shown to be especially useful for characterizing annealing processes and revealing unique residual defect patterns.

Keywords: ultra shallow junction, photoluminescence, metrology, as-implanted.
PACS: 78.55.-m

INTRODUCTION

Current generation source-drain extension implants employ relatively low energy and high implant dose to create junctions on the order of 10 nm below the substrate surface. The integrity of these so called, ultra-shallow junctions (USJ) is critical for optimum device performance where short channel effects and leakage will prevail in improperly formed devices. Typically, these junctions are characterized using electrical methods (Sheet Resistance, (Rs)) and/or physical methods (SIMS, TEM). Although effective within specific parameter windows, such methods require either electrical activation of the p-n junction – which requires a subsequent anneal – or they employ a destructive technique. Other less stringent implant steps, such as threshold voltage adjustment and channel engineering implants require similar characterization in order to develop and monitor their integrity. A nondestructive, in-line process inspection approach on monitor and patterned wafers is especially desired to reduce cost and latency associated with existing physical measurements. A specialized, non-contact, carrier lifetime-based room temperature photoluminescence (RTPL) method meets this demand. The RTPL system, which uses a novel excitation path design to achieve carrier confinement, device-suitable probing depth and submicron scanning resolution, offers a quick, non-destructive reading which is sensitive to dose, energy, leakage and the amorphous depth of the implanted layer while providing a full-wafer map image of the as-implanted and after anneal uniformity.

PHOTOLUMINESCENCE METROLOGY

Low and room temperature spectroscopic PL systems have been used for many years and are well established for defectivity characterization in silicon materials.(1, 2) However, such solutions are impractical for high volume, in-line process control. In contrast, the edge-band, carrier-lifetime-reliant RTPL system is well suited for nondestructive, high throughput damage and contamination assessment. The tool can operate in full wafer and micro mapping modes of operation. Simultaneously with the PL maps, surface reflectivity (SR) maps are collected. The RTPL measurement setup principles are illustrated in Fig. 1. The tool is equipped with two different wavelength lasers, called the Channel and the Bulk Probe. Depending on the desired probing volume (implantation energy), the Channel and Bulk Probes can be appropriately selected for evaluation. In the lateral dimension, laser beams are focused to a two micrometer spot by a system of lenses.

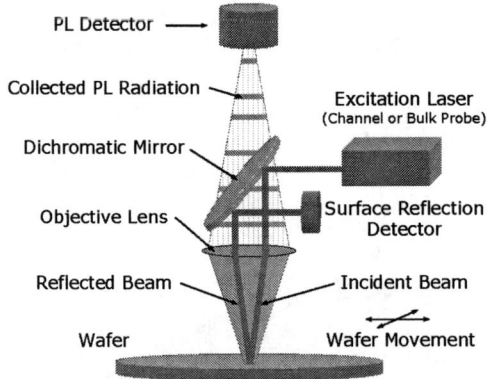

FIGURE 1. RTPL measurement setup.

Electrically active defects that are present in the bulk, such as post-implantation damage, lattice disturbances (crystal amorphization) or metal contamination are likely to be detected only in the PL image. Surface defects such as particles, growing faults or etching pits are predominantly seen in SR images. Surface defects may also be identified in PL images, if they modify the light reflection/ absorption path.

Differential and average PL signals along with their standard deviations provide useful and relevant information for quantification, inspection and control of ion implantation and annealing processes. To this goal, process limits can be defined using PL and surface reflectivity responses. In addition, results can be compared against a tabulated or otherwise pre-determined data repository (obtained via correlation) in order to quantify process excursions in terms of dose, energy or annealing condition variation.

RESULTS

A range of different implant conditions, including doses from 10^{11} to 10^{16} cm^{-2} and energy from 60 keV to 900 keV for ^{31}P species, as well as high ^{11}B doses implanted at 500 eV typical for USJ applications have been analyzed with the RTPL technique. An impact of different annealing conditions from relatively long, low temperature solid phase epitaxial (SPE) regrowth to very short, high temperature flash and laser anneals on damage removal is presented. Signatures of typical USJ annealing methods are illustrated using PL inspection (PLi). The level of damage is quantified in PLi arbitrary units.

As Implanted Samples

Figures 2 and 3 summarize the relationship between implantation dose and energy on PLi, respectively, for ^{31}P specie. Depending on the implantation conditions, the Channel and Bulk Probes were used for data collection.

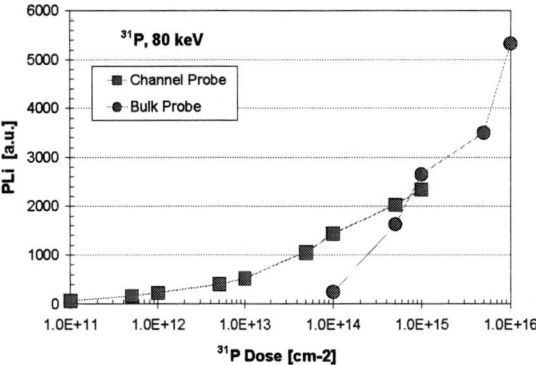

FIGURE 2. Impact of implantation dose on PLi levels for ^{31}P specie implanted at 80 keV.

Dose and energy sensitivity and detectability data are shown in Table 1. The data indicate that at applied implantation conditions, changes in dose and energy in the sub-1% range can be detected by PLi. Similar performance can be achieved for typical low energy, high dose USJ implantation conditions. High sensitivity and good measurement precision of the RTPL instrument enable imaging of damage uniformity over the entire wafer area; see Fig. 4, where implanter-specific variations in damage level over the entire wafer area are revealed. At a given equivalent implantation dose, despite differences in the implantation specie, the PLi signals can be effectively correlated to physical dimensions of the damaged layer, see Fig. 5 where PLi levels versus amorphous layer thickness, determined by TEM, are shown.

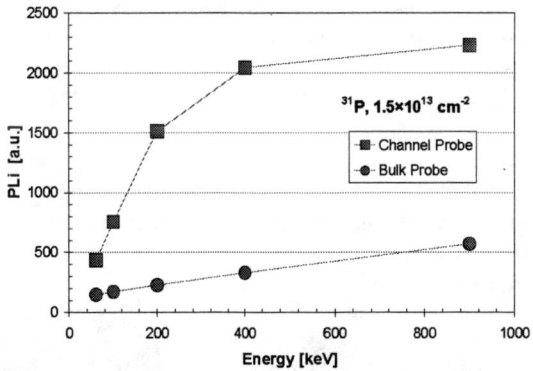

FIGURE 3. Impact of implantation energy on PLi levels for ^{31}P specie implanted at a 1.5×10^{13} cm^{-2} dose.

TABLE 1. Dose and energy sensitivity and detectability for ^{31}P specie implanted at constant energy (80 keV) or dose (10^{13} cm^{-2}), evaluated with (a) Channel or (b) Bulk Probe, using $D = 3\sigma/S$, $\sigma = 0.1\%$.

	Dose [cm^{-2}]		Energy [keV]	
	~$10^{13(a)}$	~$10^{15(b)}$	~$100^{(a)}$	~$500^{(b)}$
Sensitivity, S	0.39	0.42	1.10	0.67
Detectability, D [%]	0.76	0.71	0.27	0.45

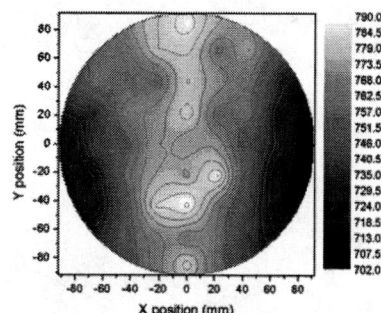

FIGURE 4. Ion implanter signature, ^{31}P, $E = 60$ keV, $C_s = 1.5 \times 10^{13}$ cm^{-3}, as-implanted sample.

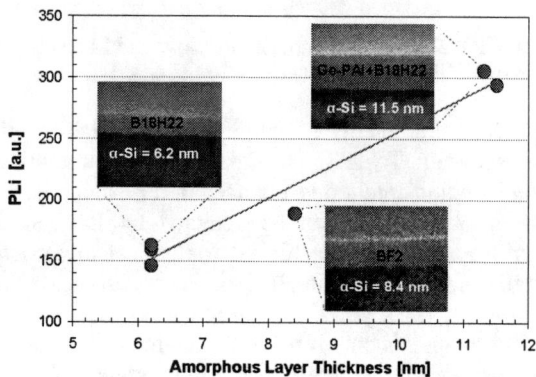

FIGURE 5. Correlation of PLi levels to amorphous layer thickness for USJ samples implanted at 10^{15} cm^{-2} equivalent dose. The thickness values were determined by TEM.

Annealed Samples

Selection of annealing conditions necessary for doping activation and damage removal at limited diffusion and relative insensitivity to pattern effects is absolutely critical for integrity of USJ formation. Suitable, high throughput and preferably non-destructive metrology is an important component of the effort. The RTPL inspection technique, with its quantifiable output, macro and micro mapping, sub-micron scanning resolution and high sensitivity to residual defects offers new possibilities in addressing these challenges.

Figure 6 shows full wafer and micro PL imaging of an ion-implanted wafer after short annealing, likely with a stripe-shaped laser beam, where stripping caused by beam overlapping can be clearly identified. Double-pass annealing reduces the residual damage in the region, as evidenced by PLi reduction from about 10.5 to 9.8 a.u. Based on the overall wafer average PLi level, it seems that the damage is not completely removed in the sample, as PLi values about 5 a.u. or lower are more typical for damage-free wafers.

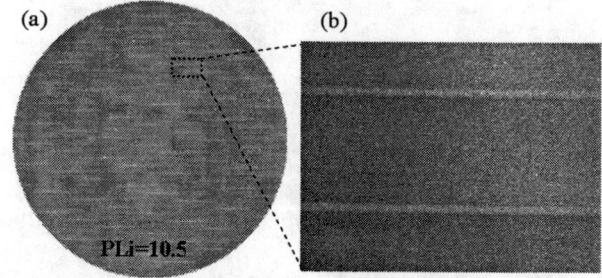

FIGURE 6. (a) PL full wafer map and (b) micromap showing close-up of stripping caused by overlap region.

The impact of typical annealing methods (SPE, spike, flash or laser) on damage removal and wafer-level uniformity is illustrated in Fig. 7 for a set of ^{11}B, 1×10^{15} cm^{-2}, 500 eV implanted samples. No preamorphization implantation was applied in this case. A large disparity in PLi levels within 5 to 110 a.u. range is observed. Note that significant uniformity problems are revealed for the annealing scenarios. For example, SPE clearly leaves behind considerable damage with PLi=110 a.u., while the spike, flash and laser anneal methods pose wafer uniformity challenges. Flash annealing led to the overall lowest (best) PLi levels of around 5 a.u., while laser annealing was not only less effective in removal of the damage, but also exhibited serious uniformity (stitching) problems.

Unique, sub-micron scale scanning resolution of RTPL enables complementary insights on damage removal capability of the annealing techniques. Figure 8 shows a series of 200 µm × 200 µm PL images, taken with 1 µm step resolution for a set of

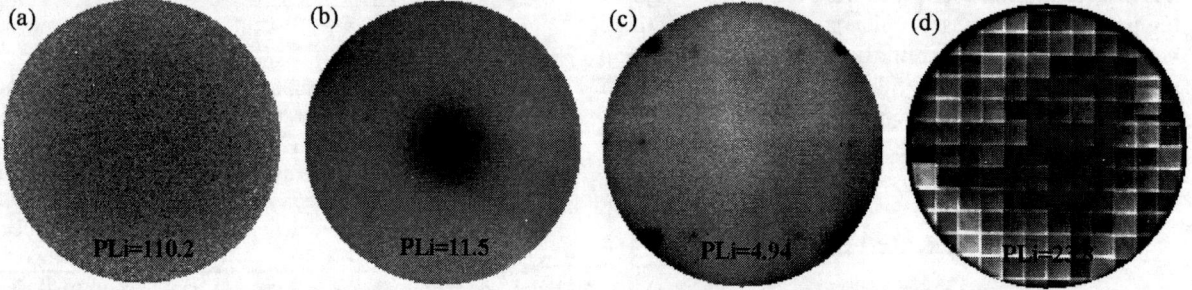

FIGURE 7. PL images of post-implantation annealed samples illustrating signatures of annealing equipment; (a) SPE, (b) spike, (c) flash, and (d) laser. All samples implanted with ^{11}B, 1×10^{15} cm^{-2}, 500 eV (without PAI).

^{11}B, 1×10^{15} cm^{-2}, 500 eV implanted samples. In this case, a germanium preamorphization step was used prior to boron implantation

The samples were subsequently subject to a variety of anneals including SPE, spike and flash. Consistently, the SPE anneal is characterized with incomplete damage removal as indicated by high PLi values. Due to the very high defect densities and small defect size, the individual defects cannot be resolved in these samples. The spike and Flash I processes led to better damage removal as compared to SPE, with PLi levels of around 12 a.u. and 29 a.u., respectively. The residual defect sizes in both cases remain too small to be individually resolved. In contrast, the more aggressive Flash II and III anneals lead to formation of large extended defects which can be easily imaged by PL. Note that the Flash II anneal produces low PLi levels on average, as the defect density is relatively small. Flash III annealing leads to a massive formation of slip defects, which are not necessarily directly attributed to the implantation process, yet result in high PLi levels indicative of residual damage. This residual, post-annealing damage deteriorates device performance due to excessive junction leakage. The damage, as assessed with PLi, is well correlated to junction leakage current, see Fig. 9.

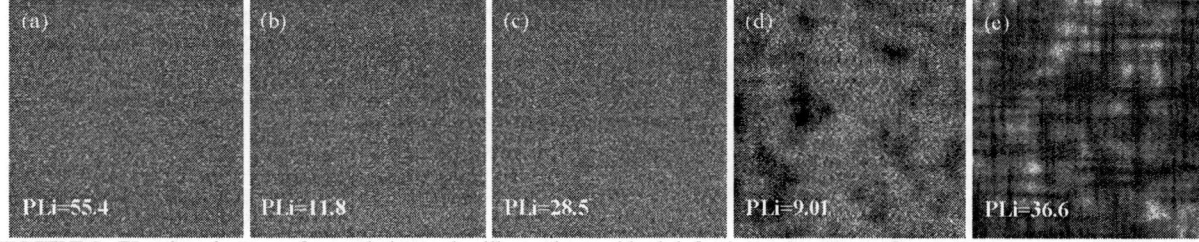

FIGURE 8. PL micro-images of annealed samples illustrating residual defectivity signatures of annealing process; (a) SPE, (b) spike, (c) flash I, (d) flash II, and (e) flash III; ^{11}B, 1×10^{15} cm^{-2}, 500 eV, with PAI implantation; 200 μm × 200 μm, 1 μm step images.

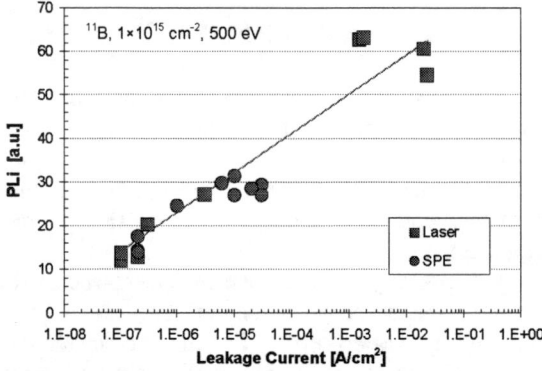

FIGURE 9. Correlation of junction leakage by RsL method (4) to residual defectivity by PLi.

SUMMARY

In this paper, a well-characterized, non-destructive, non-contact system utilizing room temperature photoluminescence (RTPL) is shown to offer significant benefits for the measurement of implant dose and energy as well as the characterization of thermal anneal processes. The following attributes are characteristic of the PL metrology system in this study:

- Ability to measure minute changes in implant conditions, enabling detection of sub-1% variations in both dose and energy
- Precise determination of residual damage following thermal annealing, such as SPE, spike, flash and laser anneals
- Detection of macro-uniformity effects using full wafer scanning, enabling troubleshooting and process optimization on a global level
- Determination of the local microstructure by using micro-scans for the detection of lattice-based defects in milli-second anneal processes

REFERENCES

1. J. Weber and M. Alonso, Physical Review B, **40**, 5683-5693 (1989)
2. M. Tajima, T. Masui, T. Abe, ECS Proceedings **90-7**, 994-1004 (1990)
3. A. Buczkowski, B. Orschel, S. Kim, S. Rouvimov, B. Snegirev, M. Fletcher and F. Kirscht, J. Electrochem. Soc., **150**, G436-G442 (2003)
4. V. Faifer, M. Current, W. Walecki, V. Souchkov, G. Mikhaylov, P. Van, T. Wong, T. Nguyen, J. Lu, S. Lau, and A. Koo, MRS. Proc., **810**, 475 (2004).

Superior Dose and Energy Monitoring Capability of the Therma-Probe System

M. Bakshi, D. Shaughnessy, L. Nicolaides, and P. Mitchell

Therma-Wave, Inc., 1250 Reliance Way, Fremont, California 94539

Abstract. The repeatability of the Therma-Probe®-XP TP630XP series ion implant monitoring system is investigated and the implications on implant monitoring capabilities of the system discussed. Improvements in tool stability enable 10-day average dose detectability of 2% 3σ using conventional on-the-fly measurements, and 0.6% 3σ using a recently introduced spatial averaging measurement mode. Dose detectability is the tool's figure of merit that combines both tool sensitivity and stability in a single parameter defining the noise-limited dose resolution of an ion implant metrology tool. The superior dose resolution of the Therma-Probe®-XP system enables implementation of in-line statistical process control for ion implant processes.

Keywords: ion implant monitoring, in-line metrology, process control.
PACS: 06.20.-f; 42.62.-b; 89.20.Bb

INTRODUCTION

As geometries continue to shrink, advanced devices require tighter implant dose control for production processes. Run-to-run control ensures that process requirements have been met, reduces scrap, and improves yield. The Therma-Probe®-XP is a non-contact non-destructive ion implant monitoring system enabling in-line metrology and eliminating costly test wafers. It is the industry leader for conventional ion implant monitoring with its current application base extended to include advanced ultra-shallow junction monitoring.[1,2]

Based on the modulated optical reflectance (MOR) technique,[3-6] Therma-Probe tools use a low intensity probe laser beam to detect small changes in the index of refraction that are induced in the semiconductor by a modulated pump laser beam. For the configuration employed with the Therma-Probe system, the change in the index of refraction is primarily a function of the local temperature and carrier density in the optically excited semiconductor,[7] which in turn will depend on the degree of crystalline damage created during the implantation process.[8] In this work, the capability of the Therma-Probe-XP system is investigated over a wide range of implant conditions. Excellent tool stability results in a superior dose resolution leading to metrology capability for in-line statistical process control.

EXPERIMENT

The dose monitoring capability of the Therma-Probe-XP system was investigated using three sets of samples: a) B^+, 10 keV, 9E11-5.5E13 ions/cm^2 (12 samples); b) B^+, 50 keV, 2E11-5E13 ions/cm^2 (8 samples); and c) As^+, 50 keV, 2E11-5E13 ions/cm^2 (12 samples).

A metrology system must provide a repeated and stable measurement to be of use in production monitoring. Short-term dynamic repeatability is evaluated here using 10 consecutive measurements on 10 keV boron implanted wafers with load/unload between successive cycles. Long-term stability is evaluated over 10 days with one measurement per day on each of the B^+ and As^+ 50 keV wafers. The wafers were offset 130 μm between cycles to ensure that no measurement overlap existed that would induce a relaxation-related trend in run-to-run results.[7,9] The measurements were performed in relatively uniform regions of the wafer in order to minimize sample influence and obtain results that are indicative of tool performance.

Traditionally, the Therma-Probe has been capable of making "on-the-fly" measurements with the stage in motion during data acquisition (area and line scans), or stationary measurements with the stage fixed in location (used for single point or template

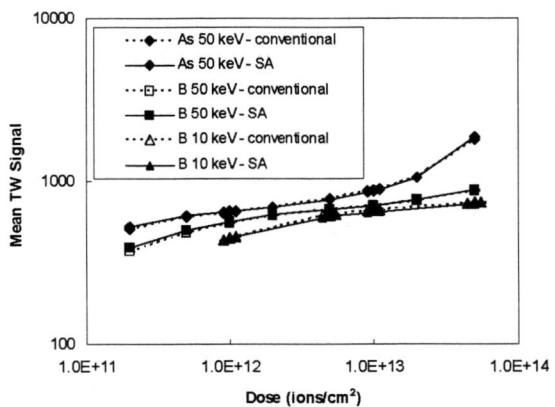

FIGURE 1. Mean TW infinity signal for all wafers using both conventional and spatial averaging (SA) measurement modes. The error bars contained within the respective data point represent signal repeatability.

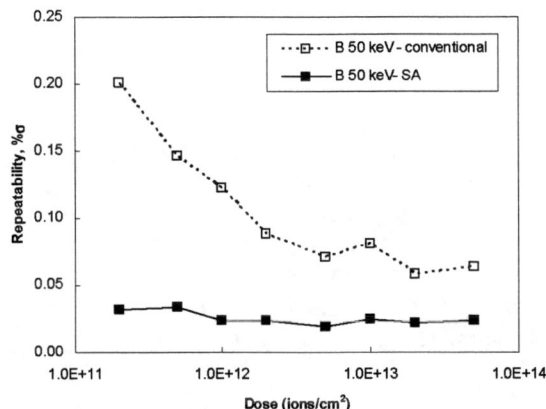

FIGURE 2. 10-day measurement repeatability for both conventional and spatial averaging modes (B⁺ 50keV).

measurements). A recently introduced proprietary measurement configuration, termed spatial averaging[10], allows the user to define a small "box" around a measurement site in single point or template mode. Measurements are taken on-the-fly over the defined lines thus allowing the user to control averaging time without over-exposing a single location on the wafer. The temporal dependence of the TW signal on laser exposure time is discussed elsewhere.[7,9]

For each case evaluated in this report, measurements were made with the conventional on-the-fly measurement ("conventional") mode, and the new spatial averaging ("SA") mode. The conventional mode is evaluated using a 5mm line scan measurement with the decay compensation feature enabled, and the spatial averaging mode is evaluated using a 50µm box at 5 sites equally distributed over 5mm length with 12s measurement time per site.

Measurement Repeatability

Overall wafer uniformity determined using area scans was found to range from 0.07% to 0.52% σ. The mean TW signal over the 10 runs for all samples is shown in Figure 1. The error bars in Figure 1 (mostly hidden within the symbols) represent the 10-run repeatability of the respective measurement.

The 10-day measurement repeatability ("long-term" or stability) using conventional and SA modes is shown in Figure 2. For clarity, the repeatability is shown for only one set of wafers (B⁺ 50keV). It is evident that the spatial averaging mode improves the long-term performance by a factor of 2-10 over the dose range of the samples. Table 1 shows a comparison for the 10-cycle repeatability (10 keV boron wafers) as well as the 10-day repeatability results (50 keV boron and arsenic implanted wafers)

TABLE 1. Mean signal and repeatability using spatial averaging measurement mode.

Boron, 10 keV			Arsenic, 50 keV			Boron, 50 keV		
Dose	Mean Signal	%σ	Dose	Mean Signal	%σ	Dose	Mean Signal	%σ
9.0E+11	438.5	0.02	2.0E+11	527.7	0.03	2.0E+11	393.0	0.03
1.0E+12	450.5	0.05	5.0E+11	612.8	0.03	5.0E+11	500.4	0.03
1.1E+12	461.0	0.02	9.0E+11	650.9	0.02	1.0E+12	568.5	0.02
4.5E+12	597.3	0.01	1.0E+12	657.7	0.02	2.0E+12	621.8	0.02
5.0E+12	604.7	0.01	1.1E+12	663.5	0.03	5.0E+12	674.3	0.02
5.5E+12	611.3	0.02	2.0E+12	699.7	0.03	1.0E+13	713.0	0.02
9.0E+12	640.3	0.01	5.0E+12	774.3	0.04	2.0E+13	765.7	0.02
1.0E+13	646.5	0.02	9.0E+12	849.8	0.03	5.0E+13	878.4	0.02
1.1E+13	651.8	0.02	1.0E+13	865.4	0.03			
4.5E+13	719.2	0.02	1.1E+13	883.5	0.04			
5.0E+13	725.2	0.01	2.0E+13	1054.7	0.04			
5.5E+13	729.8	0.01	5.0E+13	1875.1	0.07			
	Average =	0.02		Average =	0.03		Average =	0.03

TABLE 2. Average repeatability and dose detectability over entire dose range for given implant species and energy.

Implant	Conventional		Spatial Averaging	
	Repeatability, %σ	DD, %3σ	Repeatability, %σ	DD, %3σ
Dynamic (load/unload)				
B, 10keV	0.09	2.0	0.02	0.5
Long Term (10 day)				
B, 50 keV	0.10	2.1	0.03	0.6
As, 50 keV	0.10	1.9	0.03	0.6

for the case of spatial averaging. A comparison of the overall performance of 10-cycle repeatability and 10-day stability using both the conventional and the SA modes is provided in Table 2.

The 10-cycle repeatability presented in Table 1 varies from 0.01% to 0.05% giving an average short-term repeatability of 0.02% as indicated at the bottom of the table. The 10-day stability ranges from 0.02 – 0.07% for 50 keV As implants, while the same for 50 keV boron implants shows very little variation at 0.02 – 0.03%. Each stability set results in an average value of 0.03%. Though not presented in detail here for the conventional mode measurements, which had repeatability values ranging from 0.02% to 0.22% σ, the same behavior is found true: namely, that the 10-cycle repeatability effectively captures the overall tool noise figure so that it changes only slightly over a 10-day period, indicating superior tool stability.

The average measurement repeatability for each species/energy combination is shown in Table 2 for both measurement modes. The use of spatial averaging mode results in a greater than 3× improvement in measurement repeatability. This allows for tighter SPC limits for process control as demonstrated for the B^+ 50keV 1E13 ions/cm^2 wafer in Figure 3, as an example. Also of note, the short-term dynamic cycle results in better measurement repeatability than the long-term 10 day cycle, as would be expected. However, the fact the repeatability of measurements acquired during a short 10-cycle test are only slightly better than the performance of the 10-day runs is due to the excellent tool stability, as discussed above.

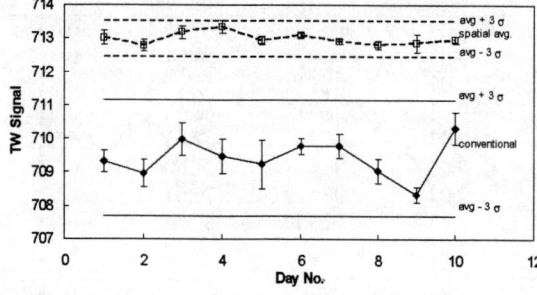

FIGURE 3. Day to day repeatability of B^+ 50keV 1E13 ions/cm^2 wafer for both conventional and spatial averaging measurement modes.

Dose Detectability

Measurement repeatability alone does not adequately characterize the performance of a tool for implant monitoring applications. Equally important is the tool sensitivity to dose. The dose sensitivity, *DS*, can be defined as the percent change in signal over the percent change in dose,

$$DS = \frac{(S_2 - S_1)/(S_2 + S_1)}{(D_2 - D_1)/(D_2 + D_1)}. \quad (1)$$

The sensitivity is essentially independent of measurement mode since the physics of the signal generation mechanisms does not change. The metric that most effectively represents the capability of a tool for SPC monitoring is the dose detectability, *DD*, expressed as %3σ parameter which can be defined as the repeatability divided by the sensitivity,

$$DD = \frac{3 \times \%\sigma}{DS}. \quad (2)$$

The detectability takes into account the tool noise and stability, which are incorporated in the repeatability measurements, as well as the tool sensitivity. Using only one of either the repeatability or the sensitivity to characterize tool performance can be misleading: very high sensitivity is not useful if poor tool noise or long-term drift compromises the measurement accuracy. On the other hand, relatively high apparent noise (including wafer uniformity) for a particular implant dose range may be due to the high dose sensitivity for that range. The dose detectability over the dose range considered for this study is shown in Figure 4 for the same set used in Figure 2 for repeatability results. Using the conventional measurement mode the detectability ranges from 0.4% to 3.3% 3σ among all the wafers. The consequences of the better repeatability of the spatial averaging mode are clearly evident with *DD* values ranging from 0.2% to 0.9%

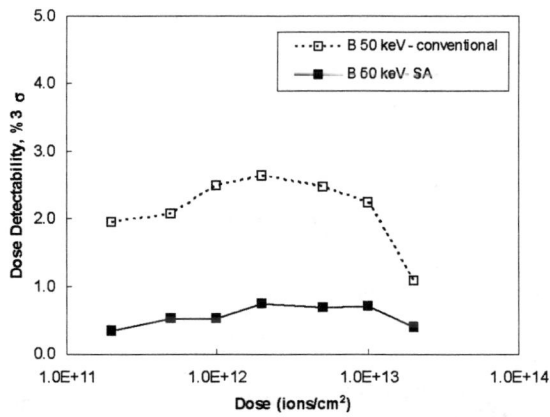

FIGURE 4. Dose detectability for both conventional and spatial averaging modes based on repeatability results of Figure 2 and sensitivity using Equation 1 (B^+ 50 keV).

3σ. The average *DD* for each species/energy combination is shown in Table 2. For a given measurement mode, the averages are nearly identical for each set of wafers. The conventional mode allows for a dose detectability of 2% 3σ on average. This represents a considerable improvement over non-XP generation Therma-Probe tools which typically exhibit dose detectability values around 6% 3σ or higher due to the relatively high tool noise.[11] Spatial averaging measurement mode results in a further improvement of more than 3× with an average dose detectability of 0.5% 3σ short term and 0.6% 3σ long-term.

Energy Detectability

Energy sensitivity and energy detectability can be defined in an analogous fashion to equations 1 and 2. The sensitivity to energy for XP generation Therma-Probe tools is unchanged from non-XP series tools. All have the same fundamental measurement parameters and thus the same underlying physics driving the signal response to implant induced crystal damage. The demonstrated enhanced stability of the Therma-Probe-XP improves energy detectability to a comparable degree as the improvements in dose detectability, since both are a consequence of improvements in overall tool noise. Typical values of energy detectability obtain an average value < 3% 3σ at medium energy conditions using the conventional mode[10]. While further improvement using the spatial averaging mode arises according to the noise reduction factor of about 2-3, just as for dose detectability.

CONCLUSION

The superior stability of the Therma-Probe-XP ion implant monitoring system has been demonstrated using a broad range of medium energy p and n-type wafers in the 2E11-5E13 cm^{-2} dose range. Average long-term dose detectability was found to be 2% 3σ for conventional mode measurements and 0.6% 3σ when using the new spatial averaging measurement choreography. The latter represents a 10× improvement in noise-limited dose resolution compared to previous generation tools. It should be noted that the noise-limited dose resolution is about the same for 10-cycles run over a period of hours as over 10-day period. Outstanding tool stability enables this level of consistent performance. A direct pay-off is thus generated in terms of a superior precision-to-tolerance metric to address the requirements of advanced device manufacturing.

ACKNOWLEDGMENTS

The authors would like to acknowledge Varian Semiconductor Equipment Associates for providing (10 keV boron implant) wafers for this study, and Alex Salnik and Jon Opsal at Therma-Wave, Inc. for helpful discussions.

REFERENCES

1. L. Nicolaides, A. Salnik, and J. Opsal, Rev. Sci. Instr. **74**, 586-588, (2003).
2. A. Salnik, L. Nicolaides, J. Opsal, A. Jain, D. Rogers, and L. Robertson, Rev. Sci. Instr. 75, 2144-2148 (2004).
3. A. Rosencwaig, J. Opsal, W. L. Smith, and D. L. Willenborg, Appl. Phys. Lett. **46**, 1013-1015 (1985).
4. W. L. Smith, A. Rosencwaig, and D. L. Willenborg, Appl. Phys. Lett. **47**, 584-586 (1985).
5. M. Guidotti and H. M. van Driel, Appl. Phys. Lett. **47**, 1336-1338 (1985).
6. J. Opsal and A. Rosencwaig, Appl. Phys. Lett. **47**, 498-500 (1985).
7. J. Opsal, M. W. Taylor, W. L. Smith, and A. Rosencwaig, J. Appl. Phys. **61**, 240-248 (1987).
8. A. Salnik and J. Opsal, J. Appl. Phys. **91**, 2874-2882 (2002).
9. J. Schurr, C. Waters, J. Maneval, N. Tripsas, A. Rosencwaig, M. Taylor, W. L. Smith, L. Golding, and J. Opsal, Nucl. Instr. and Meth. **B21**, 554-558 (1987).
10. U.S. Patent Application Ser. No.11/067,961.
10. Therma-Wave Inc. internal documentation, unpublished.

Mapping Leakages Of USJ Test Wafers

James T C Chen, Tatiana Dimitrova and Dimitar Dimitrov

Four Dimensions, Inc., 3140 Diablo Ave., Hayward, California 94545
Tel: (510) 782-1843, Fax: (510) 786-9321, Email: info@4dimensions.com

Abstract. A method for mapping ultra-shallow junction (USJ) leakage currents at a selected bias voltage for four-point probe test wafer is proposed and its measurement principle is explained. An example of using its test results as one of the references in optimizing USJ process is given. Comparison with leakage measurement of the Sheet Resistance-Leakage Current (RsL) technique is made.

Keywords: Ultra Shallow Junction, Leakage Current, Wafer Mapping.

PACS: 73.43Fj, 73.40.Lq

INTRODUCTION

It is well recognized that making the leakage current for the 45nm node and further advanced ICs to be within an acceptable limit is difficult but of vital importance to device reliability. In such devices, the leakage current between source or drain extension and the halo is dominating. Hence, monitoring the USJ leakage current for process control during IC fabrication is very important.

Methods for monitoring USJ leakage current using test wafers have been proposed by Chen and Liu[1] and Faifer et al. [2]. However, the former requires a masking step although it should result in more accurate measurement than others; the latter is based on equations derived through much simplification from the basic differential equations. In calculating the leakage current, the latter uses the ratio of two alternative surface photo voltages measured through non-contact capacitive coupling, which is not easy to be accurate, and the USJ junction capacitance, which is not directly measurable, as the parameters in the equations. Also, the leakage resistance it refers to is the leakage due to forward-biased recombination, not due to reverse bias.

We are proposing a new method for measuring and mapping the leakage currents of non-patterned USJ test wafer under a bias. It employs an automatic mercury dot-ring probe, with the dot connected to a ground potential integrator and the ring connected to ground, to probe the USJ layer on a four-point probe test wafer whose substrate is biased at a voltage Vs through the chuck, as shown in Fig.1.

PRINCIPLE OF THE METHOD

The substrate bias voltage causes current to flow from the substrate, through the USJ junction throughout the wafer, to the ring and the dot, with the junction current density decreasing further from the ring and dot. Since the mercury ring is surrounding the mercury dot and the USJ layer is very thin and conductive, the former

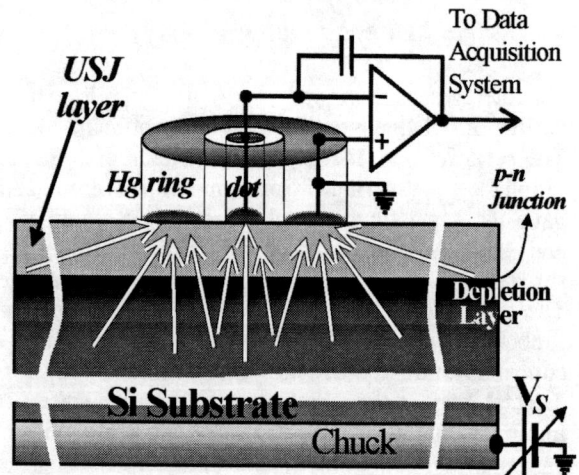

FIGURE 1. Schematic representation of the USJ leakage current distribution (white arrows) under the Hg dot-ring electrodes and how the leakage current around the dot is measured.

should collect almost all the current leaking through the whole USJ junction and only let the mercury dot collect the junction leakage current in the area close to or immediately under the dot. Thus junction leakage current under a reverse bias can be measured and mapped with good geometric resolution for a non-patterned USJ layer.

Also, under reverse bias, the bias voltage will mostly appear at the junction under the mercury dot if the junction is not too leaky because the depletion layer resistance in that small area is much higher than the dot-USJ contact resistance and the series resistances due to substrate to chuck contact and substrate bulk. Therefore, at the sites of good junctions, I-V of the USJ can be measured accurately within a reasonable range of bias voltage. In contrast to the low-leakage sites, the measured leakage currents at the leaky sites are relatively high. Consequently mapping based on multiple-site leakage current measurements identifies good junction areas and reveals the correct leakage resistance distribution in these areas.

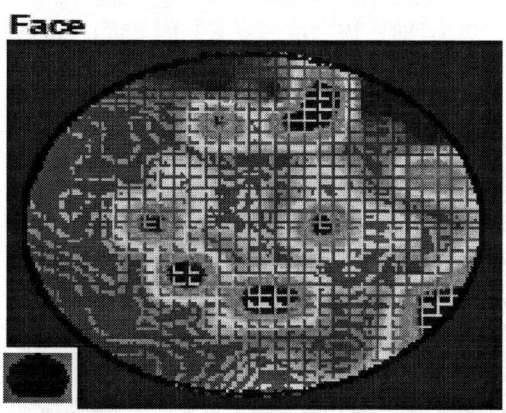

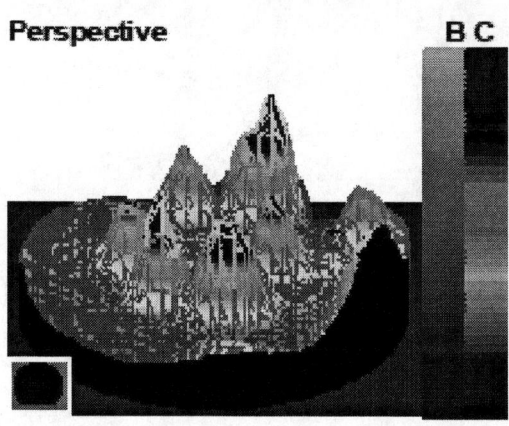

FIGURE 2. Contour and 3D maps (face and perspective) of leakage current of an USJ sample wafer. The leakage current was measured with the mercury ring-dot probe at 1V reverse bias at 116 sites.

It is generally accepted that the equivalent leakage resistance of a junction under reverse bias is mainly determined by the generation current in the space charge layer and the tunneling current across it [3]. Hence, the leakage resistance measured in this way is different from that measured with the RsL technique [2] which measures recombination current in a range of mV AC signal. In Borland's recent report on his USJ round-robin experiment [4], he shows that the leakage current extracted with the RsL technique correlates well with photoluminescence analysis of damage for the samples used in the round-robin. Since photoluminescence is based on detection of the photon radiation due to recombination of the laser excited carriers [5], its detection should be mainly concentrated near the USJ's surface, not as much reflecting the defects in the depletion layer and the junction tunnel current. This is also another way of saying that the leakage resistance extracted with the RsL technique is not the same as that measured under reverse junction bias.

EXPERIMENTS AND DISCUSSION

Samples from Borland's recent round-robin experiment [4] were used for trying out the proposed USJ leakage current measuring technique. Each of these samples was prepared in three processing steps.

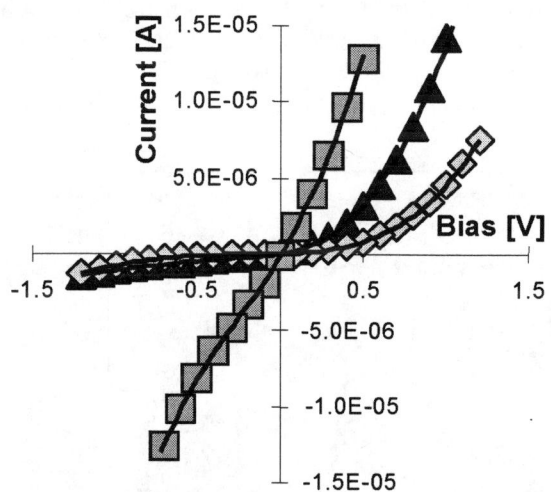

FIGURE 3. Examples of the I-V curves between the mercury dot and wafer back measured at different sites by the mercury ring-dot probe head on an USJ sample.

In each step, the process is one of those listed below:
- *1st Step PAI:* with or without Ge-5 keV/5x10^{14}/cm^2
- *2nd Step Implant:* 500 eV-B$_{11}$, 2.5 keV-BF$_2$, 5 keV-B$_{10}$H$_{14}$, or 10 keV- B$_{18}$H$_{22}$, each with 10^{15}/cm^2 dose.
- *3rd Step Dopant Activation:* Laser (Sopra), 1300 Flash (Mattson), 1080 Spike (Mattson), 1000 Spike (Mattson), or 650 SPE 5 s (Mattson).

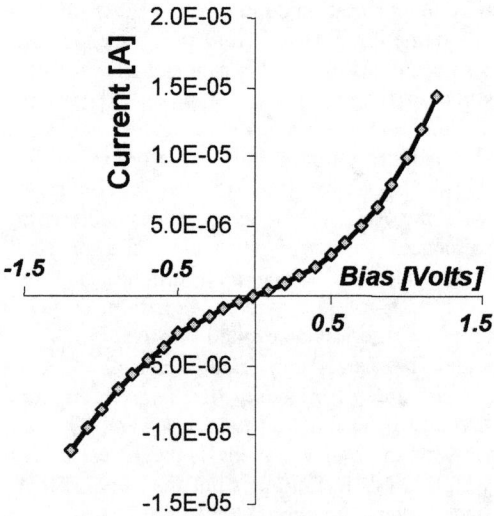

FIGURE 4. The current measured between the mercury ring and mercury dot for finding and mapping the USJ contact resistivity.

The substrates are all N type with resistivity of about 2.5 Ω.cm. Using the mercury ring-dot probe with effective dot area of 0.0388 cm^2, the leakage current at several sites in each sample was measured and some samples were mapped. A mapping example is shown in Fig. 2. This sample was processed with 5 keV-Ge-PAI/500eV-B$_{11}$/1300Flash. Measurements were made at 116 sites. A mean of 9.8x10^{-4} A/cm^2 with standard deviation of 60% was obtained. Fig. 3 shows I-V plots at three sites in another sample prepared with 5 keV-Ge-PAI/BF$_2$-2.5 keV/1300Flash processes. It also shows a large leakage current variation on the same wafer. We observe that the range of leakage current variation throughout the wafers is normally higher than 10% for all the round-robin samples.

Implanting boron of 10^{15}/cm^2 for forming an USJ layer in less than 200 Å thick will certainly result in having much non-activated boron at the layer surface even after annealing. This can greatly increase the contact resistance to the USJ layer. An example of I-V plot between the mercury ring and dot on the USJ layer of 2.5keV-10^{15}cm^{-2}-BF$_2$-PAI/1300Flash is shown in Fig. 4.

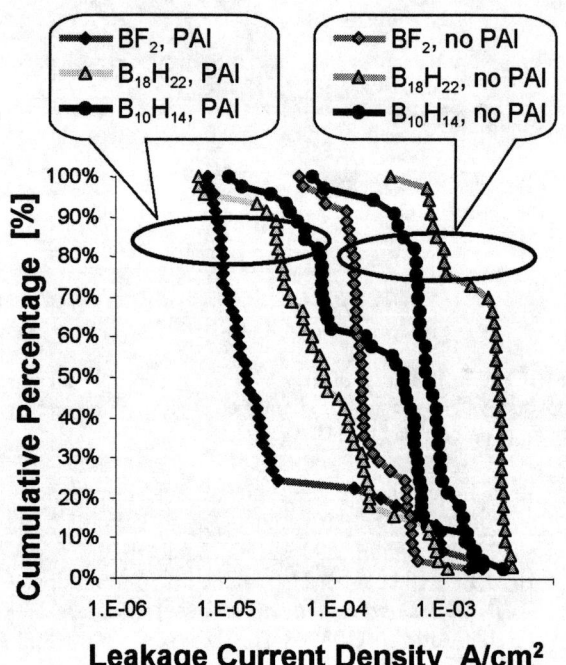

FIGURE 5. Cumulative plots of measured leakage current on molecular implanted samples group of wafers in John Borland's round robin [4].

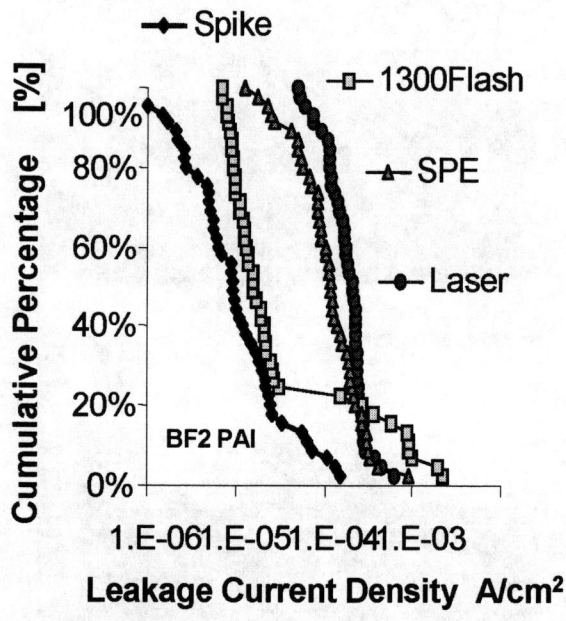

FIGURE 6. Cumulative plots of measured leakage current on BF$_2$ implanted samples with PAI annealed by different techniques.

The curve indicates a total resistance of 177 kΩ between the dot and the ring. With the sheet resistance of 660 Ω/sq and equivalent of about 0.24 square between the dot and the ring, one can see a contact resistance of about 177 kΩ under the dot of 8.04x10^{-3} cm^2, which is too high for the system to accurately measure the leakage current at the leaky sites but is not too high for accurate measurement of leakage current at not leaky site. However, it is suggested that reducing the USJ implant dosage to half of $10^{15}/cm^2$ maybe advantageous.

Fig. 5 shows cumulative plot of leakage current measured with this technique at 45 sites of each wafer from the molecular implanted group of the samples in John Borland's round robin study [4]. One can clearly see that the USJ layers processed with PAI step have lower leakage compared with those without PAI, at the low-leaky sites in each wafer. This is true especially for those samples having Spike or Flash annealing, as it is confirmed also by the cumulative plots shown in Fig. 6 for BF_2 implanted wafers. But, base on Borland's report, spike annealing drives in the USJ much deeper than Flash annealing, and yet both annealing give low enough sheet resistivity. Thus this leakage measurement helps pointing out that Flash annealing with BF_2 implant and PAI is the best process in that USJ experiment.

CONCLUSION

Using a ring-dot mercury probe system as described above, one can directly measure local leakage current of a non-patterned USJ layer under a reverse-biased voltage with good resolution. This measurement is different from the leakage current obtained by the RsL technique, which extracts the forward-biased recombination leakage current through photovoltage measurements and can be obscured by the defect density in the USJ layer itself.

Mapping the samples from the Borland et al. round-robin shows large variation in leakage current even within the same sample. This suggests that the proposed tool can be used for studying uniformity and performance of USJ processes to provide references for optimization.

Using the same system, the effect of excessive implanted boron on contact resistance of USJ layers can also be monitored.

Combining the test results reported above with John Borland's report on sheet resistivities and annealing caused depth extension for the same group of samples, it is suggested that PAI/molecular implant /Flash gets the best combined results.

REFERENCES

1. J.T.C.Chen and W. Liu, " Using Mercury Probe to Characterize USJ Layer", USJ-2005, Daytona Beach, Florida, USA, pp. 103-108, 2005.
2. V.N. Faifer, et al., "Non-Contact Electrical Measurements of Sheet Resistance and Leakage Current Density for Ultra-shallow (and Other) Junctions", MRS-Spring 2004, Symp.C, Vol. 810, p. 475, 2004.
3. S.M. Sze, *Physics of Semiconductor Devices*, Second Edition, Wiley, NY, 1981, p. 91, 537.
4. J. Borland et al., "Annealing techniques for optimizing 45nm-node USJ", Solid State Technology, May, 2006.
5. D.K. Schroder, *Semiconductor Material and Device Characterization*, Third Edition, IEEE Press, Wiley-Interscience, NY, p, 604, 2006.

Improved Techniques for Characterization and Optimization of SIMOX Implantation

R. Dolan[a], C. McKenna[a], S. Richards[a], Y. Aoki[b], T. Nakai[b], S. Nakamura[b], M. Walden[c]

[a] Ibis Technology Corporation, 32 Cherry Hill Drive, Danvers, MA 01923, USA
[b] SUMCO Corporation, 314 Nishisangao, Noda-shi, Chiba 278-0015, Japan
[c] SUMCO USA Sales Corporation, 49090 Milmont Drive, Fremont, CA 94538, USA

Abstract. Silicon-on-insulator starting wafers continue on a march toward commercial adoption. The dramatic growth in 300mm SOI is expected to continue and to outpace the growth of the bulk silicon market. The Bonded SOI type wafer currently dominates the SOI market but the competing SIMOX-SOI wafer remains a potentially simpler, higher quality and lower cost process. The SIMOX process requires a high dose (2-3E17/cm^2) implant step followed by a high temperature anneal (>1300°C). It has been found that the implant step is the most critical step in determining the final quality of the SIMOX–SOI wafer.

We will discuss improved techniques for characterization and optimization of the SIMOX implantation step. The techniques will include unique characterization methods for determining and improving the i2000 oxygen implanter product quality. We will also describe several as-implanted material characterization techniques that allow one to determine the end of line quality of the SIMOX-SOI wafer without going through the anneal and post anneal characterization cycle. We will include end of line quality data on 300mm SIMOX-SOI material produced on the i2000 implanter.

Keywords: SIMOX-SOI, Implant, Anneal, Uniformity
PACS: 78.30.Am, 61.72.Tt

INTRODUCTION

The concept of utilizing oxygen implantation to create buried insulating layers of SiO$_2$ in Silicon has been around for nearly thirty years. [1] During that time, significant effort has gone into refining the process for producing high quality buried oxides [2,3,4] at doses substantially lower [5] than the original 2E18/cm^2 requirement. As should be expected, the ability to control the characteristics of the implant process is essential to achieving the highly uniform, reproducible SOI/BOX layer combinations necessary for the most advanced devices. For conventional dopant-ion implantation steps, there has been a myriad of characterization tools [6] developed over the years for optimizing and regularly monitoring the dose uniformity, accuracy and reproducibility. SIMOX implantation involves a non-electrically-active element and the doses fall more into the category of materials modification than doping. Also, the successful application of SIMOX requires control over both the dose level and the uniformity, as well as over the characteristics of the associated damage, which in turn depends upon the combination of dose, dose rate, temperature and temperature uniformity. Unfortunately, very few of the quick turn-around tools used for characterizing conventional ion implants such as four-point probe, Therma-Wave, Optical Densitometry, etc. are useable for the high-dose SIMOX applications, and the less typical options such as SIMS and RBS/NRA are far too slow and expensive to be practical. Until relatively recently, the only measurement device available was the final SIMOX product. Since the cycle time for this feedback involves several tens of hours, it makes the optimization process tedious at best. Furthermore, control of the SIMOX process requires the ability to achieve independent measures (and ultimately control) of the implanted dose and damage distributions, as they each have significant impacts, in their own right, on the quality of the final SOI/BOX layers. In this paper, we will describe certain techniques developed for characterization and optimization of SIMOX implant steps.

IN-LINE SIMOX UNIFORMITY CONTROL

The current SIMOX process [7] consists of two implant steps, the first being the primary buried layer implant (typically 1-5 E17/cm^2 at 150-220 keV with temperatures ~ 350°C – 500°C), and the second being a damage engineering implant (~ E15/cm^2 in the same energy range as the buried layer implant at much lower temperatures). The purpose of the second implant is to introduce damage which enhances the diffusion of oxygen through the surface silicon and accelerates the internal oxidation, increasing the BOX thickness during the anneal. These two implants, in conjunction with an extended anneal at temperatures exceeding 1300°C, portions of which are done in an oxygen-rich atmosphere, create a highly stoichiometric buried oxide (BOX) ~ 1500Å thick with SOI layers between 500Å - 900Å. The uniformities of the BOX and SOI layers most strongly correlate with the uniformities of the dose distributions and of the wafer temperatures from the implant steps.

While Spectroscopic Ellipsometry (SE) or Optical Reflectometry remain valuable techiques for the characterization of SOI layers after high temperature annealing; the first order of business was to come up with a technique for measuring the as-implanted oxygen dose. On Line Technologies, working with IBIS, was able to modify their Thin Film Analyzer tool using Fourier Transform Infrared Reflectometry [8] to provide a relative measure of oxygen dose in as-implanted wafers for doses > $1E17/cm^2$ with an accuracy of ~ ±1%. The reflectance of the as-implanted layer for the longer wavelength infrared light was found to correlate with only the oxygen dose as opposed to the convolutions of the signals from the oxygen with the crystal damage that occurs for measurements in the visible portion of the spectrum.

The FTIR spectrum is empirically calibrated against dose in the following way. A model is constructed for a particular energy and dose, which predicts the positions of the front and rear of the BOX. The FTIR wavelength is long compared with the roughness of these interfaces prior to anneal so optical interference dominates the spectrum. It is assumed that the difference between the refractive index of the unimplanted silicon and the implanted region is proportional to the dose so the amplitude of the interference fringes increases with dose. Wafers are implanted with the target dose ±10%, and the FTIR spectra acquired. The amplitude of the interference fringes is plotted against dose, and the slope of this plot is then used to translate FTIR fringe amplitude to dose. Clearly this approach will work only over a limited dose range close to that of the model.

Fringes in the FTIR spectrum require an oxygen dose of at least $1E17$ cm^{-2}, and the noise of the measurement technique reduces as the dose increases, so in practice ~$2E17/cm^2$ is needed. With this dose the FTIR technique provides a total range about the mean of about 2%. Thus repeated measurements are needed to establish the desired 1% total range.

Measuring the damage engineering implant presents some additional problems for using the FTIR technique. The uniformity of this dose strongly affects that of the final SOI thickness after anneal. Using the set-up and correction for the primary dose implant is not practical since a lower beam current is preferred in order to achieve an adequate number of scans for the lower dose. One approach is simply to extend the implant time until a dose of ~$2E17/cm^2$ is built up; unfortunately, this process takes several hours. A second approach is to pre-dose a set of monitor wafers to 1 to $1.5E17/cm^2$. One of these wafers is then rotated 90 degrees from the original orientation, and again implanted so the total dose is ~$2E17/cm^2$. By rotating the wafer 90 degrees, any non-uniformity in the beam scan direction from the pre-dose implant is eliminated from the background, so the FTIR line scan reveals only the non-uniformity in the second implant. FTIR is subsequently used for both implanter scan corrections [9] and in-line process control.

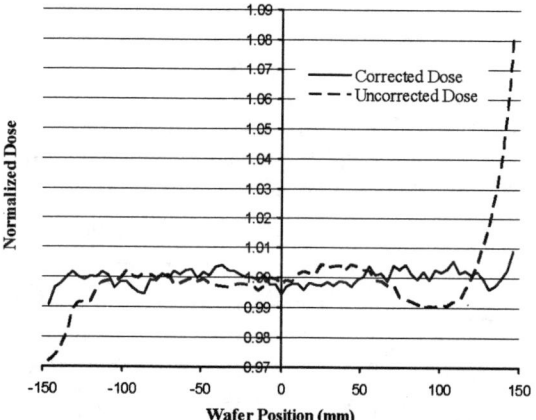

FIGURE 1. Dose Profile before and after scan correction

In addition to controlling the as-implanted dose, it is equally important to the final SOI uniformity that the damage created by this second implant be uniform. The damage uniformity is influenced not only by the scan (dose) uniformity but also by the temperature uniformity across the wafer and by the beam current density or ion beam shape. We have found that the Therma-Wave Therma-Probe tool used for conventional dopant implants can give important feedback on as-implanted wafers for both the dose uniformity and the relative temperature uniformity during the run. The Thermal-Wave (TW) technology measures surface modification (i.e., "damage") induced by ion implantation, and is well established in wafer processing to characterize and control ion implant dose accuracy and uniformity [10]. Figure 2a shows a Therma-Probe measurement of a wafer following a damage implant. The total dose uniformity is approximately 1% however there is a strong thermal signature caused by the wafer holder assembly. Figure 2b is a measurement following a redesign of the i2000 wafer holder assembly [9] that eliminated the thermal signature.

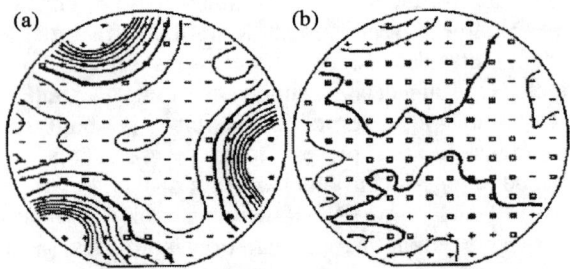

FIGURE 2a. Thermawave map showing thermal effect after "damage implant" caused by i2000 wafer holder

FIGURE 2b. Thermawave after redesign

Beam current density can also be a source of non-uniformity from the damage implant; "hot spots" in the O+ beam can result in non-uniform damage during the wafers first pass through the beam. Hot spots in the beam can lead to a condition in SIMOX known as "striping". Striping is mainly a cosmetic defect, representing less than 10Å peak to valley SOI thickness variation. The Therma-Wave tool can be used on an as-implanted wafer (must be a monitor wafer with only the damage implant) - see Fig. 3.

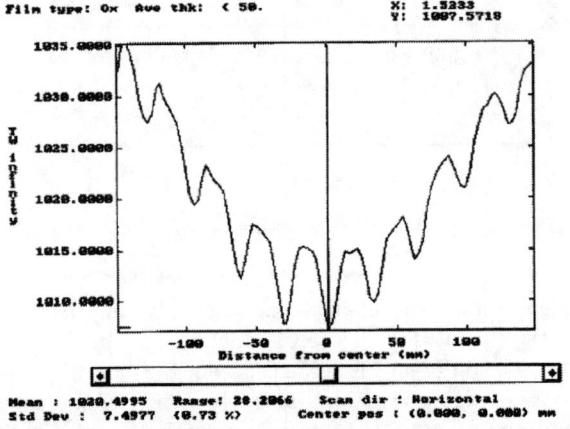

FIGURE 3 Thermawave of wafer implanted with "hot spots"

Another tool that can be used for controlling beam "hot spots" is a laser based particle counter. The main benefit is that the measurement is done as an inline inspection on real product. Ibis uses a KLA SP1-DLS particle counter to monitor implanter particle levels. The SP1 also produces a haze map. Haze is the low frequency signal caused by the scattering of laser light during darkfield inspection and can reflect minute variations in surface uniformity or roughness that are caused by wafer processing [11]. The haze map can be used to assess damage uniformity of the implant. Figure 4 is an example of a haze map of an as-implanted wafer that had 'hot spots' in the O+ beam during the damage implant portion of the two-step SIMOX implant process.

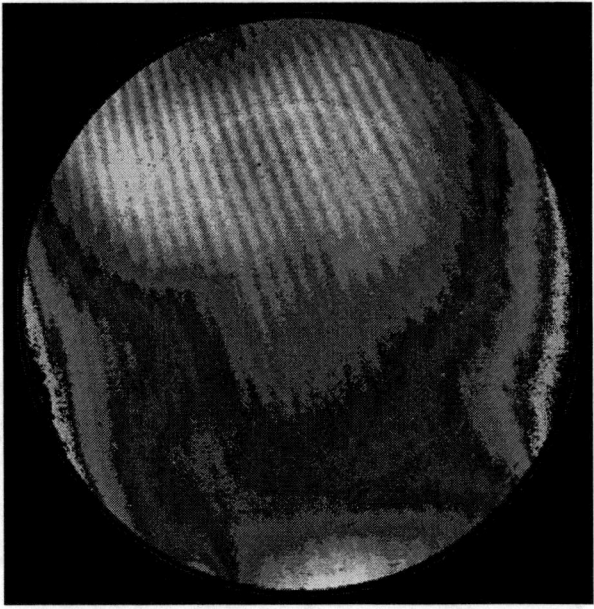

FIGURE 4 As- implanted wafer's haze map from a KLA SP1-DLS implanted with beam "hot spots"

The result of developing several in-line control techniques and the modifications to the i2000 implanter is that SIMOX –SOI final SOI thickness uniformity is routinely sub 20Å (6 sigma).

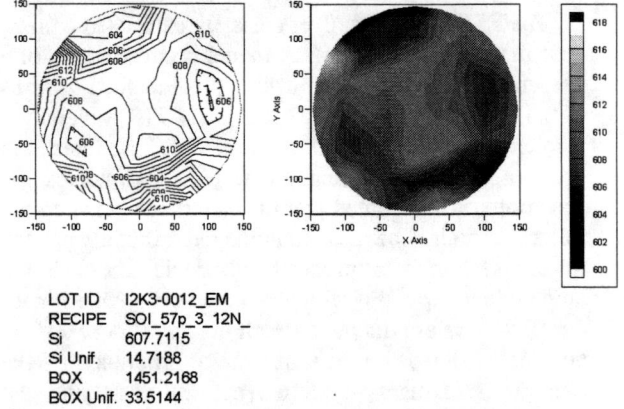

LOT ID   I2K3-0012_EM
RECIPE   SOI_57p_3_12N_
Si   607.7115
Si Unif. 14.7188
BOX   1451.2168
BOX Unif. 33.5144

FIGURE 5 Post anneal SIMOX- SOI uniformity of <15A

CURRENT SIMOX-SOI QUALITY STATUS

Another critical implanter performance area is contamination control or more specifically particulate control. The evolution of both the process (approximately 10x reduction in dose) and implanter design (the i2000 represents the 3rd generation of Oxygen implantation equipment) has reduced the typical particle levels to <200 0.2µm particle adders per 2–4 E17 dose. Today's SIMOX wafer cleaning technology is typically 80-85% efficient in particle removal, leading to typical final counts of less than 30 particles per wafer. Excessive particle contamination can lead to BOX defects, also known as BOX pinholes. The newer low dose process has achieved some immunity in the particle to BOX defect relationship, but given that this is a killer defect in SOI, implant particle level reduction will continue to be an ongoing activity. Most of the remaining screening parameters for determining wafer quality are dependent on the SIMOX process.

SOI Parameter	Typical Value
Silicon	100 - 1450Å
Silicon Uniformity	< 20Å
Box	1450Å
HF	< 0.05/cm^2
Roughness(10x10um)	< 3.5Å
Interface Roughness	< 3Å
Breakdown	> 8MV/cm
Particles	< 0.05/cm^2
Pinholes	< 0.05/cm^2
Metallics	< 1E9/cm^2
Dislocations	< 10E2/cm^2

TABLE 1. Critical SOI parameters with current SIMOX values

CONCLUSION

Many years of process and implanter development have led to a SIMOX-SOI process that produces material that is greatly improved and leads to CMOS product yields that are not affected by wafers defect problems [12]. The implanter is still responsible for the critical parameters such as SOI, BOX and particle performance but a shift has occurred whereby manufacturability and cost of ownership will likely determine the success or failure of SIMOX-SOI.

REFERENCES

[1] K. Izumi, M. Doken, H. Ariyoshi, Electron Lett. 14 (18) (1978) 593.
[2] M. Bruel, J. Margail, J. Stoemenos, P. Martin and C. Jaussaud, Vacuum 35 (1985), 589.
[3] S. Nakashima and K. Izumi, J. Mat. Res 8 (1993), 523.
[4] S. Nakashima, T. Katayama, Y. Miyamura and A.Matsuzaki, Pro IEEE SOI Conf (1994), 71.
[5] Devendra Sadana and Michael Current, in *Ion Implantation Science and Technology, 2000 Edition*, ed. J. F. Ziegler.341.
[6] C. Yarling and Michael Current, in *Ion Implantation Science and Technology, 6th Edition*, ed. J.F. Ziegler, (1996),674.
[7] Devendra Sadana, IIT School : *Ion Implantation Science and Technology*, IIT2004.(2004).
[8] U. Mantz et al., "Model-Based Infrared Spectroscopy: New Opportunities for In-Line Process Control", in *Future Fab International*, 19 (2005)
[9] C. McKenna et al.,"Enhancing the Ibis i2000™ SIMOX Oxygen Ion Implanter", these Proceedings *Proc. 2006 Int. Conf. Ion Implant Tech. (Marseilles, France)*
[10] L. Nicolaides, A. Salnik and J. Opsal, "Nondestructive Analysis of Ultrashallow Junctions Using Thermal Wave Technology", *Review of Scientific Instruments* 74, (2003), 586-588.
[11] F. Holsteyns et al., "Seeing Through the Haze, Process Monitoring and Qualification Using Comprehensive Surface Data", *Yield Management Solutions*, Spring (2004), 50-54.
[12] D. Greenlaw, "Technology Requirements for Harnessing High-Performance SOI, *Semiconductor Manufacturing*, April (2006), 44-47.

Non-contact sheet resistance and leakage current monitoring of multi-implant, ultra-shallow junctions: Doping and damage effects for ms-anneals

M.I. Current, V.N. Faifer, T.M.H. Wong, T. Nguyen and A. Koo

Frontier Semiconductor, 1631 N. 1st Street, San Jose, CA 95112 USA
Net: michaelcurrent@frontiersemi.com, www.frontiersemi.com

Abstract. Junction photo-voltage methods are used to measure sheet resistance and recombination leakage current effects in ultra-shallow junctions formed with low-energy Boron implants annealed with ms-timescale thermal cycles at ~1300 C. The impact of sub-junction doping and implant damage levels present before the ms-anneal is studied.

Keywords: Sheet resistance, junction photo-voltage, leakage current, ultra-shallow junction.
PACS: 73.40Lq, 73.50Pz, 73.61Cw

INTRODUCTION

The delays in implementation of advanced gate stage structures using high-k dielectrics and metallic gate electrodes has put strong, immediate pressures on junction formation process to accelerate SDE junction depth scaling to 10 nm and below [1]. The continued increase in the number of transistors per chip coupled with the proliferation of high-performance mobile applications has brought junction leakage current limitation to a high priority. The implant and annealing community has responded to these challenges with multi-implant processes (using multi-profile doping implants coupled with pre-amorphization and various "cocktail" ions) and development of "diffusionless" anneals with thermal cycles limited to the ms regime. Limiting dopant profile diffusion with these innovations has had the unfortunate general effect of strongly increasing junction leakage density and compromising dopant activation levels for many process combinations.

The development of non-contact metrologies for measurement of sheet resistance and leakage current (RsL) has provided a vital capability for rapid, high-resolution Rs mapping, independent of junction depth, as well as the use of junction leakage current as an in-line process control parameter. This paper will highlight the use of RsL methods to illustrate the effects of implant damage and SDE/halo doping levels on junction leakage current and dopant activation. In particular we will discuss the effects of PAI energy, halo dose and damage distributions, non-dopant species implants and the high sensitivity of ms-scale annealing to accumulated damage distributions.

JUNCTION PHOTO-VOLTAGE

The measurement of junction sheet resistance, Rs, and leakage current of USJ is based on monitoring of lateral drift carrier spreading via junction photo-voltage (JPV) signals measured inside (V_{in}) and outside (V_{out}) an illuminated area (Fig. 1) [2,3].

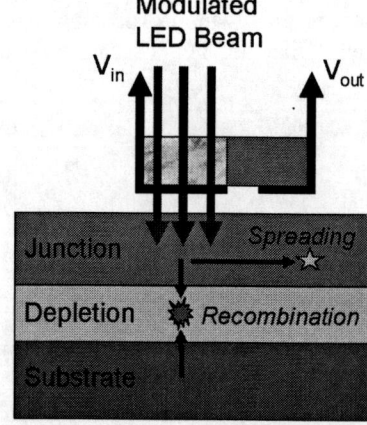

FIGURE 1. Sketch of the RsL probe electrodes showing the spreading of carriers away from the illuminated area under the inner electrode and the recombination of carriers in the depletion layer under the surface junction.

The analysis consists of measuring and modeling of the JPV signals, V_1 and V_2, captured by the transparent and non-transparent electrodes when modulated light flux produces electron – hole pairs in the semiconductor material containing a p-n junction. The dynamics of the photo-generated carriers are modeled with exact 3-dimensional solutions of charge continuity and Poisson's equations under the conditions of low light flux levels. The JPV value at a distance, r, from the illumination point is:

$$v_0(r) = Const * K_0(kr) \quad (1)$$

where: $k = [R_S * G + i\omega * R_S * C_S]^{1/2}$
$K_0(kr)$ is a modified Bessel function of the second kind
R_S = junction sheet resistance
G = junction conductivity
C_S = junction capacitance
ω = 2π*light modulation frequency.

The sheet resistance, Rs, is determined by analysis of the lateral carrier spreading based on the JPV signals from the two electrodes. The junction conductivity, G, is determined by analysis of the dependence of the JPV signals on the light modulation frequency. The junction conductivity, G, and leakage current density, I_o, are related by:

$$G = I_o * (q/kT) \quad (2)$$

where: I_o = prefactor (reverse bias leakage current) in I-V diode formula $I = I_o * [e^{(qV/kT)} - 1]$
q = electron charge
k = Boltzman's constant
T = wafer temperature (K)
V = junction bias voltage.

Mapping of sheet resistance and leakage current provides a high-spatial resolution ($\sim 10^3$ points/wafer) picture of the degree of success of the anneal process.

LEAKAGE CURRENT IN USJ

The dominant factor in the leakage current in Si under small junction photo-voltage is the recombination of electron-hole pairs created by the illumination photons in the depletion region of p-n junction. The recombination current, J_{rec}, is given by an integral of the recombination rate, U(z), across the depletion layer [3], where the recombination rate is proportional to such factors as the number and depth distribution of carrier trap sites, local doping levels and crossections for carrier recombination.

The impact of local substrate doping in the region near an USJ is two-fold. An increased doping level in the region of the junction decreases the width of the surface charge (depletion) region and moves the point of maximum recombination rate, closer to the junction edge. In addition, if the local doping is accomplished by an implantation step, the residual lattice damage after annealing, strongly increases the local concentration of recombination centers, increasing the recombination current (Fig. 2).

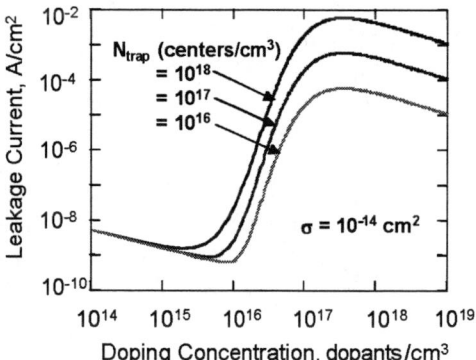

FIGURE 2. Leakage current for peak trap densities of 10^{16}, 10^{17} and 10^{18} traps/cm^3 as a function of substrate doping.

LOCAL THERMAL EFFECTS

The local heating associated with pulsed or scanned laser beams, with spot dimension of the order of ~1 cm, often results in a local variation of dopant activation and annealing which is characteristic of the laser exposure procedure. In the example shown in Fig. 3, the sheet resistance map and diameter scan of a laser annealed wafer show the effect of the linear scan tracks of the laser beam with a narrower laser spot on the upper part of the wafer. The local variations in Rs are ~2.4% according to both the Rs map and line scan.

In a "flash" anneal using a ms-timescale exposure over the entire wafer surface, Fig. 4, local variations in the peak anneal temperature resulted in an average sheet resistance of 474 Ohm/square with a variation of 3.9%. The leakage current map showed higher values in the lower part of the wafer, where the sheet resistance was also higher than the upper portions of the wafer, perhaps reflecting a cooler wafer temperature in the lower parts of the wafer.

Rs and leakage maps for a "step and expose" laser annealing scan, Fig. 5, showed strong variations in local Rs (~25 %) and leakage (>2 orders of magnitude) at different exposure sites.

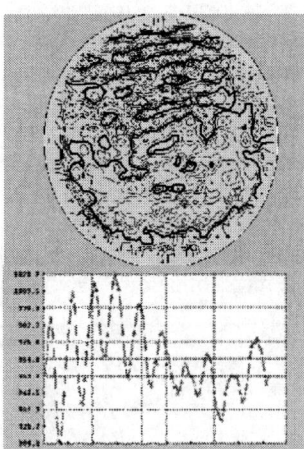

FIGURE 3. A 973-point sheet resistance map (1% contour lines) (upper) and a 121 point vertical diameter scan (lower) of a laser annealed wafer showing effects of variations in anneal beam width between the top (narrower) and bottom (wider) of the wafer. The average sheet resistance was 974 Ohm/square with a uniformity of 2.4 %.

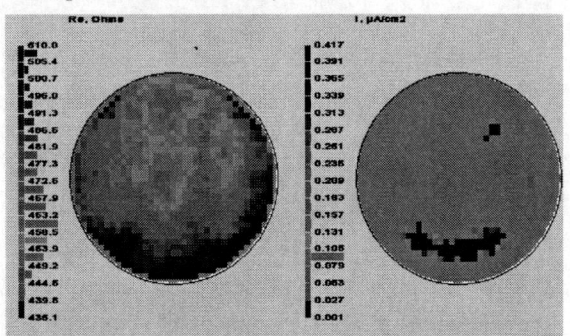

FIGURE 4. Rs (left) and leakage current (right) maps of a "flash" annealed wafer. Note high Rs and leakage in the lower part of the wafer.

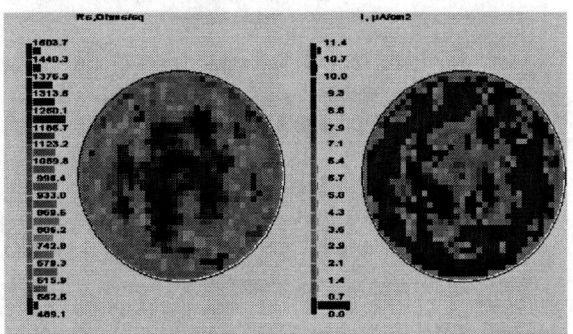

FIGURE 5. Rs (left) and leakage current (right) maps of a laser annealed wafer using a step and expose scan method. Note high Rs and leakage currents, indicated by darker colors, in similar regions across the wafer. The Rs average was 996 Ohm/square with an across wafer variation of 25.46 %. The leakage current ranged from $<10^{-7}$ to $\sim 2 \times 10^{-5}$ A/cm^2.

The variation in leakage current in ms-annealed wafers is much wider than the dopant activation. The standard deviation of the distribution of Rs values of the laser annealed wafer shown in Fig. 5 is ~25%, while the leakage current ranges from $<10^{-7}$ to $\sim 2 \times 10^{-5}$ A/cm^2 (Fig. 6).

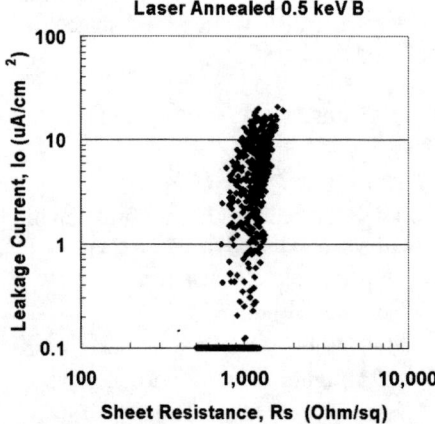

FIGURE 6. Distribution of Rs and leakage values for 973 test sites in the laser annealed wafer shown in Fig. 5.

DOPING AND DAMAGE EFFECTS

The limited annealing of implantation damage during ms-scale, "diffusion-less" anneals results in a strong sensitivity for the leakage current on the density and location of implant damage accumulated prior to anneal. For example, the junction leakage currents after ms-scale anneals are much more sensitive to the presence of deep pre-amorphization implants (PAI) than "spike" anneals, at ~1050 C for ~1 s with full wafer heating. Leakage currents of junctions formed with ms-anneals are also more sensitive to such process variations as the halo implant dose and energy and the use separate annealing cycles for halo activation and damage reduction. In addition, the use of additional "cocktail" high-dose implants with such ions as F and C, which are designed to reduce dopant diffusion through the formation of interstitial traps, are highly likely to result in additional recombination centers in the junction depletion layer, increasing recombination/generation current levels.

The progression of recombination leakage current ranges is shown in Fig. 7 for a variety of process choices associated with a flash anneal (~1300 C for ~10 ms) of a 0.5 keV B implant at a dose of 10^{15} B/cm^2. When the B junction is formed in a 10-Ohm-cm test wafer, with a doping level of ~10^{15} dopants/cm^3, the leakage current is very low, $<10^{-7}$ A/cm^2. When the B doping is preceded by a 30 keV

Ge pre-amorphization implants (PAI), the leakage current after the ms-anneal rises to the mid-10^{-6} range.

When the doping under the p-junction is increased to ~10^{19} As/cm^3 by a 40 keV As "halo" implant at a dose of 4×10^{13} As/cm^2 and when the damage created by the implant is reduced by an RTP anneal (1050 C/ 10 s) prior to the final ~1300 C/ ~10 ms anneal, the carrier recombination occurs within a much thinner (~10 nm) depletion layer and the leakage current increases to the mid-10^{-5} range. This leakage level, due primarily to the increase in sub-junction doping, is greater than 100x higher than for B implants into lightly-doped (~10^{15} n/cm^3) 10 Ohm-cm wafers with depletion layers of ~1 um.

When the B implant is preceded by As and Ge implants, with no anneals other than the ms-cycle, the leakage current ranges from ~10^{-3} to over 10^{-2} A/cm^2. At leakage currents of ~10^{-2} A/cm^2 and higher, the surface doped layer no longer functions as a p-n junction since the carrier recombination rate is so high that no carriers survive long enough to give measurable JPV signals in the outer electrode of the RsL probe (see Fig. 1). Note that this last process sequence, a combination of halo doping and pre-amorphization implants followed by SDE doping and annealing, corresponds closely to standard process for SDE formation in CMOS transistors.

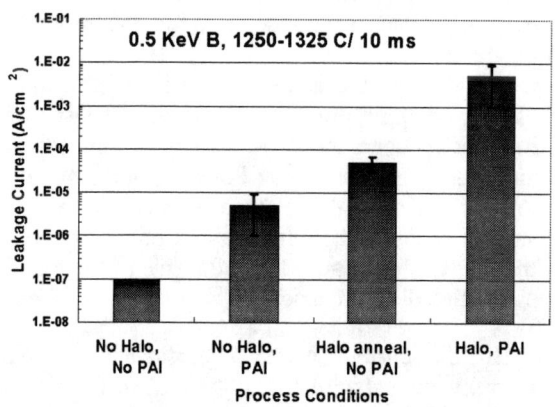

FIGURE 7. Recombination leakage current for flash annealed (~10 ms) 0.5 keV B implanted junctions with various combinations of As halo implant, post-halo, pre-Boron implant anneal and Ge PAI. Note the relation to trends shown in Fig. 2

SUMMARY

The shortening of the thermal anneal time scales into the ms-range strongly increases the impact of damage accumulation during the preceding implantation cycles on the junction leakage currents. Local variation in the thermal histories for the short (<10 ms) timescale anneals result in systematic variations in dopant activation and leakage which reflect the thermal history and scanning methods. These conditions will require increased understanding and monitoring of damage accumulation, doping and annealing processes in order to arrive at processes which provide high-activation and low leakage current for junction depths of 10 nm and less. Non-contact mapping of local Rs and leakage values provides a valuable quantitative method for monitoring the success of dopant activation and damage annealing for ultra-shallow junctions.

ACKNOWLEDGMENTS

We acknowledge many continued contributions from our colleagues and customers towards our collective understanding of process and metrology effects. In particular, we thank John Borland for expertly coordinating the round-robin processing and testing which provided some of the data in this paper and acknowledge Paul Timans of Mattson for sustained discussions of fundamental and specific issues raised by the experimental data.

REFERENCES

1. See Table 69 in the Front End Process section of the ITRS05, www.itrs.net
2. V.N. Faifer et al., MRS Proc. 810, C11.9. (2004)
3. V.N. Faifer, M.I. Current, T.M.H. Wong, V.V. Souchkov, J.Vac. Sci. Technol. B21(1) 414-420 (2006).

Characterizing Dopant Contamination Using Ion Implantation

John J. Naughton and Janet M. Towner

AMI Semiconductor, 2300 Buckskin Rd. Pocatello Id, 83201

Abstract. In the current work a method of investigating and characterizing dopant contamination was developed using ion implantation. An oxide integrity degradation was observed on both silicon gate oxide and poly-poly capacitors. Boron and phosphorus contamination within or close to the surface of the respective oxides caused an elevation in premature breakdown. Using shallow implants of boron and phosphorus, the contamination effect was reproduced and characterized. Interestingly, phosphorus was most detrimental to n-doped poly structures and boron was most detrimental to p-doped poly structures.

Keywords: Ion Implantation, Dopant Contamination, Oxide Quality.
PACS: 61.72.Tt, 85.40.Ry, 61.72.Vv, 77.22.Jp.

INTRODUCTION

Analog mixed-signal device manufacturing has many specialized modules that add to its complexity. Poly-poly capacitors for enhanced linearity, MIM capacitors for RF applications, dual gate modules for minimal die size and I/O flexibility are examples of these. As these devices are typically used in critical automotive and medical applications, they must also meet stringent reliability requirements.

Factors that are thought to contribute to oxide failure include metal contamination and incorporation of poor quality oxide [1-3]. However, it has been reported that the majority of metals do not cause problems [4]. For example, copper and zinc had no apparent effect on the defect density, as they diffuse rapidly into the Si substrate. Rather, problems are seen with metals such as Ca, Fe or Al that remain present during Si oxidation and cause an increase in the Si surface roughness or cause defects to form in the oxide

Rougher surfaces are known to degrade oxide integrity in both gates and poly-poly capacitors. Asperities at the Si-SiO$_2$ interface serve to concentrate the electric field during device operation or testing, accelerating failure. This can be seen in the polarity dependence of stressing. Lower breakdown voltages are observed when a positive bias is applied to the top electrode [5]. Interestingly, polyoxides grown on more highly doped (phosphorus) material shows an increase in electric field strength due to a larger poly grain size and a smoother surface However, with greater increases in phosphorus concentration, wearout failures increase. This increase was attributed to positive charge generation due to phosphorus atoms incorporated in the oxide

One additional mechanism for oxide degradation is plasma charging during etch and photoresist ashing [6,7]. The degradation is thought to result from electrostatic stress causing Fowler-Nordheim current to be forced through the oxide. This current results in surface states at the interface and trapped charge in the oxide. These cause latent damage in the oxide or even immediate breakdown.

In the current work, a mechanism of oxide integrity degradation was observed on both silicon gate oxide and poly-poly capacitors. Boron and phosphorus contamination within or close to the surface of the respective oxides caused an elevation in breakdown. Using ion implantation, this contamination could be simulated. Once simulated, the contamination effect on the capacitor oxide integrity could be analyzed and predicted.

BACKGROUND

In our previous work it was shown that dopant cross contamination in both a Chemical Vapor Deposition (CVD) oxide cluster tool and photo resist asher was the source of elevated oxide integrity defectivity. [8] This was observed on poly-poly capacitor structures and gate oxide structures respectively.

In order to reproduce the contamination without compromising production equipment, shallow implants of certain dopants on top of an oxide test device were considered. The most suitable test structure proved to be the poly-poly capacitor. It was chosen over a gate oxide structure as it provided more flexibility in controlling electrode polarity and manufacturability.

In this work, n+ polysilicon/oxide/n+ polysilicon and p+ polysilicon/oxide/p+ polysilicon capacitors were fabricated on active devices. First level polysilicon was 370nm thick, second level polysilicon was 250nm thick. For the standard case of n-doped poly, phosphorus doping was accomplished for both films with an implanter at 40keV energy and doses of 1.75×10^{16} at/cm^2 and 8.8×10^{15} at/cm^2, with final sheet resistances of 25Ω/ and 46Ω/ , respectively. In the case of p-poly, boron was implanted at 15keV energy and dose of 4×10^{15} at/cm^2. Capacitor oxide was deposited in an Applied Materials P5000. The oxide was deposited using TEOS and O_2 as precursors at 400° C, 300W, in a single frequency 13.56MHz at a rate of 2.6nm/sec to a thickness of 36nm.

EXPERIMENTAL

Various dopants were applied near the surface of the capacitor oxide using a Varian EHP500 medium current implanter. In order to maintain a reasonable uniformity across the surface of the wafer (<2%), three rotations were used (0° tilt, twists of 120°, 240° and 360°). Phosphorus[31] implants were performed from PH$_3$ source gases. Acceptable uniformities at low energy could not be obtained for Boron[11] so BF$_2$[49] was implanted from BF$_3$ source gas. Argon[40] was also used. Implant energies were 2keV for P, 3keV for Ar and 5keV for BF$_2$. These correspond to implant depths of 2-4 nm into the oxide. Figure 1 shows a cross sectional diagram of the poly-poly capacitor test structure used for the experiment.

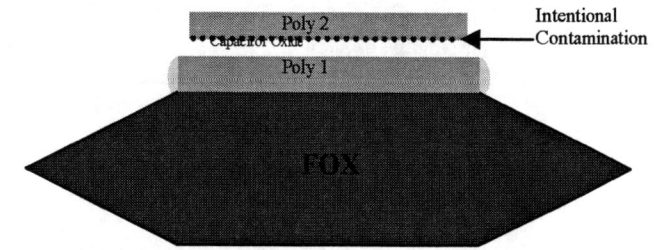

FIGURE 1. Cross sectional diagram of the Poly-Poly Capacitor test structure showing the intentional contamination experiment.

In-line oxide integrity was evaluated using a VcJox test. This test shows good correlation to ramped voltage testing. It is performed on both gate and inter-poly oxide capacitor test structures by forcing a current and measuring the voltage to determine VcJox;

$$I = 1\mu A * Area/100,000\mu m^2.$$

This test was accomplished by applying a positive current on the top plate of the capacitor and grounding the bottom plate. In this work, the reverse test was also done, applying a negative current on the top plate of the capacitor. Figure 2 summarizes the implant matrix.

Dopant	Energy	Dose (Max)
Phosphorus	2kev	1e15
		2e15
		4e15
BF2	5kev	1e15
		2e15
		4e15
Ar	3kev	2e15
Control	N/A	N/A

FIGURE 2. Intentional dopant contamination matrix.

RESULTS AND DISCUSSION

Figure 3 shows the results of the oxide integrity analysis on the n-poly capacitor. Clearly the defectivity is higher with N-type dopant contamination. What is also clear is that the Argon implant used to determine the oxide damage effect did not show any elevation in defectivity. This indicates the oxide integrity failure is the results of a like dopant.

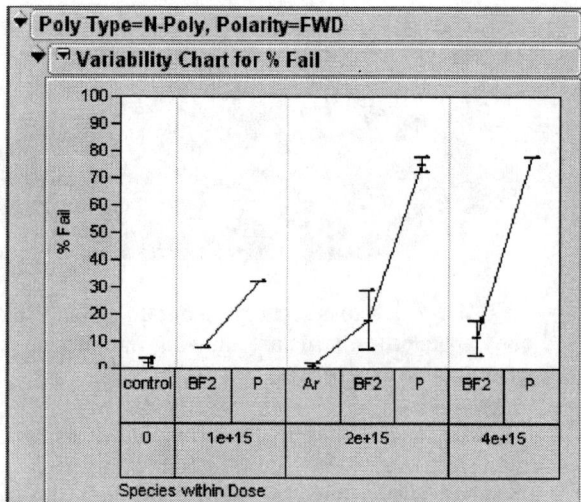

FIGURE 3. N-poly-poly capacitor defectivity results with forward polarity sweep.

Figure 4 shows the results of the N-poly capacitor oxide integrity analysis using a reverse polarity sweep on the VcJox test. It is clear that the results are consistent with the forward polarity sweep.

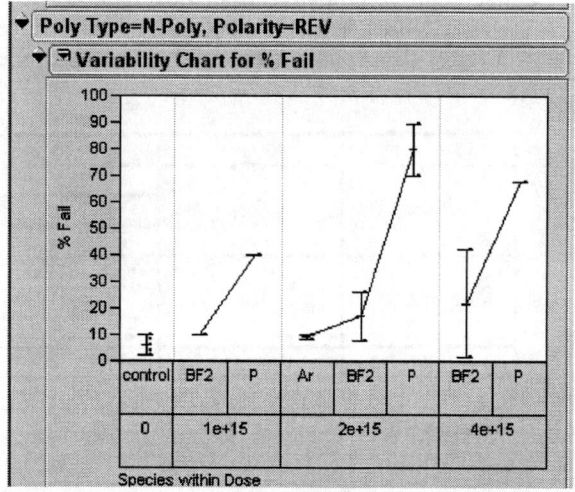

FIGURE 4. N-poly-poly capacitor defectivity results with reverse polarity sweep.

Figure 5 shows the results of the oxide integrity analysis on the p-poly capacitor. In contrast to the N-poly structure it is clear that the defectivity is higher with P-type dopant contamination. In addition, it is clear that the magnitude of the detrimental effect is less severe at the same level of dose.

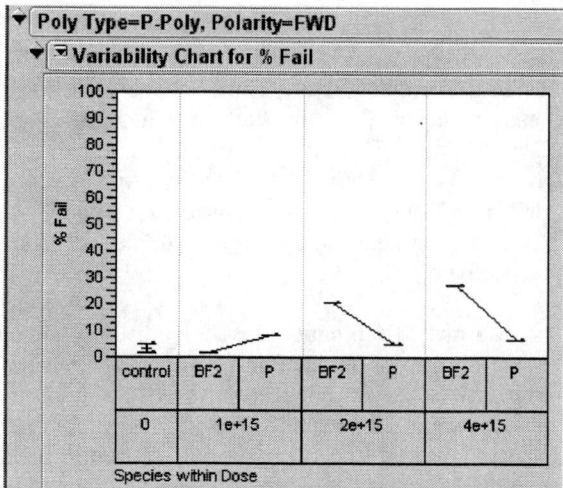

FIGURE 5. N-poly-poly capacitor defectivity results with reverse polarity sweep.

Figure 6 shows the results of the P-poly capacitor oxide integrity analysis using the VcJox test but a reverse polarity sweep. The results are consistent with the forward polarity sweep just as the case of the N-poly structure reverse polarity testing.

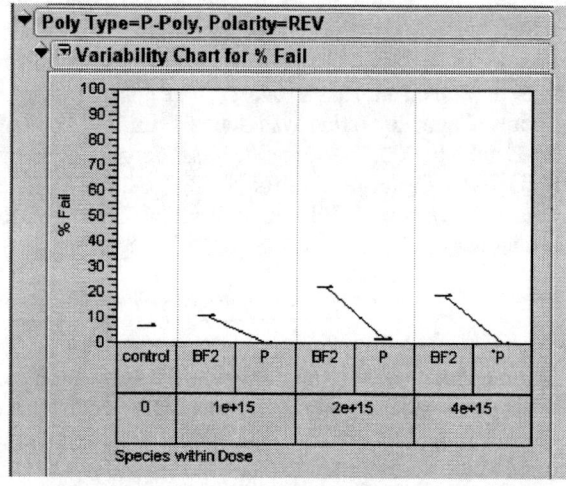

FIGURE 6. P-poly-poly capacitor defectivity results with reverse polarity sweep.

It is widely known that in a MOS or Inter-poly oxide structure where a precipitate is present at the oxide to silicon interface the electric field lines become concentrated at the point of the defect. This results in a localized concentration in the electric field. This defect will then become the location of failure of the structure and lead to device breakdown [9].

The underlying mechanism for the like polarity effect on accelerated oxide breakdown is not clearly understood. However, since the VcJox test described above stresses the oxide structures in the accumulation state, we have postulated that if this precipitate is of like dopant to the heavy doping in the poly-silicon it can also enhance the electric field at that location accelerating the point of failure. Conversely if the dopant precipitate is of opposite polarity it will not enhance the electric field. In the latter case the asperity still exists as a site of concentrated field and ultimate oxide failure but does not function in a "complimentary" mechanism.

CONCLUSION

This investigation revealed a mechanism of oxide integrity degradation on both silicon gate oxide and poly-poly capacitors. Boron and phosphorus contamination within or close to the surface of the respective oxides caused an elevation in premature breakdown. These dopants were present as a result of inadvertent cross contamination during processing. Simulation of this contamination was successfully performed using shallow implants on the surface of the oxide. A like polarity effect was observed, where phosphorus was most detrimental to n-doped poly structures and boron was most detrimental to p-doped poly structures.

ACKNOWLEDGMENTS

Acknowledgements are extended to Bruce Greenwood, Greg Scott and Muhammad Anser for their insights into oxide integrity. Also, Gail Burton, Thomas Haskett and Dan Rogers for providing extensive electrical analysis. Finally, special acknowledgements are extended to Dr. Raymond Lappan for his unwavering support of this work.

REFERENCES

1. C. Cobianu and O. Popa, IEEE Electron Device Lett, vol 14 p.213, 1993
2. C Y. Mikata, S. Mori, K. Shinada and T. Usami, Proc. 1985 Int. Reliability Physics Symp.IEEE p. 32.
3. S. Mori, N. Arai, Y. Kaneko and K. Yosikawa, IEEE Trans. Electron Devices, v 38 p. 270, 1991
4. S. Verhaverbeke, M. Meuris, P. Mertens, M. Heyna A. Philipaossian, D. Graf and A. Schnegg, Proc. 1991 Int. Electron Device Meeting, p. 71.
5. S. Wu, T. Lin, C. Lee and T. Lei, IEEE Electron Device Lett, vol. 14 p. p. 113 (1993).
6. H. Shin, C. King and C. Hu, Proc. 1992 Int. Reliability Physics Symp.IEEE p. 37.
7. T. Brozek, Y. Chan and C. Viswanathan, IEEE Electron Device Lett, vol. 17 p. 288 (1996).
8. J. Naughton and J. Towner, 2005 ECS Transactions, The Physics and Chemistry of SiO2 and the Si-SiO2 Interface, Vol. 1 No. 1 pp.243-252.
9. T. Winarski, IEEE Electrical Insulation magazine, Nov/Dec 2001,Volume: 17, Issue: 6, pp.34-47.

Manufacturing Precision Polysilicon Resistors Using Ion Implantation

Janet M. Towner

AMI Semiconductor
2300 Buckskin Road, Pocatello ID 83201

Abstract. Two techniques exist to form high resistance, precision polysilicon resistors. In both cases polysilicon films are degenerative doped with phosphorus to form the gate electrode. In the first method, phosphorus at a dose near 1×10^{15} at/cm^2 is implanted in a portion of the second level polysilicon that is masked during the previous implant. Resistors can also be formed in a portion of the first level polysilicon by counter doping with boron. This provides resistors with lower temperature coefficients of resistance and the ability to manufacture precision resistors in a single poly process. As resistors comprise only a small portion of the die (<2%), the wafer is almost entirely free of resist in the first process and entirely covered with resist in the second process. Using two types of high current, batch implanters, resistors were successfully fabricated using the first method. For the boron resistors, consistent uniformity could only be obtained using one type of implanter. In the other type of implanter, variations in resistor values of more than 300ohms were observed which far exceeded the design specification. Variation was dependent on lot size, with small lots showing better uniformity. Variation across the wafer manifested as horizontal bands with the lowest values in the center of the wafer. The cause of the nonuniformity was determined to be photoresist outgassing and ion neutralization. Photoresist breaks down under ion bombardment releasing large quantities of hydrogen gas. The released gas neutralizes incoming positive ions. This results in overdosing, which lowers the resistance most dramatically in the center of the wafer where the quantity of photoresist is greater.

Keywords: mixed signal, manufacturing, ion implantation.
PACS: 52.77.Dq, 84.32.Ff, 61.72.Ss

INTRODUCTION

Analog circuits operate in a world apart; they use signals that vary continuously from zero to the supply voltage. This is in contrast with digital circuits, where either voltage is on or voltage is off. Analog circuits manipulate precise voltages, currents and charges. They also rely on extremely precise control of the ratios of these quantities for matched pairs of resistors and capacitors. Mixed signal devices such as analog-to-digital (A/D) and digital-to-analog (D/A) converters, sensors and clocking circuitry contain both types of components. Subtleties of the manufacturing process, which have no effect on pure digital parts, become major yield issues for these designs [1]. Many process steps have to be manually adjusted to compensate for normal process variation.

Resistor linearity and precision matching performance is of vital importance in mixed signal device functions. Resistive pair
s are required with small voltage and temperature coefficients and closely matched values that do not deviate from one another over the operational range of the device. As resistance will directly affect RC timing, control is critical for these types of designs.

PROCESS MODULE

Two techniques exist to form precision polysilicon resistors with targeted values of 1000Ω/sq. In both cases polysilicon films are degenerative doped with phosphorus at doses greater than 1×10^{16} at/cm^2 to form the gate electrode and capacitor plates with a resistance of ~25 Ω/sq. In the first method, phosphorus at a much lower dose of (1×10^{15} at/cm^2) is implanted in a portion of the second level polysilicon that is protected against the previous implant. Implant energies are 40keV. A typical dose response curve for this process is shown in Figure 1.

Resistors can also be formed in a portion of the first level polysilicon by counter doping the phosphorus with boron or BF$_2$ [2]. However, the fluorine in the BF$_2$ can have a detrimental effect on resistor stability. It has a tendency to segregate to the polysilicon/oxide interface.

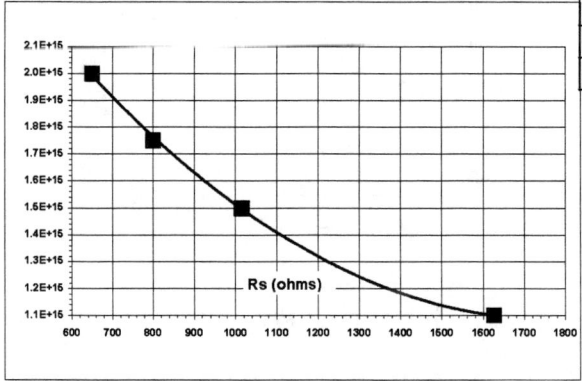

Figure 1: Dose response curve for precision resistors formed by method A. Graph shows sheet resistance versus dose.

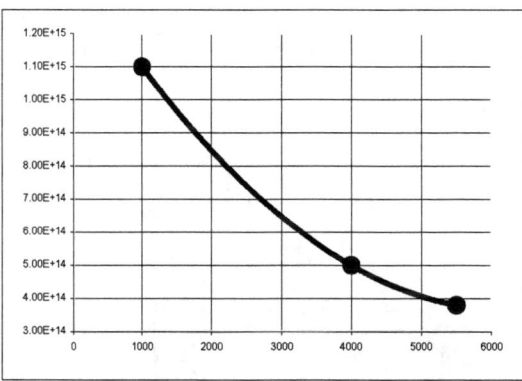

Figure 2: Dose response curve for precision resistors formed by method B. Graph shows sheet resistance versus dose.

The counter doping method allows the manufacture of precision resistors in a single layer polysilicon process. As resistors comprise only a small portion of the die in either case (~2%), during the precision implant the wafer is almost entirely free of photoresist in Process A and entirely covered with photoresist in Process B. Table I compares the two processes.

Table I: Comparison of two techniques for forming precision polysilicon resistors.

PROCESS A	PROCESS B
POLY2 DEPOSITION	POLY1 DEPOSITION
RESISTOR MASK	POLY1 IMPLANT
CAPACITOR IMPLANT	SCRUB
ASH/WET CLEAN	RESISTOR MASK
PRECISION RESISTOR	PRECISION RESISTOR
IMPLANT	IMPLANT
WET CLEAN	ASH/WET CLEAN
POLY2 MASK/ETCH	POLY1 ANNEAL
	POLY 2 MASK/ETCH

Resistors have long been successfully fabricated in our facility with Process A using two different types of high current implanters. Each implanter has an identically sized disk, rotating at 1200rpm and scanning along the vertical axis. These high current implants are batch processes with each disk requiring thirteen 200mm diameter wafers. In the event that less than thirteen product wafers are available, dedicated silicon dummy wafers are used to complete the batch.

In Process B we use B^{11} at lower energy (25-30keV) at a dose near 1×10^{15} at/cm^2. A typical dose response curve is shown in Figure 2.

Ion channeling along the columnar polysilicon grain boundaries is a serious consideration [3], particularly for elemental boron. This channeling will result in a less uniform distribution of dopant. For this reason the machine originally selected for the precision resistor implant had better capability for angular control as it could rotate about two independent axes. This will be referred to as Vendor A tool.

RESULTS

Initial process development was done with lots consisting of a few wafers so the problem was not initially apparent. As volume increased remarkably large variations in resistor values were observed across the wafer. This variation manifested as horizontal stripes with the lowest values in the middle of the wafer, shown in Figure 3. The specification for this resistor is met only at the extreme top and bottom edges of the wafer.

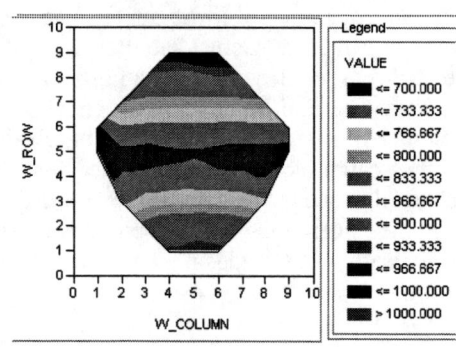

Figure 3: Variation across a wafer for Process B precision resistors run on Vendor A tool.

Remarkably large variations in resistor values were also observed as a function of lot size, with small lots exhibiting good uniformity. The underlying cause of this variation was diagnosed when the 3 lots shown in Table II were analyzed. Fortuitously, these lots each consisted of fourteen wafers, one more wafer than the implanter batch size. These lots were processed together at almost all process stages. The pattern suggested by this data was confirmed after reviewing the implant data logs. Uniformities for all wafers were poor except for the three wafers in batches 2 and 5. This corresponded to batches with a minimal amount of photoresist present during the implant.

Table II: Loading pattern of 3 fourteen-wafer lots run in Vendor A tool for process B precision resistor implant.

BATCH	LOT ID	WAFERS	BEAM CURRENT (mA)
1	C	13	3.5
2	C	1	4.7
3	B	13	3.6
4	B A	1 12	3.5
5	A	2	4.4

Beam current during setup measured at 5mA, the apparent drop during implant was the result of ions being neutralized. This drop was accompanied by a concomitant increase in the endstation pressure. Neutral atoms are not counted by the Faraday but are still electrically active. This explains the overall overdosing of the wafer and the lower average resistance.

DISCUSSION

The sequence of interactions is as follows; photoresist breaks down under ion bombardment releasing large quantities of hydrogen gas and creating a hardened carbonized crust at the top of the photoresist [4]. The released gas neutralizes incoming positive ions. The Faraday is located behind the disk and periodically sampled through slits in Vendor A tool. Thus neutral atoms are not counted though are still electrically active. Wafers are overdosed most heavily in the center of the wafer where the quantity of photoresist is greater. Overdosing with neutrals lowers the resistance.

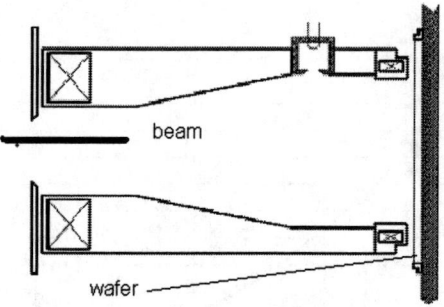

Figure 4: Schematic representation of dosimetery system in Vendor B tool. Note, wafer forms the back surface of the Faraday.

Resistance variations were expected to be reduced by moving this implant from Vendor A to Vendor B tool. This was due to two key design differences. First, the Vendor B tool dosimetry system continuously monitors the beam in a Faraday located in front of the wafers. Therefore, all outgasing occurs within the Faraday and undercounting should be minimal. The hardware for Vendor B tool is depicted in Figure 4.

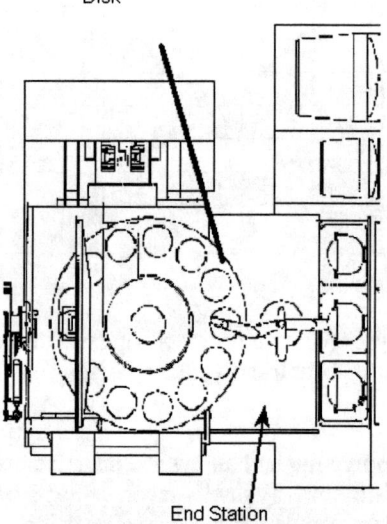

Figure 5: Schematic representation of endstation of Vendor B tool. The large volume through which the disk rotated to a vertical position allowed for effective dilution of outgasing from photoresist.

Second, the endstation vacuum on Vendor B tool is considerably better than on Vendor A tool during implant owing to the large process chamber which allowed for rapid gas dilution. The endstation is also equipped with four rather than a single cryogenic pump, so waste products are more effectively removed. This is shown in Figure 5.

Results for wafers run in Vendor B tool with a full load of photoresist coated wafers are shown in Figure 6. They confirmed our expectations of good across wafer uniformity.

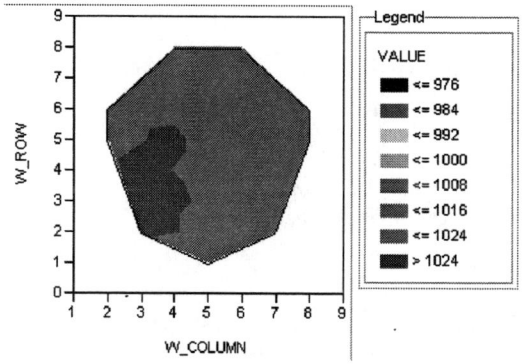

FIGURE 6: Resistance variation across wafer in Vendor B tool with a full load of photoresist wafers

It has also been reported that these counter doped precision resistors can be successfully fabricated using a third type of high current implanter [5].

CONCLUSIONS

Precision resistors can be formed in a portion of the first level polysilicon by counter doping with boron. This provides resistors with lower temperature coefficients of resistance and gives the flexibility to manufacture precision resistors in a single level polysilicon process. As resistors comprise only a small portion of the die (<2%), the wafer is almost entirely covered with photoresist. For the boron resistors, consistent uniformity could only be obtained using one of two types of high current implanters. In the other type of implanter, variations in resistor values of more than 300ohms were observed which far exceeded the design specification. Variation was dependent on lot size, with small lots showing better uniformity. Variation across the wafer manifested as horizontal bands with the lowest values in the center of the wafer. The cause of the nonuniformity was determined to be photoresist outgassing and ion neutralization. Photoresist breaks down under ion bombardment releasing large quantities of hydrogen gas. The released gas neutralizes incoming positive ions. This results in overdosing, which lowers the resistance most dramatically in the center of the wafer where the quantity of photoresist is greater.

ACKNOWLEDGMENTS

The contributions of Jeffrey Gerlach, Bruce Greenwood, Mike Thomason and Muhammad Anser are gratefully acknowledged.

REFERENCES

1. K. Kundert, H. Chang, D. Jefferies, G. Lamant, E. Malavasi and F. Sendig, Computer-Aided Design of Integrated Circuits and Systems, IEEE Transactions 19, 1561-1571 (2000)
2. S. Gupta, J. Electrochemical Society **149**, G271-G275 (2002).
3. G Fuse, S. Shibata and Y. Kato, Proc. of the 11th International Conference on Ion Implantation Technology, 642-645 (1996).
4. P. Kopalidis and S. Kondratenko, J. Electrochemical Society **152**, G375-G377 (2005).
5. Marnix Tack, AMIS Internal Communication.

Automated Dose and Dopant Level Monitoring by SIMS

Hans Maul*, Norbert Loibl[1], Ulrich Ehrke[1], Alex Merkulov[2], Paula Peres[2], Michel Schuhmacher[2]

[2]CAMECA SA 103 Blvd. St.-Denis, Courbevoie 92403, France
*[1]CAMECA GmbH Bruckmannring 40, 85764 Oberschleissheim/Munich, Germany
hans.maul@cameca.com

Abstract. Full wafer SIMS without breaking the 300mm wafer has the advantage of saving cost and speeding-up response time. A variety of performance examples for B, As and P implants in Si and for B levels in SiGe is shown for both Cameca tools, the quadrupole based SIMS 4600 and the magnet based IMS Wf. Dose mapping and dose monitoring with an RSD around 0.5% are demonstrated and the added value of dopant profiles is illustrated.

Keywords: SIMS, depth profiling, wafer, semiconductor, SiGe, dopant, dose, implant, B, As, P, Si
PACS: 68.49.Sf, 82.80.Ms, 61.72.Ss, 81.70.Jb, 61.72.Tt, 61.72.–y, 71.55.–I

INTRODUCTION

SIMS must not necessarily mean the end of life for 300mm wafers. Full wafer analysis capability allows monitoring subsequent process steps and additional metrology on the same wafer, saving cost and skilled labour and shortening the analytical response time during process development. For process monitoring easy access and timely response is of key importance. For both aspects full wafer SIMS is superior to lab type SIMS. Moreover, for metrology on patterned wafers, full wafer analysis allows a high degree of automation with straightforward access to test pad positions providing a much faster turn around time and higher throughput.

The requirements for full wafer SIMS are quite diverse ranging from daily monitoring of just one process step to running the full variety of SIMS applications. This paper discusses two approaches made to satisfy both concepts: the Cameca SIMS 4600 and the Cameca IMS Wf.

A variety of performance examples for B, P and As dose monitoring and depth profile mapping of 300mm Si wafers and for B level monitoring in SiGe is shown for both, the SIMS 4600 and the IMS Wf.

THE FULL WAFER SIMS TOOLS

The CAMECA SIMS4600 is a quadrupole spectrometer based metrology tool tailored to specific depth monitoring tasks with a high degree of simplification and automation. It provides depth profiling of implants and multi-layers and dose monitoring for any kind of doping method. It covers, with just one oxygen primary ion gun, B, As and P monitoring of EPI and implant doping process steps. An optional Cs primary ion gun provides low dose As implant monitoring. Automated stage-height positioning assures high measurement reproducibility over the wafer surface and from wafer to wafer. The SIMS4600 provides unattended cassette to cassette operation on 300mm and 200mm wafers with automated measurements on pre-selected positions. A marathon test involving 1000 wafer loads in a single unattended row was finished without any assistance.

FIGURE 1. CAMECA SIMS 4600, dose monitoring and depth profiling on full wafers, also with ultra low energy.

The IMS Wf is a magnetic sector spectrometer based tool designed for automated dose and depth profile monitoring at pre-selected positions on 300 mm wafers. It preserves the full diversity of SIMS capabilities for full wafers which is otherwise only available upon breaking the wafers. The IMS Wf covers high depth resolution for high dose ultra low energy implants as well as the lowest detection limits for low dose high energy implants, depth profiling and dose monitoring regardless of what doping method has been used and the large variety of other SIMS applications. It applies automation features such as automated stage-height positioning assisted by optical auto focus and primary ion beam centring systems to assure high measurement reproducibility over the wafer surface and from wafer to wafer.

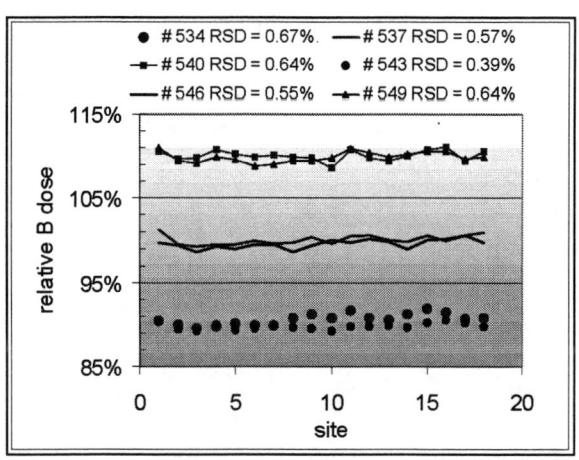

FIGURE 3. Ultra shallow, high dose, as-implanted B wafers (3 each) with implant dose differing by ± 10%. 18 sites, each wafer run twice on SIMS 4600. Dose RSD: around 0.5%.

FIGURE 2. CAMECA IMS Wf, dose monitoring and depth profiling on full wafers with the full protocol range of magnetic sector SIMS instruments.

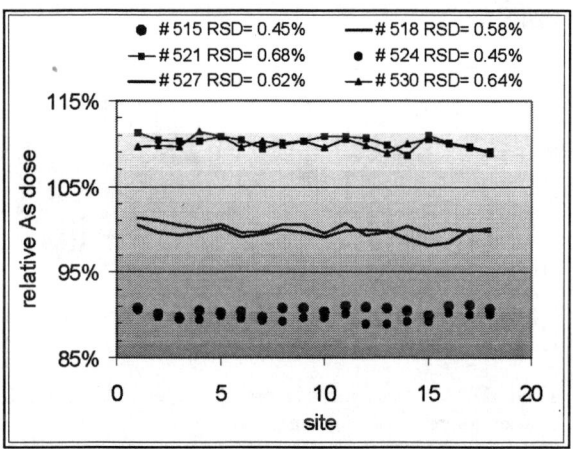

FIGURE 4. Ultra shallow, high dose, as-implanted As wafers (3 each) with implant dose differing by ± 10%. 18 sites, each wafer run twice on SIMS · 4600. Dose RSD: around 0.5%.

IMPLANT PROCESS MONITORING APPLICATIONS

Dose Mapping

The dose uniformity for a number of 300 mm B and As wafers has been determined from the SIMS depth profiles [1]. The B and As results in Fig. 3 and Fig. 4 have been obtained by SIMS 4600 using 500eV O2 at normal (0°) incidence angle.

Dose maps obtained by the IMS Wf are shown in Fig.5. The As dose map in Fig.5 has been obtained using 500eV Cs, 60° and the B dose maps have been obtained using 1 keV O2, 44° and oxygen flooding. The variation of dose values across the wafers includes both, the variation of the actual dose and the variation of the measurement. The dose RSD obtained from very uniform wafers is around 0.5%.

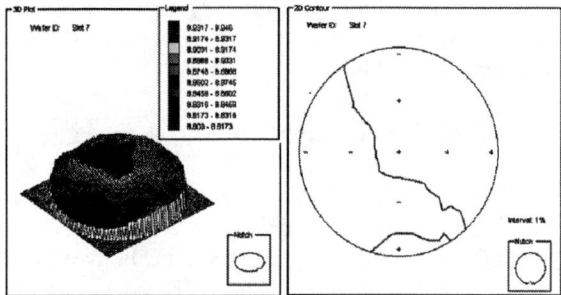

FIGURE 5. 3D and 2D plots of a high dose, low energy As implanted and annealed wafer. 7 sites dose RSD: 0.58%.

In any case the RSD is better 1%, except for the wafer shown at the right hand side of Fig. 6. In this case the RSD of 3.2% indicates non-uniform doping and the 2D map discloses that the B dose is decreasing from the wafer centre to the edges.

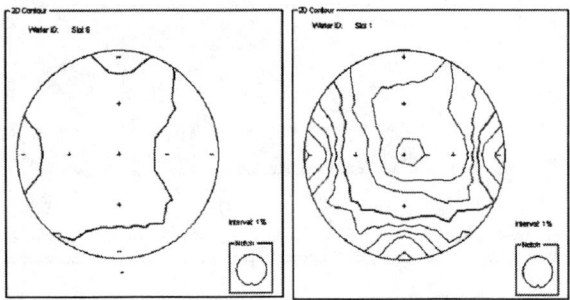

FIGURE 6. Two high dose low energy B as-implanted wafers with the same dose. However, the first wafer is uniformly doped (RSD 0.7%) while the wafer map of the second wafer on the right shows higher dose in the center (RSD: 3.2%).

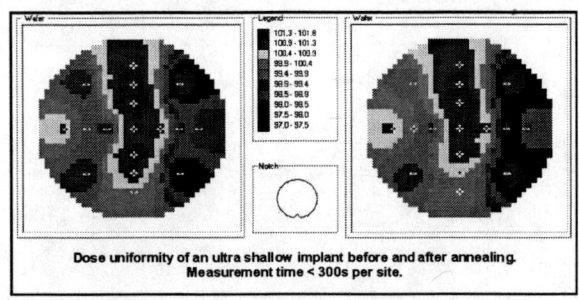

FIGURE 7. Dose map before and after RTP, the dose distribution remains unchanged.

One benefit of full wafer SIMS is that the same wafer can be analyzed before and after a particular process step. This reduces the number of process development wafers needed. Fig. 7 shows the dose map of an ultra shallow, high dose B implant before and after annealing, measured by SIMS 4600. This result proves that the dose distribution has NOT been changed by the RTP tool [2].

Dose Reproducibility and Accuracy

Reproducibility is not everything, but everything is nothing without reproducibility.

Fig. 8 and 9 show the dose reproducibility obtained by the 4600 for 5 load/unloads of 3 B and 3 As wafers, respectively (same wafers and SIMS conditions as above). 4 sites have been measured during each load. All wafers have been loaded once before starting the wafers for the 2nd, 3rd ... round.

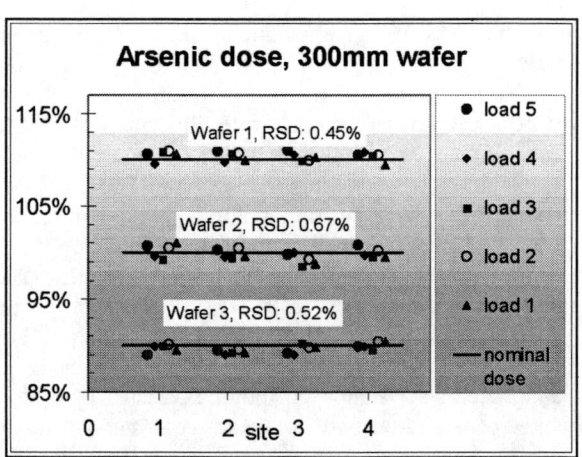

FIGURE 8. Load to load dose repeatability for arsenic, RSD around 0.5%

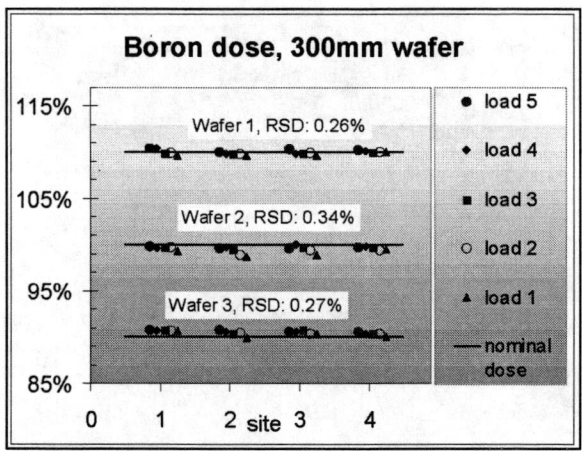

FIGURE 9. Load to load dose repeatability for boron. RSD around 0.3%

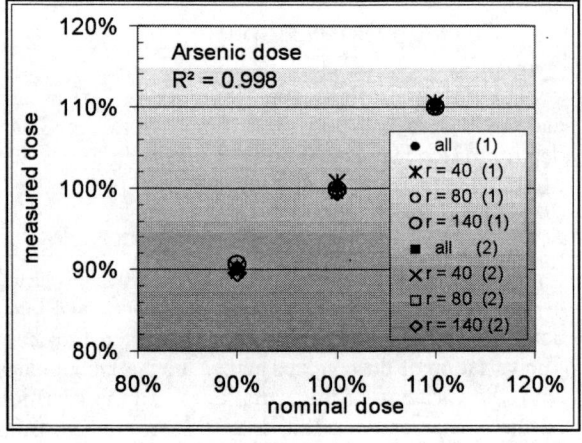

FIGURE 10. Accurate arsenic measurement of (intentionally) ± 10% wafer to wafer dose deviations

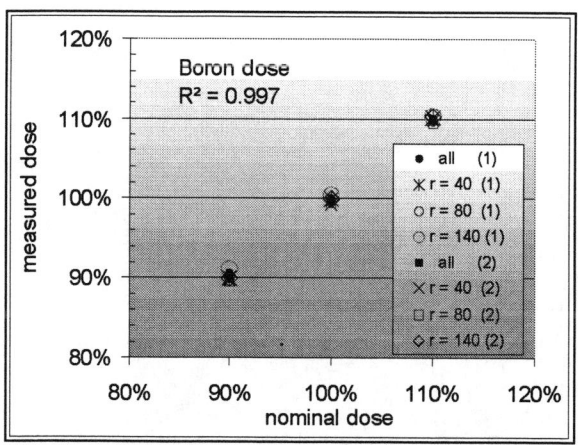

FIGURE 11. Accurate boron measurement of (intentionally) ± 10% wafer to wafer dose deviations

The RSD values are around 0.5% for the As wafers and 0.3% for the B wafers. The accuracy of dose deviations measurements can be seen from plots of the measured dose versus the nominal doses values. In fig. 10 and 11 all data from the dose uniformity measurements of fig. 3 and 4 have been included. The average values (red dotes) exactly match the nominal values. The correlation coefficients R2 0.998 for As and 0.997 for B confirm the high accuracy.

With SiGe one major subject of interest is monitoring of the B levels generated by EPI processes. The B levels on a blanket wafer have been measured by the SIMS 4600. Sets of 5 measurement on neighboring sites at R= 21mm and R = 134 mm delivered RSDs of 0.78%, 0.58%, 0.41% and 0.36%, 0.40 %, respectively. Likewise, sets of 5 measurements on test pads delivered RSD values around 0.5%. Some sets included measurements done one week apart without retuning the tool.

With EPI processed B and Ge wafers the B and Ge concentration levels vary across the wafer due to the so-called "loading effect". For the correlation of the measured data with the nominal values an R^2 of 0.98 has been obtained.

TABLE 1. The day to day dose repeatability of an ultra-shallow boron implant is around 0.5%

Wafer (slot)	1	1	1	2	2	2	3	3	3
Site on wafer	Center [at/cm²]	Edge [at/cm²]	Center / Edge	Center [at/cm²]	Edge [at/cm²]	Center / Edge	Center [at/cm²]	Edge [at/cm²]	Center / Edge
Day 1	7.53E+14	6.89E+14	1.09	6.85E+14	6.38E+14	1.07	8.12E+14	7.53E+14	1.08
Day 2	7.54E+14	6.89E+14	1.09	6.79E+14	6.43E+14	1.06	8.05E+14	7.56E+14	1.06
Day 3	7.56E+14	6.97E+14	1.08	6.78E+14	6.32E+14	1.07	8.16E+14	7.56E+14	1.08
Mean dose	7.54E+14	6.92E+14	1.09	6.81E+14	6.38E+14	1.07	8.11E+14	7.55E+14	1.07
Max-Min	3.50E+12	8.70E+12	0.01	7.10E+12	1.10E+13	0.02	1.03E+13	3.10E+12	0.01
std. dev.	1.83E+12	4.81E+12		3.76E+12	5.51E+12		5.26E+12	1.73E+12	
RSD	0.24%	0.70%		0.55%	0.87%		0.65%	0.23%	

Examples of day to day reproducibility demonstrated by the IMS Wf are shown in Table I. The B dose of 3 ultra shallow, high dose implants have been measured at two sites each (center and 5mm from the edge) at 3 consecutive days without re-tuning the instrument. The RSD of the dose values itself as well as the dose ratios between the 2 sites is better 1%.

Junction Depth

A major motivation to monitor the depth profile shape is the determination of the junction depth. Fig. 11 shows the after-anneal junction depth map with a 3% junction depth reduction from the center to the edge.

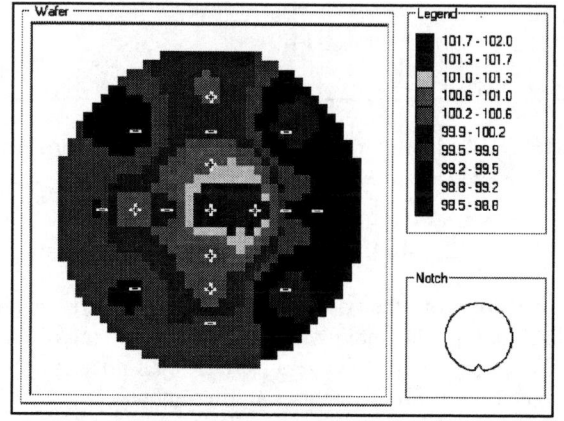

FIGURE 12. Accurate boron measurement of (intentionally) ± 10% wafer to wafer dose deviations

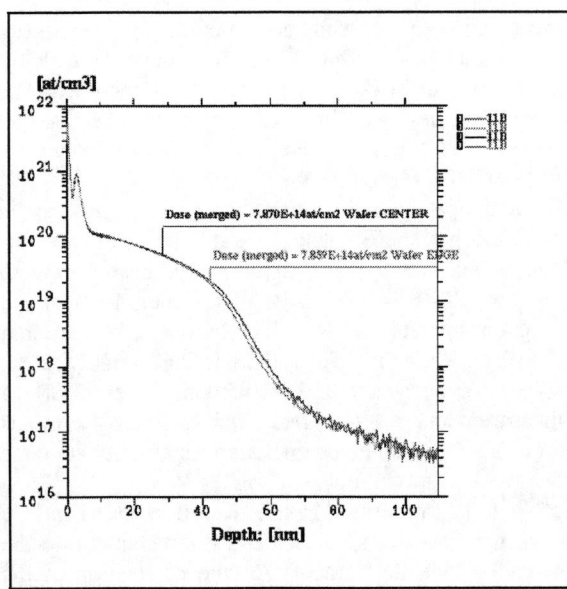

FIGURE 13. After anneal center to edge B **junction depth** difference

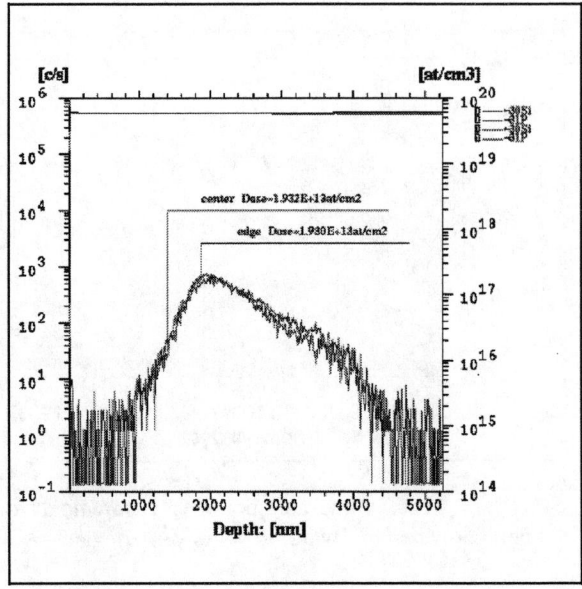

FIGURE 15. After anneal center to edge P **profile shape** difference

CONCLUSIONS

The variety of dose mapping, dose repeatability and accuracy data show that fully automated dose and depth profile monitoring is made available by both Cameca full wafer SIMS tools, the SIMS 4600 and the IMS Wf. The demonstrated dose repeatability and accuracy is better 1% (RSD). As SIMS dose values are determined from the dopant depth profiles, junction depth and profile shape changes are controlled as well.

REFERENCES

1. J.L. Maul, U. Ehrke, N. Loibl and C. Schnuerer-Patschan, Proc. SIMS XV, to be published
2. U. Ehrke, A. Sears, W. Lerch, S. Paul, G. Roters, D.F. Downey, E.A. Arevalo, J. Vac. Sci. Technol. B 22(1) (2004) p346

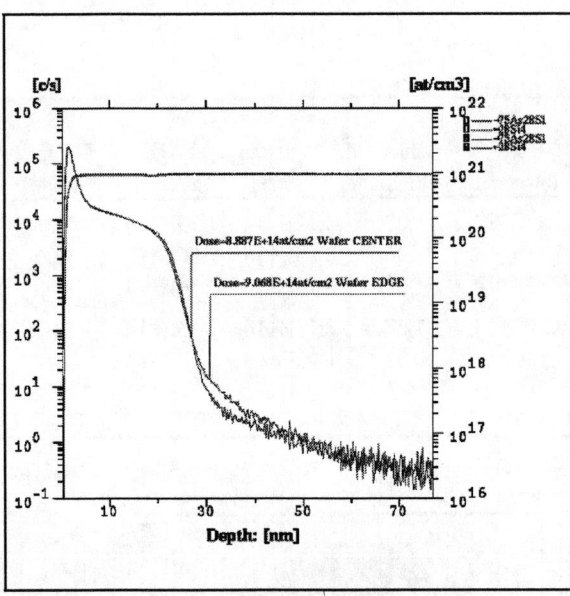

FIGURE 14. After anneal center to edge As profile **tail** difference

Depth Profiles

Details of the depth profile shapes are also disclosed as illustrated with the following figures 13 - 15 showing center to edge differences: an after anneal B junction depth difference, an after anneal As profile tail difference, and a depth profile shape change of a high energy, low dose P implant which could be caused by a slight implant angle difference.

MATERIALS

Optima HD: Single Wafer Mechanical Scan Ion Implanter

P. Splinter, M. Graf, C. Godfrey, Y. Huang, D. Polner, J. Danis and K. Ota

Axcelis Technologies Inc., 108 Cherry Hill Drive, Beverly, Ma. 01915, USA

Abstract. The trend toward single wafer high current ion implanters to support high tilt angles and to avoid damage during the implant process has led to the novel design of the Optima HD architecture. A high frequency (> 3 Hz) reciprocating pendulum coupled with a fixed spot beam has effectively extended the dose operating range without sacrificing wafer throughput. A high scan frequency enables more reliable dose uniformity by allowing a greater number of wafer passes through the ion beam in a given time period. A fixed spot beam preserves the important performance characteristics of rapid beam setup and precise angle control typical of conventional high current implanters. This paper will examine the design approach used to deliver high scan rates with a reciprocating pendulum mechanism. The challenges overcome include the minimization of vibration coupling, angle control, wafer handling and the preservation of the ion beam focal length to the wafer plane. Together with the novel wafer scan approach, the beamline is designed to deliver superior low energy performance through the implementation of a fixed spot beam which simplifies the control of beam angles, control of space charge and design of wafer neutralization components. Resulting process parameter models and empirical results will be further explored in the paper.

Keywords: Ion implantation, high current implanter, wafer scan
PACS: 85.40.Ry

INTRODUCTION

The semiconductor industry trend for single wafer process equipment used in the manufacturing of 45nm and below devices is driven by a mix of requirements in both process technology and wafer production. This trend has accelerated the adoption of single wafer high dose ion implanters which have traditionally been implemented as batch processing equipment due to their historic capability for high process utilization, high throughput and wafer uniformity. Examples of emerging process requirements include precise angle control and high tilt implants while manufacturing requirements are driven by process flexibility with a trend for decreasing number of wafers per manufactured lot.

Numerous challenges exist to the designer of single wafer high dose (> 1E14 ions/cm^2) equipment in order to achieve and even surpass the historical performance advantages of batch processing equipment. The traditional spinning disk implanter minimized engineering design problems such as wafer temperature control and wafer uniformity. Disk velocities in excess of 1000 RPM provided sufficient contact forces for good wafer cooling and the speed also enabled many passes of the wafer through the ion beam in order to average out any non-uniformities. While the high spin speed is necessary to provide strong centrifugal forces on the wafer for heat transfer, they also proved to be a detriment due to particle impingement on the wafer [1]. In addition to the advantages of the mechanical scan system, the equipment design benefited from a simple spot beam architecture for ion generation and transport. This ion source and beamline architecture led to the development of highly reliable systems with typical recipe setup times under 5 minutes. The tradeoffs during design of single wafer high dose systems attempt to maintain a majority of these types of benefits found in earlier generation equipment, while at the same time meeting the future needs of production.

SINGLE WAFER SCANNING FOR ION IMPLANTERS

In the design of ion implantation equipment one can generalize the equipment by the type of wafer scanning system or endstation used to present the wafer or work piece to the ion beam. In this regard, Table 1 categorizes all single wafer ion implanters by: one axis mechanical scan, two axes mechanical scan and no mechanical scan. The mechanical scan method is then combined with a particular design of the ion beam delivery system in order to create a complete wafer processing implanter. For example, Figure 1 further highlights the possible combinations of equipment design for a spot beam, single axis mechanical scan architecture.

Table 1: Mechanical Scan Axes and Beam Type					
Mechanical Scan Axes	0		1		2
Beam Type	Broad	Spot	Ribbon	Spot	Spot

Of the machine architectures in Table 1, the single axis spot beam scanned machines have been used in semiconductor production since the beginnings of high volume manufacturing [2]. These machines have typically been associated with low dose implants but recently high dose capability with this approach has been developed [3]. A requirement for all types of machines is to provide an ion beam that is exactly perpendicular to the surface of the wafer and maintains the same angle across the entire implant area. Therefore, scanning is required to traverse the wafer area while angle correction or collimation is needed to create an orthogonal ion beam.

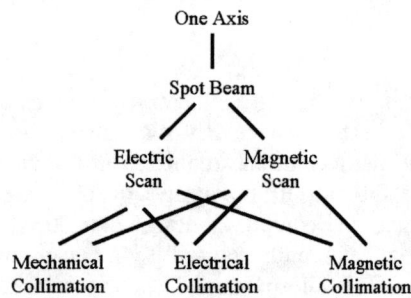

FIGURE 1. Possible combinations of active beam optical components to create a scanned beam

The spot beam, single axis ion implanter can be implemented utilizing either magnetic or electric scanning components. In addition to the scanning mechanism, the beam undergoes a collimation step which can be accomplished either magnetically or electrically, or even mechanically by adding an additional axes to the wafer scanner.

Alternatively, the single axis mechanical scan is used with a rectangular ribbon beam which is required to be greater than the width of the target wafer. Similar to scanned beam systems, angle correction and uniformity is achieved through a combination of active magnetic or electrical components to manipulate the optics and transport of the ion beam. In the case of a rectangular ribbon beam, the origin of the beam can be either a spot beam or a ribbon beam which is then projected through further optics to image itself at the wafer.

In the case of a fixed mechanical scan, the beam either encompasses the entire wafer as in plasma immersion systems or is scanned in two dimensions using the methods described for a single axis system.

Lastly, a uniform and completely orthogonal ion beam transport can be accomplished with a two axes mechanical scan. In this case, the beam transport and optics are greatly simplified by utilizing a static ion beam and therefore eliminating the need for additional active optical elements to manipulate the beam for collimation. In these designs, the focus is on the construction of the scanning systems which can be implemented in a linear [4] or rotary fashion. The Optima HD system represents a two axes, spot beam, rotary architecture for single wafer ion implantation.

OPTIMA HD ARCHITECTURE

The architecture of the Optima HD has been designed to maximize the implanted area utilization and to support doses from 5E11 ions/cm^2 to greater than 1E16 ions/cm^2 (Figure 2). Although the machine is most accurately characterized as a high dose beamline, semiconductor manufacturers often demand overlapping coverage between types of equipment in order to create the most flexibility in their production lines. To accomplish this broad range of implants and to achieve high beam utilization, the Optima HD has demonstrated turnaround times of the reciprocating pendulum of approximately ~ 30ms with overall maximum scan rates in excess of 3Hz.

The beamline design was driven by overall simplicity and to deliver high beam currents from 1keV to 60keV without energy contamination. Further, accuracy of implant angle control was accomplished through an endstation design that provides the same transport length from the ion source to every part of the wafer regardless of the implant angle.

Wafer Scanning Architecture

The mechanical scanning architecture implemented on the Optima HD uses a classical pendulum motion traversing an arc of approximately 52 degrees. The pendulum rotary action is particularly effective in this architecture by enabling high scan rates in excess of 3Hz and ensuring a reliable design. The scan drive mechanism is composed of typical components found on semiconductor production equipment such as rotary ferrofluidic vacuum seals and the reciprocating motion allows for minimal displacement of facility harnesses servicing the scan head. Of major concern in designing for the high scan rate was vibration transmitted to the rest of the system or even to the factory floor. To minimize the vibration, the patent pending rotary scan motor was designed to allow the stator of the motor to act as integral reaction mass and absorb the forces generated during the reversal of direction. The stator is allowed to move during turnaround and the motor control modifies the commutation between the rotor and stator to maintain torque. The overall position of the scan head and subsequently the target wafer is controlled by a positional encoder between the rotor and the static frame of the assembly. The technique of utilizing the stator as a reaction mass allows the system to run at high scan rates while measured vibration to the process chamber housing is typically low at .01 g's with peaks of up to .04g's.

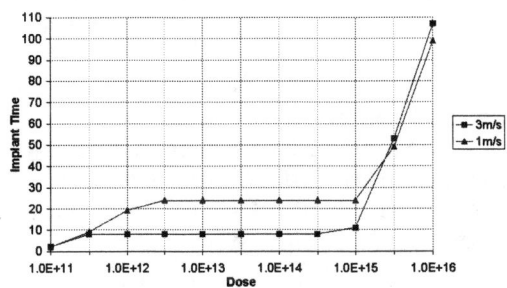

FIGURE 2. Implant time versus implanted dose for 1m/s scanning and 3m/s scanning.

The fast scan motor supports a 54cm counter-balanced wafer scan arm. The wafer is mounted by a Johnsen-Rahbeck electrostatic chuck which develops > 250 Torr clamping pressure using < 800 volts. The scan head can rotate 360 degrees to enable quad implants. The scan head rotation is also used to counter-rotate the wafer in reverse synchronous action with the scan arm. The counter rotation motion during angled implants effectively eliminates any twist error that would occur due to the arc traversed by the wafer.

The fast scan drive motor is mounted on a differentially pumped slide plate which completes the two dimensional scanning motion. The up and down motion is perpendicular to the reciprocating pendulum fast scan and reaches speeds of up to .15 meters / sec.

The entire fast and slow scan drive systems, including the process chamber, are attached to a turntable enabling implant angles of up to 45 degrees (Figure 3). The rotation along the scan axis positions the wafer such that the beam path length from the ion source to the wafer is held constant during scanning. This feature is important in maintaining the same spot size and density across all parts of the wafer.

FIGURE 3. The cutaway illustration depicts the fast scan motor and slow scan drive systems which are mounted on a turntable for implant angle capability of up to 45 degrees.

Beamline Architecture

With the shift to 300mm and an increased mix of electronics for the retail market, overall manufacturing lot sizes have shrunk. The ion transport system of the Optima HD is designed with minimal active beam optical elements in order to achieve the highest possible drift beam currents with short setup time.

The ion generation and extraction system is comprised of an indirectly heated cathode source and variable aperture extraction electrode. The electrode aperture width is geared to be proportional to the gap from the electrode to the ion source.

The beamguide through the 90 degree mass analysis dipole magnet has a 12cm opening while the magnet field is capable of greater than 1 Tesla. The beamguide and the beam tunnel immediately downstream of the magnet contain passive magnets developing a cusp confinement field in order to entrap electrons to aid in low energy transport.

Lastly, the beamline encompasses a plasma electron flood for neutralization of charging at the wafer surface.

Process Results

In spot beam ion implanters, the desired across wafer uniformity is achieved by averaging the entire beam across the area of the wafer. This averaging of the ion beam mitigates the need for precise control of the ion beam features. A minimal number of beam passes across each point of the wafer is required to produce a wafer of uniformities in the 1.0% range, one sigma. Typically the Optima HD has demonstrated that a minimum of 7 passes of the beam across every wafer location produces uniformities as shown in Figure 4.

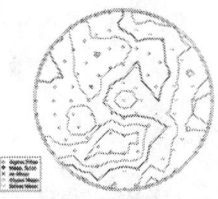

FIGURE 4. 300mm wafer uniformity map from the Optima HD of As+, 1E15 ions/cm^2, 10keV with 16mA of beam current. Mean sheet resistance of 113.7 ohms/square and uniformity of .618%, 1σ.

The unique construction of the pendulum motion allows the Optima HD to deliver high utilization and scan rates that can be used to provide implants for typical high doses and for those less than 1E14 ions/cm^2. In Figure 5 we match implants from the Optima HD and the Optima MD. The Optima MD has an architecture of a typical medium current ion implanter utilizing a 1 kHz scanned beam. In this example, the ThermaWave units are within 1% while uniformities from both machines are comparable.

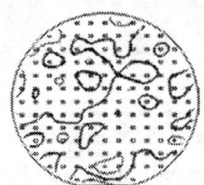 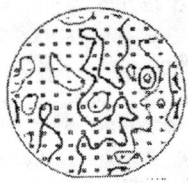

FIGURE 5. 300mm wafer uniformity map from the Optima HD left (390 TW, 0.11%) and the Optima MD right (386 TW, 0.16%) of B+, 1E12 ions/cm^2, 10keV, with ~10uA of beam current.

For control of the incident implant angle, the Optima HD incorporates a passive beam profiling scheme using the scan arm as a shadow mask across the ion beam. At the beginning of implant, the scan arm is positioned such that the ion beam is above the wafer and the beam can only strike the support arm and not the target wafer. The scan arm is then moved through the beam and beam current data samples are collected at intervals of 50ms. After a complete sweep of the arm through the beam the derivative of the samples are calculated. As shown in Figure 6, the derivative samples reconstruct the horizontal profile of the ion beam.

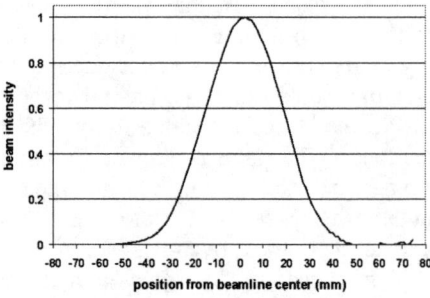

FIGURE 6. Beam profile and offset data for a 12mA beam resulting in a 0.1 degree incident angle. Profile constructed from the differential of beam current values collected using the scan arm as a shadow mask.

The beam profile is then matched against the known scan arm angle of each sample, thus providing a reference for an overall spatial position of the ion beam. A beam offset of 10mm translates to an angle offset of one degree. These angle offsets can then be corrected by use of the mass analysis magnet which provides horizontal position control of the beam. Figure 6 shows an overall horizontal angle accuracy of 0.1 degrees.

Summary

The Optima HD is a two dimensional mechanical scan machine that provides a simple and flexible means of achieving single wafer implants. The scanning mechanism reaches speeds of over 3m/s which enables implant capabilities across a wide dose range. Overall implant angle integrity is maintained by the architecture providing constant focal length from the source to the wafer and by measuring angle offsets through the use of a passive beam profiler.

REFERENCES

1. Y. Kawasaki, K. Tokunaga, et al., "The Collapse of Gate Electrode in High Current Implanter of Batch Type", International Workshop on Junction Technology, 2004, pp 39-41
2. D. W. Berrian, et al, Nucl. Instru. & Meth., B37/38 0500, 1989
3. V. Benveniste, R. Rathmell and Y. Huang, U.S. Patent No. 6881966 (19 April 2005)
4. A. Murrell, et al, "Quantum X: Single Wafer High Current Ion Implantation Using Mechanical Wafer Scan", Proceedings, IIT2004, Part II, pp 20-24.

High performance medium current ion implanter system EXCEED3000AH-G3

Shigeki Sakai, Masayasu Tanjyo, Nariaki Hamamoto, Sei Umisedo, Tomoaki Kobayashi, Takatoshi Yamashita, Takao Matsumoto, Tadashi Ikejiri, Kohei Tanaka, Yuji Koga, Satoru Yuasa, Masao Naito, Nobuo Nagai

Nissin Ion Equipment co.,ltd, 575 Kuze Tonoshiro cho Minami-ku 601-8205 Kyoto Japan

Abstract. A new medium current ion implanter has been developed based on the EXCEED3000, which is highly reliable and widely used in 300mm fabs. The ion implanter now has to be designed so that it can precisely measure and control beam characteristics. For example beam angles have to be controlled in halo implantation because high tilt angle implantation is done according to the device geometric structure. Not only horizontal beam profile system but also vertical beam profile system are implemented in EXCEED3000AH-G3 for the precise implantation control.

Keywords: Ion implantation, Angle control, Beam Size
PACS: 85.40.Ry; 41.75.-I, 41.85.Ja

INTRODUCTION

The conventional EXCEED3000 met almost all the 300mm/65nm CMOS device production requirements. Sub-65nm device manufacturing requires further upgrade of tool capabilities. The Nissin 300mm medium current ion implanter EXCEED3000AH-G3 has now been developed for such application requiring more precise implantation.

Figure1 shows a layout of EXCEED3000 series. The layout is almost the same as the EXCEED2300[1]. The beam formation components are ion source, analyzing magnet, acceleration column, final energy magnet, beam sweep magnet and collimator magnet. These components succeed the each proven properties of EXCEED2300's components and guarantee lowest risk at the transition to next generation node mass-productions. All the magnetic elements for beam transport have been remodeled with enlarging the beam acceptance This provides a boron beam current of 600uA at 5keV. A new two dimensional measurement system for detecting spatial variation of the beam current and the beam incident angle has been implemented.

Precise implantation

Importance of precise control of ion beam angular properties increases with reduction of increases with

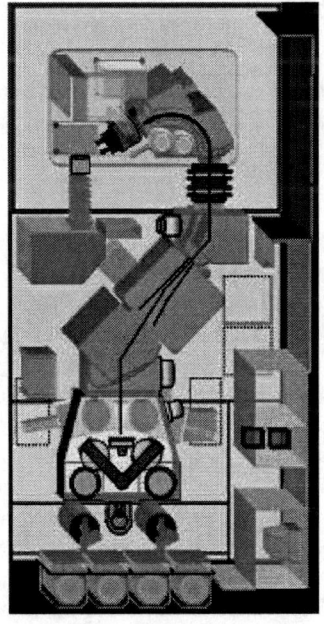

Figure1. Layout of EXCEED3000, same size to EXCEED2300.The width is 3.2m and the length is 6.3m.

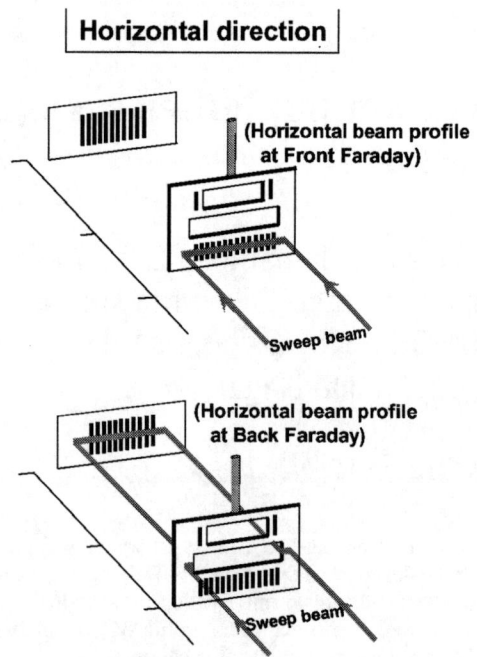
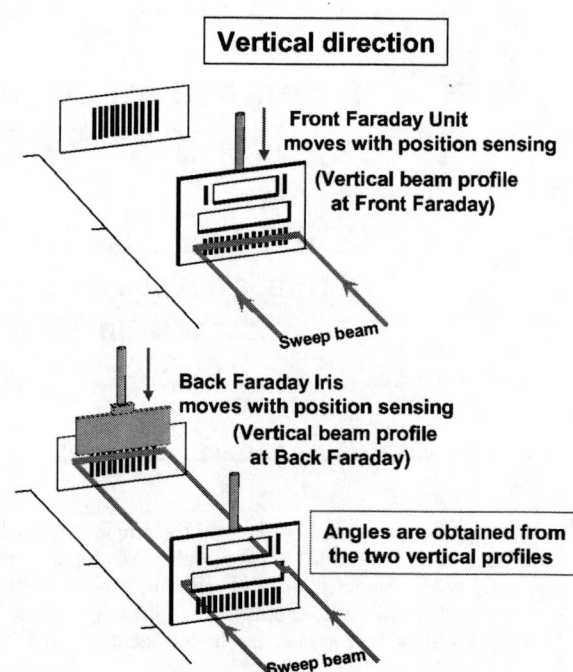

Figure 2. Horizontal beam measurement with the Front and the Back Faraday.

Figure 3. Vertical beam measurement by vertically moving the Front Faraday and the Back Faraday Iris.

reduction of intrawell isolation space and with device shrinkage[2].

EXCEED3000 has a pair of multiple Faraday systems, the Front Faraday and the Back Faraday, located at about 300 mm upstream and down stream of the platen, respectively. The Front Faraday has 16 Faraday cups and the Back Faraday has 11 Faraday cups. Using the beam current signals from the Front Faraday and the Back Faraday the beam parallelism or the spatial variation of the beam incident angle to the wafer along the beam scanning direction (horizontal direction) is calculated and tuned within an acceptable value. The configuration of this system is shown in Fig.2. Both of the mechanical setting of tilt and twist angle can be monitored by separately installed cameras as well as by encoders.

Since 1990[3] such system has been successfully used to control not only beam parallelism but also beam uniformity in horizontal direction. Now, this system provides measurement of the beam width at selected Faraday cups at the Front and the Back Faraday. This is based on the simple fact that the shape of the beam current change in time measured by a fixed Faraday cup during a beam scanning indicates information of beam spot density profile in the horizontal direction. Combination of the measurements of beam widths at the Front and the Back Faraday enables calculation of beam divergence or convergence in horizontal direction.

Figure 3 shows measurement of the beam property in the vertical direction. The Front Faraday is the three-story unit. The top story consists of the dose Faraday cups to count beam current during implantation and the implantation mask through which the beam hits wafers during implantation. The middle story is the blank mask and the bottom story is the array of 16 Faraday cups. The beam current detected by the array changes from a certain value to zero according to the vertical position of the array with respect to the beam when the vertical array position moves from where the array receives entire beam in vertical direction and to where it cannot receive any part of the beam. The vertical array position is detected by a sensors mounted in atmosphere. Calculating derivative of beam current measured by Front Faraday versus its vertical position yields vertical beam profile. It is well known that this differentiation is called as knife edge method [4]. Then to get the vertical beam profile at the Back Faraday, the Back Faraday Iris located in immediate proximity to Back Faraday moved vertically. When the beam is sweeping at Back Faraday, the iris moves down in front of the Back Faraday. The beam current signal to the Back Faraday decreases as the iris position lowers. Again this signal is differentiated by the position of the Iris; we can get the vertical beam profile at the Back Faraday. Using these two beam profile data at the Front Faraday and the Back Faraday, the vertical beam size at the wafer,

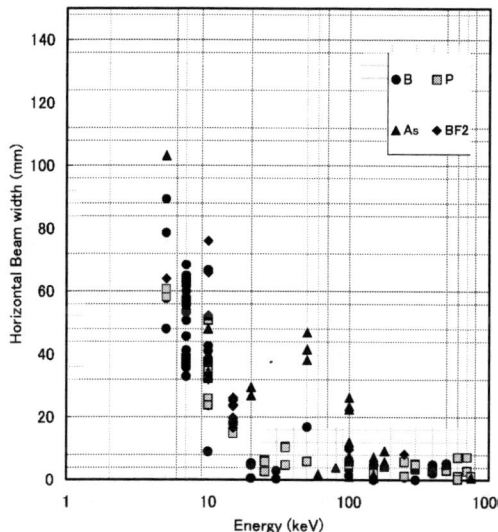

Figure 4. Horizontal beam size measured as a function of ion beam energy for various ion species

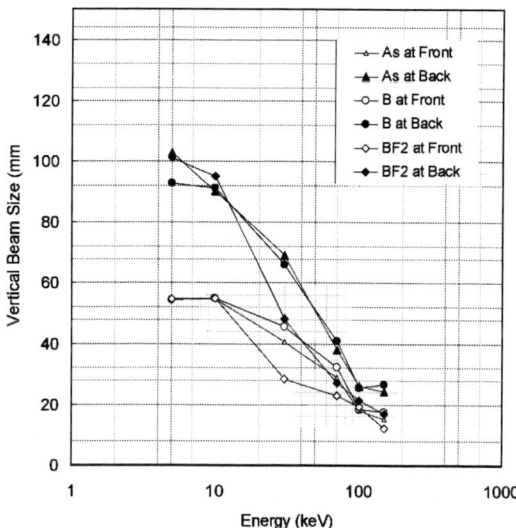

Figure 5. Vertical beam sizes measured as a function of ion beam energy for various ion species. Vertical beam sizes are measured at the Front Faraday and at the Back Faraday.

the vertical beam incident angle and beam divergence or convergence angle can be calculated.

Figure 4 shows the horizontal beam sizes as a function of beam energy for various ion species and bean current. The beam size markedly increases as the ion energy goes down below 20keV. Roughly estimating, the beam size is 10 to 20 mm at 100keV and 50 to 90mm at 5keV. Figure 5 illustrates the vertical beam size at the Front and the Back Faraday as a function of beam energy for various ion species at maximum beam current. The beam size at the wafer can be estimated from Fig.5 as about 20mm at 100keV and about 80mm at 5keV.

Using this 2 dimensional beam profile system we can get the horizontal and vertical beam divergence /convergence angle. Figure 6 shows the horizontal beam divergence angle measured as a function of ion energy for various ion species. It indicates that the beam divergence angle is almost zero degree at the energies over 100keV, while the angle changes its polarity from negative to positive by lowering the

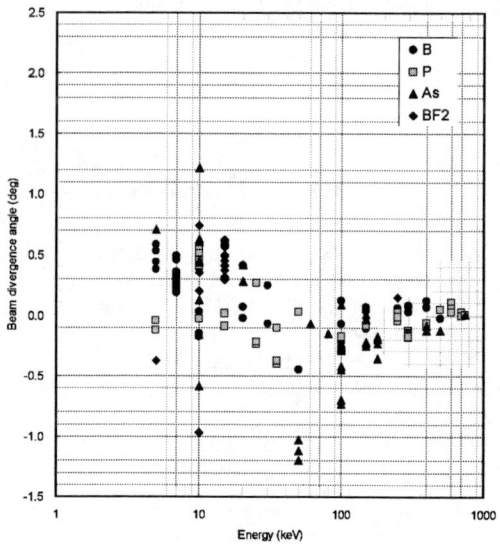

Figure 6. Horizontal beam divergence angle as a function of ion energy for various ion species

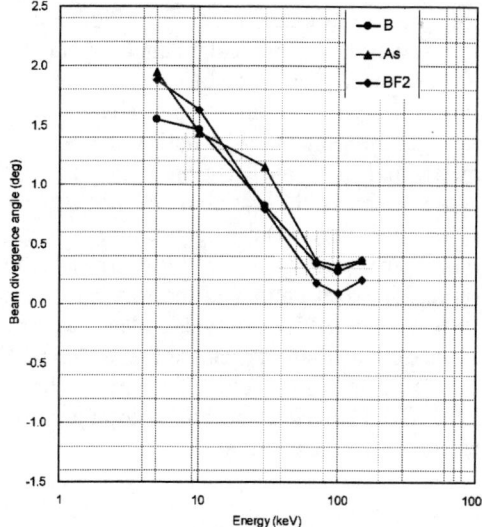

Figure 7. Vertical beam divergence angle as a function of beam energy for various ion species

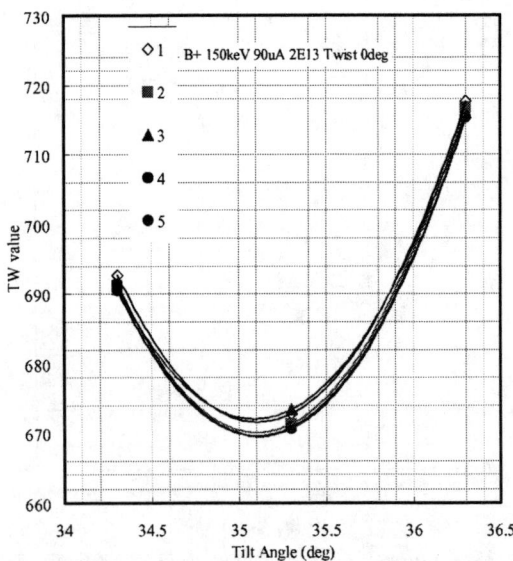

Figure 8. 5 day to day repeatability of beam incident angle using {211} plane channeling. Each data is auto tuned from cold start.

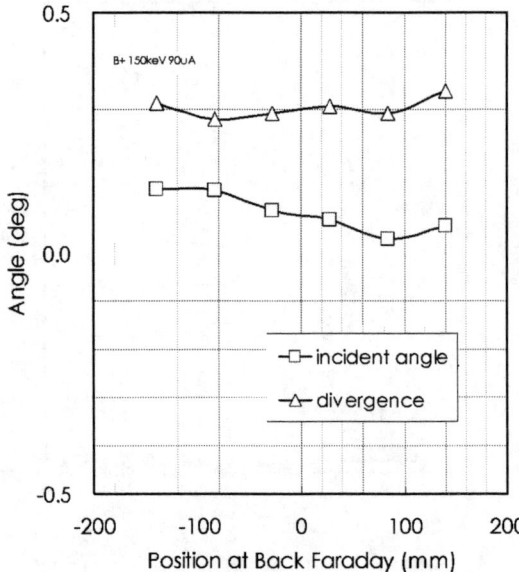

Figure 9. Vertical Beam characteristics along the horizontal position

energy below 100keV. This change of beam divergence in horizontal direction coincides with the focusing characteristics of the acceleration column.

Figure 7 shows the vertical divergence angle measured as a function of beam energy for various ion species. The vertical beam divergence angle is below 0.5 degree at the energies over 100keV and increases to 1.5 degree by lowering the energy below 10keV.

Figure 8 shows Themawave units from the wafers implanted with automatically tuned B^+ 150keV 90uA and the 2E13 cm^{-2} dose in the 5 day-to-day Marathon run for EXCEED3000AH-G3. It is known that the {211} plane channeling occurs at tilt 35.3deg and twist 0 deg wafer orientation. So Fig.7 indicates repeatability of the beam incident angle with respect to wafer. The fact that the tilt angle which gives minimum Thermawave unit did not change in 5 days shows a good repeatability of optical setting of beam parameters and mechanical setting of the wafer.

Figure 9 shows the variation of vertical incident angle and the vertical divergence angle along the horizontal position at B^+ 150keV 90uA beam. In Fig.9 the vertical incident angle and the vertical divergence are kept within 0.12deg and 0.34deg, respectively. The standard deviation of incident angle along 300mm width is estimated as 0.04deg. This result shows that the EXCEED3000AH-G3 ensures a good performance of precise angle implantation.

CONCLUSION

A new medium current ion implanter EXCEED3000AH-G3 has been developed based on the EXCEED3000 machine. The machine has been improved for precise angle control. The two dimensional beam monitor to measure the horizontal and vertical beam profile and beam angles of incidence as well as divergence. The obtained beam angle data showed a good repeatability in setting the implant angle and a very small deviation below 0.1 degree in vertical incident angle along the horizontal position.

REFERENCES

1. N.Nagai, Y.Tamura, S.Yuasa, K.Iwasawa, T.Matsumoto, M.Nakaya, M.Nakamura, and T.Nagayama ,, Proceedings *of the 2000 Int'l Conf. on Ion Implantation Tech*, pp.415-418
2. J. Weeman, J. Olson, B. N. Guo, U. Jeong, G.C. Li, S. Metha, Proceedings of the *of the 2002 Int'l Conf. on Ion Implantation Tech*, pp.276-278
3. N.Nagai, et al, Nucl. Inst. and Meth. B55 (1991) 393-397
4. E.Bauer, H. Poppa, G. Todd and F. Bonczek: J.Appl. Phys., 45, 5164 (1974)

Down to 2 nm Ultra Shallow Junctions : Fabrication by IBS Plasma Immersion Ion Implantation Prototype PULSION®

Frank TORREGROSA, Hasnaa ETIENNE, Gilles MATHIEU, Laurent ROUX

ION BEAM SERVICES, ZI Peynier-Rousset, rue Gaston Imbert Prolongée, 13790 Peynier, FRANCE

Abstract. Classical beam line implantation is limited in low energies and cannot achieve P+/N junctions requirements for <45nm node. Compared to conventional beam line ion implantation, limited to a minimum of about 200 eV, the efficiency of Plasma Immersion Ion Implantation (PIII) is no more to prove for the realization of Ultra Shallow Junctions (USJ) in semiconductor applications: this technique allows to get ultimate shallow profiles (as implanted) thanks to no lower limitation of energy and offers high dose rate. In the field of the European consortium NANOCMOS, Ultra Shallow Junctions implanted on a semi-industrial PIII prototype (PULSION®) designed by the French company IBS, have been studied. Ultra shallow junctions implanted with BF_3 at acceleration voltages down to 20V were realized. Contamination level, homogeneity and depth profile are studied. The SIMS profiles obtained show the capability to make ultra shallow profiles (as implanted) down to 2nm.

Keywords: Plasma Immersion Ion Implantation (PIII), Ultra Shallow Junctions (USJ).
PACS: 52.77.Dq ; 85.40.-e

INTRODUCTION

Due to constant shrinking of sizes in CMOS fabrication, junction depth requirements for Source / drain extension doping is more and more difficult to achieve. For bulk MPU ASICS, ITRS roadmap 2005 [1] imposes junction depth as low as 6.5nm for 45 nm node and 4.5 nm for 36 nm node. As Classical beam line implanters are reaching there limits for these specifications, Plasma Immersion Ion Implantation (PIII) or Plasma Doping (PLAD) has been proved as a possible solution for some years now [2-18]. The French Company IBS has developed its own PIII machine, named PULSION®, based on a original, simple and robust working mode. This machine has been tested for 45 nm CMOS application within the European project NANOCMOS. Metallic Contamination, homogeneity and demonstration of extremely shallow doped layers are presented bellow.

MACHINE DESCRIPTION

The structure of PULSION® is described in figure 1. The pulsed plasma is created thanks to a self designed ICP source located above a vacuum chamber.
Wafer is located on a chuck at the bottom of the chamber. The chuck is polarized through a capacitor and thanks to a current source (High voltage capacitor charging system). To improve uniformity on large substrates, the substrate holder rotates under the non centered plasma source. To avoid metallic contamination, the chamber is in aluminum alloy, plasma source in quartz with an external antenna and substrate holder is coated with a thick silicon coating.

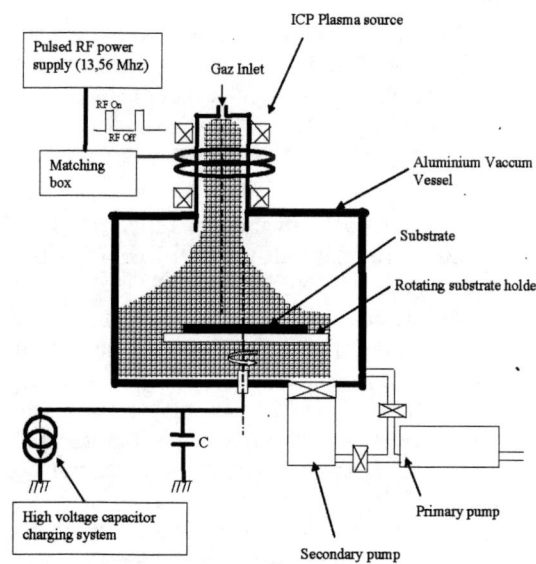

FIGURE 1. Machine structure of PULSION®

Polarization modes

PULSION® can work under two polarisation modes:
- **Mode 1**: pulsed plasma & constant polarisation voltage. This polarisation mode is very robust and easy to control. Acceleration voltage is kept constant thanks to the use of a big capacitor (> 1µF). Pulsed plasma offers the possibility to reduce thermal budget, etching and contamination. The efficiency of this mode has been proved on blank wafer [19]. Nevertheless, when an insulating layer is deposited on the wafer, charging problems occur. To avoid this drawback a new polarisation mode has been developed and patented.
- **Mode 2**: pulsed plasma and "self pulsed" acceleration voltage. The principle of this mode is the following:

1 – phase 1: plasma is off and the power supply charges the capacitor up to the desired acceleration voltage.
2 – phase 2: the power supply is inhibited when the plasma is ON. As the capacitor discharges, an implantation "pulse" occurs and the substrate voltage drops to 0V at the end of plasma pulse. Then plasma electrons are attracted towards the substrate and then can neutralize the positive charges accumulated on the substrate surface.
3 – Then, plasma is off again, and power supply inhibition is switched off to allow a new capacitor charging cycle….etc.

The capacitor value must be small to allow a rapid drop of the voltage and no irremediable charging problems.

The mode 2 has been proved to be compatible with real patterned wafers containing resists and oxides. It is used in all the following experiments.

CONTAMINATION STUDIES

Metallic contamination was checked by ToF SIMS on 200mm implanted wafers (PULSION® BF_3, 500V) and compared with the ITRS 2005 specification values. Results are presented on Figure 2. Except from Cu and Ni, the levels are all under the ITRS specifications. These results are good considering that the machine has no liner yet, and should be considerably improved soon by the setting up of a liner. Moreover, it appears that contamination at the center of the wafer is less important than at the edge, which indicates that major source of contamination may come from the substrate holder. For this reason, a new substrate holder is also under development to cope with this problem.

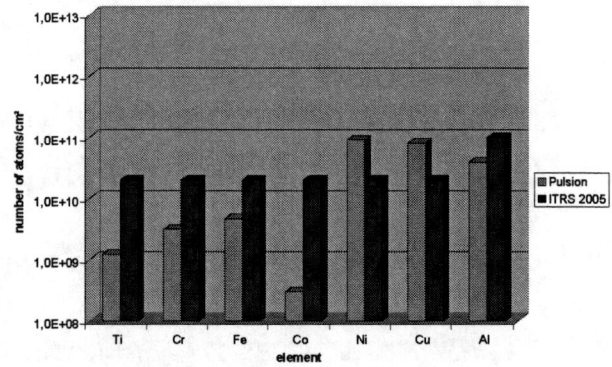

FIGURE 2. Metallic contamination level: ToF Sims analysis of a BF_3 500V implanted sample. Values (in atome/cm²) are measured at the center of the sample.

HOMOGENEITY

A non homogeneity of 3.5% on 200mm wafers implanted with PULSION® (BF_3, 500V, 1E15/cm²), activated using spike anneal (1050°C) and measured using four point probe has already been presented [20]. In the present paper, we have chosen to study non homogeneity on 200mm wafers (PULSION®, 500V, 5E15/cm²) using CAMECA LEXES tool, which allows measurement on as-implanted wafers, in such a way that we suppress possible non homogeneity due to the annealing process. Figure 3 presents the mapping obtained. As we can see on this figure, the non homogeneity looks good except from the wafer edge. Then, when we compare the implanted wafer mapping to a blank wafer mapping (see figure 4), we notice that this effect is mainly due to native oxide masking.

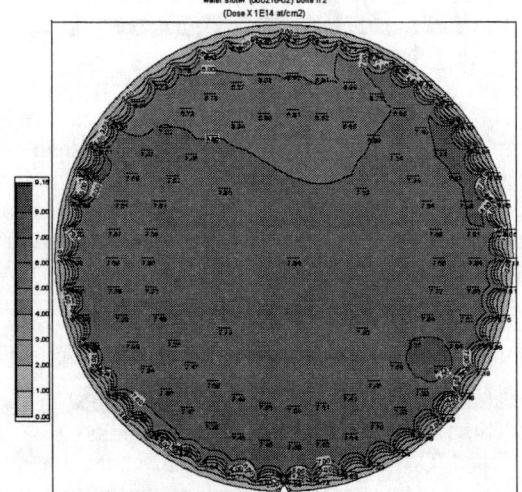

FIGURE 3. CAMECA LEXES mapping of a 200 mm wafer implanted with PULSION® using 500V polarisation voltage. Measured non homogeneity: 5%, but LEXES non reproducibility : 3.7%.

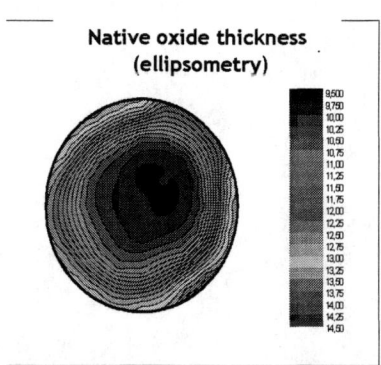

FIGURE 4. Mapping of a 200mm blank wafer measured by ellipsometry at SEMILAB.

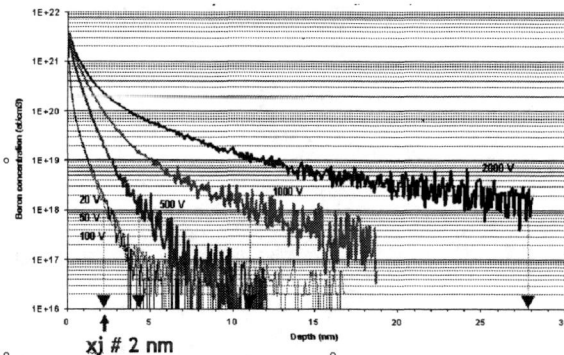

FIGURE 5. SIMS profiles of Boron for BF_3, 1E15/cm² PULSION® implantations from 20V to 2000V acceleration voltage.

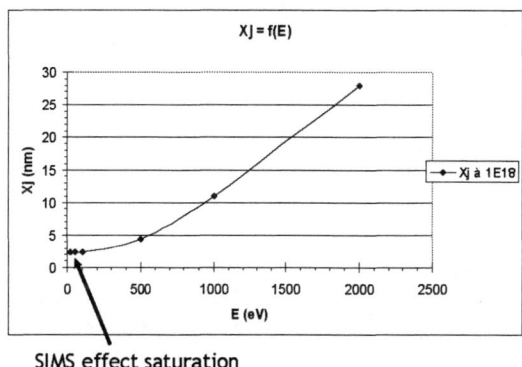

FIGURE 6. Implantation depth (at 1E18/cm3) as a function of acceleration voltage for BF_3, 1E15/cm² PULSION® implantations.

DOPING PROFILES

The measurement of implantation profile using dynamic SIMS has to be considered carefully when the ion used for the analysis (Oxygen or Cesium) has an incident energy higher than the one used to accelerate the studied specie (Boron) in the ion implanter. Indeed, due to recoil effect during SIMS analysis, the implantation depth can be bigger after analysis than the initial one.

We have started collaboration with CAMECA to use their last generation of dynamic SIMS (7F version) to characterize USJ implanted down to extremely low energies (PULSION® BF_3 from 2kV to 20V). The interest of this tool is its ability to work down to 300eV in positive mode with O_2^+ duoplasmatron source and down to 500eV in positive mode with Cs microbeam source using Cesium clusters (MCs^+ and MCs_2^+). Boron and Aluminium profiles were analysed using Oxygen beam, Fluorine profile was analysed using Cesium.

Boron profiles

The Boron profiles obtained are presented on figure 5: from 2kV to 500V, the implantation depth at 1E18/cm3 is proportional to the acceleration voltage (see figure 6). When the acceleration voltage is lower than 100V, there is no difference between the profiles due to the lack of sensitivity of SIMS at very low energy, as explained before. Recently, ToF SIMS has succeeded in separating these profiles.

Nevertheless, the possibility to implant Boron in extremely shallow layers (down to 2 nm at $1E18/cm^3$) using PULSION® has been demonstrated.

Fluorine profile

SIMS profiles of Fluorine were obtained using Cs at 400eV and 700eV depending on implantation depth.

An example of Fluorine profile compared with Boron one is showed in figure 7. As expected with a plasma which contains mainly BF_2^+ ions, Fluorine profile is almost superposed with Boron one. Figure 7 also shows Aluminium contamination profile, confirming the contamination level of 0.7% measured by ToF SIMS.

F/B ratio has been calculated from SIMS profiles integration. It varies from 2.5 to 3.5 for acceleration voltages higher than 500 V and is around 8 for lower voltages (see fig 8). F/B ratio was also measured on a sample after a thermal treatment of 750°C during 20s: exodiffusion of Fluorine is observed with a reduction of F/B ratio from 3.5 as implanted to 2 after annealing.

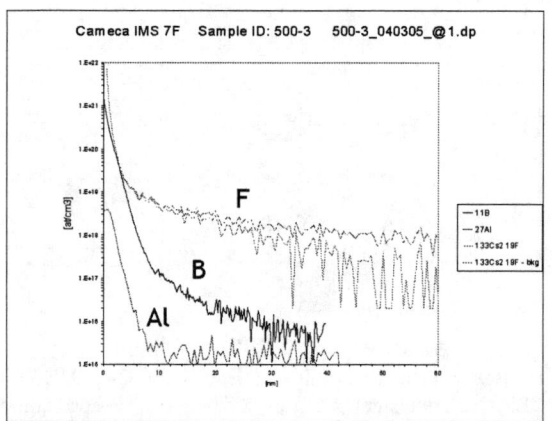

FIGURE 7. Fluorine, Boron and Aluminium profiles obtained by SIMS for BF_3, $1E15/cm^2$, 500V PULSION® implantation

FIGURE 9. TEM picture of a PULSION® BF_3 500V $1E15/cm^2$ as implanted sample.

CONCLUSION

The possibility to make PIII with a simple polarisation system avoiding any high voltage switching system has been demonstrated. A prototype named PULSION®, which allows implantation on wafers up to 300mm wafers was built. We have demonstrated that this machine was able to make Boron Ultra Shallow Junctions down to 2 nm using BF_3 and acceleration voltages down to 20V. Despite the fact that we are reaching the limits of all characterization systems, non uniformity on 200mm and 300mm has been measured on very shallow implanted layers and arises native oxide masking and edge effects, but it could be acceptable, if we take into account the high non reproducibility of the measurement and the improvements that have to be made. Then, metallic contamination is going to be improved by the setting up of a liner in the chamber and the development of a new substrate holder in order to decrease Ni and Cu levels and to be CMOS compatible. Finally, even if Boron and Fluorine concentrations at the surface of the silicium are important, no crystal defect was observed by TEM under the native oxide layer.

The next step of this study is the fabrication of a real alpha test machine scheduled to be located at LETI in 2007 thanks to the European integrated project SEA-NET.

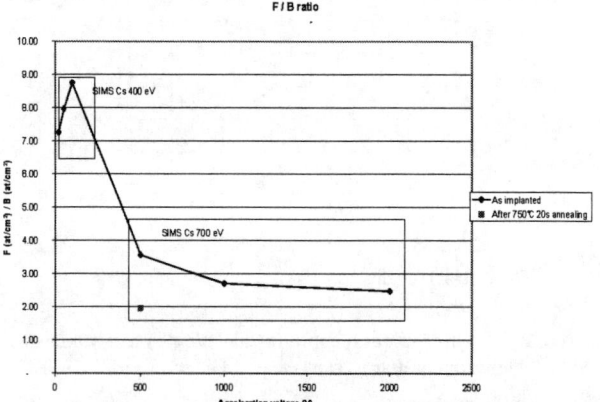

FIGURE 8. F/B ratio as a function of acceleration voltage for a sample as implanted and after thermal treatment samples.

IMPLANTATION DEFECTS

Implantation defects were studied using TEM observation on a PULSION® BF_3 500V, $1E15/cm^2$ as implanted sample. The TEM picture is presented in figure 9. As one can see no defect is observed under the native oxide layer (about 2 nm) even if the SIMS analysis gives an implantation depth of 5 nm for $1E18/cm^3$ and 10 nm for $1E16/cm^3$. It seems that for this sample, that was not etched before implantation to suppress native oxide, most of the energy was deposited within the native oxide, thus the amorphous layer is in the native oxide layer. And even if there are some Boron atoms under the native oxide layer, we can notice the absence of end of range defect, which is likely to be of great interest for the post activation process and for the final device electrical characteristics.

ACKNOWLEDGMENTS

We would like to acknowledge NANOCMOS European consortium for supporting this study, CAMECA for its collaboration in SIMS characterization studies, SEMILAB for its mappings and CNRS CEMES for TEM pictures.

REFERENCES

1. ITRS INTERNATIONAL TECHNOLOGY ROADMAP FOR SEMICONDUCTORS 2005 EDITION, FRONT END PROCESSES
2. P.K. Chu et al : Plasma immersion ion implantation – a fledging technique for semiconductor processing. Materials Science and engineering, R17 1996 207-280
3. P.K. Chu et al: Plasma doping : progress and potential Solid State Technology. Oct 1999
4. F. Le Cœur et al. : Ion implantation by plasma immersion : interest, limitations and perspectives. Surface and Coating technology 125 (2000) 71-78
5. M.J. Goeckner et al. : Plasma doping for shallow junctions, J. Vac. Sci. Technol. B 17(5), Sep/Oct 1999
6. Susan Felch et al.: Optimised BF_3 P^2LAD Implantation with Si-PAI for shallow, abrupt and high quality p+/n junctions formed using low temperature SPE annealing.
7. P.K.Chu : Recent Applications of Plasma Immersion Ion Implantation. Semiconductor international / 165, june 1996.
8. P.K. Chu , D.T.KWok et al : The importance of bias pulse rise time for determining shallow implanted dose in plasma ion implantation. Appl. Phys. Lett. Vol 82, N°12, 24 March 2003
9. S.B.Felch et al. : Plasma doping for the fabrication of ultra shallow junctions. Surface and Coatings Technology 156 (2002) 229-236.
10. Woo Sik Yoo et al. : Results from SRTF show promise for shallow junction implant anneal. Solid State Technology. Dec 2002
11. A. Anders, Handbook of Plasma Immersion Ion Implantation and Deposition, Wiley-interscience publication, 2000
12. M.J Goeckner, S.B. Felch et al : Plasma doping for shallow junctions J. Vac. Sci. Technol. B 13(5) Sept/Oct 1999,p. 2290-2293
13. M. Takase, B. Mizuno : Neaw doping technology-Plasma Doping for Next Generation CMOS process with Ultra Shallow Junction- LSI yield and surface contamination issues. IEEE 0-7803-3752-2 /97.
14. D. Lenoble et al. : The fabrication of advanced transistors with plasma doping. Surface and Coatings Technology 156(2002). 262-266.
15. D. Lenoble : Etude, réalisation et intégration de jonctions P+/N ultra-fines pour les technologies CMOS inférieures à 0.18 µm. Thèse de 3ème cycle soutenue le 13/12/2000.
16. D. Lenoble : Direct comparison of electrical performance of 0.1µm pMOSFET's doped by plasma doping or low energy ion implantation. International conference on ion implantation technology 2000.
17. M. Takase et al. : Shallow Source / Drain extensions for pMOSFET's with High Activation and low Process Damaged Fabricated by Plasma Doping. IEEE 0-7803-4100-7 /97.
18. K.S Jones et al. Appl. Phys.Lett 75 (1999) 3659.
19. F. TORREGROSA, C. LAVIRON, F. MILESI, Miguel HERNADEZ H. FAIK, Julien VENTURINI: Ultra shallow P+/N junctions using Plasma immersion ion implantation and laser annealing for sub 0.1 µm CMOS devices. *IIT 2005 Taipe*
20. F. TORREGROSA, C. LAVIRON, H. FAIK, D. BARAKEL, F. MILESI, S. BECCACCIA : Realization of ultra shallow junctions by PIII, application to solar cells. (PBII-2003), San Antonio, USA, 16-19 Sept. 2003. *Surface & Coating Technology* 186(2004) 93-98..

Advanced Single Wafer High Current Beamline Architecture for Sub-65nm

G. Redinbo, C. Campbell, J. Mullin

Varian Semiconductor Equipment Associates, Inc., 35 Dory Road, Gloucester, MA 01930, USA

Abstract. Single wafer has been established as the preferred method for high current ion implant in next-generation semiconductor processing. To deliver the required gains in leading edge device fabrication, high current implanters must be capable of higher doses at lower energies, tight beam angle control, implant dose uniformity, charge and contamination control. The fundamental architecture of the implanter is critical to achieve the productivity and ion beam quality needed for sub-65nm device applications. The design elements of the VIISta HCP dual magnet ribbon beam high current implanter that enable increased low energy productivity, tight contamination control and interlocked ion beam quality are described here. We review implanter performance characteristics for productivity (beam current, tune times, wafer handling) and device yield (beam stability, beam angle and particle control). Finally we describe vMask™, a unique platform feature which significantly reduces device development costs.

Keywords: High current implanter, ion implantation, single wafer.

INTRODUCTION

High current ion implanters are used to create the conducting paths in the transistor such as the source and drain (S/D), source or drain extension (SDE) and gate doping. These device applications require higher implant doses and, as transistor scaling reduces critical device dimensions, lower implant energies for shallower junctions. Device scaling also drives lateral control requirements of implanted profiles. In fact, the need for diffusionless activation techniques such as flash or laser annealing has shifted the focus of high current implant to include not only junction depth control but also precise as-implanted dopant placement in the lateral dimension. The combined requirements of higher doses and precise as-implanted profiles drive the need for increased productivity at lower energies and precise implant beam control. A summary of typical high current implants is shown in Table 1.

Historically, high current implanters were configured as batch systems that relied on multiple wafer passes to ensure dose uniformity. Recent yield problems on batch systems from ballistic particles and the well-known cone angle effect have shifted the focus to single wafer high current architectures. Two approaches to single wafer high current implant have emerged: single magnet spot beam architectures with two dimensional wafer scanning and the dual magnet ribbon beam architecture with one dimensional wafer scanning. Here we describe the design elements of Varian's fourth generation single wafer high current implanter, VIISta HCP, that provide increased productivity, precise dopant placement and low particle contamination.

TABLE 1. High Dose Implants

Implant	Dose Range (/cm2)	Energy Range (eV)	Tilt
S/D	8E14-5E15	1-5k	No
SDE	5E14-5E15	200-5k	.Low
Gate Doping	5E15-5E16	2k-5k	No
PAI	5E14-1E15	10-30k	No

SYSTEM DESCRIPTION

VIISta HCP, shown in Figure 1, is the latest generation single wafer high current system from Varian. The system builds on the production-proven reliability of VIISta HC. To address sub-65nm high current implant applications, innovative improvements have been incorporated into the ion beam injector optics, transport optics as well as the measurement and process control systems of VIISta HCP.

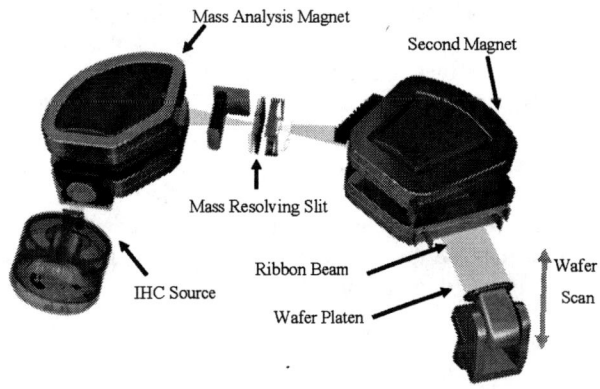

FIGURE 1. VIISta HCP

The VIISta HCP architecture incorporates high output long life injector optics, an indexed analysis magnet with focusing quadrupoles for horizontal and vertical beam control and a decel lens between the dual magnets that incorporates a state-of-the art optics design for improved low energy, high perveance ion beam transport. A new 70° collimating second magnet design provides the uniformity and angle control required for ion beam optical precision. The system delivers angle control, ion beam symmetry and excellent contamination control. Electrostatic deceleration slit lens design enhancements after beam collimation provide high perveance space-charge compensation of the beam that result in improved low energy beam currents and active vertical and horizontal focus control. These changes specifically address low energy SDE gate over-lap implant precision requirements.

In addition, the low energy transport characteristics of the VIISta HCP low density ribbon beam have been optimized. Electro-static lenses in the beamline provide effective space-charge compensation and maintain excellent ion beam optical implant quality (ion beam/wafer incidence). The dual deceleration architecture enables ion beam transport of a low energy, normally high perveance beams at higher energy. The beam is decelerated prior to the second magnet and transported at 2-10x the final energy as a well neutralized low space-charge beam; it is then decelerated to the final energy.

PRODUCTIVITY

One of the major challenges of device scaling is the requirement to maintain device manufacturability. Specifically, productivity levels for the key driver high dose implants shown in Table 1 must remain at acceptable levels in order to be production worthy and cost effective. VIISta HCP leverages the ribbon beam architecture to deliver these levels of productivity today and the extendibility to continue to meet these needs beyond 45nm. Significant improvements in beamline optics and transport efficiency have resulted in a >150% productivity improvement over four generations of Varian's single wafer high current implanter (Figure 2). Based on the VIISta common endstation with up to 500 wph wafer handling capability [1], the VIISta HCP is the industry benchmark for productivity.

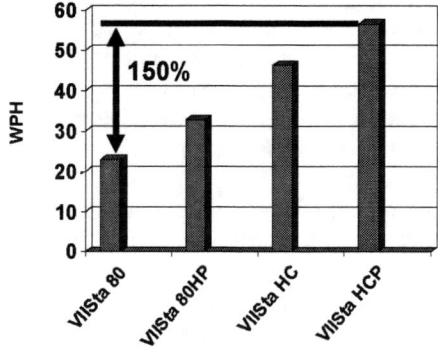

FIGURE 2. Productivity increase for four generations of VIISta single wafer high current tools. Aggregate throughput for an advanced logic recipe set.

Beam tuning performance is also key to productivity. The VIISta high current dual magnet architecture leverages the common Varian Control System (VCS) to deliver <4 minutes average tune time. Figure 3 shows tuning performance for typical applications.

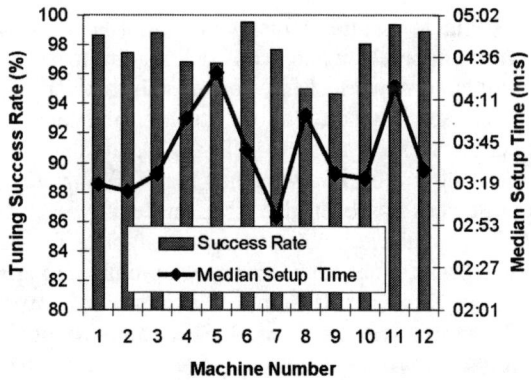

FIGURE 3. VIISta high current beam tuning success rate (%) and median beam setup time is shown for 5181 setups from 12 tools in production for a 30 day period. Overall success rate is 98% and median tune time for the entire population is 3:36.

PROCESS CONTROL

High current implant applications require control of implant dose, implant angle, particle contamination, metals, and wafer charging during implant. Detection of beam drop out during implant is a critical requirement for single wafer dose control. Beam instabilities must be detected and dose fill-in procedures designed for multiple, nested dose recovery events to assure dose integrity. Classic methods may not satisfy dose uniformity requirements when extreme beam drop-out occurs during a fill-in recovery. VIISta HCP incorporates advanced beam drop-out detection and dose fill-in capability that address all beam loss dose uniformity issues. Table 2 summarizes wafer test results from an extreme condition in which beam drop-out events are introduced in various degrees of severity on a sensitive monitor implant. Wafer #1 is the reference wafer with no glitches; wafer #2 has 2 glitches, with 1 introduced during normal recovery (requiring detection and fill-in recovery); wafers #3 - #5 have 3 to 5 glitches introduced to force execution of multiple fill-in areas during the implant. The beam drop-out detection and fill-in recovery results indicate identical process matching as the reference (no glitch wafer).

TABLE #2. Fill-in results - B 35keV 1.5E14 (7°/23°)

Wafer	Mean (ohm/sq)	% Std Dev
Wafer 1 Reference	512.8	0.62
Wafer 2: 2 glitches (1 fill-in)	513.5	0.69
Wafer 3: 3 glitches (2 fill-ins)	513	0.72
Wafer 4: 4 glitches (3 fill-ins)	513.1	0.74
Wafer 5: 5 glitches (4 fill-ins)	512.7	0.83

The angle control elements of VIISta HCP ensure the beam parallelism and beam steering are measured, controlled and interlocked prior to each implant. The ribbon beam provides a low divergence ion beam with typical measured values of < 0.5° [2]. The benefit of minimizing ion beam divergence for SDE implants in logic devices has been described in detail elsewhere [3]. The smaller beam divergence inherent in a ribbon beam enhances junction abruptness, minimizes overlap capacitance and results in faster devices. Furthermore, the VPS™ system on VIISta HCP ensures repeatable beam steering from wafer to wafer, minimizing skew of device parametrics such as I_{dsat}, I_{off} and V_t. Beam steering repeatability performance is < 0.1°.

The main motivation to shift to single wafer high current systems from traditional batch tools was yield loss due to ballistic particles [4]. There are two main contributions to particles in high current ion implant: mechanical adders from wafer handling and beam-borne adders from the beamline. VIISta HCP relies on a simple wafer scan technique that minimizes mechanical adders and the system uses a dual magnet beamline that isolates the wafer from sources of particles with a second magnet. The VIISta HCP has been recognized as one of the cleanest wafer processing systems in the fab. Particle performance is shown in Figure 4. The same design criteria also deliver metals performance to better than the specification of $2E10$ cm^{-2} for heavy metals and $2E11$ cm^{-2} for aluminum.

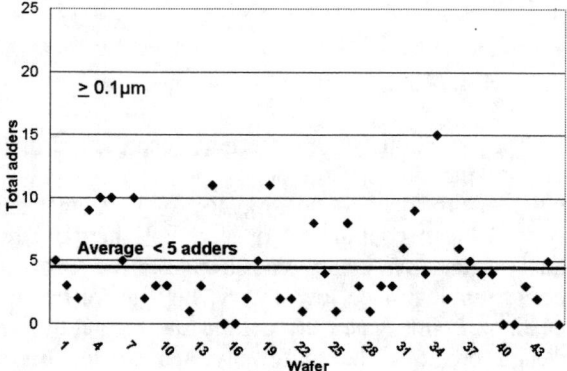

FIGURE 4. Particle adders on VIISta HCP in logic production. Particle size ≥ 0.1um over approximately 6 weeks of data collection.

A new plasma flood gun technology [5] has been incorporated into VIISta HCP. This provides both improved charge control and reduced metals contamination. It is an RF-based plasma system that provides an abundant supply of low energy electrons for charge control and eliminates tungsten as a major source of metals contamination. Lifetime has also been increased to twice that of traditional plasma flood systems.

COST ADVANTAGE

During the development of new technology nodes, ion implant is among the biggest users of test wafers. Given the dramatic increase in multiple implants to control diffusion and damage profiles combined with the lack of good TCAD predictability, the number of full flow device wafers is rising. This increase in expensive full-flow wafers along with the pressure to improve R&D efficiency has led Varian to release a new feature on its high current implanters that reduces development wafer costs by up to 75%. vMask™ is a carbon implant proximity mask that enables up to four implant splits per wafer and has been integrated into Varian's common endstation. The vMask™ feature reduces the number of split lot wafers for development of a new process, limits the effect of split lot variability and shortens time to market for new device designs. A typical split lot process result is shown in Figure 5.

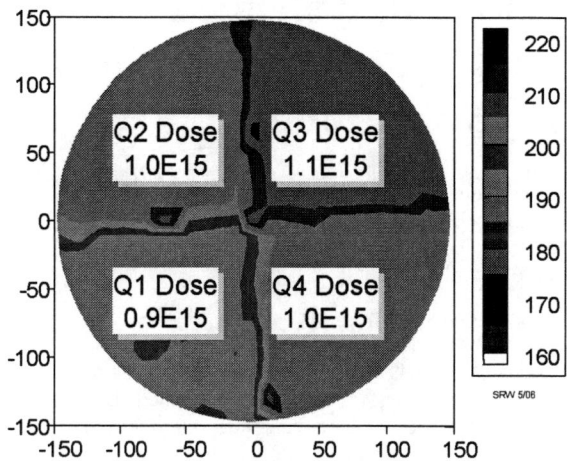

FIGURE 5. vMask process result for 5keV B implant with dose splits of 9E14 cm^{-2} to 1.1E15 cm^{-2}.

CONCLUSION

VIISta HCP provides a significant increase in productivity for all high current implant applications. Design changes have been incorporated into this fourth generation system to increase beam current delivered to the wafer. Together with average demonstrated tune times of < 4 minutes and the VIISta common endstation for high speed wafer handling, the VIISta HCP continues the track record of highest overall productivity for logic, memory and foundry applications. Improvements have also been made to the process capability of the tool, particularly in the area of dose integrity, angle performance and charge control. Finally, vMask™ enables multiple implants per wafer, significantly reducing wafer usage and development costs.

ACKNOWLEDGMENTS

We would like to thank Jay Sheuer, Alex Perel, Joe Olson, Peter Nunan and Yuri Erokhin for their contributions to this paper.

REFERENCES

1. A. Renau, "Next Generation Medium Current Product: VIISta 900XP", *Proc. 16th Int. Conf. on Ion Implantation Tech.* (2006).
2. J. Olson, G. Angel, A. Gupta, R. Mollica, D. Distaso and J. Liu, "Metrology Requirements for Single Wafer Ion Implanters," *Proc. 16th Int. Conf. on Ion Implantation Tech.* (2006).
3. Y. Erokhin and J. Liu, "High Current Precision Implant Requirements for sub-65nm Logic Devices," *International Workshop on Junction Technology.* (2006).
4. L. Pipes, M. Taylor, G. Zietz, A. Al-Bayati, M Castle, T. Marin and J. Simmons, "Characterization and reduction of a new particle defect mode in sub-0.25um semiconductor process flows," *Proc. 15th Int. Conf. on Ion Implantation Tech.* (2004).
5. P. Kurunczi, A. Perel, E. Wright, S. Kikuchi and J. Scheuer, "Advanced charge control for high current single wafer implanters," *International Workshop on Junction Technology.* (2006).

Applied Quantum X Implant System: Technology Enhancements to Enable Production-Worthy Performance at the 45 nm Node

Adrian Murrell, Peter Edwards, Richard Goldberg, Peter Banks, Bob Mitchell, Erik Collart, Sean Morley, Geoffrey Ryding, Theodore Smick, Marvin Farley, Takao Sakase, David Hacker, Peter Kindersley.

Applied Implant Technologies, Foundry Lane, Horsham, RH13 5PX, U.K.

Abstract. Mechanical scanning of the wafer in 2 dimensions is one approach that has been used to achieve single wafer processing for high current ion implantation. This approach simplifies the beamline design, compared to scanned beam or ribbon beam architectures, but has required a number of new technologies and methods in the scanner hardware and in dosimetry control. The Applied® Quantum X Implant system was designed to incorporate these new technologies, and has achieved the process performance and low energy productivity required for advanced junction formation at the 65 nm technology node. Since its introduction, extensive qualification and development work has been carried out, to extend its capability to the next technology generation. A number of further innovations and improvements to the beamline and platform have been developed, extending its throughput and process control capability to be production-worthy at 45 nm.

This paper will review the process control challenges associated with the 2d mechanical scanning approach and the new methods and hardware that have recently been implemented on the Applied Quantum X Implant system. The theory and design of the enhancements will be described and illustrated with process data in the areas of angle control, dosimetry and energy purity.

Keywords: Ion implantation, mechanical scan, beamline
PACS: 85.40.Ry

INTRODUCTION

After many years in which batch tools were used for high current implant, many makers of advanced devices have recently switched to single wafer. This has been driven by a number of factors, including yield issues from ballistic particle impact, and cone angle affects on within-wafer angle uniformity (1). There are 3 broad categories for single wafer processing, (i) scanned beam, with 1d wafer scan in the other plane, (ii) ribbon beam, again with 1d wafer scan, (iii) 2d mechanical scan of the wafer. The first two of these require long, complex ion beamlines, with magnetic or electrostatic elements to scan or collimate the beam, making it difficult to transport the low energy beams that are required for advanced devices. 2d mechanical scan allows a short simple beamline, of the same type used for many years on batch tools. However, new approaches have been required, to achieve the same level of dose uniformity and process control. These new approaches were pioneered on the Applied Quantum X implanter, which has gained rapid market acceptance. During the 2 years since its introduction much new data has been collected, and a number of new process refinements and design improvements have been developed. This paper describes some of these second generation features.

SYSTEM OVERVIEW

3 key technology areas under-pin the 2d mechanical scan architecture of Quantum X. These are (i) the mechanical scan hardware and motion control, (ii) the low energy optimized beamline, and (iii) the scan methods for achieving good process performance.

2d scanning of the wafer is carried out by the Stepscan™ assembly, which uses air-bearings and linear motors to achieve high scan frequency with no friction. This is essential to high reliability operation without particle generation. As well as advanced hardware design, a number of new methods have been developed for motion control, to allow rapid turnaround without tool vibration. Fast turnaround improves scan efficiency, increasing throughput.

The beamline is short, with a 90 degree analyzer magnet, and has a minimum of focusing or tuning elements. This allows rapid beam tune times, on the order of 3 minutes, and high transmission of low energy beams. A decel lens is used to achieve high beam currents for the lowest implant energies, down to 200 eV. The third area of technology, the scan methods, are described further in the next section.

PROCESS METHODS AND PERFORMANCE

Angle Control

The Stepscan design enables a high level of angle precision during implantation. The x and y motions are linear and independent, and this allows scan lines to be drawn that are parallel in all dimensions. Every part of the wafer sees every part of the ion beam, and this critical attribute results in very uniform and equivalent processing across the wafer. This contrasts to scanned or ribbon beam systems, where the left of the wafer sees a different part of the beam (and therefore slightly different trajectory, beam divergence and beam density) than the right side.

This uniform angle performance is demonstrated by figure 1, which shows Thermawave data for a B+/150 keV/2e13 implant, with mean and one-sigma uniformity data plotted against tilt angle. At 0 degrees the implant aligns with the (100) channeling direction, reducing the measured damage. Tilting to 1 degree changes the mean TW by 4%. Applying this sensitivity to the uniformity data, the 0.1% Thermawave can be equated to 0.025 degree one-sigma. This uniformity is significantly tighter than the angle accuracy required for 45 nm devices.

Dose Uniformity and Repeatability

As the wafer scan velocity is 2 orders of magnitude lower than on the spinning disk in batch tools, the number of passes through the beam is also fewer. The process parameters that control the number of scan lines, in particular the fast scan speed (Vx) and the spacing between the scan lines (y-pitch), are critical to achieving good uniformity, especially at low dose, and also have a strong effect on throughput.

At low dose, the best throughput is obtained using the fastest Vx and largest y-pitch that gives the required uniformity. If a smaller pitch is used, the beam current must be reduced, to achieve the same dose, and throughput drops. Knowing the optimum pitch is therefore essential. Quantum X calculates this pitch from the beam profile. Taking an accurate beam profile in the wafer plane, the vertical projection is used to calculate a plot of predicted uniformity versus y-pitch, as shown in figure 2. Large, smoothly shaped beams allow larger scan y-pitch, beams with narrow or sharp features require a smaller y-pitch to achieve good uniformity. This predictive method is essential to optimize throughput, but also to safeguard process. Day-to-day variation in the beam profile e.g. across preventative maintenance cycle, can result in changes in uniformity. Quantum X is able to adjust pitch to optimize this dynamically, and to set a recipe interlock to safe-guard against mis-process.

At high dose, there are many scan lines, and uniformity is no longer a limitation. Throughput is instead determined by (i) the beam current and (ii) the scan efficiency. Scan efficiency is affected by the over-scan distance and time. Smaller beam width in the x-plane reduces the over-scan distance, while use of lower scan speed reduces turn-around time. Therefore lower Vx is used at high dose.

As well as y-pitch and beam profile, the orientation of scan lines is also important to achieving good uniformity. Quantum X uses parallel scan lines, by carrying out the y-step at the end of each line. An alternative approach is continuous movement in y, resulting in a diagonal set of scan lines. This is easier

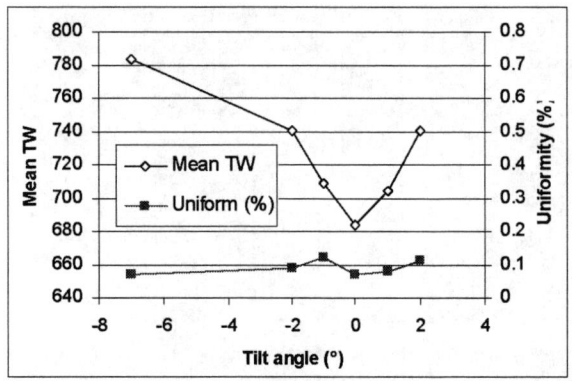

FIGURE 1. Thermawave data versus tilt angle for a B+/2e13/150 keV implant on Quantum X.

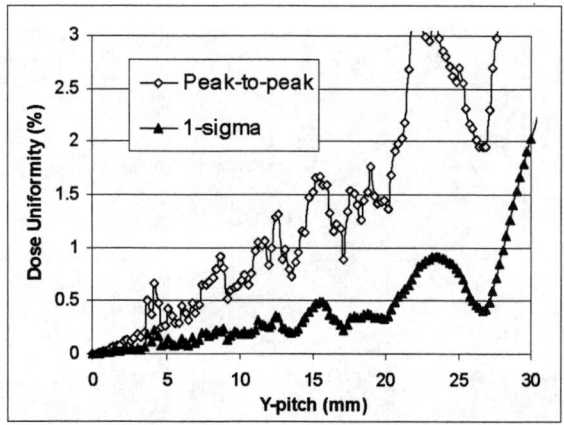

FIGURE 2. Predicted uniformity versus y-pitch, calculated from beam profile data.

to achieve in scan hardware, but the process uniformity is degraded. Figure 3 shows a simulation of scan uniformity, comparing parallel scan lines with diagonal. With the same number of scan lines, the uniformity with the parallel case is dramatically better. This is because, in regions of the diagonal scan pattern where the scan lines cross, the effective y-pitch is twice as large. Therefore, twice the scan frequency is required, using diagonal scan to achieve the same process performance.

Also shown in figure 3 is a simulation of dose uniformity in quad mode. Here a quarter of the dose is implanted at each of 4 twist angles, spaced by 90°. As with the diagonal scan, quad mode shows a much worse uniformity than the non-quad, for the same number of scan lines. Orienting half the scan lines at 90°, results again in 2x the effective pitch spacing at some points.

Historically quad mode has been used for several reasons. These include (i) use of 7° tilt to suppress channeling, (ii) averaging angle non-uniformity from cone angle on batch tools, (iii) compensating dose and angle non-uniformity in ribbon beam implanters, (iv) averaging angle alignment variation, due to day-to-day beam set-up or wafer crystal orientation repeatability. As has been demonstrated by the data in figures 1 and 3, (ii) and (iii) do not apply to the Quantum X mechanical scan approach. (i) is still relevant for high tilt implants, but for advanced devices most high current implants are now done at 0°, to avoid the shadowing effects that result from high aspect ratios on the device. For the remaining driver, (iv), beam measurement and control algorithms have recently been enhanced on Quantum X. In addition, a new scan method has been developed on Quantum X, termed dual-twist co-interlace.

With dual-twist co-interlace, two implants are carried out, one with the wafer twist rotated 180°. This achieves the same angle averaging that is seen with quad mode, reducing sensitivity to angle variations. However, with the dual implant, the 2 sets of scan lines can be interlaced to achieve optimum uniformity. For the simple case, where one set is offset by half the y-pitch, the effective y-pitch is halved. The resultant pattern is the same as that shown in figure 3 (a). However, when the wafer is rotated 180 degrees, any small asymmetry in the vertical beam profile results in a change in the uniformity vs y-pitch plot (as in figure 1). To achieve comparable uniformity to a single twist implant, the interlace alignment of the 0 degree scan lines and 180 degree scan lines must be offset. In Quantum X, this offset is again calculated from the beam profile before implant.

Table 1 shows dose uniformity data collected in dual co-interlace mode. Sheet resistance uniformity is shown for single twist implants and dual co-interlaced implants with twice the y-pitch on each half of the dual. The 1-sigma (standard deviation) uniformity is for the whole wafer map, the peak-to-peak value was collected using a diameter scan in the vertical direction, to study the line-interlace uniformity. For both types of data the uniformity for the dual co-interlaced implant is comparable to that of the single implant that used half the y-pitch.

TABLE 1. Sheet resistance uniformity data comparing single twist with dual co-interlace mode, P+/1e15/10 keV

Mode	y-pitch (mm)	Rs Uniformity data (%)	
		1-sigma	Pk-to-pk
Single	8	0.58	1.40
Dual co-int.	16	0.63	1.60
Single	4	0.48	0.40
Dual co-int.	8	0.51	0.40

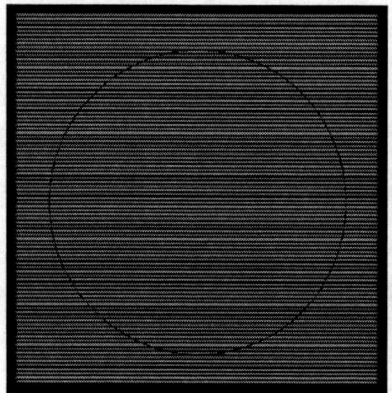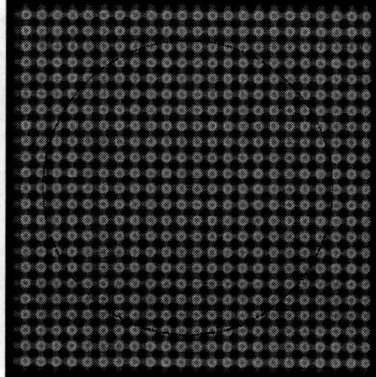

FIGURE 3. Dose uniformity simulations for the same number of scan lines, using (a) parallel scan lines, non-quad – uniformity = 0.17%, (b) diagonal scan lines, non-quad – uniformity = 1.55%, (c) parallel scan lines, quad – uniformity = 1.66%.

Energy Purity

Decel mode offers a substantial increase in productivity at low energy, but introduces a potential process risk, associated with neutralization of ions before deceleration. The neutrals formed have the pre-acceleration energy and can implant into the wafer, causing a deeper tail on the implant depth profile. The process for neutralization is charge exchange with the background gas, and the probability of this charge exchange reaction is described by the following equation:

$$\frac{I(neutral)}{I(ion)} = c.P.x.\sigma \quad (1)$$

where I = beam current, c = a constant, P = pressure, x = length, σ = charge-exchange cross-section.

Length in the above equation is the distance, with line of sight to the wafer, over which neutralization can occur. The charge exchange cross-section is the probability of the reaction, and varies as a function of ion species and energy, and background gas composition (2). To improve the energy purity of decel implants, on Quantum X, the length of beamline after the analyzer magnet has been further reduced by 150 mm, and a larger turbo pump fitted to this region.

Equation 1 only describes the % of neutrals formed before the decel stage. To predict the % neutrals that implant into the wafer, the relative efficiency of transport of the ions and the neutrals, through the decel lens must also be considered (2). Adding these two transport efficiencies (Tx) into equation 1 gives:

$$\frac{I(n-wafer)}{I(i-wafer)} = \frac{Tx(n)}{Tx(i)}.c.P.x.\sigma \quad (2)$$

The transport efficiencies also have design implication. The neutral transport is affected by ion trajectories prior to neutralization and the line of sight to the wafer. Therefore a spot beamline and narrow decel lens have an advantage compared to the open geometry of a ribbon beamline. The ion transport depends on the efficiency of the decel lens, and this can vary with energy and species. Optimization of the decel electrostatic optics, does not just influence beam current but also therefore energy purity. A new decel lens has recently been developed on the Quantum X system, called the Vortex lens, which improves ion transport efficiency significantly.

Figure 4 shows energy purity performance on Quantum X, with the beamline improvements described above. The SIMS depth profile compares a drift mode B+/0.5 keV implant with decel, using a 2->0.5 keV ratio. The effective energy contamination can be quantified by measuring the difference between the 2 curves in the tail region. From this method, the neutral contaminant is 0.09% of the total dose. This type of implant is typically used for a source-drain extension, and the background channel is typically 3-6e18. Down to 3e18 the SIMS profiles are identical.

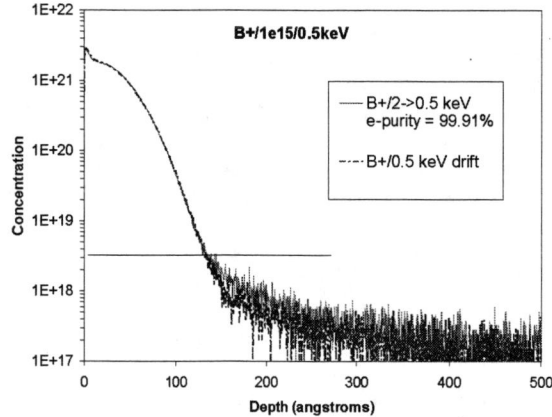

FIGURE 4. SIMS profile for B+/0.5/1e15 implant, showing energy purity in decel mode.

CONCLUSION

The 2d mechanical scan approach to high current implant offers a number of process and productivity advantages. Three key technology areas have been described that underpin the approach: (i) the mechanical scan, (ii) the low energy beamline, and (iii) the scan process methods. It has been shown how these three areas interact and influence the process performance, and data has been presented in the areas of angle control, dosimetry and energy contamination. In all these areas, new innovations in hardware and methods have been developed recently on the Applied Quantum X implanter, to enable the precision required at the 45 nm technology node.

REFERENCES

1. L. Pipes, M. Taylor, G. Zietz, A. Al-Bayati, M. Castle, T. Marin, J. Simmons, *Nucl. Inst. Meth. in Phys. Rev.* **B237** (2005) 330-335

2. D. G. Armour, J. A. Van den Berg, G. Wostenholm, A. H. Al-Bayati, A. J. Murrell, R.D. Goldberg, E.H.J. Collart, *Proc. IIT 2006*, 181-184.

Enhancing the Ibis i2000™ SIMOX Oxygen Ion Implanter

C. McKenna, R. Dolan, J. Blake, S. Richards

Ibis Technology Corporation, 32 Cherry Hill Drive, Danvers, MA 01923, USA

Abstract. The results of the Ibis i2000 uniformity enhancement program are presented. Two contributors to the uniformity of the top Silicon (SOI) and Buried Oxide (BOX) layers are identified; the implant dose uniformity and the temperature uniformity of the wafer during implant. A technique for achieving implant dose uniformities of ±1% by optimizing the scanner current waveform on the magnetically-scanned i2000 is detailed. The sparse structure of the i2000's wafer holder ensures that during the implant the wafer's thermal environment is dominated by radiative coupling to the process chamber's cooled inner wall. The principal residual non-uniformity is in the vicinity of the wafer holding pins and is strongly dependent upon the choice of pin material. A comparative survey of pin materials is presented. Combining these two approaches routinely results in SOI uniformities of <20 Å total range on all points of a 57 point map.

Keywords: SIMOX SOI Implant Uniformity Wafer Holder Pin
PACS: 61.72.Tt, 29.27.Eg

INTRODUCTION

In the four years since its initial release [1] the i2000™ has been the focus of an extensive development program aimed at both performance enhancement and reduced cost of ownership. The uniformity of the thicknesses of the top Silicon (SOI) and Buried Oxide (BOX) layers has become increasingly critical as SIMOX moves into production. Achieving the desired level of uniformity for these layers while maintaining competitive performance for other SOI parameters (such as surface roughness, BOX breakdown voltage etc.) requires a thorough understanding of the various factors that control each area of the process. One of the most basic is dose uniformity. Experience has shown that a total uniformity range of ±1% is sufficient to meet current and anticipated layer uniformity specifications.

SIMOX PROCESS

The current leading-edge SIMOX process involves a principal implant of 1-5E17 ions/cm^2 at energies of 150-220 keV performed at elevated temperatures (300 – 500 °C), and a damage engineering implant of ~E15 ions/cm^2 in the same energy range [2]. The damage engineering implant enhances the level of internal thermal oxidation (ITOX) that occurs during the anneal such that one-third or more of the final BOX thickness comes from ITOX. As such, achieving excellent final SIMOX uniformities requires control of the uniformity for both the dose and damage levels.

DOSE MEASUREMENT

Since none of the conventional metrology techniques [3] used for measuring dose levels of typical (electrically active) dopant implants are applicable for O+ implants, particularly at the high doses required, IBIS chose Fourier Transform Infrared Reflectance (FTIR) Spectroscopy, and co-developed a measurement methodology [4] together with On-line Technologies. This technique was refined to measure small spatial samples (1-2 mm diameter), permitting up to 81 points to be read for a diameter scan. With the proper calibration procedures [5], this technology is able to correlate reflectance measurements with oxygen dose to +/- 1%.

UNIFORMITY

Both the i2000™ and its predecessor, the Ibis 1000, are batch machines using Ibis's magnetic beam scanning technology [6]. The scan frequency is 150 Hz. The beamline, including the analyzer magnet, quadrupole lenses, scanner and collimator was designed by Hilton Glavish (Zimec Inc.), and fabricated by Buckley Systems.

The first step in controlling the implant uniformity was to ensure a repeatable beam trajectory. In contrast to the Ibis 1000, the i2000's analyzer magnet and scanner both deflect the beam in the horizontal plane. The i2000's extraction aperture has two degrees of freedom, permitting adjustment of the extraction gap and the lateral (horizontal) position of the aperture. As a result of this arrangement, an analyzer magnet current that will guide the beam through the mass resolving aperture can be found for any extraction aperture lateral position. There is, however, only one combination of analyzer current and extraction lateral position that will result in the beam propagating along the beamline axis after it leaves the mass resolving aperture, an essential condition for repeatable scanner operation. A beam tuning procedure was developed to ensure axial beam propagation; the procedure exploits the fact that beamline transmission is a maximum when the beam is confined to the beamline axis. With the beam tuned to the desired implant conditions, the extraction lateral position is varied in steps of 0.1 mm; the beam current is peaked by adjusting the analyzer magnet current at each position, and magnet and beam currents are plotted as a function of lateral position. A typical plot is shown in Fig. 1. The optimum extraction aperture lateral position corresponds to the peak of the beam current curve.

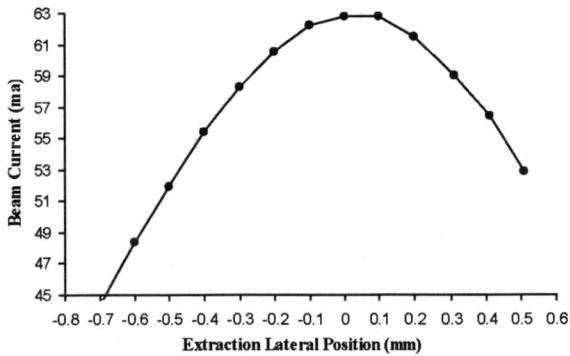

FIGURE 1. Optimizing the Extraction Lateral Position

To perform a uniform implant in a batch machine with a 1-D rotary mechanical scan, the beam's radial velocity across the wafer must follow the classic $1/r$ profile. The scan system computes the scanner current waveform required by this profile by evaluating a theoretically derived polynomial that provides scanner current as a function of beam position. A one-time beam centering calibration is performed after major beamline maintenance to compensate for tolerance accumulation. The beam is centered by replacing one or more wafers with a shield shaped like a section of an annulus with inner and outer radii corresponding to the inner and outer radii of the circle of wafers (Fig. 2).

The beam current is monitored by an extended Faraday cup located behind the wafer plane. The time (and hence the scanner current) at which the beam crosses the inner and outer radii may be determined by inspecting the beam current waveform (Fig. 3). This provides two scanner current/radius calibration points, which are used to perform a gain and offset calibration of the scan system.

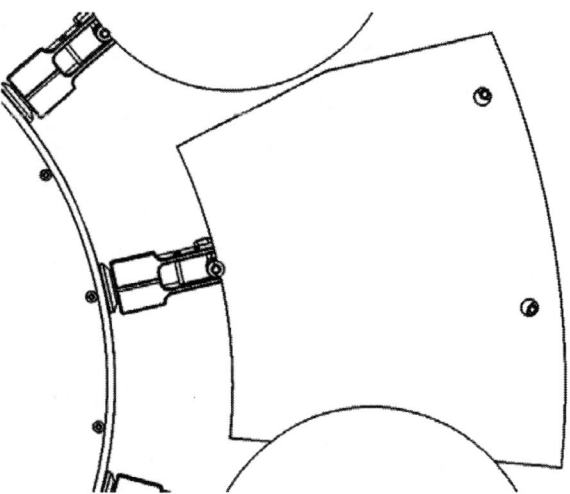

FIGURE 2. The Beam Centering Shield

The scan system also provides a mechanism for making small corrections to the dose (referred to as the Dose Correction Factor, DCF) by modulating the beam's velocity as it traverses the wafer. The corrections are specified as an arbitrary number of position/DCF pairs; intermediate values are linearly interpolated. The DCF is used to optimize implant uniformity. Since the non-uniformities may depend on the beamline's operating parameters, the decision was made to store the DCF profile in the process recipe. In practice we have found that this degree of flexibility is seldom necessary; a single DCF profile is effective over a range of implant energies and beam currents.

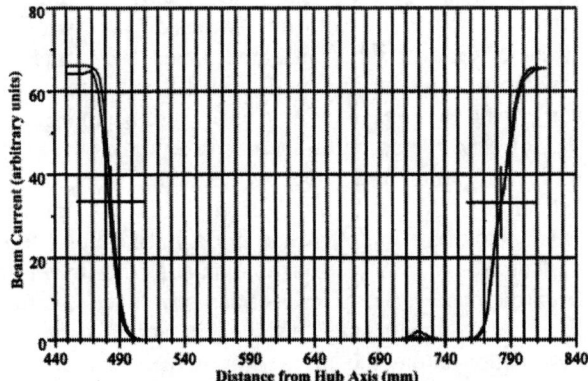

FIGURE 3. The Beam Centering Display

The next step is to determine the DCF profile that will produce a uniform dose. The i2000™ is initially run with the default (uncorrected) DCF profile [(-200, 1.0)(200, 1.0)]. The dose is measured using FTIR along the diameter of the wafer that was parallel to the magnetic scan axis during the implant. The resultant dose profile is normalized to a unity average, a polynomial of selectable order is fitted to the data to reduce noise, and the DCF profile is calculated by taking the reciprocal of the polynomial. Since the i2000's control software requires position/DCF pairs, an iterative least squares linear approximation algorithm is used to decompose the continuous DCF profile into the minimum number of linear segments needed to match the profile to a specified accuracy. The calculation is implemented in an Excel worksheet for convenience. A typical dose profile before and after correction is shown in Fig. 4.

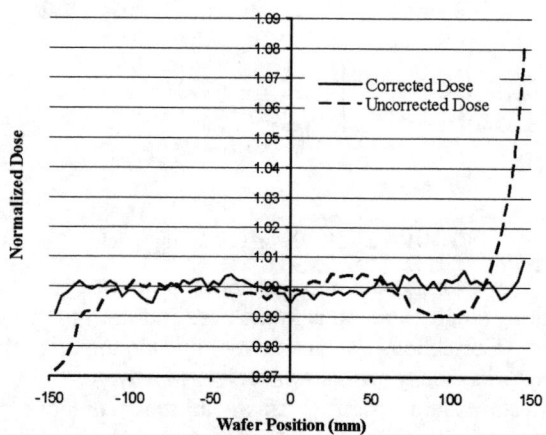

FIGURE 4. Dose Profile before and after correction

It was found that this procedure sometimes over corrected, necessitating one or (infrequently) two iterations before yielding satisfactory results (Fig. 4 is the result of one iteration). The over-correction was invariably adjacent to a region in which the higher order derivatives of the dose profile were significant. The experimentally observed dose profile is a convolution of the beam profile with an "intrinsic" dose profile characteristic of the machine and the implant conditions. This intrinsic profile would be the result of performing the implant with a zero width beam. Since the correction is based on the observed rather than the intrinsic profile, it will be less accurate in a region extending up to one beam width (typically 40 mm) on either side of an abrupt change in the intrinsic profile's first derivative. A method of extracting the intrinsic from the observed profile has been developed but has not been implemented, since it requires more information about the beam profile than is routinely available.

DAMAGE UNIFORMITY

The damage created in the silicon layer in which the SOI/BOX structures are to be formed is dependent upon the dose, dose rate and substrate temperature for the two implants involved in the process. By controlling the total dose (integrated current), dose uniformity (scan profile) and dose/damage rate (beam current), we should be able to produce both uniform dose and damage levels. As noted above, the first high-dose implant is performed at elevated temperatures achieved by a combination of the beam power and lamp arrays that are servo-controlled to maintain a thermocouple monitor wafer at a constant temperature [1]. Due to the combination of high dose and high temperature, the only measure of the damage uniformity from this implant is in the final product. Relative to damage uniformity, the damage engineering implant typically has the greatest impact. This implant starts at a relatively low temperature and is performed with the lamp arrays off such that the wafer temperature rise is driven only by beam heating. At the lower end of the dose range for this implant, it is possible to get some indication of the damage uniformity from a Thermawave (TW) measurement [3]. The TW measurements demonstrated that the principal source of damage non-uniformity came from the thermal influence of the wafer–holder pins. The final product layer thicknesses also reflected the thermal signature of the pins, with larger uniformity variations arising from the more thermally conductive pin material.

WAFER HOLDERS AND PINS

One of the areas of the i2000 that was the focus of considerable development was the wafer holder assembly, thirteen (13) of which are mounted to the hub of the rotary scan drive. A 3-D computer model of a holder is shown in Fig. 5.

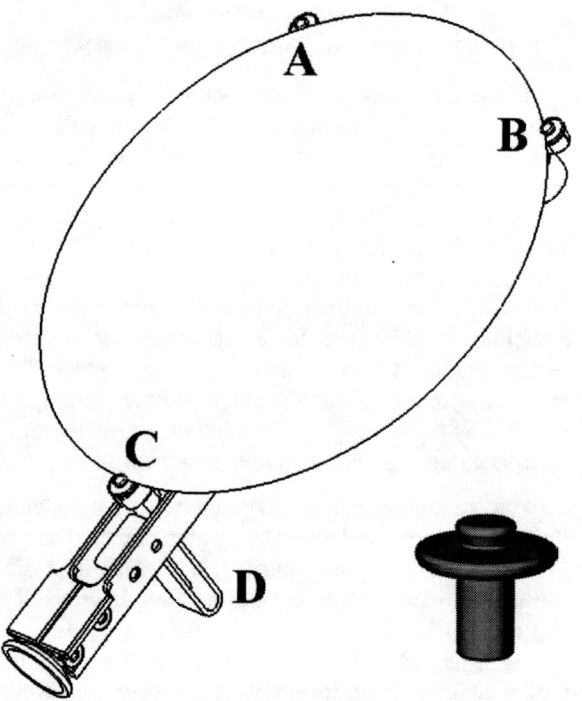

FIGURE 5. The Wafer Holder

The wafers are secured in the holder by three pins, two (A&B) on the side of the wafer farthest from the hub and one (C) as part of a spring clamp mechanism on the side of the wafer closest to the hub. The wafers are loaded by moving the innermost pin toward the hub by compressing the spring clamp (D), then inserting the wafer and gently releasing the spring clamp. Contact with the wafer is limited to a small region of the pin between the cap and the shoulder.

Due to the aggressive nature of oxygen ions and the considerable power levels of the beam (~20 KW), many materials were investigated for fabrication of the pins. While pin lifetime and electrical conductivity were the initial concerns, it became clear that the pin features that most dramatically impacted the quality of the final product were the thermal signature (influence on the damage uniformity noted above) and particle performance (including wafer edge damage).

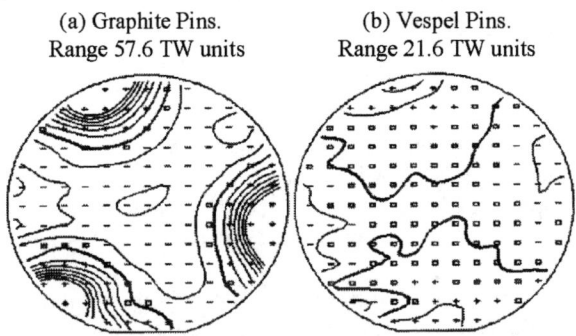

FIGURE 6a,b. Thermawave maps showing the thermal effects of Graphite (a) and Vespel (b) pins

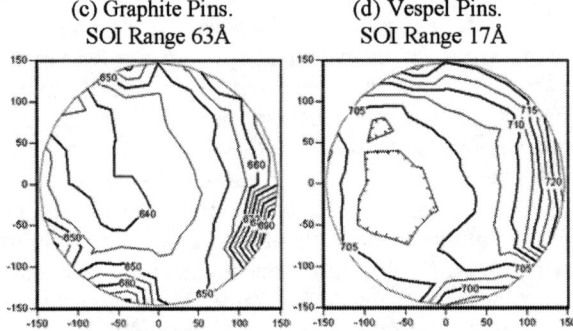

FIGURE 6c,d. Post anneal SOI thickness map

Based on the original criteria, graphite was chosen for the pin material. However, the excellent thermal conductivity of these pins resulted in signature variations in uniformity due to temperature differences between the pins and the bulk of the wafer (Fig. 6a,c). In addition, these pins, while exhibiting good particle performance over extended periods, would intermittently emit bursts of particles, most likely from oxygen buildup below the surface.

Subsequently, a polymer (Vespel) was chosen for the pin material. This material exhibited a dramatically reduced thermal signature as its temperature tracked much more closely to that of the wafer (Fig 6b,d). Implants performed with Vespel pins also demonstrated excellent particle levels.

CONCLUSIONS

By controlling system parameters to optimize both the dose and damage uniformities, the i2000 is capable of delivering implants that produce state-of-the-art SIMOX-SOI wafers with SOI total range variations <20Å for a 57 point measurement.

ACKNOWLEDGMENTS

The authors would like to thank Brian Quintal and Brian Gori for performing the experiments, and Mike Robbins and John Tricomi for processing the wafers.

REFERENCES

[1] J. Blake et al, "The Ibis i2000 SIMOX Ion Implanter" in *Proc. 2002 Int. Conf. Ion Implant Tech. (Taos, New Mexico)*, September 22-27, 2002.

[2] Devendra Sadana and Michael Current, "Fabrication of Silicon on Insulator (SOI) Wafers" in *Ion Implantation Science and Technology, 2000 Edition*, ed. J. F. Ziegler.

[3] C. Yarling and Michael Current, "Ion Implantation Process Measurement, Characterization and Control" in *Ion Implantation Science and Technology, 6th Edition*, ed. J.F. Ziegler, 1996.

[4] R. Dolan et al, "Improved Techniques for Characterization and Optimization of SIMOX Implantation", these Proceedings *Proc. 2006 Int. Conf. Ion Implant Tech. (Marseilles, France)*

[5] U. Mantz et al., "Model-Based Infrared Spectroscopy: New Opportunities for In-Line Process Control", in *Future Fab International*, 19 (2005)

[6] Hilton F. Glavish, US Patent numbers 07/843,391; 08/106,351; 08/383,422.

Note: Ibis and i2000 are trademarks and Advantox is a registered trademark of Ibis Technology Corporation. All other trademarks are the property of their respective owners

Indium Performance on the V810

Russell J. Low & Qing Zhai

Varian Semiconductor Equipment Associates, Inc., 35 Dory Road, Gloucester, MA 01930, United States

Abstract. Indium has been used as an alternative p-type dopant to boron in a number of implant applications. Typically, solids such as $InCl_3$ have been used as source feed materials, which have proven problematic for a number of reasons. Issues have included charge life, conditioning & PM recovery time (e.g. where hydroscopic materials have been used). This paper presents productivity improvements made to the V810 for running indium ion beams, along with advances made in process control.

Keywords: Ion Implantation, Indium Chloride, Indium Iodide
PACS: 81.05.Cy, 85.40.Ry, 61.72.Tt

I. INTRODUCTION

Indium continues to be used as an alternative p-type dopant to boron for applications requiring steep implant profiles, such as retrograde channel & halo implants [1,2]. The choice of source feed material for indium beam generation can have a significant impact on the tool performance and productivity [3,4]. Typically, solids such as $InCl_3$ have been used as source feed materials, which have proven problematic for a number of reasons. Issues have included charge life, conditioning and PM recovery time (e.g. because the material is hydroscopic). This paper presents advances made in running indium implants on a standard V810 EHP [5] using indium (I) iodide as the source feed material.

II. INDIUM PERFORMANCE

Indium (III) chloride ($InCl_3$) has been in use as a source material for generating indium beams for many years [6]. While providing productive beam currents and stable performance, $InCl_3$ does suffer from a couple of limitations.

The first limitation is a result of the hydroscopic nature of the pure solid. Even with careful handling, it is very difficult to keep the solid dry. Fully hydrated $InCl_3$ is nearly liquid, requiring the solid to be heated to at least 100°C to drive off the water. This outgassing operation, which involves heating the solid gently in vacuum, needs to be completed prior to being able to run indium ion beams. If heated too quickly, the escaping water vapor leads to the charge within the vaporizer blowing out into the arc chamber. This can result in significant clean up within the vacuum system.

During operation, indium and chlorine related by-products, which are also hydroscopic, form on the source or in the source area. If the source area remains vented for any significant period of time, the deposits absorb water vapor from the air. Although not difficult to remove, they do constitute an additional requirement for maintenance. Any region within the vacuum system not properly cleaned will greatly impact the vacuum recovery time, due to the time taken to outgas the water.

A second issue with $InCl_3$ is the chemical reactivity of the breakup products generated in the ion source. For every indium atom, there are three chlorine atoms, which have significant chemical reactivity. Although not an issue with a well maintained ion source, there is a known isobaric interference between $^{115}In^{++}$, $^{56}Fe^+$ and $^{58}Ni^+$ (and $^{58}Fe^+$, but with a natural abundance of only 0.33%, this is not a significant concern) [7]. In the event of iron or nickel containing materials (e.g. stainless steel) entering into the ion source environment, it is possible to get energetic iron and nickel metal contamination in the wafer. Naturally, if less halogen related material was available, or if the chemical reactivity of the halogen were reduced, this becomes even less of a concern [8].

Indium (I) Iodide (InI) has been investigated as an alternative source feed material that will maintain all the advantages of InCl$_3$, while mitigating the aforementioned issues.

III. RESULTS & DISCUSSION

A. Maintenance Recovery Performance

InI and InCl$_3$ have very similar operating temperatures, ranging from 360°C to 400°C, depending on the amount of material in the vaporizer. This allows the same preexisting optimized vaporizer hardware and control algorithms to be used.

Figure 1 presents a comparison between InCl$_3$ and InI in terms of the time taken to condition a fresh charge ready for normal operation (outgassing time). The sample was loaded into a standard vaporizer, installed into the ion source and the whole system pumped down to <1e-6 Torr. The vaporizer temperature was then ramped steadily to outgas the charge, while maintaining the system vacuum below a predetermined limit.

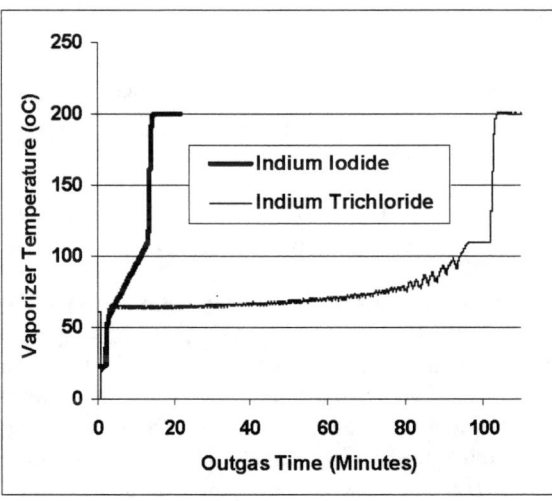

FIGURE 1. Comparison of InI and InCl$_3$ Outgassing Performance.

The vaporizer charged with 8g of InCl$_3$ required nearly two hours to reach the required stand-by temperature of 200°C, the point at which the feed material is sufficiently conditioned for use in the ion source. The 10g InI sample achieved this same operating condition within 20 minutes, which is a significant reduction in time. In both cases, care was taken with the material handling, particularly regarding the time exposed to atmosphere. In the case of InI, this is not as important, since the material is significantly less hydroscopic than InCl$_3$. Additionally, once InI has been heated to operating temperature, the material forms a solid lump in the vaporizer with a greatly reduced surface area. Again, this reduces the hydroscopic material concerns.

InI is denser and can be packed more efficiently into a standard vaporizer than InCl$_3$. This allows the largest charge to be ~30g for InI, compared to only ~8g for InCl$_3$, without any significant increase in the required outgassing time. Charge life observed with a single vaporizer running continuous indium ion beams has been measured at >160 hours. A standard VIISta V810 ion source is able to accommodate two vaporizers simultaneously, allowing >320 hours of dedicated indium operation. For a recipe mix with <50% indium operation, the charge life will not limit the source maintenance cycle.

After running the system for ~120 hours with InI, the system was vented for inspection. The source and source area were very clean, with no signs of hydroscopic deposits. No maintenance was required and vacuum recovery was comparable to a standard recovery when not operating with InCl$_3$.

B. Contamination Performance

As part of the InI characterization work, the potential for isobaric interference of iron and nickel with ^{115}In^{++} was investigated. Multiple implants were performed using a standard In^{++}/150keV beam setup at various times during the InI charge life (9.5, 65, 97 and 150 hrs of charge life). The implant dose was set at 5e15 at/cm^2 to improve the potential of detecting energetic ^{57}Fe$^+$ or ^{58}Ni$^+$ using Secondary Ion Mass Spectroscopy (SIMS). Figure 2 below shows the SIMS results for ^{57}Fe$^+$ and Figure 3 shows the SIMS results for ^{58}Ni$^+$. The SIMS spectrum for an unimplanted wafer from the same batch is included as a control.

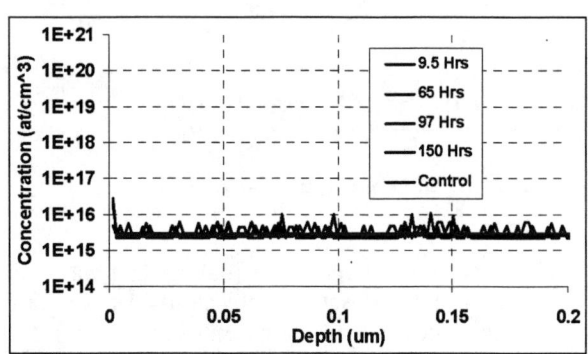

FIGURE 2. SIMS Profile of ^{57}Fe after In^{++}/150keV/5e15 Implant.

Based on calculations using SRIM [9], the expected projected range for ^{57}Fe$^+$ and ^{58}Ni$^+$ are

~0.06um. In all cases, no peak in the SIMS plot could be seen, indicating no energetic metal contamination is present, as expected.

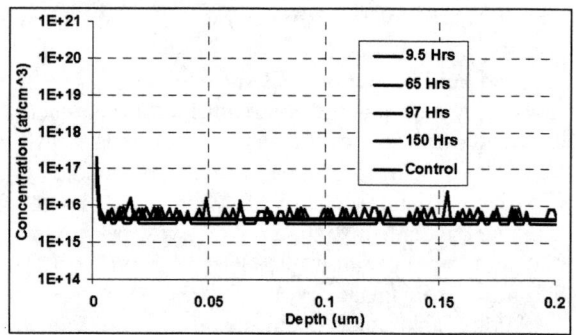

FIGURE 3. SIMS Profile of ^{58}Ni after In^{++}/150keV/5e15 Implant.

Figure 4 shows a typical mass spectrum measured in the mass/charge range of 56 to 59. Two peaks are clearly seen, representing the two isotopes of indium at 56.5 m/q & 57.5 m/q. The natural abundance of ^{113}In and ^{115}In is 4.2% and 95.8% respectively [7], which can be seen in Figure 4. Based on the full width at half maximum (FWHM) of the ^{115}In^{++} peak shown in Figure 4, the mass resolution can be calculated to be >115. This allows easy separation of indium isotopes.

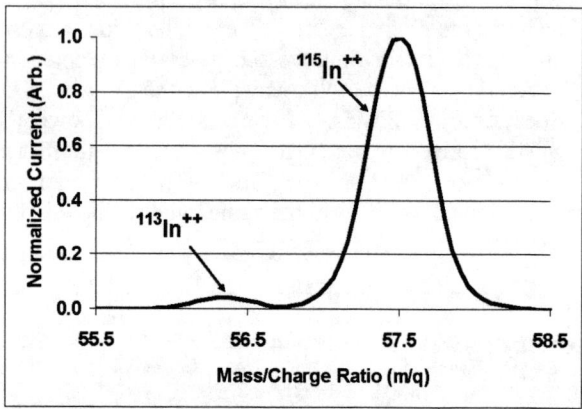

FIGURE 4. Mass Spectrum Showing In++

In the event of foreign parts finding their way into the source area, the various metal isotopes can typically be measured in the mass spectrum when using InCl$_3$ feed material. If the relative contamination information is convolved with the mass resolution, it is possible to predict the contamination level expected in the wafer. This prediction allows the Varian patent-pending interlock to be set within the software that can be used to ensure that the required metal performance is achieved for ^{115}In^{++} implants. Clearly it is optimal to utilize a less chemically reactive feed material, and use recommended source materials. In this event, there is no contamination issue.

C. Productivity

Figure 5 shows the 300mm uniform beam current as a function of beam energy for InI. The VIISta 810EHP uses In$^+$ up to 130keV and In^{++} at higher energies, irrespective of whether InCl$_3$ or InI is used as the feed material. The indium beam current specifications for the product are the same for both feed materials.

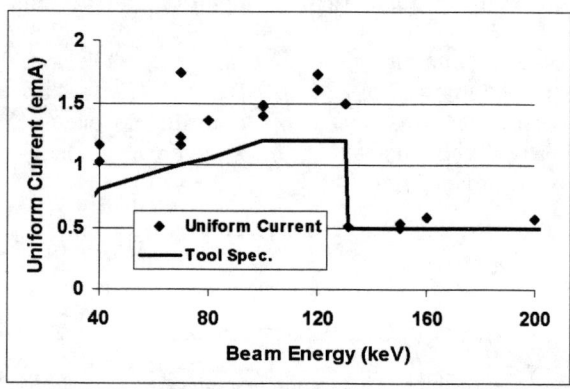

FIGURE 5. 300mm Uniform Beam Current as a Function of Beam Energy

Two implants were chose to demonstrate productivity. In$^+$/130keV/3e13at/cm^2 was chosen to represent an indium halo implant, which is performed as a quad (four position) implant at 30°/0° tilt/twist angles. In^{++}/180keV/1e13at/cm^2 represents a typical super steep retrograde (SSR) channel implant. The implant was performed as a single rotation implant at 7°/22° tilt/twist angles. The throughput results can be seen in Table 1 for the three cycles that were performed. Excellent setup-to-setup repeatability results in little variation in the measured throughput.

Recipe	Run Rate (WPH)			Tune Time (Mins)
	Min	Max	Mean	Mean
In+/130keV/3e13; Q30/0	245	247	246	12:49
In++/180keV/1e13; 7/22	255	257	256	9:02

TABLE 1. Run rate achieved for two indium recipes during three cycles.

Table 1 also shows the average beam tuning time for a gas-to-solid transition, since tune times need to be considered when discussing implanter productivity. As can be seen, high productivity can be achieved for indium implants.

IV. CONCLUSIONS

In this paper we have discussed how the choice of source feed material for indium beam generation can have a significant impact on tool performance and productivity. The medium current VIISta 810EHP offers high productivity, over a broad operating range, when using indium iodide as the source feed material. At the same time it provides superior capability with respect to dose performance, mass resolution, tune time, charge lifetime and cross contamination.

ACKNOWLEDGMENTS

The authors would like to thank Marie Welsch, Wilhelm Platow, Matt Beach and Dennis Rodier at VSEA for their useful discussions, support and hard work in collecting this data.

REFERENCES

1. G. G. Shahidi, "Indium Channel Implant for Improved Short Channel Behavior of Submicrometer NMOSFETs", IEEE Electron Device Letters, Vol. 14, No. 8, p.410-411, 1993.
2. N. Variam, S. Mehta, T. Feudel, M. Horstmann, C. Krueger & C. Ng, "Application of Indium Ion Implantation for Halo Doping", IEEE 2000 International Conference on Ion Implantation Technology Proceedings, Alpback, Austria, pp.42-45.
3. G. Luckman and R. D. Rathmell, "Indium Implantation Process Performance on the 8250_{HT}", IEEE 2000 International Conference on Ion Implantation Technology Proceedings, Alpback, Austria, pp444-447.
4. T. Yamashita, N. Miyamoto, K. Miyabayashi and T. Nagayama, "Cleaning Procedure for Indium Implantation", IEEE 2002 International Conference on Ion Implantation Technology Proceedings, New Mexico, USA, pp455-458.
5. A. Renau, J. Scheuer, D. Brennan, S. Todorov, A. Cucchetti & J. Olson, "Performance Characteristics of the Varian VIISta 810 Single Wafer Medium Current Ion Implanter", IEEE 2000 International Conference on Ion Implantation Technology Proceedings, Alpback, Austria, pp435-438.
6. J. Reyes, "$InCl_3$ as Vaporizer Feed Material for the Bernas Dual Vaporizer in the E220/500 Implanters", IEEE 1998 International Conference on Ion Implantation Technology Processings, Kyoto, Japan, pp. 284-287.
7. Chemistry Data Book, Edited by J. Stark & H. Wallace, Published by John Murry, 1982.
8. H. Fujisawa, T. Yamashita, S. Ishida, N. Hamamoto, N. Miyamoto, K. Miyabayashi & N. Nagayama, "Indium Beam Implantation", IEEE 2000 International Conference on Ion Implantation Technology Proceedings, Alpback, Austria, pp62-65.
9. J. F. Ziegler, J. P, Biersack, U. Littmark, "The Stopping and Range of Ions in Solids," vol. 1 of series "Stopping and Ranges of Ions in Matter," Pergamon Press, New York (1984).

Understanding The Calibration Methodology For The Axcelis GSD/HE Final Energy Magnet And A Means For Manipulating The Calibration Curve

Ronald Johnson, John Schuur

INNOViON Corporation, Foundry Ion Implantation, Gresham, OR 97230, USA

Abstract. In the course of monitoring and maintaining an accurate calibration of the Final Energy Magnet (FEM) located at the output end of the RF Linac structure in an Axcelis GSD/HE ion implanter, it would be useful to know exactly how the control system generates FEM setpoints for a given ion species. This paper presents the physical equations and mathematical model used by the HE control system to control the FEM magnetic field as measured by an internal gauss probe. Details of the mathematical model used by the control system to correlate actual gauss probe readings to calculated (theoretical) field strengths is presented. Included is a discussion of the relationship between ion electromagnetic rigidity and magentic field. Finally, a method for directly manipulating the calibration curve of the FEM is discussed as it relates to the specific methodology employed by the HE FEM.

Keywords: Axcelis, GSD/HE, FEM, final energy magnet, calibration
PACS: 61.72.Tt, 85.40.Ry, 29.17.+w,

INTRODUCTION

The ion energy delivered by the Axcelis GSD/HE ion implanter is determined by the setting of the Final Energy Magnet (FEM). A gauss probe situated between the FEM poles is used to measure magnetic field; this probe has high precision and repeatability, but does not measure absolute magnetic field. The location of the probe outside of the central field region leads to measurement non-linearity and to errors in actual field value. In order to translate readings from the gauss probe to values useful for controlling the FEM, a calibration curve is used which correlates the gauss probe response of a particular field to the actual field required for a given ion. The FEM is set to a value of magnetic field corresponding to the ion's rigidity and via the calibration curve, the FEM is controlled by a servo loop that matches actual gauss probe readbacks to what is required by the ion's rigidity.

The gauss probe response is linear to a measured field up to around 2500-3000 gauss, but the response becomes less sensitive at higher magnetic fields. Axcelis chose a quadratic fitting function for higher rigidity ions (up to 9000 gauss). The fitting function factors, along with the original FEM Calibration point data are stored in a file called, *cal_config.dat* on the system hard drive. The determination of the correction factors and a method to calculate the whole calibration curve is the subject of this paper.

PHYSICAL RELATIONS

Electromagnetic rigidity, ρ, is related to an ion's mass, energy and charge. For a given magnetic field, B, an ion's rigidity determines the radius of curvature, r, along which the ion will travel while subject to the field:

$$\rho = Br = \sqrt{[2mE/(eq)^2]}, \text{ Tesla-meters} \quad (1)$$

In the HE system, mass is customarily measured by amu and energy by keV; therefore some conversion constants will be required to enable accurate calculations:

$$\rho = Br = \sqrt{[2*(mk)*(e'E*1000)/(eq)^2]} \quad (2)$$

k = 1.6605E-27 kg/amu
e' = 1.602E-19 Joules/eV
e = elementary charge = 1.602E-19 Coulomb

For example, 31P+ at 1000 keV and at 500 keV, the respective quantities are:

m = 31, E = 500 and 1000, q = 1, which gives
ρ(500 keV) = 0.5666 T-m or 5666 gauss-meters, and
ρ(1000 keV) = 0.8013 T-m or 8013 gauss-meters.

An important factor in relating measured gauss to required gauss for the calibration curve calculations is the radius of curvature for the FEM. A number of measurements on the HE systems indicates this value is around 0.84 meter. Using this value with the rigidities above gives:

B(500 keV) = 0.6745 T or 6745 gauss, and B(1000 keV) = 0.9539 T or 9539 gauss.

Both of these ions have adequate rigidity(field) to require quadratic correction in the FEM calibration curve; for such processes, it is essential to have a repeatable method for matching the energy calibration among multiple HE systems.

CALIBRATION OVERVIEW

The HE system software provides four calibration points that are set by tuning specific ion beams at specific energies (with all rf cavities off). The software records ion mass, extraction voltage and gauss probe reading for each of the four points. The gauss measured by the FEM probe always reads lower than the actual gauss required for a given rigidity. All of the relevant data collected during the calibration is stored in the *cal_config.dat* file. Once the calibration data have been recorded, the software generates a linear fitting function for low to moderate rigidity ions and a quadratic fitting function for the moderate to high rigidity ions.

The field where these two functions intersect is labeled 'FEM cal quadratic cutoff B'. The linear fit results are stored in the configuration file as a slope/intercept format (y = mx + b) with the quantity 'FEM magnet scale correction' serving as the slope and the 'FEM magnet offset correction' as the intercept. For the quadratic fit, three constants are listed that correspond to a general quadratic equation of the form, $y = Ax^2 + Bx + C$. The one oddity of the organization of the constants is that 'y' corresponds to the real gauss while 'x' corresponds to the gauss probe response so that to find the internal gauss probe reading required for an ion of rigidity, y, requires one to invert the appropriate equation and solve for 'x'. This is trivial to do for the linear fit, but more involved for the quadratic portion of the FEM calibration curve. Possibly, the software may use an iterative method to calculate the uncorrected readback. Figures 1 and 2 illustrate the departure of the gauss probe readback from linearity at higher field strengths.

DISCUSSION

The calibration mechanism provided by Axcelis corrects the gauss probe reading to improve energy linearity over a wide range of electromagnetic rigidities. For the highest rigidity ions that the system is designed to accelerate, the departure from linearity of the gauss probe response can lead to errors in the measurement of B of 5%, or higher. Since a linear fit to the HE gauss probe response over-estimates B, use of such a fitting function could lead to energy errors of +10% or higher.

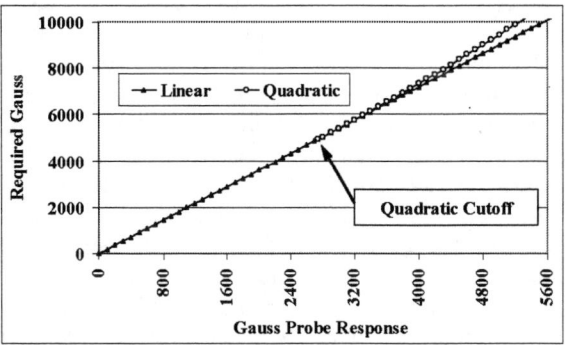

FIGURE 1. Example plots of the linear and quadratic curve fits for HE FEM Calibration

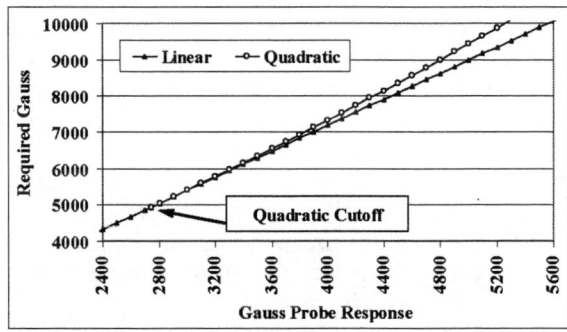

FIGURE 2. Example plots of the linear and quadratic curve fits for HE FEM Calibration - Detail

In principle, the calibration procedure offered by Axcelis is straightforward. The FEM high and low points are obtained from simple to tune boron and argon ion beams. However, the two so-called FEM ULTRA calibration beam set-ups require triply-charged Xe to be introduced into the HE linac. The FEM is then adjusted to analyze the singly charged fraction that results from collisions with the background gas over the 2 meters of the linac's length. Due to limitations in mass analysis prior to the linac, correct identification of the desired Xe isotope is not always clear. Again, energy errors may result.

The time involved in tuning the four ion beams, optimizing their transport through the linac and in gas bottle changes for the Xe source gas can be significant. Also, depending on the level of pre/post FEM Calibration verification one chooses to require, additional non-productive tool time can be consumed. A regular calibration or calibration verification program can be costly in a production environment.

MANIPULATION OF THE CALIBRATION CURVE

One can imagine a scenario where a site with multiple HE installations establishes an energy monitoring program with the intent of maintaining good agreement across all machines. If, however, one determines that the energy delivered by a particular machine has deviated from the norm (and the deviation were small), it would be useful to be able to adjust the machine's energy output with as little interruption to production as possible. Knowing the form of the FEM calibration curve fit and how to interpret the constants stored in the software system allows for direct manipulation of the FEM calibration curve without running the complete FEM calibration procedure.

There are two potential solutions to this endeavor. One is to realize that when executing the FEM Calibration procedure (or even the Analyzing Magnet Calibration procedure), that the system software doesn't actually check the quality of the beam set-up or even whether any beam current exists. The software verifies only that the extraction energy, amu setpoint and ion charge are correct for a given step – no actual ion beam need be set-up. The final, crucial information the system needs to set a calibration point is for the FEM to be set at the desired value. An approximation of the required magnetic field can be made from an estimate of the energy error:

$$\Delta B/B \approx \tfrac{1}{2}\Delta E/E \qquad (3)$$

Using the calibration constants, one can then estimate how to set the FEM to give the desired change in calibration. This method does not involve detailed calculations on the part of the user – the system software will automatically udpate the configuration constants.

The second method involves direct editing of the configuration constants using only calculated values. This method requires more detailed calculations and an understanding on the part of the user of how to apply curve-fitting routines to data sets. In order to effect this, one would need to know how to fit a parabola to an existing line given two additional points (FEM ULTRA1 and FEM UTLRA2). When mating two curves, e.g., a line and a quadratic, it is customary for the two functions to have continuous first derivatives which also have the same value at the point of intersection. In the case of the FEM Calibration curve, this requirement is what partially defines the 'Quadratic cutoff B' value. This requirement also ensures that the quadratic portion of the curve fit always lies above the linear fit for a given ordinate value. Once having calculated the desired values for ULTRA1 and ULTRA2, then one could also calculate the quadratic fitting constants and directly edit the configuration file. A system reboot would be recommended to ensure that the values are activated. While it has been verified that user chosen values can be directly entered into the configuration file and the system will use those values in setting the FEM, this approach has not been deployed in production.

CONCLUSIONS

The FEM calibration curve relies on a two part curve fit comprised of a linear portion for lower rigidity ions and a quadratic portion for higher rigidity ions. The constants used to define these curves as well as the data used to calculate them are contained in a file called *config_cal.dat*. It is possible to directly manipulate the constants to produce a calibration curve without executing the FEM Calibration procedure. This would be desirable for maintaining an energy match between mupltiple machines and also for reducing tool down time for the purpose of checking or resetting the FEM calibration curve.

ACKNOWLEDGMENTS

The work presented in this paper was developed on HE/VHE Software Revision 3.1

REFERENCES

1. Axcelis Engineering Procedure 8608043, Revision B. Final Energy Magnet New Calibration Procedure
2. Axcelis Engineering Procedure 8608049, Revision A. Final Energy Magnet: Some Questions and Answers
3. D. Halliday, R. Resnick, Physics, part 2, 3rd ed. John Wiley and Sons, 1978, pp. 730-734.

New Medium Current Ion Implanter SOPHI-200

Hidekazu Yokoo[1], Hideo Suzuki[1], Ryouta Fukui[1], Takeshi Hisamune[1], Makoto Tomita[1], Tsutomu Nishihashi[2], Kazuhiko Tonari[2] and Seiji Ogata[3]

1) ULVAC, Inc. Semiconductor Equipment Div.2, 1220-14 Suyama Susono Shizuoka 410-1231 JAPAN
2) ULVAC, Inc. Inst. Semiconductor Technology, 1220-1 Suyama Susono Shizuoka 410-1231 JAPAN,
3) ULVAC, Inc. Research & Development Div., 2500 Hagisozono Chigasaki Kanagawa 253-8543 JAPAN

Abstract. The new medium current ion implanter SOPHI-200 has been developed. By newly designed low aberration optics, high performances have been obtained at a compact implanter system with light tuning. ULVAC's wafer handling system, which is production proved, has realized the direct implantation to the ultra thin wafer for power device manufacturing. Overview of the system and the performance characteristics will be discussed.

Keywords: ion implantation; ion implanter; medium current; beam optics; power device;
PACS: 85.40.Ry

I. INTRODUCTION

ULVAC has released a new medium current ion implanter SOPHI-200. Compact footprint and light operational tuning have been achieved by exhaustive examinations of optical design through newly developed beam simulation [1]. Combined with the wafer handling system applicable for ultra thin wafers, which is production proved on ULVAC's semiconductor equipment system, the ion implantation for new generation power devices has been realized.

In this paper, overview of the system and the performance characteristics will be discussed.

II. OVERVIEW

SOPHI-200 is a parallel beam medium current ion implanter applicable for from 5 to 8inch wafers. The plane view of the system is shown in figure 1. The footprint is compact as 5.98meter length and 2.5meter width and total weight is light as 12ton. System specifications are summarized in table 1. As the standard system of SOPHI-200, the maximum energy is 200keV with the single charged ion. Optionally, the maximum energy is enhanced to 260keV.

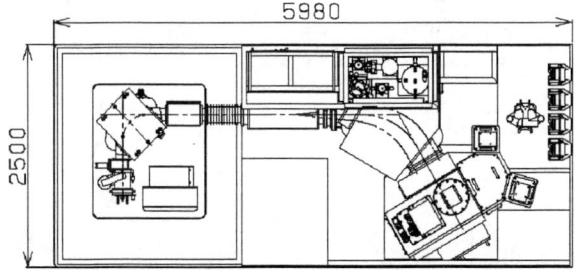

FIGURE 1. Plane view of SOPHI-200

TABLE 1. Summary of the specification

Item	Value
Beam Energy Range (Single Charged) (Optional) (Double Charged)	3 to 200keV 3 to 260keV 200 to 400keV
Throughput	210pcs/hour
Dose Uniformity	< 0.5% (S.D.)
Auto setup time	< 5 minutes
Dimensions	Length 5980mm Width 2500mm Height 2300mm
Weight	12000kg

1. Optical Design

Usually the maximum beam current will decrease as the implantation energy decrease, since the divergence of the beam due to the space charge effect is inversely proportional to the power of 3/2 of the energy. Optical system of SOPHI-200 was designed aiming at enlarging the beam current at the low energy side.

Ion beam generated from a long life ion source is mass analyzed by a 90degree low-aberration sector magnet and accelerated or decelerated to the implantation energy. Combined with a mid-electrode in the acceleration column, two pairs of electrostatic quadrupole doublet lenses arranged at both sides of acceleration column achieve wide range focusing of the ion beam. The beam current profile of SOPHI-200 contrasts with that of the conventional system in peaking at the intermediate energy as shown in figure 2. After passing through the final quadrupole lens, the beam is scanned in the horizontal plane by a one pair of electrostatic plates and parallelized by a 60degree sector magnet. Aberrations in the parallelizing magnet are drastically reduced through ray tracing simulations that utilize measurements of the fringing magnetic field in the 3-D 10mm mesh.

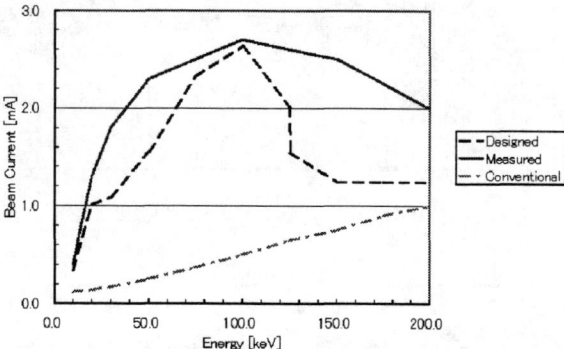

FIGURE 2. Comparison between designed and measured beam currents of SOPHI-200. The ion species is P+.

2. End Station

The wafer is chucked electro statically on the platen, which is scanned mechanically in the vertical direction. The wafer is tilted from 0 to 60degree and rotated from 2 to 32 steps. Heat transfer of the electrostatic chuck of the platen is 2.4E-3 W/cm2/deg. and this cooling power results the saturated temperature of less than 75degree at a dose of 1E15/cm2 and an implantation energy of 150KeV with implantation time of 200seconds. Positioning reproducibility of the platen is less than 0.05degree. The beam profile is measured by using a scan profile monitor incorporated in the main Faraday cup mounted just after the platen.

3. Transfer System

There are enlarging demands for power devices such as IGBT (Insulated Gate Bipolar Transistor) or MOSFET (Metal Oxide Semiconductor Field Effect Transistor). Thinner wafer thickness down to 50micron is required for these devices aiming at higher switching speed and lower power loss.

Ultra thin wafers down to 50micron thickness (see figure 3) can be transferred directory from the carrier to the carrier. Both the front side and the back side of the wafer can be processed by the identical implanter. The minimum thickness of the wafer applicable is tabulated in table 2. The WSS (Wafer Support System) such as plastic taped or glass coated wafer also can be handled. Load lock chambers and the transfer robots designed for the ultra thin wafer realize the chipping free transfer and low contamination of less than 5 particles at up to 0.20micron on the 8inch wafer. Each wafer after the process is stored back into the identical slot of the carrier, where the wafer was stored before the process. A mapping sensor incorporated in the transfer robot senses the empty slot in order to reduce the wasteful time.

FIGURE 3. Ultra thin wafer in the carrier

TABLE 2. Minimum wafer thickness applicable on SOPHI-200

Diameter	Thickness
5 inch	50 micron
6 inch	65 micron
8 inch	100 micron

III. PERFORMANCE

As key issues of the medium ion implanter, the dose uniformity, parallelism, contamination and automatic setup time are discussed below.

1. Dose Uniformity

At conventional implanters, high uniformity excessively relies on the modulation of the scanning waveform, which requires wasteful tuning time. Low aberration optics of SOPHI-200 results high uniformity of less than 0.5% at the standard deviation without invoking the modulation of the scanning waveform. Sheet resistance map of 8inch wafer implanted by SOPHI-200 are shown in figure 4. Uniformity is 0.26% at one sigma for B+ implanted and 0.47% for P+ implanted. These wafers were implanted without the modulation of the scanning waveform.

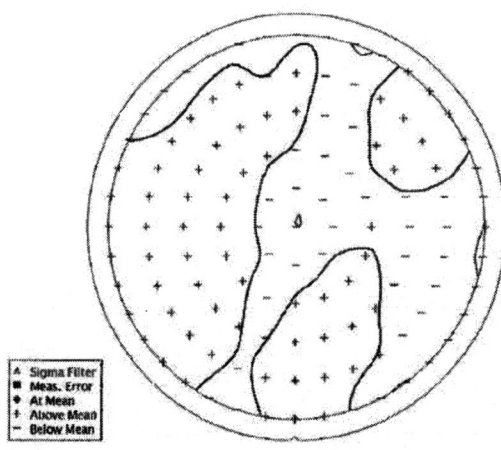

FIGURE 4. Sheet resistance map of a 8inch wafer implanted by SOPHI-200. The ion dose is 5E13/cm2 at 100Kev with B+ ions. The tilt angle is 7degree. The wafer is annealed at 950 degree for 30miniutes under N2 flow. The average sheet resistance is 863ohm/square and the standard deviation is 0.26%.

2. Parallelism

Estimated parallelism utilizing the measured magnetic field is tabulated in table 3. Contrary to the mean value of the incident angle, which is hardly affected by the beam shape, the standard deviation of the incident angle depends on the beam shape. The standard deviations in the horizontal plane and the vertical plane have been estimated as a function of the beam current. The point of view on this tradeoff seems to be neglected at traditional discussions. The minimum angle deviation reaches as low as 0.35degree due to the low aberration optics of SOPHI-200. The mean angle in the vertical direction is zero due to the symmetric structure.

TABLE 3. Beam parallelism as a function of beam current. Ion species is As+ and the implantation energy is 50keV.

Current [mA]	Mean in Hori [deg.]	S.D. in Hori [deg.]	S.D. in Vert [deg.]
0.2	0.02	0.28	0.35
0.4	0.02	0.28	0.50
0.6	0.02	0.32	0.50
0.8	0.02	0.43	0.75

3. Contamination

Energetically undesirable ions are mostly rejected at the parallelizing magnet as shown in figure 5. The breakups of molecular ions, which are not separated by magnetically with the multi charged ions, are rejected electrostatic filter mounted just after the mass analyzing slit. SIMS depth profiles with P+ and P++ at identical implantation energy are shown in figure 6. Significant difference is not observed between the depth profile with P+ and that with P++. SIMS depth profile with low energy implantation is shown in figure 7. The vacuum system, which has been designed through computer simulation of pressure distribution, keeps the end station at low pressure less than 5.0E-5Pa. The energy contamination, which may be generated at the deceleration, is hardly distinguished in higher energy tail.

Metal contaminations measured are listed in table 4. These data are obtained from ICP (Inductive Coupled Plasma) mass spectroscopy. Heavy metal species such as Ni or Mo, which causes serious defects in the device, are less than the detection limit.

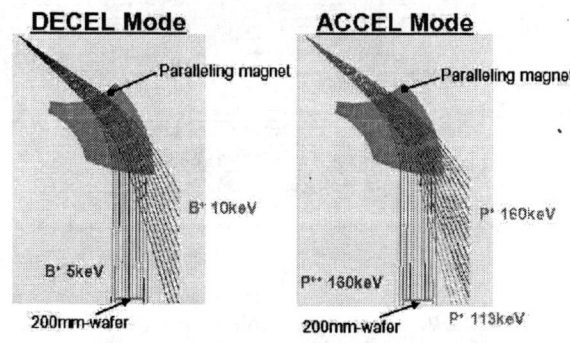

FIGURE 5. The energy contamination is rejected at the parallelizing magnet both at the deceleration mode and the acceleration mode.

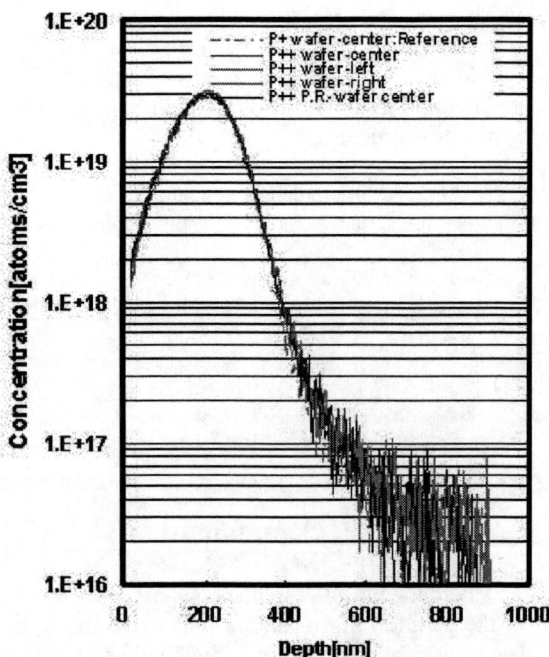

FIGURE 6. SIMS depth profiles of Si wafers implanted with P+ and P++ at an identical implantation energy of 160keV. Acceleration voltage for P+ is 160kV and that for P++ is 80kV. Energy contamination, which may be generated with the multiply charged ion, is fully removed.

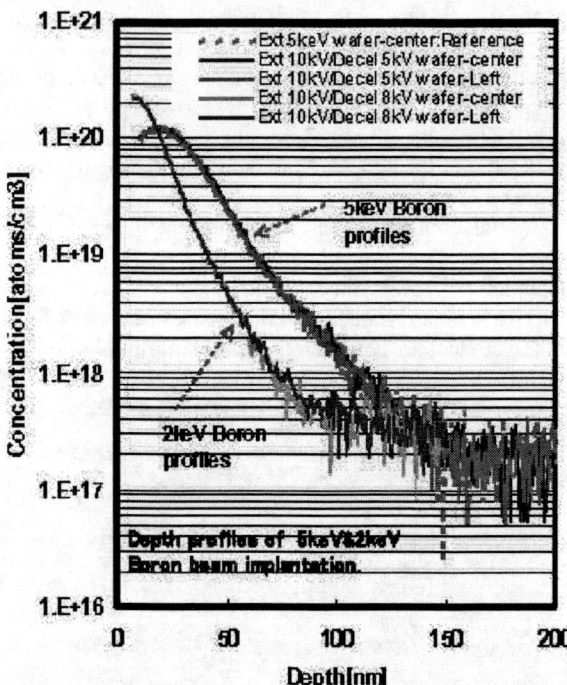

FIGURE 7. SIMS depth profile of Si wafer implanted with deceleration mode. Upper profiles are results of B+ implantation at 5keV. Lower ones are result of B+ at 2keV.

TABLE 4. Metal contamination measured by ICP Mass (VPD). The ion specie is As+ with dose of 1.0E15/cm2 at 150keV.

Metal	Concentration	Detection Limit
Al	7.1E09/cm2	1.8E09
Fe	6.0E09	3.4E09
Ni	ND	1.3E10
Mo	ND	3.7E08

4. Automatic Setup

Optimum operational conditions are easily obtained with light tuning due to the low aberration optical design. Automatic setup time is tabulated in table 5. For recipes that are not stored in the memory, virtual optimization code [1] is executed to yield the optimum lens parameters within 30 second. Since the modulation of the scan waveform is not required, the operational condition is set up automatically with light tuning. The specification of the automatic setup time is less than 5minutes.

TABLE 5. Automatic setup time for various implantation conditions.

Ion	Energy [keV]	Current [microA]	Setup Time
B+	10	70	3'45"
B+	20	200	4'51"
B+	100	500	3'43"
BF2+	30	200	4'58"
P+	20	300	4'01"
P+	70	700	4'09"
P+	100	800	4'07"

ACKNOWLEDGMENTS

The authors greatly thank Messes Y. Sakurada, J. Fujiyama and T.Oka for their continuous support and warm encouragements. We also thank the colleagues of ULVAC for their sincere collaboration in realization of SOPHI-200.

REFERENCE

1. S. Ogata et al., "Real time optimization of optical parameters of ion implanters" in these proceedings.

Optima MD: Mid-Dose, Hybrid-Scan Ion Implanter

K. W. Wenzel, A. M. Ray, B. H. Vanderberg, and R. D. Rathmell

Axcelis Technologies, Inc., 108 Cherry Hill Dr., Beverly, MA 01915

Abstract. The emergence of higher-dose and lower-energy halo and source-drain extension implants for 90 and 65-nm nodes drove the design of Axcelis' Optima MD ion implanter. The Optima MD extends the proven process performance of traditional medium-current implanters over an energy range of 1 keV to 250 keV for singly-charged ion species with beam currents ranging from 1 pµA to 4800 pµA. The Optima MD comprises a hybrid-scan beamline architecture, a new endstation, and a new control system that offers the user more flexibility in data acquisition and statistical process control. The beamline has the capability to provide milliamp beam currents in the low energy (<5 keV) range. The use of an aggressive final deceleration (decel) in combination with a final angular energy filter (AEF) enables this productivity improvement of higher beam currents while mitigating the risk of energy contamination. In this paper the new beam transport system that reduces space charge blowup and maintains focus control at low energies is described. Energy purity is quantified with SIMS analysis in several implants using comparisons with pure drift implants. For the most aggressive decel ratios with low energy beams, energy contamination remains below present detection limits. The system can be configured to process either 200-mm or 300-mm wafers with full 300-mm factory automation capability. This paper describes the system-level characteristics and performance parameters of the Optima MD.

Keywords: Ion implanter, equipment, serial
PACS: 61.72.Tt, 85.40.Ry

INTRODUCTION

Scaling of CMOS technology below 100-nm gate length has required the introduction of aggressive low-energy halo and source-drain extension implants with increasing doses [1]. The Optima MD was designed to productively cover this new range of implants in addition to providing coverage for all traditional medium current implants. It combines a hybrid-scan beamline architecture and new features to enhance low-energy beam currents with a fast end station for competitive throughputs. The beam transport relies on aggressive decel and an appropriate energy filter to enable improved beam currents with uncompromised energy purity. A new control system based on a Windows™ operating system includes automated implant angle control, charging control, faster Autotune™ times, and an advanced statistical control package that supports real-time monitoring and report generation.

BEAMLINE DESIGN

A number of improvements over traditional medium current tools [2,3] were implemented on the Optima MD beamline in order to deliver and transport higher beam currents at lower energies. Figure 1 shows a beamline system layout. The source housing has high conductance to minimize charge-exchange beam loss prior to the AMU magnet. The indirectly-heated cathode Eterna™ ELS ion source shares many parts in common with other Axcelis ion implant systems [4]. The extraction optics have been optimized, and the extraction energy is 45 kV to increase the beam perveance as it is transported through most of the beamline. These changes have resulted in a 50% increase in current over the MC3 at most energies.

Downstream of the electrostatic parallelizing lens (P-Lens) and accel/decel column is an electrostatic angular energy filter (AEF). As shown in Figure 2, the AEF deflects the beam 15° vertically, and eliminates off-energy beam contaminants.

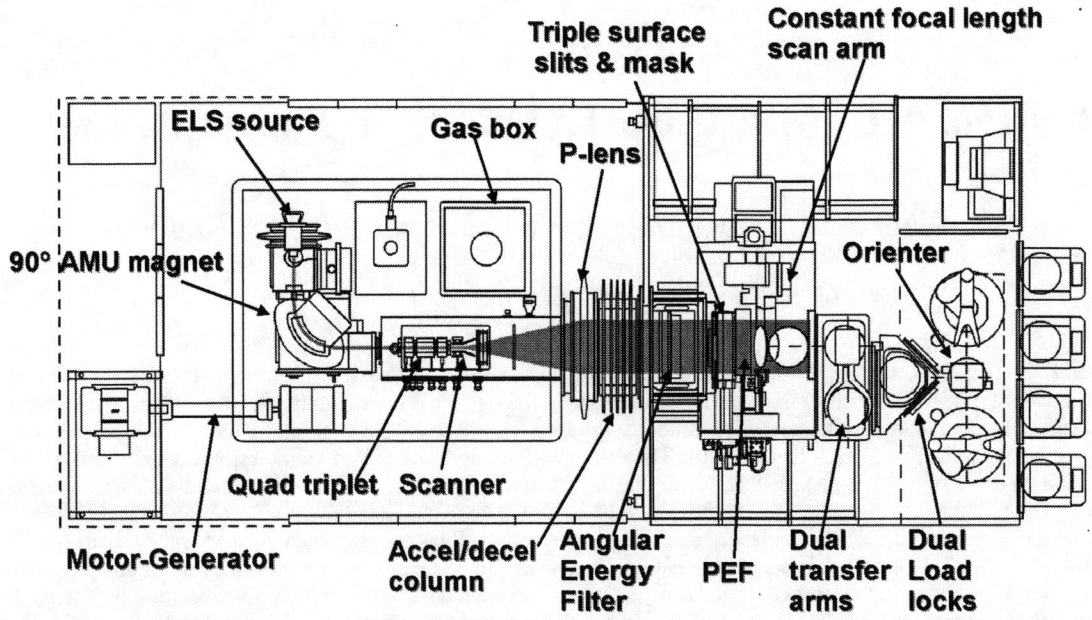

FIGURE 1. Optima MD Beamline includes an electrostatic quadrupole triplet for focusing prior to the scan vertex. The scanned beam passes through a parallelizing lens then an accel/decel column followed by an electrostatic energy filter.

Below 20 keV, for optimized low-energy beam transport, the bending plates of the AEF are negatively biased with respect to ground. This results in increased beam energy through the AEF, where the beam is fully stripped of neutralizing electrons, and reduces space-charge forces acting on the beam ions. The electric field configuration in this region is carefully designed to optimize both electrode geometries and voltages while maintaining control over beam focus, parallelism, and energy contamination. At high energies, the acceptance of the beamline has been maximized with beam optics design tools including matrix transport and ray-tracing codes. Maximum ion source output effectively limits the implanter beam current at high energies. At low energies the implanter current is transport-limited. Space-charge forces in the beam generate a radial acceleration of ions near the edge of the beam that causes the beam diameter to expand as energy decreases and as mass increases. While this dictates that current will be limited as energy is reduced, the same principle makes it clear that keeping the energy up as long as possible will improve transport. Figure 3 shows unscanned spec beam currents for Optima MD below 50 keV

The power supply configuration used to determine the final beam energy has been changed to allow a dual range acceleration supply to connect directly to the source so that below 10 keV, the beam energy is determined by a single, high stability, 10 kV power supply. The P-lens voltage is independent of

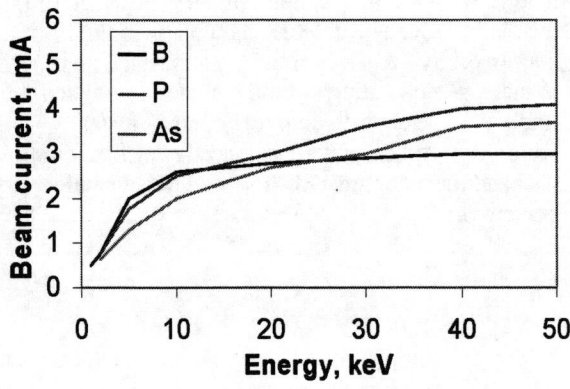

FIGURE 2. Angular energy filter assembly with deflection plates and AEF focus electrode.

FIGURE 3. Spec beam currents for Optima MD below 50 keV shows mA currents to below 5 keV.

the final beam energy, so that the lens voltage is used only for optimizing parallelism.

END STATION DESIGN

The Optima MD endstation, illustrated in Figure 4, is a new design, optimized for horizontal wafer handling through the system, high reliability, ease of set up, and improved mechanical throughput. Wafers can be delivered from four standard loadports. Dual external robots and a single in-air wafer alignment station are used to deliver wafers to dual single-wafer load-locks. In-vacuum wafer transfer is achieved with dual rotary robots with a common vertical axis and direct drive motors selected for speed and reliability. The wafer is scanned in front of the beam using a constant-focal-length scanning system with a linear drive scan arm plus direct drive motors for tilt and twist adjustments. The direct drive tilt and twist motors allow for precise angle control and closed-loop control of the vertical incident beam angle.

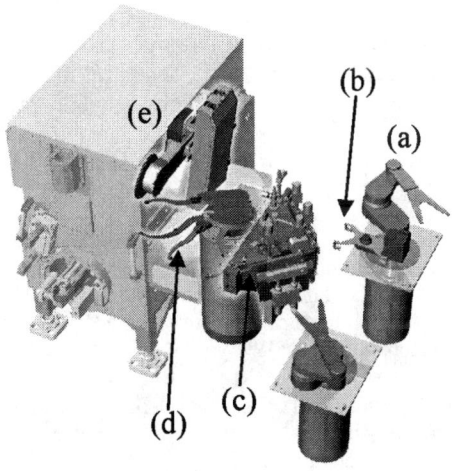

FIGURE 4. Wafer handling mechanisms of the Optima MD Endstation: (a) dual fixed in-air robots, (b) single aligner station, (c) dual load locks, (d) dual in-vacuum transfer robot, and (e) the constant-focal length wafer scan system.

CONTROL SYSTEM

The control system is based on a Windows™ operating system and introduces an object-oriented framework and configuration scheme designed to improve code maintenance and flexibility. It uses ACE / TAO "Adaptive Communication Environment" / "The Ace Orb", enabling high-performance, real-time QoS distributed applications to invoke operations without concern for application location, programming language, OS platform, communication protocols and interconnects, and hardware.

Included are infrastructure services such as events, logging, alarm, data, configuration, user administration etc. At the supervisory level a sequencer orchestrates the running of process and control jobs by utilizing these machine resources to move material and perform the work.

The control system includes a dedicated computer for collecting data at 10 Hz that can be used for diagnostics or Statistical Process Control (Profile™). Parameters can easily be assigned to a charting routine for live display of data. SPC limits can be applied by the end user for all parameters on the implanter and can protect wafers by inhibiting implants if the user defined limits are exceeded.

PROCESS PERFORMANCE

The Optima MD system was designed in order to deliver state-of-the-art control over the most critical implant parameters: energy purity, dose, angle, and charging. The Optima MD corrects the horizontal angle using quadrupole steering or optimizing the P-lens ratio for angle correction. Angle control in the vertical plane is easily achieved by determining the incident beam angle with an *in-situ* vertical beam angle measurement that is accurately referenced to the surface of the wafer and subsequent adjustment of the wafer scan tilt angle. Details of the Optima MD angle control are reported in a separate paper [5].

Energy Purity

Energy purity can be a risk in beamline systems that use decel. In the Optima MD, decel is used for all final energy implants from 0.5 keV to 120 keV. Low energy implants may use full extraction energy down to 5 keV final energy. Previous experience has shown that the 15° bend of the AEF is adequate to eliminate all neutrals created prior to the bend [6].

Below 20 keV, Optima MD maintains an elevated energy using AEF Bias to improve beam transport. The AEF Bias acts as a deflecting decel that provides the minimum risk for charge exchange that can lead to energy contamination by combining careful field configuration and minimum path length for charge exchange with differential pumping to control the pressure in the AEF region. SIMS profiles of implants of 10 keV B, which uses an AEF bias of -10 kV and 1 keV B, which uses -3.5 kV AEF bias are shown in Figure 5. Profiles compared to drift mode implants on

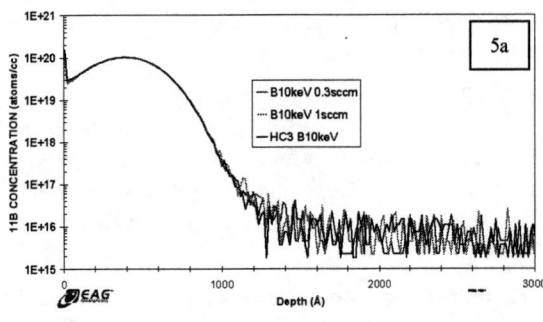

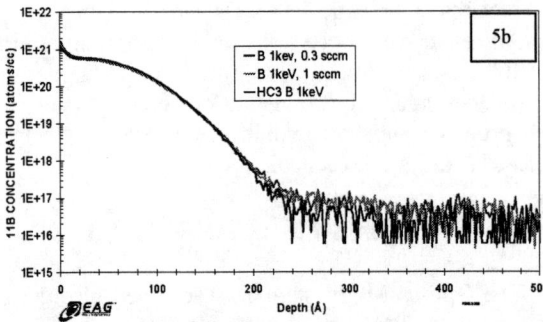

FIGURE 5. SIMS profiles of 5E14 dose implants of 10-keV B+ and 1-keV B+ with PEF Xe gas flow of 0.3 and 1.0 sccm are compared to drift implants of 10-keV B+ and 1-keV B+ from an HC3 Ultra.

the Axcelis HC3 high current implanter, where there is no risk of energy contamination, show that there is no detectable energy contamination on Optima MD.

Charge Control

Effective charge control is achieved with a multi-cusp confined plasma electron flood and *in-situ* electrical charge monitors near the plane of the wafer surface. The plasma electron flood (PEF) uses a standard arc chamber to feed plasma into a multi-cusp magnetically confined chamber to distribute the plasma across the scanned beam in the beam scanning direction.

Higher beam currents at low energies produce typical charging patterns commonly seen on high current tools when charge control is turned off. Use of the Xe PEF eliminates the charging pattern as shown in Figure 6.

The Optima MD PEF has provided adequate charge control to safely manufacture devices in the 90 nm node and below.

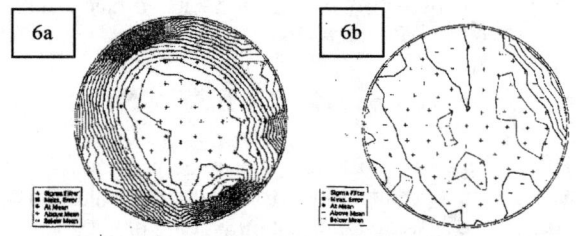

FIGURE 6. Dose uniformity for 5-keV, 2.2-mA B+ implant with no PEF, 6a, is 3.8% and with PEF ON, 6b, it is 0.53%.

SUMMARY

The new Optima MD mid-dose implanter has been proven to be a highly productive implanter with superior beam currents and process performance especially for the emerging halo implant applications. The hybrid-scan, single-wafer architecture offers state of the art beam current performance and medium current process control capabilities. The optima MD has superior performance over an extended range of implant space.

The Optima MD capabilities and advantages have already been proven at a number of advanced logic, memory, and foundry semiconductor manufacturing sites. Angle control, energy purity, and charge control are able to meet the strict demands of the advanced nodes under production today.

REFERENCES

1. T. Ghani, S. Ahmed, P. Aminzadeh, et al., "100 nm Gate Length High Performance/ Low Power CMOS Transistor Structure", IEDM, (1999).
2. J. M. Harlan and K. Petry, IEEE Proc. of Intl. Conf. on Ion Implantation Tech., Kyoto, Japan 266-269 (1998)
3. O. F. Campbell and A. M. Ray, IEEE Proc. of Intl. Conf. on Ion Implantation Tech., Kyoto, Japan 154-157 (1998)
4. T. Horsky, et al, IEEE Proceedings of 11th Intl. Conf. on Ion Implantation Technology, Austin, TX (1996) 414
5. R. D. Rathmell, et al "Implant Angle Control on Optima MD", these proceedings
6. D. E. Kamenitsa and R. D. Rathmell, Nucl. Instr. and Meth. B96 (1995) 13

ACKNOWLEDGEMENTS

We thank our colleagues Wes Huff, Anatoly Dvinov, Brian McGrath, and Patrick Heres for their input and evaluation. We also appreciate the efforts of Mark Harris and Alex Kontos to provide timely and high-quality process results.

Optimised Charging Performance On Quantum X Ion Implanters

David A. Kirkwood[1], Takao Sakase[2], Ryuichi Miura[3], Richard D. Goldberg[1] & Adrian J. Murrell[1]

[1]Applied Materials UK Ltd., Foundry Lane, Horsham, West Sussex, RH13 5PX, UK.
[2]Applied Materials Orion Group, 66 Cherry Hill Drive, Beverly, MA 01915, USA.
[3]Applied Materials Japan, 14-3 Shinizumi, Narita-Shi, Chiba 286-8516, Japan.

Abstract. A key parameter in the optimisation of CMOS device yield is the minimisation of charging-induced damage and/or breakdown of the gate dielectric material during ion implantation. In typical ion beams used for transistor doping applications, beam potentials can charge up the wafer surface if not controlled, and hence this potential must be neutralised to avoid damage to devices. MOS capacitor TEG (Test Element Group) wafers are an industry standard metric for determining the charging performance of ion implanters. By optimising the performance of the High Density Plasma Flood System (HDPFS) of the Applied Materials Quantum X ion implanter, TEG device yields of >90% at antenna ratios of 1E5:1 for a gate dielectric thickness of 3.5 nm on 300 mm wafers have been demonstrated.

Keywords: Ion Implantation, Charging.
PACS: 61.72.Tt

INTRODUCTION

The move towards single wafer (serial) ion implantation has been driven in the main by the need to eliminate the ballistic particle damage to poly-bit lines which occurred on spinning disc style batch ion implanters. The transition from batch to serial tools has the potential to provide new challenges in the control of semiconductor processes, of which wafer charging is a prime example. Although existing charging BKMs (Best Known Methods), which have successfully been used on batch tools to generate very high yields on antenna-based test structures [1] (such as TEG or SPIDER (Sematech Process Induced Damage Effect Revealer and Remover) wafers), still yield good performance when the recipes are transferred to Quantum X, it was decided to determine what had changed in the charging environment and optimise the PFS (plasma flood system) performance on the single wafer tool to exceed the performance on the batch tool. In this paper it will be demonstrated how such an approach led to very high yields on a diverse range of charge measuring tests.

THEORY

Wafer charge-up is a necessary consequence of the use of ion implantation to add dopant atoms to substrate materials. The degree to which this can be tolerated is very process dependent. One of the major reasons that it is desirable to minimize the amount of potential generated at the wafer plane is to avoid charge-up damage to the gate dielectric material in CMOS devices [2-4]. Typically charge induced breakdown of the SiO_2 in the gate will result in large shifts in V_{th}.

The wafer charges up due to interaction with the beam potential and the fact that the surface which the wafer presents to the beam is not uniformly conducting and grounded. Control of the beam potential close to the wafer plane is therefore required to minimize wafer charge up. Any device added to accomplish this must not itself contribute to charging or otherwise impact process performance.

The potential of the ion beam is dependent on the mass, charge, energy and current density of the beam. In the simplest case, neglecting space charge neutralization, this is given by equation 1 [5].

$$V = \frac{I}{4\pi\varepsilon} \cdot \left(\frac{m}{2qE}\right)^{1/2} \cdot \left[1 + 2\ln\left(\frac{R}{r}\right)\right] \quad (1)$$

Here V is the beam potential, I is the beam current, m is the mass of the ion, E is the beam energy, q the charge on the ion, R the distance to the nearest grounded surface and r the radius of the ion beam.

From this, it can be seen that for high mass, low energy beams the potential is significant, in fact it can approach the ion kinetic energy for low energy, high current As^+ beams.

Another important parameter is charge density. This is directly related to the beam density. For low energy beams this is less important however for high energy beams, which are in general well collimated and do not suffer from space charge blow-up, the beam current density could exceed $5mA/cm^2$ if not controlled. In these instances, there are two primary effects which occur, localized wafer heating due to insufficient thermal dissipation and excessive charge accumulation.

The current generation of plasma flood system used in the Applied Materials implanter, the HDPFS, has proved very successful over the years in negating the charge-up effects during ion implantation. It also has the secondary effect of enhancing low energy beam currents through space charge neutralization of these beams.

On transfer of technologies to single wafer implanters one of the main differences is the charge per pass intercepting the wafer. This is because the velocity of the wafer as it intercepts the ion beam is significantly slower than on the batch tool. This is due to the different scanning system on Quantum X which has a much better scan efficiency.

Another important difference is the way in which the wafer is grounded. On the batch tool, wafers were grounded via edge contact by the aluminium fixed restraints, whereas on the single wafer tool the wafer is grounded via six spring loaded tungsten carbide pins which impinge on the backside of the wafer.

EXPERIMENTAL

The data presented here were gathered on Applied Materials Quantum X and Quantum Xplus serial wafer ion implanters. Charge up damage was evaluated by carrying out As/50keV/16mA/5E15 implants into TEG and SPIDER test structures. This beam was chosen as it provides a tough charging environment on the wafer surface (i.e. energetic ions with a high charge density causing secondary electron emission). During the testing beam current density was controlled using the focus lens of the ion implanter and also by varying ion source conditions.

The TEG charging test structures were purchased from Hitachi. Each TEG wafer has 52 test sites with 4 capacitor structures at each site of the following collector to gate antenna ratio 1E3:1, 1E4:1, 1E5:1 and 1E6:1. The thickness of the gate oxide is 3.5 nm in all cases. This represents a stern test of the tool – thicker gate oxides can tolerate larger voltages before break down, whereas thinner oxides have sufficient leakage that charge build-up is less prevalent. Post implant, the test wafers are probed by ramping an applied voltage to the gate and monitoring leakage current. When the current flows above 1nA, the gate has broken down. This is considered a pass if more than 5V applied is required to attain this condition, and a fail if less than 5V is required.

SPIDER structures were manufactured internally by the Applied Materials MTCG unit in Santa Clara. SPIDER wafers have greater spatial resolution than TEGs, with 149 test sites and have the following antenna ratios: 2E4:1, 1E5:1, 2E5:1, 4E5:1 and 1E6:1. The gate capacitor has a surface area of $1.5\ \mu m^2$ and the gate oxide thickness is 5 nm. In these tests, a device is considered to have passed if the breakdown voltage (V_{bd}) lies within the range $-15V<V_{bd}<-10V$.

The choice of PFS gas for these experiments – argon and xenon, was dictated by the fact that both are in use in industry today. For many customers, Ar is the gas of choice for most applications due to its relative cheapness and availability with respect to Xe. Ar has other benefits over Xe – the collisional cross section and the cross section for charge exchange are both lower for Ar hence Xe is an unsuitable choice of PFS gas for use with large molecular ions with weak internal bonds (e.g. $B_{10}H_{14}$) and also is unsuitable for decel applications. Perhaps surprisingly, given the relative ease of ionization (the first IP of Xe is 12.1 eV, which is substantially lower than that of Ar, at 15.8 eV), use of Xe as a PFS gas also results in higher electron energy distributions than when using Ar. This is shown in figure 1 (internal Applied Materials data).

Here it can be seen that for a given arc current and voltage, the peak electron energy is higher for Xe than for Ar. Decreasing the gas pressure in the PFS increases substantially the emission current for Xe without significantly changing the peak of the electron energy distribution. Changing the PFS arc voltage, or current, will shift the peak position of the electron energy distribution. It should be noted that the data are from an older style of PFS and not the current HDPFS model.

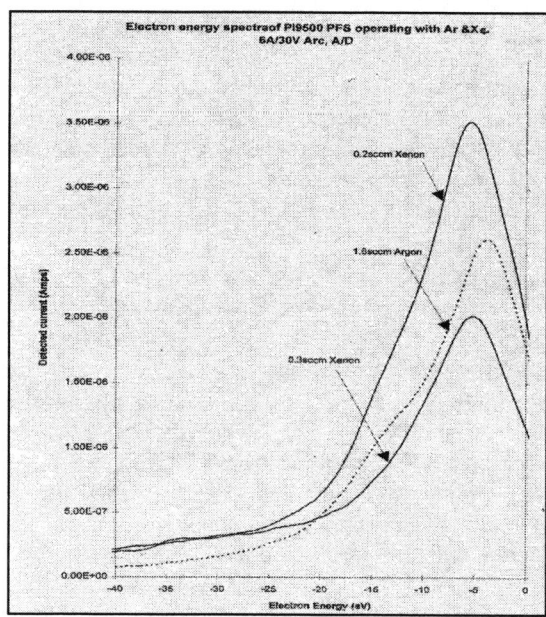

FIGURE 1. Typical electron energy distribution from an older model of PFS.

RESULTS

Data obtained from implanting into the TEG device structures are displayed graphically in figure 2, which shows the device yield following implant using firstly the existing BKM PFS conditions with Ar in the PFS and then using Xe with modified arc conditions. Each data point represents the % yield of the 52 test sites probed for each antenna ratio. From these one can immediately see that the Quantum X performs well even with the existing Ar BKM at those lower antenna ratios which correspond approximately to 65nm devices however at very large ratio the yield diminishes. On use of Xe in the PFS however, the performance at low arc volts is very encouraging – yields approaching 100% are obtained on devices with antenna ratios 1E5:1.

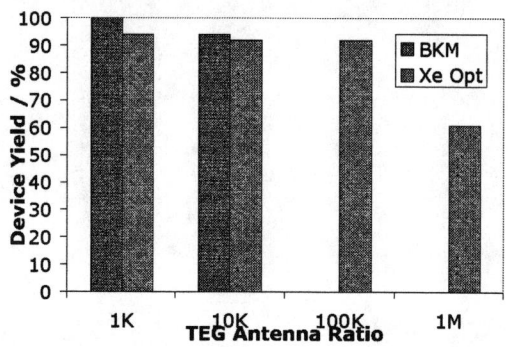

FIGURE 2. TEG device yields obtained on Quantum X for existing Ar BKM and Xe optimised BKM.

Corresponding data from the SPIDER wafers confirms these results, and are given in table 1. In general, the SPIDER wafers gave higher yields for the same antenna ratio – this is due to the slightly thicker oxide layer on the SPIDER devices. As such, they are less sensitive to charge-induced damage however they do afford better spatial resolution due to the increased number of test sites. None of the wafer maps exhibited gross failure areas – device failures appeared uniformly scattered over the wafer surface.

Test	Antenna Ratio / Yield (%)			
	100K	200K	400K	1M
Ar BKM	100	100	100	74
Xe BKM	100	100	100	99

TABLE 1. SPIDER device yields obtained on Quantum X for existing Ar BKM and Xe optimised BKMs.

DISCUSSION

From the results presented here it might be assumed these are counterintuitive, as figure one implies that Ar would always outperform Xe. However, an important parameter has been neglected – over-flooding due to energetic electrons from the PFS charging the wafer negatively. Negative charging, arising from over-flooding, can be just as damaging as under-flooding to gate oxide integrity.

The key to the improvements observed is the ability to run the PFS at much lower arc voltages when Xe is used as the feed gas, and yet still maintain sufficient emission current to neutralize the beam effectively and negate the positive potential build-up at the wafer surface.

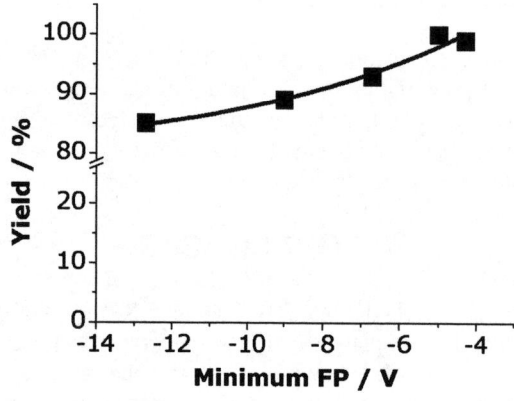

FIGURE 3. TEG yield as a function of negative floating potential measured at the wafer plane.

Measurements with a prototype charge head sensor verify these assertions – there is a clear correlation between TEG device yield for a given antenna ratio and the measured negative floating potential at the wafer plane (see figure 3). Here it can be seen that minimizing the over-flooding from the PFS by reducing the electron energy results in an increase in TEG device yield.

It is believed that the improvement in device yield arises principally from the reduction in PFS arc voltage, rather than from the use of Xe as a feed gas. It is the Xe that allows these lower voltages to be used and still maintain a stable plasma. Lowering the arc voltage results in lower negative potentials at the wafer plane and hence reduces damage due to over-flooding. Correct setting of focus electrode voltage and guide tube bias voltage are also critical to ensure optimal charging performance. Tests have shown beam current density, of which positive floating potential is a strong function, does not effect TEG yields to any significant degree within the range looked at in these experiments. This is demonstrated in figure 4.

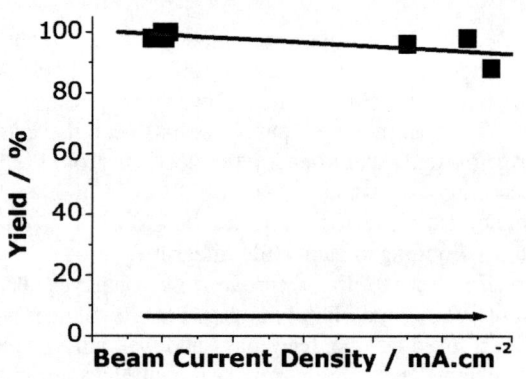

FIGURE 4. Positive floating potential at the wafer plane as a function of beam current density.

The correlation between TEG yield and positive wafer potential on the Quantum X is less clear than for negative potential due to the very effective way in which the HDPFS counteracts positive charging, for both Ar and Xe gas feeds.

CONCLUSIONS

The Applied Materials Quantum X and Quantum Xplus ion implanters offer a superior package for charge control. When very high antenna ratio devices are to be implanted, the use of Xe rather than Ar in the HDPFS increases the parameter space of the operational envelope of the system, enabling excellent TEG yields to be attained even for the highest antenna ratios.

Xe is not a blanket solution for all applications however – in addition to slightly increasing the cost of consumables of the implanter, it is also likely to impact levels of energy contamination in decel beams and should therefore only be used in conjunction with drift beams on implants into devices which are especially sensitive to charge-induced damage.

ACKNOWLEDGMENTS

The authors would like to thank both the Applied Materials Applications team and the MTCG group in Santa Clara for all their collaboration on charging work, and also the Applications lab group in Horsham for supplying tool time and system engineer resource to facilitate this project, without which none of the data presented herein would have been taken.

REFERENCES

1. H. Ito, H. Asechi, Y. Matsunaga, M. Niwayama, K. Yoneda, M. Vella, M. Reilly and W. Hacker, *1998 International Conference on Ion Implantation Technology Proceedings*, IEEE, p478
2. B. Singh, A. Elkind, M. Mack, M. Ameen, D. Marshall, P. Ring and W. Krull, *2000 International Conference on Ion Implantation Technology Proceedings*, IEEE, p561
3. M. E. Mack, in Handbook of Ion Implantation Technology, edited by J.F. Ziegler, Elsevier, (1992), p599
4. H. Ito, T. Kamata, J. England, I. Fotheringham, F. Plumb and M.I. Current, Nucl. Intr. And Meth. B **96**, (1995), p30-33
5. N. White, M. E. Mack, G. Ryding, D. H. Douglas-Hamilton, K. Steeples, M. Farley, V. Gillis, A. Wittkower and R. Lambracht, Solid State Technology **28**, 151 (1985)

65nm Device Characteristics Matching on Single and Batch System Ion Implanter

Ken-ichi Okabe, Ryuichi Miura*, Masataka Kase

Device Development Division Fujitsu Limited 50, Fuchigami, Akiruno, Tokyo 197-0833, Japan
*Applied Materials Japan, Yokoso Rainbow Tower, 3-20-20 Kaigan, Minato-ku, Tokyo 108-8444, Japan

Abstract. We examined the process matching between two types of high current implanters; a batch system and a single wafer system. In particular, we investigated the formation by ion implantation of ultra shallow junctions for device characteristics of 65nm generation CMOS logic transistors. Secondary ions mass spectroscopy (SIMS) measurements were performed to profile boron and arsenic. Sample were implanted under various conditions by both types of implanters. The device characteristics were successfully matched using the adjusted implanted conditions using the SIMS profile.

Keywords: high current implanter, batch, single, process matching.
PACS: 61.72.Tt, 85.40.Ry

INTRODUCTION

Single high current ion implanter which has the advantages on (1) the reduction of pattern damage [1,2], (2) low particle from edge-clamp-less, (3) high tilt angled implant application, (4) eliminating the cone angle issue, and (5) turn around time on a small batch, are highlighted. When the translation from a conventional batch system to a single system is performed, the most important issue has been the device characteristics matching, in particular, a dopant profile difference. For example, the difference on the dose rate and implant temperature are caused from the difference on the ion beam shape and wafer scanning system. These are causing the Si damage such as the thickness of amorphous layer and defect density into the Si crystal, therefore the device characteristics are changed by the change of dopant profile and resistance [3-5]. When the same dose is implanted both of a batch and a single implanter, it is possible that the matching of device characteristics is not successful.

In this paper, we examined the dopant profile of the Si wafers implanted using the batch and single ion implanted system of Applied Materials, and successfully performed the matching of device characteristics of 65 nm generation CMOS logic transistors. The process focused on this paper is the source drain extension (SDE) of transistor for low standby leakage power (LSTP) applications.

EXPERIMENT

Arsenic and boron ions were implanted at the ultra low energy and high dose area using the batch and single ion implanted system in a fixed scanning speed and beam density. As described on previous section, the dose rate is quite important. The dose rate of single implanter is much higher than that of batch inplanter, as shown in Table 1. For the as-implanted samples, the dose of implanted ions was calibrated with conventional dose matching scheme using the dose of secondary ions mass spectroscopy (SIMS), which had a dose accuracy of 3 %. The result is shown in Table 2. We did not use the matching scheme using the sheet resistance, because the diffusion and activation of dopant might cause the error [3]. The Si damage was evaluated by thermal wave technology.

To examine the effect of dopant diffusion, the wafers are annealed by the advanced spike annealing after the SiO_2 coating is adopted on the Si surface using LPCVD (low pressure chemical vapor deposition) technique at the temperature below 600 °C to simulated preciously the dopant profile underneath the sidewall structure of MOSFET. The spike annealing is performed at the peak temperature of 1000, 1050 and 1100 °C. After spike annealing the surface SiO_2 is carefully removed out with no Si loss, then SIMS are corrected. The implant condition is varied with the energy and dose ranging with ±10%. The device characteristics such as the gate overlap capacitance, threshold voltage, short channel effect and Ion-Ioff are measured.

TABLE 1. Typical Scan Speed comparison of a Batch implanter and a Single implanter.

Batch	Single
30000mm/s(400rpm)	300~1000mm/s

In the case of Low Spin(400rpm), Single Implanter Dose rate becomes 30-100 times.

TABLE 2. ThermaWave results.

	Ultra low Energy Boron (0.05mA/cm^2)	Low Energy Arsenic (0.13mA/cm^2)
Batch (400rpm)	728	694
Single (600mm/s)	734	700

The wafer temperature is less than 35°C @ thermal label. The implant dose is matched using SIMS.

RESULTS

The results of thermal wave are summarized on Table 2. The ThermaWave result shows a higher damage in the case of single implanter. The temperature of the wafer was less than 35°C in the measurement of the thermal label. Since this implants are implanted in the enough small power, the effect of temperature is not effective. Although the implanted dose corrected by SIMS measurement is identical, the single implant induces a higher damage. If the difference is converted into the dose, it corresponds to 10%. A possibility to make the different damage is the beam implant angle. The batch system has a cone angle effect, therefore implanted ion has small angle deviation such as ±1°. However, the large beam angle deviation that is a cause of high damage does not let channeling ions increase. We recognize that the reason of high damage on the single implanter would be high dose rate.

Figures 1 and 2 show the As profiles of as-implanted and after annealing at the implanted energy of ultra low and low, respectively. Figure 1 shows that the As profile has significantly shallow at the single implant in case of annealing temperature at 1100 °C. In the same implantation temperature, when dose rate is large, it is thought that an amorphous-crystal interface becomes sharp. In this case, it is thought that enhanced diffusion is suppressed [6]. In Fig. 2, the extended channeling tail can be seen in the batch implant, so it is consistent with the ThermaWave result that the single implant makes a higher damage which can suppresses the channeling tail relatively.

Moreover, it should be noted that the optical measurement of the thickness of amorphous layer indicate the difference that ta=10.5 nm at the batch implant and ta=10.8nm at the single implant. The diffusion of As at 1100 °C is same as ultra low energy case. As a result, the profile matching between the batch and single implanter is easy at the case of temperature of less than 1050 °C. In the case of 1100°C, since vertical abruptness of As profiles is same, we think that the matching of profiles is possible by adjustment of the dose and the energy.

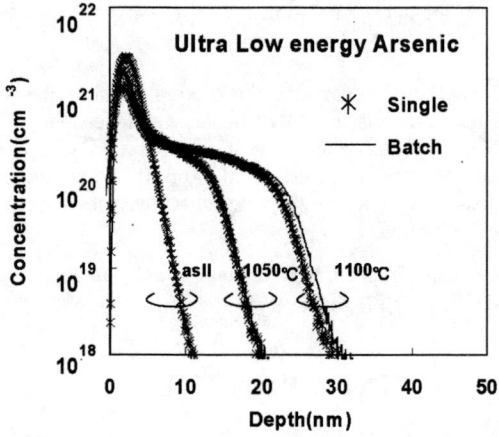

Fig.1 Ultra low energy Arsenic profile of a Batch implanter and a Single implanter

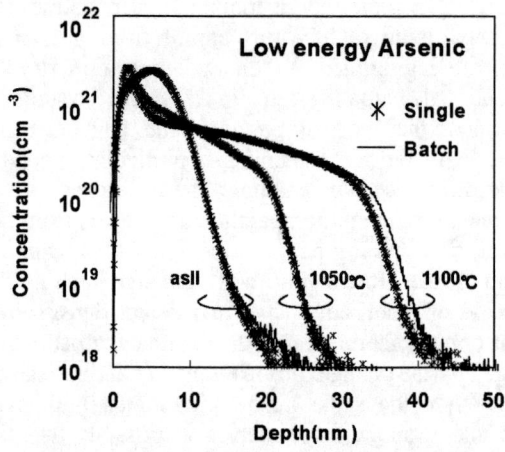

Fig.2 low energy Arsenic profile of a Batch implanter and a Single implanter

Figure 3 shows that the B profiles of as-implanted and after annealing at the ultra low energy. Although Decel implant were used for the ultra low energy boron, the energy contamination is less than 0.1%. Therefore, it is not effective in change of the profiles. The extended channeling tail of after implanted sample can be seen in the single implant, so it is not coincident with the ThermaWave result that presently we do not know what is the cause. However, since the channeling tail difference is very large, a surface damage layer may be not sensitive on a channeling tail about ultra low energy Boron. Therefore, we may need to evaluate more about the accuracy of the cone angle and the beam divergence. The profiles of after annealing has significantly deep at the single implant, reflecting the as-implanted profile. Figure 4 is as example of adjustment of dopant profile which used 1050°C spike annealing. The junction depth and vertical abruptness can be matched by decreasing 10 % of implant dose. We also try successfully to adjust the profile by tuning the implant energy (Fig.5). In a viewpoint of productivity, we can choose the decreasing dose.

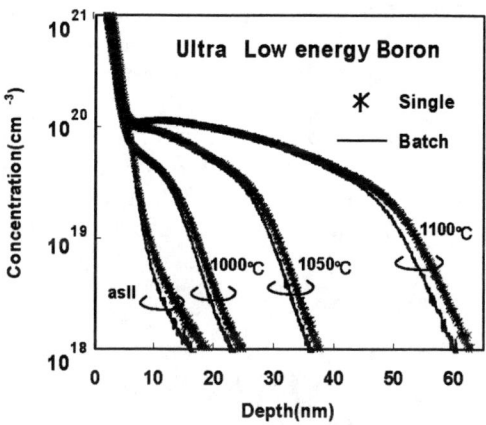

Fig.3 Ultra low energy Boron profile of a Batch implanter and a Single implanter.

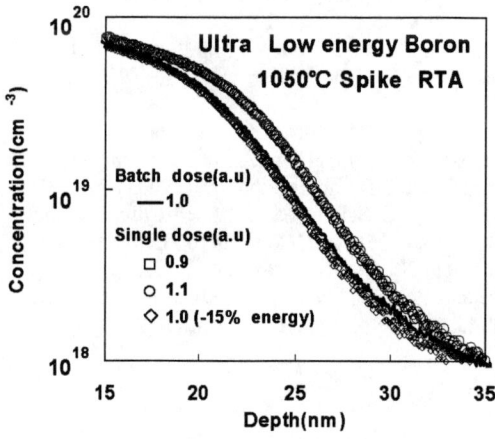

Fig.4 Boron profiles by adjustment of B doses and energy.

On the adjusted implant condition we fabricate the 65 nm generation CMOS devices of low leak transistor. The baseline process is determined using the batch implanter. The implant condition is rewritten to a new condition by the examination based on the SIMS analysis shown in previous session. This process was produced with RTA less than 1050°C. Figure 6 is the relation between the threshold voltage (Vth) and relative gate length in arbitrary unit on the PMOSFET.

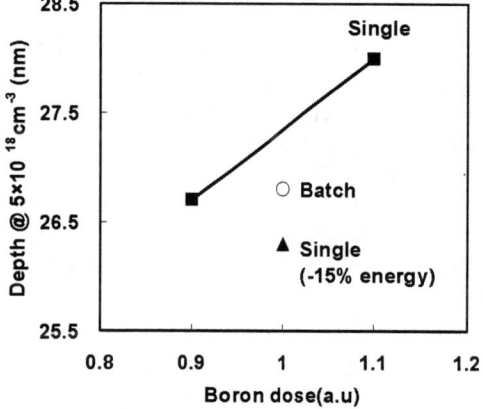

Fig.5 Junction depth of Boron at 5E18cm^{-3} in Fig.4, a Batch implanter and a Single implanter.

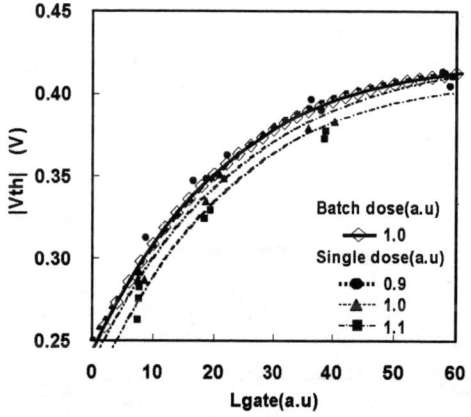

Fig.6 V_{th} roll-off characteristics of pMOS with LSTP.

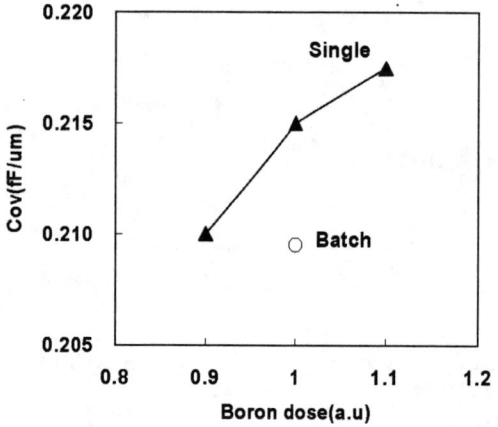

Fig.7 Dependence of Cov(Gate overlap capacitance) on implantation doses.

Figure 7 is the dependence of Cov (Gate overlap capacitance) on the implantation doses of same device of Fig. 6. Both figures results we can successfully match 10% decreased dose which is same result with the SIMS result. In the case of same dose implant the Cov is 2% larger than the batch case. In the matched Cov, extension resistance in the transistor did not change. Figure 8 is the relation between the Vth and relative gate length in arbitrary unit on the NMOSFET. No significant difference is obtained. In the case of same dose implant the Cov is same as the batch case. Figure 9 is the NMOSFET and the PMOSFET characteristics of Ion-Ioff. The device characteristics were matched decreasing 10% of total implant dose at the PMOSFET and using same total dose at the NMOSFET.

Based on the same protocol, we also achieve to match the device characteristics of high performance transistor which uses the difference halo implant condition. When heavy ion was used in the halo, the difference of extension profiles was not found in the batch and the single implanter. We think that the damage of the halo is dominant. For the matching between the batch and the single implanter, the examination of SIMS profiles underneath the side wall insulator after annealing, would be quite valid on the various situation depending on the transistor fabrication process.

SUMMARY

The source drain extension implant on the 65 nm generation CMOS logic transistors are examined depending on using the batch implanter and the single implanter. The Si damage, the relation between dopant diffusion depending on the dose rate, and the channeling tail structure are carefully compared. The device characteristics were successfully matched using the adjusted implanted conditions using the SIMS profile.

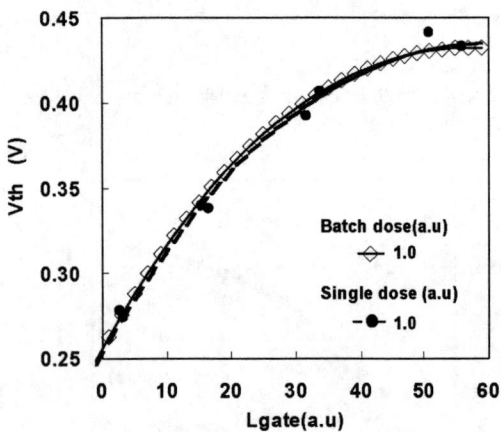

Fig.8 V_{th} roll-off characteristics of nMOS with LSTP.

REFERENCES

1. Y. Kawasaki, K. Tokunaga, et al., International Workshop on Junction Technology 2004, pp. 39-41.
2. L. Pipes, M. Taylor, G. Zietz, A. Al-Bayati, M. Castle, T. Marin, and J. Simmons, Nucl. Instr. and Meth. B237 (2005) pp330-335.
3. R. Simonton, J. Shi, T. Boden, P. Maillot, and L. Larson, ang, Mat. Res. Soc. Symp. Proc, Vol.316, 1994, pp153-158.
4. G. Fuse, M. Sano, H. Murooka, T. Yagita, M. Kabasawa, T. Shiraishi, Y.Fujino, N. Suetsugu, H. Izutani, M. Sugitani, Nucl. Instr. and Meth. B237 (2005) pp77-82.
5. D. W. Franke, M. Schmeide, H. M. Brussow, F. Hundt, E. Trepte, Proc. 15th Intl .Conf. on Ion Implantation Technology, 2004, pp102-106.
6. S.Whelan, J. A. Van den Berg, S. Zhang, D. G. Armour and R. D. Goldberg, Appl. Phys. Lett.76 (2000) pp571-573

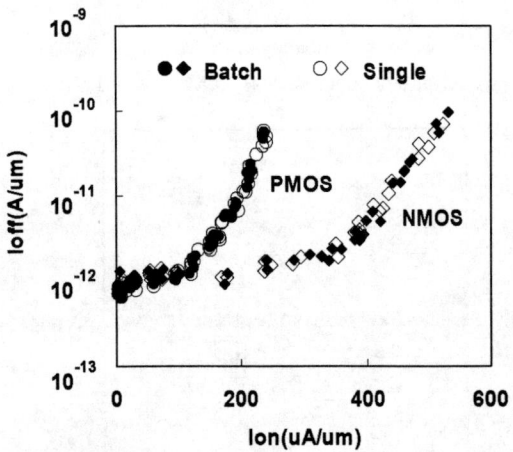

Fig.9 *Ion-Ioff* characteristics of nMOS,pMOS with LSTP. ($|V_{dd}|$=1.2V)

An Adaptive Knowledge-based Ion Source Automation Methodology To Improve Beam to Beam Switch Performance On Applied Materials Quantum® X Single Wafer Ion Implanter

Chris Burgess, Martin Keane, Robert Oliver

Applied Materials UK Limited, Foundry Lane, Horsham, West Sussex, RH13 5PX, U.K.

Abstract. In both the integrated device manufacturers (IDM) and the foundry market sectors, 300mm processing has been accompanied by decreasing size of device lots and increased variability in lot sizes. This has resulted in an increased focus on ion implant automated tune / recipe change in order to minimise the lost production time associated with this phenomenon. In parallel, significantly higher levels of automation have led to the necessity for consistency and reliability in achieving these changes.

This study demonstrates the performance of a knowledge-based control mechanism for controlling the ion source on a Quantum X single wafer ion implanter. Data is reviewed spanning multiple systems deploying multiple process recipes over the extended lifecycle associated with the indirectly heated cathode ion source. Intelligent automated recovery sequences are described along with tuning success rate and time data are presented.

Keywords: Ion Implant, Automation.
PACS: 61.72.Tt

INTRODUCTION

With the increase in fab automation there is a growing expectation increasing levels of automated processing with the highest possible process worthy throughput. There is also a greater need to reduce lost production time associated with beam to beam switching and interrupts. Described in this paper is a methodology that was adopted to improve the average tune time specifications for the Applied Materials Quantum® X Single Wafer Ion Implanters [1,2] whilst at the same time working towards increasing the success rate of the automated beam to beam switch sequence through the addition of enhanced beam switching algorithms.

METHODOLOGY

In order to better understand this methodology a brief outline of the main steps in the beam to beam switch, shown in figure 1, is required. Process recipe inputs drive the "Calculate" step and provide the starting values for the sequence of beam "defining". The "Change" step describes the migration of the existing recipe values to the values determined by the Calculate step. Having established the new conditions a further fast fixed sequence of steps is taken to ensure the resultant beam is being recorded by the dosimetry control system. The "Tune" step simply changes the starting values to acquire the correct beam current to facilitate accurate dosimetry.

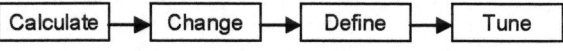

FIGURE 1. Stages of a beam switch

The time taken to switch between beams was addressed by pursuing two solutions. One is a modified "Change" step and the other being the introduction of an adaptive knowledge based system delivering better parameters from which to calculate the starting values.

The evolution of the Quantum X from its predecessors has provided for a much more robust platform for supporting a higher degree of parallel processes within the beam to beam switch. Excursions within the change step are better tolerated by the combination of improved power supplies and ancillary electronics and this has allowed for a sequence change

CP866, *Ion Implantation Technology,*
edited by K. J. Kirkby, R. Gwilliam, A. Smith, and D. Chivers

where it is now possible to remove the pre-accelerating voltage, change source conditions, extraction setting, mass analysis, plasma flood conditions quicker before turning the high voltage back on and setting it directly to the target voltage. This methodology is much quicker and additional auto recovery has made it more reliable than its predecessor.

Traditionally a beam was tuned from the same fixed set of values without any heed for the changing conditions as the source and extraction age. A knowledge based enhancement has been included within the recipe structure. When the beam is now completes the "tune" step the newly optimized parameters are retained and used in subsequent ion beam "Calculate" steps. Thus allowing the "Define" step to start with source parameters better suited to the age of the source. This eliminates the need for the beam switch sequence to search for as long and thus saves time.

Repeated use of the recipe builds up a list of evolved starting parameter for the beam to beam switch sequence. Only one value is stored for each parameter in each phase of the source life which we have defined as having 10 phases. This value is the rolling average of previous successful attempts. A rolling average with a predetermined maximum number of stored values would provide some stability to the parameter but allow it to evolve if other changes were made that would enhance the recipe performance.

In the pilot release of this feature the number of source parameters was deliberately limited to three. Those that that would most affect tune time and least affect process. Those that steer the beam laterally were not included as they are tuned over the same range in all instances of the source life during the "define" step and the auto tune sequence would undo any stored value.

Within the time gained through these two techniques it is now possible to introduce further adaptations to the beam to beam switch that enhance the quality of the information fed into the adaptive knowledge based system.

With field data now showing a more consistent Pareto it has been possible to focus the development of these additional adaptations to maximize the impact of the sequence of steps to the "Tune" step to improve the success rate of beam to beam switching. To minimize the effect of these additional sequences on the average tune time they are deployed towards the back end of the "Tune" step when a struggling beam switch requires assistance to acquire the beam current. Although the beam to beam switch has just performed an extended tune the adaptive knowledge based system will learn these new found values and use them in the future. This should then negate the need for the tune sequence to optimize for an extended period minimizing the time required to tune the next time this recipe is used at this point in the source life.

RESULTS OF INTERNAL STUDY

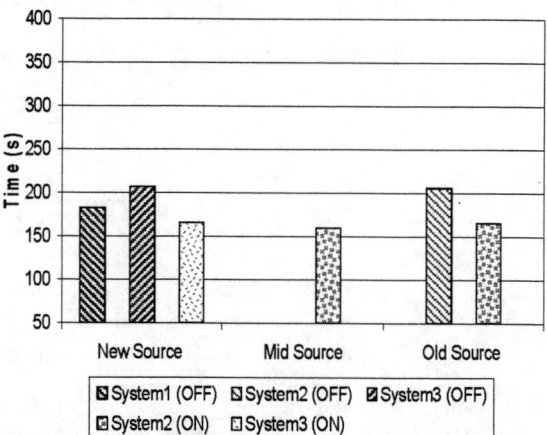

FIGURE 2. The average within species tune times obtained off three systems at different points in the source life when the adaptive knowledge-based system is On and off

A sequence of 23 typical beams was tuned and the times taken to switch from one to the other measured. The 23 beams covered a wide range of energies, the four main species (B+, As+, P+ and BF_2+) and a mix of drift and decelerated beams. Figure 2 shows the typical tune times for within species beam switches when under the influence of the adaptive knowledge-based system. It can be seen that the average tune times are better when the adaptive system is used. The effect of the adaptive system on the cross species tune times was not so apparent. This is most probably due to the effect of residual amounts of the previous species causing the tune times to be affected. However, the improvements achieved by the modified "Change" step more than compensated for this effect and in figure 3 this is demonstrated.

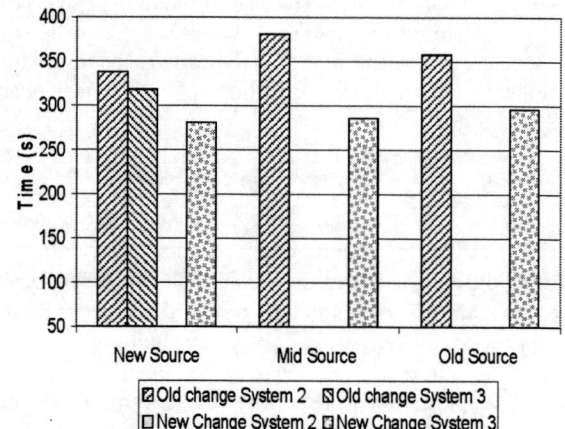

FIGURE 3. The average cross species tune times obtained

off two systems at different points in the source life when the improved "change" sequence is On and Off

When the modified "Change" sequence was enabled, tens of seconds were removed from the tune times. The benefit is most prominent in this situation as there is a need two perform two beam switches, one to the intervening inert beam and then to the target beam.

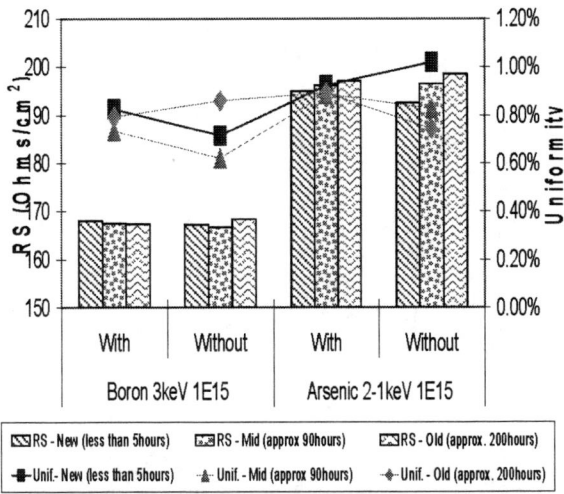

FIGURE 4. Sheet resistance and uniformity results from 2 recipes tuned at three points in a source life with and without the Adaptive Knowledge-based system enabled.

The process results provided in figure 4 shows that the values acquired by the adaptive knowledge-based system have no effect on the process. Since the learnt values were obtained by the beam to beam switch sequence, the same beam to beam switch sequence that is used in both cases it was unlikely that the process would have been affected. The sheet resistance and uniformity seen is within specified tolerances in both cases.

RESULTS OF EXTERNAL STUDY

Having now established the faster beam to beam switching and in keeping with the methodology a study has been undertaken of actual tune success rates and the reasons for failing to tune beams of systems in full production. Increasing the beam to beam switch success rate would also result in a more consistent tune times by eliminating some of the retries required to achieve the beam current. Event logs were analysed from three different customer sites off three tools spanning a total of 812hours. In Table 1 and figure 5 we see the results of this study.

TABLE 1. Tune Success Rates

Origin of data	Success rate
Customer A	90%
Customer B	96%
Customer C	96%
Internal	98%

Customer A and B are not using this adaptive knowledge-based system whilst Customer C is. The lack of tune success rate at Customer A is to some extent an artifact of the interrupt pareto shown in figure 5. Within the data obtained from Customer A there was a predominance of errors clustered around a single event and short periods of time. The majority of these Error/ Operator reasons were can be reduced by several methods and if these had not occurred the tune success rate would have been more inline with Customers B and C. With this in mind it can be seen that the tune success rate is quite consistent from one site to another.

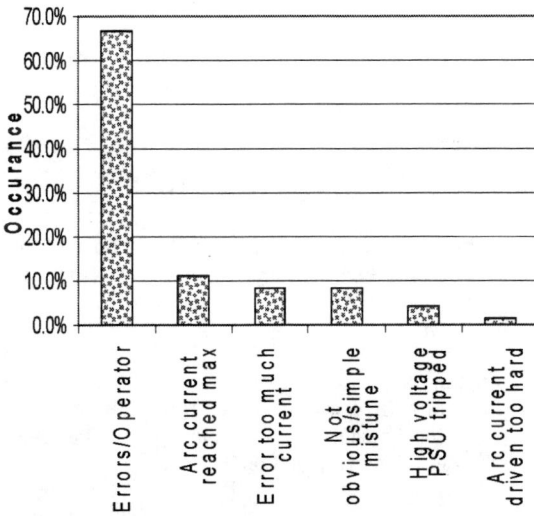

FIGURE 5. The Pareto of tune failures extracted from system event logs.

Figure 5 provides a break down of the most common beam tuning failures seen on the QuantumX™ series of implanters. Much is being done to reduce the failures occurring within the beam to beam switch sequence. These include but are not limited to the addition of tune sequences, expanding the use of tuned source parameters and auto recovery sequences. The Error/Operator events come in two types, the first are caused by the system not being ready to tune a beam, such as when something that can be shut, turned on, enabled, opened, configured etc is not in the correct state. The second is the tendency of the operator to use the retry button prior to investigating the tune failure. Events such as these can be significantly reduced if best practices are followed.

CONCLUSION

The changes made to date within the beam to beam switch algorithms within the Applied Materials QuantumX^{TM} series ion implanter has brought about a significant improvement in the average time taken to switch between both the same species and mixed species. The overall sequence of tuning is remains largely unchanged; changes have come before the tune, during the very early stages of the tune and through additional sequences to the back end of the "Tune". In the vast majority of the cases the tune sequence will appear very similar to the user.

In addition to this work it was also possible to investigate the causes of tune failures and lay the foundation for a series of improvements that will deliver improved tune success.

ACKNOWLEDGMENTS

All those who graciously shared their system event logs without which all this analysis would not have be possible.

REFERENCES

1. Murrell, A., et al., "Applied Quantum X Implant System: Technology Enhancements to Enabled Production-worthy Performance at the 45nm Node" in *These proceedings*

2. Murrell, A et al., "QuantumX: Single Wafer High Current Ion Implantation Using Mechanical Wafer Scan" in the Ion Implantation Technology. 2004. Proceedings of the 15th International Conference on

Lunch time at IIT 2006

Proceedings Editors Karen Kirkby and Andy Smith at work

Adrian Murrell at work

Conference Chairs Dave Chivers and Laurent Roux

Conference Chair Russell Gwilliam at prize giving

Conference Reception

Conference Dinner and Entertainment

Dave Chivers passes on the conference to
Sue Felch conference chair IIT 2008

Author Index

A

Abo, S., 542
Absil, P., 37
Adachi, K., 295, 317
Adams, B., 449
Adams, D., 178
Aiba, I., 524
Allen, A., 365
Alquier, D., 229
Ameen, M. S., 202
An, J., 137
Anderle, M., 113, 117
Ando, Y., 417
Andoh, Y., 429
Angel, G., 538
Aoki, Y., 578
Aoyama, T., 186
Arai, N., 295, 317
Araki, M., 433
Arevalo, E., 249, 511
Arita, Y., 300
Armour, D. G., 340
Arnó, J., 477, 489
Asaji, T., 373, 389
Augendre, E., 37, 129
Auriac, N., 133

B

Baek, S., 105
Bakshi, M., 570
Banks, P., 365, 618
Barakel, D., 253
Barozzi, M., 84
Batalin, V. A., 369
Bauer, A. J., 287
Beloto, A. F., 233
Bender, H., 21
Benistant, F., 58
Benjamin, M., 96
Bennett, N. S., 54, 73
Bernardini, J., 125
Bersani, M., 33, 73, 84, 113, 117
Bettincurt, P., 313
Bettiol, A. A., 269

Bigarella, D., 225
Bishop, S., 477, 489, 497
Biswas, S., 335, 365
Blackwood, D., 269
Blake, J., 377, 622
Bolze, D., 109
Borland, J. O., 96, 566
Bourdelle, K. K., 65
Braatz, R. D., 50
Bradley, J. D. B., 9
Breeden, T., 503
Breese, M. B. H., 269
Brown, B., 485
Bruel, M., 76
Buczkowski, A., 96, 566
Budrevich, A., 41
Bugaev, A. S., 369
Burgess, C., 649
Burle, N., 283

C

Cagnat, N., 133
Callahan, W., 381
Campbell, C., 377, 614
Carrere, M., 257
Chain, M., 46
Chalupa, Z., 353
Chan, L., 58
Chang, C.-P., 516
Charavel, R., 325
Chen, E., 460
Chen, H., 516
Chen, J., 96, 335
Chen, J. T. C., 574
Cheng, Y. C., 92
Cherkashin, N., 65
Cheung, N., 69
Chevrier, F., 283
Chien, C. C., 46
Chivers, D. J., 261
Cho, C., 137
Cho, H. J., 25, 163
Cho, H. T., 163, 171
Choi, H. W., 308
Chu, P. K., 245

Claverie, A., 65
Collart, E. J. H., 37, 73, 84, 340, 618
Colombeau, B., 58, 84
Cowern, N. E. B., 33, 37, 54, 73, 84
Cucchetti, A., 381
Cummings, J., 405
Current, M., 96
Current, M. I., 582

D

Danis, J., 601
De Almeida, A., 304
Deichler, J., 202
Delgado, S., 313
Dev, K., 50
DiMeo Jr., F., 477, 489
Dimitrov, D., 574
Dimitrova, T., 574
Distaso, D., 405, 538
Ditzler, K., 554
DiVergilio, W., 206
Doherty, R., 92
Dohi, S., 417
Dolan, R., 578, 622
Don, E., 534
Dudognon, J., 275
Duffy, R., 29, 37
Dunn, J. P., 481, 485
Dzengelaski, J., 405

E

Eddy, R., 313, 361
Edwards, P., 618
Egami, M., 493
Ehrke, U., 594
Eisner, E. C., 206
Elshot, K., 554
Erokhin, Y., 137, 353, 409, 472, 520
Erre, R., 275
Etienne, H., 229, 609

F

Faifer, V. N., 96, 582
Fakhreddine, H., 503

Falepin, A., 129
Falk, S., 409, 472
Fan, H., 460
Fang, Z., 25, 249, 511
Farley, M., 618
Felch, S. B., 21, 37, 129, 516
Ferrara, J., 453
Ferrell, C. E., 441
Feudel, T., 13
Fiala, J., 340
Foad, M. A., 33, 46
Foggiato, J., 113
Fornara, P., 125
Fortunier, R., 65
Foster, P. J., 9
Franke, D.-W., 441, 562
Freer, B. S., 507, 554
Frey, L., 121, 464
Frioulaud, L., 133, 353
Fujita, H., 421
Fujiyama, J., 401
Fukagawa, Y., 524
Fukui, R., 633
Funk, K., 182
Fuse, G., 17, 457

G

Gehre, D., 13
Gelpey, J., 96
Gennaro, S., 33, 73, 84, 113, 117
Gifford, K., 425
Gilchrist, G. F. R., 178, 198
Giubertoni, D., 33, 73, 84, 113, 117
Gladwin, S., 503
Glavish, H. F., 167
Godfrey, C., 601
Goldberg, R. D., 340, 618, 641
Gotoh, N., 295, 317
Gotoh, Y., 295, 317
Gou, S. H., 9
Graf, M. A., 507, 554, 601
Grant, J., 441
Graoui, H., 46
Grim, J. S., 481, 485
Grisenti, R., 117
Guenther, T., 441
Guo, B., 137, 353, 472, 520
Gupta, A., 249, 538

Gurhan, I., 329
Gushenets, V. I., 369
Gwilliam, R., 54, 84, 113, 335
Gwinn, M., 174

H

Hacker, D., 618
Haerting, F., 441
Hahn, S. H., 25, 137
Hamada, K., 357
Hamamoto, N., 167, 186, 421, 605
Hamilton, B., 3
Hamilton, J., 449
Hamilton, J. J., 73
Han, I. K., 62
Han, J. J., 140
Harada, M., 317
Harris, M., 206, 349
Harris, M. A., 155, 202
Hartz, C. L., 481
Hasan, M., 105
Haslam, B., 182
Häublein, V., 464
Hautala, J., 174
Hazelton, R. C., 261
Hendrix, D., 425
Henke, D., 202
Heo, S., 105, 171
Herden, M., 13
Herrmann, L., 13
Hershcovitch, A., 369
Hillard, R., 96
Hisamune, T., 433, 633
Hoffman, T., 129
Hoglund, D. E., 507, 554
Holmes, A. J. T., 340
Hong, W., 308
Horsky, T. N., 159, 167, 178, 182, 198, 206
Hsieh, T. J., 155, 206
Huang, Y., 453, 601
Hudak, C., 313
Huh, T. H., 101, 163
Hummel, S. G., 566
Hundt, F., 441
Hung, R., 516
Huntington, D., 361
Hur, N., 137

Hwang, H., 105, 171
Hwang, S. H., 25
Hwang, S.-H., 163
Hwang, Y., 137
Hyung, Y. W., 140

I

Iida, T., 218, 373, 389
Ikejiri, T., 605
Ikenaga, E., 546
Ila, D., 304, 329
Inouchi, Y., 417
Ishibashi, T., 295, 317
Ishijima, T., 300
Ishikawa, J., 295, 317
Ishikawa, M., 401
Isogai, H., 194
Ito, H., 524, 546
Ito, Y., 218
Iwai, H., 524
Izunome, K., 194
Izutani, H., 457

J

Jacobson, D., 96, 155, 167
Jakubowski, F., 202
Jank, M. P. M., 121
Janssens, T., 21, 37
Jeon, W. H., 101
Jeon, Y. B., 25
Jessop, P. E., 9
Jillson, S., 472
Jin, C. G., 524, 546
Jin, S. W., 62
Johnson, B. M., 369
Johnson, R., 630
Joo, Y. H., 163
Joung, Y. S., 62
Juang, L. S., 385
Jung, T. W., 163

K

Kabasawa, M., 393
Kaeppelin, V., 257, 353

Kagawa, T., 218
Kaim, R., 477, 489, 497
Kamenitsa, D. E., 349
Kandziora, C., 121
Kase, M., 645
Kashima, K., 194
Katahira, K., 300
Kato, Y., 218, 373, 389
Kawashita, M., 190, 321
Keane, M., 649
Keitz, M. D., 261
Kellerman, P., 377, 413
Kernevez, N., 65
Ki, Y. J., 62
Kikuchi, S., 445
Kikuchi, Y., 393
Kim, D. S., 163
Kim, E., 520
Kim, G. D., 308
Kim, J. J., 546
Kim, J. K., 308
Kim, S., 101, 163
Kim, Y. K., 140, 449, 468
Kim, Y. S., 101
Kindersley, P., 618
Kinoyama, T., 421
Kirkby, K. J., 33, 73
Kirkwood, D. A., 641
Klepper, C. C., 261
Knights, A. P., 9
Kobata, M., 546
Kobayashi, K., 546
Kobayashi, T., 605
Koga, Y., 605
Kokrot, M., 562
Kolomiets, A. A., 369
Kondratenko, S. I., 101
Konishi, M., 417, 429
Koo, A., 582
Kotaki, H., 295, 317
Krimbacher, B., 441
Kropachev, G. N., 369
Krueger, C., 531
Krull, W. A., 96, 163, 171, 182
Krytsch, G., 531
Kubo, T., 373, 389
Kudo, T., 17
Kuibeda, R. P., 369
Kulevoy, T. V., 369
Kurunczi, P. F., 445

Kwack, K.-D., 550
Kwak, N.-Y., 550
Kwok, C. T. M., 50
Kwok, D. T. K., 245

L

LaFontaine, M., 453
Laugier, F., 65
Laviron, C., 133, 534
Lazzari, J.-L., 283
Lee, B.-S., 550
Lee, C. H., 140
Lee, D., 105, 171, 472
Lee, D.-H., 550
Lee, H. D., 140
Lee, H. R., 308
Lee, J. K., 25, 163
Lee, K. S., 140
Lee, M.-K., 550
Lee, M.-Y., 62
Lee, P. S., 58
Lee, S. C., 140
Lee, S. W., 25
Lee, W., 385
Lee, Y., 137
Lendzian, H., 353
Lepienski, C. M., 233, 241
Lerch, W., 96, 109
Leterte, F., 65
Li, C. I., 46
Li, Z., 96, 566
Liebert, R., 425
Lim, T. J., 140
Litovko, I. V., 369
Liu, J., 133, 137, 520, 538
Locke, J., 503
Loh, K., 472
Loh, N., 472
Loibl, N., 594
López, P., 29
Low, R. J., 626
Luey, K., 381
Lurcott, S., 497

M

Maas, G., 29
Machado, R. M., 481
Mangaiyarkarasi, D., 269
Mangelinck, D., 125
Marganski, P., 178
Markevich, V. P., 3
Masunov, E. S., 369
Mathieu, G., 609
Mathiot, D., 133
Matsuda, Y., 417, 429
Matsumoto, T., 429, 605
Matsuo, J., 214
Matumoto, S., 17
Maul, H., 594
McComb, B., 92
McCoy, S., 96
McHugh, J., 503
McKenna, C., 578, 622
McManus, J., 497
McNally, P. J., 54
McTeer, A., 249
Mehta, S., 133, 353, 409, 449, 472
Merkulov, A., 594
Meunier-Beillard, P., 29
Milgate, R., 178
Milgate III, R. W., 198
Miller, T., 249, 511
Minamisawa, R. A., 304
Mineji, A., 96
Minotani, T., 295, 317
Miranda, J., 155, 206
Mitchell, B., 618
Mitchell, P., 425, 570
Miura, R., 641, 645
Miyabayashi, K., 421
Miyashita, F., 88, 291
Mizuno, B., 524, 546
Mollica, R., 538
Moon, J. T., 140
Moore, E., 377
Morita, K., 300
Morley, S., 618
Moschella, J. J., 261
Mulcahy, C., 58, 335, 365
Mullin, J., 614
Muntele, C., 304
Muntele, I., 304
Murrell, A. J., 618, 641

N

Na, S. K., 62
Nagai, N., 167, 605
Nagata, S., 300
Nagayama, T., 186
Naito, M., 167, 417, 605
Nakagawa, N., 542
Nakai, T., 578
Nakamura, S., 578
Nakanoya, T., 493
Nakao, K., 417
Nakase, S., 88
Nakayama, K., 190, 321
Naughton, J. J., 586
Newman, D., 554
Ng, C.-H., 531
Nguyen, T., 582
Nicolaides, L., 570
Ninan, B., 516
Nishihashi, T., 357, 401, 433, 633
Nishiwaki, T., 357
Nizou, S., 229
Noda, T., 21, 129
Noh, J. H., 140
Nouri, F., 129
Ntsoenzok, E., 283
Nukayama, M., 417
Nunan, P., 409, 520

O

Oechsner, R., 534
Oga, E., 457
Ogata, S., 433, 633
Oh, J. G., 163
Oh, J.-G., 25
Ohnishi, H., 317
Okabe, K., 645
Okada, T., 321
Okashita, K., 524, 546
Oks, E. M., 369
Okumine, T., 295, 317
Okumura, T., 524, 546
Oliveira, R. M., 233, 237, 241
Oliver, R., 649
Olson, J. C., 381, 405, 538
O'Reilly, L., 54
Orihira, K., 417

Ostrowski, B., 361
Ota, K., 601
Ottaviani, L., 253
Ozdal-Kurt, F., 329

P

Pap, A., 534
Parada, M. A., 304
Parihar, V., 21, 129
Parisi, N., 449
Park, S.-K., 550
Park, S. W., 62
Parrill, T. M., 249, 511
Pasquinelli, M., 253
Paul, S., 96, 109
Pavelka, T., 534
Pawlak, B. J., 37
Peaker, A. R., 3
Pelaz, L., 29
Peng, K. T., 385
Pepponi, G., 117
Perel, A. S., 445
Peres, P., 594
Pershin, V. I., 369
Persing, H., 249, 511
Personnic, S., 65
Petrenko, S. V., 369
Pfeffer, M., 534
Pharand, M., 453
Pineau, A., 275
Pokidov, I., 453
Polner, D., 601
Polozov, S. M., 369
Poole, H. J., 369
Portavoce, A., 125
Pyi, S. H., 25, 163

Q

Qin, S., 249

R

Ra, G. J., 101
Raissi, M., 283
Raj, D., 225
Rambach, M., 287

Raskin, J.-P., 325
Rathmell, M. A., 437
Rathmell, R. D., 349, 637
Ray, A. M., 349, 637
Redinbo, G., 377, 614
Reece, R. N., 101, 441
Regula, G., 283
Renau, A., 345
Rendon, M., 503
Reuther, H., 233, 237
Reynolds, W. P., 206
Rice, J. H., 279
Richards, S., 578, 622
Rodier, D., 468
Rodrigues, M., 329
Rolland, J. L., 481, 485
Romig, T., 520, 554
Roozeboom, F., 29
Rossi, J. O., 233, 237, 241
Rouh, K. B., 62
Roux, L., 229, 609
Roy, M., 229
Rubin, L. M., 155, 554
Rudskoy, I., 369
Ryding, G., 618
Ryssel, H., 121, 287, 464
Ryu, C.-K., 550
Ryu, S., 137

S

Saito, H., 357
Saito, J., 389
Sakai, S., 460, 605
Sakamoto, T., 542
Sakase, T., 618, 641
Sakuragi, S., 17
Sasaki, H., 373, 389
Sasaki, Y., 524, 546
Sato, F., 218, 373, 389
Sato, S., 17
Satoh, T., 317
Satta, A., 3
Sauddin, H., 524
Sauter, B., 562
Scheuer, J. T., 381, 405, 445
Scheutelkamp, R., 37
Schmeide, M., 441, 562
Schreutelkamp, R., 129

Schuhmacher, M., 594
Schuur, J., 630
Sealy, B. J., 54, 84
Seebauer, E., 50
Seki, T., 214
Seleznev, D. N., 369
Sen, B. H., 329
Senda, T., 194
Seto, S., 17
Severi, S., 37, 129
Shao, Y., 174
Sharp, J. A., 33
Shaugnessy, D., 570
Sheen, D. S., 25, 163
Shim, K., 137, 472, 520
Shim, K.-H., 353
Shin, K. I., 101
Shin, S.-W., 550
Shishiguchi, S., 96
Shon, H.-S., 550
Silva, G., 237
Simoen, E., 3
Simola, R., 125
Sinclair, F., 167, 413
Sing, D. C., 503
Singh, V., 25, 225, 249, 511
Skinner, W., 174
Smayling, M., 516
Smick, T., 618
Smith, A. J., 54, 84
Smith, G., 497
Sohn, Y. S., 25, 62, 163
Son, G. H., 140
Soncini, V., 225
Song, D. G., 140
Song, Y. W., 62
Sotta, D., 534
Splinter, P., 601
Srinivasan, M. P., 58
Steeples, K., 558
Storozhenko, P. A., 369
Suetsugu, N., 397
Sugai, H., 300
Sugitani, M., 393, 397, 457
Sun, H. L., 385
Suzuki, H., 433, 633
Suzuki, T., 17
Svarovski, A. Ya., 369
Sweeney, J., 178

T

Takai, M., 542
Takaoka, G. H., 190, 321
Takeda, D., 190
Tamura, H., 524, 546
Tan, M. C., 472
Tanaka, K., 421, 542, 605
Tanaka, Y., 17
Tanii, M., 429
Tanjyo, M., 96, 186, 421, 460, 605
Tatemachi, J., 417
Tauzin, A., 65
Taylor, M., 41
Teo, E. J., 269
Tieger, D. R., 155, 206
Timberlake, D., 405
Ting, R., 46, 92
Tokioka, H., 542
Tomita, M., 633
Tonari, K., 357, 401, 433, 633
Torregrosa, F., 229, 253, 257, 609
Toshiyuki, N., 516
Towner, J. M., 586, 590
Toyoda, E., 194
Toyoda, N., 194, 210
Tseng, H. P., 385
Tsidilkovski, E., 558
Tsuchiya, B., 300
Tsuji, H., 295, 317
Tsukihara, M., 393, 397
Tsun, H. Y., 385
Tsutsui, K., 524
Tutto, P., 534
Tzou, S. F., 46

U

Ueda, M., 233, 237, 241
Umisedo, S., 186, 421, 605

V

Vaidyanathan, R., 50
Van Daele, B., 21
Van Den Berg, J. A., 117, 340
Vanderberg, B. H., 349, 637
Vanderpool, A., 41
van der Tak, K., 29

Vandervorst, W., 21, 37
Variam, N., 377, 468
Vayer, M., 275
Venezia, V. C., 155, 202
Venturini, J., 96
Verheyden, K., 182
Vervisch, V., 229, 253
Vilela, W. A., 241
Vorrada, L., 69
Vrancken, C., 21

W

Wakaya, F., 542
Walden, M., 578
Walker, T., 96, 566
Walther, S. R., 225, 409
Wang, H. Y., 46
Wang, L., 489
Webb, R. P., 33
Welsch, M., 381
Wenzel, K. W., 637
Werner, M., 117
White, N., 335
Williams, J. M., 261
Winder, E. J., 249, 511
Wodjenski, M., 477, 497
Wong, B., 472
Wong, T. M. H., 582
Woo, H. J., 308
Wright, E., 445
Wu, C.-T., 279
Wu, K., 349
Wyon, C., 534

X

Xu, J., 520
Xu, K., 385

Y

Yamada, I., 147, 194, 210
Yamada, R., 17
Yamada, T., 397
Yamagiwa, H., 542
Yamashita, T., 421, 605
Yanagitani, T., 317
Yang, C. L., 46
Yang, J. K., 140
Yang, M. H., 361
Yano, Y., 291
Yedave, S., 477, 489
Yeong, S. H., 58
Yokoo, H., 433, 633
Yokota, K., 88, 291
Yoo, D. H., 140
Yoo, W. S., 113
Yoshino, M., 300
Yu, Y. S., 140
Yuasa, S., 605
Yuhara, J., 300
Yushkov, G. Yu., 369

Z

Zhai, Q., 468, 626
Zhao, Z., 425, 531
Zimmerman, R., 329
Ziti, M., 229